Michael Zipf
865-8 768

System Dynamics and Control

Eronini Umez-Eronini
Morgan State University

PWS Publishing

An Imprint of Brooks/Cole Publishing Company

I(T)P An International Thomson Publishing Company

Pacific Grove • Albany • Belmont • Bonn • Boston • Cincinnati • Detroit
Johannesburg • London • Madrid • Melbourne • Mexico City • New York
Paris • Singapore • Tokyo • Toronto • Washington

Sponsoring Editor: *Michael J. Sugarman*
Marketing Team: *Nathan Wilbur, Jean Thompson*
Production: *Clarinda Publication Services*
Cover Design: *Julie Gecha*
Cover Photo: *Corbis*

Photo Research: *Judith Weber*
Typesetting: *The Clarinda Company*
Cover Printing: *Phoenix Color Corporation*
Printing and Binding: *R. R. Donnelley & Sons—Crawfordsville*

COPYRIGHT © 1999 by Brooks/Cole Publishing Company
A Division of International Thomson Publishing Inc.

I(T)P The ITP logo is a registered trademark used herein under license.

For more information, contact PWS Publishing at Brooks/Cole Publishing Company:

BROOKS/COLE PUBLISHING COMPANY
511 Forest Lodge Road
Pacific Grove, CA 93950
USA

International Thomson Publishing Europe
Berkshire House 168-173
High Holborn
London WC1V 7AA
England

Thomas Nelson Australia
102 Dodds Street
South Melbourne, 3205
Victoria, Australia

Nelson Canada
1120 Birchmount Road
Scarborough, Ontario
Canada M1K 5G4

International Thomson Editors
Seneca 53
Col. Polanco
11560 México, D. F., México

International Thomson Publishing GmbH
Königswinterer Strasse 418
53227 Bonn
Germany

International Thomson Publishing Asia
60 Albert Street
#15-01 Albert Complex
Singapore 189969

International Thomson Publishing Japan
Hirakawacho Kyowa Building, 3F
2-2-1 Hirakawacho
Chiyoda-ku, Tokyo 102
Japan

Printed in the United States of America.

10 9 8 7 6 5 4 3 2 1

Library of Congress Cataloging-in-Publication Data

Umez-Eronini, Eronini,
    System dynamics and control / Eronini Umez-Eronini.
        p.   cm.
    Includes index.
    ISBN 0-534-94451-5
    1. Automatic control.   2. Dynamics—Mathematical models.
3. System analysis.   I. Title.
TJ213.U453   1998
    629.8—dc21   98-18838
                                            CIP

Contents

Chapter 3   Engineering System Models in State Space 79

Chapter 4   Other System Models in State Space 149

Chapter 10   Stability of Dynamic Systems 535

Part III   **System Design 597**

Chapter 11   Introducing Automatic Control Systems Design 599

Preface

THIS BOOK HAS THREE MAIN OBJECTIVES. The first is to provide, in one text, a unified treatment of system dynamics and automatic controls. Because control theory is almost exclusively applied to dynamic systems, and a realistic dynamic model is essential to the design of a good control strategy, it is a sound pedagogic approach to cover system dynamics and automatic controls in a single textbook. The first objective is also compatible with the trend in engineering programs to diffuse design throughout the undergraduate curriculum. The use of one book for the two related subjects should also result in considerable financial savings to students.

The second objective of the book is to emphasize the key role that dynamic system modeling plays in most engineering endeavors and to illustrate the proper balance between theory and current practice. System dynamics is one of the few engineering courses in which all the various fundamental subjects are largely integrated and applied to individual problems. Consequently, illustrative examples describing realistic and seemingly complex engineering systems as well as everyday objects are analyzed in the book and assigned as problems—printer printhead and optical disk drive dynamics, high-power stepping motor dynamics and control, micromachined microactuator and sensor subsystems, patient temperature/clinical thermometer monitoring problems; oilwell-suckerrod-walking beam systems, continuous stirred tank reactor dynamics and control, banking/economic subsystems, population and ecosystems—to mention a few. Throughout the book, several carefully executed schematic diagrams, block diagrams, and flowcharts are used not only to clarify the ideas presented but to provide practice in the use of these tools in the solution of actual problems.

The third aim of the text is to highlight certain concepts and practices that are important in professional work. Examples of these concepts and techniques include the description and application of electromechanical systems, computer-aided design and analysis of control systems using personal computers and computational packages, and hardware and software realization of digital computer control and measurement systems. Unlike most other books in the field, *System Dynamics and Control* includes a chapter on modern controller implementation, complete with sections on interfacing with external equipment, signal conditioning, data acquisition hardware and software, and a case study illustrating the actual realization of a PC-based system. The application of computers and commercially available software packages is integrated throughout the text. In addition to specific examples, in various chapters that detail software computation and simulation of time response, frequency response and stability, and synthesis of controllers, the source files for most of the illustrative material, figures, and solutions developed in the book using computer tools are available by anonymous ftp, as described below. Also, practice problems that require the use of a software package are provided at the end of each chapter. A brief segment called **On the Net,** which appears at the end of relevant chapters, itemizes the downloadable files.

These three objectives are met without sacrificing necessary depth and breadth of coverage through a judicious selection of the contents, and the structure and style of the text. *System Dynamics and Control* presents system dynamics via one consistent format: the state space approach. It covers conventional and modern control system analysis, design, and implementation for both continuous-time and discrete-time systems, including nonlinear and process control systems, classical controller and compensator design, state variable feedback control systems as well as observer design and optimal control. Common analytical techniques developed for continuous-time systems are then applied to discrete-time systems without the repetition necessary in a separate treatment. Similarly, nonlinear systems are considered wherever appropriate analytical or computational tools are introduced rather than in a separate chapter.

Throughout the book, subject material is clustered around common ideas and techniques. Electrical system modeling follows mechanical and fluid-mechanical system modeling. Thermal energy systems lead to process engineering systems, and electromechanical systems are discussed following the development of a broad view of energetic systems. The solution of the vector state equation follows the development of the response of the structurally similar but simpler scalar first-order system. There is a logical progression from sinusoids and phasors through Fourier series and Fourier transform to Laplace transform and frequency response. A conscious effort was made to ensure that the style of presentation be interesting and the material easy to comprehend. With the exception of some elements of linear algebra that are provided in an appendix, other mathematical concepts required in the book are developed within the text itself.

The book comprises three parts: Physical Modeling and Model Construction, Model Solution, and System Design. Each part contains a brief summary or overview and a list of references. The summaries, which appear at the beginning of each part, help maintain the unity of the work and provide pointers to crucial ideas in the chapters contained in that part. Answers to selected problems appearing at the end of each chapter are provided in an appendix. These problems are part of the large number of practice problems and solved examples provided in the book. The complete solutions are included in the *Instructor's Manual,* which is available to those adopting the book as a course text.

There are five chapters in each part of the book. Chapter 1, the introduction, is written in a relaxed manner that is intended to engage the reader's interest while elaborating on the underlying concepts and philosophy of the text. The chapter also introduces the physical modeling process, the actual starting point of dynamic system modeling. Rather than commence with already determined schematic diagrams and linear graphs of physical systems, we first provide a technique for developing these representations of the physical model from a rudimentary understanding of the problem and the goals of the study. Chapter 2 elaborates on the details of physical modeling and preliminary aspects of model construction, including the compromises and significant approximations that are often made in representing physical variables and describing dynamic behavior. The state space approach to system modeling is also introduced here. Chapter 3 examines the details of the state modeling method. It deals with mathematical models of mechanical, incompressible-fluid, and electrical systems. Chapter 4 continues with the models of thermal systems and process engineering systems. It also briefly covers distributed-parameter models and nonengineering system examples. This material helps to impart a balanced perspective of real systems modeling—that actual systems are neither lumped-parameter nor distributed-parameter but can be modeled usefully as such. The mathematical modeling of engineering systems concludes in Chapter 5 with examples, especially of electromechanical devices, in which different types of systems are combined in one problem. It is shown that the unifying quality of these interacting systems—power transmission at their interconnections—can be exploited to develop generalized models. Micromachined devices, another increasingly significant class of such systems, are introduced briefly at the end of the chapter.

Chapters 6 through 10 constitute the second part of the text and cover methods of solution and analysis of the systems modeled in the first part of the book. Chapter 6 considers analytical solution of the scalar first-order equation and the vector state equation, and digital computer simulation of such systems. The application of computational software is formally introduced here. Chapter 7 deals with second-order and higher-order scalar systems and considers their response to various types of input, including sinusoidal and periodic-non-sinusoidal inputs. The related subjects of forced mechanical vibrations and Fourier series are introduced here. Transform methods: Fourier-, Laplace-, and z-transform solutions, are taken up in Chapter 8. Block-diagram algebra and frequency response analysis, which are used later for single-loop feedback control systems and compensator design, are introduced in Chapter 9. Also, the relations between state models and transfer functions, phase plane description of linear and nonlinear behavior, and controllability and observability concepts are covered here. Chapter 10 concludes Part II with the stability analysis of linear and some nonlinear dynamic systems, including continuous-time and discrete-time systems.

Part III consists of Chapters 11 to 15. Automatic control system design is introduced in Chapter 11, with emphasis on classical techniques. Controller tuning for given performance criteria using available solutions and computer simulation is also included in this chapter. Chapter 12 considers frequency domain design both for specified closed-loop performance and for a given compensator. The examples include dead time and nonlinear control systems. Chapter 13 emphasizes modern control of continuous-time systems, while Chapter 14 stresses control of discrete-time systems. The topics introduced in these two chapters include feedforward and cascade control systems, direct digital control systems, state vector feedback control, observer design, adaptive control, optimal control, and the H_∞ control problem. Chapter 15 deals with the implementation of designs using the microcomputer or PC. It includes material on interfacing the computer with external equipment and the operation and programming of such interfaces, data acquisition and control instrument options, control valve characteristics, and an example illustrating the complete realization of a computer control system. These sections should help further unite theory and practice and sustain interest in the subject of automatic controls generated by earlier material.

This book is intended for mechanical engineering juniors, seniors, and first-year graduate students; and readers, including chemical and electrical engineering students, with similar backgrounds. Also, practicing engineers will find the text a good review source and reference for system dynamics and/or automatic controls. For a graduate introductory course in automatic controls, the use of assigned reading from other sources to augment the material on the more advanced topics is recommended.

Code Availability   Although application of the commercially available math software package MATLAB^{TM} is specifically described in the text, the application programs code, consisting of both complete Mathcad^{TM} MCD-files and MATLAB m-files for most illustrations and computations performed in the book, is available via anonymous ftp at pws.com. It can also be accessed through the World Wide Web at the URL http://www.pws.com/eng/ (follow the links from there).

I would like to thank the following reviewers for their helpful comments and suggestions: Brian Fabian, University of Washington; Ganiyu Hanidu, Hampton University; Thomas Kurfess, Georgia Tech; Christopher D. Rahn, Clemson University; Dan A. Tortorelli, University of Illinois at Urbana–Champaign; James Williams, MIT.

E. I. Umez-Eronini

Physical Modeling and Model Construction

Dynamic systems can be studied in four stages:

1. Description of the system physically (physical modeling)
2. Description of the system mathematically (model construction)
3. Analysis of the mathematical description (model solution)
4. Synthesis of a preferred manifestation of the system (system design).

Part I of this book considers the first two stages, that is, the physical and mathematical description of dynamic systems, including mechanical, fluid, thermal, electrical, chemical-process, and mixed systems. The analysis of the mathematical model (also called system model) is taken up in Part II, and the synthesis of automatically controlled dynamic systems (system design for automatic control) is considered in Part III.

Part I examines the definition of a dynamic system. One consequence of the nature of such a system is that it can be controlled in a particular fashion, known as *feedback control.* Because a system can be defined in terms of its components and their interconnections, its physical model can be constructed, effectively, by representing, usually graphically, the system's constituent *behavior components* and their interactions, once these have been deduced from the observed or intended overall behavior of the system. Schematic diagrams, block diagrams, and electrical networks are examples of the graphical representations of physical models considered in the first part of the text. The system models are derived using the state space approach, which is a type of formal, mathematical description of physical cause and effect in the behavior components, once these components and their interactions are exposed through a process of separation called *decomposition.* This approach is quite general and can be applied to a wide variety of systems in engineering and even to nonengineering systems.

Chapter

1

Introduction

ENGINEERING CAN BE DESCRIBED in a very general manner as the solution of problems. A slightly more specific description is that it is the application of science to the employment of available resources to mankind's benefit. *Science* is used here in a broad sense. Although at one time the major constituents of science were certain laws of physics, today, due to the growing diversity of human endeavors and the creation of new knowledge, the volume of scientific laws from which the engineer draws has greatly increased. The more noticeable additions are those involving concerns for the environment, the conservation of resources, and information and computer technology. One consequence of this trend is that greater creativity, judgment, intuition, and ability to integrate knowledge from diverse fields are necessary to solve today's engineering problems. Subjects like system dynamics and control systems design can provide experiences useful to the development of these qualities, skills, and abilities by engineers.

The important steps in the solution of an engineering problem are often summarized as follows:

1. Problem formulation
2. Problem analysis
3. Search for alternative solutions
4. Selection of preferred solution
5. Specification of selected solution

Steps 1 to 3 are the subjects of most engineering courses and are dealt with in this book. Step 4 is generally considered in economic studies due to the important role played by cost factors in the selection process. Step 5 is seldom studied independently, although it is very significant in engineering and elsewhere. A problem remains unresolved until its solution is described adequately. The reader is urged always to try to utilize available communication resources to ensure that solutions are specified in a logical, understandable, and, if possible, interesting and compelling manner.

This text deals with problems associated with *dynamic systems* and focuses on a particular aspect of problem formulation, namely, the specification of the system by means of a *physical model*. Problems will be analyzed through the solution of mathematical or *system models* constructed from the physical model. The search for alternative solutions will take the form of *design,* especially for automatic control performance.

In this chapter dynamic and static systems are defined and illustrated with examples. The concept of a dynamic system is further explained in an introductory discussion of *feedback control.* Next, the meaning of the word *system* is considered in the context of the type of problems studied in this book. Finally, physical modeling is introduced, and the chapter concludes with an illustration of *systems concept,* which is a critical first step in the construction of a physical model.

1.1   CONCEPT OF DYNAMIC AND STATIC SYSTEMS

In this book, *a dynamic system is one in which the current effects (outputs) are the result of present and previous causes (inputs).* In this definition, the notion of cause and effect, which is often used to describe both engineering and various nonengineering phenomena, is introduced. The details of this viewpoint are deferred to a later chapter. For now, it is interesting to note that the ideas that will be explored subsequently also apply to nonengineering systems. An example is the population of a nation or a species, where, in simple terms, the *present* number (output) is the result of *previous* deaths and births (inputs).

The preceding definition differs from the general notion of a dynamic system as one that is in motion, that is, changing (usually rapidly) with time. Change in most human populations generally appears to be very gradual, although the growth of certain biological cells and cultures is another matter! Similarly, although the amount of cash in a bank with no electronic teller facilities may remain fixed over the weekend, this amount is due to all the deposits and withdrawals that occurred during the preceding work week.

In order to explore further the differences between the two viewpoints of dynamic systems, we give the corresponding definition of a static system: *A static system is one in which the current effects (outputs) depend only on current causes (inputs).* By this definition, a system whose output is changing with time can be described as a static system as long as the input is also changing in a corresponding manner. Note that the time scale in which the systems are perceived can make a lot of difference. Given any system, it is possible to choose a sufficiently small measure of time such that the inherent delay between input and output will become noticeable, thereby making the system appear to be dynamic. In general, the choice of a time scale depends on the goals for the investigation of the system.

Figure 1.1 shows a water supply system that may be typical of a farm, for example. We are concerned at this time only with the tank and the valve as subsystems or components. If the pressure in the pipe immediately upstream of the valve is the gage pressure P, and the valve discharges to the atmosphere, then the pressure drop across the valve is also P.* If we ignore for now fluid velocity and friction in the pipe, P is proportional to the water density ρ, and the height of the water column

$$P = \rho g(h + H_0) \tag{1.1}$$

where g is the acceleration of gravity and h and H_0 are defined in Fig. 1.1. We have assumed that the top of the tank is at atmospheric pressure. (The tank is not completely covered.) The relation between the pressure drop across a valve and the rate of volumetric flow through the valve is generally (for *turbulent* flow) of the form

---

*If the atmospheric pressure is P_0, the absolute pressure is $(P + P_0)$, so the pressure drop is $(P + P_0) - P_0$.

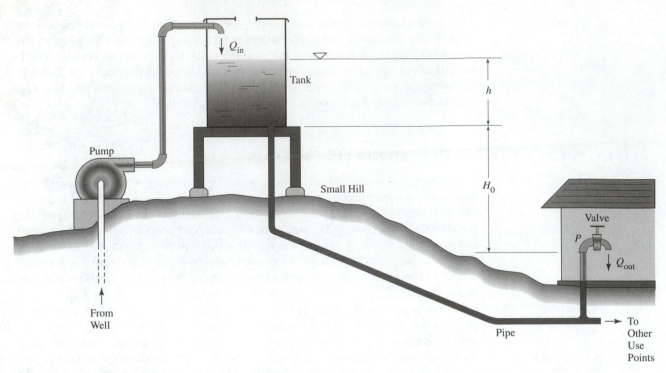

Figure 1.1

Dynamics of a Water Supply System

$$Q_{\text{out}} = C\{P\}^{1/2} \tag{1.2}$$

where C is an empirical constant representing the opening of the valve and other valve characteristics. (Valves are discussed in more detail in Sections 3.3 and 15.3.) We will assume that the valve opening is fixed. From equation (1.1),

$$Q_{\text{out}}(t) = C\{(\rho g)[h(t) + H_0]\}^{1/2} \tag{1.3}$$

Since every quantity on the right-hand side of equation (1.3)† except h is a constant, Q_{out} and h are related algebraically. In other words, any change in h will immediately be reflected by a corresponding change in Q_{out}. Thus, if we focus our attention on the valve and consider it the system, equations (1.2) and (1.3) imply that the input is h or the pressure P, the output is the flowrate Q_{out} (note that in this case pressure drop *causes* flow), and the valve is a static system. More precisely we say that the valve is described by a *static relation*. Even when the pressure drop is changing, such as is the case if the pump that lifts water from the well into the tank is turned on and the tank level h is increasing, the system will, by our definition, still be static because the current outflow $Q_{\text{out}}(t)$ will continue to depend on the current pressure drop $P(t)$, according to equation (1.2) or (1.3).

Consider now the tank as the system. If the net rate of inflow $(Q_{\text{in}} - Q_{\text{out}})$ is taken as the input, then the output is the resulting change in level, or just the level of the tank h. (The question of what to take as input or output for a given component is

---

†$Q_{\text{out}}(t)$, $h(t)$ are written in place of Q_{out}, h to emphasize the explicit involvement of time in these relations.

considered in Section 3.1.) The level of water in the tank at any time is due to all previous inflows of water into the tank and outflows of water from the tank. Even if all operations are shut down today, so that the current tank level remains constant, the present quantity of water in the tank will still be due to all the inflows and outflows that occurred previously. Thus we have a dynamic system. Mathematically, the input and output are related through the volume of water in the tank:

$$Volume\ (t) = \int_0^t \{Q_{in}(t) - Q_{out}(t)\}dt + Volume(0) \qquad (1.4)$$

The volume and level of water in the tank are algebraically related:

$$h(t) = \text{algebraic function of } Volume(t) \qquad (1.5a)$$

Note that in our present discussion the volume of water in the tank is also a logical output for the tank system. Such interchangeability of variables is often suggested, as we shall see, when the relation between them is static (equation (1.5a) for example). For the special case in which the cross-sectional area of the tank A is independent of the level,

$$Volume(t) = Ah(t) \qquad (1.5b)$$

and we have

$$h(t) = \left(\frac{1}{A}\right)\int_0^t \{Q_{in}(t) - Q_{out}(t)\}dt + h(0) \qquad (1.5c)$$

Basic Dynamic Operators

The water tank provides us with a very simple model of a basic dynamic system that we call an *integrator*. We shall also on occasion refer to a dynamic system as an *accumulator*. The tank in Fig. 1.1 "accumulates" the net input (flow), which is positive when the input flowrate exceeds the output flowrate. The accumulation is negative when the output flowrate is greater than the input flowrate, but the term still applies. The integration specified by equation (1.5c) is given the usual interpretation as the area under a curve in Fig. 1.2. In this case, it is $(1/A)$ times the difference between the area under the Q_{in} curve and that under the Q_{out} curve, from time zero to the present time t. This is the shaded area in Fig. 1.2. The dependence of the current level $h(t)$ (lower figure) on past inputs (upper figure) should be clear from these results. The present level of the tank (h at time = t) is as shown because the tank "remembers" all the past inflows and outflows. Thus we shall also refer to a dynamic system as one with *memory*. Note again that the level of water in the tank has not changed in the period identified as Today, yet the system is, in our context, dynamic.

The word *memory* also describes another basic dynamic system model that we shall call the (time) *delay*. An example of the delay operator is the hot water supply on a cold winter morning. If we take as input the act of turning on the tap or the flow of hot water *from the boiler* in response to the pressure signal at the tap,* and as output the flow of hot water *out of the faucet,* then we know from experience that hot water does not come out immediately when the tap is opened (the water out of the faucet is not hot initially). Rather, the cold water in the pipe is discharged until the hot water travels from the boiler to the tap. The time it takes for this to happen is the delay period or *transportation lag* or *dead time T*, and it is evident that we are dealing with a dynamic system. Mathematically, if the flow of hot water from the boiler is represented

---

*The pressure signal travels at the speed of sound in water, so there is a negligible time lapse between valve opening and the time hot water leaves the boiler.

Figure 1.2

Water Tank as an Integrator

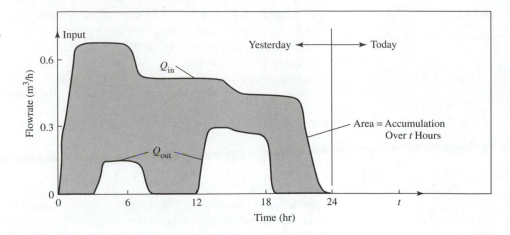

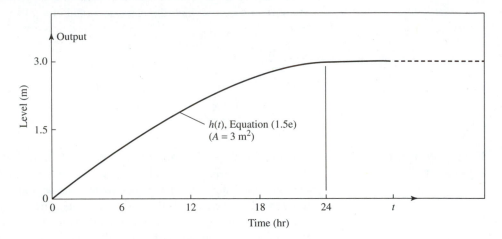

by $Q(t)$, then the flow of hot water out of the faucet is given by $Q(t - T)$ and it is understood that the function $Q(t - T)$ is the same function as $Q(t)$ but shifted T time units to the right. The two basic dynamic operators are illustrated in Fig. 1.3.

Units and Dimensions

In the equations we have written thus far, we did not attach units to the various quantities represented. In general, this will be the case throughout the book. There are two reasons for this approach. The first is that we prefer to "plug-in" numbers and units at the end, since this makes for a clearer result. The second reason is that we want our fundamental relations to be independent of any particular system of units. However, in the numerical examples where final results are given, units are introduced. Also, in graphical results such as Fig. 1.2, the units of the various quantities represented are given. This is in accord with the importance we attach to the specification of solutions. Graphs generally contain a lot of condensed information, and the potential for misinterpretation is high. Thus a graphical output should be carefully prepared and should include a title (including other relevant information), axis labels (names and dimensional units), and proper scales. A legend should also be provided when necessary.

Although the units were left out of the equations, the dimensions were not ignored. An equation must be not only arithmetically balanced but also dimensionally consistent. Dimensions such as length, time, mass, temperature, and electrical current are the quantitative descriptions of matter. These quantities can be measured in differ-

Figure 1.3

Basic Dynamic Operators

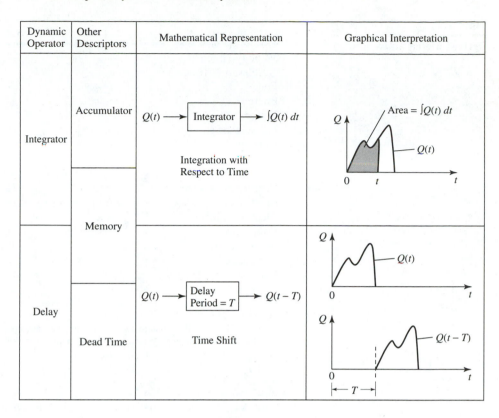

ent systems of units. In SI* or the metric system of units, for instance, length is measured in meters (m),† time in seconds (s), and mass in kilograms (kg) (see [1]). The corresponding U.S. customary units are, respectively, foot (ft), second (s), and pound (lb or lbm).

Tables 1.1 and 1.2 list the SI base units and the derived metric units. Once we fix the system of units, it is more convenient to perform the dimensional balance on the

---

*Le Système International d'Unités

†*Metre* is the British spelling and is commonly used in European countries.

Table 1.1

Basic SI Units

QUANTITY	UNIT	SYMBOL[a]
Length	meter	m
Mass	kilogram	kg
Time	second	s
Electric current	ampere	A
Temperature	kelvin[b]	K
Luminous intensity	candela	cd
Amount of matter	mole	mol

SOURCE: *ASME Orientation and Guide for Use of Metric Units,* 3d ed., ASME, 1973.
[a]Symbols are not followed by a period.
[b]Kelvin is the unit of thermodynamic temperature. Common temperature measurements are generally based on the Celsius temperature scale. The thermodynamic zero for temperature is 0.00 K = −273.15°C (0°C = 273.15 K). A change of 1°C is equivalent to 1 K change.

Table 1.2

Derived Metric Units

QUANTITY	UNIT	SYMBOL	DEFINITION
Velocity (linear)	—	—	m/s
Acceleration (linear)	—	—	m/s^2
Force	newton	N	kg $\cdot$ m $\cdot$ s^{-2}
Energy	joule	J	N $\cdot$ m
Power	watt	W	J/s
Pressure or stress[a]	pascal	Pa	N/m^2
Density	—	—	kg/m^3
Electric charge	coulomb	C[b]	A $\cdot$ s
Electric potential	volt	V	W/A
Electric resistance	ohm	Ω	V/A
Electric capacitance	farad	F	C/V
Electric inductance	henry	H	V $\cdot$ s/A
Frequency	hertz	Hz	1/s or s^{-1}
Magnetic flux	weber	Wb	V $\cdot$ s
Conductance	siemens	S	A/V
Thermal flux density	—	—	W/m^2
Entropy	—	—	J/K

SOURCE: *ASME Orientation and Guide for Use of Metric Units,* 3d ed., ASME, 1973.
[a]Care should be taken to differentiate between absolute and gage pressure where this is important.
[b]The Celsius symbol is written °C to avoid confusion with the coulomb, C.

basis of units. This procedure also causes the appropriate units to be assigned to the results automatically (see **On the Net** at the end of the chapter). Consider equation (1.3), for example. Suppose we are using the metric system. The units of density ρ, acceleration g, and height h or H_0, are, from Table 1.1, kg/m^3, m/s^2 and m, respectively. Flowrate Q is volume per unit of time; hence the unit is m^3/s. Substituting these units into equation (1.3) gives

$$\text{m}^3/\text{s} = [\text{units of } C][(\text{kg/m}^3)(\text{m/s}^2)(\text{m})]^{1/2} \tag{1.6}$$

or

$$\text{m}^3 \cdot \text{s}^{-1} = [\text{units of } C] \, \text{kg}^{1/2} \cdot \text{m}^{-1/2} \cdot \text{s}^{-1} \tag{1.7a}$$

so that the units of C must be kg$^{-1/2}$m$^{7/2}$. If we already had units for C and they were different from kg$^{-1/2}$m$^{7/2}$, then equation (1.7a) would not "balance," implying an error somewhere in the results. Thus a dimensional check can be useful in determining if a result is wrong. However, dimensional consistency by itself does imply a correct solution. In terms of U.S. customary units, equation (1.3) gives

$$\text{ft}^3/\text{s} = [\text{units of } C][\text{lb/ft}^3)(\text{ft/s}^2)(\text{ft})]^{1/2} \tag{1.7b}$$

so that the units of C are lb$^{-1/2}$ft$^{7/2}$.

With regard to the quantities force and weight, the units of force are derived from the basic unit of mass according to Newton's second law of motion, which we shall simply state for the present as force = mass × acceleration.

In the metric system of units, the product of one kilogram of mass and one meter per second squared of acceleration is termed a *newton:*

$$(1)\text{N} = [(1)\text{kg}][(1)\text{m/s}^2] = (1)\text{kg} \cdot \text{m/s}^2 \tag{1.8}$$

Thus 5.0 N is required to maintain a quantity of matter of mass 2.0 kg at an acceleration of 2.5 m/s^2.

In U.S. customary units, the unit of force is the pound-force, which is the force with which a standard pound-mass is attracted to the earth at a location where the acceleration due to gravity is 32.174 ft/s². That is,

$$(1)\text{lbf} = [(1)\text{lbm}][(32.174)\text{ft/s}^2] = 32.174 \text{ lbm} \cdot \text{ft/s}^2 \qquad (1.9)$$

However, weight is the measure of the gravitational force acting on a quantity of matter at a specific location, so that force is measured in terms of weight in the U.S. customary system.* A mass of one pound will weigh at standard gravity (32.174 ft/s²), one pound-force. The force required to maintain a quantity of matter of mass 7.7 lbm at an acceleration of 11.3 ft/s² is

$$\frac{(7.7)(11.3)}{(32.174)} \text{ lbf} = 2.7 \text{ lbf}$$

In SI, the international standard gravity value is 9.80665 m/s², so that the weight of one kilogram mass is approximately 9.807 N (the exact value will depend on the location).

Since the earth's gravitational pull is ubiquitous and does not change very significantly for the same object at any point on the globe, the weight is sometimes used in place of the mass to represent a quantity of matter. This practice should be avoided. To further confuse matters, the pound-force (lbf) is at times simply referred to as the pound (lb), so that one has to determine from the context whether the mass or the weight is implied. (Use lbm and lbf, respectively.)

This book preferentially uses SI units, which have become a worldwide standard. A table of selected conversion factors between SI units and U.S. customary units has been provided in Appendix A.

1.2 FEEDBACK CONTROL CONCEPT

If for a dynamic system the present output depends on past inputs, then by introducing suitable inputs in the present, one can influence the output in the immediate future. The key words are *suitable* and *influence*. How do we determine a suitable input? Is the influence substantial, negligible, good, or bad? It is reasonable to expect that specification of the proper input will require at least some knowledge of the workings of the system and that desirable results (subsequent output) can be achieved if the present output is known. Generally, the operation of the system is represented by some model of its behavior, and the most direct determination of the present output is a measurement.

Consider again the water supply system of Fig. 1.1. It is evident from our earlier discussion that the operation of the water tank system is fairly well determined. As was shown by equation (1.3), the flow at a faucet, and indeed at any point where the water is utilized, is a function of the tank water level h. Some of the applications of the water supply (again take a farm as an example) could be fairly sensitive to the water flow or pressure at the point where the water is used. Jet nozzles for scrubbing or cleaning equipment or produce, and sprinkler irrigation components are two good examples. How can we ensure that the tank water level will remain high enough and fairly steady for these applications? A simple approach, assuming that over significant periods of

---

*In the British gravitational system of units, the lbf is an independently defined unit. The derived unit for mass is the slug. The product of one slug of mass and one foot per second squared of acceleration is one pound-force.

time the aggregate volume of water used at the various sites is fairly uniform and predictable, and that some variation in pressure and therefore performance can be tolerated in the applications, is to operate the pump and fill the tank from the well on some sort of time schedule.

Figure 1.4a shows the way the actual tank water level depends on the scheduled pump operation or the desired tank level. Although the pump operating schedule could have been deduced from, say, statistical analysis of past data on water use on the farm, there is nothing in the control scheme of Fig. 1.4a that prescribes any relationship between one variable of the tank and the other (the variables in this case being Q_{in} and h. The term Q_{out} is related to h through the valve opening parameter C, which is taken here as a perturbation of the system or a disturbance input). Indeed, if at any time the pattern of water use (equivalent valve opening) should deviate significantly from that on which the pump schedule is based, the control system could break down or perform very poorly. This type of control scheme is referred to as *open-loop control.*

The structure of an open-loop control system is shown in Fig. 1.4b. The specific characteristic or variable of the system, the output, that we wish to control is referred to as the *controlled variable,* whereas the characteristic or variable that is determined by the control action is called the *control input.* The control elements can collectively be called the *controller,* although, as we shall see later, the control elements can be contained in other subgroups. The *reference input* generally implies the desired value of the controlled variable. In Fig. 1.4b the loop is open in the sense that there is no path through which the control input can be determined in terms of the controlled variable, that is, the controller does not prescribe a relationship between the control input and the controlled variable.

Figure 1.4

Open-loop Control of Water Supply System

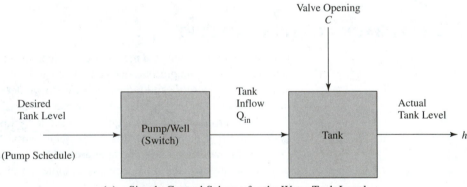

(a)   Simple Control Scheme for the Water Tank Level

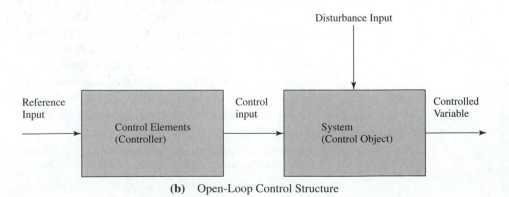

(b)   Open-Loop Control Structure

Feedback Control Structure

Another approach to the control of the water tank level is to have the p___ turn the pump on whenever the tank water level drops below the refere___ to turn it off again when the tank fills beyond the reference level. To faci___ of the operator, a level measuring/transmitting and indicating system (LT and LI) can be connected to the tank and used to display the tank level at a location where it can conveniently be observed (see Fig. 1.5a). The result is a *closed-loop control* system, since there is a prescribed (albeit switching) relationship between the tank level and input flow. The operator observing the tank level provides the path that closes the loop. However, a control system with a human operator as an element of the controller is *manual,* that is, not automatic. On a busy farm or where it is not feasible to dedicate an employee to operation of the pump only, such manual control would clearly be in-

Figure 1.5

Manual and Automatic Control of Water Tank Level

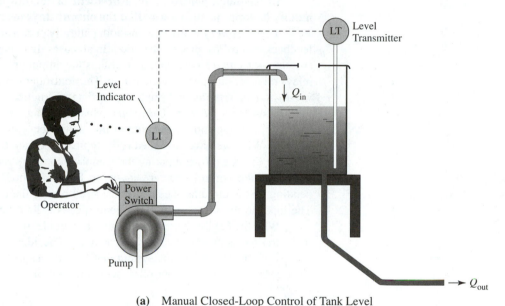

(a) Manual Closed-Loop Control of Tank Level

(b) Automatic Closed-Loop Control of Tank Level

adequate or unsatisfactory. An example of an *automatic* (closed-loop) control system for the tank level is shown in Fig. 1.5b. Here the comparison of the tank level signal with the desired value of the tank level (entered into the system as a setpoint setting) and the turning of the pump on or off are all performed by appropriate hardware in the controller. However, note that both the manual and automatic level control systems, just described, involve

1. the maintenance (in spite of external disturbances) of some prescribed relationship of one system variable to another using the result of a comparison of these variables;
2. the transmission of some signal or information from a later (output) stage to an earlier (input) stage of the system, that is, loop closure through *feedback*.

The concept involved here is that of *feedback control*.

   In feedback control, a measurement of the output of the system is used to modify its input in such a way that the output stays near the desired value (see [2]).

   There are very many systems both naturally occurring and man-made that rely on feedback control for proper operation. In all cases there is usually a sensory instrument *(sensor)* or *feedback element* that measures the output or system variable of interest and relays this measurement to a *controller*. The controller compares the signal with the *desired value* or *setpoint* and sends appropriate instructions to the *actuator* or "effector" mechanism (or *final control element*), which then acts on the system or *control object* (or *plant*) to bring subsequent outputs into agreement (prescribed relationship) with the setpoint. What we have described is the typical feedback (or closed-loop) control structure (see Fig. 1.6). In comparing the feedback control structure shown in Fig. 1.6 with the open-loop control structure shown in Fig. 1.4b, note that the actuator could, depending on the circumstances, be considered part of the controller or part of the plant. The input to the actuator from the controller is called the *manipulated variable*.

   With the feedback control structure as a guide, we can analyze systems suspected of operating as feedback control processes. The idea is to identify components or groups of components of the system that function as the various elements in the feedback control structure—controller, actuator, control object, and sensor—as well as to

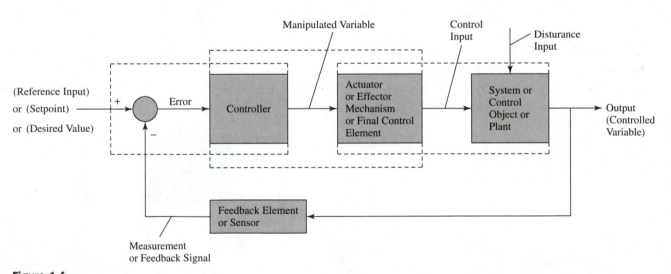

Figure 1.6

**Typical Feedback
Control Structure**

establish the controlled variable and the setpoint. If all this can be done, then the system under study is a feedback control system. Consider the case of the so-called human thermostat. This remarkable system maintains the body temperature at around 37°C (98.6°F) give or take about one degree, whether one is in Alaska in the middle of winter or in Florida in the middle of summer. In the latter case, that is, prevention of overheating, the actuation (control input) is the dilation of the blood vessels in the skin, which increases heat transport from the interior of the body to the surface; and sweating, which increases the rate at which this heat is lost from the surface to the environment, through evaporation (see [3]). The controller is located in the hypothalamus, an area at the base of the brain stem above the crossing of the optic nerves. But where are the sensors? For a while it was thought that the skin was the predominant measurement device for overheat protection. An experiment was performed in which (1) the skin temperature, (2) the internal temperature (as measured by a thermocouple held against the ear drum, which is near the hypothalamus), and (3) the rate of sweating were all recorded for a subject while at rest and during exercise. The results showed that the main sensors for overheat protection are located in the controller (in the hypothalamus) itself! In Fig. 1.7a the skin temperature (as a candidate feedback signal) shows no consistent relationship with the rate of sweating (a control input). Figure 1.7b, on the other

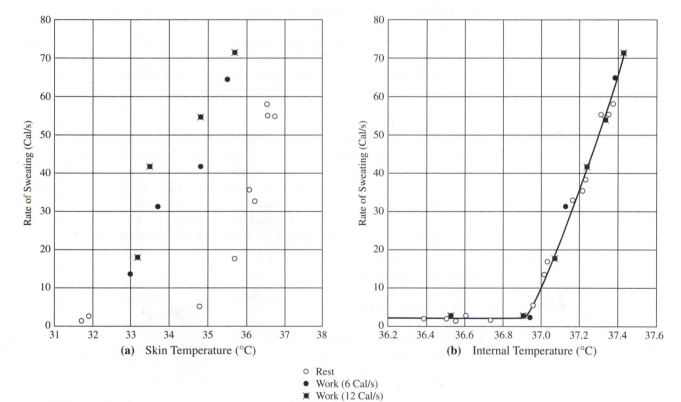

○ Rest
● Work (6 Cal/s)
✹ Work (12 Cal/s)

Figure 1.7

Heat Dissipation as a Function of Skin and Internal Temperatures. From "The Human Thermostat," by T. H. Benzinger. Copyright © 1960 by Scientific American, Inc. All rights reserved.

hand, indicates that the rate of sweating is relatively fixed when the internal temperature (the other candidate feedback signal) as measured near the controller is below about 36.9°C (98.4°F). Above this temperature, the rate of sweating increases dramatically with internal temperature. These results define rather sharply the setpoint for this subject as 36.9°C (98.4°F) at the time of the experiment. Further, the results indicate that there is indeed a sensor located close to or inside the controller and that the overheat protection system is a feedback control system.

An example of a man-made system in which the controller and the sensor are combined is the water clock invented by Ktesibios of Alexandria, Egypt, who is presumed to have lived during the reign of King Ptolemy II (Philadelphus), 285–247 B.C. (see [4]). This clock is believed to be the earliest known feedback device. The essential element of the water clock (see Fig. 1.8a) is the float valve or controller, which ensures a constant rate of water trickle out of the regulating vessel (control object) by maintaining a constant water level (controlled variable) in the vessel. The level of water in the regulating vessel depends on the rate of external water supply to the vessel. As shown in Fig. 1.8a, the same float that determines the valve opening and hence the rate of water inflow to the vessel is also responding to the height of water in the vessel. Thus the controller is controlling and "measuring" the water level simultaneously. The water clock function is achieved by collecting the water trickling out of the regulating vessel in another vessel (not shown), whose levels are calibrated to read time.

Fig. 1.8b illustrates the feedback control structure of the Egyptian water clock. Note that if the desired water level in the regulating vessel is that at which the valve (inverted cone and float) is almost closed (by the float), then the pressure p in the water supply line represents essentially a disturbance to the system.

Other Types of Control

Our discussion thus far has suggested a number of classifications or types of control systems. We have seen that control systems can be open-loop or closed-loop, manual or automatic, and in a universal sense, naturally occurring or man-made. There are many more types of control systems that can be classified on the basis of various aspects of the subject. For example, naturally occurring versus man-made, and manual versus automatic can be considered classifications based on environmental or social factors. Table 1.3 summarizes types of control systems classified on the following bases:

1. Environmental or social
2. Nature of loop
3. Nature of controller
4. Nature of plant
5. Type of controlled variable
6. Nature of setpoint
7. Type of control law
8. Analytical or design technique

It is not our intent, in this section, to discuss in detail all the control system types represented in Table 1.3, since the various underlying principles are yet to be developed. Rather, the purpose of Table 1.3 is to illustrate the large number of possible types of control systems and to convey a feeling for the wide spectrum of activities that can involve control functions. As you progress through the relevant principles in subsequent chapters, you may refer back to Table 1.3 to obtain a perspective on how the material fits into the general framework of control systems.

Figure 1.8

**Components and Structure
of the Water Clock**

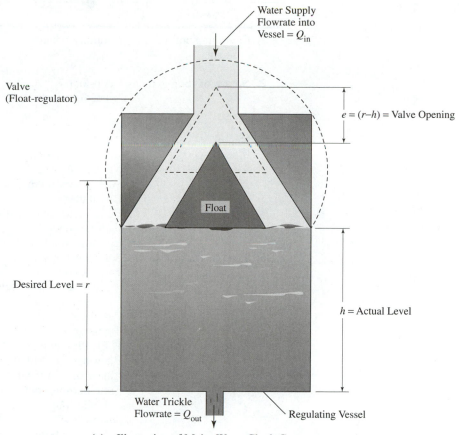

Water Supply
Flowrate into
Vessel = Q_{in}

Valve
(Float-regulator)

$e = (r - h)$ = Valve Opening

Float

Desired Level = r

h = Actual Level

Water Trickle
Flowrate = Q_{out}

Regulating Vessel

(a)   Illustration of Major Water Clock Components

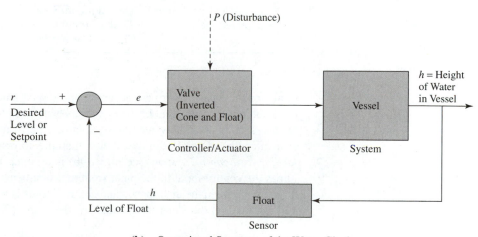

P (Disturbance)

r
Desired
Level or
Setpoint

$+$

e

Valve
(Inverted
Cone and Float)

Controller/Actuator

Vessel

System

h = Height
of Water
in Vessel

$-$

h

Level of Float

Float

Sensor

(b)   Operational Structure of the Water Clock

**Nature of Loop: Feed-
forward Control
System**

The open-loop control of the water tank level illustrated in Fig. 1.4a is considered inadequate for significant perturbations in the anticipated valve opening. This situation
can be improved significantly if the control action is determined from a measurement
of the valve opening. (See Fig. 1.9a). Indeed, if our knowledge of the plant is exact

Table 1.3

Types of Control Systems

BASIS	TYPE	
1. Environmental or social	Naturally occurring	Man-made Manual Automatic

2. Nature of loop	Open-loop
	    Feedforward
		Closed-loop
		    Single or Multiple
		    Cascade
		    Multivariable

3. Nature of controller	Continuous or discontinuous Analog or digital Numerical Supervisory/hierarchical Distributed	
4. Nature of plant	Continuous-time or discrete-time Sampled-data Linear or nonlinear	

5. Type of controlled variable	Servomechanism
	Process control
	    Attitude control, level control, temperature
	       control, etc.	

6. Nature of setpoint	Regulator Tracking	

7. Type of control law	P, PI, PID, compensator, state vector feedback, etc.
	Bang-bang (on-off)
	Linear or nonlinear
	Time-optimal, optimal
	    Finite-time settling, optimal regulator
	Adaptive
	    Model reference adaptive control, self-tuning
	       regulator
	Robust	

8. Analytical or design technique	Time-domain/State space or frequency domain Stochastic control system, H_∞, etc. Classical or modern	

(equations (1.3) and (1.5c) are accurate, and the parameter values, such as A, are precise) and our measurement of the disturbances is complete (the equivalent C at all water use sites is accurately measured), then the performance of the open-loop control system can be as good as or even better than the corresponding closed-loop system. A *feedforward control system* (see Fig. 1.9b) is an open-loop control system that determines the control input to the plant from measurements of the plant disturbance input. The path between the disturbance and control inputs is a feedforward loop because both variables are on the same input "side" of the plant. (See equation (1.5c).)

Nature of Controller: Discontinuous versus Control Control

All the controllers considered thus far are *discontinuous* in their operation. In particular, the control action (pump operation) or control law is *on-off*, also called *bang-bang*. The pump is either on (presumably at full speed and maximum displacement) or off (zero flow). This is in contrast with *continuous* control, in which the controller output (manipulated variable) is a continuous function of the controller input (error variable).

Figure 1.9

Feedforward Control System

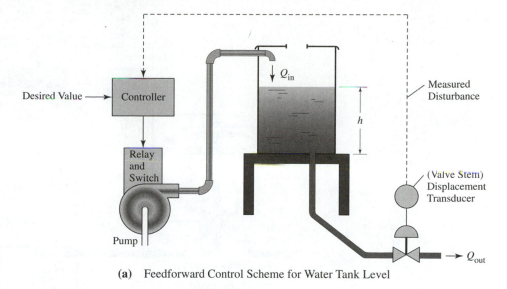

(a)   Feedforward Control Scheme for Water Tank Level

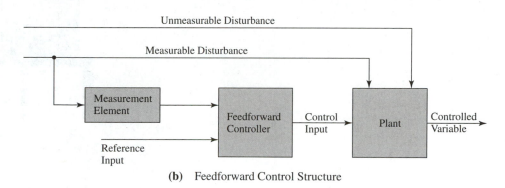

(b)   Feedforward Control Structure

Figure 1.10 shows an example of a continuous tank level control system. In this case the manipulated variable is the inlet flow control valve stem position. Thus the rate of water flow into the tank Q_{in} is continuously variable. A steady source of water supply is presumed.

Type of Controlled Variable

On the basis of the controlled variable, the water tank level control systems of Figs. 1.4a, 1.5, 1.9a, and 1.10 could be called *liquid-level control systems*. Other examples of controlled variables include attitude (as for a spacecraft attitude control system), pitch (as for an aircraft pitch control system), speed (of a DC motor, for example), position, force, torque, pressure, flowrate, current, voltage, magnetic flux, frequency, temperature, humidity, density, concentration, and chemical composition. In Fig. 1.10, if the flow into the tank is the controlled variable of interest, and the tank is viewed as part of the flow measurement system, then the control system illustrated in the figure can be described as a *flow control* system. Liquid-flow, liquid-level, and pressure and temperature controls are common in reactors, distillation columns, heat exchangers and liquid storage systems. These are typical components in what are often called process industries. More general classifications of control systems can be made on the basis of the type of controlled variable—for example, *process control* systems and *servomechanisms*.

Figure 1.10

**Continuous Control
of Tank Level**

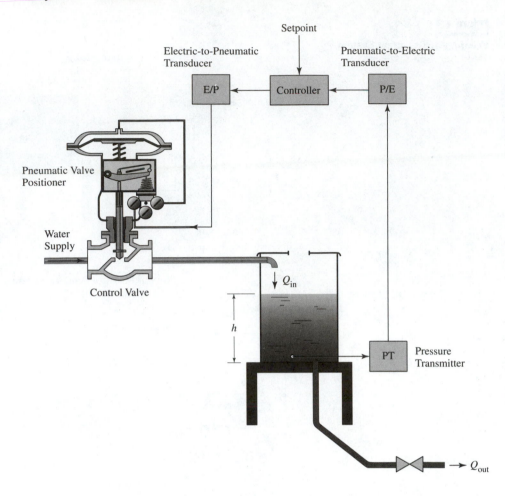

Servomechanism versus Process Control.   Strictly speaking, a servomechanism is a
feedback control system in which the controlled variable is a (mechanical) position; for
example, the linear position of a hydraulic piston or angular position of a motor shaft.
More generally, mechanical position is closely associated with velocity, acceleration,
force, torque, and (considering electromechanical devices) frequency, voltage, and current. In this context, process control generally refers to control systems with such other
controlled variables as liquid level, pressure, temperature, density, concentration, or
chemical composition.

An Example of a Servomechanism.   The printhead positioning system of a dot
matrix impact printer (see Fig. 1.11a) is an example of a servomechanism that should
be familiar to most of you. If you have at any time removed the cover of such a
printer to, say, change the ribbon, you must have observed most of the mechanical
components of the carriage transport system illustrated in Fig. 1.11b. The printhead
positioning system actually consists of both the carriage (lateral) positioning system
(the actual servomechanism) and a paper (longitudinal) positioning system (which
can be open loop, except perhaps for a paper-out sensor input). A common paper
transport system is the friction-feed system (not shown), in which the paper is
pressed against the platen by pressure rollers so that as the platen rotates, the paper
is carried along between the platen and the rollers. The printhead is mounted on a

Figure 1.11

Dot Matrix Printer and Print Head Servomechanism

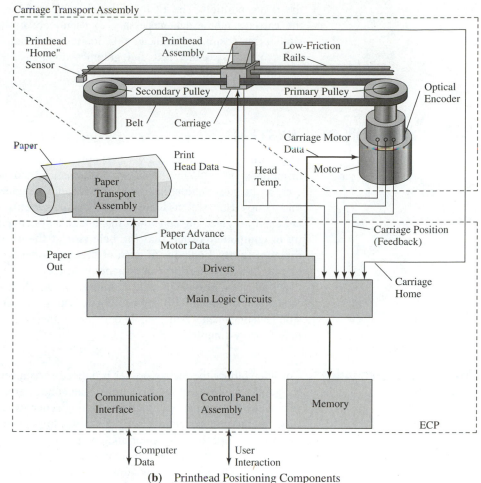

Carriage Transport Assembly

Printhead "Home" Sensor

Printhead Assembly

Low-Friction Rails

Secondary Pulley

Primary Pulley

Optical Encoder

Belt

Carriage

Carriage Motor Data

Paper

Print Head Data

Head Temp.

Motor

Paper Transport Assembly

Paper Advance Motor Data

Carriage Position (Feedback)

Paper Out

Drivers

Carriage Home

Main Logic Circuits

Communication Interface

Control Panel Assembly

Memory

ECP

Computer Data

User Interaction

(a) Dot Matrix Impact Printer

(b) Printhead Positioning Components

carriage that rides on low-frictionrails and is carried back and forth across the page by a slide (usually a timing belt) that constitutes the drive linkage between two pulleys and a stepping motor drive train.

The desired printhead position and the dot pattern to print at that position are part of the data sent by the computer to the printer's electronic control package (ECP). Typically the ECP is made up of the five major sections shown in Fig. 1.11b (see [5]): communication interface, drivers, control panel, memory, and main logic. Data interchange between the computer and the printer is handled by the communication interface. The control panel is where a person interacts with the printer (for example, to change operating modes such as fonts). Drivers are necessary to convert low-power digital signals (control output) from the ECP into voltage and current levels that can operate the actuators—motors and solenoids. The main logic unit includes one or more microprocessors that make the control decisions. For carriage positioning, the controller (ECP) output, that is, carriage motor data, is sent to the actuator (stepping motor) through the carriage motor driver. An optical encoder (usually integral with the stepping motor shaft) sends distance and direction (and even speed) information back to the ECP. There is also a printhead "home" sensor that informs the ECP when the carriage has reached the home position. The ECP logic unit utilizes these feedback signals to compute the carriage motor data necessary to accurately position the printhead at the desired print position. Printhead positioning resolutions of 0.0028 in. or better are now common, so that text and graphics dot densities of 360 dots per inch (dpi) are routinely realized.

1.3 DEFINITION OF A SYSTEM

The word *system* is used in a very general manner not only in engineering but in other disciplines as well—in chemistry, mathematics, and politics, for example. Usually, there is the presumption that the word's meaning is understood, as was the case before this section. However, as it applies to dynamic systems studied in this book in particular, and to engineering constructs in general, a system can be defined as **a thing made up of components such that the behavior of the overall combination can be predicted if (a) the behavior of each of the components can be predicted, and (b) the interaction between the components is known.**

This is a very important definition, as we shall see. The formation of system models, and practically all engineering activities, exploit this definition by dealing with simpler components and their interactions rather than with the entire system, which may be very complex.

Purpose of a Dynamic Study

Consider further the matrix printer printhead servomechanism of Fig. 1.11. The need for good knowledge of the behavior of the control object is even more critical in this case given the desired very high positioning accuracy. To design the controller for the carriage transport system, we need to study it so as to be able to predict in advance (using paper and pencil or monitor and computer) and with sufficient accuracy the printhead's response to a variety of inputs (such as the stepping motor shaft angular displacement pattern, the corresponding lateral displacements of the slide, and the carriage motion). In other words, *we need to know the dynamic characteristics or the dynamics* of the carriage transport. Thus if we consider the print head, the slide, and the carriage as components of the system (problem) under study, then the purpose of a dynamic investigation of this system is to satisfy items (a) and (b) in the definition, namely, to be able to predict the behavior of each component and the interactions among these components. This somewhat circuitous statement says, in effect, that the object of a dynamic study of a system is to be able to define that system. An important implication of this explanation is that through dynamic analysis we can define a system without the physical benefit of it, in other words, without having the matrix printer physically present, or if we do, not necessarily as one unit (perhaps the printer is yet to be built or assembled).

Stages of a Dynamic Study

How do we conduct a dynamic investigation? What do we need? These questions can be answered by outlining the various steps in a dynamic study. The first step is to *specify the system* or component to be studied.* Is it the entire matrix printer or just the carriage transport assembly? If it is the carriage transport, do we consider the relative motion of the printhead assembly or even that of the print wires? Thus the objective of the study or the problem (design of a printhead positioning controller) influences the

---

*Note that we can isolate a component of a system and regard it as another system, which may also have its own components.

specification of the system. In this book the product of system specification will be an abstract and simplified description of the system called a *physical model.**

The second step in a dynamic study is the *construction of a system model.* If one has the requisite resources, ability, and time, building or procuring the specified system is ideal. Usually a mathematical model consisting of a number of equations is constructed. The theory of system dynamics and *system decomposition*—the ability to break up a system into simpler components to which this theory can be applied—are utilized in this stage.

The third step is to *solve the model* or the equations. This process consists of subjecting the model to various inputs and examining the outputs in terms of the expected behavior of the system we specified. In other words, we study the *dynamic behavior* of the model. Is the behavior the desired one? Did we model a bidirectionally rigid (like a screw) drive in place of a one-way rigid (belt) drive? If we did and this is inadequate, then we have to go back and reconstruct the model.

The last step is *design.* Here we mold the model (choose parameters, set certain values, etc.) to achieve our objective—accurate printhead positioning. For instance, it may be that the settling time of the present stepping motor shaft at the command angular position is too long, thus requiring electronic damping of the motor or even a DC servomotor drive instead of the stepping motor system. The last alternative may require that we go all the way back to a new specification of the system or at least to the model construction stage.

Introducing the Block Diagram

The stages of a dynamic study can be summarized as shown in Fig. 1.12. The figure represents, in a general sense, a *block diagram.* The fact that we can represent the ideas contained in the preceding paragraphs with a diagram illustrates the universal applicability of this important tool! We could have been talking about the steps for baking a fruitcake or how to start your own business (see the practice problems in Section 1.5).

---

*The physical model itself can be a schematic diagram, a block diagram, a circuit diagram, a bondgraph, a flowchart, or a verbal description, for example.

Figure 1.12

Block Diagram (flowchart) of Stages of a Dynamic Study

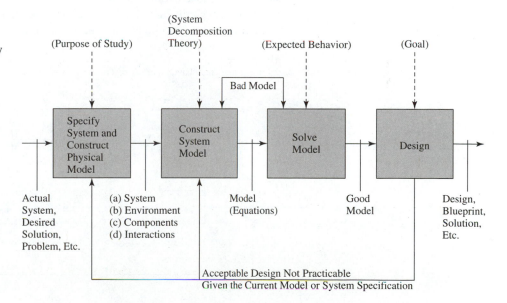

Figure 1.13

**Generalized Format
for a Block**

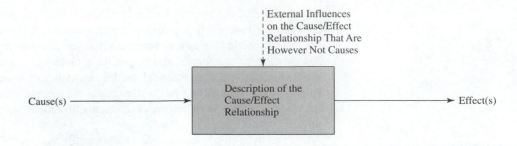

A block diagram is a sequence of causes and effects. It partitions whatever it is that is under investigation into subsystems or *blocks,* each of which can be studied more easily. The format for individual blocks in a block diagram is illustrated schematically in Fig. 1.13. The causes are shown as solid lines with arrows pointing to the block, and the effects are represented by solid lines with arrows pointing away from the block.* The actual causes and/or effects can be described on these lines if this makes the diagram clearer. The description of the cause-effect relationship is entered into the block and can take many forms depending on the application. For example, it can be a verbal description consisting of several statements, or an equation, or just a word such as DESIGN. In our current application, use of DESIGN in Fig. 1.12 relies on the prior description of the design stage, which connects the cause (a good model) to the effects (final design, blueprint, etc., or return to either system specification or model construction stages). It is also good practice to show other external influences on the cause-effect relationship on the block diagram, particularly if these are important but are not obvious from the causes and the description of the cause-effect relationships. In Fig. 1.12 the design activities depend on the goal or objective of the study, although this goal or objective is not by itself the cause for going into the design stage. Dashed lines are used to emphasize the fact that this is not a cause but a consideration in going from the cause to the effect.

We have given a general description of block diagrams. Specific applications of the block diagram are usually given special names. For instance, Fig. 1.12, which depicts a logical process or procedure (a computer program is another example) is more appropriately referred to as a flowchart. In these specific applications, special symbols may be used to represent particular operations such as a decision (typically a diamond-shaped block) in a flowchart or a comparison (summation) in a block diagram (see Fig. 1.14,

---

*Note that these lines may have branches that also represent the same effect(s) or cause(s) for other blocks.

Figure 1.14

**Block Diagram of Ancient
Water Clock**

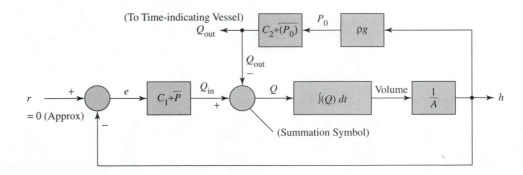

for example). The term *block diagram* is normally used to refer to an application for representing mathematical models. Here the causes and effects are actual variables or signals, and the blocks contain the operations that are performed on the input variables. Signals meeting at a junction (which is represented by a small circle) are conventionally added with their indicated signs but proceed unaltered along regular branches. Such an operational block diagram or (simply) block diagram is illustrated in Example 1.1, which considers the water clock that we introduced previously.

| Example 1.1 | Block Diagram of the Ancient Water Clock |

Draw a block diagram for the water clock device illustrated in Fig. 1.8. The major components are the valve and the regulating vessel (tank), whose characteristics were discussed in section 1.1. Let the pressure above the water in the regulating vessel be atmospheric, and let P be the gage pressure in the water supply line. Assume that the empirical valve parameter C of equation (1.2) is proportional to the valve opening e and that the regulating vessel is a uniform tank with cross-sectional area A.

Solution: See Fig. 1.14.

Note that the indicated relationships were determined as follows:

From equation (1.2),

$$Q_{in} = C\{P^{1/2}\} = C_1 e\{P^{1/2}\}$$

where C_1 is a proportionality factor

From equation (1.4),

$$Volume(t) = \int_0^t Q \, dt + Volume(0); \quad Q = Q_{in} - Q_{out}$$

From equation (1.5b),

$$h = \left(\frac{1}{A}\right) Volume$$

From equations (1.2) and (1.1),

$$Q_{out} = C_2\{P_0\}^{1/2}; \quad P_0 = (\rho g)h$$

| Example 1.2 | Automobile Engine Cooling System |

Most automobile engines are cooled by water circulating through channels in the engine block. The heated water rejects its heat to the atmosphere in the radiator, which functions essentially as a forced convection heat exchanger. The engine temperature is affected by the load on the engine. A component of this load is the engine speed, which is related to the throttle input. Assume that the water pump and the radiator fan are driven by the automobile engine, so that they are also affected by the throttle input. Further, most cooling system thermostatic controls work by recirculating some of the hot water leaving the engine block without letting it cool in the radiator. This allows the engine to warm up quickly under cold starting conditions. Draw a block diagram to describe in detail your understanding of the cooling system for a water-cooled automobile engine.

Figure 1.15

Automobile Engine Cooling System

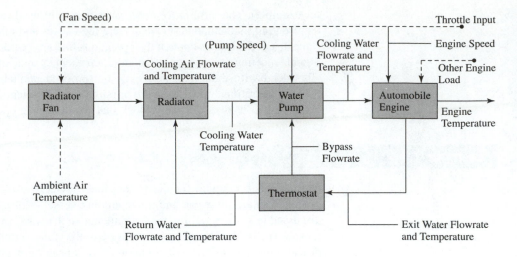

Solution: See Fig. 1.15.

Note that

1. The significant quantity in this system is the engine temperature. It is the object of the cooling system to maintain this temperature at tolerable levels at all times.
2. Although there are feedback loops within the system, the *overall* engine cooling system as shown in the figure is not a feedback control system—there is no controller and no genuine feedback of the engine temperature, and there are no independent actuators. (It is for this reason that many older model cars overheat in rush hour traffic—the stop-and-go driving puts heavy loads on the engine, thus raising its temperature. However, since the driving speeds are also low, the cooling water cannot reject the engine heat fast enough to the environment, especially on hot summer days.
3. The block diagram for a feedback-controlled automobile engine cooling system, typical of some later model cars is given as an exercise (see Practice Problem 1.9).

We shall now consider individually and in detail the stages of a dynamic investigation. Since this is also the framework of system dynamics, the rest of the book is divided into parts according to these stages. Stages one and two are considered in the rest of Part I, while *model solution* or stage three is taken up in Part II. Part III deals mostly with (control system) *design,* the last stage.

1.4 STAGE ONE: PHYSICAL MODELING

The object of system specification and physical modeling is to provide a description of the system that is as accurate as possible yet simple enough to allow subsequent analysis and design. In general, **a physical model is constructed by isolating a portion of the universe as the system of interest and then conceptually partitioning the behavior of this system into known components.** Two themes dominate the physical modeling and system modeling stages: *compromise* and *approximation*. The compromises are made on the original intent or goal of the study in order to specify a simpler system or an uncomplicated set of interactions at the boundary between the system and the rest of the world. The approximations are made in the descriptions of the behavior of the components of the physical model in order to take advantage of existing and well-developed theories as well as to minimize the level of complexity of required analytical tools. The important considerations at all times are how much compromise should be made and what approximations are valid. The ability to resolve the first consideration comes from practice and the exercise of mature judgment. The approximations are validated through an iterative process of model construction and model solution as indicated in Fig. 1.12. If the system model is constructed on the basis of invalid assumptions, the solution of the model when matched with the expected behavior will indicate a bad model, necessitating new approximations and the construction of a new model.

The first step in isolating the system from the rest of the world is inherent in *systems concept,* which is discussed subsequently. The second step is a matching process. One must have a sufficient array of theoretical behaviors with which to match the conceptual behavior components. Again, experience (developed through practice) is important here. We begin to build up our array of theoretical behaviors in Chapter 2.

Systems Concept: Illustration

We shall illustrate the preceding ideas with the dot matrix printer printhead servomechanism problem. First we need to specify the printhead positioning system. A broad interpretation of our goal would be that we want the printhead to move quickly to the commanded position on the page and remain there with very little oscillation in both the lateral and longitudinal directions. This requires not only that our system include both carriage and paper transport systems but that we consider the relative motion between the printhead assembly and its carriage. However, recognizing the high stiffness of a friction-fed paper drive system, we can assume exact longitudinal positioning by a discrete and fairly stiff actuator such as a stepping motor. Also we can assume that the carriage drive motor can faithfully reproduce the input angular step commands without introducing any significant dynamics of its own. With these compromises, we can isolate the carriage and belt (with pulley) as our system. (See Fig. 1.16.) Once we do this, everything else outside of the carriage transport becomes the environment.

Systems concept involves the process of isolating conceptually a part of the universe that is of interest and calling it the system while specifying the interactions between this system and the rest of the world, which is called the environment. The boundary between a system and its environment is imaginary, and as we saw above, it is a compromise. The interactions between our system and its environment are the angular motion of the primary pulley as derived from the angular step commands to the carriage motor, and some frictional resistance to the motion of the carriage, due to the rails. These particular interactions are due in part to the choice of the system boundary. With our choice of system, the relative motion between the printhead assembly and the carriage represents an internal interaction. In general, simplification of the interactions between the system and its surroundings, through the choice of the boundary, is often achieved at the expense of increased complexity of the system enclosed by the boundary. This brings up again the notion of compromise and the necessity to exercise good engineering judgment. Suppose, for example, we did not neglect the effect of the carriage motor dynamics on the printhead positioning. Then our system boundary would have enclosed the carriage motor as well. In place of angular motion of the primary pulley, the new interaction with the surroundings would be the step pulses to the carriage motor. However this minor simplification of the boundary interactions requires that we now deal with the problem of describing how a stepping motor behaves. Such

Figure 1.16

**Illustration of
Systems Concept**

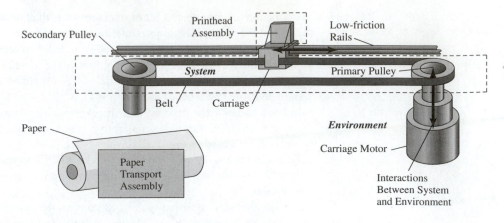

(a)   Systems Concept

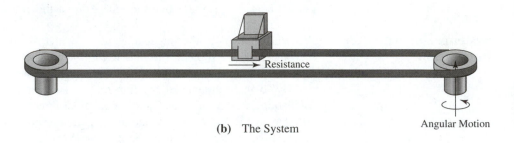

(b)   The System

an additional problem should not be undertaken unless it is determined to be necessary
for realizing the goals of the study adequately.

O n   t h e   N e t

- SC1.MCD: Some computational software packages such as Mathcad have unit capabilities. This file gives a quantitative example for equation (1.3) or (1.7a).
- Auxiliary exercise: Practice Problem 1.7.

1.5 PRACTICE PROBLEMS

1.1 Yes or no? A system whose output does not change with
time must be a static system. Explain your answer.

1.2* Hooke's law for a lumped spring can be written as
$F = k(x_2 - x_1)$, where F, x_1, x_2 are variables, and k is a parameter for the spring.

(a) Is this a static or a dynamic relation?

(b) As an element that can store potential energy, is a spring
a static or a dynamic element?

1.3 Consider a gun duel in which two types of guns are
available. Gun type A produces a visible flash on being fired
but is otherwise totally silent. Gun type B produces an
audible report but no light of any kind when fired. The com-
batants are to be separated by a distance of 300 m. The
speed of a bullet fired from either gun is the same and is 300

m/s. Consider now the life-and-death situation in which time
is measured in terms of human reflex periods (on the order
of ¼ second), and "cause" is the firing of a gun, and "effect"
is the reaction of the target combatant.

(a) What type gun will more likely produce a dynamic
cause-effect relationship? Explain your answer in detail.

(b) For the dynamic situation, which of the two basic
dynamic operators given in Table 1.1 would you say is
involved?

(c) What type gun would you rather your opponent had if
you were in such a duel?

1.4* In a rack-and-pinion steering system, the steering wheel
turns the pinion, which in turn moves the rack. The rack
is attached to the car wheels through some linkages so that
the car wheels essentially mimic the driver's steering
wheel.

Backlash as defined in Fig. 1.17 is necessary to allow the rack and pinion to mesh.

(a) Is the relation between the motions of the steering wheel and the vehicle wheels static or dynamic?

(b) Could backlash have a role in some accidents that result from loss of vehicle control at high speeds? If so, can you explain the role?

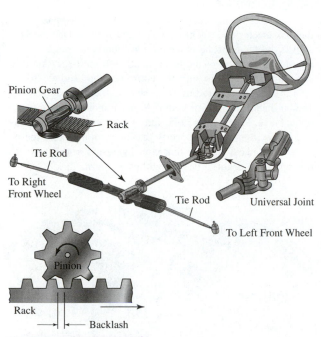

Figure 1.17 (Problem 1.4)

1.5 Consider a constant-speed conveyor onto which a compacting type material (bread dough for example) is fed from a hopper with adjustable gate height $u(t)$. The thickness of the material is measured at two stations as shown in Fig. 1.18. Determine the relation between the measurements h_1 and h_2 and the gate height u. Is this a static or a dynamic relation? If it is dynamic, what type of dynamic operator is involved?

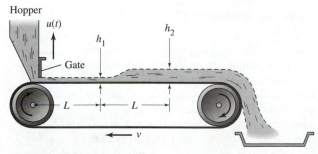

Figure 1.18 (Problem 1.5)

1.6* The number of marketable apples available each day in a supermarket noted for high-grade apples depends on the rate (apples per day) at which shoppers purchase the apples r_p, the spoilage rate of the apples r_s, and the rate at which new apples are brought in from the warehouse r_n. Derive an expression for the number of apples in the supermarket each day. Indicate whether the relation is dynamic or static, and if dynamic, what is the dynamic operator.

1.7 (a) If the unit of C is F (farads; [Mathcad: coul/volt]; see Table 1.2), R is Ω (ohm), J is kg · m^2, K_r is N · m, K_t is N/m, m is kg, and a = linear acceleration, f = force, i = current, x = displacement, T = torque, V = voltage, and θ = angle, determine whether the following equations are dimensionally correct:

(i) $m\dfrac{d^2x}{dt^2} + K_t x = f$

(ii) $f = J \cdot a$

(iii) $J\left(\dfrac{d\theta}{dt}\right)^2 + K_r\theta^2 = T$

(iv) $\dfrac{1}{C}\dfrac{dV}{dt} + \dfrac{V}{R} = i$

(b) What are the units of β in the following equations:

(v) $J\dfrac{d^2\theta}{dt^2} + \beta\dfrac{d\theta}{dt} + K_r\theta = T$

(vi) $m\dfrac{d^2x}{dt^2} + \beta\left(\dfrac{dx}{dt}\right)^2 + K_r x = f$

1.8 (a) Give one example of a system process/device found within a household that satisfies the following classification:

(i) Manual open-loop control system

(ii) Manual closed-loop control system

(iii) Automatic open-loop control system

(iv) Automatic closed-loop control system

(v) Feedforward control system

(vi) Discontinuous control system

(vii) Continuous control system

(b) In each of the closed-loop control examples you gave in part a, indicate the *controlled variable,* and then determine to which of the categories, *process control system* or *servomechanism,* broadly interpreted, the example belongs.

1.9 Employ feedback control to modify the automobile cooling system of Example 1.2 so as to prevent engine overheating in slow traffic.

1.10* Draw a block diagram of the water tank manual closed-loop control system shown in Fig. 1.5a. Include as many potential disturbance inputs to the system as you can think of, and identify the *controller,* the *actuator,* and the *measurement element.*

1.11 Repeat Problem 1.10 for the automatic closed-loop control system of Fig. 1.5b.

1.12 Consider the process of spooning food into the mouth.

(a) Identify in your own terms the components of the feedback control system involved in this process, that is, indicate the *controller,* the *control object,* the *controlled variable,* the *feedback element,* and the *setpoint.*

(b) Put the components in a block diagram to explain the dynamics of eating.

1.13 Most long distance runners practice self-suggestion as a means of improving or maintaining their speed when they sense physical exhaustion. Assuming that it is possible to increase running speed by autosuggestion and also considering muscle fatigue, describe in detail (using a block diagram for example) the attempt of a runner to break an existing long distance record as a feedback control process.

1.14 Consider the airline company AIRBROKE and its management in a price war situation. The airline has promised its customers to set its fares on all its routes at 5% below lowest market price. The management intends to keep this promise even when the airline is losing money, in order to retain its clients and to remain in business. The passengers on their part generally fly the airline with the lowest fares. AIRBROKE has the capacity to carry 2000 passengers per day (total for all routes). Due to other investments, AIRBROKE can remain operational as long as its losses remain below $1 million per quarter. Draw a block diagram to explain the operations of AIRBROKE in the competitive environment, and show whether or not this is a feedback control system.

1.15 Draw a functional block diagram that is descriptive of each of the following:

(a) Your favorite recipe such that others can prepare it successfully.

(b) An assembly robot that is programmed to pick up bolts from a supply tray and insert them in threaded holes in a large piece of machinery. The robot picks up a bolt by feeling along the tray from one end to the other. Similarly, it identifies a threaded hole by remembering the location of the last bolted hole and dragging the bolt along the surface of the machine piece starting from that location.

(c) A thermostat-controlled home heating system.

(d) The registration process at your university or college such that a new student will have little difficulty in getting into all her (his) classes.

1.16* For the water supply system illustrated in Fig. 1.1, suppose there is interest in the day-to-day level of water in the tank. Partition this part of the universe into a system and its environment, and specify the interactions across the system boundary.

1.17 Fig. 1-19 shows components typical of a read-only optical disk drive (used with compact disk read-only memory (CD-ROM), for example). A narrow beam of light from a laser is focused to a small spot (about 1 μm in diameter) on the surface of the spinning disk through a series of lenses and other optical elements. The data that are encoded on the disk surface as a pattern of reflectivity cause varying amounts of light to be reflected back along the original path to the beam splitter, where the light is diverted through another set of lenses and optical elements to a photo detector. The laser beam is kept locked onto the desired part of the disk (sector and track) by voice-coil motors. One motor adjusts the focus of the objective lens, while the other adjusts the position of the beam relative to the track. Another servo motor is used to shift the entire reading head along rails to different desired tracks. The track density can be 16,000 tracks per inch or better. A fourth motor spins the optical disk. Partition this unit into a not too complex system and its environment and specify the interactions across the system boundary. Assume that your interest is in the positioning of the beam to desired addresses on the disk.

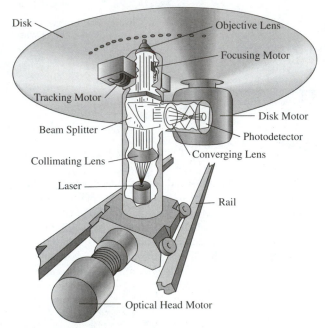

Figure 1.19 (Problem 1.17)

1.18 Consider a nuclear power plant, its structure, and its interaction with the community in which it is situated: electric power delivery and demand, cooling water discharge to the river, fishes, pollution, reactor, control rods, feed water and steam flows, condenser, turbine, generator, and the like. Apply the systems concept to this power plant, that is, isolate the system from the rest of the world and identify the internal and external interactions.

1.19* Consider a car with some passengers, sitting on car seats and being driven along a bumpy road. There is a driver controlling the car (speed and heading). The steering wheel is connected by linkages to the car wheels. The engine vibrates on its mounting, and the vibration is related to the engine speed. The car frame, passengers, seats, etc., constitute the mass supported above the shock absorbers and springs. The engine, wheels, axles, etc., represent the mass supported below. The air in the tires also provides some damping and compliance. Your goal is to evaluate the ride quality of the car for this particular road. Starting with Fig. 1.20, make a sketch of this part of the universe and isolate a proper system for study. Indicate the interactions between the system and its surroundings. Assume the car has identical front and rear suspension units.

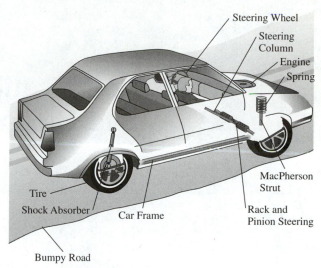

Figure 1.20 (Problem 1.19)

Chapter
2
Specification of Dynamic Systems and Behavior

TO CONTINUE WITH THE INTRODUCTION of *physical modeling* and subsequently *system modeling,* we first need to develop knowledge of some theoretical behaviors that will allow us to recognize the corresponding physical behavior components, including potential interactions across the boundary separating the components from the environment or other components, in systems of interest. In this chapter we shall look at theoretical behaviors typical of mechanical and electrical components and consider simple examples of physical modeling of systems containing such components. We shall then introduce the mathematical modeling of dynamic systems by exploring the process of *system decomposition* and some of the approximations of behavior that are often made in system modeling.

2.1 INTRODUCING MECHANICAL BEHAVIOR COMPONENTS

In the discussion of the dot matrix impact printer servomechanism and in some of the practice problems of Chapter 1 dealing with systems concept, we referred to such notions as rigidity and stiffness, friction in guide rails, supported masses, and air damping and compliance, with the assumption that the types of component behaviors associated with these terms were at least intuitively understood if not known theoretically. In this section we shall review simple but formal descriptions of the behavior of such elementary mechanical components as springs, masses, and frictional resistance.

Compliance and Inertia

The material properties of compliance (represented by such quantities as modulus of elasticity, modulus of rigidity, and bulk modulus, or derived spring constant) and inertia (represented by mass or derived moment of inertia) are fundamental to the behavior of mechanical system components, especially when such systems are in motion. Since we usually differentiate between translatory motion and rotational motion, we consider two types of each of these behavior components.

The Translational Spring

The (tensile) force exerted by a spring is a function of the extension (stretch) of the spring:

$$F = f(\Delta x) \tag{2.1a}$$

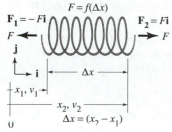

$$F = f(\Delta x)$$

$F_1 = -F\mathbf{i}$ $F_2 = F\mathbf{i}$

$$\Delta x = (x_2 - x_1)$$

(a)   Coil Spring Under
Tensile Load

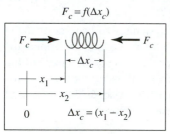

$$F_c = f(\Delta x_c)$$

$$\Delta x_c = (x_1 - x_2)$$

(b)   Symbolic Spring
in Compression

Figure 2.1

Ideal Translational Spring

As shown in Fig. 2.1a, given a consistent sign convention for forces and displacements,* the tensile force F is positive when the relative displacement of the ends of the spring is positive (extension). The behavior of a spring can also be described in terms of the compressive force and the compression (see Fig. 2.1b):

$$F_c = f(\Delta x_c) \tag{2.1b}$$

To avoid confusion these equations should always be accompanied by the corresponding symbolic representation of the spring force (free body diagram of the spring) such as shown in Fig. 2.1b.

In this book translational springs will be represented symbolically as in Fig. 2.1, although only coil springs actually look like that. Examples of other mechanical components that function as springs are shown in Fig. 2.3. The equivalent spring constant (discussed below) is given in each case. Most elastic materials obey (to some degree) Hooke's law, which (considering the case of a single dominant (nonzero) component of stress) states that stress is proportional to strain:

$$\sigma = E\epsilon \quad (E = \text{modulus of elasticity}) \tag{2.2a}$$

$$\tau = G\gamma \quad (G = \text{modulus of rigidity}) \tag{2.2b}$$

where σ is the (normal) stress, ϵ is the corresponding strain, τ is the shear stress, and γ is the corresponding shear strain. Consider the simple case of a homogeneous material with a constant but very small (compared with the length) cross-sectional area A (a rod or cable for instance). If we assume that the strain is uniform along the length L, equation (2.2a) gives the normal force F,

$$\left(\frac{F}{A}\right) = E\left(\frac{\Delta x}{L}\right) \tag{2.2c}$$

or

$$F = \left(\frac{AE}{L}\right)\Delta x = f(\Delta x) \tag{2.2d}$$

where (F/A) is the normal (average) stress, $(\Delta x/L)$ is the strain, and Δx is the stretch in the material. The *spring constant* k is defined by

$$F = k\Delta x \tag{2.3a}$$

so that for the rod or cable,

$$k = AE/L \tag{2.3b}$$

Since the changes in length and in the quantity (A/L), under normal loads, are small for a rod of the usual materials, L and indeed k are essentially constants, and equation (2.3a) is directly applicable. For long strips or cables, or rods of highly deformable material, the change in length or in (A/L) is normally substantial, so that equation (2.2c) should be used. A procedure for determining the instantaneous cable length and stretch is presented in Section 3.2.

---

*Figure 2.1 assumes the free length L of the spring is zero (ideal *point* spring). Otherwise, consider the position of the spring's right end as $(L + x_2)$ instead of x_2. The expressions $\Delta x = x_2 - x_1$ and $\Delta x_c = x_1 - x_2$ then remain valid.

In terms of force and displacement, the behavior of a spring as given by equations (2.1) through (2.3b) is static. However, we shall see that the spring is often encountered in situations in which the input to the component can be regarded as velocity. (The matter of what constitutes input and output for a given component is discussed in Section 3.1.) Thus given the spring end velocities v_1 and v_2 (refer to Fig. 2.1a), equation (2.1a) becomes

$$F = f\left(\int [v_2 - v_1]\, dt\right) \tag{2.4a}$$

and equation (2.3a) can be replaced by

$$F = k \int (v_2 - v_1)\, dt \tag{2.4b}$$

From our discussion of time integration and dynamic systems, we can see that the spring behavior is dynamic when the input is velocity. In the preceding developments, F, x, and v represent the magnitudes of parallel vectors or the same components of the force, displacement, and velocity vectors. In general, this relationship will be true for the other behavior components to be discussed subsequently. Also, in Fig. 2.1a and throughout the text, boldface symbols are used to represent vectors.

The Torsional Spring

The translational spring discussed in the previous section is associated with translatory motion. For rotary motion we have the torsional spring, for which the torque T is a function of the angular twist $\Delta\theta$:

$$T = f(\Delta\theta) \tag{2.5}$$

For a linear torsional spring, the torque is given by

$$T = k_r \Delta\theta \tag{2.6a}$$

where k_r is the torsional spring constant. In terms of the rotary speeds of the spring ends (refer to Fig. 2.2 and note the implied sign convention*), the linear torsional spring behavior is

$$T = k_r \int (\omega_2 - \omega_1)\, dt \tag{2.6b}$$

Note that the dimensions of the torsional spring constant are different from those of the translational spring. For instance, in metric units, k in equation (2.3a) has units of (N/

---

*The double arrow and directed partial circle notations used for the torque in Fig. 2.2 are redundant, since one can be obtained from the other by the right-hand rule. Only the double-arrow notation will be used in the rest of the book.

Figure 2.2

Ideal Torsion Spring

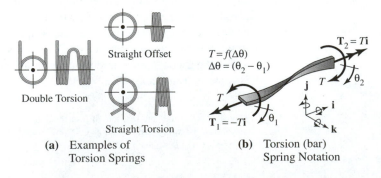

(a) Examples of
      Torsion Springs

(b) Torsion (bar)
      Spring Notation

m), whereas k_r in equation (2.6a) has units of (N · m). Although the torsional spring can be an actual torsion spring (Figs. 2.2a and 2.3a), in general, it is a flexible (compliant) shaft or link (see Figs. 2.2b, 2.3b, 2.3c, 2.3e, and 2.3f) within a mechanical arrangement.

The Translational Mass

Because all matter has some mass, so do components of engineering systems. A mechanical system component is generally regarded as a mass element if the behavior of the component due to its *inertia* is significant relative to other characteristics. Thus, in most applications, a spring is considered only in terms of its compliance unless it is also a very heavy spring. We can describe *inertia* as the reluctance of a body to change its current state of motion. An object at rest begins to move only when a force is applied to it. Similarly, a mass moving at a constant speed in a straight line will change its direction and/or its speed only if a new force is applied to it. For instance, a flywheel is a large mass that once set in motion will continue to spin for a long time, thereby storing energy for later use. The behavior of a mass is described by Newton's second law of motion, which states that *the (vector) change in momentum of a particle due to an applied force is proportional to the impulse of that force*. The *momentum* of a particle of mass m with velocity $\mathbf{v}$ is $m\mathbf{v}$, and the *impulse* of a force is equal to the integral of the force with respect to time. Thus the mathematical statement of Newton's second law is

$$m\mathbf{v} - (m\mathbf{v})_0 = \int_0^t \mathbf{F}\, dt \tag{2.7}$$

By differentiating equation (2.7), we can write the second law as

$$\mathbf{F} = \frac{d}{dt}(m\mathbf{v}) \tag{2.8a}$$

If the mass of the particle is constant (an assumption we shall generally make)* equation (2.8a) yields the popular expression

$$\mathbf{F} = m\frac{d}{dt}(\mathbf{v}) = m\mathbf{a} \tag{2.8b}$$

where $\mathbf{a}$ is the acceleration of the particle. It is important to note that equation (2.8b) is only a special case of equation (2.8a). An example of a situation in which it is not appropriate to use equation (2.8b) is in describing the dynamics of a rocket, since much of the mass of the rocket at any time is made up of the fuel that is constantly being burned to produce the thrust that propels the rocket forward. Even here, equation (2.8a) cannot be applied directly, since this is effectively a system of particles that is exchanging mass with its environment (see Practice Problem 2.18). Also note that $\mathbf{F}$ is the resultant of all the forces applied to the particle, and Newton's second law applies to motion with respect to an *inertial (fixed) frame* of reference. For a rigid body, $\mathbf{v}$ is the velocity of the mass center measured with respect to a the fixed reference point. The reader is referred to a dynamics text such as reference [6] for detailed explanations of the more complex applications of Newton's second law.

We shall be concerned mostly with motion in one coordinate direction, so that the quantities in equations (2.7) or (2.8) do not have to be vectors but can represent the same components of the respective vectors. If we regard the force on a mass as the input and the resulting change in motion or the velocity as the output, then according to equation (2.7) the behavior of a mass is dynamic.

---

*We shall ignore mass changes due to relativistic effects.

Coil Spring

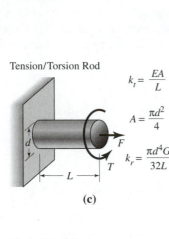

$$k_t = \frac{Gd^4}{64R^3N}$$

$$k_r = \frac{Ed^4}{128RN}$$

N = Number of Active Coils

(a)

Cantilever Beam/Rectangular Bar

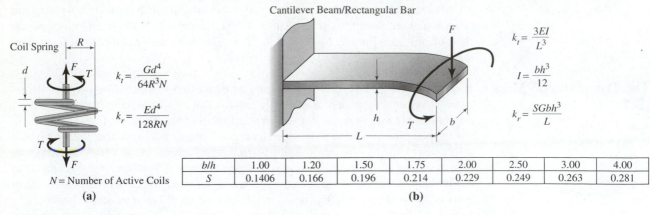

$$k_t = \frac{3EI}{L^3}$$

$$I = \frac{bh^3}{12}$$

$$k_r = \frac{SGbh^3}{L}$$

b/h	1.00	1.20	1.50	1.75	2.00	2.50	3.00	4.00
S	0.1406	0.166	0.196	0.214	0.229	0.249	0.263	0.281

(b)

Tension/Torsion Rod

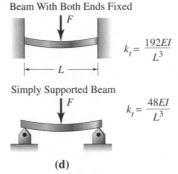

$$k_t = \frac{EA}{L}$$

$$A = \frac{\pi d^2}{4}$$

$$k_r = \frac{\pi d^4 G}{32L}$$

Beam With Both Ends Fixed

$$k_t = \frac{192EI}{L^3}$$

Simply Supported Beam

$$k_t = \frac{48EI}{L^3}$$

Cylindrical Rubber Spring in Shear

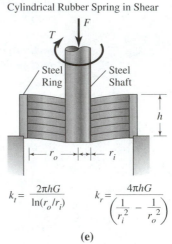

Steel Ring   Steel Shaft

$$k_t = \frac{2\pi hG}{\ln(r_o/r_i)}$$

$$k_r = \frac{4\pi hG}{\left(\dfrac{1}{r_i^2} - \dfrac{1}{r_o^2}\right)}$$

(c)   **(d)**   **(e)**

Clock (Spiral) Spring

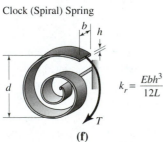

$$k_r = \frac{Ebh^3}{12L}$$

Rubber (Elastomeric) Spring in Compression

Plate Area A

$$k_t = f(F, A, h, H)$$

$$H = (\text{Durometer Hardness})$$

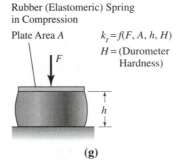

Oil (Hydraulic) Spring

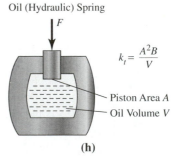

$$k_t = \frac{A^2 B}{V}$$

Piston Area A
Oil Volume V

(f)   **(g)**   **(h)**

Definitions:

K_t = translational spring constant
K_r = torsional spring constant
E = modulus of elasticity
G = modulus of rigidity
B = bulk modulus

Air (Pneumatic) Spring

$$k_t = \frac{A^2 P}{V}$$

Bellows   Piston

Volume V
Pressure P

Volume V
Pressure P

Piston Area A

Area A Diaphragm

(i)

Figure 2.3

Examples of Practical Springs

Rotary Inertia   Newton's second law also applies to the rotation of particles and rigid bodies. In this case we talk about angular momentum and the resultant applied torque or moment. Although a more detailed treatment of the material covered here is available in most dynamics books (see [6] for example), a brief review of this subject should help provide the proper perspective on the applications discussed in this text.

The resultant moment about a fixed point (inertial reference) due to all forces with resultant **F** on a body is

$$\mathbf{T} = \mathbf{r} \times \mathbf{F} \quad \text{(where × implies vector cross product; see Appendix B)} \tag{2.9}$$

r is the position vector from the fixed point to the point of application of the resultant force **F**. From equation (2.8a),

$$\mathbf{T} = \mathbf{r} \times \left\{ \frac{d}{dt}(m\mathbf{v}) \right\} = \frac{d}{dt}(\mathbf{r} \times m\mathbf{v}) \tag{2.10a}$$

since

$$\left(\frac{d}{dt}\mathbf{r} \right) \times m\mathbf{v} = \mathbf{0}$$

If we define the angular momentum as (the moment of momentum),

$$\mathbf{H} = \mathbf{r} \times m\mathbf{v} \tag{2.10b}$$

then the corresponding Newton's second law for angular motion is

$$\mathbf{T} = \frac{d}{dt}\mathbf{H} \tag{2.11}$$

For a rigid body rotating about a fixed point on the body or about its mass center with angular velocity **ω**, the instantaneous velocity of a mass particle m_i with respect to the fixed point or the center of mass (Fig. 2.4) is

$$\mathbf{v}_i = \boldsymbol{\omega} \times \mathbf{r}_i \tag{2.12a}$$

where the position vector $\mathbf{r}_i$ is measured relative to the fixed point or mass center. Substitution of equation (2.12a) into equation (2.10b) and integration over all the particles making up the body yields the angular momentum for a rigid body (see [6] for example):

$$\mathbf{H} = \int \{\mathbf{r} \times (\boldsymbol{\omega} \times \mathbf{r})\}\, dm \tag{2.12b}$$

Expansion of equation (2.12b) into its components in the x-y-z frame that is attached to the body at the mass center or at the fixed point (see Fig. 2.4) gives

$$\mathbf{H} = \mathbf{i}\left\{ \omega_x \int (y^2 + z^2)\, dm - \omega_y \int (xy)\, dm - \omega_z \int (xz)\, dm \right\}$$
$$+ \mathbf{j}\left\{ -\omega_x \int (yx)\, dm + \omega_y \int (z^2 + x^2)\, dm - \omega_z \int (yz)\, dm \right\} \tag{2.13a}$$
$$+ \mathbf{k}\left\{ -\omega_x \int (zx)\, dm - \omega_y \int (zy)\, dm + \omega_z \int (x^2 + y^2)\, dm \right\}$$

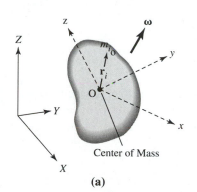

Center of Mass

(a)

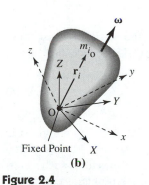

Fixed Point

(b)

Figure 2.4

Definition of Angular Momentum

or defining the (mass) moments of inertia J_{xx}, J_{yy}, J_{zz}, and the products of inertia J_{xy}, J_{xz}, J_{yz}, we have

$$
\begin{aligned}
\mathbf{H} = \ &\mathbf{i}(J_{xx}\omega_x - J_{xy}\omega_y - J_{xz}\omega_z) \\
&+ \mathbf{j}(-J_{xy}\omega_x + J_{yy}\omega_y - J_{yz}\omega_z) \\
&+ \mathbf{k}(-J_{xz}\omega_x - J_{yz}\omega_y + J_{zz}\omega_z) \\
= \ &\mathbf{i}H_x + \mathbf{j}H_y + \mathbf{k}H_z
\end{aligned}
\tag{2.13b}
$$

In the general case in which the x-y-z coordinate system is rotating with angular velocity $\boldsymbol{\Omega}$, which may be different from the angular velocity of the body $\boldsymbol{\omega}$, the rate of change of the angular momentum is given by

$$
\frac{d}{dt}\mathbf{H} = \left(\frac{d}{dt}\mathbf{H}\right)_{xyz} + \boldsymbol{\Omega} \times \mathbf{H}
$$

so that for the rigid body equation (2.11) becomes

$$
\mathbf{T} = \left(\frac{d}{dt}\mathbf{H}\right)_{xyz} + \boldsymbol{\Omega} \times \mathbf{H}
\tag{2.14a}
$$

In terms of components, equation (2.14a) gives

$$
\begin{aligned}
T_x &= \dot{H}_x - H_y\Omega_z + H_z\Omega_y \\
T_y &= \dot{H}_y - H_z\Omega_x + H_x\Omega_z \\
T_z &= \dot{H}_z - H_x\Omega_y + H_y\Omega_x
\end{aligned}
\tag{2.14b}
$$

Note again that these are the sums of all the applied torques in a given direction. The components of $\mathbf{H}$ are given by equation (2.13b).

For any fixed point of rotation, there is one set of axis directions attached to the body at this point, called the *principal axes*, for which the products of inertia are zero. If the x-y-z axes coincide with the principal axes, then equation (2.14b) becomes

$$
\begin{aligned}
T_x &= J_{xx}\dot{\omega}_x - J_{yy}\Omega_z\omega_y + J_{zz}\Omega_y\omega_z \\
T_y &= J_{yy}\dot{\omega}_y - J_{zz}\Omega_x\omega_z + J_{xx}\Omega_z\omega_x \\
T_z &= J_{zz}\dot{\omega}_z - J_{xx}\Omega_y\omega_x + J_{yy}\Omega_x\omega_y
\end{aligned}
\tag{2.15}
$$

Further, if motion takes place about only one of the axes, say, the z-axis, and we choose $\boldsymbol{\Omega} = \boldsymbol{\omega}$, then $\omega_x = \Omega_x = 0$, and $\omega_y = \Omega_y = 0$, and equation (2.15) reduces to

$$
T_z = J_{zz}\dot{\omega}_z \quad \text{or} \quad T = J\frac{d}{dt}\omega
\tag{2.16}
$$

Mass moments of inertia that can be used with equation (2.15) or (2.16) are given in Fig. 2.5 for some solids of fairly regular or common shapes, along with the mass center for such bodies. More extensive data can be found in some dynamics texts such as reference [6]. Although equation (2.16) is applicable to many of the problems considered in this book, a three-dimensional analysis (equation (2.11) or (2.14)) must be used whenever it is necessary to do so.

Figure 2.5

Mass Properties of Some Homogeneous Solids

Body of Mass m	Mass Center G	Mass Moments of Inertia
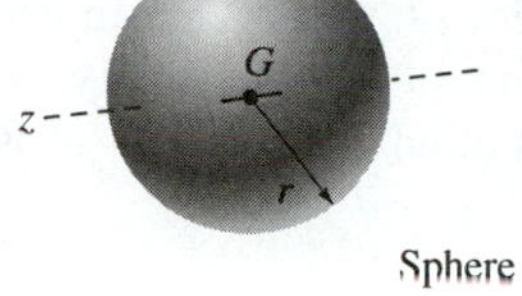 Sphere	–	$J_{zz} = \frac{2}{5}mr^2$
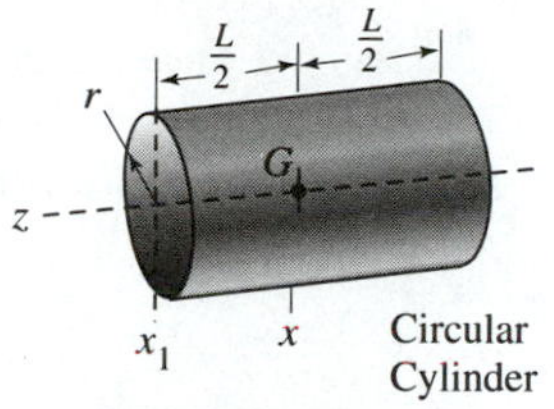 Circular Cylinder	–	$J_{xx} = \frac{1}{4}mr^2 + \frac{1}{12}mL^2$ $J_{x_1x_1} = \frac{1}{4}mr^2 + \frac{1}{3}mL^2$ $J_{zz} = \frac{1}{2}mr^2$
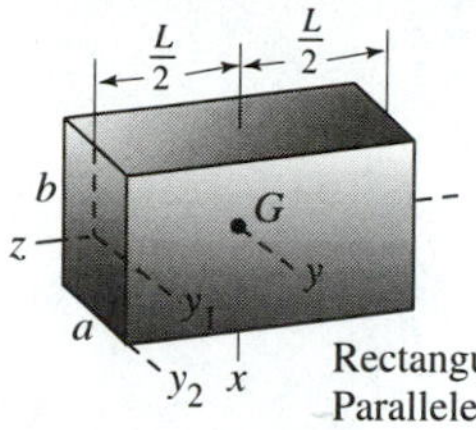 Rectangular Parallelepiped	–	$J_{xx} = \frac{1}{12}m(a^2 + L^2)$ $J_{yy} = \frac{1}{12}m(b^2 + L^2)$ $J_{zz} = \frac{1}{12}m(a^2 + b^2)$ $J_{y_1y_1} = \frac{1}{12}mb^2 + \frac{1}{3}mL^2$ $J_{y_2y_2} = \frac{1}{3}m(b^2 + L^2)$
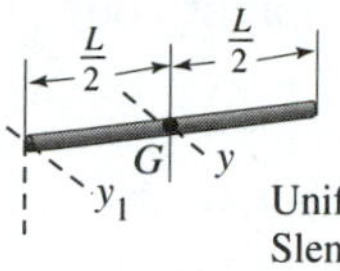 Uniform Slender Rod	–	$J_{yy} = \frac{1}{12}mL^2$ $J_{y_1y_1} = \frac{1}{3}mL^2$

Mechanical Damping

In mechanical systems, damping generally refers to any resistance to motion that results in its attenuation. Unlike masses and springs that store energy, elements that cause damping dissipate energy (and are static elements). We shall consider the energy viewpoint for the behavior of components later.

Ideal Damper/Viscous Friction

An example of a physical damper is the automobile shock absorber cylinder (usually used in combination with a large coil spring in the front of the vehicle and a leaf spring in the rear (see Fig. 2.6a). We shall represent such translational dampers symbolically as shown in Fig. 2.6b, by a dashpot. The automobile cylinder develops a force (F) that *opposes* the motion of the car end (v_2) relative to the tire end (v_1). By this means, large motions in the wheels result in only small oscillations of the car frame (they are attenuated or damped out) and, ultimately, of the car occupants.

Mathematically the relation for the ideal damper is

$$F = \beta v_{rel} \tag{2.17}$$

where β is the damper constant, and v_{rel} is the relative velocity between the two ends of the damper. For the notation of Fig. 2.6b, $v_{rel} = (v_2 - v_1)$. The majority of elements described as dampers are not physical dampers but are called dampers because

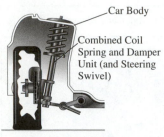

(a)   Automobile Shock Absorber

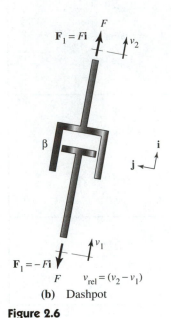

(b)   Dashpot

Figure 2.6

Translational Physical Damper and Dashpot

they obey relation (2.17). This type of damping often appears when two bodies separated by a viscous medium move relative to each other.

In Fig. 2.7a body B is moving with velocity v_2, while body A has velocity v_1. The part of the medium right next to A will have velocity equal to v_1, whereas the part next to B will move along with B with velocity v_2. In between, there is velocity distribution from v_1 to v_2. Because of friction or the viscosity of the medium μ, a shear stress tending to oppose the relative motion of the layers of the medium is set up. This stress is given by

$$\tau = \frac{F}{A} = \mu \frac{\partial v}{\partial y} \tag{2.18}$$

If the velocities of the bodies are not too high, the velocity gradient $(\partial v / \partial y)$ is linear, as shown in Fig. 2.7a, so that

$$\frac{\partial v}{\partial y} = \frac{v_2 - v_1}{h}$$

Then

$$\frac{F}{A} = \frac{\mu}{h}(v_2 - v_1) \text{ or } F = \left(\frac{\mu A}{h}\right)(v_2 - v_1) = \beta v_{\text{rel}} \tag{2.19}$$

This force opposes the motion of body B. For body A, the relative velocity is $(v_1 - v_2)$, so that the force opposing the motion of body A is $-F$, which is in agreement with Newton's third law.

Note that the phenomenon just described also represents the behavior of the physical damper. If we consider the opposite ends of the damper as the two bodies A and B that are subject to the applied forces F_A and F_B, respectively, as shown in Fig. 2.7b, then the force opposing body A is $F_{dA} = \beta(v_1 - v_2)$, while the force opposing body B is $F_{dB} = \beta(v_2 - v_1)$. Because an ideal damper has no mass, the resultant force on each body must be zero by Newton's second law. Thus $F_A = -F_{dA}$ and $F_B = F_{dB}$. But $F_{dA} = -F_{dB}$. Hence $F_A = F_B = F$, as shown in Fig. 2.6b.

Corresponding expressions can also be developed for the frictional resistive torque in rotary motion. An example of a physical rotary viscous damper is a fluid clutch (Fig. 2.8), which is used to transmit rotary motion from one shaft to another without mechanical coupling. In simple terms, rotation of the input shaft sets up shear stresses in the fluid, as described above, which drive the output shaft. The frictional torque resisting the relative motion of the shafts (see Fig. 2.8) is given by

$$T = \beta_r(\omega_2 - \omega_1) \tag{2.20}$$

Other Types of Mechanical Resistance

Viscous friction is always present to some degree whenever two bodies move relative to each other. Note that the medium can be air, a liquid, or even a solid; and the thickness h can be very small (such as a fluid film) or quite large. When the relative velocity between a fluid medium and a body is high, the relation between the viscous *drag* force on the body and the relative velocity is not linear. Rather, the force depends on the relative velocity raised to some (usually the second) power.

$$F = \beta_h v_{\text{rel}} |v_{\text{rel}}| \tag{2.21}$$

This type of damping, which we call *hydrodynamic damping,* is actually more common than the damper behavior given by equation (2.19), but it is described by a more com-

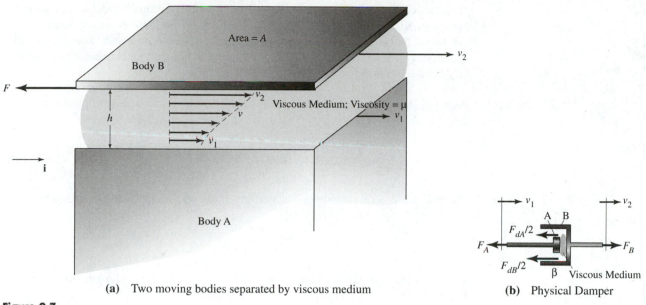

(a)   Two moving bodies separated by viscous medium   **(b)**   Physical Damper

Figure 2.7

**Viscosity and
Viscous Damping**

plex expression. Other types of frictional resistance found in mechanical systems include dry friction or Coulomb damping, magnetic damping, and internal or structural
damping. The Coulomb damping force F on a body moving with respect to another
with relative velocity v_{rel} can be represented by

$$F = fN\ \text{sign}(v_{rel}) \tag{2.22}$$

where f is the coefficient of (dynamic) friction, and N is the normal (perpendicular to
the direction of motion) reaction force between the bodies.

Figure 2.8

Simple Fluid Clutch

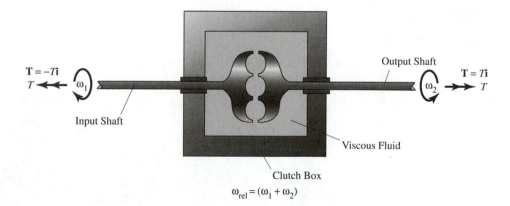

$$\omega_{rel} = (\omega_1 + \omega_2)$$

2.2 SOME ELECTRICAL BEHAVIOR COMPONENTS

The behavior of components of electrical systems is often described in terms of the
voltage-current characteristics of these components. An electrical circuit is a closed
path or combination of paths through which current can flow. It is generally composed

of circuit elements interconnected by wires. Thus the physical model of an electrical system is often represented by a circuit diagram. A wire here is an idealized perfect conductor, that is, all points along the wire are presumed to be exactly at the same voltage. Figure 2.9 summarizes the symbolic representation and ideal behavior of elementary circuit components.

Ideal Resistor   A circuit element recognized as a resistor (Fig. 2.9b) is characterized by the voltage-current relationship known as Ohm's law, a static expression:

$$i_{A-B} = \frac{1}{R}(V_A - V_B); \quad [\text{Ohm's law}] \tag{2.23a}$$

The resistance R of a conductor of uniform cross section is given approximately by

$$R = \rho\frac{l}{A} \quad (\text{ohms})$$

where ρ (ohm-meter) is the resistivity of the material, l (m) is the length of the conductor, and A (m^2) is the uniform cross-sectional area.

Voltage and Current Sources

For current to flow indefinitely in an electrical circuit, at least one electric-energy source must be present. Voltage and current sources are examples of electric-energy sources. As shown in Fig. 2.9c, the voltage drop across an ideal *voltage source* (a battery with zero internal resistance for example) is prescribed, whereas the current through it is not specified—it can have any value consistent with the overall circuit containing the voltage source. In the case of a battery, the voltage drop is constant with time. In contrast AC generators produce voltages that are periodic.

As described in Fig. 2.9d, the current through an ideal *current source* is prescribed, whereas the voltage drop across it is not specified. We shall not be concerned here with the actual realizations of voltage and current sources.

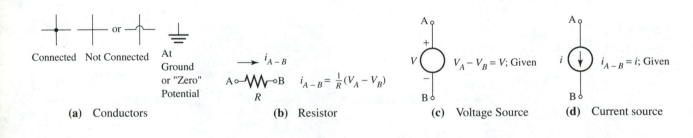

(a) Conductors   (b) Resistor   (c) Voltage Source   (d) Current source

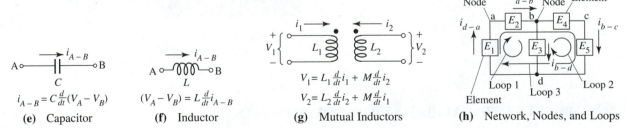

(e) Capacitor   (f) Inductor   (g) Mutual Inductors   (h) Network, Nodes, and Loops

Figure 2.9

Symbols for Some Ideal Electrical Circuit Elements

Ideal Capacitor and Inductor

With the notation in Fig. 2.9e, the constitutive behavior of an ideal electrical *capacitor* is

$$V_A - V_B = \frac{1}{C} \int i_{A-B} \, dt; \quad \text{(C is the capacitance)} \tag{2.24a}$$

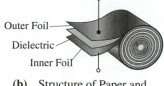

(a) Parallel-plate Capacitor

The dynamic relation (2.24a) is implied when the mathematically equivalent expression, equation (2.24b), is written:

$$C \frac{d}{dt} (V_A - V_B) = i_{A-B} \tag{2.24b}$$

The capacitance of a simple flat-plate capacitor (Fig. 2.10a or 2.10b) is given by

$$C = \frac{\epsilon A}{d} \quad \text{(farads)} \tag{2.24c}$$

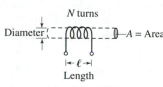

(b) Structure of Paper and Plastic Dielectric Capacitors

where A (m^2) is the area of either plate, d (m) is the distance between the plates, and ϵ (F/m) is the dielectric constant.

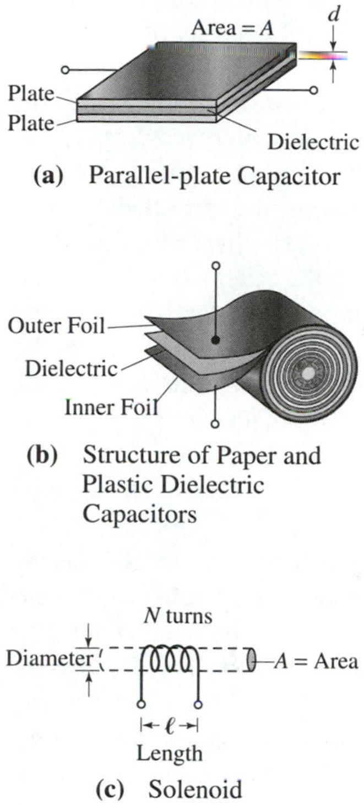

The Inductor

An ideal inductor (a real inductor always has a certain resistance) with inductance L (see Fig. 2.9f) is a dynamic component with the behavior

$$i_{A-B} = \frac{1}{L} \int (V_A - V_B) \, dt \tag{2.25a}$$

The mathematical model is often written as

$$L \frac{d}{dt} i_{A-B} = (V_A - V_B) \tag{2.25b}$$

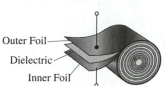

(c) Solenoid

Figure 2.10

Simple Realizations of Two Capacitors and an Inductor

The inductance of a solenoid-type coil (see Fig. 2.10c) is given approximately by

$$L = \frac{\mu N^2 A}{l} \quad \text{(henrys)} \tag{2.25c}$$

where μ (H/m) is the permeability of the material in the core of the coil, N is the number of turns, A (m^2) is the cross-sectional area of the solenoid, and l (m) is the solenoid length.

When two current-carrying wires are physically close together (within each other's magnetic field), a variation of current in one wire will, in addition to inducing a voltage drop in that wire (self-inductance), cause a voltage drop in the other wire (mutual inductance). If the mutual inductance is M, the total voltage drop in each wire (see Fig. 2.9g) is given by

$$V_1 = L_1 \frac{d}{dt} i_1 + M \frac{d}{dt} i_2$$

$$V_2 = L_2 \frac{d}{dt} i_2 + M \frac{d}{dt} i_1 \tag{2.26}$$

Equation (2.26) applies for the sign convention represented by the dots in Fig. 2.9g, that is, the positive sides of the voltages are toward the dots, whereas the current is positive when it flows into the terminal marked by the dot.

2.3 PHYSICAL MODELING EXAMPLES

As was discussed previously, construction of a physical model consists of two steps. The first step, which is to partition the universe into the system of interest and its environment, was shown to be part of the systems concept and was illustrated with the matrix printerhead positioning problem. The second step, which involves decomposition of the system into behavior components, requires that we have available an array of theoretical component behaviors. In the last several sections, the ideal behaviors of a few mechanical and electrical components were developed. The physical model itself is the representation in some form (detailed sketches, symbolic schematics, block diagrams, flowcharts, circuit diagrams) of these behavior components interconnected into an arrangement whose overall characteristics simulate the system's behavior. For instance, the schematic diagram of Fig. 1.8a, the generalized block diagram of Fig. 1.8b, and the operational block diagram in Fig. 1.14 are all physical models of the Egyptian water clock.

Simple Mechanical Systems

Care must be exercised in interconnecting components of mechanical systems to ensure that observed *causal* (see Section 3.1) relationships are maintained. Consider, for example, the automobile front suspension of Fig. 2.6a. The proper interconnection of the components is as shown in Fig. 2.11a. The arrangement represented by Fig. 2.11b is incorrect because the tire end of the damper and spring should "see" the same velocity (and displacement).* In both figures, m represents the sprung mass or the mass of the car (frame, engine, occupants, etc.). The unsprung mass (the mass of the tire for example) is ignored in Fig. 2.11.

We shall now consider in more detail some examples in physical modeling of mechanical systems.

_____

*As will be explained in Section 3.1, there is also a causal conflict for the mass in Fig. 2.11b.

Figure 2.11

Two Candidate Interconnections of Three Components of the Automobile Suspension System

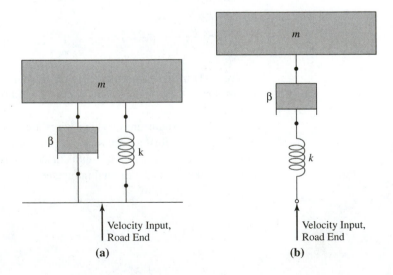

Velocity Input, Road End

(a)   **(b)**

Example 2.1	The Bow and Arrow Problem

Consider the part of the universe or the general system consisting of an archer shooting an arrow from a bow, as illustrated in Fig. 2.12a. Develop, as a schematic diagram, the physical model of the system with the goal of describing the horizontal (component of) motion of the arrow.

Solution:

Step 1—Isolating the system: As shown in Fig. 2.12b, the system consists of the string-bow (combined) and the arrow. The archer is considered only to the extent of determining the input motion of the bow and the displacement of the middle region of the string and does not have to be included in the system. The boundary interactions are thus: the frame of the string-bow is fixed (input displacement here is zero—archers normally hold the bow fairly steady while extending the string); the middle of the string is initially displaced by an mount x_{in}; and there is air resistance to the motion of the arrow when it is in flight. Also, a gravitational force acts on the arrow, although this force is not involved in the horizontal motion.

Some compromises involved:

1. Although slight movements of the bow frame will affect the dynamics of the arrow, this aspect can be ignored in the present problem.
2. The arrow and the string are always perfectly centered with respect to the bow frame, which is also held in a vertical attitude so that a vector description of the string forces will not be required.
3. The friction between the arrow and the bow frame is small and only momentary. The air resistance in flight is more important.

Note that some of these compromises may not be appropriate, given a more detailed objective for the study, but are listed for illustrative purposes only.

Step 2—Conceptual behavior partitioning and matching: From our knowledge of the behavior of certain mechanical elements and from intuition, we can make the following matches:

> *Spring* for string-bow
> *Mass* for arrow
> *Hydrodynamic damping* for air resistance

Physical model: Figure 2.12c shows the interconnection, in a consistent manner, of the components isolated in step 1 but represented by schematic symbols according to the behavior partitioning obtained in step 2.

Some approximations that could be involved:

1. The change in length of the string is sufficiently small to allow the string to be treated as an ideal spring rather than as a cable.
2. The arrow is rigid (no compliance) and does not rotate or spin.
3. The air resistance is neither hydrodynamic nor viscous, but the velocities involved are such that viscous damping would not be a good approximation.

Notes:

1. The approximations are normally made during the construction of the mathematical model but are presented here to illustrate the continuity between the physical model and the system model.

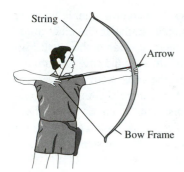

(a) The General System

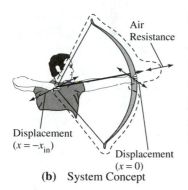

Displacement
$(x = -x_{in})$

Displacement
$(x = 0)$

(b) System Concept

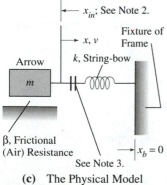

x_{in}; See Note 2.

x, v

Fixture of Frame

k, String-bow

Arrow

m

β, Frictional (Air) Resistance

$x_b = 0$

See Note 3.

(c) The Physical Model

Figure 2.12

The Bow and Arrow Problem

2. The initial displacement $x = -x_{in}$ is negative.
3. The interconnection between the mass and the spring is a special one. The mass is connected to the spring only as long as x is less than zero. For x greater than or equal to zero, the model consists of only the mass, the frictional resistance, and the gravitational force. However, the arrow now has a finite velocity. Thus in model solution, the full model will be solved until $x = 0$. Then the value of the mass velocity at this point is used to solve the reduced model until the arrow comes to rest.

Example 2.2	Motor-Actuated Control Valve

A control valve is often used to regulate flow in response to a control signal, which is typically an electrical voltage (converted into a pneumatic signal in an electric-to-pneumatic transducer, as in the case of the pneumatic valve positioner shown in Fig. 1.10, or converted into a current signal in an amplifier circuit, as in the case of a motor- or solenoid-actuated valve). In most motor-driven valves, the valve stem displacement is proportional to the angle of twist of the motor shaft. This can be accomplished by a number of means, including using a power screw, a torque arm, a cam, or a gear train. The current through the motor windings (armature coil for permanent magnet motors) is essentially proportional to the motor torque. Construct a simple physical model of a motor-actuated control valve system that will allow the determination of the flow through the valve, given the pressure drop across the valve and the current into the motor.

Note that no visual information has been provided, so this is a case where the physical model is to be developed without the physical benefit of the system. This situation is typical of new product design or product redesign where what is available is mostly the idea of what the product is to accomplish, or a similar product may exist that needs to be improved.

Solution: Considering the problem description, there are obviously many feasible solutions. The solution we have adopted is one that exploits the theoretical behaviors of mechanical components that we have developed so far. It is not necessarily a solution that will be found in many commercial motor-actuated valves. Specifically, we have used the combination of a torque arm and a flexible shaft to transfer the motor angular motion to the valve stem travel (see Fig. 2.13).

System isolation: The system consists of a rotor, including the armature coils and integral shaft, and a valve. The current input results in an applied torque T_{in} to the rotor. One end of the shaft is fixed, while the other end is free to rotate within bearings. The valve stem is linked to the rotor (the actual attachment could be to the motor shaft rather than the rotor).

Compromises:

1. The electrical details of the motor are not very significant to the present problem. The applied torque is simply proportional to the input current.
2. The details of the linkage between the rotor and the valve stem are also not very significant to the problem. The arrangement in Fig. 2.13 is adequate.

Conceptual behavior partitions:

>   *Mass* (rotary inertia) for rotor
>   (Torsion) *spring* for motor shaft
>   *Valve*
>   *Viscous damping* for valve stem

Figure 2.13

Physical Model of Control Valve System

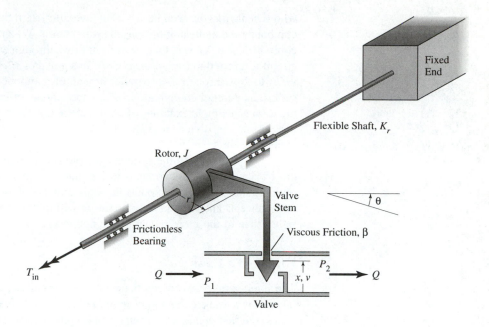

Possible approximations:

1. The masses of the rotor, motor shaft, and moving parts in the valve can all be combined (lumped) in the rotor mass.
2. There is no friction in the motor bearings; however, friction in the valve stem due to valve packing is significant.
3. The frictional resistance in the valve stem is of the viscous (damper) type.
4. There is no compliance in the valve stem. Also, the force on the valve stem due to the pressure in the flowing fluid is negligible (or is included in the viscous damping force on the valve stem).

Other possible approximations:

5. Bending and hence lateral vibration of the motor shaft is insignificant because the shaft is not long enough.
6. The valve opening is directly proportional to the valve stem position, and the flow through the valve is proportional to the product of the valve stem position and the square root of the pressure drop across the valve.
7. The valve behavior is static. There is negligible fluid storage in the valve.

Notes:

1. See note 1 in Example 2.1
2. The frictional force in the valve stem is proportional to the linear velocity of the valve stem v. The force acts through the moment arm r to cause a damping torque on the rotor. Similarly, the linear velocity of the valve stem is related to the angular velocity of the rotor through the moment arm r.

Some Examples of Electrical Networks

Electrical systems are common in engineering, especially in actuation, measuring and testing devices, input/output components, and man-machine interface units. Although such systems are often complex, they are usually composed of subsystems or circuits whose behaviors are well known. The overall system behavior can then be obtained by combining the models of the various subsystems. When the components of a circuit are combined in such a way that the current flows through each component in sequence, the circuit is referred to as a *series circuit*. In a *parallel circuit,* the current in the circuit divides to flow through each parallel branch of components. A circuit with a number of series and parallel combinations of components is often referred to as a *network*. Figure 2.9h is a simple example. Here the elements E_1 and E_2 are in series ($i_{d-a} = i_{a-b}$); the current i_{a-b} divides into i_{b-d} and $i_{b-c} = i_{c-d}$ to flow through the parallel branches containing the elements E_3 and E_4 (in series with E_5), respectively.

Our approach to the synthesis of physical models of electrical systems, that is, electrical circuits or networks, is to accumulate experience of the behavior of potential building blocks of these models through analysis of such blocks. In order to illustrate new concepts and methods, our attention will mainly be directed to the simpler circuits rather than to the more recognizable applications.

Kirchhoff's Laws

The interaction between components of electrical circuits is described by very simple laws known as the *Kirchhoff current law* and the *Kirchhoff voltage law*. The current law states: *the sum of all currents entering a node is zero*. The current law thus implies that charge cannot accumulate at a pure junction. This is similar to the law of conservation of matter discussed in Section 3.3. Since currents are signed quantities, the current law can also be stated as *the sum of all currents leaving a node is zero*. For node b in Fig. 2.9h, for example, we can write

$$i_{a-b} + i_{d-b} + i_{c-b} = 0 \text{ or } i_{b-a} + i_{b-d} + i_{b-c} = 0 \tag{2.27a}$$

The Kirchhoff voltage law states that *the sum of all the voltage drops around a complete loop is zero*. The same law can also be stated using *voltage increases* in place of *voltage drops*. In either case, a loop is any closed path (starting from a node and back to the same node) in an electrical circuit. Three loops are shown for Example in Fig. 2.9h: loops abda, bcdb, and abcda. The voltage law for loop 3, for instance, would be

$$(V_a - V_b) + (V_b - V_c) + (V_c - V_d) + (V_d - V_a) = 0 \text{ or}$$

$$(V_b - V_a) + (V_c - V_b) + (V_d - V_c) + (V_a - V_d) = 0 \tag{2.27b}$$

Application of Kirchhoff's Laws   Traditional electrical circuit analysis is based on one or the other of the two Kirchhoff's laws. If the current law is used, the procedure is often called the *node method*. Analysis based on the voltage law is termed the *loop method*. In the node method, the current law is written for each *independent* node, and the behavior of the circuit is described by a set of independent voltage variables. If the total number of elements (or branches) in a circuit is b_t, and the total number of nodes is n_t, then the number of independent nodes, which is the same as the number of independent voltages in the circuit, is given by

$$n = n_t - 1 \tag{2.28a}$$

The number of independent loops is equal to the number of independent currents and is given by

$$l = b_t - n = b_t - n_t + 1 \tag{2.28b}$$

However, when an independent variable is prescribed (for example, by a voltage or current source), the number of independent variables for the circuit analysis must be reduced accordingly. To facilitate analysis, it is normal practice to designate one point (node) on the circuit as being at *ground* potential. The voltage values at all other points are then determined relative to the actual value of the voltage at that ground point, with no loss of generality.

Example 2.3 — Application of Node and Loop Methods

Use the node and loop methods to obtain sets of equations sufficient in each case to determine the voltage drop for the inductor in the electrical circuit shown in Fig. 2.14a.

Solution—Node Method:

1. **Free body circuit diagram:** Label the nodes in the circuit (this also identifies the node voltages), and assume the sense of each branch current (see Fig. 2.14b). Choose one of the nodes (node d) as ground point.
2. **Number of independent nodes (voltages):** The nodes are a, b, c, and d, that is, $n_t = 4$. Equation (2.28a) then gives $n = n_t - 1 = 4 - 1 = 3$.
3. **Current law for n nodes:**
   Node a: Sum currents into node a:

$$i_s + i_1 = 0$$

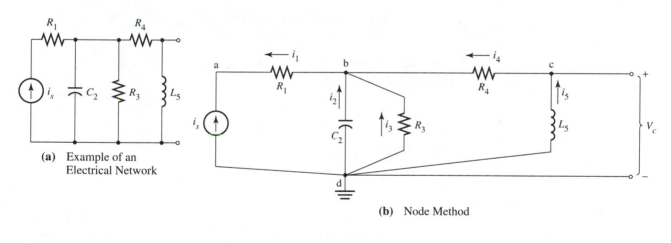

(a) Example of an Electrical Network

(b) Node Method

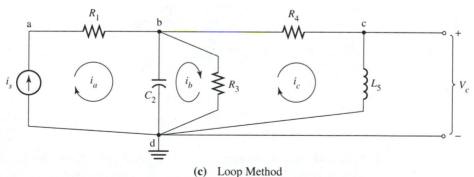

(c) Loop Method

Figure 2.14

Application of Node and Loop Methods

Equation (2.23a), that is, Ohm's law gives

$$i_1 = \frac{(V_b - V_a)}{R_1}$$

Hence

$$i_s + \frac{(V_b - V_a)}{R_1} = 0 \qquad \text{(a)}$$

Node b: Sum currents into node b:

$$-i_1 + i_2 + i_3 + i_4 = 0$$

By equation (2.23a),

$$i_3 = \frac{(0 - V_b)}{R_3} = \frac{-V_b}{R_3}, \qquad i_4 = \frac{(V_c - V_b)}{R_4}$$

The ideal capacitor model, equation (2.24b) gives

$$i_2 = C_2 \frac{d}{dt}(0 - V_b) = -C_2 \frac{d}{dt} V_b$$

Thus

$$\frac{-(V_b - V_a)}{R_1} - C_2 \frac{d}{dt} V_b - \frac{V_b}{R_3} + \frac{(V_c - V_b)}{R_4} = 0 \qquad \text{(b)}$$

Node c: Sum currents into node c: $-i_4 + i_5 = 0$.
The inductor model, equation (2.25b) gives

$$i_5 = \frac{1}{L_5} \int_0^t (0 - V_c)\, dt + i_5(0) = i_5(0) - \frac{1}{L_5} \int_0^t V_c\, dt$$

so that

$$-\frac{(V_c - V_b)}{R_4} + i_5(0) - \frac{1}{L_5} \int_0^t V_c\, dt = 0 \qquad \text{(c)}$$

4. **Set of equations:** From equation (a), V_b and V_a are dependent, since i_s is prescribed. The number of independent variables (voltages) therefore reduces to two, that is, V_b and V_c. The solution is given by equations (d) and (e), which are obtained from equations (a), (b), and (c). The desired quantity is $V_c - V_d = V_c$:

$$C_2 \frac{d}{dt} V_b + \frac{V_b}{R_3} - \frac{(V_c - V_b)}{R_4} = i_s \qquad \text{(d)}$$

$$\frac{(V_c - V_b)}{R_4} + \frac{1}{L_5} \int_0^t V_c\, dt = i_5(0) = \frac{1}{R_4}[V_c(0) - V_b(0)] \qquad \text{(e)}$$

Solution—Loop Method:

1. **Free-body circuit diagram:** Label the nodes and choose the ground point, as before. Indicate the loop currents and their directions (see Fig. 2.14c).
2. **Number of independent loops (currents):** The total number of branches $b_t = 6$. From equation (2.28b), $l = b_t - n_t + 1 = 6 - 4 + 1 = 3$. However, one loop current, $i_a = i_s$, is specified. hence $l = 2$.

3. **Voltage law for loops b and c:** The independent currents are i_b and i_c.
   Loop b: Sum voltage drops around loop b:

   $$(V_b - V_d) + (V_d - V_b) = 0 \tag{f}$$

   Loop c: Sum voltage drops around loop c:

   $$(V_c - V_d) + (V_d - V_b) + (V_b - V_c) = 0 \tag{g}$$

   From equation (2.24a),

   $$(V_b - V_d) = V_b = \frac{1}{C_2} \int_0^t (i_s - i_b)\, dt + V_b(0)$$

   Equation (2.23a) gives

   $$(V_d - V_b) = R_3(i_c - i_b) \quad \text{and} \quad (V_b - V_c) = R_4 i_c$$

   From equation (2.25b),

   $$(V_c - V_d) = L_5 \frac{d}{dt} i_c$$

4. **Set of equations:** Substituting the last three equations into equations (f) and (g), we obtain the following two equations in the two unknowns i_b and i_c:

   $$\frac{1}{C_2} \int_0^t (i_s - i_b)\, dt + R_3(i_c - i_b) = -V_b(0) \tag{h}$$

   $$L_5 \frac{d}{dt} i_c + R_3(i_c - i_b) + R_4 i_c = 0 \tag{i}$$

   where the desired quantity is given in terms of i_b and i_c by

   $$(V_c - V_d) = -(V_d - V_b) - (V_b - V_c) = -R_3(i_c - i_b) - R_4 i_c \quad \text{or}$$
   $$(V_c - V_d) = R_3 i_b - (R_3 + R_4) i_c \tag{j}$$

Notes:
1. The set of variables found in the results is different for each method: (V_b, V_c); and (i_b, i_c). However the number of independent variables (two) is the same in each case.
2. For certain problems, an advantageous choice between the two traditional techniques—node and loop methods—exists. For instance, in the present problem, three node equations were required in the node method (V_a was actually eliminated by utilizing one of these equations), whereas only two loop equations were necessary in the loop method.
3. Although it was not illustrated in this example, where the freedom to do so exists, a judicious choice of the ground point on a circuit may substantially simplify the analysis of that circuit.

....

Series and Parallel Combinations of R, C, and L Elements

A combination of components, which occurs often in electrical circuits, can be replaced by a single equivalent element in order to reduce the analytical effort. It is necessary however to first recognize the presence of such a combination of elements in the circuit. Figure 2.15 illustrates the series and parallel combinations of n elements without specifying the type of element. Using Kirchhoff's laws, we can easily derive the expression for the equivalent element E_e for each case of combined resistors, capacitors, and inductors. We leave it as an exercise for the reader to confirm the following results.

Figure 2.15

**Series and Parallel Resistors,
Capacitors, and Inductors**

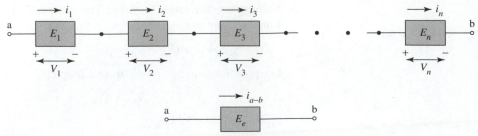

(a)   Series Combination of n Elements

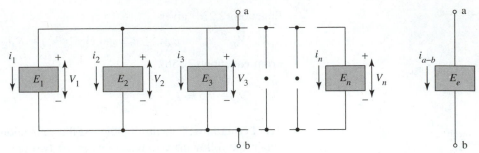

(b)   Parallel Combination of n Elements

Series resistors (Fig. 2.15a, $E_j = R_j$, $E_e = R_e$):

$$(V_a - V_b) = i_{a-b} \sum_{j=1}^{n} R_j = i_{a-b}R_e; \qquad R_e = R_1 + R_2 + R_3 + \ldots + R_n \tag{2.29}$$

Series capacitors (Fig. 2.15a, $E_j = C_j$, $E_e = C_e$):

$$(V_a - V_b) = \left(\sum_{j=1}^{n} \frac{1}{C_j} \right) \int i_{a-b}\, dt = \frac{1}{C_e} \int i_{a-b}\, dt; \tag{2.30a}$$

$$\frac{1}{C_e} = \frac{1}{C_1} + \frac{1}{C_2} + \frac{1}{C_3} + \cdots + \frac{1}{C_n}$$

For $n = 2$, for example,

$$C_e = \frac{C_1 C_2}{C_1 + C_2} \tag{2.30b}$$

Series inductors (Fig. 2.15a, $E_j = L_j$, $E_e = L_e$):

$$(V_a - V_b) = \sum_{j=1}^{n} V_j = \sum_{j=1}^{n} L_j \frac{d}{dt} i_{a-b} = L_e \frac{d}{dt} i_{a-b}; \tag{2.31}$$

$$L_e = L_1 + L_2 + L_3 + \ldots + L_n$$

Parallel resistors (Fig. 2.15b, $E_j = R_j$, $E_e = R_e$):

$$(V_a - V_b) = i_{a-b}R_e; \qquad \frac{1}{R_e} = \sum_{j=1}^{n} \frac{1}{R_j}, \qquad \left[n = 2, \quad R_e = \frac{R_1 R_2}{R_1 + R_2} \right] \tag{2.32}$$

Parallel capacitors (Fig. 2.15b, $E_j = C_j$, $E_e = C_e$):

$$(V_a - V_b) = \frac{1}{C_e} \int i_{a-b}\, dt; \qquad C_e = C_1 + C_2 + C_3 + \ldots + C_n \qquad \textbf{(2.33)}$$

Parallel inductors (Fig. 2.15b, $E_j = L_j$, $E_e = L_e$):

$$(V_a - V_b) = L_e \frac{d}{dt} i_{a-b}; \qquad \frac{1}{L_e} = \sum_{j=1}^{n} \frac{1}{L_j}; \qquad \left[n = 2, \quad L_e = \frac{L_1 L_2}{L_1 + L_2} \right] \textbf{(2.34)}$$

Practical R, RC, and RCL Networks

The number of practical circuits that can also serve as potential building blocks in the synthesis of electrical systems is very large, so we can examine here only a very small number of such circuits. Specifically, we shall consider one example from each of the three categories of such circuits: *R networks,* which are *resistive* circuits composed of combinations of resistors; *RC networks,* which are *capacitive* circuits containing resistors and capacitors; and *RCL networks,* which are circuits containing resistance, capacitance, and inductance. Additional examples including *inductive* circuits or *RL networks* are given as practice problems at the end of the chapter.

Practical resistor networks include the *voltage divider,* a series resistor network that can be used to provide a specific potential difference or a number of potential differences from one voltage source; the *range multiplier,* commonly found in meters, where it is used in series with a current-measuring subcircuit to permit measurement of potential difference; and the *bridge circuit,* one application of which is the *Wheatstone bridge,* used often in resistance measuring devices. Some applications of *RC* networks include the *capacitive coupling* circuit (or *high-pass filter*), which is used to transfer higher frequency AC (alternating current) signals from one subcircuit to another while attenuating or blocking lower frequency AC or DC (direct current) signals; *capacitive bypass* circuit (or *low-pass filter*), which is often used to remove high-frequency components of AC signals from some point in a circuit by providing a bypass to ground; and *phase shifting* and *wave shaping* networks. Major applications of *RCL* networks are in *resonant* or *tuned* circuits and in *filters.*

Example 2.4   The Voltage Divider

Derive the expression for the output potential difference V_{out} of the simple voltage divider shown in Fig. 2.16a. Extend the results to determine the voltage drop between each of the output terminals and ground in the multisection voltage divider shown in Fig. 2.16b.

Solution: Under the open circuit condition shown in Fig. 2.16a, that is, in which no current is flowing out the output terminal from node w, resistors R_1 and R_2 are in series, so that the current flowing to ground through R_2 is given by equation (2.29):

$$i_{w-g} = \frac{V_{in}}{(R_1 + R_2)}$$

But Ohm's law, equation (2.23a), also gives

$$i_{w-g} = \frac{V_{out}}{R_2}$$

Figure 2.16

Unloaded Voltage Divider Circuits

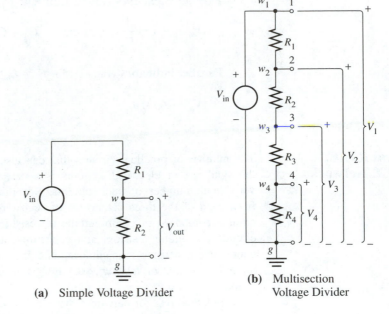

(a)   Simple Voltage Divider

(b)   Multisection Voltage Divider

so that

$$V_{out} = \left(\frac{R_2}{R_1 + R_2} \right) V_{in} \quad \text{(voltage divider)} \tag{2.35a}$$

In the same way, in Fig. 2.17b,

$$i_{w_1-g} = \frac{V_{in}}{(R_1 + R_2 + R_3 + R_4)} = \frac{V_j}{\sum\limits_{k=j}^{4} R_k} \tag{2.35b}$$

so that $V_1 = V_{in}$ and

$$V_2 = \left(\frac{R_2 + R_3 + R_4}{R_1 + R_2 + R_3 + R_4} \right) V_{in}; \qquad V_3 = \left(\frac{R_3 + R_4}{R_1 + R_2 + R_3 + R_4} \right) V_{in};$$

$$V_4 = \left(\frac{R_4}{R_1 + R_2 + R_3 + R_4} \right) V_{in}$$

Example 2.5	Low-Pass RC Filter

a. Obtain the differential equation relating the output voltage drop to the input voltage drop for the bypass circuit shown in Fig. 2.17a. Note that in a low-pass filter or bypass circuit, the output is taken across the capacitor.

b. Let the input potential difference be an AC voltage, $V_{in} = A \sin(\omega t)$. Assuming that the particular solution for the output potential difference is of the form $V_{out} = B \sin(\omega t + \phi)$, determine the constants B and ϕ by substituting this trial solution into the differential equation obtained in (a).

Solution a: Figure 2.17b shows the bypass circuit redrawn with the nodes labeled, the assumed sense of the current, and the chosen ground point (that is, the free body circuit

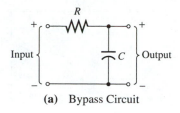

(a)   Bypass Circuit

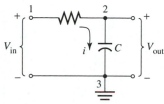

(b)   Free Body Circuit Diagram

Figure 2.17

Low-pass *RC* Filter

diagram). Note that $V_{in} = V_1$, and $V_{out} = V_2$. Using equations (2.23a) and (2.24b), we have

$$i_{1-2} = i_{2-3} = \frac{1}{R}(V_1 - V_2) = C\frac{d}{dt}V_2 = i_{2-3}$$

Thus the differential equation for V_{out} (that is, V_2) in terms of V_{in} (or V_1) is

$$\frac{d}{dt}V_2 = \frac{1}{RC}(V_1 - V_2) \tag{2.36a}$$

Solution b:

$$\frac{d}{dt}[B\sin(\omega t + \phi)] = B\omega\cos(\omega t + \phi) = \frac{1}{RC}[A\sin(\omega t) - B\sin(\omega t + \phi)]$$

But

$$\cos(\omega t + \phi) = \cos(\omega t)\cos\phi - \sin(\omega t)\sin\phi \quad \text{and} \quad \sin(\omega t + \phi)$$
$$= \sin(\omega t)\cos\phi + \cos(\omega t)\sin\phi$$

Hence

$$B\omega[\cos(\omega t)\cos\phi - \sin(\omega t)\sin\phi]$$

$$= \frac{1}{RC}[A\sin(\omega t) - B\{\sin(\omega t)\cos\phi + \cos(\omega t)\sin\phi\}]$$

Equating coefficients of $\cos(\omega t)$ and of $\sin(\omega t)$ on both sides, we have

$$B\omega\cos\phi = \frac{-B}{RC}\sin\phi \tag{a}$$

and

$$-B\omega\sin\phi = \frac{A - B\cos\phi}{RC} \tag{b}$$

Equation (a) gives ϕ:

$$\tan\phi = -RC\omega, \text{ or } \phi = -\tan^{-1}(RC\omega) \tag{2.36b}$$

Solving equations (a) and (b) simultaneously for B, we obtain $B = A\cos\phi$, or $B/A = \cos\phi$. Squaring both sides of equation (a) and using $\sin^2\phi = 1 - \cos^2\phi$ gives us $[1 + (RC\omega)^2]\cos^2\phi = 1$. Thus

$$r = \frac{B}{A} = \frac{1}{\sqrt{1 + (RC\omega)^2}} = \frac{1}{\sqrt{1 + f^2}}, \quad \text{where } f = RC\omega \tag{2.36c}$$

It follows that the output potential difference is given by

$$V = \frac{A\sin(\omega t - \tan^{-1}[RC\omega])}{\sqrt{1 + (RC\omega)^2}} = \frac{A\sin\left(\frac{1}{RC}ft - \tan^{-1}(f)\right)}{\sqrt{1 + f^2}} \tag{2.36d}$$

The quantity $r = B/A$ is known as the amplitude ratio. Figure 2.18 shows a plot of r as a function of dimensionless frequency f, for $0.001 \leq f \leq 1000$ (see **On the Net**).

Figure 2.18

Frequency Response of a Low-pass Filter

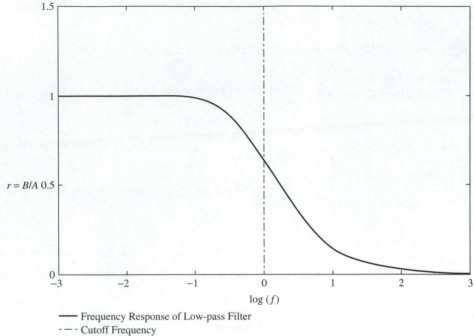

— Frequency Response of Low-pass Filter
-–-· Cutoff Frequency

The behavior of the low-pass *RC filter* is as the name suggests. At low frequencies, the input signal is transferred to the output with little or no amplitude attenuation (r = 1 approximately). At high frequencies, the circuit provides a low resistance path or bypass to ground, thus severely attenuating the signal transferred to the output. The demarcating frequency appears to be $\omega = 1/RC$ (that is $f = 1$; $\log(f) = 0$), which is known as the *cutoff frequency* or *corner frequency* (see Section 9.2)

Example 2.6   Series Resonant Circuit

Consider the series *RLC* resonant circuit of Fig. 2.19a. Let the input potential difference be $V_{in} = V \sin(\omega t)$, and define the dimensionless frequency $f = \omega\sqrt{(LC)}$ and parameter $Q = \sqrt{L/C}/R$. If the resulting current in the circuit is of the form $i = I \sin(\omega t + \phi)$,

a. Obtain the second-order differential equation relating i to V_{in} or its derivative.
b. Determine ϕ and the dimensionless amplitude ratio

$$r = \frac{I\sqrt{(L/C)}}{V}$$

Solution a: Applying the voltage law to the free body circuit diagram shown in Fig. 2.19b gives us

$$L\frac{d}{dt}i + Ri + V_3 = V_{in} \tag{a}$$

Equation (2.24b) yields

$$C\frac{d}{dt}V_3 = i \tag{b}$$

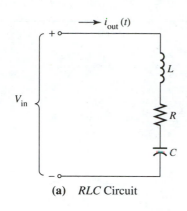

(a) *RLC* Circuit

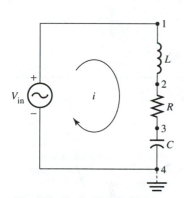

(b) Free Body Circuit Diagram

Figure 2.19

Series *RLC* Circuit

Differentiating (a) with respect to time and substituting (b) gives us the desired result:

$$\frac{d^2i}{dt^2} + \frac{R}{L}\frac{di}{dt} + \frac{1}{LC}i = \frac{1}{L}\frac{d}{dt}V_{in} \tag{2.37a}$$

Solution b: Substituting $i = I\sin(\omega t + \phi)$, and $V_{in} = V\sin(\omega t)$ into equation (2.37a), we obtain

$$-I\omega^2\sin(\omega t + \phi) + \frac{IR\omega}{L}\cos(\omega t + \phi) + \frac{I}{LC}\sin(\omega t + \phi) = \frac{V\omega}{L}\cos(\omega t)$$

or expanding, we have

$$I\left(\frac{1}{LC} - \omega^2\right)[\sin(\omega t)\cos\phi + \cos(\omega t)\sin\phi]$$

$$+ \frac{IR\omega}{L}[\cos(\omega t)\cos\phi - \sin(\omega t)\sin\phi] = \frac{V\omega}{L}\cos(\omega t)$$

Equating coefficients of $\cos(\omega t)$ and $\sin(\omega t)$ on both sides of the preceding equation, gives us

$$I\left(\frac{1}{LC} - \omega^2\right)\sin\phi + \frac{IR\omega}{L}\cos\phi = \frac{V\omega}{L} \tag{c}$$

$$I\left(\frac{1}{LC} - \omega^2\right)\cos\phi - \frac{IR\omega}{L}\sin\phi = 0 \tag{d}$$

From equation (d), we obtain the result

$$\phi = \tan^{-1}\left(\frac{1 - \omega^2 LC}{\omega RC}\right) \tag{2.37b}$$

Also, equation (d) gives

$$\left(\frac{1}{LC} - \omega^2\right)^2\cos^2\phi = \left(\frac{R\omega}{L}\right)^2\sin^2\phi, \text{ or } \cos\phi = \frac{\omega RC}{\sqrt{(1 - \omega^2 LC)^2 + (\omega RC)^2}} \tag{e}$$

Dividing equation (c) by $\cos\phi$, then substituting for $\tan\phi$ from equation (2.37b) and for $\cos\phi$ from equation (e), and solving for the dimensionless amplitude ratio $r = [I\sqrt{(L/C)}]/V$ gives us the result

$$r = \frac{I\sqrt{L/C}}{V} = \frac{\omega\sqrt{LC}}{\sqrt{(1 - \omega^2 LC)^2 + (\omega RC)^2}} \tag{2.37c}$$

In terms of f and Q, as defined in this problem,

$$r = \frac{I\sqrt{L/C}}{V} = \frac{f}{\sqrt{(1 - f^2)^2 + (f/Q)^2}} \tag{2.37d}$$

Fig. 2.20 shows a plot of the amplitude ratio r as a function of frequency f, with Q as parameter.

....

The main information that should be obtained from the last two examples is (1) the process of analyzing the circuit to obtain its description (differential equation) and (2) the type of performance that can be obtained with these circuits, given the proper

Figure 2.20

**Frequency Response of a
Series-resonant Circuit**

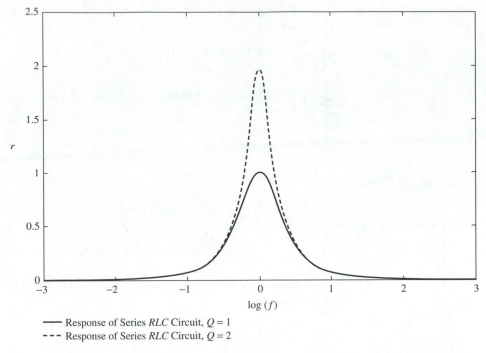

— Response of Series *RLC* Circuit, $Q = 1$
- - - Response of Series *RLC* Circuit, $Q = 2$

The series *RLC* circuit exhibits the phenomenon of resonance, a condition in which the output (the current amplitude) becomes arbitrarily large at some frequency of the input. The frequency at which resonance occurs, that is the resonant frequency is apparently given by $f = 1$. The parameter Q, known as the quality factor, seems to control the sharpness (the peak amplitude, and the frequency range over which the amplitude is large) of the current-frequency response. This type of circuit is the basis of tuning a radio or television receiver: The resonant frequency is "tuned" close to the frequency of the desired signal; the undesired signals with sufficiently different frequencies are rejected by the tuned circuit.

choice of the values of the parameters. With such knowledge, one should be able to decide when and how to incorporate similar circuits as building blocks of an overall model. The purpose of solving the differential equations now is to illustrate the peculiar performance of the circuits. The solution of differential equations is a subject of its own, much of which is reviewed later in the text as needed (see Chapters 6, 7, and 8).

2.4 STAGE TWO—MODEL CONSTRUCTION: PRELIMINARIES

The process of *system decomposition* is necessary for the systematic construction of mathematical models of dynamic systems. Although it is possible, with experience, to go from the physical model to the system equations with only a rudimentary decomposition of the problem, a disciplined and formal application of system decomposition in many cases represents the major and more significant part of mathematical modeling. Once a system is decomposed, the behavior of each of the components and their interactions can be approximated using known theories. As we showed in earlier examples, the specification of a system and construction of its physical model involve compromises about the goal of the dynamic study, whereas approximations of component behavior are involved in the development of the mathematical model. The approximations we make in going from the physical model to the system model have some rather definite characteristics, so that we can study them individually. We do so in this section, following the discussion and illustration of system decomposition.

System Decomposition

System decomposition—the process of breaking up a system into simpler components and analyzing the individual parts, can be performed formally as follows:

a. **Identify the components:** Name them and draw free body diagrams showing inputs and outputs and the internal interactions (at component interconnections) and external interactions (for boundary components). Indicate all necessary parameters and variables and their senses; adopt a sign convention.

b. **Separate the components into static and dynamic elements.**

c. **Write the input-output relations:** Describe the behavior of these components. Note that the relations describing static elements are quite different from dynamic expressions (integration or differentiation, delay or difference).

In identifying the components, draw the free body diagrams in an arbitrary configuration so that all relevant variables can be defined. As an illustration, consider the water supply system of Fig. 1.1. The main components are:

- Pump
- Pipes (pump to tank, and tank to valve)
- Tank
- Valve

Figure 2.21 shows the free body diagram of each of the components. The valve, we know, is a static element from our previous discussions, whereas the tank is a dynamic element. The pipes, depending on their lengths, can be considered either static elements or dynamic components, as we shall see later. If we consider the pump as simply converting an electrical input such as current into a fluid mechanical output—flowrate, with a one-to-one correspondence or algebraic relation between the current and the flowrate, then we can classify the pump as a static element. The behavior of the tank and valve were derived previously (see equations (1.2) and (1.5c)). For the tank we had, assuming a tank of uniform cross-sectional area,

Figure 2.21

Components of the Water Supply System

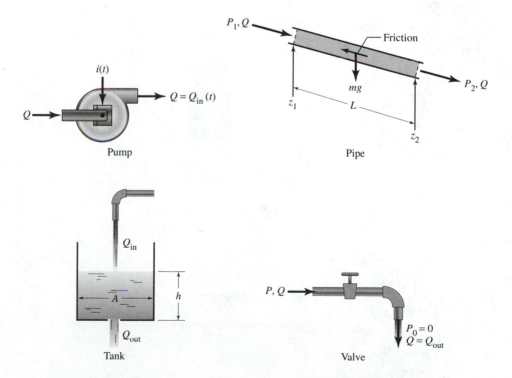

Pump

Pipe

Tank

Valve

$$h(t) = \left(\frac{1}{A}\right) \int_0^t (Q_{in} - Q_{out})\, dt + h(0)$$

and for the valve,

$$Q_{out}(t) = CP^{1/2}$$

If i is the current supplied to the pump motor to cause the flow Q_{in}, then we can write for the pump,

$$Q_{in}(t) = K_p i(t) \tag{2.38}$$

where K_p is a constant for the pump. Equation (2.38) is, of course, a static relation. We shall skip the description of the pipe behavior at this time.

Example 2.7	The Bow and Arrow Problem: Decomposition

The physical model developed in Example 2.1 (see Fig. 2.12c) already shows the separate components. This is the nature of most schematic diagrams and electrical circuit diagrams. Figure 2.22c shows the free body diagram of each of these components.

The nature and behavior of each component are as follows.

Hydrodynamic damping (static): From equation (2.21),

$$F_d = \beta v_{rel} |v_{rel}|$$

which gives

$$F_d = \beta v |v| \tag{a}$$

Figure 2.22

The Bow and Arrow System and Component Free Body Diagrams

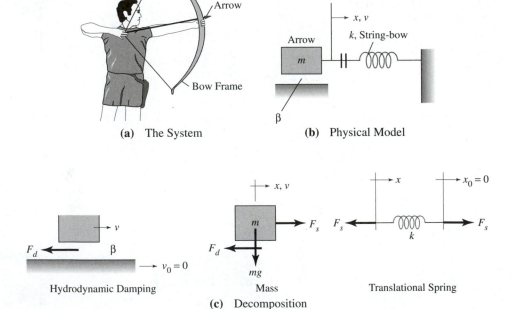

(a) The System

(b) Physical Model

Hydrodynamic Damping

Mass

Translational Spring

(c) Decomposition

Mass (dynamic): Newton's second law, equation (2.8b), for the horizontal motion is

$$m \frac{dv}{dt} = \begin{cases} F_s - F_d; & (x \le 0) \\ -F_d; & (x > 0) \end{cases}; \qquad v(0) = 0 \tag{b}$$

Translational spring (dynamic): Assuming that the string is stretched by only a very small amount when it is plucked, Hooke's law (equation (2.3a) and Fig. 2.1a (for interpretation of Δx), and equation (2.4b)), give

$$F_s = k(x_0 - x) = -kx$$

$$\frac{dx}{dt} = v; \qquad x(0) = -x_{\text{in}} \tag{c}$$

Notes:

1. Although the dynamic operator involved here is integration, differential equations have been written as the preferred *format* for the mathematical models. This subject is discussed in more detail in Section 3.1. Meanwhile, there should be no confusion as long as the appropriate inputs and outputs are recognized in these relations. For example, in the model for the mass, the output is velocity, and a solution for the velocity requires integration of the input force.

2. Recall that only the horizontal component of the travel of the arrow is of interest. To describe the motion in the vertical direction, we must include the weight of the arrow in the vertical forces.

3. The interaction between the mass and the spring is that the forces acting on each at their interconnection are equal and opposite—Newton's third law—and that they have a common displacement. These interactions are incorporated directly in the free body diagrams as trivial relations.

4. $x_0 = 0$ is an external interaction on the spring. Similarly, $v_0 = 0$ (the wind speed is neglected) is a boundary interaction on the hydrodynamic damper.

Example 2.8   Motor-Actuated Control Valve: Decomposition

The physical model of the motor-actuated control valve that was developed in Example 2.2, and the free body diagrams of the components, are shown in Fig. 2.23.

The assumed behaviors of the components are as follows:

Viscous damping (static): From equation (2.17),

$$F_f = \beta v \tag{a}$$

Valve (static): Equation (1.2), modified by introducing the sign function to ensure that the flow direction is always the correct one, gives

$$Q = C \, \text{sign}(\Delta P)(|\Delta P|)^{1/2}; \qquad \Delta P = P_1 - P_2 \text{ is given} \tag{b}$$

Assuming that C is proportional to the valve opening,

$$C = cx; \quad c = \text{constant} \tag{c}$$

Mass (dynamic): This is a rotary inertia behavior component. Rotation takes place only about the axis of the shaft. Hence (see note 1 of Example 2.7), equation (2.16) gives

$$J \frac{d\omega}{dt} = T_{\text{in}} - T_f - T_s \tag{d}$$

Figure 2.23

**Motor-actuated Valve
and Component Free
Body Diagrams**

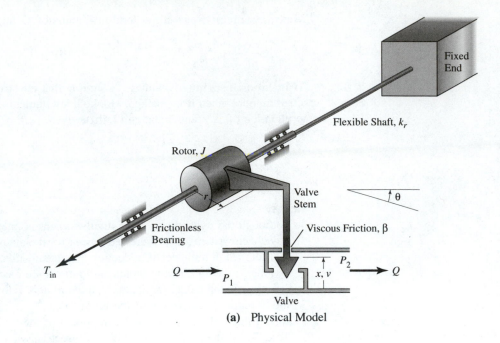

(a) Physical Model

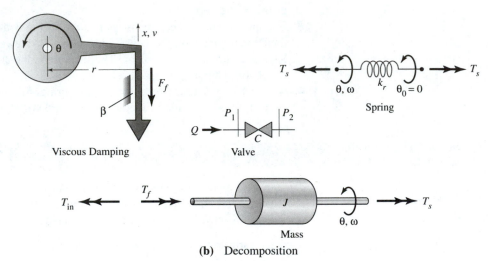

(b) Decomposition

Spring (dynamic): From equations (2.6a) and (2.6b)

$$T_s = k_r \theta$$

$$\frac{d\theta}{dt} = \omega \qquad\qquad\qquad \textbf{(e)}$$

Internal interactions: The following interactions between components that can be deduced from the physical model are implied in the free body diagrams:

Rotor-shaft: The rotor and shaft are integral bodies and have the same rotation.

Rotor–valve stem: Approximately,

$$T_f = rF_f \qquad\qquad\qquad \textbf{(f)}$$

Further, for small θ, which is assumed,

$$x = r\theta, \quad \text{approximately} \tag{g}$$

so that

$$v = r\omega \tag{h}$$

Example 2.9   Example of an Electrical Circuit

The free body circuit diagram that we have used for the traditional analysis of electrical circuits is in effect an informal decomposition of the circuit diagram. The labeling of nodes, indication of branch or loop currents and their directions, and so forth, constitute an isolation of components without physically breaking them apart. This is sufficient in most cases. Let us however, for completeness, illustrate the formal process of system decomposition for the electrical circuit of Example 2.3.

The free body circuit diagram of Fig. 2.14b is shown again in Fig. 2.24a, and the free body diagrams of the separated components are shown in Fig. 2.24b. Accordingly, the behavior of each component is follows:

> **Ground point** (actually a boundary interaction): $V_d = 0$ **(a)**
> **Current source** (static input element): $i_1 = -i_s$; i_s is given   **(b)**

Figure 2.24

Example of an Electrical Circuit and Its Decomposition

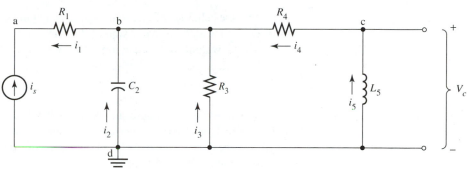

(a)   Example of an Electrical Circuit

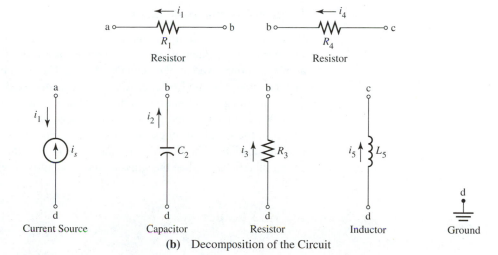

(b)   Decomposition of the Circuit

Resistors (all static elements): From Ohm's law, equation (2.23a),

$$R_2: i_1 = \frac{(V_b - V_a)}{R_1} \tag{c}$$

$$R_3: i_3 = \frac{(0 - V_b)}{R_3} = \frac{-V_b}{R_3} \tag{d}$$

$$R_4: i_4 = \frac{(V_c - V_b)}{R_4} \tag{e}$$

Capacitor (dynamic): Equation (2.24b) gives

$$i_2 = C_2 \frac{d}{dt}(0 - V_b) = -C_2 \frac{d}{dt} V_b \tag{f}$$

Inductor (dynamic): From equation (2.25b),

$$L_5 \frac{d}{dt} i_5 = (0 - V_c) = -V_c \tag{g}$$

Internal interactions: At nodes b and c, Kirchhoff's current law requires that

$$-i_1 + i_2 + i_3 + i_4 = 0 \tag{h}$$

and

$$-i_4 + i_5 = 0 \Rightarrow i_4 = i_5 \tag{i}$$

Note that the behavior of a current source is that the current in the branch is prescribed. Equation (b) is also an expression of the internal interaction at node a. The present results are equivalent to the equations developed using the node method in Example 2.3.

....

Lumping, Linearity, and Stationarity

The equations we use to describe the various behavior components, whether mechanical, electrical, or otherwise, are ideal or approximate models rather than the actual behavior of the components. We shall now consider the nature of some of the particular approximations that are made when system components are represented by ideal springs, masses, dashpots, resistors, capacitors, inductors, and the like.

Lumped-System Assumption

Recall that the stretch in a spring is specified in terms of the displacement of the ends. In contrast, the stretch of an actual spring varies along its length. Consider, for example, a spring fixed at one end (Fig. 2.25a). The stretch near the fixed end is smaller than the stretch at the free end. Also, although the ideal spring has no mass, the mass of a real spring is *distributed* along the length of the spring.

We can analyze the spring from elementary principles as follows: Isolate an element of the spring of length dx, as shown in Fig. 2.25b. If the stretch of the element is u (a function of both position x and time t), and the modulus of elasticity for the spring material is E, then the stress on the element is given by Hooke's law or equation (2.2c):

$$\left(\frac{F_s}{A}\right) = E\frac{\partial u}{\partial x} \tag{2.39a}$$

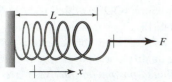

(a) Distributed-parameter Model

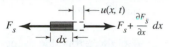

(b) Spring and a Differential Spring Element

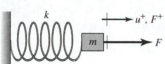

(c) Lumped-parameter Model

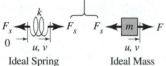

(d) Ideal Spring and Ideal Mass

Figure 2.25

Distributed- and Lumped-parameter Models for a Real Spring with Mass

where A is the cross-sectional area and $(\partial u/\partial x)$ is the strain. (Note that the force will also vary with position and time). Newton's second law for the element gives

$$-F_s + F_s + \frac{\partial F_s}{\partial x}\,dx = (dm)\frac{\partial v}{\partial t} \tag{2.39b}$$

But if the material density is ρ, then, $(dm) = \rho A\,dx$. Also,

$$v = \frac{\partial u}{\partial t} \Rightarrow \frac{\partial v}{\partial t} = \frac{\partial^2 u}{\partial t^2}$$

so that equation (2.39b) can be written as

$$\frac{\partial F_s}{\partial x}\,dx = (\rho A\,dx)\frac{\partial^2 u}{\partial t^2} \tag{2.39c}$$

From equation (2.39a),

$$\frac{\partial F_s}{\partial x} = \frac{\partial}{\partial x}\left(AE\frac{\partial u}{\partial x}\right)$$

so that, for A = constant, equation (2.39c) becomes

$$\frac{\partial^2 u}{\partial t^2} = \left(\frac{1}{\rho A}\right)\frac{\partial}{\partial x}\left(AE\frac{\partial u}{\partial x}\right) = \left(\frac{E}{\rho}\right)\frac{\partial^2 u}{\partial x^2} \tag{2.40a}$$

The boundary conditions are

$$u\Big|_{x=0} = 0, \qquad \frac{\partial u}{\partial x}\Big|_{x=L} = \frac{F}{(EA)} \tag{2.40b}$$

Equations (2.40a) and (2.40b) represent the *distributed-parameter* (the spring and mass effects are mixed or distributed over all the element) model of the fixed-free spring. This is a partial differential equation that is more difficult to solve than the ordinary differential equation given by the alternative *lumped-parameter* analysis (see below). A lumped system assumption for the above spring "lumps" all the spring effects into a "point" or ideal spring with no mass, and all the mass effects into a "point" or ideal mass, as shown in Fig. 2.25c. In other words, we *separate the mixed effects into ideal components*. Hooke's law for the ideal spring, equation (2.3a), gives

$$F_s = ku$$

and Newton's law for the ideal mass, equation (2.8a), is

$$-F_s + F = m\frac{d^2 u}{dt^2}$$

Combining the two equations, we obtain

$$\frac{d^2 u}{dt^2} = \left(\frac{1}{m}\right)F - \left(\frac{k}{m}\right)u \tag{2.41}$$

Equation (2.41) is a second-order ordinary differential equation (ODE), which is easier to solve than the second-order partial differential equation (PDE), as mentioned earlier.

When is a lumped approximation justified? We shall answer this question in rather general terms using the spring example. First, we qualify our answer somewhat by stating that *both the lumped-parameter and distributed-parameter analyses are mathematical approximations to the real thing,* so that the proper question is perhaps, When is lumped-parameter approximation better than distributed-parameter approximation? If the parameters of the element, such as the stretch per unit length, the density, and the area for the spring example vary spatially within the element, and this variation is large, a lumped model may not be justified.* If the variation is small, the advantages of a simpler lumped model may be sufficient to offset the errors due to the lumped-parameter approximation. Another consideration, which is somewhat related to the first, has to do with the frequency or rate of change of inputs or disturbances applied to the component. In equation (2.40a) the quantity (E/ρ) has dimensions or units of velocity squared, and $(E/\rho)^{1/2}$ gives a characteristic velocity for the material (for propagation of (elastic) strain waves). For a given length L of the spring, the corresponding "characteristic" frequency will be

$$f_c = \frac{1}{L}\left(\frac{E}{\rho}\right)^{1/2}$$

(2.42)

If the rate at which things are "happening" to the spring is much less than f_c, (for example, for steel, $f_c \approx 2 \times 10^5$ Hz for $L = 0.0254$ m), then there is sufficient time for one disturbance to propagate throughout the element before the next one occurs, and so on, so that the element appears to behave as one (lumped) unit. Under these circumstances, lumping may be adequate. Note also that in equation (2.42), if we let L tend to zero, which is the physical description for our ideal point spring, then f_c tends to infinity, and the frequency of any disturbance is always less than the characteristic frequency of an ideal or point spring.

Linearity Assumption

Consider the damper of Fig. 2.26 with the power law relation:† $F = Bv^2$. If this component is operating in our system about the point (v_0, F_0), the behavior of the damper can be approximated by the linear relation $F_d = \beta v_d$, where F_d and v_d are measured relative to F_0 and v_0 respectively. As long as the operating point does not shift away from (v_0, F_0) by a large amount, this approximation may be adequate, and it simplifies our analysis a lot. How large a deviation from the operating point is acceptable depends on the goal and application of the analysis.

A system is linear when its behavior is described by linear relations. The relation between two variables F and v, $[F = f(v)]$ is linear if for $F_1 = f(v_1)$, $F_2 = f(v^2)$, . . . , $F_n = f(v_n)$, $F_1 + F_2 + . . . + F_n = f(v_1) + f(v_2) + . . . + f(v_n)$ for all $v_1, v_2, . . . ,$ v_n. This definition is easily verified in Fig. 2.26. For the real damper, it is evident that for the given (v_1, F_1) and (v_2, F_2), the point (F, v), with $v = v_1 + v_2$, results in $F \neq F_1 + F_2$. The preceding definition also illustrates another important implication and application of linearity besides simplifying the analysis. This is the principle of *su-*

---

*A lumped-parameter "equivalent" to the distributed model can be constructed by separating the mixed effects into a large number of lumped ideal elements connected together. The order of the ODE however increases with the number of dynamic elements.

†Written as $F = Bv|v|$ to account for the change in the sign of F with a sigh change in v.

Figure 2.26

Real and Ideal Damper Behavior

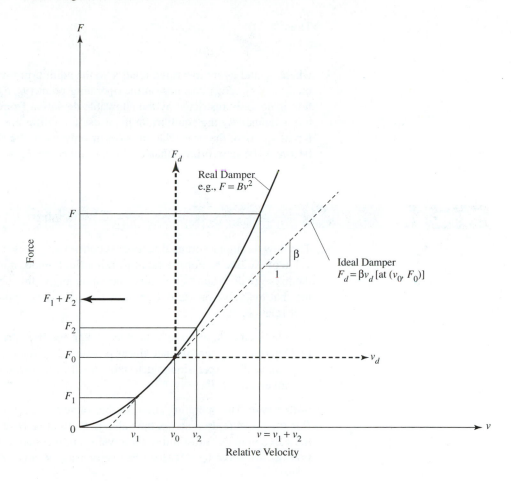

perposition. We can find the effect that a number of inputs (causes) acting together will have on our system by adding together (superposing) all the separate effects of the individual inputs, as long as our system is linear. We can now explain linear control systems in Table 1.3 (classification on the basis of nature of plant) as those control systems that are composed entirely of linear elements.

All physical systems are essentially nonlinear in their behavior. However, as demonstrated in Fig. 2.26, a large number of engineering elements in common use can be considered linear in the neighborhood of an operating (usually equilibrium) point. The process of linearization can be illustrated mathematically: Suppose

$$F = Bv^2 = f(v) \tag{2.43}$$

Expanding $f(v)$ in a Taylor series about v_0, gives

$$f(\Delta v + v_0) = f(v_0) + \left.\frac{df}{dv}\right|_{v=v_0} \Delta v + \frac{1}{2!}\left.\frac{d^2 f}{dv^2}\right|_{v=v_0}(\Delta v)^2 + \cdots \tag{2.44a}$$

$$= F_0 + 2Bv_0\Delta v + \text{higher order terms in } \Delta v$$

The slope of the linear approximation β is the same as the tangent to the nonlinear curve at $v = v_0$, that is,

$$\beta = \frac{f(\Delta v + v_0) - F_0}{\Delta v} \approx \left.\frac{df}{dv}\right|_{v=v_0} = 2Bv_0 \tag{2.44b}$$

Thus

$$(F - F_0) \approx \beta(v - v_0), \quad \text{or } F_d \approx \beta v_d \tag{2.45}$$

where F_d and v_d are measured relative to the equilibrium values. Note that the approximation ($\beta = 2Bv_0$) depends on the operating point (v_0, F_0). Thus a linear approximation is not only restricted by the allowable deviation from the given equilibrium point but is defined by the equilibrium point itself—a different β will be obtained for a different v_0. Also, the new relation is linear only when the equilibrium (operating) point becomes the new origin (that is, the variables are F_d and v_d, as defined in equation (2.45)).

Example 2.10	Linearized Valve Pressure Drop–Flow Relation

The flow through a certain directional control valve varies only slightly from a nominal value of 0.02 m³/s. For the usual working fluid, under normal operating temperatures, the pressure drop across the valve corresponding to this flow rate is 0.64 MPa. Assuming that the valve pressure drop–flow behavior is adequately described by the square root relation, equation (1.2),

a. What is the flow through the valve when the pressure drop is 0.7 MPa?
b. Determine an approximate linear pressure drop–flow expression for this valve and evaluate the expression to determine the flow rate when the pressure drop across the valve is 0.7 MPa.

Solution a: The operating point for this valve is $(P_0, Q_0) = (0.64 \times 10^6, 0.02)$, where P is the pressure drop across the valve in Pa (that is, N/m²), and Q is the flow through the valve in m³/s. Substituting these values into equation (1.2), repeated here as equation (a), we solve for C, which can be considered a constant for a directional control valve:

$$Q = C\{P\}^{1/2} \tag{a}$$

$$0.02 = C\{0.64 \times 10^6\}^{1/2} \Rightarrow C = 2.5 \times 10^{-5} \text{ N}^{-1/2}\text{m}^4\text{s}^{-1}$$

Substituting for C and $P = 0.7$ MPa in equation (a), we obtain the required flow rate:

$$Q = (2.5 \times 10^{-5})(0.7 \times 10^6)^{1/2} = 0.02092 \text{ m}^3/\text{s}$$

Solution b: Evaluate the slope ($1/R_f$) of the Q versus P curve at the operating point:

$$\left(\frac{1}{R_f}\right) = \frac{\partial}{\partial P}(C\{P\}^{1/2})\bigg|_{P = P_0} = \frac{1}{2}C\{P\}^{-1/2}\bigg|_{P = 0.64 \times 10^6}$$

$$= 1.5625 \times 10^{-8} \text{ N}^{-1}\text{m}^5\text{s}^{-1}$$

The linear pressure drop–flow relation for the valve is therefore

$$Q^* = \left(\frac{1}{R_f}\right)P^*; \quad \text{where } Q^* = (Q - Q_0) \quad \text{and} \quad P^* = (P - P_0) \tag{b}$$

Evaluating equation (b) at $P^* = (0.7 - 0.64) = 0.06$ MPa, we have

$$Q^* = (1.5625 \times 10^{-8})(0.06 \times 10^6) = 0.00094 \Rightarrow$$

$$Q = 0.02 + 0.00094 = 0.02094 \text{ m}^3/\text{s}$$

The error in the linear approximation at this pressure drop is only about 0.1%. At a pressure drop of 1.28 MPa (double the operating point pressure drop), the error in the linear approximation increases to about 6%, which could still be tolerable for many applications.

Stationarity Assumption

A *stationary* or *time invariant* model of a dynamic system results when the parameters in the component relations do not vary with time. For the behavior components we have studied thus far, the associated parameters include the spring constant, the mass, the moment of inertia, the damping constant, the electrical resistance, the capacitance, and the inductance. In real life, these quantities do change with time (for example, due to environmental changes), so that there are no truly stationary systems. However, if the period of change of the system parameters is large compared with the time frame† or time measure for the study, or the effect of the change in the parameter value on the results of the dynamic study is negligible relative to the goals of the study, then stationarity is a good assumption, since the analysis of stationary systems is in general simpler than that of nonstationary systems.

Example 2.11   Temperature Coefficient of Resistivity

The resistivity of conducting material is a function of the temperature of the material. According to experimental data, for a substantial temperature range around typical room temperatures, the resistance (or resistivity) of a conductor at a temperature T_2 can be found, fairly accurately, in terms of the resistance (or resistivity) at temperature T_1, using the formula

$$R_2 = R_1[1 + \alpha(T_2 - T_1)] \tag{2.46}$$

where R_1 and R_2 are the resistances (or resistivities) corresponding to the temperatures T_1 and T_2, respectively, and α is the temperature coefficient of resistance (or resistivity).

Consider now the voltage divider circuit of Fig. 2.16a in use in a piece of electronic equipment whose enclosure attains a steady operating temperature of 30°C after power is applied to the circuit, at room temperature of, say, 20°C. As shown in Fig. 2.27, let $R_L = 40\ \Omega$ represent the equivalent resistance of the rest of the circuit (that is, the load). Determine the error, if any, in the estimated current through the load i_L, based on the room temperature values of the resistors: $R_1 = 7\ \Omega$ and $R_2 = 5\ \Omega$, assuming the load resistance remains the same, input voltage to the circuit is 20 V, and the resistors are made of nickel, which has a temperature coefficient of resistance at 20°C of 0.006 ohm/(ohm-°C).

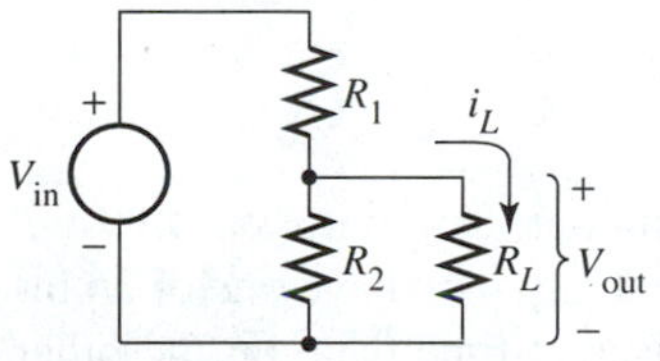

Solution: Figure 2.27 represents a loaded simple voltage divider. If we replace the two resistors R_2 and R_L with the parallel combination (see equation (2.32)), equation (2.35a) gives the estimated output voltage:

$$V_{out} = \left[\frac{(R_2 R_L)/(R_2 + R_L)}{R_1 + (R_2 R_L)/(R_2 + R_L)} \right] V_{in} \tag{2.47}$$

Figure 2.27

Loaded Simple Voltage Divider Circuit

---

†Recall that whether a system is considered static or dynamic can also depend on the time frame for the analysis.

Substituting the nominal (20°C) values of the resistors, we obtain V_{out} = 7.767 V, so that the estimated load current is

$$i_L = \frac{V_{out}}{R_L} = \frac{7.767}{40} = 0.194 \text{ A}$$

According to equation (2.46), the steady values of the resistances after the temperature rises from 20°C to 30°C are: R_1 = 7[1 + 0.006(30 − 20)] = 7[1.06] = 7.42 Ω; R_2 = 5[1.06] = 5.3 Ω.

Substitution of these values into equation (2.47) gives the final value of the output voltage:

$$V_{out} = \left\{ \frac{[(5.3)(40)]/(5.3 + 40)}{7.42 + [(5.3)(40)]/(5.3 + 40)} \right\}(20) = 7.7354 \text{ V}$$

Therefore the steady-state load current is

$$i_L = \frac{V_{out}}{R_L} = \frac{7.7354}{40} = 0.193 \text{ A}$$

Thus there is an error in the estimated load current of about 0.5%.

....

Uncertainty, Continuous and Sampled Data

Another assumption that is usually made to simplify analysis is that the system is *deterministic*. For instance, in exponential decay or sinusoidal variation, the variables change in a deterministic pattern. There is no *uncertainty* in the value of the variable at any time. This is in contrast with a *random* pattern of variable change wherein the value of the variable at any time is given on a probabilistic basis. It is sometimes necessary in engineering to consider random patterns of variables. For instance the presence of noise either due to the measurement device (a loose connection in an electronic instrument for example) or due to the measurement environment (atmospheric disturbances in communication signals for example) very often causes an otherwise deterministic variable pattern to become completely random or random to some degree *(stochastic)*. A deterministic assumption may still be justified if, for instance, the magnitude of the noise is negligible relative to the measured signal and the desired precision in the analysis, or if it is known a priori that the noise will be effectively removed by conditioning (frequency filtering for example).

In Table 1.3, under classification of control systems on the basis of analytical or design technique, we can now explain stochastic control systems as those control systems that are developed from a probabilistic representation of the measured plant output and/or the plant behavior. This text considers only deterministic systems or deterministic models of dynamic systems.

Continuous versus Discrete Data Assumption

Variable changes may also be characterized as continuous (in time) or discrete. Continuous-time systems are described by variables that at any time have one of an infinite number of possible values over a given range. Discrete-time data, on the other hand, consist of variables that exist or are measured (read, recognized) only at discrete intervals of time rather than continuously. Signals measured in this manner are referred to as *sampled data,* and the measurement process is called *sampling*. The time intervals may be regular or irregular (of varying lengths). However only sampling (at regular intervals of time) with a fixed period will be considered in this text.

There are many natural and man-made systems that are characterized by variables that are discrete in time *(discrete-time systems)*. This is especially true if we allow arbitrarily small resolutions in our time measure. Some perhaps more familiar examples include electronic activity in, say, a quartz crystal (governed by energy level transitions); the heart and the human circulatory system (various events may be correlated with the heartbeat); interest payments on bank savings accounts (calculated daily, monthly, or quarterly); and a digital computer (all operations take place on some internal clock cycle). In Table 1.3, under classification by nature of plant, a discrete-time control system would imply a control system characterized by discrete-time variables because the plant is a discrete-time system or is described by discrete-time data.

Discrete-time expressions are often constructed as approximations to continuous-time system models in order to obtain numerical solutions of specific problems or to allow integration of a discrete-time device in a continuous-time environment. However, more general results should be obtained if the option of a discrete-time system model is considered right from the start. Some factors that could influence the choice of the type of model include the nature of the system components, the time frame for the analysis, the goals of the study, and perhaps the necessity for a later numerical solution.

O n   t h e   N e t
- SC2.MCD (Mathcad), fig2_18.m (MATLAB): source file for Fig. 2.18
- SC3.MCD (Mathcad), fig2_20.m (MATLAB): source file for Fig. 2.20
- Auxiliary exercise: Practice Problems 2.27 through 2.30

2.5 PRACTICE PROBLEMS

2.1 The deflection at the end of a cantilever beam due to a load on the end (see Fig. 2.3b) is given by $x = FL^3/3EI$, where $I = bh^3/12$ is the area moment of inertia. Derive the spring constant shown in the figure.

2.2 (a) The hydrostatic stress on an elastic material is proportional to the volumetric strain (change in volume per unit volume). The bulk modulus is the proportionality constant. Using this definition of the bulk modulus, derive the expression for the spring constant of a hydraulic (oil) spring given in Fig. 2.3h.

2.2 (b)* Consider the viscous damper function for the same device. If the equivalent internal radius of the oil cylinder is R, the radius of the piston (plunger) is r, and the viscosity of the oil is μ, derive an expression for the damping force on the piston at a given depth of piston penetration.

2.3 (a) The force shown in Figure 2.28 is applied to a pure mass of 1 kg that is initially at rest at $x = 0$. Calculate and sketch as a function of time, the (i) acceleration, (ii) velocity, (iii) displacement, (iv) transmitted power, and (v) stored ~kinetic energy. Indicate in each case whether the relation between the force and the required quantity is static or dynamic.

2.3 (b) Repeat 2.3a if the force is applied to a spring with the opposite end fixed, in place of a mass. The spring constant is 1 N/m.

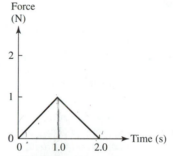

Force
(N)

Figure 2.28   (Problem 2.3)

2.3 (c) Repeat 2.3a except substitute a viscous damper with the opposite end fixed, for the mass. Also replace (v) stored energy, with (v) dissipated energy. The viscous damping constant is 1 N · s/m.

2.4* Obtain the relation between the level of gasoline in the funnel h, and the input flow rate Q_{in} (see Fig. 2.29). Is the funnel a static or dynamic system when it is described in terms of its level? Does your expression apply to very small values of h? Explain briefly.

2.5 In a certain design project, the retarding force on a prototype race car is represented by

$$F = (500 + 6v + 0.1v^2) \text{ N},$$

where v is the speed of the car in km/h. Sketch each term in this formula as a function of velocity and indicate what type of damping is represented by each term.

2.6 The rubber spring in Fig. 2.3g has the force displacement relation shown in Fig. 2.30. If the spring is used as the shock mount for heavy rotating equipment and the force on

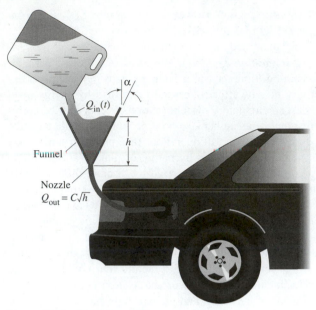

Figure 2.29 (Problem 2.4)

the spring is given by $F = (4000 + 3000 \sin 2t)$ N, determine approximately and plot the displacement $x(t)$ over one period of the force. If vibrations with amplitude greater than 1 cm are considered detrimental, would you recommend

Figure 2.30 (Problem 2.6)

this spring as an effective shock mount for the rotating equipment?

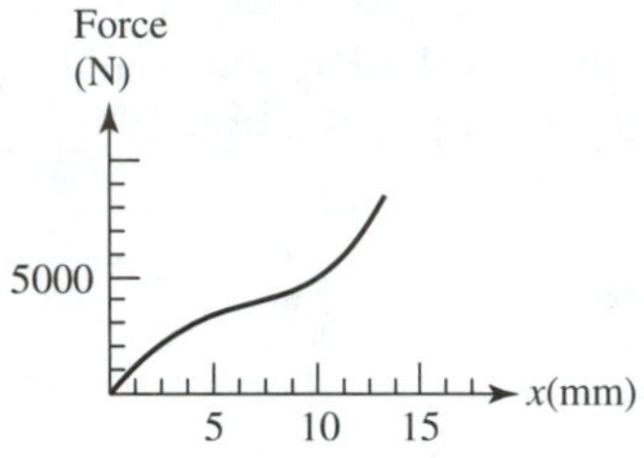

2.7* A space vehicle of mass m (ignore the size of the vehicle) is traveling with constant velocity v in a zero-gravity environment. If the rate of fuel consumption $me = -dm/dt$ is constant, find the thrust (force) developed by the vehicle exhaust, and derive an expression for the average velocity at which mass is expelled from the vehicle exhaust. Assume that there are no environmental resistances to the motion of the vehicle and that the force on the vehicle due to the average static pressure P in the exhaust gas is PA, where A is the cross-sectional area of the nozzle exit.

2.8 Two round wires of the same conductive material and length are used as resistors in a circuit. If the diameters of the wires are 1.2 mm and 0.8 mm, respectively, which wire will have more current flow through it when the wires have the same potential difference across them?

2.9 Describe three ways by which the capacitance of a parallel-plate capacitor could be increased.

2.10* Find the spacing between the plates of the parallel-plate capacitor shown in Fig. 2.10a if a capacitance of 750

pF is desired and the area of the plates is 0.02 m². Assume the dielectric is paper, which has $\epsilon = 3.1 \times 10^{-11}$ F/m.

2.11 How many turns, approximately, are needed to produce a solenoid-type coil of length 100 mm and core diameter 32 mm, if the inductance of the coil is to be 2.84 mH? Assume the permeability of the air in the core is 1.257 μH/m.

In problems 2.12 through 2.18, construct the physical model of the system described. Indicate any major compromises and potential assumptions.

2.12* A hydroelectric power plant supplied from a dam on a large river (see Fig. 2.31). The objective is to predict the electrical power output from the plant. Assume that the turbine-generator combination simply converts fluid power (pressure times volume flowrate) to electrical power (voltage times current). Include plant controls such as a surge tank and a wicket gate (valve) in the main conduit leading from the dam to the turbine.

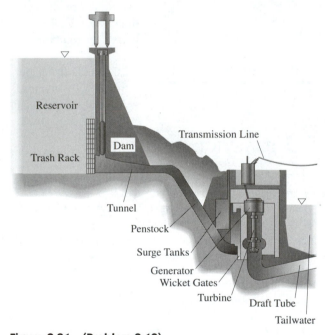

Figure 2.31 (Problem 2.12)

2.13* A printhead drive mechanism (say, for a typewriter; see Fig. 2.32). The printhead is driven across the page by a motor that reels in a tow cable. The angular velocity of the motor is known. The printhead rides on a platform that is lubricated with a highly viscous fluid. The objective is to determine the settlingtime of the printhead—a parameter that influences print quality.

2.14 The front suspension system of the car system of Problem 1.19. (See Fig. 2.6a, for example.) The goal is to evaluate ride quality by determining the vibrations of the car for various road profiles. The unsprung mass should be included.

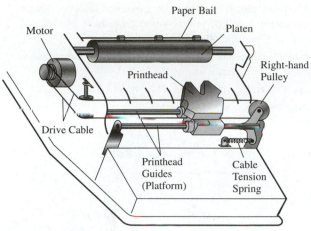

Figure 2.32 (Problem 2.13)

2.15* A ship propulsion system consisting of a turbine, a drive shaft, and propeller. The drive torque developed at the turbine rotor is prescribed. The goal is to determine the dynamic behavior of the system at high speeds.

2.16 A fluid clutch. The objective is to determine the angular velocity of the load. The drive torque and the load torque are given.

2.17 A cam and follower used in the (tappet) valve opening mechanism of an automobile engine cylinder. The cam

rotates with constant angular velocity ω, and the cam profile is given by $y_1 \cdot (t) = R\sin\theta$, where R is a constant and $\theta = \omega$. Note that the follower moves relative to the engine block, which vibrates with a known velocity $v_e(t)$. The motion of the follower is of interest. Consider the valve stem (see Fig. 2.33) the ultimate follower.

2.18 A drill column in a mile-long inclined oil well. The drill column consists of several jointed segments of drill pipe running from the surface to near the bottom of the well, followed by a set of very heavy drill pipes (collars) and a small rotary drill bit (see Fig. 2.34). During drilling, a tensile force and rotation are applied to the drill pipe at the surface. The subsequent rotation of the bit coupled with the weight above it causes the bit to chew into the rock. The motion of the bit

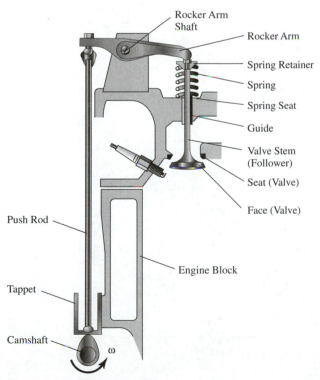

Figure 2.33 (Problem 2.17)

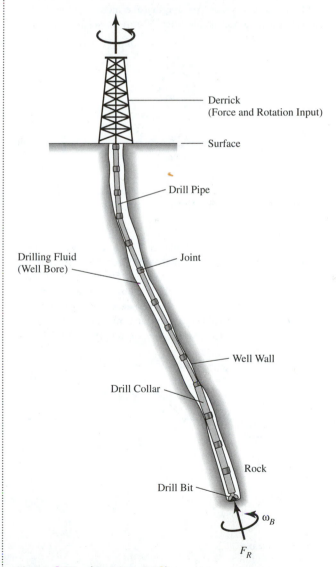

Figure 2.34 (Problem 2.18)

can be described simply as $v_B = v_0 \sin \omega_B t$, where v_B is the linear velocity of the last drill collar segment. The reaction of the rock is described by a force on the bit F_R and a torque on the drill collar $T_R = f(F_R)$, which is a known function of F_R. The objective is to determine F_R and ω_B. There is viscous fluid in the annular region between the drill column and the well wall, and in some sections, the drill pipes rub against the walls of the well bore.

2.19* A current meter with a resistance of 5 Ω shows a full-scale reading when the current through it is 50 mA. The meter is to be used to measure current flow into a circuit, as shown in Fig. 2.35. If the largest circuit current to be measured is 1.0 A, determine the resistance of the shunt (a *shunt* is a resistor placed in parallel with a load (the meter in this case) in order to reduce the current through the load) so that the meter will indicate full scale when the circuit current is at this maximum value. Assume that the meter will be damaged if a current greater than 50 mA passes through it.

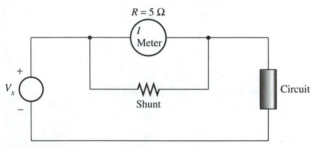

Figure 2.35   (Problem 2.19)

2.20 The current meter of Problem 2.19 is to be used as a voltmeter to measure source potential differences. Since the current through the meter cannot exceed 50 mA, the voltage drop across the meter must at all times be limited to 0.25 V (full-scale deflection), because the resistance of the meter is 5 Ω. Determine the values of two range multipliers R_1 and R_2 (see Fig. 2.36; a *range multiplier* or simply *multiplier* is a

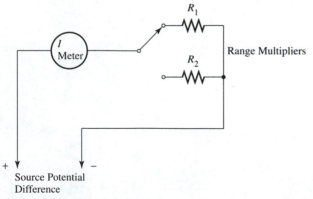

Figure 2.36   (Problem 2.20)

resistance placed in series with a meter to limit the circuit current) that will permit the voltmeter to have two ranges, one 15 V and the other 150 V.

2.21 Calculate the unknown voltage drop or current indicated in each circuit in Fig. 2.37.

(a) The voltage drop across R_2

(b) The output voltage drop V_{out}

(c) The current through R_3

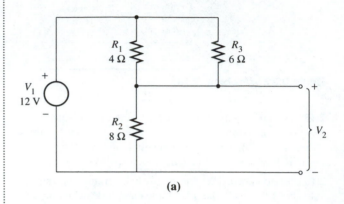

(a)

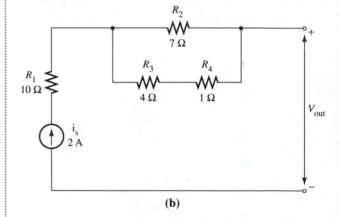

(b)

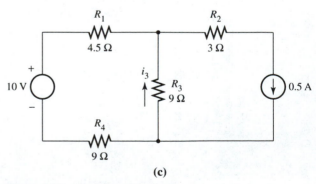

(c)

Figure 2.37   (Problem 2.21)

(d) The current through R_2

(e) The output voltage drop V_{out}

(f) The current through R_5

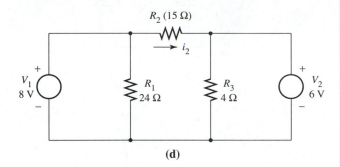

(d)

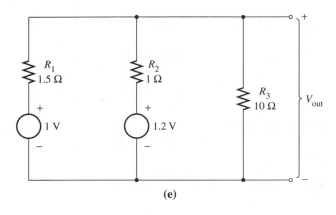

(e)

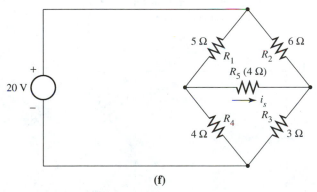

(f)

Figure 2.37   (Problem 2.21)

2.22* Figure 2.38 shows a *Wheatstone bridge* circuit used to measure an unknown resistance R_3. R_2 is a variable resistor or potentiometer. When the value of R_2 is adjusted such that the current measured by the meter is zero, the bridge is said to be *balanced*. Show that for a balanced bridge, $R_1/R_4 = R_2/R_3$, so that $R_3 = R_2R_4/R_1$.

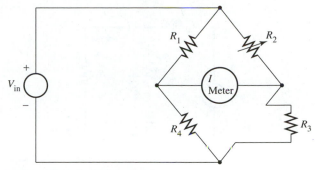

Figure 2.38   (Problem 2.22)

2.23 Calculate the value of the equivalent element E_e that can replace the combination of elements shown in the network of Fig. 2.39 if

(a) E stands for resistance R.

(b) E stands for capacitance C.

(c) E stands for inductance L.

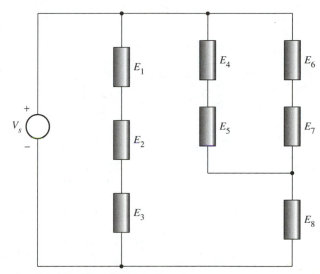

Figure 2.39   (Problem 2.23)

2.24 For the networks shown in Fig. 2.40, derive using either the node method or the loop method, the complete set of equations necessary to determine

(a) the voltage at node b relative to ground potential.

(b) the output potential difference V.

(c) the voltage drop V_2.

(d)* the voltage drop across the capacitor.

(e) the voltage at any node relative to ground potential, and the current through any branch of the circuit.

(f) the potential difference V_3.

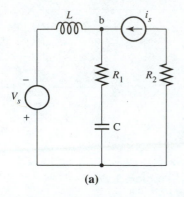

(a)

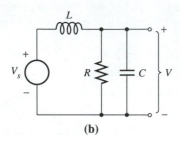

(b)

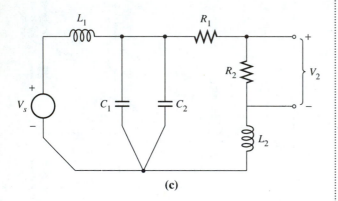

(c)

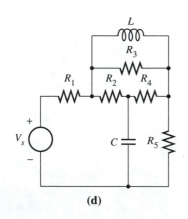

(d)

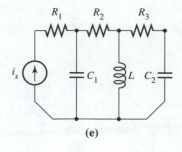

(e)

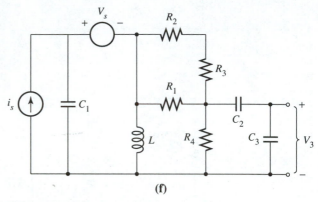

(f)

Figure 2.40   (Problem 2.24)

2.25 Determine the set of equations that could be used for predicting the voltage output V of the indicated circuit in Fig. 2.41.

(a) The bridge circuit

(b) The circuit with mutual inductors. Assume that V_s is a time-varying voltage source.

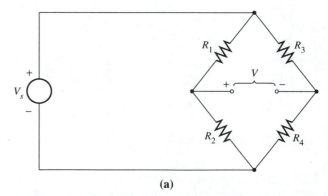

(a)

Figure 2.41   (Problem 2.25)

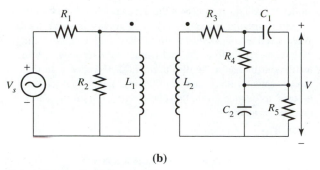

(b)

Figure 2.41 (Problem 2.25)

2.26 Electronic (game) equipment that dissipates 0.025 W when the operating potential difference is 5 V DC is picking up 60 Hz noise from the power line. Figure 2.42a shows a typical solution, which is to connect a bypass capacitor across the (5 V) DC+ inlet point of the "DC" subcircuit to ground. This gives, in effect, a low-pass filter as shown in Fig. 2.42b. A rule of thumb for selecting a value for the bypass capacitor is to let the virtual resistance have a value $R_e = R_L/20$. Determine C for the present problem by positioning the cutoff frequency of the effective low-pass filter at 60 Hz.

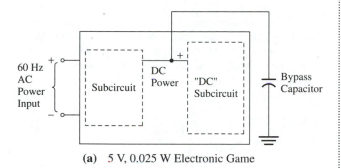

(a) 5 V, 0.025 W Electronic Game

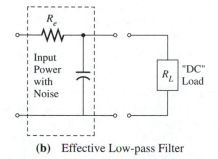

(b) Effective Low-pass Filter

Figure 2.42 (Problem 2.26)

2.27 Repeat Example 2.5 using the high-pass RC filter shown in Fig. 2.43. Note that in a high-pass filter the output

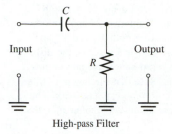

High-pass Filter

Figure 2.43 (Problem 2.27)

is taken across the resistor. The dimensionless frequency and amplitude ratio have the same definition as in the example. Plot r versus f, as in Fig. 2.18.

2.28 Repeat Example 2.5 using the following inductive circuits. Let the dimensionless frequency be defined by $f = \omega L/R$. Plot the amplitude ratio r (as defined here) versus f as in Fig. 2.18:

(a) The series LR circuit shown in Fig. 2.44a, $r = B/A$.

(b) The parallel LR circuit shown in Fig. 2.44b, $r = B/RA$

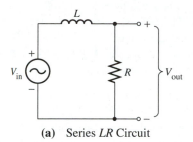

(a)   Series LR Circuit

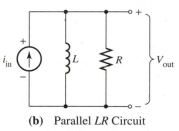

(b)   Parallel LR Circuit

Figure 2.44 (Problem 2.28)

2.29* The high-pass filter of Fig. 2.45 functions as an RC waveshaping network known as a *differentiator* when the parameters are chosen such that $RC \leq T/10$, where T is the

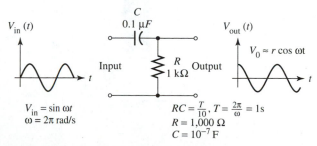

$V_{in} = \sin \omega t$
$\omega = 2\pi$ rad/s

$RC = \frac{T}{10}, T = \frac{2\pi}{\omega} = 1s$
$R = 1{,}000\ \Omega$
$C = 10^{-7}\ F$

Figure 2.45 (Problem 2.29)

period of the highest frequency component of the input signal. Using your results from Problem 2.27 and the data given in Fig. 2.45, show by plotting both V_{in} and V_{out} as functions of time that, as illustrated in Fig. 2.45, the differentiator circuit would shape the input sine wave into an approximate cosine wave output. Note that the low-pass filter, on the other hand, would give an *integrator* network. The network parameters in this case should be chosen such that $RC \geq 10T$.

2.30 Repeat Example 2.6 using the parallel *RLC* circuit shown in Fig. 2.46. In this case, consider the current $i_{in} = I \sin(\omega t)$, as the input, and the potential difference across the terminals, $V_{out} = V \sin(\omega t + \phi)$, as the output. Let the dimensionless amplitude ratio be given by $r = V[\sqrt{(C/L)}]/I$. The dimensionless frequency f and the quality factor Q remain as defined in the example. Plot r versus f as in Fig. 2.20.

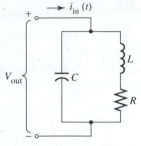

Parallel Resonant Circuit
Figure 2.46   (Problem 2.30)

In problems 2.31 through 2.44 decompose the systems whose physical models are described, that is, identify components, draw their free body diagrams indicating inputs/outputs, interactions, parameters, and their senses; identify static and dynamic elements; and write the input-output relations describing the behavior of the components. Describe all major assumptions.

2.31 The matrix printerhead positioning system of Fig. 1.16b. Describe the horizontal motion of the printhead on the rails. Assume that the belt does not slip on the pulleys.

2.32 The water clock (float-regulator and regulating vessel) of Example 1.1.

2.33 The *seismic* instrument shown in Fig. 2.47 is used to sense and record inertial motions of the case. When used as

an *accelerometer* it indicates the acceleration of the case with respect to inertial space (the effect of gravity is superposed). When used as a *velocimeter* it indicates the velocity of the case with respect to an inertial frame. As a *seismograph* the instrument indicates the displacement of the case x. In this mode, which is illustrated in the figure, the seismograph can be used to deduce ground displacement $x_g(t)$, during earthquakes for example.

2.34* The loading system shown in Fig. 2.48. The mass of the cable is negligible. The pulley is heavy and experiences viscous resistance in its pivot. The mass is subject to (dry) sliding friction. The motion of the tow truck $x(t)$ is given.

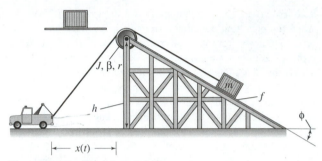

Figure 2.48   (Problem 2.34)

2.35 The compressor model shown in Fig. 2.49. There is no leak in the chambers, and the gas behavior can be considered elastic. Coulomb friction is present in the cylinders, and there is viscous friction in the flywheel pivot. The torque developed by the drive motor (not shown) is given.

Figure 2.49   (Problem 2.35)

2.36* The hydroelectric power plant whose physical model was constructed in Problem 2.12.

2.37 The typing head drive mechanism of Problem 2.13.

2.38 The car front suspension system of Problem 2.14.

2.39 A ship propulsion system given in Problem 2.15. Assume that the rotor motion is viscously damped and that the load on the propeller, due to the sea, is proportional to the square of the propeller speed.

2.40 The fluid clutch of Problem 2.16.

2.41 A cam and follower mechanism as given in Problem 2.17.

2.42 (a)* The experimental sled system shown in Fig. 2.50. Note that the no-slip condition implies that m_2 rolls on m_1.

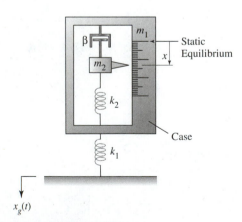

Figure 2.47   (Problem 2.33)

2.42 (b) Repeat Problem 2.42a but assume that m_2 slips on m_1.

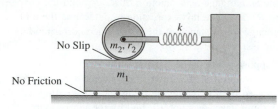

Figure 2.50 (Problem 2.42)

2.43 A model of an elevator system as shown in Fig. 2.51. The cable makes several turns around the drive pulley. The motor speed is given.

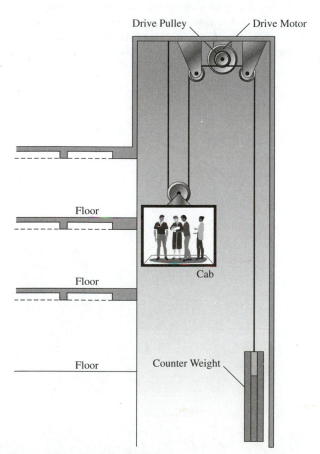

Figure 2.51 (Problem 2.43)

2.44* The hydraulic arrangement for lifting heavy weights remotely (see Fig. 2.52). No long fluid lines are involved, and the mass of the small cylinder is negligible.

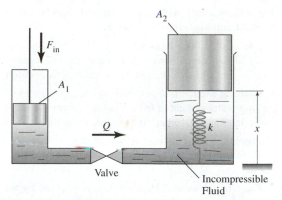

Figure 2.52 (Problem 2.44)

2.45 The system model for the bow and arrow problem (decomposed in Example 2.7) for viscous damping can be shown to be

$$\frac{dx}{dt} = v$$

$$m\frac{dv}{dt} = \begin{cases} -kx - \beta v, & x \le 0 \\ -\beta v, & x > 0 \end{cases}$$

Is this model linear or nonlinear? Explain your answer.

2.46 Obtain a linearized damping coefficient β for the total frictional force on the race car of Problem 2.5 if it operates at around 100 km/h.

2.47* Obtain a linear relation for the gasoline level in the funnel of Problem 2.4. Assume that operation is around $Q_{in} = 5$cc/s, $Q_{out} = 5$cc/s, and $h = 10$ cm.

2.48 The behavior of a certain linear system can be described by $x = f(u)$. If $x = 7.5$ when $u = 5.7$ and $x = 6.0$ when $u = 4.2$, what will be the value of x when

(a) $u = 1.5$? **(b)** $u = 9.9$? **(c)** $u = 2.7$?

2.49 Is the system of Problem 2.18 best described by a distributed-parameter model or a lumped-parameter model? Justify your answer.

2.50 Which of the assumptions discussed in Section 2,4—lumped system, linearity, stationarity, and deterministic—are made in each of the following situations?

(a) In analyzing the auto suspension system of Problem 2.14, a road profile resulting in the displacement input to the tires, $x = x_0\sin \omega t$, is used.

(b) In the motor-actuated control valve problem of Example 2.8, the frictional torque is taken as $T_f = -rF_f$, independent of angular position of the rotor.

(c) The thermal resistance of heat transfer surfaces in a heat exchanger is assumed fixed in spite of fouling (corrosion, deposition, erosion, etc.).

(d) The area of a tank with slightly tapered sides is assumed constant and independent of height.

(e) The seismograph of Problem 2.33, although calibrated while in California, is assumed to give the exact readings of ground displacement in an Antarctic expedition.

(f) In studying the thermal pollution in a lagoon due to cooling water discharge from a nearby power plant, the entire lagoon is assumed to have one temperature.

(g) In Problem 1.6 the number of apples that spoil each day is given as 5% of the total number of apples in the supermarket.

2.51* The strain measured by a strain gage is given by the unit change in resistance of the gage divided by the *gage factor,* a constant for the gage. When employed in strain measurement, the unknown resistor R_3 of the Wheatstone bridge circuit of Fig. 2.38 is replaced with the active gage that is (attached to the test specimen) subject to strain. Usually the resistor R_4 is also an identical strain gage but is not subject to strain (dummy gage). In an experiment, let the value of R_1 be fixed at 1000 Ω, and let the intrinsic resistance of the gages be 1.0 Ω. With no strain, the bridge is balanced with $R_2 = 1000$ Ω. Suppose when the active gage is subjected to strain the bridge is balanced with $R_2 = 1010$ Ω

(a) derive the expression for the unit change in resistance of the active gage as a function of R_1, R_2, and the change in R_2. Evaluate this expression to determine the unit change in resistance of the active gage corresponding to the test strain.

(b) Suppose while the test specimen was under strain the temperature (of the active strain gage environment) increased by 5°C, while the bridge area, including the dummy gage, was not affected by the temperature change. Determine the error in the estimated unit change in resistance of the active gage if the temperature coefficient of resistivity of the gage material is 0.0005 Ω/Ω-°C.

(c) Show that if the dummy gage had been placed in the same environment as the active gage so that it was subject to the same temperature effects but not subject to the test strain, the error in the estimated unit change in active gage resistance would have been negligible, that is, *temperature compensation* of the bridge circuit would have been achieved.

Chapter

3

Engineering System Models in State Space

F OR MANY DYNAMIC ENGINEERING SYSTEMS, the construction of the mathematical model from a given physical model can be relatively straightforward. This is because the behaviors of the components of most engineering systems are, as we saw in Chapter 2, governed by well-known natural laws (Newton's laws, Hooke's law, Ohm's law, etc.), and the problem of recognizing the presence of these behavior components and properly interconnecting them must have already been solved during the process of developing the physical model. However, various procedures are available and are in use for deriving system equations from a given physical model. In this chapter, the state modeling procedure will be explained and used to derive system models from the physical models of various engineering problems. The theoretical mechanical and electrical behavior components developed in Chapter 2 will be used where applicable, and new constitutive relations will be introduced when needed. In particular, lumped-parameter models of (strictly) mechanical and electrical systems will be considered in addition to incompressible fluid systems and simple electrical-electronic system examples. The discussion will continue in Chapter 4 with thermal energy and chemical engineering systems and a few examples illustrating distributed-parameter modeling.

3.1 THE STATE SPACE APPROACH

In general most modeling approaches can be viewed as belonging to one of two groups of methods: those that are based on formal considerations of physical cause and effect in the system components and those that consider other phenomena. In the latter group are the energy balance methods (Lagrange's method for example), which are based on the law of conservation of energy. The *state space approach* used in this book is a member of the first group. This method is considered particularly attractive because, in our opinion, it provides the user with sufficient opportunity to develop and to apply physical intuition and because it is essentially transparent to the type of system being modeled—mechanical, electrical, chemical-process, energetic/nonenergetic, conservative/nonconservative, and so forth.

The Concept and Definition of State

The idea of *state* is a fundamental concept in the state space method. Let us examine this concept from the already established viewpoint of a dynamic system. Recall that a dynamic system was described as one with *memory*. The output (or, more generally, the behavior or disposition or configuration or whatever quantity of interest) of a dynamic system at any instant is determined by the effects of the inputs acting on it at that very instant, as well as by the "stored" effects of inputs that acted on it in the past. Imagine this store or memory of effects to be like a digital computer memory. Each unit or cell of the memory holds a number. The set of numbers at any instant, such that given the inputs acting on the system at that instant and for all future time, is sufficient to determine the behavior or disposition of the system for that instant and for all future time, constitutes the state of the system at that instant. The state of a system is thus like a small code that is sufficiently compact so that given the system inputs, it can predict the future behavior of the system, and it can update itself.

More formally, we define the state of a dynamic system as follows:*

*The **state** of a dynamic system at any time t_0 is the **smallest set of numbers** that is sufficient to determine the behavior or disposition of the system for all time $t \geq t_0$, given these numbers and the inputs at time t_0 as well as the inputs for all time $t \geq t_0$.*

In the preceding definition, the idea of a minimal set of numbers is convenient but not necessary. A larger set of numbers that satisfies all the other requirements could also constitute the state. It will simply contain some superfluous or redundant data. Another factor in the definition is that the time t_0 is arbitrary. The time we commence our observation of the system becomes the beginning time t_0, which demarcates the times constituting the *past* from the times referred to as the *future*. Further, by *time* we mean continuous time t or discrete time $(\ldots, t_{-2}, t_{-1}, t_0, t_1, t_2, \ldots)$, so that the definition applies to both continuous-time and discrete-time systems.

Finally, it is apparent from the foregoing that *state* is a quantitative or mathematical concept. However, since physical phenomena are the subject of our mathematical models, the state of a system is also a physical concept. **The state of a dynamic system represents the minimum amount of information about the system at time t_0, that is necessary to determine its future behavior from its inputs at t_0 and its future inputs** (that is, without reference to its inputs before time t_0). It follows that one way to construct a system model is to write equations describing the state of the physical model with respect to time, since the behavior of the system can be obtained by solving this model for the state at any instant. This is the state modeling concept or the state space approach. A logical question to ask at this point is, How do we mathematically describe the state of a physical system?

State Variables and the State Equation

From the definition of state, we can infer that the state of a physical system can be described by any set of variables whose values at any instant represent the minimal set of numbers referred to in the definition. We shall call the set of **linearly independent** variables used to specify the state of a system the *state variables*.

A set of n variables $x_1, \ldots, x_n$ is said to be *linearly independent* if there is no set of constants $c_1, c_2, \ldots, c_n$, not all zero, that satisfy the equation

$$c_1 x_1 + c_2 x_2 + \cdots + c_n x_n = 0 \qquad \textbf{(3.1)}$$

If a set of constants other than one of all zero constants satisfies the equation, then the set of variables is said to be *linearly dependent* (see Appendix B, equation (B1.6b)). The number of (scalar) state variables determines the *order* of a system. A system

---

*More rigorous definitions of state are available. See for example, Zadeh, L. A., and Desoer, C. A., *Linear System Theory—the State Space Approach.* McGraw-Hill, 1963.

whose state is described by a finite number of state variables is called a *finite order system*. As we shall see, system models derived using lumped-parameter analysis are finite order, whereas distributed-parameter systems require theoretically an infinite number of state variables to describe their state.

The State Equation.   The state variables must be formulated in such a way that if one knows their values at a given instant, together with the values of the input variables for that time and all future time, then the disposition of the system and these variables is completely determined for that time and all future time. A mathematical construct that particularly satisfies this requirement is the initial-value problem of a first-order ordinary differential (for continuous-time systems) or difference (for discrete-time systems) equation. Given the value of the dependent variable at any instant (initial condition) and the values of the input variables for that instant and all subsequent time, the differential (or difference) equation can be solved for the value of the dependent variable for all time from that instant.

Suppose we have n state variables: $x_1, x_2, \ldots, x_n$, and suppose our system is subject to r (external) inputs: $u_1, u_2, \ldots, u_r$ and is characterized by a number of parameters and constants (values), all of which we shall generally refer to as parameters (p's). Then the set of n first-order ordinary differential equations,

$$\frac{d}{dt} x_1 = f_1(x_1, x_2, \ldots, x_n, u_1, u_2, \ldots, u_r, p\text{'s}, t); \qquad x_1(t_0) = x_{1-0} \qquad \textbf{(3.2a)}$$

$$\frac{d}{dt} x_2 = f_2(x_1, x_2, \ldots, x_n, u_1, u_2, \ldots, u_r, p\text{'s}, t); \qquad x_2(t_0) = x_{2-0}$$

$$\vdots$$

$$\frac{d}{dt} x_n = f_n(x_1, x_2, \ldots, x_n, u_1, u_2, \ldots, u_r, p\text{'s}, t); \qquad x_n(t_0) = x_{n-0}$$

where t refers to time and t_0 is some initial time, are *state equations*. For discrete-time systems (with constant sampling period Δt), we can write the corresponding difference equations,

$$x_{i,k+1} = f_{i,k}(x_{1,k}, x_{2,k}, \ldots, x_{n,k}, u_{1,k}, u_{2,k}, \ldots, u_{r,k}, p\text{'s}, t_k); \qquad t_k = k\,\Delta t$$

$$x_{i,k_0} = x_{i-0}; \qquad t_0 = k_0\Delta t \qquad\qquad \textbf{(3.2b)}$$

$$i = 1, 2, \ldots, n$$

where the subscript i refers to a particular state variable x_i, and the subscript k is a discrete-time index, defined in equation (3.2b).

It is important to note carefully the format of the state equation. The left-hand side (LHS) is the time **derivative** (or time shift-**advance**) of the state variable, whereas the right-hand side (RHS) is an **algebraic** function of state variables, inputs, parameters, and time, **only.** The right-hand function does not have to contain all the state variables, inputs, parameters, or time but **it cannot contain any other variable that is not one of these** unless a supplemental **algebraic** (not a differential or difference) equation is written defining that variable in terms of only state variables, inputs, parameters, and time. Finally, there is a state equation for each state variable (with no exception). Since an incomplete or improperly formulated model will often yield undefined variables in the state equations, consistent and strict observance of the state equation format described above gives the state space approach to system modeling a self-checking character.

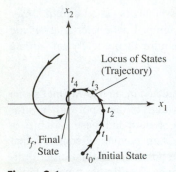

Figure 3.1

State Space for a
Second-order System

The State Space. We describe the *state space* as the mathematical (*n*-dimensional) space whose coordinates (see Appendix B) are the state variables. The state of the system at any instant is thus represented by a "point" in the state space. This description explains the use of the term *state space approach* to describe system modeling on the basis of state equations. The state space is easily visualized when it consists of only two state variables (a plane in this case), as shown in Fig. 3.1. For higher order systems, graphical visualization of the state space is difficult. However as we shall see in Chapter 9, certain concepts in dynamic system analysis could be explained perhaps more vividly when interpreted in state space. As illustrated in Fig. 3.1, the solution of the state equation, for instance, can be represented in state space by a locus of states or a *trajectory*. Trajectory patterns for various initial states collectively can reveal a great deal about the dynamic behavior of a system (second-order system trajectories are discussed in Section 9.4).

| **Example 3.1** | Bow and Arrow Problem: Trajectory of Arrow |

Consider the physical model of the bow and arrow system that was decomposed in Example 2.8.

a. Determine the state-equations model for the system.

b. Use a computational mathematics package such MATLAB to solve the state equations obtained in (a), and display the results in state space. Let $x_{in} = 0.1$ m, and assume the following parameter values: $m = 0.12$ kg, $k = 0.12$ N/m, and $\beta = 0.6$ N $\cdot$ s^2/m^2.

Solution a: Although we have yet to suggest how to select state variables, the state of motion of a free mass element (the arrow in flight) should be adequately represented by its velocity, considering the property of inertia and Newton's second law. Equation (b) of Example 2.7 gives the velocity of the arrow as

$$m \frac{d}{dt} v = \begin{cases} F_s - F_d; & (x \le 0) \\ -F_d; & (x > 0) \end{cases}; \qquad v(0) = 0 \tag{a}$$

Substituting for F_d and F_s from the results of Example 2.7, we have

$$m \frac{d}{dt} v = \begin{cases} -kx - \beta v|v|; & (x \le 0) \\ -\beta v|v|; & (x > 0) \end{cases}; \qquad v(0) = 0 \tag{b}$$

According to the state equation format, equation (3.2a), x in the preceding equation must either be a state variable or be algebraically related to v. However, equation (c) of Example 2.7 confirms that x is a (second) state variable for this system. Thus the state model is as follows:

$$\frac{d}{dt} x = v; \qquad x(0) = -x_{in}$$

$$\frac{d}{dt} v = \begin{cases} -\dfrac{k}{m} x - \dfrac{\beta}{m} v|v|; & (x \le 0) \\[2mm] -\dfrac{\beta}{m} v|v|; & (x > 0) \end{cases}; \qquad v(0) = 0 \tag{3.3}$$

where there are two state variables: $x_1 = x$, and $x_2 = v$.

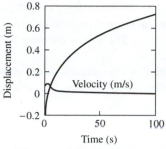

(a) Motion in Time Domain

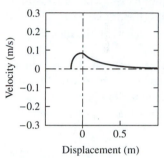

(b) Motion in State Space

Figure 3.2

Horizontal Motion of Arrow

Solution b: First, create a function M-file, called **arrow.m,** containing the differential equations (3.3):

```
function xdot = arrow(t,x)
km = 0.12/0.12; bm = 0.6/0.12;
xdot = zeros(2,1);
xdot(1) = x(2);
xdot(2) = - bm.*x(2).*abs(x(2));
if x(1) <= 0
    xdot(2) = xdot(2) - km.*x(1);
end
```

To solve the differential equation defined in **arrow.m** over the time interval $0 \leq t \leq 100$ and generate the state space trajectory, we invoke the fourth-/fifth-order Runge-Kutta-Fehlberg method (ode45) in the following script M-file called **fig3_2.m:**

```
% Script to generate Figure 3.2
% Uses function M-file, arrow.m
%
t0 = 0; tf = 100;        % time interval
x0 = [-.1 0]';            % initial conditions
[t,x] = ode45('arrow',t0,tf,x0);
%
axis('square');                  % square aspect ratio
subplot(211);                    % split graph window
plot(t,x); title(' (a) Motion in Time Domain');
ylabel('DISPLACEMENT (m)');
xlabel('TIME (sec)'); text(17,.02,'VELOCITY (m/sec)');
%
axis([-.3 .9 -.3 .3]);    % directly scale axes
plot(x(:,1),x(:,2)); title (' (b) Motion in State Space');
ylabel('VELOCITY (m/sec)'); xlabel('DISPLACEMENT (m)');
hold on; plot([-.3 .9],[0 0],'-.');
plot([0 0],[-.3 .3],'-.'); hold off;
```

Figure 3.2a shows the response of the arrow, that is, the horizontal displacement and velocity as a function of time. The horizontal trajectory in state space is shown in Figure 3.2b. These are the results generated by the MATLAB script and function files described above.

The State Modeling Procedure

Given the physical model of a dynamic system, we recommend the following method for deriving the state model:

1. **Perform system decomposition** (as described in Section 2.4, that is, identify components—draw free body diagrams, show all variables, inputs, interactions, sign convention, separate dynamic and static elements, and write relations governing the behavior of each element).

2. **Assign state variables** *(to the dynamic components as a first attempt).*
3. **Write the state equation for each independent state variable** (using the relations from step 1 and any additional relations between variables.) *The format for the state equations must be followed.*
4. **From considerations of the objectives of the model, write output equations and/or modify the state equations.** *Specify what constitutes the final system model.*

Output Equations and State Model

In the final step in the state modeling method, the output equations are concise expressions relating output quantities (which represent the modeling objectives) with the state variables. It is good practice always to write an output equation even when it is a simple statement setting a variable representing the output quantity (the *output variable*) equal to one of the state variables. This ensures completeness and clarity in the modeling effort.

In the state space approach, an output equation is necessarily an algebraic (not a differential or difference) equation, since otherwise new state variables would have to be defined. Thus it is sometimes necessary in step 4 of the modeling process to substitute state variables or define additional ones. The format for the output equation (for continuous-time systems) can in general be represented as follows:

$$y_1 = g_1(x_1, x_2, \ldots, x_n, u_1, u_2, \ldots, u_r, p\text{'s}, t) \tag{3.4}$$

$$y_2 = g_2(x_1, x_2, \ldots, x_n, u_1, u_2, \ldots, u_r, p\text{'s}, t)$$

$$\vdots$$

$$y_m = g_m(x_1, x_2, \ldots, x_n, u_1, u_2, \ldots, u_r, p\text{'s}, t)$$

where there are *m* output variables: $y_1, \ldots, y_m$, and $g_1, \ldots, g_m$ are (algebraic) functions of the state variables and parameters, and possibly of inputs and time. Thus a complete state model consists of the following set of equations:

$$\frac{d}{dt} x_i = f_i(x_1, x_2, \ldots, x_n, u_1, u_2, \ldots, u_r, p\text{'s}, t); \qquad x_i(t_0) = x_{i-0}$$

$$i = 1, 2, \ldots, n \tag{3.5a}$$

$$y_j = g_j(x_1, x_2, \ldots, x_n, u_1, u_2, \ldots, u_r, p\text{'s}, t); \qquad j = 1, 2, \ldots, m$$

For discrete-time systems, the corresponding form of the state model is

$$x_{i,k+1} = f_{i,k}(x_{1,k}, x_{2,k}, \ldots, x_{n,k}, u_{1,k}, u_{2,k}, \ldots, u_{r,k}, p\text{'s}, k\Delta t); \qquad x_{i,k_0} = x_{i-0}$$

$$i = 1, 2, \ldots, n \tag{3.5b}$$

$$y_{j,k} = g_{j,k}(x_{1,k}, x_{2,k}, \ldots, x_{n,k}, u_{1,k}, u_{2,k}, \ldots, u_{r,k}, p\text{'s}, k\Delta t);$$

$$j = 1, 2, \ldots, m$$

Assigning State Variables and Causality

Step 2 of the modeling procedure provides a beginning or a "handle" to state variable assignment, which is to choose the **output** variable (the *dependent variable* of the variable pair) in the relation describing the behavior of each of the dynamic components in the system. The state variables so assigned are considered a first choice, since subsequent activity in steps 3 and 4 of the modeling procedure could result in the selection of additional state variables (recall Example 3.1 and the introduction of the displacement x as a state variable) or even the replacement of some of them.

Which variable in an expression of component behavior is the output? The explicit identification of input and output or independent and dependent variables is called *causality*. Causality can be *natural* (that is physical) or *assigned*. It can be *reversible* or *nonreversible*. By studying certain aspects of natural causality we shall derive a definite criterion for assigning causality to dynamic components and hence the basis for assigning state variables.

Causality: Natural and Assigned

In general, the variable we assign as input or output is a reflection of the way we see the behavior of a system component. Consider the water supply system example of Chapter 1 (see Fig. 1.1). The level or amount (volume) of water in the tank is a logical quantity of interest in describing a water tank. This leaves the rates of water flow into and out of the tank as the inputs. However, for the valve, the rate of water flow through the valve, which is also the water flowrate out of the tank, is of particular interest within the context of the farm water supply. Thus depending on the circumstances, a given signal may be input or output for different components or systems. However, for a given element under a specified set of circumstances, the causality for the signal can be fixed and may not be reversed. If the causality for the variables describing a component is fixed no matter the circumstances, we say that the component has *nonreversible* causality. As we shall see shortly, the causality assignment to the water tank as a dynamic component is nonreversible. In contrast, the valve can be found in situations where the pressure drop caused by the flow through the valve is the proper output. Such will be the case for instance for a valve connected immediately downstream of a power pump.* Thus the valve is an example of a component with *reversible* causality. In general the causality assignments to static elements are reversible.

Natural Causality.   For certain elements the causality for the usual variables is fixed by the physical nature of these systems. A good example is a pure (with negligible internal resistance) electrical capacitor. If such a capacitor (initially discharged) is suddenly connected across a battery that also has a very small internal resistance (an ideal voltage source), the result will likely be a small explosion and destruction of the capacitor! (See Fig. 3.3a.) A similar set of circumstances may have been created inadvertently, through shorting of wires, by some startled hobbyists who at one time or another have poked carelessly inside an older radio or amplifier circuit. The electrical capacitor is characterized by a *natural causality* that demands that the input to the capacitor be a current rather than a voltage. This is shown in Fig. 3.3b, in which the same capacitor and battery are connected together through a series resistor. A resistor, as a static element, has reversible causality, so that with a voltage input to the resistor, its output and therefore the input to the capacitor is a current. In the discussion of physical modeling examples in Chapter 2 (see Fig. 2.11), it was pointed out that behavior components must be properly interconnected to prevent causal conflicts. Figure 3.3a represents a connection with a causal conflict. Similarly in Fig. 2.11b, with a velocity or displacement input from the road specified, the output of the spring and hence input to the damper is the force. This implies a velocity output for the damper and thus a velocity

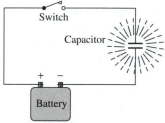

(a)   Casual Conflict: Voltage Input (Attempted)

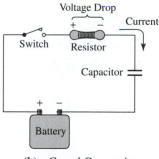

(b)   Casual Connection: (Input is Current)

Figure 3.3

Natural Causality and an Electrical Capacitor.

---

*The *Standard Handbook for Mechanical Engineers,* 7th ed., McGraw-Hill, describes power pumps as "positive displacement machines which, at constant speed, deliver essentially the same capacity at any pressure within the capability of the driver and the strength of the pump."

input for the mass. As shown in Example 3.2, the *causal* input for a mss is force, so the physical model of Fig. 2.11b is also in causal conflict.

Assigned Causality for Dynamic Components.   The constitutive behavior of the ideal electrical capacitor in terms of the current i and the voltage drop in the direction of the current V is equation (2.24a) repeated:

$$V = \frac{1}{C} \int^t i \, dt \tag{3.6a}$$

which is a dynamic relation. If we differentiate equation (3.6a) with respect to time, the (mathematically) alternative expression, equation (2.24b), is obtained:

$$i = C \frac{d}{dt} V \tag{3.6b}$$

As we have already seen, the causality implied in equation (3.6b)—voltage is input and current is output—is not natural and is incorrect for the capacitor. There is an even more general rational that we can apply to other dynamic elements as well. In equation (3.6b) the current is obtained by differentiating the voltage (drop). We recall the definition of the derivative operator:

$$\frac{d}{dt} V = \lim_{\Delta t \to 0} \left[\frac{V(t + \Delta t) - V(t)}{\Delta t} \right] \tag{3.7}$$

A proper interpretation of definition (3.7) shows that we need *knowledge of the future,** $V(t + \Delta t)$, in order to differentiate! Because this is not possible physically (unless one is a magician), we can say that *differentiation is unrealizable.* Integration, that is, equation (3.6a), on the other hand, is realizable, since it depends only on present and past inputs (see Section 1.1).

The preceding arguments also apply to the delay operator. A time delay is realizable, whereas the time advance is not. The *assigned causality* for dynamic elements should now be clear. *The input is the variable that is integrated or time-delayed.* Note again that the time derivative and time advance used in state equations to represent dynamic behavior are mathematical expressions. The implied behavior is actually described by a time integral or a time delay.

Example 3.2	Assigning State Variables

Assign causalities to the behavior of an ideal translatory linear spring, mass, and damper using the variables resultant force F and velocity v.

Solution: Refer to the free body diagrams shown in Fig. 3.4.

Spring behavior $F = k \int^t v \, dt$

State equation: $\dfrac{d}{dt} F = kv$ or $\dfrac{d}{dt} x = v$ and $F = kx$

---

*In numerical analysis the derivative operator may be *approximated* by expressions that require only present and previous data, such as the backward difference: $[V(t) - V(t - \Delta t)]/\Delta t$.

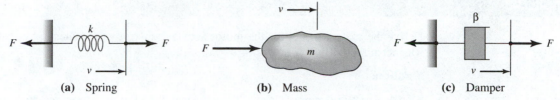

(a) Spring   **(b)** Mass   **(c)** Damper

Figure 3.4

Free Body Diagrams for Spring, Mass, and Damper

Causality:	output = F or x; input = v
Mass behavior	$mv = \int^t F\,dt$
State equation:	$\dfrac{d}{dt}v = \left(\dfrac{1}{m}\right)F$
Causality:	output = v; input = F
Damper behavior	$F = \beta v$ or $v = \left(\dfrac{1}{\beta}\right)F$ (static element)
Causality:	(output = F, input = v) or (output = v, input = F)

"The proof of the pudding is in the eating." In the examples that follow, the state space approach to system modeling will be tested using mechanical, electrical, and other engineering systems. The reader is urged to adopt a similar philosophy to all significant ideas introduced in this book—*test them!* Familiar examples and practice problems at the end of each chapter should be used for this purpose.

3.2 MECHANICAL SYSTEMS

We begin with systems that can be viewed as (strictly) mechanical. Many of the modeling concepts discussed in preceding chapters were illustrated with examples of this type. The usual behavior components are springs, masses, and frictional elements. In one-dimensional mechanical systems, the interconnection of these components and the constraints on their motion are such that the system configuration can be adequately described in one-dimensional space. This makes the mathematical modeling process relatively simple, since vectorial analysis is then not necessary. The behavior of systems whose configurations are specified in two or three dimensions (particularly the latter) of space is most conveniently described in terms of vector relations. However, the state variables in the state space method are scalar quantities, so the component expressions of the vector relations should be used with the state space approach.

One-Dimensional Examples

In the Examples that follow it should be understood that the state modeling procedure outlined in Section 3.1 is being utilized even when the individual steps of the method are not identified explicitly. Also, certain constitutive behaviors developed earlier will often be used without further reference.

Example 3.3   Motor-Actuated Control Valve: System Model

Derive the system model for the motor-actuated control valve whose physical model was decomposed in Example 2.8.

Solution:

Step 1—system decomposition: The physical model of the motor-actuated valve and the free body diagrams of the components were shown in Fig. 2.23. The results of the system decomposition carried out previously are equations (a) through (h) of Example 2.8. The dynamic elements are the spring and the rotor (rotary inertia).

Step 2—state variables: The output variables of the dynamic components are θ and ω. Hence let $x_1 = \theta$, and $x_2 = \omega$.

Step 3—state equations: Beginning with equations (e) and (d) of Example 2.8, we have

$$\frac{d}{dt}\theta = \omega \tag{a}$$

$$\frac{d}{dt}\omega = \left(\frac{1}{J}\right)[T_m - T_f - k_r\theta]$$

Using equations (f), (a), and (h) of Example 2.8, we can eliminate T_f in terms of ω:

$$T_f = rF_f = r\beta v = r^2\beta\omega$$

According to the explanation in Example 2.2, T_{in} can be assumed to be directly proportional to the input current: $T_{in} = K_T i_{in}$. Thus the state equations are

$$\left.\begin{array}{l} \dfrac{d}{dt}\theta = \omega \\[2mm] \dfrac{d}{dt}\omega = \left(\dfrac{1}{J}\right)[K_T i_{in} - r^2\beta\omega - k_r\theta] \end{array}\right\}; \quad \text{state equations} \tag{b}$$

Step 4—output equations: Recall from Example 2.2 that the objective of the model is to be able to determine the flow through the valve given the pressure drop across the valve and the current into the motor. Thus the output should be $y = Q$, and from equations (b) and (c) of Example 2.8,

$$Q = cx\,\text{sign}(P_1 - P_2)[(|P_1 - P_2|)^{1/2}]$$

However, x needs to be eliminated in favor of state variables, inputs $[u_1 = i_{in}, u_2 = (P_1 - P_2)]$, and parameters $[c, r, J, k_r, \beta, K_T]$. From equation (g) of Example 2.8: $x = r\theta$, the output equation is

$$Q = cr\,\text{sign}(P_1 - P_2)[(|P_1 - P_2|)^{1/2}]\theta; \quad \text{output equation} \tag{c}$$

The final system model is thus

$$\left.\begin{array}{l} \dfrac{d}{dt}\theta = \omega \\[2mm] \dfrac{d}{dt}\omega = \left(\dfrac{1}{J}\right)[K_T i_{in} - r^2\beta\omega - k_r\theta] \\[2mm] Q = cr\,\text{sign}(P_1 - P_2)[(|P_1 - P_2|)^{1/2}]\theta \end{array}\right\}; \quad \text{system model} \tag{3.8}$$

Figure 3.5

**Physical Model and Free
Body Diagram of Components
of Grain Scale**

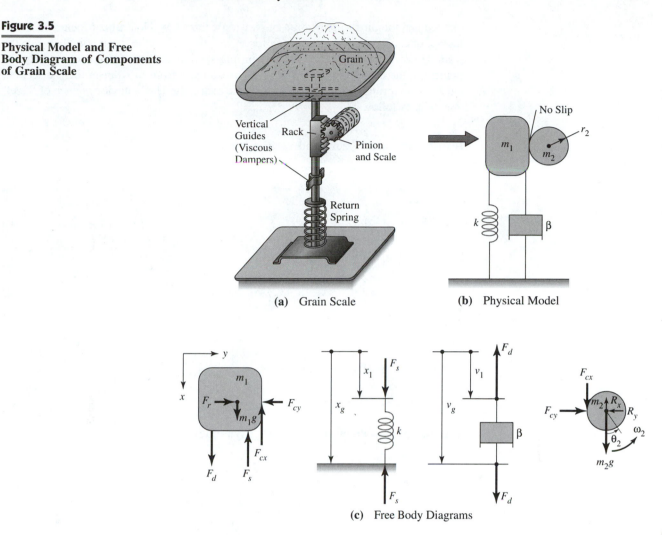

(a) Grain Scale

(b) Physical Model

(c) Free Body Diagrams

| Example 3.4 | Grain Scale |

A particular grain scale operates by means of a rack and pinion gear arrangement. The rack is attached to the weighing platform and moves with it. A calibrated dial is attached to the gear and forms an integral part of the gear. The motion of the rack is constrained by greased vertical guides. A return spring connecting the rack to the base of the scale is used to bring the platform back to neutral position. Construct a model for this system that can be used to determine how long to wait before reading the weight of a measure of grain after placing it on the platform.

Solution: On the basis of the system description, the physical model shown in Fig. 3.5a is constructed and is further reduced to the model in Fig. 3.5b. Here the mass of the typical quantity of grain has been combined with the mass of the rack, giving m_1. Backlash in the meshing gears has been ignored, and the radius r_2 is the pitch radius.

Step 1: We have a translatory mass m_1, a rotating mass m_2, a spring k, and a dashpot β, which are shown in arbitrary configurations in Fig. 3.5c. Note the sign convention adopted and the direct application of Newton's third law. Note in particular the introduction of equal and opposite contact forces, F_{cy} and F_{cx}, at the mesh point between the gears. However m_1 is idealized as a point mass, so that all the forces are actually coincident at the center of mass. The rack is restrained from moving in the y-direction, and F_r is the resultant force developed by the restraints. The behavior of each of the elements is as follows:

Translatory mass—grain and rack m_1 (dynamic):

$$m_1 \frac{d}{dt} v_1 = m_1 g - F_s + F_d - F_{cx} \tag{a}$$

$$F_r - F_{cy} = 0 \tag{b}$$

Return spring k (dynamic):

$$F_s = k(x_1 - x_g) = kx_1 \qquad (x_g = 0; \text{ boundary interaction}) \tag{c}$$

$$\frac{d}{dt} x_1 = v_1 \tag{d}$$

Dashpot—greased guides β (static):

$$F_d = \beta(v_g - v_1) = -\beta v_1 \tag{e}$$

Rotating mass—pinion and scale m_2 (dynamic):

$$J_2 \frac{d}{dt} \omega_2 = F_{cx} r_2; \qquad J_2 = \frac{1}{2} m_2 r_2^2 \tag{f}$$

$$F_{cy} - R_y = 0 \tag{g}$$

$$F_{cx} - R_x + m_2 g = 0 \tag{h}$$

Step 2: The first state variable assignment is v_1, x_1, and ω_2, that is, the outputs of the three dynamic components.

Step 3: The state equations that can be derived from equations (a), (d), and (f) involve the unknown variable F_{cx}, which cannot be eliminated using any of the relations described thus far. Hence more equations are required.

Additional relations between variables: We note that

$$\frac{d}{dt} \theta_2 = \omega_2 \tag{i}$$

and for no slip between the rack and pinion,

$$r_2 \omega_2 = v_1 \quad \text{(internal interaction)} \tag{j}$$

Hence

$$r_2\theta_2 = x_1 \tag{k}$$

v_1 and ω_2 are therefore dependent (so are x_1 and θ_2), and the number of state variables is reduced to 2. ω_2 and θ_2 will be used because they are more directly related to the scale indication. F_{cx} can now be eliminated using equations (a), (f), and (j). Substituting (j) in (a) and comparing with (f), we obtain

$$m_1 r_2 \frac{d}{dt}\omega_2 = m_1 g - F_s + F_d - F_{cx} = 2\frac{m_1}{m_2}F_{cx} \Rightarrow \left(1 + 2\frac{m_1}{m_2}\right)F_{cx} =$$

$$m_1 g - F_s + F_d$$

$$F_{cx} = \left(\frac{m_2}{2m_1 + m_2}\right)(m_1 g - F_s + F_d) \tag{l}$$

Thus the state equations are

$$\frac{d}{dt}\theta_2 = \omega_2 \tag{3.9a}$$

$$\frac{d}{dt}\omega_2 = \left(\frac{2}{r_2}\right)\left(\frac{1}{2m_1 + m_2}\right)(m_1 g - kr_2\theta_2 - \beta r_2\omega_2)$$

Step 4: The object of this model is actually to predict the angular oscillations in the scale. Thus the output is

$$y = \theta_2 \tag{3.9b}$$

The state model then is given by equations (3.9a) and (3.9b). Note that this example illustrates a situation in which the eventual state variables are different from the initially assigned set of variables. Finally, observe that equations (b), (g), and (h) were not needed in the development, thus making this essentially a one-dimensional problem.

....

Cables: Length and Stretch

The following suggestions are offered for the analysis of problems involving extensible cables:

1. Establish a reference point of zero velocity, preferably (if the problem allows) on the cable.
2. If the reference point is on the cable, then the length of the cable at any time is given by the distance between the free end of the cable and the reference point, **measured along the cable.**
3. If the reference point is outside the cable, the length of the cable at any time is given by the difference of the distances between the two free ends of the cable and the reference point, **measured along the cable.**
4. To determine the stretch in the cable, assume one end of the cable is fixed and consider the displacement of the other end. This displacement will contribute positively to the stretch in the cable if it tends to elongate (that is, stretch) the cable. The displacement is subtracted from the stretch (negative contribution) if it tends to relax (that is, buckle) the cable. Repeat this process for the other end of the cable to obtain the total stretch in the cable.

Example 3.5   Loading Arrangement

Consider the physical model of a loading system shown in Fig. 3.6a. The mass of the cable is negligible. The pulley is heavy and experiences viscous resistance in its pivot. There is no slip between the pulley and the cable. The load is subject to dry sliding friction. If the displacement of the tow vehicle x is prescribed, derive a model of this system that could be used to determine at any time the progress of the load up the incline.

Solution:

Step 1: The components of this system include cables, a translational mass (load), a rotating mass (pulley), viscous friction (in pulley pivot), and dry friction between the load and the surface of the incline. Because of the no-slip condition the points of contact between the cable and the pulley are points of instantaneous zero velocity, and the cable can thus be considered as two cables with reference points as shown in Fig. 3.6b.

Figure 3.6

Physical Model and Free Body Diagram of a Loading System

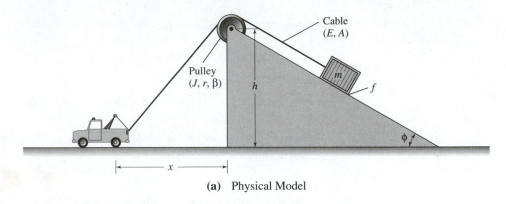

(a) Physical Model

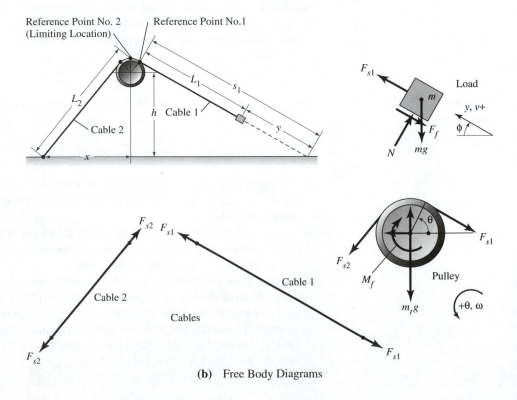

(b) Free Body Diagrams

Cables (dynamic):

a. The length of cable 1, $L_1(t)$, is the distance between the mass end of the cable and reference point no. 1. That is,

$$L_1(t) = s_1 - y(t) \approx h \operatorname{cosec} \phi - y(t) \tag{a}$$

where the approximation is based on neglecting the radius of the pulley relative to s_1. The exact value of s_1 can of course be calculated if necessary.

b. If we fix the pulley end of cable 1, the displacement of the mass end y tends to relax the cable. If the mass end is considered fixed, the displacement of the pulley end $r\theta$, (note the sign convention adopted) tends to stretch the cable. Hence the stretch in cable 1 $\Delta L_1(t)$ is

$$\Delta L_1(t) = r\theta(t) - y(t) \tag{b}$$

c. The length of cable 2, $L_2(t)$, is the distance between the tow vehicle end and reference point no. 2. This is given by geometry (again neglecting the contribution due to the curvature of the pulley and the variation with x of the actual point of tangency between the cable and the pulley) as

$$L_2(t) \approx \sqrt{\{x(t)^2 + (h+r)^2\}} \tag{c}$$

d. Consider the pulley end of cable 2 fixed. The displacement along the cable due to x is $L_2(t) - L_2(0)$, where $L_2(0)$ is the value of $L_2(t)$ when $x = 0$, that is, $L_2(0) = (h+r)$. This displacement tends to stretch the cable. If the vehicle is fixed, the rotation of the pulley tends to cause the cable to relax by the amount $r\theta$. Hence the stretch in cable 2 is

$$\Delta L_2(t) = \sqrt{\{x(t)^2 + (h+r)^2\}} - (h+r) - r\theta(t) \tag{d}$$

e. From equation (2.2c),

$$F_{s1} = (EA)\Delta L_1(t)/L_1(t) \tag{e}$$

$$F_{s2} = (EA)\Delta L_2(t)/L_2(t) \tag{f}$$

where E is the cable material modulus of elasticity, and A is the cross-sectional area.

Translatory mass (dynamic):

$$m\frac{d}{dt}v = F_{s1} - mg \sin \phi - F_f \tag{g}$$

$$0 = N - mg \cos \phi \Rightarrow N = mg \cos \phi; \qquad \text{(boundary interaction)} \tag{h}$$

Rotary inertia (pulley) (dynamic):

$$J\frac{d}{dt}\omega = (F_{s2} - F_{s1})r - M_f \tag{i}$$

Viscous friction (static):

$$M_f = \beta\omega \tag{j}$$

where β is the viscous damper constant

Dry friction (static):

$$F_f = fN \operatorname{sign}(v) = fmg \cos \phi \operatorname{sign}(v) \tag{k}$$

where f is the coefficient of dry friction.

Step 2: The initial set of state variables is y for cable 1, θ for cable 2, v for the translatory mass, and ω for the pulley. Note that the cable force and stretch are not convenient state variables to use in this problem. The stretch in each cable is, however, completely specified by y, θ, and x.

Step 3: The state equations are

$$\frac{d}{dt}y = v \tag{3.10a}$$

$$\frac{d}{dt}\theta = \omega \tag{3.10b}$$

From equations (g) and (h), (e), (a), and (b), and (k),

$$\frac{d}{dt}v = \frac{1}{m}\left\{\left[\frac{(EA)(r\theta - y)}{(h \operatorname{cosec} \phi - y)}\right] - mg \sin \phi - fmg \cos \phi \operatorname{sign}(v)\right\} \tag{3.10c}$$

From equations (i), (e) and (f), (c) and (d), (a) and (b), and (j)

$$\frac{d}{dt}\omega = \frac{1}{J}\left\{(EA)\left[\frac{\sqrt{x^2 + (h + r)^2} - (h + r) - r\theta}{\sqrt{x^2 + (h + r)^2}} - \frac{(r\theta - y)}{(h \operatorname{cosec} \phi - y)}\right] - \beta\omega\right\} \tag{3.10d}$$

Step 4: The desired output is y, and the state model is given by equations (3.10a) through (3.10d).

....

Multidimensional Examples

We consider now a few simple mechanical system examples that require two or more dimensions to describe their behavior in space. The relatively more complex examples such as the material on gyroscopic motion and Example 3.8 are presented here for completeness and as further evidence of the applicability of the state space method to such problems. The reader seeking more detailed treatment of the subject is again referred to a text in dynamics (such as reference [6]).

Example 3.6   Stone Crusher

The stone crusher consists of a large thin disk of mass m that is pin-connected to a horizontal axle. If this axle is turned by the application of the torque T_0 as shown in Fig. 3.7, derive a model of the system that will allow the determination of the normal force exerted by the disk on the stones. Assume that the disk rolls without slipping, and ignore the mass of the drive column and the axle.

Solution: The only relevant components are the disk (see free body) and the resistance to the rolling of the disk that results in the force F_f. The disk is an inertial behavior component—a dynamic component to which we tentatively assign the state variables

Figure 3.7

Simple Stone Crusher and Free Body Diagram

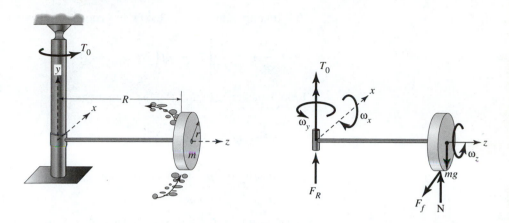

ω_x, ω_y, and ω_z (these being the x-y-z components of the angular velocity of the disk). Since the x-y-z axes are also principal axes of the disk, equation (2.15) is applicable:

$$T_x = J_{xx}\dot{\omega}_x - J_{yy}\Omega_z\omega_y + J_{zz}\Omega_y\omega_z$$
$$T_y = J_{yy}\dot{\omega}_y - J_{zz}\Omega_x\omega_z + J_{xx}\Omega_z\omega_x \qquad \textbf{(3.11)}$$
$$T_z = J_{zz}\dot{\omega}_z - J_{xx}\Omega_y\omega_x + J_{yy}\Omega_x\omega_y$$

The x-y-z coordinates have been chosen such that $\Omega_x = \Omega_z = 0$, and $\Omega_y = \omega_y$. The arrangement of the axle suggests $\omega_x = 0$, and the no-slip condition on the disk implies

$$R\omega_y = -r\omega_z \qquad \textbf{(a)}$$

Thus the number of independent state variables reduces to 1, ω_y. The state equation is given by the middle equation (3.11),

$$\frac{d}{dt}\omega_y = \frac{1}{J_{yy}}T_y \qquad \textbf{(b)}$$

From the free body diagram,

$$T_y = T_0 - F_f R \qquad \textbf{(c)}$$

Assuming that the rolling resistance is due to dry friction, with a coefficient of friction f, then we have

$$F_f = fN \qquad \textbf{(d)}$$

Further, from the free body diagram,

$$T_x = (mg - N)R, \text{ or } N = mg - \frac{1}{R}T_x \qquad \textbf{(e)}$$

From equation (3.11),

$$T_x = J_{zz}\omega_y\omega_z = -J_{zz}\frac{R}{r}\omega_y^2 \qquad \textbf{(f)}$$

Substituting equations (c) to (f) into equation (b), we obtain for the state equation,

$$\frac{d}{dt}\omega_y = \frac{1}{J_{yy}}\left[T_0 - Rf\left(mg + \frac{J_{zz}}{r}\omega_y^2\right)\right]$$ (3.12a)

From equations (e) and (f), the output equation is

$$N = mg + \frac{J_{zz}}{r}\omega_y^2$$ (3.12b)

Note that by the *parallel-axis theorem*,

$$J_{yy} = J_{y_1y_1} + mR^2$$ (3.12c)

and from Fig. 2.5 (circular cylinder with $L \approx 0$),

$$J_{yy} = \tfrac{1}{4}mr^2 + mR^2$$ (3.12d)

and

$$J_{zz} = \tfrac{1}{2}mr^2$$ (3.12e)

The remaining component of the torque T_z, if required, can be found from equations (3.11) and (3.12a):

$$T_z = J_{zz}\dot{\omega}_z = -J_{zz}(R/r)\dot{\omega}_y$$ (3.12f)

$$T_z = \frac{J_{zz}R}{J_{yy}r}\left[Rf\left(mg + \frac{J_{zz}}{r}\omega_y^2\right) - T_0\right]$$

Equations (3.12a) and (3.12b), however, constitute the solution of the present problem.

....

Gyroscopic Action and Gyroscopes

Gyroscopic motion occurs when the axis about which a body is *spinning* is itself also rotating about another axis. The disk of Example 3.6 executes gyroscopic motion because the spin axis z rotates about the y-axis.

Consider the symmetric rotor of Fig. 3.8, which is spinning about the z-axis at a high steady rate p. If a torque **M** about the x-axis is applied on the rotor through, for example, the equal and opposite forces **F,** the rotor shaft will not rotate about the x-axis as would be the case if it were not spinning, rather the shaft will rotate *(precess)* in the x-z plane about the y-axis, at a relatively slow but steady rate $\dot{\phi}$. Conversely, if instead of the couple **M**, the rotor shaft is given a steady precession $\dot{\phi}$ about the y-axis while the rotor is spinning about the z-axis, the gyroscopic torque **M** will be developed by the rotor against the rotor supports. These results can be illustrated using equation (2.15). Here, we know that $\Omega_z = 0$, $\omega_z = p$, and $\dot{\omega}_z = \dot{\omega}_y = 0$. Thus

$$T_x = J_{xx}\dot{\omega}_x + J_{zz}\Omega_y p = M$$

$$T_y = -J_{zz}\Omega_x p$$

$$T_z = -J_{xx}\Omega_y\omega_x + J_{yy}\Omega_x\omega_y$$

T_y and T_z are both zero, since the y- and z-axes are free of torque. It follows that $\Omega_x = 0$, as stated above. Also $\omega_x = \dot{\omega}_x = 0$. Hence

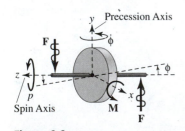

Figure 3.8

Spinning Symmetric Rotor

$$J_{zz}\Omega_y p = M = J_{zz}\dot{\phi}p$$ (3.13)

The symmetric rotor demonstrates *gyroscopic action* for the simple case in which the spin axis is perpendicular to the precession axis. There are many important engineering applications of the *gyroscope* (also called *gyro*). In directional control devices such as *inertial guidance* systems, airplane *automatic pilots* and space vehicle *attitude control* systems, the gyro is mounted on frictionless supports called *gimbals,* as shown in Fig. 3.9. Such a gyroscope, called a *free gyro,* is free from external moments applied to its base and thus retains its orientation while the base (or vehicle) may tilt through various angles. Other applications of the gyroscope include the *gyro compass* and vehicle stabilizing devices. In the gyro compass, the arrangement of the gyro is such that the rotation of the earth causes the gyro to precess so that the spin axis always points north.

Example 3.7	Spin-Stabilized Projectile

The velocity **v** of (the center of mass of) a projectile makes a small angle θ with its geometric axis. The projectile is subject to an aerodynamic resistance force **R** that is directed opposite **v** and passes through a point A, r units ahead of the mass center G, as shown in Fig. 3.10a. To stabilize the projectile, it is given a constant spin p, which causes it to precess at a rate $\dot{\phi}$ about the axis of the velocity. Derive a model that can be used to determine the transient behavior of the projectile over a short segment of its flight path. Also determine the expression for the minimum spin velocity for which the projectile will be spin-stabilized (with $\dot{\theta} = 0$), at steady-state precession ($\ddot{\phi} = 0$).

Solution: The only component is the projectile (mass). The x-y-z axes are chosen, as shown in Fig. 3.10b, so that the velocity axis is contained in the y-z plane. It follows that the total torque on the projectile (at the center of mass) is

$$\mathbf{T} = Rr \sin \theta \mathbf{i}$$

Figure 3.9

Application of Gyroscope in Directional Guidance

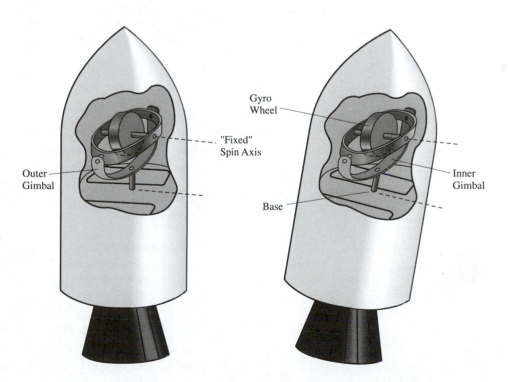

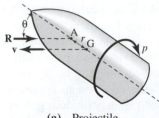

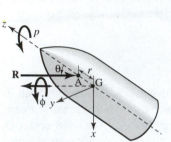

(a) Projectile

(b) Free Body Diagram

Figure 3.10

Spin-stabilized Projectile and Free Body Diagram

That is, $T_y = T_z = 0$, and $T_x = Rr \sin \theta$. Equation (2.15), repeated here, is applicable:

$$T_x = J_{xx}\dot{\omega}_x - J_{yy}\Omega_z\omega_y + J_{zz}\Omega_y\omega_z$$
$$T_y = J_{yy}\dot{\omega}_y - J_{zz}\Omega_x\omega_z + J_{xx}\Omega_z\omega_x \tag{3.14}$$
$$T_z = J_{zz}\dot{\omega}_z - J_{xx}\Omega_y\omega_x + J_{yy}\Omega_x\omega_y$$

Recalling that Ω_x, Ω_y, Ω_z are the components of the angular velocity of the x-y-z coordinates, and ω_x, ω_y, ω_z are the components of the angular velocity of the projectile, we see that

$$\Omega = \omega_x = -\dot{\theta}; \ \dot{\omega}_x = -\ddot{\theta}$$
$$\Omega_y = \omega_y = \dot{\phi} \sin \theta; \qquad \dot{\omega}_y = \ddot{\phi} \sin \theta + \dot{\phi}\dot{\theta} \cos \theta \tag{3.15}$$
$$\Omega_z = \dot{\phi} \cos \theta; \qquad \omega_z = \dot{\phi} \cos \theta + p; \qquad \dot{\omega}_z = \frac{d}{dt}(\dot{\phi} \cos \theta + p)$$

Substituting equations (3.15) in equations (3.14), and noting that due to symmetry, $J_{xx} = J_{yy} = J_T$, we obtain

$$T_x = -J_T(\ddot{\theta} + \dot{\phi}^2 \sin \theta \cos \theta) + J\dot{\phi} \sin \theta(\dot{\phi} \cos \theta + p)$$
$$T_y = J_T(\ddot{\phi} \sin \theta) + J\dot{\theta}(\dot{\phi} \cos \theta + p) = 0 \tag{3.16}$$
$$T_z = J\frac{d}{dt}(\dot{\phi} \cos \theta + p) = J(\ddot{\phi} \cos \theta - \dot{\phi}\dot{\theta} \sin \theta) = 0$$

where $J = J_{zz}$.

State variables: As a first attempt, we should assign $(\omega_x, \omega_y, \omega_z)$ the components of the angular velocity of the mass element (projectile) as the state variables. However, from equations (3.15), $\dot{\theta}$ and $\dot{\phi}$ appear to be an alternative set of **independent** state variables that describe the projectile behavior. Equation (3.16) yields

$$\frac{d}{dt}\dot{\theta} = \frac{-Rr \sin \theta}{J_T} - \dot{\phi}^2 \sin \theta \cos \theta + \frac{J}{J_T}\dot{\phi} \sin \theta(\dot{\phi} \cos \theta + p) \tag{3.17a}$$

$$\frac{d}{dt}\dot{\phi} = -\left(\frac{J}{J_T}\right)\frac{\dot{\theta}(\dot{\phi} \cos \theta + p)}{\sin \theta} = \frac{\dot{\phi}\dot{\theta} \sin \theta}{\cos \theta} \tag{3.17b}$$

where the last term on the RHS of (3.17b) is given by the last of equations (3.16). However, from equation (3.17b),

$$\dot{\phi} = \frac{(J/J_T)p \cos \theta}{\{[1 - (J/J_T)]\cos^2\theta - 1\}} \tag{3.18a}$$

This implies a further reduction of the state variables to $\dot{\theta}$ only. However, the relation between θ and $\dot{\theta}$ is a dynamic one, so that θ is also a state variable. Thus the necessary set of state equations and hence the model is given by equations (3.18b) and (3.18c):

$$\frac{d}{dt}\theta = \dot{\theta} \tag{3.18b}$$

$$\frac{d}{dt}\dot{\theta} = \frac{J}{J_T}\dot{\phi}p \sin \theta - \frac{1}{J_T}Rr \sin \theta - \left(1 - \frac{J}{J_T}\right)\dot{\phi}^2 \sin \theta \cos \theta \tag{3.18c}$$

where $\dot{\phi}$ is given in terms of the state variables θ and $\dot{\theta}$ by equation (3.18a), and p is input. To determine the minimum spin velocity for $\dot{\theta} = 0$, we substitute $\dfrac{d}{dt}\dot{\theta} = 0$ in equation (3.18c) to get the quadratic in $\dot{\phi}$:

$$\dot{\phi}^2(J_T - J)\cos\theta - \dot{\phi}Jp + Rr = 0$$

The solution for which is

$$\dot{\phi} = \frac{Jp \pm \sqrt{(Jp)^2 - 4Rr(J_T - J)\cos\theta}}{2(J_T - J)\cos\theta}$$

The minimum p for which a real rate of precession will exist is given by $(Jp)^2 > 4Rr(J_T - J)\cos\theta$, or

$$p > (2/J)\sqrt{Rr(J_T - J)\cos\theta} \tag{3.19}$$

Example 3.8   Sliding Ladder

The uniform ladder AB of mass m with center of mass at G slides along the smooth walls under the action of gravity (see Fig. 3.11). Derive a dynamic model for the instantaneous position of the ladder.

Solution: The ladder is approximated with a uniform slender rod whose free body diagram is shown. Equation (2.8a) for the mass center gives

$$m\frac{d}{dt}v_{Gx} = N_B = ma_{Gx} \tag{a}$$

$$m\frac{d}{dt}v_{Gy} = N_A - mg = ma_{Gy} \tag{b}$$

Also, from equation (2.16),

$$J_G\frac{d}{dt}\omega = N_A(l\cos\theta) - N_B(l\sin\theta) = J_G\alpha \tag{c}$$

Note that from Fig. 2.5,

$$J_G = (\tfrac{1}{12})ml^2 \tag{d}$$

Also,

$$\frac{d}{dt}\theta = -\omega \tag{e}$$

Because the slender rod is a mass element, the state variables should be velocity components, v_{Gx}, v_{Gy}, and ω. However the involvement of θ in equation (c) suggests θ as a fourth state variable. Although the linear position of the center of mass is desired, x_G and y_G do not have to be made state variables as well because the constraint that the ladder slides along the walls implies that

$$x_G = l\cos\theta \quad \text{and} \quad y_G = l\sin\theta \tag{f}$$

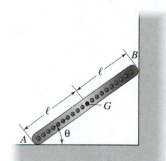

(a)   Sliding Ladder

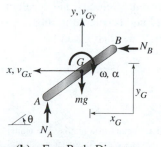

(b)   Free Body Diagram

Figure 3.11

Sliding Ladder and Free Body Diagram

Equation (f) further implies that v_{Gx} and v_{Gy} are dependent on ω and θ, so that the number of independent state variables reduces to two: θ and ω.

N_A and N_B represent unknowns that must be eliminated by writing additional relations. A possibly simpler approach is to begin with equation (f) and determine a_{Gx} and a_{Gy} as functions of θ, ω, and α; substitute these into equations (a) and (b) and then with α given by equation (c); solve the resulting equations (a) and (b), simultaneously, for N_A and N_B. This is left as an exercise for the reader (see Practice Problem 3.27). We opt however, for an alternative but possibly more tedious approach, using relations from kinematics on rigid body rotation and relative motion (see [6]), since this will permit us to explore the latter concept, which was not discussed directly when the behavior component of inertia was considered in Section 2.1.

Rigid body rotation: From equation (2.12a), the velocity of a point P at the position $\mathbf{r}_{P/G}$ from the center of mass G of a rigid body rotating about the mass center with angular velocity ω when the mass center is **fixed** is

$$\mathbf{v}_{P/G} = \dot{\mathbf{r}}_{P/G} = \omega \times \mathbf{r}_{P/G} \tag{3.20a}$$

so that the acceleration is

$$\mathbf{a}_{P/G} = \dot{\omega} \times \mathbf{r}_{P/G} + \omega \times \omega \times \mathbf{r}_{P/G} \tag{3.20b}$$

For the present ladder, $\omega = \omega\mathbf{k}$, and $\dot{\omega} = \alpha\mathbf{k}$. If we let $\mathbf{r}_{P/G} = r_x\mathbf{i} + r_y\mathbf{j}$, then the components of equation (3.20b) are given by*

$$\mathbf{a}_{P/G} = \begin{vmatrix} \mathbf{i} & \mathbf{j} & \mathbf{k} \\ 0 & 0 & \alpha \\ r_x & r_y & 0 \end{vmatrix} + \omega \times \begin{vmatrix} \mathbf{i} & \mathbf{j} & \mathbf{k} \\ 0 & 0 & \omega \\ r_x & r_y & 0 \end{vmatrix} \tag{g}$$

$$= -r_y\alpha\mathbf{i} + r_x\alpha\mathbf{j} + \begin{vmatrix} \mathbf{i} & \mathbf{j} & \mathbf{k} \\ 0 & 0 & \omega \\ -r_y\omega & r_x\omega & 0 \end{vmatrix} = -(r_y\alpha + r_x\omega^2)\mathbf{i} + (r_x\alpha - r_y\omega^2)\mathbf{j}$$

Further, for points A and B,

$$r_{Ax} = l\cos\theta; \qquad r_{Ay} = -l\sin\theta; \qquad r_{Bx} = -l\cos\theta; \qquad r_{By} = l\sin\theta \tag{h}$$

Relative motion: If the point of reference (the center of mass of the ladder) is not fixed, then the above velocity and acceleration are not *absolute* velocity and acceleration but *relative* velocity and acceleration (that is, velocity and acceleration relative to point G). The absolute velocity and acceleration of point P are then given by

$$\mathbf{v}_P = \mathbf{v}_G + \mathbf{v}_{P/G} \tag{3.21}$$

$$\mathbf{a}_P = \mathbf{a}_G + \mathbf{a}_{P/G}$$

---

*We use here the interpretation of the cross or vector product as the determinant of a matrix. The reader who is not familiar with matrix methods may refer to Appendix B or otherwise use the scalar interpretation: $\mathbf{P} \times \mathbf{Q}$ is a vector perpendicular to the plane formed by $\mathbf{P}$ and $\mathbf{Q}$. It has magnitude $PQ\sin\beta$, where β is the angle between $\mathbf{P}$ and $\mathbf{Q}$, and direction given by the right-hand rule (the fingers curling from $\mathbf{P}$ to $\mathbf{Q}$).

where $\mathbf{v}_{P/G}$ and $\mathbf{a}_{P/G}$ are given by equations (3.20a) and (3.20b), respectively. For the present ladder, the acceleration of points A and B on the ladder is thus given by

$$\mathbf{a}_A = \mathbf{a}_G + \mathbf{a}_{A/G}, \text{ and } \mathbf{a}_B = \mathbf{a}_G + \mathbf{a}_{B/G} \tag{i}$$

where $\mathbf{a}_{A/G}$ and $\mathbf{a}_{B/G}$ are given by equations (g) and (h). Using these equations and expanding equation (i) into the components, we obtain

$$a_{Ax} = a_{Gx} + \alpha l \sin\theta - \omega^2 l \cos\theta \tag{j}$$

$$a_{Ay} = 0 = a_{Gy} + \alpha l \cos\theta + \omega^2 l \sin\theta \tag{k}$$

$$a_{Bx} = 0 = a_{Gx} - \alpha l \sin\theta + \omega^2 l \cos\theta \tag{l}$$

$$a_{By} = a_{Gy} - \alpha l \cos\theta - \omega^2 l \sin\theta \tag{m}$$

a_{Ay} and a_{Bx} are equal to zero because the ends of the ladder are constrained to move only along the walls. Substituting for a_{Gx} from equation (a), a_{Gy} from equation (b), and α from equation (c), in equations (k) and (l), we have

$$g - \frac{N_A}{m} = \frac{l^2}{J_G}\cos\theta(N_A\cos\theta - N_B\sin\theta) + \omega^2 l \sin\theta \tag{n}$$

$$\frac{N_B}{m} = \frac{l^2}{J_G}\sin\theta(N_A\cos\theta - N_B\sin\theta) - \omega^2 l \cos\theta$$

The solution of which is (see **On the Net**):

$$N_A = \frac{mg}{ml^2 + J_G}(J_G + ml^2\sin^2\theta) - m\omega^2 l \sin\theta \tag{3.22a}$$

$$N_B = \frac{m\cos\theta(mgl^2\sin\theta - J_G\omega^2 l - m\omega^2 l^3)}{ml^2 + J_G}$$

Finally, from equations (e), (c), (d), and (f), the state model is

$$\left. \begin{aligned} \frac{d}{dt}\theta &= -\omega \\ \frac{d}{dt}\omega &= \left(\frac{12}{ml}\right)(N_A\cos\theta - N_B\sin\theta) \end{aligned} \right\}; \text{ state equations} \tag{3.22b}$$

$$\left. \begin{aligned} x_G &= l\cos\theta \\ y_G &= l\sin\theta \end{aligned} \right\}; \quad \text{output equations} \tag{3.22c}$$

where N_A and N_B are given by equation (3.22a).

3.3 INCOMPRESSIBLE FLUID SYSTEMS

The underlying assumption in this section is that the fluid (system) is incompressible. One way to view compressibility is to look at the pressure-density relation for a fluid of constant mass. An assumption of incompressibility implies that the density of the fluid is independent of the pressure. This is not true, for instance, for a gas. Recall the perfect gas relation, which can be written as

$$\frac{P_1}{\rho_1 T_1} = \frac{P_2}{\rho_2 T_2} \tag{3.23}$$

where ρ, P, T are density, absolute pressure, and absolute temperature, respectively, of the gas. Thus the material in this section does not apply to gases and other compressible fluids.

The Law of Conservation of Matter.   Just as Newton's third law was always invoked, most of the time without formal reference, at internal and external boundary sections of mechanical components, the law of conservation of matter is similarly significant and applicable to junctions or nodes of fluid elements. For instance, for the flow of an incompressible fluid into an elastic junction, the law of conservation of matter implies that

$$\frac{d}{dt}m = w_{in} - w_{out} = 0 \tag{3.24a}$$

where m is the mass of fluid in the junction, and w_{in} and w_{out} are the mass rate of flow of fluid into and out of the junction, respectively. Thus equation (3.24a) simply states that what flows into the junction must flow out; this could be construed as a trivial relation. In terms of a general rigid junction with n flow points (ports), equation (3.24a) implies that

$$w_1 + w_2 + w_3 + \ldots + w_n = 0 \tag{3.24b}$$

In equation (3.24b) any flow *out* of the junction simply has the opposite sign. The reader may recall Kirchhoff's current law, equation (2.27a), which is a similar law governing the flow of electric current into a conductor junction. For a single incompressible fluid system the volume flowrate Q can be substituted for the mass flow rate w in equations (3.24a) and (3.24b), as will generally be the case.

Gravitational Body Force.   Another preliminary matter to consider is the body force on a quantity of fluid. An element of fluid of mass m is always subject to the force due to gravity mg, where g is the gravitational acceleration. It follows that the free body diagram for a fluid element whose mass is not negligible (relative to other behavior components) must include the gravitational force (weight) on the element.

Pressure.   Finally, recall that the static pressure at any point in a fluid is the same in all directions at that point and acts normally to any surface at that point.

Short and Long Constrictions

A *constriction* is used here in a general sense to refer to the geometric condition of a fluid channel that introduces a resistance to the flow of the fluid. A fluid resistance is characterized in terms of its pressure drop–flow relation. An easily visualized fluid resistance is a porous element or plug. For instance, the slow filtration of liquid through a sand pack or even filter paper illustrates the resistance of the porous medium to fluid flow.

Porous Plug

Incompressible flow of fluid through a porous plug is governed by Darcy's law, which was formulated from the results of experimental studies on the flow of water through unconsolidated sand filter beds. The law states,

$$\frac{Q}{A} = -\left(\frac{k}{\mu}\right)\frac{dP}{dl} \tag{3.25a}$$

where Q is the volumetric flow rate, A is the apparent or total cross-sectional area of the channel, μ is the fluid *viscosity,* and k is a measure of the fluid conductivity of the porous medium called the *permeability.* The expression dP/dl is the pressure gradient in the direction of flow. In terms of the porous plug illustrated in Fig. 3.12, equation (3.25a) gives

$$\frac{1}{A}Q_{1-2} = \frac{-(k/\mu)(P_2 - P_1)}{L}$$

Thus the pressure drop–flow relation (Q_{1-2} implies volumetric flow from (points) 1 to 2) is simply

$$(P_1 - P_2) = \left(\frac{\mu L}{kA}\right)Q_{1-2} = R_f Q_{1-2} \tag{3.25b}$$

where the fluid flow resistance R_f has been defined in equation (3.25b). We note that the porous plug (and, as we shall see, any fluid resistor) is a static element. Also observe the similarity between the behavior of the (porous plug) fluid resistor and Ohm's law, equation (2.23a).

Short Constriction

Other examples of actual physical constrictions of flow channels include *orifices,* valves, and *nozzles* (see Fig. 3.13). There are many types of orifices, nozzles, and valves (a few valve types are illustrated in Fig. 3.13), each with its own detailed pressure drop–flow characteristics, which are often determined by experiment. We consider these elements in general as short constrictions. An ideal short constriction is characterized by negligible fluid storage and by short length (extent). It follows that body forces are negligible and there is insufficient length (surface area) to develop predominant frictional forces. The general behavior of such a component was given in equation (1.2) (see also equation (b) of Example 2.8), which is repeated here:

$$Q_{1-2} = C\,\text{sign}(P_1 - P_2)(|P_1 - P_2|)^{1/a} \tag{3.26}$$

where α is experimentally established but is usually close to 2.0, and C is a characteristic quantity for the constriction that depends on flow cross-sectional area (orifice area, valve opening, etc.), and constructional details of the constriction. Note that equation (3.26) also describes the behavior of the porous plug, for which $\alpha = 1.0$. The relation (3.26) is otherwise nonlinear, although linearization in the vicinity of an operating point can be carried out (see Section 2.4).

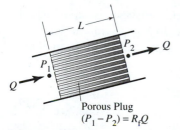

Porous Plug
$(P_1 - P_2) = R_f Q$

Figure 3.12

Resistance of a Porous Plug

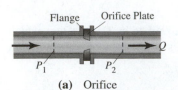

(a)   Orifice

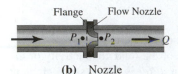

(b)   Nozzle

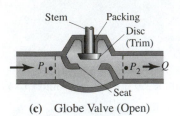

(c)   Globe Valve (Open)

(d)   Gate (Slide) Valve (Closed)

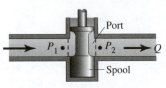

(e)   Spool Valve (Open)

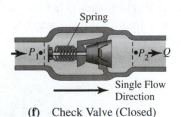

(f)   Check Valve (Closed)

Figure 3.13

Examples of Short Constrictions

Long Constrictions

An ideal long constriction is essentially a fluid channel of uniform cross-sectional area with appreciable length but also with very small (relative to the fluid resistance function) fluid storage. Thus body forces are neglected. However because of the "long" length, channel roughness, size, geometry, fluid velocity, and fluid viscosity are all important. These quantities determine the inertial and (viscous) friction forces experienced by the fluid. The nature of the pressure drop–flow behavior of the constriction depends on whether the viscous or inertial forces are dominant. The Reynolds number $\mathcal{R}$, which is given by

$$\mathcal{R} = v(4R_h)\rho/\mu \tag{3.27a}$$

where v, ρ and μ are the fluid velocity, density and viscosity respectively, and

$$R_h = (\text{wetted area})/(\text{wetted perimeter}) \tag{3.27b}$$

where R_h is the fluid-channel hydraulic radius, is a dimensionless flow characteristic representing the relative magnitudes of the inertia forces to the viscous forces on the fluid. When the Reynolds number is greater than a threshold value (about 2000 for most of the systems considered here), the inertial forces are considered dominant, and the nature of the fluid flow is described as *turbulent*. For $\mathcal{R}$ less than the threshold value, the viscous frictional forces are presumed to dominate the fluid flow, which is then described as *laminar*. Detailed discussions of these topics are available in most fluid mechanics texts such as reference [7].

Fluid resistance in long constrictions can be represented by the frictional force resisting the flow of the fluid F_f, consistent with the viscous and hydrodynamic damping forces for mechanical systems discussed in Section 2.1. In general, the frictional force for long constrictions is given by

$$F_f = B \, \text{sign}(v)v^\alpha; \qquad \alpha \cong \begin{cases} 1 \text{ for laminar flow} \\ 2 \text{ for turbulent flow} \end{cases} \tag{3.28a}$$

where B is a constant depending on fluid properties and geometric parameters of the flow channel, and $v = Q/A$ is the average velocity of the fluid. Assuming one-dimensional flow along the channel and uniform fluid velocity over the length of the resistance element, application of Newton's second law to the body of fluid in the long constriction shown in Fig. 3.14 gives the frictional pressure drop

$$F_f = A(P_1 - P_2)_f \tag{3.28b}$$

where A is the channel cross-sectional area. Equations (3.28a) and (3.28b) are consistent with equation (3.26).

In traditional analysis of fluid flow in pipes, fluid resistance is represented by yet another quantity, the *headloss* h_L. For the component shown in Fig. 3.14, application of the Bernoulli equation (see [7]) for incompressible fluid motion in pipes will give

$$h_{L_{1-2}} = \left(\frac{1}{\rho g}\right)(P_1 - P_2)_f \tag{3.29a}$$

where g is the acceleration due to gravity. In equations (3.28b) and (3.29a) the subscript f has been appended to the pressure drop to emphasize that this is the component of

Figure 3.14

Long Constriction and Fluid Resistance

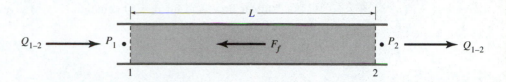

the total drop in pressure that is due to fluid resistance only. The Darcy-Weisbach proposal on the basis of experimental evidence for circular pipes is

$$h_L = f\frac{Lv^2}{2dg} \tag{3.29b}$$

where d is the (internal) pipe diameter, L is the length of the channel, and f is a coefficient of proportionality called the friction factor. f is primarily a function of the Reynolds number and the channel roughness. For turbulent flow, f is usually obtained from charts prepared from experimental data (Moody diagram for commercial pipes for example). For laminar flow, an analytical value for head loss is given by the Hagen-Poiseuille law:

$$h_L = \frac{32\,\mu Lv}{\rho g d^2}; \quad \text{(laminar flow)} \tag{3.29c}$$

Thus the friction factor for laminar flow using equations (3.29b) and (3.29c) is

$$f = \frac{64\,\mu}{\rho d v} = 64/\mathbf{R}; \quad \text{(laminar flow, Hagen-Poiseuille)} \tag{3.29d}$$

Note that in equations (3.29b) through (3.29d), for other than a circular pipe that is flowing full, d can be replaced by $4R_h$. Also from equations (3.28b) and (3.29a),

$$F_f = \rho A g h_L \tag{3.30}$$

Example 3.9   Pressure Drop-Flow Relations for Long Constrictions

Derive the pressure drop–flow behavior given by each of the two representations of fluid resistance in long constrictions, that is, equations (3.28a) and (3.29b). What is the relation between these two representations?

Solution: The pressure drop–flow behavior given by equation (3.28a), using (3.28b), is

$$F_f = A(P_1 - P_2)_f = B\left(\frac{Q}{A}\right)^{\alpha} \tag{3.31a}$$

while the relation given by equations (3.29b) and (3.30) is

$$F_f = A(P_1 - P_2)_f = f\frac{\rho L Q^2}{2\,Ad}; \quad \text{(Darcy-Weisbach)} \tag{3.31b}$$

For laminar flow (and for $\alpha = 1$) equation (3.31a) gives

$$F_f = \frac{BQ}{A}; \quad \text{(laminar flow, } \alpha = 1\text{)} \tag{3.31c}$$

whereas equation (3.31b) with $f = 64\,\mu/(vd\rho)$ yields

$$F_f = \frac{32\,\mu L Q}{d^2}; \quad \text{(laminar flow, Darcy-Weisbach/Hagen-Poiseuille)} \tag{3.31d}$$

so that

$$B = \frac{32\,\mu A L}{d^2}; \quad \text{(laminar flow, } \alpha = 1\text{, Darcy-Weisbach/Hagen-Poiseuille)} \tag{3.31e}$$

(Recall the viscous damper constant β dependence on surface length and viscosity, equation (2.19).)

For turbulent flow (and for $\alpha = 2$), equation (3.31a) gives

$$F_f = \frac{B}{A^2} Q^2; \quad \text{(turbulent flow, } \alpha = 2\text{)} \tag{3.31f}$$

hence

$$B = \frac{f\rho A L}{2d}; \quad \text{(turbulent flow, } \alpha = 2\text{, Darcy-Weisbach)} \tag{3.31g}$$

Note again that d can be replaced by $4R_h$ as appropriate. Equation (3.31a) suggests that equation (3.26) is applicable to both long and short constrictions, provided the factor C is properly interpreted.

Fluid Storage and Fluid Inertia

An example of a fluid storage element is a tank or reservoir. The behavior of a tank was derived in Section 1.1 but will be considered again. The assumption for an ideal fluid storage element is that the fluid velocities inside it are negligible. It follows that the pressure difference between the top of the reservoir and any other point in the reservoir is due to the weight of fluid above that point. In particular, for the fluid storage element shown in Fig. 3.15,

$$P_B = P_T + \rho g h \tag{3.32a}$$

Also, the law of conservation of matter, equation (3.24a), requires that

$$\frac{d}{dt}(m) = \frac{d}{dt}(\rho Volume) = w_{in} - w_{out} = \rho_{in} Q_{in} - \rho_{out} Q_{out} \tag{3.32b}$$

where ρ is the fluid density. For an incompressible fluid, $\rho = \rho_{in} = \rho_{out} = $ constant. Hence

$$\frac{d}{dt}(Volume) = Q_{in} - Q_{out} \tag{3.32c}$$

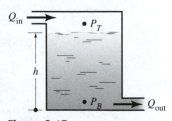

Figure 3.15

Fluid Storage Element

Fluid Inertia

A fluid channel segment is considered an inertia element if the mass of fluid in the channel is large enough so that a finite force is necessary to accelerate it. Since this implies a long channel or substantial internal surface area, **fluid friction in long constrictions should be considered along with fluid inertia.** However the volume rate of flow into and out of the channel segment is assumed to be the same, so that fluid storage is not considered. Finally, since changes in elevation of the channel segment will introduce a component of the body force in the direction of fluid flow, this aspect of fluid system behavior will also be introduced along with fluid inertance.

Consider the fluid pipe segment shown in Fig. 3.16. The mass of fluid in the pipe is $m = \rho A L$, so that Newton's second law applied to the element gives

$$\rho A L \frac{d}{dt} v = A(P_1 - P_2) - F_f + mg \sin \theta$$

since

$$A \frac{d}{dt} v = \frac{d}{dt} Q \quad \text{and} \quad \sin \theta = \frac{(z_1 - z_2)}{L}$$

Then

$$\rho L \frac{d}{dt} Q = A(P_1 - P_2) + \rho A g(z_1 - z_2) - F_f \tag{3.33}$$

For a particular problem, the appropriate expression for F_f (equations (3.31a) through (3.31g)) should be used.

Examples in Fluid Systems

| Example 3.10 | Water Reservoir–Hydraulic Generator |

The water reservoir (as formed by a dam for example) is connected by a long pipe line to a hydraulic generator system. The valve at the end of the pipe line is controlled by

Figure 3.16

General Fluid Inertance and Resistance

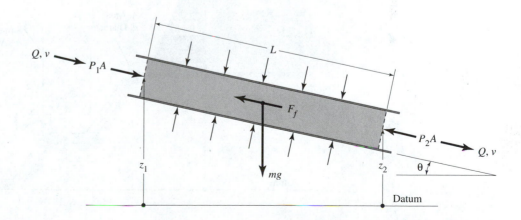

a turbine governor (speed control) and may rapidly stop the water flow if the generator loses its load.

a. Derive the dynamic model for the level of the surge tank, assuming that the height of the tailwater h_3 is fixed and that the turbine-generator is simply an energy converter that changes fluid power into electrical power.

b. Explain the role of the surge tank in the system.

Solution a:

1. Decomposition: As was the case for electrical circuit diagrams, the free body diagram of some fluid circuits can to a large extent be realized without formally breaking up the system. From Fig. 3.17, the components of the present system are the reservoir and tailwater, the long pipe of length L, the surge tank, the constriction at the inlet to the surge tank, the valve, and the turbine-generator subsystem. The length of pipe between the surge tank and the valve is assumed to be short, and its resistance can be considered lumped into the valve resistance (that is, $P_2 = P_4$).

The reservoir and tailwater. Although otherwise a dynamic component (see equation (3.32c)), the rate of change of the volume or level of a large reservoir (a river with a dam for instance) due to flow to the turbine-generator system is negligible. Hence the reservoir is simply a source of pressure (energy) input to this system. From equation (3.32a)

$$P_1 = \rho g h_1; \quad h_1 = \text{known input} \tag{a}$$

Similarly, the tailwater is like a means of removing energy from the system. That is, the instantaneous pressure head (potential energy) available for conversion in the plant is $(h_1 - h_3)$.
Thus

$$P_6 = \rho g h_3; \quad h_3 = \text{known (fixed control) input} \tag{b}$$

Long pipe (dynamic): This actually comprises two components: fluid inertance (dynamic) and long constriction (static). From equations (3.33) and (3.31b), assuming a

Figure 3.17

Water Reservoir–Hydraulic Generator System

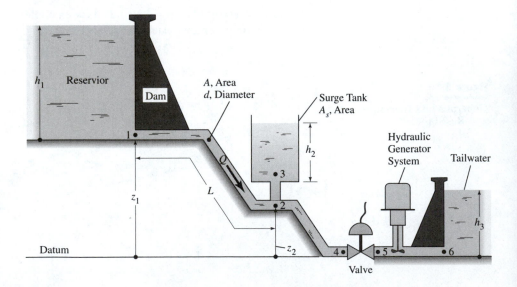

circular pipe (of internal diameter d and cross-sectional area A), which is flowing full, we have

$$\rho L \frac{d}{dt} Q = A(P_1 - P_2) + \rho A g(z_1 - z_2) - F_f; \qquad F_f = B\left(\frac{Q}{A}\right)^{\alpha} \qquad \text{(c)}$$

Tank inlet (static): This is a fluid resistance element, that is, a short constriction. From equation (3.26)

$$Q_{2-3} = C_3 \text{sign}(P_2 - P_3)(|P_2 - P_3|)^{1/\alpha_3} \qquad \text{(d)}$$

Surge tank (dynamic): From equations (3.32a) through (3.32c)

$$\left.\begin{array}{l} A_s \dfrac{d}{dt} h_2 = Q_{2-3} \\[2mm] P_3 = \rho g h_2 \end{array}\right\}; \quad A_s = \text{tank cross-sectional area} \qquad \text{(e)}$$

Valve (static): From equation (3.26),

$$Q_{2-4} = C_4 \text{sign}(P_4 - P_5)(|P_4 - P_5|)^{1/\alpha_4} \qquad \text{(f)}$$

Turbine-generator system (assumed to be a static transducer):

$$(P_5 - P_6)Q_{2-4} = IV/e; \quad \tfrac{IV}{e} = \text{known input} \qquad \text{(g)}$$

where I is the current, V is the voltage and e is the conversion efficiency.

2. Static variables: Assign Q and h_2.

3. State equations: First, considering equations (c) and (e), we need the relation between Q and Q_{2-3}. This is given by equation (3.24b):

$$Q_{2-3} = Q - Q_{2-4} \qquad \text{(h)}$$

From equations (a) and (c), and (d) and (e), the state equations are

$$\frac{d}{dt} Q = \frac{A}{\rho L}(\rho g h_1 - P_2) + \frac{A}{L} g(z_1 - z_2) - \frac{B}{\rho L}\left(\frac{Q}{A}\right)^{\alpha} \qquad \text{(3.34a)}$$

$$\frac{d}{dt} h_2 = \frac{C_3}{A_s} \text{sign}(P_2 - \rho g h_2)(|P_2 - \rho g h_2|)^{1/\alpha_3} \qquad \text{(3.34b)}$$

where, according to equations (f), (h), (d), (g), and (b), P_2 is obtained by solving simultaneously the following four equations in four unknowns (P_2, Q_{2-4}, P_4, and P_5):

$$P_2 = P_4 \qquad \text{(3.34c)}$$

$$Q_{2-4} = C_4 \text{sign}(P_4 - P_5)(|P_4 - P_5|)^{1/\alpha_4} \qquad \text{(3.34d)}$$

$$Q = Q_{2-4} + C_3 \text{sign}(P_4 - \rho g h_2)(|P_4 - \rho g h_2|)^{1/\alpha_3} \qquad \text{(3.34e)}$$

$$P_5 = \frac{IV}{(eQ_{2-4})} + \rho g h_3 \qquad \text{(3.34f)}$$

and α in equation (3.34a) is approximately equal to 1 or 2, according to whether the magnitude of Q is such that the flow in the long pipe is laminar or turbulent:

$$\alpha = \begin{cases} 1; & \dfrac{4\rho Q}{\pi d \mu} < 2000 \quad \text{(laminar flow)} \\[2ex] 2; & \dfrac{4\rho Q}{\pi d \mu} \geq 2000 \quad \text{(turbulent flow)} \end{cases}$$

(3.34g)

3. Output: This is simply h_2.

Solution b. The opening or closing of the valve by the turbine governor results in a pressure variation that travels as a pressure wave at the speed of sound in water along the pipe. The pressure wave is reflected at the reservoir (to a wave of opposite type: *compression wave* to *expansion wave* and vice versa) and (to a like wave) at the valve, which represents a solid boundary. The instantaneous pressure at the valve is the static pressure plus the variation. If the valve movement is sufficiently rapid, the rise in pressure at the valve may be high enough *water hammer)* to destroy the valve. The surge tank is used to absorb the excess (or deficit) pressure (through flow into (or out of) the surge tank) and hence to protect the valve and other equipment.

Note that the coupling (feedback loop) between the turbine-generator speed and the valve opening (see Example 4.5) was ignored in the preceding analysis. The valve opening was assumed to be normally constant.

Example 3.11	Simplified Oil Well Lift System

The rig shown in Fig. 3.18a is used to pump oil from wells in which the reservoir pressure is not sufficient to naturally lift the oil to the surface at an economic rate. A physical model of the system for pumping oil from such a well is shown in Fig. 3.18b. For simplicity, the drive machinery (see Fig. 3.18c1) has been replaced by the input torque $T_{in}(t)$. The system consists of the *walking beam,* which has mass m_1 and moment of inertia J_1 about the pivot point O; a long cable (see free body diagram, Fig. 3.18c2) that has a no-slip reference on the cam surface of the walking beam, as shown; the *sucker rod* of mass m_2 (Fig. 3.18c3); and the well bore above and below the sucker rod (Figs. 3.18c4 and 3.18c5). The well bore below the sucker rod includes the surrounding rock formation from which oil flows into the well bore through perforations in the casing and up the bore hole as the sucker rod moves up. The pressure in the surrounding rock (which is called *reservoir* or *formation*) is fixed at P_f.

Derive a model for this system that applies during the upstroke of the pumping rod. Assume that the walking beam moves through small angles, that the oil is incompressible, and that the flow of oil up the well bore is laminar. Assign parameters as needed and note that the well is quite deep.

Solution:

Decomposition and relations: This system has combined (strictly) mechanical and fluid elements. The free body diagram of the components is shown in Fig. 3.18c.

Mass m_1 (dynamic): Newton's second law gives

$$J_1 \frac{d}{dt} \omega_1 = T_{in} - m_1 g L_1 \cos \theta_1 - F_s L_2$$

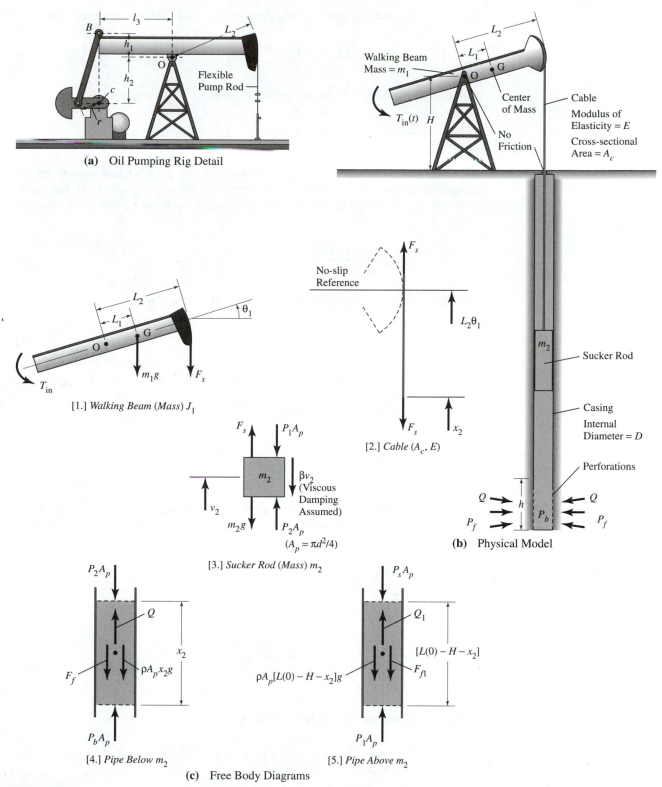

(a) Oil Pumping Rig Detail

[1.] *Walking Beam (Mass) J_1*

[2.] *Cable (A_c, E)*

[3.] *Sucker Rod (Mass) m_2*

$(A_p = \pi d^2/4)$

(b) Physical Model

[4.] *Pipe Below m_2*

[5.] *Pipe Above m_2*

(c) Free Body Diagrams

Figure 3.18

Oil Well Lift System

We assume the walking beam rotates through the small angles ($\cos \theta \cong 1$); hence

$$J\frac{d}{dt}\omega_1 = T_{in} - m_1 g L_1 - F_s L_2 \tag{a}$$

Cable (dynamic): We use the procedure suggested for handling cables,

$$F_s = \frac{A_c E \Delta L}{L}; \qquad \Delta L = L_2\theta_1 - x_2; \qquad L = L(0) - x_2$$

where $L(0)$ is the distance between the level of the pivot O and essentially the bottom of the well, but excluding the length of the sucker rod, A_c is the cross-sectional area of the cable, and E is the modulus of elasticity for the cable material. Thus

$$F_s = (A_c E)\left[\frac{L_2\theta_1 - x_2}{L(0) - x_2}\right] \tag{b}$$

Mass m_2 (dynamic): Applying Newton's second law, we have

$$m_2\frac{d}{dt}v_2 = P_2 A_p + F_s - P_1 A_p - \beta v_2 - m_2 g \tag{c}$$

where A_p is the cross-sectional area of the well bore, and β is the assumed viscous damping constant for the pump rod.

Long pipe below m_2 (dynamic): From equation (3.33),

$$\frac{d}{dt}Q = \frac{A_p}{\rho x_2}(P_b - P_2) + \frac{A_p g}{x_2}(0 - x_2) - \frac{F_f}{\rho x_2}$$

where P_b is the pressure at the bottom of the well bore, and d is the well bore diameter. Using Darcy-Weibach/Hagen-Poiseuille, this time for laminar flow, we obtain from equation (3.31d)

$$F_f = \frac{32\,\mu x_2 Q}{d^2}$$

Hence

$$\frac{d}{dt}Q = \frac{A_p}{\rho x_2}(P_b - P_2) - A_p g - \frac{32\,\mu Q}{\rho d^2} \tag{d}$$

Long pipe above m_2 (dynamic): Similarly,

$$\frac{d}{dt}Q_1 = \frac{A_p}{\rho[L(0) - H - x_2]}(P_1 - P_s) - A_p g - \frac{32\,\mu Q_1}{\rho d^2} \tag{e}$$

where P_s, the pressure at the surface or surface valve, is assumed to be known.

State variables: We assign ω_1 for m_1, v_2 for m_2, x_2 for the cable, and Q and Q_1 for the long pipes.

Other relations: If we assume no leakage through the sucker rod (valve) during the upstroke, then, for incompressible flow, equation (3.32c) gives

$$Q = A_p v_2 = Q_1 \tag{f}$$

If we also assume that the rock and the perforated casing act like a porous plug fluid resistance, equation (3.25b) yields

$$R_f Q = P_f - P_b; \qquad P_f = \text{constant} \tag{g}$$

For radial flow of oil in the formation, the resistance to fluid flow R_f is obtained from the cylindrical form of equation (3.25a):

$$\frac{Q_r}{2\pi r h} = -\left(\frac{k}{\mu}\right)\frac{d}{dr}P \tag{3.35a}$$

where r is radial distance, h is height (length) of the cylinder, and Q_r, is a radial flow, positive in the outward sense. Integrating equation (3.35a) between an inner radius r_i and outer radius r_o, obtain

$$\frac{\mu \ln(r_o/r_i)}{2\pi k h} Q_r = (P_i - P_o) \Rightarrow R_f = \frac{\mu \ln(r_o/r_i)}{2\pi k h} \tag{3.35b}$$

For the present problem, r_i is the radius of the well bore ($d/2$), and r_o is the assumed radius (radial extent) of the reservoir r_e. Also $Q, = -Q_r$, hence

$$\frac{\mu \ln(2r_e/d)}{2\pi k h} Q_r = (P_f - P_b) \Rightarrow R_f = \frac{\mu \ln(2r_e/d)}{2\pi k h} \tag{3.36a}$$

From equation (f), Q, Q_1, and v_2 are all dependent. Using equations (f) and (g) and solving equations (c) through (e) simultaneously for v_2, P_1, and P_2, we get

$$P_1 = \frac{\left\{\dfrac{A_p}{m_2}(P_f - R_f A_p v_2) + \left(\dfrac{A_p}{m_2} + \dfrac{1}{\rho x_2}\right)\left(\dfrac{x_2 P_s}{L(0) - H - x_2}\right) + \left(\dfrac{32\mu}{\rho d^2} - \dfrac{\beta}{m_2}\right)v_2 + \dfrac{A_c E}{m_2}\left[\dfrac{L_2 \theta_1 - x_2}{L(0) - x_2}\right]\right\}}{\left\{\dfrac{A_p}{m_2} + \left(\dfrac{A_p}{m_2} + \dfrac{1}{\rho x_2}\right)\left(\dfrac{x_2}{L(0) - H - x_2}\right)\right\}} \tag{3.36b}$$

$$P_2 = P_f - R_f A_p v_2 + \left(\frac{x_2}{L(0) - H - x_2}\right)(P_s - P_1) \tag{3.36c}$$

Note that there is no static relation for eliminating θ_1. Hence θ_1 must be assigned as another state variable while eliminating Q and Q_1 in favor of v_2.

The state equations:

$$\frac{d}{dt}x_2 = v_2 \tag{3.36d}$$

$$\frac{d}{dt}v_2 = \frac{1}{m_2}\left\{(P_2 - P_1)A_p + (A_cE)\left[\frac{L_2\theta_1 - x_2}{L(0) - x_2}\right] - \beta v_2 - m_2g\right\} \tag{3.36e}$$

$$\frac{d}{dt}\theta_1 = \omega_1 \tag{3.36f}$$

$$\frac{d}{dt}\omega_1 = \frac{1}{J_1}\left\{T_{\text{in}} - m_1gL_1 - (L_2A_cE)\left[\frac{L_2\theta_1 - x_2}{L(0) - x_2}\right]\right\} \tag{3.36g}$$

where P_1 and P_2 are given by equations (3.36a) through (3.36c). If we are interested in the oil flow rate out of the pump, then the following output equation completes the solution:

$$Q = A_pv_2 \tag{3.36h}$$

3.4 ELECTRICAL SYSTEMS

This section considers physical and system models of electrical and electronic engineering systems. Although a few common electronic behavior compartments *(diodes, transistors,* and *operational amplifiers)* are introduced here, the primary basis of analysis is still the electrical circuit concepts developed in Chapter 2. However, the unifying theme of dynamic and static behavior is maintained, and the state modeling procedure will be found to be compatible with the traditional method of circuit analysis employed in Chapter 2.

Example 3.12	Application of State Space Approach

Consider again the electrical circuit of Example 2.3 Derive a state model for this circuit that is sufficient to determine the voltage drop across the inductor.

Solution:

Step 1—system decomposition: The circuit of Example 2.3 was decomposed in Example 2.9. The circuit and component free body diagrams in that example were shown in Fig. 2.24. The results of Example 2.9 (equations (a) through (i) and Fig. 2.24 of that example) are utilized in the following.

Step 2—state variables: The dynamic elements are the capacitor and the inductor, and the corresponding output variables are V_b and i_5.

Step 3—state equations: We start with equations (f) and (g) of Example 2.9:

$$\frac{d}{dt}V_b = -\frac{1}{C_2}i_2 \tag{j}$$

$$\frac{d}{dt}i_5 = -\frac{1}{L_5}V_c \tag{k}$$

From equations (h), (b), (d), and (i) of Example 2.9,

$$i_2 = i_1 - i_3 - i_4 = -i_s - i_3 - i_4 = -i_s + \frac{V_b}{R_3} - i_4 = -i_s + \frac{V_b}{R_3} - i_5 \qquad \text{(l)}$$

From equations (e) and (i), $V_c = R_4 i_4 + V_b = R_4 i_5 + V_b$ **(m)**
    Thus the state equations become

$$\frac{d}{dt} V_b - \frac{1}{C_2}\left(\frac{V_b}{R_3} - i_3 - i_s\right) \qquad \textbf{(3.37a)}$$

$$\frac{d}{dt} i_5 = -\frac{1}{L_5}(V_b + R_4 i_5) \qquad \textbf{(3.37b)}$$

Step 4—output equation: The voltage drop across the inductor V_c is given by equation (m) of Example 2.9. That is, the output equation is

$$V_c = V_b + R_4 i_5 \qquad \textbf{(3.37c)}$$

Equations (3.37a) through (3.37c) constitute the state or system model. These results are compatible with the results of Example 2.3. In Example 2.3, with the node method, V_c was given by the differential equation (d) and integral equation (e) of that example, with V_b and V_c as the independent variables. With the loop method, the solution was given by equations (h) and (i) of that example, with i_b and i_c $(= -i_5)$ as the independent variables. However, equation (3.37c) shows that V_c is not independent of V_b and i_5, and

$$i_3 = i_c - i_b \Rightarrow i_b = i_c - i_3 = -i_5 + \frac{V_b}{R_3}$$

so that (V_b, i_5), (V_b, V_c) and (i_b, i_c) are simply alternative sets of independent state variables for this circuit. For the state variable set (V_b, V_c) the reader can easily verify that the state model is

$$\left.\begin{aligned}
\frac{d}{dt} V_b &= -\frac{1}{C_2}\left[\left(\frac{1}{R_3} + \frac{1}{R_4}\right)V_b - \frac{1}{R_4}V_c - i_s\right] \\
\frac{d}{dt} V_b &= -\frac{1}{C_2}\left(\frac{1}{R_3} + \frac{1}{R_4}\right)V_b - \left(\frac{R_4}{L_5} - \frac{1}{R_4 C_2}\right)V_c + \frac{i_s}{C_2}
\end{aligned}\right\}; \text{ state equations}$$

Output $= V_c$

Similarly, for the state variable set (i_b, i_c) the state model is

$$\frac{d}{dt} i_b = \left(\frac{R_3}{L_5} - \frac{1}{R_3 C_2}\right)i_b - \left(\frac{R_3}{L_5} + R_4\right)i_c + \frac{i_s}{R_3 C_2}$$

$$\frac{d}{dt} i_c = \frac{1}{L_5}[-(R_3 + R_4)i_c + R_3 i_b]$$

$$V_c = R_3 i_b - (R_3 + R_4)i_c$$

Example 3.13   Passive Filter Circuit

A circuit that can be used to realize a fourth-order *Butterworth* low-pass *(passive)** fil-ter (see Section 8.1: Signal Processing) is shown in Fig. 3.19. Obtain the state model for this network that can be used to predict the output voltage $V_{out}(t)$ for given input voltage $V_{in}(t)$.

Solution:

Decomposition: Figure 3.19 is also a free body (decomposed) circuit diagram. The components are

1. Voltage source, input element:

$$V_a = V_{in}(t) \tag{a}$$

2. Source resistance R_s (static): from equation (2.23a),

$$V_a - V_b = R_s i_1 \tag{b}$$

3. Inductor L_1 (dynamic): from equation (2.25b),

$$L_1 \frac{d}{dt} i_1 = (V_b - V_c) \tag{c}$$

4. Capacitor C_1 (dynamic): from equation (2.24b),

$$C_1 \frac{d}{dt} V_c = i_1 - i_2 \tag{d}$$

5. Inductor L_2 (dynamic): from equation (2.25b),

$$L_2 \frac{d}{dt} i_2 = (V_c - V_d) \tag{e}$$

---

*An *active* (Chebycheff) filter is given in Practice Problem 3.51d.

Figure 3.19

Circuit for a Fourth-order Butterworth-type Low-pass Passive Filter

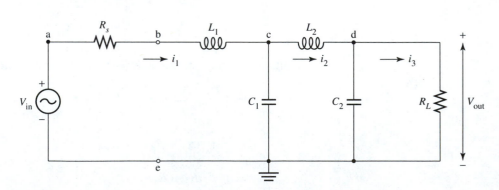

6. Capacitor C_2 (dynamic): from equation (2.24b),

$$C_2 \frac{d}{dt} V_d = i_2 - i_3 \tag{f}$$

7. Load resistance R_L (static): from equation (2.23a),

$$i_3 = \frac{V_d}{R_L} \tag{g}$$

State variables: Given the dynamic components, we assign i_1, V_c, i_2, and V_d as state variables.

State equations: Substitution of equations (a) and (b) for V_a and V_b, respectively, in (b) gives equation (3.38a). We rewrite equation (d) as equation (3.38b); we rewrite equation (e) as equation (3.38c); and substitute (g) in (f) to obtain equation (3.38d):

$$\frac{d}{dt} i_1 = \frac{1}{L_1}(V_{\text{in}}(t) - R_s i_1 - V_c) \tag{3.38a}$$

$$\frac{d}{dt} V_c = \frac{1}{C_1}(i_1 - i_2) \tag{3.38b}$$

$$\frac{d}{dt} i_2 = \frac{1}{L_2}(V_c - V_d) \tag{3.38c}$$

$$\frac{d}{dt} V_d = \frac{1}{C_2}\left(i_2 - \frac{1}{R_L} V_d\right) \tag{3.38d}$$

State Model: Equations (3.38) constitute the state mode, with the desired output given by

$$V_{\text{out}} = V_d \tag{3.38e}$$

Some Electronic Circuit Components

Semiconductor-based components such as diodes, transistors, and various types of integrated circuits are the key elements in modern electronic circuits. Unlike the *passive* circuit elements introduced in Section 2.2, which either *store* (capacitor and inductor) or *dissipate* (resistor) a circuit's energy (the energy viewpoint for components of engineering systems is taken up formally in Section 5.1), most semiconductor components are *active* circuit elements that *utilize* a circuit's energy to perform specific sets of functions. The detail of how these functions are realized within the component is not our concern here. Rather, as was the case in Section 2.2, we shall describe the behavior of each element in terms of its voltage-current characteristics. Further, because of the inherent complexity of the voltage current characteristics of semiconductor-based components, our description of the behavior of each such component will concentrate on the *essential property* of the element in the given application or mode of use. Finally, we shall limit our attention to low-frequency applications and only a few combinations of electronic components and modes of their use.

Junction Diodes

Junction *diodes* are *two-terminal* semiconductor devices with the typical characteristics shown in Fig. 3.20. The figure also illustrates the circuit symbol we use for the (rectifier) diode. Note that the applied voltage axis scale for positive voltages has been expanded severalfold relative to the scale for negative voltages; however, the quantitative values shown in the figure are for illustrative purposes only. Also, we have introduced in Fig. 3.20 a convention for representing voltage drops across semiconductor device terminals: V_{AC} implies $(V_A - V_C)$. For applied voltages more negative than V_B, the *breakdown* voltage, the voltage drop across the diode V_{AC} remains essentially constant at V_B, independent of the current. This is the mode of operation identified in Fig. 3.20 as *zener/avalanche* mode. *Zener diodes* are diodes specially manufactured to exhibit specific breakdown voltages. Such diodes are used in voltage regulators and signal level clamps.

The *essential property* of the diode when the applied voltages are greater than V_B is that when the diode is *forward biased*, that is, $V_{AC} > 0$, some appreciable current can pass through the diode, whereas when it is *reverse biased*, $V_{AC} < 0$, only a negligible current can pass through the diode. This property is known as *rectification*, and the region $V_{AC} > V_B$ is identified in Fig. 3.20 as rectifier mode. A good mathematical description of the rectifier diode characteristics is

$$i_{A-C} = I_r(e^{V_{AC}/V_t} - 1); \quad \text{(rectifier diode equation)} \tag{3.39}$$

where I_r (amperes) is the *reverse saturation current*, and V_t (volts) is the *voltage equivalent of temperature* (a function of temperature). For a typical diode, such as the silicon pn junction diode with the characteristics shown in Fig. 3.20, I_r would have a value in the nanoampere range (50×10^{-9} A, for example), and V_t would have a value of about 25 mV at room temperature. If we substitute these values ($I_r = 50 \times 10^{-9}$ A, $V_t = 25$ mV) in the diode equation (3.39), we see that the for practical currents within the rating of this diode (1.0 A), the voltage drop acrossthe diode does not exceed about 0.42 V. This voltage drop would be considered negligible in a circuit, such as some mo-

Figure 3.20

Typical Junction Diode Characteristics

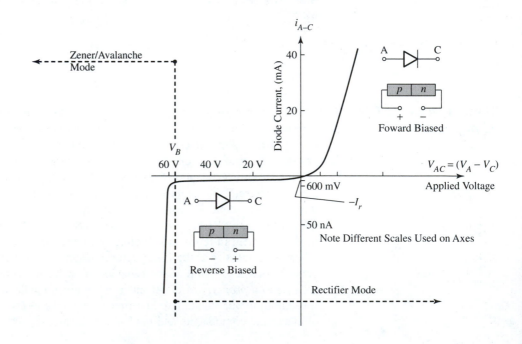

tor and other inductive-device drivers, in which the operating voltages are all greater than, say, 8.4 V (0.42 is only 5% of 8.4 V).

We shall thus use two *approximate* models for the rectifier diode. As shown in Fig. 3.21a, for large voltage circuits we shall represent the diode by a special switch that is open (hence no current passes through the diode) as long as the voltage drop across the diode is negative. When the switch is closed, the value of the current through the diode is determined by the rest of the circuit. It is assumed that the balance of the circuit is such that this current is always within the diode current rating. As a way to represent analytically such special switches, we introduce a switching function that has a value of 1 when the switch is closed and a value of 0 when the switch is open:

$$\text{switch}(\eta) = \begin{cases} 0; & \eta < 0 \\ 1; & \eta \geq 0 \end{cases} \tag{3.40a}$$

Thus for the large-voltage model switch, the switch argument is $\eta = (V_A - V_C)$.

The small-voltage model shown in Fig. 3.21b represents the diode by a switch in series with a voltage source and a resistor. The values of the voltage source V_d and the resistor R_d depend on the assumed nominal operating point. As shown in the figure, for a given operating point P, V_d, which is known as the *cut-in voltage,* is given by the intersection of the tangent line with the diode characteristic curve at P with the voltage axis, while R_d is defined in terms of V_d and the operating point (i_o, V_o) by,

$$R_d = (V_o - V_d)/i_o \tag{3.40b}$$

Thus the voltage-current relationship for the rectifier diode for small voltages can be expressed as

$$i_{A-C} = \text{switch}(V_{AC} - V_d)\left[\frac{(V_{AC} - V_d)}{R_d}\right] \tag{3.40c}$$

Figure 3.21
_____

Approximate Models for the Rectifier Diode

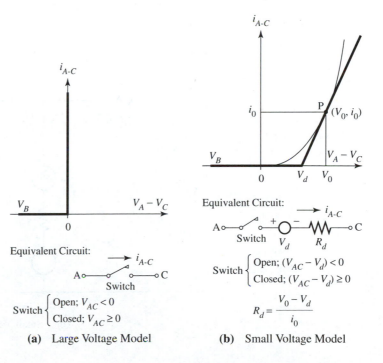

(a) Large Voltage Model

(b) Small Voltage Model

Example 3.14   Diode As a Voltage Limiter

Figure 3.22a shows a diode being used as a voltage limiter in a circuit. For the input voltage limits $+V$ and $-V$, it is assumed that the output voltage without the diodes would be excessively high. Diode D_1 used alone in the circuit would prevent the output from going much higher than the bias voltage V_H, whereas diode D_2 by itself would

Figure 3.22

Application of the Diode as a Voltage Limiter

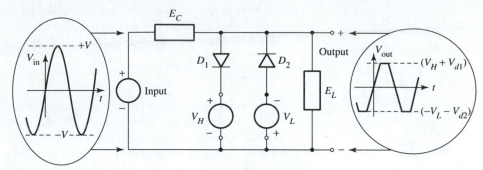

(a)   Voltage Limiting Diode Circuit

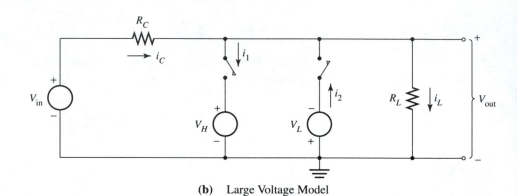

(b)   Large Voltage Model

(c)   Small Voltage Model

limit the output to a low of about $-V_L$. With both diodes, the output voltage will always lie, approximately, between V_H and $-V_L$.

a. Demonstrate these predictions using the large-voltage diode model and resistors $R_C = 5\ \Omega$ and $R_L = 25\ \Omega$ in place of the equivalent circuits or elements E_C and E_L. Assume that the input voltage is given by $V_{in} = 24\sin(2\pi t)$ and that the bias voltages are $V_H = 12$ V and $V_L = 9$ V.

b. Show by substituting the small-voltage diode model for the large-voltage model in (a) that the range of the output voltage is more like $(V_H + V_{d1}) \geq V_{out} \geq (-V_L - V_{d2})$. Use, for this purpose, the small-voltage model parameters $R_{d1} = R_{d2} = 1.5\ \Omega$, and $V_{d1} = V_{d2} = 0.7$ V.

Solution a: Figure 3.22b shows the equivalent circuit for the large-voltage diode model. Without the diodes, the circuit is just a voltage divider, and equation (2.35a) gives

$$V_{out} = \left(\frac{R_L}{R_L + R_C}\right)V_{in} = V_0 \tag{a}$$

As shown in Fig. 3.23a, without the diodes and for the given values of R_L and R_C, V_0 is just another sine curve with an amplitude of 20 V.

With diode D_1 in place, any time switch D_1 closes V_{out} is constrained to be equal to V_H; the currents in the circuit merely adjust to permit this. When the switch is open, V_{out} is given by equation (a). Thus using the switching function defined earlier, we can write

$$V_{out} = V_1 = [1 - \text{switch}(V_0 - V_H)]V_0 + \text{switch}(V_0 - V_H)V_H \tag{b}$$

If diode D_2 was in place instead, then a similar analysis would give, for this case,

$$V_{out} = V_2 = [1 - \text{switch}(-V_0 - V_L)]V_0 - \text{switch}(-V_0 - V_L)V_L \tag{c}$$

If both diodes are in place, we note that the arguments of the two switching functions are such that both switches are never closed at the same time. At any rate, the logical combination of V_1 and V_2 is

$$V_{out} = V_3 = [1 - \text{switch}(V_0 - V_H)][1 - \text{switch}(-V_0 - V_L)]V_0 +$$
$$\text{switch}(V_0 - V_H)V_H - \text{switch}(-V_0 - V_L)V_L \tag{d}$$

The simulation of equations (b), (c), and (d) and the resulting outputs are shown in Fig. 3.23.

Solution b: The equivalent circuit for the small-voltage model is shown in Fig. 3.22c. Kirchhoff's current law, equation (2.27a), and Ohm's law, equation (2.23a), give

$$i_L = i_C - i_1 + i_2 \quad \text{and} \quad i_C = \frac{V_{in} - V_{out}}{R_C}; \qquad V_{out} = R_L i_L$$

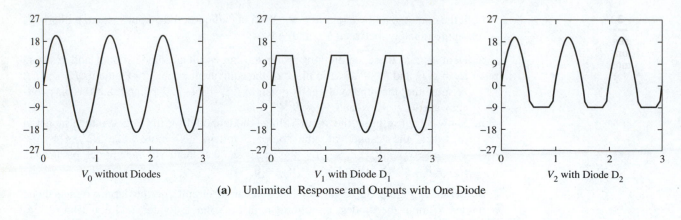

(a)   Unlimited Response and Outputs with One Diode

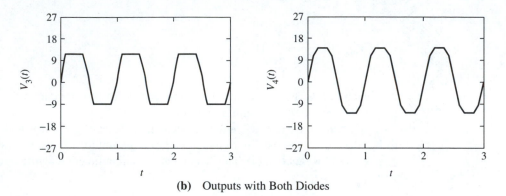

(b)   Outputs with Both Diodes

Figure 3.23

Response of Voltage Limiter

so that

$$V_{\text{out}} = V_4 = \left(\frac{R_L}{R_C + R_L}\right)V_{\text{in}} - R_C\left(\frac{R_L}{R_C + R_L}\right)(i_1 - i_2) \tag{e}$$

$$= V_0 - R_C\left(\frac{R_L}{R_C + R_L}\right)(i_1 - i_2)$$

However, by equation (3.40c), the diode currents are given by

$$i_1 = \text{switch}(V_0 - V_H - V_{d1})\left(\frac{V_{\text{out}} - V_H - V_{d1}}{R_{d1}}\right) \tag{f}$$

$$= \text{Ss1}\frac{V_{\text{out}}}{R_{d1}} - \text{Ss1}\frac{(V_H + V_{d1})}{R_{d1}}$$

$$i_2 = \text{switch}(-V_0 - V_L - V_{d2})\left(\frac{-V_{\text{out}} - V_L - V_{d2}}{R_{d2}}\right) \tag{g}$$

$$= -\text{Ss2}\frac{V_{\text{out}}}{R_{d2}} - \text{Ss2}\frac{(V_L + V_{d2})}{R_{d2}}$$

Figure 3.24

Representation of the Bipolar Junction Transistor (Common-base Configuration)

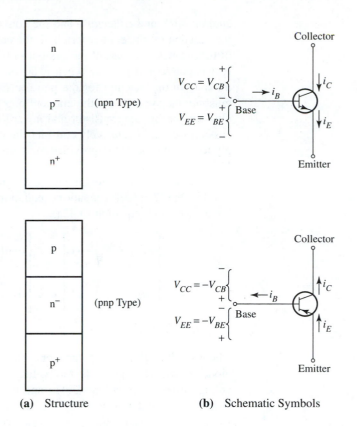

(a) Structure   (b) Schematic Symbols

where

$$Ss1 = \text{switch}(V_0 - V_H - V_{d1}); \qquad Ss2 = \text{switch}(-V_0 - V_L - V_{d2}) \qquad \textbf{(h)}$$

Substituting equations (f) and (g) into equation (e), we obtain finally

$$V_{\text{out}} = V_4 = \left[\frac{R_{d1}R_{d2}(R_C + R_L)}{R_{d1}R_{d2}(R_C + R_L) + Ss1(R_{d2}R_C R_L) + Ss2(R_{d1}R_C R_L)} \right] \qquad \textbf{(i)}$$

$$\left\{ V_0 + \frac{R_C R_L}{R_C + R_L} \left[\frac{Ss1(V_H + V_{d1})}{R_{d1}} - \frac{Ss2(V_L + V_{d2})}{R_{d2}} \right] \right\}$$

The simulation of equation (i) gives the expected result. As shown in Fig. 3.23b, V_4 lies, approximately, in the range -10 to 13, which is wider than the range -9 to 12 given by the large-voltage model of the diode.

....

Bipolar Junction Transistors

A transistor is a three-terminal semiconductor device whose *essential property* is that the current through it can be controlled at one of the terminals (the *base*). Figure 3.24 shows the schematic structure and circuit symbols we use for npn and pnp transistors. The figure also identifies the relative locations of three terminals of the transistor—

emitter, base, and *collector*—and the convention we have adopted for current flow in the junction (at the emitter terminal, the current is considered positive in the direction of the arrow that is part of the symbol of the transistor. At the other terminals, the current is considered positive when it flows *into* the junction for the npn transistor and *away* from the junction for the pnp transistor). Also note that the convention distinguishing the symbols for the npn and pnp junction transistors is that the arrow points outward for the npn transistor and inward for the pnp transistor. In the following discussion we shall exclusively consider the npn transistor. However, the results also apply to the pnp transistor, provided the current directions and the polarities for V_{EE} and V_{CC} are reversed as shown in Fig. 3.24.

A good mathematical model of the junction transistor characteristics is given by (see [8]) the *Ebers-Moll* equations (equations (3.41a and b) and Kirchhoff's current law for the junction, equation (3.42a):

$$i_E = k_1 \left[\exp\left(\frac{V_{BE}}{V_t} \right) - 1 \right] + k_2 \left[\exp\left(\frac{V_{CB}}{V_t} \right) - 1 \right] \tag{3.41a}$$

$$i_C = k_3 \left[\exp\left(\frac{V_{BE}}{V_t} \right) - 1 \right] + k_4 \left[\exp\left(\frac{V_{CB}}{V_t} \right) - 1 \right] \tag{3.41b}$$

$$i_E = i_B + i_C \tag{3.42a}$$

where k_1 through k_4 are parameters related to reverse saturation currents (recall the diode equation), and V_t is the voltage equivalent of temperature. Note that the voltage drops in equations (3.41a and b) follow the convention introduced with the diode. In normal transistor operation, the base-emitter junction is forward biased, whereas the collector-base junction is reverse biased. Thus (recall the junction diode characteristics) the voltage between the base and the emitter V_{BE} is limited to a small positive value (say, about 0.5 V), while V_{CB} is relatively large and negative (say, -5 or more volts), so that the second term in each of the equations (3.41a) and (3.41b) can be ignored (note, for example, $\exp(-5/.025) \cong 0.0$) relative to the first term. Thus we can let

$$i_C \cong \alpha i_E \tag{3.42b}$$

where α is a parameter of the junction transistor (the *DC short-circuit gain in the common-base configuration;* configurations are explained below) and has a value very close to, but always less than, 1 (typically between 0.980 and 0.999). Further, the base-emitter junction, being forward biased, behaves like the pn diode junction, so that the base current is given by an equation similar to the rectifier diode equation (3.39): $i_B = I_{rB}(e^{V_{BE}/V_t} - 1)$. Thus we can also let

$$i_C \cong \beta i_B \tag{3.42c}$$

However, β does not represent another parameter. Substituting equations (3.42b) and (3.42c) into equation (3.42a), we see that

$$\beta = \frac{\alpha}{1 - \alpha} \tag{3.42d}$$

For the typical values of α mentioned earlier, β would have a range of about 50 to 1000. Thus, in principle, a small current at the transistor base can be used to control relatively very large collector currents. Equations (3.42a) through (3.42d) represent our working model for the junction transistor.

Common-Emitter Configuration. In most applications of the junction transistor, the input is applied between two terminals while the output is obtained at another pair of terminals. With only three terminals on the transistor, one terminal is common to both input and output. Of the possible configurations of common terminal and input/output terminal pairs, we shall consider only the case when the emitter terminal is common, and output is taken between the collector and emitter terminals—the *common-emitter* (CE) *configuration* (see Fig. 3.25b). As before, the base-emitter junction is normally forward biased, so that, as shown in Fig. 3.25a, the CE input characteristics are similar to the rectifier diode characteristics. The typical output characteristics are shown in Fig. 3.25c. Observe that for practical values of V_{CE}, the collector current depends on the base current, according to equation (3.42c), and is essentially independent of the collector-emitter voltage drop. However the CE output characteristic is quite sensitive to temperature, and a significant part of transistor circuit design involves introducing external features that minimize the effect of temperature.

Example 3.15	Design Aspects of a Simple Junction Transistor Amplifier

Figure 3.26a shows the circuit diagram of a prototype single-stage bipolar junction transistor amplifier. The design problem is to determine the biasing and external components (resistors in this case) to ensure that the transistor's operating point is within a restricted desired region and that this operating point is stable with respect to temperature and minor variations in the power supply. The major restrictions on the operating region are that (1) the (maximum continuous) collector current I_C, maximum collector-emitter voltage when the base is open-circuited ($i_B = 0$) V_{CEO}, and the maximum (continuous device) power dissipation P_D must all be within the rating of the transistor; (2) the base current should be sufficiently positive to forward-bias the base-emitter junction and place the transistor operation in regions where our design equations are applicable; and similarly, (3) the collector-emitter voltages must exceed about 0.2 to 0.5 V (see Fig. 3.25c) to ensure that operation is in the "constant" current region. Assuming a supply voltage of 15 V, as shown, and that the transistor sample has the same characteristics as given in Fig. 3.25 (including $I_C = 800$ mA, $V_{CEO} = 30$ V, and $P_D = 3$ W), determine good values of R_{bias} and R_L that would yield reasonable amplification.

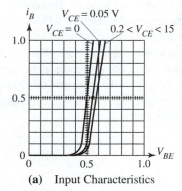

(a) Input Characteristics

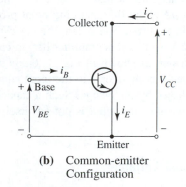

(b) Common-emitter Configuration

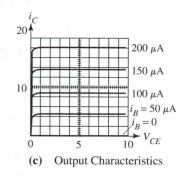

(c) Output Characteristics

Figure 3.25

Typical Common-emitter Input and Output Characteristics

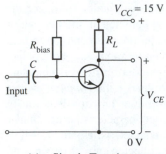

(a) Simple Transistor Amplifier Circuit

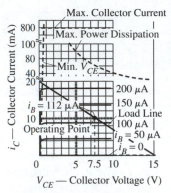

(b) Common-emitter Output Characteristics of a Silicon npn Transmitter

Figure 3.26

Elementary Design of a Simple Amplifier

Solution: We observe that the transistor in Fig. 3.26a is in the common-emitter configuration. We also note that the supply voltage is well within the V_{CEO} rating.

$$P_D \cong i_C V_{CE} \tag{3.43a}$$

Figure 3.25c has been reproduced in Fig. 3.26b with off-scale extensions of the current axis. The boundaries given by equation (3.43a), the maximum collector current rating, and the requirement on the minimum collector-emitter voltages are shown by dashed lines on the figure. These dashed lines (the bottom boundary is the $i_B = 0$ line, and the right boundary, which is not shown, is the collector-emitter junction zener breakdown voltage) enclose the restricted operating region discussed above.

Considering the given supply voltage of 15 V, and assuming the output should be able to swing through positive and negative values, we can choose the operating point at $V_{CE} = 7.5$ V. If we choose 10 mA as a reasonable collector current at the operating point (this decision could also be influenced by actual load considerations), the required base current, from Fig. 3.26b, is about 112 μA. From Fig. 3.25a, we note that at this current level, the base-emitter voltage is about 0.51 V and that the base-emitter junction is sufficiently forward biased as required.

We can now determine the values of the external resistors, using the following equations:

$$R_{\text{bias}} = \frac{(V_{CC} - V_{BE})}{i_B} \tag{3.43b}$$

$$R_L = \frac{(V_{CC} - V_{CE})}{i_C} \tag{3.43c}$$

Equation (3.43c) is known as the *DC load line* equation (see Fig. 3.26b). Given that the base-emitter is forward biased by about 0.51 V at the operating point, equation (3.43b) gives $R_{\text{bias}} = (15 - 0.51)(\text{V})/112(\mu\text{A}) = 129.375$ kΩ. The nearest standard resistor value consistent with ensuring that the base-emitter is forward biased (120 kΩ) should be used. Similarly, equation (3.43c) yields $R_L = (15 - 7.5)(\text{V})/10(\text{mA}) = 750$ Ω. Given these resistor values, the actual operating point would shift but only very slightly. Finally, the design in Fig. 3.26b shows that a variation of the base current of around 63 μA about the operating point produces approximately a 4.0-V variation in output voltage. From Fig. 3.25a, the corresponding variation in input voltage is only about 0.02 V, so that this amplifier design can achieve amplifications of up to 200. An input capacitor, as shown in Fig. 3.26a, is usually necessary to isolate the base from zero-frequency components in the input signal, which could upset the DC bias condition established by the design. Such a capacitor should be sized so that the useful frequency content of the input is not adversely attenuated.

Operational Amplifiers

The design of a transistor amplifier described in the preceding paragraph should convey an idea of the variety in type and complexity (including number of stages) of discrete amplifier circuits needed for different types of signals and desired amounts of amplification (called *gain*) in various areas of application. Among the types of amplifiers available are current amplifiers, voltage amplifiers, power amplifiers (both current and voltage are amplified), DC amplifiers (steady- or zero-frequency signals are amplified), and AC amplifiers (time-varying signals are amplified). Generally, the practice is to minimize as much as possible the design effort by utilizing prepackaged circuits in which a number of requirements in a given application area are already implemented in families of such circuits. This trend has been aided by continuing developments in *large-scale integration* of electronic circuits. Amplifiers belong to the *analog integrated circuit* category, as opposed to digital integrated circuits. Analog circuits operate on continuous data, while digital circuits handle discrete data. The two types of signals were explained in Section 2.4.

The *operational amplifier* (*op-amp* for short) is a very important group of analog integrated circuits that are used as building blocks in various applications. It is a DC-coupled amplifier in which the output voltage, over the operating range and in a manner analogous to a mathematical operation, can be made to follow linearly the input voltage (hence the name operational amplifier). Figure 3.27a shows the simple and convenient model we shall use for the op-amp. The + terminal of the op-amp is known as the *noninverting input,* while the − terminal is called the *inverting input.* A signal applied to the inverting input (with the + terminal grounded) gives rise to an output signal of opposite sign, whereas a signal applied to the + terminal remains unchanged. The op-amp model has the following properties:

1. **Wide bandwidth** and ability to respond down to zero frequency (DC)
2. **Very high gain** (amplification). The quantity A in the model is of the order 10^4 to 10^9. This makes the op-amp without a *passive feedback circuit* very sensitive to noise and difficult to regulate.
3. **Active operation.** The op-amp has an external power source ($\pm V_s$ in the model). It is an active device and it draws very little current from the circuit in which it is placed. R_i in the model is quite high, and the current flowing into the op-amp input terminal is of the order of 10^{-9} A.

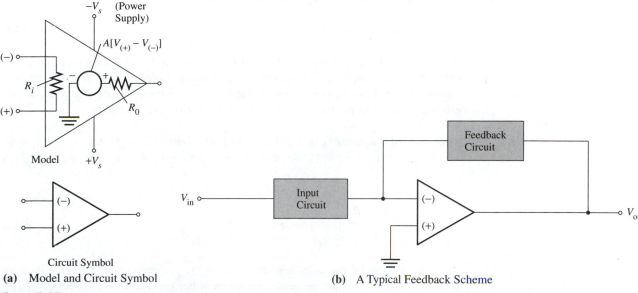

(a) Model and Circuit Symbol

(b) A Typical Feedback Scheme

Figure 3.27

The Operational Amplifier

With passive feedback elements such as shown in the typical scheme in Fig. 3.27b, the overall gain is reduced considerably, but the op-amp acquires the following two very useful properties:

4. **Unilateral operation.** As a *unilateral coupling* device, the state of the component upstream of the op-amp does not affect the state of the downstream component.
5. **Precisely regulated operation.** The (linear) relation between op-amp input and output can be specified by the user and can be as accurate as the user can make it.

Op-amp circuits can be analyzed very simply by utilizing two approximate rules that are derived from the listed properties:

1. The voltage difference between the two op-amp input terminals, $V(+) - V(-)$, equals zero.
2. The currents flowing into the two op-amp input terminals equal zero.

As a practical matter the absolute value of the output voltage of the op-amp cannot exceed that of the power supply voltage. Thus we must have $|V(+) - V(-)| \leq 2V_s/A$, and since A is so large, rule 1 follows. Also, given the first rule and the fact that R_i is so large, rule 2 follows.

| **Example 3.16** | Some Useful Op-Amp Circuits |

Use the two op-amp rules to determine the functions performed by the op-amp circuits shown symbolically in Fig. 3.28. These circuits are based on the use of the inverting input. Note that only a single output terminal is necessary on the circuit symbols because by convention the output voltage is measured with respect to ground. Also, the power supply connections are understood to be present but are not shown.

Solution:

Voltage follower (Fig. 3.28a): $V_0 = V(-)$, and by rule 1, $V(-) = V(+)$, hence

$$V_0 = V(+); \quad \text{voltage follower} \tag{3.44}$$

The voltage follower has no amplification but it has a high input resistance and a low output resistance (see [8]). This makes it a very useful *isolator* (separator or *buffer* interposed between two parts of a circuit that should not interact) especially when a low-resistance load is to be driven by a source with high source resistance.

Noninverting amplifier (Fig. 3.28b): Kirchhoff's current law gives $i_f + i_1 + i_b = 0$, and from rule 1, $V(-) = V_1$, so that $(V_0 - V_1)/R_f - V_1/R_1 + i_b = 0$. Rule 2 gives $i_b = 0$, hence

$$V_0 = \left(\frac{R_1 + R_f}{R_1} \right) V_1; \quad \text{noninverting amplifier} \tag{3.45}$$

The amplification of the amplifier $(R_1 + R_f)/R_1$ is determined by the user through choice of R_1 and R_f. Since the noninverting amplifier possesses, to a sufficient extent, the high input resistance and low output resistance typical of the voltage follower, it can also be used as an isolator.

Figure 3.28

Some Operational Amplifier Circuits

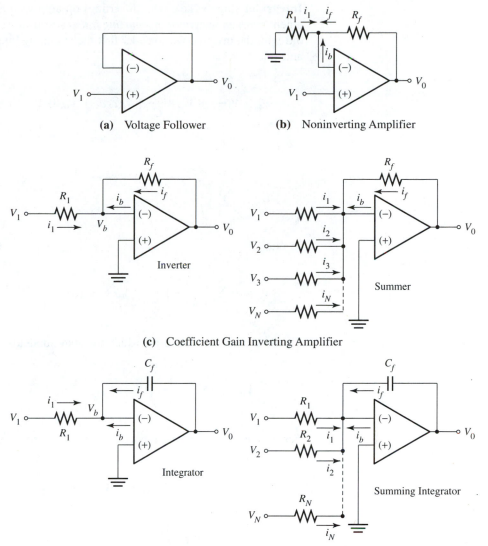

(a) Voltage Follower

(b) Noninverting Amplifier

(c) Coefficient Gain Inverting Amplifier

(d) Integrating Amplifier

Coefficient gain/inverter (Fig. 3.28c): This inverting amplifier circuit is the basis of the *summer* or summing amplifier, which is also shown. Since we can consider the *inverter* a special case of a summer with one input, it is sufficient to obtain first the results for the summer. Kirchhoff's current law at the *summing* junction is $i_1 + i_2 + i_3 + \ldots + i_N + i_b + i_f = 0$. From rule 1 $V(-) = 0$, so that Ohm's law gives $i_i = V_i/R_i$. Also, $i_f = V_0/R_f$, and from rule 2 $i_b = 0$. Thus the current law yields

$$V_0 = -\sum_{i=1}^{N}\left(\frac{R_f}{R_i}\right)V_i; \text{ summer} \tag{3.46a}$$

Note that the summer is actually a summing inverter and that each input can have a separate user-determined gain. The inverter relation follows directly:

$$V_0 = -(R_f/R_1)V_1; \text{ inverter} \tag{3.46b}$$

The minus sign gives the inversion, and amplification is possible as before.

Integrator (Fig. 3.28d): The (inverting) op-amp with a capacitor as the feedback element gives an integrator or *summing integrator* if there is more than one input. As we did with the inverter, we consider first the summing integrator. From equation (2.24a) and rule 1,

$$V_0 - V(-) = V_0 = \frac{1}{C_f} \int_0^t i_f(t)\, dt + V_0(0)$$

Further, rules 1 and 2 and Ohm's law give

$$i_f = -\sum_{i=1}^{N} \frac{V_i}{R_i}$$

so that

$$V_0 = -\sum_{i=1}^{N} \left(\frac{1}{R_i C_f} \right) \int_0^t V_i(t)\, dt + V_0(0) \tag{3.47a}$$

This is a dynamic system for which the state model is

$$\frac{d}{dt} V_0 = -\sum_{i=1}^{N} \left(\frac{1}{R_i C_f} \right) V_i \tag{3.47b}$$

For the integrator, the corresponding equations are

$$V_0 = -\left(\frac{1}{R_1 C_f} \right) \int_0^t V_1(t)\, dt + V_0(0) \tag{3.47c}$$

$$\frac{d}{dt} V_0 = -\left(\frac{1}{R_1 C_f} \right) V_1 \tag{3.47d}$$

....

Further Examples in Electrical Engineering

The examples of electrical system models considered next are in addition to those already presented in this chapter and those analyzed in Section 2.3. However, the systems modeled include behavior components introduced in this chapter.

Example 3.17   DC Level Adjuster

Recall that one of the applications of *RC* networks is as DC coupling of a time-varying signal from one stage of a circuit to the other. It is generally required that the circuit time constant *(RC)* be much greater than the period of the highest frequency component of the signal being transferred. As shown in Fig. 3.29a, the DC component of the input signal is lost when it passes through such an *RC* network.

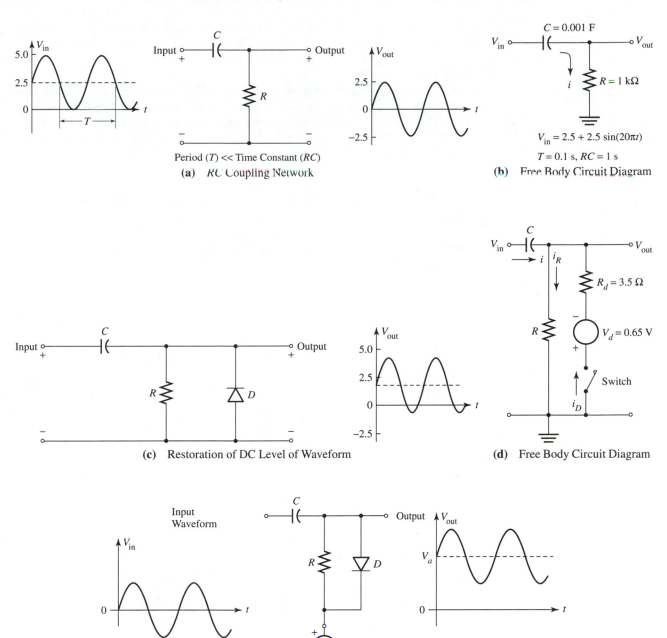

(a) RC Coupling Network

Period $(T) \ll$ Time Constant (RC)

(b) Free Body Circuit Diagram

$V_{in} = 2.5 + 2.5 \sin(20\pi t)$

$T = 0.1$ s, $RC = 1$ s

(c) Restoration of DC Level of Waveform

(d) Free Body Circuit Diagram

(e) Adjustment of DC Level of Waveform

Figure 3.29

Use of the Diode as a DC Level Adjuster

a. Derive the model for the *RC* network shown in Fig. 3.29a, and using a computational package such as MATLAB, solve this model for the output voltage assuming the input voltage and the circuit parameters are as given in the free body circuit diagram in Fig. 3.29b.

b. An application of the diode to restore the DC level of the input waveform is shown in Fig. 3.29c. Derive the model for this circuit and demonstrate, using MATLAB or a similar package, and the input and circuit parameters given in Fig. 3.29b, that the DC component is indeed restored to approximately within the diode cut-in voltage. Use the diode small-voltage model parameters given in Fig. 3.29d. More generally, the DC level of an input waveform can be adjusted to any desired value, using a diode and a voltage source, as shown in Fig. 3.29e. It is left as an exercise (see Practice Problem 3.43a) for the reader to demonstrate this result.

Solution a: From Fig. 3.29b, the circuit components are a resistor (static) and a capacitor (dynamic):

  Resistor: Ohm's law gives

$$V_{out} = iR. \tag{a}$$

Capacitor: Equation (2.24b) gives

$$C\frac{d}{dt}(V_{in} - V_{out}) = i \tag{b}$$

Considering equation (b), let the state variable be $x = V_{in} - V_{out}$. Then the state model is

$$\frac{d}{dt}x = \left(\frac{1}{RC}\right)(V_{in} - x) \tag{c}$$

$$V_{out} = V_{in} - x \tag{d}$$

To solve these equations, we first create a MATLAB M-file, called **highpf.m,** containing the differential equation (c). The script file shown in the box called **fig3_30.m,** uses MATLAB's ode45 to solve the differential equation over the time interval $0 \le t \le 3$ (about three times the time constant of the circuit). However only the last half-second of the output is displayed. Figure 3.32a shows the result (solid curve), which is as expected. The dashed curve is the input waveform. Note that the script file also solves the model derived for the solution of part (b) of this example. The initial condition on x in both cases is obtained by assuming a zero initial value for the output voltage (that is, that the capacitor is initially discharged), so that, by equation (d), the consistent initial value of x is $x(0) = V_{in}(0) = 2.5$.

Solution b: Consider the free body circuit diagram of Fig. 3.29d. Equation (b) still applies. However, the current law gives $i = i_R - i_D$, and Ohm's law gives $i_R = V_{out}/R$. The small-voltage diode model, equation (3.40c), yields

$$i_D = \text{switch}(-V_{out} - V_d)\left(\frac{-V_{out} - V_d}{R_d}\right) \tag{e}$$

Figure 3.30

Transferred Waveforms by *RC* Coupling Network With and Without DC Restoration

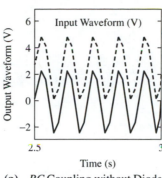

(a) *RC* Coupling without Diode

(b) Diode Restoration of DC Level

so that the state model is

$$\frac{d}{dt}x = \left(\frac{V_{in} - x}{RC}\right) - \text{switch}(-V_{in} + x - V_d)\left(\frac{-V_{in} + x - V_d}{R_d C}\right) \tag{f}$$

$$V_{out} = V_{in} - x \tag{g}$$

Equation (f) is contained in the MATLAB M-file called **highpf2.m** (see below). The script file, **fig3_30.m,** is used to generate the output for both cases—with and without a diode. As shown in Fig. 3.30b, the DC level of the waveform has been restored, although not completely. The difference is in the range of the diode cut-in voltage, as anticipated. The M-files follow:

```
function xdot = highpf(t,x)
R = 1000; C = .001; iRC = 1/(R*C);
dc = 2.5; w = 20*pi; a = 2.5;
xdot = zeros(1,1);
xdot(1) = (a*sin(w*t) + dc - x(1))*iRC;
end
```

```
function xdot = highpf2(t,x)
R = 1000; C = .001; iRC = 1/(R*C); Rd = 3.5; Vd = .65; iRdC = 1/(Rd*C);
dc = 2.5; w = 20*pi; a = 2.5;
xdot = zeros(1,1);
xdot(1) = (a*sin(w*t) + dc - x(1))*iRC;
if (x(1) - a*sin(w*t)-dc-Vd) >= 0
      xdot(1) = xdot(1) - ((x(1)-a*sin(w*t)-dc-Vd)*iRdC);
end
```

Example 3.18 Transistor Switching of Inductive Loads

Figure 3.31a shows the application of a transistor as a switch for inductive loads. Common examples include relay coils, stepping motors, and similar inductive components of motion control systems. Figure 3.31b, for instance, illustrates a

```
% fig3_30.m
% Script to generate Figure 3.30
% Uses function M-files, highpf.m and highpf2.m
%
disp(' ')
disp('          Please wait, this will take a little while!')
%
t0 = 0; tf = 3;        %time interval
x0 = [2.5];            % initial condition
%
[t1,x1] = ode45('highpf',t0,tf,x0); % case of no diode
Vin1 = 2.5*sin(20*pi*t1)+2.5;      % regenerate Vin
Vo1 = Vin1 - x1;                    % compute output, eq. (d)
%
[t2,x2] = ode45('highpf2',t0,tf,x0); % case with diode
Vin2 = 2.5*sin(20*pi*t2)+2.5;      % regenerate Vin
Vo2 = Vin2 - x2;                    % compute output, eq. (g)
%
disp(' ')
disp('   done!');
%
axis('square');.............% square aspect ratio
axis([2.5 3 -3 7]);         % scale axes, range of t is 2 to 3
subplot(211);               % split graph window
plot(t1,Vo1);               % plot the output waveform
title('(a) RC Coupling Without Diode');
ylabel('OUTPUT WAVEFORM (V)'); xlabel('TIME (sec)'); hold on;
plot(t1,Vin1,'--');         % plot input waveform (dashed line)
text(2.52,5.005,'INPUT WAVEFORM (V)'); hold off;
%
axis([2.5 3 -3 7]);         % scale axes for part (b)
plot (t2,Vo2);              % plot the output waveform
title('(b) Diode Restoration of dc Level');
ylabel('OUTPUT WAVEFORM (V)'); xlabel('TIME (sec)');
```

possible application of the transistor switch in the closed-loop control of the water tank level (see Fig. 1.5b) discussed in Section 1.2. A characteristic of inductive systems with moving armatures is the opposing *electromotive force* (voltage) (or back *emf*) which is produced (see Section 5.3). Thus, as shown in Fig. 3.31c, the inductive load model is not just an inductor but a series combination of an inductor, a velocity-dependent voltage source (representing the back emf), and a resistor (representing the actual resistance of the coil wire, which may not be negligible). The presence of the dependent voltage source implies that V_{CE} could exceed V_{CC}, normally on switching the load off. The purpose of the diode in Fig. 3.31a is therefore to protect the transistor from the resulting high voltage by providing a lower resistance (leakage) path around the coil. Such diodes are called *flyback* diodes in some applications. The flyback diode, in turn, is protected from excessive current by an appropriately sized resistor R_F placed in series with it.

By developing the model for the system of Fig. 3.31c, show that, in this application, (1) the transistor indeed acts like a switch, (2) the current supplied to the load

Figure 3.31

Use of a Transistor Switch with Flyback Diode as Driver for an Inductive Load

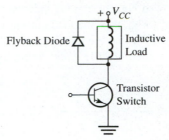

(a) Transistor Switch for Inductive Load

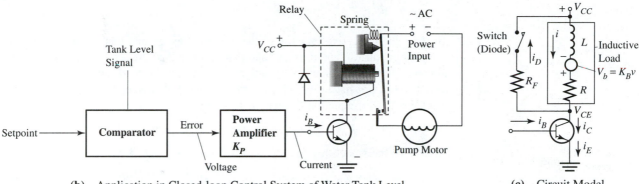

(b) Application in Closed-loop Control System of Water Tank Level　　　**(c)** Circuit Model

while the load is switched on is essentially constant, and (3) the diode does prevent voltage overload of the transistor when the load is suddenly switched off.

Solution: We note that the transistor configuration is the common-emitter mode. We assume that the transistor is matched to the load or that biasing components (not shown) are in place to ensure that our transistor model is applicable and that operation is in the constant-current region. From Fig. 3.31c, the inductor behavior along with Ohm's law yields:

$$L\frac{d}{dt}i = V_{CC} - iR + K_B v - V_{CE} \qquad \textbf{(a)}$$

The current law and the large-voltage diode model give

$$i_C = i - i_D = i - \text{switch}(V_{CE} - V_{CC})(V_{CE} - V_{CC})/R_F \qquad \textbf{(b)}$$

With the stated assumptions, our transistor model gives

$$i_C = \beta i_B \qquad \textbf{(c)}$$

where i_B is the control input. Substituting equation (c) in equation (b), we have

$$i = \beta i_B + \text{switch}(V_{CE} - V_{CC})(V_{CE} - V_{CC})/R_F \qquad \textbf{(d)}$$

1. According to equation (d), as long as $(V_{CE} - V_{CC})$ is less than zero, which is the normal situation, $i = \beta i_B$, so that by switching i_B between zero and a sufficiently positive value, the load (current) can be turned on and off.

2. Since the transistor base current can be independently controlled, if a constant-current source is used for i_B, then the load current will be essentially constant. The implicit assumption here is that the transistor switching time is very short and that the inductor dynamics (the *inductive effect*) is negligible. If we rewrite equation (a) as follows

$$\left(\frac{L}{R}\right)\frac{d}{dt}i = \frac{(V_{CC} - V_{CE}) + K_B v}{R} - i \tag{e}$$

then the quantity (L/R) represents the inductive effect, so that the assumption is

$$\left(\frac{L}{R}\right)\frac{d}{dt}i \cong 0 \tag{f}$$

3. To show the protective function of the diode, let us assume that the inductive load is a spring-return soleniod (relay) and that as a result of the action of the spring, the armature (core) instantaneously attains the velocity $v = v_A$ once the current is switched off. Consistent with the assumption in (2), that is, equation (f), we can neglect the LHS of equation (a) and write

$$V_{CE} = V_{CC} + K_B v_A - \beta R i_B - \text{switch}(V_{CE} - V_{CC})(V_{CE} - V_{CC})R/R_F$$

Given that the switch has just been turned off, $\beta R i_B = 0$, leaving

$$V_{CE} = (V_{CC} + K_B v_A) - \text{switch}(V_{CE} - V_{CC})(V_{CE} - V_{CC})R/R_F \tag{g}$$

Thus without the diode,

$$V_{CE} = (V_{CC} + K_B v_A) > V_{CC}$$

would be applied to the transistor during this period, no matter how brief. However, with the diode present, switch $(V_{CE} - V_{CC})$ becomes equal to 1, and the diode conducts the instant the overvoltage occurs, so that the excess voltage is "bled" off:

$$V_{CE} = (V_{CC} + K_B v_A) - (V_{CE} - V_{CC})R/R_F$$

These predictions can be verified by the simulation of the full equations developed here, using a computational mathematical package, for example.

Example 3.19	An Electronic PID Controller

A common and important closed-loop control law (see Table 1.3) in industry is the *PID* (or proportional-plus-integral-plus-derivative) controller. As the name implies, the *manipulated* variable (see Fig. 1.6) is the sum of three terms: the first term is proportional to the error, the second term is proportional to the integral of the error, and the third term is proportional to the derivative of the error. Analytically,

$$m(t) = K_c \left\{ e(t) + \frac{1}{T_i}\left[\int_0^t e(t)\,dt + e(0)\right] + T_d \frac{d}{dt}e(t) \right\} \tag{3.48}$$

where $m(t)$ is the manipulated variable, $e(t)$ is the error, and K_c, T_d, and T_i are proportionality "constants" or (adjustable) parameters of the controller. Figure 3.32 shows the use of the electronic op-amp circuits introduced in Example 3.16 to realize the function in equation (3.48). The *comparator* (subtraction of the feedback signal from the set-

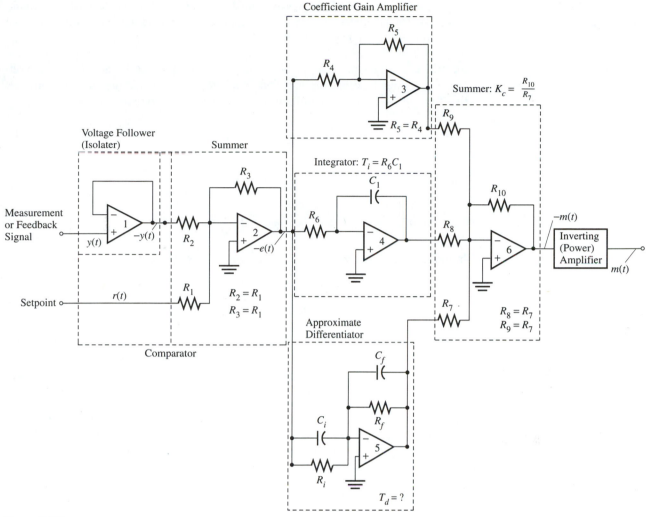

Figure 3.32

An Op-Amp Circuit Realization of the PID Control Law

point signal) is accomplished with a *voltage follower* (which doubles as an isolator of the controller from the rest of the system) and a *summer.* Note that the output of the summer is $-e(t)$. This output passes through (1) a *coefficient gain amplifier* (with gain of 1) to produce the proportional term $e(t)$; (2) an *integrating amplifier* with the values of the input resistor and feedback capacitor chosen (or set) such that $R_3C_1 = T_i$, to give the integral term; and (3) an op-amp circuit described as an *approximate differentiator.* Presumably, the output of this circuit is approximately equal to the derivative term. Finally, a *summing amplifier* with the ratio of the value of the feedback resistor to the value of each input resistor set equal to K_c is used to add the three terms and amplify the result by the amount K_c, as required by equation (3.48). Note that the output of the summing amplifier is $-m(t)$, so that a further inversion is required. This is accomplished, in this case, by an *inverting (power) amplifier* so that the controller output has the requisite strength and form for connection to the downstream device(s).

Derive the model of the approximate differentiator circuit and determine the conditions on the parameters that will ensure a close approximation. Why is there no exact differentiator for use in this application?

Solution: Let the voltage input to the differentiator circuit be V_{in} and the output voltage, V_{out}. Using the capacitor behavior, Ohm's law, and op-amp rules 1 and 2, and setting the sum of currents entering the $(-)$ terminal equal to zero, we have

$$C_i \frac{d}{dt} V_{in} + \frac{V_{in}}{R_i} + \frac{V_{out}}{R_f} + C_f \frac{d}{dt} V_{out} = 0 \tag{a}$$

which gives the model

$$R_f C_f \frac{d}{dt} V_{out} = -V_{out} - \frac{R_f}{R_i} V_{in} - R_f C_i \frac{d}{dt} V_{in} \tag{b}$$

Introducing the *time constants* $\tau = R_f C_f$ and $T = R_i C_i$, we have

$$\tau \frac{d}{dt} V_{out} = -V_{out} - \frac{\tau}{T}\left(\frac{C_i}{C_f}\right) V_{in} - \left(\frac{R_f}{R_i}\right) T \frac{d}{dt} V_{in} \tag{3.49a}$$

If we select $T \gg \tau$ and $\tau \approx 0$, then equation (3.49a) gives the approximate result

$$V_{out} \cong \left(\frac{R_f}{R_i}\right) T \frac{d}{dt} V_{in} = T_d \frac{d}{dt} V_{in} \tag{3.49b}$$

True differentiation (see equation (3.7)) was shown in Section 3.1 to be physically unrealizable. Thus a physical op-amp circuit that is an exact differentiator does not exist. Another op-amp circuit that is used less often than the present circuit but also functions as an approximate differentiator is given in Practice Problem 3.50.

....

On the Net

- arrow.m and fig3_2.m (MATLAB): source file for Example 3.1 and Fig. 3.2
- SC4.MCD (Mathcad): solution of equation (n) of Example 3.8 by symbolic matrix inversion
- SC5.MCD (Mathcad), fig3_23.m (MATLAB): source file for Fig. 3.23
- highpf.m, highpf2.m, and fig3_30.m (MATLAB): source file for Example 3.17 and Fig. 3.30
- Auxiliary exercise: Practice Problems 3.4, 3.19, 3.39 through 3.45.

3.5 PRACTICE PROBLEMS

3.1 Is the state of a system sufficiently described by giving only its configuration? Explain your answer.

3.2 Suppose that the behavior of a component of a system can be described in terms of two variables, e and f. If this behavior is given by a relation of the form $f = \int_0^t F(e)\, dt$ (where $F(e)$ means function of e), indicate the causality for e

and f and state what variable, if any, should be assigned as the state variable for the component.

3.3 The charge buildup q on the plate of a capacitor following its connection across the terminals of a current source is $q = \int^t i\, dt$, where i = current, and t = time since connection (see Fig. 3.33). The voltage across the capacitor V is $V = (1/C)q$, where C is a constant termed the capacitance. Indicate which of the following statements is a correct and appropriate state variable assignment to the capacitor and explain your choice:

(i) any one of i, q, and V

(ii) either q or V

(iii) either i or q

(iv) either i or V

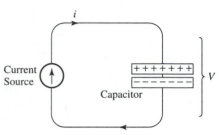

Figure 3.33 (Problem 3.3)

3.4 (a) Derive a model for the dynamic behavior of the bucket in Fig. 3.34 over long periods of time. Assume that the leak rate is proportional to the pressure drop across the hole and ignore the mass of the bucket material. Is your model linear or nonlinear?

(b) Use a computational software to explore the relation between x and h in a phase plane with those variables as coordinates by solving your system equations, assuming the following parameter values and initial conditions: $k = 1961$ N/m, $A = 0.1$ m^2, $\rho = 1000$ kg/m^3, $R = 10^6$

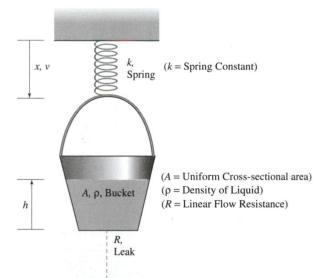

(A = Uniform Cross-sectional area)
(ρ = Density of Liquid)
(R = Linear Flow Resistance)

Figure 3.34 (Problem 3.4)

N · s/m^5, $x(0) = 0.2$ m, $v(0) = 0$ m/s, and $h(0) = 0.4$ m. Plot $x(t)$ versus $h(t)$ for $0 \le t \le 10$s.

3.5* A certain radioactive substance is known to decay at a rate proportional to the amount present. When a quantity of this substance, originally 8 kg, was placed under observation, its mass after 45 minutes was reduced to 6 kg. Develop a model for predicting the mass of the substance at any time. Indicate all measurement units.

3.6 A passenger train departs a station along a straight, level track. Assuming simply that the train is subject to only two forces, a thrust proportional to the rate of consumption of fuel and a retarding force proportional to the speed of the train, and that the rate of fuel consumption is a control input, derive a model that will allow determination of the distance the train will travel before its fuel runs out.

In Problems 3.7 through 3.12, derive the mathematical model for the systems whose physical models are shown.

3.7 The force $F_1(t)$ is prescribed, and the objective of the model construction is to predict x_3 (see Fig. 3.35).

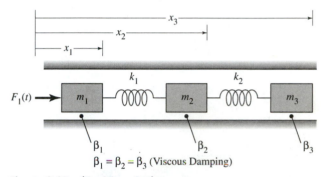

$\beta_1 = \beta_2 = \beta_3$ (Viscous Damping)

Figure 3.35 (Problem 3.7)

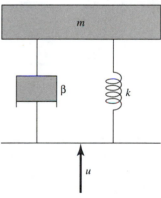

Figure 3.36 (Problem 3.8)

3.8 Simplified car suspension system without the unsprung mass. u is the velocity input due to the road profile (see Fig. 3.36).

3.9* The cylinder (mass m, moment of inertia about center J, and radius r) rolls without slipping. The objective is the linear position of the cylinder x due to the translation of the cylinder (see Fig. 3.37).

3.10 The cable has unstretched length L, modulus of elasticity E, and cross-sectional area A. The lever has mass m and moment of inertia J about pin O. Assume initially that the

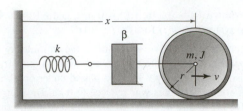

Figure 3.37   (Problem 3.9)

lever moves through small angles θ. Then repeat the problem without this assumption. The goal of the model is to determine the angular position of the lever (see Fig. 3.38).

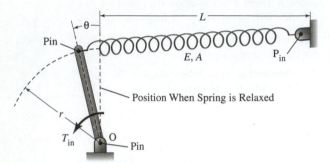

Figure 3.38   (Problem 3.10)

3.11 The printhead drive mechanism of Practice Problem 2.13 (see Fig. 3.39).

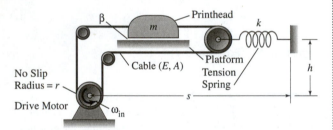

Figure 3.39   (Problem 3.11)

3.12 The ship propulsion system of Practice Problem 2.15 (see Fig. 3.40).

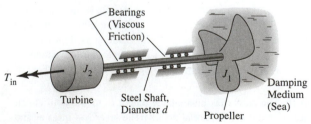

Figure 3.40   (Problem 3.12)

In problems 3.13 through 3.17 derive the state model for the systems whose physical models were assigned for decomposition in Chapter 2.

3.13* The seismograph model of Practice Problem 2.33.

3.14 The compressor model of Practice Problem 2.35. The objective is the motion of the cylinders.

3.15 The sled system of Practice Problem 2.42a. The position of mass m_2 relative to the sled (m_1) is desired.

3.16 The simple elevator system of Practice Problem 2.43. The position, velocity, and acceleration of the elevator are of interest.

3.17 The hydraulic lift arrangement of Practice Problem 2.44. The goal of the model is the motion of the large piston.

3.18* Derive the model for the ship stepped–propeller shaft system shown in Fig. 3.41. The torsional spring constant for each portion of the shaft can be determined from the given data. The moment of inertia of the propeller is J. The coefficient for the damping on the propeller due to the water is β. Assume that the angular vibration of the engine $\theta_e(t)$ is known. The angular displacement θ of the propeller is required.

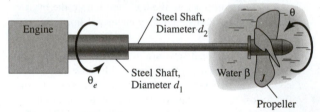

Figure 3.41   (Problem 3.18)

3.19 A simple physical model of a vehicle/flywheel system is shown in Fig. 3.42. During braking or acceleration of the vehicle, energy is transferred to or from the flywheel via the clutch, whose equation is $F_f = B(\omega_2 - \omega_3)$, where F_f is the average tangential frictional force in the opposing clutch faces that resists the motion of the vehicle when $\omega_2 > \omega_3$ (during braking) and aids this motion during acceleration (when $\omega_3 > \omega_2$). Assume that F_f acts at the radius r_c. B is a parameter of the clutch.

(a) Obtain a model for this system, assuming that the system input is the torque developed by the vehicle engine T_e. Neglect the inertia of the gear (wheels).

(b) Use a computational mathematics package to solve the state equations obtained in (a) and display the results in a state space with ω_2 and ω_3 as coordinates, for the braking case. Assume that the initial velocities are $\omega_2(0) = 80 \text{ s}^{-1}$ and $\omega_3(0) = 50 \text{ s}^{-1}$, the applied torque for braking, $T_e = -1500 \text{ N} \cdot \text{m}$; and the following parameter values: $J_1 = 100$ kg $\cdot$ m^2, $J_2 = 200$ kg $\cdot$ m^2, $J_3 = 125$ kg $\cdot$ m^2, $r_c = 0.2$ m, $B = 5000$ N $\cdot$ s, and gear ratio $r_2/r_1 = 0.5$.

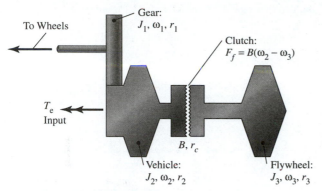

Figure 3.42 (Problem 3.19)

3.20 A simple physical model of the draw works, elevators, and derrick used in pulling up drill pipes in an oil well is shown in Fig. 3.43. The angular velocity of the draw works motor and cable reel ω_{in} is given. The pipes and swivel can be represented by a lumped mass m_2. There is viscous friction between the pipes and the well as shown. Assuming that the angle α remains essentially constant, derive a model of the system for an interest in the vibrations (displacement) of the derrick.

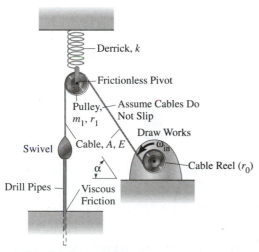

Figure 3.43 (Problem 3.20)

3.21* The physical model of the engine valve opening mechanism of Practice Problem 2.17 is shown in Fig. 3.44. The valve is opened by the action of a cam on a slightly compliant push rod whose motion is transmitted to the valve stem through the rocker arm. The moment of inertia of the

rocker arm about the pivot is J. The cam profile is given by $y_1(t) = R \sin \theta$, where R is a constant, and the cam is driven at a known constant speed $\dot{\theta} = \omega$. Derive the model for this system and state any crucial assumptions that you make. Ignore the mass of the push rod but note that the motion of the valve is subject to some damping.

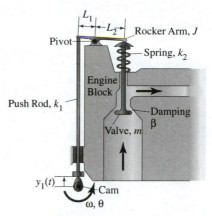

Figure 3.44 (Problem 3.21)

3.22 Obtain the state equations describing the motion of the mechanical system whose physical model is shown in Fig. 3.45. Assume that the magnitude of the force F_{in} is given but that the value of the angle α is always such as to prevent the rotation of the mass m_2. *Hint:* No slip implies that the point of contact between m_3 and ground is an instantaneous center of zero velocity, so that the angular velocity of m_3 is related to the linear velocity of m_2. Also, there is a frictional force at this interface.

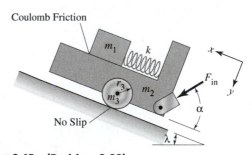

Figure 3.45 (Problem 3.22)

3.23 The gyroscopic moment produced by the controlled precession of a large gyro is used to counteract the rolling of a ship. In the arrangement shown in Fig. 3.46, a motor with a pinion turns the large gear, which causes the gyro to precess about a horizontal transverse axis y in the ship. The rotor has a mass m_2 and radius of gyration r and is turning counterclockwise (as seen from the top) with a constant speed p. Ignoring the mass of the pinion and assuming that the ship is subject to a disturbance torque tending to roll it to port, derive a model to predict the angular (roll) displacement of the ship θ_x. In what direction should the motor turn (as seen looking from starboard to port) to counteract the ship roll to port?

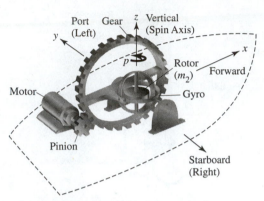

Figure 3.46   **(Problem 3.23)**

3.24 The turbine rotor in a ship's power plant is spinning at a constant rate p, in the direction shown in Fig. 3.47. The rotor shaft is mounted on frictionless bearings A and B that are supported on the ship with the rotor axis in the horizontal fore-and-aft direction as shown. If the ship is making a turn to port (left) of radius R with speed s, derive a model to predict the angle of rise or fall of the ship's bow due to the gyroscopic action. The turbine rotor has mass m, radius of gyration r, and center of mass G as indicated in the figure. Assume that the ship's center of mass without the rotor is at C and that its moment of inertia about a horizontal port-starboard axis through C is J.

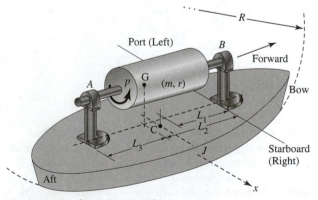

Figure 3.47   **(Problem 3.24)**

3.25* A vibration-producing device consists of two wheels with eccentric masses (m_e) as shown in Fig. 3.48. The wheels rotate in opposite directions but are synchronized so that the unbalanced masses always reach the same vertical position at the same time. If the total mass of the device is m, derive a model for the vibration of the device.

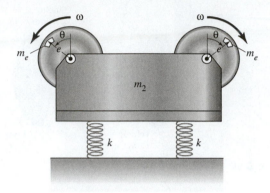

Figure 3.48   **(Problem 3.25)**

3.26 A physical model of a loading situation on an unsupported slender beam is shown in Fig. 3.49. The beam of mass m is initially at rest when a prescribed force F is applied at the position shown. Derive a model for the subsequent motion (of the center of mass G) of the beam if the coefficients of static and kinetic friction between the end of the beam and the surface are both f. Assume that the magnitude of F is such that the beam slips at O.

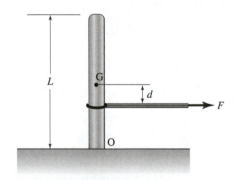

Figure 3.49   **(Problem 3.26)**

3.27 Determine N_A and N_B by the alternative approach suggested in Example 3.8, that is, using equations (e) and (c) of the example, differentiate equations (f) twice and substitute the results in equations (a) and (b); then solve the resulting equations simultaneously, for N_A and N_B.

3.28 The spacecraft in Fig. 3.50 is rotating with constant speed Ω about the z-axis while its center is moving through space at constant velocity. If the solar panels are also rotating about the y-axis at a constant rate $\dot{\theta}$, develop a model for the components of the absolute angular and linear velocities of a point A on the solar panel. *Hint:* See equations (3.20a) and (3.20b) and discussion of relative motion in Example 3.8.

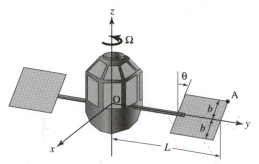

Figure 3.50 (Problem 3.28)

3.29* The solid circular disk of mass m and small thickness spins freely on its shaft at a steady rate p (see Fig. 3.51). Derive the horizontal components of the reaction forces at A and B exerted by the respective bearings on the horizontal shaft as the assembly falls after release from the vertical position ($\theta = 0$) with $\dot{\theta} = 0$. Neglect the mass of the two shafts compared with m. Also ignore all friction.

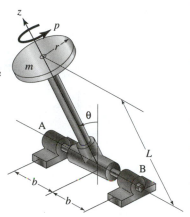

Figure 3.51 (Problem 3.29)

3.30 A system of cascading reservoirs is given as shown in Fig. 3.52. The highest reservoir is very large, while the lower two are of finite dimensions. Derive the state model for this system. Assign any necessary parameters.

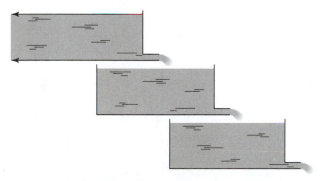

Figure 3.52 (Problem 3.30)

3.31 The liquid levels in the two tanks shown in Fig. 3.53 are initially equal, and the intermediate valve is open while the outflow valve is closed. If the flow through both valves is laminar, develop the mathematical model relating the level of tank 2 with time after the outflow valve is opened. The intermediate valve remains open throughout.

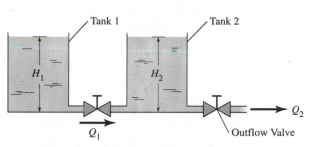

Figure 3.53 (Problem 3.31)

3.32 Write the dynamic equations necessary to determine $h(t)$ for the system shown in Fig. 3.54. State any assumptions that you make. Assign parameters as needed.

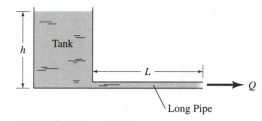

Figure 3.54 (Problem 3.32)

3.33* Derive a model for the system shown in Fig. 3.55 if the piston displacement and velocity x_2 and v_2 are prescribed inputs. The desired output is the cylinder displacement x_1. Assume that the mass of the cylinder is negligible.

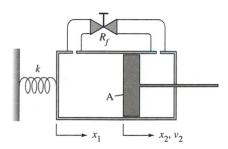

Figure 3.55 (Problem 3.33)

3.34 Complete, if necessary, the decomposition of the water supply system of Fig. 1.1 that was started in Section 2.4.

Then derive the model for the system assuming the goal is to predict the level of water in the tank at the end of each day's operations.

3.35* Derive a dynamic model for the arrangement for lifting heavy weights from a remote location as shown in Fig. 3.56. The tube of internal diameter d is quite tall (h is large, and laminar flow is anticipated). The shaft of the large piston passes inside the spring so that the top end of the spring pushes against the chamber while the lower end pushes against the large piston. Assume that the behavior of the valve is linear.

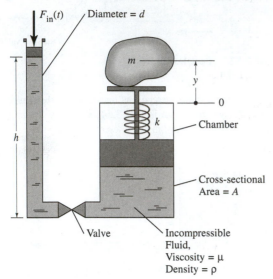

Figure 3.56   **(Problem 3.35)**

3.36 A fish from a fish tank is accidentally swept into the discharge line of the tank as shown in Fig. 3.57. Water flows

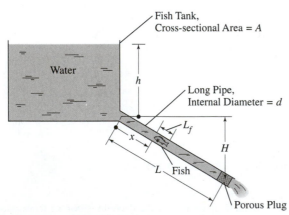

Figure 3.57   **(Problem 3.36)**

slowly out of the tank through a porous plug as shown. The mass of the fish is m, but the fish is small enough to pass through the pipe (assume that the fish can be represented by a cylinder with cross-sectional area a_f and length L_f as shown). However, the water flowing around the fish dampens its motion, and the coefficient of viscous friction is β. Develop a model for this system that can predict the displacement x of the fish from the tank.

3.37 Derive complete models for each circuit in Fig. 3.58 using the state space approach.

(a) The voltage at node b is desired.

(b) The objective is V.

(c) The model should be able to predict V_2.

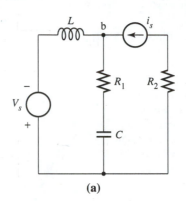

(a)

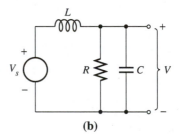

(b)

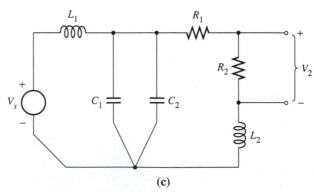

(c)

Figure 3.58   **(Problem 3.37)**

(d) The voltage drop across the capacitor is of interest.

(e) A general model that can be used to determine any desired voltage or current in the circuit is required.

(f) The needed model output is V_3.

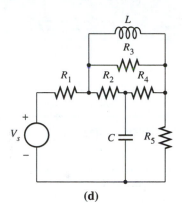

(d)

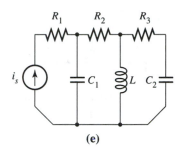

(e)

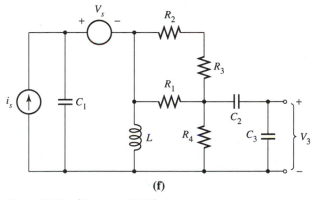

(f)

Figure 3.58   (Problem 3.37)

3.38* The circuit shown in Fig. 3.59 (enclosed in broken lines) is known as a *lead network*. It is often applied in the *frequency domain* design of feedback controllers (the subject of Chapter 12). Assuming that the input voltage V_1 is prescribed, find the expression for the output voltage V_2 that includes the effect of the system (represented by the load resistance R_L) connected to the output leads of the lead

network. *Hint:* Consider using point a for ground, and combining the resistors R_2 and R_L.

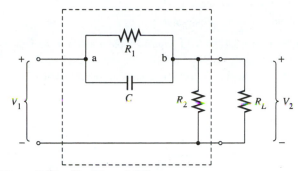

Figure 3.59   (Problem 3.38)

3.39 Figure 3.60 shows a diode voltage limiter circuit, similar in operation to the circuit shown in Fig. 3.22a (case of diode D2 only) and analyzed in Example 3.14. Using a computational mathematics package, show as was done in Example 3.14, that the output voltage is limited to a low of about $-V_L$. Assume the same input and circuit parameter values as in Example 3.14, and consider both the large- and small-voltage diode models.

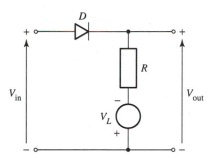

Figure 3.60   (Problem 3.39)

3.40 The voltage limiter circuit shown in Fig. 3.61 is known as a *positive comparator* circuit. Using the approach of

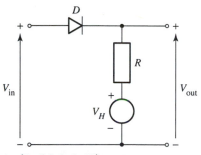

Figure 3.61   (Problem 3.40)

Example 3.14, and the input function and circuit parameter values given in that example, determine the behavior of the present circuit, assuming a

(a) large-voltage diode model.

(b) small-voltage diode model.

In Problems 3.41 to 3.43, repeat Problem 3.40 for the indicated circuits:

3.41 The alternative positive comparator circuit shown in Fig. 3.62.

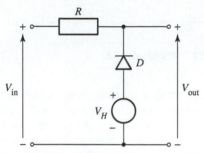

Figure 3.62 (Problem 3.41)

3.42a* The negative comparator circuit shown in Fig. 3.63.

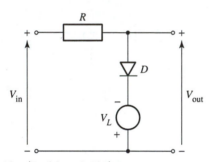

Figure 3.63 (Problem 3.42a)

3.42b The alternative negative comparator circuit shown in Fig. 3.64.

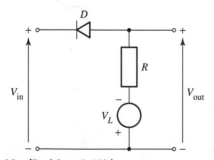

Figure 3.64 (Problem 3.42b)

3.43a The general diode DC level adjuster circuit mentioned in Example 3.17 and shown in Fig. 3.65.

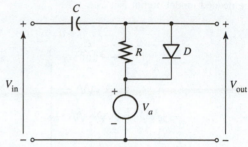

Figure 3.65 (Problem 3.43a)

3.43b The circuit shown in Fig. 3.66, which is used for clipping at two positive voltages.

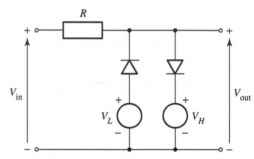

Figure 3.66 (Problem 3.43b)

3.43c The circuit shown in Fig. 3.67, which is used for clipping at two negative voltages.

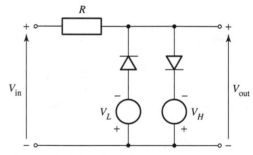

Figure 3.67 (Problem 3.43c)

3.44 Figure 3.68 shows a diode-capacitor filter circuit for supplying uniform and steady voltage required by some electronic apparatus.

(a) Using the diode small-voltage model, derive the model of the circuit for the voltage drop across the load.

(b) Solve the model obtained in (a) and display the result assuming the input function and circuit parameter values given in the figure

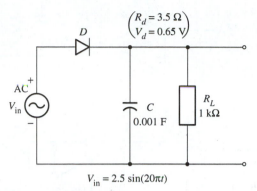

$$V_{in} = 2.5 \sin(20\pi t)$$

Figure 3.68   (Problem 3.44)

3.45 Repeat Problem 3.44 for the circuit shown in Fig. 3.69.

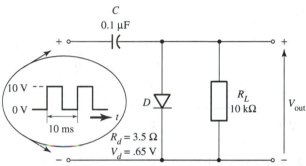

Figure 3.69   (Problem 3.45)

3.46a Determine approximately the operating point for the *fixed-bias* circuit shown in Fig. 3.70 if $R_B = 13$ kΩ, and $R_L = 9.1$ kΩ Assume a silicon transistor with $\beta = 50$ and a V_{CC} of 18 V.

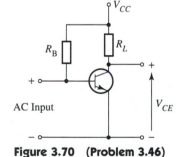

3.46b For the fixed-bias circuit shown, determine the values of R_B and R_L
Figure 3.70   (Problem 3.46)
that would permit an operating point of approximately $i_C = 5$ mA; $V_{CE} = 10$ V, given that $V_{CC} = 20$ V, and the transistor is a silicon transistor with $\beta = 50$.

3.47a* Figure 3.71 shows a *self-bias* transistor circuit. The operating point for such a circuit is less dependent on the transistor parameter β and more stable with respect to tem-

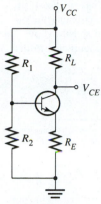

perature effects. Determine approximately the operating point for this circuit if $R_1 = 18$ kΩ, $R_2 = 3.6$ kΩ, $R_E = 330$ Ω, and $R_L = 1.2$ kΩ. Assume a silicone transistor with $\beta = 80$ and that the base-emitter junction is forward biased at the operating point with 0.6 V. Take $V_{CC} = 12$ V.

3.47b Determine R_1, R_2, and R_E for an operating point ($i_C = 3$ mA, $V_{CE} = 8.5$ V) for the self-bias transistor circuit shown, assuming a silicon transistor with $\beta = 50$ and a supply voltage $V_{CC} = 12$ V. Assume a load $R_L = 2.2$ kΩ.

Figure 3.71
(Problem 3.47)

3.48 The circuit shown in Fig. 3.72 is known as a *voltage-to-current converter.* Show that $i_{out} = (1/R_i)V_i$.

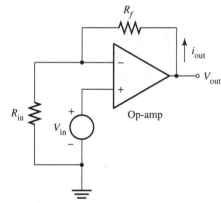

Figure 3.72   (Problem 3.48)

3.49 Figure 3.73 shows a prototype *logarithmic amplifier.* Show that the output is given approximately by the expression $V_{out} = -K \log\left(\dfrac{V_{in}}{R}\right)$. What does the parameter K consist of?

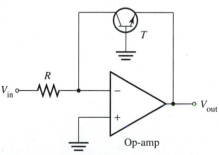

Figure 3.73   (Problem 3.49)

3.50* The operational amplifier (op-amp) circuit shown in Fig. 3.74 is an approximate differentiator. Assuming the op-amp can be modeled as indicated for small signals, show that the output voltage V_{out} is approximately proportional to the time derivative of the input voltage V_{in}.

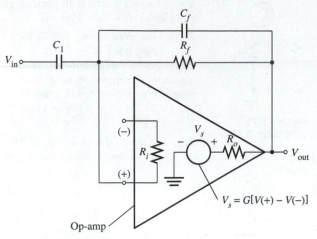

Figure 3.74   (Problem 3.50)

3.51 Derive the model for the indicated circuits shown in Fig. 3.75.

(a) The *average detector* circuit

(b) The *peak detector* circuit

(c) The *second-order low-pass filter* circuit

(d) The *second-order Chebyshev filter* (active) circuit. K is a design constant that is greater than 1.

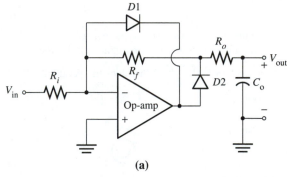

(a)

Figure 3.75   (Problem 3.51)

Chapter

4

Other System Models in State Space

W E CONTINUE IN THIS CHAPTER with the development of models of engineering systems, particularly those systems that can be considered as predominantly *thermal energy* systems and *process* (or *chemical*) engineering systems. As in previous chapters, we shall introduce behavior components typical of these systems and later use them in the synthesis of the system models. We shall also briefly explore system modeling based on the assumption that component behavior is *distributed parameter*. The equivalent model based on a *lumped-parameter* analysis will also be developed in each case to demonstrate the perspective that actual system behavior is neither lumped parameter nor distributed parameter but can usefully be modeled based on an assumption of distributed- or lumped-parameter behavior. Finally, examples of models of non-engineering, but familiar, systems will be developed in state space. This material, which can be omitted on first reading without loss of continuity in the text, reinforces the idea that the state space approach is transparent to the type of system being modeled.

4.1 THERMAL SYSTEMS

Energy that is transferred as a result of a *temperature gradient* is sometimes called *thermal energy* or *heat*. Heat transfer takes place from a region of higher temperature to one of lower temperature (just as electric current is said to flow from a region of higher potential (voltage) to one of lower potential.

Conduction, Convection, and Radiation

There are three modes of heat transfer—*conduction, convection,* and *radiation*. In general, a specific heat exchange situation may involve simultaneous heat transfer by two or more of these modes.

Conduction

The transfer of thermal energy by molecule-to-molecule interaction is referred to as conduction. The relation governing heat conduction through a homogenous material with constant properties and negligible energy storage (in other words, through a lumped ideal resistor to heat conduction) is Fourier's law of heat conduction:

$$q = -kA\frac{d}{dx}T \tag{4.1a}$$

where dT/dx is the temperature gradient in the direction of heat flow, A is the area normal to the heat flow direction, and k is the *thermal conductivity*—a parameter of the material. Note that the (rate of) heat transfer q has the same units as power.

For one-dimensional heat conduction as shown in Fig. 4.1a, the component behavior from integration of equation (4.1a) is

$$q_{1-2} = \left(\frac{kA}{\Delta x}\right)(T_1 - T_2) \tag{4.1b}$$

This relation assumes that the temperature profile is linear with x and that the area is not a function of x or that Δx is sufficiently small to justify these assumptions.

For radial heat flow such as through the cylinder of Fig. 4.1b, we note that the area at any radius r is $A = 2\pi rL$, hence

$$q_{i-o}\int_{r_i}^{r_o}\frac{1}{r}\,dr = -2\pi kL\int_{T_i}^{T_o}dT$$

which gives

$$q_{i-o} = \frac{2\pi kL}{\ln(r_o/r_i)}(T_i - T_o) \tag{4.1c}$$

Convection

Convection involves the transfer of heat by conduction as well as by *mass motion*. In *natural convection* (as when hot air rises from the earth's surface to be replaced by colder air) the mass motion is due to natural phenomena. In *forced convection* (as when a room is cooled by blowing cold air into it, using a fan) the mass motion is artificially induced. In both cases the heat must initially be transferred to the mobile medium by molecular interaction (conduction).

Detailed consideration of heat convection can be found in most heat transfer texts (such as [9]). We consider here the simple case of (one-dimensional) convection of heat from a wall or body at temperature T_w to a *free stream* that is at a uniform temperature T_∞. As shown by the velocity profile in Fig. 4.2, the free stream is the region where the fluid velocity has become a constant, $v = v_\infty$. At the wall, the fluid (air, water, oil, etc.) velocity is zero, so that heat transfer from the wall to the fluid is entirely by conduction. However, the amount of heat conducted depends on how much heat is carried away by the fluid. Thus we speak of the overall effect of convection, which is given by Newton's law of cooling:

$$q_c = hA(T_w - T_\infty) \tag{4.2}$$

where A is the surface area normal to the direction of heat flow, and h is the convection *heat transfer coefficient*. In general, h is determined by experiment and depends on several quantities including the fluid properties and fluid velocity.

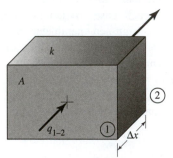

(a)   Linear, $q_{1-2} = \dfrac{kA}{\Delta x}(T_1 - T_2)$

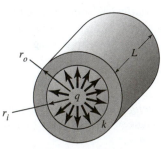

(b)   Radial,
$$q_{i-o} = \frac{2\pi kL}{\text{Ln}(r_o/r_i)}(T_i - T_o)$$

Figure 4.1

Example of Ideal Heat Conductance

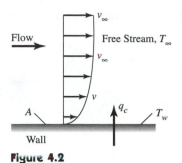

Flow

Free Stream, T_∞

A

Wall

Figure 4.2

Illustration of Overall Heat Convection

Radiation

Heat transfer through the propagation of *electromagnetic waves* is termed *thermal radiation*. The heat exchange between two directly opposed ideal radiators or *black-bodies* (they completely absorb all incident radiation without reflection) is given by the Stefan-Boltzmann law of thermal radiation:

$$q_{1-2} = \sigma A(T_1^4 - T_2^4) \tag{4.3a}$$

where A is the "directly opposed" area (mutually intersected by parallel rays from each body) normal to the heat flow; σ is the Stefan-Boltzmann constant, which has a value of 5.669×10^{-8} W/m$^2 \cdot$ K^4; and T_1 and T_2 are the absolute temperatures of the two bodies.

Nonideal Radiators   An approximate treatment of radiative heat exchange between nonblackbodies that also accounts for configurations in which the bodies are not directly opposed is given by

$$q_{1-2} = \left[\frac{1}{(1 - \epsilon_1)/\epsilon_1 A_1 + 1/A_1 F_{1-2} + (1 - \epsilon_2)/\epsilon_2 A_2} \right] \sigma(T_1^4 - T_2^4); \tag{4.3b}$$

$$A_1 F_{1-2} = A_2 F_{2-1}$$

where ϵ_1 and ϵ_2 are the *emissivities* of the respective bodies, A_1 and A_2 are the (surface) areas of the respective bodies, and F_{1-2} and F_{2-1} are the *shape* (or geometric) *factors* (also *view factors*) for radiant energy interchange between the bodies. The determination of these functions and other details are usually given in heat transfer texts such as reference [9]. However, a particularly useful interpretation of equation (4.3b) is to treat the denominator term as a series "resistance": $R_1 + R_2 + R_3$, where $R_1 = (1 - \epsilon_1)/(\epsilon_1 A_1)$ is a "surface" resistance that accounts for the reduction of the *emissive power* that a blackbody would have had at temperature T_1, $(E_{p1} = \sigma T_1^4)$, to the actual total radiation that leaves the surface of (a nonblack) body 1 per unit time and per unit area (that is, the *radiosity* of body 1, $J_1 = E_{p1} - R_1 q_{1-2}$). $R_3 = (1 - \epsilon_2)/(\epsilon_2 A_2)$ is a similar "surface" resistance for body 2 $(J_2 = E_{p2} - R_3[-q_{1-2}])$; and $R_2 = 1/(A_1 F_{1-2})$, is a "space" resistance between the two radiosity "potentials," so that $q_{1-2} = (J_1 - J_2)/R_2$. Thus

$$(R_1 + R_2 + R_3)q_{1-2} = E_{p1} - E_{p2} = \sigma T_1^4 - \sigma T_2^4.$$

Heat Energy Storage

Thus far we have considered ideal *thermal resistance* (to either conduction, convection, or radiation) with no heat *storage* or *generation* within the element. In practice the assumption of an ideal thermal resistance is reasonable for a body of negligible mass such as very thin materials or bodies with small volumes. The behavior of materials that store energy is described by the first law of thermodynamics, which states that *the change in the total energy of a system is equal to the heat transferred to it less the work done by the system:*

$$dE = \delta Q - \delta W \tag{4.4a}$$

The total energy of a system consists of its *internal energy U, potential energy PE,* and *kinetic energy KE.* Thus

$$dE = dU + d(KE) + d(PE) \tag{4.4b}$$

In the typical ideal thermal energy system component, the quantities δW, $d(KE)$, and $d(PE)$ are all assumed to be negligible (or have been separated out into other components). Hence

$$dU = \delta Q \tag{4.5a}$$

For steady flow of heat ($\dot{Q} = q$), equation (4.5a) gives

$$\frac{d}{dt}U = q \tag{4.5b}$$

The *specific heat* (at constant volume)* c is a material property that is defined as

$$c = \frac{1}{m}\frac{dU}{dT} \tag{4.5c}$$

where m is the mass of the material and T is its temperature. It follows from equations (4.5b) and (4.5c) that

$$\frac{d}{dt}U = mc\frac{d}{dt}T = q \tag{4.6a}$$

or

$$C\frac{d}{dt}T = q; \qquad C = mc; \text{ (heat capacitance)} \tag{4.6b}$$

The quantity $mc = C$ is known as the *heat capacitance,* and the dynamic relation (4.6b) describes the behavior of a thermal capacitor or a heat storage element. It is important to note that q in equations (4.6a) and (4.6b) represents the **total** heat flow **into** the component. Also, a lumped system assumption is inherent in equations (4.6) because the entire material is assumed to be at the same temperature. An example of a thermal capacitor is a large brick wall in a home. During the day, while the sun is out, the wall absorbs heat by radiation and conduction, and its temperature rises slowly (large mass: large thermal capacitance). At night, while the temperature outside the house is low, the wall loses the stored heat (gradually) to the room (as well as to the outside), thus keeping the inside of the house warm.

Examples in Thermal Systems

Example 4.1   Quenching in Oil

Hot forging is quenched in the specially designed vat shown in Fig. 4.3. The vat has a circular cross section and a bottom of thickness w. The vat is also surrounded on the side (but not the bottom) by an insulating material (not a perfect insulator). Heat can be conducted to the atmosphere from the sides and bottom of the vat. The oil in the vat

---

*For solids and most liquids there is little difference between the value of the specific heat at constant volume and the specific heat at constant pressure.

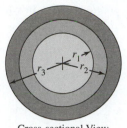

Cross-sectional View

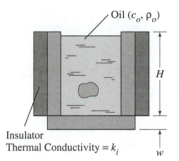

Oil (c_o, ρ_o)

H

Insulator
Thermal Conductivity = k_i

w

(a) Forging in Quenching Vat

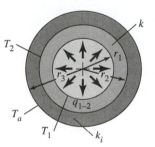

T_2
r_1
r_3 r_2
q_{1-2}
T_a
T_1
k_i
k

1. Vat Sides

q_b
T_1
Area = πr_1^2
w
T_a
k

2. Vat Bottom

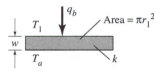

T_f
T_1
q_f

3. Forging

(b) Free Body Diagrams

Figure 4.3

Hot Forging and Quenching Vat

can also lose heat to the atmosphere from its surface; the surface (convective) heat transfer coefficient is h. The vat is made of material whose thermal conductivity is k. The specific heat c_o and density ρ_o of the oil are given. The mass of the forging is m and its specific heat is c. The initial temperatures of the forging and oil are known but not their subsequent temperatures. Derive a model for this system that will allow determination of the oil temperature $T_1(t)$. The ambient temperature T_a is given. The model should contain only known parameters. Assume the vat material and the insulator do not store heat and that the heat transfer coefficient h also applies to the surface (interface with oil) of the forging. The surface area of the forging A is known.

Solution:

Decomposition: The free body diagrams of the components are shown in Fig. 4.3b.

thermal resistance (static):

1. Conduction (radially) through cylindrical vat sides and insulation, assuming there is no heat flow along the walls. Equation (4.1c) gives

$$q_{1-2} = \frac{2\pi k H}{\ln(r_2/r_1)}(T_1 - T_2) \tag{a}$$

$$q_{2-a} = \frac{2\pi k_i H}{\ln(r_3/r_2)}(T_2 - T_a) \tag{b}$$

$$q_{1-2} = q_{2-a}; \quad \text{(series resistance, see note)} \tag{c}$$

where T_1 and T_2 are the temperatures at the surfaces as shown. The entire oil surface and the inner face of the vat are assumed to be at the same temperature T_1.

2. Conduction through vat bottom—equation (4.1b):

$$q_b = \frac{k}{w}(\pi r_1^2)(T_1 - T_a) \tag{d}$$

3. Convection at oil surface and at face of forging—equation (4.2):

$$q_s = h(\pi r_1^2)(T_1 - T_a) \tag{e}$$

$$q_f = hA(T_f - T_1) \tag{f}$$

where the entire forging including its surface is assumed to be at temperature T_f.

thermal capacitance (dynamic): Equation (4.6a) is applicable:

1. Forging:

$$mc\frac{d}{dt}T_f = -q_f \tag{g}$$

2. Oil:

$$m_o c_o \frac{d}{dt}T_1 = \sum q_{\text{into}} = q_f - q_{1-2} - q_b - q_s \tag{h}$$

where

$$m_o = \text{mass of oil} = \pi r_1^2 H \rho_o \tag{i}$$

State variables and state equations: The state variables are T_f and T_1. In equation (h), q_{1-2} involves the unknown T_2 from equation (a).

Using (b) and (c), we have

$$T_2 = \frac{\ln(r_3/r_2)}{2\pi k_i H} q_{1-2} + T_a$$

so that

$$q_{1-2} = \left\{ \frac{2\pi kH/\ln(r_2/r_1)}{1 + [2\pi kH/\ln(r_2/r_1)][\ln(r_3/r_2)/2\pi k_i H]} \right\}(T_1 - T_a) \tag{j}$$

$$= \left\{ \frac{1}{\ln(r_2/r_1)/2\pi kH + \ln(r_3/r_2)/2\pi k_i H} \right\}(T_1 - T_a)$$

The model: From equations (g) through (j), the state equations are

$$\frac{d}{dt}T_f = -\left(\frac{1}{mc}\right)hA(T_f - T_1); \qquad T_f(0) = T_{f-0} \tag{4.7a}$$

$$\frac{d}{dt}T_1 = \left(\frac{1}{\pi r_1^2 H \rho_o c_o}\right)\left\{ hA(T_f - T_1) - \left[\left(\frac{2\pi H}{\ln(r_2/r_1)/k + \ln(r_3/r_2)/k_i}\right)\right.\right. \tag{4.7b}$$

$$\left.\left. + (\pi r_1^2)\left(\frac{k}{w}\right) + (\pi r_1^2 h)\right](T_1 - T_a)\right\}; \qquad T_1(0) = T_{1-0}$$

The state model consists of equations (4.7a), (4.7b), and the output equation

$$y = T_1 \tag{4.7c}$$

Note:

Equation (j) could have been easier to obtain by recognizing that the vat and the insulation constitute resistances in series. Analogies between engineering systems (thermal and electrical in this case) are discussed in Chapter 5.

Example 4.2	Electronic Package Cooling

Most active electronic packages such as power supplies generate a lot of heat and must be housed within instruments in such a way that the heat they generate is quickly dissipated. Figure 4.4 shows a power supply P mounted inside an enclosure E that is installed in a computer. Part of the frame of the computer F is shown. Both the frame F and the enclosure E are connected to the heat sink S, which dissipates heat to the atmosphere by convection. Derive a two-lump model (with P and E as heat storage elements) that will give the temperature of the power supply following the turning on of electric power to the system. Assume that the rate of heat generation within P is known and that it is constant at q_P. Also assume that heat transfer from P to E is by radiation as well as by conduction and that the frame F is a perfect insulator. Note that the temperature of the heat sink is essentially constant, as is the case with the atmospheric temperature T_a. Assign any needed parameters.

Figure 4.4

**Computer Power Supply
Thermal Package System**

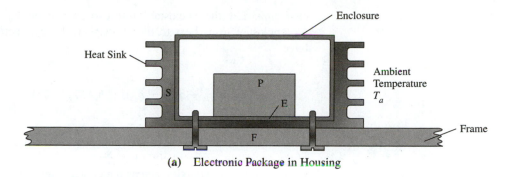

(a)   Electronic Package in Housing

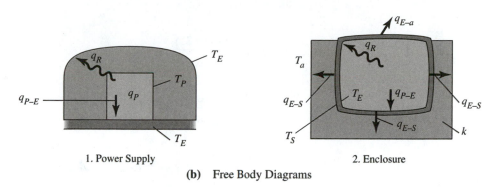

1. Power Supply

2. Enclosure

(b)   Free Body Diagrams

Solution: Let the uniform temperatures of the power supply and the enclosure be T_P and T_E, respectively, and let the temperature of the external surface of the heat sink be T_S.

power supply: If the heat transfer from P to E by radiation and by conduction are q_R and q_{P-E}, respectively, then from equations (4.6b) and (4.3b),

$$C_P \frac{d}{dt} T_P = \sum q_{\text{into}} = q_P - q_R - q_{P-E}; \qquad q_R = \frac{\sigma(T_P^4 - T_E^4)}{R_{\epsilon P} + 1/(A_P F_{P-E}) + R_{\epsilon E}} \qquad \textbf{(a)}$$

where C_P is the effective heat capacitance of P, and, as explained under equation (4.3b), $R_{\epsilon P}$ and $R_{\epsilon E}$ are "surface resistance" terms due to the emissivities of the power supply package and the enclosure, respectively, while F_{P-E} is the view factor for radiation from the power supply to the enclosure. All these parameters are appropriately computed for the material and configuration of the power supply and enclosure shown in Fig. 4.4a. A_P is the surface area of the power supply excluding the base area A_b where P is attached to the enclosure. Equation (4.1b) gives

$$q_{P-E} = \frac{(T_P - T_E)}{R_E} \qquad \textbf{(b)}$$

where R_E is the interface thermal resistance or the contact resistance between P and E over the contact area A_b.

enclosure: Let the enclosure heat capacitance be C_E. Assuming that the top of the enclosure loses heat to the atmosphere by convection (see Fig. 4.4b), we then have

$$C_E \frac{d}{dt} T_E = q_R + q_{P-E} - q_{E-S} - h_t A_t (T_E - T_a) \tag{c}$$

where h_t and A_t are the convective heat transfer coefficient and area of the top surface of the enclosure, respectively. Also, given R_S, that is, the effective interface thermal resistance to heat conduction across the contact area between the enclosure and the heat sink, we have

$$q_{E-S} = \frac{(T_E - T_S)}{R_S} \tag{d}$$

We note that the temperature T_S is an unknown.

heat sink: For average heat sink temperature T_{Sa} we can write

$$C_S \frac{d}{dt} T_{Sa} = 0 = q_{E-S} - q_{S-a}; \qquad q_{S-a} = h_S A_S (T_S - T_a) \tag{e}$$

where A_S and h_S are the exposed heat sink surface area and convective heat transfer coefficient, respectively. There is no heat transfer through the computer frame, since it is considered a *perfect* insulator. Thus from equations (d) and (e),

$$\frac{(T_E - T_S)}{R_S} = h_S A_S (T_S - T_a)$$

and

$$T_S = \left(\frac{1}{1 + R_S h_S A_S} \right)(T_E + R_S h_S A_S T_a) \tag{f}$$

The model: the state variables are T_P and T_E, and the state model is given by equations (a), (b), (c), (d), and (f), as follows:

$$\frac{d}{dt} T_P = \left(\frac{1}{C_P} \right) \left[q_P - \frac{\sigma(T_P^4 - T_E^4)}{R_{\epsilon P} + 1/(A_P F_{P-E}) + R_{\epsilon E}} - \frac{1}{R_E}(T_P - T_E) \right] \tag{4.8a}$$

$$\frac{d}{dt} T_E = \left(\frac{1}{C_E} \right) \left\{ \frac{\sigma(T_P^4 - T_E^4)}{R_{\epsilon P} + 1/(A_P F_{P-E}) + R_{\epsilon E}} + \frac{1}{R_E}(T_P - T_E) \right.$$
$$\left. - \left[\left(\frac{h_S A_S}{1 + R_S h_S A_S} \right) + h_t A_t \right](T_E - T_a) \right\} \tag{4.8b}$$

Output $= T_P$ \hfill (4.8c)

4.2 PROCESS ENGINEERING SYSTEMS

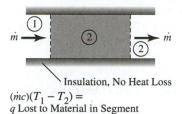

Insulation, No Heat Loss

$(\dot{m}c)(T_1 - T_2) =$
q Lost to Material in Segment

Figure 4.5

Material Transport Through Containing Segment

A process engineering system is a purposeful arrangement of *processing units* such as tanks, pumps, compressors, conveyors, heat exchangers, absorbers, evaporators, dryers, reactors, boilers, combustion chambers, distillation columns, digestors, fermentors, mixers, precipitators, concentrators, or receivers. Usually the ultimate objective of such a system is to convert raw materials into other useful products. A few examples include (petroleum) refineries, (electric) power generating plants, air separation plants, and polyethylene plants. The "processing units" of process engineering systems by themselves are actually full-fledged engineering systems, and many of the behavior components already introduced with respect to electrical, mechanical, and especially incompressible fluid and thermal energy systems are applicable. However, additional behavior components are necessary to accurately describe even some of the simplest units. In this section we consider some of these theoretical behaviors as we develop system models of process units that contain such behavior components. The overall system model is thus determined from the arrangement of the models of the individual units of which a process engineering system is composed.

Simple Material Transport

Many processing units of process engineering systems involve *material transport* in one form or another, although the primary objective can be something else, such as the transfer of heat to or from a process fluid. An important example is a *heat exchanger.* The analysis in Section 4.1 of heat energy storage is easily extended to apply to what we call *ideal mass transport.*

Consider the heat gain or loss for a material moving within a containment of some sort, as shown in Fig. 4.5, at a steady rate of mass flow $\dot{m}$. Heat loss from the container to the environment is assumed negligible. The entering material is at a temperature T_1, while the exit material temperature is T_2. Complete mixing within the containing segment is also assumed, so that the entire material within the container is at a uniform temperature, which is the same as the exit temperature T_2. If the containing volume is sufficiently large so that each material particle dm spends a finite amount of time dt within the container, then we can speak of the change in internal energy of the material. Under the usual assumptions of no work and negligible changes in potential and kinetic energies, this change is given by equations (4.5a) and (4.5c):

$$dU = (dm)c \, dT = dQ \tag{4.9}$$

Dividing equation (4.9) by dt, and noting that $dT = T_2 - T_1$, we see that under the assumed conditions the rate of heat energy gain by the material is given by

$$(\dot{m}c)(T_2 - T_1) = q_{\text{gained}}; \quad \text{(simple material transport)} \tag{4.10}$$

Example 4.3   Heat Exchanger System

Consider the *single-lump* model of a prototype (single tube and shell) *double-pipe* heat exchanger (Fig. 4.6a) shown in Fig. 4.6b. In this example we consider a possible application for utilization of solar heat. Hot water from the roof flat-plate collector (process fluid B) is circulated at (mass-flow) rate $\dot{m}_b$ through the inner cylinder (tube) while air (process fluid A) is drawn from the outside and blown at the rate $\dot{m}_a$ through the outer cylinder (shell). This air could eventually be used for, say, clothes drying in the laundry room. Derive a model to predict the temperature of the air leaving the heat exchanger using the information supplied in Fig. 4.6b. Assume that the cylinders are both of length L, and that the external surface of the outer cylinder is covered with a perfect insulator. Also, the air inlet temperature (the ambient temperature in this case) T_a and the temperature of the water leaving the collector T_b are prescribed quantities.

Solution: Seven behavior components can be recognized: (1) the inner-cylinder fluid thermal capacitance C_B, (2) the convective heat transfer coefficient between the water and the inner surface of the tube h_i, (3) the tube material thermal conductivity k_1, (4) the convective heat transfer coefficient between the outer surface of the tube and the air h_o, (5) the tube mass transport, (6) the annular chamber fluid thermal capacitance C_A, and (7) the shell mass transport. Let

T_B = temperature of water in inner cylinder
T_A = temperature of air in annular chamber
q_{b-a} = heat transfer from water in inner cylinder to air in annular region
q_b = heat gained by the water flowing from the collector into the inner cylinder
q_a = heat gained by the air blown into the annular chamber
A_i = effective inner surface area of tube $\sim 2\,\pi r_i L$
A_o = effective outer surface area of tube $\sim 2\,\pi r_o L$
ρ_B = density of water at temperature T_B
ρ_A = density of air at temperature T_A

Figure 4.6

Heat Exchanger System

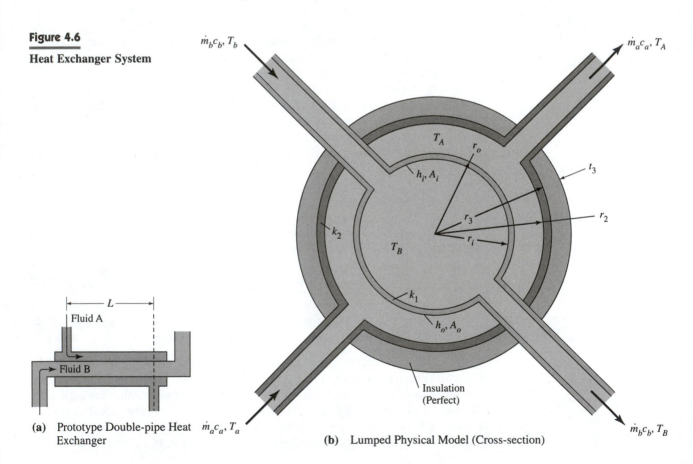

(a) Prototype Double-pipe Heat Exchanger

(b) Lumped Physical Model (Cross-section)

The behaviors of the seven components are as follows:

1. *Tube fluid thermal capacitance:* From equation (4.6b),

$$C_B \frac{d}{dt} T_B = -q_b - q_{b-a}; \qquad C_B \cong \rho_B \pi r_i^2 L c_b \tag{a}$$

2, 3, and 4. Tube-to-shell thermal resistance: The tube inner-surface heat transfer coefficient, the tube material thermal conductivity, and the tube outer-surface heat transfer coefficient represent thermal resistances in series. From equations (4.2) and (4.1c),

$$q_{b-a} = \left[\frac{1}{\frac{1}{h_i A_i} + \frac{\ln(r_o/r_i)}{2\pi k_1 L} + \frac{1}{h_o A_o}} \right] (T_B - T_A) \tag{b}$$

5. *Tube material transport:* From equation (4.10),

$$q_b = \dot{m}_b c_b (T_B - T_b) \tag{c}$$

6. *Shell fluid thermal capacitance:* From equation (4.6b),

$$C_A \frac{d}{dt} T_A = -q_a + q_{b-a}; \qquad C_A = \rho_A \pi (r_2^2 - r_o^2) L c_a \tag{d}$$

7. *Shell material transport:* From equation (4.10),

$$q_a = \dot{m}_a c_a (T_A - T_a) \tag{e}$$

The **state variables** are T_B and T_A, and from equations (a) through (e), the model is

$$\frac{d}{dt} T_B = -\frac{1}{C_B} \left\{ \dot{m}_b c_b (T_B - T_b) + \left[\frac{1}{\frac{1}{h_i A_i} + \frac{\ln(r_o/r_i)}{2\pi k_1 L} + \frac{1}{h_o A_o}} \right] (T_B - T_A) \right\} \tag{4.11a}$$

$$= -\frac{1}{C_B} \{ \dot{m}_b c_b (T_B - T_b) + [U_{\text{all}} A](T_B - T_A) \}$$

$$\frac{d}{dt} T_A = \frac{1}{C_A} \left\{ \left[\frac{1}{\frac{1}{h_i A_i} + \frac{\ln(r_o/r_i)}{2\pi k_1 L} + \frac{1}{h_o A_o}} \right] (T_B - T_A) - \dot{m}_a c_a (T_A - T_a) \right\} \tag{4.11b}$$

$$= \frac{1}{C_A} \{ [U_{\text{all}} A](T_B - T_A) - \dot{m}_a c_a (T_A - T_a) \}$$

$$\text{Output} = T_A \tag{4.11c}$$

We have introduced the idea of the *overall heat transfer coefficient* in equations (4.11a) and (4.11b). The overall heat transfer coefficient is defined as follows:

$$q = U_{\text{all}} A \Delta T_{\text{overall}} \tag{4.12a}$$

At the discretion of the user of equation (4.12a), U_{all} can be based on either the inside or outside area of the tube. For example, in our present application, U_{all} is given by either equation (4.12b) or (4.12c):

$$[U_{all}][A](\Delta T_{overall}) \qquad (4.12b)$$

$$= \left[\cfrac{1}{1/h_i + A_i \ln(r_o/r_i)/2\,\pi k_1 L + \cfrac{A_i}{A_o}\cfrac{1}{h_o}} \right][A_i](T_B - T_A); \quad \text{(using inside tube area)}$$

$$[U_{all}][A](\Delta T_{overall}) \qquad (4.12c)$$

$$= \left[\cfrac{1}{\cfrac{A_o}{A_i}\cfrac{1}{h_i} + \cfrac{A_o \ln(r_o/r_i)}{2\,\pi k_1 L} + 1/h_o} \right][A_o](T_B - T_A); \quad \text{(using outside tube area)}$$

The advantage of using the overall heat transfer coefficient is that the result need not be specific to a geometric configuration of the heat exchanger. For instance, by using the appropriate U_{all} and corresponding area A, we could apply equations (4.11a) through (4.11c) to a configuration of concentric parallelopipeds instead of circular pipes, or a multiple-tubes-in-shell heat exchanger, or a double-pipe heat exchanger with multiple passes of the tube.

The implication of the present *lumped-parameter* analysis should be noted. In practice, the temperature of the fluids and rate of heat exchange between them vary along the length of the heat exchanger. Consequently, there is a difference between the behavior of the heat exchanger for the *parallel-flow* configuration shown in Fig. 4.6a and the behavior of a *counterflow* (the two fluids enter from opposite ends of the heat exchanger) configuration. With the lumped-system assumption there is no difference between such flow configurations.

Compressible, Mixing, and Reacting Systems

Process engineering systems generally involve a variety of physical phenomena. For instance, mass transport can be the result of *forced mass motion* (where the driver can be a turbomachine) or it can be due to *natural convection* (where the driving effect is a density gradient) or to *diffusion* (where the driving effect is a *concentration* gradient). Density changes can occur in *compressible fluids*. Variations in concentration can occur in a drying process, for example, which usually also involves *phase changes* and equilibrium of phases, or *in mixing processes*. A mixing process can also involve *phase equilibrium* or *reacting components*, and it can be *exothermic* (heat is released in the process) or *endothermic* (heat is absorbed in the process). Chemical *reaction rates* usually depend on the concentration of the reacting substances, and a *rate equation* or a *rate law* is necessary to relate the concentration of the reactants to the reaction rate. The possibilities are many. We consider next four illustrative problems: an air heater with flow-rate control for batch moist product drying in a through-circulation drying chamber, a centrifugal-flow air compressor, a continuous mixing process in a well-stirred tank, and an endothermic first-order reaction process in a *continuous stirred tank reactor* (CSTR).

Steady Compressible Flow in a Duct

A major behavior component associated with heating/heated channels, narrow combustion chambers, fluid turbines, and similar units is the *steady flow of a compressible fluid in a duct*. We use the term *duct* to emphasize that the flow can be assumed to be one dimensional (the duct defines a *stream tube* that is coaxial with a *streamline* of the flow). In general, the flow of a given fluid has to satisfy four fundamental physical principles:

1. Conservation of mass (matter), expressed by the *continuity equation*
2. Newton's second law of motion, expressed by the *momentum equation*
3. Conservation of energy, expressed by the *energy equation,* and
4. Constitutive relation for the fluid, which is an expression of the *material behavior*

Consider the segment of a duct shown in Fig. 4.7. Since we are interested in the gross behavior of the unit (lumped-system assumption), a volume element of the fluid occupying the entire duct (known as a *control volume*) has been isolated to facilitate the development of the duct equations. The *control surface* is identified by the dashed line. Although the results to be obtained are identified by general names such as continuity equation and momentum equation, these results are specific to the system and assumptions inherent in Fig. 4.7 and are not general as such.

Continuity Equation   Because the duct is considered inelastic, it represents an inelastic fluid junction for which the conservation of mass is given by equation (3.24a):

$$w_{in} = w_{out} = \dot{m} \text{ or } \dot{m} = \rho A v = \text{constant}; \quad \text{(continuity equation, duct)} \qquad \text{(4.13)}$$

Note that $\dot{m}$ in the preceding equations is a variable, that is, the mass rate of flow of fluid through the duct, as indicated in Fig. 4.7, and not the rate of change of mass in the junction, written dm/dt. A represents cross-sectional area, ρ is density, and v is the fluid velocity.

Momentum Equation   Newton's second law is given by equation (2.8a). In words, *force = time rate of change of momentum.* In general, the force consists of *body forces* (such as gravitational and electromagnetic forces) and *surface forces* (such as those due to pressure and shear stress). Body forces are ignored in this development. The surface effects identified in Fig. 4.7 yield for the duct

$$\text{Surface forces} = P_1 A_1 - P_2 A_2 - \tau_w \ell L \qquad \text{(4.14a)}$$

where P_1 and P_2 are the pressures at the duct inlet and exit, τ_w is the average shear stress, ℓ is the circumference or perimeter of the cross section, and L is the length of the duct. The shear stress term can be used to represent general frictional effects.

For a control volume, the change in momentum consists of the *net momentum carried out by the flow* and the *change of momentum of fluid contained in the control volume.* That is,

$$\frac{d}{dt}(mv) = -(\rho_1 v_1 A_1)v_1 + (\rho_2 v_2 A_2)v_2 + \frac{d}{dt}\int_0^L \rho v A \, dx$$

For *steady flow,* the last term is zero. The continuity equation can be substituted in the first two terms to give

$$\frac{d}{dt}(mv) = \dot{m}(v_2 - v_1) \qquad \text{(4.14b)}$$

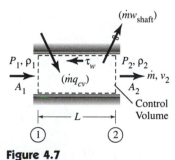

Figure 4.7

Duct Unit and Control Volume

Newton's second law then becomes

$$\dot{m}(v_2 - v_1) + P_2 A_2 - P_1 A_1 = -\tau_w \ell L; \quad \text{(momentum equation, duct)} \tag{4.15}$$

Energy Equation   The energy equation is given by the rate form of the first law of thermodynamics, equation (4.4a). That is,

$$\frac{d}{dt} E = \dot{Q} - \dot{W} \tag{4.16a}$$

In general, the rate of heat transfer to the fluid in the duct can include direct *volumetric heating* of the fluid, and heat conduction through the duct walls. We assume there is no heat loss or gain through the walls of the duct. The volumetric heating is assumed to be given by the product of the mass flowrate and an average rate of heat input per unit mass of the system q_{cv}(J/kg). Thus

$$\dot{Q} = \dot{m} q_{cv} \tag{4.16b}$$

The work done by the system can include *shaft* (or *mechanical*) work (done **by** the fluid), flow work due to the pressure forces, work (done on the system) due to body forces, and work (done **on** the fluid) due to shear forces. Again, the body forces are ignored here. Hence

$$\dot{W} = [\dot{m} w_{\text{shaft}}] + [P_2 A_2 v_2 - P_1 A_1 v_1] - \left[\tau_w \ell L \left(\frac{v_1 + v_2}{2} \right) \right]$$

The shaft- or mechanical-power output is also assumed to be the product of the mass flowrate and an average mechanical work output per unit mass of the system w_{shaft}. Note that the mechanical work can be done **on** the fluid, as in a compressor (a negative w_{shaft} in this case). Using the continuity equation, we can write the rate of work term as

$$\dot{W} = \dot{m} \left\{ [w_{\text{shaft}}] + \left[\frac{P_2}{\rho_2} - \frac{P_1}{\rho_1} \right] - \left[\left(\frac{1}{\rho_1 A_1} + \frac{1}{\rho_2 A_2} \right) \left(\frac{\tau_w \ell L}{2} \right) \right] \right\} \tag{4.16c}$$

The rate of change of total energy consists of the rate of change of total energy in the control volume and the net total energy flow across the control surfaces. For the steady-flow, steady-state assumption, the first component is zero, so that (recall that total energy is the sum of internal energy and kinetic and potential energies)

$$\frac{d}{dt} E = \dot{m} \left[\left(u_2 + \frac{v_2^2}{2} + g z_2 \right) - \left(u_1 + \frac{v_1^2}{2} + g z_1 \right) \right] \tag{4.16d}$$

The duct of Fig. 4.7 does not include changes in elevation, so the gz terms cancel out. Finally, the resulting energy equation (4.16a) is

$$\left[\left(u_2 + \frac{v_2^2}{2} \right) - \left(u_1 + \frac{v_1^2}{2} \right) \right] \tag{4.16e}$$

$$= [q_{cv}] - [w_{\text{shaft}}] + \left[\frac{P_1}{\rho_1} - \frac{P_2}{\rho_2} \right] + \left[\left(\frac{1}{\rho_1 A_1} + \frac{1}{\rho_2 A_2} \right) \left(\frac{\tau_w \ell L}{2} \right) \right]$$

Note that the mass flowrate cancels out of the equation. Applying the definition of the *enthalpy* (per unit mass), we have

$$h = u + \frac{P}{\rho}; \quad \text{(enthalpy)} \tag{4.17a}$$

$$\left[\left(h_2 + \frac{v_2^2}{2}\right) - \left(h_1 + \frac{v_1^2}{2}\right)\right] \tag{4.17b}$$

$$= [q_{cv}] - [w_{\text{shaft}}] + \left[\left(\frac{1}{\rho_1 A_1} + \frac{1}{\rho_2 A_2}\right)\left(\frac{\tau_w \ell L}{2}\right)\right]; \quad \text{(energy equation, duct)}$$

Material Behavior   In general, material behavior is described by experimental data. If the working fluid can be assumed to be an ideal gas, simplified expressions are possible. Ideal gas behavior means that the internal energy (per unit mass) is a function of the (absolute) temperature only: $u = f(T)$, and the pressure, density, and (absolute) temperature are related according to

$$\frac{P}{\rho} = RT \quad \text{or} \quad \frac{P}{\rho T} = \text{constant}; \quad \text{(ideal gas)} \tag{4.18a}$$

where R (J/kg · K) is the *gas constant*. It follows from equation (4.17a) that the enthalpy is also a function of temperature only:

$$h = u + RT \tag{4.18b}$$

Further, over reasonable temperature ranges, the temperature functions for u and h can be approximated by straight line segments, so that we can assume

$$u_2 - u_1 = c_v(T_2 - T_1) \tag{4.18c}$$

$$h_2 - h_1 = c_P(T_2 - T_1) \tag{4.18d}$$

where c_v and c_P are the constant-volume and constant-pressure specific heats, respectively. Equation (4.18b) requires that the specific heats are related according to

$$c_P - c_v = R \tag{4.18e}$$

For an ideal gas, we can substitute equations (4.18a) and (4.18d) into the energy equation to get

$$c_P(T_2 - T_1) + \frac{v_2^2 - v_1^2}{2} \tag{4.18f}$$

$$= q_{cv} - w_{\text{shaft}} + \left(\frac{R}{2}\right)\left(\frac{T_1}{P_1 A_1} + \frac{T_2}{P_2 A_2}\right)\tau_w \ell L; \quad \text{(energy equation, ideal gas)}$$

We note that overall, and for the usual variables, the duct unit is a static component. However, units connected upstream or downstream of the duct can involve dynamic behavior.

Example 4.4   Regulated Air Heater for a Moist Product Dryer

Figure 4.8 shows the schematic model of a heating duct with flow regulation. The duct supplies heated air at a controlled flow rate to a through-circulation moist product drying chamber to which it is coupled. Given the state model of the drying process, as discussed next, determine the necessary description of the air heater so that the model for the entire system is complete. Assume that the properties of the air at the inlet to the heater, and the power input at the heating coils, are specified.

Solution: A model of the dryer is necessary to determine the required inputs from the heater. The drying of a moist product generally involves (1) diffusion of moisture from the inside of the product to the product surface, and (2) vaporization of the moisture from the surface of the product into the drying air. The quality of the dried product in a once-through dryer can be controlled by regulating the drying temperature and the drying rate. These are, in turn, determined to a large extent by the properties and flow rate of the air entering the dryer.

Fick's law: The rate of diffusion of moisture within the product is governed by *Fick's law of diffusion,* which states that the *mass flux* of a constituent is proportional to the concentration gradient (see [9], for example):

$$\frac{\dot{m}_A}{A} = -D\frac{\partial C_A}{\partial x} \tag{4.19}$$

where $[\dot{m}_A/A]$ (kg/m^2 · s) is the mass flux of component A, D (m^2/s) is a proportionality constant known as the *diffusion coefficient,* C_A (kg/m^3) is the mass concentration of component A per unit volume, and x (m) is linear distance along which the concentration profile is defined. Application of equation (4.19) is complicated by a number of factors. The diffusion coefficient is not easy to determine for many products. Most practical products have complex and nonuniform geometry. The drying process itself significantly alters the properties and hence the behavior of many products while drying is in progress. For instance, the drying of most moist products follow a two-stage process: an initial high drying rate period during which the vaporization of moisture from the surface regions of the product is the limiting process, and a second reduced

Figure 4.8

Air Heater and Dryer

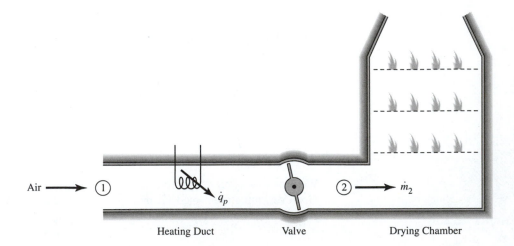

Air ① $\dot{q}_p$ Valve ② $\dot{m}_2$

Heating Duct   Valve   Drying Chamber

drying rate period when moisture diffusion from the interior of the product is the limiting process.

A third-order model of the drying of moist material in an insulated through-circulation drying chamber (see [10]) is given by the following:

$$\frac{d}{dt}x_1 = \left(\frac{-1}{2m}\right)x_2$$

$$\frac{d}{dt}x_2 = (8\,\lambda^2 m)x_1 - (5\lambda)x_2 - (8\lambda^2 m)q_e \qquad (4.20)$$

$$\frac{d}{dt}x_3 = \frac{1}{C_e}\left[-\left(\frac{H}{2}\right)x_2 + \left(\frac{c_a + w_2 c_v}{1 + w_2}\right)\dot{m}_2(T_2 - x_3) + \left(\frac{c_w - c_v}{2}\right)x_2 x_3\right]$$

where the state variables x_1, x_2, and x_3 are the average moisture content (on a dry basis) of the product in the chamber, twice the rate of moisture loss from the material, and the temperature of the moist material, respectively; m is the dry mass of the moist material; q_e is the equilibrium moisture content (on a dry basis) for the particular product; H is the latent heat of (water) vaporization; C_e is the equivalent humid heat capacity of the material; c_a, c_v, and c_w are the (constant-pressure) specific heats of the dry air, moisture, and water, respectively; and $\dot{m}_2$, w_2, and T_2 are the flowrate, humidity ratio, and temperature of the inlet air, respectively. The model assumes an initial high moisture loss (moisture vaporization limited) rate (4λ) that is four times the moisture loss rate (λ) limited by moisture diffusion within the solid material, so that the overall moisture loss process follows a second-order *overdamped* (see Section 7.1) behavior:

$$\frac{d^2 x_1}{dt^2} + 5\lambda\frac{dx_1}{dt} + 4\lambda^2 x_1 = 4\lambda^2 q_e$$

It follows that the required inputs from the air heater description are $\dot{m}_2$, w_2, and T_2. The humidity ratio of an air-water vapor mixture is defined by

$$w = \frac{\text{mass of water vapor}}{\text{mass of dry air}} = 0.622\frac{P_v}{P_a} \qquad (4.21a)$$

and the relative humidity is given by

$$\phi = \frac{P_v}{P_g} \qquad (4.21b)$$

where P_v, P_g, and P_a are the partial pressure of the vapor as it exists in the mixture, the saturation pressure of the vapor at the same temperature, and the partial pressure of the dry air, respectively. Since the properties of the air at the inlet to the heater are known, both ϕ and P_g are known at this point. P_v is also determined, and from *Dalton's law of partial pressures*,

$$P = P_v + P_a \qquad (4.21c)$$

P_a is determined as well. Hence the humidity ratio of the air entering the heater is known. It is assumed that no processes occur within the heater that change the relative mass of vapor and dry air, so that w_2 is also determined.

To establish the other two quantities, we need to modify the duct model to include the flow control valve. Note that $\dot{m}q_{cv} = \dot{q}_p$ and that $w_{shaft} = 0$. Further, we can lump all frictional effects within the heater at the exit valve. Also, the flow resistance presented by the drying chamber, if not negligible, can be considered lumped with the valve resistance as well. Equation (4.18f) becomes

$$\dot{m}_2\left[c_p(T_2 - T_1) + \frac{v_2^2 - v_1^2}{2} \right] \tag{4.21d}$$

$$= \dot{q}_p + \dot{m}_2 R\left(\frac{T_2}{P_2 A_2} \right)\tau_w \ell L = \dot{q}_p + \dot{m}_2\left(\frac{\tau_w \ell L}{\rho_2 A_2} \right)$$

We shall assume that the heater is operated under turbulent flow conditions at the exit. According to our previous valve model, equation (3.28a) or (3.31f), we can let

$$\tau_w \ell L = B v_2^2 \tag{4.21e}$$

where B depends on the valve characteristics, including the valve opening. Note that the valve opening is now represented by A_2, which is the means of regulating the flow. Substituting equation (4.21e) into equations (4.15) and (4.21d) gives us the air heater description:

continuity equation:

$$\dot{m}_2 = \rho_2 A_2 v_2 \tag{4.22a}$$

momentum equation:

$$\dot{m}_2(v_2 - v_1) + B v_2^2 + P_2 A_2 = P_1 A_1 \tag{4.22b}$$

energy equation:

$$\dot{m}_2\left[c_p(T_2 - T_1) + \frac{v_2^2 - v_1^2}{2} - B\left(\frac{v_2^2}{\rho_2 A_2} \right) \right] = \dot{q}_p \tag{4.22c}$$

ideal gas behavior:

$$P_2 = R\rho_2 T_2 \tag{4.22d}$$

Equations (4.22a) through (4.22d) constitute a system of four algebraic equations in four unknowns, $\dot{m}_2$, v_2, P_2, and T_2. Therefore $\dot{m}_2$ and T_2 can be obtained as functions of the heater inlet conditions and valve opening A_2.

....

Second Law of Thermodynamics and Ideal Turbomachinery

The *second law of thermodynamics* for a simple compressible substance undergoing a process can be written on a per unit mass basis (see [11], for example) as

$$T\,ds \geq \delta q; \quad \text{(second law of thermodynamics)} \tag{4.23a}$$

where s is the *entropy* per unit mass, a property of the substance (J/kg · K); and T is the absolute temperature. For an (ideal) *reversible* process, the second law is

$$T\, ds = \delta q; \quad \text{(second law, reversible process)} \tag{4.23b}$$

If we can neglect changes in kinetic and potential energies of the substance, the *first law of thermodynamics* for the (unit mass of) substance as given by equations (4.4a) and (4.4b) is

$$du = \delta q - \delta w; \quad \text{(first law of thermodynamics)} \tag{4.23c}$$

Recall the definition of the property enthalpy, equation (4.17a). For the process we can write

$$dh = du + P\, d\!\left(\frac{1}{\rho}\right) + \frac{1}{\rho}\, dP \tag{4.23d}$$

But for a simple compressible substance,

$$\delta w = P\, d\!\left(\frac{1}{\rho}\right) \tag{4.23e}$$

Substituting equations (4.23e) and (4.23d) in equation (4.23c) and then in equation (4.23b) gives us the relation between properties:

$$T\, ds = dh - \frac{1}{\rho}\, dP = du + P\, d\!\left(\frac{1}{\rho}\right); \quad \text{(reversible process)} \tag{4.24}$$

Isentropic Process   Consider now a *reversible adiabatic* (no heat transfer, $\delta q = 0$) *process*. This is an *isentropic* (constant entropy) process. By equation (4.23b), $T\, ds = 0$, so that equation (4.24) gives

$$dh - \frac{1}{\rho}\, dP = du + P\, d\!\left(\frac{1}{\rho}\right) = 0; \quad \text{(reversible and adiabatic = isentropic)} \tag{4.25a}$$

Equation (4.25a) has the following implication for an ideal gas for which we can assume constant specific heat:

$$dh = c_p\, dT = \frac{1}{\rho}\, dP; \quad \text{(isentropic, constant specific heat)} \tag{4.25b}$$

From the ideal gas equation (4.18a),

$$T = \left(\frac{1}{R}\right)\frac{P}{\rho} \Rightarrow dT = \frac{1}{R}\left[\frac{1}{\rho}\, dP + P\, d\!\left(\frac{1}{\rho}\right)\right]$$

Substituting for dT in equation (4.25b), and introducing the ratio of specific heats $k = c_p/c_v$, we obtain

$$c_p\left[\frac{1}{\rho}\,dP + P\,d\!\left(\frac{1}{\rho}\right)\right] = R\left[\frac{1}{\rho}\,dP\right] = (c_p - c_v)\left[\frac{1}{\rho}\,dP\right] \Rightarrow$$

$$c_p P\,d\!\left(\frac{1}{\rho}\right) = -c_v\left[\frac{1}{\rho}\,dP\right]$$

or

$$k\left[\rho\,d\!\left(\frac{1}{\rho}\right)\right] = -\frac{dP}{P}$$

Integration of the last equation gives

$$P\left(\frac{1}{\rho}\right)^k = constant; \quad \text{(ideal gas, isentropic process)} \tag{4.25c}$$

For an ideal gas undergoing an isentropic process between two states, equation (4.25c) yields

$$\frac{P_2}{P_1} = \left(\frac{\rho_2}{\rho_1}\right)^k; \quad \text{(ideal gas, isentropic process)} \tag{4.25d}$$

and equation (4.18a) gives

$$\frac{T_2}{T_1} = \left(\frac{\rho_2}{\rho_1}\right)^{k-1} = \left(\frac{P_2}{P_1}\right)^{(k-1)/k}; \quad \text{(ideal gas, isentropic process)} \tag{4.25e}$$

Turbomachines   The process that takes place in most ideal turbomachinery can sometimes be approximated by a *reversible adiabatic process in a duct* with negligible changes in fluid kinetic and potential energies. The energy equation (4.17b) gives for this situation

$$w_{\text{shaft}} = h_1 - h_2; \quad \text{(adiabatic turbomachine)} \tag{4.26a}$$

Substituting equation (4.24), we obtain

$$w_{\text{shaft}} = -\int_1^2 \left(\frac{1}{\rho}\right) dP; \quad \text{(turbomachine)} \tag{4.26b}$$

Sometimes the approximation *reversible isothermal* (constant temperature) *process in a duct* with negligible changes in fluid kinetic and potential energies is applicable to the turbomachine. The energy equation (4.17b) and the property relation equation (4.24) yield for this case

$$w_{\text{shaft}} = q_{cv} + (h_1 - h_2) = [q_{cv} - \int_1^2 T\,ds] - \int_1^2 \left(\frac{1}{\rho}\right) d P$$

But by equation (4.23b) the term in brackets is zero. Hence equation (4.26b) applies to both *ideal adiabatic* and *ideal isothermal* turbomachinery.

Example 4.5	Centrifugal-Flow Compressor

Figure 4.9a shows the end view and cross section typical of a *centrifugal-flow compressor*. The impeller is rotated by the shaft, while the diffuser guide vanes are fixed to the casing. The fluid comes in axially and is discharged through a scroll casing (volute) surrounding the diffuser. Figure 4.9b shows the velocity diagram for the fluid leaving the impeller. The absolute velocity **v** is the sum of the velocity of the impeller $\mathbf{v}_i$ and the velocity of the fluid relative to the impeller $\mathbf{v}_{rel}$. As shown in Fig. 4.9c, the fluid velocity can also be resolved into tangential and radial components $\mathbf{v}_t$ and $\mathbf{v}_r$ respectively. The entire work of compression is essentially performed in the impeller. A parameter of the centrifugal compressor is the *slip factor* (SF) defined as the ratio of $\mathbf{v}_t$ to $\mathbf{v}_i$. Assuming the fluid is an ideal gas, determine the required work input per unit mass, the isentropic efficiency, and the pressure ratio for the compressor, given the slip factor, the radius of the impeller tip, the rotating speed of the impeller, and the inlet temperature.

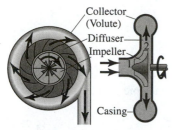

Collector (Volute)
Diffuser
Impeller
Casing

(a)   End View and Section

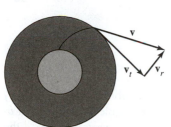

(b)   Absolute Fluid Velocity at Impeller Tip

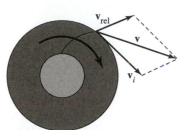

(c)   Tangential and Radial Components of Fluid Velocity at Impeller Tip

Figure 4.9

Centrifugal-flow Compressor

Solution: Consider a spiral duct segment bounded by the impeller surfaces. The fluid velocity along this duct is, in effect, the tangential component. Ignoring wall friction, we can write the momentum equation (4.15) in terms of angular momentum:

$$\dot{m}(r_2 v_{t2} - r_1 v_{t1}) = P_1 A_1 r_1 - P_2 A_2 r_2 = T_{net} \tag{4.27a}$$

where T_{net} is the net torque applied to the shaft. Because the flow enters the impeller in the axial direction, v_{t1} may be taken as zero, so that equation (4.27a) becomes

$$\dot{m}(r_2 v_{t2}) = T_{net} \tag{4.27b}$$

The power input to the compressor is $T_{net}\omega$, where ω is the impeller rotating speed. Introducing the slip factor, we have

$$SF = \frac{v_{t2}}{v_{i2}} \Rightarrow v_{t2} = SF v_{i2} = SF r_2 \omega$$

Equation (4.27b) yields

$$\dot{m} SF (r_2 v_{i2}) \omega = \dot{m} SF r_2^2 \omega^2 = T_{net}\omega = \dot{m}(-w_{shaft}) \tag{4.27c}$$

so that

$$(-w_{shaft}) = SF r_2^2 \omega^2 \tag{4.28a}$$

The *isentropic efficiency* is the ratio of the work input required for isentropic compression to the actual work input, for the same inlet and exit pressures:

$$\eta_s = \frac{(-w_s)}{(-w_{shaft})} \tag{4.28b}$$

From equation (4.25d),

$$P = P_1 \left(\frac{\rho}{\rho_1} \right)^k$$

To find w_s we can substitute this P into equation (4.26b) and integrate or, more easily, use equations (4.26a) and (4.25e) as follows:

$$-w_s = h_2 - h_1 = c_p(T_2 - T_1) = c_p T_1 \left(\frac{T_2}{T_1} - 1 \right) \tag{4.28c}$$

$$= c_p T_1 \left[\left(\frac{P_2}{P_1} \right)^{(k-1)/k} - 1 \right]$$

Substituting equations (4.28a) and (4.28c) into equation (4.28b), we obtain

$$\eta_s = \frac{c_p T_1}{SF r_2^2 \omega^2} \left[\left(\frac{P_2}{P_1} \right)^{(k-1)/k} - 1 \right] \tag{4.28d}$$

Finally, we can find the pressure ratio in terms of the isentropic efficiency:

$$\frac{P_2}{P_1} = \left(\frac{\eta_s SF r_2^2 \omega^2}{c_p T_1} + 1 \right)^{k/(k-1)} \tag{4.28e}$$

....

Thermochemical Analysis

In problems involving change of composition (due to chemical reaction, for example), certain concepts and standards facilitate analysis.

Heat of Reaction   Every pure substance is assumed to have an enthalpy, and the energy (heat) liberated or absorbed in a reaction (or solution process), which is known as the *heat of reaction* (ΔH), is given by

$$\Delta H = \sum H_{\text{products}} - \sum H_{\text{reactants}} \tag{4.29a}$$

or

$$\sum H_{\text{products}} = \sum H_{\text{reactants}} + \Delta H \tag{4.29b}$$

Enthalpy of Formation.   The heat of reaction in which one mole (abbreviated mol) of a substance is prepared from its constituents (in their most stable forms) at a standard pressure P_s and reference temperature T_0, is known as the *enthalpy of formation* ($\Delta \tilde{H}_f$). The tilde ($\sim$) implies enthalpy per mole or *molar enthalpy*. In SI units the standard pressure and reference temperature are 0.1 MPa and 25°C, respectively. Given the enthalpy of formation (ΔH_f), the enthalpy of a substance at any pressure (P) and temperature (T) is

$$H_{T,P} = \Delta H_f + \Delta H_{(T_0,P_s) \rightarrow (T,P)} \tag{4.29c}$$

Thus **the enthalpies of different substances can be added or subtracted, since they are all given relative to the same base.** For solids and liquids, a good approximation for the enthalpy change from the reference state is

$$\Delta H_{(T_0,P_s)\rightarrow(T,P)} = c_p(T - T_0) \tag{4.29d}$$

where c_p is the constant-pressure specific heat.

Example 4.6	Continuous Mixing in a Well-Stirred Tank

Figure 4.10 shows a continuous well-stirred tank mixing process. The input, stream 1, enters the tank at the volumetric rate Q_1 and has two components, A and B, with molar concentrations c_{A1}, and c_{B1}. Stream 2 consists entirely of component B (this could be a thickening/lightening or blending process) and enters the tank at the volumetric rate Q_2. Cooling or heating fluid circulates through a coil (heat exchanger) immersed in the tank in order to cool or heat the contents of the tank, as required by the process. The rate of heat transfer to the tank from the cooling/heating fluid $\dot{q}_p$ is prescribed. Stream 3, the output stream, is composed of the same two components as stream 1, with molar concentrations c_{A3} and c_{B3}. The production rate, as determined by the volumetric rate of the output stream Q_3, is also a controlled (prescribed) quantity. Determine a model for this system that will permit determination of the instantaneous level of material in the tank and the molar concentration of component A in the product. Although the density of each stream will, in general, depend on the component concentrations and temperature, assume that the effect of stream 2 on the product density is negligible and that the density of the product can be considered the same as that of stream 1.

Solution: If we assume a tank of uniform area, the level of material in the tank is proportional to the volume:

$$L = \frac{V}{A} \tag{a}$$

Conservation of matter: The accumulation of material in the tank is given by equation (3.32b):

$$\frac{d}{dt}(\rho_3 V) = \rho_1 Q_1 + \rho_2 Q_2 - \rho_3 Q_3 \tag{b}$$

where V is the volume of material in the tank, and the density of the product is assumed to be the same as the density in the tank. Indeed, since the tank is well mixed, all the properties of the product are assumed to be the same as those of the tank. Substituting $\rho_3 = \rho_1$, as given, in equation (b), we have

$$\frac{d}{dt}V = Q_1 + rQ_2 - Q_3; \qquad r = \frac{\rho_2}{\rho_1} \tag{c}$$

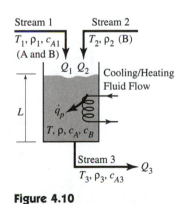

Figure 4.10

Continuous Well-stirred Tank Mixing Process

Each component must satisfy the material balance: *rate of accumulation of component = net rate of flow of component **into** tank.* Thus

component A:

$$\frac{d}{dt}(c_{A3}V) = c_{A1}Q_1 - c_{A3}Q_3 \tag{d}$$

component B:

$$\frac{d}{dt}(c_{B3}V) = c_{B1}Q_1 + \frac{\rho_2}{M_2}Q_2 - c_{B3}Q_3 \tag{e}$$

where M_2 is the molecular weight of stream 2.

Conservation of energy: Consider a control volume coincident with the material in the tank. From equation (4.16a), $\dot{Q} = \dot{q}_p$, and $\dot{W} = 0$ (or $\dot{Q} - \dot{W} = \dot{q}_p$ if the stirring of the material in the tank must be accounted for). The rate of change of total energy consists of the rate of change of total energy in the tank and the net total energy flow out of the tank. If we use the approximation, reasonable for solids and liquids, $U \cong H$, and neglect kinetic and potential energies everywhere, equation (4.16a) becomes

$$\frac{d}{dt}E = \frac{d}{dt}(H_3) + \rho_3 Q_3 h_3 - [\rho_1 Q_1 h_1 + \rho_2 Q_2 h_2] = \dot{q}_p$$

Thus

$$\frac{d}{dt}(\rho_3 V h_3) = [\rho_1 Q_1 h_1 + \rho_2 Q_2 h_2] - \rho_3 Q_3 h_3 + \dot{q}_p \tag{f}$$

From equations (4.29c) and (4.29d),

$$h_1 = h_{1f} + c_{p1}(T_1 - T_0) \tag{g}$$

$$h_2 = h_{2f} + c_{p2}(T_2 - T_0) \tag{h}$$

$$h_3 = h_{3f} + c_{p3}(T_3 - T_0) \tag{i}$$

And from equation (4.29b),

$$\rho_1 h_{1f} = c_{A1}\Delta\tilde{H}_{Af} + c_{B1}\Delta\tilde{H}_{Bf} + c_{A1}\Delta\tilde{H}_{1(A)} \tag{j}$$

$$\rho_3 h_{3f} = c_{A3}\Delta\tilde{H}_{Af} + c_{B3}\Delta\tilde{H}_{Bf} + c_{A3}\Delta\tilde{H}_{3(A)} \tag{k}$$

$$h_{2f} = \frac{\Delta\tilde{H}_{Bf}}{M_2} \tag{l}$$

where $\Delta\tilde{H}_{Af}$ and $\Delta\tilde{H}_{Bf}$ are molar enthalpies of formation, and $\Delta\tilde{H}_{1(A)}$ and $\Delta\tilde{H}_{3(A)}$ are molar heats of (solution) reaction per mole of component A for streams 1 and 3, respectively, at standard pressure and reference temperature. Substituting equations (g) through (l) into equation (f), we obtain

$$
\frac{d}{dt}\{V[c_{A3}\Delta\tilde{H}_{Af} + c_{B3}\Delta\tilde{H}_{Bf} + c_{A3}\Delta\tilde{H}_{3(A)}] + V\rho_3 c_{p3}(T_3 - T_0)\}
$$

$$
= Q_1[c_{A1}\Delta\tilde{H}_{Af} + c_{B1}\Delta\tilde{H}_{Bf} + c_{A1}\Delta\tilde{H}_{1(A)} + \rho_1 c_{p1}(T_1 - T_0)]
$$

$$
+ Q_2\rho_2\left[\frac{\Delta\tilde{H}_{Bf}}{M_2} + c_{p2}(T_2 - T_0)\right] - Q_3[c_{A3}\Delta\tilde{H}_{Af} + c_{B3}\Delta\tilde{H}_{Bf}
$$

$$
+ c_{A3}\Delta\tilde{H}_{3(A)} + \rho_3 c_{p3}(T_3 - T_0)] + \dot{q}_p
$$

or, expanding and grouping terms, we have

$$
\left[\frac{d}{dt}(c_{A3}V\Delta\tilde{H}_{Af}) - c_{A1}\Delta\tilde{H}_{Af} + c_{A3}Q_3\Delta\tilde{H}_{Af}\right]
$$

$$
+ \left[\frac{d}{dt}(c_{B3}V\Delta\tilde{H}_{Bf}) - c_{B1}Q_1\Delta\tilde{H}_{Bf} - \frac{\rho_2}{M_2}Q_2\Delta\tilde{H}_{Bf} + c_{B3}Q_3\Delta\tilde{H}_{Bf}\right] \qquad \text{(m)}
$$

$$
+ \left[\frac{d}{dt}(c_{A3}V\Delta\tilde{H}_{3(A)}) + c_{A3}Q_3\Delta\tilde{H}_{3(A)}\right] + \left\{\frac{d}{dt}[V\rho_3 c_{p3}(T_3 - T_0)]\right\}
$$

$$
= Q_1[c_{A1}\Delta\tilde{H}_{1(A)} + \rho_1 c_{p1}(T_1 - T_0)] + Q_2[\rho_2 c_{p2}(T_2 - T_0)]
$$

$$
- Q_3[\rho_3 c_{p3}(T_3 - T_0)] + \dot{q}_p
$$

The first term of equation (m) is zero as a result of equation (d). The second term is also zero due to equation (e). By equation (d), the third term is equal to $c_{A1}Q_1\Delta\tilde{H}_{3(A)}$. Thus equation (m), with expansion of the fourth term, becomes

$$
\left[\rho_3 c_{p3}(T_3 - T_0)\frac{d}{dt}V\right] + \left[\rho_3 c_{p3}V\frac{d}{dt}T_3\right]
$$

$$
= Q_1[c_{A1}(\Delta\tilde{H}_{1(A)} - \Delta\tilde{H}_{3(A)}) + \rho_1 c_{p1}(T_1 - T_0)]
$$

$$
+ Q_2[\rho_2 c_{p2}(T_2 - T_0)] - Q_3[\rho_3 c_{p3}(T_3 - T_0)] + \dot{q}_p
$$

Substituting equation (c) in the first term of the preceding equation and rearranging the result gives us

$$
\rho_3 c_{p3}V\frac{d}{dt}T_3 = Q_1[c_{A1}(\Delta\tilde{H}_{1(A)} - \Delta\tilde{H}_{3(A)}) + \rho_1 c_{p1}(T_1 - T_0) - \rho_3 c_{p3}(T_3 - T_0)] \quad \text{(n)}
$$

$$
+ Q_2[\rho_2 c_{p2}(T_2 - T_0) - \rho_3 c_{p3}r(T_3 - T_0)] + \dot{q}_p
$$

If we also expand equation (d),

$$
V\frac{d}{dt}c_{A3} + c_{A3}\frac{d}{dt}V = c_{A1}Q_1 - c_{A3}Q_3
$$

and substitute equation (c), we get

$$V\frac{d}{dt}c_{A3} = (c_{A1} - c_{A3})Q_1 - rc_{A3}Q_2 \tag{o}$$

The state variables are V, T_3, and c_{A3}. Note that c_{B3} is not independent of c_{A3}, hence it is not assigned as one of the state variables. The corresponding state equations are from equations (c), (n), and (o),

$$\frac{d}{dt}V = Q_1 + rQ_2 - Q_3 \tag{4.30a}$$

$$\frac{d}{dt}T_3 = \frac{1}{\rho_3 c_{p3} V}\{Q_1[c_{A1}(\Delta\tilde{H}_{1(A)} - \Delta\tilde{H}_{3(A)}) + \rho_1 c_{p1}(T_1 - T_0) \tag{4.30b}$$

$$- \rho_3 c_{p3}(T_3 - T_0)] + Q_2[\rho_2 c_{p2}(T_2 - T_0) - \rho_3 c_{p3}r(T_3 - T_0)] + \dot{q}_p\}$$

$$\frac{d}{dt}c_{A3} = \frac{1}{V}[(c_{A1} - c_{A3})Q_1 - rc_{A3}Q_2] \tag{4.30c}$$

One of the desired outputs is given by equation (a). The other is c_{A3}. Hence the output equation is

$$y_1 = L = \left(\frac{1}{A}\right)V \tag{4.30d}$$

$$y_2 = c_{A3}$$

Equations (4.30a) through (4.30d) constitute the system model. Note that in equation (4.30b) the magnitude of $\dot{q}_p$ is positive or negative depending on whether the tank is being heated or cooled.

Example 4.7 First-Order Reaction in a Continuous Stirred Tank Reactor

Consider the well-mixed continuous-flow chemical reactor shown in Fig. 4.11, in which the component A *(reactant)* is consumed to produce the component B *(product)*. The reaction is first order. That is, in the reaction A ⟶ B, the reaction rate, or the rate of disappearance of A (expressed as the negative of the rate of change (increase) of concentration of A), is proportional to the amount of A present (the concentration of A). That is,

$$-\frac{d}{dt}c_A = kc_A \tag{4.31a}$$

The rate constant k varies with temperature according to the *Arrhenius equation:*

$$k = k_0 e^{-E_a/\tilde{R}T} \tag{4.31b}$$

where k_0 is a constant [s^{-1}] called the *frequency factor,* E_a is the *energy of activation* [J/mol] needed for a successful reaction, $\tilde{R}$ is the molar gas constant [8.3143

Figure 4.11

Continuous Stirred Tank Reactor

Simplified Model Nomenclature

w = Reactor Feed Rate (kg/s)

T_1 = Feed Temperature (K)

u_1 = Concentration of Reactant A in Feedstream (kg/m^3)

u_2 = Rate of Heat Input (kW)

x_1 = Reactor Concentration of A (kg/m^3)

x_2 = Reactor Temperature (K)

ρ = Density of Reacting Mixture (kg/m^3)

c = Specific Heat of Reacting Mixture (kJ/kg · K)

H = Heat of (Endothermic) Reaction (of A) (kJ/kg)

k_b = Reaction Rate Constant (s^{-1})

V = Reactor Volume (m^3)

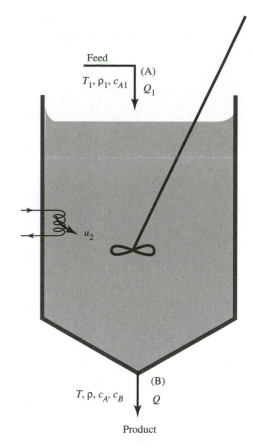

J/(K · mol)], and T is the absolute temperature [K]. Given the input stream properties as shown in Fig. 4.11, the rate of direct heat supply to the reactor (the reaction is endothermic and heating of the reactor is presumed) and the production rate, derive a model to determine the concentration of reactant A in the product and the temperature in the reactor.

Solution: Having demonstrated a detailed procedure in Example 4.6, we opt for a very simplified solution in this case. The detailed problem is left as an exercise in Practice Problem 4.16. For the approximate solution, we make the following assumptions: (1) the rate constant is independent of temperature, (2) the densities and specific heats of the reacting mixtures are constant and the same in the feed stream and in the reactor, (3) the feed rate (in terms of mass flowrate) w is the same as the production rate, and (4) the enthalpy of formation is the same for each stream. With these assumptions, we can utilize the alternative quantities shown in the box in Fig. 4.11 to determine the model. As in Example 4.6, we perform material and energy balances:

Material balance:

tank material: Since the feed rate and the production rate are equal and mass is not created in the reactor, the rate of accumulation of material in the reactor is zero. That is, the mass of material in the reactor is constant.

component B: On the basis of the given quantities, equation (4.31a) is

$$\text{Rate of creation of product} = \frac{d}{dt}x_1 \bigg|_{\text{reaction}} = -k_b x_1 \tag{a}$$

However, the total rate of change of the concentration of A includes the net rate of flow of A into the tank. Hence

$$\frac{d}{dt}x_1 = -k_b x_1 + \frac{w}{\rho V}(u_1 - x_1) \tag{b}$$

Energy balance: If we neglect again kinetic and potential energies, the rate of change of total energy in the tank is equal to the net total energy flow into the tank plus the rate of direct heat supply to the tank minus the *rate of heat absorption by the reaction:*

$$\frac{d}{dt}(\rho Vh) = w(h_1 - h) + u_2 - Hk_b V x_1 \tag{c}$$

Assumptions (2) and (4) imply $(h_1 - h) = c(T_1 - x_2)$, so that

$$\frac{d}{dt}(\rho Vh) = h\frac{d}{dt}(\rho V) + \rho V \frac{d}{dt}h = \rho Vc\frac{d}{dt}x_2 = wc(T_1 - x_2) + u_2 - Hk_b V x_1 \tag{d}$$

State model: The state variables are x_1 and x_2. The state model is therefore

$$\frac{d}{dt}x_1 = -k_b x_1 + \frac{w}{\rho V}(u_1 - x_1) \tag{4.32a}$$

$$\frac{d}{dt}x_2 = -\frac{Hk_b}{\rho c}x_1 + \frac{w}{\rho V}(T_1 - x_2) + \frac{u_2}{\rho Vc} \tag{4.32b}$$

$$y_1 = x_1 \tag{4.32c}$$

$$y_2 = x_2 \tag{4.32d}$$

4.3 EXAMPLES OF DISTRIBUTED-PARAMETER MODELS

In Section 2.4 we showed that consideration of the spatial variation of the spring and mass effects in a real spring leads to a distributed-parameter model of the spring behavior. In this section we consider distributed-parameter models of some simple but common engineering components or systems. In state space, the system models are systems of first-order *partial* differential equations as opposed to systems of first-order *ordinary* differential equations that describe the lumped-parameter model.

Longitudinal and Torsional Vibrations in Thin Rods

We begin with a rod of thin but constant cross section that is subject to longitudinal disturbance. Such a situation is typical of many earthworking or construction tools (a pneumatic jackhammer, for example) in which a long narrow steel tool (bit) is driven percussively against the work surface. Consider the differential segment of the rod shown in Fig. 4.12a. The longitudinal displacement $u(x, t)$, the force $F(x, t)$, and the body force (such as the force of gravity) per unit volume $N(x, t)$, at any section, are

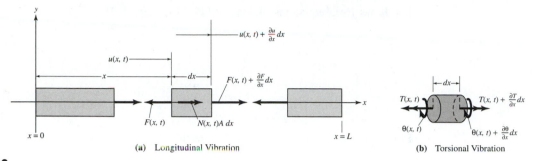

Figure 4.12

Thin Rod Segment and Variables

functions of both time t and position x. Thus partial derivatives are required as shown in the figure. Applying Newton's second law to this element, we have

$$\rho A \, dx \frac{\partial^2 u}{\partial t^2} = \left[F(x, t) + \frac{\partial F}{\partial x} dx \right] - F(x, t) + NA \, dx = \frac{\partial F}{\partial x} dx + NA \, dx \qquad \textbf{(4.33a)}$$

where ρ is the density of the rod material and A is the cross-sectional area. The strain at each cross section is given by $\partial u/\partial x$, and from Hooke's law (neglecting any contribution by bending to the average stress distribution in the thin cross section),

$$F(x, t) = EA \frac{\partial u}{\partial x} \qquad \textbf{(4.33b)}$$

where E is the material modulus of elasticity. From equation (4.33b) we see that

$$\frac{\partial F}{\partial x} dx = \frac{\partial}{\partial x} \left(E \frac{\partial u}{\partial x} \right) A \, dx = EA \frac{\partial^2 u}{\partial x^2} dx; \quad \text{(for uniform } E\text{)}$$

Substituting in equation (4.33a) we obtain the distributed-parameter model:

$$\frac{\partial^2 u}{\partial t^2} = \frac{1}{\rho} \left[\frac{\partial}{\partial x} \left(E \frac{\partial u}{\partial x} \right) + N \right] = \frac{E}{\rho} \frac{\partial^2 u}{\partial x^2} + \frac{N}{\rho} \qquad \textbf{(4.34)}$$

Equation (4.34) requires two boundary conditions and two initial conditions for solution. For instance, for the boundary values, the displacement at both ends of the rod or the force and displacement at some point on the rod could be specified. The initial conditions normally consist of the initial displacement (or force) and velocity profiles throughout the rod.

It is possible to obtain this model in a format that includes only first-order terms and that is therefore compatible with the state space procedure. The (state) variables here are force F and velocity v. We note from the definition of the strain ϵ that

$$\frac{\partial \epsilon}{\partial t} = \frac{\partial}{\partial t} \left(\frac{\partial u}{\partial x} \right) = \frac{\partial}{\partial x} \left(\frac{\partial u}{\partial t} \right) = \frac{\partial v}{\partial x}$$

It follows from equations (4.33a) and (4.33b) that

$$\frac{\partial v}{\partial t} = \frac{1}{\rho A} \frac{\partial F}{\partial x} + \frac{N}{\rho} \tag{4.35a}$$

$$\frac{\partial F}{\partial t} = EA \frac{\partial v}{\partial x} \tag{4.35b}$$

Equations (4.35a) and (4.35b) constitute the alternative model.

Torsional Vibration of a Thin Rod

A similar model for the torsional vibration of a thin rod is easily derived. Again, if we neglect bending stresses, the shear strain γ (for a circular cross section) is given by

$$\gamma = r\frac{\partial \theta}{\partial x} = \frac{r}{IG} T; \qquad I = \frac{\pi}{2} r^4 \tag{4.36}$$

where r is the radius of the rod cross section, I is the polar moment of inertia of the cross-sectional area, G is the shear modulus (modulus of rigidity), θ is the angular twist, and T is the torque. Both θ and T vary along the length of the rod. Introducing the angular velocity ω, we have

$$\frac{\partial}{\partial t}\left(\frac{\partial \theta}{\partial x}\right) = \frac{\partial}{\partial x}\left(\frac{\partial \theta}{\partial t}\right) = \frac{\partial \omega}{\partial x}$$

Equation (4.36) yields

$$\frac{\partial T}{\partial t} = IG\frac{\partial \omega}{\partial x} \tag{4.37a}$$

Newton's second law, equation (2.16), applied to the segment of rod shown in Fig. 4.12b, in the absence of body torques, results in

$$\frac{\partial \omega}{\partial t} = \left(\frac{dx}{J}\right)\frac{\partial T}{\partial x} = \left(\frac{1}{\rho I}\right)\frac{\partial T}{\partial x} \tag{4.37b}$$

where J is the (mass) moment of inertia of the rod segment about the x-axis. Equations (4.37a) and (4.37b) constitute the distributed-parameter model.

One-Dimensional Heat Conduction

Consider one-dimensional heat conduction in a uniform (thin) rod. The temperature $T(x, t)$ and heat flow $q(x, t)$ vary in both time and space. For the segment of the rod shown in Fig. 4.13, if A, k, c, and ρ are the cross-sectional area, the thermal conductivity, the specific heat, and density, respectively, Fourier's law of heat conduction (equation (4.1a)) and an energy balance over the element (see equation (4.6a)) yield

$$q(x, t) = -\frac{kA}{dx}\left[T(x, t) + \frac{\partial T}{\partial x}dx - T(x, t)\right] \tag{4.38a}$$

$$q(x, t) - \left[q(x, t) + \frac{\partial q}{\partial x}dx\right] = (\rho A\,dx)c\frac{\partial T}{\partial t} \tag{4.38b}$$

Figure 4.13

Heat Conduction in a Uniform Rod

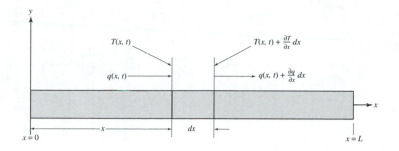

Thus the model is

$$\frac{\partial T}{\partial t} = -\frac{1}{\rho A c}\frac{\partial q}{\partial x} \tag{4.39a}$$

$$q = -(kA)\frac{\partial T}{\partial x} \tag{4.39b}$$

An alternative model, the one-dimensional unsteady heat conduction equation, can be obtained by taking the partial derivative with respect to x of equation (4.39b) and substituting the result in (4.39a):

$$\frac{\partial T}{\partial t} = \left(\frac{k}{\rho c}\right)\frac{\partial^2 T}{\partial x^2} = \alpha\frac{\partial^2 T}{\partial x^2} \tag{4.40}$$

Thus the state model is first order (in time, second order in the spatial variable). The quantity α is known as the (thermal) *diffusivity*. Equation (4.40) requires the initial temperature distribution and two boundary conditions (for instance, the temperature at both ends of the rod or the temperature and heat flux at one end) for solution. Equations (4.39) or (4.40) can also be considered the distributed-parameter model of a thermal conduction resistance element (which allows for the mixed effects of thermal conductivity and thermal capacitance).

Lumped-Parameter Alternatives

In Section 2.4 it was suggested that a lumped-parameter alternative to the distributed-parameter model can be constructed by separating the mixed effects into a large number of ideal lumped elements. Consider the thin rod subject to torsional vibration as an example. The mixed effects are torsional compliance (spring) (equation (4.36)) and rotary inertia (equation (4.37b)). Figure 4.14 shows a causal linking of a large number of rotary springs and masses. Each spring-mass pair represents the (uniform) properties and behavior of a segment of the rod of length dx. The length dx is assumed to be sufficiently short so that these properties can be considered uniform throughout the segment. With respect to the general nth segment whose decomposition is shown,

$$k_n = \frac{I_n G_n}{dx_n} \tag{4.41a}$$

$$J_n = (\rho_n A_n \, dx_n \, r_n^2)/2 = \rho_n I_n \, dx_n; \text{ (cylinder)} \tag{4.41b}$$

Figure 4.14

**Lumped-parameter
Model Structure**

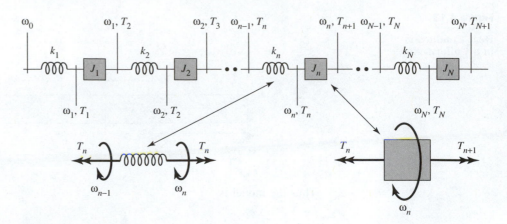

and

$$\frac{d}{dt} T_n = k_n(\omega_n - \omega_{n-1}) \tag{4.42a}$$

$$\frac{d}{dt} \omega_n = \frac{1}{J_n}(T_{n+1} - T_n) \tag{4.42b}$$

Equations (4.42a) and (4.42b) applied to all $n(n = 1$ to $N)$ yield the lumped model. The structure of Fig. 4.14 assumes that the boundary conditions include the angular velocity at the left end and the torque at the right end. Also, the state variables are defined at the ends of each segment.

The first-order formats of the distributed-parameter models of equations (4.35), (4.37), and (4.39) suggest a straightforward lumped-parameter alternative that can be viewed as a discretization in space. This is accomplished by applying these equations to each of N spatial segments and replacing the partial derivatives with respect to the spatial variable with a difference expression. Since the spatial dependence is thus removed, the partial derivatives with respect to time become absolute derivatives.

Consider the longitudinal vibration example. Equations (4.35a) and (4.35b) become

$$\left. \begin{aligned} \frac{d}{dt} v_n &= \left(\frac{1}{\rho A\, dx}\right)_n (F_n - F_{n-1}) + \left(\frac{N}{\rho}\right)_n \\ \frac{d}{dt} F_n &= \left(\frac{EA}{dx}\right)_n (v_{n+1} - v_n) \end{aligned} \right\}, \qquad n = 1, 2, \ldots, N \tag{4.43}$$

As a state equation, the model has $2N$ state variables (v_n, F_n, $n = 1, 2, \ldots, N$). F_0 and v_{N+1} are boundary values. Hence the system is indeed of order $2N$. Note that the pattern of the space index n in each of the preceding equations is causal. In the first equation, force is the input or independent variable, and for the first (boundary) element the boundary value is a force. In the second equation, velocity is the independent variable, and the boundary input (last element) is a velocity.

Example 4.8	Continuous Pizza Oven

Consider the oven unit in a pizza parlor as shown in Fig. 4.15a. The pizza patties are removed from cold storage at the beginning of the day and placed in the kitchen, so that

Figure 4.15

Commercial Pizza Oven System

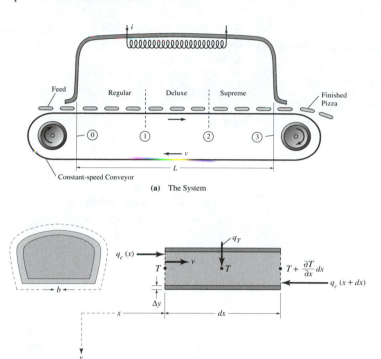

(a) The System

(b) Heat Percolation Process

the temperature of the pizza feedstock to the oven is variable. The pizzas are prepared in batches (three pans deep) of different types—regular, deluxe, and supreme—but of the same size. The oven can contain three of these batches at any time, as shown in Fig. 4.15. Assuming that a uniform temperature can be maintained inside the oven, for instance, by electrical resistance heating, derive a distributed-parameter model as well as a lumped-parameter model for the system that can predict the temperature attained by each batch of pizza in the oven at any time.

Solution—Distributed-Parameter Analysis:

heat percolation process: We shall derive the results without regard to the discontinuous nature of the pizza stream. Consider Fig. 4.15b, which shows a short (length = dx) segment of channel (oven) with flowing material (pizza). We assume that the temperatures of the channel and of the material in the channel are uniform at u and T, respectively, over the length dx, although T is evidently a function of x and time t. Equation (4.39a) thus gives

$$(\rho A c \, dx)\frac{\partial T}{\partial t} = \hat{C} dx \frac{\partial T}{\partial t} = q_c(x) + q_T + q_c(x + dx) \tag{4.44a}$$

where $\hat{C}$ is the thermal capacitance of the material per unit length of the channel. The heat transferred from the channel to the material in the segment q_T is given by equation (4.1b):

$$q_T = \frac{kA}{\Delta y}(u - T) = \left(\frac{kb}{\Delta y}\right) dx \, (u - T) = \left(\frac{1}{R}\right)^{\hat{}} dx \, (u - T) \tag{4.44b}$$

where $(1/R)^\wedge$ is the surface conductance per unit length of channel. According to equation (4.10), the heat gained by the material in the segment (lost by the flowing material) is, at x,

$$q_c(x) = 0 \tag{4.44c}$$

and at $(x + dx)$,

$$q_c(x + dx) = \left(\rho A c \frac{\partial x}{\partial t}\right)\left(T - \left[T + \frac{\partial T}{\partial x} dx\right]\right) = -C^\wedge \frac{\partial x}{\partial t} \frac{\partial T}{\partial x} dx \tag{4.44d}$$

Substituting equations (4.44b) through (4.44d) in equation (4.44a), we obtain

$$\frac{\partial x}{\partial t} = v \tag{4.45a}$$

$$\frac{\partial T}{\partial t} = \left(\frac{1}{R}\right)^\wedge \frac{1}{C^\wedge}(u - T) - v\frac{\partial T}{\partial x} \tag{4.45b}$$

Equations (4.45a) and (4.45b) describe a *heat percolation process*. Given the boundary and initial conditions, they can be solved for the position x and temperature T of any pizza.

Solution—Lumped-Parameter Model: For a constant belt speed and the requirement for pizza temperature at specific locations inside the oven, a discrete-time lumped-parameter model seems most appropriate. Figure 4.15a shows a possibly adequate, computationally economical discretization in space (numbered 0 to 3). A more accurate division could be to use each pizza location (that is 10 points, 0–9). The desired temperatures are T_1, T_2, and T_3. The heat capacitance equation (4.6b) and the material transport equation (4.10) can be applied in discrete-time form to the pizza **at** n according to

$$\frac{C_n}{\Gamma}(T_{n,k+1} - T_{n,k}) = \sum q_{\text{into}} = \frac{1}{R_n}(u_k - T_{n,k}) + (\dot{m}c)_n(T_{n-1,k} - T_{n,k}) \tag{4.46a}$$

where Γ is the transportation time as well as the sampling period:

$$\Gamma = \frac{L}{3v} \tag{4.46b}$$

and

$$(\dot{m}c)_n = \rho A c v = (\rho A\, dx\, c)\left(\frac{v}{dx}\right) = \frac{C_n}{\Gamma} \tag{4.46c}$$

The first term of the RHs of equation (4.46a) represents heat supplied by the oven, whereas the second term is the heat carried in from the previous location. Note that the energy balance is with respect to the material already at n, not the material arriving at n. Further, the position in the oven of pizza batch type j is given by

$$x_{j,k+1} = x_{j,k} + \Gamma v \tag{4.46d}$$

Considering the three positions and the corresponding pizza types, equations (4.46) give us the following model:

$$T_{1,k+1} = -\frac{\Gamma}{(RC)_1}T_{1,k} + T_{0,k} + \frac{\Gamma}{(RC)_1}u_k \tag{4.47a}$$

$$T_{2,k+1} = -\frac{\Gamma}{(RC)_2}T_{2,k} + T_{1,k} + \frac{\Gamma}{(RC)_2}u_k \tag{4.47b}$$

$$T_{3,k+1} = -\frac{\Gamma}{(RC)_3}T_{3,k} + T_{2,k} + \frac{\Gamma}{(RC)_3}u_k \tag{4.47c}$$

$$x_{1,k+1} = x_{1,k} + \Gamma v \tag{4.47d}$$

$$x_{2,k+1} = x_{2,k} + \Gamma v \tag{4.47e}$$

$$x_{3,k+1} = x_{3,k} + \Gamma v \tag{4.47f}$$

where $T_{0,k}$ is actually an input—the temperature of the feed. Equations (4.47a) through (4.47f) constitute a system of first-order difference equations that are the discrete-time state equations. The temperature of any pizza batch is determined by first obtaining its position (equations (4.47d) to (4.47f)) and then the temperature at that position (equations (4.47a) to (4.47c)).

4.4 NONENGINEERING SYSTEM EXAMPLES

In Section 3.1 we expressed the opinion that one of the attractions of the state space approach is its transparency to the type of system being modeled. The engineering system examples we have considered thus far have generally supported this viewpoint. This section considers only two examples of models of nonengineering systems obtained using the state space approach (additional examples are provided as practice problems). The material in this section, however, is further evidence of the wide applicability of the state model. But being applicable does not necessarily imply being easy to apply. The problem with most nonengineering systems is that the components are usually hidden, and the behavior of certain components can sometimes be contrary to normal intuition. Thorough familiarity with the system under study is thus often necessary for adequate decomposition and description of the behavior components.

Example 4.9	Savings-Only Commercial Bank

A simplified view of the operations of a savings-only bank is as follows:

1. Money comes into the bank as deposits paid into savings accounts by the customers and as dividends or profits made by the bank from its investments (which include loans, revenue-yielding property, etc.).
2. Money leaves the bank through withdrawals from savings accounts by customers, overhead expenditure (salaries, supplies, etc.), and investments by the bank.
3. The interest rate on savings accounts is usually fixed for relatively long periods of time, say, at $100\alpha_1\%$, (a typical value is $100\alpha_1\% = 3\%$, or $\alpha_1 = 0.03$), and compounding is (assumed here to be) quarterly. The bank makes on the average, say,

$100\alpha_3$% profit each quarter on all existing investments (a typical value may well be over 10%, $\alpha_3 = 0.1$, for some banks).

4. The bank's current policy is to invest each quarter $100\alpha_2$% (for example 70%, $\alpha_2 = 0.7$) of the money left in the bank at the end of a previous quarter.

Develop a dynamic model of the operations of this bank that will assist the bank's policy makers. Assume that the amounts of deposits, withdrawals, and overhead expenditure each quarter are known.

Solution: Although savings and withdrawals are continuous, all major events connected with policy decision occur at distinct points in time. A discrete-time system assumption therefore seems appropriate. The time period is 3 months, that is, one quarter.

Step 1: Three distinct components (see Fig. 4.16) can be identified: savings and interest, investments, and total account or money in the bank. These components are suggested by the describe activities of the bank and do not necessarily correspond to any physical departments into which the bank can be divided physically.

savings and interest (dynamic): Let D_k, W_k, and I_k represent the total deposits, the total withdrawals, and the total interest to be paid, respectively, during quarter k. Then the total amount in the savings account during any quarter is given by

$$S_{k+1} = S_k + I_k + D_k - W_k; \qquad I_k = \alpha_1 S_k \tag{a}$$

Note that the interest is not subtracted from the savings, as we might expect. The interest on savings accounts represents an earning by the bank until payments are made. Let

$$u_{1,k} = D_k - W_k \tag{b}$$

Then

$$S_{k+1} = (1 + \alpha_1)S_k + u_{1,k} \tag{c}$$

investments (dynamic): Let A_k represent the total amount in the bank at the end of period k, the total account. Then the total amount already invested by period k is given by

$$V_{k+1} = V_k + \alpha_2 A_k \tag{d}$$

Figure 4.16

**Savings Bank
System Components**

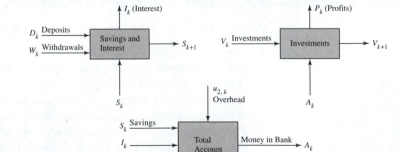

The profit from investments for the period k is

$$P_k = \alpha_3 V_k \tag{e}$$

total account (static): The total account is just the summation of all accounts, plus receipts (to the total account) less disbursements (from the total account) during the period,

$$A_k = S_k + P_h - I_k - u_{2,k} \tag{f}$$

Observe that the interest that was added to the savings account is subsequently paid from the total account. The implication is that any investment by the bank is considered a separate operation from savings account interest generation, even though the money paid out as interest could eventually come from profits on investments. Put differently, savings account interest is just money borrowed by the bank and paid when it is due.

Step 2: Let the state variables be the amount of savings $x_{1,k} = S_k$ and investments $x_{2,k} = V_k$. The inputs are the net payments into the savings accounts $u_{1,k}$ and the overhead expenditure during the quarter $u_{2,k}$.

Step 3: From equations (c), (d), (e), and (f), the state equations are

$$x_{1,k+1} = (1 + \alpha_1)x_{1,k} + u_{1,k} \tag{4.48a}$$

$$x_{2,k+1} = x_{2,k} + \alpha_2[(1 - \alpha_1)x_{1,k} + \alpha_3 x_{2,k} - u_{2,k}] \tag{4.48b}$$

$$= (1 - \alpha_1)\alpha_2 x_{1,k} + (1 + \alpha_2\alpha_3)x_{2,k} - \alpha_2 u_{2,k}$$

Step 4: It would appear that the more significant quantities for the bank's policy makers are the money in the bank at the beginning of a quarter, after investments, and the profits made from investments in any quarter. Thus the outputs can be taken as

$$y_{1,k} = (1 - \alpha_2)A_k = (1 - \alpha_2)(1 - \alpha_1)x_{1,k} \tag{4.48c}$$

$$+ (1 - \alpha_2)\alpha_3 x_{2,k} - (1 - \alpha_2)u_{2,k}$$

$$y_{2,k} = \alpha_3 x_{2,k} \tag{4.48d}$$

The state model is thus given by equations (4.48a) through (4.48d).

Example 4.10   Prediction Model for a Buyers' Group

A buyers' group in a large fruit-producing region of the country desires to develop a model to predict the daily price of cantaloupes as a decision tool for purchasing this fruit for its members during the summer months. From carefully kept records, the group has deduced the following about the cantaloupe industry:

1. Consumer demand for cantaloupes increases substantially with the average daily temperature and decreases somewhat with the retail price of cantaloupes in the grocery stores. Daily temperatures for the summer months can be predicted fairly well.

2. The price that the grocery stores demand for cantaloupes decreases as the amount of cantaloupes available for the market increases but increases with consumer demand.
3. Cantaloupe farmers in the region tend to ship more of their cantaloupes elsewhere when the retail price of cantaloupes in the region is rising in order to further drive up prices.* However, they respond to increasing consumer demand by providing more cantaloupes for the region's market.

Derive a state model for this consumer group using the given information. Assume that the group can supply all necessary correlation coefficients (c's, see below). Identify all state variables, inputs, and parameters.

Solution:

Step 1: The components are cantaloupe consumers, grocery stores, and cantaloupe farms (or farmers) (see Fig. 4.17).

consumers (static): Let D = consumer demand (cantaloupe/day), P = retail price of cantaloupes:

$$D = c_1 T - c_2 P \tag{a}$$

grocery stores (pricing) (dynamic): Let S = supply of cantaloupes (cantaloupes/day) (see notes):

$$P = \int^t (c_3 D - c_4 S)\, dt \tag{b}$$

grocery stores (purchasing) (dynamic): Let Q = total quantity of cantaloupes in the market:

$$Q = \int^t (S - D)\, dt \tag{c}$$

cantaloupe farms (static): (see notes):

$$S = c_5 D - c_6 P \tag{d}$$

---

*This observation is probably contrary to the reader's first impressions of market dynamics.

Figure 4.17

**Buyers' Group
System Components**

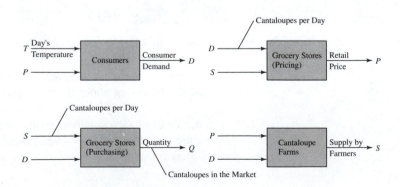

Notes:
1. It is assumed that the grocery store fixes prices after analyzing the cumulative quantities bought and sold, in other words, after an inventory. Cantaloupe farmers, on the other hand, are assumed to be able to respond instantaneously to prices and demand, since they need merely to divert more or less of their already harvested cantaloupes to other markets.
2. The unit of time measure for this problem is a day.

Step 2: The state variables are $x_1 = P$ and $x_2 = Q$. The input is $u = T$

Step 3: From equations (a) through (d), the state equations are

$$\frac{dP}{dt} = c_3 D - c_4 S = c_3(c_1 T - c_2 P) - c_4(c_5 D - c_6 P) \qquad \text{(e)}$$

$$= c_3(c_1 T - c_2 P) - c_4 c_5(c_1 T - c_2 P) + c_4 c_6 P$$

$$\frac{d}{dt} Q = S - D = c_7 D - c_6 P = c_7(c_1 T - c_2 P) - c_6 P; \qquad c_7 = c_5 - 1 \qquad \text{(f)}$$

Step 4: The objective for the model is the daily price P of cantaloupes. Hence let the output be $y = P$. The state model then is

$$\frac{d}{dt} x_1 = a_{11} x_1 + b_1 u \qquad \text{(4.49a)}$$

$$\frac{d}{dt} x_2 = a_{21} x_1 + b_2 u \qquad \text{(4.49b)}$$

$$y = x_1 \qquad \text{(4.49c)}$$

where the input is $u = T$, that is, each day's temperature, and the parameters are

$$a_{11} = c_4 c_6 - c_2(c_3 - c_4 c_5); \qquad a_{21} = -c_2 c_7 - c_6 = -c_2(c_5 - 1) - c_6 \quad \text{(4.50a)}$$

$$b_1 = c_1(c_3 - c_4 c_5); \qquad b_2 = c_1 c_7 = c_1(c_5 - 1) \qquad \text{(4.50b)}$$

....

4.5 PRACTICE PROBLEMS

4.1 The structural detail for heat transfer from a heat engine is shown in Fig. 4.18. The combustion chamber temperature is steady at the known value T_c. The temperature of the environment outside the engine is also fixed at T_a. It is desired to predict the temperature of the protective coating in order to establish its integrity and capacity to perform its protective functions. Derive a model for this system that meets this requirement. Explain your assumptions.

4.2 Part of the mission of a space shuttle is the opening of the cargo bay doors to release heat trapped inside the vehicle. Assume simply that the shuttle has the cylindrical configuration shown in Fig. 4.19 and that the tiles covering the entire external surface are insulators of very low thermal conductivity. Assume also that the heat generation inside the

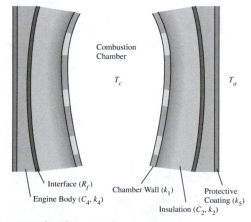

Combustion Chamber

T_c

T_a

Interface (R_f)

Engine Body (C_4, k_4)

Chamber Wall (k_1)

Protective Coating (k_5)

Insulation (C_2, k_2)

Figure 4.18 (Problem 4.1)

shuttle is due mainly to resistive electrical parts and active electronic components that consume power at a steady rate q_e. Obtain a model of the shuttle that can be used to determine how long the vehicle can remain in orbit should the cargo bay doors fail to open. Assume that heat is lost from the shuttle mainly by radiation.

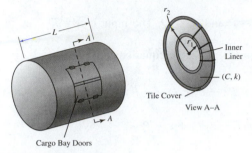

Figure 4.19 (Problem 4.2)

4.3* A thermometer with thermal heat capacitance C is placed inside a liquid bath in which its surface convective heat transfer coefficient is h (see Fig. 4.20). if the density and specific heat of the liquid in the bath are ρ and c, respectively, write dynamic equations for the temperature indicated by the thermometer and for the temperature of the liquid. The liquid is initially at a temperature T_0, while the thermometer was initially outside the bath at the ambient temperature T_a.

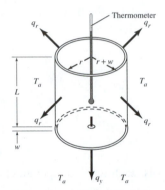

Figure 4.20 (Problem 4.3)

4.4 A building consists of two rooms that are heated by one furnace (see Fig. 4.21). The furnace delivers equal quantities of heat q_f to each room. The outside temperature is T_a. One wall of room 2 is made of glass, which loses heat by convection. Assuming that there is no heat storage in the walls and the roof and that the floors and ducts are perfect insula-

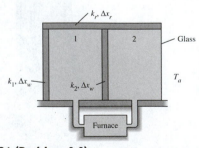

Figure 4.21 (Problem 4.4)

tors, derive a set of equations from which the temperature of each room can be found after the furnace is turned on. State your assumptions and assign parameters as needed.

4.5 Consider the constant-temperature bath system often used in chemical and heat transfer experiments to induce a constant-temperature boundary condition T_b on the experiment (see Fig. 4.22). Cooling fluid is pumped at a steady rate $\dot{m}$ through the bath. The fluid is introduced at a temperature equal to the desired boundary temperature. Derive a model for this system and state the circumstances under which the desired boundary condition is approached in the experiment. Assume simply that the experimental package generates heat at a known rate q.

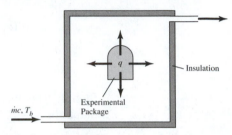

Figure 4.22 (Problem 4.5)

4.6* Consider the following mixing tank problem. Hot water and cold water flow at predetermined rates into a mixing tank where the mixture is thoroughly stirred (see Fig. 4.23). It is desired to control the temperature T of the output stream from the tank by adjusting the flowrate of the hot water into the tank. Derive a model for the system that can be used to design the controller. Identify the control input. Assume that the enthalpy of formation is the same for the hot water, the cold water, and the mixture.

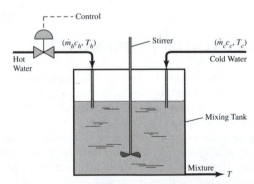

Figure 4.23 (Problem 4.6)

4.7 Repeat Problem 4.4 with the following modifications: The heat delivery to each room is not known. Rather, the furnace heats the rooms by circulating air clockwise from room 1, through a duct in the separating wall, to room 2 and back into the furnace. The mass flowrate of the air and the temperature in the furnace are prescribed. Assume that there are no open windows or other holes through which air can escape to the outside. State all your assumptions.

4.8 A two-room building has one of its rooms cooled by an air conditioner (see Fig. 4.24). The air blower in the air conditioner blows cold air into the room at a known rate $\dot{m}$. The air leaving the air conditioner can be assumed to be at the temperature of the evaporator coils, and the evaporator temperature is effectively constant. Half of the first wall is made of glass of convective heat transfer coefficient h. The wall between the rooms is covered with decorative wallboard on both sides as shown. The floor of each room can be considered a perfect insulator, and there is negligible heat storage in the walls and roof of the building. If the outside temperature is T_a, derive a model for the temperature of each room following the turning on of the air conditioner.

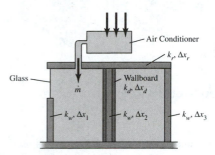

Figure 4.24 (Problem 4.8)

4.9* A storage tank with a parabolic shape and circular cross section, that is, the radius of the cross section r is related to the distance y (measured from the top) by $r = b\sqrt{y}$, is being filled with liquid from a pump connected to the tank through a long fluid line (see Figure 4.25). There is a check valve at the entrance to the tank to prevent backflow of liquid. The

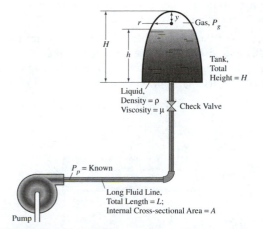

Figure 4.25 (Problem 4.9)

output pressure at the pump P_p is known, and the air trapped above the liquid in the tank behaves like a perfect gas. Since the tank can sustain only a finite pressure, the trapped air pressure during the filling process is desired. Derive a model for this system that will allow this pressure to be determined.

4.10 Consider the arrangement for launching a projectile. The projectile of mass m and volume V_m is initially held in position by a pin in a tube of length L, as shown in Fig. 4.26. To the left of the projectile is a gas (assume it is an ideal gas) initially at a high temperature and pressure. The pin is released to launch the projectile. Derive a model for this system that can be used to predict the exit velocity v_e of the projectile. Neglect friction in the walls of the tube.

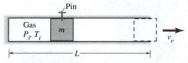

Figure 4.26 (Problem 4.10)

4.11 Consider again the moist product dryer of Example 4.4. Suppose as shown in Fig. 4.27, the air heater is a *natural-convection* type, inclined at an angle α to the horizontal. Derive the description of this heater that will provide all required inputs to the dryer. Note the following differences from Example 4.4: (1) the inlet air velocity is an unknown, (2) gravitational body force is present, (3) the heater is inclined at an angle and is not horizontal, and (4) the pressure gradient is due to the change in elevation and is given by $dP/dx = -\rho_\infty g \cos \alpha$, where x is distance along the channel, and ρ_∞ may be taken as the density at the entrance to the channel (actually it is the density in the free stream, and in this case the free stream velocity is zero). Assume that the fluid behaves like an ideal gas. Can you identify the *buoyancy force* term in your momentum equation?

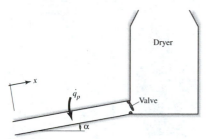

Figure 4.27 (Problem 4.11)

4.12 Repeat Problem 4.9 with the following modifications: the output pressure at the pump is not known; however, the pump is a centrifugal-flow pump; and the pump inlet pressure and impeller rotary speed are known. Assume any needed geometric data on the pump. Also assume incompressible behavior for the liquid.

4.13* A large and deep water tank discharges water at the bottom through a linear valve and centrifugal-flow water turbine connected to the valve, as shown in Fig. 4.28. Assuming any needed geometric data of the turbine, determine the model of the system that can be used to predict the angular velocity of the turbine shaft. *Hint:* Assume that the turbine is essentially a centrifugal-flow compressor, operated in reverse (see Example 4.5) and that the water is incompressible.

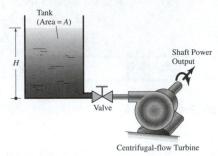

Figure 4.28 (Problem 4.13)

4.14 Consider the portable power supply system shown in Fig. 4.29. The gas in the pressure-cylinder is initially at a high pressure. A regulating valve at the exit of the cylinder is used to regulate the inlet pressure to the turbine as the cylinder discharges. Assuming that the electrical power output from the generator is simply proportional to the mechanical power output from the turbine, develop a model to predict the instantaneous electric power output of the generator. You may take the turbine behavior as similar to a centrifugal-flow compressor, operated in reverse (see Example 4.5). Also assume that the gas behaves like an ideal gas and that the initial pressure in the pressure-cylinder and the volume of the cylinder are known.

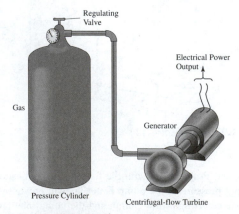

Figure 4.29 (Problem 4.14)

4.15 Repeat example 4.6, by making the following assumption: The densities and specific heats of each stream are constant and the same.

4.16 Repeat Example 4.7 without the simplifying assumptions. Rather, follow the procedure of Example 4.6. Use the variables and parameters identified in the Fig. 4.30.

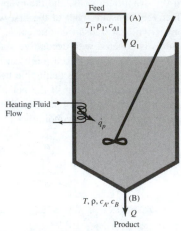

Figure 4.30 (Problem 4.16)

4.17* A continuous well-stirred tank mixer is used to mix three components, A, B, and C. The components are delivered in three feed streams, each stream consisting entirely of one component only. Assume that means is available for direct cooling or heating of the tank. Given the temperature, density, and volumetric rate for each feed stream, and the volumetric rate of the output stream, develop a model for the mixing tank that can be used to predict the concentration of components B and C in the product.

4.18 Consider again the continuous well-stirred tank mixing process modeled in Example 4.6. Suppose stream 2 is also composed of the two components A and B. Determine the model of the stirred tank in this case.

4.19 **(a)** Develop a distributed-parameter model for the parallel-flow, double-pipe heat exchanger of Fig. 4.31, considering thermal capacitance effects, material transport and thermal conductivity of the pipe walls.

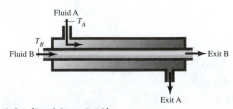

Figure 4.31 (Problem 4.19)

4.19 (b) Obtain a lumped-parameter model of the heat exchanger system, using three spatial divisions along the length of the heat exchanger.

4.20 (a)* The bit used with a pneumatic jackhammer has the shape of a long tapered circular shaft (see Fig. 4.32). The behavior of the rock to which the tool is applied can be described as a nonlinear spring response with spring constant $k(y) = cy^{-1/4}$. Derive a distributed-parameter model for the longitudinal vibrations of the bit, assuming that the displacement-time history of the top of the bit $u_0(t)$ is known.

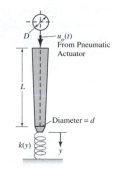

Figure 4.32 (Problem 4.20)

4.20 (b) Derive a three-lumped model for the system of Problem 4.20a.

4.21 (a) Obtain a distributed-parameter model for the torsional motion of a screwdriver used in driving a screw nail into wood (see Fig. 4.33). Assume that the screwdriver consists of a rigid handle of diameter D and a uniform elastic rod of length L_2 and diameter d. The screw nail should be considered short, and the resistance to its rotation within the wood is proportional to the product of the rotating speed and the angle of twist of the screw.

4.21 (b) Construct an equivalent lumped-parameter model for the screwdriver system using a minimum number of essential lumped components. Explain your choice of lumped elements.

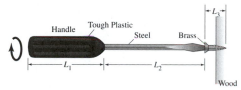

Figure 4.33 (Problem 4.21)

4.22* A heat sink design for a circular transistor consists of a close-fitting metal tube that is closed at both ends (see Fig. 4.34). If the transistor generates heat at a known rate q, and if all the heat generated is transferred into the cylindrical region through the bottom end, derive a distributed-parameter model of the heat sink operation.

4.23 (a) Obtain a distributed-parameter model for flow in an uncased circular tunnel connecting the bottom of a reservoir

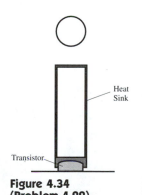

Figure 4.34 (Problem 4.22)

and a free discharge point B, as shown in Fig. 4.35. Assume that the tunnel flows full and that the pressure in the formation (surrounding rock) is uniform. Note that the radial form of Darcy's law for linear flow was given in Example 3.11, equation (3.35a).

4.23 (b) Derive an equivalent three-lump model for the problem.

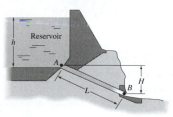

Figure 4.35 (Problem 4.23)

4.24* Derive the distributed-parameter model for the oil well drilling column described in Problem 2.18.

4.25 Construct a discrete-time state model of the adult dog population in a certain city that predicts the number of rabid adult dogs in that city in any year. The following are known: (1) 10% of female adult dogs get rabies, whereas only 5% of male adult dogs are rabid; (2) the number of baby dogs born each year is 80% of the female adult dog population; (3) 40% of all baby dogs are male and the rest are female; (4) a baby dog will grow into an adult dog in 2 years; (5) 20% of adult dogs die by the end of each year. *Hints:* (a) Let the number of male and female dogs in any year k be $x_{1,k}$ and $x_{2,k}$. (b) The output equation should be the number of rabid dogs in year k. (c) Define an intermediate age group, 1 year between baby and adult.

4.26* In Problem 1.6 if the spoilage rate of apples and the rate at which new apples are brought in from the warehouse are fixed and predictable quantities, and if the rate at which shoppers purchase apples is proportional to the amount of marketable apples in the supermarket, derive a model for this system stating what is the state variable, the input, and the output in the model.

4.27 In a certain developing economy, the government's expenditure in the agricultural sector at year k, E_k, can be considered to be made up of expenditure on acquisition, maintenance and operation of farm machines T_k, expenditure on welfare and alternative employment for farmers and farmworkers who are not utilized in farming G_k, and the subsidy and other assistance to farmers and farmworkers actually engaged in farming F_k. In other words,

$$E_k = T_k + G_k + F_k, \ k = 0, 1, 2, \ldots$$

Suppose the expenditure on mechanized farming is proportional to the total agricultural expenditure in the previous year (this could be government policy), that is, $T_k = aE_{k-1}$,

$0 < a < 1$, and that the expenditure on displaced and unutilized farmers and farmworkers is increased or decreased by the change in the mechanized farming expenditure over the previous year, that is,

$$G_k = G_{k-1} + b(T_k - T_{k-1}), \, 0 < b < 1$$

If the sum of government expenditure on the farmers and farmworkers utilized in farming and those not so employed is fixed (there is very little change in this combined population) over the years, $G_k + F_k = u$ (all k), derive a state model that can be used to study the agricultural policy in this economy.

4.28* Consider a certain region inhabited by foxes and rabbits. Suppose that without the rabbits to eat, the mortality rate for the foxes is greater than the birthrate, while in the absence of foxes, the birthrate for rabbits exceeds their mor-

tality rate. Suppose further that the effect of the foxes on the rabbit population is to increase the mortality rate of the rabbits by an amount proportional to the number of foxes, while the presence of rabbits to eat results in an increase in the birthrate of the foxes by an amount proportional to the number of rabbits. Derive a model for this predator-prey system that will permit the determination of the number of each species at any time. Consider the use of the following parameters in your model:

$B_1, B_2 =$ birthrate: of rabbits in the absence of foxes, of foxes in the absence of rabbits

$M_1, M_2 =$ mortality rate: of rabbits in the absence of foxes, of foxes in the absence of rabbits

$x_1, x_2 =$ number: of rabbits at time t, of foxes at time t

Chapter

5

Generalized System Models and Analogs

THE DISCUSSION AND PROBLEMS in the preceding chapter showed that many practical engineering systems have combined in them mechanical, fluid, thermal, chemical, and electrical components. We are able to develop unified models of these systems because the behaviors of the various dynamic components are described by the same class of differential (difference) equations. In this chapter we exploit this aspect of engineering elements to develop generalized models and analogs for such systems. In particular, we use the concept of *energy and power,* which is applicable to all such systems, as the basis for a new set of common components and variables discussed in Section 5.1.

An important class of generalized system examples—*electromechanical* systems—is considered in this chapter (Section 5.2). Electromechanical devices are ubiquitous in engineering. They find applications as actuators and sensors in virtually every field of endeavor. Some well-known examples of electromechanical systems include electric motors and generators, stepping motors, solenoids and relays, and various other types of actuators, electrodynamometers, differential transformers, load cells, accelerometers, flowmeters, and various other displacement and pressure transducers and motion detectors.

A significant trend in electromechanical systems especially is the miniaturization of such systems to *micron* and *submicron* sizes—*microactuators* and *microsensors*—through the application of *microfabrication* techniques developed from *microelectronics* (particularly silicon integrated-circuit fabrication technology) and other special processing methods. These *micromachined* systems are the eyes and ears and sometimes muscle of many of the extremely sophisticated instruments and very efficient control systems required in some of today's machines and processes. Micromachined devices are introduced briefly in Section 5.4 as a follow-up to a more general coverage, in Section 5.3, of other hybrid and integrated systems.

5.1 THE CONCEPT OF ENERGETIC SYSTEMS

We consider as *energetic* those systems whose components interact with their environment (including other systems) through the transmission of power. Consider, for example, the elements shown in Fig. 5.1 an interconnected spring and mass will transmit power at their interface in the form of force times linear velocity (or torque times an-

193

Figure 5.1

Engineering Systems and Power Transmission

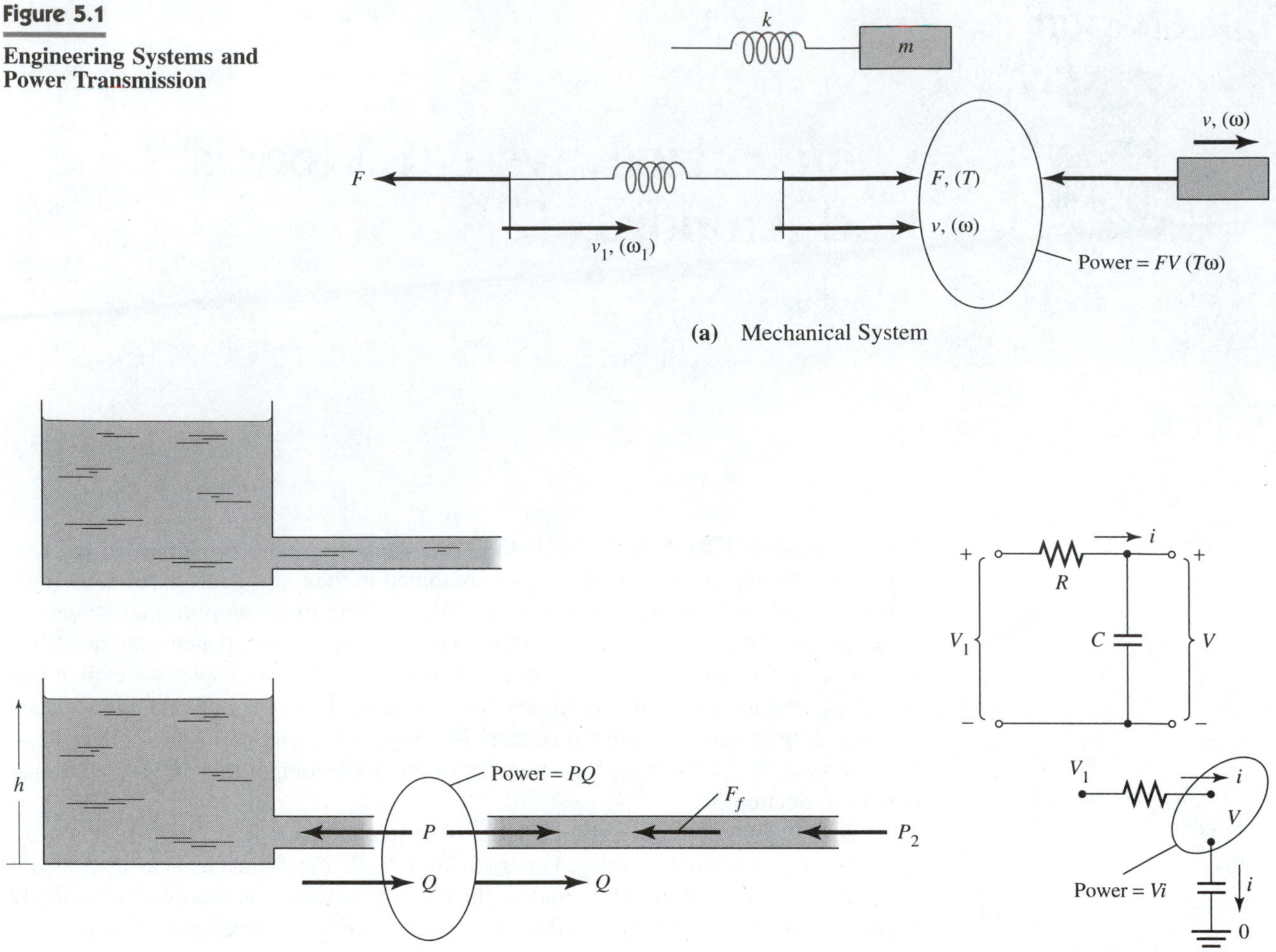

(a)   Mechanical System

(b)   Fluid (Hydraulic) System

(c)   Electrical System

gular velocity); two fluid system components such as a tank and a long fluid line will interact with each other through power transmission in the form of pressure times volume flowrate; and power in the form of current times voltage is transmitted at the interface of an electrical resistor and a capacitor. Similarly, a heat storage element will interact with surrounding thermal resistance material through the flow of heat. Thus energetic system components either *store* energy, that is, accumulate over time the input power, or *dissipate* (or *transfer*) energy, that is, transmit all input power, without storage, to (other connected components) or to the *environment* (or, in the case of *source* components, *generate* energy).

Generalized Signal Variables and Elements

It follows from the relation between the concepts of "state" and "memory" (recall our discussion in Section 3.1) that the state of an energetic system at any instant can be related to the amount of energy contained (stored) in its elements. Further, to specify state, a variable pair (one input and the other the output, or one independent and the other the dependent variable) whose product at any instant gives the instantaneous power transmitted at the interface between the component and its environment should be sufficient to describe the "energetic" behavior of that component. Such variable pairs, one variable of which we term *effort* or *potential* (*e*), while the other we call *flow*(*f*), have already been employed in describing engineering components. For example, force and velocity, pressure and volumetric flowrate, and voltage and current. If more than one form of power is transmitted between an element and its surroundings, a different interface or (energy) *port* with the necessary variable pair is implied for each type of power transmission. We should note that the product of the variable pair temperature and heat flowrate, which we have used to describe thermal systems, is not power, although heat flowrate by itself is power (the product of temperature and rate of entropy flow is power). However, since the known behaviors of relevant thermal system components are conveniently described in terms of temperature and heat flow, we shall continue to use these variables.

System Analogs

The concept of *analogs* arises in the context of generalized signal variables and elements, in terms of defining the flow and effort for an element in a particular type of system (mechanical, electrical, etc.). Thus a *force-flow/velocity-potential* analogy is often used for mechanical systems, just as is the *velocity-flow/force-effort* analogy. A velocity-flow/force-effort analogy for a mechanical component implies that given the behavior of a nonmechanical component, the behavior of the *analogous* mechanical component can be obtained from the description of the behavior of the nonmechanical element by replacing the flow variable with velocity and the effort variable with force, and vice versa. The velocity-flow/force-effort analogy is preferred here because it results in relations that are consistent with what can be considered generally accepted concepts (see, for example, the discussion below on kinetic and potential energy).

Generalized Elements

Five ideal behavior components or *generalized elements*—R, C, I, S_e, and S_f—can be described in terms of the effort and flow variable pair (e, f). Without loss of generality, we shall consider mostly linear continuous-time models.

Resistance   The behavior of an R (resistance) element is given by

$$e = Rf \tag{5.1a}$$

or

$$f = \frac{e}{R} \tag{5.1b}$$

Such an element is nonlinear if the parameter R depends on f or e.

Capacitance and Inertance C (capacitance) and I (inertance) elements have the behavior

$$e = \int^t \left(\frac{1}{C}\right) f \, dt \tag{5.2}$$

$$f = \int^t \left(\frac{1}{I}\right) e \, dt \tag{5.3}$$

Note that equations (5.2) and (5.3), as written, apply to both linear and nonlinear components. We can see that C and I elements represent generalized dynamic components, whereas the resistance element is a static component. Further, for state space models the flow variable f should be assigned the state variable for an I element, whereas e should be the state variable for the C element.

Effort and Flow Sources. The behavior of effort and flow sources, respectively, is

$$e = S_e \quad \text{(independently of } f) \tag{5.4}$$

$$f = S_f \quad \text{(independently of } e) \tag{5.5}$$

Table 5.1 summarizes the analogy between these generalized behaviors and some of the components encountered in previous chapters. For instance, a constant-displacement pump can be considered a flow source. A good example is a reciprocating piston pump. For incompressible fluid flow, the volume flowrate per piston depends on the piston or liner diameter D, the piston rod diameter d, the stroke L, and the specified rotary speed of the crank ω and is independent of the pressure difference $(P_2 - P_1)$. For a double-acting piston pump (see Fig. 5.2) which is typical of oilfield operations, we have

$$Q_s = \frac{1}{4}\,[\pi D^2 + (\pi D^2 - \pi d^2)]\,L\left(\frac{\omega}{2\pi}\right) e_{ff}; \quad e_{ff} = \text{volumetric efficiency} \tag{5.6}$$

Kinetic and Potential Energy

In the earlier chapters, part of the ease with which engineering systems could be modeled was attributed to the structural simplicity of those systems whose behavior components were often found already to be separated into the physical parts. For energetic systems there is the additional tendency for different energy storage modes to be associated with individual physical parts of the system. For instance, as is well known in mechanical systems, springs store potential energy, while masses store kinetic energy. With respect to the generalized elements, we shall consider the energy stored by a C-element as potential energy, and the energy stored by an I element as kinetic energy.

Kinetic Energy

Energy is the integral of power over time. That is,

$$E = \int^t ef\,dt \tag{5.7}$$

Considering linear components, let us define for an I-element a new generalized variable p,

$$p = \int^t e\,dt \tag{5.8}$$

so that $dp = e\,dt$. Equation (5.3) implies that

$$f = \frac{p}{I} \tag{5.9}$$

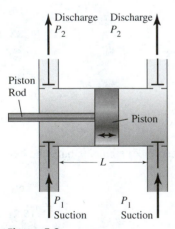

Discharge P_2 Discharge P_2

Piston Rod

Piston

L

P_1 Suction P_1 Suction

Figure 5.2

Operating Schematic of a Double-acting Piston Pump

Table 5.1 Generalized Elements and Analogs

System Type	Generalized Elements					Analogy
	R	C	I	S_e	S_f	
Mechanical (translation)	1. $F = \beta V$ 2. $F = \beta\|V\|V$ 3. $F = fN\,\text{sign}(V)$ $R \equiv \beta$ or $f_R(V)$	$F = k\int_0^t v\,dt$ $C \equiv \dfrac{1}{k}$	$v = \dfrac{1}{m}\int_0^t F\,dt$ $I \equiv m$	$F = F_s$ $S_e \equiv F_s$	$v_a = v_s$ $S_f \equiv v_s$	$e \Rightarrow F$ $f \Rightarrow v$ $F = $ Force $v = $ Linear Velocity
Mechanical (rotation)	1. $T = \beta\omega$ 2. $T = \beta\|\omega\|\omega$ 3. $T = fNr\,\text{sign}(\omega)$ $R \equiv \beta$ or $f_R(\omega)$	$T = k\int_0^t \omega\,dt$ $C \equiv \dfrac{1}{k}$	$\omega = \dfrac{1}{J}\int_0^t T\,dt$ $I \equiv J$	$T = T_s$ $S_e \equiv T_s$	$\omega_a = \omega_s$ $S_f \equiv \omega_s$	$e \Rightarrow T$ $f \Rightarrow \omega$ $T = $ Torque $\omega = $ Angular Velocity
Fluid (incompressible)	$Q = C(\|P_a - P_b\|)^{\frac{1}{\alpha}}$ C depends on L and Flow Regime. $R \equiv f_R(Q)$	$h = \dfrac{1}{A}\int_0^t Q\,dt$ or $(P_a - P_b) = \dfrac{\rho g}{A}\int_0^t Q\,dt$ $C \equiv A/(\rho g)$	$A = $ Area of Cross Section $Q = \dfrac{A}{\rho L}\int_0^t (P_a - P_b)\,dt$ $I \equiv \rho L/A$	Large Reservoir $h_s \approx$ Constant $(P_a - P_b) = P_s = \rho g h_s$ $S_e \equiv P_s$	Pump $\omega = $ Given Displacement $= $ Constant $Q = Q_s$ $S_f \equiv Q_s \equiv f_Q(\omega)$	$e \Rightarrow (P_a - P_b)$ $f \Rightarrow Q$ $P = $ Pressure $Q = $ Volume Flowrate
Thermal	$q = \dfrac{kA}{\Delta x}(T_a - T_b)$ $R \equiv \Delta x/(kA)$ Cylindrical $R \equiv \dfrac{\text{Ln}(r_o/r_i)}{2\pi kL}$	$T = \dfrac{1}{C_T}\int_0^t q\,dt$ $C \equiv C_T$		$T = T_s$ Cooling Fluid $\dot{m} \to \infty$ $S_e \equiv T_s$	Heat Generator q_s e.g., Radioactive Source $S_f \equiv q_s$	$e \Rightarrow T$ $f \Rightarrow q$ $T = $ Temperature $q = $ Heat Flowrate
Electricial	$(V_a - V_b) = Ri_{a-b}$ $R \equiv R$	$(V_a - V_b) =$ $\dfrac{1}{C}\int_0^t (i_{a-b})dt$ $C \equiv C$	$(i_{a-b}) =$ $\dfrac{1}{L}\int_0^t (V_a - V_b)dt$ $I \equiv L$	$(V_a - V_b) = V_s$ $S_e \equiv V_s$	$(i_{a-b}) = i_s$ $S_f \equiv i_s$	$e \Rightarrow (V_a - V_b)$ $f \Rightarrow i_{a-b}$ $V = $ Voltage $i = $ Current

so that equation (5.7) becomes

$$E = \frac{1}{I} \int^{p} p \, dp; \quad \text{(kinetic energy)} \tag{5.10a}$$

This is the energy stored in a linear I-element. If in particular $p(t = 0)$ is zero, then from equation (5.10a),

$$E(t) = \frac{1}{2} \left(\frac{1}{I} \right) p(t)^2; \quad \text{(kinetic energy)} \tag{5.10b}$$

Substituting equation (5.9) into equation (5.10b), we obtain the kinetic energy in terms of f:

$$E(t) = \frac{1}{2} I f(t)^2; \quad \text{(kinetic energy)} \tag{5.10c}$$

Consider the case of the mechanical mass. If we use the force-potential analogy of Table 5.1, f represents the velocity and I represents the mass, so that equation (5.10c) gives the familiar

$$E = \frac{1}{2} m v^2 \tag{5.10d}$$

Potential Energy

Let us define another new generalized variable q (not to be confused with heat flow-rate) for a C-component:

$$q = \int^{t} f \, dt \tag{5.11}$$

Equation (5.2) then gives for the linear C-element,

$$e = \frac{q}{C} \tag{5.12}$$

Again the stored energy is given by equation (5.7), so that the energy stored in the C-element is

$$E = \frac{1}{C} \int^{q} q \, dq; \quad \text{(potential energy)} \tag{5.13a}$$

or for $q(t = 0) = 0$,

$$E(t) = \frac{1}{2} \left(\frac{1}{C} \right) q(t)^2; \quad \text{(potential energy)} \tag{5.13b}$$

Substituting equation (5.12) into equation (5.13b), we obtain the potential energy in terms of the effort variable:

$$E(t) = \frac{1}{2} C e(t)^2; \quad \text{(potential energy)} \tag{5.13c}$$

For a mechanical spring element and force-potential analogy, $(1/C) = k$. Since the integral of velocity is displacement x, then $q = x$, and equation (5.13b) implies the easily recognized result

$$E = \frac{1}{2} kx^2 \tag{5.13d}$$

Energy Dissipated in a Resistance Element

An R-component does not store energy. The energy dissipated in an R-element is given by equations (5.1) and (5.7):

$$E(t) = \int^t Rf^2 \, dt = \int^t \frac{1}{R} e^2 \, dt; \quad \text{(dissipated energy)} \tag{5.14}$$

Note that in all cases the energy stored or dissipated is a positive quantity.

Transformers and Gyrators, Transducers, and Example Systems

Some additional generalized elements with special ideal constitutive relations can be identified for engineering systems. These components—*transformers, gyrators,* and *transducers*—in general transmit power at two energy ports (two-port elements) without accumulating the power (that is, without storing energy). These elements are therefore *static* components. However, unlike the situation with the resistance element, all the available power at one port is transmitted to the other port, and none is dissipated to the environment. Physical models of energetic systems result when two or more components are interconnected somehow. The physical modeling process has already been considered in chapters 1 and 2, so that our objective here is simply to present the generalized nature of component junctions and a few system models illustrating the ideas of this section.

Transformers and Gyrators

In *transformers* and *gyrators,* the **same type of power is transmitted** at the two ports. There is merely a change in the relation between the flows and efforts at the two ports. For the transformer, the relation is between the effort at one port and the effort at the other port, and between one flow and the other. However, for the gyrator, the variables are *crossed,* that is, the relation is between effort and flow at each port.

The Transformer Element   The behavior of a transformer or a *TF*-component is given by

$$e_1 = me_2; \quad \text{(transformer)} \tag{5.15a}$$

$$f_1 = \frac{1}{m} f_2; \quad \text{(transformer)} \tag{5.15b}$$

The second factor is $(1/m)$ and not something else because the power input to the *ideal* transformer must equal the power output, since there is no energy storage and no energy dissipation. To see this, we express the equality of power at the two ports:

$$e_1 f_1 = e_2 f_2; \quad \text{(power balance)} \tag{5.16}$$

If we substitute equation (5.15a) into equation (5.16), we recover equation (5.15b).

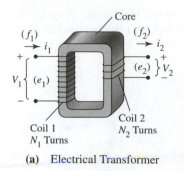

(a)   Electrical Transformer

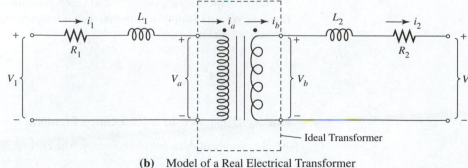

(b)   Model of a Real Electrical Transformer

Figure 5.3

Examples of Electrical Transformers

Thus the transformer can be described either by both equations (5.15a) and (5.15b) or by equation (5.16) and either equation (5.15a) or (5.15b). Note that equations (5.15) and (5.16) presume a sign convention for variables such that power is input at one port and output at the other port.

Examples of Transformers   There are many examples of transformers in engineering. The electrical transformer of Fig. 5.3a usually comes to mind when transformers are mentioned. Typically, it consists of two coils of N_1 and N_2 turns wound on the same core of large (magnetic) *permeability,* such as one made from a *ferromagnetic* material. One coil is termed the *primary* coil, and the other is called the *secondary* coil. Ideally the *flux linkage* λ in each coil is proportional to the number of turns N in the coil, so that

$$\frac{\lambda_1}{\lambda_2} = \frac{N_1}{N_2} = m; \quad \text{(turns ratio)} \tag{5.17a}$$

But the flux linkage in an element of inductance L through which current i is flowing is defined by (see also equation (5.41) of Section 5.2)

$$\lambda = Li; \quad \text{(flux linkage)} \tag{5.17b}$$

so that by equations (2.25b) and (5.17a),

$$\frac{V_1}{V_2} = \frac{N_1}{N_2} \quad \text{or} \quad V_1 = \left(\frac{N_1}{N_2}\right) V_2 = mV_2 \tag{5.18a}$$

For the ideal transformer being described, the power balance equation (5.16) applies, so that

$$i_1 = \left(\frac{N_2}{N_1}\right) i_2 \tag{5.18b}$$

Thus m for an ideal electrical transformer is the *turns ratio* (N_1/N_2).

In a real transformer, not all the flux that links coil 1 also links coil 2, as there is some flux leakage from the core. Also, the coils have finite resistances. A real transformer can be modeled as shown in Fig. 5.3b, by separating the nonideal effects into representative elements: coil resistances R_1 and R_2, and inductances L_1 and L_2, to simulate flux leakages. The ideal transformer is represented in Fig. 5.3b by the mutual inductors model of Fig. 2.9g. However, note that the current direction at the right port has been reversed, so that, consistent with equation (5.16), power is output at this port.

Perhaps the example that is closest to an ideal transformer is a light pivoted mechanical lever (Fig. 5.4a). Considering the rotation about the pivot O, we see that $v_1/a = v_2/b$, so that

$$v_1 = \left(\frac{a}{b}\right) v_2 \tag{5.19a}$$

Also, since the lever is assumed massless (light), a moment balance about O gives

$$F_1 a = F_2 b \quad \text{or} \quad F_1 = \left(\frac{b}{a}\right) F_2 = m F_2 \tag{5.19b}$$

Hence for the mechanical lever, $m = (b/a)$ (that is, the arm-length or lever ratio).

The simple gear train of Fig. 5.4b is another mechanical system transformer. If the inertia of the gears, friction between the gear tooth surfaces, and backlash can all be neglected or separated into other components, as was done for the electrical transformer, then the behavior of the ideal gear train is given by

$$T_1 = \left(\frac{r_1}{r_2}\right) T_2 = \left(\frac{N_1}{N_2}\right) T_2 = m T_2 \tag{5.20a}$$

and

$$\omega_1 = \left(\frac{N_2}{N_1}\right) \omega_2 \tag{5.20b}$$

where r_1 and r_2 are the pitch radii of the gears, and N_1 and N_2 are the numbers of teeth in the gears. The quantity $m = (N_1/N_2)$ this time is called the *gear ratio* or the *velocity ratio*.

The Gyrator Element   The constitutive relation for the gyrator (*GY*-element) is

$$e_1 = r_G f_2 \quad \text{(gyrator)} \tag{5.21a}$$

$$f_1 = \frac{1}{r_G} e_2 \quad \text{(gyrator)} \tag{5.21b}$$

Equations (5.21a) and (5.21b) also imply a power balance for the ideal gyrator. Because of the "crossed" form of the relation between the variables in equations (5.21), different forms of behavior can, in principle, be obtained when gyrators are combined with other elements or other gyrators. For example, two gyrators in cascade

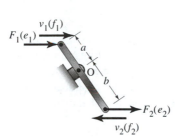

(a)   Mechanical Lever

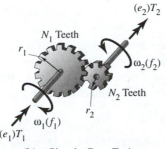

(b)   Simple Gear Train

Figure 5.4

Examples of Mechanical Transformers

$$e_1 = r_{G1} f_2; \qquad f_2 = \frac{1}{r_{G2}} e_3; \qquad \Rightarrow e_1 = \frac{r_{G1}}{r_{G2}} e_3 = m e_3:$$

$$f_1 = \frac{1}{r_{G1}} e_2; \qquad e_2 = r_{G2} f_3; \qquad \Rightarrow f_1 = \frac{r_{G2}}{r_{G1}} f_3 = \frac{1}{m} f_3$$

yield, overall, the transformer behavior. A shunt capacitance element and a gyrator,

$$e_1 = \int^t \left(\frac{1}{C}\right) f_1 \, dt; \qquad e_1 = rf_2; \qquad \Rightarrow rf_2 = \int^t \left(\frac{1}{C}\right) f_1 \, dt$$

$$f_1 = \frac{1}{r} e_2; \quad \Rightarrow rf_2 = \int^t \left(\frac{1}{rC}\right) e_2 \, dt; \quad \Rightarrow f_2 = \int^t \left(\frac{1}{r^2 C}\right) e_2 \, dt = \int^t \left(\frac{1}{I}\right) e_2 \, dt$$

give a relation between the variables at the output port equivalent to that of a shunt inertance.

Example of a Gyrator   The spinning symmetric rotor of Fig. 3.8 is perhaps the best known example of a gyrator. Consider the gyroscope shown in Fig. 5.5 with a prescribed steady spin p, rotor moment of inertia about the spin axis J, and frictionless bearings. If a force F or moment $T_1 = Fd$, is applied as shown, the resulting steady precession ω_2 is given by equation (3.13), restated,

$$T_1 = (Jp)\omega_2 \tag{5.22a}$$

Thus from equation (5.21a),

$$r_G = Jp; \quad \text{(ideal gyroscope)} \tag{5.22b}$$

Equation (5.21b) is automatically satisfied, since, in this ideal example, ω_1 and T_2 are both zero (see the development leading to equation (3.13)).

Transducers

The term *transducer* is often used in general to mean energy converter. Transducers here refer to two-port elements in which different forms of power are transmitted at the ports. Since power is the rate of change of energy, it follows that the energy forms associated with each port or with the "input" and "output" ports of a transducer are different, hence the term *energy converter*. The constitutive behavior of the transducer (*TD*-element) is given by

$$e_1 = \frac{1}{r_T} f_2; \quad \text{(transducer)} \tag{5.23a}$$

$$f_1 = r_T e_2; \quad \text{(transducer)} \tag{5.23b}$$

Mathematically, equations (5.21) and (5.23) represent the same behavior, so that the *essential* difference between the transducer and the gyrator is that the power transmitted at the two ports of a transducer is *understood* to be of different forms, otherwise the generalized variables e and f, by themselves, do not differentiate between system types or power forms.

Examples of Transducers   Perhaps the best known transducers are the electric motor and generator for which the associated power forms are electrical (Vi) and mechanical (usually $T\omega$). the electrical motor is said to convert electric energy into (usually rotational) mechanical energy, while the electrical generator produces electric power from mechanical power. The *direct current* (DC) motor whose circuit diagram is shown in Fig. 5.6a is considered in more detail in Section 5.2, along with some other electrome-

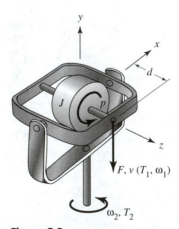

Figure 5.5

Example of a Gyrator: An Ideal Gyroscope

Figure 5.6

Two Examples of Transducers

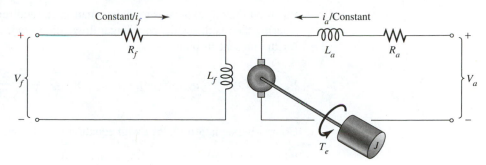

(a) DC Motor under Armature/Field Control (Circuit Diagram)

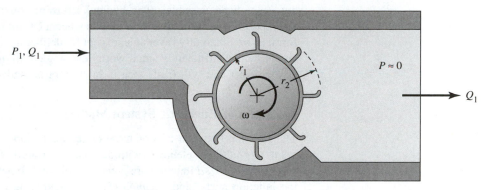

(b) Hydraulic Turbine (Schematic Diagram)

chanical transducers. The torque T_e developed by the ideal DC motor (excluding damping and inertia) can be written,

$$T_e = (K_1 i_f) i_a; \quad \text{(armature control)} \tag{5.24}$$

or

$$T_e = (K_1 i_a) i_f; \quad \text{(field control)} \tag{5.25}$$

where K_1 is a constant, and i_f and i_a are, respectively, the constant field current and armature current (equation 5.24)) or the field current and constant armature supply current (equation 5.25)), depending on whether *armature* or *field control* is being used. Thus by equation (5.23b), with $T_e = e_2$, $r = (1/K_1 i_f)$ for armature control and $r = (1/K_1 i_a)$ for field control.

The *hydraulic turbine* shown schematically in Fig. 5.6b converts fluid-mechanical power (PQ) into rotary-mechanical (or shaft) power ($T\omega$). Consider the case of an ideal turbine and incompressible working fluid. We shall assume, without loss of generality, complete expansion in the turbine, and fluid entry tangent to the rotor blades. Thus the exit fluid pressure is $P_2 = 0$ (gage), and the tangential component of the fluid velocity at the impeller entrance is $v_{t1} = v_1$. The work output per unit mass of fluid is given by equation (4.26b):

$$w_{\text{shaft}} = -\int_{P_1}^{0} \left(\frac{1}{\rho}\right) dP = \left(\frac{1}{\rho}\right) P_1$$

so that the mechanical power output is

$$\dot{m} w_{\text{shaft}} = \rho Q_1 w_{\text{shaft}} = P_1 Q_1 = T_2 \omega_2 \tag{5.26a}$$

Equation (5.26a) expresses a power balance, as required for the transducer. Also, considering the result for the centrifugal-flow compressor, equation (4.27c), we can write for the present turbine,

$$\dot{m}SFr_2^2\omega_2^2 = \rho Q_1 SFr_2^2\omega_2^2 = T_2\omega_2 \quad \text{or} \quad Q_1 = \left(\frac{1}{\rho SFr_2^2\omega_2}\right)T_2 \tag{5.26b}$$

Comparing equation (5.26b) with equation (5.23b), we can set $r_T = \left[\dfrac{1}{\rho SFr_2^2\omega_2}\right]$. The three generalized elements—transformers, gyrators, and transducers—are characterized by *reversible causality* (recall the discussion in Section 3.1). This should be expected, since the behavior of these elements is in each case described by static relations among the generalized variables. The hydraulic turbine, on reversal, yields a pump of the *rotary-vane* type; and a generator results when a DC motor is operated in reverse. Transformers are inherently reversible, except by deliberate design. A worm gear set, for example, with a sufficiently small worm lead angle will normally operate with the worm as the driver and will (self) lock if an attempt is made to use the gear as the driver.

Examples of Energetic System Models

The junction between two or more components can itself be viewed as another element—a multiport element (with at least two ports). Two models of the behavior of such a *generalized* junction are *common-flow* or 1-Junction (the flow is the same at all the junction ports) and *common-effort* or 0-Junction (the effort is identical at all the ports). That is, for a three-port junction, for example,

$$f_1 = f_2 = f_3; \quad \text{(1-junction)} \tag{5.27}$$

$$e_1 = e_2 = e_3; \quad \text{(0-junction)} \tag{5.28}$$

In addition to equation (5.27) or (5.28), power conservation is fundamental to all junctions:

$$e_1f_1 + e_2f_2 + e_3f_3 = 0; \quad \text{(power balance; power input at all three ports)} \tag{5.29}$$

Equation (5.29) assumes that power is input at all the ports. Actually, since a junction neither creates nor consumes energy, power must be input on at least one port and output on at least one other port. In general, the representation of a junction should include an indication of the assumed power flow directions. Power flow at a port is in the assumed direction if the effort and flow are both positive.

| **Example 5.1** | Model of a Real Electrical Transformer |

Determine a model for the real electrical transformer of Fig. 5.3b that can predict the voltage V_2, given the voltage V_1 and the (ideal transformer) turns ratio m.

Solution: The circuit diagram is an adequate physical representation of an electrical system. For completeness, note that in Fig. 5.3b the current directions are the assumed power flow directions at every port where the voltage is above the reference potential. V_1 represents an effort source S_e with $e = V_1$; an 0-Junction can be assumed between this source and the resistance R_1. Between R_1 and the inertance L_1 a 1-Junction is implied. The right half of the circuit can be interpreted similarly, except that for the purpose of the present exercise, a load R_L or energy sink can be assumed as the appropriate right-end element.

System model: Equations (5.1) and (5.3) or the appropriate equations from Chapter 2 give

$$R_1 i_a + L_1 \frac{di_a}{dt} = V_1 - V_a \tag{a}$$

$$R_2 i_b + L_2 \frac{di_b}{dt} = V_b - V_2 \tag{b}$$

The ideal transformer equation (5.15a) and the power balance equation (5.16) imply

$$V_a = mV_b \quad \text{and} \quad i_b = mi_a \tag{c}$$

so that equation (b) can be solved for V_2 in terms of i_a and V_a:

$$V_2 = \frac{V_a}{m} - mR_2 i_a - mL_2 \frac{di_a}{dt} \tag{d}$$

However, V_a is still unknown. With the new direction of i_b, the mutual inductors model (Fig. 2.9g) is

$$V_a = L_a \frac{d}{dt} i_a - M \frac{d}{dt} i_b \tag{e}$$

$$V_b = -L_b \frac{d}{dt} i_b + M \frac{d}{dt} i_a \tag{f}$$

Using equations (c), we can determine M in terms of m, L_a, and L_b from equations (e) and (f):

$$M = \frac{(L_a + m^2 L_b)}{2m} \tag{5.30}$$

so that V_a can be given by equation (e). The final model equations obtained from these results are

$$\text{System:} \quad \frac{d}{dt} i_a = \left[\frac{2}{2L_1 + L_a - m^2 L_b} \right] (V_1 - R_1 i_a) \tag{5.31a}$$

$$\text{Output:} \quad V_2 = \left[\frac{L_a - m^2 L_b - 2m^2 L_2}{m(2L_1 + L_a - m^2 L_b)} \right] (V_1 - R_1 i_a) - mR_2 i_a \tag{5.31b}$$

Example 5.2	Analogous Model of a Mechanical Grain Scale

Develop an analogous circuit diagram or generalized physical model of the grain scale of Example 3.4 (see Fig. 3.5a). Assume, as in that example, that the indicating wheel is driven by a rack and pinion gear arrangement and that the motion of the weigh platform is viscously damped. Show that the resulting system model is the same as was determined in Example 3.4.

Solution: As in Example 3.4, we shall assume a coordinate system in which displacement is positive downward and the rotation of the pinion is positive counterclockwise.

Decomposition: From Fig. 3.5a (refer also to Table 5.1), we can directly identify the following elements: the lumped mass (m_1) of material being weighed, the platform, and the rack constitute an inertance I_1-element. Viscous damping (β) in the platform column gives an R-element. The return spring ($1/k$) is a C-element. The indicating wheel ($J_2 = \frac{1}{2}m_2r_2^2$) constitutes a second inertance I_2-element. The rack and pinion gear set (excluding the inertia components that have been separated out) constitute a transformer or TF-element. The gear set transforms the linear motion of the platform column v_1 into the rotary motion of the indicating wheel ω_2. Since $v_1 = r_2\omega_2$, by equation (5.15b), $m = 1/r_2$ for the transformer.

The physical model shown in Fig. 5.7 is developed as follows: The absence of any restraint to motion at the top of the platform can be viewed as a zero-force input (boundary). However, the gravitational body force on m_1 gives an effort (voltage) source S_e (with $e = m_1g$ relative to ground) at this end. The input energy is reduced by the viscous damping, for which the relative velocity is v_1, the velocity of m_1 and of the rack (the damped motion of m_1 is relative to fixed guides), hence R is (in series with) connected by 1-Junction [d] to I_1, which is in series (1-Junction [a]) with the rack (transformer input segment). m_1 is also subject to (opposed by) a spring force. The velocity of the upper end of the spring is the same as v_1, hence the 1-Junction [c]. However the lower end the spring is fixed, giving the shunt-C-to-ground arrangement shown. The "voltage" drop (force change) across the transformer input port is identified as ΔF_{rack}, and the corresponding torque change at the output port is ΔT_{pinion}.

System model: Directly using the system variables, also indicated in Fig. 5.7, rather than those of the analogous circuit, we easily obtain the following loop equations:

$$m_1g - \beta v_1 - m_1 \frac{d}{dt} v_1 - \Delta F_{rack} - kx_1 = 0; \qquad \frac{d}{dt} x_1 = v_1 \tag{a}$$

$$\Delta T_{pinion} - J_2 \frac{d}{dt} (\omega_2) = 0 \tag{b}$$

Figure 5.7

Analogous Physical Model of a Mechanical Grain Scale

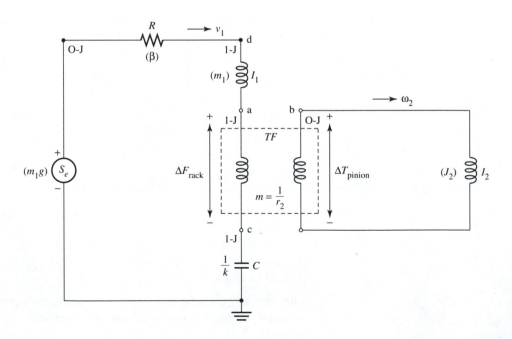

Note that we write the spring force kx_1 directly rather than use the integral formula for the voltage drop across the capacitor. The ideal transformer equations (5.15a) and (5.15b) yield

$$\Delta F_{\text{rack}} = \frac{1}{r_2} \Delta T_{\text{pinion}}; \qquad v_1 = r_2\omega_2 \tag{c}$$

Equations (b) and (c) further give

$$\Delta F_{\text{rack}} = \frac{1}{r_2} J_2 \frac{d}{dt} \omega_2 = \frac{1}{2} m_2 r_2 \frac{d}{dt} \omega_2 \tag{5.32}$$

Substituting equation (5.32) in (a) and using $v_1 = r_2\omega_2$ and $x_1 = r_2\theta_2$ to replace v_1 and x_1, we obtain the state model:

$$\frac{d}{dt}\theta_2 = \omega_2 \tag{5.33}$$

$$\frac{d}{dt}\omega_2 = \left[\frac{2}{r_2(2m_1 + m_2)} \right] (m_1 g - \beta r_2\omega_2 - kr_2\theta_2)$$

Which is the same as equation (3.9a) of Example 3.4. The output is $y = \theta_2$, as before.

Note:

The transformer in the preceding physical model could have been replaced by considering only the left or translational circuit with a dependent effort source

$$\Delta F_{\text{rack}} = \frac{1}{2} m_2 r_2 \frac{d}{dt}\omega_2 = \left(\frac{1}{2} m_2 \right) \frac{d}{dt} v_1$$

or an additional inductor I_r in series with I_1:

$$I_r = \frac{1}{2} m_2$$

In *motion control* practice, one of the traditional ways of accounting for this sort of influence across a transformer is through what is known as *reflected inertia*. For instance, in Example 3.4, equation (a), the contact force F_{cx} would have been omitted, the base inertia would be m_1, and the reflected inertia, that is, the contribution across the transformer by m_2 would have been $\frac{1}{2}m_2$. This approach presumes that the presence of the reflected inertia is understood and the process of determining it without constructing a model is known. Also nonlinear effects such as backlash and stiction may not be easily "reflected" without considerable insight into the problem.

....

5.2 ELECTROMECHANICAL SYSTEMS

Electromechanical systems are a special class of energetic systems in which the two predominant energy or power forms are electrical (that is, voltage times current) and mechanical (that is, force times velocity). Such electromechanical *energy converters* or transducers (recall transducers transform energy or power from one form into another) are widely distributed in engineering, particularly in instrumentation and control systems. Their applications range from a simple loudspeaker to sophisticated motion control by a stepping or linear motor.

Electrical Transducers

Although the term *electromechanical system* is, in practice, usually applied to devices in which *mechanical motion* takes place, either because of applied *mechanical forces* or due to *forces* of electrical (usually *electromagnetic*) origin, electromechanical system can be broadly interpreted to include more general forms of mechanical power—pressure times volume flowrate *(electrohydraulic systems)* and heat flowrate *(electrothermal systems)*. There are many such devices or electrical transducers in which one of the energy or power forms is electrical while the other is mechanical. For instance, the potentiometer (which is discussed below) can be used to obtain a voltage that is proportional to a mechanical displacement. A strain gage is another example of a transducer based on the resistive phenomenon. Capacitance probes also measure displacement, by allowing the displacement to either change the separation of the two plates of a parallel-plate capacitor or the relative proportion of two dielectric media with different dielectric constants while the plates remain stationary. In the latter case, the component suffering the displacement does not have to be conductive. Piezoelectric crystals develop a voltage when subjected to mechanical stress. Piezoelectric vibrators are made from plates cut from piezoelectric crystals such that when an alternating voltage of the proper frequency is applied to the arrangement, it vibrates at resonance. Similarly, an electric motor (DC motors in particular are discussed below) will convert electric power into mechanical power. A thermocouple is an application of thermoelectric phenomena in which heat is transformed to electric power. The particular behavior involved here is the Seebeck effect (see [13]). Other physical effects exploited by electrical transducers include electromagnetic (discussed further in this section), inductive, photoelectric, photovoltaic, and ionization. Figure 5.13 presents a summary of some of the electrical transducers along with the idealized models. Note that in the thermocouple circuit, an electronic circuit can be used in place of the potentiometer to dynamically generate the balancing voltage V so that a continuous measurement is possible.

The Potentiometer Example

The potentiometers shown in Figs. 5.8a and 5.8b (examples of physical potentiometers are also shown in Fig. 5.9) are applications of the *voltage divider* behavior (see Fig. 2.16a) determined in Example 2.4, equation (2.35a). The linear slide potentiometer (Figs. 5.8a and 5.9a), for example, is used to measure linear displacement (*x*). A wiper arm that is mechanically coupled to the moving component is in sliding but electrical contact with a resistor that is often a long winding of bare wire or a strip of conductive plastic. Because the resistance of the material is proportional to its length, equation (2.35a) gives

$$V_{\text{out}} = \left(\frac{R_2}{R_1 + R_2}\right) V_{\text{in}} = \left(\frac{R_2}{R_T}\right) V_{\text{in}} = \left(\frac{x}{L}\right) V_{\text{in}} = K V_{\text{in}} \qquad \textbf{(5.34a)}$$

where R_T is the total resistance. The motion of the wiper thus results in a variation of *x* (or R_2) and hence V_{out} as given by equation (5.34a). If the output voltage can be approximately differentiated (why approximately?), then the potentiometer can also indicate velocity. The arrangement shown in Fig. 5.8b (see also Figs. 5.9b and 5.9c) operates on the same principle as described except that the wiper rotates over a circular resistance element. Thus the rotary potentiometer is used to measure angular displacement θ. Potentiometers or *pots* used in analog circuits (some radio sets, for example)

Figure 5.8

Examples of Electrical Transducers

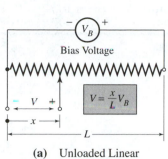

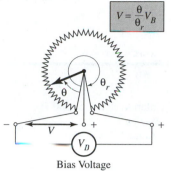

$$V = \frac{\theta}{\theta_r} V_B$$

Bias Voltage

(a) Unloaded Linear Potentiometer

(b) Unloaded Rotary Potentiometer

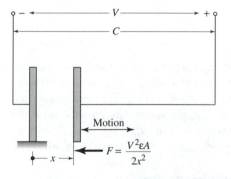

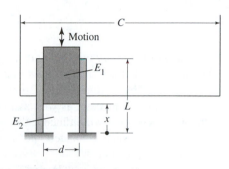

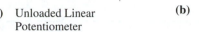

$$F = \frac{V^2 \varepsilon A}{2x^2}$$

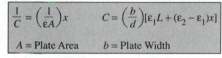

$$\frac{1}{C} = \left(\frac{1}{\varepsilon A}\right)x \qquad C = \left(\frac{b}{d}\right)[\varepsilon_1 L + (\varepsilon_2 - \varepsilon_1)x]$$

$A =$ Plate Area $\qquad b =$ Plate Width

(c) Parallel-plate Probe Examples

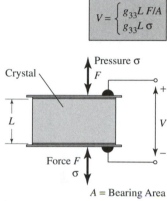

$$V = \begin{cases} g_{33} L\, F/A \\ g_{33} L\, \sigma \end{cases}$$

Pressure σ

Crystal

$A =$ Bearing Area

(d) Piezoelectric Element

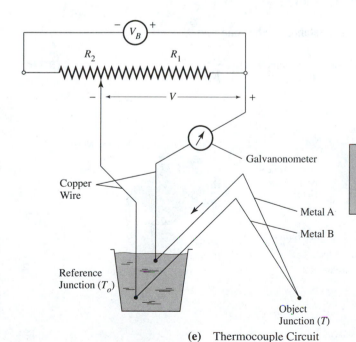

Galvanonometer

Copper Wire

Metal A

Metal B

Reference Junction (T_o)

Object Junction (T)

(e) Thermocouple Circuit

Potentiometer at Balance

$$V = (\alpha_A - \alpha_B)(T - T_o)$$
$$= \alpha_{AB}(T - T_o)$$
$$\alpha = \text{Seeback Coefficient}$$

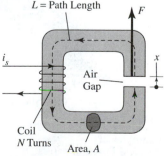

$$F = -\frac{1}{2}i_s^2 \frac{dL}{dx}$$

$$L = AN^2\left[\frac{\mu_c \mu_o}{L\mu_o + x\mu_c}\right]$$

$$L \approx AN^2\left(\frac{\mu_o}{x}\right)$$

$$F \approx \frac{A}{2}\mu_o\left(\frac{Ni_s}{x}\right)^2$$

$L =$ Path Length

Air Gap

Coil N Turns

Area, A

(f) Electromagnetic Element

EXAMPLE MATERIAL	ϵ PERMITTIVITY (F/m)	α SEEBECK COEFFICIENT (V/K) $\times 10^{-3}$	G_{33} COUPLING COEFFICIENT (V $\cdot$ m/N) $\times 10^{-3}$	μ PERMEABILITY (H/m)
Air	1.0006			1.257×10^{-6}
Wax Paper	1.75			
Polyvinyl Chloride (PVC, at 100 Hz)	6.5			
Iron − Constantan		0.05		6.53×10^{-3} (Iron)
Chromel − Alumel		0.04		
Steel				8.8×10^{-3}
Ceramic "B"			14	
PZT-4 Ceramic			24.9	

Figure 5.8, continued

Examples of
Electrical Transducers

are usually of the rotary type but with a scale on the knob. Here the purpose is to obtain a variable V_{out} that is a fraction (less than 1) of another variable, V_{in}. The desired fraction is obtained by "setting the pot," that is, rotating the knob (wiper) to the appropriate scale position. In this case, as well as in the other applications, any circuit connected to the wiper terminals will influence the actual position of the knob for the desired fraction. In practice, such a "loaded" pot is set by monitoring V_{out} without regard to the eventual knob position.

The behavior of a "loaded" voltage divider circuit was determined in Example 2.12, equation (2.47). In terms of the total resistance of the potentiometer R_T and the pot knob scale setting K, equation (2.47) can be written as

$$V_{\text{out}} = \left(\frac{K}{1 + (R_2/R_L)(1 - K)} \right) V_{\text{in}} \tag{5.34b}$$

where R_L is the equivalent resistance of the load and

$$K = \frac{x}{L} \text{ (linear pot)} \quad \text{or} \quad \frac{\theta}{\theta_r} \text{ (rotary pot)} = \frac{R_2}{R_T} \tag{5.34c}$$

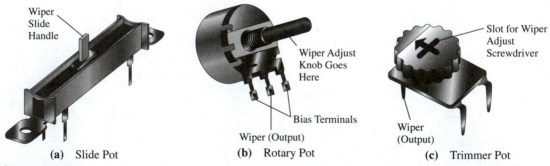

Wiper
Slide
Handle

(a) Slide Pot

Wiper Adjust
Knob Goes
Here

Bias Terminals

Wiper (Output)

(b) Rotary Pot

Slot for Wiper
Adjust
Screwdriver

Wiper
(Output)

(c) Trimmer Pot

Figure 5.9

Examples of
Physical Potentiometers

It is clear from equation (5.34b) that substantial error can be made by just setting the pot (K) to the desired fraction especially when R_L is finite (of the order of R_2).

The potentiometer as just described is not suitable for measuring displacement where contact with the moving element is undesirable (gas bearings for example) or where high frequencies (or velocities) are involved, such as in mechanical vibration studies. Here, the inertia of the wiper arm and possible frictional forces can severely distort the measured signal. In both cases the capacitance probe or even a fiber optic (given proper moving surface texture) probe can be more useful.

In the balance of this section we shall consider strictly electromechanical transducers, particularly those involving electromagnetic phenomena. We shall briefly examine the nature of electromagnetic forces and torques, and we shall develop the dynamic model of a simple electromagnetic relay system that represents a typical application of such forces. The control of DC motors will be explored in this context, and the operation of stepping motors and their controllers will be discussed.

Introducing Electrome-chanical Energy Con-version Principles

Our approach to modeling electromechanical systems is to separate the system into an essential *energy converter* or transducer that is lossless, and electrical and mechanical subsystems coupled by the ideal transducer. Since we already know how to model each of the subsystems, the entire electromechanical system can be described once the transducer relation is determined. By handling nonideal effects (losses) in the subsystems, we somewhat simplify the description of the transducer. This approach is valid as long as any loss mechanism associated with the conversion process itself can be neglected or transferred to the subsystems. It is also consistent with our concept of the generalized transducer or gyrator element, provided any energy storage feature or dynamic behavior in the converter is negligible or appropriately accounted for in another component.

Figure 5.10 shows a model of an electromechanical system based on this approach. Recall the first law of thermodynamics, equation (4.16a). To apply it to the electromechanical system, we expand the work term to include mechanical *(shaft)* work and electrical work. As illustrated in Fig. 5.10, the electrical power input to the conservative subsystem or transducer is Vi. This is the negative of the rate of change of the electrical work done by the transducer $(V_b i)$. Similarly, the mechanical power input to the transducer is Fv, and it is the negative of the rate of change of mechanical

Figure 5.10

Energy Model of an Electromechanical System

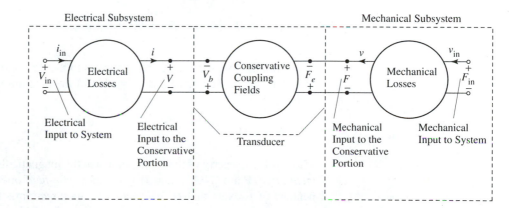

work done by the transducer $F_e v$. For the ideal transducer implied, the heat transfer is zero. Thus the first law for the transducer is

$$\frac{d}{dt} E_{\text{em}} = -V_b i - F_e v = Vi - F_e v = Vi + Fv = -V_b i + Fv \qquad \textbf{(5.35a)}$$

or

$$\frac{d}{dt} E_{\text{em}} = Vi - F_e v \qquad \textbf{(5.35b)}$$

where E_{em} is the energy stored in the coupling fields. In terms of changes in stored energy (that is, equation (4.4a) applied to the system), equation (5.35b) gives

$$dE_{\text{em}} = Vi \, dt - F_e \, dx; \qquad \frac{dx}{dt} = v \qquad \textbf{(5.35c)}$$

Equation (5.35c) will be our basis for determining the relation for the transducer (or the gyrator that we shall use to represent it). Thus the ability to describe forces (F_e) caused by electrical (especially electromagnetic) effects and (magnetically) generated voltages (V_b) due to mechanical motion is essential to modeling electromechanical systems.

Magnetic Fields and Reluctance

Two electrical behaviors are fundamental to the description of electromagnetic forces and magnetically induced voltages. The first is the concept of magnetic fields and circuits and *Lorentz force law,* and the second is *Faraday's law.* Magnetic fields are produced by moving electric charges and by currents in conductors. A measure of the strength of the magnetic field around a conductor is the *magnetic-flux density* **B** (Wb/m^2 or A $\cdot$ H/m^2 or V $\cdot$ s/m^2). The magnetic-flux density is a vector in a plane perpendicular to the conductor. Its direction can be determined by the following *right-hand rule* (#1):

> Grasp the conductor with the right hand so that the thumb points along the conductor in the direction of the current. The magnetic flux encircles the conductor in the direction of the fingers

The *magnetic flux* ϕ (Wb) is the integral of magnetic-flux density over the area through which the flux passes:

$$\phi = \int_A \boldsymbol{B} \cdot d\boldsymbol{A} = \int_A \boldsymbol{B} \cdot \boldsymbol{n} \, dA; \quad \text{(magnetic flux)} \qquad \textbf{(5.36a)}$$

where **n** is an (outward) unit vector normal to the differential surface area dA. The effective magnetic-flux density is a function of the medium through which the magnetic flux passes:

$$\mathbf{B} = \mu \mathbf{H} \qquad \textbf{(5.36b)}$$

where μ is a property of the region in which the magnetic field exists known as the *permeability* (H/m or N/A^2), and **H** is the *magnetic-field intensity* (A/m), which is independent of μ. Note that **H** is a vector and its direction is the same as that of **B**.

According to **Ampere's law,** the increment of magnetic-flux density $d\mathbf{B}$ at a point P due to the differential length dl of a conductor carrying current i is inversely proportional to the square of the distance r between dl and P and directly proportional to the current:

$$d\mathbf{B} = \left(\frac{\mu i}{4\pi r^3}\right) d\mathbf{l} \times \mathbf{r} \tag{5.36c}$$

where **r** is a position vector directed from dl to point P, and **dl** is a directed line segment (vector) along the conductor in the direction of the current.

Concept of Magnetic Circuit and Reluctance   The *toroid* with N turns of a current-carrying winding distributed around it, as shown in Fig. 5.11a, is our prototype simplified model of a structure composed, in the main, of high-permeability material. The effect of high permeability of the structural material is that the magnetic field is contained almost totally in the core, that is, the magnetic flux is confined to the path defined by the structure. Equation (5.36c) can be used to determine the flux density at any point within the toroid. The result (see [14]) is

$$B = \frac{\mu}{2\pi a}[Ni] = \frac{\mu}{l}[Ni] \tag{5.37a}$$

$$\text{or } H = \frac{1}{l}[Ni] \tag{5.37b}$$

where l is the mean length of the flux path. The direction of B or H is as indicated in Fig. 5.11a (that is, for ϕ; on account of equation (5.36a), we shall generally attribute a direction to ϕ that is the same as that of B, although ϕ is a scalar quantity).

If the cross-sectional area of the core is A, equation (5.36a) gives for the toroid $\phi = BA$, hence

$$\phi = \left(\frac{\mu A}{l}\right)[Ni] = \frac{1}{\Re}[Ni] \tag{5.37c}$$

The quantity Ni is known as the *magnetomotive force* (mmf). Further, the closed path followed by the magnetic field around the coil can be considered a *magnetic circuit.* By analogy with an electrical circuit, if ϕ and $[Ni]$ are analogous to the

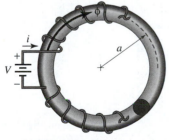

(a)   Toroid with N Turns

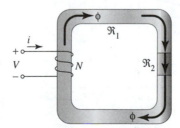

(b)   Series Magnetic Circuit

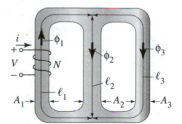

(c)   Parallel Magnetic Circuit

Figure 5.11

Magnetic Circuit Concepts

current and voltage, respectively, then equation (5.37c) represents "Ohm's law" for the magnetic circuit and defines a "resistance" to magnetic flux, known as the *reluctance* of the magnetic circuit $\Re$. Figures 5.11b and 5.11c show some of the implications of this analogy.

Example 5.3   Magnetic Circuit with Air Gap

The reluctance of many magnetic circuit structures is unfortunately not a constant because the permeability of most materials is a function of the magnetomotive force. However, this is not the case for air (the permeability of free space is constant at $4\pi \times 10^{-7}$ H/m). Also, as illustrated in Fig. 5.12a, for some materials of interest here (sheet steel, for example) if the circuit conditions are such that the material is *not saturated*, then the permeability and hence the reluctance (for fixed cross section and path length) may be taken as constants. We shall generally assume unsaturated conditions and ignore other nonlinear effects in the magnetic properties, such as *hysteresis* (not shown in the figure).

Consider now the magnetic circuit with air gap shown in Fig. 5.12b. The core material is laminated steel. The cross-sectional area of each section is the same, $A_c = 10$ cm^2, and the lengths of the sections are as indicated in the figure. The air gap is $x_g = 0.5$ cm. The winding of 250 turns is carrying 2 A. Find the air-gap magnetic flux.

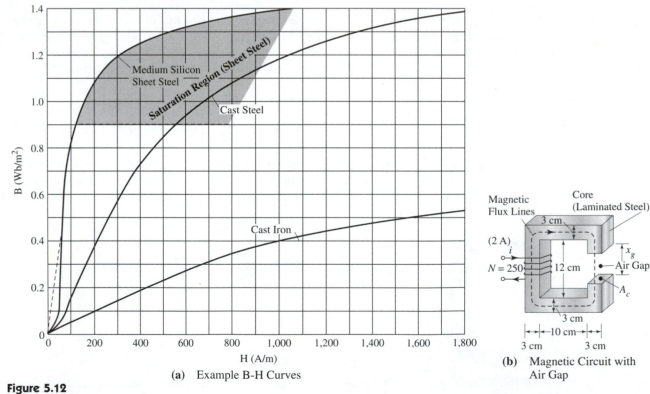

Figure 5.12

Example B-H Curves and Magnetic Circuit

(a)   Example B-H Curves

(b)   Magnetic Circuit with Air Gap

Solution: We assume that the air gap length is small enough that the magnetic flux remains continuously constrained inside the core and through the air gap. Thus the air gap constitutes a series reluctance of the same (neglecting fringing effects, see [15]) cross-sectional area A_c. The mean length of the flux path in the gap is the length of the gap x_g. Within the core, the mean length is

$$l_c = \{2\,[10 + 3) + (12 + 3)] - 0.5\} \times 10^{-2} \text{ m} = 0.555 \text{ m}$$

The total mmf is

$$[Ni] = 250(2) = 500 \text{ A}$$

Equation (5.37c) gives

$$\phi\mathfrak{R} = [Ni] = \phi\,(\mathfrak{R}_c + \mathfrak{R}_g) = \phi\left(\frac{l_c}{\mu_c A_c} + \frac{x_g}{\mu_o A_c}\right) \tag{5.38a}$$

Assuming medium silicon sheet steel property values and operation in the unsaturated region (this will be checked), from Fig. 5.12a, we take

$$\mu_c \cong \frac{0.8}{100} = 0.008 \text{ (Wb/A} \cdot \text{m)}$$

Substituting in (5.38a), we obtain

$$\phi\left[\frac{0.555}{(0.008)(0.001)} + \frac{0.005}{(4\pi \times 10^{-7})(0.001)}\right] = \phi(6.9375 \times 10^4 + 3.9789 \times 10^6)$$

$$= \phi(4.0483 \times 10^6) = [Ni] = 500$$

which gives $\phi = 1.235 \times 10^{-4}$ Wb.

The mmf drop in the core is $(1.235 \times 10^{-4})(6.9375 \times 10^4) \doteq 8.57$ A. The corresponding magnetic field intensity is $H = 8.57/0.555 = 15.44$ A/m, only, so that the assumption of unsaturated operation is acceptable (see Fig. 5.12a).

Observe that the air gap reluctance (3.9789×10^6) is very large relative to that of the high-permeability material (6.9375×10^4). It is about 57 times larger in this case. Thus it is possible in many problems to neglect the reluctance of most high-permeability components relative to the reluctance of the air gap component. Under this circumstance, equation (5.38a) becomes

$$\phi\left(\frac{x_g}{\mu_o A_c}\right) \approx Ni \tag{5.38b}$$

....

Lorentz Force Law

The force **F** (N) on a point charge q (C) moving in a magnetic field (the electric field is ignored here) is given by the *Lorentz law:*

$$\mathbf{F} = q\,(\mathbf{E} + \mathbf{v} \times \mathbf{B}) = q\,(\mathbf{v} \times \mathbf{B}) \tag{5.39a}$$

where **v** (m/s) is the velocity of the point charge relative to the magnetic field. From equation (5.39a), the differential (magnetic) force (dF) acting on a differential element (dl) of a conductor in which current i is flowing is

$$d\mathbf{F} = i(\mathbf{dl} \times \mathbf{B})$$ (5.39b)

where **dl** is a directed line segment (vector) along the conductor in the direction of the current, and **B** is, as before, the magnetic-flux density (Wb/m²). The direction of the force **F** can be established by another *right-hand rule* (#2):

> Align the thumb, index finger, and third finger of the right hand so that they are mutually at right angles with each other, the index finger is in the direction of the current, and the third finger is in the direction of the magnetic-flux density. The thumb will then be in the direction of the force

Equation (5.39b) gives the force acting on the conducting material itself and explains the nature of magnetic field forces. For example, if the magnetic-flux density is zero, then no force is produced. Equation (5.39b) is otherwise not used directly in the following developments.

Faraday's Law for Electromotive Force

An electric field is produced in space when a magnetic field varies with time (see [15]). The *induced voltage* or the *electromotive force* (emf) across the terminals of a coil due to a time-varying magnetic field in the core is given by *Faraday's law:*

$$V = N \frac{d}{dt} \phi = \frac{d}{dt} \lambda; \qquad \lambda = N\phi \quad \text{[Faraday's law]}$$ (5.40)

where λ is known as the *flux linkage* of the coil and N is the number of turns of the coil, as before. Equation (5.40) applies only when every winding is linked by flux ϕ. The induced voltage V is in a direction that opposes the change in flux linkage. For a *linear* magnetic circuit (constant permeability or operation in unsaturated region), the relation between the flux linkage and the current can be established from the inductive behavior of the circuit. From equation (2.25b),

$$L \frac{d}{dt} i = V = \frac{d}{dt} \lambda$$

so that

$$L = \frac{\lambda}{i} = \frac{N\phi}{i}$$ (5.41)

Faraday's law equation (5.40) is applicable to the voltage drop across the electrical input terminals of a magnetic-circuit-based transducer. Thus in equation (5.35c) we can take

$$V\, dt = i \left(\frac{d}{dt} \lambda \right) dt = i\, d\lambda$$

so that equation (5.35c) becomes

$$dE_{em} = i\,d\lambda - F_e\,dx \tag{5.42a}$$

Equations (5.42a) and (5.41) imply that E_{em} is a unique function of two independent variables, either (i, x) or (λ, x). Because these quantities are independent ("thermodynamic") state properties, E_{em} is a point function, and its value at any state is independent of the path to that state (see [11] or [15]). We can thus integrate equation (5.42a) in the following way, for example:

$$\int_{(0,0)}^{(\lambda, x)} dE_{em} = \left[\int\int_{(0,0)}^{(0,x)} (i\,d\lambda - F_e\,dx) \bigg|_{\lambda = \text{constant} = 0} \right] \tag{5.42b}$$
$$+ \left[\int\int_{(0,x)}^{(\lambda, x)} (i\,d\lambda - F_e\,dx) \bigg|_{x = \text{constant} = x} \right] = \int_0^\lambda i\,d\lambda$$

The final result in equation (5.42b) is obtained because the first term in the middle is zero, since for constant λ, $d\lambda = 0$ and $\lambda = 0$ implies $F_e = 0$ (recall equation (5.39b)); and in the second term, $dx = 0$, since x is held constant. Substituting for i from equation (5.41) gives us the result

$$E_{em}(\lambda, x) = \int_0^\lambda i\,d\lambda = \int_0^\lambda \frac{\lambda}{L}\,d\lambda = \frac{1}{2}\frac{\lambda^2}{L} = \frac{1}{2}\lambda i = \frac{1}{2}Li^2 \tag{5.43a}$$

An alternative expression for E_{em} (see [15]) is in terms of the energy density of the magnetic field integrated over the volume Vol of the magnetic field:

$$E_{em} = \int_{Vol} \frac{1}{2}(\mathbf{H} \cdot \mathbf{B})\,dVol \tag{5.43b}$$

which for a magnetic medium with constant permeability reduces to

$$E_{em} = \int_{Vol} \frac{1}{2}\left(\frac{B^2}{\mu}\right)\,dVol \tag{5.43c}$$

We can also differentiate E_{em} as a function of λ and x, as follows:

$$dE_{em} = \frac{\partial E_{em}}{\partial \lambda}\,d\lambda + \frac{\partial E_{em}}{\partial x}\,dx \tag{5.44}$$

Since equations (5.42a) and (5.44) should be equal for all λ and x, we must have

$$i = \frac{\partial E_{em}}{\partial \lambda} \tag{5.45a}$$

$$F_e = -\frac{\partial E_{em}}{\partial x} \tag{5.45b}$$

For rotary (mechanical power) systems, the equation for the torque T_e, corresponding to equation (5.45b) is simply

$$T_e = -\frac{\partial E_{em}}{\partial \theta} \tag{5.45c}$$

where θ is angular displacement. Note that the differentiation in equation (5.45a) is carried out with x (or θ) held constant, whereas in that of equation (5.45b) or (5.45c), λ is held constant. Equations (5.45a) through (5.45c) represent the transducer model. They are directly applicable to (magnetic-circuit-based) electromechanical energyconversion devices having only one winding. The present results can be extended, in a fairly straightforward manner (see [15], for example), to *multiply excited systems* (that is, those having multiple windings), but this will not be done here.

| **Example 5.4** | Model of a Simple Electromagnetic Relay System |

The electromagnetic relay, although structurally quite simple, contains most of the essential ingredients of a magnetic-circuit-based electromechanical system and is therefore a good vehicle for illustrating the model construction procedure developed above. Consider again Example 3.18 on the switching of an inductive load by a transistor. As shown in Fig. 5.13a, let the inductive load be an electromagnetic relay. The relay armature parameters are as indicated in the figure. The resistance to the armature motion at the pivot is of the viscous type. The flyback diode circuit with *suppression resistance* R_F, described in Example 3.18, has been omitted in order to limit the focus of the present problem. Let the resistance of the relay winding be R_m. Assume that the value of R_m is large enough to limit the current in the relay winding to the rated value. Derive a state model for this system that will yield the instantaneous value of the transistor

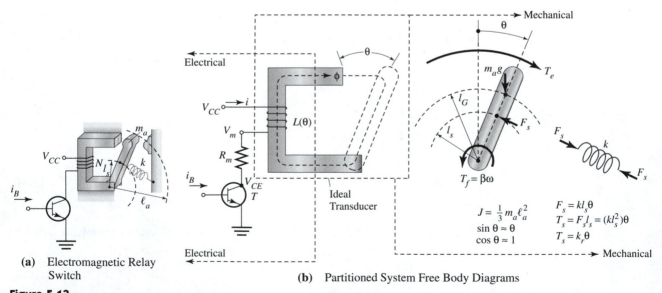

(a) Electromagnetic Relay
        Switch

(b) Partitioned System Free Body Diagrams

Figure 5.13

Simple Electromagnetic Relay
and Transistor Driver

collector-emitter voltage V_{CE}. Assume that the magnetic core and the armature material are highly permeable.

Solution: Figure 5.13b shows the partitioning of the system of Fig. 5.13a into electrical and mechanical subsystems and an ideal transducer. The coil within the transducer has no resistance. The resistance is in the electrical subsystem. Similarly, the armature within the transducer has no mass and no friction in the pivot. These effects are all in the mechanical subsystem. To the electrical subsystem, the transducer is apparently just an inductor within the circuit. For the mechanical subsystem, the transducer represents an external torque (force) on the armature (mass). We consider first the transducer equations.

Transducer model: The transducer magnetic circuit is equivalent to the magnetic circuit of Example 5.3. Thus the gap flux is given by equation (5.38a). From our experience in Example 5.3, we can neglect the reluctance of the core and armature without significant loss of accuracy. That is, we can use equation (5.38b) for the gap flux:

$$\phi \left(\frac{l_c}{\mu_c A_c} + \frac{x_g}{\mu_o A_c} \right) = [Ni] \cong \phi \left(\frac{x_g}{\mu_o A_c} \right) \Rightarrow \phi = \frac{(\mu_o A_c N)i}{x_g} \tag{5.46a}$$

The inductance is given by equation (5.41). Since $x_g \cong l_a \theta$, we take

$$L = \frac{N\phi}{i} \approx \frac{(\mu_o A_c N^2)}{x_g} \cong \frac{(\mu_o A_c N^2)}{l_a \theta} \tag{5.46b}$$

Note that when the reluctance of the core is neglected, the detailed geometry of the core is lost from the problem, so that the expressions summarized in Fig. 5.8f can be used directly.

The energy stored in the transducer is given by equation (5.43a), and the torque output is given by equation (5.45c). Thus

$$T_e = -\frac{\partial}{\partial \theta} \left(\frac{1}{2} \frac{\lambda^2}{L} \right) = -\frac{\lambda^2}{2} \frac{\partial}{\partial \theta} \left(\frac{1}{L} \right) = -\frac{\lambda^2}{2} \frac{\partial}{\partial \theta} \left(\frac{l_a \theta}{\mu_o A_c N^2} \right) \tag{5.46c}$$

$$= -\left(\frac{l_a}{2\mu_o A_c N^2} \right) \lambda^2$$

From equation (5.41), $\lambda = Li$, so that

$$T_e = -\left(\frac{l_a}{2\mu_o A_c N^2} \right) L^2 i^2 = -\left(\frac{\mu_o A_c N^2}{2 l_a \theta^2} \right) i^2 \tag{5.46d}$$

Note that equation (5.46d) agrees with the entry in Fig. 5.8f with $T_e = -F l_a$ and $\theta = x/l_a$. If we compare equation (5.46d) with equation (5.21b), with $T_e \equiv e_2$ and $i \equiv f_1$, we see that we have a nonlinear gyrator.

Electrical subsystem: The circuit diagram of the electrical subsystem is already implied in Fig. 5.13b. Note that this circuit actually includes within it the transducer (that is, the inductive aspect).

Starting with the dynamic component, "the inductor," we note that equation (2.25b) is not applicable, since the inductance here is a function of θ and hence of time. Another interpretation of this observation is that this circuit element is not a normal inductor but a special element—an electromechanical energy storage component! Hence equation (5.35b) is the applicable dynamic relation

$$\frac{d}{dt} E_{em} = Vi - T_e \omega \tag{5.47a}$$

If we now substitute for

$$E_{em} = \frac{1}{2} \lambda^2$$

$$L = \frac{1}{2} L(\theta)i^2$$

and

$$T_e = \frac{-\mu_0 A_c N^2}{2l_a \theta^2} i^2 = \frac{-L(\theta)i^2}{2\theta}$$

and use equation (5.46b) for $L(\theta)$, we obtain

$$\frac{dE_{em}}{dt} = \frac{d}{dt}\left[\frac{1}{2} L(\theta)i^2\right] = i\left[L(\theta)\frac{di}{dt} - \frac{i\omega}{2}\frac{L(\theta)}{\theta}\right] = Vi + \frac{i^2\omega}{2}\frac{L(\theta)}{\theta} \tag{5.47b}$$

Finally, noting that the voltage drop across the "inductor" is $V = V_{CC} - V_m$, we take

$$L(\theta)\frac{d}{dt} i = \left[\frac{L(\theta)i}{l_a\theta}\right]l_a\omega + V_{CC} - V_m \tag{5.47c}$$

But

$$V_m - V_{CE} = iR_m$$

so that

$$L(\theta)\frac{d}{dt} i = \left[\frac{L(\theta)i}{l_a\theta}\right]l_a\omega + V_{CC} - iR_m - V_{CE} \tag{5.48a}$$

To compare equation (5.48e) with equation (a) of example 3.18, we note that the inductance is assumed constant in Example 3.18. Suppose we assume nominal or equilibrium values about which i and θ vary only slightly, that is, i_0 and θ_0. Then we can take

$$L(\theta) \approx L(\theta_0) = L$$

and

$$\left[\frac{L(\theta)i}{l_a\theta} \right] \approx \left[\frac{L(\theta_0)i_0}{l_a\theta_0} \right] = K_B$$

We see that equation (5.48a) and equation (a) of Example 3.18 are the same under these circumstances, since $l_a\omega = $ velocity $= v$. Thus the idea of a "back emf" in switched electromagnetic (inductive) loads is explained.

To complete the present model of the electrical system we note that the collector-emitter voltage is given by the transistor's output characteristics (Fig. 3.25c):

$$V_{CE} = f_{To}(i_c, i_B) = f_{To}(i, i_B) \tag{5.48b}$$

Equations (5.48a) and (5.48b) along with equation (5.46b) constitute the model for the electrical subsystem.

Mechanical subsystem: The free body diagram of the mechanical subsystem is shown in Fig. 5.13b. In the free body diagram note that the input torque, T_e, is drawn so that it agrees with the implied sign convention for positive rotation and moment. This way, $T_e d\theta$ gives a positive differential work done on the mechanical subsystem by the transducer. Similarly, in the determination of the spring force, the positive direction of linear displacement should be consistent with the sign convention on angular displacement. The equations for the mechanical subsystem can be developed from the free body diagram using the state modeling approach of chapters 2 through 4, or from the generalized methods of this chapter.

Rotary inertia (inertance):

$$J \frac{d}{dt} \omega = (m_a g l_G \theta + T_e) - \beta\omega - T_s \tag{5.48c}$$

Note that from Fig. 2.5 (uniform slender rod),

$$J = \frac{1}{3} m_a l_a^2$$

Spring (capacitance):

$$\frac{1}{k_r} \frac{d}{dt} T_s = \omega; \qquad k_r = k l_s^2$$

However, as before, the capacitor equation can be replaced by the following two equations:

$$\frac{d}{dt} \theta = \omega \tag{5.48d}$$

$$T_s = k_r \theta$$

Electromechanical torque output (transducer or gyrator): From equations (5.46b) and (5.46d)

$$T_e = -\left(\frac{\mu_o A_c N^2}{2 l_a \theta^2}\right) i^2 = \frac{-L(\theta) i^2}{2\theta} \tag{5.48e}$$

Note that the power balance gives

$$V_b i = T_e \omega \Rightarrow V_b = \frac{-L(\theta) i}{2\theta} \, \omega$$

Overall system: The total system model is given by equations (5.48a) through (5.48e), as follows

$$\frac{d}{dt}\, \theta = \omega \tag{5.49}$$

$$\frac{d}{dt}\, \omega = \frac{3}{m_a l_a^2}\left[(m_a g l_G - k l_s^2)\, \theta - \beta\omega - \frac{L(\theta)}{2\theta}\, i^2\right]$$

$$\frac{d}{dt}\, i = \frac{i\omega}{\theta} + \frac{1}{L(\theta)}\, [V_{CC} - iR_m - f_{To}(i, i_B)]$$

$$V_{CE} = f_{To}(i, i_B)$$

$$L(\theta) = \mu_o A_c N^2/(l_a \theta)$$

....

DC Motor Control

Most motors used in control systems are direct current machines. This is because of the ease with which the speed of a DC motor can be varied. Also, the direction of rotation can be reversed by merely changing the polarity of the applied voltage. A DC motor is basically a multiply excited electromechanical system consisting of a current-carrying rotor (*armature* coil) in a magnetic *(stator)* field. The current is introduced into the armature in such a way that the direction of the current relative to the magnetic field remains fixed as the motor rotates. Also, the rotation of the armature windings in the magnetic field induces a counter emf *(back emf)* that opposes the applied voltage in the armature.

As shown in Figs. 5.14a to 5.14d, the field and armature windings may be *shunt wound, series connected, compounded,* or *separately excited.* Control system motors are usually separately excited and are typically controlled either by varying the field voltage V_f while the armature current i_a is kept constant (that is, *field control*) or by varying the armature voltage V_a while keeping the field current i_f constant (that is, *armature control*).

The torque T developed by a motor is essentially proportional to the product of the armature and field currents as long as the field current is not large enough to saturate the field iron:

$$T = K_1 i_a i_f; \qquad K_1 = \text{constant} \tag{5.50a}$$

Similarly, the armature back emf V_b under this condition is proportional to the product of the armature (rotary) speed ω and the field current:

$$V_b = K_2 i_f \omega; \qquad K_2 = \text{constant} \tag{5.50b}$$

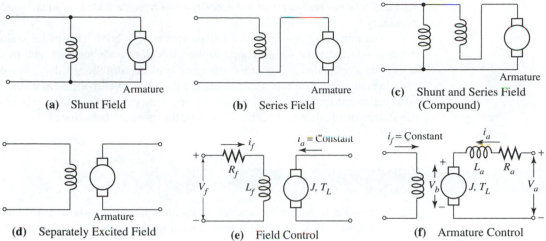

Figure 5.14

DC Motor Windings and Control

Field Control

For *field control* (see Fig. 5.14e), if the resistance and inductance of the windings are R_f and L_f, respectively, then the following equations describe, approximately, the operation of the motor:

$$\frac{d}{dt}\theta = \omega \tag{5.51a}$$

$$\frac{d}{dt}\omega = \frac{1}{J}(T - T_L) = \frac{1}{J}(K_f i_f - T_L); \qquad K_f = (K_1 i_a) = \text{constant} \tag{5.51b}$$

$$\frac{d}{dt}i_f = \frac{1}{L_f}(V_f - R_f i_f) \tag{5.51c}$$

where θ is the angular displacement of the armature (motor shaft), J is the armature moment of inertia, and T_L is the assumed load on the motor. Frictional resistance to the rotor motion has been ignored in the model but is easily included. Note the qualitative similarity between these results and the behavior of the electromagnetic relay of Example 5.4.

Armature Control

For *armature control* (see Fig. 5.14f), if the armature winding resistance and inductance are R_a and L_a, respectively, then the following model obtains approximately,

$$\frac{d}{dt}\theta = \omega \tag{5.52a}$$

$$\frac{d}{dt}\omega = \frac{1}{J}(K_a i_a - T_L); \qquad K_a = (K_1 i_f) = \text{constant} \tag{5.52b}$$

$$\frac{d}{dt}i_a = \frac{1}{L_a}(V_a - R_a i_a - K_b \omega); \qquad K_b = (K_2 i_f) = \text{constant} \tag{5.52c}$$

Again, frictional resistance to the rotor motion may be added to equation (5.52b), if necessary.

The preceding description of DC motors is very brief. The reader seeking more detail is referred to books dedicated to the subject, to trade journals, and to manufacturers' catalogs. Also, AC motors were not considered although they are sometimes used in control applications. In Example 2.2 the motor was modeled as a static element with the motor torque proportional to the armature current (armature control). This is the statement of equation (5.50a). However, the dynamic behavior of the motor (given by equations (5.52) was ignored in that example.

| **Example 5.5** | Armature-Controlled DC Motor |

Derive the state model for the armature-controlled DC motor whose schematic diagram is shown in Fig. 5.15a and hence verify equations 5.52. Include the inertia of the rotor and friction in the bearings.

Solution: The free body diagram of the partitioned system is shown in Fig. 5.15b. First note the following observations: The armature control voltage V_a can be considered an effort source. The same current circulates through the armature voltage source and the armature windings. The angular velocity in the bearing is the same as the angular velocity of the rotor, which for no compliance is also the velocity of the load. The load is actually an effort sink (negative source). Finally the coupling of the mechanical

Figure 5.15

Model for Armature-controlled DC Motor

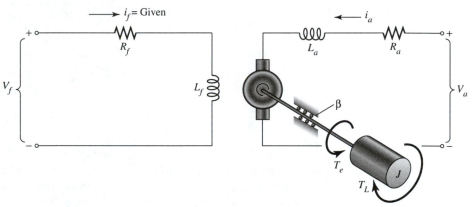

(a)   Schematic (Circuit) Diagram of System

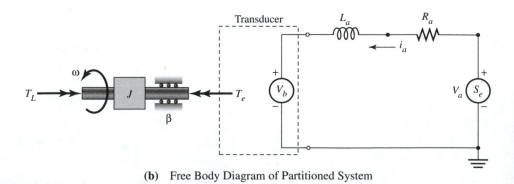

(b)   Free Body Diagram of Partitioned System

and electrical segments through the transducer (or gyrator) was previously discussed. The electrical power (input) $V_b i_a$ is converted into the mechanical power (output)$T_e \omega$.

Transducer: Let the angular velocity of the rotor be ω and the generated torque be T_e. Equations (5.23) and (5.24) give

$$i_u = r_T T_e; \qquad r_T = \frac{1}{K_1 i_f} \Rightarrow T_e = K_1 i_f i_a \qquad (5.53)$$

$$V_b = \frac{1}{r_T} \omega \Rightarrow V_b = K_1 i_f \omega$$

Electrical:

$$V_a - R_a i_a - L_a \frac{d}{dt} i_a = V_b \qquad (5.54a)$$

Mechanical:

$$\frac{d}{dt} \theta = \omega \qquad (5.54b)$$

$$J \frac{d}{dt} \omega = T_e - \beta \omega - T_L$$

Overall system: Depending on the application, the output for this problem will normally be the rotor (shaft) angular displacement θ, or speed ω. The state variables are i_a, θ, and ω. Using equations (5.53) to substitute for V_b in (5.54a) and T_e in (5.54b), we obtain the state equations:

$$\frac{d}{dt} \theta = \omega \qquad (5.55a)$$

$$\frac{d}{dt} \omega = \frac{1}{J} (K_1 i_f i_a - \beta \omega - T_L) \qquad (5.55b)$$

$$\frac{d}{dt} i_a = \frac{1}{L_a} (V_a - R_a i_a - K_1 i_f \omega) \qquad (5.55c)$$

which are the same as equations (5.52), except for the damping term $(\beta \omega)$ that was added to the problem. Equations (5.55a) through (5.55c) also require the coupling constants K_a and K_b to be equal ($K_a = K_b = K_1 i_f$), while equations (5.52a) through (5.52c) imply no such restriction. One way to model such a nonlinear coupling is to separate the system into two subsystems, one having the R_a and L_a elements but with the armature voltage V_a and the back emf ($K_b \omega$) as sources, and the other containing the TY or GY, β, and J elements, but with $K_a i_a$ and $(-T_L)$ as sources. Such an approach is necessary for a field-controlled DC motor, since there is no back emf in the field circuit (see Practice Problem 5.13). It is nonetheless interesting that the back emf came up *naturally* in a generalized system view of the armature-controlled DC motor.

Introducing Stepping Motors and Drives

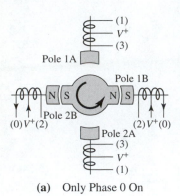

(a)   Only Phase 0 On

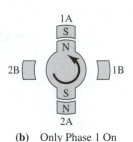

(b)   Only Phase 1 On

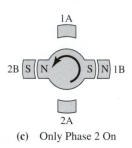

(c)   Only Phase 2 On

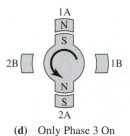

(d)   Only Phase 3 On

Figure 5.16

Stepping Motor
Stator-rotor Model

Stepping motors are designed to rotate in finite steps of a given number of degrees each time the stator is excited by a DC pulse. Since the angle of rotation per pulse is fixed, the motor shaft angular position is predictable with a high degree of accuracy and is repeatable. Because each angular step of the output shaft requires only a dc pulse input, the stepping motor can be driven directly from a digital signal in which the desired shaft positions and speed are encoded. Such a signal can be taken directly off a computer data bus. These characteristics make it possible to use stepping motors effectively in open-loop mode and in closed-loop direct digital control systems. Other advantages of stepping motors over comparable DC motors include low cost, ruggedness, and reliability. Consequently, stepping motors are in widespread use in such applications as printers, disk drives and similar computer peripherals, computer numerically controlled (CNC) machine tools, robots, and other advanced motion control systems. Some of the disadvantages of stepping motors include resonance effects, long settling times, and noisy and often rough performance.

Operation of Permanent Magnet–Type Stepping Motors

There are three main types of stepping motors, *permanent magnet* motors, *variable reluctance* motors, and *hybrid* motors, which combine the operating principle of the other two types of stepping motors. A common stepping motor is the four-phase permanent magnet–type motor containing two *(bifilar)* winding pairs that are center-tapped to provide four (stator) windings. As shown in Fig. 5.16a, the equivalent electrical model for the mechanical stator-rotor combination for such a motor consists of four stator poles and windings located 90° apart and a single north-south pole unit representing the rotor. The pattern of application and removal of current to each winding is known as *coil energization sequence*. Consider the simple case of *unipolar* coil energization. The switching of the current in the stator phases causes changes in the polarity of the stator poles and consequently rotation of the rotor (in *bipolar* coil energization, stator operation is determined by the direction of the winding current).

Figures 5.16a through 5.16d illustrate a single-step unipolar coil energization sequence in which only one phase (winding) is on (current applied) at a time. This is the *full-step one-phase-on sequence*. In Fig. 5.16a, with phase 0 on (see the current flow arrows shown on the $V+$ and (0) terminals) the stator poles 1B and 2B have polarities S (south) and N (north), respectively, and the rotor will be aligned, as shown, to minimize the reluctance. If next the current in phase 0 is switched off while phase 1 is switched on, the 1A and 2A stator poles will have the polarities shown in Fig. 5.16b. The rotor in realigning itself in the magnetic field will rotate counterclockwise, one-quarter of an (equivalent) revolution. Figures 5.16c and 5.16d show the sequential energization of the two remaining phases. It is evident that at the end of this sequence the rotor will have completed one (equivalent) revolution. The process is then repeated for continuous rotation. The actual step angle that is obtained each time the polarity of one winding is changed is determined by the number of teeth on the rotor and stator. With 50 teeth on the rotor and the appropriate number of teeth on each stator pole ($A = 50$, which is a typical value), the step angle of the four-phase motor is 1.8° or 200 steps in one actual revolution.

Mechanical Subsystem Model

As was the case for the DC motor and indeed for most electromechanical components, the main requirement in describing their behavior is the mechanical torque T_e (or force F_e) and the electrical back emf V_b, that is, the outputs of the electromechanical transducer (see Fig. 5.10). The electromechanical torque output for any of the three types of stepping motors can be taken as

$$T_e = - \sum_{n=0}^{N-1} K_T i_n \sin\left(A\,\theta - \frac{1}{2}\,n\pi\right) \tag{5.56a}$$

where K_T is the *torque constant* (N · m/A), i_n is the current in phase n, N is the total number of phases in the motor, and A is the number of rotor teeth. Equation (5.56a) assumes that the phases are numbered starting from zero (see Fig. 5.16a). The corresponding expression for the back emf in each phase circuit is

$$V_b = K_b \omega \sin\left(A\,\theta - \frac{1}{2}\,n\pi\right) \tag{5.56b}$$

where K_b is *back-emf constant,* ω is the angular velocity of the shaft, and n is the particular phase under consideration. As we saw for the armature-controlled DC motor, the values of K_T and K_b should be equal in the ideal case but are generally different for practical motors.

The stepping motor mechanical subsystem model can be described as follows:

$$\frac{d}{dt}\,\theta = \omega \tag{5.56c}$$

$$\frac{d}{dt}\,\omega = \frac{1}{J_T}\left(-\sum_{n=0}^{N-1} K_T i_n \sin\left(A\,\theta - \frac{1}{2}\,n\pi\right) - \beta_m \omega - T_L \right)$$

where $\beta_m \omega$ is the friction torque (assumed in this equation to be of a viscous nature) and T_L is the load torque. J_T is the rotor inertia when there is no load on the motor, otherwise it is the sum of the rotor inertia, the inertia of gears and or pulleys, etc., of the transmission system, and the *reflected* inertia of loads that are included in the transmission chain, which in some cases can include the weight of the motor itself.

Stepping Motor Drives

The electrical subsystem model for the stepping motor depends on the drive circuit. The stepping motor drive delivers electrical power to the motor in response to low-level signals from the control system. The current switching in motor windings, discussed earlier, occurs whenever the motor drive circuit receives a pulse input step command; the performance of stepping motor systems depends a great deal on the power amplifier drive scheme employed. The major drive types are *unipolar* and *bipolar* drives, and within each drive type are *voltage* or *resistance-limited (R-L)* drives and current drives. There are also *chopper* drives, which are usually of the bipolar type.

Unipolar Drive Circuit   Figure 5.17 shows a circuit model that is typical of unipolar *R-L* drives for a four-phase motor. Only the final stage of the power amplifier (that is, the transistor switch) is considered. The series resistance R_s is used to limit the steady-state current to the motor windings to the rated value and decrease the current rise time following switching. L_m, R_m, and V_b represent, respectively, each stator winding inductance, resistance, and (back) emf voltage that is induced in the stator winding as a result of the magnetic rotor motion. The polarity is generally such as to oppose the torque-producing current. The flyback or voltage suppression diode circuit, which was

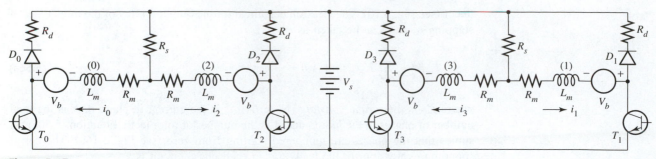

Figure 5.17

**Model of a Unipolar
Drive Circuit**

discussed in Example 3.18, includes suppression resistance R_d, which controls the rate of decay of the inductive current when the drive transistor cuts off. Assuming the transistors are indeed operating like switches (see Example 3.18) with switching times that can be taken as relatively instantaneous, we can write the loop equations for each phase circuit (see [16]) as follows:

$$\frac{d}{dt} i_0 = \frac{1}{L_m} \left[S_{p0}V_s + K_b\omega \sin (A\,\theta) - R_s (i_0 + i_2) \right.$$
$$\left. - \{R_m + (1 - S_{p0})\,R_d\}\, i_0 \right] S_{d0}$$
(5.57a)

$$\frac{d}{dt} i_1 = \frac{1}{L_m} \left[S_{p1}V_s + K_b\omega \sin \left(A\,\theta - \frac{1}{2}\pi \right) - R_s (i_1 + i_3) \right.$$
$$\left. - \{R_m + (1 - S_{p1})\,R_d\}\, i_1 \right] S_{d1}$$
(5.57b)

$$\frac{d}{dt} i_2 = \frac{1}{L_m} \left[S_{p2}V_s + K_b\omega \sin (A\,\theta - \pi) - R_s (i_0 + i_2) \right.$$
$$\left. - \{R_m + (1 - S_{p2})\,R_d\}\, i_2 \right] S_{d2}$$
(5.57c)

$$\frac{d}{dt} i_3 = \frac{1}{L_m} \left[S_{p3}V_s + K_b\omega \sin \left(A\,\theta - \frac{3}{2}\pi \right) - R_s (i_1 + i_3) \right.$$
$$\left. - \{R_m + (1 - S_{p3})\,R_d\}\, i_3 \right] S_{d3}$$
(5.57d)

where S_{pn} and S_{dn} are switching variables defined as follows:

$$S_{pn} = \begin{cases} 1 & \text{if phase } n \text{ (transistor) is on} \\ 0 & \text{if phase } n \text{ (transistor) is of} \end{cases}$$
(5.57e)

$$S_{dn} = \begin{cases} 1 & \text{if } i_n \geq 0 \\ 0 & \text{if } i_n < 0 \end{cases}$$
(5.57f)

S_{dn} simulates the conduction of the suppression diode during forward bias only. Note that the large-voltage model of the diode is being used.

Table 5.2 shows the phase-energizing sequence that can be used with the preceding drive circuit to move a four-phase motor through one complete *(electrical)* revolution. For a motor with 50 rotor teeth, one full step is 1.8°, so that in one electrical revo-

Table 5.2

Unipolar Phase-Energizing Sequences

FULL STEPPING (1.8°) COIL ENERGIZING SEQUENCE				HALF-STEPPING (0.9°) SEQUENCE	
One-Phase-On		Two-Phase-On			
Phase Current (n)	Poles	Phase Current (n)	Poles	Phase Current (n)	Poles
1	1A	1 & 2	1A + 2A	1 & 0	1A + 2B
2	2A	3 & 2	1B + 2A	1	1A
3	1B	3 & 0	1B + 2B	1 & 2	1A + 2A
0	2B	1 & 0	1A + 2B	2	2A
1	1A	1 & 2	1A + 2A	2 & 3	2A + 1B
				3	1B
				3 & 0	1B + 2B
				0	2B
				1 & 0	1A + 2B

lution, which requires four current switches in *full-stepping* mode, the motor shaft will rotate 7.2°. Both *one-phase-on* and *two-phase-on* energizing schemes are shown. There are subtle but usually important differences in motor performance depending on the energizing scheme used. However, many factors are involved, some specific to the motor in use, so that we can only refer the interested reader to dedicated books on the subject and manufacturers' catalogs. When the energizing scheme is an alternation between one-phase-on and two-phase-on *(half-stepping)*, the motor rotates in increments of half the basic step (that is, 0.9° for a 1.8° step angle motor) angle. This energizing scheme, which is also shown in Table 5.2, is quite effective in minimizing resonance effects and rough operation.

Recirculating Chopper Drive   The *recirculating chopper* is a method of current control used in many stepping motor drives. It is based on a four-transistor bridge together with recirculating diodes and a sense resistor. Flyback diodes and series resistors are not needed, as such, (although they can be added) because of the direct control of the current level in each winding by *switching* or *chopping*.

This ability directly to control the current level is a significant advantage for chopper drives. It allows very complex drive schemes such as *microstepping* to be achieved much more easily. In microstepping, the current level during switching is maintained at a value that causes the motor to rotate only a fraction of the basic step angle (as in half-stepping). However, this fraction can be as low as $\frac{1}{50}$, so that instead of, say, 200 steps per revolution, 10,000 steps per revolution can be accomplished. Also, in most cases the time interval during which the current is on is controlled (pulse width modulation) as well, causing the torque level to average out. Thus extremely fine resolutions and smooth operation are possible with such drive schemes. However for moderate- to high-speed applications, the microstepping drive must also be capable of operating at extremely high frequencies, hence they are generally more expensive than other drives.

Figure 5.18 shows a circuit model for a *single-supply parallel bipolar recirculating chopper drive*. It is assumed that bifilar winding pairs if present have been connected in *parallel* so that there are only two windings for a four-phase motor. The inductance and winding resistance indicated are, in this case, the parallel combination values. Current is injected into the first winding by turning on the top switch (S_{p1}) and

Figure 5.18

Model of a Bipolar Recirculating Chopper Drive Circuit

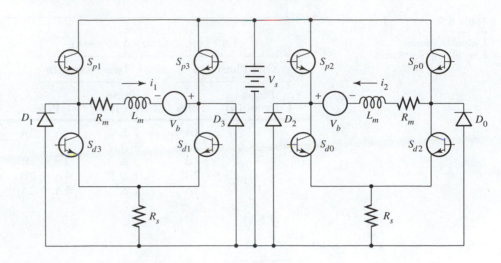

the bottom switch (S_{d1}). The full power supply voltage is thus applied across the winding, and current flow is in the direction indicated. The sense resistor (R_s) provides a feedback voltage signal that is proportional to the current. This signal is monitored by the drive controls, and when the current level reaches the required value, the top switch (S_{p1}) is turned off. The current will begin to decay (as a result of losses in the system) while still circulating (maintained by the back emf or, if you like, the energy stored in the transducer) through the bottom switch and the diode D1. When the current level reaches a preset lower threshold, the top switch is turned back on. In this way the current in the winding is maintained very closely at the desired average value. The bottom switch (S_{d1}) remains on as long as it is desired that the corresponding phase (say, phase 1) be on. Thus the switching or chopping function is more or less independent of the phase-switching function and usually proceeds along at a high but fixed frequency.

If the top switch (S_{p3}) and the bottom switch (S_{d3}) are used instead, the current flows through the winding in the reverse direction, which is equivalent to another phase (say, phase 3). Thus switching between two phases can be accomplished by reversing the current. This is the meaning of a bipolar drive. The present drive is a single-supply drive because the current reversal is accomplished using only one power supply instead of two. The other switch pairs shown in the figure provide the remaining two phases. Chopper drives are normally used with the two-phase-on energizing scheme. The two-phase-on sequence shown in Table 5.2 for unipolar drives applies to the bipolar chopper drive if we interpret the phase current column as indicative of the state of the lower switches (recall that the top switches are not really part of phase switching). It should be evident from the preceding description that the chopper drive equations can also be developed along the same lines as equations (5.57) by using the facility of switching functions.

5.3 OTHER HYBRID AND INTEGRATED SYSTEM EXAMPLES

In the preceding section we consider electrical transducers in general and electromechanical transducers in particular. Such energetic devices are *hybrid* systems in the sense that different forms of power transmission (or energy store)—electrical and mechanical, for example—are combined in the same component. In this section we continue to look at hybrid systems, especially some examples of energetic systems having mechanical-thermal and mechanical-hydraulic power combinations. These types of systems are referred to generally as *mechanical transducers*. We shall also illustrate the modeling of *integrated* systems, that is, energetic systems in which the different forms of power although possibly present in different components strongly interact because of the close coupling of these components within a functional entity. Process engineering systems that were considered in Section 4.2 are usually integrated systems.

Transducers In Mechanical Systems

There are also many common or significant transduction devices in which the forms of power are mechanical in the general sense of the term. Two examples are discussed next: the bimetallic temperature gage, in which heat is transformed to pure mechanical power, and the hydraulic amplifier, actually the spool valve component, which can transform power between fluid (hydraulic) and mechanical forms.

The Bimetallic Temperature Gage

When a solid is heated, it expands by an amount that is proportional to the change in its temperature. The coefficient of linear (thermal) expansion or the linear expansivity α is defined by

$$\alpha = \frac{(L_2 - L_1)/L_1}{(T_2 - T_1)} \tag{5.58a}$$

where L_1 and L_2 are the lengths of the solid at temperatures T_1 and T_2, respectively. Equation (5.58a) implies for small ΔL,

$$(L_2 - L_1) = \alpha L_1 (T_2 - T_1) \quad \text{or} \quad \Delta L = \alpha L \Delta T \tag{5.58b}$$

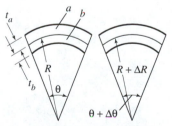

(a)   Temperature Gage

The bimetallic element is often made by welding or riveting two thin strips of metal of different coefficients of expansion. Because one of the strips expands more than the other, a change in temperature causes the element to twist or bend. In the temperature gage application, the sensitivity of the instrument is further improved by winding a long element into a bimetallic helix or spiral. This device is often used in home thermostats, not only to indicate the temperature but also to actuate automatic temperature control systems using the movement of the bimetallic element.

Consider a segment of the helix or spiral that includes the angle θ and has a radius of curvature R. With the notation of Fig. 5.19b,

$$\Delta L_a = \left(R + \Delta R + \frac{t_a}{2} \right) (\theta + \Delta\theta) - \left(R + \frac{t_a}{2} \right) \theta$$

$$\Delta L_b = \left(R + \Delta R - \frac{t_b}{2} \right) (\theta + \Delta\theta) - \left(R - \frac{t_b}{2} \right) \theta$$

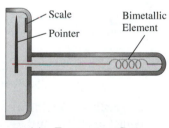

(b)   Segment of Helix or Spiral Before and After Change in Temperature

Figure 5.19

Bimetallic Temperature Gage

also

$$\Delta L_a = \alpha_a L_a \Delta T = \alpha_a \left(R + \frac{t_a}{2} \right) \theta \Delta T$$

$$\Delta L_b = \alpha_b L_b \Delta T = \alpha_b \left(R - \frac{t_b}{2} \right) \theta \Delta T$$

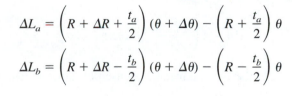

Figure 5.20

Hydraulic Amplifier

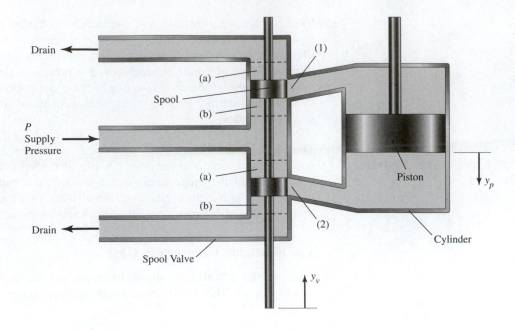

From the preceding equations, neglecting second-order terms in ΔR and $\Delta \theta$, we obtain

$$\Delta L_a - \Delta L_b = \left(\frac{t_a + t_b}{2}\right) \Delta \theta = \left[(\alpha_a - \alpha_b) R + \left(\frac{\alpha_a t_a + \alpha_b t_b}{2}\right)\right] \theta \Delta T$$

Hence

$$\Delta \theta = \left[\frac{2(\alpha_a - \alpha_b)}{(t_a + t_b)} R + \left(\frac{\alpha_a t_a + \alpha_b t_b}{t_a + t_b}\right)\right] \theta \Delta T \qquad \textbf{(5.59a)}$$

From equation (5.59a) it is apparent that for high sensitivity we need

$$\alpha_a \gg \alpha_b \quad \text{and} \quad t_a + t_b \ll 1 \qquad \textbf{(5.59b)}$$

Under these conditions and considering that α is quite a small quantity—for instance, the values for diamond (at 40°C), aluminum (at 20°C), and sodium (at 40°C) are 1.18, 25.5, and 40.4 (10^{-6} per °C), respectively—the second term on the RHS of equation (5.59a) can be neglected, so that

$$\Delta \theta = 2 \left(\frac{\alpha_a - \alpha_b}{t_a + t_b}\right) R \theta \Delta T \qquad \textbf{(5.59c)}$$

For the entire bimetallic element of uncoiled length L,

$$\Delta \theta = 2 \left(\frac{\alpha_a - \alpha_b}{t_a + t_b}\right) L \Delta T \qquad \textbf{(5.59d)}$$

The Spool Valve and the Hydraulic Amplifier

The hydraulic amplifier shown in Fig. 5.20 consists of a spool valve and a large piston-cylinder unit. Such an arrangement is capable of producing very large piston forces with the application of relatively very small forces to the spool valve. Since power cannot be created from nothing, a high-pressure supply (power source) is an integral part of the spool valve. The power source is a common feature of all types of amplifiers or, more generally, *active* components (recall the transistor and op-amp components of Section 3.4).

The operation of the amplifier is easily explained with the aid of the broken outlines showing alternative positions of the valve spools. When the valve is moved upward to, say, position (a), high-pressure fluid from the supply passes through the now-open port (1) to the top of the piston while fluid drains from below the piston through the also opened port (2). The piston thus moves downward. Similarly if the valve is lowered, say, to (b), high-pressure fluid is admitted to the cylinder below the piston, and the piston is raised. In either direction, the piston movement is powered by the high-pressure supply, and large forces can be developed in the piston.

For the situation shown in Fig. 5.20 in which there is no external load on the piston (and approximately for the case of a loaded piston), the pressure drop across the valve remains essentially constant, and the flow through the ports is therefore proportional to the open area of the ports. This is the area uncovered by the valve spools, hence

$$Q = Cy_v \tag{5.60a}$$

where Q is the flow rate of fluid into the cylinder, and C is a proportionality constant (linear valve behavior is assumed). This is the equation for the spool valve transduction element. For incompressible fluid flow,

$$Q = A \frac{dy_p}{dt} \tag{5.60b}$$

where A is the internal cross-sectional area of the large cylinder, so that

$$\frac{d}{dt} y_p = \frac{C}{A} y_v \tag{5.60c}$$

Integrated System Examples

In addition to the process engineering systems analyzed in Chapter 4, some problems studied in Chapter 3, such as Examples 3.10, 3.11, and the application suggested in Example 3.18 (Fig. 3.31b) can be viewed as integrated systems. We consider here two additional illustrations.

Example 5.6   Speed Control System

This problem, which is adapted from a similar example in reference [17], relates to the speed control of engines such as gas or steam turbines and diesel engines using a hydraulic servomotor (see Fig. 5.21). As illustrated by the physical model shown, the desired speed is set by the throttle lever. The flyweight is driven by the engine whose speed is being controlled, and the differential displacement of the spring due to the flyweight and the throttle lever determines the input to the hydraulic servomotor (*servomechanisms* were defined in Section 1.2 (see Table 1.3)). We desire a simple model for the entire system. Determine the model in state space using system decomposition.

Solution a:

Decomposition: Figure 5.22 shows the free body diagrams of the system components

Figure 5.21

Engine Speed Control System

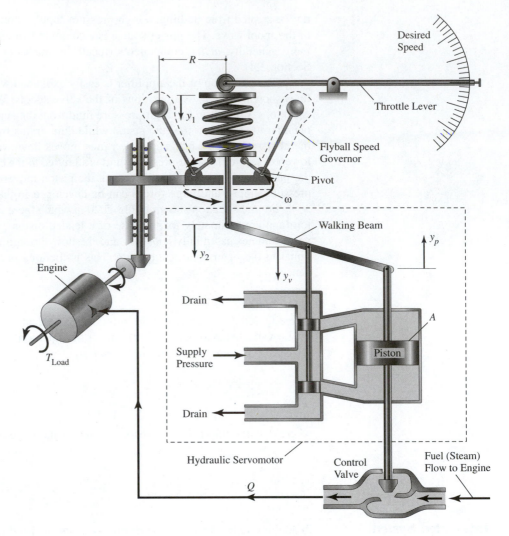

The hydraulic servomotor: This component consists of the floating lever or walking beam and the hydraulic amplifier that was analyzed previously. Because of the floating nature of the beam, the displacement y_v is due in part to y_2 as well as y_p. Assuming that the displacements are small, we find the contribution to y_v by y_2, that is, y_{v2}, by fixing end B while end A is displaced by y_2, that is,

$$y_{v_2} = \left(\frac{b}{a + b} \right) y_2$$

Similarly, the contribution of the displacement of B by y_p while A is fixed is

$$y_{v_p} = \left(\frac{a}{a + b} \right) y_p$$

The walking beam (static): Hence

$$y_v = y_{v_2} + y_{v_p} = \left(\frac{by_2 - ay_p}{a + b} \right)$$

(5.60d)

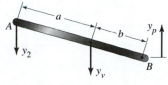

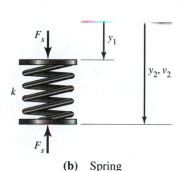

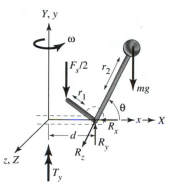

(a)   Walking Beam

(b)   Spring

(c)   Flyweight and Lever

Figure 5.22

**Free Body Diagrams of
Engine Speed Control System**

The hydraulic amplifier (dynamic): From equation (5.60c)

$$\frac{d}{dt} y_p = \frac{C_2}{A} y_v \tag{5.60e}$$

Spring (dynamic):

$$F_s = k \, (y_1 - y_2) \tag{5.61a}$$

$$\frac{d}{dt} y_2 = v_2$$

Throttle lever (static): Let the desired speed (change from reference speed) be ω_d:

$$y_1 = c_1 \omega_d; \qquad c_1 = \text{constant} \tag{5.61b}$$

The flyweight and lever (dynamic): This component has a complex model, which is given by equation (2.14b). However, for the choice of flyweight and lever segment and system of coordinates, the problem is somewhat simplified. Referring to the free body diagram, we note that

$$\omega_y = \Omega_y = \omega$$
$$\omega_x = \Omega_x = 0 \tag{5.62a}$$
$$\omega_z = \dot\theta; \qquad \Omega_z = 0$$

$$J_{xx} = m r_2^2 \sin^2\theta; \qquad J_{xy} = m \, (r_2\cos\theta + d)(r_2\sin\theta)$$
$$J_{yy} = m \, (r_2\cos\theta + d)^2; \qquad J_{yz} = J_{zx} = 0 \tag{5.62b}$$
$$J_{zz} = m \, [(r_2\cos\theta + d)^2 + r_2^2\sin^2\theta] = m \, [r_2^2 + d^2 + 2r_2 d \cos\theta]$$

Equation (2.13b) reduces to

$$H_x = -J_{xy}\omega$$
$$H_y = J_{yy}\omega \tag{5.62c}$$
$$H_z = J_{zz}\omega_z$$

Equation (2.14b) becomes

$$T_x = -\dot{J}_{xy}\omega - J_{xy}\dot\omega + J_{zz}\omega_z\omega$$
$$T_y = \dot{J}_{yy}\omega + J_{yy}\dot\omega \tag{5.62d}$$
$$T_z = \dot{J}_{zz}\omega_z + J_{zz}\dot\omega_z + J_{xy}\omega^2 = J_{zz}\dot\omega_z + J_{xy}\omega^2 - 2m r_2 d \sin\theta\omega_z^2$$

Again from the free body diagram,

$$T_z = R_y d - \frac{F_s}{2}\,(d - r_1\sin\theta) - mg \, (r_2\cos\theta + d)$$

and

$$\sum F_y = R_y - \frac{F_s}{2} - mg = ma_y = m \, (r_2\cos\theta\dot\omega_z - r_2\sin\theta\dot\omega_z^2)$$

so that

$$\dot{\omega}_z = \left[\frac{1}{J_{zz} - mr_2 d \cos \theta} \right]$$

$$\left[mr_2 d \sin \theta \omega_z^2 - J_{xy} \omega^2 + \frac{F_s}{2} r_1 \sin \theta - mgr_2 \cos \theta \right] \tag{5.63a}$$

Without loss of generality, the angle included by the lever at the pivot can be taken as 90°, so that

$$-y_2 = r_1 \cos \theta \tag{5.63b}$$

The engine (dynamic): This is also a complex dynamic system. However, a very simple model can be derived if we assume that the torque due to the ingestion of fuel is produced instantaneously. For typical engine characteristics, this torque T_i is evidently proportional to the rate of fuel input and inversely proportional to the engine speed. Finally, assuming that the engine is subject to a frictional torque T_f that is proportional to the engine speed, we have

$$J \frac{d}{dt} \omega = T_i - T_f - T_L = \frac{c_4 Q}{\omega} - \beta \omega - T_L \tag{5.64a}$$

Valve (static): See Example 2.2, for instance,

$$Q = \left(C \sqrt{\Delta P} \right) y_p = c_3 y_p \tag{5.64b}$$

State variables: We assign θ for the spring (y_2 is an alternative), y_p for the hydraulic amplifier, ω for the engine, and ω_z for the flyball governor.

State equations: First,

$$\frac{d}{dt} \theta = \omega_z \tag{5.65a}$$

From equations (5.60d), (5.60e), and (5.63b),

$$\frac{d}{dt} y_p = -\frac{c_2}{A} \left[\frac{br_1 \cos \theta + ay_p}{a + b} \right] \tag{5.65b}$$

From equations (5.64a) and (5.64b),

$$\frac{d}{dt} \omega = \frac{1}{j} \left[\frac{c_3 c_4 y_p}{\omega} - \beta \omega - T_L \right] \tag{5.65c}$$

and equations (5.63) and (5.61) give

$$\frac{d}{dt} \omega_z = \frac{1}{(r_2^2 + d^2 + r_2 d \cos \theta)} \left[r_2 d \sin \theta \, (\omega_z^2 - \omega^2) - r_2^2 \cos \theta \sin \theta \omega^2 \right. \tag{5.65d}$$

$$\left. - gr_2 \cos \theta + \frac{k}{2m} (c_1 \omega_d + r_1 \cos \theta) \, r_1 \sin \theta \right]$$

The model: The object of the model is the actual speed of the engine. Assuming, as shown in Fig. 5.21, that the speed governor is driven directly by the engine through some gear arrangement, then we have

$$\text{Output} = c_5 \omega \tag{5.65e}$$

Equations (5.65a) through (5.65e) constitute the overall system model.

Example 5.7	Temperature Control System

A possible bimetallic element–based temperature control system for a room air conditioning unit is shown in Fig. 5.23. The desired temperature is set by a slide lever. A wiper attached to the bimetallic spiral and the slide lever form a switch for a solenoid that actuates another switch for the compressor. Note that the current through the bimetallic element has to be small to prevent resistance heating of the element, which could interfere with the temperature-sensing function. The resistance R_i therefore serves to limit this current. The compressor operates at a fixed rotary speed as determined by the constant-current source i_c. Air from the room is circulated at a fixed rate (by a blower)

Figure 5.23

Room Temperature Control System

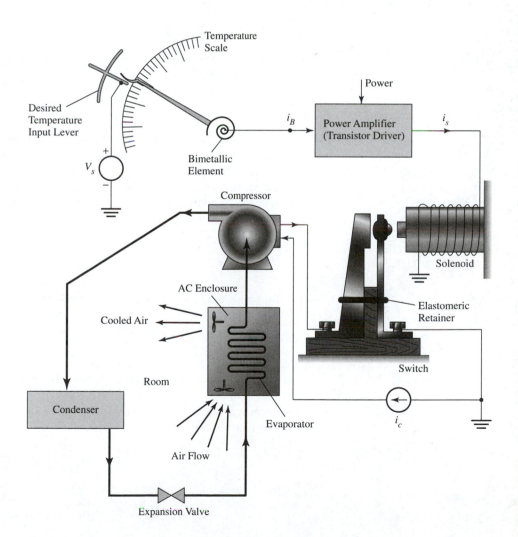

through the air conditioner enclosure containing the evaporator coils. Assuming that the room gains heat by conduction from the outside and by generation, at a fixed rate, by the room occupants, derive a simple overall model for this system. Note that the bimetallic element is inside the air conditioner enclosure.

Solution: A significant feature of temperature processes that do not involve reactive or explosive conditions is the relatively slow rate of most events. Thus the dynamics of such components as electric motors and relays can be neglected relative to the significant thermal capacitances in the process. Assuming that the outside or ambient temperature is fairly constant, the evaporator temperature is usually approximately constant. The refrigerant flow rate and the heat gain by the evaporator are effectively fixed as long as the compressor is operative. Given such steady conditions, and the on-off operation of the air conditioning unit, a discrete-time model may provide an adequate but simple description of this system. Therefore, we let k be the (running) time index, and Γ, the sampling period. We also assume that the constants used in the following expressions are known quantities: $c_1, c_2, \ldots, c_5$. The free body diagrams of the various components are shown in Fig. 5.24.

Bimetallic element and lever (static): From equation (5.59d) we deduce

$$\theta_k = c_1 T_{e,k} + c_2 \tag{5.66a}$$

The desired temperature lever setting θ_d can be assumed to be proportional to the desired temperature T_d. Hence

$$\theta_d = c_3 T_d \tag{5.66b}$$

Compressor (static, assumed): The compressor is driven by an electric motor. If we neglect the motor dynamics, then according to equation (5.24) or equation (5.50a) (for a series field), the torque input to the compressor T_{net} is proportional to i_c^2. Since the control mode is on-off, we can assume that the compressor is operated at a constant rotating speed ($\omega = $ constant), determined by the input current i_c, whenever it is on. From equation (4.27c) then, we take the refrigerant mass flow rate as

$$\dot{m}_{r,k} = \frac{T_{\text{net}}\omega}{\text{SF}r_2^2\omega^2} = c_4 i_{c,k}^2 \tag{5.66c}$$

Also, for an ideal adiabatic compressor, equations (4.26) and (4.28a) give

$$\text{SF}r_2^2\omega^2 = h_{2,k} - h_{e,k} = c_{rv}(T_{2,k} - T_{e,k}) = c_5 \tag{5.66d}$$

where c_{rv} is the specific heat of the refrigerant vapor. It is assumed that the refrigerant enters the compressor as (probably superheated) vapor.

Solenoid switch (static, assumed): As in Example 3.18, if we neglect the dynamics of this relay, then we can take

$$i_{s,k} = \beta i_{B,k}$$

$$i_{c,k} = \begin{cases} i_c; & i_{s,k} = 0 \\ 0; & i_{s,k} \neq 0 \end{cases} \tag{5.66e}$$

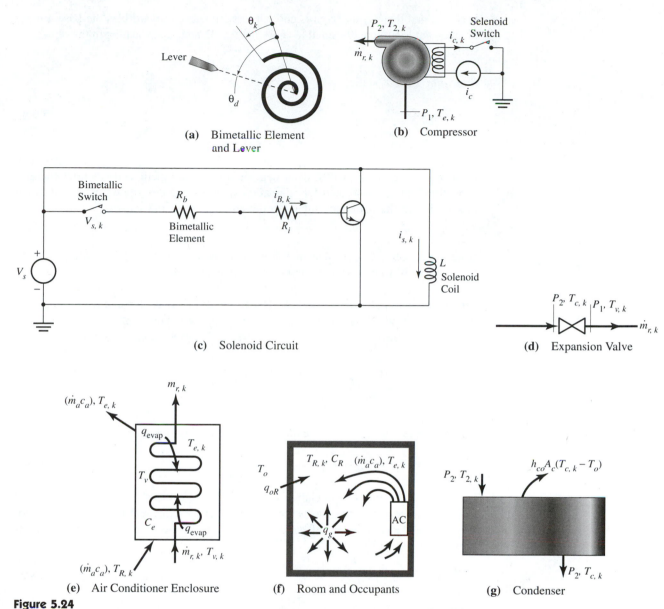

Figure 5.24

Free Body Diagrams for Temperature Control System

Solenoid circuit (static, assumed): The relay is the potential dynamic element in the solenoid circuit. Because this component is assumed to be static, the entire circuit is also static. R_b (the equivalent electrical resistance of the bimetallic element) is in series with R_i. For the typically small forward bias voltage V_{BE} we can take

$$i_{B,k} = \left(\frac{1}{R_b + R_i}\right) V_{s,k} \tag{5.66f}$$

It is assumed that $i_{B,k}$ as given is enough to adequately forward-bias the base-emitter junction but is sufficiently small to cause negligible resistance heating in the bimetallic strip.

Bimetallic switch (static):

$$V_{s,k} = \begin{cases} V_s; & \theta_k \geq \theta_d \\ 0; & \theta_k < \theta_d \end{cases} \tag{5.66g}$$

Expansion valve (static): The refrigerant leaves the condenser as a liquid and undergoes a *throttling* process in the expansion valve, leaving it as part vapor and part liquid. In the energy equation (4.17b) if we neglect changes in kinetic and potential energies as well as wall friction, then for adiabatic expansion the description of the throttling process is that the outlet and inlet specific enthalpies are the same:

$$h_{v,k} = h_{c,k} \tag{5.66h}$$

Note that equation (5.66h) does not imply that the inlet and exit temperatures are the same. There is a phase change within the expansion valve.

Air conditioner enclosure (dynamic): The air conditioning enclosure includes within it the evaporator, which is assumed to have negligible heat storage capacity. Considering the thermal capacitance C_e, of the enclosure, if T_R is the room temperature, then we have

$$T_{e,k+1} = T_{e,k} + \frac{\Gamma}{C_e}(q_{\text{air}} - q_{\text{evap}}) \tag{5.66i}$$

The heat given up by the circulating air q_{air} is given by equation (4.10):

$$q_{\text{air}} = \dot{m}_a c_a (T_{R,k} - T_{e,k}) \tag{5.66j}$$

where $\dot{m}_a$ is the rate of circulation of the air through the air conditioner. We shall assume that the refrigerant leaves the evaporator at the same temperature as the enclosure. The heat gained by the refrigerant is then given by the change in enthalpy (see energy equation (4.17b)):

$$q_{\text{evap}} = \dot{m}_{r,k} (h_{e,k} - h_{v,k}) \tag{5.66k}$$

Room (dynamic): If T_o is the outside or ambient temperature, then

$$T_{R,k+1} = T_{R,k} + \frac{\Gamma}{C_R}(q_g + q_{oR} - q_{\text{air}}) \tag{5.66l}$$

where q_g is the heat generation by room occupants, assumed known, q_{oR} is the heat transfer into the room from outside by conduction:

$$q_{oR} = \frac{1}{R_w}(T_o - T_{R,k}); \qquad R_w = \sum_{j=1}^{4} \frac{\Delta x_j}{k_j A_j} \tag{5.66m}$$

where R_w is the total thermal resistance of the walls, Δx is the wall thickness, k is the thermal conductivity, and A is the wall normal area. Heat transfer through the roof and floor is ignored. Finally, the heat given up to the enclosure is given in equation (5.66j).

Condenser (static, assumed): We shall also neglect thermal capacitance effects within the condenser. If we assume heat loss from the condenser to the environment is mainly by convection, then the energy equation applied to the condenser yields

$$\dot{m}_{r,k}(h_{2,k} - h_{c,k}) = h_{co}A_c(T_{c,k} - T_o) \tag{5.66n}$$

where h_{co} is the convective heat transfer coefficient, and A_c is the condenser heat transfer surface area. Because the refrigerant leaving the condenser is liquid, we can take (consistent with 5.66d))

$$h_{c,k} = c_{rl}T_{c,k} \tag{5.66o}$$

where c_{rl} is the specific heat of the refrigerant liquid.

State variables: We assign T_e and T_R.

State equations: From equations (5.66i), (5.66j), and (5.66k),

$$T_{e,k+1} = T_{e,k} + \frac{\Gamma}{C_e}[\dot{m}_a c_a(T_{R,k} - T_{e,k}) - \dot{m}_{r,k}(h_{e,k} - h_{v,k})] \tag{5.67a}$$

From equations (5.66l), (5.66m), and (5.66j),

$$T_{R,k+1} = T_{R,k} + \frac{\Gamma}{C_R}\left[q_g + \frac{1}{R_w}(T_o - T_{R,k}) - \dot{m}_a c_a(T_{R,k} - T_{e,k})\right] \tag{5.67b}$$

where from equations (5.66c) and (5.66e),

$$\dot{m}_{r,k} = \begin{cases} c_4 i_c^2; & i_{B,k} = 0 \\ 0; & i_{B,k} \neq 0 \end{cases}$$

and from (5.66f) and (5.66g),

$$i_{B,k} = \begin{cases} \left(\dfrac{V_s}{R_b + R_i}\right); & c_1 T_{e,k} + c_2 \geq c_3 T_d \\ 0; & c_1 T_{e,k} + c_2 < c_3 T_d \end{cases}$$

so that

$$\dot{m}_{r,k} = \begin{cases} c_4 i_c^2; & c_1 T_{e,k} + c_2 < c_3 T_d \\ 0; & c_1 T_{e,k} + c_2 \geq c_3 T_d \end{cases} \tag{5.67c}$$

We still need to eliminate $h_{e,k}$ and $h_{v,k}$ from equation (5.67a). From equation (5.66d),

$$h_{e,k} = c_{rv}T_{e,k} \quad \text{and} \quad h_{2,k} = c_{rv}T_{2,k} = c_5 + h_{e,k} = c_5 + c_{rv}T_{e,k}$$

From equations (5.66h), (5.66n), and (5.66o),

$$h_{v,k} = h_{c,k} = h_{2,k} - \frac{h_{co}A_c}{\dot{m}_{r,k}}(T_{c,k} - T_o) = c_5 + c_{rv}T_{e,k} - \frac{h_{co}A_c}{\dot{m}_{r,k}}\left(\frac{1}{c_{rl}}h_{c,k} - T_o\right)$$

hence

$$h_{v,k} = \left(\frac{\dot{m}_{r,k}c_{rl}}{\dot{m}_{r,k}c_{rl} + h_{co}A_c}\right)\left(c_5 + c_{rv}T_{e,k} + \frac{h_{co}A_c}{\dot{m}_{r,k}}T_o\right)$$

so that

$$-\dot{m}_{r,k}(h_{e,k} - h_{v,k}) = \left(\frac{\dot{m}_{r,k}}{\dot{m}_{r,k}c_{rl} + h_{co}A_c}\right) \tag{5.67d}$$

$$[c_5\dot{m}_{r,k}c_{rl} + h_{co}A_c(c_{rl}T_o - c_{rv}T_{e,k})]$$

Model: The model consists of equations (5.67a) through (5.67d) and the equation

$$\text{Output}_k = T_{R,k} \tag{5.67e}$$

5.4 INTRODUCING MICROMACHINED DEVICES

We conclude this chapter with a cursory look at *micromachined* systems. These devices are again mostly electromechanical systems, that is, *microelectromechanical systems* or MEMS. However, due to fabrication factors, electromagnetic systems are not yet predominant in microsensors and actuators, as they are in *macro* systems. In contrast, *electrostatic* (especially capacitive) and piezoelectric systems were common in first-generation practical micromachined devices.

Microsensors and Actuators

The demands for instrumentation and control in the fields of manufacturing and other processing industries, aerospace, defense, and (especially recently) communications have always increased as technology has advanced. For some time, developments in microprocessors and very large scale integrated circuit (VLSI) technology have allowed engineers to meet these demands through improvements in the speed and resolution of existing instruments and the implementation of more sophisticated control systems. With the availability today of new types of instruments, *microsensors* and *microactuators* may well present a new challenge—that of finding applications for these devices that can utilize their inherent characteristics. One of the heavy users of *micromachined* products is the automobile industry. Applications here range from micromachined silicon pressure sensors used in manifold absolute pressure measurement as a means of monitoring air intake in automobile engines, to micromachined accelerometers with on-chip self-testing features that are used in the deployment of air bags in collisions. An example of a more common place application is the micromachined nozzle used in commercially available ink-jet printers, thousands of which are used to rapidly and accurately place tiny ink droplets on the material being printed. A significant and growing area of application is in communications, where the twin technologies of *micromachining* and *fiber optics* are merging to further accelerate the growth of the communications industry.

Methods of Micromachining

There are three distinct *microfabrication* processes (see [18] and [19]): *bulk micromachining, surface micromachining,* and the *LIGA* process.

Bulk Micromachining   Early silicon-based devices such as mass-produced pressure- and acceleration-related sensors, valves, nozzles for ink-jet printers, switches, fluidic amplifiers, and gas chromatographs were produced using the bulk micromachining process. Silicon is the preferred material not only for its semiconductor properties, which recommend it for electronic components, but also for its strength and crystal structure.

Although silicon is brittle and allows little or no plastic deformation, its elastic modulus is higher than that of steel, it has very little mechanical hysteresis, and it is highly tolerant of fatigue loading. Bulk micromachining makes use of the well-established and highly refined fabrication techniques and equipment of microelectronics, including the highly economical *batch fabrication* approach; *photolithography; wet etching* and *etch-stops;* deposition and patterning equipment, as well as developments in *anisotropic etchants,* which attack different crystallographic planes of a crystal at different rates; *electrostatic bonding;* and *fusion bonding.*

Figure 5.25 illustrates the typical three-dimensional structures that can be formed by anisotropic etching of single-crystal silicon. Membrane and other structures with thin diaphragms, for example, are used to make pressure, strain, and displacement sensors, load cells, and pressure switches, while cantilevers and other suspended-mass structures are the basis of vibration and flow sensors and accelerometers. The bulk micromachining process as well as the other two processes summarized next are explained in more detail in references [18] and [19]. Descriptions of the bulk machining of some of the early silicon-based structures are provided in reference [20].

Surface Micromachining In surface micromachining, the desired structure is defined using microelectronics-type techniques and equipment in a surface-layer-by-layer process in which one or more layers are *sacrificial.* Polysilicon in particular and other dielectric materials, metals, and alloys that can be freed from their supported substrate by selective etching of dissolvable material, in general, are used in surface micromachining. Bridges and other suspended-mass components are common structures in surface-micromachined systems. Slidable members can also be produced by surface micromachining. Polysilicon grown by chemical vapor deposition has a high surface mobility, which allows it to refill undercut regions in a structure to form restraining flanges, as necessary, for a sliding member. Figure 5.26 illustrates the process sequence for a linear actuator *(microgripper)* made by a combination of bulk- and surface-micromachining. The phosphosilicate glass layer deposited in sequence (b) represents a sacrificial layer. The alignment window and V-groove are bulk-micromachined.

LIGA Process LIGA is an acronym for the German term for lithography, electroforming, and plastic molding. In the LIGA process, the pattern is formed by deep-etching X-ray lithography on an X-ray resist (deposited over a plating base for follow-up electroforming), usually based in poly(methyl methacrylate), PMMA. The resulting structure is either the final part or the *form* (mold) for producing the part by electroforming or plastic molding. The LIGA process produces well-defined, thick microstructures with parallel walls and flat surfaces. Electroplated products are permanently anchored to a substrate because of the plating base deposited on the substrate. However, if a sacrificial layer is first deposited on the substrate before the plating base is applied, components that are partially or completely free from the substrate can be produced. This so-called *sacrificial LIGA* (SLIGA) process has been used to make micro gear trains, turbines and micromotors and other products that require spinning parallel parts in contact. Indeed, the LIGA process and its derivatives were the basis for the fabrication of most of the early *micromagnetic* actuators that utilized relatively standard coil configurations.

The major disadvantage of the LIGA process is the availability and cost of the required highly collimated and very bright X-ray source, such as a synchroton. Considerable effort has therefore been expended to develop fabrication technology that can use conventional equipment and yet achieve results that are more or less comparable

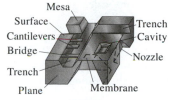

(a) Crystal Planes Normal to the Surface Produce Perpendicularly Intersecting Sidewalls

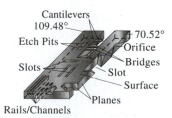

(b) Sloping Sidewalls are Formed by Self-limiting Etch Planes

Figure 5.25

Typical Three-dimensional Structures Formed by Anisotropic Etching of Single-crystal Silicon (Monosilicon). After Jerome Rosen, reference [18]

Figure 5.26

Process Sequence for a Linear Actuator (Microgripper) Made by Combination of Bulk- and Surface-micromachining Processes. After Lee O'Connor, reference [19]

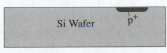

(a) Boron is Diffused into a Silicon Wafer

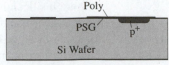

(b) A 2-μm-thick layer of phosphosilicate glass (PSG) is deposited on the wafer. This is followed by the deposition of a 2.5-μm-thick polysilicon layer. The shape of the polysilicon layer is defined through chemical etching of the regions that are not needed.

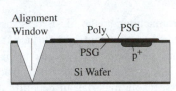

(c) PSG is deposited, the wafer is anneled, and an alignment window is etched.

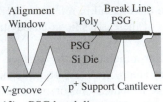

(d) PSG break lines are patterned around the gripper. A V-groove and the silicon die are etched into the wafer.

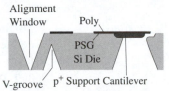

(e) PSG is etched away except where polysilicon is anchored to the wafer.

to the LIGA process. One such fabrication method based on polyimide electroplating molds and a standard UV radiation source is described in reference [21]. Figure 5.27 summarizes both this approach (b) and the LIGA method (a) for the fabrication of electroplated microstructures.

Modeling Micromachined Devices

Micromachined devices are typically *hybrid* systems that often include electrical and mechanical power forms (electromechanical component). However, because of the compatibility between most micromachining techniques and the fabrication of microelectronic circuits, many micromachined devices also have on-chip integrated circuitry for electronic signal processing and/or conditioning. Such devices are therefore *integrated* systems. But since we can develop adequate models of the integrated circuits (particularly *analog integrated circuits*) using the methods of Section 3.4, we shall focus our attention on the hybrid element in the micromachined device. Typical power forms involved in the hybrid component of micromachined devices include the following:

1. **Electrical and mechanical**—*electromechanical;* an example is the accelerometer, in which the vibration of a mass element at the end of a cantilever structure is sensed by a micromachined capacitor (see Example 5.8)
2. **Electrical and fluid**—*electrohydraulic* (incompressible fluid system) or *electropneumatic* (air system); as in the microturbine and the vibrating diaphragm pump, in which a diaphragm microstructure is vibrated electrostatically. Note that electromechanical energy conversion is also implied.

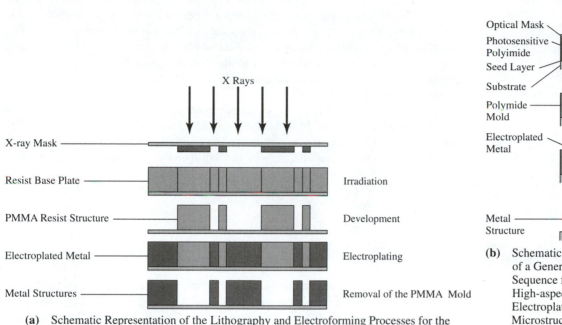

(a) Schematic Representation of the Lithography and Electroforming Processes for the LIGA Method.

(b) Schematic Representation of a Generic Process Sequence for Fabricating High-aspect-ratio Electroplated Microstructures Using Photosensitive Polyimide

Figure 5.27

Schematic Representations of the LIGA Method and a Comparable Generic Process for the Fabrication of Electroplated Microstructures. After Bruno Frazier and Mark Allen, reference [21]

3. **Electrical and thermal** *electrothermal;* an example is a temperature sensor utilizing a micromachined *thermistor* (an element, usually a semiconductor, whose resistance changes with temperature)

4. **Electrical and chemical**—*electrochemical;* an example is an electrochemical sensor array small enough to be inserted into a human artery that is used to monitor levels of oxygen, carbon dioxide, and pH in the blood. An acoustic wave device incorporating an acoustically thin micromachined silicon nitride membrane sputtered with zinc oxide functions as an electromechanical multisensor when acoustic waves sensitive to different parameters such as density, temperature, and pressure are produced in the device.

5. **Electrical and optical**—*electroptical;* an example is an array of torsional mirrors used in a fiber-optic communication system. The torsion bars that support the mirrors and are driven electrostatically function as electromechanical-optical amplifiers

6. **Thermal and fluid**—*thermofluid;* as in a low-flow sensor employing convective heat transfer between two microbridge structures to determine mass flowrate. Since the bridges are actually thermistors, the device also utilizes electrical power.

Micro-Behavior Components. The foregoing examples suggest that micromachined devices can be modeled in the same way as regular hybrid devices, provided appropriate *micro*-behavior components are substituted wherever the corresponding *macro*-behavior component is inadequate. One such element of behavior is friction. It has been found that high friction forces exist at the microscale, resulting in the need for large input forces or torques to produce relative motion between contacting microstructures. Better understanding of the mechanical property of friction at the micro level is still required for adequate quantification of this micro-behavior component. Another area in which macro behavior is an insufficient guide is in the effect of residual and thermal stresses. Because of the planar nature of most micromachined components, minor warpage due to residual or thermal stresses can cause interference or misalignment in moving elements. Consequently, the required separation between components relative to the size of the components is generally much larger for micromachined devices with moving members than for macro systems. Alternatively, additional processing steps may be needed to relieve residual stresses that build up during normal fabrication. Neither of the two examples considered next involves contact between a moving microstructure and another microfeature, and we shall assume that such microstructures are sufficiently separated.

| **Example 5.8** | Micromechanical Accelerometer with Capacitive Sensor |

Figure 5.28a shows the schematic illustration of the accelerator and integrated MOS detection circuitry of the micromachined device described in reference [22]. The sensing element is a small cantilever, beam with plated end-mass. The beam is a composite structure consisting of (see Fig. 5.34b) a silicon dioxide (SiO_2) beam, 0.46 μm thick by 25 μm wide by 108 μm long, coated with a chromium-gold (Cr-Au) (20 nm Cr, 40 nm Au) metal layer that can be assumed to be equivalent (reference [22]) to a 50-nm Au metal film only. The end-mass weighs about 0.35 μg and is about 16 μm thick by 34 μm wide and occupies about the final quarter of the beam. The metal coating on top of the beam and the p^+ layer below the beam constitute a variable capacitance C_B. The average spacing between the beam metal layer and the p^+ layer is 7 μm with about 1.5° initial upward bend of the beam due to residual stresses.

Assuming the description of the detection circuitry is well determined (see circuit diagram, Fig. 5.28c) so that the current in the variable capacitance branch of the circuit can be presumed, derive an approximate but adequate model for the voltage drop across C_B as a function of the current in that branch of the circuit, the relative displacement of the beam metal layer from the p^+ layer, and the acceleration of the surface on which the accelerometer is attached. Determine the conditions under which the relative displacement, as measured by the voltage drop across C_B, is generally a good indication of the acceleration of the accelerometer base.

Solution: The densities of the SiO_2 and Au are approximately 2250 kg/m^3 and 19320 kg/m^3, respectively. For the given dimensions, the mass of the composite beam is approximately 0.054 μg. This is negligible relative to the mass of the plated end-mass, so that the mass of the beam can be ignored in the following analysis. First, we shall determine an effective spring constant for the cantilever beam. We shall then use this value in developing the equation of motion of the end-mass and the variation of the capacitance of C_B.

Deflection of cantilever beam with mass on end. Although the mass of the end-plate is distributed over a significant (one-quarter) length of the cantilever beam, given the short overall length of the beam, the deflection characteristics of the beam due to a distributed model of the mass will not be much different from the behavior

Figure 5.28

Schematic of Accelerometer with Detection Circuitry, Mechanical Model of Cantilever Beam, and the Detection Circuit Diagram. After Kurt Petersen et al., reference [22]

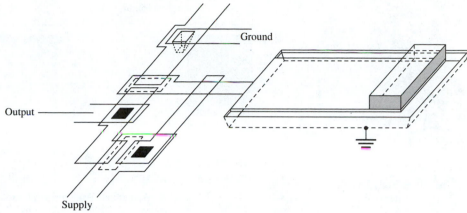

(a) Schematic of Completed Accelerometer Indicating Detection Circuitry, Etched Ground Contact Via, Free-standing Beam, and Plated End-mass

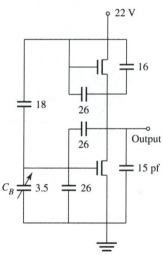

(c) Circuit Diagram, Including Parasitics and Output Capacitance, Numerically Modeled for Output Response as a Function of dc Capacitance Variation in C_B. Capacitance Values are in Femtofarads Unless Indicated Otherwise. Because the MOS Devices are p-type, the Supply Voltage and V_{out} are Both Negative.

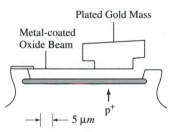

(b) Mechanical Model of the Cantilever Beam

given by an appropriate lumped-mass assumption. Figure 5.29a shows the free body diagram of the cantilever beam with the end-mass replaced by a point load at the position corresponding to the center of the end-plate. The assumed boundary conditions are that the left end of the beam is *fixed,* while the right end is *free* (both with respect to the p$^+$ layer). Given that our objective is the vibration of the plated mass, Fig. 2.3b applies to this case provided we use *a* in Fig. 5.29a as the length of

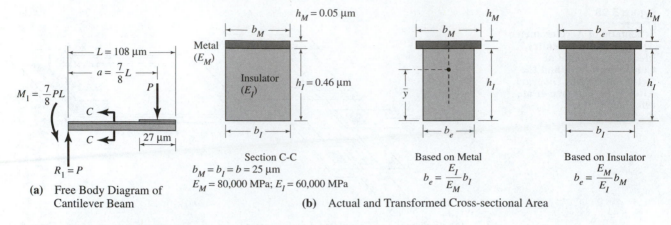

(a)   Free Body Diagram of Cantilever Beam

Section C-C
$b_M = b_I = b = 25 \ \mu m$
$E_M = 80,000 \ MPa; \ E_I = 60,000 \ MPa$

(b)   Actual and Transformed Cross-sectional Area

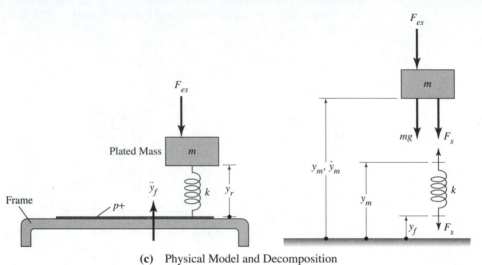

(c)   Physical Model and Decomposition

Figure 5.29

Free Body Diagram of Cantilever Beam with End-mass and Physical Model of the Accelerometer

the beam. Thus the equivalent spring constant for the vertical vibration of the plated mass can be taken as

$$k = \frac{3 \ EI}{a^3} \tag{5.68a}$$

Since the present beam has a composite cross section, the quantities E and I must be interpreted appropriately. As shown in Fig. 5.29b, for a cross section composed of two materials, the *equivalent* or *transformed cross-sectional area* can be based on either material. If we use the metal coating as our basis, then E in equation (5.68a) is the modulus of elasticity of gold, that is, E_M. The area moment of inertia for the composite cross section is given by the *parallel-axis theorem*:

$$I = I_0 + Ad^2 \tag{5.68b}$$

where I is the (area) moment of inertia about an axis that is parallel to the centroidal axis but separated from it by the perpendicular distance d, A is the area of the cross section, and I_0 is the moment of inertia of the cross section about the centroidal axis. The centroid of the composite cross section $\bar{y}$ is given by

$$\bar{y} = \frac{\sum_j A_j \bar{y}_j}{\sum_j A_j} = \frac{A_M \bar{y}_M + A_I \bar{y}_I}{A_M + A_I} = \frac{(bh_M)\left(\frac{1}{2}h_M + h_I\right) + (b_e h_I)\left(\frac{1}{2}h_I\right)}{bh_M + b_e h_I} \tag{5.68c}$$

$$= 0.2623 \ \mu m$$

where subscript e refers to the equivalent cross section of the insulator shown in the middle figure in Fig. 5.29b. For a rectangular cross section with width b and height h,

$$I_0 = \frac{bh^3}{12}; \quad \text{(rectangle)} \tag{5.68d}$$

Applying equation (5.68b) to the two areas, we obtain

$$I = I_{0M} + A_M \left(\frac{1}{2}h_M + h_I - \bar{y}\right)^2 + I_{0e} + A_e \left(\frac{1}{2}h_I - \bar{y}\right)^2 \tag{5.68e}$$

$$= 0.2233 \ (\mu m)^4$$

Substituting this value of I and the value of E_M into equation (5.68a), we have $k = 0.0635$ N/m.

Vertical vibration of plated mass: Figure 5.29c shows the equivalent physical model for the vertical motion of the plated mass. Note that vertical displacements have been defined as positive upward to agree with the implied spring behavior in which the restoring force on the mass is downward. F_{es} is the electrostatic force. It is assumed that the frame to which the accelerometer is attached is subject to the acceleration $\ddot{y}_f$, where y_f is the absolute vertical displacement of the frame and hence the p$^+$ base of the accelerometer (the accelerometer is assumed to be rigidly attached to the frame). The absolute vertical displacement of the plated mass is y_m. Hence the relative displacement of the beam metal layer from the p$^+$ layer is

$$y_r = y_m - y_f \tag{5.69a}$$

Newton's second law and Hooke's law (see decomposition, Fig. 5.29c) give

$$F_s = k\,(y_m - y_f) = ky_r \tag{5.69b}$$

$$m\,\frac{d}{dt}\,\dot{y}_m = -mg - F_s - F_{es} = -mg - ky_r - F_{es} \tag{5.69c}$$

Our interest is in the equation for y_r. From equations (5.69a) and (5.69c), we see that

$$m\,\frac{d}{dt}\,\dot{y}_r = m\,\frac{d}{dt}\,\dot{y}_m - m\,\frac{d}{dt}\,\dot{y}_f = -\,[mg + ky_r + F_{es}] - m\ddot{y}_f$$

or

$$\frac{m}{k}\frac{d}{dt}\dot{y}_r = -\left(y_r + \frac{F_{es}}{k} + \frac{mg}{k}\right) - \frac{m}{k}\ddot{y}_f$$

(5.69d)

Electrostatic force and variable capacitance: Although the cantilever beam is, in general, deflected into a curve, the change in capacitance between the beam metal layer and the p^+ layer with respect to an equilibrium value can be determined quite adequately still assuming a parallel-plate configuration. Similarly, the electrostatic force can be taken from Fig. 5.8c as

$$F_{es} = \frac{V_b^2 \epsilon_0 bt}{2y_r^2}$$

(5.69e)

where ϵ_o is the permitivity of free space and V_b is the voltage drop. If we assume calibration of the detection circuitry such that $V_b = 0$ and hence $F_{es} = 0$ at equilibrium ($\ddot{y}_f = 0$ and $\ddot{y}_r = \dot{y}_r = 0$), then the equilibrium displacement of the beam at the plated mass is the negative of the static deflection there:

$$y_E = -\frac{P}{k} = -\frac{mg}{k} \approx -0.054 \ \mu m$$

(5.69f)

The equilibrium separation between the "parallel" plates (also at the plated mass) x_E is the difference between the average spacing, the static deflection, and the deflection due to residual stress. The deflection of the beam at the center of the plated mass region due to residual stress is approximately

$$\delta_S \approx a\pi \frac{(-1.5°)}{180°} = -2.474 \ \mu m$$

(5.69g)

We note that the magnitude of the deflection due to residual stress is significant relative to the average spacing, $x_A = 7 \ \mu m$. However, this deflection is in a direction that increases the average spacing, so that there is no possibility of contact between the vibrating beam and its base due to the effect of the residual stress. For the equilibrium separation,

$$x_E = x_A + y_E - \delta_S = (7 - 0.054 + 2.474) \approx 9.42 \ \mu m$$

(5.69h)

From Fig. 5.8c, the corresponding nominal or equilibrium capacitance is given by

$$\frac{1}{C_{B0}} = \frac{x_E}{\epsilon_0 bL}$$

(5.69i)

State model: We consider now the relative displacement of the plates with respect to the equilibrium point x_r, or the change in separation of the plates from the nominal value:

$$x_r = y_r - y_E = y_r + \frac{mg}{k}; \qquad \frac{dx_r}{dt} = \frac{dy_r}{dt}; \qquad \frac{d\dot{x}_r}{dt} = \frac{d\dot{y}_r}{dt} = \frac{dv_r}{dt}$$

(5.70a)

The instantaneous parallel-plate capacitance is given by

$$\frac{1}{C_B} = \frac{x_E + x_r}{\epsilon_0 bL} = \frac{1}{C_{B0}} + \frac{x_r}{\epsilon_0 bL} \tag{5.70b}$$

However, for the approximate model desired, the instantaneous electrostatic force, equation (5.69e), may be considered negligible, given the assumed calibration, $V_b(y_E) = 0$. Substituting $F_{es} = 0$ and equation (5.70a) in equation (5.69d), we obtain

$$\frac{d}{dt} x_r = v_r \tag{5.70c}$$

$$\frac{d}{dt} v_r = -\frac{k}{m} x_r - \ddot{y}_f \tag{5.70d}$$

Finally, given the current i through the beam capacitor, the voltage drop is approximately given by equation (2.24b) is

$$\frac{d}{dt} V_b = \left(\frac{1}{C_{B0}} + \frac{x_r}{\epsilon_0 bL} \right) i \tag{5.70e}$$

$$\text{Output} = V_b \tag{5.70f}$$

Equations (5.70c) through (5.70f) constitute the model for the accelerometer. The assumption implicit in equation (5.70e) is that the current i is independently determined and is unaffected by changes in the beam capacitance. To explore the condition for x_r to provide, in general, a good measure of the frame acceleration, we substitute equation (5.70c) into (5.70d) and consider the resulting second-order linear ordinary differential equation:

$$\frac{d^2 x_r}{dt^2} + \omega_0^2 x_r = -\ddot{y}_f; \qquad \omega_0 = \sqrt{\frac{k}{m}} \tag{5.71a}$$

Note that due to the definition of x_r as relative displacement from the equilibrium displacement, the initial conditions of equation (5.71a) are zero. Suppose the input acceleration can be represented by a sinusoid of frequency ω, then as shown in Section 7.3 (see also Example 2.6 and Section 6.2), the ratio of the relative displacement amplitude to the input acceleration amplitude is given by

$$\frac{|x_r|}{|\ddot{y}_f|} = \frac{1}{\sqrt{(\omega_0^2 - \omega^2)^2}} \tag{5.71b}$$

where ω_0 is known as the natural frequency of the accelerometer. If we consider only the amplitude ratio or the sensitivity (other measures of performance are considered in Section 7.3), this quantity as given by equation (5.71b) will become relatively independent of the input acceleration if the natural frequency of the device is much higher than the frequency of the input acceleration. In practice, accelerometers are designed to have natural frequencies that are at least twice the highest frequency of acceleration to be measured. For the present accelerometer, the natural frequency is approximately

$$\omega_0 = \sqrt{\frac{k}{m}} = \sqrt{\frac{0.0635}{0.35 \times 10^{-9}}} = 13,470 \text{ rad/s, or } 2144 \text{ Hz}$$

The experimentally determined natural frequency of the accelerometer in reference [22] was 2200 Hz. Note that the effect of the plated end-mass is to lower the natural frequency of the cantilever beam structure. If this mass is reduced, the natural frequency of the accelerometer can be made quite high. Indeed, because of their size and material of manufacture, micromachined structures normally have very high natural frequencies, so these structures are inherently good accelerometer components.

Example 5.9   Surface-Micromachined Magnetic Microactuator

Considerable effort has gone toward realizing magnetic-based microactuators that can be fabricated on the same substrate with an integrated circuit. Such a fully integrated electromagnetic microactuator fabricated using surface micromachining technique is described in reference [23]. Figures 5.30a and 5.30b show, respectively, the *scanning electron micrograph* (SEM) and the schematic diagram of the microactuator. The mag-

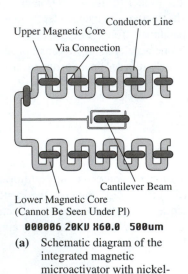

(a)   Schematic diagram of the integrated magnetic microactivator with nickel-iron cantilever beam.

(b)   Analogous Schematic Diagram

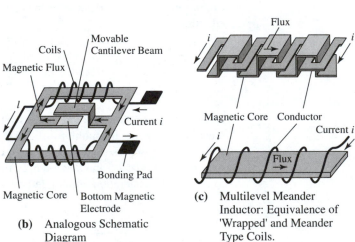

(c)   Multilevel Meander Inductor: Equivalence of 'Wrapped' and Meander Type Coils.

Figure 5.30

Scanning Electron Micrograph, Schematic Diagrams, and Processing Sequence for a Fully Integrated Magnetic Microactuator. After Chong Ahn and Mark Allen, reference [23]

netic circuit in Fig. 5.30b consists of a *meander*-type coil, represented in the figure by equivalent "wrapped" solenoid type coils, and a meander magnetic core incorporating an air gap formed by a movable surface-micromachined nickel-iron cantilever beam, 2.5 μm thick, 25 μm wide, and 780 μm long.

The actual meander inductor is illustrated schematically in Fig. 5.30c. It contains 17 turns of a meander planar conductor, interlaced with a multilevel meander magnetic core (the core is wrapped or woven around the conductor—a reversal of the normal configuration found in macroinductors; see lower figure). By switching in this way the roles of conductor and core, the wrapping effect of a regular coil is realized in a meander coil. The processing sequence for the microactuator is illustrated in Fig. 5.30d. The conductor for the present example is copper, and the magnetic core is a (81%) nickel-(19%) iron alloy. Derive a state model for this system that will yield the cantilever beam tip deflection as a function of the voltage applied to the coils.

Solution: We make the following assumptions in developing the model:

1. The magnetic flux across the air gap (between the movable cantilever beam and the bottom magnetic electrode, see Fig. 5.30b) is distributed uniformly, and fringing effects are negligible.
2. The magnetic core does not saturate, and the permeability has a constant value.
3. The magnetic force generated in the air gap is also uniformly distributed along the length of the cantilever beam.

Tip deflection and strain energy of cantilever beam with uniformly distributed load: Because of assumption (3), the free body diagram of the cantilever beam is as shown in Fig. 5.31a. Although Fig. 2.3b does not apply in this case, the deflection of a *beam with uniform load* is one of the standard beam deflection problems whose solutions are usually tabulated in solid mechanics or machine design texts. From reference [24], for example, entry (3) of Table A-9 gives the tip deflection y_t, the deflection y at any point x, a lateral location from the built-in end, and the moment M at x, as follows:

$$y_t = \frac{w l_b^4}{8EI} \tag{5.72a}$$

$$y = \frac{w}{24EI}(6l_b^2 x^2 + x^4 - 4l_b x^3) \tag{5.72b}$$

$$M = -\frac{w}{2}(l_b - x)^2 \tag{5.72c}$$

where E is the modulus of elasticity of the beam material, I is the (area) moment of inertia of the beam cross section (a rectangle in this case), and w is the (distributed) uniform load per unit length of the beam. The deflections in equations (5.72a) and (5.72b) are positive in the direction of the distributed load. The external work done on an elastic member in deforming it is transformed into strain or potential energy. The strain energy stored in a beam due to bending is given by (see reference [24], for example)

$$U = \int \frac{M^2 \, dx}{2EI}; \quad \text{(strain energy, bending)} \tag{5.72d}$$

Polyimide Layer

Plated Bottom Core

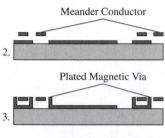

1.

Si Wafer

Titanium Seed Layer

Meander Conductor

2.

Plated Magnetic Via

3.

Cavity

4.

Sacrificial Layer

5.

Top Core   Cantilever Beam

6.

Released Cantilever Beam

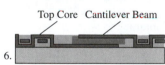

7.

(d)   Processing Sequence:
1. Multi-spincoating of Polyimide, Dry Etching, and Bottom Core Plating
2. Patterning of Conductor
3. Magnetic Via Plating
4. Dry Etching of Cavity
5. Patterning of Sacrificial Layer
6. Top Core and Cantilever Beam Plating
7. Release of Cantilever Beam

Figure 5.30, *continued*

Figure 5.31

Free Body Diagram of
Cantilever Beam, Magnetic
Circuit, and Air Gap

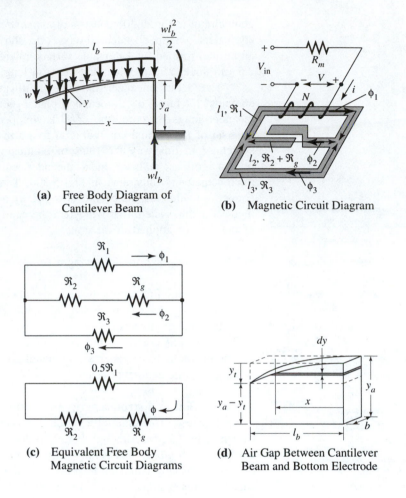

(a)   Free Body Diagram of
Cantilever Beam

(b)   Magnetic Circuit Diagram

(c)   Equivalent Free Body
Magnetic Circuit Diagrams

(d)   Air Gap Between Cantilever
Beam and Bottom Electrode

Substituting equation (5.27c) into equation (5.27d), we obtain

$$U = \int_0^{l_b} \frac{w^2}{8EI} (l_b - x)^4 \, dx = \frac{w^2 l_b^5}{40EI} \tag{5.72e}$$

From equation (5.72a), in terms of the beam tip deflection, the stored strain energy is

$$U = \frac{8EI}{5l_b^3} y_t^2 \tag{5.72f}$$

Magnetic circuit and reluctance of air gap: Figure 5.31b shows the magnetic circuit diagram of the microactuator. As was illustrated in Fig. 5.11c, the present circuit is a parallel magnetic circuit. Because the magnetic core lengths l_1 and l_3 are equal, and the core material and cross-sectional areas are the same, the reluctance of both paths $\mathfrak{R}_1$ and $\mathfrak{R}_3$ (see equivalent "free body" magnetic circuit diagrams, Fig. 5.31c) are also the same, so that their parallel combination is equal to $0.5\,\mathfrak{R}_1$. As shown in Fig. 5.31c, the resulting circuit is a series magnetic circuit with the total reluctance

$$\mathfrak{R} = 0.5\mathfrak{R}_1 + \mathfrak{R}_2 + \mathfrak{R}_g \tag{5.73a}$$

where $\mathfrak{R}_2$ is the total reluctance of the central magnetic core of total length l_2 but excluding the air gap that interrupts this core path and has the reluctance $\mathscr{R}_g$. From the definition, equation (5.37c), and equation (5.73a), the magnetic flux in the series circuit, which is also the magnetic flux in the air gap, is given by

$$\phi = \frac{Ni}{0.5\mathfrak{R}_1 + \mathfrak{R}_2 + \mathfrak{R}_g} = \frac{Ni}{0.5\, l_1/\mu_c A_c + l_2/\mu_c A_c + \mathfrak{R}_g} \tag{5.73b}$$

where N is the number of turns ($N = 17$), i is the current in the meander conductor, μ_c is the permeability of the magnetic core, and A_c is the core cross-sectional area, which is assumed to be the same for all the core paths. Because of the assumed distributed nature of the magnetic force and the relatively large base cross-sectional area A_g of the air gap path, a determination of $\mathfrak{R}_g$ that takes into account the deflected curve of the cantilever beam is appropriate. As shown in Fig. 5.31d, for a given average spacing between the movable cantilever beam and the fixed bottom magnetic electrode y_a and tip deflection y_t, the air gap can be divided into a constant-height segment and a variable-height portion, so that the reluctance is given by

$$\mathfrak{R}_g = \frac{y_a - y_t}{\mu_o A_g} + \frac{1}{\mu_o} \int_0^{y_t} \frac{dy}{bx} \tag{5.73c}$$

Substituting for dy from equation (5.72b), we have

$$\mathfrak{R}_g = \frac{y_a - y_t}{\mu_o A_g} + \frac{w}{24EI\mu_o b} \int_0^{l_b} (12l_b^2 + 4x^2 - 12l_b x)\, dx \tag{5.73d}$$

$$= \frac{y_a - y_t}{\mu_o A_g} + \frac{22l_b^3 w}{72EI\mu_o b}$$

Using equation (5.72a) and $A_g = bl_b$, we finally obtain

$$\mathfrak{R}_g = \frac{y_a - y_t}{\mu_o A_g} + \frac{22y_t}{9\mu_o A_g} = \frac{y_a + \dfrac{13}{9} y_t}{\mu_o A_g} \tag{5.73e}$$

Equation (5.73b) then becomes

$$\phi = \frac{Ni}{0.5\, \dfrac{l_1}{\mu_c A_c} + \dfrac{l_2}{\mu_c A_c} + \dfrac{y_a + \dfrac{13}{9} y_t}{\mu_o A_g}} \tag{5.73f}$$

$$= \frac{Ni\mu_c A_c \mu_o A_g}{(0.5l_1 + l_2)\, \mu_o A_g + \left(y_a + \dfrac{13}{9} y_t \right) \mu_c A_c}$$

Electromechanical energy converter (inductor): By equation (5.41), the inductance is

$$L = \frac{N\phi}{i} = \frac{N^2 \mu_c A_c \mu_o A_g}{(0.5l_1 + l_2)\mu_o A_g + \left(y_a + \frac{13}{9} y_t\right)\mu_c A_c} \tag{5.74a}$$

Also, from equations (5.43a) and (5.74a), the stored electromechanical energy is given by

$$E_{em}(\lambda, y_t) = \frac{1}{2}\frac{\lambda^2}{L} = \frac{1}{2}\frac{\lambda^2\left[(0.5l_1 + l_2)\mu_o A_g + \left(y_a + \frac{13}{9} y_t\right)\mu_c A_c\right]}{N^2 \mu_c A_c \mu_o A_g} \tag{5.74b}$$

Consequently, the equations corresponding to equations (5.42a) and (5.44) are as follows:

$$dE_{em} = i\, d\lambda - dU \tag{5.74c}$$

$$dE_{em} = \frac{\partial E_{em}}{\partial \lambda} d\lambda + \frac{\partial E_{em}}{\partial y_t} dy_t \tag{5.74d}$$

Thus

$$\frac{dU}{dt} = -\frac{\partial E_{em}}{\partial y_t}\frac{dy_t}{dt} = -\frac{1}{2}\frac{\lambda^2\left[\frac{13}{9}\mu_c A_c\right]v_t}{N^2 \mu_c A_c \mu_o A_g}; \qquad v_t = \frac{dy_t}{dt} \tag{5.74e}$$

which gives us, using $\lambda = Li$ and equation (5.74a),

$$\frac{dU}{dt} = -\frac{1}{2}\frac{\frac{13}{9} N^2 \mu_c^2 A_c^2 \mu_o A_g i^2 v_t}{\left[(0.5l_1 + l_2)\mu_o A_g + \left(y_a + \frac{13}{9} y_t\right)\mu_c A_c\right]^2} \tag{5.74f}$$

Equation (5.43a) also gives the stored electromechanical energy as

$$E_{em} = \frac{1}{2} Li^2 \tag{5.74g}$$

which implies

$$\frac{dE_{em}}{dt} = \frac{1}{2} i^2 \frac{dL}{dt} + Li\frac{di}{dt} \tag{5.74h}$$

Also, for this case equation (5.35b) is

$$\frac{dE_{em}}{dt} = Vi - \frac{dU}{dt} \tag{5.74i}$$

where V is the voltage drop across the inductor (transducer) only. Setting equation (5.74h) equal to equation (5.74i), substituting for dL/dt using equation (5.74a), and using equation (5.74f) for dU/dt, we obtain

$$L\frac{d}{dt} = V + \frac{\frac{13}{9}N^2\mu_c^2A_c^2\mu_o A_g}{\left[(0.5l_1 + l_2)\mu_o A_g + \left(y_a + \frac{13}{9}y_t\right)\mu_c A_c\right]^2}iv_t \qquad \text{(5.74j)}$$

Motion of the beam tip: The mass of the movable cantilever beam is assumed to be negligible. This means that the beam behaves like a flexural spring only. The equivalent spring constant, for example, can be obtained by comparing the potential energy as given by equation (5.72f) with the generalized expression, equation (5.13b) or (5.13d), with y_t as the coordinate variable. That is,

$$k = \frac{16EI}{5l_b^3} \qquad \text{(5.75a)}$$

To find the beam tip velocity, we differentiate the potential energy equation (5.72f) and set the result equal to equation (5.74e):

$$\frac{dU}{dt} = \left[\frac{16EI}{5l_b^3}y_t\right]v_t = -\frac{1}{2}\frac{\lambda^2\left[\frac{13}{9}\mu_c A_c\right]v_t}{N^2\mu_c A_c\mu_o A_g} \qquad \text{(5.75b)}$$

We solve for y_t:

$$y_t = -\frac{65l_b^3}{288EI}\frac{\lambda^2}{N^2\mu_o A_g} \qquad \text{(5.75c)}$$

An alternative approach to equation (5.75c) is illustrated in Practice Problem 5.29.

Taking the derivative of both sides and substituting for $\lambda = Li$ from equation (5.74a), we get

$$v_t = -\frac{65l_b^3}{144EI}\frac{\lambda}{N^2\mu_o A_g}\frac{d\lambda}{dt} \qquad \text{(5.75d)}$$

$$= -\frac{65l_b^3}{144EI}\frac{\mu_c A_c i}{\left[(0.5l_1 + l_2)\mu_o A_g + \left(y_a + \frac{13}{9}y_t\right)\mu_c A_c\right]}\frac{d\lambda}{dt}$$

We apply equation (5.40) to obtain finally

$$v_t = -\frac{65l_b^3}{144EI}\frac{\mu_c A_c}{\left[(0.5l_1 + l_2)\mu_o A_g + \left(y_a + \frac{13}{9}y_t\right)\mu_c A_c\right]}Vi \qquad \text{(5.75e)}$$

State model: The major difference between the present problem and Example 5.4 is that the mechanical subsystem here consists of a spring only, and the electrical circuit does not include a transistor junction. Beginning with the inductor, the state equation is given by equations (5.74a) and (5.74j). That is,

$$\frac{di}{dt} = \frac{\left[(0.5l_1 + l_2)\mu_o A_g + \left(y_a + \frac{13}{9}y_t\right)\mu_c A_c\right]}{N^2\mu_c A_c \mu_o A_g}V$$

$$+ \frac{\frac{13}{9}\mu_c A_c}{\left[(0.5l_1 + l_2)\mu_o A_g + \left(y_a + \frac{13}{9}y_t\right)\mu_c A_c\right]}iv_t$$

But $V = V_{in} - iR_m$ (see Fig. 5.31b), where R_m is the resistance of the meander conductor. Substituting for V and for v_t from equation (5.75e), we obtain

$$\frac{di}{dt} = \left\{\frac{\left[(0.5l_1 + l_2)\mu_o A_g + \left(y_a + \frac{13}{9}y_t\right)\mu_c A_c\right]}{N^2\mu_c A_c\mu_o A_g} - \frac{845l_b^3}{1296EI}\right.$$

$$\left.\frac{\mu_c^2 A_c^2}{\left[(0.5l_1 + l_2)\mu_o A_g + \left(y_a + \frac{13}{9}y_t\right)\mu_c A_c\right]^2}i^2\right\}\{V_{in} - iR_m\} \quad \text{(5.76a)}$$

Finally, for the spring, we write

$$\frac{dy_t}{dt} = -\frac{65l_b^3}{144EI}\frac{\mu_c A_c}{\left[(0.5l_1 + l_2)\mu_o A_g + \left(y_a + \frac{13}{9}y_t\right)\mu_c A_c\right]}\{V_{in} - iR_m\} \quad \text{(5.76b)}$$

Equations (5.76a) and (5.76b) constitute the state model of the magnetic microactuator. The output is y_t

5.5 PRACTICE PROBLEMS

5.1(a) Derive the transformer equations (5.15a) and (5.15b) for the simple gear train shown in Fig. 5.4b, assuming ideal behavior.

5.1(b) Derive the system model for a real gear train, which includes the inertia of the two gears, viscous friction in the shaft bearings, and backlash in the teeth. Assume that backlash can be modeled effectively in a continuous-time analysis by slippage (nonlinear shunt resistance).

5.2 Derive the state equations for the motion of the gear train shown in Fig. 5.32, assuming the members are rigid and that friction between the gears is negligible.

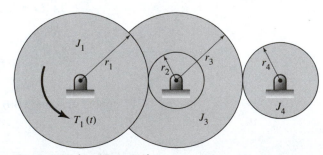

Figure 5.32 (Problem 5.2)

5.3* Derive the model of the vehicle/flywheel system of Problem 3.19 using a generalized or partitioned system approach. Explain your model of the clutch described in Problem 3.19.

5.4 Use a generalized system approach to obtain the equations for

(a) the electrical circuit shown in Fig. 5.33a.

(b) the transistor circuit shown in Fig. 5.33b.

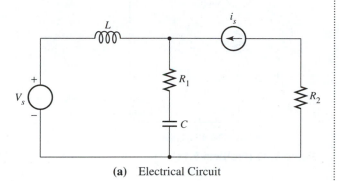

(a) Electrical Circuit

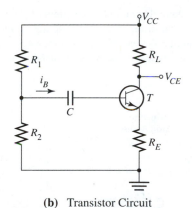

(b) Transistor Circuit

Figure 5.33 (Problem 5.4)

5.5 Derive the system model for the electrical system whose circuit diagram is shown in Fig. 5.34, assuming that the desired output is the voltage across R_4.

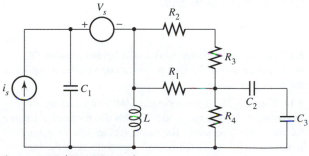

Figure 5.34 (Problem 5.5)

5.6* Repeat Problem 4.13 from a generalized system viewpoint. Assume that the working fluid is incompressible.

5.7 Derive the state model for the system shown in Fig. 5.35. Assume that the shape and material of the movable member are identical with those of the fixed member.

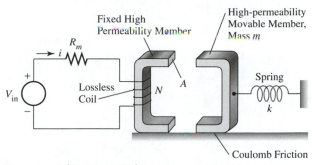

Figure 5.35 (Problem 5.7)

5.8 For the given relay system in Fig. 5.36, derive the equations necessary to determine the motion x of the plunger. The magnetic core and the plunger material can be assumed to be infinitely permeable. Assume that their air gap y_g is small compared with the thickness of the plunger.

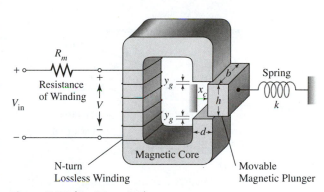

Figure 5.36 (Problem 5.8)

5.9* Figure 5.37 illustrates a rudimentary reluctance machine having a single winding on the stator. Assuming the rotor and stator are shaped in such a way that the variation of the inductance of the winding is sinusoidal with respect to the rotor angular position as shown, find the instantaneous torque if the rotor angular velocity is ω_m.

5.10 Figure 5.38 shows the cross section of a loudspeaker and a schematic diagram of an electrodynamic speaker. Derive a model for this system, assuming the output sound wave is due to the pressure variation in (compressible) air within the speaker cone.

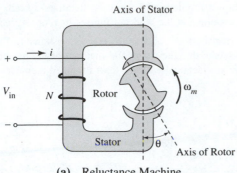

(a)   Reluctance Machine

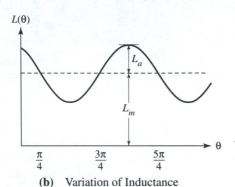

(b)   Variation of Inductance

Figure 5.37 (Problem 5.9)

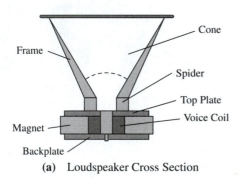

(a)   Loudspeaker Cross Section

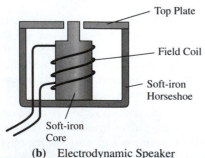

(b)   Electrodynamic Speaker

Figure 5.38 (Problem 5.10)

5.11* Develop a model for the capacitor microphone shown in Fig. 5.39 that will predict the output AC voltage due to a sinusoidal sound (pressure) wave incident on the diaphragm (see Fig. 5.8c).

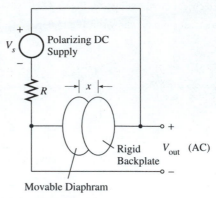

Figure 5.39 (Problem 5.11)

5.12 A schematic diagram of a real single-pole–single-throw relay is as shown in Fig. 5.40. Develop a state model for the system, assuming the use of a transistor switch. State the simplifying assumptions you made in arriving at your model.

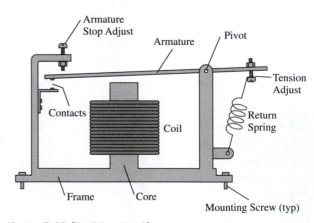

Figure 5.40 (Problem 5.12)

5.13 Derive the dynamic model of a field-controlled DC motor (discussed in Section 5.2). Include viscous damping of the rotor motion.

5.14 Figure 5.41 shows a simple $L/4R$ (unipolar) stepping motor drive. Derive the model for this drive circuit assuming the transistors operate in the cutoff region. Use switching functions to represent the control inputs. Include the back emf and resistance for each winding (inductor).

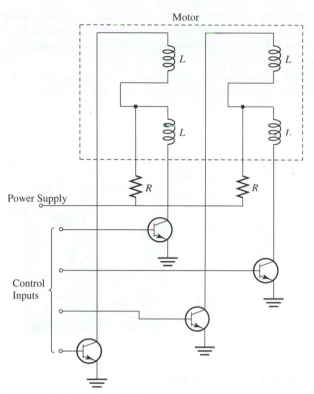

Figure 5.41 (Problem 5.14)

5.15 Employing switching functions and the model of the unipolar drive circuit as a guide, develop the model of the bipolar chopper drive described in Section 5.2 and illustrated in Fig. 5.18. Explain your assumptions.

5.16* Find the model for the temperature T of the liquid bath shown in Fig. 5.42. The effect of the stirrer is to give a

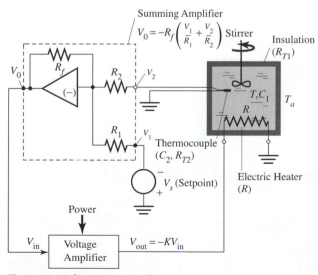

Figure 5.42 (Problem 5.16)

uniform distribution of temperature throughout the bath. Assume that the voltage drop developed across the thermocouple leads is approximately the Seebeck voltage.

5.17 Derive the mathematical model for the simple rolling mill shown in Fig. 5.43. The desired thickness is set by the thumbscrew, which alters the reference position of a walking beam in the hydraulic servomotor.

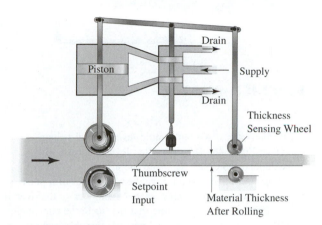

Figure 5.43 (Problem 5.17)

5.18* Derive the model for the hydraulic brake system shown in Fig. 5.44. Assume that the gas above the oil reservoir behaves like a perfect gas, and the frictional force between the brake shoe and drum is of the Coulomb type. State any other assumptions you make in arriving at your solution.

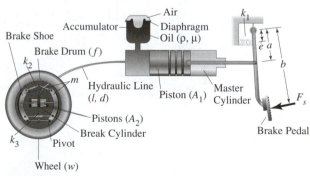

Figure 5.44 (Problem 5.18)

5.19 Consider control of the pumping system of Problem 4.9. A piezoelectric crystal pressure transducer is used to switch the pump off when the pressure in the tank above the liquid exceeds a certain value. Derive the overall model for the arrangement shown in Fig. 5.45.

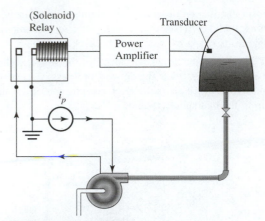

Figure 5.45 (Problem 5.19)

5.20* Consider a two-tank liquid-level control system as shown in Fig. 5.46. The controlled level is measured by a capacitance probe. The flow into the upper tank, which supplies the controlled tank, is regulated by means of a pneumatic amplifier control valve arrangement. Derive the overall model for this system. Assume that the air in the pneumatic amplifier behaves like a perfect gas and that the discharge through the nozzle is given by

$$\dot{m} = c_1 A_n \left\{ P\left[\left(\frac{P_a}{P}\right)^{1.4} - \left(\frac{P_a}{P}\right)^{1.7} \right] \right\}^{1/2}$$

for unchoked flow, that is, for $(P_a/P) > 0.53$, and for choked flow $[(P_a/P) < 0.53]$ by

$$\dot{m} = c_2 A_n P$$

where c_1 and c_2 are constants, A_n is the nozzle area, P_a is the absolute (ambient) pressure outside the nozzle, and P is the absolute pressure inside the nozzle and valve diaphragm chambers.

5.21 Repeat Problem 5.20 but replace the capacitance probe with the float-potentiometer arrangement shown in Fig. 5.47.

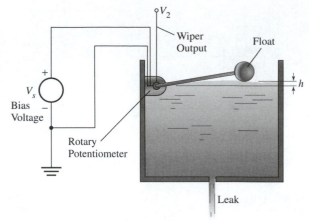

Figure 5.47 (Problem 5.21)

5.22 Develop a dynamic model for the laboratory water clock display unit shown in Fig. 5.48. Two tanks in cascade are utilized in the control of the flow into the clock display tube. However, only the level of the first tank is measured

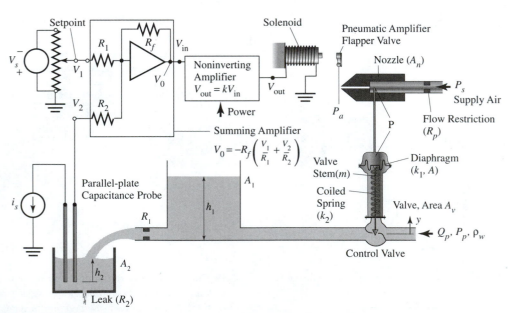

Figure 5.46 (Problem 5.20)

(using a capacitance probe). For simplicity, assume that the controller uses only proportional control, that is, the voltage-supply to the motor actuator is proportional to the error in the tank level. Also assume that the motor operates under armature control.

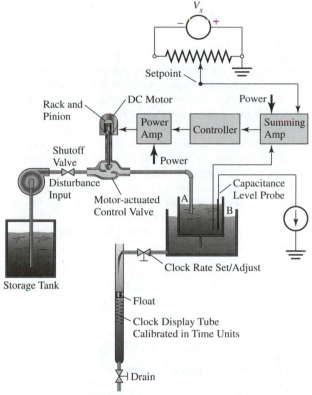

Figure 5.48 (Problem 5.22)

5.23* Derive continuous-time and discrete-time models for the configuration shown in Fig. 5.49 of a Control System Laboratory Reconfigurable Double Pendulum Unit. The actuator is an electroproportional solenoid that operates by pulse width modulation (see Section 15.2). However, its operation can be described for the present purposes simply as follows: the piston displacement u is proportional to the command input V_m. Also, ignore for now the details of the signal conditioning and converter circuits, but simply assume that the control algorithm implemented in the digital computer results in a command input V_m that is given by

$$V_m = K_c[(\text{error}) + I_d(\text{rate of change of error})]$$

where the error is the difference between the setpoint entered into the computer program and the measured pendulum position, and K_c and I_d are controller parameters. Assume that the motion of the pendulum is undamped.

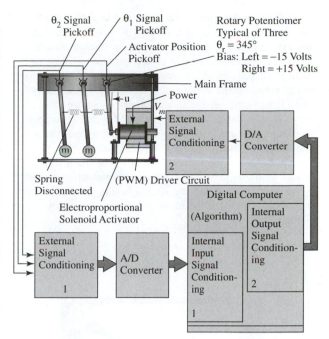

Figure 5.49 (Problem 5.23)

5.24* Figure 5.50 shows a schematic diagram of a micromachined check valve based on the cantilever beam with plated end-mass microstructure. Determine the equations describing the operation of the valve based on the voltage input to the capacitive actuator. The electrostatic force as a function of the voltage drop is given in Fig. 5.8c. Assume any needed parameters.

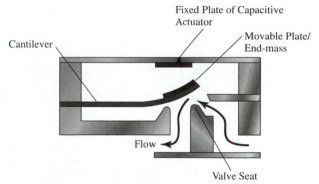

Figure 5.50 (Problem 5.24)

5.25 Repeat Problem 5.24 assuming a micromachined piezoelectric device is used as the actuator.

5.26 The micromachined cantilever beam device of Problem 5.24 is to be used as flow sensor. As shown in Figure 5.51, the sensing device is a micromachined strain gage. A strain measurement system utilizing strain gages and a

Wheatstone bridge circuit was described in Problem 2.51. Assuming that an electronic version of such a signal reduction circuit is integrated with the present microflowmeter, derive a state model that can predict the volume flowrate of an incompressible fluid through the device. Assume that two micromachined strain gages, oriented parallel to the axis of the cantilever beam, are utilized as shown. The strain gage at the tip of the beam is a dummy or compensator gage. It suffers no strain because the normal stress (the bending moment) is zero here. Because of the orientation of the gages, the one at the root of the beam measures the strain due to the maximum normal stress caused by bending. Assume that the simple bending stress formula is applicable and that the material behavior is linearly elastic.

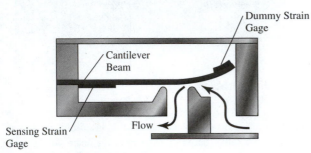

Figure 5.51 (Problem 5.26)

5.27 Determine the model of the micromachined diaphragm pressure sensor shown in Fig. 5.52. The sensing element is a parallel-plate capacitor. Assume that the moving plate of the capacitor constitutes a plated mass on the diaphragm, which has a circular surface area, is thin (a membrane), and otherwise has negligible mass. The displacement at the center of a circular membrane y due to a uniform pressure over the entire surface area P is given by

$$\frac{P}{E}\left(\frac{r}{h}\right)^4 = \frac{5.33}{1-v^2}\left(\frac{y}{h}\right) + \frac{2.60}{1-v^2}\left(\frac{y}{h}\right)^3$$

where r is the radius of the circular surface, h is the thickness of the membrane, and E and v are the modulus of elasticity and Poisson's ratio, respectively, of the membrane material. Assume that the pressure being measured is uniformly distributed over the entire surface of the diaphragm. Explain any other simplifying assumptions that you make.

5.28* Figure 5.53 illustrates an application of the micromachined diaphragm device of Problem 5.27 as an electrostatically actuated control valve. Derive the model for the system to predict the flowrate of an essentially incompressible fluid through the valve as a function of the applied voltage drop

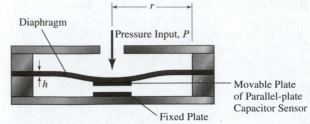

Figure 5.52 (Problem 5.27)

across the capacitor plates. Assume that the relation between pressure drop and flow across the valve is linear when based on the annular flow area between the diaphragm and the valve seat. Also assume that the diaphragm behaves like a thin membrane, so that the electrostatic force is uniformly distributed over the entire surface of the diaphragm. The relation between the electrostatic force and the voltage drop across a parallel-plate capacitor is given in Fig. 5.8c. The relation between the uniform pressure on the diaphragm surface and the displacement of the center of the diaphragm was given in Problem 5.27. Assume as in Problem 5.27 that the moving plate of the capacitor constitutes a plated mass on the diaphragm.

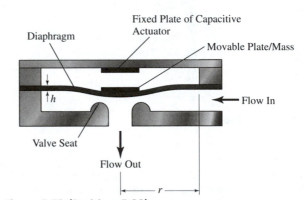

Figure 5.53 (Problem 5.28)

5.29 Show that in example 5.9, for the motion of the beam tip, equation (5.75c) can also be obtained from the electromechanical transducer force output F_e and Hook's law: $F_e = -ky_t$, where k is given by equation (5.75a) and F_e is given by equation (5.45b): $F_e = -\dfrac{\partial E_{em}}{\partial y_t}$.

5.30 Repeat Example 5.9 for the case in which the cantilever beam has a plated end-mass. What are the major differences between the behavior of the magnetic microactuator with and without a plated mass on the end of the cantilever beam microstructure?

REFERENCES FOR PART I

1. *ASME Orientation and Guide for Use of Metric Units,* 3rd ed., ASME, 1973.

2. Auslander, D. M., et al., *Introducing Systems and Control,* McGraw-Hill, New York, 1974.

3. Benzinger, T. H., "The Human Thermostat," *Scientific American,* January 1961.

4. Mayr, O., *The Origins of Feedback Control* (trans. from German), MIT Press, Cambridge, Mass., 1970.

5. Bigelow, S. J., *Troubleshooting and Repairing Computer Printers,* Windcrest/McGraw-Hill, New York, 1992.

6. Meriam, J. L., and Kraige, L. G., *Engineering Mechanics,* Vol. 2: *Dynamics,* John Wiley, New York, 1992.

7. Janna, W. S., *Introduction to Fluid Mechanics,* Brooks/Cole Engineering Division, Monterey, Calif., 1983.

8. Oldham, W. G., and Schwarz, S. E., *An Introduction to Electronics,* Holt, Rinehart and Winston, New York, 1972.

9. Holman, J. P., *Heat Transfer,* 5th ed., McGraw-Hill, New York, 1981.

10. Umez-Eronini, E. I., "Automatic Control and Simulation of a Small-Scale Low-Cost Solar Dryer," *The Nigerian Engineer,* Vol. 25, No. 4, pp. 10–22, 1990.

11. Van Wylen, G. J., and Sonntag, K. E., *Fundamentals of Classical Thermodynamics,* 3rd ed. John Wiley, New York, 1986.

12. Rosenberg, R. C., and Karnopp, D. C., *Introduction to Phyiscal System Dynamics,* McGraw-Hill, New York, 1983.

13. Angrist, S. W., *Direct Energy Conversion,* Allyn and Bacon, Boston, 1971.

14. Matsch, L. W., *Electromagnetic and Electromechanical Machines,* 2nd ed., Thomas Y. Crowell, New York, 1977.

15. Fitzgerald, A. E., et al., *Electric Machinery,* 4th ed., McGraw-Hill, New York, 1983.

16. Umez-Eronini, I. E., "Dynamics of a Stepping Motor Driven Conventional Machine Slide," *Proceedings of the 1988 ASME Design Automation Conference,* DE-Vol. 14, pp. 345–52, 1988.

17. Raven, F. H., *Automatic Control Engineering,* McGraw-Hill, New York, 1978.

18. Rosen, J., "Machining in the Micro Domain," *Mechanical Engineering,* Vol. 111, No. 3, pp. 40–46, March 1989.

19. O'Connor, L., "MEMS: Microelectromechanical Systems," *Mechanical Engineering,* Vol. 114, No. 2, pp. 40–47, February 1992.

20. Cochin, I., and Plass, H., Jr., *Analysis and design of dynamic systems,* Harper and Row, New York, 1990.

21. Frazier A. B., and Allen, M. G., "Metallic Microstructures Fabricated Using Photosensitive Polyimide Electroplating Molds," *Journal of Microelectromechanical Systems,* Vol. 2, No. 2, pp. 87–94, June 1993.

22. Petersen, K. E., Shartel, A., and Raley, N. F., "Micromechanical Accelerometer Integrated with MOS Detection Circuitry," *I.E.E.E. Trans. Electron Devices,* Vol. ED-29, No. 1, pp. 23–27, January 1982.

23. Ahn, C. H., and Allen, M. G., "Fully Integrated Surface Micromachined Magnetic Microactuator with a Multilevel Meander Magnetic Core," *Journal of Microelectromechanical Systems,* Vol. 2, No. 1, pp. 15–22, March 1993.

24. Shigley, J. E., and Mischke, C. R., *Mechanical Engineering Design,* 5th ed., McGraw-Hill, New York, 1989.

Model Solution

THE THIRD STAGE IN A DYNAMIC INVESTIGATION is *model solution*. The solution of the various system models constructed in Part I of the text is developed in Part II by subjecting classes of differential/difference equations that are representative of these models to various types of inputs and deriving the mathematical expressions for the system response. The results are therefore general and are applicable to any physical system described by the equations. Indeed, some of the subject matter of Chapters 6 and 7 parallels similar material in elementary texts on ordinary differential equations (see [1], for example).

We begin in Chapter 6 with the *time-domain* solution of the *first-order* system, which represents the simplest case. However, a system of more than one first-order differential/difference equation, of the general higher order state model, is structurally similar to a single first-order equation, so the higher order system or *vector state equation* is also considered in Chapter 6, following some preliminary material on *complex numbers*. Nonetheless, closed-form solutions are available for only a few simple and or linear cases. For the majority of models developed in Part I, numerical simulation by digital computer represents an effective means of analysis. Chapter 6 concludes with a discussion of digital computer solution of system models, especially the application of computational mathematics software, which in most cases can be used with little or no additional programming.

Because of their preponderance in the models of dynamic engineering systems, second- and higher-order *scalar ordinary differential equations* (ODEs) are treated separately, beginning with Chapter 7, and the results are applied especially to the response of *tuned RCL circuits* and to *forced mechanical vibrations*. Depending on the type of input, special methods of solution such as the *phasor transform* solution for models subject to *sinusoidal* inputs, and the *Fourier series* solution for systems subject to *periodic* inputs, are traditional. The solution of dynamic engineering system models by transformation is treated in Chapter 8. Transformation from the *time domain* to the *Laplace* and *frequency domains* and to the *z-domain* (for discrete-time systems) particularly reveals the structure and behavior of these systems. *Identification* and *frequency response* and the analysis of system behavior in *state space* are considered in this context in Chapter 9. Part II concludes with a detailed look, in Chapter 10, at *stability* of both *continuous-* and *discrete-time* systems, which is an important aspect of behavior that is determined by the structure of the system.

6

Response of Lumped-Parameter Systems

W E SAW IN PART I THAT LUMPED-PARAMETER SYSTEMS generally are described adequately by ordinary differential/difference equations. We consider in this chapter the particular case of linear ODEs and difference equations, especially of the first order. We shall exploit the special nature of these equations—*linear* and *ordinary*—to develop general solutions to this class of dynamic models. We shall do so by taking advantage of the well-established theory for ordinary differential equations, including the application of the *principle of superposition,* which allows the response of a *linear* system to be obtained as the sum of individual responses. For the general case or nonlinear state model, we shall present the solution of such systems by *digital computer simulation.* In particular, we shall consider the application of computational mathematics packages such as MATLAB and Mathcad to such problems.

6.1 STAGE THREE: MODEL SOLUTION

In stage three of a dynamic study we solve the equations representing the system model as a way of studying the dynamic behavior of the system. In so doing we determine whether the system model is good or bad, that is, whether the behavior is as expected. Also, we often use the same technique (in an iterative process) to arrive at an acceptable design that satisfies the purpose of the investigation. Model solution is therefore critical to the study of dynamic systems. In particular, we require solution techniques that will allow us to determine the system behavior under a variety of inputs and for different configurations or values of the system parameters. The natural response of a system initially displaced from its equilibrium state and in the absence of external inputs is known as *free response.* This is distinguished from the *forced response* or the response from the equilibrium state under the influence of an external input or disturbance.

Free Response, Time Constant, and Stability of First-Order Systems

We begin our consideration of solutions of dynamic system equations with the simplest case:

$$\frac{d}{dt} x(t) = ax(t); \qquad x(0) = x_0; \quad \text{(first order, free, and stationary)} \tag{6.1}$$

This is a *first-order* (there is only one state variable x) *free* (there are no input terms) system. Further, the coefficient a is a constant (that is, a *stationary* system) and represents some parameter of the system. The initial state of the system is $x(t) = x_0$ at time $t = 0$ (a more general initial condition is for $t = t_0$, the solution for which is considered later).

Example 6.1   Uniform Tank with a Leak

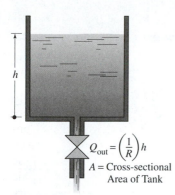

Figure 6.1

Leaking Tank

Consider the tank (of uniform cross section) with a linear valve or hole such that the flow out of the tank is directly proportional to the level of liquid in the tank (see Fig. 6.1). Determine the state model and relate it to equation (6.1). Assume the tank initially contains some quantity of liquid.

Solution: Equation (3.32b) gives

$$\frac{d}{dt}[\text{Volume}] = \frac{d}{dt}[Ah(t)] = -Q_{\text{out}} = -\left(\frac{1}{R}\right)h(t)$$

which further yields

$$\frac{d}{dt}h(t) = \left(\frac{-1}{AR}\right)h(t); \qquad h(0) = \frac{\text{Volume}(0)}{A} \tag{6.2}$$

Comparing equations (6.1) and (6.2), we can see that the state variable is $x = h$, the parameter is $a = -1/(AR)$, and the initial state is $x_0 = h(0) = (\text{Volume}(0)/A)$.

Example 6.2   Low-Pass RC Filter

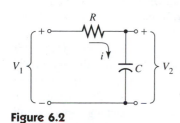

Figure 6.2

Low-pass *RC* Filter

Obtain the differential equation relating the output voltage drop to the input voltage drop for the bypass circuit shown in Fig. 6.2. For the condition of zero input and an initially charged capacitor, compare the resulting model with the system of equation (6.1).

Solution: The differential equation was derived in Example 2.5. The result, equation (2.36a) repeated, is

$$\frac{d}{dt}V_2 = \frac{1}{RC}(V_1 - V_2) \tag{6.3a}$$

For zero input ($V_1 = 0$, that is the input terminal is grounded) and initial charge resulting in the potential drop V_0 across the capacitor, the charge on the capacitor will leak off according to the equation

$$\frac{d}{dt}V_2 = \left(\frac{-1}{RC}\right)V_2; \qquad V_2(0) = V_0 \tag{6.3b}$$

This compares with equation (6.1) as follows: state variable $x = V_2$, system parameter $a = -1/(RC)$, and initial state $x_0 = V_0$.

Free Response

We can solve equation (6.1) by direct integration using separation of variables:

$$\int \frac{dx}{x} = \int a \, dt \tag{6.4a}$$

Carrying out the indicated integrations, we have

$$\log_e x = at + C_1; \qquad C_1 = \text{constant of integration} \tag{6.4b}$$

Taking the exponential power of both sides, we obtain

$$\exp(\log_e x) = x = \exp(at + C_1) = C_2 \exp(at); \quad \text{where } C_2 = \exp(C_1) \tag{6.4c}$$

Applying the initial condition, we have

$$x(0) = C_2 \exp(0) = C_2 = x_0 \tag{6.4d}$$

so that the solution is

$$x(t) = e^{at} x_0; \quad \text{(first-order free response)} \tag{6.5}$$

Equation (6.5) is the *free response* (also described as *natural* response) of the first-order system given by equation (6.1). The response is **free in the sense that there are no (forcing) inputs.**

Time Constant and Stability Characteristics

Consistent with the relation between the coefficient a and the physical parameters of the system (recall Examples 6.1 and 6.2), the free response of a first-order system is completely characterized by the quantity a. Figure 6.3 shows the nature of the free response for the three possible ranges in which the value of a may lie: $a = 0$, $a < 0$, and $a > 0$.

If $a = 0$, then $x(t) = e^{0t} x_0 = x_0$; $x(t)$ remains constant (at the initial value) for all time. As a dynamic system with memory, it has a complete remembrance of the past (the initial state in this case). Physically, $a = 0$ implies that the corresponding system parameter function has a very (negligibly) small value. For instance, in Example 6.2, if the resistance value is very large (approaching open-circuit conditions), then $1/(RC) \approx 0$, and the charge on the capacitor will remain almost indefinitely. Eventually, of course, it will gradually decrease and the capacitor will become discharged after a long time has elapsed, since the condition is not quite open-circuit and a leakage path to ground exists. This is the response shown in Fig. 6.3 under $a < 0$, $|a| \equiv$ small.

If $a < 0$, evidently the time it takes the capacitor to become discharged (that is the time for the state to approach zero) depends on the magnitude of a. As discussed above, the time is very long when the magnitude of a is very close to zero. For very large (but negative) values of a, the rate of decay can be quite rapid.

Time Constant   A very important measure for this type of dynamic behavior, that is, *exponential decay,* is the *time constant* of the response T. The time constant can be defined in various ways. The one closest to our present discussion is: *the time constant*

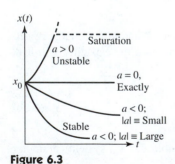

Figure 6.3

Graph of e^{at} for Ranges of a

is the time it takes the exponential decay to reach (1/e) of its initial value. Mathematically, we have

$$x(T) = e^{aT}x_0 = \left(\frac{1}{e}\right)x_0 \quad \text{or} \quad e^{aT} = \frac{1}{e} = e^{-1}$$

Hence

$$T = \left(\frac{-1}{a}\right) \tag{6.6}$$

Also, the tangent to the response curve at any time intersects the time axis (the final value of the response) T time units later. Considering, for instance, the tangent to the response curve at $t = 0$, we have

$$\frac{d}{dt}x(t)\bigg|_{t=0} = \frac{d}{dt}(e^{at}x_0)\bigg|_{t=0} = ae^{at}x_0\big|_{t=0} = \frac{-x_0}{T} \tag{6.7a}$$

The slope of the tangent line at $t = 0$ is given by

$$\text{Slope} = -\frac{x_0}{\Delta t} \tag{6.7b}$$

Hence $T = \Delta t$, where Δt is the time at which the tangent line intersects the time axis. Although the preceding two interpretations of the time constant are demonstrated here at $t = 0$, they are applicable at any t, as shown in Fig. 6.4 (the case for any t is given as an exercise in Practice Problem 6.1).

Figure 6.4

The Time Constant in Exponential Decay

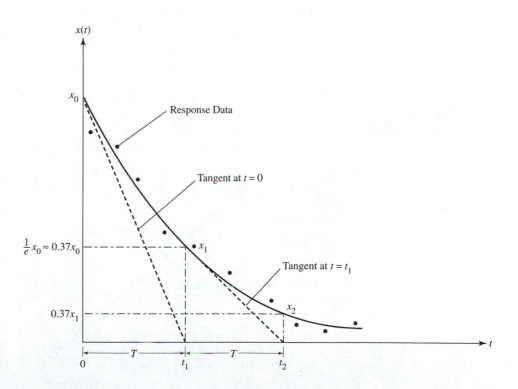

For a measured response, the interpretation of the time constant based on the semilog plot of the response, as shown in Fig. 6.5, may be more practical, since in general a smooth (exponential) curve through the data points representing the measured response is not available a priori. On the other hand, it is easy to fit a representative straight line on a semilog (in general, log to the base 10) plot of the response data:

$$\log_e x(t) = \log_e (e^{-t/T} x_0) = \frac{-t}{T} + \log_e (x_0) \tag{6.7c}$$

since

$$\log_e y = 2.3 \log_{10} y$$

$$\log_{10} x(t) = \left(\frac{-1}{2.3T}\right) t + \log_{10} (x_0) \tag{6.7d}$$

Equations (6.7c) and (6.7d) represent straight lines on semilogarithmic coordinates, and the time constant can be obtained from the slope of these straight lines. For example, in equation (6.7d),

$$\text{Slope} = \frac{-1}{(2.3T)} \tag{6.7e}$$

Alternatively, for any two times t_1 and t_2,

$$\log_{10} x(t_1) - \log_{10} x(t_2) = \frac{(t_2 - t_1)}{(2.3T)} \quad \text{or} \quad 2.3 \log_{10} \left[\frac{x(t_1)}{x(t_2)}\right] = \frac{(t_2 - t_1)}{T} \tag{6.7f}$$

If we select t_1 and t_2 such that $x(t_2)$ is $x(t_1)/2.73$, then

$$2.3 \log_{10} \left[\frac{x(t_1)}{x(t_2)}\right] = 1, \quad \text{for } x(t_2) = \frac{x(t_1)}{2.73} \tag{6.7g}$$

Figure 6.5

Semi-log Plot of an Exponential Decay Data

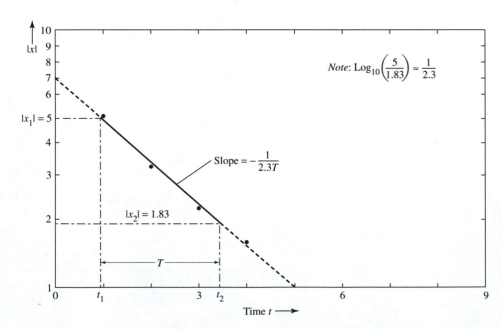

Note: $\text{Log}_{10}\left(\dfrac{5}{1.83}\right) \approx \dfrac{1}{2.3}$

so that

$$T = (t_2 - t_1)$$

This result is illustrated in Fig. 6.5. Note that the magnitude of the response $|x(t)|$ is plotted on the semilog coordinates, since the logarithm is undefined for nonpositive numbers.

Identification of System   The preceding results imply that given the response of a first-order system (or a system that can be approximated with a first-order model), one can construct the model of that system, that is, write the first-order equation, by estimating the time constant (and hence a) from the free response data without actually seeing or analyzing the internal structure of the system (see Practice Problem 6.2). In terms of the time constant, the first-order model of a free system is

$$\frac{d}{dt} x(t) = \left(-\frac{1}{T} \right) x(t) \tag{6.8a}$$

Considering the two previous examples, we see that the time constant for the leaking tank is $T = (AR)$, and for the filter circuit, $T = (RC)$.

Stability and Eigenvalue   To complete the discussion of the nature of the free response, we consider the case $a > 0$. As is evident from the two examples of the tank and the filter, this situation is not very common in normal physical systems. A system with $a > 0$ represents an *unstable* system, as shown in Fig. 6.3. The magnitude of the response increases indefinitely (much like the membership of an unrestrained group of gangsters!). This is an *exponential growth*. Certain chain reactions in physics are also governed by this type of behavior. Similarly, the growth of many biological systems when unrestricted is close to exponential. As a practical matter, the growth of unstable systems, particularly man-made devices, is not unlimited. At some point something generally gives (fails) or reaches its limit, and thereafter the response remains at this *saturation* value or ceases altogether.

Recall that another approach to the solution of a homogeneous equation such as equation (6.1) is to substitute the trial solution $x = e^{pt}$. The result,

$$p - a = 0 \tag{6.8b}$$

is known as the characteristic equation because the values of p that satisfy the equation—the root (or *eigenvalue;* see Section 6.3) determine the character of the solution, equation (6.4c):

$$x = C_2 e^{pt}$$

Thus we see that for stability of the first-order system, the eigenvalue $p = a$ must not be positive. Further, for a stable first-order system, the time constant and the eigenvalue are related by equation (6.6).

Forced Motion, Linearity, and Superposition

Recall the low-pass RC filter with input model, equation (6.3a):

$$\frac{d}{dt} V_2 = \frac{1}{RC} (V_1 - V_2) = \left(\frac{-1}{RC} \right) V_2 + \left(\frac{1}{RC} \right) V_1 \tag{6.9}$$

In general, the first-order system with forcing input will be written as

$$\frac{d}{dt} x(t) = ax(t) + bu(t); \qquad x(0) = x_0; \quad \text{(first order, forced, stationary)} \tag{6.10}$$

where $u(t)$ is the input or forcing term. From equation (6.9) we can identify $a = -1/(RC)$, as before; $b = 1/(RC)$; and $u(t) = V_1(t)$. Thus it is sufficient to derive the response for the system (6.10) rather than for specific cases as represented by equation (6.9).

Equation (6.10) is a *first-order nonhomogeneous ordinary differential equation with constant coefficients*. The **forced motion is the response due only to the (forcing) input $u(t)$, with zero initial conditions.** This is similar to but not the same as the *particular* solution in mathematics. (The difference between the *homogeneous* and free response, and the particular and *forced response,* is in the handling of the initial conditions. This will become clear in subsequent developments.) One of the most general of the mathematical techniques for finding the particular solution is the *method of variation of parameters*. In simple terms, the particular solution is assumed to be a variant of the homogeneous solution. In the present case, we shall use a product of the homogeneous solution $(C_2 e^{at})$ and another function (with zero initial value) that is to be determined. That is, we let

$$x_f(t) = e^{at} v(t) \tag{6.11a}$$

where $x_f(t)$ is the forced response, $v(t)$ is a function to be determined, and the constant C_2 is absorbed in $v(t)$. Substituting equation (6.11a) in the differential equation (6.10), we have

$$\frac{d}{dt} x_f(t) = ae^{at} v(t) + e^{at} \frac{d}{dt} v(t) = ax_f(t) + bu(t) = ae^{at} v(t) + bu(t) \tag{6.11b}$$

so that

$$\frac{d}{dt} v(t) = e^{-at} bu(t) \tag{6.11c}$$

Integrating, we obtain

$$v(t) = \int_0^t e^{-a\tau} bu(\tau) \, d\tau + [v(0) = 0] \tag{6.11d}$$

Hence

$$x_f(t) = e^{at} v(t) = e^{at} \int_0^t e^{-a\tau} bu(\tau) \, d\tau; \quad \text{(first-order forced response)} \tag{6.12a}$$

For analytical evaluation of the forced response, form (6.12a) is probably preferable. However, a perhaps familiar and more general form of the same result is obtained by combining the two exponential terms to get

$$x_f(t) = \int_0^t e^{a(t-\tau)} bu(\tau) \, d\tau; \quad \text{(first-order forced response)} \tag{6.12b}$$

Equation (6.12b) is the *convolution integral* for the "homogeneous solution" $[e^{at}]$ and the nonhomogeneous term $[bu(t)]$. In general, the convolution (*) of two functions $g(t)$ and $f(t)$ is given by

$$g(t)*f(t) = \int_0^t g(t - \tau)f(\tau)\, d\tau; \quad \text{(convolution integral)} \tag{6.13}$$

The method of variation of parameters is general in the sense that equation (6.12b) is not limited to any particular form of the forcing $u(t)$. However, analytical evaluation of the convolution integral may not be possible for some $u(t)$, but a numerical solution for the response to any input can be obtained as discussed in Section 6.5.

Linearity and Superposition

Now that we know the free and forced responses of the first-order system, what is the total or overall response? In other words, what is the *general solution* of equation (6.10)? Recall that in mathematics the general solution is that sum of the homogeneous and particular solutions that satisfies the initial conditions. Here the **total response is the sum of the free and forced responses.** The initial conditions are already accounted for in the individual solutions. We explore the reason for this from our own viewpoint: Consider the separate responses of the system (6.10) under two different conditions. In the first, there is no input ($u(t) = u_A(t) = 0$ for all t) and the initial state is $x(0) = x_A = x_0$. In the second, the input is the actual input: $u(t) = u_B(t) = u(t)$, and the initial state is zero, $x(0) = x_B = 0$. The solution under the first condition is the same as the free response that we obtained previously, (equation (6.5)). The solution for the second condition, is the forced response, or equation (6.12b). The general or actual condition of a system subject to initial state x_0, and forcing input $u(t)$ is the same as the sum of the two conditions given above, that is,

$$x_A + x_B = x_0$$

and

$$u_A(t) + u_B(t) = u(t)$$

Since the system of equation (6.10) is *linear* (the coefficient a is not a function of x, that is, a can be a constant or a function of t only), it satisfies the *principle of superposition,* so that the response to the general case is the *sum* of the responses under the two individual conditions:

$$x(t) = x_{\text{free}} + x_{\text{forced}} = e^{at}x_0 + \int_0^t e^{a(t-\tau)}bu(\tau)\, dt; \quad \text{(first-order system response)} \tag{6.14a}$$

If the initial value is specified at $t = t_0$, that is, $x(t_0) = x_0$, then

$$x(t) = e^{a(t-t_0)}x_0 + \int_{t_0}^t e^{a(t-\tau)}bu(\tau)\, dt; \quad \text{(general first-order system response)} \tag{6.14b}$$

The superposition principle can also be applied to good advantage when the input $u(t)$ is complicated but can be split into the sum of less complex individual input functions.

To elaborate on this consequence of linearity, we consider the general nth order linear ODE:

$$\frac{d^n x}{dt^n} + a_{n-1} \frac{d^{n-1} x}{dt^{n-1}} + \cdots + a_1 \frac{dx}{dt} + a_0 x = u(t) \tag{6.15}$$

Again, system (6.15) is linear when the coefficients a_i, $i = 0, \ldots, n - 1$ are not functions of x. The free response of (6.15), in view of our previous discussion, can be described as the response of the system due only to the initial conditions, and the forced response can be termed the system response due only to the forcing input (that is, with zero initial conditions). Suppose the forcing input $u(t)$ is made up of several functions: $u(t) = u_1(t) + u_2(t) + u_3(t) + \ldots$. The forced response when the input is only $u_i(t)$ can be described as the system response due to u_i. The superposition principle applied to (6.15) under these circumstances is

$$x(t) = [\text{response due to initial conditions}] + [\text{response due to } u_1]$$

$$+ [\text{response due to } u_2] + \ldots \tag{6.16}$$

$$= [\text{response due to initial conditions}] + [\text{response due to } u]$$

Linear Operations.   Consider now the forced response only or the *total response under zero initial conditions*. In either case, equation (6.16) reduces to

$$x(t) = [\text{response due to } u_1] + [\text{response due to } u_2] + \ldots \tag{6.17a}$$

$$= [\text{response due to } u]$$

Suppose the forced response or the total response under zero initial conditions of the system (6.15) to a specific input $u_A(t)$ is known and it is $x_A(t)$. The forced response (or total response under zero initial state) to any other input $u_B(t)$ that can be obtained from $u_A(t)$ by a linear operation on $u_A(t)$ is the solution given by applying the same linear operation on $x_A(t)$. In other words, given that

$$x_A(t) = [\text{response due to } u_A(t)] \tag{6.17b}$$

and

$$u_B(t) = Ł_S[u_A(t)] \tag{6.17c}$$

where $Ł_S[\ ]$ stands for a specific linear operation, then

$$x_B(t) = [\text{response due to } u_B(t)] = Ł_S[x_A(t)] \tag{6.17d}$$

This result, which is as a consequence of the linearity of (6.15), is easily proved for a particular system by substituting x_A in equation (6.15) and performing the specific linear operation on the resulting equation (see Practice Problem 6.12). To sketch a proof in a somewhat general manner, let us first explain the term *linear operator*. A linear operator $Ł$ can be defined (see [1], for example) as one for which

$$Ł[c_1 f_1(t) + c_2 f_2(t)] = c_1 Ł[f_1(t)] + c_2 Ł[f_2(t)] \tag{6.18}$$

where f_1 and f_2 are any two functions to which $Ł$ can be applied and c_1 and c_2 are two arbitrary constants. We note from this definition that the operations represented in equation (6.15), that is, ordinary differentiation, multiplication by a constant, and addition, are all linear operations and hence equation (6.15) itself involves the linear differential operator, which is defined by

$$\mathcal{L}_{Dn}[f] = \frac{d^n f}{dt^n} + a_{n-1} \frac{d^{n-1} f}{dt^{n-1}} + \cdots + a_1 \frac{df}{dt} + a_0 f \tag{6.19a}$$

so that equation (6.15) becomes

$$\mathcal{L}_{Dn}[x] = u(t) \tag{6.19b}$$

Substituting $x_A(t)$ in equation (6.19b) and applying the specific linear operation of interest, we obtain

$$\mathcal{L}_S [\mathcal{L}_{Dn} [x_A(t)]] = \mathcal{L}_S [u_A(t)] = u_B(t) \tag{6.20a}$$

For a specific linear operation, one can show, using definition (6.18) and the condition of zero initial conditions, that

$$\mathcal{L}_S [\mathcal{L}_{Dn} [x_A(t)]] = \mathcal{L}_{Dn} [\mathcal{L}_S [x_A(t)]] \tag{6.20b}$$

Hence $\mathcal{L}_S[x_A(t)] = x_B(t)$ satisfies the differential equation (6.15) with $u = u_B$, and the proof is complete. From the present exercise we can see that addition/subtraction, multiplication by a constant, differentiation, and integration (from zero to t) are all linear operations, whereas multiplication of two functions, for example, is not a linear operation.

As was mentioned previously, the combination of equation (6.17a) and equation (6.17d) can be employed to simplify complicated problems as well as to save a great deal of time. For the latter case, suppose the forced response to a certain input $u_B(t)$ is desired. If $u_B(t)$ is recognized to be, for instance, the integral (from zero to t) of another forcing function for which the forced response of the system is available, the solution for the new forcing input can be obtained quickly by integrating (from zero to t) the known response. This approach is illustrated in Examples 6.4 and 6.5.

Forced Response to Some Special Input Functions

Since the free response is known for any given initial state, and the general response is the sum of the free and forced motions, it is sufficient to concentrate our attention now on the forced component.

Example 6.3   The Step Input

We define the **_unit_** *step function* as follows:

$$U_S(\alpha) = \begin{cases} 0, & \alpha \leq 0 \\ 1, & \alpha > 0 \end{cases}; \text{ (unit step)} \tag{6.21}$$

That is, the magnitude is zero for all α except for $\alpha > 0$, for which the magnitude is unity. The general step of magnitude u_0 is simply obtained by multiplying the unit step by u_0. The argument α has been used instead of time t so as to give the general definition. Thus in Figs. 6.6a and 6.6b, the same expression $u_0 U_S(\alpha)$ can be used to represent both functions provided that α is given the proper value as shown. An example of a physical system whose output approximates a step function is an on-off switch (connected to a constant-voltage source, for instance; see Fig. 6.6c). The output of a

Figure 6.6

Step Input Function and a
Switch as Function Generator

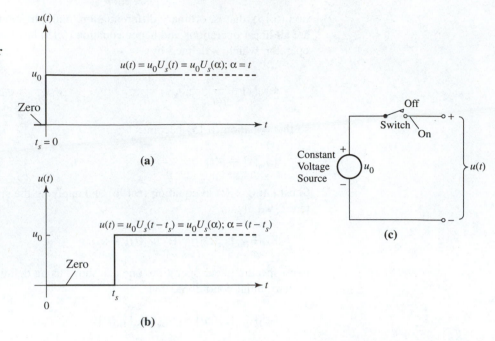

(a)

(b)

(c)

switch is zero as long as the switch is off. The output instantaneously becomes nonzero when the switch is turned on and remains at the new value as long as the switch is left on. The sudden opening or closing of a valve is another example of an approximate step input.

Find the forced response of the system (6.10) to the step input $u_0 U_S(t)$. Repeat the problem for a step input of the same magnitude but commencing at time $t = t_s$.

Solution: Using the definition of the unit step and equation (6.12a), we have

$$x_f(t) = e^{at}\int_0^t e^{-a\tau} bu_0 U_S(\tau)\,d\tau = \begin{cases} 0, & t \le 0 \\ bu_0 e^{at}\int_0^t e^{-a\tau}\,d\tau, & t > 0 \end{cases} \tag{a}$$

Performing the indicated integration and substituting $a = -(1/T)$, we obtain

$$bu_0 e^{at}\int_0^t e^{-a\tau}\,d\tau = \frac{bu_0}{a}\left[e^{at} - 1\right] \tag{b}$$

$$x_f(t) = \begin{cases} 0, & t \le 0 \\ (bu_0 T)[1 - e^{-t/T}], & t > 0 \end{cases} \tag{6.22a}$$

We can also employ the definition of the unit step to write the same result as

$$x_f(t) = (bu_0 T)[1 - e^{-t/T}]U_S(t) \tag{6.22b}$$

Equations (6.22a) and (6.22b) are equivalent. The response represented by these equations is easily sketched (see Fig. 6.7a) if we note that the value of the exponential term is 1 at $t = 0$ and approaches zero as t becomes large. Also, the definition of the time constant helps fill in the details of the sketch as shown.

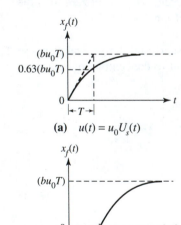

(a) $u(t) = u_0 U_s(t)$

(b) $u(t) = u_0 U_s(t - t_s)$

Figure 6.7

Sketch of a Forced
Step Response

If the step input occurs at time t_s instead of at zero, we can use the general definition of the step input to write the forced response as

$$x_f(t) = (bu_0 T)[1 - e^{-\alpha/T}]U_S(\alpha); \quad \text{(forced response to a step)} \tag{6.23}$$

Thus when $\alpha = t$, the response is exactly the same as Fig. 6.7a. With $\alpha = (t - t_s)$,

$$x_f(t) = (bu_0 T)[1 - e^{-(t-t_s)/T}]U_S(t - t_s) \tag{6.24}$$

This response is shown in Fig. 6.7b. It is the same response as in Fig. 6.7a but shifted t_s time units to the right.

Example 6.4	The Ramp Input

A *ramp* input starting at zero is a straight line with slope u_0, that is,

$$u(t) = u_0 t = u_0 U_R(t)$$

where $U_R(t)$ is a ramp of slope $= 1$, starting at zero (see Fig. 6.8). Determine the forced response of the system (6.10) to the ramp input.

Solution: The forced response to a ramp input can be found by direct substitution in equation (6.12a):

$$x_f(t) = bu_0 e^{at} \int_0^t \tau e^{-a\tau} \, d\tau \tag{a}$$

and integrating by parts to obtain

$$x_f(t) = (bu_0 e^{at}) \left. \frac{e^{-a\tau}}{a^2} (-aT - 1) \right|_0^t \tag{b}$$

$$f(t) = \frac{bu_0}{a^2} (e^{at} - at - 1) = bu_0 T^2 \left(e^{-t/T} + \frac{t}{T} - 1 \right) \tag{6.25}$$

We note that the forced response becomes unbounded as t gets very large, as should be expected for this particular forcing input.

The solution could perhaps have been obtained more easily if it was recognized that the ramp $u_0 t$ results when a step of magnitude u_0 occurring at $t = 0$ is integrated from zero to t:

$$\int_0^t u_0 U_S(t) \, dt = u_0 t$$

Thus by equation (6.17d), we simply integrate equation (6.22b):

$$x_f(t) = (bu_0 T) \int_0^t [1 - e^{-\tau/T}] \, d\tau = \left. bu_0 T (\tau + Te^{-\tau/T}) \right|_0^t$$

$$x_f(t) = bu_0 T [t + T(e^{-\tau/T} - 1)] = bu_0 T^2 (e^{-\tau/T} + t/T - 1)$$

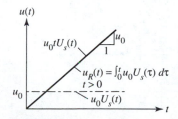

Figure 6.8

Ramp Input Function

as before. For the general ramp function specified in terms of the unit ramp, which is defined as

$$U_R(\alpha) = \begin{cases} 0, & \alpha \le 0 \\ \alpha, & \alpha > 0 \end{cases}; \quad \text{(unit ramp)} \tag{6.26}$$

the forced response due to $u(t) = u_0 U_R(\alpha)$ is

$$x_f(t) = bu_0 T^2 (e^{-\alpha/T} + \alpha/T - 1)\, U_S(\alpha); \quad \text{(forced response to a ramp)} \tag{6.27}$$

The simplification offered by the use of the utility α can be appreciated by deriving (6.27) longhand. The reader is urged to verify equation (6.27). Figure 6.9 shows the response to the ramp, equation (6.25), and how it can be sketched as the sum of the three terms in equation (6.25).

Example 6.5	The Pulse Input

We define the ***unit pulse function*** of duration T_p as follows:

$$U_P(\alpha) = \begin{cases} 0, & \alpha \le 0 \\ 1, & 0 < \alpha \le T_p; \quad \text{(unit pulse)} \\ 0, & \alpha > T_p \end{cases} \tag{6.28}$$

That is, the magnitude of the pulse is zero for all α except for $0 < \alpha \le T_p$, for which the magnitude is unity. We obtain the general pulse of magnitude u_0 (see Fig. 6.10) by multiplying the unit pulse by u_0. the argument α is used for the same reason as for the

Figure 6.9

Forced Response to a Ramp Input

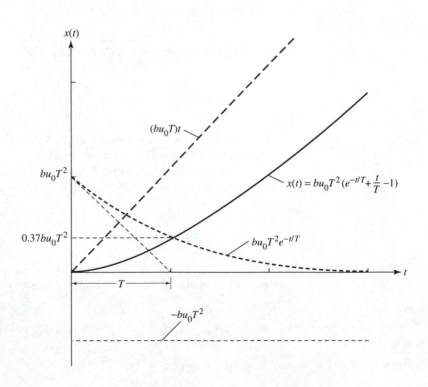

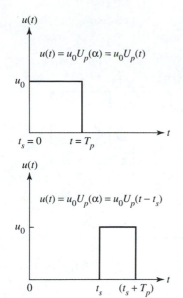

Figure 6.10

Pulse Input Function

unit step. A pulse function can be approximated physically by switching, as was the case for the step input. At time t_s the switch (electric switch, a valve, etc.) is turned on and left on for a time duration equal to T_p. Then the switch is turned off again. Determine the forced response of the system (6.10) to a pulse input.

Solution: The forced response to a pulse (occurring at $t = 0$) can be found using the definitions (6.28) and the convolution integral (6.12b), as follows:

$$\text{For } t \leq 0, \quad x_f(t) = 0$$

$$\text{For } 0 < t \leq T_p, \quad x_f(t) = bu_0 e^{at} \int_0^t e^{-a\tau}\, d\tau \tag{6.29a}$$

which gives, considering the step response result, equation (6.22a),

$$\text{For } 0 < t \leq T_p, \quad x_f(t) = bu_0 T\,[1 - e^{-t/T}] \tag{6.29b}$$

The situation $t > T_p$, deserves careful attention. Although according to the definition (6.28) the input will no longer be present, the state of the system has already been altered by the forcing during the period $0 < t \leq T_p$. Suppose the state of the system at $t = T_p$ is $x(T_p) = x_{0p}$. Then if we consider a new time axis t', with origin at T_p, the response is the same as a free motion (no forcing) with initial condition equal to x_{0p}, that is,

$$x_f(t') = e^{-t'/T} x_{0p} \tag{6.29c}$$

We find x_{0p} by evaluating equation (6.29b) at $t = T_p$:

$$x_{0p} = bu_0 T\,[1 - e^{-T_p/T}] \tag{6.29d}$$

since t' and t are related by

$$t' = t - T_p \tag{6.29e}$$

We can substitute equations (6.29e) and (6.29d) into equation (6.29c) to get, for $t > T_p$:

$$\text{For } t > T_p, \quad x_f(t) = bu_0 T\,[1 - e^{-T_p/T}]\, e^{-(t-T_p)/T} \tag{6.29f}$$

We can summarize our results for $\alpha = t$ as follows:

$$x_f(t) = \begin{cases} 0, & \alpha \leq 0 \\ bu_0 T\,[1 - e^{-\alpha/T}]\,U_S(\alpha), & 0 < \alpha \leq T_p; \\ bu_0 T\,[1 - e^{-T_p/T}]\,e^{-(\alpha-T_p)/T}\,U_S(\alpha), & \alpha > T_p \end{cases} \quad \begin{array}{l}\text{(forced}\\\text{response}\\\text{to a pulse)}\end{array} \tag{6.30}$$

Equation (6.30) applies as well to the general pulse occurring at $t = t_s$, for which $\alpha = t - t_s$.

Alternatively, we can derive the pulse response by using superposition and the step response. As shown in Fig. 6.11, the pulse input of magnitude u_0 is equal to the sum of two step inputs, one of magnitude u_0 and the other of magnitude $-u_0$. The step inputs are separated in time by an amount equal to T_p. The solution to the pulse input then is the sum of the solutions to the two step inputs $u_1(t) = u_0 U_S(\alpha)$ and

Figure 6.11

Pulse Response as Sum of Step Responses

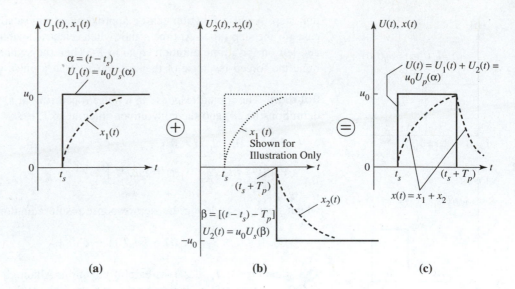

$u_2(t) = -u_0U_S(\alpha - T_p)$. From equation (6.23) the individual responses to $u_1(t)$ and $u_2(t)$ are, respectively,

$$x_1(t) = (bu_0T)[1 - e^{-\alpha/T}]U_S(\alpha)$$

and

$$x_2(t) = -(bu_0T)[1 - e^{-\beta/T}]U_S(\beta) = -(bu_0T)[1 - e^{-(\alpha-T_p)/T}]U_S(\alpha - T_p)$$

so that the forced response to the pulse input is, by equation (6.17a),

$$x_f(t) = x_1(t) + x_2(t)$$

$$= (bu_0T)\{[1 - e^{-\alpha/T}]U_S(\alpha) - [1 - e^{-(\alpha-T_p)/T}]U_S(\alpha - T_p)\} \qquad \textbf{(6.31)}$$

Equations (6.30) and (6.31) are equivalent, and both results are plotted in Fig. 6.11c.

Impulse Response and Convolution

Another special input function is the *impulse*. One way to visualize the impulse is by considering a degenerate pulse. Take, for example, the pulse of width T_p and magnitude u_0 occurring at $t = t_s$ (Fig. 6.12). The area under the pulse is $A_0 = u_0T_p$. An impulse input of strength (area) A_0 occurring at $t = t_s$, $[u(t) = A_0\delta(t - t_s)]$, will result in the limit that the width of the pulse becomes arbitrarily small while the area remains constant at A_0. Obviously, this requires that the magnitude approach infinity as shown. Thus an impulse is not physically realizable, although it can be approximated, for instance, by a very narrow but very large pulse. Also, from the preceding development, the only quantity characterizing an impulse is its strength or area A_0. By comparison, a pulse is characterized by its magnitude and width, while a step is characterized only by its magnitude. the *unit* impulse $\delta(\alpha)$ is an impulse

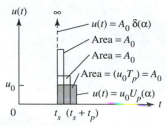

Figure 6.12

Impulse Input as the "Limit" of a Pulse

whose strength is unity. Thus we adopt the notation of describing an impulse as the product of the strength and the unit impulse. In mathematics that unit impulse function is often called the *Dirac delta function*.

We define the unit impulse as follows:

$$\delta(\alpha) = \begin{cases} \int_{-\epsilon}^{+\epsilon} \delta(\alpha)\, d\alpha = 1, & \epsilon > 0 \\ 0 & \text{for all } \alpha \neq 0; \quad \text{(unit impulse)} \\ \infty & \text{for } \alpha = 0 \end{cases} \qquad (6.32)$$

From equation (6.12a), the response to an impulse input of strength A_0 occurring at $t = 0$ is

$$x_f(t) = e^{at} \int_0^t e^{-a\tau} b A_0 \delta(\tau)\, d\tau \qquad (6.33)$$

An important rule concerning the integration of products of other functions with the unit impulse (see [2]) can be deduced from the definition of equation (6.32) as follows: *The (time) integral of the product of any function and the unit impulse is equal to the value of the function evaluated at the instant of time (within the domain of integration) when the argument of the unit impulse function is zero.* When we apply this rule to equation (6.33), the function multiplying the Dirac delta function is $(e^{-a\tau} b A_0)$, while the argument of the delta function is τ. Thus we have

$$x_f(t) = e^{at} e^{-a \cdot 0} b A_0 = b A_0 e^{at} \qquad (6.34a)$$

or

$$x_f(t) = b A_0 e^{-t/T} \qquad (6.34b)$$

For the general impulse occurring at time t_s the forced response is given by

$$x_f(t) = b A_0 e^{-\alpha/T} U_S(\alpha); \quad \text{(forced response to an impulse)} \qquad (6.34c)$$

The impulse response for $t_s = 0$ is sketched in Fig. 6.13. It can be seen from this figure, as well as from equation (6.34b), that the impulse response is very much like the free response. Indeed, *the impulse response is the same as the free response if the initial condition is $x_0 = b A_0$.* As we shall explore later in Section 8.2, the impulse response of a linear system can be used to identify the system (obtain its model, that is, differential equation or input-to-output transfer function) without the benefit of the system's physical model. (Recall the first-order free system model, equation (6.8a), which was constructed using the time constant obtained from the free response.)

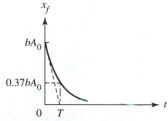

Figure 6.13

Forced Response to an Impulse Input

Forced Response via Convolution with Impulse Response   The general forced response of equation (6.12b) can be viewed as the convolution of the unit impulse response be^{at} and the forcing input $u(t)$. It follows that if we know the impulse response of a first-order system, we can determine its response to any other input without constructing a model of the system or performing further experiments on the system. We shall see later that this is also true for second-order systems and indeed for any linear system.

Singularity Functions   The special forcing functions that we have considered here—the impulse, the step, and the ramp—form a family of functions with unique properties. These functions have at times been referred to as *singularity functions*. We saw that integration of a step of magnitude u_0 results in a ramp of slope u_0. Also, differentiating a ramp of slope u_0 results in a step of magnitude u_0. The same relations apply between the step and the impulse if we ignore the fact that derivatives are not strictly defined at discontinuities. Thus "differentiating" a step will yield a quantity with infinite magnitude, or an impulse. Similarly, by the definition, equation (6.32), of the unit impulse and the rule for integrating its product with other functions, integrating an impulse of strength A_0 will yield a step of magnitude A_0.

Thus we should expect that the impulse response we just derived can also be obtained by differentiating the forced response to a step. Performing the indicated operation on equation (6.23) and noting that $d/dt = d/d\alpha$, we have

$$x_f(t) = \left(\frac{d}{dt} \{(bu_0 T)\,[1 - e^{-\alpha/T}]\} \right) U_S(\alpha)$$

$$= (bu_0 T) \left[0 + \frac{1}{T}\, e^{-\alpha/T} \right] U_S(\alpha) = bu_0 e^{-\alpha/T} U_S(\alpha)$$

which is the same as equation (6.34c) if $u_0 = A_0$.

Example 6.6 — Response of a Linear First-Order Stationary System

A preliminary study of ways to slow down a car under specific road conditions is to be made. The angular velocity of the rear wheels is considered the critical variable, and a viscously damped rotor ($J = 40.8$ kg $\cdot$ m^2, $\beta = 0.408$ N $\cdot$ m $\cdot$ s) model of the rear-wheel-brake assembly (see Fig. 6.14) is considered adequate for the present analysis. The braking torque is of the Coulomb type and is independent of the angular speed of the rear wheels. The initial velocity of the car (rear wheel) is 50 rad/s.

Obtain the rear wheel angular velocity-time behavior for the following braking (full-brake is equivalent to 800 N $\cdot$ m braking moment) schemes:

a. Brake is applied gradually, increasing linearly from zero to half full-brake over a period of 8 s and then released.
b. One-quarter full-brake is suddenly applied and maintained for 8 s and then released.
c. Full-brake is suddenly applied and maintained for 2 s only and then released.
d. Braking moment is increased linearly from zero to half full-brake in 4 s and then linearly decreased to zero in another 4 s.

Note that the braking strength, as given by the area under the braking moment–time curve, is the same in each case.

Solution: The model of the system of Fig. 6.14 is given by equation (2.16):

$$\frac{d}{dt}\,\omega = \left(-\frac{\beta}{J} \right) \omega + \left(\frac{1}{J} \right) M_{\text{in}}; \qquad \omega(0) = 50\ \text{s}^{-1}$$

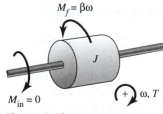

$M_f = \beta\omega$

$M_{\text{in}} = 0$

ω, T

Figure 6.14

Viscously Damped Rotor

or

$$\frac{d}{dt} x = ax + bu; \qquad x \equiv \omega, x_0 = 50 \text{ s}^{-1}; \qquad u \equiv M_{in}$$

$$a = \left(-\frac{\beta}{J}\right) = -\frac{(0.408) \, (\text{kg} \cdot \text{m} \cdot \text{s}^{-2}) \, (\text{m} \cdot \text{s})}{(40.8) \, (\text{kg} \cdot \text{m}^2)} = -0.01 \text{ s}^{-1}; \qquad \text{(a)}$$

$$T = \frac{-1}{a} = 100 \text{ s}$$

$$b = \left(\frac{1}{J}\right) = \frac{1}{(40.8) \, (\text{N} \cdot \text{s}^2 \cdot \text{m}^{-1}) \, (\text{m}^2)} = 0.0245 \text{ N}^{-1} \cdot \text{m}^{-1} \cdot \text{s}^{-2}$$

Free response: The free response is the same for all cases and is given by equation (6.5)

$$\omega_{\text{free}} = e^{at}\omega(0) = 50e^{-0.01t} \qquad \text{(b)}$$

Total response:

Case 1: The input is given by

$$(-)M_{in}(t) = \begin{cases} (50t) \text{ N} \cdot \text{m}, & 0 < t \le 8 \quad \Leftarrow \text{a ramp} \\ 0 \text{ N} \cdot \text{m}, & 8 < t \end{cases} \qquad \text{(c)}$$

For $0 < t \le 8$: from equation (6.27),

$$\omega_{\text{forced}} = \frac{bu_0}{a^2}(e^{at} - at - 1) = -12250 \, [e^{-0.01t} + 0.01t - 1] \text{ s}^{-1} \qquad \text{(d)}$$

$$\omega(t) = \omega_{\text{free}} + \omega_{\text{forced}} = [50e^{-0.01t} - 12250 \, (e^{-0.01t} + 0.01t - 1)] \text{ s}^{-1}$$

At $t = 8$, $\omega(8) = 7.98 \text{ s}^{-1}$.
  For $8 < t$: since the input is zero, we have a free response with initial ($t = 8$) condition $\omega_0 = \omega(8)$. That is,

$$\omega(t) = 7.98e^{-0.01(t-8)} \text{ s}^{-1} \qquad \text{(e)}$$

Case 2: The input is

$$(-)M_{in}(t) = \begin{cases} 200 \text{ N} \cdot \text{m}, & 0 < t \le 8 \quad \Leftarrow \text{a step} \\ 0, & 8 < t \end{cases} \qquad \text{(f)}$$

For $0 < t \le 8$: from equation (6.23),

$$\omega_{\text{forced}} = (0.0245) \, (-200) \, (100) \, [1 - e^{-0.01t}] \text{ s}^{-1}$$

and

$$\omega(t) = [50e^{-0.01t} - 490 \, (1 - e^{-0.01t})] \text{ s}^{-1} \qquad \text{(g)}$$

$$\omega(8) = 8.48 \text{ s}^{-1} \qquad \text{(h)}$$

For $8 < t$: as for case 1,

$$\omega(t) = 8.48e^{-0.01(t-8)} \text{ s}^{-1} \tag{i}$$

Note:

The same results can be obtained using equation (6.31) (the reader should check this), since the input is also a pulse (see case 3).

Case 3: First, we approximate the input with an impulse. The strength of the impulse is (-1600) N · m · s. Using equation (6.34c), we have

$$\omega_{\text{forced}} = (0.0245)\,(-1600)\,e^{-0.01t} = -39.2e^{-0.01t} \text{ s}^{-1}$$

and

$$\omega(t) = (50 - 39.2)\,e^{-0.01t} = 10.8e^{-0.01t} \text{ s}^{-1} \tag{j}$$

Second, the input is actually a narrow pulse with magnitude $(-)u_0 = 800$ N · m and duration $T_p = 2$ s. By equation (6.31), the forced response is

$$\omega_{\text{forced}} = (0.0245)\,(800)\,(100)\,\{[1 - e^{-0.01t}] - [1 - e^{-0.01(t-2)}]\,U_S\,(t-2)\} \text{ s}^{-1}$$

and

$$\omega(t) = (50e^{-0.01t} - 1960\,\{[1 - e^{-0.01t}] - [1 - e^{-0.01(t-2)}]\,U_S\,(t-2)\}) \text{ s}^{-1} \tag{k}$$

Case 4: The input is

$$(-)M_{\text{in}}(t) = \begin{cases} 100t \text{ N} \cdot \text{m}, & 0 < t \leq 4 \Leftarrow \text{ramp } (+\text{slope}) \\ 800 - 100t \text{ N} \cdot \text{m}, & 4 < t \leq 8 \Leftarrow \text{ramp } (-\text{slope}) + \text{offset} \\ 0 \text{ N} \cdot \text{m}, & 8 < t \end{cases} \tag{l}$$

As shown in Fig. 6.15, this input function can be obtained as the sum of three functions (there are other possibilities).

$$M_{\text{in}}(t) = \{-100t\} + \{200\,(t-4)\,U_S\,(t-4)\} + \{-100\,(t-8)\,U_S\,(t-8)\} \tag{m}$$

Figure 6.15

Case 4 Input as the Sum of Three Functions

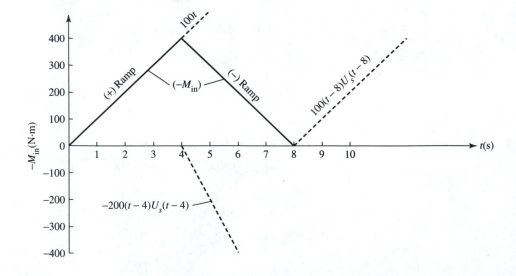

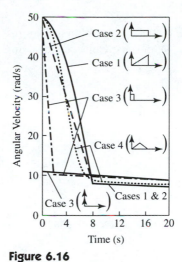

Figure 6.16

**Total Response of a
First-order Dynamic System**

By superposition, the forced response is the sum of the forced responses to each input. Using the previous result for the ramp, we find

$$\omega_{\text{forced}} = \{2\,(-12250)\,[e^{at} - at - 1]\}$$
$$+ \{4\,(12250)\,[e^{a\,(t-4)} - a\,(t - 4) - 1]\,U_S\,(t - 4)\}$$
$$+ \{2\,(-12250)\,[e^{a\,(t-8)} - a\,(t - 8) - 1]\,U_S\,(t - 8)\}\,\text{s}^{-1}$$

Thus

$$\omega(t) = 50e^{-0.01t} - 24500\,\{[e^{-0.01t} + 0.01t - 1] \qquad \text{(k)}$$
$$- 2\,[e^{-0.01(t-4)} + 0.01\,(t - 4) - 1]\,U_S\,(t - 4)$$
$$+ [e^{-0.01(t-8)} + 0.01\,(t - 8) - 1]\,U_S\,(t - 8)\}\,\text{s}^{-1}$$

The angular velocities calculated for the various cases are plotted in Fig. 6.16 as a function of time. Each case is identified by a sketch of the input braking moment $(-M_{\text{in}})$. Note the differences in the time it takes to reduce the speed of the car in each case. Also note the rather abrupt changes in angular velocity that will give rise to severe angular accelerations for most cases except case 4. Which braking strategy would you prefer?

6.2 REVIEW OF COMPLEX NUMBERS AND THEIR REPRESENTATIONS

The solution of first-order systems was treated in detail in the preceding section. In this section we prepare to handle higher-order systems whose solutions usually involve the analysis of quantities with complex values.

The square of a *real* number is always a positive quantity. The family of numbers whose squares are negative are *imaginary* numbers. An imaginary number can be represented by $c(-1)^{1/2}$ or cj, where c is a real number and j is a symbol defined as

$$j = (-1)^{1/2} \qquad \textbf{(6.35)}$$

so that

$$jj = j^2 = -1 \qquad \textbf{(6.36)}$$

Rectangular Representation

A complex number is the sum of a real number and an imaginary number. The complex number z can be written as

$$z = x + jy \qquad \textbf{(6.37)}$$

where x and y are both real, and x is termed the *real part* of z, and y the *imaginary part* (Some texts refer to jy as the imaginary part. This is a matter of definition. It is essential however to be consistent.) Equation (6.37) is a representation of the complex number in *rectangular form*. Other representations of the complex number will be considered subsequently.

Complex Conjugate

Every complex number ($z = x + jy$) has its *conjugate* ($z^* = x - jy$), obtained by reversing the algebraic sign before every term in which (j) appears. Consider the complex number (see Example 6.7)

$$z = \frac{1 + 3j}{3 - 5j}$$

The complex conjugate is given by

$$z^* = \frac{1 - 3j}{3 + 5j}$$

The product of z and z^* is a real number, that is, the *absolute square* of z is

$$z \cdot z^* = (x + jy)(x - jy) = x^2 + y^2 \tag{6.38a}$$

Note that multiplication is defined in the usual manner when supplemented by equation (6.36). Similarly, two complex numbers are added together (or subtracted) by separately adding (subtracting) the real parts and the imaginary parts. The square root of the absolute square of z is the *absolute value* of z or the *magnitude* of z:

$$|z| = (x^2 + y^2)^{1/2} \tag{6.38b}$$

Example 6.7   Real and Imaginary Parts

Find the real and imaginary parts of $z = \dfrac{1 + 3j}{3 - 5j}$.

Solution: Make the denominator real by multiplying with its complex conjugate:

$$z = \frac{(1 + 3j)(3 + 5j)}{(3 - 5j)(3 + 5j)} = \frac{(1 + 3j)(3 + 5j)}{9 + 25} = \left(\frac{-12}{34}\right) + \left(\frac{14}{34}\right)j$$

Real part $= -12/34$;   Imaginary part $= 14/34$

The Complex Plane

The real and imaginary parts of a complex number can be viewed as coordinates on two mutually perpendicular axes—the *real axis* and the *imaginary axis,* respectively. The "geometric" plane defined by these two axes is known as the *complex plane.*

Another useful representation of a complex number in the complex plane is the *polar form.* In this representation the complex number is defined by its magnitude M and a *(polar or phase)* angle ϕ **measured counterclockwise from the real axis** as shown in Fig. 6.17. The relation between the rectangular and polar forms is given by the following trigonometric identities, which are evident in Fig. 6.17.

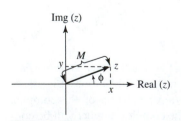

Figure 6.17

The Complex Plane

$$x = M \cos \phi = \text{real part} \tag{6.39a}$$

$$y = M \sin \phi = \text{imaginary part} \tag{6.39b}$$

$$\phi = \tan^{-1}\left(\frac{y}{x}\right) = \tan^{-1}\left(\frac{\text{imaginary part}}{\text{real part}}\right) \tag{6.39c}$$

Thus,

$$z = M (\cos \phi + j \sin \phi) \tag{6.40}$$

Exponential Representation

Recall the identities

$$\cos \phi = \frac{e^{j\phi} + e^{-j\phi}}{2} \tag{6.41a}$$

$$\sin \phi = \frac{e^{j\phi} - e^{-j\phi}}{2j} \tag{6.41b}$$

When combined, these relations give Euler's formula:

$$e^{j\phi} = \cos \phi + j \sin \phi \tag{6.41c}$$

Equations (6.40) and (6.41c) give the *exponential representation,* which is an alternative polar form:

$$z = Me^{j\phi}; \quad \text{(exponential representation)} \tag{6.42}$$

It is important to note that ϕ in equation (6.42) must be dimensionless (that is, in radians). Also, since trigonometric functions are periodic with angular period of 2π, then

$$z = Me^{j(\phi + 2n\pi)}, \quad n = 0, 1, 2, \ldots; \quad \text{(exponential representation)} \tag{6.43a}$$

The major advantage of the exponential representation is due to the properties of the exponential function that are also applicable to the representation.

Example 6.8	Conjugates and Products in Exponential Form

Show that in exponential form the complex conjugate is obtained by changing the sign of the angle and that the product of two complex numbers is given by the product of their magnitudes and the sum of their angles.

Solution: Let $z = M (\cos \phi + j \sin \phi)$, then $z^* = M (\cos \phi - j \sin \phi) = M [\cos(-\phi) + j \sin(-\phi)]$

Using Euler's formula, we have

$$z^* = Me^{-j\phi} \tag{6.43b}$$

Next, we consider

$$z_1 = M_1 e^{j\phi_1} = M_1 (\cos \phi_1 + j \sin \phi_1)$$

and

$$z_2 = M_2 e^{j\phi_2} = M_2 (\cos \phi_2 + j \sin \phi_2)$$

$$z_1 \cdot z_2 = M_1 M_2 [(\cos \phi_1 + j \sin \phi_1)(\cos \phi_2 + j \sin \phi_2)]$$

$$z_1 \cdot z_2 = (M_1 M_2)[(\cos \phi_1 \cos \phi_2 - \sin \phi_1 \sin \phi_2)$$
$$+ j (\sin \phi_1 \cos \phi_2 + \cos \phi_1 \sin \phi_2)]$$

$$z_1 \cdot z_2 = (M_1 M_2)[(\cos (\phi_1 + \phi_2) + j \sin (\phi_1 + \phi_2)]$$

by equation (6.41c),

$$z_1 \cdot z_2 = M_1 M_2 e^{j(\phi_1 + \phi_2)} \tag{6.43c}$$

Example 6.9

a. Convert $z = -3.6 + 2.9j$ to exponential form.

b. Find the cube root of $z = -3.6 + 2.9j$.

Solution A:

$$M = \sqrt{(x^2 + y^2)} = \sqrt{(3.6^2 + 2.9^2)} = 4.62$$

$$\phi = \tan^{-1}\left(\frac{y}{x}\right) = \tan^{-1}\left(\frac{-2.9}{3.6}\right) = -0.678 \text{ or } 2.463 = (-0.678 + \pi)$$

Only one of the two angles is correct. **The sure way to determine the angle of a complex number is to plot the complex number in the complex plane.** Such a plot will show this complex number in the second quadrant, so that $\phi = 2.463$ is the correct angle. Thus

$$z = 4.62e^{j2.463}$$

Solution B: Given the exponential representation of z, the result (6.43c) of Example 6.8 suggests that one cube root should be

$$z^{1/3} = 4.62^{1/3}e^{j(2.463/3)}$$

However, there should be three cube roots. Equation (6.43a) allows the determination of all the necessary roots:

$$z^{1/3} = M^{1/3}e^{j(\phi + 2n\pi)/3}, \quad n = 0, 1, 2, \ldots \tag{6.43d}$$

giving,

$$z^{1/3} = 1.666e^{j0.821}; \qquad 1.666e^{j2.916}; \qquad 1.666e^{j5.010}$$

Sinusoids and Phasors

Another special class of forcing inputs is the *sinusoidal function*. By sinusoids we mean functions of the type

$$f(t) = A\cos(\omega t + \phi) \tag{6.44}$$

Three parameters (all constants) characterize this function:

1. the *amplitude* (A), which is the value of the sinusoid at its maximum
2. The *phase* or *phase angle* (ϕ), which specifies the position of the sinusoid with respect to time and particularly with respect to other sinusoids. Hence ϕ is also called the *phase difference* or *phase shift*. Note that ϕ and (ωt) must have the same units—radians.
3. The *(circular or angular) frequency* (ω), which defines the period of the function

A sinusoid is completely specified when the listed three parameters are given (see Fig. 6.18). Although sinusoids can also be defined in terms of sines, the cosine function is preferred here. Sinusoidal input functions are important for various reasons, including the following:

1. Sinusoids are easy to generate (just as are step inputs), and their parameters can be conveniently measured.
2. Many important physical signals are sinusoidal or nearly sinusoidal—for example, radio waves.
3. Most signals of nonsinusoidal form can be expressed as the sum of sinusoids.

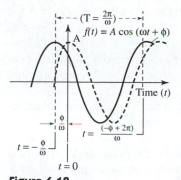

Figure 6.18

A Sinusoid

4. A linear operation on a sinusoid (differentiation, integration, addition, etc., but not multiplication and division by other sinusoids) results in another sinusoid of the same frequency.

The last item provides a key to analyzing linear systems that are subject to sinusoidal forcing: The forced response is also a sinusoid of the same frequency (recall the method of undetermined coefficients in mathematics). Because for stable systems the free response decays with time, the *steady-state* solution is dominated by the forced response. The steady-state response under sinusoidal input is thus sometimes called *sinusoidal steady state*.

   Phasors.   Because the input and the (forced) output frequencies are the same under sinusoidal forcing, a representation of sinusoids in which the frequency and hence the time information are suppressed is convenient for analyzing linear systems subject to such inputs. A *phasor is a complex number that represents a sinusoid in terms of its magnitude and phase.* For instance, in exponential form, the phasor representing the sinusoid of equation (6.44) is

$$z = Ae^{j\phi} \tag{6.45a}$$

Note that z is not equal to the function but merely represents it. However, from Euler's formula, equation (6.41c), it is evident that

$$A \cos (\omega t + \phi) = \text{Real part}[ze^{j\omega t}] \tag{6.45b}$$

Example 6.10   Sum of Sinusoids

Find the sum of the following sinusoids using complex numbers:

$$f_1 = 7 \sin 5t; \qquad f_2 = 3\sqrt{2} \cos (5t + \pi/4)$$

Solution: Note that both sinusoids have the same frequency. Also, $\sin \theta = \cos(\theta - \pi/2)$, hence

$$7 \sin 5t = 7 \cos \left(5t - \frac{\pi}{2} \right)$$

$$z_1 = 7e^{-j\pi/2} = 0 - 7j$$

$$z_2 = 3\sqrt{2}e^{j\pi/4} = 3 + 3j$$

$$z_1 + z_2 = 3 - 4j = 5e^{-j0.9273}$$

Note that $3 - 4j$ is in the fourth quadrant in the complex plane. Finally,

$$f_1 + f_2 = \text{Real} [(z_1 + z_2) e^{j5t}] = \text{Real} [5e^{j(5t-0.9273)}]$$

$$7 \sin 5t + 3\sqrt{2} \cos (5t + \pi/4) = 5 \cos (5t - 0.9273)$$

....

Damped Sinusoids

The preceding developments can be generalized slightly to include a fairly wide class of functions, namely, *damped sinusoids*. A damped sinusoid is a function of the form

$$f(t) = A \cos (\omega t + \phi) e^{-\sigma t} \tag{6.46a}$$

The only additional information is the parameter (constant) σ, which we can call the *decay rate*. It is assumed to be positive. As shown in Fig. 6.19, the function (6.46a) also represents an *undamped sinusoid* (discussed previously) if $\sigma = 0$. Further, with both the frequency ω and the phase ϕ equal to zero, the damped sinusoid represents a pure exponential decay. Even a step function can be described as a damped sinusoid ($\omega = \phi = \sigma = 0$).

Because of the property of exponential functions, in particular, the integral or derivative of an exponential function is an exponential function of the same argument; A phasor representation of the damped sinusoid is also possible. Such a representation suppresses not only the frequency information but also the decay rate and can be very useful analytically. If $z = Ae^{j\phi}$ is the phasor representation of the damped sinusoid, then

$$A \cos(\omega t + \phi)e^{-\sigma t} = \text{Real part } [ze^{(-\sigma + j\omega)t}] \tag{6.46b}$$

In Section 7.2 we shall explore a transform solution for the forced response of linear systems that is based on the phasor concept. For now, we consider the response of systems of first-order ordinary differential equations. As we shall see, these systems are characterized by modes of behavior that are represented, generally, by damped sinusoids.

Figure 6.19

Damped Sinusoids

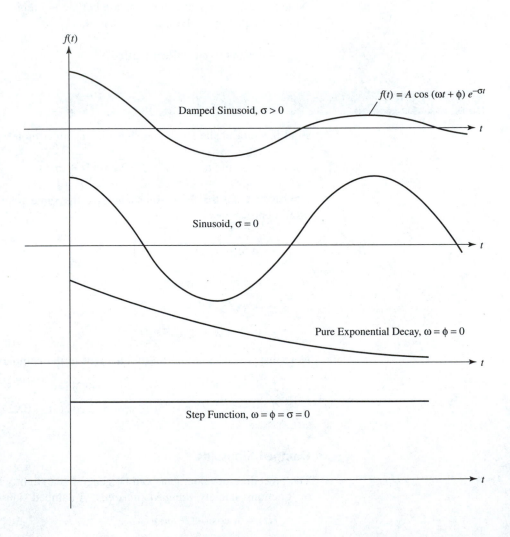

6.3 TIME-DOMAIN SOLUTION OF THE VECTOR STATE EQUATION

We continue in this section with model solution, in particular, the solution of the linear *vector state equation*. Such a model is a direct result of the state space method of system modeling developed in the first part of the text.

State Vector and Vector Differential (or Difference) Equation

Recall from Section 3.1 that the system model in state space consists of a set of first-order differential or difference equations. The set of equations (3.5a) can be represented in compact form* as follows:

$$\frac{d}{dt} \mathbf{X}(t) = \mathbf{f}\,(\mathbf{X}, \mathbf{U}, t); \qquad \mathbf{X}(t_0) = \mathbf{X}_0 \tag{6.47a}$$

$$\mathbf{Y}(t) = \mathbf{g}\,(\mathbf{X}, \mathbf{U}, t) \tag{6.47b}$$

where $\mathbf{f}$ and $\mathbf{g}$ are vector functions of the state variables, inputs, and time, whose elements are given by equation (3.5a); $\mathbf{X}$ is a column vector of n state variables, known as the *state vector;* $\mathbf{U}$ is a column vector of r input variables; and $\mathbf{Y}$ is a column vector of m output variables:

$$\text{State vector: } \mathbf{X}(t) = \begin{bmatrix} x_1 \\ x_2 \\ \cdot \\ \cdot \\ \cdot \\ x_n \end{bmatrix}; \qquad \text{input vector: } \mathbf{U}(t) = \begin{bmatrix} u_1 \\ u_2 \\ \cdot \\ \cdot \\ \cdot \\ u_r \end{bmatrix}; \qquad \text{output vector: } \mathbf{Y}(t) = \begin{bmatrix} y_1 \\ y_2 \\ \cdot \\ \cdot \\ \cdot \\ y_m \end{bmatrix}$$

Thus equation (6.47a) represents a *vector state equation* or a vector differential equation. For the discrete-time system described by the set of difference equations (3.5b), the corresponding vector state model is

$$\mathbf{X}_{k+1} = \mathbf{f}_k\,(\mathbf{X}_k, \mathbf{U}_k, k); \qquad \mathbf{X}_{k_0} = \mathbf{X}_0 \tag{6.48a}$$

$$\mathbf{Y}_k = \mathbf{g}_k\,(\mathbf{X}_k, \mathbf{U}_k, k) \tag{6.48b}$$

where k is a (running) time index, $\mathbf{X}_k$, $\mathbf{U}_k$, and $\mathbf{Y}_k$ are the state vector, the input vector, and the output vector at the kth sampling instant, respectively, and the elements of the vectors $\mathbf{f}_k$ and $\mathbf{g}_k$ are given by equation (3.5b). We shall return to discrete-time systems in Section 6.4.

Linear Vector Differential Equation

The vector state equation (6.47a) will represent a linear system if the vector function $\mathbf{f}$ contains only terms linear in $\mathbf{X}$ and $\mathbf{U}$ (the state variables and inputs). For example, the functions $f_1 = -3.7x_1 + x_2 + 2.3u_1 - \sin(3t)u_2$ and $f_2 = x_1 + (1 - e^{-t})\,x_2 - u_1$ contain only terms that are linear in the variables x_1, x_2, u_1, and u_2. In contrast, a function such as $f_3 = x_1^2 - x_1x_2 + 2x_2 + u_1 - x_1u_2$ contains the terms x_1^2, x_1x_2, and x_1u_2, which are nonlinear in these variables. A linear vector differential equation and output equation, as described, can be written as follows:

$$\frac{d}{dt} \mathbf{X}(t) = \mathbf{A}\mathbf{X}(t) + \mathbf{B}\mathbf{U}(t); \qquad \mathbf{X}(t_0) = \mathbf{X}_0 \tag{6.49a}$$

$$\mathbf{Y}(t) = \mathbf{C}\mathbf{X}(t) + \mathbf{D}\mathbf{U}(t) \tag{6.49b}$$

---

*Readers who are not familiar with matrix methods may refer to Appendix B for a review of this material.

where **A** is an $n \times n$ matrix all of whose elements are either constants or functions of time only, and **B, C,** and **D** are, respectively, $n \times r$, $m \times n$, and $m \times r$ matrices whose elements are also either constants or time-varying only. Recall the stationarity assumption described in Section 2.4: The system represented by equations (6.49a) and (6.49b) is a *stationary* one if no element of any of the matrices in the equations is time-varying, that is, they are all constants.

| **Example 6.11** | Composition of System Matrices |

Determine the elements of the state vector, the input vector, the output vector, and the **A, B, C,** and **D** matrices applicable to the armature-controlled DC motor modeled in Example 5.5. Assume the application is control of shaft speed.

Solution: From the state model equations (5.55a) through (5.55c) repeated,

$$\frac{d}{dt}\theta = \omega$$

$$\frac{d}{dt}\omega = \frac{1}{J}(K_1 i_f i_a - \beta\omega - T_L)$$

$$\frac{d}{dt}i_a = \frac{1}{L_a}(V_a - R_a i_a - K_1 i_f \omega)$$

let

$$\mathbf{X} = \begin{bmatrix} x_1 \\ x_2 \\ x_3 \end{bmatrix} = \begin{bmatrix} \theta \\ \omega \\ i_a \end{bmatrix}$$

There is (see Fig. 6.20) one control input $u_1 = V_a$ and a disturbance input, the load torque $u_2 = T_L$. Hence

$$\mathbf{U} = \begin{bmatrix} u_1 \\ u_2 \end{bmatrix} = \begin{bmatrix} V_a \\ T_L \end{bmatrix}$$

For speed control, the output or controlled variable is given by $y = x_2 = \omega$. Thus the system matrices are

$$\mathbf{A} = \begin{bmatrix} 0 & 1 & 0 \\ 0 & -\dfrac{\beta}{J} & \dfrac{K_1 i_f}{J} \\ 0 & -\dfrac{K_1 i_f}{L_a} & -\dfrac{R_a}{L_a} \end{bmatrix}; \quad \mathbf{B} = \begin{bmatrix} 0 & 0 \\ 0 & -\dfrac{1}{J} \\ \dfrac{1}{L_a} & 0 \end{bmatrix}; \quad \mathbf{C} = [0 \ \ 1 \ \ 0]; \quad \mathbf{D} = 0$$

Linearization of the Vector State Equation

In Section 2.4 we were able to come up with linear approximations to nonlinear component behavior in the neighborhood of an operating point. Given a nonlinear vector state equation (6.47a) and an operating "point," ($\mathbf{X} = \mathbf{X}_0, \mathbf{U} = \mathbf{U}_0$), what is the ap-

Figure 6.20

**Armature-controlled
DC Motor**

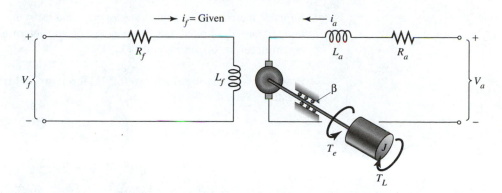

proximate but linear vector state equation (6.49a) that is valid in a small neighborhood of this operating point?

The Taylor series expansion utilized in Section 2.4 was for a single variable. A similar but multivariable Taylor series expansion of the components of f in equation (6.47a) about the operating or equilibrium point $(\mathbf{X}_0, \mathbf{U}_0)$ yields in this case the system

$$\frac{d}{dt}\hat{\mathbf{X}}(t) = \mathbf{A}\hat{\mathbf{X}}(t) + \mathbf{B}\hat{\mathbf{U}}(t); \qquad \hat{\mathbf{X}}(t_0) = \hat{\mathbf{X}}_{t=t_0} \tag{6.50a}$$

$$\mathbf{A} = \frac{\partial \mathbf{f}}{\partial \mathbf{X}}(\mathbf{X}, \mathbf{U}, t)\bigg|_{\substack{\mathbf{U}=\mathbf{U}_0 \\ \mathbf{X}=\mathbf{X}_0}} \quad \text{and} \quad \mathbf{B} = \frac{\partial \mathbf{f}}{\partial \mathbf{U}}(\mathbf{X}, \mathbf{U}, t)\bigg|_{\substack{\mathbf{U}=\mathbf{U}_0 \\ \mathbf{X}=\mathbf{X}_0}} \tag{6.50b}$$

which is linear in the deviation variables

$$\hat{\mathbf{X}} = \mathbf{X} - \mathbf{X}_0; \qquad \hat{\mathbf{U}} = \mathbf{U} - \mathbf{U}_0 \tag{6.50c}$$

and the interpretation of an $(n \times m)$ *Jacobian* matrix is

$$\frac{\partial \mathbf{f}}{\partial \mathbf{L}} = \begin{bmatrix} \dfrac{\partial f_1}{\partial z_1} & \cdots & \dfrac{\partial f_1}{\partial z_m} \\ \vdots & & \vdots \\ \dfrac{\partial f_n}{\partial z_1} & \cdots & \dfrac{\partial f_n}{\partial z_m} \end{bmatrix} \tag{6.50d}$$

Example 6.12   Linearization of the Moist Material Drying Model

Consider the state model for drying of a moist material in an insulated through-circulation drying chamber, equation (4.20) repeated:

$$\frac{d}{dt}x_1 = -\left(\frac{1}{2m}\right)x_2 \tag{6.51a}$$

$$\frac{d}{dt}x_2 = (8\lambda^2 m)\,x_1 - (5\lambda)\,x_2 - (8\lambda^2 m)\,q_e \tag{6.51b}$$

$$\frac{d}{dt}x_3 = \frac{1}{C_e}\left[-\left(\frac{H}{2}\right)x_2 + \left(\frac{c_a + w_2 c_v}{1 + w_2}\right)\dot{m}_2(T_2 - x_3) + \left(\frac{c_w - c_v}{2}\right)x_2 x_3\right] \tag{6.51c}$$

Suppose the adjustable inputs are $u_1 = q_e$ and $u_2 = \dot{m}_2$. Assuming that the dryer inlet temperature T_2 is fixed and independent of $\dot{m}_2$, determine an approximate linear state equation for an operating point $(\mathbf{X}_0, \mathbf{U}_0)$.

Solution: Comparing equations (6.51a) through (6.51c) with equation (6.47a), we can identify

$$f_1(x_1, x_2, x_3, u_1, u_2, t) = -\left(\frac{1}{2m}\right) x_2 \tag{6.52}$$

$$f_2(x_1, x_2, x_3, u_1, u_2, t) = p_1 x_1 - p_2 x_2 - p_1 u_1; \qquad p_1 = 8\lambda^2 m, \quad p_2 = 5\lambda \tag{6.53}$$

$$f_3(x_1, x_2, x_3, u_1, u_2, t) = e_1 x_2 + e_2 u_2 (T_2 - x_3) + e_3 x_2 x_3 \tag{6.54}$$

$$e_1 = -\left(\frac{H}{2C_e}\right), \quad e_2 = \frac{1}{C_e}\left(\frac{c_a + w_2 c_v}{1 + w_2}\right), \quad e_3 = \left(\frac{c_w - c_v}{2C_e}\right)$$

Substituting equations (6.52) to (6.54) into equation (6.50b) and applying equation (6.50d), we obtain

$$\mathbf{A} = \begin{bmatrix} \dfrac{\partial f_1}{\partial x_1} & \dfrac{\partial f_1}{\partial x_2} & \dfrac{\partial f_1}{\partial x_3} \\[2mm] \dfrac{\partial f_2}{\partial x_1} & \dfrac{\partial f_2}{\partial x_2} & \dfrac{\partial f_2}{\partial x_3} \\[2mm] \dfrac{\partial f_3}{\partial x_1} & \dfrac{\partial f_3}{\partial x_2} & \dfrac{\partial f_3}{\partial x_3} \end{bmatrix} = \begin{bmatrix} 0 & -\left(\dfrac{1}{2m}\right) & 0 \\[2mm] p_1 & -p_2 & 0 \\[2mm] 0 & (e_1 + e_3 x_{3,0}) & (e_3 x_{2,0} - e_2 u_{2,0}) \end{bmatrix} \tag{6.55}$$

$$B = \begin{bmatrix} \dfrac{\partial f_1}{\partial u_1} & \dfrac{\partial f_1}{\partial u_2} \\[2mm] \dfrac{\partial f_2}{\partial u_1} & \dfrac{\partial f_2}{\partial u_2} \\[2mm] \dfrac{\partial f_3}{\partial u_1} & \dfrac{\partial f_3}{\partial u_2} \end{bmatrix} = \begin{bmatrix} 0 & 0 \\[2mm] -p_1 & 0 \\[2mm] 0 & e_2(T_2 - x_{3,0}) \end{bmatrix} \tag{6.56}$$

The linear vector state equation is given by equation (6.50a) with **A** and **B** given by equations (6.55) and (6.56), respectively, and from equation (6.50c),

$$\hat{\mathbf{X}} = \begin{bmatrix} x_1 - x_{1,0} \\ x_2 - x_{2,0} \\ x_3 - x_{3,0} \end{bmatrix} \quad \text{and} \quad \hat{\mathbf{U}} = \begin{bmatrix} u_1 - u_{1,0} \\ u_2 - u_{2,0} \end{bmatrix} \tag{6.57}$$

First-Order System Analogy and State Transmission Matrix

We shall now concentrate on the solution of the nth-order linear continuous-time vector state equation (6.49a). It is understood that equation (6.49a) also represents (approximately, and in a small region about an operating or equilibrium point) a nonlinear system described by deviation variables **X** and **U** from the operating or equilibrium point.

Except for the matrices and vectors, equation (6.49a) is very much like equation (6.10), with $t_0 = 0$. By analogy with the first-order ($n = 1$) case (whose solution is equation (6.14a)), we can write the general solution to (6.49a) for the case $t_0 = 0$ as

$$\mathbf{X}(t) = e^{\mathbf{A}t}\mathbf{X}_0 + \int_0^t e^{\mathbf{A}(t-\tau)}\mathbf{B}\mathbf{U}(\tau)\, d\tau \tag{6.58}$$

The first term of the RHS is the free response, and the second term is the forced solution. This approach is in the tradition of solving new differential equations, provided we can show that equation (6.58) satisfies equation (6.49a). The matrix $e^{\mathbf{A}t}$ is known as the solution matrix and is defined in a manner similar to that used for the scalar exponential (see Appendix B, equation (B4.3d)),

$$e^{\mathbf{A}t} = \mathbf{I} + \mathbf{A}t + \frac{\mathbf{A}^2 t^2}{2!} + \frac{\mathbf{A}^3 t^3}{3!} + \cdots \tag{6.59}$$

where $\mathbf{I}$ is the identity matrix of the same order as $\mathbf{A}$. The infinite matrix series (6.59) converges (see [3] and Appendix B) for all finite t and all $n \times n$ matrices $\mathbf{A}$ with finite elements. The validity of the solution given by equations (6.58) and (6.59) can be established by substituting these results into the vector differential equation (6.49a). This is left as an exercise (see Practice Problem 6.30). Note that using (6.59), we can show that

$$\frac{d}{dt} e^{\mathbf{A}t} = \mathbf{A}e^{\mathbf{A}t} = e^{\mathbf{A}t}\mathbf{A} \tag{6.60}$$

and

$$e^{\mathbf{A}(t+\tau)} = e^{\mathbf{A}t}e^{\mathbf{A}\tau} \tag{6.61}$$

However, all the properties of the scalar exponential do not transfer to the matrix exponential. For example, $e^{(\mathbf{A}+\mathbf{B})} \neq e^{\mathbf{A}}e^{\mathbf{B}}$ unless $\mathbf{A}$ and $\mathbf{B}$ commute, that is, $\mathbf{AB} = \mathbf{BA}$ (see Practice Problem 6.31).

As regards using equation (6.58) to compute the total response, we note that the integration in this equation is to be carried out term by term, thus removing any difficulty from that quarter. This leaves computation of the matrix exponential as the stumbling block. Indeed, equation (6.59) may not be computationally very useful except for very special forms of the matrix $\mathbf{A}$ that we shall consider subsequently. For systems whose overall behavior can be characterized by a number of distinct first-order-type behaviors, another interpretation of equation (6.59) is available and is developed in the forthcoming material. Also, in Chapter 8, Section 8.3, we shall provide another solution to equation (6.49a) and hence an alternative interpretation of the matrix exponential using *inverse Laplace transforms*.

Example 6.13   **A Matrix Is Diagonal**

Determine the matrix exponential for the matrix

$$\mathbf{A} = \text{diag}\,[a_{11}, a_{22}, \ldots, a_{nn}] = \begin{bmatrix} a_{11} & 0 & \cdots & 0 \\ 0 & a_{22} & & \vdots \\ \vdots & & \ddots & 0 \\ 0 & \cdots & 0 & a_{nn} \end{bmatrix}$$

Solution: Since $\mathbf{A}$ is diagonal, then

$$\mathbf{A}^m = \text{diag}\,[a_{11}^m, a_{22}^m, \ldots, a_{nn}^m] \tag{6.62a}$$

Substituting equation (6.62a) into equation (6.59), we obtain the solution*

$$e^{\mathbf{A}t} = \text{diag } [e^{a_{11}t}, e^{a_{22}t}, \dots, e^{a_{nn}t}] = \begin{bmatrix} e^{a_{11}t} & 0 & \cdots & 0 \\ 0 & e^{a_{22}t} & & \vdots \\ \vdots & & \ddots & 0 \\ 0 & \cdots & 0 & e^{a_{nn}t} \end{bmatrix} \qquad (6.62b)$$

Although few **A** matrices are diagonal, the result of Example 6.13 is significant, as we shall see, since most nondiagonal systems can be transformed into the diagonal form by a suitable *transformation matrix*. The solution given by equations (6.62b) and (6.58) in the new transformed variables can then be converted into the original state variables by performing the inverse transformation.

Consider

$$\mathbf{A} = \begin{bmatrix} -1 & 0 \\ 0 & -2 \end{bmatrix}$$

We have

$$\mathbf{A}^2 = \begin{bmatrix} 1 & 0 \\ 0 & 4 \end{bmatrix}; \qquad \mathbf{A}^3 = \begin{bmatrix} -1 & 0 \\ 0 & -8 \end{bmatrix}; \qquad \mathbf{A}^4 = \begin{bmatrix} 1 & 0 \\ 0 & 16 \end{bmatrix}, \text{ etc.}$$

so that

$$e^{\mathbf{A}t} = \begin{bmatrix} 1 & 0 \\ 0 & 1 \end{bmatrix} + \begin{bmatrix} -t & 0 \\ 0 & -2t \end{bmatrix} + \begin{bmatrix} t^2/2! & 0 \\ 0 & 4t^2/2! \end{bmatrix} + \begin{bmatrix} -t^3/3! & 0 \\ 0 & -8t^3/3! \end{bmatrix} + \cdots$$

$$= \begin{bmatrix} e^{-t} & 0 \\ 0 & e^{-2t} \end{bmatrix}$$

Example 6.14   **A** Matrix Is in a Special Form

Consider

$$\mathbf{A} = \begin{bmatrix} 1 & 1 \\ 0 & 1 \end{bmatrix}$$

Determine the matrix exponential.

Solution: It can be shown by induction that

$$\mathbf{A}^n = \begin{bmatrix} 1 & n \\ 0 & 1 \end{bmatrix}$$

_____

*Equation (6.62b) suggests that an nth-order system has at most n time constants or modes. There is more discussion of this later.

Substituting in equation (6.59), we obtain

$$e^{\mathbf{A}t} = \begin{bmatrix} e^t & te^t \\ 0 & e^t \end{bmatrix}$$

Example 6.15   A Matrix Is Simple and of Low Order

Consider

$$\mathbf{A} = \begin{bmatrix} -1 & 0 \\ 1 & -2 \end{bmatrix}$$

a. Determine the matrix exponential.
b. If a system with the given **A** matrix is subject to a step input of magnitude 2, and if the initial condition ($t_0 = 0$) and the **B** matrix are as given below, what is the total response?

$$\mathbf{X}(0) = \begin{bmatrix} 2 \\ 3 \end{bmatrix}; \qquad \mathbf{B} = \begin{bmatrix} 1 \\ 0 \end{bmatrix}; \qquad \mathbf{U}(t) = 2U_S(t)$$

Solution A: For the given **A** matrix, we have

$$\mathbf{A}^2 = \begin{bmatrix} 1 & 0 \\ -3 & 4 \end{bmatrix}; \qquad \mathbf{A}^3 = \begin{bmatrix} -1 & 0 \\ 7 & -8 \end{bmatrix}; \qquad \mathbf{A}^4 = \begin{bmatrix} 1 & 0 \\ -15 & 16 \end{bmatrix}, \text{etc.}$$

Let

$$e^{\mathbf{A}t} = \begin{bmatrix} e_{11} & e_{12} \\ e_{21} & e_{22} \end{bmatrix}$$

Then by equation (6.59),

$$e_{11} = 1 - t + \frac{t^2}{2!} - \frac{t^3}{3!} + \frac{t^4}{4!} - \ldots = e^{-t}$$

$$e_{12} = 0 + 0 + 0 + 0 + 0 + \ldots = 0$$

$$e_{22} = 1 - 2t + \frac{4t^2}{2!} - \frac{8t^3}{3!} + \frac{16t^4}{4!} - \ldots = e^{-2t}$$

At most two time constants are expected for the second-order system and hence two exponential characteristics. Since two have already been found, other exponential terms can only be *linear combinations* of these two:

$$e_{21} = 0 + t - \frac{3t^2}{2!} + \frac{7t^3}{3!} - \frac{15t^4}{4!} + \ldots = [c_1 e^{-t} + c_2 e^{-2t}]$$

$$= \left[1 - t + \frac{t^2}{2!} - \frac{t^3}{3!} + \frac{t^4}{4!} - \ldots \right] - \left[1 - 2t + \frac{4t^2}{2!} - \frac{8t^3}{3!} + \frac{16t^4}{4!} - \ldots \right]$$

Hence

$$e_{21} = e^{-t} - e^{-2t}$$

and

$$e^{\mathbf{A}t} = \begin{bmatrix} e^{-t} & 0 \\ (e^{-t} - e^{-2t}) & e^{-2t} \end{bmatrix}$$

Solution B: By equation (6.61b) we can write the general solution as

$$\mathbf{X}(t) = e^{\mathbf{A}t}\mathbf{X}(0) + 2e^{\mathbf{A}t}\int_0^t e^{-\mathbf{A}\tau}\mathbf{B}\,d\tau$$

The free response is

$$e^{\mathbf{A}t}\mathbf{X}(0) = \begin{bmatrix} e^{-t} & 0 \\ (e^{-t} - e^{-2t}) & e^{-2t} \end{bmatrix}\begin{bmatrix} 2 \\ 3 \end{bmatrix} = \begin{bmatrix} 2e^{-t} \\ (2e^{-t} + e^{-2t}) \end{bmatrix}$$

For the forced response,

$$e^{-\mathbf{A}\tau}\mathbf{B} = \begin{bmatrix} e^{\tau} & 0 \\ (e^{\tau} - e^{2\tau}) & e^{2\tau} \end{bmatrix}\begin{bmatrix} 1 \\ 0 \end{bmatrix} = \begin{bmatrix} e^{\tau} \\ (e^{\tau} - e^{2\tau}) \end{bmatrix}$$

so that

$$\int_0^t e^{-\mathbf{A}\tau}\mathbf{B}\,d\tau = \begin{bmatrix} \int_0^t e^{\tau}\,d\tau \\ \int_0^t (e^{\tau} - e^{2\tau})\,d\tau \end{bmatrix} = \begin{bmatrix} (e^t - 1) \\ \left(e^t - \dfrac{1}{2}e^{2t} - \dfrac{1}{2}\right) \end{bmatrix}$$

$$2e^{\mathbf{A}t}\int_0^t e^{-\mathbf{A}\tau}\mathbf{B}\,d\tau = \begin{bmatrix} 2e^{-t} & 0 \\ 2(e^{-t} - e^{-2t}) & 2e^{-2t} \end{bmatrix}\begin{bmatrix} (e^t - 1) \\ \left(e^t - \dfrac{1}{2}e^{2t} - \dfrac{1}{2}\right) \end{bmatrix}$$

$$= \begin{bmatrix} 2(1 - e^{-t}) \\ (1 - 2e^{-t} + e^{-2t}) \end{bmatrix}$$

Hence the total response is

$$\mathbf{X}(t) = \begin{bmatrix} 2e^{-t} \\ (2e^{-t} + e^{-2t}) \end{bmatrix} + \begin{bmatrix} 2(1 - e^{-t}) \\ (1 - 2e^{-t} + e^{-2t}) \end{bmatrix} = \begin{bmatrix} 2 \\ (1 + 2e^{-2t}) \end{bmatrix}$$

....

The State Transition Matrix

Equation (6.58) was suggested by the similarity between equations (6.49a) and (6.10), that is, the nth-order linear stationary vector state equation and the general first-order linear stationary system. Let us then explore the nature of solutions to (6.49a) along lines similar to those in our discussion of first-order systems.

We consider first the homogeneous equation,

$$\frac{d}{dt}\mathbf{X} = \mathbf{A}\mathbf{X}; \qquad \mathbf{X}(t_0) = \mathbf{X}_0 \tag{6.63}$$

The free response of the system (6.49a) is the solution of (6.63) that satisfies the given initial condition. We found previously that

$$\mathbf{X}_{\text{free}}(t) = e^{\mathbf{A}t}\mathbf{X}_0 \quad (t_0 = 0) \tag{6.64a}$$

More generally,

$$\mathbf{X}_{\text{free}}(t) = e^{\mathbf{A}(t-t_0)}\mathbf{X}_0 = \Phi(t, t_0)\mathbf{X}_0 \tag{6.64b}$$

The matrix exponential

$$\Phi(t, t_0) = e^{\mathbf{A}(t-t_0)} \tag{6.64c}$$

is called the *state transition matrix* because it relates the state at any other time t to the state at t_0. We also found previously, for a fair number of examples, that the elements of the solution matrix $e^{\mathbf{A}t}$ are generally combinations of exponential terms (like e^{at}). This is consistent with linear ODE theory, which suggests that equations (6.64a) through (6.64c) may be expressed in component form as follows:

$$x_1(t) = v_{11}e^{p_1 t} + v_{12}e^{p_2 t} + \ldots + v_{1n}e^{p_n t}$$
$$x_2(t) = v_{21}e^{p_1 t} + v_{22}e^{p_2 t} + \ldots + v_{2n}{}^{p_n t} \tag{6.65a}$$
$$\vdots$$
$$x_n(t) = v_{n1}e^{p_1 t} + v_{n2}e^{p_2 t} + \ldots + v_{nn}e^{p_n t}$$

assuming the exponential arguments are all distinct.

Eigenvalues, Eigenvectors, and Response Modes

In vector form, equation (6.65a) is

$$\mathbf{X}(t) = \mathbf{V}_1 e^{p_1 t} + \mathbf{V}_2 e^{p_2 t} + \cdots + \mathbf{V}_n e^{p_n t} \tag{6.65b}$$

where

$$\mathbf{V}_i = \begin{bmatrix} v_{1i} \\ v_{2i} \\ \vdots \\ v_{ni} \end{bmatrix} \tag{6.65c}$$

The vectors $\mathbf{V}_i$ are, as we shall see, the *eigenvectors* of the system. Since equation (6.65b) must satisfy equation (6.63), we differentiate equation (6.65b) and substitute in (6.63) to obtain

$$\frac{d}{dt}\mathbf{X}(t) = p_1\mathbf{V}_1 e^{p_1 t} + p_2\mathbf{V}_2 e^{p_2 t} + \cdots + p_n\mathbf{V}_n e^{p_n t} = \mathbf{A}\mathbf{X} \tag{6.66}$$

$$= \mathbf{A}\left[\mathbf{V}_1 e^{p_1 t} + \mathbf{V}_2 e^{p_2 t} + \cdots + \mathbf{V}_n e^{p_n t}\right]$$

which yields

$$p_i\mathbf{V}_i = \mathbf{A}\mathbf{V}_i \tag{6.67a}$$

Equation (6.67a) defines an *eigenvalue problem* (see Appendix B, equation (B3.1a)), and its solution, V_1, is the eigenvector associated with a given p_i. Further, equation (6.67a) can be put in the form

$$(p_i\mathbf{I} - \mathbf{A})\mathbf{V}_i = 0 \tag{6.67b}$$

which is a homogeneous system of n linear algebraic equations for which the condition for nontrivial solutions to exist is (see Appendix B)

$$|p_i\mathbf{I} - \mathbf{A}| = 0; \quad \text{(characteristic equation)} \tag{6.68}$$

Equation (6.68) is the *characteristic equation* of the system (recall equation (6.8b) for the first-order system). The roots of the characteristic equation, the p_i's, are the *eigenvalues* of the system. From equations (6.65a), (6.64), and (6.58), the eigenvalues are the arguments of exponential *modes,* which completely characterize the response of the system. This is the same result we found for the first-order case.

Response Modes

Let us consider all the possible values of the eigenvalues:

a. If the eigenvalues are *all real and distinct,* then the exponential modes are $e^{p_1 t}$, $e^{p_2 t}$, . . . , $e^{p_n t}$. If, in addition, all the p_i's have negative values, then only *exponential decay modes* (see Fig. 6.21a) are present, and the behavior of the system is completely stable (recall the discussion of stability for the first-order response). As shown in Fig. 6.21a, the exponential decay can be said to be *fast* or *slow,* depending on the relative magnitude of the real and negative eigenvalue.

b. If the roots of the characteristic equation are *all real but some repeat,* the modes for the nonrepeating roots are as given in (a). For a repeating root p_k with *multiplicity* n_k (that is, number of the same roots p_k, which also means that the eigenvalue repeats $(n_k - 1)$ times), the modes corresponding to these roots are

$$e^{p_k t}, te^{p_k t}, \left(\frac{t^2}{2!}\right)e^{p_k t}, \ldots, \left[\frac{t^{(n_k-1)}}{(n_k-1)!}\right]e^{p_k t}$$

For example, if an eigenvalue repeats once (multiplicity = 2), and the value of the root is real and negative, then the two modes corresponding to the two equal roots are as shown in Fig. 6.21b. From this figure we see that once the value of the repeating root is negative, then both modes (and indeed all modes due to such repeating root, no matter the multiplicity) are stable.

c. If the eigenvalues are *complex-valued,* then they occur in *conjugate pairs.* We can represent each such eigenvalue pair (p_k, p_{k-1}) as a pair of complex numbers, as follows:

$$p_k = -\sigma + j\omega; \qquad p_{k-1} = -\sigma - j\omega \tag{6.69}$$

where $(-\sigma)$ is defined as the real part of the complex-valued root. The linear combination of the pair of modes due to the pair of roots actually represents one mode. To see this, let the combination of the modes be $c_1 e^{(-\sigma+j\omega)t} + c_2 e^{(-\sigma-j\omega)t} = e^{-\sigma t}[c_1 e^{j\omega t} + c_2 e^{-j\omega t}]$, where c_1 and c_2 are arbitrary but possibly complex-valued

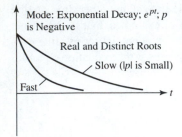

Mode: Exponential Decay; e^{pt}; p is Negative

Real and Distinct Roots

Slow ($|p|$ is Small)

Fast

(a)   All Real and Distinct Eigenvalues

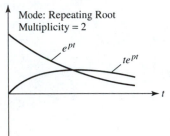

Mode: Repeating Root
Multiplicity = 2

e^{pt}

te^{pt}

(b)   All Real with Some Repeating Eigenvalues

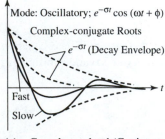

Mode: Oscillatory; $e^{-\sigma t} \cos(\omega t + \phi)$

Complex-conjugate Roots

$e^{-\sigma t}$ (Decay Envelope)

Fast

Slow

(c)   Complex-valued (Conjugate Pairs of) Eigenvalues

Figure 6.21

Response Modes (of stable systems)

constants. Now, the sum of the following two sinusoids of the same frequency: $A \cos \omega t$ and $B \cos \omega t$, where A and B are arbitrary are real constants, is another sinusoid of the same frequency: $A \cos \omega t + B \cos \omega t = C \cos(\omega t + \phi)$, where C and ϕ are real constants (see Example 6.10). Also, by Euler's formula, equations (6.41a) through (6.41c),

$$A \cos \omega t + B \sin \omega t = \frac{1}{2} \left[A \left(e^{j\omega t} + e^{-j\omega t} \right) - jB \left(e^{j\omega t} - e^{-j\omega t} \right) \right]$$

$$= \frac{1}{2} [A - jB] e^{j\omega t} + \frac{1}{2} [A + jB] e^{-j\omega t} = c_1 e^{j\omega t} + c_2 e^{-j\omega t}$$

Thus the equivalent single mode is a "damped" sinusoid: $C \cos(\omega t + \phi) e^{-\sigma t}$. If the real part of the eigenvalues is negative, then as shown in Fig. 6.21c, the system behavior is stable. Further, depending on the magnitude of the (real part of the) eigenvalues, the response mode can also be described as *fast* (magnitude or absolute value is large) or *slow* (magnitude is small). These are relative terms. The slow modes dominate the long-term response of the system, since they remain for longer periods of time, while the fast modes decay (become negligibly small) very quickly.

d. If the value of the real part of **any** eigenvalue is positive (note that this applies to any of the three cases listed: real and distinct, real and repeated, and complex-valued), for instance, p_k, then the mode contributed by this eigenvalue will result in an exponential growth mode (which for the case of complex-valued eigenvalues will also be oscillatory). Such a mode will always dominate the response of the system (other response modes, being stable, decay with time). Consequently, the system behavior will be unstable in the same way as a first-order system with a positive was unstable.

Note that the representation of the eigenvalue given in equation (6.69) can apply to all four cases above, depending on the values of σ and ω. Hence we can say, in general, that the response of systems described by nth-order vector state equations is characterized by modes that are damped sinusoids.

Returning to the problem of computing the eigenvectors and the free response, we recall that the homogeneous system of algebraic equations (6.67b) under the condition of equation (6.68) has an infinite number of nontrivial solutions. Thus equation (6.67a) or (6.67b) determines only the direction of the eigenvectors and not their magnitudes. Obviously if we fix the magnitudes of the eigenvectors by considering the initial condition of the system, then equation (6.65a) will give the free response directly for the system with *distinct eigenvalues*. If the characteristic roots are complex-valued but distinct, then the eigenvectors, if computed from equation (6.67a) or (6.67b), will contain complex numbers.

Example 6.16 Computation of Eigenvectors and Free Response

Consider again the problem of Example 6.15, for which

$$\mathbf{A} = \begin{bmatrix} -1 & 0 \\ 1 & -2 \end{bmatrix} \quad \text{and} \quad \mathbf{X}_0 = \begin{bmatrix} 2 \\ 3 \end{bmatrix}$$

Determine the eigenvectors and the free response.

Solution: We first determine the eigenvalues from the characteristic equation (6.68):

$$\det\left[\begin{bmatrix} p & 0 \\ 0 & p \end{bmatrix} - \begin{bmatrix} -1 & 0 \\ 1 & -2 \end{bmatrix}\right] = \det\begin{bmatrix} p+1 & 0 \\ -1 & p+2 \end{bmatrix} = (p+1)(p+2) = 0$$

which gives, $p_1 = -1$ and $p_2 = -2$

The eigenvalues are real and distinct (recall that these were the arguments of the exponential terms obtained in the expansion of the solution matrix in Example 6.14). We can compute the direction of the first eigenvector from equation (6.67a) or (using the reduction method given in Appendix B) for solving homogeneous systems of linear algebraic equations) from equation (6.67b):

$$(-1)\begin{bmatrix} v_{11} \\ v_{21} \end{bmatrix} = \begin{bmatrix} -1 & 0 \\ 1 & -2 \end{bmatrix}\begin{bmatrix} v_{11} \\ v_{21} \end{bmatrix} = \begin{bmatrix} -v_{11} \\ v_{11} - 2v_{21} \end{bmatrix}$$

which gives $v_{21} = v_{11}$ (geometrically,* a 45° line in two-dimensional space).

For the second eigenvector,

$$(-2)\begin{bmatrix} v_{12} \\ v_{22} \end{bmatrix} = \begin{bmatrix} -1 & 0 \\ 1 & -2 \end{bmatrix}\begin{bmatrix} v_{12} \\ v_{22} \end{bmatrix} = \begin{bmatrix} -v_{12} \\ v_{12} - 2v_{22} \end{bmatrix}$$

yields $v_{12} = 0$ (geometrically, the horizontal axis if v_{12} is the coordinate on the vertical axis).

Substituting these results and the initial condition into equation (6.65a), we have

$$2 = v_{11} + v_{12} = v_{11}$$
$$3 = v_{21} + v_{22} = v_{11} + v_{22} = 2 + v_{22}$$

Hence $v_{11} = 2$ and $v_{22} = 1$, so that the eigenvectors are

$$\mathbf{V}_1 = \begin{bmatrix} 2 \\ 2 \end{bmatrix} \text{ and } \mathbf{V}_2 = \begin{bmatrix} 0 \\ 1 \end{bmatrix}$$

Again, from equation (6.65a), the free response is

$$x_1(t) = v_{11}e^{-t} + v_{12}e^{-2t} = 2e^{-t}$$
$$x_2(t) = v_{21}e^{-t} + v_{22}e^{-2t} = 2e^{-t} + e^{-2t}$$

which is the solution obtained in Example 6.14.

To fix ideas, it is instructive to explore the preceding developments relative to a system with repeating roots. In Example 6.14 the solution matrix was found for the following **A** matrix:

$$\mathbf{A} = \begin{bmatrix} 1 & 1 \\ 0 & 1 \end{bmatrix}; \qquad e^{\mathbf{A}t} = \begin{bmatrix} e^t & te^t \\ 0 & e^t \end{bmatrix}$$

-----

*Geometric interpretations of this type are generally not possible especially for higher order systems.

The free response of a system with this $\mathbf{A}$ matrix and an initial state ($t_0 = 1$) of, say,

$$\mathbf{X}(1) = \begin{bmatrix} 1 \\ 1 \end{bmatrix}$$

will be

$$\mathbf{X}(t) = e^{\mathbf{A}(t-1)}\mathbf{X}(1) = \begin{bmatrix} e^{(t-1)} + (t-1)\,e^{(t-1)} \\ e^{(t-1)} \end{bmatrix} = \begin{bmatrix} te^{(t-1)} \\ e^{(t-1)} \end{bmatrix}$$

The eigenvalues of this system are given by equation (6.68):

$$\det[p\mathbf{I} - \mathbf{A}] = \det \begin{bmatrix} (p-1) & -1 \\ 0 & (p-1) \end{bmatrix} = (p-1)^2 = 0$$

giving $p_1 = p_2 = 1$. Thus the response modes are e^t and te^t. It follows that the free response should consist of a linear combination of these modes:

$$x_1(t) = c_{11}e^t + c_{12}te^t \tag{6.70}$$

$$x_2(t) = c_{21}e^t + c_{22}te^t$$

Note the difference between equation (6.70) and equation (6.65a). Equation (6.70) must satisfy equation (6.63) with the $\mathbf{A}$ matrix as given. That is,

$$\frac{d}{dt} \begin{bmatrix} c_{11}e^t + c_{12}te^t \\ c_{21}e^t + c_{22}te^t \end{bmatrix} = \begin{bmatrix} 1 & 1 \\ 0 & 1 \end{bmatrix} \begin{bmatrix} c_{11}e^t + c_{12}te^t \\ c_{21}e^t + c_{22}te^t \end{bmatrix} \tag{6.71}$$

$$c_{11}e^t + c_{12}(e^t + te^t) = (c_{11} + c_{12})\,e^t + (c_{12} + c_{22})\,te^t$$

$$c_{21}e^t + c_{22}(e^t + te^t) = c_{21}e^t + c_{22}te^t$$

Equation (6.71) gives $c_{22} = 0$ and $c_{12} = c_{21}$. Next, equation (6.70) must satisfy the initial conditions:

$$1 = c_{11}e^1 + c_{12}e^1$$

$$1 = c_{21}e^1$$

which gives $c_{11} = 0$, and $c_{12} = c_{21} = e^{-1}$. Hence the free response is

$$x_1(t) = e^{-1}te^t = te^{(t-1)}$$

$$x_2(t) = e^{-1}e^t = e^{(t-1)}$$

which is the same as the result given by the solution matrix!

Magnitudes of Eigenvectors   For general use, the magnitudes of the eigenvectors are arbitrarily fixed, as indicated in Appendix B. Consider, for instance, the system of Example 6.16. The directions of the eigenvectors were determined as

$$\mathbf{V}_1 = \begin{bmatrix} v_{11} \\ v_{11} \end{bmatrix}; \qquad \mathbf{V}_2 = \begin{bmatrix} 0 \\ v_{22} \end{bmatrix}$$

Using the method of setting the first nonzero element of the eigenvector to 1, we obtain

$$\mathbf{V}_1 = \begin{bmatrix} 1 \\ 1 \end{bmatrix}; \qquad \mathbf{V}_2 = \begin{bmatrix} 0 \\ 1 \end{bmatrix}$$

Note that these vectors are collinear with those determined using the initial conditions. If we fix the magnitude of the eigenvectors by *normalization,* that is,

$$|\mathbf{V}_i| = \sum_{k=1}^{n} (v_{ki})^2 = 1 \tag{6.72}$$

then for the same system of Example 6.16, we obtain the normalized eigenvectors

$$\mathbf{V}_1 = \begin{bmatrix} \sqrt{2}/2 \\ \sqrt{2}/2 \end{bmatrix}; \qquad \mathbf{V}_2 = \begin{bmatrix} 0 \\ 1 \end{bmatrix}$$

The eigenvectors in this case are *unit vectors* in *state space* and may be viewed as establishing coordinate directions along which the free response is described (see State Trajectories for Second-Order Systems, Section 9.4).

Forced Response and Application of Linear Transformations

The forced response of the system (6.49a) as given by equation (6.58) is

$$\mathbf{X}_{\text{forced}}(t) = \int_{t_0}^{t} e^{\mathbf{A}(t-\tau)} \mathbf{B} \mathbf{U}(\tau)\, d\tau = e^{\mathbf{A}t} \int_{t_0}^{t} e^{-\mathbf{A}\tau} \mathbf{B} \mathbf{U}(\tau)\, d\tau \tag{6.73}$$

or in terms of the state transition matrix,

$$\mathbf{X}_{\text{forced}}(t) = \Phi(t, t_0) \int_{t_0}^{t} \Phi(t_0, \tau) \mathbf{B} \mathbf{U}(\tau)\, d\tau \tag{6.74}$$

Equation (6.73) is easily demonstrated. Consider (6.49a) premultiplied on both sides by $e^{-\mathbf{A}t}$:

$$e^{-\mathbf{A}t}\left(\frac{d}{dt}\mathbf{X}\right) = e^{-\mathbf{A}t}\mathbf{A}\mathbf{X} + e^{-\mathbf{A}t}\mathbf{B}\mathbf{U} \tag{6.75a}$$

Noting from equation (6.60) that

$$\frac{d}{dt}(e^{-\mathbf{A}t}\mathbf{X}) = -e^{-\mathbf{A}t}\mathbf{A}\mathbf{X} + e^{-\mathbf{A}t}\left(\frac{d}{dt}\mathbf{X}\right)$$

We see that equation (6.75a) becomes

$$\frac{d}{dt}(e^{-\mathbf{A}t}\mathbf{X}) = -e^{-\mathbf{A}t}\mathbf{B}\mathbf{U} \tag{6.75b}$$

Integrating equation (6.75b) between the time limits t_0 and t we obtain

$$e^{-\mathbf{A}t}\mathbf{X}(t) = \int_{t_0}^{t} e^{-\mathbf{A}\tau} \mathbf{B} \mathbf{U}(\tau)\, d\tau \tag{6.75c}$$

which gives equation (6.73).

Application of Linear Transformations

As we saw in Example 6.15, the forced response is easily computed if the solution matrix or the state transition matrix is known. From equation (6.62b) the solution matrix is trivial for the case when **A** is diagonal. It follows that one way to find the forced response (or for that matter the free response and hence the total solution) is to *diagonalize* (transform to diagonal form) the system (6.49a). The solution in the diagonal domain can then be transformed back to the original domain to obtain the forced (or free or total) response.

Specifically, we want to make a change of "coordinates" from **X** to $\hat{\mathbf{X}}$ using the *linear transformation*

$$\mathbf{X}(t) = \mathbf{T}\hat{\mathbf{X}}(t); \quad \text{(linear transformation; T = transformation matrix)} \tag{6.76}$$

such that the resulting system in $\hat{\mathbf{X}}$ (equations (6.78a) through (6.78e)) is diagonal. To find the new system, we substitute (6.76) in equations (6.49a) and (6.49b):

$$\frac{d}{dt}\mathbf{X} = \mathbf{T}\frac{d}{dt}\hat{\mathbf{X}} = \mathbf{AX} + \mathbf{BU} = \mathbf{AT}\hat{\mathbf{X}} + \mathbf{BU}; \quad \mathbf{X}(0) = \mathbf{T}\hat{\mathbf{X}}(0) \tag{6.77a}$$

Output: $\mathbf{Y} = \mathbf{CT}\hat{\mathbf{X}}$ \hfill (6.77b)

Assuming the inverse of the transformation matrix exists, we premultiply equations (6.77) on both sides by $\mathbf{T}^{-1}$ to get

$$\frac{d}{dt}\hat{\mathbf{X}} = (\mathbf{T}^{-1}\mathbf{AT})\,\hat{\mathbf{X}} + (\mathbf{T}^{-1}\mathbf{B})\,\mathbf{U}; \quad \hat{\mathbf{X}}(0) = \mathbf{T}^{-1}\mathbf{X}(0)$$

or

$$\frac{d}{dt}\hat{\mathbf{X}} = \hat{\mathbf{A}}\hat{\mathbf{X}} + \hat{\mathbf{B}}\mathbf{U}; \quad \hat{\mathbf{X}}(0) = \mathbf{T}^{-1}\mathbf{X}(0) \tag{6.78a}$$

Output: $\mathbf{Y} = \hat{\mathbf{C}}\hat{\mathbf{X}}$ \hfill (6.78b)

where

$$\hat{\mathbf{A}} = \mathbf{T}^{-1}\mathbf{AT} \tag{6.78c}$$

$$\hat{\mathbf{B}} = \mathbf{T}^{-1}\mathbf{B} \tag{6.78d}$$

$$\hat{\mathbf{C}} = \mathbf{CT} \tag{6.78e}$$

As given in Appendix B, if the eigenvectors $\mathbf{V}_i$, $i = 1, \ldots, n$, associated with the n eigenvalues p_i of **A** are linearly independent, then the $n \times n$ transformation matrix having the eigenvectors as its columns (the *eigenvector matrix*) is *not singular* (that is, $\mathbf{T}^{-1}$ exists) and diagonalization is possible. Further, the diagonal matrix has the eigenvalues of **A** (in the same order as the associated eigenvectors are in **T**) as its main diagonal elements. That is,

$$\mathbf{T} = [\mathbf{V}_1, \ldots, \mathbf{V}_n]; \quad \text{(eigenvector matrix)} \tag{6.79a}$$

$$\hat{\mathbf{A}} = \mathbf{T}^{-1}\mathbf{AT} = \begin{bmatrix} p_1 & & & 0 \\ & p_2 & & \\ & & \ddots & \\ 0 & & & p_n \end{bmatrix}$$

The Modal Matrix as the Eigenvector Matrix

Since

$$e^{\hat{A}t} = \text{diag} \left[e^{\hat{a}_{11}t}, e^{\hat{a}_{22}t}, \ldots, e^{\hat{a}_{nn}t} \right] = \text{diag} \left[e^{p_1 t}, e^{p_2 t}, \ldots, e^{p_n t} \right]$$

we see that the transformation matrix $\mathbf{T}$ causes a *decoupling of the modes* of the system. This can be demonstrated by writing equation (6.78a) in component form (assuming m inputs):

$$\frac{d}{dt} \hat{x}_1(t) = p_1 \hat{x}_1 + \sum_{j=1}^{m} \hat{b}_{1j} u_j \tag{6.79c}$$

$$\frac{d}{dt} \hat{x}_2(t) = p_2 \hat{x}_2 + \sum_{j=1}^{m} \hat{b}_{2j} u_j$$

$$\vdots$$

$$\frac{d}{dt} \hat{x}_n(t) = p_n \hat{x}_n + \sum_{j=1}^{m} \hat{b}_{nj} u_j$$

Each equation is a first-order system that depends on only one eigenvalue and that is not related (not coupled) to the other equations. The solution to each equation is of course characterized by the exponential mode for the associated eigenvalue. For these reasons, the transformation matrix $\mathbf{T}$ is called the *modal matrix*. What, then, is the relation between the eigenvalues of the new system and those of the original system? This question can be answered by substituting equation (6.78c) in the characteristic equation for the diagonal system and using properties of the inverse matrix (see Appendix B):

$$\det [p\mathbf{I} - \hat{\mathbf{A}}] = \det [p\mathbf{I} - \mathbf{T}^{-1}\mathbf{A}\mathbf{T}] = \det [\mathbf{T}^{-1} (p\mathbf{I} - \mathbf{A}) \mathbf{T}] = 0 \quad \text{or}$$

$$\det [\mathbf{T}^{-1}] \det [p\mathbf{I} - \mathbf{A}] \det [\mathbf{T}] = 0$$

giving (since $\mathbf{T}$ is not singular)

$$\det [p\mathbf{I} - \mathbf{A}] = 0 \tag{6.80}$$

Equation (6.80) says that the eigenvalues for the two systems are the same! This result should have been expected. The eigenvalues are determined by the physical parameters of the system. Physically we still have the same system; we have merely made a mathematical change in the variables with which we describe it!

From Appendix B, we note that *the eigenvectors are linearly independent when the eigenvalues are all distinct.* This does not imply that nondistinct eigenvalues may not yield linearly independent eigenvectors. Rather, a transformation to diagonal form is always possible for all systems with distinct eigenvalues.

Example 6.17   Solution by Diagonalization

Solve the system (6.49a) with

$$\mathbf{A} = \begin{bmatrix} -3 & 4 & 4 \\ 1 & -3 & -1 \\ -1 & 2 & 0 \end{bmatrix}; \qquad \mathbf{B} = \begin{bmatrix} 2 \\ -\dfrac{1}{2} \\ 1 \end{bmatrix}; \qquad u(t) = U_S(t); \qquad \mathbf{X}_0 = \begin{bmatrix} 0 \\ 1 \\ 2 \end{bmatrix} = \mathbf{X}(0)$$

Solution: The characteristic equation (6.68) is

$$\det [p\mathbf{I} - \mathbf{A}] = \det \begin{bmatrix} (p+3) & -4 & -4 \\ -1 & (p+3) & 1 \\ 1 & -2 & p \end{bmatrix} = 0$$

Or

$$(p + 3) [p (p + 3) + 2] + 4 [-p - 1] - 4 [2 - p - 3] = 0$$
$$= (p + 3) (p^2 + 3p + 2) = (p + 3) (p + 2) (p + 1) = 0$$

Thus the eigenvalues are $p_1 = -3$, $p_2 = -2$, $p_3 = -1$, all distinct.

Using equation (6.67a) or (6.67b) and $v_{11} = 1$, $v_{22} = 1$, $v_{13} = 1$, we obtain the eigenvectors

$$\mathbf{V}_1 = \begin{bmatrix} 1 \\ -1 \\ 1 \end{bmatrix}; \qquad \mathbf{V}_2 = \begin{bmatrix} 0 \\ 1 \\ -1 \end{bmatrix}; \qquad \mathbf{V}_3 = \begin{bmatrix} 1 \\ 1 \\ 2 \\ 0 \end{bmatrix}$$

The transformation matrix is then

$$\mathbf{T} = \begin{bmatrix} 1 & 0 & 1 \\ -1 & 1 & \dfrac{1}{2} \\ 1 & -1 & 0 \end{bmatrix}$$

Also,

$$\mathbf{T}^{-1} = \begin{bmatrix} 1 & -2 & -2 \\ 1 & -2 & -3 \\ 0 & 2 & 2 \end{bmatrix}$$

We note the check:

$$\mathbf{T}^{-1}\mathbf{A}\mathbf{T} = \begin{bmatrix} -3 & 0 & 0 \\ 0 & -2 & 0 \\ 0 & 0 & -1 \end{bmatrix} = \hat{\mathbf{A}}$$

The diagonal system from equations (6.78a) through (6.78e) is

$$\frac{d}{dt} \hat{x}_1 = -3\hat{x}_1 + U_S(t); \qquad \hat{x}_1(0) = -6$$

$$\frac{d}{dt} \hat{x}_2 = -2\hat{x}_2; \qquad \hat{x}_2(0) = -8$$

$$\frac{d}{dt} \hat{x}_3 = -\hat{x}_3 + U_S(t); \qquad \hat{x}_3(0) = 6$$

The solution to the preceding equations is obtained from equation (6.14a):

$$\hat{x}_1(t) = -6e^{-3t} + \frac{1}{3}(1 - e^{-3t}); \qquad \hat{x}_2(t) = -8e^{-2t};$$

$$\hat{x}_3(t) = 6e^{-t} + (1 - e^{-t})$$

or

$$\hat{x}_1(t) = -\frac{1}{3}(19e^{-3t} - 1); \qquad \hat{x}_2(t) = -8e^{-2t}; \qquad \hat{x}_3(t) = 5e^{-t} + 1$$

Note that this solution could also have been obtained using equation (6.58), since $e^{\hat{A}t}$ is known. The solution to the original problem is given be equation (6.76), that is,

$$\mathbf{X}(t) = \begin{bmatrix} 1 & 0 & 1 \\ -1 & 1 & \frac{1}{2} \\ 1 & -1 & 0 \end{bmatrix} \begin{bmatrix} -\frac{1}{3}(19e^{-3t} - 1) \\ -8e^{-2t} \\ 5e^{-t} + 1 \end{bmatrix} = \begin{bmatrix} 5e^{-t} - \frac{1}{3}(19e^{-3t} - 4) \\ \frac{5}{2}e^{-t} - 8e^{-2t} + \frac{19}{3}e^{-3t} + \frac{1}{6} \\ -\frac{1}{3}(19e^{-3t} - 1) + 8e^{-2t} \end{bmatrix}$$

Nondistinct Eigenvalues and Jordan Form   Although diagonalization may not be possible for some systems with repeated eigenvalues, a transformation to a form that is still relatively easy to solve is always possible. Any square matrix is *similar* to (can be transformed to) the *Jordan form* (see [3]):

$$\hat{\mathbf{A}} = \text{diag}\,[\mathbf{J}_{k_i,1}(p_i),\,\mathbf{J}_{k_i,2}(p_i),\,\ldots\,,\,\mathbf{J}_{k_i,m_i}(p_i)], \quad i = 1, 2, \ldots, q \qquad \textbf{(6.81a)}$$

where m_i is the number of Jordan blocks for eigenvalue i, q is the number of repeated and distinct eigenvalues, and $\mathbf{J}_k(p)$ is the $k \times k$ **Jordan block:**

$$\mathbf{J}_k(p) = \begin{bmatrix} p & 1 & 0 & \cdot & \cdot & 0 & 0 \\ 0 & p & 1 & \cdot & \cdot & 0 & 0 \\ \cdot & \cdot & \cdot & \cdot & \cdot & \cdot & \cdot \\ \cdot & \cdot & \cdot & \cdot & 0 & p & 1 \\ 0 & 0 & \cdot & \cdot & 0 & 0 & p \end{bmatrix} \qquad \textbf{(6.81b)}$$

For example,

$$\hat{\mathbf{A}} = \begin{bmatrix} -2 & 1 & 0 & 0 & 0 & 0 \\ 0 & -2 & 0 & 0 & 0 & 0 \\ 0 & 0 & -2 & 0 & 0 & 0 \\ 0 & 0 & 0 & -1 & 1 & 0 \\ 0 & 0 & 0 & 0 & -1 & 1 \\ 0 & 0 & 0 & 0 & 0 & -1 \end{bmatrix}; \qquad \begin{array}{l} q = 2, p_1 = -2, p_2 = -1 \\ m_1 = 2, k_{1,1} = 2, k_{1,2} = 1 \\ m_2 = 1, k_{2,1} = 3 \end{array}$$

can be represented as

$$\hat{\mathbf{A}} = \text{diag}\,[\mathbf{J}_2\,(-2),\,\mathbf{J}_1\,(-2),\,\mathbf{J}_3\,(-1)]$$

Also,

$$\hat{\mathbf{A}} = \begin{bmatrix} -3 & 0 & 0 \\ 0 & -2 & 0 \\ 0 & 0 & -1 \end{bmatrix} = \text{diag } [\mathbf{J}_1 (-3), \mathbf{J}_1 (-2), \mathbf{J}_1 (-1)]$$

Thus the diagonal matrix is only a special case of the general Jordan form. When a system has some repeated roots, a transformation into the Jordan form can be obtained in which the distinct roots give rise to diagonal blocks, and the repeated eigenvalues give Jordan blocks of size (k) equal to the multiplicity of the repeating root. The solution to a system described by a Jordan block is easily found due to the "relay line"–type coupling between the component equations—once the equation at the end of the line (which is completely decoupled) is solved, the rest are determined. Consider the following system:

$$\frac{d}{dt} \hat{\mathbf{X}} = \mathbf{J}_k(p)\hat{\mathbf{X}} + \hat{\mathbf{B}}\mathbf{U}; \qquad \hat{\mathbf{X}}(0) = \mathbf{T}^{-1}\mathbf{X}(0) \tag{6.82a}$$

where $\mathbf{J}_k(p)$ is defined by equation (6.81b). The component equations are

$$\frac{d}{dt} \hat{x}_1 = p\hat{x}_1 + \hat{x}_2 + \sum \hat{b}_{1,j} u_j \tag{6.82b}$$

$$\frac{d}{dt} \hat{x}_2 = p\hat{x}_2 + \hat{x}_3 + \sum \hat{b}_{2,j} u_j$$

$$\vdots$$

$$\frac{d}{dt} \hat{x}_{k-1} = p\hat{x}_{k-1} + \hat{x}_k + \sum \hat{b}_{k-1,j} u_j$$

$$\frac{d}{dt} \hat{x}_k = p\hat{x}_k + \sum \hat{b}_{k,j} u_j$$

We see that the last equation is decoupled from the rest and is easily solved as a first-order system. Once $\hat{x}_k$ is known, it can be substituted in the equation for $\hat{x}_{k-1}$, where it becomes part of the forcing function in this equation. Thus the $(k-1)$th equation becomes a first-order system and is easily solved. With $\hat{x}_{k-1}$, known, $\hat{x}_{k-2}$ can be found, and so on up the line. The solution matrix for the Jordan block is also easily found (see Practice Problem 6.40) and it can be used instead to obtain the solution.

To illustrate how the column vectors of the transformation matrix corresponding to a Jordan block can be found, we consider the case of an $n \times n$ Jordan block—that is, the system has only one eigenvalue and it has a multiplicity equal to the order of the system, n. Equation (6.78c) premultiplied on both sides by $\mathbf{T}$ gives

$$\mathbf{TJ}_n(p) = \mathbf{AT} = \mathbf{T} \begin{bmatrix} p & 1 & 0 & \cdot & \cdot & 0 & 0 \\ 0 & p & 1 & \cdot & \cdot & 0 & 0 \\ \cdot & \cdot & \cdot & \cdot & \cdot & \cdot & \cdot \\ \cdot & \cdot & \cdot & \cdot & 0 & p & 1 \\ 0 & 0 & \cdot & \cdot & 0 & 0 & p \end{bmatrix} \tag{6.83a}$$

In terms of column vectors we have

$$\mathbf{AT}_1 = p\mathbf{T}_1$$

$$\mathbf{AT}_2 = \mathbf{T}_1 + p\mathbf{T}_2$$ (6.83b)

$$\vdots$$

$$\mathbf{AT}_n = \mathbf{T}_{n-1} + p\mathbf{T}_n$$

Note that the equation for the first column vector (the only eigenvector) is the same as equation (6.67a). The equations for the remaining column vectors are coupled by the 1's in the Jordan block. This is evident in equation (6.83b). However, the set of column vectors is linearly independent.

Example 6.18   Case with a Jordan Block

Solve the system (6.49a) with

$$\mathbf{A} = \begin{bmatrix} -3 & 2 & 0 \\ -1 & -4 & 1 \\ -2 & 0 & -1 \end{bmatrix}; \quad \mathbf{B} = \begin{bmatrix} 2 \\ 0 \\ 2 \end{bmatrix}; \quad \mathbf{X}_0 = \begin{bmatrix} 1 \\ 0 \\ 1 \\ 2 \end{bmatrix}; \quad u(t) = \delta(t)$$

Solution:

$$\det[p\mathbf{I} - \mathbf{A}] = \det \begin{bmatrix} (p+3) & -2 & 0 \\ 1 & (p+4) & -1 \\ 2 & 0 & (p+1) \end{bmatrix} = \begin{matrix} (p+3)(p+4)(p+1) + 2(p+3) \\ = (p+3)^2(p+2) = 0 \end{matrix}$$

Eigenvalues:

$$p_1 = p_2 = -3; \qquad p_3 = -2$$

We expect a 2×2 Jordan block for the repeated eigenvalue and a 1×1 diagonal block for the distinct eigenvalue. That is,

$$\hat{\mathbf{A}} = \begin{bmatrix} -3 & 1 & 0 \\ 0 & -3 & 0 \\ \hline 0 & 0 & -2 \end{bmatrix}$$

The eigenvector for the repeating root is given by (6.67a) or the first line of equation (6.83b):

$$-3 \begin{bmatrix} T_{11} \\ T_{21} \\ T_{31} \end{bmatrix} = \begin{bmatrix} -3 & 2 & 0 \\ -1 & -4 & 1 \\ -2 & 0 & -1 \end{bmatrix} \begin{bmatrix} T_{11} \\ T_{21} \\ T_{31} \end{bmatrix} = \begin{bmatrix} -3T_{11} + 2T_{21} \\ -T_{11} - 4T_{21} + T_{31} \\ -2T_{11} - T_{31} \end{bmatrix}$$

Solving, we obtain

$$T_{21} = 0; \ T_{11} = T_{31}$$

Let

$$T_{11} = 1 = T_{31}$$

The second linearly independent vector is given by (6.83b):

$$-3 \begin{bmatrix} T_{12} \\ T_{22} \\ T_{32} \end{bmatrix} + \begin{bmatrix} 1 \\ 0 \\ 1 \end{bmatrix} = \begin{bmatrix} -3T_{12} + 2T_{22} \\ -T_{12} - 4T_{22} + T_{32} \\ -2T_{12} - T_{32} \end{bmatrix}$$

Solving, we have

$$T_{22} = \frac{1}{2}; \qquad T_{32} = \frac{1}{2} + T_{12}$$

Let

$$T_{12} = 1; \qquad T_{32} = \frac{3}{2}$$

The eigenvector for the distinct root is also given by (6.67a):

$$-2 \begin{bmatrix} T_{13} \\ T_{23} \\ T_{33} \end{bmatrix} = \begin{bmatrix} -3T_{13} + 2T_{23} \\ -T_{13} - 4T_{23} + T_{33} \\ -2T_{13} - T_{33} \end{bmatrix}$$

which yields

$$T_{13} = 2T_{23} = \left(\frac{1}{2} \right) T_{33}$$

Let

$$T_{13} = 1; \qquad T_{23} = \frac{1}{2}; \qquad T_{33} = 2$$

Thus the transformation matrix is

$$\mathbf{T} = \begin{bmatrix} 1 & 1 & 1 \\ 0 & \frac{1}{2} & \frac{1}{2} \\ 1 & \frac{3}{2} & 2 \end{bmatrix} \quad \text{and} \quad \mathbf{T}^{-1} = \begin{bmatrix} 1 & -2 & 0 \\ 2 & 4 & -2 \\ -2 & -2 & 2 \end{bmatrix}$$

Check: $\mathbf{T}^{-1}\mathbf{A}\mathbf{T} = \hat{\mathbf{A}}$; $\qquad \hat{\mathbf{B}} = \mathbf{T}^{-1}\mathbf{B} = \begin{bmatrix} 2 \\ 0 \\ 0 \end{bmatrix}$; $\qquad \hat{\mathbf{X}}(0) = \mathbf{T}^{-1}\mathbf{X}(0) = \begin{bmatrix} 1 \\ 1 \\ -1 \end{bmatrix}$

The transformed system of equations is

$$\frac{d}{dt}\hat{x}_1 = -3\hat{x}_1 + \hat{x}_2 + 2\delta(t); \qquad \hat{x}_1(0) = 1$$

$$\frac{d}{dt}\hat{x}_2 = -3\hat{x}_2; \qquad \hat{x}_2(0) = 1$$

$$\frac{d}{dt}\hat{x}_3 = -2\hat{x}_3; \qquad \hat{x}_3(0) = -1$$

Solving, beginning with the last equation, we obtain

$$\hat{x}_3(t) = -e^{-2t}$$

$$\hat{x}_2(t) = e^{-3t}$$

$$\hat{x}_1(t) = e^{-3t} + e^{-3t} \int_0^t e^{3\tau} [\hat{x}_2(\tau) + 2\delta(\tau)] \, d\tau$$

$$= e^{-3t} + e^{-3t} \int_0^t e^{3\tau} [e^{-3\tau} + 2\delta(\tau)] \, d\tau = 3e^{-3t} + te^{-3t}$$

Thus

$$\mathbf{X}(t) = \mathbf{T}\hat{\mathbf{X}}(t) = \begin{bmatrix} 1 & 1 & 1 \\ 0 & \dfrac{1}{2} & \dfrac{1}{2} \\ 1 & \dfrac{3}{2} & 2 \end{bmatrix} \begin{bmatrix} (3+t)\,e^{-3t} \\ e^{-3t} \\ -e^{-2t} \end{bmatrix} = \begin{bmatrix} (4+t)\,e^{-3t} - e^{-2t} \\ \dfrac{1}{2}\,(e^{-3t} - e^{-2t}) \\ \left(\dfrac{9}{2} + t\right) e^{-3t} - 2e^{-2t} \end{bmatrix}$$

6.4 SOLUTION OF THE LINEAR DISCRETE-TIME MODEL

A parallel development of much of the material in Section 6.3 is possible for discrete-time systems. Consider the first-order homogeneous difference equation

$$x_{k+1} - px_k = 0; \qquad x_0 = \text{given} \tag{6.84a}$$

or

$$x_{k+1} = px_k; \qquad x_0 = \text{given} \tag{6.84b}$$

Equation (6.84b) represents an iteration problem. Thus beginning with x_0, we easily find x_k:

$$x_1 = px_0$$

$$x_2 = px_1 = p^2 x_0 \tag{6.85}$$

$$\vdots$$

$$x_k = p^k x_0$$

Response Modes

We note that the solution or response, equation (6.85), is characterized by the term p^k, which is thus the response mode. For $p > 1$, the progression $p^1, p^2, \ldots, p^k$ grows with k, the rate of growth depending on the magnitude of p. For $p < -1$, the response mode is an oscillatory growth with k. Other values of p yield stable modes, as summarized in Fig. 6.22.

We consider now the nth-order homogeneous discrete-time system:

$$\mathbf{X}_{k+1} = \mathbf{P}\mathbf{X}_k; \qquad \mathbf{X}_0 = \text{given} \tag{6.86}$$

By analogy with the continuous-time case, we can expect that the solution to (6.86) consists of a linear combination of modes $p_1^k, p_2^k, \ldots, p_n^k$, that is,

$$x_{1k} = v_{11}p_1^k + v_{12}p_2^k + \ldots + v_{1n}p_n^k$$
$$x_{2k} = v_{21}p_1^k + v_{22}p_2^k + \ldots + v_{2n}p_n^k, \text{ or}$$
$$\vdots$$
$$x_{nk} = v_{n1}p_1^k + v_{n2}p_2^k + \ldots + v_{nn}p_n^k$$
$$\mathbf{X}_k = \mathbf{V}_1p_1^k + \mathbf{V}_2p_2^k + \ldots + \mathbf{V}_np_n^k \tag{6.87}$$

where $\mathbf{V}_i$ is the vector

$$\mathbf{V}_i = \begin{bmatrix} v_{1i} \\ v_{2i} \\ \vdots \\ v_{ni} \end{bmatrix}$$

Note that the p_i's are not the elements of $\mathbf{P}$. Substituting equation (6.87) into equation (6.86), we obtain

$$\mathbf{V}_1p_1^{k+1} + \mathbf{V}_2p_2^{k+1} + \ldots + \mathbf{V}_np_n^{k+1} = \mathbf{P}\,[\mathbf{V}_1p_1^k + \mathbf{V}_2p_2^k + \ldots + \mathbf{V}_np_n^k]$$
$$= p_1\mathbf{V}_1p_1^k + p_2\mathbf{V}_2p_2^k + \ldots + p_n\mathbf{V}_np_n^k$$

Figure 6.22

**Response Modes
of a First-order
Discrete-time System**

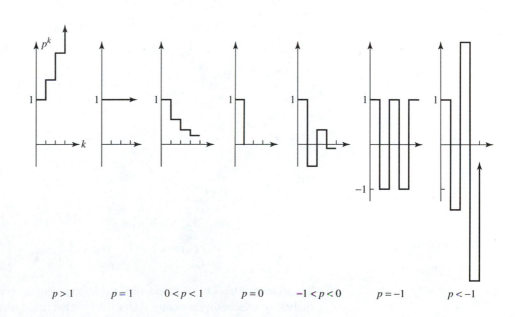

| $p > 1$ | $p = 1$ | $0 < p < 1$ | $p = 0$ | $-1 < p < 0$ | $p = -1$ | $p < -1$ |

which yields on equating terms,

$$p_i \mathbf{V}_i = \mathbf{P} \mathbf{V}_i \tag{6.88a}$$

or

$$(p_i \mathbf{I} - \mathbf{P}) \, \mathbf{V}_i = 0 \tag{6.88b}$$

As was the case for continuous-time systems, equations (6.88a) and (6.88b), define an eigenvalue problem, and the p_i's are the eigenvalues or the roots of the characteristic equation

$$\det [p_i \mathbf{I} - \mathbf{P}] = 0; \quad \text{(characteristic equation)} \tag{6.89}$$

Also, for the nth-order system, real and complex-conjugate eigenvalues are possible, so that in addition to the modes described for the first-order case, other oscillatory modes of various patterns are also obtained. These are considered in more detail in Section 10.3.

Computation of Discrete-Time System Response

For a system with distinct eigenvalues, equation (6.88a) or (6.88b) determines the directions of the vectors $\mathbf{V}_i$, which in this case are eigenvectors, and the given initial state $\mathbf{X}_0$ determines the magnitudes of the vectors. The homogeneous system solution equation (6.87) is then complete.

The general state model for a linear discrete-time system is given by

$$\mathbf{X}_{k+1} = \mathbf{P}\mathbf{X}_k + \mathbf{Q}\mathbf{U}_k \tag{6.90a}$$

$$\mathbf{Y}_k = \mathbf{C}\mathbf{X}_k + \mathbf{D}\mathbf{U}_k \tag{6.90b}$$

For example, the model of the savings-only commercial bank developed in Example 4.9, equations (4.48a) through (4.48d) repeated,

$$x_{1,k+1} = (1 + \alpha_1) \, x_{1,k} + u_{1,k}$$

$$x_{2,k+1} = (1 - \alpha_1) \, \alpha_2 x_{1,k} + (1 + \alpha_2 \alpha_3) \, x_{2,k} - \alpha_2 u_{2,k}$$

$$y_{1,k} = (1 - \alpha_1) \, (1 - \alpha_2) \, x_{1,k} + (1 - \alpha_2) \, \alpha_3 x_{2,k} - (1 - \alpha_2) \, u_{2,k}$$

$$y_{2,k} = \alpha_3 x_{2,k}$$

gives

$$\mathbf{X}_k = \begin{bmatrix} x_{1,k} \\ x_{2,k} \end{bmatrix}; \qquad \mathbf{U}_k = \begin{bmatrix} u_{1,k} \\ u_{2,k} \end{bmatrix}; \qquad \mathbf{Y}_k = \begin{bmatrix} y_{1,k} \\ y_{2,k} \end{bmatrix}$$

and

$$\mathbf{P} = \begin{bmatrix} (1 + \alpha_1) & 0 \\ (1 - \alpha_1) \, \alpha_2 & (1 + \alpha_2 \alpha_3) \end{bmatrix}; \qquad \mathbf{Q} = \begin{bmatrix} 1 & 0 \\ 0 & -\alpha_2 \end{bmatrix}$$

$$\mathbf{C} = \begin{bmatrix} (1 - \alpha_1) \, (1 - \alpha_2) & (1 - \alpha_2) \, \alpha_3 \\ 0 & \alpha_3 \end{bmatrix}; \qquad \mathbf{D} = \begin{bmatrix} 0 & -(1 - \alpha_2) \\ 0 & 0 \end{bmatrix}$$

Equation (6.90a) is the discrete-time vector state equation that corresponds to equation (6.49a) of the continuous-time system. As for the first-order discrete-time case, (6.90a) is an iteration problem that can be solved directly, given $\mathbf{X}_0$ and the input sequence $\mathbf{U}_0$, $\mathbf{U}_1, \ldots, \mathbf{U}_k$:

$$
\begin{bmatrix}
\mathbf{X}_1 = \mathbf{PX}_0 + \mathbf{QU}_0 \\
\mathbf{X}_2 = \mathbf{PX}_1 + \mathbf{QU}_1 \\
\mathbf{X}_3 = \mathbf{PX}_2 + \mathbf{QU}_2 \\
\vdots \\
\mathbf{X}_k = \mathbf{PX}_{k-1} + \mathbf{QU}_{k-1}
\end{bmatrix}
\begin{aligned}
&= \mathbf{P}^2\mathbf{X}_0 + \mathbf{PQU}_0 + \mathbf{QU}_1 \\
&= \mathbf{P}^3\mathbf{X}_0 + \mathbf{P}^2\mathbf{QU}_0 + \mathbf{PQU}_1 + \mathbf{QU}_2 \\
&\vdots \\
&= \mathbf{P}^k\mathbf{X}_0 + \sum_{j=0}^{k-1} \mathbf{P}^{k-1-j}\mathbf{QU}_j
\end{aligned}
\qquad (6.91)
$$

Thus the general solution to equation (6.90a) is

$$
\mathbf{X}_k = \mathbf{P}^k\mathbf{X}_0 + \sum_{j=0}^{k-1} \mathbf{P}^{k-1-j}\mathbf{QU}_j \qquad (6.92)
$$

In practice, equation (6.92) is computationally tedious, even for a digital computer. The recurrence solution, shown within brackets in equation (6.91), from which equation (6.92) was deduced, represents the more practical solution approach.

Example 6.19	Discrete-Time System Response

Compute the response of the second-order system

$$
\mathbf{X}_{k+1} = \begin{bmatrix} -1 & 0 \\ 1 & -1 \end{bmatrix} \mathbf{X}_k, \text{ for } k = 1, 2, \ldots,
$$

if the initial state is

$$
\mathbf{X}_0 = \begin{bmatrix} 1 \\ 1 \end{bmatrix}
$$

What are the eigenvalues and response modes of this system?

Solution: Using the general response, equation (6.92), we have

$$
\mathbf{X}_k = \begin{bmatrix} -1 & 0 \\ 1 & -1 \end{bmatrix}^k \begin{bmatrix} 1 \\ 1 \end{bmatrix}
$$

The problem is to compute $\mathbf{P}^k$.

$$
k = 2, \mathbf{P}^2 = \begin{bmatrix} -1 & 0 \\ 1 & -1 \end{bmatrix} \begin{bmatrix} -1 & 0 \\ 1 & -1 \end{bmatrix} = \begin{bmatrix} 1 & 0 \\ -2 & 1 \end{bmatrix}
$$

$$
k = 3, \mathbf{P}^3 = \begin{bmatrix} -1 & 0 \\ 1 & -1 \end{bmatrix} \begin{bmatrix} 1 & 0 \\ -2 & 1 \end{bmatrix} = \begin{bmatrix} -1 & 0 \\ 3 & -1 \end{bmatrix}
$$

$$
k = 4, \mathbf{P}^4 = \begin{bmatrix} -1 & 0 \\ 1 & -1 \end{bmatrix} \begin{bmatrix} -1 & 0 \\ 3 & -1 \end{bmatrix} = \begin{bmatrix} 1 & 0 \\ -4 & 1 \end{bmatrix}
$$

$$
\vdots
$$

$$
k = k, \mathbf{P}^k = \begin{bmatrix} (-1)^k & 0 \\ -k(-1)^k & (-1)^k \end{bmatrix}; \qquad k \geq 1
$$

Thus

$$\mathbf{X}_k = \begin{bmatrix} (-1)^k & 0 \\ -k(-1)^k & (-1)^k \end{bmatrix} \begin{bmatrix} 1 \\ 1 \end{bmatrix} = \begin{bmatrix} (-1)^k \\ (1-k)(-1)^k \end{bmatrix}$$

The **P** matrix is especially simple, and a closed-form solution is possible. The recurrence solution is as follows:

$$k = 1, \mathbf{X}_1 = \mathbf{P}\mathbf{X}_0 = \begin{bmatrix} -1 & 0 \\ 1 & -1 \end{bmatrix} \begin{bmatrix} 1 \\ 1 \end{bmatrix} = \begin{bmatrix} -1 \\ 0 \end{bmatrix}$$

$$k = 2, \mathbf{X}_2 = \mathbf{P}\mathbf{X}_1 = \begin{bmatrix} -1 & 0 \\ 1 & -1 \end{bmatrix} \begin{bmatrix} -1 \\ 0 \end{bmatrix} = \begin{bmatrix} 1 \\ -1 \end{bmatrix}$$

$$k = 3, \mathbf{X}_3 = \mathbf{P}\mathbf{X}_2 = \begin{bmatrix} -1 & 0 \\ 1 & -1 \end{bmatrix} \begin{bmatrix} 1 \\ -1 \end{bmatrix} = \begin{bmatrix} -1 \\ 2 \end{bmatrix}$$

$$k = 4, \mathbf{X}_4 = \mathbf{P}\mathbf{X}_3 = \begin{bmatrix} -1 & 0 \\ 1 & -1 \end{bmatrix} \begin{bmatrix} -1 \\ 2 \end{bmatrix} = \begin{bmatrix} 1 \\ -3 \end{bmatrix}$$

$$\vdots$$

$$k = k, \mathbf{X}_k = \mathbf{P}\mathbf{X}_{k-1} = \begin{bmatrix} (-1)^k \\ (1-k)(-1)^k \end{bmatrix}$$

as before.

The eigenvalues are given by equation (6.89). That is,

$$\det \begin{bmatrix} (p+1) & 0 \\ -1 & (p+1) \end{bmatrix} = (p+1)(p+1) = 0$$

which gives the repeated root $p_1 = p_2 = -1$.

One of the modes is thus $(-1)^k$. From the solutions shown, we conclude that the second response mode must be $k(-1)^k$. Recall that for a continuous-time system with the **A** matrix equal to the **P** matrix shown, the response modes will be e^{-t} and te^{-t}.

....

6.5 DIGITAL COMPUTER SIMULATION OF DYNAMIC SYSTEMS

The use of digital computers for computation is widespread. An aspect of this application of computers is digital *computer simulation,* whose power in control system design will be demonstrated in Part III. Computer simulation is an experiment using a *computer model,* which is one (usually convenient) form of implementation of a system model. Thus given the variety of system models and the possible types of experiments to which the computer model can be subjected, the subject of computer simulation can indeed be very broad and open ended. Our discussion here is limited to the simulation of linear and nonlinear dynamic systems described in state space. The experimentation with the computer model will take the form of obtaining the time response of the model to a specific set of inputs. As we shall see later for controlled systems, such response when matched with the specified (desired) performance gives clues to the beneficial adjustment of relevant system parameters, either in a systematic manner or on an entirely ad hoc basis. This is control system design through simulation. Such design activity presumes that the computer model is a sufficiently accurate representation of the actual system.

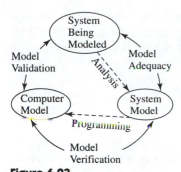

Figure 6.23

The Modeling Environment

Figure 6.23 illustrates the modeling environment and the interrelationships among the actual system, the system model, and the computer model. The process of ensuring that the computer model is an accurate representation of the actual system is known as model *validation*. In general, validation is performed by simulation! The model response in this case is matched with the known (or measured) response of the actual system under the same experimental conditions. Where there is lack of sufficient agreement between these two responses, the computer model is considered not validated and may need to be reconstructed. The computer models to be developed here will not be formally validated but will nonetheless be considered sufficiently accurate representations of the actual systems. The reason for this is that in our present activity the system model is already determined and must be presumed to be an appropriate (adequate) description of the actual system. Thus according to Fig. 6.23, we merely need to carry out model *verification,* that is, ensure that the computer model is an accurate implementation of the system model.

Programming is simply the means of producing the computer model, which is thus made up of *programs*. In addition to the overall verification of the computer model, which ensures that simulation results can in confidence be related to the system being modeled, module-by-module verification of the programs that make up the computer model is necessary to ensure that any numerical analysis used or any computer-dependent input to the model does not itself introduce significant errors. A computer model can be verified by giving the computer program a problem with a known analytical solution and comparing the computer solution with the analytical solution. A computer model can also be verified by solving the same system model but with a (reduced) manageable size, using the computer and another means, such as a calculator, and comparing the two results. An example of module-by-module verification for numerical analysis or computer-derived errors is the repeated application of a numerical integration scheme (module) to a given problem using different (time) step sizes in order to determine the maximum step size for which the algorithm is stable or convergent. The modules of commercially available computer programs are generally considered verified and are used on the *faith* that they are actually so. Programming with such packages simply involves putting together in some logical sequence a collection of relevant modules available in the package. The ease with which this can be done varies with the software package as well as with the simulation problem.

Introducing Some Math Packages

Commercial software that can be categorized as *computational packages* or *mathematical toolboxes* is available. Such software, which is designed to solve a broad range of problems, includes Derive, IMSL, Macsyma, Maple, Mathcad, Mathematica, MATLAB, and TK Solver, among many others. The eight programs listed were featured in a survey of such math packages described in reference [4]. Math packages can further be divided into *numeric* packages and *symbolic* packages. Numeric packages, in general, evaluate given expressions numerically and return numeric answers. For instance, to evaluate an integral, most numerical integration routines sample the integrand at many points in the interval of integration and use these samples to approximate the integral. In contrast, symbolic processors are able to return closed-form solutions. To integrate on a variable, the symbolic processor will attempt to find a closed-form expression that is equal to the indefinite integral and return a symbolic result. IMSL, Mathcad, MATLAB, and TK Solver are predominantly numeric packages, while Derive, Macsyma, Maple, and Mathematica are predominantly symbolic packages. However, the trend in numeric packages is to include some symbolic capabilities, and symbolic packages, in general, include some capacity for numerical evaluation of expressions. For purposes of illustration only, we give next brief descriptions of the two math packages featured in this book: Mathcad (version 3.1 for Windows), whose applications are available mainly as electronic files for downloading, and MATLAB (student edition version S3.5), whose applications are described in this text and are available as downloadable electronic files as well.

Mathcad

A product of Mathsoft Inc. (Cambridge, Massachusetts), Mathcad combines the live document interface of a spreadsheet with a graphical "what you see is what you get" interface typical of a word processor. This interface allows users to work with equations, text, and graphics, all combined on the screen in real time, much like with a calculator. There is essentially no hidden information; everything appears on the screen in familiar mathematical notation. For example, to create simple expressions, one just types them. As with a spreadsheet, as soon as a change is made anywhere in the document, the program immediately updates results and redraws graphs to reflect the change. Mathcad guides the user through the creation of plots, integrals, and other more complex mathematical expressions by providing the framework for the user to fill in the blanks. Mathcad can run on Macintosh and Sun workstations and on IBM PC-compatible computers. There are also DOS, Windows, and UNIX versions, and files created with the program are portable among all platforms.

The computational features of Mathcad 3.1 for Windows include matrix arithmetic, unit conservation and conversion, real and complex Fourier transforms, complex variables and roots, solution of simultaneous linear or nonlinear equations, cubic splines, automatic or manual plot scaling, three-dimensional plots, etc. It also allows user-defined functions. A variety of useful formulas, constants, and graphic images are available in the form of *electronic handbooks* that can be purchased separately from the software. Information from an electronic handbook can be inserted directly in a Mathcad document by clicking with the mouse. The inserted information is immediately "live" and part of the Mathcad document. A standard handbook, which includes some mathematical and engineering formulas, physical constants, properties of various materials, and some other useful information, is shipped with Mathcad 3.1 as part of the program package. There are also *Application Packs,* which are specially written Mathcad documents that utilize Mathcad features to carry out common design calculations and other applications in some specific area. Application packs are available, for example, for electrical engineering, chemical engineering, mechanical engineering, and statistics. The collection of downloadable (as described in the Preface) MCD-files for illustrations and computations performed in this book constitute an application pack (albeit of a specific scope) for system dynamics and control.

MATLAB

The basic building blocks of MATLAB are matrices, differential equations, arrays of data, plots, and graphs. MATLAB from The Mathworks Inc. (Natick, Massachusetts) is available in two versions: a *professional* version and a low-cost *student edition.* The student edition, which is available for MS-DOS–compatible personal computers and Macintoshes, is identical with the professional version except for limitation in matrix size (1024 elements, 32 × 32 array) and the absence of metafile support and the graphics postprocessor (the student edition can print only through screen dump procedures). The student edition also comes bundled with The Signals and Systems Toolbox, a special collection of signal processing, linear systems, and controls functions that expand MATLAB's problem-solving and modeling capabilities.

Other toolboxes that can be purchased at extra cost include Optimization, Control System, System Identification, MMLE3 State-Space Identification, Mu-Analysis and Synthesis, Signal Processing, Robust Control, Spline, and Chemometrics. These application-specific toolboxes add specific functionality to MATLAB's more than 300 functions, which cover linear equation solving, differential equations, polynomial arithmetic, eigenvalues, matrix math, convolution, one- and two-dimensional fast Fourier transforms, curve-fitting and cubic splines, two- and three-dimensional graphics, and many other areas. MATLAB includes its own matrix-based programming language. However, MATLAB is also an interactive and interpretive environment. In addition to the normal command-driven mode in which single-line commands are processed and the results displayed immediately when the command is entered, MATLAB can also execute sequences of commands that are stored in files. Disk files that contain MATLAB statements are called M-files. Many of the application functions such as the Runge-Kutta solution of differential equations, which has already been utilized in some earlier examples in this text, are actually *function* M-files that allow new functions to be created from the existing ones.

Digital Computer Solution of Continuous-Time Systems

The computation of discrete-time system response has already been considered in Section 6.4. The problem we wish to solve here is the following: Given a vector function $\mathbf{f}(\mathbf{X},t)$, find a vector function $\mathbf{X}(t)$ that is an approximate solution to the ordinary vector differential equation

$$\frac{d}{dt}\mathbf{X} = \mathbf{f}(\mathbf{X}, t) \tag{6.93a}$$

with the initial condition

$$\mathbf{X}(t_0) = \mathbf{X}_0 \tag{6.93b}$$

and for the time interval

$$t_0 \leq t \leq t_f \tag{6.93c}$$

In Section 6.3 we learned to solve this problem **exactly** for special types of $\mathbf{f}(\mathbf{X}, t)$ such as the low-order, linear, and time-invariant case. In the present discussion no particular restriction is applied to $\mathbf{f}(\mathbf{X}, t)$, which can in fact represent a nonlinear function. A continuing problem in dynamic system analysis is how to deal with nonlinear systems. Whereas most of the available solutions to this problem are ad hoc in nature, the simulation approach is generally applicable.

Direct or Explicit Methods

Consider the time interval $[t_0, t_f]$ divided into N subintervals of length T, such that

$$t_k = t_0 + kT \tag{6.94a}$$

With $\mathbf{X}_0$ given by the initial condition, we can obtain the approximate values $\mathbf{X}_1, \mathbf{X}_2, \ldots, \mathbf{X}_N$ to the exact values $\mathbf{X}(t_1), \mathbf{X}(t_2), \ldots, \mathbf{X}(t_N)$ by using the simplest direct approximation of the derivative at the point $(\mathbf{X}_k, t_k)$ that is the difference quotient:

$$\mathbf{f}(\mathbf{X}_k, t_k) \approx \frac{1}{T}(\mathbf{X}_{k+1} - \mathbf{X}_k) \tag{6.94b}$$

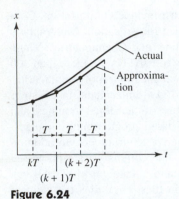

Figure 6.24

Euler's or Tangent Method

This approximation gives the recursion formula known as *Euler's* (or *tangent*) method:

$$\mathbf{X}_{k+1} = \mathbf{X}_k + T\mathbf{f}(\mathbf{X}_k, t_k); \qquad t_k = t_0 + kT; \ \text{(Euler's method)} \qquad (6.94c)$$

The solution concept given by equation (6.94c) is illustrated in Fig. 6.24 for the first-order case, where it is evident how the name *tangent method* can apply to the process. We can also observe in Fig. 6.24 the disadvantage of this simple method, which is that the step length T must be very short for reasonable accuracy. This is a general problem for all direct methods, and the precise size of T depends on the function $\mathbf{f}$.

To explore this problem further, note that equation (6.94c) and other direct methods (see Runge-Kutta methods below) are difference equations:

$$\mathbf{X}_{k+1} = \mathbf{f}(\mathbf{X}_k, t_k) \qquad (6.95a)$$

whose stability behavior, generally, depends on T. Consider, for example, the first-order linear and homogeneous state equation

$$\frac{d}{dt} x = ax; \qquad x(0) = x_0 \qquad (6.95b)$$

The exact solution,

$$x(t) = x_0 e^{at} \qquad (6.95c)$$

we know is stable for all $a < 0$. However, the approximate solution according to Euler's method is given by the difference equation (6.95a), which is, for this case,

$$x_{k+1} = x_k + Tax_k \qquad (6.95d)$$

or equation (6.84b) repeated,

$$x_{k+1} = (1 + Ta)\,x_k = px_k; \qquad x_0 = \text{given} \qquad (6.95e)$$

From Section 6.4 the eigenvalue of (6.95e) is p, and the mode for its stability behavior is p^k or $(1 + Ta)^k$. Thus the system (6.95e) is unstable for all $|1 + Ta| > 1$. This gives the stability condition for Euler's explicit method for the problem (6.95b) as $-2 < Ta < 0$ or, since $a < 0$ is assumed,

$$T < \left| \frac{2}{a} \right|; \ \text{(step size)} \qquad (6.95f)$$

Equation (6.95f) gives an upper limit on the step size T for Euler's method to even work (be stable). For reasonable accuracy, T should be substantially less than what is given by equation (6.95f) but not so small that random errors due to the limited resolution of the computer in use become predominant. In general, it is not always convenient or even possible to obtain an explicit condition on the magnitude of T. Rather, the common practice is to start with a large but reasonable value for T and progressively decrease this value until successive simulation results are satisfactory and appear to remain the same.

Shannon's Sampling Theorem   Another consideration in the determination of the step size is based on the fact that T is the *sampling period* for all input processes involved in the simulation. It follows that the value of the step size must be small enough to adequately represent such input information. The upper bound on the step size, in this regard, is given by *Shannon's sampling theorem*, which can be stated thus: *For any signal containing one or more oscillatory components, the maximum sampling interval* *that can be used without losing fundamental information is half the period of the fastest oscillatory component* (see [2]), that is,

$$T \leq 0.5 T_{\min} \tag{6.96}$$

where $T_{\min}$ is the period of the fastest (highest-frequency) oscillatory component. In practice, values of the time step much less than the theoretical limit given by equation (6.96) are used in order to gain more information and/or to satisfy stability or convergence (implicit methods) conditions.

Runge-Kutta Methods   The problem of improving the explicit solution of (6.93) can be stated as follows:

Given $\mathbf{X}(t_k)$, how can we best compute the approximation $\mathbf{X}_{k+1}$ to $\mathbf{X}(t_k + T)$?

The Runge-Kutta approach to the solution of this problem is to compute $\mathbf{f}(\mathbf{X}, t)$ at several strategically chosen points near the solution curve in the interval $[t_k, t_k + T]$ and to combine these values in such a way that good accuracy is obtained in the computed increment $\mathbf{X}_{k+1} - \mathbf{X}_k$ (see [5]). Many Runge-Kutta methods are available with different virtues. The best known Runge-Kutta method is the fourth-stage Runge-Kutta with the following formula:

$$\mathbf{R}_1 = T\mathbf{f}(\mathbf{X}_k, \mathbf{t}_k)$$

$$\mathbf{R}_2 = T\mathbf{f}\left(\mathbf{X}_k + \frac{1}{2}\mathbf{R}_1, \mathbf{t}_k + \frac{1}{2}T\right)$$

$$\mathbf{R}_3 = T\mathbf{f}\left(\mathbf{X}_k + \frac{1}{2}\mathbf{R}_2, \mathbf{t}_k + \frac{1}{2}T\right) \tag{6.97}$$

$$\mathbf{R}_4 = T\mathbf{f}(\mathbf{X}_k + \mathbf{R}_3, \mathbf{t}_k + T)$$

$$\mathbf{X}_{k+1} = \mathbf{X}_k + \frac{1}{6}(\mathbf{R}_1 + 2\mathbf{R}_2 + 2\mathbf{R}_3 + \mathbf{R}_4)$$

This formula is *fourth order,* which essentially means that the global truncation error in the approximating polynomial is of order T^4. although the function $\mathbf{f}$ is computed four times in each time step, the formulas in equation (6.97) are relatively easy to program, and many commercial programs for the numerical solution of ordinary differential equations are often based on the Runge-Kutta technique. MATLAB offers two such programs implemented as M-files in its MATLAB Toolbox: **ode23,** which is for medium accuracy and is based on a simple second- and third-order pair of Runge-Kutta formulas, and **ode45,** which uses a fourth- and fifth-order pair of Runge-Kutta formulas for higher accuracy. The **ode45** program has already been applied to some problems in Chapter 3 (Examples 3.1 and 3.17). Both programs determine the adequate step size automatically. The important things to note in this regard are that (1) the user can control the measurement of adequacy of the step size by specifying a tolerance value (*tol*) that represents the desired accuracy of the solution, and (2) the step size at each time step (the automatic step size procedure results in a variable step size) is determined to lie between a minimum value and a maximum value according to

$$\left(\frac{t_f - t_0}{20000}\right) < T \leq \left(\frac{t_f - t_0}{5}\right) \tag{6.98}$$

If no adequate step size is found in this range, the program aborts with the error condition "SINGULARITY LIKELY." Since the domain of the solution $[t_0, t_f]$ is often determined by considerations that are unrelated to the desired accuracy of the solution, these MATLAB ODE solution functions can fail to solve an otherwise well-behaved or nonsingular problem unless effective use is also made of the *tol* input argument, and/or a workable solution domain other than the desired domain is utilized. The default value of *tol* (if not specified) is 0.001 for **ode23** and 1×10^{-6} for **ode45.**

Example 6.20   Response of a Third-Order System: Runge-Kutta Method

Consider the system

$$\frac{d}{dt}\mathbf{X} = \begin{bmatrix} -3 & 4 & 4 \\ 1 & -3 & -1 \\ -1 & 2 & 0 \end{bmatrix} \mathbf{X} + \begin{bmatrix} 2 \\ -\dfrac{1}{2} \\ 1 \end{bmatrix} u; \qquad u(t) = U_S(t); \qquad \mathbf{X}(0) = \begin{bmatrix} 0 \\ 1 \\ 2 \end{bmatrix}$$

The exact solution was given in Example 6.16 as

$$\mathbf{X}(t) = \begin{bmatrix} 5e^{-t} - \dfrac{1}{3}(19e^{-3t} - 4) \\ \dfrac{5}{2}e^{-t} - 8e^{-2t} + \dfrac{19}{3}e^{-3t} + \dfrac{1}{6} \\ -\dfrac{1}{3}(19e^{-3t} - 1) + 8e^{-2t} \end{bmatrix}$$

Using the **ode45** function of MATLAB, obtain a solution of this problem in the interval $0 \le t \le 1$, and display the results graphically for two values of desired accuracy of the solution:

a. $tol = 1 \times 10^{-6}$, that is, the default tolerance
b. $tol = 5 \times 10^{-5}$, that is, 50 times the default value.

Solution: The MATLAB ODE solution algorithms require a function M-file that contains the differential equations to be integrated. The function file named **r3dsys.m** is as follows:

```
function xdot = r3dsys (t,x)
A = [−3 4 4; 1 −3 −1; −1 2 0]; B = [2 −0.5 1]';
xdot = zeroes (3,1);
u = 1;
xdot = A * x + B * u;
end
```

The function is utilized by the **ode45** algorithm according to the script named **fig6_25.m** and shown in the box. The **ode45** function is invoked twice. In the first case no value is entered for the desired accuracy, so that the default value of 1×10^{-6} is used by the program, as required. In the second case the *tol* input argument is set to the appropriate value. The script file generates both the computed response and the exact solution for each desired accuracy of the solution. The corresponding graphical output produced by the script is shown in Fig. 6.25. Note that the exact solution is indicated

```
% fig6_25.m
% Script to generate Figure 6.25
% Uses function M-file r3dsys.m
% x(t) is the computed result using default accuracy
% v(s) is the computer result using tol = 0.00005
%
t0 = 0; tf = 1;                          % time interval
x0 = [0 1 2]';                           % Initial state
tol = 0.00005;                           % desired accuracy for (b)
%
% Case of default accuracy (tol = 0.000001)
[t,x] = ode45 ('r3dsys',t0,tf,x0);
% Generate exact solution over the same time interval
e1 = 5*exp(-t); e2 = 8*exp(-2*t); e3 = (19/3)*exp(-3*t);
y1 = e1 - e3 + (4/3);
y2 = 0.5*e1 - e2 + e3 + (1/6);
y3 = (1/3) - e3 + e2;
%
% Case of desired accuracy: tol = 0.00005
[s,v] = ode45('r3dsys',t0,tf,x0,tol);
% Generate exact solution over the same time interval
e1 = 5*exp(-s); e2 = 8*exp(-2*s); e3 = (19/3)*exp(-3*s);
w1 = e1 - e3 + (4/3);
w2 = 0.5*e1 - e2 + e3 + (1/6);
w3 = (1/3) - e3 + e2;
%
axis ([0 1 0 3.125]);                    % scale axes
subplot (211)                            % split graph window
plot (t,x,'-');                          % plot computed data (a)
title ('(a) Desired Accuracy = 1E-6');
ylabel ('Elements of X'); xlabel ('Time');
text (.38,2.8,'x1'); text (.7,1.55,'x3');
text (.2,.4,'x2'); hold on;
plot (t,y1,'*',t,y2,'*',t,y3,'*');       % plot exact solution (a)
plot ([.15 .3], [1.25 1.25],'-');
text (.35,1.15,'Computed Result');       % write legend
plot ([.15 .2],[.9 .9],'*');
plot ([.25 .3],[.9 .9],'*');
text (.35,.8,'Exact Solution'); hold off;
%
axis ([0 1 0 3.125]);                    % scale axes
plot (s,v,'-');                          % plot computed data (b)
title (' (b) Desired Accuracy = 5E-5');
ylabel ('Elements of X'); xlabel ('Time');
text (.38,2.8,'x1'); text (.7,1.55,'x3');
text (.2,.4,'x2'); hold on;
plot (s,w1,'*',s,w2,'*',s,w3,'*');       % plot exact solution (b)
plot ([.15 .3],[1.25 1.25],'-');
text (.35,1.15,'Computed Result');       % write legend
plot ([.15 .2], [.9 .9],'*');
plot ([.25 .3],[.9 .9],'*');
text (.35,.8,'Exact Solution'); hold off;
```

by *star* point type. Since the exact solution is computed at each time step, the star points also demarcate the step length at each time step. The variable step size feature of the ODE solution functions should be evident. A remarkable advantage of automatic step size is that for a well-behaved system such as the present case, the desired accuracy in the result can be achieved with fewer time steps and hence within a shorter computation time than would be required for a fixed time step process. In Fig. 6.25 a desired accuracy of solution of 1×10^{-6} is accomplished in only 16 time steps, whereas on the basis of the shortest step size obtained in that simulation, a total of 80 or more time steps would have been required for a fixed step size procedure.

Finally, note that in Fig. 6.25b the sample points given by the desired accuracy of the solution are not close enough to define a sufficiently smooth curve. Interpolation (using the *spline* function) can be used to obtain a smoother plot. Alternatively, the specification for *tol* can be made smaller than is otherwise required.

Example 6.21   Stiff Problem and Runge-Kutta Method

Consider the first-order system

$$\frac{d}{dt} x = -100x + 100 \sin t; \quad x(0) = 0$$

Figure 6.25

Computation of a Third-order System Response Using a Runge-Kutta Method

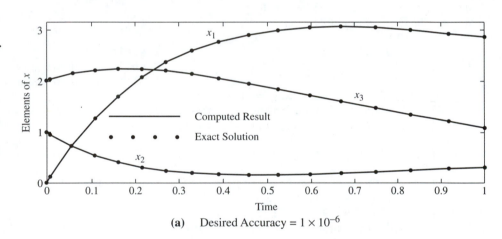

(a)   Desired Accuracy $= 1 \times 10^{-6}$

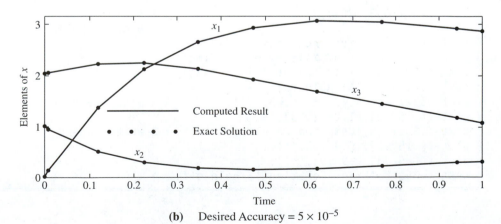

(b)   Desired Accuracy $= 5 \times 10^{-5}$

which has the exact solution

$$x(t) = \frac{\sin t - 0.01 \cos t + 0.01 e^{-100t}}{1.0001}$$

Use the **ode45** Runge-Kutta function to find the approximate solution to the system for $0 \leq t \leq 3$, using a desired accuracy of solution of 5×10^{-5}. Display the results in the form of a table of about 22 rows, showing for each time represented, the exact solution, the computed result, and the applicable step size.

Solution: The required function M-file (named **stiff.m**) is as follows:

```
function xdot = stiff(t,x)
xdot = 100*(sin(t) −x);
end
```

The script file used to compute the results (**tab6_1.m**) is shown in the box. Observe that the step size at each time step is computed using the MATLAB *diff* function. To display only about 22 of the data points, *subscripting* is used to assign output points to the display matrix y in increments of 6 points. Also, note that the output exact solution and computed solution are actually 10 times the actual values. This is done so that under the default MATLAB "short" format of 5-decimal digits, the full desired accuracy of the solution can be observed. Table 6.1 shows the result of execution of the script file.

It can be observed that at an average step size of, say, 0.025, the full desired accuracy of the algorithm is obtained. However the earliest step sizes for this problem are quite small. The beginning six step sizes, of which only the very first one is shown in Table 6.1, are 0.0090, 0.0099, 0.0118, 0.0143, 0.0179, and 0.0236. This type of behavior is typical of some *stiff problems* (differential equations that present difficulties in their numerical solution).

```
% tab6_1.m
% Script to generate Table 6.1
% Uses function M-file stiff.m
%
t0 = 0; tf = 3;                              % time interval
x0 = 0;                                      % initial condition
tol = .00005;                               % desired accuracy
%
[t,x] = ode45 ('stiff',t0,tf,x0,tol);
dx = diff(t);                               % compute step sizes
y(1:21,1) = t(1:6:121); y(1:21,3) = 10*x(1:6:121);
y(2:21,4) = dx(1:6:115)
y(:,2) = (sin(y(:,1)) − 0.01*cos(y(:,1))+0.01*exp(−100*y(:,1)))/.10001;
%
disp (' Table 6.1 Performance of Runge-Kutta Method');
disp ('    Time Exact (E-1) Result (E-1) StepSize');
disp (y(1:21,:));
```

Table 6.1

**Performance of
Runge-Kutta Method**

TIME	EXACT (10^{-1})	RESULT (10^{-1})	STEP SIZE
0	0	0	0
0.0866	0.7647	0.7647	0.0090
0.2892	2.7556	2.7557	0.0370
0.4657	4.4010	4.4011	0.0308
0.6272	5.7872	5.7873	0.0276
0.7807	6.9660	6.9661	0.0260
0.9291	7.9502	7.9503	0.0250
1.0740	8.7428	8.7429	0.0244
1.2165	9.3434	9.3435	0.0239
1.3574	9.7510	9.7511	0.0236
1.4973	9.9646	9.9647	0.0234
1.6367	9.9839	9.9840	0.0233
1.7762	9.8091	9.8092	0.0232
1.9164	9.4417	9.4417	0.0233
2.0577	8.8836	8.8837	0.0235
2.2009	8.1375	8.1376	0.0237
2.3468	7.2062	7.2063	0.0241
2.4966	6.0913	6.0913	0.0247
2.6520	4.7908	4.7909	0.0254
2.8161	3.2922	3.2922	0.0266
2.9968	1.5413	1.5414	0.0286

A common type of stiff differential equation is one that incorporates processes of significantly different time scales. This is the case with the problem of Example 6.21, which includes an oscillatory process with a period of 2π (that is approximately 6.28) and a first-order exponential decay process with a time constant of 0.01. To solve such problems successfully with a fixed step size procedure, we must consider the fastest process while determining the time step, even though the solution component corresponding to that process can die out very quickly (see [5]). Obviously, a variable step size scheme is advantageous for such problems. For instance, the stability criterion equation (6.95f) for the first-order system equation (6.95b) for Euler's method requires for the fastest process ($a = 100$) $T < 0.02$. As shown in Example 6.22 for the present problem, Euler's method is stable although not very accurate at a time step of 0.020 but unstable at a time step of 0.021. Another approach is to try an implicit method (discussed subsequently), since such methods have been most successful for stiff problems. However, the convergence condition for some implicit techniques can also limit their application.

Implicit Methods: The Trapezoidal Approach

Although explicit methods have error propagation or stability problems that can severely limit the permissible values of the step size, *implicit* techniques are **unconditionally stable,** but this does not necessarily guarantee that they will be more accurate for a particular problem. Also, convergence problems can arise, especially for nonlinear systems.

The *trapezoidal method* is obtained by integrating equation (6.93a) over the interval (t_k, t_{k+1}) with $\mathbf{f}(\mathbf{X}, t)$ approximated in this interval by a straight line $\hat{\mathbf{f}}(\mathbf{X}, t)$ of arbitrary slope passing through the end points $\mathbf{f}(\mathbf{X}_k, t_k)$ and $\mathbf{f}(\mathbf{X}_{k+1}, t_{k+1})$. That is,

$$\mathbf{X}_{k+1} - \mathbf{X}_k = \int_{t_k}^{t_{k+1}} \hat{\mathbf{f}}(\mathbf{X}, t) \, dt$$

Because the value of the integral is the area of the **trapezoid** with sides equal to $\mathbf{f}(\mathbf{X}_k, t_k)$ and $\mathbf{f}(\mathbf{X}_{k+1}, t_{k+1})$ and width equal to T, the trapezoidal method is given by

$$\mathbf{X}_{k+1} - \mathbf{X}_k = \frac{1}{2}\,[\mathbf{f}(\mathbf{X}_k, t_k) + f(\mathbf{X}_{k+1}, t_{k+1})]; \qquad k = 0, 1, 2, \ldots, N \qquad \textbf{(6.99)}$$

The method is an *implicit* one in the sense that $\mathbf{X}_{k+1}$, which is to be computed (in the kth step), appears (implicitly) on the RHS of equation (6.99). Problems of this sort are normally solved using some *iterative* technique. The starting value of $\mathbf{X}_{k+1}$ for the iteration scheme must be estimated (*predicted*) somehow. Thus the iterative solution for $\mathbf{X}_{k+1}$ can be viewed as a correction of this initial estimate. The whole procedure is thus called a *predictor-corrector* method.

An Euler-predictor trapezoidal-corrector scheme, for example, will be given by the following when applied to the problem of equations (6.93a) through (6.93c): Beginning at time step k, we have

Euler predictor:

$$\mathbf{X}_{k+1}^{(0)} = \mathbf{X}_k + T\mathbf{f}(\mathbf{X}_k, t_k) \qquad \textbf{(6.100a)}$$

Trapezoidal corrector:

$$\mathbf{X}_{k+1}^{(j+1)} = \mathbf{X}_k + \frac{1}{2}\,T[\mathbf{f}(\mathbf{X}_k, t_k) + \mathbf{f}(\mathbf{X}_{k+1}^{(j)}, t_{k+1})]; \qquad j = 0, 1, 2, \ldots, l \ \textbf{(6.100b)}$$

The absence of an iteration index (j) above a variable in the scheme indicates that the value has been iterated and accepted in an earlier solution step. A sufficient criterion for the convergence of the preceding iteration is

$$\frac{1}{2}\,T \left\| \frac{\partial \mathbf{f}}{\partial \mathbf{X}_k} \right\| < 1 \qquad \textbf{(6.100c)}$$

where the indicated norm (see [5]) is the magnitude of the elements of the Jacobian matrix (see equation (6.50d) and Example 6.12). However the stopping of the iterations may be controlled either by comparing the difference $|\mathbf{X}_{k+1}^{(j+1)} - \mathbf{X}_{k+1}^{(j)}|$ with some preset tolerance or by predetermining the number of iterations. The later method is often used, and two or three iterations are usually sufficient.

Example 6.22   Stiff Problem and an Implicit Trapezoidal Mathcad Program

a. Develop a Mathcad file implementing an implicit trapezoidal algorithm. Minimize duplicate computations, and make the program easily adaptable to other problems.
b. 1. Apply the problem developed in (a) to the problem of Example 6.21. Execute the program for a step size of $T = 0.021$ and display the results in a table.
   2. Use Euler's method to solve the same problem for two step sizes: $T = 0.021$ and $T = 0.02$, and also display the results in tabular form.

Solution a: An implicit trapezoidal program that uses the previously iterated value (instead of an Euler or other approximation) as the prediction for the next time step can be developed on the basis of the *solve block* feature of Mathcad. Also Mathcad's stopping method for the iteration of a solve block, which is to consider the block solved

TRAP.MCD:IMPLICIT TRAPEZOIDAL PROGRAM FOR SOLUTION OF ODEs

By E.I. Umez-Eronini, 05-17-93

The program uses the solve block feature of Mathcad™ to implement an implicit predictor-trapezoidal-corrector scheme in which the predicted value is simply the previous solution, to solve

$$\frac{dx}{dt} = f(x, t) \text{ subject to } x(t_0) = x_0 \text{ for } t_0 \leq t \leq t_f$$

to use the program, *substitute* your right-hand (RH) functions $f(x, t)$ as indicated, *provide* initial values and parameters, including a specification for *TOL*, whose value (between 0 and 1) determines the desired accuracy of convergence in the corrector, and *expand* or *contract* all references to state variables x or w according to the number of state variables of the system and in line with the existing formats.

Notes:

1. Subscripts on initial time t_0, final time t_f, and RH function and solve block state variables x or w are all literal subscripts, and these variables are all scalar quantities.

2. Solution vector x and all other vectors (arrays) have subscripts that begin at 1 (see statement ORIGIN $\equiv$ 1). Hence the solution vector x subscripts match the literal subscripts of the state variables x or w.

3. Convergence problems can result in the error: **did not find solution.** Make sure that your RH functions have been entered correctly and all expansions/contractions with respect to state variables have been carried out properly before considering increasing the value of *TOL*. (The default value of *TOL* is 0.001.)

Provide initial conditions and parameters: ORIGIN $\equiv$ 1 (All array subscripts begin at 1)

$t_0 := 0$ $t_f := 1$ $T := 0.01$ Expand or contract $TOL := 0.001$

$t_1 := t_0$ $h := 0.5 \cdot T$ $Np1 := \left(\dfrac{t_f - t_0}{T}\right) + 1$ $x^{<1>} := \begin{pmatrix} 0 \\ 1 \\ 2 \end{pmatrix}$ $\begin{bmatrix} x_1 \\ x_2 \\ x_3 \end{bmatrix} := x^{<1>}$ ◄——Expand or contract

┗━━ This is an array subscript

Substitute user-defined RH functions: $f_1(x_1, x_2, x_3, t) := -3 \cdot x_1 + 4 \cdot x_2 + 4 \cdot x_3 + 2$
                    Expand or contract ——► $f_2(x_1, x_2, x_3, t) := x_1 - 3 \cdot x_2 - x_3 - 0.5$
                    $f_3(x_1, x_2, x_3, t) := -x_1 + 2 \cdot x_2 + 1$

Trapezoidal corrector: Given $x_1 = w_1 + h \cdot (f_1(w_1, w_2, w_3, s) + f_1(x_1, x_2, x_3, t))$
            Expand or contract ——► $x_2 = w_2 + h \cdot (f_2(w_1, w_2, w_3, s) + f_2(x_1, x_2, x_3, t))$
                    $x_3 = w_3 + h \cdot (f_3(w_1, w_2, w_3, s) + f_3(x_1, x_2, x_3, t))$

Expand or contract _____
                                    ▼   ▼
Iteration: $k := 1 .. Np1$ $Trap(w_1, w_2, w_3, s, t) := Find(x_1, x_2, x_3)$

$x^{<k+1>} := Trap[(x^{<k>})_1, (x^{<k>})_2, (x^{<k>})_3, t_k, t_{k+1}]$ $t_{k+1} := t_0 + k \cdot T$

Expand or contract ——┛

Illustrations of computed results:

$i := 1 .. 3$

	t_i	$(x^{<i>})_1$	$(x^{<i>})_2$	$(x^{<i>})_3$
Tabular	0	0	1	2
display ——►	0.01	0.137	0.946	2.029
	0.02	0.27	0.895	2.055

Graphical display ——►

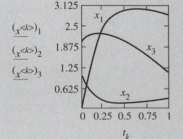

$(x^{<k>})_1$
$(x^{<k>})_2$
$(x^{<k>})_3$

SC7.MCD: STIFF PROBLEM SOLUTION USING TRAP.MCD AND EULER

Example 6.22, solution continued:

We shall use the vectors x and X to represent the implicit (TRAP) and explicit (Euler) solutions, respectively, for the step sizes $T = 0.021$, and V to represent the Euler solution for $T = 0.02$. The corresponding time variables and exact answers for the two step sizes are represented by t and s and y and z, respectively.

Initial conditions and parameters: $\qquad$ ORIGIN $\equiv 1 \quad TOL := 0.00005 \quad t_0 := 0 \quad t_f := 3 \quad T := 0.021$

$$h := 0.5 \cdot T \qquad t_1 := t_0 \qquad x^{<1>} := \begin{pmatrix} 0 \\ 0 \end{pmatrix} \qquad X^{<1>} := x^{<1>} \qquad \begin{pmatrix} x_1 \\ x_2 \end{pmatrix} := x^{<1>} \qquad Npl := \left(\frac{t_f - t_0}{T} \right) + 1$$

Solution Using Implicit Trapezoidal (TRAP) Prgram

User-defined RH functions: $\qquad f_1(x_1, x_2, t) := 100 \cdot (\sin(t) - x_1) \qquad f_2(x_1, x_2, t) := 0$

Trapezoidal corrector: Given $\qquad x_1 = w_1 + h \cdot (f_1(w_1, w_2, s) + f_1(x_1, x_2, t))$

$$x_2 = w_2 + h \cdot (f_2(w_1, w_2, s) + f_2(x_1, x_2, t))$$

Trap $(w_1, w_2, s, t) := $ Find (x_1, x_2)

Iteration for T = 0.021: $\quad k := 1 .. Np1 \qquad t_{k+1} := t_0 + k \cdot T \qquad$ **Exact Solution T = 0.021**

$$x^{<k+1>} := \text{Trap} [x^{<k>})_1, (x^{<k>})_2, t_k, t_{k+1}] \qquad y_k := \frac{\sin(t_k) - 0.01 \cdot \cos(t_k) + 0.01 \cdot \exp(-100 \cdot t_k)}{1.0001}$$

Solutions Using Explicit (Euler's) Method $\quad X^{<k+1>} := X^{<k>} + T \cdot \begin{bmatrix} 100 \cdot [\sin(t_k) - (X^{<k>})_1] \\ 0 \end{bmatrix}$

Case of T = 0.02: $\quad T := 0.02 \quad s_1 := t_0$

$$Np1 := \left(\frac{t_f - t_0}{T} \right) + 1 \qquad V^{<1>} := \begin{pmatrix} 0 \\ 0 \end{pmatrix} \qquad \text{Iteration:} \quad k := 1 .. Np1 \qquad s_{k+1} := t_0 + k \cdot T$$

$$V^{<k+1>} := V^{<k>} + T \cdot \begin{bmatrix} 100 \cdot [\sin(s_k) - (V^{<k>})_1] \\ 0 \end{bmatrix} \qquad \begin{array}{c} \textbf{Exact Solution T = 0.020} \\ z_k := \frac{\sin(s_k) - 0.01 \cdot \cos(s_k) + 0.01 \cdot \exp(-100 \cdot s_k)}{1.0001} \end{array}$$

Results: $i := 1, 12 .. 132$ (TRAP)

t_i	y_i (Exact)	$(x^{<i>})_1$	(Euler) $(X^{<i>})_1$ (T = 0.021)	s_i	z_i (Exact)	(Euler) $(V^{<i>})_1$ (T = 0.02)
0	0	0	0	0	0	0
0.231	0.21919	0.21919	0.19069	0.22	0.20845	0.19847
0.462	0.43674	0.43674	0.5182	0.44	0.41685	0.42689
0.693	0.63109	0.63109	0.3989	0.66	0.60516	0.59522
0.924	0.79191	0.79191	1.45466	0.88	0.76429	0.77437
1.155	0.91066	0.91066	−0.9799	1.1	0.88658	0.87667
1.386	0.98104	0.98104	6.37542	1.32	0.96614	0.97623
1.617	0.99929	0.99929	−14.3911	1.54	0.99912	0.98922
1.848	0.96446	0.96446	44.87545	1.76	0.98394	0.99404
2.079	0.8784	0.8784	−124.40438	1.98	0.92132	0.91142
2.31	0.74567	0.74567	358.19239	2.2	0.8143	0.82438
2.541	0.57332	0.57332	$-1.01926 \cdot 10^3$	2.42	0.66802	0.65809

whenever the difference between successive approximations to the solution becomes less than the value of the predefined variable *TOL*, is compatible with the first stopping criterion, discussed above, for the corrector.

The program **TRAP.MCD**, shown in the box, uses the third-order system of Example 6.21 to illustrate a Mathcad Implicit Trapezoidal Program developed by the author. Observe that the trapezoidal corrector, equation (6.100b), appears under the *Given* statement in the solve block and that the iterated value which is also the prediction at any time step, is represented by the state variables $w_1, w_2, \ldots$, etc. When the program is executed, the system of equations constituting the corrector is solved iteratively until the difference between successive approximations to the solution, that is, $\left| \mathbf{X}_{k+1}^{(j+1)} - \mathbf{X}_{k+1}^{(j)} \right|$ becomes less than *TOL*. Therefore *TOL* is a specification of the desired accuracy of solution for this program. If no value is entered for *TOL*, the default value or 0.001 automatically applies.

The corresponding MATLAB program developed by the author, **trap.m (On the Net)** uses a specified number of iterations of the corrector instead.

Note that to use **TRAP.MCD**, the user is expected to make a copy of the file and then modify the indicated sections in the copy to reflect the actual system to be solved.

Solution b: The file **SC7.MCD** which follows **TRAP.MCD**, shows the application of the implicit trapezoidal program (TRAP) to the stiff problem of Example 6.21.

The computed results are remarkably close to the exact solution. Note that according to equation (6.100c) the time step for the implicit trapezoidal scheme should be less than 0.02 for this problem. However, equation (6.100c) is only a *sufficient condition* for convergence; it is not also a *necessary condition*. The present algorithm is convergent even for a time step of 0.021. The performance of **TRAP.MCD** both for the illustrative problem and for the present stiff problem implies the verification of this program.

The situation for the explicit Euler scheme is another matter. At a step size of 0.021, which is close to but greater than the stability value of 0.02, the algorithm explodes. An unstable oscillation of increasing magnitude is evident in the computed result shown in the first table in SC7.MCD. The computed data bear no resemblance to the exact solution. On the other hand, as shown in the second table, for a step size of 0.02, the algorithm is stable but not sufficiently accurate. These results support the previous discussions above about stability and stiff problems.

The TRAP program developed here is used with the Mathcad applications that require the solution of a differential equation. The file **SC7.MCD** is presented here to illustrate TRAP. Although only MATLAB application programs are subsequently discussed in the text, the corresponding Mathcad MCD files are available for downloading, as discussed in the Preface.

On the Net
- SC6.MCD (Mathcad): Response of third-order system of example 6.20 by Eulers method
- fig6_25.m and r3dsys.m (MATLAB): source file for Fig. 6.25
- tab6_1.m and stiff.m (MATLAB): source file for Table 6.1
- TRAP.MCD (Mathcad): Implicit Trapezoidal program for solution of ODEs
- trap.m and rhf.m (MATLAB): Implicit Trapezoidal program for solution of ODEs
- SC7.MCD (Mathcad): Stiff problem solution using TRAP program
- Auxiliary exercise: Practice Problems 6.10, and 6.51 through 6.58

6.6 PRACTICE PROBLEMS

6.1(a) Prove that the tangent to an exponential response $(x_0 e^{at} + x_f)$ at any instant of time intersects the asymptotic final value of the response x_f, T time units after the tangent point, where T is the time constant.

6.1(b) Prove that the value of a first-order free response at any instant of time is $1/e$ times the value of the response T time units after that instant; $e = \exp(1)$, $T =$ time constant.

6.1(c) How many time constants elapse before the free response of a first-order system decays to 25% of its original value? How far does it decay in two time constants? Would you consider the motion of a first-order free system essentially complete after two time constants? Explain your answer.

6.19(d) A certain radioactive substance is known to decay at a rate proportional to the amount present. A quantity of this substance, originally of mass M_0 kg, was observed to experience a 20% reduction in mass after 50 min. Derive an expression for the mass of the substance at any time. Find the time that it will take for a quantity of this substance to decay to half of its original mass. This time period is known as the *half-life* of the radioactive material. Rewrite the expression you derived above in terms of the half-life. Do you consider the half-life and the time constant equivalent concepts? If so, what is the relation between them? If not, why not?

6.2* An unknown mechanical device gave the measured free response data shown in Fig. 6.26. It is desired to model the device as a first-order system. Determine this mathematical model from the given data. What is your estimate of the initial state of the device?

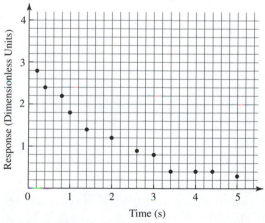

Figure 6.26 (Problem 6.2)

6.3 If the data plotted in Fig. 6.27 are the response of a first-order system to the impulse $2\delta(t)$, identify the first-order model.

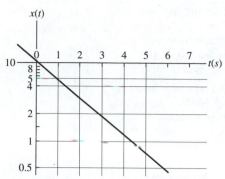

Figure 6.27 (Problem 6.3)

6.4 Sketch the function (i) $2(1 - e^{-t/3})$; (ii) $7e^{-2t}$, and indicate the time constant and the asymptotic final value of the function.

6.5 Find and sketch (roughly) the forced response of the first-order system $dx/dt = ax + u$ when the input is $u(t) = e^{\omega t}$. Consider all possible values of ω and a, that is, (i) $\omega < a < 0$; $a < \omega < 0$ (ii) $a < 0$, $\omega > 0$; $a < 0$, $\omega < 0$ (ii) $\omega = a$ (iv) $a > \omega > 0$; $\omega > a > 0$. Discuss the character of the response in each case.

6.6* Find and sketch the total response of the system $dx/dt = -x + u$; $x(0) = 1$

(a) when the input is as shown in Fig. 6.28;

(b) when $u(t) = 2t$.

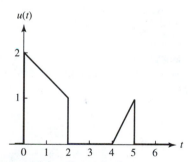

Figure 6.28 (Problem 6.6)

6.7a A first-order linear system $(dx/dt)x = ax + bu$ is subject to the input shown in Fig. 6.29a

(a) If this input can be approximated by an impulse at $t = 5$, find the response of the system when $x(0) = 10$, $a = -0.1$, and $b = 1.0$.

(b) if $a = -1.0$ (instead of -0.1), will the approximation of the input by an impulse still be a good idea? Explain your answer.

(c) Repeat (a) for the exact input.

6.7b Find the response of the system of Problem 6.7a for the input shown in Fig. 6.29b when $x(0) = 10$, $a = -0.1$, and $b = 1.0$.

6.8* An oil tanker weighing 9.807×10^7 N is initially at rest in the terminal. A tugboat is then used to push it (on the stern) with a force of 4×10^4 N for 10 s before the tugboat backs away. Find and sketch the resulting motion (velocity) of the tanker (a) for the exact tugboat input; (b) assuming the input to be an impulse. (c) Compare the two results. For what initial velocity of the tanker will the free response of the tanker be the same as the forced response for (b)? Assume that the motion of the tanker is viscously damped with a damping constant of 5×10^5 N · s/m.

6.9(a) For the network shown in Fig. 6.30, find the forced response $V_{out}(t)$ (i) if $V_{in}(t) = V_a \sin(\omega t)$; (iii) if $V_{in}(t) = V_a \cos(\omega t)$.

6.9(b) Repeat Problem 6.9a if $V_{in}(t) = V_a e^{-\omega t}$.

6.10 The system of Problem 6.6 subject to the input shown in Fig. 6.31. Find approximately

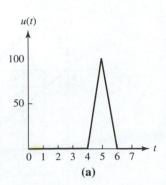

(a)

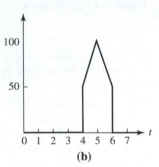

(b)

Figure 6.29 (Problem 6.7)

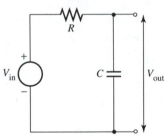

Figure 6.30 (Problem 6.9)

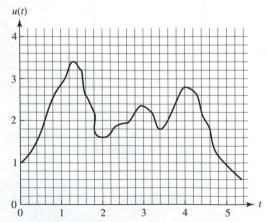

Figure 6.31 (Problem 6.10)

the value of the forced response at $t = 3$ by numerically evaluating the convolution integral, equation (6.12b).

6.11 A clinical thermometer with the model $dT/dt = (1/\lambda)(T_s - T)$, where T = temperature (reading) of thermometer, T_s = surrounding temperature, λ = time constant = 20 s, is initially in a cooler where the temperature is maintained at 10°C. To take the temperature of a patient, a nurse places the thermometer underneath the patient's tongue and leaves it there for 20 s. The nurse then withdraws the thermometer and waits for another 10 s before reading the thermometer.

(a) Sketch carefully the forcing input that the thermometer experiences. Also sketch the response you expect for the thermometer (on the same graph).

(b) Express the thermometer's response analytically.

(c) What will be the nurse's reading or the patient's temperature?

(d) If the nurse cannot replace this thermometer, how can she improve the performance?

(e) If the nurse can buy a new thermometer, what do you recommend?

6.12 Show that if $y_1(t)$ is a forced response to the system

$$\frac{d^2y}{dt^2} + 2\zeta\omega_0 \frac{dy}{dt} + \omega_0^2 y = f_1(t); \qquad y(0) = \dot{y}(0) = 0.$$

(a) $\dfrac{dy_1}{dt}$ is the forced response to the input $f_2(t) = \dfrac{d}{dt} f_1(t)$.

(b) $\displaystyle\int_0^t y_1 \, dt$ is the forced response to the input

$$f_2(t) = \int_0^t f_1(\tau) \, d\tau.$$

6.13* A linear system has the following response under zero initial conditions to an impulse of strength 3.0: $y_{I3}(t) = 1 - e^{-3t}$. If this system is subjected to the input $f(t) = te^{-2t}$,

(a) Find and sketch its response to this forcing under zero initial conditions.

(b) What will be the (zero initial state) response to the input $f(t) = (1 - 2t) e^{-2t}$?

6.14 Find

$$\text{Real}\left[\frac{5 + j2}{1 + 7j} + \frac{2 - j3}{1 - j4} \right]$$

6.15 Find (a) $\sqrt{j}$ (two roots); (b) $\sqrt[5]{-1}$ (five roots), (c) Show that $(-1 + j)^7 = -8(1 + j)$.

6.16 Find the magnitude and angle of (a) $z = 5e^{j0.5} + 2e^{j0.25}$; (b) $\exp(2 + j\,\pi/4) + \exp(5 - j\,\pi/2)$.

6.17 Find (a) $\ln(-1)$; (b) $\ln(-3)$.

6.18* Find the magnitude and angle of $z = \dfrac{(1 + j2)(2 - j3)}{(2 + j)(1 + j3)}$.

6.19 If $z = 0.6e^{-0.8j}$, find (a) $|z|$; (b) z^*; (c) x and y for $z = x + jy$.

6.20* Show that

$$\frac{d}{dt}[\text{Real}\{z\}] = \text{Real}\left\{\frac{d}{dt}[z]\right\}$$

where $z = x(t) + jy(t)$.

6.21 Use complex numbers to find the sum of the following sinusoids:

(a) $f_1 = -4\cos 2t$; $f_2 = 3\cos(2t + \pi/2)$

(b) $f_1 = 2\sin 5t$; $f_2 = 3\sin(5t - \pi/6)$

6.22 Express each of the following sinusoids as the real part of a complex number:

(a) $f = A\sin(\omega t + \phi)$;   (b) $f = 5e^{-2t}\cos(3t + \pi/4)$;

(c) $f = \sin(\omega t) - 2\cos(\omega t)$

6.23* Prove that $(\cos\phi + j\sin\phi)^n = \cos(n\phi) + j\sin(n\phi)$, where $z = Me^{j\phi}$.

6.24 Indicate whether each of the following ordinary differential equations is (a) nonlinear, (b) linear and nonstationary; or (c) linear and time-invariant. For the nonlinear systems, explain the nonlinearity:

(i) $\dfrac{d^3x}{dt^3} + 5t\dfrac{d^2x}{dt^2} + 2\left(\dfrac{dx}{dt}\right)^2 + tx = t^2$

(ii) $2\dfrac{d^3x}{dt^3} + 4\dfrac{d^2x}{dt^2} + 6\dfrac{dx}{dt} + 8x = \sin 2t$

(iii) $\dfrac{d^2x}{dt^2} + 7x\dfrac{dx}{dt} = 0$ (iv) $\dfrac{d^2x}{dy^2} + 4\dfrac{dx}{dy} + 6x = y$

(v) $\dfrac{d^2y}{dt^2} + t\dfrac{dy}{dt} + 2y = 7t$ (vi) $\dfrac{dx}{dt} + 3x = 4tx$

6.25 Determine the state vector, the input vector, the output vector, and the corresponding component functions of $f(\mathbf{X}, \mathbf{U}, t)$ and $g(\mathbf{X}, \mathbf{U}, t)$, as defined in equations (6.47a) and (6.47b) for the following:

(a) The sliding ladder model determined in Example 3.8, equation (3.22a) through (3.22c)

(b) The spin-stabilized projectile of Example 3.7, equations (3.18a) through (3.18c)

(c) The micromechanical accelerometer with capacitive sensor of Example 5.8, equations (5.70c) through (5.70f)

6.26 The system model for the bow and arrow problem (Example 3.1) is as follows:

$$\frac{dx}{dt} = v$$

$$m\frac{dv}{dt} = \begin{cases} -kx - \beta|v|v, & x \le 0 \\ -\beta|v|v, & x > 0 \end{cases}$$

Is this model linear or nonlinear? Explain your answer.

6.27 Consider the following predator-prey population model:

$$\frac{d}{dt}x_1 = x_1 - 2x_1x_2$$

$$\frac{d}{dt}x_2 = -2x_2 + x_1x_2$$

Determine

(a) The equilibrium point (other than the origin) of the system.

(b) The linear model that is valid around the equilibrium point from (a).

6.28* A dual pendulum-spring model is given by

$$\frac{d}{dt}\begin{bmatrix} \theta_1 \\ \theta_2 \\ \omega_1 \\ \omega_2 \end{bmatrix} = \begin{bmatrix} \omega_1 \\ \omega_2 \\ -\sin 2\theta_1 + \cos\theta_1\sin\theta_2 - g\sin\theta_1 + \cos\theta_1 u \\ \sin\theta_1\cos\theta_2 - \dfrac{1}{2}\sin^2\theta_2 - g\sin\theta_2 \end{bmatrix}$$

Linearize this model about an operating point at the origin and $u = 0$. That is, determine the $\mathbf{A}$ and $\mathbf{B}$ matrices.

6.29 Determine the elements of the vectors and matrices in the linear vector state model equations (6.49a) and (6.49b) as applicable to the following:

(a) The grain scale model of Example 3.4, equations (3.9a) and (3.9b)

(b) The fourth-order Butterworth passive filter circuit of Example 3.13, equations (3.38a) through (3.38c)

(c) The first-order CSTR model of Example 4.7, equations (4.32a) through (4.32d)

6.30 Using the definition, equation (6.59), of the matrix exponential, and equations (6.60) and (6.61), show that equation (6.58) satisfies the state equation (6.49a).

6.31(a) * Prove using the expansion for the matrix exponential that $\exp(\mathbf{A} + \mathbf{B})$ is not equal to $\exp(\mathbf{A})\exp(\mathbf{B})$ for two square matrices $\mathbf{A}$ and $\mathbf{B}$ unless $\mathbf{A}$ and $\mathbf{B}$ commute with each other (that is, $\mathbf{AB} = \mathbf{BA}$).

6.31(b) Using the results from part (a), determine the inverse of $\exp(\mathbf{A})$ and deduce that $\exp(\mathbf{A})$ is nonsingular for every square matrix $\mathbf{A}$. Note that if $\mathbf{D}$ is an $n \times n$ matrix and there is an $n \times n$ matrix $\mathbf{E}$ such that $\mathbf{DE} = \mathbf{ED} = \mathbf{I}_n$, $\mathbf{D}$ is said to be invertible or nonsingular, and $\mathbf{E}$ is said to be the inverse of $\mathbf{D}$; $\mathbf{I}_n = n \times n$ identity matrix.

6.32 Find the output of the following system when the input is a unit step:

$$\frac{d}{dt}\mathbf{X} = \begin{bmatrix} -1 & 0 & 0 \\ 0 & -2 & 0 \\ 0 & 0 & -3 \end{bmatrix}\mathbf{X} + \begin{bmatrix} 1 \\ 2 \\ 0 \end{bmatrix}u;$$

$$\mathbf{X}(0) = \mathbf{0}; \qquad y = \begin{bmatrix} 1 & 1 & 1 \end{bmatrix}\mathbf{X}$$

6.33 Obtain the characteristic equation, modes, and eigenvectors for the second-order linear system with the **A** matrix

(a) $\mathbf{A} = \begin{bmatrix} 1 & 2 \\ 4 & 3 \end{bmatrix}$; **(b)** $\mathbf{A} = \begin{bmatrix} 0 & -1 \\ 1 & -1 \end{bmatrix}$

Describe the modes in each case and identify the fast and slow modes.

6.34* Determine the local response of the following system by direct application of equations (6.58) and (6.59):

$$\frac{d}{dt}\mathbf{X} = \begin{bmatrix} 1 & 2 \\ 4 & 3 \end{bmatrix} \mathbf{X} + \begin{bmatrix} 1 \\ -1 \end{bmatrix} \sin 2t; \qquad \mathbf{X}(0) = \begin{bmatrix} 1 \\ 4 \end{bmatrix}$$

6.35 Diagonalize the following third-order system:

$$\frac{d}{dt}\mathbf{X} = \begin{bmatrix} -10 & -12 & 6 \\ 2 & 0 & -2 \\ -1 & 3 & 1 \end{bmatrix} \mathbf{X} + \begin{bmatrix} 1 & 0 \\ 0 & 1 \\ 1 & 1 \end{bmatrix} \mathbf{U};$$

$$\mathbf{X}(0) = \begin{bmatrix} 1 \\ 2 \\ 1 \end{bmatrix}; \qquad \mathbf{Y} = \begin{bmatrix} 1 & 1 & 0 \\ 1 & 0 & 1 \end{bmatrix} \mathbf{X}$$

6.36 Determine the solution to the vector state equation

$$\frac{d}{dt}\mathbf{X} = \begin{bmatrix} 2 & -4 \\ 3 & -6 \end{bmatrix} \mathbf{X} + \begin{bmatrix} 1 & 0 \\ 0 & 1 \end{bmatrix} \mathbf{U}, \quad \text{if } \mathbf{X}(0) = \begin{bmatrix} 1 \\ 2 \end{bmatrix}; \quad \mathbf{U} = \begin{bmatrix} 1 \\ 1 \end{bmatrix}$$

6.37* Find the response of the following third-order system when the input is a unit step:

$$\frac{d}{dt}\mathbf{X} = \begin{bmatrix} -3 & 0 & 1 \\ -2 & -3 & 5 \\ 0 & -1 & 1 \end{bmatrix} \mathbf{X} + \begin{bmatrix} 1 \\ 0 \\ 1 \end{bmatrix} u; \qquad \mathbf{X}(1) = \begin{bmatrix} -1 \\ 1 \\ 2 \end{bmatrix}$$

6.38 Obtain the characteristic equation, modes, eigenvectors and a similar diagonal matrix for

$$\mathbf{A} = \begin{bmatrix} 0 & 2 & 0 \\ -2 & -4 & 2 \\ -4 & 0 & 0 \end{bmatrix}$$

6.39 Find the free response of the system

$$\frac{d}{dt}\mathbf{X} = \begin{bmatrix} -3 & 4 & 4 \\ 1 & -3 & -1 \\ -1 & 2 & 0 \end{bmatrix} \mathbf{X}; \qquad \mathbf{X}(0) = \begin{bmatrix} 1 \\ -2 \\ 2 \end{bmatrix}$$

given the modal matrix

$$\mathbf{T} = \begin{bmatrix} 1 & 0 & 1 \\ -1 & 1 & \frac{1}{2} \\ 1 & -1 & 0 \end{bmatrix}$$

6.40(a) A third-order free system given by

$$\frac{d}{dt}\mathbf{X} = \begin{bmatrix} p & 1 & 0 \\ 0 & p & 1 \\ 0 & 0 & p \end{bmatrix} \mathbf{X}; \qquad \mathbf{X}(0) = \mathbf{X}_0$$

consists of a first-order free system

$$\frac{d}{dt}x_3 = px_3$$

and another first-order system if x_3 is regarded as the input:

$$\frac{d}{dt}x_2 = px_2 + bu; \qquad bu = x_3$$

and yet another first-order system when x_2 is known:

$$\frac{d}{dt}x_1 = px_1 + bu; \qquad bu = x_2$$

Solve the third-order system as three first-order equations as suggested, and obtain the solution for the original system.

6.40(b) Deduce from your results in part (a) the solution matrix for the general $n \times n$ Jordan block:

$$\mathbf{J}_n(p) = \begin{bmatrix} p & 1 & 0 & \cdot & \cdot & 0 & 0 \\ 0 & p & 1 & \cdot & \cdot & 0 & 0 \\ \cdot & & & & & & \cdot \\ \cdot & & & & \cdot & p & 1 \\ 0 & 0 & \cdot & \cdot & 0 & 0 & p \end{bmatrix}$$

6.41 Determine the elements of the vectors and matrices in the linear discrete-time vector state model, equations (6.90a) and (6.90b), as applicable to the continuous pizza oven model of Example 4.8, equations (4.47a) through (4.47f).

6.42(a) * Find by iteration, the general solution of the system

$$\mathbf{X}_{k+1} = \begin{bmatrix} 0 & -1 \\ 1 & 0 \end{bmatrix} \mathbf{X}_k + \begin{bmatrix} 1 \\ 1 \end{bmatrix} u_k;$$

$$u_k = 1, \quad k = 0, 1, 2, \ldots; \qquad \mathbf{X}_0 = \begin{bmatrix} 0 \\ 1 \end{bmatrix}$$

6.42(b) * Find $\mathbf{X}_4$ for the system using the result of (a) and by using equation (6.91).

6.43 A second-order linear discrete-time system is described by

$$\mathbf{X}_{k+1} = \begin{bmatrix} 1 & 1 \\ 0 & 1 \end{bmatrix} \mathbf{X}_k + \begin{bmatrix} 2 \\ 1 \end{bmatrix} u_k, \qquad \mathbf{X}_0 = \begin{bmatrix} 1 \\ 1 \end{bmatrix}$$

Given that $u_k = k$,

(a) Show by use of equation (6.92) that the total solution to the system is

$$\mathbf{X}_k = \begin{bmatrix} (1 + k) + k(k-1)(k+4)/6 \\ 1 + k(k-1)/2 \end{bmatrix}$$

(b) Verify the result of (a) for $k = 3$ by the iterative method.

6.44 Find the eigenvalues, the response modes, and the eigenvectors of the discrete-time system described in Problem 6.43.

6.45 Calculate the eigenvalues and describe and sketch the modes (the combined modes for complex-conjugate eigenvalues) for the discrete-time systems with the following **P** matrices:

(a)*
$$\begin{bmatrix} 0 & -0.5 \\ 0.5 & 0 \end{bmatrix}$$

(b)
$$\begin{bmatrix} -2 & 1 \\ -1 & -2 \end{bmatrix}$$

(c)
$$\begin{bmatrix} 1 & -2 & 2 \\ -1 & 2 & -1 \\ 1 & 1 & 0 \end{bmatrix}$$

6.46 Describe two ways by which a computer model can be verified.

6.47* Determine the stability condition for the magnitude of the step size T such that the two-stage algorithm

$$R_1 = Tf(x_k, t_k)$$

$$R_2 = Tf(x_k + R_1, t_k + T)$$

$$x_{k+1} = \frac{1}{2}(R_1 + R_2)$$

is stable for the system

$$\frac{dx}{dt} = ax + b$$

where a and b are constants with a negative.

6.48 Explain Shannon's sampling theorem. Describe one type of stiff problem.

6.49 In a modeling environment,

(a) What are the relations among the actual system, the system model, and the computer model?

(b) How can validation of a computer model be carried out?

(c) Why is computer model verification a necessity?

6.50 Determine the recursive formula or difference equation for Euler's method with step length T applied to the problem $dy/dt = -y$, $y(0) = 1$. For what values of T is the sequence $\{y_k\}_0^\infty$ bounded?

6.51 Solve the following system:

$$\frac{d}{dt}\mathbf{X} = \begin{bmatrix} -1 & 0 \\ 1 & -2 \end{bmatrix}\mathbf{X} + \begin{bmatrix} 1 \\ 0 \end{bmatrix} 2U_S(t); \qquad \mathbf{X}(0) = \begin{bmatrix} 2 \\ 3 \end{bmatrix}$$

using

(a) Euler's shooting method with a step size $T = 0.05$.

(b) MATLAB's Runge-Kutta (**ode45**) method.

(c) The implicit trapezoidal method—(**TRAP**) with step size $T = 0.05$.

6.52 Using the indicated method, solve approximately the following systems and compare the result with the exact solution:

(a) System of Example 6.17 by MATLAB (**ode45**).

(b) System of Example 6.18 by (**TRAP**). Approximate the impulse with a pulse of one time step duration. Experiment with the magnitude of the pulse relative to accuracy of the computed results and any numerical problems.

6.53 Given the differential equation $dy/dx = x + y$ with $y(0) = 1$, compute an approximation to $y(0.2)$ (use 5 correct decimals)

(a) By the Runge-Kutta method (MATLAB **ode45**).

(b) By the trapezoidal method (**TRAP**) with step size $T = 0.1$.

(c) Compute $y(0.2)$ exactly.

6.54* Given $dy/dx = 1 + x^2y^2$, $y(0) = 0$, compute $y(0.5)$ using a step size of 0.25 and the implicit trapezoidal method.

6.55 Use any numerical technique (software) to simulate the response of the system of Problem 6.34, subject to the initial state $x_1(0) = x_2(0) = 0$.

6.56 Use any available software package to simulate the free response of both the original dual pendulum system given in Problem 6.28 and the linearized model obtained in that problem. Assume the initial state is $\theta_1(0) = \theta_2(0) = 0$, $\omega_1(0) = 2$, $\omega_2(0) = 0$, and a unit step input. Compare the results for the linear and nonlinear models.

6.57 A three-tank system is described by the model

$$\frac{d}{dt}\mathbf{X} = \begin{bmatrix} -0.5 & 0.5 & 0 \\ 1 & -1.5 & 0.5 \\ 0 & 0.2 & -0.36 \end{bmatrix}\mathbf{X} + \begin{bmatrix} 1 \\ 2 \\ 0 \end{bmatrix} u$$

Under a control scheme, the input is given by $u = 27.8x_1 - 19.22x_2 - 200.9x_3$. Determine by simulation the controlled behavior of the system given the initial state $x_1(0) = x_2(0) = x_3(0) = 1$.

6.58 A controlled system has the following fifth-order model:

$$\dot{x}_1 = x_2$$

$$\dot{x}_2 = -411.73x_1 + 25.73x_3$$

$$\dot{x}_3 = 0.09(u - 50)$$

$$\dot{x}_4 = -\frac{122.3}{T_1}x_1$$

$$\dot{x}_5 = \frac{1}{\tau}\left[122.3\left(\frac{T_2}{\tau} - 1\right)x_1 - x_5\right]$$

If the output and control are given by

$$y = x_5 - \frac{122.3T_2}{\tau}x_1 \quad \text{and} \quad u = K[0.1k_c(x_4 + y) - 17.73x_3]$$

use an available computational software package to simulate the response of this system from the initial state $x_1 = -0.03$, $x_2 = x_3 = x_4 = x_5 = 0$. Assume the following parameter values: $K = 50$, $k_c = 2$, $T_1 = 0.01$, $T_2 = 1.01$, and $\tau = 0.01$. Display graphically the output y for $0 \le t \le 1$.

7

Solution of Higher-Order Scalar Systems

THIS CHAPTER CONTINUES WITH THE RESPONSE of lumped-parameter systems. However, the emphasis is on systems described by *scalar* ordinary differential equations of second or higher order. Although the result of the state space approach to system modeling is typically a system of first-order ODEs or a vector state equation, such a model is equivalent to and is reducible to a single scalar ordinary differential equation, provided the state model is a single-input, single-output (SISO) system. This equivalence between the two types of dynamic system models will be considered formally in Section 9.3.

The free and forced responses of systems described by second-order time-invariant linear scalar ODEs are considered first. This material is important not only for the necessary background it provides for the general analysis of scalar ODE problems but also because the behaviors of a large number of engineering systems are often described, conveniently, by a second-order ordinary differential equation. The development of the subject is therefore, like the first-order case in Chapter 6, fairly complete and detailed. Forced solution techniques that take advantage of the nature of the scalar linear ODE system model and its inputs are considered next. Two such methods, the phasor transform technique, which is useful for stable linear systems subject to sinusoidal forcing inputs, and the more general Fourier series solution for systems subject to periodic inputs are discussed. The performance of tuned *RCL* networks or radio frequency (RF) circuits and forced mechanical vibration systems are two examples of significant applications of these analytical methods that are introduced briefly in this chapter.

7.1 RESPONSE OF SECOND-ORDER SYSTEMS

We begin our analysis of multiple-order scalar systems by considering separately, in this section, second-order systems. As just stated, there are two main reasons for this approach. The first is that many dynamic systems in engineering can be described adequately with a second-order model. Although physically these systems are of much higher order (actually infinite order) their behavior may be similar to the response of second-order systems. The second reason is that certain important analytical concepts and design considerations are more convenient to introduce here, and they have traditionally been associated with systems described by second-order ordinary differential equations.

| **Example 7.1** | Model of an Auto Suspension System |

Consider the physical model of a car suspension system shown in Fig. 7.1. Determine a scalar ordinary differential equation model for the vertical displacement of the car relative to the road surface.

Solution: Using the methods of Chapters 2 and 3, we can derive the following state model:

$$\frac{dr}{dt} = u \tag{7.1a}$$

$$\frac{dh}{dt} = v \tag{7.1b}$$

$$\frac{dv}{dt} = -\left(\frac{k}{m}\right)h - \left(\frac{\beta}{m}\right)v + \left[\frac{\beta}{m}u + \frac{k}{m}r - g\right] \tag{7.1c}$$

where h is the vertical displacement of the car, v is the vertical velocity of the car, r is the vertical displacement of the road surface from the datum level (road profile), and u is the velocity input at the wheels due to the road profile. We shall assume that the acceleration input at the wheels due to the road profile $a(t)$ is actually the specified input, from which $u(t)$ and $r(t)$ can be derived.

Given our interest in the vertical motion of the car, we can differentiate (7.1b) with respect to time and substitute (7.1c) into the result. After subtracting $a(t)$ from both sides of the result, we get the second-order ordinary differential equation:

$$\left(\frac{d^2h}{dt^2} - \frac{d^2r}{dt^2}\right) + \frac{\beta}{m}\left(\frac{dh}{dt} - \frac{dr}{dt}\right) + \frac{k}{m}(h - r) = -g - a(t)$$

Defining $y(t) = h(t) - r(t) =$ displacement of the car relative to the road surface, we finally obtain

$$\frac{d^2y}{dt^2} + \frac{\beta}{m}\frac{dy}{dt} + \frac{k}{m}y = -g - a(t) \tag{7.1d}$$

We have placed all the assumed inputs together on the RHS. In general, we shall write the second-order scalar model in the following form:

$$\frac{d^2y}{dt^2} + 2\zeta\omega_0\frac{dy}{dt} + \omega_0^2 y = f(t) \tag{7.2}$$

Thus for the suspension system, the output variable y is the relative displacement of the car, and

$$\omega_0^2 = \frac{k}{m} \Rightarrow \omega_0 = \sqrt{\frac{k}{m}} \tag{7.3a}$$

$$2\zeta\omega_0 = \frac{\beta}{m} \Rightarrow \zeta = \frac{\beta}{2\sqrt{mk}} \tag{7.3b}$$

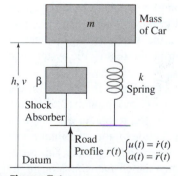

Figure 7.1

Auto Suspension System

Free Response: Natural Frequency and Damping

In equation 7.2 ω_0 is termed the *natural frequency* and ζ is called the *damping ratio*. The significance of these quantities will be discussed subsequently. $f(t)$ is the *nonhomogeneous* term or forcing function (input). Equation (7.2) is a nonhomogeneous second-order ordinary differential equation with constant coefficients. The solution of the homogeneous equation $[f(t) = 0]$ is given by

$$y_H(t) = C_1 E_1(t) + C_2 E_2(t); \qquad C_1, C_2 = \text{constant} \tag{7.3c}$$

where the functions $E_1(t)$ and $E_2(t)$ are linearly independent solutions that can be found by substitution of the trial solution $e^{\lambda t}$ into the differential equation (7.2) with $f(t)$ set to zero. This yields the characteristic equation

$$\lambda^2 e^{\lambda t} + 2\zeta\omega_0 \lambda e^{\lambda t} + \omega_0^2 e^{\lambda t} = 0$$

or

$$\lambda^2 + 2\zeta\omega_0 \lambda + \omega_0^2 = 0; \quad \text{(characteristic equation)} \tag{7.4a}$$

The two roots of this equation,

$$\lambda_1, \lambda_2 = -\zeta\omega_0 \pm \sqrt{\zeta^2 \omega_0^2 - \omega_0^2}$$

or

$$\lambda_1, \lambda_2 = -\omega_0 (\zeta \pm \sqrt{\zeta^2 - 1}) \tag{7.4b}$$

yield (recall the discussion of response modes in Section 6.3)

$$\left.\begin{array}{l} E_1(t) = e^{\lambda_1 t} \\ E_2(t) = e^{\lambda_2 t} \end{array}\right\} \lambda_1, \lambda_2; \quad \text{real and distinct} \tag{7.5a}$$

$$\left.\begin{array}{l} E_1(t) = e^{\lambda t} \\ E_2(t) = t e^{\lambda t} \end{array}\right\} \lambda_1 = \lambda_2 = \lambda = -\zeta\omega_0; \quad \text{real and repeated} \tag{7.5b}$$

$$\left.\begin{array}{l} E_1(t) = e^{-\sigma t}\cos(\omega t) \\ E_2(t) = e^{-\sigma t}\sin(\omega t) \end{array}\right\} \lambda_1, \lambda_2 = -\sigma \pm j\omega; \quad \text{complex conjugate} \tag{7.5c}$$

where,

$$\sigma = \omega_0 \zeta; \qquad \omega = \omega_0 \sqrt{1 - \zeta^2} \tag{7.5d}$$

It is evident that the arguments of the exponential terms E_1 and E_2 completely characterize the response of the system (recall the role of eigenvalues in the vector state model). It is for this reason that equation (7.4a) is known as the *characteristic equation*. Further, the roots of the characteristic equation are the *eigenvalues* of the system. Also, the exponential functions $E_1(t)$ and $E_2(t)$ represent the response modes for a second-order system. In general, the eigenvalue is a complex number (recall equation (6.69) and see equation (7.5c), and it can be displayed in the complex plane like any other complex number (see Fig. 7.2). Note from Fig. 7.2 that given the eigenvalues of the system, the parameters in equation (7.2) can be determined, that is, the system can be identified (see Practice Problem 7.2).

The time functions represented by $E_1(t)$ and $E_2(t)$ are sketched in Fig. 7.3 for the different values of the characteristic roots given by equations (7.5a) through (7.5d). Note the method used to sketch the product functions. For example, to obtain $e^{-\sigma t}\cos(\omega t)$, the curves for $e^{-\sigma t}$ and $\cos(\omega t)$ are first sketched (dashed curves), then the final result is obtained by multiplying together the values of these curves at a few points. This product is zero each time the cosine is zero and is $\pm e^{-\sigma t}$ when the cosine is plus or minus 1. Similarly, sums of functions can be sketched individually and the values at a few points added together to produce the final sketch. As was the case for the first-order system, the *free response of a second- or multiple-order scalar system is that homogeneous*

Figure 7.2

Complex Plane Display of an Eigenvalue

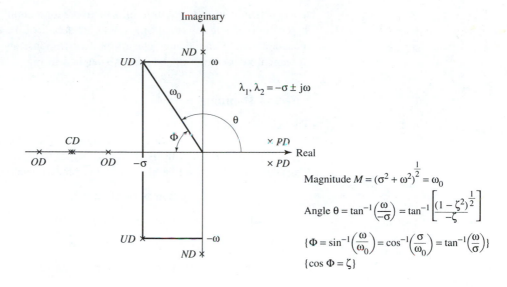

$$\lambda_1, \lambda_2 = -\sigma \pm j\omega$$

$$\text{Magnitude } M = (\sigma^2 + \omega^2)^{\frac{1}{2}} = \omega_0$$

$$\text{Angle } \theta = \tan^{-1}\left(\frac{\omega}{-\sigma}\right) = \tan^{-1}\left[\frac{(1-\zeta^2)^{\frac{1}{2}}}{-\zeta}\right]$$

$$\left\{\Phi = \sin^{-1}\left(\frac{\omega}{\omega_0}\right) = \cos^{-1}\left(\frac{\sigma}{\omega_0}\right) = \tan^{-1}\left(\frac{\omega}{\sigma}\right)\right\}$$

$$\{\cos\Phi = \zeta\}$$

solution that satisfies the initial condition. According to equation (7.3c) the free response is made up of functions of the type sketched in Fig. 7.3. This solution is sometimes erroneously regarded as the transient response. Strictly speaking, *the transient response is the total solution (free plus forced) for a short time after the response is initiated but before the solution reaches a steady value.* For the usual inputs, the forced response con-

Figure 7.3

Sketch of Second-order System Characteristic Modes

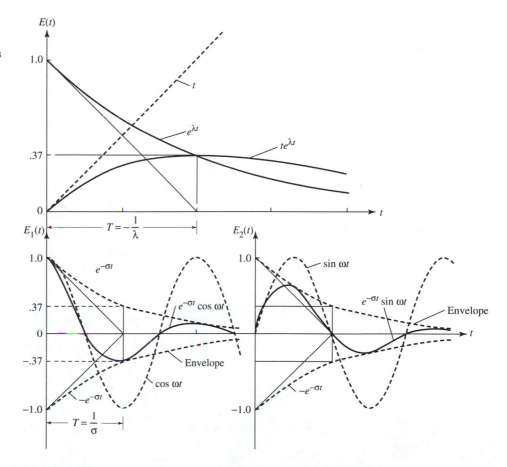

tains very little system dynamic information, (for example the particular solution for a step input is also a step function) so that the details of the transient response depend almost completely on the character of the free response and this character is determined by which combination of the modes sketched in Fig. 7.3 appear in equation (7.3c).

Effect of Damping

The frequency of oscillation ω and the rate of decay σ depend on the value of the damping ratio ζ. From equation (7.3b) the value ζ is related to the amount of the damping in the system. But why is it called the damping ratio? The reason is as follows: Consider the following five classes of the range of values of ζ,

$\zeta > 1$;   System is overdamped

$\zeta = 1$;   System is critically damped

$0 < \zeta < 1$;   System is underdamped

$\zeta = 0$;   System is undamped (not damped)

$\zeta < 0$;   System is unstable

The value of the damping constant corresponding to the *critically damped* case, for the auto suspension system for example, is given by

$$\zeta = 1 = \frac{\beta_c}{2\sqrt{mk}} \Rightarrow \beta_c = 2\sqrt{mk} \tag{7.6a}$$

This value is termed *critical* because the system's response is oscillatory for values of $\zeta < 1$. In terms of the critical damping constant, equation (7.3b) becomes

$$\zeta = \frac{\beta}{\beta_c} \tag{7.6b}$$

Thus damping ratio is the ratio of the value of the actual damping to the critical damping value. Let us now examine in turn the system behavior associated with each of the classes of values of the damping ratio and their correlation with the system eigenvalues.

Stable and Overdamped Behavior   When the damping ratio is greater than 1, we see from equation (7.4b) that all the eigenvalues are real and negative, so that the response is given by equations (7.3c) and (7.5a). In other words, the solution is a combination of two purely exponential decay modes, one of which always decays faster than the other (recall Fig. 6.21a). The behavior is *stable,* and on the complex plane, Fig. 7.2 (see example points labeled *OD*), both eigenvalues are located *on* the *negative* real axis.

Example 7.2   Free Response of an Overdamped Second-Order System

Suppose the car suspension system of Fig. 7.1 has the following parameter values:

$$\left(\frac{k}{m}\right) = 4 \text{ s}^{-2}; \qquad \left(\frac{\beta}{m}\right) = 5 \text{ s}^{-1}$$

For the initial conditions $y(0) = 3$ cm, $\dot{y}(0) = 0.0$ cm $\cdot$ s^{-1}, determine the free response.

Solution:

$$\omega_0 = 2; \qquad \zeta = \frac{(\beta/m)}{2\omega_0} = \frac{5}{4} = 1.25$$

$\zeta > 1 \Rightarrow$ overdamped $\Rightarrow y(t) = C_1 e^{\lambda_1 t} + C_2 e^{\lambda_2 t}$

$$\lambda_1, \lambda_2 = -2\left(\frac{5}{4} \pm \sqrt{\frac{25}{16} - 1}\right) = -4 - 1; \qquad \text{eigenvalues} \atop (\textit{both real and negative})$$

C_1 and C_2 are determined from the initial conditions:

$$C_1 + C_2 = y(0)$$
$$\lambda_1 C_1 + \lambda_2 C_2 = \dot{y}(0)$$

Solving by matrix inversion (see Appendix B), we have

$$\begin{bmatrix} C_1 \\ C_2 \end{bmatrix} = \begin{bmatrix} \left(\dfrac{\lambda_2}{\lambda_2 - \lambda_1}\right) & \left(\dfrac{-1}{\lambda_2 - \lambda_1}\right) \\ \left(\dfrac{-\lambda_1}{\lambda_2 - \lambda_1}\right) & \left(\dfrac{1}{\lambda_2 - \lambda_1}\right) \end{bmatrix} \begin{bmatrix} y(0) \\ \dot{y}(0) \end{bmatrix} \qquad (7.7)$$

$$C_1 = -1; \qquad C_2 = 4; \qquad y(t) = -e^{-4t} + 4e^{-t}$$

The response is sketched in Fig. 7.4a. Note that equation (7.7) is a general result for the *overdamped* case.

Stable and Critically Damped Behavior   In equation (7.4b) the only time the eigenvalues are equal to each other and real is when the damping ratio equals 1. This situation corresponds to the critically damped case for which the solution is given by equations (7.3c) and (7.5b). The modes are again exponential decay type, with no oscillatory components. The two eigenvalues, although equal, are still on the negative real axis of the complex plane (see example point labeled *CD* in Fig. 7.2), and the system behavior is stable.

Figure 7.4

**Overdamped and
Critically Damped
Second-order Response**

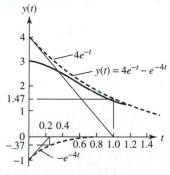

(a)   Overdamped Free Response

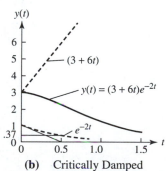

(b)   Critically Damped
Free Response

Example 7.3   Free Response of a Critically Damped System

Repeat Example 7.2 but with

$$\left(\frac{k}{m}\right) = 4 \text{ s}^{-2}; \qquad \left(\frac{\beta}{m}\right) = 4 \text{ s}^{-1}$$

Solution:

$$\omega_0 = 2; \qquad \zeta = \frac{(\beta/m)}{2\omega_0} = 1 \Rightarrow \text{critically damped}$$

Hence

$$y(t) = (C_1 + C_2 t)e^{\lambda t}$$

$$\lambda = -(1)\omega_0 = -2; \qquad \text{repeated eigenvalue (}real \text{ and } negative\text{)}$$

Applying the initial conditions, we have

$$C_1 = y(0) \tag{7.8a}$$

$$\lambda C_1 + C_2 = \dot{y}(0) \Rightarrow C_2 = \dot{y}(0) - \lambda y(0) \tag{7.8b}$$

$$C_1 = 3; \qquad C_2 = 6; \qquad y(t) = (3 + 6t)e^{-2t}$$

This result is shown in Fig. 7.4b. Again, equations (7.8a) and (7.8b) are general for the *critically damped* second-order system. In practice, a system with damping ratio in the range 0.95–1.05 will normally be regarded as being critically damped, since this quantity is never known precisely.

....

Stable and Underdamped Behavior   The *underdamped* condition gives the most interesting response pattern and it is usually an adequate representation of the damping naturally present in many physical systems. In equation (7.4b), $0 < \zeta < 1$ implies that the eigenvalues are complex (conjugate). However the real part is negative (see example points labeled *UD* in Fig. 7.2), and the system is stable. The response given by equations (7.3c) and (7.5c) consists of damped oscillatory modes.

Example 7.4   Free Response of an Underdamped Second-Order System

In the automobile suspension example, let

$$\left(\frac{k}{m}\right) = 10 \text{ s}^{-2}; \qquad \left(\frac{\beta}{m}\right) = 1 \text{ s}^{-1}, \text{ and take the initial conditions as} \atop y(0) = 1 \text{ cm}; \dot{y}(0) = 0.0 \text{ cm} \cdot \text{s}^{-1}$$

Solution:

$$\omega_0 = \sqrt{10}; \zeta = \frac{(\beta/m)}{2\omega_0} = \frac{1}{\sqrt{10}} < 1 \Rightarrow \text{underdamped}$$

Hence

$$y(t) = (C_1 \cos \omega t + C_2 \sin \omega t)e^{-\sigma t} \tag{7.9a}$$

From equation (7.5d),

$$\sigma = \omega_0 \zeta = 1 \quad \text{and} \quad \omega = \omega_0 \sqrt{(1 - \zeta^2)} = 3$$

so that the eigenvalues are

$$-1 \pm j3 \quad \text{(complex conjugate with negative real part)}$$

Introducing the initial conditions, we have

$$C_1 = y(0) = 1 \tag{7.9b}$$

$$-\sigma C_1 + \omega C_2 = \dot{y}(0)$$

$$C_2 = \frac{\dot{y}(0) + \sigma y(0)}{\omega} = \frac{1}{3}$$

Recall from the discussion following equation (6.69) that the two sinusoids given by the equations (7.3c) and (7.5c) can be combined into one oscillatory mode: $C \cos(\omega t + \phi)e^{-\sigma t}$, which is our preferred representation for a damped sinusoid. By the trigonometric identity

$$C[\cos(\omega t + \phi)] = C[\cos(\omega t) \cos \phi - \sin(\omega t) \sin \phi] \tag{7.10}$$

$$= C_1 \cos(\omega t) + C_2 \sin(\omega t)$$

we can write the response (7.9a) as

$$y(t) = C \cos(\omega t + \phi)e^{-\sigma t}; \quad \text{(underdamped second-order free response)} \tag{7.11a}$$

where

$$C \cos \phi = C_1$$

$$-C \sin \phi = C_2$$

so that

$$C = \sqrt{C_1^2 + C_2^2} \quad \text{and} \quad \phi = \arctan\left(\frac{-C_2}{C_1}\right) : C = \frac{\sqrt{10}}{3}, \phi = -0.32 \tag{7.11b}$$

or given the solution (7.11a), we can obtain C and ϕ directly from the initial conditions:

$$C \cos \phi = y(0) \tag{7.11c}$$

$$C\omega \sin \phi = -\dot{y}(0) - \sigma y(0) \tag{7.11d}$$

....

The response given by equation (7.11a) is easily plotted as the product of a cosine and an exponential curve. In the sketch shown in Fig. 7.5, some important quantities characterizing the response of an underdamped second-order dynamic system are indicated. The solution is a damped oscillation with circular (angular) frequency ω. We shall refer to ω as the *damped circular frequency* to distinguish it from the natural circular frequency ω_0. From equation (7.4d),

$$\omega = \omega_0 \sqrt{1 - \zeta^2} \tag{7.12a}$$

Figure 7.5

Underdamped Second-order Free Response

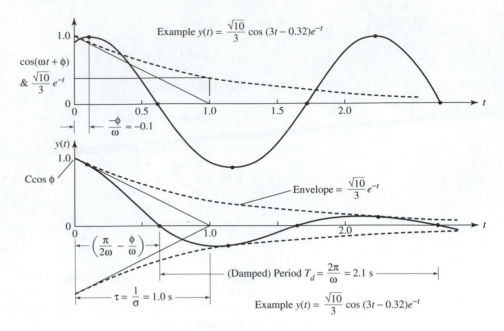

Note that the damped circular frequency is always less than the natural circular frequency. The damped frequency and the period of oscillation (the damped period) are given by

$$f = \frac{\omega}{2\pi} = \frac{\omega_0 \sqrt{1 - \zeta^2}}{2\pi} \tag{7.12b}$$

$$T_d = \frac{1}{f} = \frac{2\pi}{\omega} = \frac{2\pi}{\omega_0 \sqrt{1 - \zeta^2}} \tag{7.12c}$$

In equation (7.11a) and in Fig. 7.5, ϕ is called the phase angle or phase shift. A positive ϕ implies a leftward shift of the response curve by an amount ϕ/ω relative to zero on the time axis. A negative ϕ gives the opposite effect. The exponential term, $C \exp(-\sigma t)$, in equation (7.11a) provides an envelope on the oscillation as shown. This *decay envelope* is characterized by the damping time constant, which is given by equation (6.6):

$$\tau = \frac{1}{\sigma} = \frac{1}{\omega_0 \zeta} \tag{7.12d}$$

Thus the decay rate is determined by the amount of damping. The response dies out very quickly for large values of σ (see Fig. 7.7). Another way to represent the decay rate is in terms of the number of cycles of oscillations required for the response to be damped to $1/e$ of its initial value. That is,

$$n = \frac{\text{time to damp to } 1/e}{\text{time for one cycle}} = \frac{\tau}{T_d} = \frac{\sqrt{1 - \zeta^2}}{2\pi\zeta} \tag{7.12e}$$

for very small damping ratios, $\zeta^2 \ll 1$,

$$n \approx \frac{1}{2\pi\zeta} \tag{7.12f}$$

Yet another representation of the decay rate, which is often used in the reduction of mechanical vibration data, is the *logarithmic decrement*. If y_k is the amplitude (peak value) of the response in cycle k and at time t_k, then y_{k+1} is the amplitude in the $k + 1$ cycle and is the value at time $t_{k+1} = t_k + T_d$. It follows that

$$\frac{y_k}{y_{k+1}} = \frac{Ce^{-\sigma t_k}}{Ce^{-\sigma(t_k+T_d)}} = e^{\sigma T_d} = \text{constant} \tag{7.12g}$$

The logarithmic decrement δ is defined by

$$\delta = \ln\left(\frac{y_k}{y_{k+1}}\right) = \ln e^{\sigma T_d} = \sigma T_d$$

Substituting for σ and T_d from equations (7.5c) and (7.12c), respectively, we obtain

$$\delta = \frac{2\pi\zeta}{\sqrt{1 - \zeta^2}} \tag{7.12h}$$

From equations (7.12e) and (7.12h), we see that the logarithmic decrement and the number of cycles to decay to $1/e$ are related: $\delta = 1/n$.

Also, if ζ is small, $\delta \approx 2\pi\zeta$. For a given experimental data, it may be more accurate or representative to measure the vibration amplitudes at other than successive cycles. In this case,

$$\ln\left(\frac{y_k}{y_{k+N}}\right) = N\delta \tag{7.12i}$$

where N is the number of cycles separating the two amplitudes. An average equivalent (viscous) damping ratio for such data is then given by

$$\zeta = \frac{N\delta}{\sqrt{(2\pi N)^2 + N^2\delta^2}} \tag{7.12j}$$

Undamped Behavior and Natural Frequency   Recall the micromechanical accelerometer with capacitive sensor of Example 5.8. The mechanical model of the cantilever beam accelerometer is shown again in Fig. 7.6. The system model was such that the behavior of the mechanical subsystem could be explored autonomously according to the second-order scalar differential equation (5.71a), here repeated:

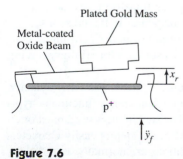

Figure 7.6

Model of Cantilever Beam Micromechanical Accelerometer

$$\frac{d^2x_r}{dt^2} + \omega_0^2 x_r = -\ddot{y}_f; \qquad \omega_0 = \sqrt{\frac{k}{m}} \tag{7.13}$$

where x_r is the relative displacement of the plates with respect to the equilibrium configuration, and $\ddot{y}_f$ is the (upward) acceleration of the frame to which the accelerometer is attached.

Comparing equation (7.13) with equation (7.2), we can see that $\zeta = 0$, that is, the micromechanical accelerometer is an undamped second-order system. In equation (7.12a), if the damping ratio is zero, then the frequency of oscillation is the same as the natural frequency. Thus *the natural frequency is the frequency characterizing the free response of the system with no damping*. Also, in equation (7.12d) $\zeta = 0$ implies an infinite time constant. That is, the decay envelope consists of two parallel lines at $\pm C$. The system thus will oscillate for all time with constant amplitude once it is disturbed. (It follows that an undamped system has no transient response in the strict sense of that term). The response neither decays nor grows, and the system can be considered *marginally stable*. The eigenvalues are purely imaginary or complex (conjugates) with zero real part (see example points located right on the imaginary axis and labeled *ND* in Fig. 7.2). These remarks suggest that the undamped motion can be viewed as a special case or a limiting case of the underdamped system. In practice, most physical systems have some finite amount of damping, so that no damping is assumed when the degree of damping is very low. For the accelerometer, some damping may be necessary to facilitate the reading of the measurement (accelerometer design is considered further in Section 7.3). For energetic systems such as the car suspension system, the undamped free response corresponds to motion with no mechanism for energy loss. The sum of the potential and kinetic energies remains constant at the initial value, so that the system oscillates indefinitely from one extreme of total kinetic energy with zero potential energy to the other extreme of total potential energy with zero kinetic energy.

Negative Damping and Instability   A positive nonzero damping ratio implies the existence of a mechanism for energy loss in the system. Thus the total energy level in the system constantly decreases when there is some damping, and the motion eventually dies out. A negative damping ratio implies the presence of a means for continually adding energy into the system. The energy level will therefore constantly increase and the amplitude of the response will grow continuously (exponentially). The system is *unstable*. Physically this state of affairs is quite uncommon. From equation (7.4b) we see that at least one of the eigenvalues is real and positive, or the eigenvalues are complex (conjugates) with positive real part. In the complex plane (see example points labeled *PD* in Fig. 7.2), the system has one or more eigenvalues located in the right half of the plane—a condition for instability (stability of general dynamic systems is discussed formally in Chapter 10). The characteristic modes therefore include one or two pure exponential growths or an oscillatory exponential growth.

Forced Response to Special Inputs

We have thus far discussed the free response of second-order scalar systems. However, the general solution or the transient response consists of both the free and forced motions. For a given input function, available mathematical techniques such as the *method of undetermined coefficients* or the *method of variation of parameters* can be used to find the particular solution that together with the homogeneous solution can be used to determine the total response. Also the phasor- and Laplace-transform methods, developed in later sections, can be used to obtain the forced response. In the present section the forced response to some specific inputs (the total response under zero initial conditions) will be developed in the time domain. A universal solution for arbitrary forcing functions will also be specified.

In design work the transient response of second-order systems and other dynamic models is usually represented by the response to one of the singularity functions—typically the step or the impulse—under zero initial state. The step input is used most often because of the ease with which it can be generated. Also, the step is easily characterized—only the magnitude is needed. Zero initial conditions are normally assumed for lack of a better knowledge of the initial state of the physical system. In addition, most stable systems come to rest at their equilibrium point after a disturbance, and this point is often the origin in any analytical description of the system.

| **Example 7.5** | Unit Step Response of a Scalar Second-Order System |

Determine the response of the system of equation (7.2) subject to zero initial conditions and a unit-step forcing input.

Solution: By the method of undetermined coefficients, the trial solution to equation (7.2) when the nonhomogeneous term is the unit step $f(t) = U_s(t)$ is (see Table 7.1) $y_p(t) = a_0$. Substituting this solution in equation (7.2) we determine the coefficient a_0 so that the particular solution is

$$y_p(t) = \frac{1}{\omega_0^2}; \qquad t > 0$$

Thus depending on the value of the damping ratio (we consider only stable systems), the general solution will be one of the following:

$$\zeta > 1: \qquad y_s(t) = C_1 e^{\lambda_1 t} + C_2 e^{\lambda_2 t} + \frac{1}{\omega_0^2}; \qquad t > 0 \tag{7.14a}$$

$$\zeta = 1: \qquad y_s(t) = (C_1 + C_2 t)e^{\lambda t} + \frac{1}{\omega_0^2}; \qquad t > 0 \tag{7.14b}$$

$$0 \leq \zeta < 1: \qquad y_s(t) = C \cos(\omega t + \phi)e^{-\sigma t} + \frac{1}{\omega_0^2}; \qquad t > 0 \tag{7.14c}$$

where λ_1, λ_2, λ, σ, and ω are given by equations (7.4) and (7.5). The constants C_1 and C_2 or C and ϕ are determined from the initial conditions (or from conditions at another point in time) such that the general solution satisfies these conditions. Consider, for instance, the *underdamped* case with zero initial conditions:

$$y(0) = \dot{y}(0) = 0 \tag{7.15a}$$

Then

$$C \cos(\phi) + \frac{1}{\omega_0^2} = 0$$

$$-\sigma C \cos(\phi) - C\omega \sin(\phi) = 0$$

Table 7.1 Trial Solutions in Method of Undetermined Coefficients

FORM OF $f(t)$[a]	TRIAL SOLUTION[b]
ce^{pt}	ae^{pt}
$c_n t^n + c_{n-1}t^{n-1} + \ldots + c_0$	$a_n t^n + a_{n-1}t^{n-1} + \ldots + a_0$
$e^{pt}(c_n t^n + c_{n-1}t^{n-1} + \ldots + c_0)$	$e^{pt}(a_n t^n + a_{n-1}t^{n-1} + \ldots + a_0)$
$c_1 \cos(\omega t) + c_2 \sin(\omega t)$	$a_1 \cos(\omega t) + a_2 \sin(\omega t)$
$e^{pt}[c_1 \cos(\omega t) + c_2 \sin(\omega t)]$	$e^{pt}[a_1 \cos(\omega t) + a_2 \sin(\omega t)]$
$e^{pt}(c_n t^n + c_{n-1}t^{n-1} + \ldots + c_0)\cos(\omega t)$	$e^{pt}(a_n t^n + a_{n-1}t^{n-1} + \ldots + a_0)\cos(\omega t)$
$+ e^{qt}(D_n t^n + D_{n-1}t^{n-1} + \ldots + D_0)\sin(\omega t)$	$+ e^{qt}(b_n t^n + b_{n-1}t^{n-1} + \ldots + b_0)\sin(\omega t)$

[a]If $f(t)$ is a sum of some of these functions, the trial solution is a sum of the corresponding trial functions.
[b]If a term of the assumed trial function appears in the homogeneous solution, the trial solution is obtained by multiplying the assumed trial function with the smallest positive integral power of t that gives a function no term of which appears in the homogeneous solution.

which gives

$$C = \frac{-1}{\omega_0^2 \cos(\phi)} \tag{7.15b}$$

$$\phi = \tan^{-1}\left(\frac{-\sigma}{\omega}\right) \tag{7.15c}$$

Also,

$$\sigma^2 C^2 \cos^2(\phi) = \omega^2 C^2 \sin^2(\phi) = \omega^2 C^2 [1 - \cos^2(\phi)]$$

so that

$$(\sigma^2 + \omega^2)\cos^2(\phi) = \omega^2 \Rightarrow \cos(\phi) = \frac{\omega}{\sqrt{\sigma^2 + \omega^2}}$$

or

$$\cos \phi = \sqrt{1 - \zeta^2} \tag{7.15d}$$

Thus the **underdamped unit-step response** is

$$y_s = C \cos(\omega t + \phi)e^{-\sigma t} + \frac{1}{\omega_0^2} = C \cos\left(\omega_0 \sqrt{1 - \zeta^2}\, t + \phi\right)e^{-\zeta\omega_0 t} + \frac{1}{\omega_0^2}$$

$$y_s = \frac{1}{\omega_0^2}\left[1 - \frac{1}{\cos\phi}\cos(\omega t + \phi)e^{-\zeta\omega_0 t}\right] \tag{7.15e}$$

$$= \frac{1}{\omega_0^2}\left[1 - \frac{1}{\sqrt{1 - \zeta^2}}\cos\left(\omega_0\sqrt{1 - \zeta^2}\,t + \phi\right)e^{-\zeta\omega_0 t}\right]$$

where

$$\phi = \tan^{-1}\left(\frac{-\sigma}{\omega}\right) = \tan^{-1}\left(\frac{-\zeta}{\sqrt{1 - \zeta^2}}\right) \tag{7.15f}$$

In Fig. 7.7a, $\omega_0^2 y_s$ (which is dimensionless) is shown plotted against normalized time $\omega_0 t$ for various values of the damping ratio. The effect of damping on the transient step response is clearly illustrated by this figure.

··

Underdamped Second-Order Transient Step Response Characteristics

The typical performance measures that characterize this type of behavior are shown in Fig. 7.7b on an isolated second-order underdamped transient step response curve. To simplify subsequent analysis, the indicated output variable is $y_d = \omega_0^2 y_s$, which from (7.15e) is given by

$$y_d = 1 - \frac{1}{\sqrt{1 - \zeta^2}}\cos\left(\omega_0\sqrt{1 - \zeta^2}\,t + \phi\right)e^{-\zeta\omega_0 t} \tag{7.16a}$$

$$= 1 - \frac{1}{\cos\phi}\cos(\omega t + \phi)e^{-\sigma t}$$

Figure 7.7

**Transient Response
of an Under-damped
Second-order System Subject
to a Unit Step Input**

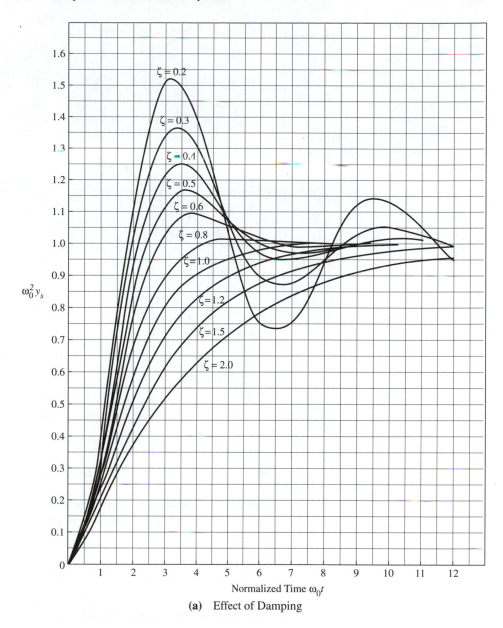

(a) Effect of Damping

1. **Rise time** (t_r) is the time for the response to first reach the final value. Since according to equation (7.16a), the final value is $B = 1$,

$$1 - \frac{1}{\cos \phi}\cos(\omega t_r + \phi)e^{-\sigma t_r} = 1 \Rightarrow \cos(\omega t_r + \phi) = 0$$

$$\Rightarrow \omega t_r + \phi = \frac{\pi}{2} + n\pi; \quad n = 0, 1, \ldots,$$

so that

$$t_r = \frac{1}{\omega}\left(\frac{\pi}{2} - \phi\right) = \frac{\pi/2 - \phi}{\omega_0\sqrt{1 - \zeta^2}} \tag{7.16b}$$

Figure 7.7 continued

**Transient Response
of an Under-damped
Second-order System Subject
to a Unit Step Input**

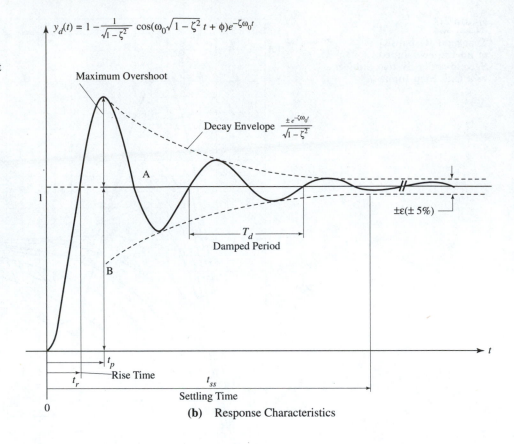

$$y_d(t) = 1 - \frac{1}{\sqrt{1-\zeta^2}} \cos(\omega_0\sqrt{1-\zeta^2}\, t + \phi)e^{-\zeta\omega_0 t}$$

(b)   Response Characteristics

Thus the *rise time* is a function of the damped frequency ω or a function of both ω_0 and ζ.

2. **Overshoot (A)** is the amount by which the maximum value of the response exceeds the final value. It is normally expressed as a fraction or percentage of the final value. Thus

$$\% \text{ Overshoot} = 100\left(\frac{A}{B}\right)$$

The maximum value of the response occurs when

$$\frac{dy_d}{dt} = 0 = \left[\frac{\omega}{\cos\phi}\sin(\omega t + \phi) + \frac{\sigma}{\cos\phi}\cos(\omega t + \phi)\right]e^{-\sigma t} \quad \text{or}$$

$$\omega\sin(\omega t + \phi) + \sigma\cos(\omega t + \phi) = 0$$

Expanding the sine and cosine terms, and noting that $\tan\phi = -\sigma/\omega$ gives us

Time at the peak response:

$$t_p = \frac{\pi}{\omega} = \frac{\pi}{\omega_0\sqrt{1-\zeta^2}} \tag{7.16c}$$

We obtain the peak value of the response by substituting t_p into equation (7.16a):

$$Y_{t_p} = 1 - \frac{1}{\cos \phi} \cos(\pi + \phi) e^{-\sigma \pi / \omega} = 1 + e^{-\sigma \pi / \omega}; \quad \text{(peak value of response)} \quad \textbf{(7.16d)}$$

so that the *overshoot* (A) and % overshoot (100(A/B); note that the final value is B = 1) are

$$\text{Overshoot} = e^{-\sigma \pi / \omega} = \exp\left(\frac{-\pi \zeta}{\sqrt{1 - \zeta^2}} \right) \qquad \textbf{(7.16e)}$$

$$\%\text{Overshoot} = 100 \exp\left(\frac{-\pi \zeta}{\sqrt{1 - \zeta^2}} \right) \qquad \textbf{(7.16f)}$$

The overshoot is a function of the damping ratio only. The smaller the damping ratio, the higher the overshoot (see Fig. 7.7a).

3. **Period of oscillation** is the *damped period*, which was given in equation (7.12c).
4. **Settling time** is the time it takes the response to settle within some prescribed range of values ($\pm \epsilon$ of the final value). This is also the time at which the magnitude of the *decay envelope* becomes smaller than the specified value. Hence for 5% settling time, for example,

$$\frac{1}{\sqrt{1 - \zeta^2}} e^{-\zeta \omega_0 t_{ss}} = 0.05; \; or \; t_{ss} = \left(\frac{-1}{\zeta \omega_0} \right) \ln(0.05 \sqrt{1 - \zeta^2}) \qquad \textbf{(7.16g)}$$

The settling time is thus a function of the two characteristic parameters, ω_0 and ζ, of the system.

The preceding step-input transient response characteristics are, as we shall see, often used in the specification of design performance for controlled systems. It should be evident from these results that given the transient step response of a presumed second-order system, the characteristic parameters ζ and ω_0 can be determined from measurements on the response data, and hence the system model, equation (7.2), can be determined (see Practice Problem 7.13).

Example 7.6	Unit Impulse Response

Determine the response of the system of equation (7.2) subject to zero initial conditions and a unit impulse as the forcing input.

Solution: Recall that the unit impulse can be defined as the derivative of the unit step, although this derivative does not exist in the strict mathematical sense; thus the response to a unit impulse for zero initial conditions can be obtained by differentiating the unit step response, equations (7.14a) through (7.14c). That is,

For $\zeta > 1$:

$$y_I(t) = \frac{1}{\omega_0^2} \left(\frac{\lambda_1 \lambda_2}{\lambda_2 - \lambda_1} \right) (e^{\lambda_2 t} - e^{\lambda_1 t}) = \frac{e^{\lambda_2 t} - e^{\lambda_1 t}}{\lambda_2 - \lambda_1} \qquad \textbf{(7.17a)}$$

where in equation (7.14a) for the unit step response,

$$C_1 = \frac{-\lambda_2}{\omega_0^2(\lambda_2 - \lambda_1)}; \qquad C_2 = \frac{\lambda_1}{\omega_0^2(\lambda_2 - \lambda_1)}; \qquad \lambda_1\lambda_2 = \omega_0^2 \tag{7.17b}$$

For $\zeta = 1$:

$$y_I(t) = \frac{\lambda^2}{\omega_0^2} te^{\lambda t} = te^{-\omega_0 t} \tag{7.17c}$$

where in equation (7.14b) for the unit step response,

$$C_1 = \frac{-1}{\omega_0^2}; \qquad C_2 = \frac{\lambda}{\omega_0^2}; \qquad \lambda = -\omega_0 \tag{7.17d}$$

For $0 \leq \zeta < 1$:

$$y_I(t) = [-\sigma C \cos(\omega t + \phi) - \omega C \sin(\omega t + \phi)]e^{-\sigma t}$$

Using trigonometric expansions for $\cos(\omega t + \phi)$ and $\sin(\omega t + \phi)$, and substituting for C, $\tan \phi$ and $\cos \phi$ from equations (7.15b), (7.15c), and (7.15d), respectively, we obtain

$$y_I(t) = \left(\frac{\sigma^2 + \omega^2}{\omega \omega_0^2}\right) \sin(\omega t)e^{-\sigma t} = \frac{1}{\omega}\sin(\omega t)e^{-\sigma t} \tag{7.17e}$$

Convolution Integral

According to the results developed for the first-order system and as will be shown in Section 8.2 for linear systems in general, the forced response to any forcing input $f(t)$ is given in terms of the (zero initial state) *unit impulse response* by the *convolution integral:*

$$y_f(t) = \int_0^t y_I(t - \tau)f(\tau)\, d\tau \tag{7.18}$$

Thus although practically it is not easy to generate an impulse input, analytically the impulse response if known is very significant in the sense that it represents the "signature" of the system because the input/output behavior (the forced response) can be obtained from it.

Example 7.7   Unit Ramp Response of a Critically Damped System

Determine the forced response of the system of Example 7.3 to a unit ramp.

Solution by Method of Undetermined Coefficients: The system is critically damped, so that the homogeneous solution is

$$y_H(t) = (C_1 + C_2 t)e^{\lambda t}$$

By the method of undetermined coefficients, for a nonhomogeneous term $f(t) = t$, the trial solution (see Table 7.1) is $a_1 t + a_0$, and the particular solution is

$$y_p(t) = \frac{1}{\omega_0^2}\left(t - \frac{2\zeta}{\omega_0}\right)$$

Thus the general solution is

$$y(t) = (C_1 + C_2 t)e^{\lambda t} + \frac{1}{\omega_0^2}\left(t - \frac{2\zeta}{\omega_0}\right); \qquad \lambda = -\omega_0$$

Applying zero initial conditions, we obtain

$$C_1 = \frac{2\zeta}{\omega_0^3} = \frac{2}{\omega_0^3}; \qquad C_2 = \frac{-1}{\omega_0^2}\left(\frac{2\lambda\zeta}{\omega_0} + 1\right) = \frac{1}{\omega_0^2}$$

Hence

$$y_f(t) = \frac{1}{\omega_0^2}\left[\left(\frac{2}{\omega_0} + t\right)e^{-\omega_0 t} + t - \frac{2}{\omega_0}\right]$$

Solution from Unit Step Response: The unit ramp forced response can be obtained by integrating the unit step forced response. From equation (7.14b),

$$y_s(t) = (C_1 + C_2 t)e^{-\omega_0 t} + \frac{1}{\omega_0^2}$$

Zero initial state gives

$$y_s(t) = \frac{1}{\omega_0^2}[1 - (1 + \omega_0 t)e^{-\omega_0 t}]$$

Integrating from zero to t, we have

$$y_f(t) = \frac{1}{\omega_0^2}\int_0^t [1 - (1 + \omega_0\tau)e^{-\omega_0\tau}]\,d\tau$$

$$= \frac{1}{\omega_0^2}\left[t + \frac{e^{-\omega_0 t} - 1}{\omega_0} - \frac{e^{-\omega_0 t}(-\omega_0 t - 1) + 1}{\omega_0}\right]$$

which gives

$$y_f(t) = \frac{1}{\omega_0^2}\left[\left(\frac{2}{\omega_0} + t\right)e^{-\omega_0 t} + t - \frac{2}{\omega_0}\right]$$

as before.

Solution by Convolution Integral: The forced response to any input can always be found by convolution with the unit impulse response. From equation (7.17c),

$$y_f(t - \tau) = (t - \tau)e^{-\omega_0(t - \tau)}$$

and from equation (7.18),

$$y_f(t) = \int_0^t (t - \tau)e^{-\omega_0(t-\tau)}\tau \, d\tau = te^{-\omega_0 t}\int_0^t \tau e^{\omega_0 \tau} \, d\tau - e^{-\omega_0 t}\int_0^t \tau^2 e^{\omega_0 \tau} \, d\tau$$

$$= te^{-\omega_0 t}\left[\frac{e^{\omega_0 t}(\omega_0 t - 1) + 1}{\omega_0^2} - \frac{e^{-\omega_0 t}t^2 e^{\omega_0 t}}{\omega_0} + \frac{2e^{-\omega_0 t}}{\omega_0}\int_0^t \tau e^{\omega_0 \tau} \, d\tau\right]$$

$$= t\left[\frac{(\omega_0 t - 1) + e^{-\omega_0 t}}{\omega_0^2}\right] - \frac{t^2}{\omega_0} + \frac{2e^{-\omega_0 t}}{\omega_0}\left[\frac{e^{\omega_0 t}(\omega_0 t - 1) + 1}{\omega_0^2}\right]$$

$$+ \frac{te^{-\omega_0 t} - t}{\omega_0^2} + \frac{2}{\omega_0}\left[\frac{(\omega_0 t - 1) + e^{-\omega_0 t}}{\omega_0^2}\right]$$

which gives

$$y_f(t) = \frac{1}{\omega_0^2}\left[\left(\frac{2}{\omega_0} + t\right)e^{-\omega_0 t} + t - \frac{2}{\omega_0}\right]$$

as before.

··

7.2 PHASOR TRANSFORM SOLUTION AND SINUSOIDAL STEADY STATE

In this section the *phasor transform* method is applied to the solution of second-order *single-degree-of-freedom systems*. Single degree of freedom implies that the configuration of the system can be specified by a single (independent) coordinate. For mechanical systems in rectilinear motion, the configuration is normally determined by the position of the system's masses. The configuration of a multiple-degree-of-freedom system is described by a number of independent coordinates that, according to Section 3.1, can be represented by a set of state variables. It follows that the general approach of Chapter 6 applies to such systems. However, the mechanical vibration of a two-degrees-of-freedom-system is considered as a special case in Section 7.3. The reader interested in general analyses of multiple-degree-of-freedom mechanical systems is referred to texts on mechanical vibrations such as reference [6]. For electrical systems, reference [7] or a similar text may be consulted.

The Phasor Transform   As we discussed in Section 6.2, sinusoids constitute an important class of special input functions. In alternating current (AC) systems, for instance, essentially all forcing inputs are sinusoidal. Consequently, a technique that is especially compatible with such input signals is often used in the solution of these dynamic systems. Some of the properties of sinusoids and complex numbers that form the basis of the phasor transform technique were introduced in Section 6.2. To these we now add the phasor representation of the derivative of a sinusoid.

Consider the sinusoid represented by the phasor $z = Ae^{j\phi}$, that is, $f(t) = A\cos(\omega t + \phi)$:

$$\frac{d}{dt}f(t) = -\omega A \sin(\omega t + \phi) \tag{17.19a}$$

But $\sin \theta = \cos\left(\theta - \dfrac{\pi}{2}\right)$, hence

$$\frac{d}{dt} f(t) = -\omega A \cos\left[\omega t + \left(\phi - \frac{\pi}{2}\right)\right] \tag{17.19b}$$

By equation (6.45a), the phasor representing the derivative (with respect to time) of $f(t)$ is

$$z_d = -\omega A e^{j(\phi - \pi/2)} = -\omega e^{-j\,\pi/2} A e^{j\phi} = j\omega A e^{j\phi} \tag{7.19c}$$

or

$$z_d = j\omega z \tag{7.19d}$$

Similarly, for the second derivative, differentiation of equation (7.19a) gives

$$\frac{d^2}{dt^2} f(t) = -A\omega^2 \cos(\omega t + \phi) \tag{7.20a}$$

which has the phasor representation

$$z_d = -\omega^2 A e^{j\phi} = (j\omega)^2 A e^{j\phi} = (j\omega)^2 z = -\omega^2 z \tag{7.20b}$$

In general, it can be shown that the phasor representation for the nth derivative of the sinusoid is

$$z_d = (j\omega)^n z \tag{7.21}$$

Introducing the symbol $\mathscr{P}$ for the phasor transform (representation), such that

$$\mathscr{P}[f_1(t)] = z_1 \tag{7.22a}$$

and

$$\mathscr{P}\left[\frac{d^n}{dt^n} f_1(t)\right] = \mathscr{P}[f_2(t)] = (j\omega)^n z_1 = z_2 \tag{7.22b}$$

and so on, we see from equation (6.45b) that the inverse phasor transform ($\mathscr{P}^{-1}$) is given by

$$\mathscr{P}^{-1}(z) = \text{Real}[e^{j\omega t} z] \tag{7.22c}$$

Equations (7.22a) through (7.22c) can be applied to the determination of the forced response of stationary and stable linear systems subject to sinusoidal input. Stability of the system is essential, since the idea of a steady-state response is meaningful only for stable systems.

Consider, for example, the analysis of *radio frequency* (RF) circuits, which are used in communication systems. Two such circuits are shown in Fig. 7.8. The resonant circuit is used in combination with other circuits in a radio, for instance, to "tune" to a given broadcast station. The signal is a sinusoid whose frequency is the same as the broadcasting (carrier) frequency of the radio station. The station is tuned when the output of the resonant circuit [$v_0(t)$ or $i_0(t)$] due to this signal is sufficiently large relative to the output caused by other input signals corresponding to noise or other radio stations broadcasting at other frequencies.

Both resonant circuits in Fig. 7.8 are second-order single-degree-of-freedom systems (recall the model for the series circuit shown in Fig. 7.8b was determined in Ex-

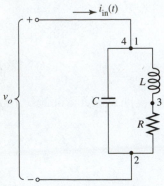

(a) Parallel Resonant Circuit

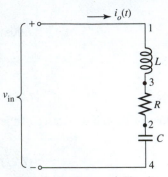

(b) Series Resonant Circuit

Figure 7.8

Examples of Radio Frequency Circuits

ample 2.6, equation (2.37a)). Consider now the parallel circuit of Fig. 7.8a. Because the input is a sinusoid and the system is linear, we can analyze the circuit directly in terms of phasors. Recall from Section 2.2 the relations for resistors, capacitors and inductors. Capital letters are used in the following for the phasor transform, while lowercase letters represent the sinusoids.

Resistor:

$$i_{3-2} = \frac{v_3 - v_2}{R} = i_{1-3}; \qquad I_{3-2} = \frac{V_3 - V_2}{R} \tag{7.23a}$$

Capacitor:

$$C \frac{d}{dt}(v_1 - v_2) = i_{4-2}; \qquad j\omega C(V_1 - V_2) = I_{4-2} \tag{7.23b}$$

Inductor:

$$L \frac{d}{dt} i_{1-3} = v_1 - v_3; \qquad j\omega L I_{1-3} = V_1 - V_3 \tag{7.23c}$$

We are interested in the voltage drop $v_0 = v_1 - v_2$:

$$j\omega C V_0 = I_{4-2} = I_{\text{in}} - I_{1-3} = I_{\text{in}} - \frac{V_3 - V_2}{R}$$

But

$$V_3 - V_2 = (V_1 - V_2) - (V_1 - V_3) = V_0 - j\omega L I_{1-3}$$

Thus

$$I_{\text{in}} - I_{1-3} = I_{\text{in}} - \frac{V_0}{R} + \frac{j\omega L}{R} I_{1-3} \quad \text{and} \quad I_{1-3} = \left(\frac{1}{R + j\omega L}\right) V_0$$

$$\text{and} \quad j\omega C V_0 = I_{\text{in}} - \left(\frac{1}{R + j\omega L}\right) V_0$$

Hence

$$V_0 = \left[\frac{R + j\omega L}{(1 - \omega^2 LC) + j\omega RC}\right] I_{\text{in}} \tag{7.24a}$$

The voltage is then given by

$$v_0(t) = \text{Real}\left\{\left[\frac{e^{j\omega t}(R + j\omega L)}{(1 - \omega^2 LC) + j\omega RC}\right] I_{\text{in}}\right\} = A_0 \cos(\omega t + \phi_0) \tag{7.24b}$$

Thus $v_0(t)$ represents the particular solution to the circuit equation for a sinusoidal forcing input

$$i_{\text{in}}(t) = A_i \cos(\omega t + \phi_i)$$

Before looking at other examples or applications, let us illustrate several concepts using these developments.

Phasor Transfer Functions: Impedance/Admittance

A transfer function is, in general, a relation between input and output (under zero initial conditions). The phasor transfer function $G(j\omega)$ therefore expresses the relation between the output phasor and the input phasor:

$$\text{Output phasor} = G(j\omega) \cdot \text{input phasor} \tag{7.25}$$

Thus for the parallel resonant circuit, the phasor transfer function may be taken as

$$G(j\omega) = \frac{V_0(j\omega)}{I_{in}(j\omega)} = \frac{R + j\omega L}{(1 - \omega^2 LC) + j\omega RC} \tag{7.26}$$

Recall that energetic systems can, in general, be described in terms of two variables, effort e and flow f, whose product gives the instantaneous power.

Impedance   The transfer function relating the *output effort phasor* to the *input flow phasor* for a system or component is termed the *impedance*. Hence $G(j\omega)$ in equation (7.26) is the impedance for the entire circuit. Similarly, the impedance for a resistor (equation (7.23a) is $Z_R(j\omega) = R$, and the impedance for an inductor (equation (7.23c) is $Z_L(j\omega) = j\omega L$. Note that the impedance concept as defined also applies to nonelectrical systems.

Admittance   The transfer function relating the *output flow phasor* to the *input effort phasor* is called the *admittance*. The admittance for a resistor is $Y_R(j\omega) = 1/R$ and that of a capacitor is $Y_C(j\omega) = j\omega C$. Note that the admittance or impedance is not necessarily defined for every system or component. For instance, an overall admittance cannot be defined for the parallel resonant circuit if we insist that the input is always a current (flow). Similarly, a voltage source by itself does not have either an admittance or an impedance, since there is no functional relationship between the voltage produced by the source and the current through the source. The voltage drop across an ideal voltage source is fixed and is independent of the current flowing through the source.

Driving-Point and Transfer Impedance/Admittance   When the input and output ends (terminals) of a system are distinguishable (see Fig. 7.9), the impedance is termed *driving point impedance* if the output effort is measured at the input terminal (same end or branch where the input flow is applied). Similarly, the concept of *driving point admittance* applies when the output flow is measured at the same input terminal or branch where the input effort is supplied. If the output effort or flow is recognized at the output end then *transfer impedance* and *transfer admittance,* respectively, are the applicable terms.

Figure 7.9

Driving-point and Transfer Impedance/Admittance

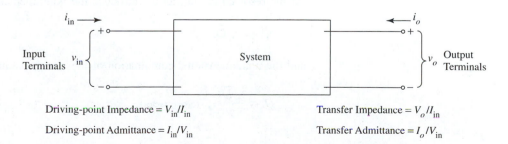

Driving-point Impedance = V_{in}/I_{in} Transfer Impedance = V_o/I_{in}

Driving-point Admittance = I_{in}/V_{in} Transfer Admittance = I_o/V_{in}

Magnitude Ratio and Phase Difference

By definition, the phasor transfer function determines completely the forced behavior of a linear system when subject to sinusoidal input. Because this transfer function is a complex number, the behavior is also completely characterized by specifying the magnitude and angle of the transfer function. The relations in such specifications may generally be nondimensionalized. The ratio of the output amplitude (magnitude) to the input amplitude (magnitude) is equal to the magnitude of the transfer function. Similarly, the difference between the output phase (angle) and the input phase (angle) gives the angle of the phasor transfer function. All these follow from the definition, equation (7.25), that is,

$$G(j\omega) = Me^{j\phi} = \frac{A_0 e^{j\phi_0}}{A_i e^{j\phi_i}} \tag{7.27a}$$

$$M = \frac{A_0}{A_i} = |G(j\omega)| \tag{7.27b}$$

and

$$\phi = \phi_0 - \phi_i = \angle G(j\omega) \tag{7.27c}$$

Consider the parallel resonant circuit example. We can nondimensionalize the result using (as it sometimes done) the quantity $\sqrt{C/L}$. Equation (7.24a) becomes

$$\sqrt{\frac{C}{LV_0}} = \left[\frac{R\sqrt{C/L} + j\omega\sqrt{CL}}{(1 - \omega^2 LC) + j\omega RC} \right] I_{\text{in}} = G(j\omega)I_{\text{in}} \tag{7.28a}$$

From the methods of Section 6.2 and equation (7.27b), the ratio of the output amplitude $\sqrt{C/L}|V_0|$ to the input amplitude $|I_{\text{in}}|$ is

$$\sqrt{\frac{C}{L}} \frac{|V_0|}{|I_{\text{in}}|} = \frac{A_0}{A_i} = M = \sqrt{\frac{R^2 C/L + \omega^2 LC}{(1 - \omega^2 LC)^2 + (\omega RC)^2}} \tag{7.28b}$$

Also, the difference in phase between the output $\sqrt{C/L}v_0(t)$ and the input $i_{\text{in}}(t)$ is

$$\phi_0 - \phi_i = \phi = \tan^{-1}\left(\frac{\omega L}{R} \right) - \tan^{-1}\left(\frac{\omega RC}{1 - \omega^2 LC} \right) \tag{7.28c}$$

The output is said to *lead* the input if the phase shift is positive. Otherwise, the output *lags* the input.

Resonance, Quality Factor, and Bandwidth

We note that both the *magnitude ratio* and the *phase shift* are functions of the input frequency. To examine this frequency dependence (*frequency response; see Section 9.2*) for the resonant circuit, let us introduce the nondimensional frequency

$$\eta = \omega\sqrt{LC} \tag{7.29a}$$

and the nondimensional combination of the circuit parameters

$$Q = \frac{\sqrt{L/C}}{R}; \quad \text{(quality factor)} \tag{7.29b}$$

Q is known as the *quality factor,* and its (actually its inverse's), significance for electrical systems is similar to that of the damping ratio for mechanical systems. In terms of the above two quantities, the magnitude ratio, equation (7.28b), becomes

$$M = \sqrt{\frac{(1/Q)^2 + \eta^2}{(1 - \eta^2)^2 + (\eta/Q)^2}} \tag{7.29c}$$

A plot of M as a function of η with Q as a parameter is shown in Fig. 7.10.

Resonance   An undamped system will vibrate indefinitely at the natural frequency with no further energy input once disturbed. (Recall the discussion of undamped second-order behavior.) It follows that continued disturbance of the system at a frequency equal to its natural frequency will cause energy to accumulate in the system. The resulting large output in the presence of a small input is known as *resonance.* If damping is present, the energy accumulation is reduced by energy dissipation. However, resonance still occurs when the frequency of the disturbance is the same as the natural frequency except that the large output is diminished according to the magnitude of damping present. In Fig. 7.10, small values of Q are associated with large damping (high values of R).

Resonant Frequency   The *resonant frequency* can thus be defined as the **input** frequency at which the magnitude ratio becomes arbitrarily large under **zero damping**

Figure 7.10

Resonant Circuit Frequency Behavior and Q Factor

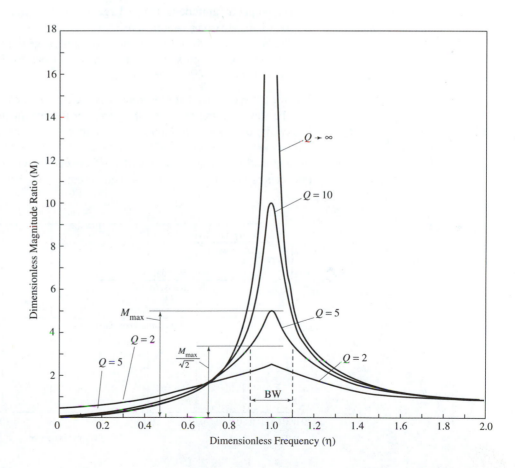

conditions. For the resonant circuit example, the no-damping ($R = 0$; $Q \longrightarrow \infty$) magnitude ratio is given by equation (7.28b):

$$M_0 = \frac{\omega\sqrt{LC}}{\sqrt{(1 - \omega^2 LC)^2}} = \frac{\eta}{\sqrt{(1 - \eta^2)^2}} \qquad \text{(7.30a)}$$

This magnitude ratio is infinite when $\omega^2 LC = 1$. This gives the resonant (circular) frequency

$$\omega_R = \frac{1}{\sqrt{LC}}; \qquad (\eta = 1) \qquad \text{(7.30b)}$$

Although it is not apparent in Fig. 7.10, when damping is present, the peak output amplitude may occur at frequencies other than at $\eta = 1$. For instance, for the present parallel resonant circuit, for $Q = 2$, M is a maximum at about $\eta = 0.987$. However, the *peak output frequency* is still very close to the resonant frequency, so that the magnitude–frequency relation is often used to obtain a good estimate of the natural frequency of the system, especially when the amount of damping is small (Q is large).

Quality Factor and Bandwidth

In applications such as radio tuning circuits, it is desirable that Q be very large for two reasons:

1. The output magnitude is then as large as possible (that is, the received station signal level will be strong—more sensitivity).
2. The larger the Q, the narrower the frequency range over which the output magnitude is large (noise from stations broadcasting at nearby frequencies will be rejected, giving better resolution or sharpness).

It is for these reasons that Q is known as the *quality factor.* The second item listed can also be stated as, the larger Q, the narrower the *bandwidth* of the tuning circuit. The bandwidth of a system is that range of frequencies (of the input) over which the system will respond satisfactorily. What is satisfactory varies with the type of system and with the application, and sometimes with the frequency range—low frequencies or high frequencies or midrange. Thus in hi-fi equipment, a wide bandwidth covering much of the

Figure 7.11

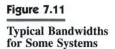

Typical Bandwidths for Some Systems

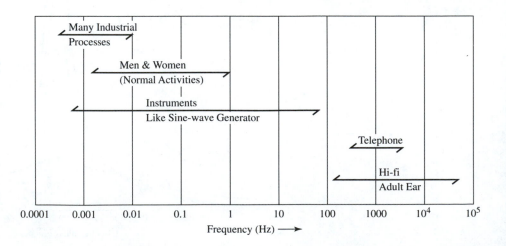

audio frequency spectrum (see Fig. 7.11) is desirable for accurate reproduction of all the details of the music (whether it is a piece of jazz or a symphony). On the other hand, we desire a narrow bandwidth for a tuning circuit. In energetic systems, the bandwidth (BW) is generally defined quantitatively by

$$\text{BW} = \text{frequency range for which } M > \frac{1}{\sqrt{2}} M_{\text{max}} \tag{7.30c}$$

For the parallel resonant circuit example, it can be shown that

$$\text{BW} \approx \frac{\omega_R}{Q} \tag{7.30d}$$

which agrees with the preceding discussion. As another illustration of divergent needs in different areas, larger damping is preferred over very low resistance in the automobile suspension system, to facilitate riding comfort. Typical bandwidths for various systems are illustrated in Fig. 7.11.

Example 7.8 — Tuned Resonant Circuit

Consider the series resonant circuit of Fig. 7.8b. If the input to the circuit is the sinusoidal voltage $v_i^n(t) = A_i \cos(\omega t)$, what is

a. The overall admittance of the circuit?
b. The phasor transfer function for the circuit?
c. If the frequency range of the stations to be tuned is 500–1500 kHz, and the desired quality factor for the circuit must be greater than 10 for this range of frequencies, determine good values for R and L that will allow the circuit to tune these stations effectively. Assume that the tuning is accomplished by a variable condenser (capacitor) with a range of 1–10 (10^{-9} F)
d. If the circuit from (c) is used to tune a station operating at 1000 kHz, obtain (i) the value of C under these conditions, (ii) the resonant frequency of the circuit for the value of C from (i), (iii) the amplitude and phase of the current output if $A_i = 0.5$ V. (iv) What is the bandwidth of the circuit under these conditions?

Solution a: The second-order differential equation relating i_0 to v_{in} was derived in Example 2.6, equation (2.37a). We can take the phasor transform of this equation and solve for the admittance, $Y(j\omega) = I_0/V_{\text{in}}$. This approach is followed in Section 7.3. Alternatively, we can consider the circuit subunits. The admittance of the inductor, resistor, and capacitor are $1/j\omega L$, $1/R$, and $j\omega C$, respectively. These are admittances in series so that the overall admittance is

$$\frac{1}{Y(j\omega)} = j\omega L + R + \frac{1}{j\omega C}$$

or

$$Y(j\omega) = \frac{j\omega C}{(1 - \omega^2 LC) + j\omega RC} \tag{7.31a}$$

Solution b: Because the overall input to the circuit is a voltage, then equation (7.31a) also represents the phasor transfer function for the circuit. That is, using $\sqrt{L/C}$ to non-dimensionalize the function, we obtain

$$G(j\omega) = \frac{I_0 \sqrt{LC}}{V_{in}} = \frac{j\omega\sqrt{LC}}{(1 - \omega^2 LC) + j\omega RC} \tag{7.31b}$$

The magnitude of the transfer function is

$$M = \frac{\omega\sqrt{LC}}{\sqrt{(1 - \omega^2 LC)^2 + (\omega RC)^2}} \tag{7.31c}$$

This is the same result obtained in Example 2.6, equation (2.37c). The phase angle is

$$\phi = \frac{\pi}{2} - \tan^{-1}\left(\frac{\omega RC}{1 - \omega^2 LC}\right) \tag{7.31d}$$

From equation (7.31b), setting $R = 0$ and considering the frequency for which M becomes unbounded, gives us the resonant frequency,

$$\omega_R = \frac{1}{\sqrt{LC}} \tag{7.31e}$$

Solution c: In terms of the quality factor $Q = \sqrt{L/C}/R$ and the normalized frequency $\eta = \omega\sqrt{LC}$,

$$M = \frac{\eta}{\sqrt{(1 - \eta^2)^2 + (\eta/Q)^2}} \tag{7.31f}$$

M is plotted versus η in Fig. 7.12 (see also equation (2.37d) and Fig. 2.20) for $Q = 10$ and 20.584 (a value determined below). From this figure (it can also be shown analytically that), the maximum value of M for the series resonant circuit occurs at $\eta = 1$ (that is, at the natural frequency, independent of the value of Q). Consider the radio operation at the extremes of its frequency range. From equation (7.31e), we can require

$$2\pi(5 \times 10^5) = \frac{1}{\sqrt{LC_{max}}}, \qquad C_{max} = 10^{-8}\text{F} \Rightarrow L = 0.01013 \text{ mH}$$

$$2\pi(15 \times 10^5) = \frac{1}{\sqrt{LC_{min}}}, \qquad C_{min} = 10^{-9}\text{F} \Rightarrow L = 0.01126 \text{ mH}$$

Thus a choice of $L = 0.0107 \times 10^{-3}$ H would be adequate. Similarly, since the minimum Q is determined by the maximum C, for $Q = 10$, equation (7.29b) gives

$$R = 0.1\sqrt{\frac{L}{C_{max}}} = 3.27 \ \Omega$$

Figure 7.12

Series Resonant Circuit Design Example

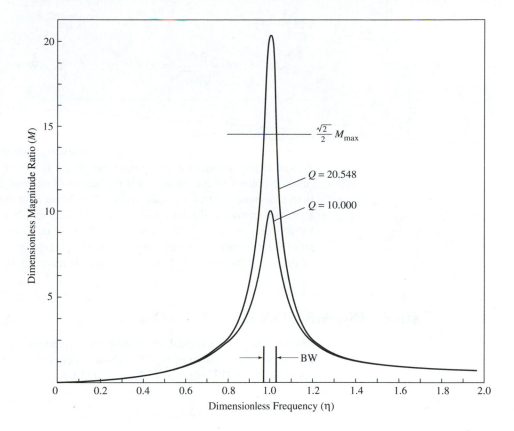

Solution d: It has been established that the response amplitude for this circuit is a maximum at the natural frequency. Assuming precise tuning, if the input frequency is also 1000 Hz, then we have

$$2\pi(10^6) = \frac{1}{\sqrt{LC}}$$

which gives $C = 2.37(10^-F)$.

The resonant frequency is also 1.0 MHz (that is $\eta = 1$). From (7.31f), $M = Q$, so that the magnitude of the current output, $y_0\cos(\omega t + \phi)$, is

$$y_0 = \sqrt{\frac{C}{L}}QA_i = \frac{1}{R}A_i = 0.153 \text{ A}$$

From (7.31d), $\phi = 0$, so that the output phase is equal to the input voltage phase, which is zero. The quality factor with the above value of C is

$$Q = \frac{1}{R}\sqrt{\frac{L}{C}} = 20.548$$

The plot of M versus η for this value of Q is also shown in Fig. 7.12. From this figure, the bandwidth is given by

$$\Delta\eta = 1.029 - 0.976 = 0.053$$

which gives

$$\Delta\omega = \text{BW} = \frac{0.053}{\sqrt{LC}} = 333 \times 10^3 \text{ s}^{-1}, \quad \text{or BW} = 53 \text{ kHz}$$

Note that part of the requirements in the preceding example (determination of values for R and L) can be considered a design problem. Indeed, the entire example can be viewed as the design of a series resonant circuit for a radio. In general, the model solution problems of this chapter and the next will normally involve some design (other than automatic control) considerations. This is because the design and model solution processes cannot really be clearly separated. The reader is urged to be on the lookout for this design aspect in the various problems in order to derive the maximum benefit from the examples. The blending of solution and design will continue even in Part III of the text, where the predominant subject is model design for automatic control.

7.3 INTRODUCING MECHANICAL VIBRATIONS

The vibrations of a typical single-degree-of-freedom mechanical system (such as the auto suspension system of Fig. 7.1) about its equilibrium level is also described by the second-order ODE equation (7.2), repeated here:

$$\frac{d^2y}{dt^2} + 2\zeta\omega_0\frac{dy}{dt} + \omega_0^2 y = f(t) \tag{7.32a}$$

If $f(t)$ is a sinusoid, that is,

$$f(t) = A_i\cos(\omega t + \phi_i) \tag{7.32b}$$

then the forced response should also be a sinusoid of the same frequency:

$$y(t) = A_0\cos(\omega t + \phi_0) \tag{7.32c}$$

Thus our task is to determine A_0 and ϕ_0. To apply phasor transforms to equations (7.32), we let

$$\mathscr{P}(f) = F = A_i e^{j\phi_i} \tag{7.32d}$$

$$\mathscr{P}(y) = Y = A_0 e^{j\phi_0} \tag{7.32e}$$

Then (7.32a) transforms to

$$-\omega^2 Y + (2\zeta\omega_0)j\omega Y + \omega_0^2 Y = F$$

Introducing the nondimensional frequency, or frequency ratio, we have

$$\eta = \frac{\omega}{\omega_0} \tag{7.33a}$$

The phasor transfer function for (7.32a) can be written as

$$G(j\omega) = \left(\frac{Y}{F}\right) = \left[\frac{1/\omega_0^2}{\sqrt{(1 - \eta^2) + j^2\zeta\eta}}\right] \tag{7.33b}$$

The magnitude ratio is

$$M = |G(j\omega)| = \frac{1/\omega_0^2}{\sqrt{(1 - \eta^2)^2 + (2\zeta\eta)^2}} \tag{7.33c}$$

The phase difference is

$$\phi = \angle G(j\omega) = -\tan^{-1}\left(\frac{2\zeta\eta}{1 - \eta^2}\right) \tag{7.33d}$$

The output amplitude is thus

$$A_0 = \frac{A_i}{\omega_0^2}\left[\frac{1}{\sqrt{(1 - \eta^2)^2 + (2\zeta\eta)^2}}\right] \tag{7.33e}$$

The quantity in brackets is often called the *magnification factor* (MF), since for a typical mechanical system, the forcing input $f(t)$ in equation (7.32a) represents acceleration (force divided by mass) when y is displacement. If this input is static, $f(t) = A_i$, then the forced response $Y_0 = A_i/\omega_0^2$ is the *static deflection,* so that MF is the ratio of the "dynamic" and static deflections:

$$\text{MF} = \frac{|Y|}{|Y_0|} = \frac{1}{\sqrt{(1 - \eta^2)^2 + (2\zeta\eta)^2}} = \omega_0^2 M \tag{7.33f}$$

where Y is the displacement from the equilibrium position due to a sinusoidal force input of magnitude F_i, while Y_0, the static deflection, is the displacement from the equilibrium position due to a static force input, also of magnitude F_i. The magnification factor gives the effect of frequency and damping on the forced (sinusoidal steady-state) response. This can be seen in Fig. 7.13, which displays equation (7.33f) as a the plot of MF versus η, with ζ as parameter.

Finally, the solution (that is, the forced response) is given by equation (7.32c), where

$$A_0 = MA_i \quad \text{and} \quad \phi_0 = \phi_i + \phi \tag{7.33g}$$

Note from equation (7.33c) that the magnitude ratio under zero damping ($\zeta = 0$) becomes infinite when $\eta = 1$, that is,

$$\omega = \omega_R = \omega_0 \tag{7.34}$$

Thus the resonant frequency is the same as the natural frequency of the system.

Single-Degree-of-Freedom Examples

The free vibrations of single-degree-of-freedom systems were considered in detail in Section 7.1.

Equations (7.33a) through (7.33g) are general results for the *forced vibrations* of second-order *single-degree-of-freedom* systems when the forcing input can be represented by equation (7.32b). As we discussed in Section 6.2, the forced vibration also implies the **steady-state response** (*sinusoidal steady state*) after any free motion must have become insignificant. We consider next some design-type examples. Additional design applications are given in the Practice Problems.

Figure 7.13

**Effect of Frequency
and Damping on the
Magnification Factor**

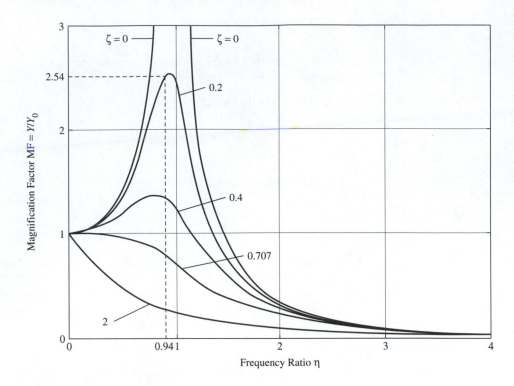

Example 7.9   Accelerometer Design

Acceleration in mechanical systems is often measured (1) in its own right as a useful quantity, (2) as a close relative to force (it is usually proportional to force), and (3) as a means of obtaining velocity (which is given by (realizable) integration of acceleration). Consider the general purpose accelerometer that consists of an arrangement of mass, damper, and spring, as shown in Fig. 7.14a. The relative displacement y between the mass and the frame can be measured by a transducer (such as an unbonded strain gage, linear voltage differential transformer (LVDT), or capacitance gage, which is not shown. The indication y is a measure of the acceleration of the frame, which is assumed to be the same as that of the body to which the frame is attached.

a. If the accelerometer is used to measure steady acceleration up to a maximum of 10 m/s² and the measurement (y scale) is limited to a physical length of 2 cm, determine the value of the mass m and the viscous damping constant β such that the instrument will function properly. Assume that the spring constant is fixed at 250 N/m.
b. Discuss the corresponding design considerations if the accelerometer is to be used to measure dynamic acceleration, such as the acceleration of a vibrating piece of mechanical equipment.

Solution: From the free body diagrams shown in Fig. 7.14b, we obtain the following relation:

$$m \frac{d}{dt} v_m = k(x - x_m) - \beta(v_m - v)$$

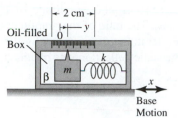

Oil-filled Box

(a)   Schematic Diagram

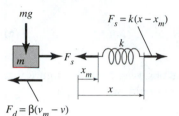

$F_d = \beta(v_m - v)$

(b)   Free Body Diagrams

Figure 7.14

A General-purpose Accelerometer

Since

$$y = x_m - x, \quad \dot{y} = v_m - v, \quad \frac{d}{dt} v_m = \ddot{y} + \ddot{x}$$

and

$$\frac{d^2 y}{dt^2} + \frac{\beta}{m}\frac{dy}{dt} + \frac{k}{m} y = -\ddot{x} = a_f \tag{a}$$

where a_f is taken as the frame acceleration. Thus as before (see equation (7.2)), the natural frequency and damping ratio are given by

$$\omega_0 = \sqrt{\frac{k}{m}} \quad \text{and} \quad 2\zeta\omega_0 = \frac{\beta}{m} \longrightarrow \zeta = \frac{\beta}{2\sqrt{mk}} \tag{b}$$

a. Static case: With damping in the system, the free response of the accelerometer dies out with time, leaving the forced response. If the acceleration is static ($a_f = $ constant), the method of undetermined coefficients gives the forced response

$$y = \frac{a_f}{\omega_0^2} \tag{c}$$

so that the measurement is proportional to the acceleration. The important considerations are thus: (1) How quickly does the free response die out, and especially, what is the settling pattern of the measurement so that a quick and accurate reading will be possible. (2) Is the sensitivity of the instrument sufficiently large?

A good solution for the first consideration is to fix the damping ratio around critical damping ($\zeta \approx 1$) as a compromise between quick response and minimal or no oscillation. With respect to response speed, the decay rate for critical damping $\sigma = \omega_0$ implies that ω_0 should be as large as possible to improve speed (decrease time constant). For the second consideration, from equation (c), the sensitivity is $S = 1/\omega_0^2$, which can be improved by decreasing the natural frequency. This is in conflict with the first speed requirement. However, there is a physical requirement on S: The indication y should be 2 cm or less for $a_f = 10$ m/s^2. That is, from equation (c),

$$0.02 \geq \frac{10}{\omega_0^2} \longrightarrow \omega_0^2 \geq \frac{10}{0.02} = 500 \tag{d}$$

As a practical matter the natural frequency should not be much larger than given by (d) if the available space for the scale is to be utilized efficiently. Thus as a first attempt, we can let $\omega_0 = 22.4$. The corresponding critical damping time constant $T = 0.045$ s is sufficiently small. Finally, we take

$$m = \frac{k}{\omega_0^2} = \frac{250}{22.4^2} = 0.498 \text{ kg}; \qquad \beta = 2\sqrt{mk} = 22.32 \text{ kg/s}$$

The present design could be the starting point for optimization based on simulating the behavior of the instrument, incorporating fine details not included in the first analysis while subjecting it to acceleration input of various values within the given range.

b. Dynamic case: We consider the input acceleration as a sinusoid of amplitude $|a_f|$ and frequency ω and define the frequency ratio $\eta = \omega/\omega_0$. The ratio of the amplitude of the indication $|y|$ to the amplitude of the frame acceleration $|a_f|$ is given by the magnitude of the transfer function, equation (7.33c). The static deflection was obtained previously:

$$y_0 = \frac{|a_f|}{\omega_0^2}$$

The *magnification factor,* or the ratio of the amplitude of y to the static deflection, is

$$\text{MF} = \frac{|y|}{|y_0|} = \omega_0^2 \frac{|y|}{|a_f|} = \frac{1}{\sqrt{(1 - \eta^2)^2 + (2\zeta\eta)^2}} \qquad (e)$$

which is the same result plotted in Fig. 7.13. It can be observed from this figure that for small η, MF tends to 1 for all values of the damping ratio, which implies

$$y \longrightarrow \frac{|a_f|}{\omega_0^2}$$

as in the static case. The design considerations are as follows:

Natural frequency: For dynamic measurements, the natural frequency of the device is often chosen to be at least two times higher than the *highest* frequency acceleration to be measured. That is, $\eta \leq 0.5$, which is considered "small."

Error in the measurement: If we assume static calibration of the accelerometer, the fractional error in the measurement is the deviation of MF from 1, that is, error = $|\text{MF} - 1|$.

Damping ratio: For zero or very small damping, the possibility exists that the acceleration being measured can have higher harmonics that can cause resonance, or give MF $>> 1$. Further, for $\eta \leq 0.5$, the deviation of MF from 1 depends on the damping ratio. Generally, the damping ratio can be chosen in the range $0.5 \leq \zeta \leq 0.707$. Indeed, with the damping ratio in this range, and $\eta \leq 0.5$, the deviation of MF from 1 is very small—the largest value of the deviation within this range of η is 0.1094.

Bandwidth: Observe from Fig. 7.13 that for any damping ratio in the range recommended, MF is reasonably close to 1 for a frequency range of 0 to $\eta = 0.75$. This is considered the acceptable frequency range, hence the bandwidth is taken as $\Delta\omega = 0.75\omega_0$.

Example 7.10	Rotating Unbalance

Consider the simple model of a mechanical equipment such as a turbine or a pump with an unbalanced rotating component shown in Fig. 7.15a. The unbalance is represented by the eccentric mass m_e (or the rotor or impeller). The mass of the equipment is m, and the effective vertical stiffness and damping of the equipment on its support are simulated by the spring k and damper β. If representative values of m_e, m, k, and β are

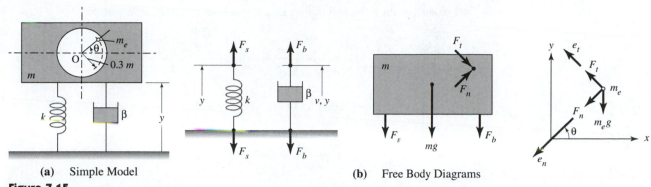

(a) Simple Model

(b) Free Body Diagrams

Figure 7.15

Mechanical Equipment with Rotating Unbalance

20 kg, 1500 kg, 60×10^6 N/m, and 120000 N · s/m, respectively, and the eccentric mass is rotating about frictionless bearings at O at a constant rate of 1800 rpm, determine

a. A model for the vertical motion of the system about its equilibrium displacement in the same format as equation (7.32a).
b. A plot of the magnification factor as a function of the frequency ratio.
c. The value of the amplitude of the response y at the given angular velocity of m_e.

Solution a: Using the techniques of Chapter 3 and with reference to the free body diagrams shown in Fig. 7.15b, we obtain from Newton's second law for the equipment,

$$m \frac{dv}{dt} = F_n \sin \theta - F_t \cos \theta - F_s - F_b - mg \tag{a}$$

where,

$$F_s = ky \quad \text{and} \quad F_b = \beta v \tag{b}$$

For the eccentric mass, in the normal and tangenital coordinate directions,

$$F_t - m_e g \cos \theta = m_e a_t \quad \text{and} \quad F_n + m_e g \sin \theta = m_e a_n \tag{c}$$

From plane rigid body rotation about a fixed point (or see equations (3.20a) and (3.20b) and discussion of relative motion in Example 3.8), we have

$$a_t = r\alpha = 0 \quad \text{and} \quad a_n = r\omega^2; \quad r = 0.3 \tag{d}$$

where α and ω are the angular acceleration and angular velocity, respectively, of the eccentric mass about O. Substituting equations (c) and (d) in equation (a) gives us

$$m \frac{dv}{dt} = m_e r\omega^2 \sin \theta - m_e g \sin^2\theta - m_e g \cos^2\theta - F_s - F_b - mg$$

and using equation (b) and noting that $\theta = \omega t$ for constant ω, we obtain

$$\frac{d^2y}{dt^2} + \left(\frac{\beta}{m}\right)\frac{dy}{dt} + \left(\frac{k}{m}\right)y = \frac{m_e}{m}r\omega^2\sin\,\omega t - \left(\frac{m + m_e}{m}\right)g \qquad \textbf{(7.35a)}$$

Comparing equations (7.35a) and (7.32a), we see that

$$\omega_0 = \sqrt{\frac{k}{m}} = \sqrt{\frac{60 \times 10^6}{1500}} = 200.00\ \text{s}^{-1} \qquad \textbf{(7.35b)}$$

$$\zeta = \frac{\beta}{2\omega_0 m} = \frac{120{,}000}{2(200)1500} = 0.200 \qquad \textbf{(7.35c)}$$

$$f(t) = \frac{m_e}{m}r\omega^2\sin\,\omega t - \left(\frac{m + m_e}{m}\right)g = 0.004\omega^2\sin\,\omega t - 9.94 \qquad \textbf{(7.35d)}$$

$$= f_1(t) + f_2(t)$$

Solution b: By the method of undetermined coefficients, the forced response to the constant input $(f_2(t) = -9.94)$ is a constant displacement y_e, which is given by

$$y_e = \frac{-9.94}{\omega_0^2} = -0.00025\ \text{m}$$

This is the equilibrium displacement—the absolute displacement of the system in the absence of external forcing input. Note the difference between the equilibrium displacement and the static deflection. By the method of undetermined coefficients, the forced response (about the equilibrium displacement) of the system (7.37a) due to a static input

$$f_1(t) = \frac{F_i}{m} = \frac{m_e r\omega^2}{m} = 0.004\omega^2$$

is $y_0 = 0.004\omega^2/\omega_0^2$. This is the *static deflection*.

However, we are interested in the frequency-dependent oscillations y_1 of the equipment about y_e, that is, the forced response due only to

$$f_1(t) = 0.004\omega^2\sin\,\omega t = 0.004\omega^2\cos(\omega t - \pi/2) = A_i\cos(\omega t + \phi_i)$$

From equations (7.32a) through (7.32c) and equations (7.33),

$$y_1(t) = A_0\cos(\omega t + \phi_0)$$

and

$$A_0 = MA_i = MFy_0; \qquad \phi_0 = \phi + \phi_i$$

where

$$\phi = -\tan^{-1}\left(\frac{2\zeta\eta}{1 - \eta^2}\right)$$

$$MF = \frac{1}{\sqrt{(1 - \eta^2)^2 + (2\zeta\eta)^2}}; \qquad \eta = \frac{\omega}{\omega_0}$$

as in the last example.

The plot of MF versus η in Fig. 7.13 includes $\zeta = 0.2$ as well as other values of the damping ratio. Note that for $\zeta = 0$, MF tends to infinity at $\eta = 1$ (the resonant frequency), as before. Also, the maximum MF occurs at values of η progressively less than 1 as ζ increases. Indeed, for a fixed ζ, the maximum magnification factor occurs at η given by

$$\frac{d}{dt} MF = 0 = \frac{-2\eta(1 - \eta^2 - 2\zeta^2)}{[(1 - \eta^2)^2 + (2\zeta\eta)^2]^{3/2}} \quad \text{or} \quad 1 - \eta^2 - 2\zeta^2 = 0$$

which implies

$$MF_{max}: \eta = \sqrt{1 - 2\zeta^2} \quad \text{for } \zeta \leq \frac{\sqrt{2}}{2} = 0.707 \qquad (7.35e)$$

For $\zeta \geq \sqrt{2}/2$, MF is less than 1 for all $\eta > 0$. The condition $\eta = 0$ represents the displacement of the system due to a statically applied load (zero frequency), and MF = 1 for each curve at this frequency ratio. This situation agrees with the earlier explanation of the magnification factor.

Solution c: For the given frequency,

$$\omega = 2\pi\left(\frac{1800}{60}\right) = 188.496 \text{ rad/s} \quad \text{and} \quad \eta = \frac{188.496}{200} = 0.94$$

The value of MF for this frequency is 2.543. Thus

$$A_0 = (0.004)(0.94)^2(2.543) = 0.009 \text{ m } (= 9 \text{ mm})$$

Eliminating Excess Vibration

The vibration amplitude of 9 mm obtained in the last example may represent an unacceptable vibration level for many types of mechanical equipment. This could result in damage to the equipment or other systems, particularly where the vibration-induced forces are transmitted through the supporting structures to these systems. Although there are a few situations such as in certain instruments and screening devices where *excess vibration* may be desirable, in general, the reduction or even elimination of vibration is often required to protect equipment and to reduce maintenance costs.

Depending on the source of the disturbances, various techniques are employed to reduce vibrations in equipment. However, from our previous discussions and the last example, we can see that excess vibrations will in general result when the *operating frequency* is close to the system's resonant frequency. Thus one solution to destructive vibrations is to shift, where feasible, the operating frequency to a value at which the vibration amplitude is acceptable. In Fig. 7.13, vibrations are severely attenuated for frequency ratios greater than about 2, no matter the amount of damping. However, if the operating frequency must be fixed, it may be possible to modify the system in order to change its natural frequency. For a given operating frequency, the frequency ratio increases with decreasing natural frequency. Equations (6.35b) and (7.3a) imply that the natural frequency can be modified by changing either the mass or the stiffness of the system or both. For small devices such as various instrument components (see Practice Problems 7.33 and 7.35) this is easily accomplished. For heavy equipment, though, the redesign necessary to modify the mass and spring constant may not be practical.

Vibration Isolators

In this section we explore the situation in which the excess vibration is being transmitted either from the equipment or to it. We seek first an expression for the transmitted force. From Fig. 7.15b, the force transmitted to the foundation is the sum of the spring force F_s and the damper force F_b:

$$f_T(t) = ky + \beta \frac{dy}{dt}$$

Employing phasors, we find that the magnitude of the transmitted force and hence the peak transmitted force is given by

$$F_{T_{max}} = |kY + j\omega\beta Y| = A_0\sqrt{k^2 + (\beta\omega)^2} = A_0 k\sqrt{1 + (2\zeta\eta)^2}$$

The ratio of the peak transmitted force to the amplitude of the applied force is defined as the *transmissibility T*,

$$T = \frac{F_{T_{max}}}{F_i} = \frac{F_{T_{max}}}{mA_i} = \frac{(M)F_{T_{max}}}{mA_0}$$

and from equations (7.33a) through (7.33g),

$$T = \frac{A_0 k}{A_0 m\omega_0^2} \sqrt{\frac{1 + (2\zeta\eta)^2}{(1 - \eta^2)^2 + (2\zeta\eta)^2}} = \sqrt{\frac{1 + (2\zeta\eta)^2}{(1 - \eta^2)^2 + (2\zeta\eta)^2}} \tag{7.35f}$$

The relation between the transmissibility and the frequency and damping ratio, given by equation (7.35f), is illustrated in Fig. 7.16. In general, for $\eta \leq \sqrt{2}$, T increases with decreasing damping, while the transmissibility is less than 1 for all $\eta > \sqrt{2}$.

Thus we see that the transmitted forces between the equipment and the vibrating foundation or between the vibrating equipment and the foundation can be reduced adequately by interposing between the equipment and the foundation a compliant support whose parameters are so chosen that the lowest frequency ratio is greater than $\sqrt{2}$. Such a support subsystem is known as a *vibration isolator*.

Figure 7.16

**Effect of Frequency
and Damping
on the Transmissibility**

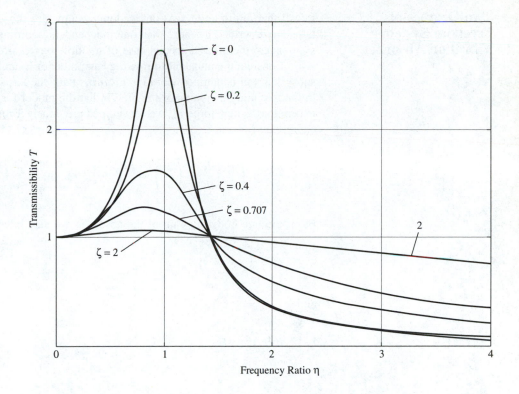

Figure 7.16

**Effect of Frequency
and Damping
on the Transmissibility**

Suppose we wish to isolate the vibration of the rotating equipment (m) of Fig. 7.15a from the foundation by a proper choice of the spring constant k and the damping factor β. The spring and damper subsystem constitute the vibration isolator in this case. For the given operating frequency of 1800 rpm (188.5 rad/s), a frequency ratio of, say, 2 requires that the natural frequency of the combined system ω_0 be 94.25 rad/s. From equation (7.35b), ω_0 is determined by both the mass of the system and the spring constant. Because the mass of the equipment is assumed to be not adjustable in this case, the required value of the spring constant is

$$k = m(94.25)^2 = 1500(94.25)^2 = 13.3 \times 10^6 \text{ N/m}$$

Note that a softer spring (relative to the previous one of spring constant $k = 60 \times 10^6$ N/m) is required, which may be contrary to the reader's physical intuition. As for the damping constant, observe from Fig. 7.16 that for $\eta > \sqrt{2}$, the transmissibility decreases with a decrease in damping, so that our design damping factor should approach zero. In other words, the best vibration isolation is given by an isolator with only spring elements. Note also that with this design of the isolator, the transmissibility (equation (7.35f)) is reduced from a previous value of 2.718 to a new value of 0.333. In practice, vibration isolators vary from simple rubber sleeves to complex pneumatic circuits.

Another means of reducing excess vibration in a piece of equipment without extensively modifying it is to use a *vibration absorber*. This is an oscillating subsystem with parameters so chosen that when it is attached to the major equipment the equipment's vibrations at a particular frequency or range of frequencies are highly attenuated. The vibration absorber is discussed next.

Two-Degrees-of-Freedom Systems: Vibration Absorber

Recall that multiple-degrees-of-freedom systems are those systems whose configurations are described by more than one independent coordinate. Two-degree-of-freedom systems are therefore a special case of multiple-degrees-of-freedom systems.

Consider the unbalanced rotating equipment of Example 7.10, with a spring-mass subsystem (an undamped vibration *absorber*) attached as shown in Fig. 7.17a. We see in the free body diagrams of Fig. 7.17b that the model for the unbalanced equipment, except for the new force F_a, was developed previously. From equation (7.35a), the new expression for y is

$$\frac{d^2y}{dt^2} + \left(\frac{\beta}{m}\right)\frac{dy}{dt} + \left(\frac{k}{m}\right)y = \frac{F_a}{m} + \frac{m_e}{m}r\omega^2\sin\omega t - \left(\frac{m+m_e}{m}\right)g \qquad (7.36a)$$

From the free body diagram of the absorber, we also have

$$F_a = k_a(y_a - y)$$

and

$$\frac{d^2y_a}{dt^2} = -\frac{F_a}{m_a} - g = -\frac{k_a}{m_a}(y_a - y) - g$$

or

$$\frac{d^2y_a}{dt^2} + \left(\frac{k_a}{m_a}\right)y_a = \frac{k_a}{m_a}y - g \qquad (7.36b)$$

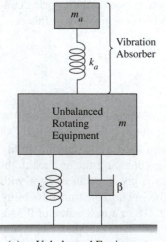

(a)   Unbalanced Equipment and Absorber

Figure 7.17

Application of Vibration Absorber

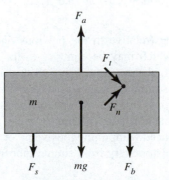
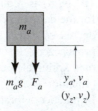
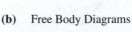
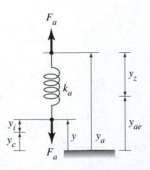

(b)   Free Body Diagrams

We have two coupled second-order systems to solve. As before, we are interested in the alternating component of the displacements, to which the constant-gravity forcing terms do not contribute. The differential equations for the alternating components can be obtained by merely dropping the gravity forcing in equations (7.36a) and (7.36b). More formally, the equilibrium displacements of the equipment and the absorber are given by

$$\left(\frac{k}{m}\right)y_e = \frac{k_a}{m}(y_{a_e} - y_e) - \left(\frac{m + m_e}{m}\right)g \quad \text{and} \quad \left(\frac{k_a}{m_a}\right)y_{a_e} = \left(\frac{k_a}{m_a}\right)y_e - g$$

or

$$y_e = -\left(\frac{m + m_a + m_e}{k}\right)g \quad \text{and} \quad y_{a_e} = -\left(\frac{m + m_a + m_e}{k} + \frac{m_a}{k_a}\right)g$$

If we define the displacements of the equipment and absorber relative to their respective equilibrium displacements, that is, the alternating components of the displacements, y_1 and y_2, by

$$y = y_1 + y_e \quad \text{and} \quad y_a = y_2 + y_{a_e} \tag{7.36c}$$

then substituting (7.36c) in equations (7.36a) and (7.36b), we obtain

$$m\frac{d^2 y_1}{dt^2} + \beta\frac{dy_1}{dt} + (k + k_a)y_1 - k_a y_2 = m_e r\omega^2 \sin \omega t \tag{7.36d}$$

$$m_a\frac{d^2 y_2}{dt^2} + k_a(y_2 - y_1) = 0$$

Equation (7.36d) represents a two-degrees-of-freedom system. The two independent coordinates are y_1 and y_2. The equations in the two coordinates are coupled, in this case, through the terms $(-k_a y_2)$ in the first equation and $(-k_a y_1)$ in the second equation. Because these terms result physically from the spring connection between the masses, the system (7.36d) is said to be *elastically coupled*. The coupling can also be through acceleration terms, resulting in *inertial coupling*. A method of analysis of these coupled systems, not discussed here (see [6], for example) is to make a change of variables to a coordinate system known as *principal* or *natural* coordinates in which the resulting equations are decoupled.

Damped Free Vibrations

To illustrate the general analysis of two-degrees-of-freedom systems, we consider first the homogeneous part of equation (7.36d), that is,

$$m\frac{d^2 y_1}{dt^2} + \beta\frac{dy_1}{dt} + (k + k_a)y_1 - k_a y_2 = 0 \tag{7.37a}$$

$$m_a\frac{d^2 y_2}{dt^2} + k_a(y_2 - y_1) = 0$$

From our previous experience, we can try solutions of the form

$$y_1(t) = C_1 e^{\lambda t} \tag{7.37b}$$

$$y_2(t) = C_2 e^{\lambda t}$$

Substituting (7.37b) in (7.37a), we obtain the homogeneous algebraic system in matrix form:

$$\begin{bmatrix} (m\lambda^2 + \beta\lambda + k + k_a) & (-k_a) \\ (-k_a) & (m_a\lambda^2 + k_a) \end{bmatrix} \begin{bmatrix} C_1 \\ C_2 \end{bmatrix} = 0 \tag{7.37c}$$

for which nontrivial solutions will exist only if the determinant of the coefficient matrix is zero. That is, the characteristic equation is

$$\begin{vmatrix} (m\lambda^2 + \beta\lambda + k + k_a) & (-k_a) \\ (-k_a) & (m_a\lambda^2 + k_a) \end{vmatrix} = 0 \tag{7.37d}$$

or

$$mm_a\lambda^4 + \beta m_a\lambda^3 + [mk_a + m_a(k + k_a)]\lambda^2 + \beta k_a\lambda + kk_a = 0 \tag{7.37e}$$

Equation (7.37e) is fourth order in λ, with all the coefficients (which represent physical parameters in the system) positive. The general conditions on these coefficients for the stability of a system having such a (higher than second-order) characteristic equation can be investigated without first solving for the roots, by methods such as the Routh test, which is discussed in Chapter 10. For stable behavior there are three possibilities for the system eigenvalues:

1. All four eigenvalues are real and negative: λ_1, λ_2, λ_3, λ_4.
2. All four eigenvalues are two pairs of complex conjugate roots: λ_1, $\lambda_2 = -\sigma_1 \pm j\omega_1$, and λ_3, $\lambda_4 = -\sigma_2 \pm j\omega_2$.
3. Two eigenvalues are real and negative, and two are a complex conjugate pair: λ_1, λ_2 and λ_3, $\lambda_4 = -\sigma \pm j\omega$.

For each of the four roots λ_i, equation (7.37c), as a homogeneous equation, can determine only the relative values of C_1 and C_2. That is, we can obtain β_i for each λ_i such that $C_{2i} = \beta_i C_{1i}$.

The homogeneous solution for each of the three cases is a combination of modes:

$$y_1(t) = C_{11}e^{\lambda_1 t} + C_{12}e^{\lambda_2 t} + C_{13}e^{\lambda_3 t} + C_{14}e^{\lambda_4 t} \tag{7.37f}$$

$$y_2(t) = \beta_1 C_{11}e^{\lambda_1 t} + \beta_2 C_{12}e^{\lambda_2 t} + \beta_3 C_{13}e^{\lambda_3 t} + \beta_4 C_{14}e^{\lambda_4 t}$$

where again, $\beta_i = C_{2i}/C_{1i}$ is determined by substituting $\lambda = \lambda_i$ in equation (7.37c), and the four remaining constants C_{1i}, $i = 1, \ldots 4$, can be found from the four initial conditions of (7.37a). Recall that for the case of complex conjugate eigenvalues, the pair of exponential modes in (7.37f) combine into a damped sinusoidal mode. Thus in general, the possibility exits for the free response to consist of purely exponential decay modes, damped harmonic (oscillatory) modes, or combinations of exponential and damped harmonic modes.

Undamped Free Vibrations and Natural Frequencies

For an undamped system, the eigenvalues given by case 2 will be $\lambda_1, \lambda_2 = \pm j\omega_1$ and $\lambda_3, \lambda_4 = \pm j\omega_2$, and ω_1 and ω_2 are, in this case, the natural frequencies of the system. Thus the two-degrees-of-freedom system has two natural frequencies. The smaller of the two frequencies is traditionally denoted by ω_1 and is called the *fundamental* or *first* frequency of the system. Consider the present problem, for example. For $\beta = 0$, the characteristic equation (7.37e) is

$$\lambda^4 + \left[\frac{mk_a + m_a(k + k_a)}{mm_a}\right]\lambda^2 + \frac{kk_a}{mm_a} = 0 \qquad (7.38a)$$

Letting $\Omega = \lambda^2$, we have the quadratic in Ω:

$$\Omega^2 + \left[\frac{mk_a + m_a(k + k_a)}{mm_a}\right]\Omega + \frac{kk_a}{mm_a} = 0$$

Using the same parameter values given in Example 7.10: $m = 1500$ kg and $k = 60 \times 10^6$ N/m, and assuming for this exercise, $m_a = 450$ kg and $k_a = 15.99 \times 10^6$ N/m, we obtain the roots

$$\Omega_1, \Omega_2 = -2.2216 \times 10^4, -6.3977 \times 10^4 \Rightarrow; \lambda_1, \lambda_2 = \pm j149.05$$

$$\text{and } \lambda_3, \lambda_4 = \pm j252.94$$

Thus the fundamental frequency is $\omega_1 = 149.05$, and the second natural frequency is $\omega_2 = 252.94$.

Continuing with the given parameter values, we substitute each of the eigenvalues, in turn, into either of the two equations in (7.37c) and find $\beta_1 = \beta_2 = 2.668 = \alpha_1$, and $\beta_3 = \beta_4 = -1.249 = \alpha_2$. Thus the free undamped vibrations from (7.37f) are

$$y_1(t) = (C_{11}e^{j\omega_1 t} + C_{12}e^{-j\omega_1 t}) + (C_{13}e^{j\omega_2 t} + C_{14}e^{-j\omega_2 t})$$

$$y_2(t) = \alpha_1(C_{11}e^{j\omega_1 t} + C_{12}e^{-j\omega_1 t}) + \alpha_2(C_{13}e^{j\omega_2 t} + C_{14}e^{-j\omega_2 t})$$

or

$$y_1(t) = A_1 \cos(\omega_1 t + \phi_1) + A_2 \cos(\omega_2 t + \phi_2) \qquad (7.38b)$$

$$y_2(t) = \alpha_1 A_1 \cos(\omega_1 t + \phi_1) + \alpha_2 A_2 \cos(\omega_2 t + \phi_2)$$

The constants to be determined from the initial conditions are thus A_1, A_2, ϕ_1, and ϕ_2. We can see that, in general, depending on the initial conditions, the free vibrations of an undamped two-degrees-of-freedom system can be one of the following: a first-mode harmonic oscillation ($A_1 \neq 0, A_2 = 0$); a second-mode harmonic oscillation ($A_1 = 0, A_2 \neq 0$); or a combination of the fundamental and second modes of vibration (equation (7.38b)), which is not harmonic (the sum of two sinusoids of different frequencies is not a sinusoid).

Principal Mode Vibrations and Mode Shapes The solutions for the harmonic oscillations, which are known as *principal mode vibrations,* are usually of particular interest. For the present system and parameter values, the first and second principal mode vibrations are

$$y_1(t) = A_1\cos(\omega_1 t + \phi_1) \tag{7.38c}$$

$$y_2(t) = \alpha_1 A_1\cos(\omega_1 t + \phi_1) = 2.668 A_1\cos(\omega_1 t + \phi_1); \quad \text{(first principal mode vibration)}$$

$$y_1(t) = A_2\cos(\omega_2 t + \phi_2) \tag{7.38d}$$

$$y_2(t) = \alpha_2 A_2\cos(\omega_2 t + \phi_2) = -1.249 A_2\cos(\omega_2 t + \phi_2); \quad \text{(second principal mode vibration)}$$

A plot of the vibration amplitude at each point in the system is known as the *mode shape.* Figure 7.18 shows the mode shapes depicting the principal modes of the present system. Figure 7.18b reveals that for the second mode of vibration, there will be a location in the system, called a *node,* where the deflection will always be zero. Such information can be very useful in determining the location of sensors and actuators for effective control action. In a large space structure such as some satellites (with typical large and flexible arrays of solar panels), a displacement sensor will not ordinarily be placed at or near known nodes for the satellite's principal modes. On the other hand, such locations should be ideal for the actuators, since minimum actuator travel would be required to realize significant changes in the configuration of the satellite. The general problems of being able to completely determine system configuration from the available measurements *(observability)* and the ability of the given control inputs to affect all the coordinates of the system *(controllability)* are subjects discussed briefly in Chapter 9.

Forced Vibrations

The forced vibration of stable two-degrees-of-freedom systems is relatively straightforward and can be considered as part of a design problem.

Example 7.11   Application of Vibration Absorber

Consider again the unbalanced rotating equipment of Example 7.10.

a. Design an undamped vibration absorber (see Fig. 7.15) such that the vibration amplitude of the equipment at the operating frequency of 1800 rpm is negligible. Assume that there is a practical limit on the largest possible value of m_a of 450 kg.

Figure 7.18

Mode Shapes from Principal Mode Analysis of a Two-degrees-of-freedom System

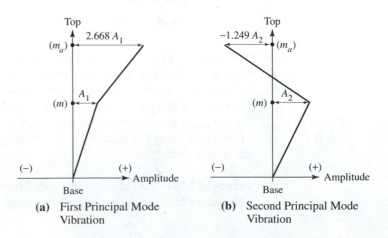

(a) First Principal Mode Vibration

(b) Second Principal Mode Vibration

b. Obtain a general plot of the dimensionless amplitude of vibration of the equipment as a function of the ratio of the input frequency to the unbalanced equipment natural frequency. Discuss the resonant behavior of the system.

Solution a: The system model, equation (7.36d), was derived previously. From our previous experience with single-degree-of-freedom systems, we can assume that the forced response is of the form

$$y_1(t) = A_1\cos(\omega t + \phi_1); \qquad \mathscr{P}[y_1(t)] = A_1 e^{j\phi_1}$$ (7.39a)

$$y_2(t) = A_2\cos(\omega t + \phi_2); \qquad \mathscr{P}[y_2(t)] = A_2 e^{j\phi_2}$$

that is, $y_1(t)$ and $y_2(t)$ are both sinusoids of the same frequency as the forcing input sinusoid:

$$m_e r\omega^2\sin\omega t = m_e r\omega^2\cos(\omega t - \pi/2); \qquad \mathscr{P}[m_e r\omega^2\cos(\omega t - \pi/2)] = m_e r\omega^2 e^{-j\pi/2}$$

Applying then the phasor transform to equation (7.36d), we obtain the following equation:

$$(k + k_a - m\omega^2 + j\omega\beta)A_1 e^{j\phi_1} - k_a A_2 e^{j\phi_2} = (m_e r\omega^2)e^{-j\,\pi/2}$$ (7.39b)

$$-k_a A_1 e^{j\phi_1} + (k_a - m_a\omega^2)A_2 e^{j\phi_2} = 0$$

Solving equation (7.39b) simultaneously for the output phasors, we obtain

$$A_1 e^{j\phi_1} = \left[\frac{m_e r\omega^2(k_a - m_a\omega^2)}{(k + k_a - m\omega^2 + j\omega\beta)(k_a - m_a\omega^2) - k_a^2}\right]e^{-j\,x/2}$$ (7.39c)

$$A_2 e^{j\phi_2} = \left[\frac{m_e r\omega^2 k_a}{(k + k_a - m\omega^2 + j\omega\beta)(k_a - m_a\omega^2) - k_a^2}\right]e^{-j\,x/2}$$

The corresponding expressions for the vibration amplitudes A_1 and A_2 are

$$A_1 = \sqrt{\frac{[m_e r\omega^2(k_a - m_a\omega^2)]^2}{[(k + k_a - m\omega^2)(k_a - m_a\omega^2) - k_a^2]^2 + (k_a - m_a\omega^2)^2\omega^2\beta^2}}$$ (7.40a)

$$A_2 = \sqrt{\frac{[m_e r\omega^2 k_a]^2}{[(k + k_a - m\omega^2)(k_a - m_a\omega^2) - k_a^2]^2 + (k_a - m_a\omega^2)^2\omega^2\beta^2}}$$ (7.40b)

From equation (7.40a), we observe that the amplitude of vibration of the equipment will be zero if the absorber mass and spring constant are chosen such that

$$k_a - m_a\omega^2 = 0 \quad \text{or} \quad \frac{k_a}{m_a} = \omega^2$$ (7.40c)

With this condition, the amplitude of vibration of the absorber itself becomes

$$A_2 = \frac{m_e r\omega^2}{k_a} = \frac{rm_e}{m_a}$$ (7.40d)

Thus we see that we must choose m_a as large as possible in order to limit the vibration of the absorber subsystem and not destroy it. From the constraints of the problem, we select $m_a = 450$ kg. Equation (7.40c) then fixes k_a:

$$k_a = (450)(188.5)^2 = 15.99 \times 10^6 \text{ N/m}$$

The absorber vibration amplitude corresponding to this design is by equation (7.40d)

$$A_2 = \frac{(0.3)(20)}{450} = 0.0133 \text{ m} = 13.3 \text{ mm}$$

Although this result can represent severe vibration in the absorber, the actual equipment that is of concern will theoretically not vibrate at all!

Solution b: In terms of a frequency ratio based on the natural frequency of the unbalanced rotating equipment,

$$\eta = \omega \sqrt{\frac{m}{k}}$$

and using

$$\zeta^2 = \frac{\beta^2}{(4km)}$$

we obtain for the ratio of the amplitude of vibration to the radius of the unbalance for the equipment and the absorber,

$$\left(\frac{A_1}{r}\right) = \frac{\eta^2 k (m_e/m) \sqrt{(k_a - \eta^2 k m_d/m)^2}}{\sqrt{[(k + k_a - \eta^2 k)(k_a - \eta^2 k m_d/m) - k_a^2]^2 + (k_a - \eta^2 k m_d/m)^2 (2\eta k \zeta)^2}}$$

$$\left(\frac{A_2}{r}\right) = \frac{\eta^2 k (m_e/m) k_a}{\sqrt{[(k + k_a - \eta^2 k)(k_a - \eta^2 k m_d/m) - k_a^2]^2 + (k_a - \eta^2 k m_d/m)^2 (2\eta k \zeta)^2}}$$

$$(7.40e)$$

Figure 7.19 shows a plot of the dimensionless amplitude of vibration versus the frequency ratio for both the equipment and the vibration absorber. The absorber design conditions are evident in the figure. At $\eta = 0.94$, which corresponds to the operating speed, the amplitude of vibration of the equipment is zero. However, Fig. 7.19 also exhibits two resonant conditions, which is typical of oscillatory two-degrees-of-freedom systems and has implications for system design.

Recall that the two-degrees-of-freedom system has two natural frequencies. For the present problem with the parameters as determined, these frequencies are $\omega_1 = 149.05$ and $\omega_2 = 252.94$. the values of the dimensionless forcing frequency corresponding to these frequencies, $\eta = 0.745$ and $\eta = 1.265$, are close to the peak output or resonant frequencies in Fig. 7.19. To establish the value of the resonant

Figure 7.19

Response of Unbalanced Equipment with Vibration Absorber

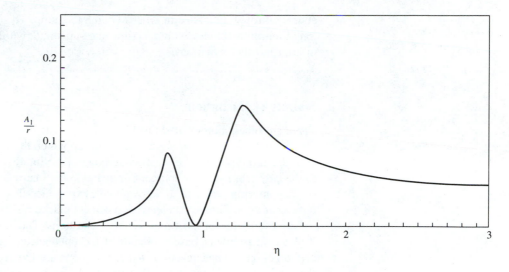

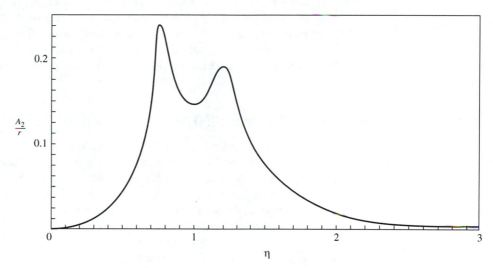

frequencies, we consider the vibration amplitudes, equations (7.40a) and (7.40b), for no damping in the system. The condition for arbitrarily large amplitudes under these circumstances is

$$(k + k_a - m\omega^2)(k_a - m_a\omega^2) - k_a^2 = 0 \qquad \textbf{(7.40f)}$$

$$\text{or} \quad \omega^4 - \left[\frac{mk_a + m_a(k + k_a)}{mm_a}\right]^2 \omega^2 + \frac{kk_a}{mm_a} = 0$$

which is the same as the characteristic equation (7.38a) with $\lambda = j\omega$; and the positive (physically meaningful) roots are $\omega = 149.05$ and $\omega = 252.94$, as before. Thus similar to the case for single-degree-of-freedom systems, the resonant frequencies of a multiple-degrees-of-freedom system are the natural frequencies.

With respect to the design, we note that the actual operating frequency is between the two resonant frequencies. Consequently, during start-up or shutdown, the equipment must accelerate past the first resonant region to reach the operating speed or decelerate through this zone in order to come to a stop. This might not be a problem for long-running machines with very few start-up/shutdown periods, but it could be significant in other situations.

···

Sources of Excitation

There are other sources of sinusoidal forcing input for mechanical equipment besides rotating unbalance. The motion of the base on which the equipment is mounted can be considered the source of excitation, although the vibration may have been transmitted to the base from another source. For example, an instrument package that is mounted on the fluttering wings of an airplane will experience this disturbance, which is called *base excitation*. Similarly, a television camera placed inside a race car will be subjected to base excitation for most road surface conditions. The vibration absorber of Example 7.11 would be under base excitation if the unbalanced equipment was viewed as its support or base. The concept of *self-excitation* is used to describe the situation in which the forcing input is proportional to the displacement or to one of the time derivatives of the displacement of the object that is being disturbed. The fluttering aircraft wing mentioned above is itself experiencing self-excitation. Other examples of self-excitation include galloping electrical transmission lines, whirling shafts, squealing brakes, and chattering machine tools (see the problem analyzed in Section 9.1).

7.4 FORCED RESPONSE TO NONSINUSOIDAL PERIODIC INPUTS

We consider in this section the forced response of linear and stationary scalar ODE systems to nonsinusoidal periodic inputs via the *Fourier series* representation of such inputs. Although a large number of periodic forcing inputs of interest can be considered sinusoidal functions, this is not always an accurate or adequate representation of the input in many cases. Various natural phenomena and contrived events such as noise patterns, acoustic and electromagnetic waves of all kinds, and vibrations of various structures and components are periodic but nonsinusoidal. A *periodic* phenomenon typically contains a pattern that repeats itself over and over again in time or space. In a periodic wave phenomenon, the pattern itself moves in time. Analytically, the function $f(t)$ is *periodic* (in time) with *period* Γ if for all t,

$$f(t + \Gamma) = f(t) \tag{7.41a}$$

It follows from this definition that if Γ is a period of $f(t)$, then 2Γ and indeed any integral multiple of Γ is also a period of $f(t)$. The smallest positive value of Γ for which equation (7.41a) holds is called (see [1]) the *fundamental period* of $f(t)$.

Introducing Fourier Series

The sine and cosine functions $\sin(\omega_0 t)$ and $\cos(\omega_0 t)$, are periodic, each with a fundamental period of $2\pi/\omega_0$:

$$\sin(\omega_0[t + \Gamma]) = \sin \omega_0 t \cos \omega_0 \Gamma + \cos \omega_0 t \sin \omega_0 \Gamma = \sin \omega_0 t \tag{7.41b}$$

$$\Rightarrow \Gamma = \frac{2n\pi}{\omega_0}; \qquad n = \pm 1, \pm 2, \ldots$$

$$\cos(\omega_0[t + \Gamma]) = \cos \omega_0 t \cos \omega_0 \Gamma - \sin \omega_0 t \sin \omega_0 \Gamma = \cos \omega_0 t$$

$$\Rightarrow \Gamma = \frac{2n\pi}{\omega_0}; \qquad n = \pm 1, \pm 2, \ldots$$

If $g_1(t)$ and $g_2(t)$ are periodic functions with common period Γ, then the sum $f(t) = g_1(t) + g_2(t)$ is also periodic with period Γ:

$$f(t + \Gamma) = g_1(t + \Gamma) + g_2(t + \Gamma) = g_1(t) + g_2(t) = f(t) \tag{7.41c}$$

Thus the sinusoid $C_1\cos(\omega_0 t) + C_2\sin(\omega_0 t) = C\cos(\omega_0 t + \phi)$ is periodic with a fundamental period of $2\pi/\omega_0$ and ω_0 is the *fundamental* (circular) *frequency*. In Section 6.2, one of the reasons given for the importance of sinusoids in linear systems analysis was that most signals of nonsinusoidal form can be expressed as the sum of sinusoids. A significant feature of many periodic functions is that they can be expanded in terms of a series of sine and cosine terms (sinusoids) known as a *Fourier series*:

$$f(t) = \frac{A_0}{2} + \sum_{n=1}^{\infty} (A_n\cos n\omega_0 t + B_n\sin n\omega_0 t) \tag{7.42a}$$

$$f(t) = \frac{A_0}{2} + \sum_{n=1}^{\infty} C_n\cos(n\omega_0 t + \phi_n) \tag{7.42b}$$

where

$$A_0 = \frac{2}{\Gamma} \int_{t_0}^{t_0+\Gamma} f(t)\, dt \tag{7.43a}$$

$$A_n = \frac{2}{\Gamma} \int_{t_0}^{t_0+\Gamma} f(t)\cos n\omega_0 t\, dt \tag{7.43b}$$

$$B_n = \frac{2}{\Gamma} \int_{t_0}^{t_0+\Gamma} f(t)\sin n\omega_0 t\, dt \tag{7.43c}$$

$$C_n = (A_n^2 + B_n^2)^{1/2} \tag{7.44a}$$

$$\phi_n = -\tan^{-1}\left(\frac{B_n}{A_n}\right) \tag{7.44b}$$

Also, t_0 is a real number, and $\Gamma = 2\pi/\omega_0$ is the period. Note that the amplitude C_n and phase ϕ_n or the coefficients (A_n and B_n) of the component terms in the Fourier series are determined from $f(t)$ (see [8]). Correspondingly, $f(t)$ is determined from knowledge of C_n and ϕ_n (or A_n and B_n). Thus the pair of equations (7.42b) and (7.44) (or (7.42a) and (7.43)) constitute a *Fourier series transform pair*. The sufficient conditions for a function to be fourier series transformable, and for the series to converge to the function itself, are known as the *Dirichlet conditions* (see [9]): If

1. $f(t)$ is defined and single-valued except possibly at a finite number of points in the interval $(t_0, t_0 + \Gamma)$
2. $f(t)$ is periodic outside $(t_0, t + \Gamma)$ with period Γ
3. $f(t)$ and $\dot{f}(t)$ are piecewise continuous in $(t_0, t_0 + \Gamma)$

then the series (7.42) with coefficients (7.43) converges to

a. $f(t)$ if t is a point of continuity.
b. $\frac{1}{2}[f(t+) + f(t-)]$ (that is, the average of the *right* and *left hand limits* of $f(t)$ at t) if t is a point of discontinuity.

Figure 7.20 illustrates the meaning of a *piecewise continuous* function. Also note that

$$f(t+) = \lim_{\epsilon \to 0^+} f(t + \epsilon); \quad \text{(ϵ approaches zero through positive values)} \tag{7.45a}$$

$$f(t-) = \lim_{\epsilon \to 0^+} f(t - \epsilon); \quad \text{(ϵ approaches zero through positive values)} \tag{7.45b}$$

The Dirichlet conditions are only *sufficient* conditions. It is not necessary that a function satisfy these conditions before it can be Fourier series transformable. However, any function that satisfies the Dirichlet conditions, and essentially all periodic functions of practical interest do, is Fourier series transformable.

Example 7.12   Amplitude and Phase Spectra

Find the Fourier series expansion of the *sawtooth* type function shown in Fig. 7.21. Determine the amplitude and phase coefficients of the expansion as given by equations (7.44a) and (7.44b) and plot these coefficients as functions of n.

Solution: We choose the interval t_0 to $t_0 + \Gamma$ as -1 to 3, so that $t_0 = -1$, $\Gamma = 2\pi/\omega_0 = 4$ ($\omega_0 = \pi/2$), and $f(t) = t + 1$ within the interval $-1 \le t \le 2$ ($f(t) = 0$ for $2 < t \le 3$). Then from equations (7.43),

$$A_0 = \frac{2}{4}\int_{-1}^{3} f(t)\, dt = \frac{1}{2}\int_{-1}^{2} (t+1)\, dt + \frac{1}{2}\int_{2}^{3} 0.\, dt = \frac{1}{2}\left(\frac{t^2}{2} + t\right)\Big|_{-1}^{2} = \frac{9}{4} \tag{7.46a}$$

$$A_n = \frac{1}{2}\int_{-1}^{3} (t+1)\cos n\omega_0 t\, dt$$

$$= \frac{1}{2}\left\{\frac{1}{(n\omega_0)^2}\cos n\omega_0 t + \frac{t}{n\omega_0}\sin n\omega_0 t + \frac{1}{n\omega_0}\sin n\omega_0 t\right\}\Big|_{-1}^{2}$$

Figure 7.20

Example of a Piecewise Continuous Function

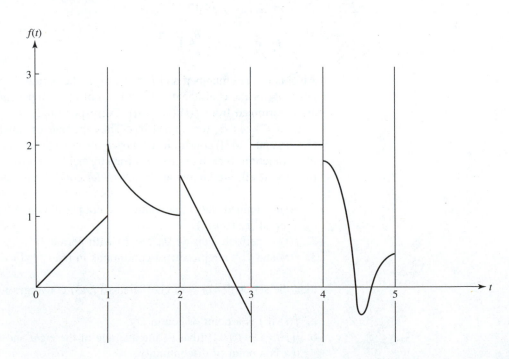

Figure 7.21

Example of a Periodic Function

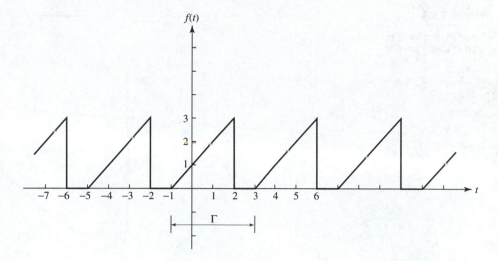

or

$$A_n = \frac{1}{2(n\omega_0)^2}(\cos 2n\omega_0 - \cos n\omega_0) \tag{7.46b}$$

$$B_n = \frac{1}{2}\int_{-1}^{3} (t + 1)\sin n\omega_0 t \, dt$$

$$= \frac{1}{2}\left\{ \frac{1}{(n\omega_0)^2}\sin n\omega_0 t - \frac{t}{n\omega_0}\cos n\omega_0 t - \frac{1}{n\omega_0}\cos n\omega_0 t \right\}\Bigg|_{-1}^{2}$$

or

$$B_n = \frac{1}{2(n\omega_0)^2}(\sin n\omega_0 - 3n\omega_0\cos 2n\omega_0) \tag{7.46c}$$

With the coefficients A_0, A_n, and B_n given by equations (7.46a) through (7.46c), the Fourier series is given by equation (7.42a) or equations (7.42b) and (7.44). Figure 7.22 shows these results. Note that in computing the phase ϕ_n, we implement equation (7.44b) in a manner that avoids a *singularity* error whenever A_n is zero (see source files in On the Net). Also, the calculation of the function $f(t)$ can make use of the periodicity of $f(t)$.

Figure 7.22a shows the original function $f(t)$ (dash-dot line) and the Fourier series expansion—one truncated at only the term $n = 3$ (dotted line), the other truncated at the $n = 15$ term (solid line). The results imply that whereas $n = 3$ represents an insufficient number of terms of the infinite series, a fairly accurate representation of this function can be obtained with a small number of terms, such as 15. However, note the overshoot of the representation at the points of discontinuity. An overshoot of at least 10% will remain at all such points, no matter the number of terms included in the Fourier series expansion. This is known as the *Gibbs phenomenon* (see [8]).

Figures 7.22b and 7.22c show, respectively, the amplitude and phase coefficients, equations (7.44a) and (7.44b) as functions of n. Such displays of coefficients versus n (harmonic frequency $n\omega_0$; $n = 1 \Rightarrow$ *fundamental* or *first harmonic*; $n = 2 \Rightarrow$ *second*

Figure 7.22

Fourier Series Amplitude and Phase Spectra

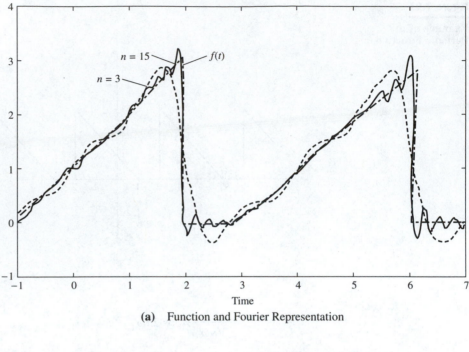

(a)   Function and Fourier Representation

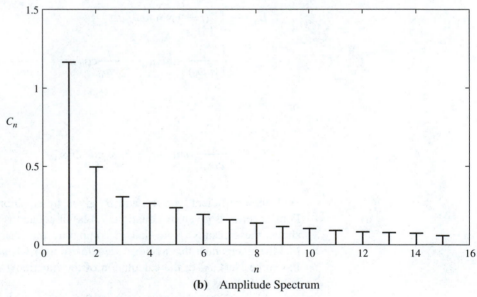

(b)   Amplitude Spectrum

harmonic; and so on) are known as *spectrums* or spectra. A pair of amplitude and phase spectra actually constitute an alternative and complete representation of the periodic function.

Effects of Symmetry

Simplifications of the formulas for the Fourier series coefficients result from the various types of symmetry that are often possessed by practical periodic waveforms. Fig-

Figure 7.22 *continued*

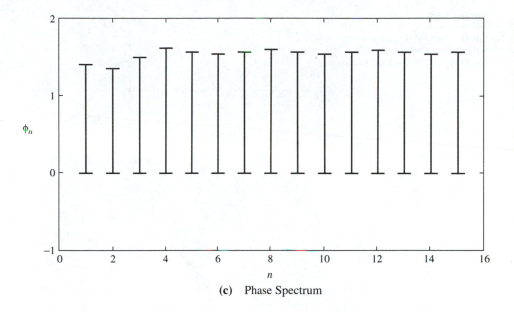

(c) Phase Spectrum

ure 7.23 illustrates the major types of symmetry with a summary of their effect on the Fourier coefficients.

Zero-Average Value A symmetry in the waveform with respect to the time axis that has the effect that, over each period, the area enclosed by the positive part of the function is equal to the area enclosed by the negative part (see Fig. 7.23a) results in *zero-average value* of the function. Because according to equation (7.43a), the coefficient A_0 is twice the *mean* of $f(t)$ over a period, the effect of this type of symmetry is that A_0 is zero.

Zero-Axis Symmetry: Even Function A waveform that is symmetric about the axis $t = 0$ (see Fig. 7.23b) is said to possess *zero-axis symmetry*. This means that $f(-t) = f(t)$. A function $f(t)$ with this property is called an *even* function. From equation (7.43c), since $\sin(-n\omega_0 t) = -\sin(n\omega_0 t)$, considering the interval t_0 to $t_0 + \Gamma = -\Gamma/2$ to $\Gamma/2$, the coefficients B_n are all zero, and the Fourier series of an even function consists of only cosine terms and possibly the constant term A_0, which can be considered a cosine term ($n = 0$) also.

Zero-Point Symmetry: Odd Function A waveform for which $f(-t) = -f(t)$ (see Fig. 7.23c) possesses *zero-point symmetry*. Such a function is also called an *odd* function. As for the case of the even function, since $\cos(-n\omega_0 t) = \cos(n\omega_0 t)$, equation (7.43b) implies that all the coefficients A_n are zero, so that in a Fourier series corresponding to an odd function, only sine terms can be present.

Half-Range Fourier Sine or Cosine Series A Fourier series in which only sine terms or cosine terms are present is known as a *half-range Fourier* (sine or cosine, respectively) *series*. To obtain the half-range Fourier series expansion for a given function, we

Figure 7.23

Illustration of Types of Symmetry and Their Effect on Fourier Series Coefficients

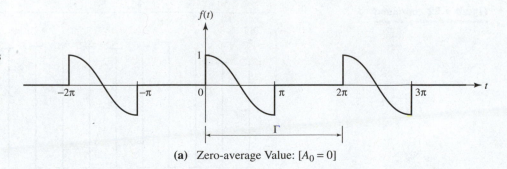

(a) Zero-average Value: $[A_0 = 0]$

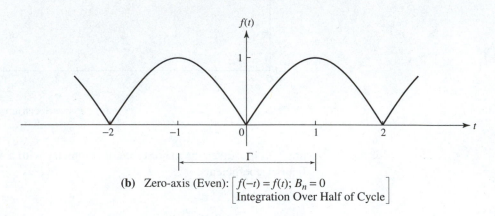

(b) Zero-axis (Even): $\left[\begin{array}{l} f(-t) = f(t); B_n = 0 \\ \text{Integration Over Half of Cycle} \end{array} \right]$

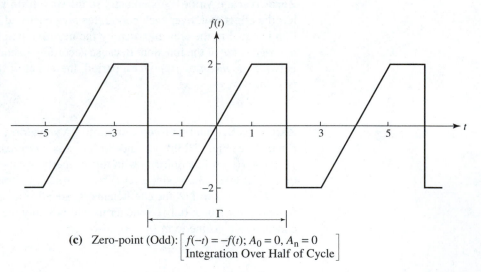

(c) Zero-point (Odd): $\left[\begin{array}{l} f(-t) = -f(t); A_0 = 0, A_n = 0 \\ \text{Integration Over Half of Cycle} \end{array} \right]$

usually define the function in the interval 0 to $\Gamma/2$, which is half the interval $-\Gamma/2$ to $\Gamma/2$, and then also specify it as odd or even, thereby defining the function in the other half of the interval, that is, $-\Gamma/2$ to 0 (see [8]). Then

$$A_n = 0; \qquad B_n = \frac{4}{\Gamma} \int_0^{\Gamma/2} f(t)\sin n\omega_0 t \, dt; \quad \text{(half-range sine series)} \qquad (7.47a)$$

$$B_n = 0; \qquad A_n = \frac{4}{\Gamma} \int_0^{\Gamma/2} f(t)\cos n\omega_0 t \, dt; \quad \text{(half-range cosine series)} \qquad (7.47b)$$

Representation of a Nonperiodic Function

Given a continuous function $f(t)$, defined over a finite interval $t_0 \leq t \leq t_f$, although the function is not periodic, a Fourier series representation of the function, valid within the original interval, can be determined for a specified period by extending the definition of the function into a periodic function. If in the original interval $t_0 = 0$, then extensions into odd or even or neither-odd-nor-even periodic functions are possible. Figure 7.24 illustrates this idea for the parabola $f(t) = t^2$, using a basic period of 2π.

Response of a Linear Time-Invariant System

Given the Fourier series representation, the forced response of a **linear** time-invariant system to a periodic forcing input is the sum of the responses of the system to the individual (sinusoidal) terms in the Fourier series expansion of the input function. From

Figure 7.24

Extensions of a Nonperiodic Function for Fourier Series Representation

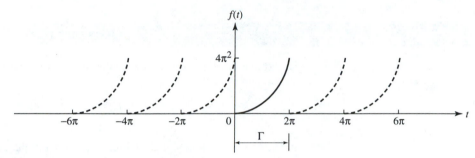

(a)   Neither Even nor Odd Extension; Period = 2π

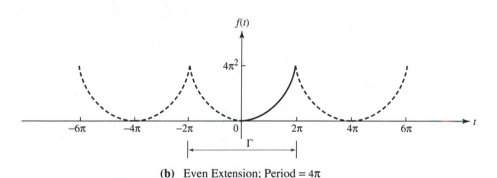

(b)   Even Extension; Period = 4π

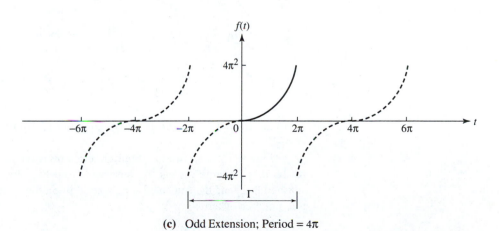

(c)   Odd Extension; Period = 4π

our experience with the phasor transform solution and sinusoidal steady state, each of the response terms is also a sinusoid of the same (harmonic) frequency as the corresponding term in the Fourier series of the input but of different amplitude and phase. Specifically, if the Fourier series representation of the input is

$$f(t) = \frac{A_0}{2} + \sum_{n=1}^{\infty} A_n \cos(n\omega_0 t + \psi_n); \quad \text{(input series)} \tag{7.48a}$$

then the forced response of the system is given by

$$y(t) = \frac{C_0}{2} + \sum_{n=1}^{\infty} C_n \cos(n\omega_0 t + \phi_n); \quad \text{(output series)} \tag{7.48b}$$

Thus the forced response is also periodic and is represented by a Fourier series.

Example 7.13   Application of Fourier Series Representation

Determine the forced response of the system of equation (7.2), repeated here,

$$\frac{d^2 y}{dt^2} + 2\zeta\omega_N \frac{dy}{dt} + \omega_N^2 y = f(t); \qquad \omega_N = \text{natural frequency} \tag{7.49}$$

to the finite parabola signal of Fig. 7.24 over the interval $0 \le t \le 2\pi$ using a Fourier series representation of the even extension shown in Fig. 7.24b. Compare the result with the solution given by the method of undetermined coefficients. Assume the second-order system (7.49) is underdamped with the following parameter values: $\omega_N = 2$, and $\zeta = 0.5$.

Solution: The Fourier series representation of an even function consists of only cosine terms. The coefficients of the half-range cosine series are determined on the basis of the half-period as follows:

$$A_0 = \frac{4}{\Gamma} \int_0^{\Gamma/2} f(t)\, dt = \frac{1}{\pi} \int_0^{2\pi} t^2\, dt = \frac{8\pi^2}{3}$$

$$A_n = \frac{4}{\Gamma} \int_0^{\Gamma/2} f(t)\cos(n\omega_0 t)\, dt = \frac{1}{\pi} \int_0^{2\pi} t^2 \cos(n\omega_0 t)\, dt$$

$$= \frac{1}{\pi}\left\{ \frac{t^2}{n\omega_0}\sin(n\omega_0 t) - \frac{2}{n\omega_0}\left[\frac{1}{(n\omega_0)^2}\sin(n\omega_0 t) - \frac{t}{n\omega_0}\cos(n\omega_0 t) \right] \right\}\Bigg|_0^{2\pi}; \quad \omega_0 = \frac{1}{2}$$

$$A_n = \frac{1}{\pi}\left\{ \frac{2}{n\omega_0}\left[\frac{2\pi}{n\omega_0}\cos(2n\pi\omega_0) \right] \right\} = \frac{4}{(n\omega_0)^2}\cos(2n\pi\omega_0)$$

Thus the input function is given by equation (7.48a) with A_0 and A_n as given, and $\psi = 0$.

     Figure 7.25a shows this Fourier series representation of the input, computed up to the $n = 15$ term. As can be observed from the plot of both the original function (dashed line) and the truncated series (solid line), the approximation is almost exact for this number of terms.

Figure 7.25

Input Function and Forced Response

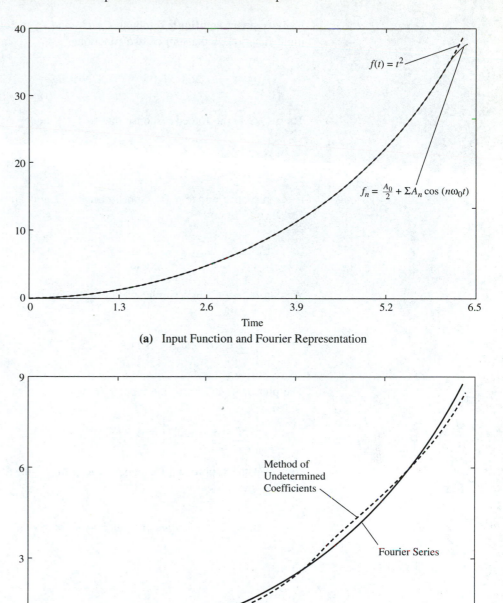

$f(t) = t^2$

$f_n = \dfrac{A_0}{2} + \Sigma A_n \cos{(n\omega_0 t)}$

Time

(a) Input Function and Fourier Representation

Method of
Undetermined
Coefficients

Fourier Series

Time

(b) Forced Response by Fourier Series and by Method of Undetermined Coeffecients

Fourier series solution: To obtain C_0, C_n, and ϕ_n in equation (7.48b) we consider the input function as the sum of two functions,

$$f_1(t) = \frac{A_0}{2} \quad \text{and} \quad f_2(t) = \sum_{n=1}^{\infty} A_n \cos(n\omega_0 t)$$

Then $C_0/2$ is the forced response due to $f_1(t)$, and

$$\sum_{n=1}^{\infty} C_n \cos(n\omega_0 t + \phi_n)$$

is the response due to $f_2(t)$. Substituting $C_0/2$ and $f_1(t)$ into the differential equation (7.49), we obtain

$$\frac{\omega_N^2 C_0}{2} = \frac{A_0}{2} \Rightarrow C_0 = \frac{A_0}{(\omega_N^2)}$$

We can obtain C_n in a similar way. Employing phasor transforms, we let

$$\mathscr{P}\left[\sum_{n=1}^{\infty} C_n \cos(n\omega_0 t + \phi_n)\right] = Y$$

The phasor transform of the input $f_2(t)$ is

$$\mathscr{P}\left[\sum_{n=1}^{\infty} A_n \cos(n\omega_0 t)\right] = A_n$$

Note that the suppressed frequency is $\omega = n\omega_0$. Substituting in the differential equation, we have

$$-(n\omega_0)^2 Y + 2\zeta\omega_N j n\omega_0 Y + \omega_N^2 Y = A_n \Rightarrow Y = \frac{A_n}{[\omega_N^2 - (n\omega_0)^2] + j2\zeta\omega_N n\omega_0}$$

so that

$$C_n = |Y| = \frac{A_n}{\sqrt{[\omega_N^2 - (n\omega_0)^2]^2 + (2\zeta\omega_N n\omega_0)^2}}$$

$$\text{and} \quad \phi_n = \angle Y = -\tan^{-1}\left[\frac{2\zeta\omega_N n\omega_0}{\omega_N^2 - (n\omega_0)^2}\right]$$

Method of undetermined coefficients: From Table 7.1, for the forcing input $f(t) = t^2$, we try the solution

$$y(t) = a_2 t^2 + a_1 t + a_0; \qquad \frac{dy}{dt} = 2a_2 t + a_1; \qquad \frac{d^2 y}{dt^2} = 2a_2$$

Substituting into equation (7.49), we obtain

$$2a_2 + 2\zeta\omega_N(2a_2 t + a_1) + \omega_N^2(a_2 t^2 + a_1 t + a_0) = t^2$$

Equating corresponding coefficients of powers of t, we have

$$\omega_N^2 a_2 = 1 \Rightarrow a_2 = \frac{1}{\omega_N^2}$$

$$4\zeta\omega_N a_2 + \omega_N^2 a_1 = 0 \Rightarrow a_1 = \frac{-4\zeta}{\omega_N^3}$$

$$2a_2 + 2\zeta\omega_N a_1 + \omega_N^2 a_0 = 0 \Rightarrow a_0 = \frac{-2(1 - 4\zeta^2)}{\omega_N^4}$$

Hence

$$y(t) = \frac{1}{\omega_N^2}\left(t^2 - \frac{4\zeta}{\omega_N} t - \frac{2(1 - 4\zeta^2)}{\omega_N^2} \right)$$

Figure 7.25b shows the Fourier series solution (for $n = 15$ terms; solid line) and the response obtained by the method of undetermined coefficients (dashed line). The correspondence between the two solutions is good but can be improved to a desired degree of accuracy by including additional terms of the infinite Fourier series, equation (7.48b). Although the present example was used to illustrate both the computation of half-range Fourier series for an even (or odd) function resulting from symmetry in the periodic function, and the representation of a nonperiodic function using Fourier series, the primary demonstration was the application of Fourier series to the solution of a linear time-invariant system subject to periodic input.

....

The Complex Fourier Series

Recall the Euler identities equation (6.41c):

$$e^{j\phi} = \cos \phi + j \sin \phi, \qquad e^{-j\phi} = \cos \phi - j \sin \phi$$

where $j = \sqrt{-1}$. Using these results, we can write the Fourier series for the function $f(t)$ that satisfies the Dirichlet conditions as follows:

$$f(t) = \sum_{n=-\infty}^{\infty} c_n e^{jn\omega_0 t} = \sum_{n=-\infty}^{\infty} |c_n| e^{j(n\omega_0 t + \phi_n)}; \qquad t_0 \leq t \leq t_0 + \Gamma \tag{7.50a}$$

where $\omega_0 = 2\pi/\Gamma$ (Γ = period) as before, and c_n is a complex-valued constant:

$$c_n = \frac{1}{\Gamma} \int_{t_0}^{t_0+\Gamma} f(t) e^{-jn\omega_0 t}\, dt = |c_n| e^{j\phi_n} = |c_n| \cos \phi_n + j|c_n| \sin \phi_n \tag{7.50b}$$

$$\phi_n = \tan^{-1}\left(\frac{\text{Img}\{c_n\}}{\text{Real}\{c_n\}} \right) \tag{7.50c}$$

If $f(t)$ is discontinuous at t, then the left side of equation (7.50a) should be replaced by the arithmetic mean of the right-hand and left-hand limits: $\frac{1}{2}[f(t+) + f(t-)]$, in accordance with the Dirichlet conditions. As before, equations (7.50a) and (7.50b) constitute a *Fourier series transform pair*. Note that equations (7.50b) and (7.50c) yield directly the amplitude and phase spectra. The correspondence between the *exponential* form

(7.50a) and the *trigonometric* form (7.42a), for real $f(t)$, can be reestablished as follows: From equation (7.50b),

$$c_{-n} = \frac{1}{\Gamma} \int_{t_0}^{t_0+\Gamma} f(t) e^{jn\omega_0 t} \, dt = \left[\frac{1}{\Gamma} \int_{t_0}^{t_0+\Gamma} f(t) e^{-jn\omega_0 t} \, dt \right]^* = c_n^* \qquad \textbf{(7.51a)}$$

Substituting equation (7.51a) into equation (7.50a) and using Euler's formula gives us see ([8])

$$f(t) = c_0 + \sum_{n=1}^{\infty} [(c_n + c_n^*)\cos n\omega_0 t + j(c_n - c_n^*)\sin n\omega_0 t] \qquad \textbf{(7.51b)}$$

Example 7.14   Representation in Complex Exponential Form

Determine using the exponential form, the Fourier series expansion of the sawtooth type function of Fig. 7.21 that was considered in Example 7.12. Obtain the amplitude and phase spectra and relate the results to those determined in example 7.12.

Solution: The Fourier series is given by equation (7.50a) where, from equation (7.50b), using $t_0 = -1$, and $\Gamma = 4$ ($\omega_0 = \pi/2$), as before,

$$c_n = \frac{1}{\Gamma} \int_{t_0}^{t_0+\Gamma} f(t) e^{-jn\omega_0 t} \, dt = \frac{1}{4} \int_{-1}^{3} (t+1) e^{-jn\omega_0 t} \, dt$$

$$c_n = \frac{1}{4} \int_{-1}^{2} (t+1) e^{-jn\omega_0 t} \, dt + \frac{1}{4} \int_{2}^{3} 0 . e^{-jn\omega_0 t} \, dt$$

$$= \frac{1}{4} \left\{ \frac{e^{-jn\omega_0 t}}{-(n\omega_0)^2} (-jn\omega_0 t - 1) + \frac{e^{-jn\omega_0 t}}{-jn\omega_0} \right\} \Bigg|_{-1}^{2}$$

Applying the limits and Euler's identities, we have

$$c_n = \frac{1}{4(n\omega_0)^2} \{ [\cos(2n\omega_0) - \cos(n\omega_0)] + j[3n\omega_0 \cos(2n\omega_0) - \sin(n\omega_0)] \} \quad \textbf{(7.52a)}$$

Equation (7.52a) gives the complex Fourier coefficients. However, note that the value of c_0 cannot be found by putting $n = 0$ in equation (7.52a). Rather, we substitute $n = 0$ in equation (7.50b) to get

$$c_0 = \frac{1}{\Gamma} \int_{t_0}^{t_0+\Gamma} f(t) \, dt = \frac{1}{4} \int_{-1}^{2} (t+1) \, dt + \frac{1}{4} \int_{2}^{3} 0 . \, dt = \frac{1}{4} \left\{ \frac{t^2}{2} + t \right\} \Bigg|_{-1}^{2} = \frac{9}{8}$$
$$\textbf{(7.52b)}$$

The amplitude spectrum as given directly by equation (7.52a) is

$$|c_n| = \frac{1}{4(n\omega_0)^2} \{ [\cos(2n\omega_0) - \cos(n\omega_0)]^2 \qquad\qquad \textbf{(7.52c)}$$
$$+ [3n\omega_0 \cos(2n\omega_0) - \sin(n\omega_0)]^2 \}^{1/2}$$

Note that there is a scaling of one-half between equation (7.52c) and the result of Example 7.12. The phase spectrum, however, is the same:

$$\phi_n = -\tan^{-1}\left\{\frac{-3n\omega_0\cos(2n\omega_0) + \sin(n\omega_0)}{\cos(2n\omega_0) - \cos(n\omega_0)}\right\} \tag{7.52d}$$

Finally, substituting equations (7.52a) and (7.52b) into equation (7.51b), we obtain

$$f(t) = \frac{9}{8} + \sum_{n=1}^{\infty}\frac{1}{2(n\omega_0)^2}\{[\cos(2n\omega_0) - \cos(n\omega_0)]\cos n\omega_0 t \tag{7.52e}$$

$$+ [-3n\omega_0\cos(2n\omega_0) + \sin(n\omega_0)]\sin n\omega_0 t\}$$

which is the same as the result obtained in Example 7.12.

On the Net

- SC8.MCD (Mathcad): source file for Fig. 7.22
- Fig 7_22.m (MATLAB): source file for Fig. 7.22
- SC9.MCD (Mathcad): source file for Fig. 7.25
- Fig7_25.m (MATLAB): source file for Fig. 7.25
- Auxiliary exercise: Practice Problems 7.39 through 7.46

7.5 PRACTICE PROBLEMS

7.1 Choose the correct statement: the transient response is (a) the free response only; (b) the forced response only; (c) the total response before steady state; (d) the total response over the entire period of interest.

7.2 The eigenvalues of a second-order system are as shown in Fig. 7.26 in the complex plane.

(a) What is the homogeneous ordinary differential equation describing this system?

(b) If the initial state of the system is $y(0) = 1$ and $\dot{y}(0) = 0$, find its free response.

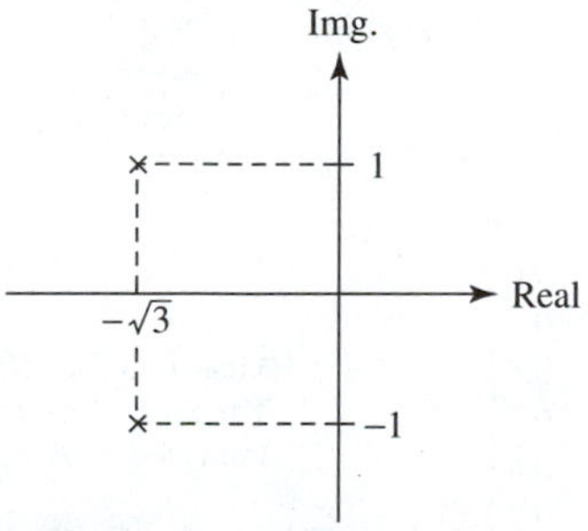

Figure 7.26 (Problem 7.2)

7.3 Sketch the unit-step response of the damped second-order system whose physical model is shown in Fig. 7.27. Label on the sketch as appropriate the magnitudes of the damped period, decay rate, phase shift, and peak displacement. Plot the eigenvalues of the system in the complex plane and label where applicable the natural frequency, damped frequency, and damping ratio, if

(a) $(\beta/2m) = 1$ s^{-1}, $(k/m) = 10$ s^{-2}, $y(0) = 1$ cm, $\dot{y}(0) = 0$ cm/s

(b) $(\beta/2m) = 1$ s^{-1}, $(k/m) = 10$ s^{-2}, $y(0) = 1$ cm, $\dot{y}(0) = 2$ cm/s

(c) $(\beta/2m) = 3$ s^{-1}, $(k/m) = 5$ s^{-2}, $y(0) = 1$ cm, $\dot{y}(0) = 2$ cm /s

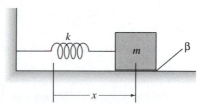

Figure 7.27 (Problem 7.3)

7.4* The system shown in Fig. 7.28 is initially at rest. If the input is a unit step, $v_{in} = U_S(t)$,

(a) Find the velocity and acceleration immediately after the application of the input, that is, $v(0+)$ and $\dot{v}(0+)$.

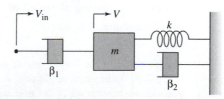

Figure 7.28 (Problem 7.4)

(b) Sketch the solution and estimate the maximum (absolute value) force in the damper β_2 if $\beta_1 = \beta_2 = 175$ N · s/m, $k = 350$ N/m, $m = 350$ kg.

(c) Find and sketch the force in the spring [from (b)] as a function of time, and estimate the maximum (absolute value) force in the spring.

7.5 Show the results given in equations (7.17a), (7.17c), and (7.17e).

7.6 Find the impulse response of the system

$$\frac{d^2y}{dt^2} + 3\frac{dy}{dt} + 2y = f(t)$$

initially at rest

7.7 Find the response of the system of Problem 7.6 when $f(t) = U_s(t) = $ unit step.

7.8* Find the solution to

(a) $\dfrac{dy}{dt} + 2y = 5$

(b) $\dfrac{d^2y}{dt^2} + 2\dfrac{dy}{dt} = 0$

   using the result from (a).

7.9 (a) The equation of motion of a mechanical system is given by

$$\frac{d^2y}{dt^2} + 2\frac{dy}{dt} + 5y = f(t)$$

(i) What is the damping ratio? The natural frequency (Hz)? Is the system overdamped critically damped, under-damped, or undamped?

(ii) Find the free response of the system from an initial state $y(0) = 1$ cm, $\dot{y}(0) = 0$.

(iii) Find and sketch the forced response when the forcing input is $f(t) = \sin(t)$.

(iv) For the same initial state given in (ii), what is the general response of the system under the forcing input in (iii)?

7.9 (b) Repeat parts (iii) and (iv) if the forcing input is $f(t) = \sin(t)e^{-t}$.

7.10 An overdamped linear second-order system with the equation of motion

$$\left(\frac{m}{k}\right)\frac{d^2y}{dt^2} + \left(\frac{\beta}{k}\right)\frac{dy}{dt} + y = u(t)$$

is initially at rest ($y(0) = \dot{y}(0) = 0$). Find the response of this system when it is subject to the forcing $u(t) = u_0 t$ by

(a) Convolution with the impulse response.

(b) Another method.

7.11* Sketch the response V_{out} of the system whose physical model is shown in Fig. 7.29. Indicate, if applicable, the magnitude of the damped period, number of cycles to damp to $1/e$ of initial value, the phase angle, the frequency of the response (Hz), and the damping ratio if the switch was for a

long time at a, and then at $t = 0$ it is switched from a to b and if $2RC = 1$ s, $LC = 0.1$ s^2, deduce the initial conditions from physical reasoning.

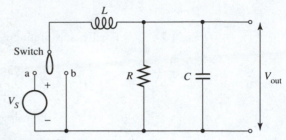

Figure 7.29 (Problem 7.11)

7.12 Repeat Problem 7.11 if the switch was for a long period initially at b and then is switched from b to a at $t = 0$.

7.13 (a) An unknown oscillatory device gave the measured free response shown in Fig. 7.30. It is desired to model the device as an underdamped second-order dynamic system. From the given data,

(i) Find the damped period, damped circular frequency, decay time constant, natural (circular) frequency, damping ratio, and logarithmic decrement.

(ii) Sketch the eigenvalues of this system in the complex plane and identify or represent on this sketch all the applicable quantities determined in (i).

(iii) Find the initial conditions for the measured response.

(iv) Write the time function representing the data and the second-order ordinary differential equation describing the system.

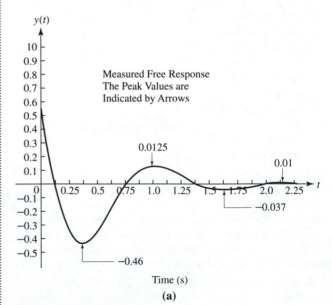

Figure 7.30 (Problem 7.13)

7.13 (b)* For the unit-step response of a second-order system given in Fig. 7.30b, determine

(i) The damped circular frequency.

(ii) The damping ratio.

(iii) The percent overshoot.

(iv) The rise time and.

(v) The settling time for 5% tolerance on the steady-state error.

Unit-step Response

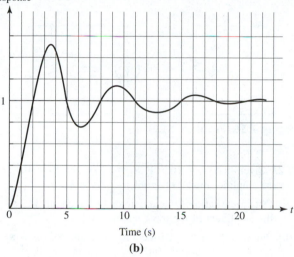

Time (s)

(b)

Figure 7.30, *cont'd* **(Problem 7.13)**

7.13 (c) You have a second-order instrument that you have determined has a maximum output four times the DC response. What is the damping ratio? *Hint:* the DC response can be viewed as the output at zero frequency.

7.14 (a) What is the number of degrees of freedom of the system described by the following model?

(i) $\dfrac{d^4y}{dt^4} + 2\dfrac{d^2y}{dt^2} + y = f(t)$

(ii) $\begin{cases} \dfrac{d^2y_1}{dt^2} + 2\dfrac{dy_1}{dt} + 450y_1 - 50y_2 = f_1(t) \\[2mm] \dfrac{d^2y_2}{dt^2} + 100(y_2 - y_1) = 0 \end{cases}$

7.14 (b) The governing equations for a linear elastic system can also be represented by the matrix model $\mathbf{m\ddot{x}} + \mathbf{c\dot{x}} + \mathbf{kx} = \mathbf{F}$.
Determine for (i) each of the systems in Problem 7.14(a), for (ii) the unbalanced equipment with vibration absorber equation (7.36d), the matrices $\mathbf{m}$, $\mathbf{c}$, and $\mathbf{k}$, which are known as the mass, damping, and stiffness matrices, respectively.

7.15 Prove by induction that

$$\mathscr{P}\left[\dfrac{d^n}{dt^n}f(t)\right] = (j\omega)^n F$$

where

$$F = Ae^{j\phi} = \mathscr{P}[f(t)] \quad \text{and} \quad f(t) = A\cos(\omega t + \phi)$$

In Problems 7.16 through 7.18, find (a) the driving-point impedance; (b) the driving-point admittance; (c) the transfer impedance; and (d) the transfer admittance.

7.16* The electrical (lag-lead) compensation network shown in Fig. 7.31.

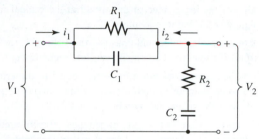

Figure 7.31 (Problem 7.16)

7.17 The electrical network as illustrated in Fig. 7.32.

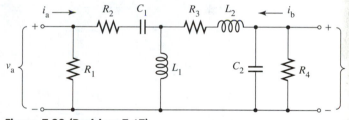

Figure 7.32 (Problem 7.17)

7.18 The mechanical compensation component shown in Fig. 7.33.

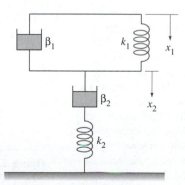

Figure 7.33 (Problem 7.18)

7.19* Find the amplitude and phase, that is, A and ϕ for the following transfer functions:

(a) $\dfrac{10 + j5\omega}{(1 + j\omega/4)(1 + j\omega)} = Ae^{j\phi}$

(b) $\dfrac{3}{(j\omega)(4 + j\omega)(1 + j3\omega)} = Ae^{j\phi}$

7.20 For the two parallel *RLC* circuits shown in Fig. 7.34, let the critical frequency be defined as the frequency of the input at which the magnitude ratio is as small or as large as possible.

(a) What is the critical frequency of each circuit.

(b) Assuming that $R = 0$ for each combination, what is the value of the impedance of the circuit at the critical frequency if the input is a sinusoid?

(c) Obtain the amplitude and phase of the voltage V for each circuit if the current i is a sinusoid of frequency ω.

(d) If the same current with a frequency equal to the critical frequency is input to both circuits, which circuit output will lead the other and by what amount?

(e) Sketch the magnitude of the impedance of each circuit as a function of frequency and for a quality factor of 5. At what frequency is the impedance a maximum/minimum? What is the bandwidth in each case? Explain your answer.

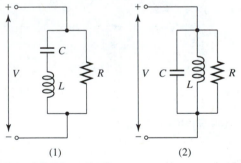

(1)   (2)

Figure 7.34 (Problem 7.20)

7.21* Develop a design (that is, determine R and L) for the series resonant circuit shown in Fig. 7.35 so that it can be used as a filter to remove a signal of undesirable frequency of 1 MHz. Assume that a signal is considered effectively removed if its output signal amplitude is less than the amplitude of a signal of useful frequency by a factor greater than $\sqrt{2}$. Also assume that the useful

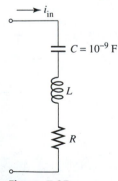

Figure 7.35 (Problem 7.21)

frequencies closest to 1 MHz are 0.9 MHz and 1.1 MHz. The input signals are all sinusoidal currents. Also, a finite value for R is necessary, since circuit components normally have intrinsic resistances.

7.22 A model of an electronic amplifier is shown in Fig. 7.36. Note the dependent current source whose value is determined by the current in R_1.

(a) What are the driving-point impedance and admittance of the circuit?

(b) What are the transfer impedance and admittance of the circuit?

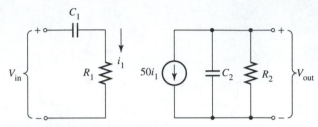

Figure 7.36 (Problem 7.22)

7.23 Given that $C_1 = 10^{-6}$ F, $R_1 = 10\ \Omega$, and $R_2 = 10^5\ \Omega$ for the amplifier model of Problem 7.22,

(a) Obtain the ratio of the amplitude of the output V_{out} to the amplitude of the input sinusoid V_{in}.

(b) Plot on a log-log scale the amplitude ratio of (a) as a function of frequency for three values of C_2: 10^{-11} F, 5×10^{-10} F, and 10^{-6} F.

(c) Which value of C_2 will you select if the amplifier is to be used in a high-fidelity amplifier, a telephone receiver, and a regular commercial function generator. Explain your answer in each case.

7.24 Show that a self-excited second-order single-degree-of-freedom mechanical system which is subject to a velocity-proportional force disturbance is equivalent to either a damped (overdamped or underdamped) or undamped second-order system or a second-order system with negative damping (that is, an unstable system).

7.25* Suppose the motor-actuated control valve of Example 3.3 is supplied by a single-acting piston pump, and there is interest in the effect of the ripple in the flow due to the pump, on the motor actuator operation,

(a) Derive a second-order model for the angular displacement of the motor rotor, assuming simply that pulsations in the flow to the valve give rise to an additional (harmonic) force $F_0 \sin 2\pi f t$ on the valve stem, where f is the pump speed in revolutions per second, and that no external torque is applied on the system ($T_{\text{in}} = 0$).

(b) What type of excitation (base or self) does the valve experience?

(c) Given the following parameter values: $\beta r^2 = 20$ N · s · m, $J = 0.2$ kg · m², $K = 2600$ N · m, and $rF_0 = 30$ N · m, what is the phasor transfer function for the angular position of the motor rotor?

(d) Draw a graph of the rotor angular displacement as a function of the pump speed (in rpm).

7.26 Consider the automobile suspension system model of Example 7.1. Suppose this model is to be used to assist in the preliminary development of the suspension system for a new vehicle whose suspension must be such that the amplitude of vehicle vibration relative to the road surface does not exceed 0.05 m for most road surface conditions. If from accumulated data the developers have established that (a) most road surface profiles are effectively represented by a sine wave as shown in Fig. 7.37, $r = r_0 \sin\dfrac{\pi x}{L}$; (b) the maximum average road surface excursion is about 0.2 m, and the range for L is 1.0 m $\le L \le$ 10.0 m; and (c) available vehicle suspension springs and dampers are limited to the following values for the spring constant and viscous damper constant: $10^7 \le K$ (N/m) $\le 3 \times 10^7$; $7500 \le \beta$ (N · s/m) $\le 10{,}000$. Obtain preliminary values of K and β that will meet the design specification at a constant horizontal vehicle speed of 10 m/s. Assume that the average mass of the car is 1000 kg.

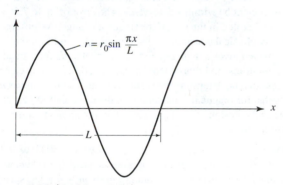

Figure 7.37 (Problem 7.26)

7.27* The vibration-producing device of Problem 3.25 has $m_e = 1$ kg and $e = 0.1$ m. If $m_2 = 100$ kg and $K = 50{,}000$ N/m, find

(a) The natural frequency of the system.

(b) The steady-state amplitude of vibration of m_2.

(c) The time lag between the peak force in the springs and the peak displacement of m_e.

7.28 A machine that is known to cause unacceptable vibrations in nearby equipment is to be mounted on available vibration isolators as shown in Fig. 7.38. The isolators collectively have an effective spring constant $K = 200{,}000$ N/m and viscous damping constant $\beta = 2000$ N · s/m. If the exciting force on the machine is given by $F = 3000 \cos 60t$ N, determine

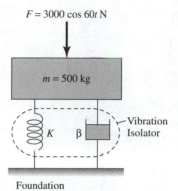

Figure 7.38 (Problem 7.28)

(a) The maximum force that the machine will transmit to the foundation.

(b) The amount by which the mass of the machine can be increased in order to reduce the maximum transmitted force by 50%.

7.29 If in Problem 7.28 other isolators with the same spring constant but different damping properties are available, select an isolator (determine β) that can reduce the maximum force transmitted to the foundation by the machine to below 390 N.

7.30 As a result of torsional vibrations in the prime mover for the overhead cable of a tramway in a ski resort, the support for a tram that has stopped during a sightseeing run reciprocates horizontally with a small amplitude, as shown in Fig. 7.39. Assuming that the response of the tram can be simulated using a pendulum that oscillates through small angles,

(a) Determine the transfer function from x to θ and the steady-state angular vibration of the tram as a function of the excitation frequency f.

(b) For what range of perturbation frequencies is the assumption of small angular oscillations valid?

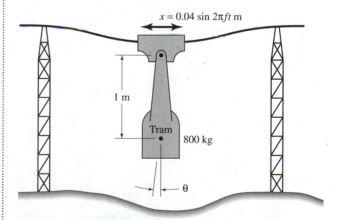

Figure 7.39 (Problem 7.30)

7.31 In Example 7.11, if the mechanical equipment is to be operated at 2400 rpm rather than 1800 rpm,

(a) Show that the vibration absorber design of that example will not be adequate at this new speed, and obtain a new design for the same absorber illustrated in Fig. 7.17. Assume that the mass of the absorber is limited to the same maximum value as before.

(b) Consider as an alternative the viscous vibration absorber shown in Fig. 7.40. Determine a design of this absorber that will permit the equipment to operate satisfactorily (as defined in Example 7.11) over a speed range of 1800–2400 rpm. Assume that the absorber mass is still limited to a maximum value of 450 kg.

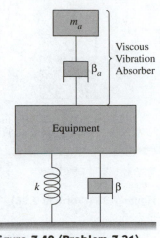

Figure 7.40 (Problem 7.31)

7.32* The reservoir-level monitoring system shown in Fig. 7.41 is used to record remotely (in the control room) the fluctuations in reservoir level in a large water treatment facility. The recorder is subject to a constant high-frequency–low-amplitude disturbance caused by level variations (ripples) due to periodic diversion of water into the reservoir, from a multiplexed pump, through a single large-capacity discharge line near the recorder. It is desired that the level recorder filter out the high-frequency variations while measuring to 2.5% accuracy the normal variations in reservoir level. Available experimental data show that normal reservoir level variation due to water use in the community served by the reservoir during the day has a (minimum) period of about 6 h with an amplitude of 1 m (see insert). The ripples have on average an amplitude of 0.04 m and a (maximum) period of 30 minutes. Assuming that the disturbance will be considered effectively filtered out if the changes in the measurement y that are due to the inlet flow are less than 2.5% of the changes in y caused by the water demand, design the level recorder, that is, find satisfactory values for d and A. The conduit length L is fixed at 20 m, and the conduit diameter and tank cross-sectional area are limited to those shown in the Table 7.2 for off-the-shelf units. Assume that a low-pressure drop in the conduit is desirable because of other instruments placed inside the conduit.

7.33* Consider the general-purpose accelerometer of Example 7.9. The accelerometer is to be used to measure dynamic accelerations with (circular) frequencies in the range 50 to 150 s^{-1}. The critical measurements have circular frequencies around 100 s^{-1}, and it is required that at this frequency the error in the measurement be within 1.5%, whereas an error of up to 7.5% can be tolerated in the rest of the frequency range. Assuming the only physical constraint is the type of oil used to fill the case, which gives a damping constant $\beta = 50$ kg/s for typical configurations, and this cannot be changed, design the accelerometer, that is, determine the spring constant k and the mass m. Note that for practical considerations, the mass should be as large as the design will allow, since other components contribute to the total mass.

7.34 An accelerometer (see Fig. 7.14) uses an LVDT to indicate the displacement of the vibrating mass. If the relevant instrument parameters are: vibrating mass $m = 0.8$ kg, spring constant $k = 0.5$ N/mm, and damping constant $\beta = 0.20$ N · s/m, and if the peak acceleration due to be measured at any frequency will not exceed 5 m/s^2 in amplitude, determine the desirable size (in centimeters) of the linear displacement range for the LVDT.

7.35 (a) A poorly designed accelerometer has a damping ratio of 0.34. If the input frequency is identical with the natural frequency of the instrument, what maximum percentage error will be made in the measurement? Assume the accelerometer is described by an underdamped second-order scalar ODE.

7.35 (b)* An accelerometer having a natural frequency of 1500 Hz and damping ratio of 0.6 was used to measure an acceleration signal of known magnitude but of uncertain frequency. If an error of 25% was made in the measurement, what is the probable frequency of the acceleration?

Table 7.2	
d(m)	A(m^2)
0.041	0.10
0.052	0.16
0.063	0.22

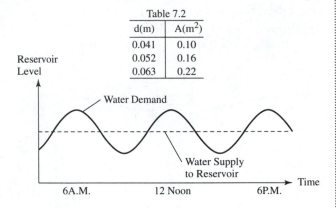

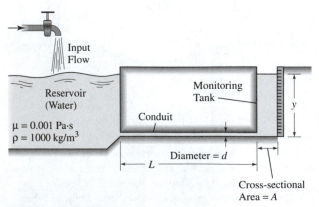

Input Flow

Reservoir (Water)

$\mu = 0.001$ Pa·s
$\rho = 1000$ kg/m^3

Monitoring Tank

Conduit

y

Diameter $= d$

L

Cross-sectional Area $= A$

Figure 7.41 (Problem 7.32)

7.36 Repeat Problem 7.9(a) (iii) using phasor transformation.

7.37 The equation of motion of a mechanical system is given by

$$\frac{d^2y}{dt^2} + \left(\frac{\beta}{m}\right)\frac{dy}{dt} + \left(\frac{k}{m}\right)y = u(t); \qquad y(0) = \dot{y}(0) = 0$$

where y is displacement. Given that $m = 2$ kg, $\beta = 8$ N·s/m, and $k = 200$ N/m,

(a) Find the total response of the system to the input $u(t) = \sin(9t)e^{-t}$.

(b) What is the ratio of the amplitude of y to the amplitude of u at steady state?

(c) What is the phase shift between y and u at steady state?

(d) The mechanical system is known to function best when the damping ratio is 0.1 and the variations in y due to potential disturbances are a minimum. However, the damping in the system β is fixed, although the mass m and the spring constant k are adjustable. If the input from (a) $u(t) = \sin(9t)e^{-t}$ represents the only disturbance (due to a nearby piece of machinery) that the system experiences, find a value of m and k that will give good performance. *Note:* Consider good performance to mean that the variations are reduced by a factor of about 5 or more over the present situation.

7.38 Find (a) the input/output phasor transfer function including its magnitude and phase, and (b) the steady-state response of the following stable linear systems:

(i) $\dfrac{d^3y}{dt^3} + 4\dfrac{d^2y}{dt^2} + 5\dfrac{dy}{dt} + 2y = 3\sin 3t$

(ii) $\dfrac{d^4y}{dt^4} + 10\dfrac{d^3y}{dt^3} + 35\dfrac{d^2y}{dt^2} + 50\dfrac{dy}{dt} + 24y = \cos t$

(iii) $\dfrac{d^3y}{dt^3} + 3\dfrac{d^2y}{dt^2} + 3\dfrac{dy}{dt} + y = \cos(2t)e^{-t}$

7.39 Determine and plot (using computational software) the trigonometric form of the Fourier series representation, including the amplitude and phase spectra of the following periodic functions shown in Fig. 7.42:

(a)* The square wave shown

(b) The indicated pulse train

(c) The zero-average function shown

(d)* The regular pulse train shown

(e) The indicated sawtooth waveform

(f) The indicated triangular wave

(g) The zero-point symmetric function shown

(h)* The impulse train of the indicated strength

7.40 Repeat Problem 7.39 using the exponential form of the Fourier series

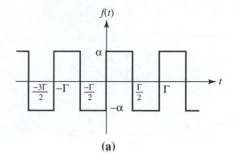

(a)

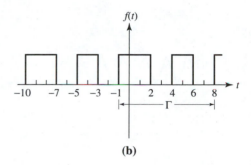

(b)

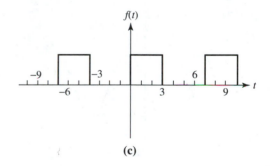

(c)

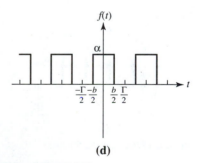

(d)

Figure 7.42 (Problem 7.39)

7.41 Find the trigonometric Fourier series expansion for the following functions:

(a) The zero-average value function shown in Fig. 7.23a. Assume that

$$f(t) = \begin{cases} \cos t, & 0 < t < \pi \\ 0, & \pi < t < 2\pi \end{cases}$$

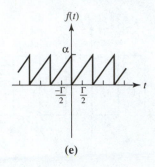

(e)

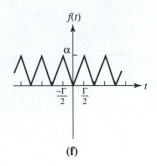

(f)

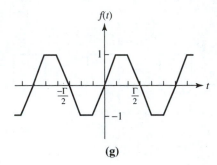

(g)

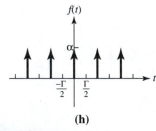

(h)

Figure 7.42 cont'd (Problem 7.39)

(b) The even function shown in Fig. 7.23b. Let $f(t) = t(2 - t)$, $0 < t < 2$.

(c) The odd function shown in Fig. 7.23c

(d)* The neither even nor odd extension of $f(t) = t^2$; $0 < t < 2\pi$ shown in Fig. 7.24a

(e) The odd extension of $f(t) = t^2$; $0 < t < 2\pi$ shown in Fig. 7.24c

7.42 Repeat Problem 7.41 using the complex exponential form.

7.43 Rework Example 7.13 using the complex exponential form of the Fourier series.

7.44 Determine the forced response of the system of Example 7.13 if the forcing input is the sawtooth waveform of Fig. 7.21. (The Fourier series representation was determined in Example 7.12.)

7.45 Find the forced response of the system of Problem 7.6 if the forcing input is

(a) The periodic function of Fig. 7.21 whose Fourier series representation was given in example 7.12.

(b) The periodic function shown in Fig. 7.23a and described in Problem 7.41a.

7.46 Find and sketch the forced response of the system of Problem 7.9 when the forcing input $f(t)$ is the period function shown in Fig. 7.23b and described in Problem 7.41b.

7.47 Find using Fourier series, the steady-state response of the following stable linear systems:

(a) The system of Problem 7.38i

(b)* The system of Problem 7.38ii

(c) The system of Problem 7.38iii

7.48 (a) Given the periodic function $f(t)$ with the Fourier series

$$f(t) = \sum_{n=-\infty}^{\infty} c_n e^{jn\omega_0 t}$$

show that the mean value of the square of the function is given by

$$\frac{1}{\Gamma} \int_{t_0}^{t_0+\Gamma} |f(t)|^2 \, dt = \sum_{n=-\infty}^{\infty} |c_n|^2$$

7.48 (b) Show the following result for the trigonometric (Fourier series) representation of $f(t)$:

$$\frac{1}{\Gamma} \int_{t_0}^{t_0+\Gamma} |f(t)|^2 \, dt = \frac{A_0^2}{4} + \sum_{n=1}^{\infty} \left(\frac{A_n^2 + B_n^2}{2} \right)$$

$$= \frac{A_0^2}{4} + \sum_{n=1}^{\infty} \frac{C_n^2}{2}; \quad \text{(Parseval's formula)}$$

8

Further Solution by Transformation

I N CHAPTER 7 WE INTRODUCED THE PHASOR TRANSFORM method as an effective *shorthand* approach to the analysis of linear stationary ODE systems that are subject to sinusoidal forcing. We consider in this chapter generalizations of this type of technique for much more arbitrary input functions and even certain nonstationary ODE systems and partial differential equation systems. Specifically, we examine the *Fourier* and *Laplace transforms,* which are usually applied to continuous-time systems, and the *z-transform* for linear discrete-time systems. As was the case with the phasor transform, a significant advantage of these transform methods is that a **differential or difference equation** problem in the *time domain* is converted into an **algebraic** problem in the *transform domain.* Difficult analytical procedures in the time domain, such as those involving complex interactions between individual dynamic systems or their components, can be carried out more easily in the transformed domain provided the inverse transform of the results back to the time domain is available. Indeed, in some cases, useful information and conclusions can be derived directly from results in the transform domain without having to perform an inverse transformation. Consequently, transform techniques are widely applied to such tasks as representation of functions and performance, solution of dynamic systems, and analysis and processing of signals.

We begin with the *Fourier integral transform,* which follows directly from the Fourier series considered in the last chapter. For signals derived experimentally or otherwise in discrete form and applied to continuous-time systems, an equivalent *discrete Fourier transform* (DFT) is available. The use of the DFT is particularly facilitated by the *fast Fourier transform* (FFT), which is a very efficient computational algorithm commonly available in many commercial mathematics software packages including MATLAB and Mathcad. The Laplace transform, which is applicable to an even wider class of functions than is the Fourier transform, and which, unlike the Fourier transform, can be used in transient (nonzero initial state) analysis, is considered next, including, very briefly, its application to nonstationary systems and partial differential equation problems. The direct application of the Laplace transform to the continuous-time vector state equation is also discussed as an alternative solution approach. The chapter concludes with a corresponding look at the interpretation and application of the *z-transform* to discrete-time systems.

8.1 THE FOURIER TRANSFORM SOLUTION

The Fourier (series) transform pair, equations (7.50a) and (7.50b) or equations (7.42a) and (7.43), represents a periodic function $f(t)$ with an infinite sum of undamped sinusoids of different frequencies that are discrete multiples of the fundamental frequency given by the least period of the function:

$$\omega = n\omega_0 = n\left(\frac{2\pi}{\Gamma}\right)$$

For $t_0 = -\Gamma/2$, equations (7.50a) and (7.50b) can be written in the form

$$f(t) = \sum_{n=-\infty}^{\infty} c_n e^{jn\omega_0 t} = \sum_{n=-\infty}^{\infty}\left[\frac{1}{\Gamma}\int_{-\Gamma/2}^{\Gamma/2} f(t)e^{-jn\omega_0 t}\,dt\right]e^{jn\omega_0 t} \qquad \textbf{(8.1)}$$

$$= \sum_{n=-\infty}^{\infty} e^{jn\omega_0 t}\left(\frac{2\pi}{\Gamma}\right)\frac{1}{2\pi}\left[\int_{-\Gamma/2}^{\Gamma/2} f(t)e^{-jn\omega_0 t}\,dt\right]$$

One way to deduce the generalization of this transform pair to cover arbitrary or nonperiodic functions is to regard such a general function $f(t)$ as a "periodic" function with an arbitrarily large period: $\Gamma \longrightarrow \infty$. The fundamental frequency in this limit approaches a differential (infinitesimally small) frequency, whereas the frequencies of the sinusoids range continuously (no longer in discrete values) from $-\infty$ to $+\infty$:

$$\lim_{\Gamma\to\infty}\left(\omega_0 = \frac{2\pi}{\Gamma}\right) \longrightarrow d\omega \qquad \textbf{(8.2a)}$$

$$\lim_{\substack{\Gamma\to\infty \\ n\to\infty}}\left(\omega = n\frac{2\pi}{\Gamma}\right) \longrightarrow \omega; \quad \text{(continuous band)} \qquad \textbf{(8.2b)}$$

In equation (8.1), after $(n\omega_0)$ is replaced by $(\omega;$ continuous variety$)$ and $(2\pi/\Gamma)$ by $(d\omega)$, the limits of the integral in brackets $(-\Gamma/2$ and $\Gamma/2)$ approach $(-\infty$ and $\infty)$, and, by the definition of the definite integral, the summation over n approaches an integral. That is,

$$f(t) = \int_{-\infty}^{\infty} e^{j\omega t}\,d\omega\frac{1}{2\pi}\left[\int_{-\infty}^{\infty} f(t)e^{-j\omega t}\,dt\right] \qquad \textbf{(8.2c)}$$

Equation (8.2c) gives the *Fourier (integral) transform pair* applicable to an arbitrary function $f(t)$ if we define the quantity in brackets as the *Fourier (integral) transform* of $f(t)$. In other words, the *Fourier transform* of a function $f(t)$ is a function of ω and is defined by the following *integral* relations:

$$\mathscr{F}[f(t)] = F(\omega) = \int_{-\infty}^{\infty} f(t)e^{-j\omega t}\,dt; \quad \text{(Fourier transform)} \qquad \textbf{(8.3a)}$$

$$\mathscr{F}^{-1}[F(\omega)] = f(t) = \frac{1}{2\pi}\int_{-\infty}^{\infty} F(\omega)e^{j\omega t}\,d\omega; \quad \text{(inverse Fourier transform)} \qquad \textbf{(8.3b)}$$

Observe that the Fourier transform $F(\omega)$ is, like the phasor transform, a complex number.

Example 8.1   Calculation of Fourier and Inverse Fourier Transforms

a. Show that the Fourier transform of the exponential function $Ce^{-a^2t^2}$ is

$$C\frac{\sqrt{\pi}}{a}e^{-(\omega/2a)^2}$$

b. Show that the inverse Fourier transform of the function

$$C\frac{\sqrt{\pi}}{a}e^{-(\omega/2a)^2}$$

is $Ce^{-a^2t^2}$.

Solution a: From equation (8.3a),

$$\mathcal{F}[Ce^{-a^2t^2}] = F(\omega) = C\int_{-\infty}^{\infty} e^{-a^2t^2}e^{-j\omega t}\,dt = C\int_{-\infty}^{\infty} e^{-a^2t^2}(\cos\omega t - j\sin\omega t)\,dt$$

$$F(\omega) = C\left\{\left[\int_{-\infty}^{\infty} e^{-a^2t^2}(\cos\omega t)\,dt\right] - j\left[\int_{-\infty}^{\infty} e^{-a^2t^2}(\sin\omega t)\,dt\right]\right\}$$

Figure 8.1a shows the sketch of the time-domain function $f(t)$, which has a *zero-axis symmetry,* that is, it is an *even* function. The product of *two even functions* is another *even function;* the product of *two odd functions* is an *even function;* and the product of an *even function and an odd function* is an *odd function.* Because cos ωt is an even function, the term under the first integral in brackets above is an even function. In contrast, the term under the second integral in brackets is an odd function, since sin ωt is odd. The integral from $-\infty$ to ∞ of an even function is twice its integral from 0 to ∞, while the integral from $-\infty$ to ∞ of an odd function is zero. Consequently,

$$F(\omega) = 2C\int_{0}^{\infty} e^{-a^2t^2}(\cos\omega t)\,dt = 2C\left[\frac{1}{2}\sqrt{\frac{\pi}{a^2}}e^{-\omega^2/4a^2}\right] = C\frac{\sqrt{\pi}}{a}e^{-(\omega/2a)^2}$$

Solution b: From equation (8.3b),

$$\mathcal{F}^{-1}\left[C\frac{\sqrt{\pi}}{a}e^{-(\omega/2a)^2}\right] = f(t) = \frac{1}{2\pi}\int_{-\infty}^{\infty}\left(C\frac{\sqrt{\pi}}{a}e^{-(\omega/2a)^2}\right)e^{j\omega t}\,d\omega$$

$$= \frac{C}{2a\sqrt{\pi}}\int_{-\infty}^{\infty} e^{-(\omega/2a)^2}(\cos\omega t + j\sin\omega t)\,d\omega$$

$$f(t) = \frac{C}{2a\sqrt{\pi}}\left\{\left[\int_{-\infty}^{\infty} e^{-(\omega/2a)^2}(\cos\omega t)\,d\omega\right] + j\left[\int_{-\infty}^{\infty} e^{-(\omega/2a)^2}(\sin\omega t)\,d\omega\right]\right\}$$

The present $F(\omega)$ happens to be entirely *real* (no imaginary part), otherwise it could have been broken into the real and imaginary parts, which could then be considered separately under the present solution approach. Figure 8.1b shows the sketch of the real

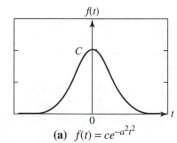

$f(t)$

C

0

(a) $f(t) = ce^{-a^2t^2}$

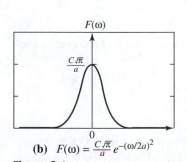

$F(\omega)$

$\dfrac{C\sqrt{\pi}}{a}$

0

(b) $F(\omega) = \dfrac{C\sqrt{\pi}}{a}e^{-(\omega/2a)^2}$

Figure 8.1

Sketch of $f(t)$ and $F(\omega)$

part of $F(\omega)$, which is also the entire function. Since this function is even, the second integral in brackets above is zero, and

$$f(t) = \frac{C}{a\sqrt{\pi}} \int_0^\infty e^{-\omega^2/4a^2}(\cos \omega t)\, d\omega = \frac{C}{a\sqrt{\pi}}\left[\frac{1}{2}\sqrt{\frac{\pi}{1/4a^2}}e^{-t^2/(4/4a^2)}\right] = Ce^{-a^2t^2}$$

The sufficient conditions for a function $f(t)$ to be Fourier transformable are the Dirichlet conditions. In particular, $f(t)$ should satisfy the condition of *absolute convergence:*

$$\int_{-\infty}^{\infty} |f(t)|\, dt < \infty; \quad \text{(absolute convergence condition)} \tag{8.4}$$

It follows that not all functions are Fourier transformable. Also, there are some functions that are Fourier transformable but do not satisfy the Dirichlet conditions. Although the $f(t)$ in Example 8.1 satisfies the absolute convergence condition, the Fourier transform pair of that example was especially chosen because of the rather straightforward manner with which it can be calculated. In general, use of equations (8.3a) and (8.3b) to calculate $f(t)$ and $F(\omega)$ can be mathematically quite involved, even for apparently simple functions. Fortunately, once calculated, Fourier transform pairs can subsequently be utilized, as required, without having to be recalculated. Such transform data are usually available in the form of a table of transform pairs that can be referred to whenever the input function ($f(t)$ or $F(\omega)$) matches an entry in the table. Table 8.1 is an abbreviated example of a *table of Fourier transform pairs*. Note that some functions in Table 8.1, such as the *singularity functions* $A_0\delta(t)$ (impulse function) and $u_0 U_S(t)$ (step function), do not satisfy the Dirichlet conditions. However, these functions are Fourier transformable.

Properties of the Fourier Transform

The application of the Fourier transform in dynamic systems analysis or signal analysis is facilitated by various transform properties, most of which are summarized in Table 8.2. The combination of a table of Fourier transform pairs and some of the properties of Fourier transforms permits the handling of some other functions not directly listed in the table of transform pairs but that can be related to available functions through operations involving such properties. In Example 8.2 we illustrate the application of a few of these properties and the use of a table of Fourier transform pairs such as Table 8.1.

Energetic Signals and Parseval's Formula

Suppose the signal $f(t)$ represents the voltage drop (volts) across a 1-Ω resistor. The instantaneous power transmission (watts) is given by

$$p(t) = f(t)[i(t)] = f(t)\left[\frac{f(t)}{R}\right] = f(t)^2$$

where $i(t)$ is the current (amperes) and $R = 1\ \Omega$ is the resistance. It follows that the energy dissipated in the resistor during the time interval $[t_0, t_f]$ is given by

$$E(t_0, t_f) = \int_{t_0}^{t_f} f(t)^2\, dt \tag{8.5a}$$

Table 8.1 Some Fourier Transform Pairs

ITEM	$f(t) = \dfrac{1}{2\pi} \displaystyle\int_{-\infty}^{\infty} F(\omega)e^{j\omega t}\, d\omega$	$F(\omega) = \displaystyle\int_{-\infty}^{\infty} f(t)e^{-j\omega t}\, dt$
1	$f(t) = u_0 U_s(t) = \begin{cases} u_0, & t > 0 \\ 0 & \text{otherwise} \end{cases}$	$F(\omega) = u_0\left[\pi\delta(\omega) - j\dfrac{1}{\omega}\right]$
2	$f(t) = \begin{cases} u_0, & t > 0 \\ 0, & t = 0 \\ -u_0, & t < 0 \end{cases}$	$F(\omega) = -j2u_0\dfrac{1}{\omega}$
3	$f(t) = u_0$	$F(\omega) = 2\pi u_0 \delta(\omega)$
4	$f(t) = Ce^{-\sigma t}U_s(t)$	$F(\omega) = \dfrac{C}{\sigma + j\omega}$
5	$f(t) = Ce^{-a^2 t^2}$	$F(\omega) = C\dfrac{\sqrt{\pi}}{a}e^{-(\omega/2a)^2}$
6	$f(t) = C\sin \omega_0 t$	$F(\omega) = j\pi C[\delta(\omega + \omega_0) - \delta(\omega - \omega_0)]$
7	$f(t) = C\cos \omega_0 t$	$F(\omega) = \pi C[\delta(\omega - \omega_0) + \delta(\omega + \omega_0)]$
8	$f(t) = \begin{cases} C(\sin \omega_0 t)e^{-\sigma t}, & t \geq 0 \\ 0, & t < 0 \end{cases}$	$F(\omega) = \dfrac{C\omega_0}{(\sigma + j\omega)^2 + \omega_0^2}$
9	$f(t) = \begin{cases} C(\cos \omega_0 t)e^{-\sigma t}, & t \geq 0 \\ 0, & t < 0 \end{cases}$	$F(\omega) = C\dfrac{\sigma + j\omega}{(\sigma + j\omega)^2 + \omega_0^2}$
10	$f(t) = A_0\delta(t)$	$F(\omega) = A_0$
11	$f(t) = A_0\displaystyle\sum_{n=-\infty}^{\infty} \delta(t - n\Gamma)$	$F(\omega) = \dfrac{2\pi A_0}{\Gamma}\displaystyle\sum_{n=-\infty}^{\infty}\delta\left(\omega - \dfrac{2n\pi}{\Gamma}\right)$
12	$f(t) = \begin{cases} u_0, & |t| \leq T_p/2 \\ 0, & \text{otherwise} \end{cases}$	$F(\omega) = 2u_0\dfrac{\sin(\omega T_p/2)}{\omega}$
13	$f(t) = \begin{cases} u_0\left(1 - \dfrac{2|t|}{T_p}\right), & |t| \leq \dfrac{T_p}{2} \\ 0 & \text{otherwise} \end{cases}$	$F(\omega) = \dfrac{1}{2}u_0 T_p\left[\dfrac{\sin(\omega T_p/4)}{(\omega T_p/4)}\right]^2$

Analogously, for a general signal $f(t)$, the energy content of the signal is proportional to the time integral of the square of the signal and is given by

$$E = \int_{-\infty}^{\infty} f(t)^2\, dt \tag{8.5b}$$

By the Fourier transform property Parseval's formula (item 5 of Table 8.2), E is also given in terms of the Fourier transform of the signal:

$$E = \frac{1}{2\pi}\int_{-\infty}^{\infty} |F(\omega)|^2 d\omega \tag{8.5c}$$

The quantity $|F(\omega)|^2$ is known as the *power density spectrum* of the function $f(t)$, while E as given by equation (8.5b) is known as the *energy spectrum density*.

Table 8.2 Some Properties of the Fourier Transform

DESCRIPTION	$f(t)$	$F(\omega)$
1. Linearity	$af_1(t) + bf_2(t)$	$aF_1(\omega) + bF_2(\omega)$
2. Scaling	$f(at)$	$\dfrac{1}{|a|}F\left(\dfrac{\omega}{a}\right)$
3. Derivative	$\dfrac{d^n}{dt^n}f(t)$	$(j\omega)^n F(\omega)$
4. Convolution	$f(t)*g(t) = \displaystyle\int_{-\infty}^{\infty} f(\tau)g(t-\tau)\,d\tau$	$F(\omega)G(\omega)$
5. Parseval's formula	$E = \displaystyle\int_{-\infty}^{\infty} f(t)^2\,d\tau$	$E = \dfrac{1}{2\pi}\displaystyle\int_{-\infty}^{\infty} |F(\omega)|^2\,d\omega$
6. Autocorrelation	$\displaystyle\int_{-\infty}^{\infty} f(\tau)f^*(\tau - t)\,d\tau$	$F(\omega)F^*(\omega) = |F(\omega)|^2$
7. Frequency shift	$e^{\pm j\omega_0 t}f(t)$	$F(\omega \mp \omega_0)$
8. Time shift	$f(t \pm t_0)$	$e^{\pm j\omega t_0}F(\omega)$
9. Time reversal	$f(-t)$	$F(-\omega)$
10. Symmetry	$F(t)$	$2\pi f(-\omega)$

In terms of the Fourier series representation of $f(t)$, equations (7.42) or (7.50), and energy dissipation over a time interval of one period, Parseval's formula (see Practice Problem 7.48) is

$$\frac{1}{\Gamma}\int_{t_0}^{t_0+\Gamma} |f(t)|^2\,dt = \sum_{n=-\infty}^{\infty} |c_n|^2; \quad \text{(complex exponential form)} \tag{8.6a}$$

or

$$\frac{1}{\Gamma}\int_{t_0}^{t_0+\Gamma} |f(t)|^2\,dt = \frac{A_0^2}{4} + \sum_{n=1}^{\infty}\left(\frac{A_n^2}{2} + \frac{B_n^2}{2}\right) \tag{8.6b}$$

$$= \left(\frac{A_0}{2}\right)^2 + \sum_{n=1}^{\infty}\frac{C_n^2}{2}; \quad \text{(trigonometric form)}$$

Equation (8.6a) implies that the energy of any signal is the sum of the energies of its spectral (frequency) components. An interpretation of equation (8.6b) is that the total power of a signal is the power of the DC component plus the (root-mean-square (rms) value) power of all the AC components. These results illustrate a practical significance of the *amplitude spectrum* of a signal, which was introduced in the last chapter. The amplitude spectrum contains information about the energy content of a signal and its distribution among the different frequencies represented in the signal.

Example 8.2   Tuned Resonant Circuit

Consider again the tuned resonant circuit of Example 7.8 (see Fig. 7.9b). Assuming that the input $v_{in}(t)$ is Fourier transformable, determine

a. The transformed input-output relationship.
b. The power density spectrum of the output of the circuit.

c. The (time-domain) unit impulse (forced) response.

d. The Fourier domain output *under zero initial state* if $v_{in}(t) = \delta(t) + e^{-3t^2}$. Plot the amplitude spectrum and the power density spectrum of the output for this case, assuming the following parameter values: $L = 0.1$ H, $C = 0.001$ F, and $R = 10$ Ω.

Solution a: As was the case with the phasor transform and Example 7.8, the Fourier transform of the component behaviors can be taken and combined according to the series network, using the *linearity* property of the transform. The result is the Fourier transform of the combination because the component behaviors are linear expressions.

Inductor: The voltage drop across the inductor is given by

$$L\frac{di_0}{dt} = v_{1-3}$$

The Fourier transform of this relation is given by the *derivative* property:

$$Lj\omega I_0(\omega) = V_{1-3}(\omega)$$

Thus the admittance (Fourier domain) is

$$\frac{I_0(\omega)}{V_{1-3}(\omega)} = \frac{1}{j\omega L}$$

Resistor and capacitor: Similarly, the admittance of the resistor is $1/R$, and that for the capacitor is given by the capacitor relation and the time differentiation (derivative) property:

$$C\frac{d}{dt}v_{2-4} = i_0 \Rightarrow Cj\omega V_{2-4}(\omega) = I_0(\omega) \Rightarrow \frac{I_0(\omega)}{V_{2-4}(\omega)} = j\omega C$$

Input-output relationship: The overall admittance is given as before by the series combination:

$$\frac{1}{Y(\omega)} = j\omega L + R + \frac{1}{j\omega C} \Rightarrow Y(\omega) = \frac{j\omega C}{(1 - \omega^2 LC) + j\omega RC}$$

so that the transformed input-output relationship or the (Fourier domain) *transfer function* is given by

$$I_0(\omega) = G(\omega)V_{in}(\omega) \text{ or } G(\omega) = Y(\omega) = \frac{j\omega C}{(1 - \omega^2 LC) + j\omega RC} \qquad \textbf{(8.7a)}$$

Solution b: The transformed output is given, in general, by

$$\text{Output}(\omega) = G(\omega)\text{Input}(\omega) \qquad \textbf{(8.7b)}$$

By equation(8.5c), the power density spectrum of the output is also given, generally by

$$|\text{Output}(\omega)|^2 = |G(\omega)|^2|\text{Input}(\omega)|^2 \qquad \textbf{(8.8a)}$$

The transformed input is, in general, a complex number with magnitude and phase:

$$\text{Input}(\omega) = M_i e^{j\phi_i(\omega)} \tag{8.8b}$$

Thus

$$|\text{Output}(\omega)|^2 = M_i^2 |G(\omega)|^2 \tag{8.8c}$$

Equation (8.8c) says that the power density spectrum of the response of a linear time-invariant system is the product of the power density spectrum of the input function (M_i^2) and the square of the magnitude of the system transfer function, that is, the square of the amplitude ratio $(|G(\omega)|^2)$. For the present system,

$$|I_0(\omega)|^2 = |V_{\text{in}}(\omega)|^2 \left[\frac{\omega^2 C^2}{(1 - \omega^2 LC)^2 + (\omega RC)^2} \right] \tag{8.8d}$$

Solution c: The Fourier transform of the unit impulse at $t = 0$ is by item 10 of Table 8.1, $\mathcal{F}[\delta(t)] = 1$. Substituting this value in equation (8.7a), we obtain the Fourier transform of the unit impulse response:

$$I_0(\omega) = G(\omega)(1) = G(\omega) = \frac{j\omega C}{(1 - \omega^2 LC) + j\omega RC}; \quad \text{(unit impulse response)} \tag{8.9}$$

The implication of this result is discussed later under Laplace domain transfer functions. To obtain the time-domain response, we can write $I_0(\omega)$ as

$$I_0(\omega) = j\omega \left[\frac{C}{(1 - \omega^2 LC) + j\omega RC} \right] = j\omega \left[\frac{1/L}{(1/LC - \omega^2) + j\omega R/L} \right]$$

and compare the quantity in brackets with the Fourier transform of the damped sine function, item 8 of Table 8.1, which can be written as

$$\mathcal{F}[A \sin(\omega_0 t) e^{-\sigma t}] = \frac{A\omega_0}{(\sigma + j\omega)^2 + \omega_0^2} = \left[\frac{A\omega_0}{(\omega_0^2 + \sigma_2 - \omega^2) + j2\sigma\omega} \right]$$

If we take

$$\frac{R}{L} = 2\sigma \quad \text{(note that } \sigma \text{ is a positive quantity)}$$

$$\frac{1}{LC} = \omega_0^2 + \sigma^2 = \omega_0^2 + \frac{R^2}{4L^2} \Rightarrow \omega_0^2 = \frac{1}{LC} - \frac{R^2}{4L^2}$$

$$A\omega_0 = \frac{1}{L} \Rightarrow A = \frac{1}{\sqrt{L/C - R^2/4}}$$

we can write $I_0(\omega)$ as

$$I_0(\omega) = j\omega \mathcal{F} \left[\frac{1}{\sqrt{L/C - R^2/4}} \sin\left(\sqrt{\frac{1}{LC} - \frac{R^2}{4L^2}} t \right) e^{-Rt/2L} \right]$$

By the derivative property, item 3 of Table 8.2, the time domain unit impulse response is given by the first time-derivative of the above term in brackets, that is,

$$
i_0(t) = \frac{1}{\sqrt{L/C - R^2/4}} \left[\sqrt{\frac{1}{LC} - \frac{R^2}{4L^2}} \cos\left(\sqrt{\frac{1}{LC} - \frac{R^2}{4L^2}} t \right) \right.
$$
$$
\left. - \frac{R}{2L} \sin\left(\sqrt{\frac{1}{LC} - \frac{R^2}{4L^2}} t \right) \right] e^{-Rt/2L}
$$

Solution d: The given input is the sum of two functions: a unit impulse (which has just been considered) and an exponential function (item 5 of Table 8.1). Thus using the entries from Table 8.1 and the linearity property (item 1 of Table 8.2), we obtain the Fourier transform of the input function:

$$
V_{in}(\omega) = 1 + \sqrt{\frac{\pi}{3}} e^{-(\omega/2\sqrt{3})^2}
$$

Note that $V_{in}(\omega)$ is a real quantity (zero imaginary part). Substituting in equation (8.7b), we have the Fourier domain output:

$$
I_0(\omega) = \frac{j\omega C}{(1 - \omega^2 LC) + j\omega RC} \left[1 + \sqrt{\frac{\pi}{3}} e^{-(\omega/2\sqrt{3})^2} \right]
$$

The power density spectrum is given by equation (8.8c) or (8.8d), that is,

$$
|I_0(\omega)|^2 = \left[\frac{\omega^2 C^2}{(1 - \omega^2 LC)^2 + (\omega RC)^2} \right] \left[1 + \sqrt{\frac{\pi}{3}} e^{-(\omega/2\sqrt{3})^2} \right]^2
$$

The amplitude spectrum is simply

$$
|I_0(\omega)| = \sqrt{|I_0(\omega)|^2} = \sqrt{\frac{\omega^2 C^2}{(1 - \omega^2 LC)^2 + (\omega RC)^2} \left[1 + \sqrt{\frac{\pi}{3}} e^{-(\omega/2\sqrt{3})^2} \right]^2}
$$

These functions are plotted in Fig. 8.2. Observe that the energy content is largest for frequencies near the resonant or natural frequency:

$$
\omega_R = 1/\sqrt{(LC)} = 100 \text{ s}^{-1}
$$

The similarity between the phasor transform and Fourier transform solutions is evident in the preceding example. Indeed, the transfer function is the same in both transform domains. The difference, which is implicit, is the meaning of ω. For the phasor transform, ω is a parameter with a specific value that is the same as the input frequency, whereas for the Fourier transform, ω is a variable that can take on all values between $-\infty$ and ∞. Note also that this example gives some indication of a shortcoming of the Fourier transform in dynamic system analysis. The system to be solved must have zero

Figure 8.2

Amplitude and Power Density Spectra

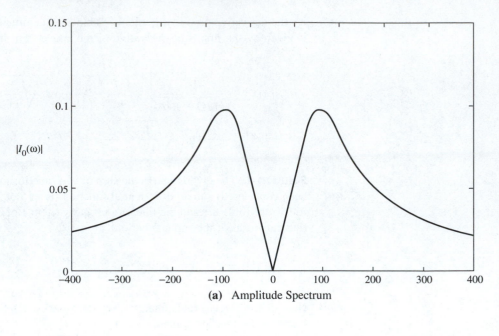

(a)   Amplitude Spectrum

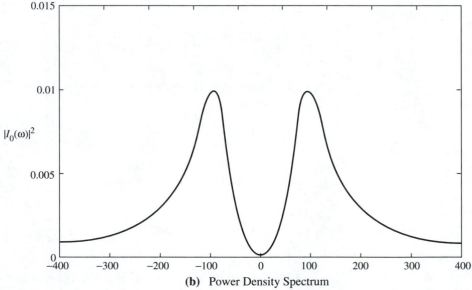

(b)   Power Density Spectrum

initial conditions (or only the forced response can be sought), since no means is otherwise provided in the definition of the transform pairs or their properties for incorporating initial conditions.

Signal Processing

The manipulation of signals is usually accomplished by using special circuits with known spectral characteristics called *filters*. Such filters can be used to remove noise (undesirable components) from a signal or to extract a desired signal from noise, to detect whether a particular signal is present *(matched filtering)*, or to equalize spectra (a procedure used to recover a signal from the output of a previous process), as well as in other signal processing applications (see [8]). The Fourier transform has important uses in the analysis of the input-output relationships and the determination of frequency spectra of linear time-invariant filters.

The processing of a signal can, in general, be represented in the Fourier domain by equation (8.7b), repeated here:

$$Y(\omega) = G(\omega)U(\omega) \tag{8.10a}$$

where $Y(\omega)$ is the Fourier transform of the output from the filter, $U(\omega)$ is the Fourier transform of the input signal, and $G(\omega)$ is the Fourier domain transfer function of the filter or the Fourier transform of the unit impulse response of the filter (recall equation (8.9)). Since all the quantities in equation (8.10a) are, in general, complex numbers (each having a magnitude and a phase), equation (8.10a) can be written as

$$Y(\omega) = |Y(\omega)|e^{j\phi_y(\omega)} = |G(\omega)||U(\omega)|e^{j[\phi_g(\omega)+\phi_u(\omega)]} \tag{8.10b}$$

where $(|Y(\omega)|, \phi_y(\omega))$ and $(|U(\omega)|, \phi_u(\omega))$ are the amplitude spectrum and phase spectrum for, respectively, the output and input signals, and $|G(\omega)|$ and $\phi_g(\omega)$ are the amplitude and phase transfer functions of the filter. Thus the output amplitude spectrum is equal to the product of the spectrum of the filter amplitude transfer function and the input signal amplitude spectrum, whereas the output phase spectrum is equal to the sum of the filter transfer phase function and the input signal phase spectrum.

According to the *convolution* property, item 4 of Table 8.2, the signal processing represented by equation (8.10a) is given in the time domain by

$$y(t) = u(t)*g(t) = \int_{-\infty}^{\infty} u(\tau)g(t - \tau)\,d\tau \tag{8.10c}$$

Often the *filtering* problem is to determine $g(t)$ (the filter construction) such that the output $y(t)$ will have the desired characteristics. A practical problem is ensuring that the resulting filter or a close approximation is *realizable,* that is, it can be physically constructed. Consider, for example, the problem of *extracting a signal from noise.* If we let the desired signal and the noise be $s(t)$ and $n(t)$, respectively, then the output signal from the filter is

$$y(t) = [s(t) + n(t)]*g(t) \tag{8.10d}$$

If the extraction is exact, then $y(t) = s(t)$, so that

$$s(t) = [s(t) + n(t)]*g(t) \tag{8.10e}$$

This gives the output Fourier spectrum

$$S(\omega) = [S(\omega) + N(\omega)]G(\omega)$$

so that the desired filter transfer function is

$$G(\omega) = \frac{S(\omega)}{S(\omega) + N(\omega)} = \frac{S(\omega)/N(\omega)}{1 + S(\omega)/N(\omega)} \tag{8.10f}$$

Equation (8.10f) suggests that $G(\omega)$ should approach unity in frequency regions where the signal spectrum is large relative to the noise spectrum, while in spectral regions where the noise spectrum is predominant, $G(\omega)$ should be very small.

A filter is *distortionless* when the form of the output is exactly identical with the input, except, possibly, for a uniform change in amplitude and/or a time lag. Equation (8.10c) for such a filter can be written as

$$y(t) = G_0 u(t - t_0) \quad \text{(distortionless filtering)} \tag{8.11a}$$

where G_0 and t_0 are the constant amplitude and time lag of the filter response, respectively. By the *time shift* property, item 8 of Table 8.2, the corresponding Fourier domain relation is

$$Y(\omega) = G_0 e^{-j\omega t_0} U(\omega); \quad \text{(distortionless filtering)} \tag{8.11b}$$

so that the transfer function of a distortionless filter is

$$G(\omega) = G_0 e^{-j\omega t_0}; \quad \text{(distortionless filter)} \tag{8.11c}$$

Various other classifications such as *amplitude-distorted, phase-distorted, low-pass, high-pass,* and *band-pass* (see [8]) are employed in describing the spectral characteristics of filters. For many of these categories of filters, well-developed procedures exist for approximating the desired response characteristics using known functions such as *elliptic, Butterworth,* or *Chebyshev.*

Butterworth Function in Low-Pass Filtering

For a brief illustration, consider the use of the Butterworth function for frequency-selective low-pass filters. An nth-order Butterworth function (see [8] or a related text for more detailed discussion) has the amplitude response shown in Fig. 8.3:

$$|H_n(j\omega)| = \frac{1}{\sqrt{1 + (\omega/\omega_c)^{2n}}} \tag{8.12a}$$

where ω_c is the cutoff frequency. It can be seen from Fig. 8.3 that the approximation of the Butterworth function to an ideal low-pass filter improves as the order n increases. For a specification of cutoff frequency and desired attenuation at some other frequency, the order of the Butterworth filter that meets these requirements can be obtained from equation (8.12a).

The realization of the Butterworth filter is another matter. It can be shown (see [8]) that a transfer function (to obtain the Fourier domain transfer function, substitute $s = j\omega$) of the form

$$H(s) = \frac{1}{a_0 + a_1 s + a_2 s^2 + \cdots + a_{n-1} s^{n-1} + a_n s^n} = \frac{k}{D(s)} \tag{8.12b}$$

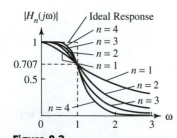

Figure 8.3

***n*th-order Butterworth Amplitude Response**

can be realized with passive circuit elements if and only if $D(s)$ is a *Hurwitz polynomial,* that is, $D(s)$ has complex-conjugate roots with negative real parts, and if $D(s)$ is partitioned into the sum of two polynomials, one containing even powers of s, and the other odd powers, then (1) the roots of each component polynomial are purely imaginary (a zero value being allowed), and (2) the roots alternate.

The transfer function corresponding to the Butterworth amplitude function (8.12a) satisfies

$$H_n(s)H_n(-s) = |H_n(j\omega)|^2 = \frac{1}{1 + [(j\omega/\omega_c)^2/j^2]^n} = \frac{1}{1 + (-[s/\omega_c]^2)^n} \tag{8.12c}$$

The $2n$ roots of the denominator polynomial $[1 + (-[s/\omega_c]^2)^n]$ are given by

$$(-1)^n[s/\omega_c]^{2n} = -1 = e^{j(2k-1)\pi}, \quad k = 1, 2, \ldots, 2n$$

$$\Rightarrow [s/\omega_c]^{2n} = e^{j(2k-1)\pi}e^{jn\pi}, \quad k = 1, 2, \ldots, 2n$$

or

$$[s/\omega_c]_k = e^{j(2k+n-1)\pi/2n}, \quad k = 1, 2, \ldots, 2n \tag{8.12d}$$

The roots $[s/\omega_c]$ all have a magnitude of 1 and are complex conjugates. Half of these roots are in the left-half complex plane, and the other half are in the right-half plane. The n(stable) roots in the left-half complex plane correspond to $H_n(s)$ in equation (8.12c), so that the Butterworth transfer function can be written as

$$H(s) = \frac{1}{1 + a_1[s/\omega_c] + a_2[s/\omega_c]^2 + \cdots + a_{n-1}[s/\omega_c]^{n-1} + [s/\omega_c]^n} \tag{8.12e}$$

where the coefficients $a_1, \ldots, a_{n-1}$ are such that the roots of the denominator polynomial in (8.12e) are the same n left-half complex-plane roots given by (8.12d). For example, for $n = 2$ (and $\omega_c = 1$), the roots are

$$s_1 = e^{j3\pi/4} = \cos\left(\frac{3\pi}{4}\right) + j\sin\left(\frac{3\pi}{4}\right) \quad \text{and}$$

$$s_2 = e^{j5\pi/4} = \cos\left(\frac{5\pi}{4}\right) + j\sin\left(\frac{5\pi}{4}\right)$$

and the denominator polynomial

$$D(s) = (s - s_1)(s - s_2) = s^2 + 1.4142s + 1$$

so that $a_1 = 1.4142$. Note that this polynomial satisfies the conditions for a Hurwitz polynomial. The two-port LC circuit shown in Fig. 8.4 can be used to realize the corresponding second-order Butterworth transfer function. The transfer function for the circuit can be shown (see also Practice Problem 8.8b) to be

$$\frac{V_{out}(s)}{V_{in}(s)} = \frac{1}{CLs^2 + (C + L)s + 2} \tag{8.12f}$$

Figure 8.4

Circuit for a Second-order
Butterworth-type Low-pass
Passive Filter

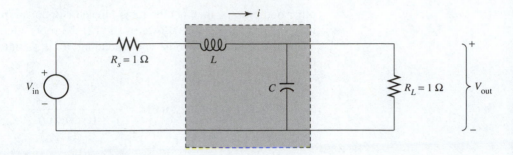

where, without loss of generality, the values of the source and load resistances (R_s and R_L) have been taken as 1 Ω. To account for the factor of 2 in the denominator of (8.12f), we can take

$$\frac{H(s)}{2} = \frac{1}{2([s/\omega_c]^2 + 1.4142[s/\omega_c] + 1)} = \frac{1}{2\left(\frac{1}{2}CLs^2 + \frac{1}{2}(C + L)s + 1\right)}$$

which for a given ω_c can be solved for C and L.

Example 8.3   Design and Realization of a Butterworth Passive Filter Circuit

Design and realize with passive circuit elements, a Butterworth-type low-pass filter having a cutoff frequency of 1 kHz and attenuation of 0.1 or less at twice the cutoff frequency.

Solution—Design: The Butterworth filter order is given by equation (8.12a), with $\omega = 2\omega_c$:

$$\frac{1}{\sqrt{1 + 2^{2n}}} \le 0.1 \Rightarrow 1 + 2^{2n} \ge 100 \Rightarrow n \ge 3.3147$$

Thus a fourth-order filter is required. The coefficients in the denominator polynomial of the transfer function (8.12e) can be found as previously described or by using the *buttap*(n) and *poly*(p) functions available in MATLAB's Signals and Systems Toolbox (see script file fig8_3.m, **On the Net**):

$$D(s) = \left[\frac{s}{\omega_c}\right]^4 + 2.6131\left[\frac{s}{\omega_c}\right]^3 + 3.4142\left[\frac{s}{\omega_c}\right]^2 + 2.6131\left[\frac{s}{\omega_c}\right] + 1 \qquad \textbf{(a)}$$

Solution—Realization: Equation (8.12f) suggests we can cascade two *LC* networks as shown in Fig. 3.19. The state model for this circuit was developed in Example 3.13, equations (3.38a) through (3.38e). The transfer function (given as an exercise in Practice Problem 8.7), which can be obtained by Fourier transforming these equations or by combining the transforms of the circuit component behaviors, is (again assuming $R_s = R_L = 1$ Ω):

$$\frac{V_{out}(s)}{V_{in}(s)} = \frac{1}{C_1C_2L_1L_2s^4 + (C_1L_1 + C_1C_2)L_2s^3 + (C_1 + C_2)(L_1 + L_2)s^2 + (C_1 + C_2 + L_1 + L_2)s + 2} \qquad \textbf{(b)}$$

Comparing this equation with the fourth-order Butterworth transfer function, given by (a):

$$\frac{H(s)}{2} = \frac{1}{2(D(s))}$$

we obtain the following nonlinear system of four algebraic equations in C_1, C_2, L_1, L_2:

$$C_1 C_2 L_1 L_2 = \frac{2}{(1000\pi)^4}$$

$$C_1 L_1 L_2 + C_1 C_2 L_2 = \frac{5.2262}{(1000\pi)^3}$$

$$(C_1 + C_2)(L_1 + L_2) = \frac{6.8284}{(1000\pi)^2}$$

$$C_1 + C_2 + L_1 + L_2 = \frac{5.2262}{1000\pi}$$

where the value of the cutoff (circular) frequency $\omega_c = 2000\pi$ has been substituted into $D(s)$.

This nonlinear system of equations can be solved using MATLAB's *fsolve* function. (See again script file fig8_3.m and M-file butta4.m, **On the Net.** Note that a good set of initial estimates of the variables is helpful in obtaining meaningful results.) Such a set of results is the following: $C_1 = 4.3399 \times 10^{-4}$ F, $C_2 = 3.9782 \times 10^{-4}$ F, $L_1 = 1.7052 \times 10^{-4}$ H, and $L_2 = 6.6122 \times 10^{-4}$ H.

Discrete Fourier Transform and Computation with Fast Fourier Transform

In a situation where the desired filter characteristic is realized using a digital filter, or when a computer is used in signal processing, it is necessary to represent functions in *sampled* form. A parallel development of Fourier series and Fourier transform for signals that are discrete-time periodic and discrete-time nonperiodic functions is available. In general, to obtain the discrete-time approximation from the corresponding continuous-time result, we consider finite sums in place of infinite sums, approximate integrals by finite sums, and use the values of the function only at the sampled points. For proof of the validity of the resulting expressions, the interested reader is referred to reference [8] or similar texts.

Discrete-Time Fourier Series

Consider the periodic function $f(t)$ expanded in the (exponential) Fourier series, equations (7.50a) and (7.50b) repeated ($t_0 = 0$):

$$f(t) = \sum_{n=-\infty}^{\infty} c_n e^{jn\omega_0 t} \tag{8.13a}$$

$$c_n = \frac{1}{T} \int_0^T f(t) e^{-jn\omega_0 t}\, dt \tag{8.13b}$$

The discrete-time approximation is

$$f(kT) = f\left(\frac{k\Gamma}{N}\right) = \sum_{n=0}^{N-1} \tilde{c}_n e^{jn\omega_0 kT} = \sum_{n=0}^{N-1} \tilde{c}_n e^{j2\pi nk/N} \qquad \text{(8.14a)}$$

$$\tilde{c}_n = \frac{1}{N}\sum_{k=0}^{N-1} f(kT)e^{-jn\omega_0 kT} = \frac{1}{N}\sum_{k=0}^{N-1} f\left(\frac{k\Gamma}{N}\right)e^{-j2\pi nk/N} \qquad \text{(8.14b)}$$

where $\Gamma = 2\pi/\omega_0$ is the fundamental period of $f(t)$, $T = \Gamma/N$ is the sampling period (sampling at equidistant points is implied), and k is the running time (sampling point) index. For regular functions, the approximation becomes exact, $\tilde{c}_n = c_n$, in the limit $N \longrightarrow \infty$. Ordinarily, the difference

$$\epsilon_n = \tilde{c}_n - c_n \qquad \text{(8.15a)}$$

is known as the *aliasing error.*

However, most engineering systems respond to some upper frequency limit, so that engineering system signals usually have a finite bandwidth. Such signals can be accurately represented by equidistant sample sequences of sufficiently short sampling period. Suppose in equation (8.13a) M is the highest harmonic in the Fourier series expansion for such a *band-limited* function. Then the truncated series representation

$$f(t) = \sum_{n=-M}^{M} c_n e^{jn\omega_0 t} \qquad \text{(8.15b)}$$

is exact, since terms beyond $|n| = M$ are zero. Thus we need to determine only $2M + 1$ terms, that is, $c_{-M}, c_{-M+1}, \ldots c_{-2}, c_{-1}, c_0, c_1, c_2, \ldots, c_M$, using $2M + 1$ values of $f(t)$ (say, $f(t_1), f(t_2), \ldots, f(t_{2M+1})$) to completely specify $f(t)$. It follows that the approximation (8.14a) will also be exact (zero aliasing error), that is, the coefficients $\tilde{c}_n$ can be found exactly, if the sampling period is such that

$$N \geq 2M + 1 \quad \text{or} \quad \frac{N}{2} > M \qquad \text{(8.15c)}$$

Note that in equation (8.14b), since $e^{-j2\pi k} = 1$ (for $k = $ integer), $e^{-j2\pi nk/N} = e^{-j2\pi(n+N)k/N}$.

Hence

$$\tilde{c}_n = \tilde{c}_{n+N} \qquad \text{(8.15d)}$$

and both $\tilde{c}_n$ and $f(kT)$ are periodic functions. Also, by substitution of $(-n)$ for (n) in equation (8.14b), we find that

$$\tilde{c}_n = \tilde{c}_n^* \qquad \text{(8.15e)}$$

Thus when we say that the coefficients can be found exactly, we mean:

$$\tilde{c}_0 = c_0 \text{ and } [\tilde{c}_n = c_n, \ \tilde{c}_{-n+N} = c_{-n} = \tilde{c}_n^*; \ n = 1, \ldots, M]$$

The frequency corresponding to M, that is,

$$\omega_M = \frac{N}{2}\omega_0 = \frac{\pi}{T} \tag{8.15f}$$

is known as the *fold-over* or *Nyquist* frequency, and it is the maximum frequency content that can be detected in the Fourier series representation.

Example 8.4	Discrete-Time Fourier Series of a Band-Limited Function

a. Find the Fourier series coefficients (equations (8.13a) and (8.13b)) for the periodic function

$$f(t) = 2[\cos(2\pi t) + \cos(4\pi t)]$$

b. Determine the corresponding discrete-time Fourier series coefficients (equations (8.14a) and (8.14b)) for the same function for $N = 3$ and $N = 5$.

Solution a: Note that the fundamental frequency of the given function is $\omega_0 = 2\pi$, so that $\Gamma = 1$ (verify this by sketching the function), and the maximum frequency is 4π; hence $M = 2$. Using Euler's identity, equation (6.41a), we have

$$f(t) = 2\left[\frac{e^{j\omega_0 t} + e^{-j\omega_0 t}}{2} + \frac{e^{j2\omega_0 t} + e^{-j2\omega_0 t}}{2}\right] = e^{-j2\omega_0 t} + e^{-j\omega_0 t} + e^{j\omega_0 t} + e^{j2\omega_0 t}$$

Comparing this representation with equation (8.13a) or (8.15b), we see that $c_{-2} = c_{-1} = 1$, $c_0 = 0$, and $c_1 = c_2 = 1$.

Solution b: For the discrete-time Fourier series, the coefficients are given by equation (8.14b).

Case of $N = 3$: This case violates the condition of equation (8.15c).

$$\tilde{c}_n = \frac{1}{3}\sum_{k=0}^{2} f\left(\frac{k}{3}\right)e^{-j2\pi nk/3}$$

yields

$$\tilde{c}_0 = \frac{1}{3}\left[f(0) + f\left(\frac{1}{3}\right) + f\left(\frac{2}{3}\right)\right] = \frac{1}{3}[4 - 2 - 2] = 0$$

$$\tilde{c}_1 = \frac{1}{3}\left[4(1) - 2\left(-0.5 - j\frac{\sqrt{3}}{2}\right) - 2\left(-0.5 + j\frac{\sqrt{3}}{2}\right)\right] = 2$$

$$\tilde{c}_2 = \frac{1}{3}\left[4(1) - 2\left(-0.5 + j\frac{\sqrt{3}}{2}\right) - 2\left(-0.5 - j\frac{\sqrt{3}}{2}\right)\right] = 2$$

We see that the coefficients are not determined exactly: $\tilde{c}_1 \neq c_1$ and $\tilde{c}_2 \neq c_2$.

Case of $N = 5$: This case satisfies the condition of equation (8.15c).

$$\tilde{c}_n = \frac{1}{5} \sum_{k=0}^{4} f\left(\frac{k}{5}\right) e^{-j2\pi nk/5}$$

yields

$$\tilde{c}_0 = \frac{1}{5}\left[f(0) + f\left(\frac{1}{5}\right) + f\left(\frac{2}{5}\right) + f\left(\frac{3}{5}\right) + f\left(\frac{4}{5}\right) \right]$$

$$= \frac{1}{5}[4 - 1 - 1 - 1 - 1] = 0$$

$$\tilde{c}_1 = \frac{1}{5}[4(1) - 1(0.309 - j0.951) - 1(-0.809 - j0.588)$$

$$- 1(-0.809 + j0.588) - 1(0.309 + j0.951)] = 1$$

$$\tilde{c}_2 = \frac{1}{5}[4(1) - 1(-0.809 - j0.588) - 1(0.309 + j0.951)$$

$$- 1(0.309 - j0.951) - 1(-0.809 + j0.588)) = 1$$

$$\tilde{c}_3 = \frac{1}{5}[4(1) - 1(-0.809 + j0.588) - 1(0.309 - j0.951)$$

$$- 1(0.309 + j0.951) - 1(-0.809 - j0.588)] = 1$$

$$\tilde{c}_4 = \frac{1}{5}[4(1) - 1(0.309 + j0.951) - 1(-0.809 + j0.588)$$

$$- 1(-0.809 - j0.588) - 1(0.309 - j0.951)] = 1$$

This time, $\tilde{c}_0 = c_0$, $\tilde{c}_1 = c_1$, $\tilde{c}_2 = c_2$, $\tilde{c}_{-1+5} = \tilde{c}_4 = \tilde{c}_1^* = c_{-1}$, and $\tilde{c}_{-2+5} = \tilde{c}_3 = \tilde{c}_2^* = c_{-2}$, as required.

....

Discrete Fourier Transform

Although the Fourier (integral) transform pair, equations (8.3a) and (8.3b) repeated,

$$\mathscr{T}[f(t)] = F(\omega) = \int_{-\infty}^{\infty} f(t)e^{-j\omega t}\, dt \tag{8.16a}$$

$$\mathscr{T}^{-1}[F(\omega)] = f(t) = \frac{1}{2\pi} \int_{-\infty}^{\infty} F(\omega)e^{j\omega t}\, d\omega \tag{8.16b}$$

applies to certain arbitrary continuous functions $f(t)$ that may not be periodic, in practice, we can process only data sequences that are available over a finite time window. For such a finite-length signal $f(t)$ we need a truncated rather than an infinite series approximation of the Fourier transform. The *discrete Fourier transform* (DFT) pair of a sequence of N samples

$$[f(kT)]; \qquad 0 \leq k \leq N - 1$$

is given by (see [8])

$$\mathscr{D}[f(kT)] = F\left(n\frac{2\pi}{NT}\right) = F(n\Omega) = \sum_{k=0}^{N-1} f(kT)e^{-j2\pi nk/N} \tag{8.17a}$$

$$= \sum_{k=0}^{N-1} f(kT)e^{-jn\Omega kT}; \qquad n = 0, 1, \ldots, N-1$$

$$\mathscr{D}^{-1}[F(n\Omega)] = f(kT) = \frac{1}{N}\sum_{n=0}^{N-1} F(n\Omega)e^{j2\pi nk/N} \tag{8.17b}$$

$$= \frac{1}{N}\sum_{n=0}^{N-1} F(n\Omega)e^{jn\Omega kT}; \qquad k = 0, 1, \ldots, N-1$$

where T is the sampling time interval, so that the signal length is $(N-1)T$, and $\Omega = \omega_s/N = 2\pi/NT$ is the frequency sampling interval (ω_s is the sampling frequency), so that N transformed values are computed. As was the case for the discrete-time Fourier series (DFS), both the time- and frequency-domain sequences above are periodic (recall equation (8.15d)). Note that the definition (8.17a) requires that $F(n\Omega)$ by multiplied by (NT) before it can be compared directly with $F(\omega)$ as given by equation (8.16a).

Note further the identical form of equations (8.14) and (8.17). Although the DFS applies to a periodic function with the number of samples N equivalent to the number of terms in the expansion over a specific period of the function only, and the DFT applies to a function that is not necessarily periodic, with N equal to the number of functional terms in the representation of a finite-length signal, the DFT can be viewed as a sampled (at ω_s) form of the DFS, so that the coefficients of a DFS can be obtained from the DFT by interpolation. This viewpoint brings up an important practical consideration in the application of the DFT. Signal contents of frequencies above the sampling frequency ω_s cannot be detected by the DFT (recall that for the DFS, accurate truncation of the series is determined by the highest harmonic frequency, according to equation (8.15c)). In order to correctly recover a signal from the DFT, **the signal must be sampled with a frequency ω_s of at least twice its bandwidth** (see equation 8.15f)).

Fast Fourier Transform

As was the case for the Fourier integral transform, the DFT $F(n\Omega)$, is, in general, a complex number. Examination of equation (8.17a) reveals that a direct solution of the DFT requires for each (kth) sample, access to N coefficients $e^{-jn\Omega kT}$, N complex multiplications, and N complex additions. For most practical problems the total number of mathematical operations and storage requirements implied for even a relatively small number of samples is exorbitant. The development of a computationally highly efficient technique for evaluating the DFT known as the *fast Fourier transform* (FFT) represented a very significant advance in digital signal processing. A number of FFT algorithms have been developed. The most efficient ones generally require the restriction of the number of samples to some particular factor (usually $N = 2^m$, where m is an integer); however, FFT algorithms are also available for arbitrary N. Other variables in FFT programs are the *normalizing factor* and the sign of the exponent. The definition of equations

(8.17a) and (8.17b) in which the normalizing factor is $(1/N)$, and it is applied to the inverse transform while the exponent is negative in the transform and positive in the inverse transform is one among many definitions. In some definitions, the same normalizing factor is applied instead to the transform and/or the signs of the exponents are reversed. Other definitions use symmetric normalizing factors such as $\left(1/\sqrt{N}\right)$ applied to both the transform and the inverse transform. The different conventions are not significant as long as the functions are used in transform-inverse transform *pairs*. We consider in Example 8.5 some particular implementations in MATLAB.

Example 8.5 — Computation of the Fast Fourier Transform

Compute, using MATLAB FFT functions, the discrete Fourier transform of a sequence of N samples of the function $f(t) = 2[\cos(2\pi t) + \cos(4\pi t)]$ defined on the time interval $0 \leq t \leq 1$, for the following cases: $T = \frac{1}{3}$, $N = 3$, $T = \frac{1}{3}$, $N = 9$ and $T = \frac{1}{8}$, $N = 16$. Obtain in each case the amplitude spectrum and the coefficients corresponding to a discrete Fourier series representation of the function.

Solution a: The results are shown in Table 8.3. The script file used to obtain these results, **tab8_3.m,** is listed in the box. Two FFT functions, fft(**x**) and fft(**x**, N), are used in the script file. The function pair fft(**x**) and ifft(**x**) compute the discrete Fourier transform pair of the vector **x** or of each column of the matrix **x** according to equations (8.17a) and (8.17b) except for the index of the input or computed sequence, which, consistent with MATLAB vectors, runs from 1 to N instead of from 0 to $N - 1$. The function fft(**x**, N) is the same as fft(**x**), except that the sequence is either truncated or padded with zeros, as necessary, so that the length of **x** is equal to the specified N. For example, the sequence $f1k$ is exactly nine samples in length so that fft(**x**) was used for the case of $N = 9$.

Table 8.3 Computation of Fast Fourier Transform by MATLAB

FREQUENCY f(Hz)	F(f)	|F(f)|	|F(f)|/N	FREQUENCY f(Hz)	F(f)	|F(f)|	|F(f)|/N
	Case of $T = \frac{1}{3}$, $N = 3$; $\Omega = 1$ Hz				*Case of* $T = \frac{1}{8}$, $N = 16$; $\Omega = \frac{1}{2}$ Hz		
0	0.0000	0.0000	0.0000	0	-0.0000	0.0000	0.0000
1.0000	$6.0000 - 0.0000j$	6.0000	2.0000	0.5000	$-0.0000 + 0.0000j$	0.0000	0.0000
2.0000	$6.0000 + 0.0000j$	6.0000	2.0000	1.0000	$16.0000 + 0.0000j$	16.0000	1.0000
				1.5000	$-0.0000 + 0.0000j$	0.0000	0.0000
	Case of $T = \frac{1}{3}$, $N = 9$; $\Omega = \frac{1}{3}$ Hz			2.0000	$16.0000 + 0.0000j$	16.0000	1.0000
				2.5000	$0.0000 + 0.0000j$	0.0000	0.0000
0	0	0	0	3.0000	$0.0000 + 0.0000j$	0.0000	0.0000
0.3333	$0.0000 + 0.0000j$	0.0000	0.0000	3.5000	$0.0000 + 0.0000j$	0.0000	0.0000
0.6667	$0.0000 + 0.0000j$	0.0000	0.0000	4.0000	0.0000	0.0000	0.0000
1.0000	$18.0000 - 0.0000j$	18.0000	2.0000	-3.5000	$0.0000 - 0.0000j$	0.0000	0.0000
1.3333	$-0.0000 - 0.0000j$	0.0000	0.0000	-3.0000	$0.0000 - 0.0000j$	0.0000	0.0000
-1.3333	$-0.0000 + 0.0000j$	0.0000	0.0000	-2.5000	$0.0000 - 0.0000j$	0.0000	0.0000
-1.0000	$18.0000 + 0.0000j$	18.0000	2.0000	-2.0000	$16.0000 - 0.0000j$	16.0000	1.0000
-0.6667	$0.0000 - 0.0000j$	0.0000	0.0000	-1.5000	$-0.0000 - 0.0000j$	0.0000	0.0000
-0.3333	$-0.0000 - 0.0000j$	0.0000	0.0000	-1.0000	$16.0000 - 0.0000j$	16.0000	1.0000
				-0.5000	$-0.0000 - 0.0000j$	0.0000	0.0000

```
% tab8_3.m
% Script to generate Table 8.3
%
N1=3; N2=9; N3=16;                             % number of samples
T1=1/3; T3=.125;                               % sampling time intervals
df2=1/(N2*T1); df3=1/(N3*T3);                  % frequency sampling intervals
ny2=1+(N2/2); ny3=1+(N3/2);                    % Nyquist frequency points
n=0:1:N3;                                      % frequency index
%
% Case of T=1/3, N=3
t1=0:T1:3-T1;                                  % (9) time sampling points
f1k=2*(cos(2*pi*t1) + cos(4*pi*t1));           % sequence of time functions
F1n=fft(f1k,N1);                               % compute FFT with N1 samples
y1(1:3,1)=n(1:3)';                             % sample frequencies (Hz)
y1(1:3,2)=F1n(1:3)'; y1(1:3,3)=abs(F1n(1:3))'; y1(1:3,4)=y1(1:3,3)/N1;
%
% Case of T=1/3, N=9
F2n=fft(f1k);                                  % compute FFT with 9 samples
for i=1:N2, y2(i,1)=df2*n(i)'; if i>ny2,
y2(i,1)=-y2(2*ny2-i,1); end; end;              % sample frequencies (Hz)
y2(1:9,2)=F2n(1:9)'; y2(1:9,3)=abs (F2n(1:9))'; y2(1:9,4)=y2(1:9,3)/N2;
%
% Case of T = .125
t3=0:T3:2;                                     % time sampling points
f3k=2*(cos(2*pi*t3)+cos(4*pi*t3));             % sequence of time functions
%
F3n=fft(f3k,N3);                               % compute FFT with N3 points
for i=1:N3, y3(i,1)=df3*n(i)'; if i>ny3,
y3(i,1)=-y3(2*ny3-i,1); end; end;              % sample frequencies (Hz)
y3(1:16,2)=F3n(1:16)'; y3(1:16,3)=abs(F3n(1:16))';
y3(1:16,4)=y3(1:16,3)/N3;
clc;                                           % clear command window
disp(' Table 8.3 Computation of Fast Fourier Transform by MATLAB');
disp(' ');
disp(' Case of T=1/3, N=3; Frequency Sampling Interval = 1 Hz');
disp(' f=Frequency(Hz)   F(f)    IF(f)I    IF(f)I/N');
disp(y1(1:3,:));
disp('   press any key to continue'); pause; disp(' ');
disp(' Case of T=1/3, N=9; Frequency Sampling Interval = 1/3 Hz');
disp(' f=Frequency(Hz)   F(f)    IF(f)I    IF(f)I/N');
disp(y2(1:9,:));
disp('   press any key to continue'); pause; disp(' ');
disp(' Case of T=1/8, N=16; Frequency Sampling Interval = 1/2 Hz');
disp(' f=Frequency(Hz)    F(f)    IF(f)I    IF(f)I/N');
disp(y3(1:16,:));
disp(' '); disp('end of program');
```

The frequency sampling interval is given by

$$\Omega = \frac{2\pi}{NT} \text{ rad/s} = \frac{1}{NT} \text{ Hz}$$

Thus the sample frequencies (Hz) are given by

$$f_i = \frac{i - 1}{NT}; \qquad i = 1, 2, \ldots$$

up to the index that is less than or equal to the Nyquist frequency "index" ($ny2$ or $ny3$ in the program). Since i starts from 1, equation (8.15f) for the Nyquist frequency point is

$$ny = \frac{N}{2} + 1$$

Beyond the index ny, the sampling frequencies are considered negative (consistent with equations (8.15d) and (8.15e)):

$$f_i = -f_{2ny - i}$$

Observe from Table 8.3 that, in accord with equation (8.15e), the Fourier transform or the series coefficients for these frequencies are complex conjugates of the coefficients for the corresponding (mirror-image) positive frequencies. Actually, the data above the Nyquist frequency are redundant and not necessary.

To interpolate for the DFS coefficients, recall from Example 8.4 that the frequency sampling interval for the DFS was $\omega_0 = 2\pi$ (rad/s) or 1 Hz. Thus the coefficients in each case can be found at the frequencies 0, 1, and 2 Hz. For $T = \frac{1}{3}$, we recover the corresponding inaccurate results of Example 8.4, ($c_0 = 0$, $c_1 = 2$, $c_2 = 2$); no matter the number of samples, $N = 3$ or 9. Note that the coefficients from the DFT have to be scaled by $(1/N)$ as required by the correspondence between equations (8.14) and (8.17). Thus the accuracy of the DFT is not directly controlled by the number of samples N, or the sample length $(N - 1)T$, but by the sampling time interval T or the sampling frequency $\omega_s = 2\pi/T$. Also, observe that for the case of $T = \frac{1}{8}$, the accurate coefficients of Example 8.4, ($c_0 = 0$, $c_1 = 1$, $c_2 = 1$), are recovered even though $N = 16$ and not 5. Clearly, as long as the sampling frequency is at least twice the bandwidth, an accurate DFT is obtained (recall the bandwidth of $f(t)$ is 4π). However, for this DFT to approximate closely the (continuous) Fourier integral transform, the sampling frequency should be as large as possible. Finally note that for this case $\left(T = \frac{1}{8}\right)$, the coefficients for frequencies beyond 2 Hz are all zero. These frequencies are not present in the input function.

8.2 INTRODUCING THE LAPLACE TRANSFORM METHOD

In Section 7.3 the vibration absorber problem was solved as two simultaneous second-order scalar differential equations. However, the two equations of (7.36d) actually constitute one fourth-order ordinary differential equation in a single variable, y for example. The phasor and other transform techniques are equally applicable to such higher-order scalar ODE system models.

Example 8.6	Frequency Transform Solution of a Third-Order Scalar ODE

Consider again the armature-controlled DC motor (see Fig. 8.5) in an application such as the motor-actuated control valve described in Example 2.2. That is, the controlled variable is shaft displacement θ (position control), and the load torque $T_L = T_s = k\theta$. Show that this control system can be described by a third-order ODE:

$$\frac{d^3y}{dt^3} + a_2\frac{d^2y}{dt^2} + a_1\frac{dy}{dt} + a_0y = f(t) = b_0u(t) \tag{8.18}$$

where $y = \theta$, $a_0 = 10^5$, $a_1 = 6.4 \times 10^4$, $a_2 = 90$, $b_0 = 10^5$, and $u(t) = V_a$. For the parameters values: $\beta = 0.04$ N $\cdot$ m $\cdot$ s, $J = 0.001$ kg $\cdot$ m^2, $R_a = 0.3$ Ω, $L_a = 0.006$ H, $K_1i_f = K_T = 0.6$ V $\cdot$ s, and $k = 2$ N $\cdot$ m.

a. If a test control input is the *undamped sinusoid* $u(t) = 10 \sin 50t$, find (i) the phasor (frequency) transfer function between the input and output, and (ii) the amplitude and phase shift of the *steady-state response*.

b. Repeat (a) if the test input is the *damped sinusoid* $u(t) = 10 \sin(50t)e^{-4t}$.

Solution: The armature-controlled DC motor model, with $y = \theta$ and $T_L = k\theta = ky$, is given by equations (5.55a) through (5.55c), repeated here:

$$\frac{d}{dt}y = \omega \tag{a}$$

$$\frac{d}{dt}\omega = \frac{1}{J}(K_Ti_a - \beta\omega - ky) \tag{b}$$

$$\frac{d}{dt}i_a = \frac{1}{L_a}(V_a - R_ai_a - K_T\omega) \tag{c}$$

Using $\omega = \dot{y}$ and $\dot{\omega} = \ddot{y}$, then from (b) we have

$$R_ai_a = \frac{R_a}{K_T}(J\ddot{y} + \beta\dot{y} + ky)$$

Figure 8.5

Armature-controlled DC Motor Actuator

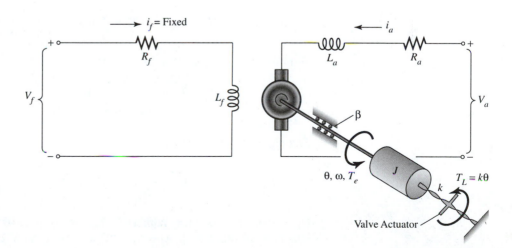

Valve Actuator

and

$$\frac{d^2}{dt^2}\omega = \frac{1}{J}\left(K_T\frac{d}{dt}i_a - \beta\dot{y} - k\dot{y}\right)$$ (d)

Substituting for $R_a i_a$ in (c) and then using the resulting (c) to substitute for $\frac{d}{dt}i_a$ in (d), we obtain

$$\frac{d^2}{dt^2}\omega = \frac{1}{J}\left\{\frac{K_T}{L_a}\left[V_a - \frac{R_a}{K_T}(J\ddot{y} + \beta\dot{y} + ky) - K_T\dot{y}\right] - \beta\ddot{y} - k\dot{y}\right\}$$

Finally, we use (a) to get

$$\frac{d^3}{dt^3}y + \left(\frac{\beta}{J} + \frac{R_a}{L_a}\right)\frac{d^2}{dt^2}y + \left(\frac{k}{J} + \frac{R_a\beta}{JL_a} + \frac{K_T^2}{JL_a}\right)\frac{d}{dt}y + \left(\frac{kR_a}{JL_a}\right)y = \left(\frac{K_T}{JL_a}\right)V_a$$ (e)

We substitute the given parameter values in (e) to get

$$\frac{d^3}{dt^3}y + 90\frac{d^2}{dt^2}y + 6.4 \times 10^4\frac{d}{dt}y + 10^5 y = 10^5 V_a$$ (f)

Solution a: Since the input is a sinusoid, $u(t) = 10 \sin 50t = 10 \cos(50t - \pi/2)$, and the system is stable (stability can be shown by the Routh test; for example, see Section 10.2), the steady-state response can be found by phasor or fourier transformation of equation (8.18). That is,

$$-j\omega^3 Y - a_2\omega^2 Y + ja_1\omega Y + a_0 Y = b_0 U$$

i. Hence the frequency transfer function is given by

$$G(j\omega) = \frac{Y(j\omega)}{U(j\omega)} = \frac{b_0}{(a_0 - a_2\omega^2) + j(a_1\omega - \omega^3)}$$

ii. The amplitude of the response is given by

$$|Y(j\omega)| = |G(j\omega)||U(j\omega)| = 10(b_0)\sqrt{\frac{1}{(a_0 - a_2\omega^2)^2 + (a_1\omega - \omega^3)^2}}$$

Setting $\omega = 50$, gives us

$$|Y(j\omega)| = 10^6\sqrt{\frac{1}{(-1.25 \times 10^5)^2 + (3.075 \times 10^6)^2}} = 0.3249$$

The phase shift is given by

$$\phi_y = \phi_u - \tan^{-1}\left(\frac{30.75}{-1.25}\right) = \phi_u - 1.611$$

Note that the complex number $-1.25 \times 10^5 + j3.075 \times 10^6$ is in the second quadrant, so that the angle is 92.328° and not −87.672°. Since $\phi_u = -\pi/2(-90°)$,

$\phi_y = -3.182(-182.328°)$. It follows that the steady-state response is $y_{ss}(t) = 0.3249 \cos(50t - 3.182)$.

Solution b: Recall from Section 6.2 that the phasor $z = Ae^{j\phi}$ can also represent the damped sinusoid:

$$f(t) = A \cos(\omega t + \phi)e^{-\sigma t} = \text{Real}[ze^{(-\sigma + j\omega t)}] \tag{8.19a}$$

Introducing the complex number

$$-s = -\sigma + j\omega \tag{8.19b}$$

we can show, as we found for undamped sinusoids (see also the time differentiation property, item 3 of Table 8.2, for the case of the Fourier transform), that if z is the phasor transform of the damped sinusoid, then

$$\mathscr{L}\left[\frac{d^n}{dt^n}f(t)\right] = s^n z \tag{8.19c}$$

Consider the case $n = 1$, for example,

$$\frac{d}{dt}[A \cos(\omega t + \phi)e^{-\sigma t}] = -A\omega \sin(\omega t + \phi)e^{-\sigma t} - A \cos(\omega t + \phi)e^{-\sigma t}$$

$$= [-\omega A \cos(\omega t + \phi - \pi/2) - \sigma A \cos(\omega t + \phi)]e^{-\sigma t}$$

By equation (8.19a),

$$\mathscr{L}\left[\frac{d^n}{dt^n}f(t)\right] = -\omega Ae^{j(\phi - \pi/2)} - \sigma Ae^{j\phi} = (-\omega e^{-j\pi/2} - \sigma)Ae^{j\phi} = (-\sigma + j\omega)Ae^{j\phi}$$

which agrees with equation (8.19c). The completion of the proof by induction is left as an exercise (see Practice Problem 8.20). It follows that we can extend the earlier phasor transform technique to damped sinusoidal forcing inputs by merely replacing $j\omega$ with s.

i. The transfer function ($G(s)$ this time) is

$$G(s) = \frac{b_0}{s^3 + a_2 s^2 + a_1 s + a_0}$$

ii. The response "amplitude" and phase angle are given by

$$|Y(s)| = |G(s)||U(s)|; \qquad \phi_y = \angle G(s) + \phi_u$$

But for the given input function,

$$s = -4 + j50; \qquad s^2 = -2484 - j400; \qquad s^3 = 29936 - j122600$$

so that

$$s^3 + a_2 s^2 + a_1 s + a_0 = -349624 + j3041400$$

Thus

$$|Y(s)| = \frac{10^6}{\sqrt{349624^2 + 3041400^2}} = 0.3266$$

and

$$\phi_y = \frac{-\pi}{2} - \tan^{-1}\left(\frac{3041400}{-349624}\right) = \frac{-\pi}{2} - 1.685 = -3.256$$

The forced response is therefore

$$y(t) = 0.3266\cos(50t - 3.256)e^{-4t}$$

As we discussed in Section 6.2, damped sinusoids represent a much larger class of functions than do undamped sinusoids. However, when an infinite number of undamped sinusoids of different frequencies were combined, a remarkable increase was obtained in the number of forcing functions for which the response of linear systems could be analyzed by a transform method. The Fourier integral transform and related functions that represented formalizations of this process were discussed in Sections 7.4 and 8.1. The preceding example suggests that an even more powerful transform for linear systems analysis could be developed on the basis of combinations of damped sinusoids of frequencies covering an infinite band. Further, the formal definition of such a transform should be close to or even coincident with the function obtained by replacing $(j\omega)$ in the definition of the Fourier transform, equation (8.16a), with $s = (-\sigma + j\omega)$. This is indeed the case for the *Laplace transform,* which is given by (see [8])

$$F(s) = \int_{-\infty}^{\infty} f(t)e^{-st}dt \tag{8.20a}$$

$$f(t) = \frac{1}{2\pi j}\int_{\sigma-j\infty}^{\sigma+j\infty} F(s)e^{st}\,ds \tag{8.20b}$$

Equations (8.20a) and (8.20b) define the bilateral or *two-sided* Laplace transform pair. In practice, we usually consider the behavior of physical systems for times $t > 0$, given the initial state of the system (defined as $t = 0$) and the input functions that are also determined for $t > 0$. Consequently, the bilateral Laplace transform is often replaced in engineering analysis by the *one-sided* Laplace transform, which has the definition

$$\mathscr{L}[f(t)] = F(s) = \int_{0}^{\infty} f(t)e^{-st}\,dt;\ \ \text{(Laplace transform)} \tag{8.20c}$$

However, the inverse transform is still given by equation (8.20b), repeated here:

$$\mathscr{L}^{-1}[F(s)] = f(t) = \frac{1}{2\pi j}\int_{\sigma-j\infty}^{\sigma+j\infty} F(s)e^{st}\,ds;\ \ \text{(inverse Laplace transform)} \tag{8.20d}$$

The calculation of the inverse Laplace transform, a *line integral,* requires its conversion to a *contour integral* (see [7] and [9]), and the evaluation of such integrals is outside the scope of this text. In practice, the inverse Laplace transform of $F(s)$ is normally obtained by referring to a table of Laplace transform pairs, previously computed using equation (8.20c), and matching an entry in the table with $F(s)$. The matching process is facilitated by the application of some of the properties of Laplace transforms and the expansion of $F(s)$ into partial fractions, both of which are discussed next. Table 8.4 is an abbreviated table of Laplace transform pairs. Some important properties of Laplace transforms are given in Table 8.5.

Table 8.4

Some Common Laplace Transform Pairs

ITEM	$F(s) = \int_0^\infty f(t)e^{-st}\,dt$	$f(t) = \dfrac{1}{2\pi j}\int_{\sigma-j\infty}^{\sigma+j\infty} F(s)e^{st}\,ds$
1	$\dfrac{1}{s}$	1, also unit step $U_s(t)$
2	$\dfrac{1}{s^2}$	t
3	$\dfrac{n!}{s^{n+1}}$	$t^n, \quad n = 0, 1, 2, \ldots$
4	$\dfrac{1}{s+a}$	e^{-at}
5	$\dfrac{1}{(s+a)^2}$	te^{-at}
6	$\dfrac{\omega}{s^2 + \omega^2}$	$\sin(\omega t)$
7	$\dfrac{s}{s^2 + \omega^2}$	$\cos(\omega t)$
8	$\dfrac{\omega}{(s+a)^2 + \omega^2}$	$e^{-at}\sin(\omega t)$
9	$\dfrac{s+a}{(s+a)^2 + \omega^2}$	$e^{-at}\cos(\omega t)$
10	$\dfrac{2\omega s}{(s^2 + \omega^2)^2}$	$t\sin(\omega t)$
11	$\dfrac{s^2 - \omega^2}{(s^2 + \omega^2)^2}$	$t\cos(\omega t)$
12	$\ln\left(\dfrac{s+a}{s+b}\right)$	$\dfrac{e^{-bt} - e^{-at}}{t}$
13	1	$\delta(t)$ (unit impulse)
14	$\dfrac{e^{-t_s s}(1 - e^{-T_p s})}{s}$	$U_p(t - t_s)$ (unit pulse)

Example 8.7   Inverse Laplace Transform by Table Lookup

Find the inverse Laplace transform of

$$F(s) = \frac{s+3}{s^2 + 6s + 12}$$

Solution:

$$F(s) = \frac{s+3}{s^2 + 6s + 12} = \frac{s+3}{s^2 + 6s + 9 + 3}$$

Comparing this result with entry 9 in Table 8.4, we see that

$$F(s) = \frac{s+3}{(s+3)^2 + \left(\sqrt{3}\right)^2}$$

Table 8.5 Some Important Properties of Laplace Transforms

DESCRIPTION	THEOREM	EXAMPLE
1. Linearity	If c_1 and c_2 are constants, $\mathscr{L}[c_1 f_1(t) + c_2 f_2(t)] = c_1 F_1(s) + c_2 F_2(s)$	$\mathscr{L}[t - 3e^{-2t} + 4\sin t] =$ $\dfrac{1}{s^2} - \dfrac{3}{s+2} + \dfrac{4}{s^2+1}$
2. Scaling	$\mathscr{L}[f(at)] = \dfrac{1}{a} F\left(\dfrac{s}{a}\right)$	$f(t) = te^{-t}, \mathscr{L}[ate^{-at}] =$ $\dfrac{1}{a}\dfrac{1}{(s/a+1)^2} = \dfrac{a}{(s+a)^2}$
3. Derivative	$\mathscr{L}\left[\dfrac{d}{dt}f(t)\right] = sF(s) = f(0).$ By repeated application of (3), $\mathscr{L}\left[\dfrac{d^n}{dt^n}f(t)\right] = s^n F(s) - s^{n-1} f(0) -$ $s^{n-2}\left(\dfrac{df}{dt}\bigg|_{t=0}\right) - \ldots - s\left(\dfrac{d^{n-2}f}{dt^{n-2}}\bigg|_{t=0}\right)$ $-\left(\dfrac{d^{n-1}f}{dt^{n-1}}\bigg|_{t=0}\right)$	$\mathscr{L}\left[\dfrac{d}{dt}\sin 2t\right] = s\mathscr{L}[\sin 2t] - 0$ $= \dfrac{2s}{s^2+4} = \mathscr{L}[2\cos 2t]$
4. Integral	$\mathscr{L}\left[\displaystyle\int_0^t f(\tau)\,d\tau\right] = \dfrac{F(s)}{s}$	$\mathscr{L}\left[\displaystyle\int_0^t \sin 2\tau\,d\tau\right] = \dfrac{\mathscr{L}[\sin 2t]}{s}$ $= \dfrac{2}{s(s^2+4)} = \mathscr{L}\left[\dfrac{1-\cos 2t}{2}\right]$
5. Multiplication by t^n	$\mathscr{L}[t^n f(t)] = (-1)^n \dfrac{d^n}{ds^n} F(s);\quad n \geq 0$	$\mathscr{L}[te^{-3t}] = \mathscr{L}[tf(t)]$ $= (-1)\dfrac{d}{ds}\left(\dfrac{1}{s+3}\right) = \dfrac{1}{(s+3)^2}$
6. s-Domain shift	$\mathscr{L}[e^{-at}f(t)] = F(s+a)$	$\mathscr{L}[e^{-at}\cos \omega t] = \dfrac{s+a}{(s+a)^2+\omega^2}$
7. Time shift	$\mathscr{L}[f(t+T)U_s(t+T)] = e^{Ts}F(s)$	$f(t) = \begin{cases}(t-3)^2, & t>3 \\ 0, & t<3\end{cases}$ $\mathscr{L}[f(t)] = \dfrac{2e^{-3s}}{s^3}$
8. Initial-value	$\lim_{t\to 0+}[f(t)] = \lim_{s\to\infty}[sF(s)]$ if these limits exist	$\lim_{t\to 0+}[e^{-2t}\cos 3t] =$ $\lim_{s\to\infty}\left[\dfrac{s(s+2)}{(s+2)^2+9}\right] = 1$
9. Final-value	$\lim_{t\to\infty}[f(t)] = \lim_{s\to 0}[sF(s)]$ if these limits exist	$\lim_{t\to\infty}[te^{-t}] = \lim_{s\to 0}\left[\dfrac{s}{(s+1)^2}\right] = 0$
10. Convolution	$\mathscr{L}^{-1}[F(s)G(s)] = \displaystyle\int_0^t f(\tau)g(t-\tau)\,d\tau$ $= f(t)*g(t)$	$\mathscr{L}^{-1}\left[\dfrac{1}{s+2}\right] = f(t) = e^{-2t}; \mathscr{L}^{-1}\left[\dfrac{1/2}{(s+1)^2}\right]$ $= g(t) = \dfrac{te^{-t}}{2}$ $\mathscr{L}^{-1}\left[\dfrac{1}{2}\left(\dfrac{1}{s+2}\right)\dfrac{1}{(s+1)^2}\right] =$ $\dfrac{1}{2}\displaystyle\int_0^t (t-\tau)e^{-(t-\tau)}e^{-2\tau}\,d\tau = \dfrac{1}{2}[(t-1)e^{-t} + e^{-2t}]$

which gives

$$f(t) = e^{-3t}\cos\left(\sqrt{3}t\right)$$

Example 8.8   Evaluation of Laplace Transform

Using equation (8.20c), find the Laplace transform of

a. e^{-at}

b. $\delta(t)$

Solution A: $f(t) = e^{-at}$; $\quad F(s) = \int_0^\infty e^{-at}e^{-st}\, dt = \int_0^\infty e^{-(s+a)t}\, dt = \left.\frac{e^{-(s+a)t}}{-(s+a)}\right|_0^\infty$

$$= \frac{1}{s+a}, \text{ for } s > -a$$

This is entry 4 in Table 8.4.

Solution B:

$$f(t) = \delta(t); \quad F(s) = \int_0^\infty \delta(t)e^{-st}\, dt$$

Using the rule given in Section 6.1 for the integration of products of other functions with the unit impulse, we see that the Laplace transform of $\delta(t) = e^{-s(0)} = 1$.

Existence of Laplace Transforms

Although the preceding introduction of the Laplace transform is only a heuristic argument, the Fourier integral transform background suggests that functions that satisfy the Dirichlet conditions are Laplace transformable. In equation (8.20c) the range of values of s is limited to those for which the real part of s ensures that the integral (8.20c) converges. In general (see for example, [1]), we shall assume that if

1. $f(t)$ is piecewise continuous on the interval $0 \le t \le T$ (see Fig. 7.16), and
2. there is a real constant q such that

$$\int_0^\infty |f(t)|e^{-qt}\, dt = \lim_{\substack{T \to \infty \\ \epsilon \to 0}} \int_\epsilon^T |f(t)|e^{-qt}\, dt < +\infty; \qquad 0 < \epsilon < T \qquad \textbf{(8.21)}$$

3. then $f(t)$ is Laplace transformable for Real$(s) > q$.

The exponential decay factor e^{-qt} in the convergence condition (8.21) makes it a much less stringent condition than the absolute convergence condition of equation (8.4). Thus many more functions are Laplace transformable than are Fourier integral transformable. However, the listed conditions are, as before, only sufficient conditions. We shall also assume that $f(t)$ is zero for all t less than zero, as needed for the nonsymmetric interval of the one-sided Laplace transform definition (8.20c).

| **Example 8.9** | Existence of a Laplace Transform |

Determine the conditions under which the following functions are Laplace transformable:

a. $f(t) = te^{-at}$
b. $f(t) = t^{-1}$
c. $f(t) = t^{-1/2}$

Solution a: $f(t)$ satisfies condition (1), and

$$\int_0^\infty te^{-(a+q)t}\, dt = \frac{e^{-(a+q)t}}{(a+q)^2}[-(a+q)t - 1]\Big|_0^\infty$$

$$= \frac{1}{(a+q)^2} < +\infty, \quad \text{if } (a+q) > 0 \quad \text{or} \quad q > -a$$

Thus te^{-at} has a Laplace transform for $\text{Re}(s) > -a$.

Solution b: $f(t) = t^{-1}$ is singular at $t = 0$. The function t^{-1} is not Laplace transformable. Note that

$$\int_\epsilon^T t^{-1}e^{-qt}\, dt = \int_q^{qT} x^{-1}e^{-x}\, dx = \ln(x)e^{-x}\Big|_q^{qT} + \int_q^{qT} e^{-x}\ln(x)\, dx$$

diverges in the limit $T \longrightarrow \infty$ and $\epsilon \longrightarrow 0+$.

Solution c: However, the function $f(t) = t^{-1/2}$, which is also singular at $t = 0$ and hence does not satisfy condition (1), is Laplace transformable. For this case, note that

$$\int_0^\infty t^{-1/2}e^{-qt}\, dt = \frac{1}{\sqrt{q}} \int_0^\infty x^{-1/2}e^{-x}\, dx = \sqrt{\frac{\pi}{q}}; \quad q > 0$$

converges and the Laplace transform of $t^{-1/2}$ is $\sqrt{\pi/s}$.

···

Transform Properties: Initial- and Final-Value Theorems

The Laplace transform is a powerful technique for solving dynamic engineering models not only because differential equations are converted to algebraic equations in the Laplace domain but also because the transform has some very useful properties. Of particular significance is the *derivative* property. The Laplace transform of the nth derivative of a function incorporates all n initial values of the function and its first $(n - 1)$ derivatives, so that unlike the phasor and Fourier transforms, the Laplace transform can be used in the analysis of transient response. Proofs of the theorems in Table 8.5 can be found in most texts on Laplace transforms (see [7] for example).

| **Example 8.10** | Time- and s-Domain-Shift Property |

Find the time function corresponding to the following Laplace transform expressions:

(a) $\dfrac{5}{(s+1)(s+7)}$ **(b)** $\dfrac{3e^{-2s}}{s+5}$ **(c)** $\dfrac{4s+12}{[(s+3)^2+4]^2}$

Solution a:

$$F(s) = \frac{5}{(s+1)(s+7)} = \frac{5}{6}\frac{(-1+7)}{(s+1)(s+7)} = \frac{5}{6}\left[\left(\frac{1}{s+1}\right) - \left(\frac{1}{s+7}\right)\right]$$

From Table 8.4, entry 4,

$$f(t) = \frac{5}{6}(e^{-t} - e^{-7t})$$

Solution b:

$$F(s) = e^{-2s}G(s); \qquad G(s) = \frac{3}{s+5}$$

From Table 8.4, entry 4,

$$g(t) = 3e^{-5t}$$

From the time-shift property (Table 8.5),

$$f(t) = g(t-2)U_S(t-2) = 3e^{-5(t-2)}U_S(t-2)$$

Solution c:

$$F(s) = \frac{4s+12}{[(s+3)^2+4]^2} = \frac{2(2)(s+3)}{[(s+3)^2+2^2]^2}$$

From Table 8.4, entry 10, if $g(t) = t\sin(2t)$, then

$$G(s) = \frac{2(2)s}{(s^2+2^2)^2}$$

Thus $F(s) = G(s+3)$ and from the s-domain-shift property (Table 8.5),

$$f(t) = te^{-3t}\sin(2t)$$

Example 8.11 Multiplication by t^n Property

Derive the Laplace transform of the following time functions:

a. $te^{-3t}\sin(2t)$, using the multiplication by t^n property
b. $f(t)$ as illustrated in Fig. 8.6

Solution a: $f(t) = tg(t)$, where $g(t) = e^{-3t}\sin(2t)$. From Table 8.4, entry 8,

$$G(s) = \frac{2}{(s+3)^2+4}$$

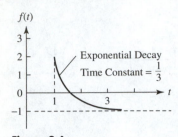

Figure 8.6

Forcing Input

Using the multiplication by t^n property (Table 8.5), we have

$$F(s) = (-1)\frac{d}{ds}\left[\frac{2}{(s+3)^2+4}\right] = \frac{(-1)(2)(-1)(s+3)}{[(s+3)^2+4]^2} = \frac{4s+12}{[(s+3)^2+4]^2}$$

Solution b: By inspection of Fig. 8.6, we can express $f(t)$ as

$$f(t) = [3e^{-3(t-1)} - 1]U_S(t-1)$$

We consider $g(t) = 3e^{-3t} - 1$, and from Table 8.4,

$$G(s) = \frac{3}{s+3} - \frac{1}{s} = \frac{2s-3}{s(s+3)}$$

and by the time shift property, Table 8.5,

$$F(s) = \frac{e^{-s}(2s-3)}{s(s+3)}$$

Example 8.12   Linearity and Derivative Properties

Solve the following differential equation using the Laplace transform method:

$$\frac{d^2y}{dt^2} + 2\frac{dy}{dt} + 5y = 2; \qquad y(0) = 1, \quad \dot{y}(0) = -1 \tag{8.22}$$

Solution: Using the linearity property, the derivative property, and Table 8.4, we can transform equation (8.22) to

$$[s^2Y(s) - sy(0) - \dot{y}(0)] + 2[sY(s) - y(0)] + 5Y(s) = \frac{2}{s}$$

where $Y(s)$ is the Laplace transform of the total response

$$[s^2 + 2s + 5]Y(s) = \frac{2}{s} + sy(0) + \dot{y}(0) + 2y(0) = \frac{2}{s} + s + 1$$

$$\Rightarrow Y(s) = \frac{2}{s(s^2+2s+5)} + \frac{s+1}{s^2+2s+5}$$

Thus

$$y(t) = \mathscr{L}^{-1}\left[\frac{2}{s(s^2+2s+5)}\right] + \mathscr{L}^{-1}\left[\frac{s+1}{s^2+2s+5}\right]$$

noting that $s^2 + 2s + 5 = (s+1)^2 + 2^2$, we see (using Table 8.4) that

$$y(t) = \mathscr{L}^{-1}\left[\frac{1}{s}F(s)\right] + e^{-t}\cos(2t)$$

where

$$F(s) = \frac{2}{(s + 1)^2 + 2^2}$$

From Table 8.4

$$f(t) = e^{-t}\sin(2t)$$

Using the integral property from Table 8.5, we obtain

$$\mathcal{L}^{-1}\left[\frac{1}{s}F(s)\right] = \int_0^t e^{-\tau}\sin(2\tau)d\tau = \frac{e^{-\tau}}{5}\left(-\sin(2\tau) - 2\cos(2\tau)\right)\Big|_0^t$$

$$= \frac{1}{5}[2 - e^{-t}(\sin 2t + 2\cos 2t)]$$

Finally

$$y(t) = \frac{1}{5}[2 - e^{-t}(\sin 2t - 3\cos 2t)]$$

Initial- and Final-Value Theorems

The initial- and final-value properties of the Laplace transform allow the determination of a system's behavior at very early times and at much later times without solution of the system's model. This represents a useful shortcut in design problems that require initial or final response data. Such data can include, for a mechanical system for instance, values of the position, velocity, and acceleration. The equilibrium state of most stable engineering system models often corresponds to the *final value* of their response. In control systems design, the final or steady-state value of the error is often used as a performance measure for the controller. These are but a few examples. The notation $\text{Lim}(t \longrightarrow 0+)$ in Table 8.5, entry 8 implies that time approaches zero from positive values as the limit is taken. This is essential, since for the *one-sided* Laplace transform the function is presumed to remain zero for all negative time until $t = 0$, where a jump discontinuity may occur.

Validity and Applicability   The final-value theorem presumes that $f(t)$ and $df(t)/dt$ are both Laplace transformable, that $F(s)$ is the Laplace transform of $f(t)$, and that the indicated limits in Table 8.5 exist. It is also valid if $sF(s)$ is *analytic* on the imaginary axis and in the right half-complex plane. This means that no *poles* of $sF(s)$ (that is, values of s at which $|sF(s)|$ becomes infinite) are located on the imaginary axis or in the right half-plane (note that the origin is excluded). Thus the final-value theorem is not applicable to a pure sinusoid, $f(t) = \cos(\omega t)$

(Table 8.4: $F(s) = \dfrac{s}{s^2 + \omega^2}$)

for example, since $sF(s) = \dfrac{1}{s^2 + \omega^2}$ has poles $(s = \pm j\omega)$ on the imaginary axis; moreover, the $\underset{t \to \infty}{\text{Lim}} \cos(\omega t)$ does not exist.

Although the initial-value theorem also presumes that $f(t)$ and $df(t)/dt$ are both Laplace transformable, that $F(s)$ is the Laplace transform of $f(t)$, and that the limits indicated in Table 8.5 exist, its application is not limited by the locations of the poles of $sF(s)$.

Example 8.13   Application of Initial- and Final-Value Theorems

The equation of motion for an accelerometer is

$$m\frac{d^2y}{dt^2} + \beta\frac{dy}{dt} + ky = ma$$

If the input is a step, $a = 9.8U_S(t)$ m/s^2, and $m = 3.0$ kg, $\beta = 10$ N $\cdot$ s/m, and $k = 250$ N/m, determine

a. The relation between the initial position $y(0)$ and the initial velocity $\dot{y}(0)$, such that the early acceleration $\ddot{y}(0+)$, is zero.

b. The final position $y(\infty)$, velocity $\dot{y}(\infty)$, and acceleration $\ddot{y}(\infty)$.

Solution: Using the given data, the equation of motion gives us

$$3\frac{d^2y}{dt^2} + 10\frac{dy}{dt} + 250y = 29.4U_S(t) \tag{8.23a}$$

Applying a Laplace transform we obtain

$$3[s^2Y(s) - sy(0) - \dot{y}(0)] + 10[sY(s) - y(0)] + 250Y(s) = 29.4/s$$

or

$$(3s^2 + 10s + 250)Y(s) = 29.4/s + 3sy(0) + 3\dot{y}(0) + 10y(0)$$

hence,

$$Y(s) = \frac{29.4/s + 3sy(0) + 3\dot{y}(0) + 10y(0)}{3s^2 + 10s + 250} \tag{8.23b}$$

(a): The transform of the acceleration is

$$\mathscr{L}\left[\frac{d^2y}{dt^2}\right] = s^2Y(s) - sy(0) - \dot{y}(0) = \mathscr{L}[\ddot{y}(t)] = A(s)$$

substituting equation (8.23b) we obtain

$$A(s) = \frac{29.4s + 3s^3y(0) + [3\dot{y}(0) + 10y(0)]s^2}{3s^2 + 10s + 250} - sy(0) - \dot{y}(0)$$

that is,

$$A(s) = \frac{[29.4 - 10\dot{y}(0) - 250y(0)]s - 250\dot{y}(0)}{3s^2 + 10s + 250}$$

(8.23c)

Using the initial-value theorem, we have

$$\operatorname*{Lim}_{t \to 0+} [\ddot{y}(t)] = \operatorname*{Lim}_{s \to \infty}[sA(s)]$$

$$= \operatorname*{Lim}_{s \to \infty}\left[\frac{(29.4 - 10\dot{y}(0) - 250y(0))s^2 - 250\dot{y}(0)s}{3s^2 + 10s + 250} \right]$$

Dividing numerator and denominator by s^2, we obtain

$$\ddot{y}(0+) = \operatorname*{Lim}_{s \to \infty}\left[\frac{29.4 - 10\dot{y}(0) - 250y(0) - 250\dot{y}(0)/s}{3 + 10/s + 250/s^2} \right]$$

$$= \frac{29.4 - 10\dot{y}(0) - 250y(0)}{3}$$

For $\ddot{y}(0+) = 0$,

$$29.4 - 10\dot{y}(0) = 250y(0) \quad \text{or} \quad y(0) = 0.1176 - 0.04\dot{y}(0)$$

(b): Equations (8.23b) and (8.23c) give the Laplace transforms of the position and acceleration, respectively. For the velocity, note that

$$\mathscr{L}\left[\frac{dy}{dt} \right] = sY(s) - y(0) = V(s)$$

so that from (8.23b),

$$V(s) = \frac{3s\dot{y}(0) + 29.4 - 250y(0)}{3s^2 + 10s + 250}$$

Using the final-value theorem, we get

$$y(\infty) = \operatorname*{Lim}_{s \to 0}[sY(s)] = \operatorname*{Lim}_{s \to 0}\left[\frac{29.4 + 3s^2y(0) + 3s\dot{y}(0) + 10sy(0)}{3s^2 + 10s + 250} \right]$$

$$= \frac{29.4}{250} = 0.1176$$

$$\dot{y}(\infty) = \operatorname*{Lim}_{s \to 0}[sV(s)] = \operatorname*{Lim}_{s \to 0}\left[\frac{s(29.4 + 3s\dot{y}(0) - 250y(0))}{3s^2 + 10s + 250} \right] = 0$$

$$\ddot{y}(\infty) = \operatorname*{Lim}_{s \to 0}[sA(s)] = \operatorname*{Lim}_{s \to 0}\left[\frac{(29.4 - 10\dot{y}(0) - 250y(0))s^2 - 250s\dot{y}(0)}{3s^2 + 10s + 250} \right] = 0$$

The last two results should be expected for a stable system, since all variation with time should cease after a sufficiently long time has elapsed.

Transfer Functions, Impulse Response, Convolution

The idea of the transfer function was first introduced in Section 7.2 in terms of the phasor transfer function for sinusoidal steady-state $G(j\omega)$. This was followed by the Fourier domain transfer function $G(\omega)$, which turned out to be the same expression as $G(j\omega)$. The Laplace domain transfer function $G(s)$, with $s = -\sigma + j\omega$, represents the most general "phasor" transfer function, and it includes the previous definitions for which $s = j\omega$. Formally, the transfer function $G(s)$ expresses the relation, in the Laplace domain, between the input and output of a *system under zero initial conditions:*

$$Y(s) = G(s)U(s) \tag{8.24}$$

where $Y(s)$ and $U(s)$ are the Laplace transforms of the system output and input, respectively. The restriction to zero initial state is essential; otherwise, a different input-output relation would be required for each set of initial conditions. Thus equation (8.24) is a unique definition of the behavior of a system. Indeed, it is the model of the system if, as is often the case in the design of stable linear systems, attention is focused on the forced response or steady-state behavior (recall that the response to a forcing input under zero initial state is the forced response and that for stable systems the free response due to the initial state has died out prior to the onset of steady-state conditions).

Consider, for example, the general second-order linear system equation (7.2), repeated here:

$$\frac{d^2y}{dt^2} + 2\zeta\omega_0\frac{dy}{dt} + \omega_0^2 y = u(t) \tag{8.25}$$

Using the derivative property of Table 8.5 and assuming zero initial conditions, we obtain the Laplace transform of (8.25):

$$s^2Y(s) + 2\zeta\omega_0 sY(s) + \omega_0^2 Y(s) = U(s) \quad \text{or} \quad Y(s) = \frac{1}{s^2 + 2\zeta\omega_0 s + \omega_0^2}U(s)$$

Thus the transfer function for the system (8.25) is

$$G(s) = \frac{1}{s^2 + 2\zeta\omega_0 s + \omega_0^2} \tag{8.26}$$

Recalling the characteristic equation for the system, that is, equation (7.4a), we see that the *denominator of $G(s)$ equal to zero* **is in fact the characteristic equation.** This is a general result for linear dynamic systems:

$$\text{Denominator } G(s) = 0; \quad \text{(characteristic equation)} \tag{8.27}$$

which reinforces the idea that the transfer function is a system model or a representation of system behavior.

Impulse Response

The transfer of the forced response of a linear system to an arbitrary input can be computed with relative ease from the system's impulse response. To understand this statement, note that the Laplace transform of an impulse is analytically very simple: $\mathscr{L}[\delta(t)] = 1$ (see Table 8.4 and Example 8.8). Consider a linear system whose input-output behavior is given by equation (8.24). Suppose the input is an impulse of strength A, $u(t) = A\delta(t)$, then the forced response is

$$Y(s) = G(s)U(s) = G(s)(A) \tag{8.28a}$$

For a unit impulse, $A = 1$, so that

$$Y(s) = G(s) \quad \text{[unit impulse response]} \tag{8.28b}$$

Thus *the Laplace transform of the forced response of a linear system to a unit impulse is the system's transfer function!*

Impulse and Step Response Identification   The foregoing result suggests the following experimental procedure for the identification (development of the model) of a system without gaining access to its internals (black box approach): With the system at its equilibrium state, apply an impulse input and measure the system's output relative to the equilibrium value. This measurement is $Ag(t)$, where A is the strength of the impulse, and $G(s) = \mathcal{L}[g(t)]$.

In practice, this experiment is applied to both linear and nonlinear systems so that a linear model (which assumes linear behavior) of the system is obtained. However, there are realization difficulties with the impulse input. Although an approximation by a large and narrow pulse is often used, such a large amplitude of the input will normally excite the nonlinearities (saturation, for example) in the system, so that a normal or average response is not realized. The method that is used extensively in industry is the step response identification, because of the relative ease with which a step input can be generated—a sudden switching-on or -off of an input valve, for instance. Although the measured output is not the transfer function, the system is easily identified.

Consider without loss of generality, a unit step input. From equation (8.24),

$$Y_s(s) = G(s)\mathcal{L}[U_s(t)] = G(s)/s$$

where $Y_s(s)$ is the Laplace transform of the system unit step input response. Thus

$$G(s) = sY_s(s) \tag{8.28c}$$

$$g(t) = \mathcal{L}^{-1}[sY_s(s)]$$

Using the derivative property (Table 8.5), and since zero initial state is presumed, we see that

$$g(t) = \frac{d}{dt}y_s(t) \tag{8.28d}$$

where $y_s(t)$ is the measured unit step response. For a general step input, the unit step response is the measured step response divided by the magnitude of the step.

The Convolution Theorem

We consider now equation (8.24) in general terms, that is, without restriction to a particular form of input (impulse, step, etc.). We seek the time-domain response $y(t)$, given $g(t)$ and $u(t)$, without first transforming the problem to the Laplace domain. The solution to this problem is given by the convolution property of Laplace transforms. The convolution theorem (see [7] for proof) states: If $\mathscr{L}^{-1}[F(s)] = f(t)$ and $\mathscr{L}^{-1}[G(s)] = g(t)$, then

$$\mathscr{L}^{-1}[F(s)G(s)] = \int_0^t f(\tau)g(t - \tau)\, d\tau = f(t)*g(t) \tag{8.29a}$$

This is item 10 of Table 8.5. If we let $t - \tau = \lambda$ or $\tau = t - \lambda$, we see that

$$\int_0^t f(\tau)g(t - \tau)\, d\tau = -\int_t^0 f(t - \lambda)g(\lambda)\, d\lambda = \int_0^t f(t - \lambda)g(\lambda)\, d\lambda$$

so that

$$f * g = g * f \tag{8.29b}$$

Since

$$y(t) = \mathscr{L}^{-1}[Y(s)] = \mathscr{L}^{-1}[G(s)U(s)]$$

the solution to the problem posed above is

$$y(t) = g(t)*u(t) = \int_0^t g(t - \tau)u(\tau)\, d\tau \tag{8.29c}$$

This is the same expression given in equation (6.13) as well as the confirmation of the general result given by equation (7.18).

Example 8.14   Implications of Transfer Function

Given a system's transfer function,

$$G(s) = \frac{1}{(s + 1)(s + 2)}$$

a. What are the eigenvalues of the system?
b. What is the differential equation representing the system model?
c. Find the forced response of this system to a step input $u(t) = 3U_s(t)$.

Solution a: From equation (8.27), the characteristic equation is

$$(s + 1)(s + 2) = 0$$

so that the eigenvalues are -1 and -2. The transient response modes would be e^{-t} and e^{-2t}.

Solution b:

$$Y(s) = G(s)U(s) = \frac{U(s)}{(s + 1)(s + 2)} = \frac{U(s)}{s^2 + 3s + 2}$$

or

$$(s^2 + 3s + 2)Y(s) = U(s)$$

Noting that $G(s)$ is defined for zero initial conditions, the derivative property suggests to us the model

$$\frac{d^2y}{dt^2} + 3\frac{dy}{dt} + 2y = u(t)$$

Solution c: Again,

$$Y(s) = G(s)U(s) = \frac{1}{(s+1)(s+2)} \frac{3}{s} - \frac{3}{s(s+1)(s+2)}$$

By *partial fraction expansion* (this is considered in the forthcoming material)

$$\frac{3}{s(s+1)(s+2)} = \frac{3}{2s} - \frac{3}{s+1} + \frac{3}{2(s+2)}$$

so that Table 8.4 yields

$$y(t) = \frac{3}{2}U_s(t) - 3e^{-t} + \frac{3}{2}e^{-2t}$$

Example 8.15	Application of Transfer Function

The output of a first-order system with the model

$$\frac{dx}{dt} + 4x = u(t)$$

is input to a second system whose transfer function is

$$G_2(s) = \frac{s}{(s+1)(s+2)}$$

Find the magnitude and phase of the output from the second system when the input to the first-order system is $u(t) = \cos(t)e^{-3t}$.

Solution: The transfer function of the first-order system is

$$G_1(s) = \frac{1}{s+4}$$

Using equation (8.24) repetitively, we obtain the overall transfer function for the combined system:

$$Y(s) = G(s)U(s) = G_2(s)G_1(s)U(s)$$

where $Y(s)$ is the transform of the overall output, which is the output of the second system. That is,

$$G(s) = G_2(s)G_1(s) = \frac{s}{(s+1)(s+2)(s+4)}$$

The phasor representation of the input is e^{st}, $s = -3 + j$. Since the input amplitude is 1, the output magnitude is

$$|G(s)| = \left| \frac{-3 + j}{(-2 + j)(-1 + j)(1 + j)} \right| = \left| \frac{-3 + j}{4 - 2j} \right| = \left| \frac{(-3 + j)(4 + 2j)}{16 + 4} \right|$$

$$= \left| \frac{-7 - j}{10} \right| = \sqrt{\frac{49 + 1}{100}} = \sqrt{\frac{1}{2}} = 0.707$$

Since the input phase angle is zero, the output phase is

$$\angle G(s) = \tan^{-1}\left(\frac{-1}{-7} \right) = 3.28 \quad \text{(or 188.13°)}$$

Example 8.16   Step Response Identification

If the forced response of a system to the step input $u(t) = 2U_s(t)$ is $-e^{-t}(t + 1)$, what will be the system's response to an input

a. $u(t) = e^{-2t}$?
b. $u(t) = 3\delta(t)$?

Solution a: The unit step input response is

$$y_{us}(t) = -\frac{1}{2}e^{-t}(t + 1)$$

and by equation (8.28d)

$$g(t) = \frac{d}{dt}\left[-\frac{1}{2}e^{-t}(t + 1) \right] = -\frac{1}{2}[-e^{-t}(t + 1) + e^{-t}] = \frac{1}{2}te^{-t}$$

Using the convolution integral equation (8.29c), we obtain the response when the input is $u(t) = e^{-2t}$:

$$y(t) = \frac{1}{2}\int_0^t (t - \tau)e^{-(t-\tau)}e^{-2\tau}\, d\tau = \frac{1}{2}e^{-t}\left[t\int_0^t e^{-\tau}d\tau - \int_0^t \tau e^{-\tau}\, d\tau \right]$$

$$= \frac{1}{2}e^{-t}\left[t(-e^{-\tau})\Big|_0^t - e^{-\tau}(-\tau - 1)\Big|_0^t \right]$$

or

$$y(t) = \frac{1}{2}e^{-t}[t(1 - e^{-t}) - 1 + e^{-t}(t + 1)] = \frac{1}{2}[(t - 1)e^{-t} + e^{-2t}]$$

Alternatively,

$$G(s) = \mathcal{L}[g(t)] = \mathcal{L}\left[\frac{1}{2}te^{-t} \right] = \frac{1}{2}\left[\frac{1}{(s + 1)^2} \right];$$

$$U(s) = \mathcal{L}[u(t)] = \mathcal{L}[e^{-2t}] = \frac{1}{s + 2}$$

and

$$Y(s) = G(s)U(s) = \frac{1}{2} \frac{1}{(s + 1)^2(s + 2)}$$

By partial fraction expansion (see next section):

$$\frac{1}{(s + 1)^2(s + 2)} = -\frac{1}{(s + 1)} + \frac{1}{(s + 1)^2} + \frac{1}{(s + 2)}$$

so that

$$y(t) = \frac{1}{2}[-e^{-t} + te^{-t} + e^{-2t}] = \frac{1}{2}[(t - 1)e^{-t} + e^{-2t}]$$

Solution b: The response to a unit impulse input is the transfer function (equation (8.28b)):

$$y_1(t) = g(t) = \frac{1}{2} te^{-t}$$

The response to an impulse of strength 3 is therefore

$$y_1(t) = \frac{3}{2} te^{-t}$$

....

The Inverse Transform and Partial Fraction Expansions

The inverse Laplace transform as given by equation (8.20d) is not easy to compute. As must have become evident in the last few examples, the practical method of inverse Laplace transformation, which is by table lookup, is facilitated by the expansion of $F(s)$ into its partial fractions. Actually, any reduction of $F(s)$ into functions than can be recognized in the available table of Laplace transforms is fair game, and this has been the approach until now. This section also considers briefly, for illustrative purposes, the application of Laplace transforms to ODEs with variable coefficients and to simple partial differential equations subject to boundary conditions.

Partial Fraction Expansions

Consider

$$F(s) = \frac{b_m + b_{m-1}s + b_{m-2}s^2 + \ldots + b_0 s^m}{a_n + a_{n-1}s + a_{n-2}s^2 + \ldots + a_0 s^n}, \quad n \geq m \tag{8.30a}$$

the equation $a_n + a_{n-1}s + a_{n-2}s^2 + \ldots + a_0 s^n = 0$ has n roots, some of which may be repeated. For example, $s^3 + 4s^2 + 5s + 2$ has three roots ($n = 3$): $-1, -1, -2$ (-1 is repeated once). This can be verified by back substitution: $(s + 1)^2 (s + 2) = s^3 + 4s^2 + 5s + 2$.

Suppose the n roots are n_1 roots $= -\lambda_1, n_2$ roots $= -\lambda_2, \ldots, n_r$ roots $= -\lambda_r$, where $n_1 + n_2 + \ldots + n_r = n$, then $a_n + a_{n-1}s + a_{n-2}s^2 + \ldots + a_0 s^n = (s + \lambda_1)^{n_1}(s + \lambda_2)^{n_2} \cdots (s + \lambda_r)^{n_r}$. For instance, in the preceding example, $n_1 = 2, \lambda_1 = 1,$

and $n_2 = 1$, $\lambda_2 = 2$. It follows that $F(s)$ as a ratio of polynomials (8.30a) can be written as

$$F(s) = \frac{b_m + b_{m-1}s + b_{m-2}s^2 + \ldots + b_0 s^m}{(s + \lambda_1)^{n_1}(s + \lambda_2)^{n_2} \cdots (s + \lambda_r)^{n_r}} \tag{8.30b}$$

The partial fraction representation of $F(s)$ is

$$F(s) = B + \sum_{k=1}^{r} \sum_{i=1}^{n_k} \frac{c_{ki}}{(s + \lambda_k)^i} \tag{8.31a}$$

where

$$B = \begin{cases} b_0 & \text{if } m = n \\ 0 & \text{otherwise} \end{cases} \tag{8.31b}$$

and

$$c_{ki} = \frac{1}{(n_k - i)!} \frac{d^{(n_k - i)}}{ds^{(n_k - i)}} \left[(s + \lambda_k)^{n_k} F(s) \right] \Bigg|_{s=-\lambda_k} \tag{8.31c}$$

For the special case in which there are no repeated roots, equations (8.31a) and (8.31c) reduce to

$$F(s) = B + \sum_{i=1}^{n} \frac{c_i}{(s + \lambda_i)}; \qquad c_i = (s + \lambda_i)F(s) \Bigg|_{s=-\lambda_i} \tag{8.31d}$$

The application of equations (8.31a) through (8.31d) is best explained with examples.

Example 8.17   Expansion into Partial Fraction

Find the partial fractions expansion of

a. $F(s) = \dfrac{1}{s^3 + 4s^2 + 5s + 2}$

b. $F(s) = \dfrac{s^2 + 2s + 2}{s^2 + 3s + 2}$

c. $F(s) = \dfrac{s + 2}{s^2 + 2s + 4}$

Solution a: It has already been shown that

$$F(s) = \frac{1}{(s + 1)^2(s + 2)}$$

and $n_1 = 2$, $\lambda_1 = 1$, and $n_2 = 1$, $\lambda_2 = 2$, so that $r = 2$.
From equation (8.31a), (with $B = 0$),

$$F(s) = \left[\frac{c_{11}}{(s+1)} + \frac{c_{12}}{(s+1)^2} \right] + \frac{c_{21}}{(s+2)}$$

Using equation (8.31c), we have

$$k = 1, i = 1: \quad c_{11} = \frac{1}{1!} \frac{d}{ds}\left[\frac{(s+1)^2}{(s+1)^2(s+2)} \right]\Bigg|_{s=-1} = \frac{-1}{(s+2)^2}\Bigg|_{s=-1} = -1$$

$$k = 1, i = 2: \quad c_{12} = \frac{1}{0!}\left[\frac{(s+1)^2}{(s+1)^2(s+2)} \right]\Bigg|_{s=-1} = \frac{1}{(s+2)}\Bigg|_{s=-1}$$

$$= 1; \quad \text{(Note that 0! = 1)}$$

$$k = 2, i = 1: \quad c_{21} = \frac{1}{0!}\left[\frac{(s+2)}{(s+1)^2(s+2)} \right]\Bigg|_{s=-2} = \frac{1}{(s+1)^2}\Bigg|_{s=-2} = 1$$

Thus

$$F(s) = \frac{-1}{(s+1)} + \frac{1}{(s+1)^2} + \frac{1}{(s+2)}$$

which is the result used in Example 8.16.

Solution b:

$$F(s) = \frac{s^2 + 2s + 2}{s^2 + 3s + 2} = \frac{s^2 + 2s + 2}{(s+1)(s+2)}$$

We have $n = m = 2$, $n_1 = n_2 = 1$, $B = 1$. From equation (8.31d),

$$c_1 = (s+1)\frac{s^2 + 2s + 2}{(s+1)(s+2)}\Bigg|_{s=-1} = 1;$$

$$c_2 = (s+2)\frac{s^2 + 2s + 2}{(s+1)(s+2)}\Bigg|_{s=-2} = -2$$

Thus

$$F(s) = 1 + \frac{1}{s+1} - \frac{2}{s+2}$$

Solution c:

$$F(s) = \frac{s+2}{s^2 + 2s + 4} = \frac{s+2}{\left[s - \left(1 + j\sqrt{3}\right)\right]\left[s - \left(-1 - j\sqrt{3}\right)\right]}$$

We have a pair of complex-conjugate roots, which are otherwise distinct. Equations (8.31a) through (8.31d) are applicable to both real- and complex-valued roots. With no repeated roots, equation (8.31d) is applicable; $n > m$ implies $B = 0$. Then

$$c_1 = \frac{s + 2}{s - \left(-1 - j\sqrt{3}\right)}\bigg|_{s = -1 + j\sqrt{3}} = \frac{1 + j\sqrt{3}}{-1 + j\sqrt{3} + 1 + j\sqrt{3}} = \frac{1 + j\sqrt{3}}{2j\sqrt{3}}$$

$$= \frac{1}{2} - \frac{j\sqrt{3}}{6}$$

$$c_2 = \frac{s + 2}{s - \left(-1 + j\sqrt{3}\right)}\bigg|_{s = -1 - j\sqrt{3}} = \frac{1 - j\sqrt{3}}{-1 - j\sqrt{3} + 1 - j\sqrt{3}} = \frac{1 - j\sqrt{3}}{-2j\sqrt{3}}$$

$$= \frac{1}{2} + \frac{j\sqrt{3}}{6}$$

and

$$F(s) = \frac{1/2 - \left(j\sqrt{3}/6\right)}{\left[s - \left(-1 + j\sqrt{3}\right)\right]} + \frac{1/2 + \left(j\sqrt{3}/6\right)}{\left[s - \left(-1 - j\sqrt{3}\right)\right]}$$

The coefficients are complex conjugates, as should be expected. What is the inverse Laplace transform? By entry 4 of Table 8.4,

$$f(t) = \left(\frac{1}{2} - \frac{j\sqrt{3}}{6}\right)e^{(-1 + j\sqrt{3})t} + \left(\frac{1}{2} + \frac{j\sqrt{3}}{6}\right)e^{(-1 - j\sqrt{3})t}$$

$$= \frac{1}{2}e^{-t}\left[e^{j\sqrt{3}t} + e^{-j\sqrt{3}t}\right] - \frac{j\sqrt{3}}{6}e^{-t}[e^{j\sqrt{3}t} - e^{-j\sqrt{3}t}]$$

$$= e^{-t}\left[\left(\frac{e^{j\sqrt{3}t} + e^{-j\sqrt{3}t}}{2}\right) + \frac{\sqrt{3}}{3}\left(\frac{e^{j\sqrt{3}t} - e^{-j\sqrt{3}t}}{2j}\right)\right]$$

Euler's identities, equations (6.41a) through (6.41c), yield

$$f(t) = \left[\cos\sqrt{3}t + \frac{\sqrt{3}}{3}\sin\sqrt{3}t\right]e^{-t} = C\cos\left(\sqrt{3}t + \phi\right)e^{-t}$$

and according to equation (7.11b),

$$C = \sqrt{(1)^2 + \left(\sqrt{3}/3\right)^2} = \frac{2\sqrt{3}}{3} \quad \text{and} \quad \phi = \tan^{-1}\left(\frac{-\sqrt{3}}{3}\right) = -0.5236$$

The partial fractions expansion procedure as just described and illustrated in the examples assumes that the order of the denominator polynomial is greater than or equal to the order of the numerator polynomial and that the roots of the denominator polynomial are known. The first assumption is not restrictive, since the numerator can be divided by the denominator to obtain a remainder term that satisfies $n \geq m$. The second assumption can be a problem especially for high-order polynomials, although a numerical solution can also be used (MATLAB's *residue* function is applicable to $F(s)$ as a ratio of polynomials; see **On the Net**). Finally, in computing the coefficients c_{ki}, an informal but direct alternative procedure is obtained through clearing of fractions and equating of like powers of s on both sides of the resulting equation. In this regard the denominator of the partial fractions need not be linear provided the degree of the numerator is less than that of the denominator and there are a total of n coefficients. The choice of the partial fractions such as

$$\frac{A}{(ps + q)} \quad \text{and} \quad \frac{As + B}{ps^2 + qs + r}$$

can, in this case, facilitate both the factoring of the denominator polynomial and the matching of Laplace transform pairs to the partial fractions.

Example 8.18   Alternative Approach

Find

$$\mathscr{L}^{-1}\left[\frac{3s + 1}{(s + 1)(s^2 + 1)}\right]$$

Solution: We let

$$F(s) = \frac{3s + 1}{(s + 1)(s^2 + 1)} = \frac{A}{(s + 1)} + \frac{Bs + C}{(s^2 + 1)}$$

We clear the fractions:

$$A(s^2 + 1) + (Bs + C)(s + 1) = A(s^2 + 1) + B(s^2 + s) + C(s + 1) = 3s + 1$$

and equate like powers of s:

$$A + B = 0, \qquad B + C = 3; \qquad A + C = 1$$

Thus $A = -1$, $B = 1$, and $C = 2$, so that

$$F(s) = \frac{-1}{s + 1} + \frac{s + 2}{(s^2 + 1)} = \frac{-1}{s + 1} + \frac{s}{s^2 + 1} + \frac{2}{s^2 + 1}$$

From Table 8.4,

$$\mathscr{L}^{-1}[F(s)] = -e^{-t} + \cos t + 2 \sin t$$

Application to Nonstationary and Distributed Systems

Recall from Section 2.4 that although the assumption of stationarity is often justified, there are no truly time-invariant systems. We now briefly illustrate the application of Laplace transforms to the solution of ODEs with time-varying coefficients, especially when the variable coefficients are simple polynomial functions of time.

Example 8.19   A Time-Varying System

Consider the mass-expelling space vehicle of Practice Problem 2.7.

a. Find the thrust and average velocity at which mass is expelled from the vehicle exhaust.

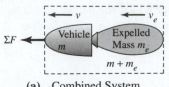

(a)   Combined System

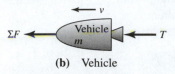

(b)   Vehicle

Figure 8.7

Combined System and Vehicle Free Bodies

b. Suppose the vehicle velocity is not constant. If at some initial state ($t = 0$) the vehicle mass and velocity are m_0 and v_0, respectively, find the displacement of the vehicle from this initial point after t time units. Assume that during this period the pressure P in the exhaust gas is given by $P = k\dot{m}_e. -2\,\dot{m}_e(V - V_0)/A$

Solution a: Consider first the general situation in which a vehicle that is moving with velocity v is simultaneously expelling mass at a prescribed rate $\dot{m}_e = -\dot{m}$, such that the expelled mass flow velocity v_e is a constant. If we assume that the expelled mass continues to move at this velocity, then the vehicle and expelled mass can be combined into one system of mass $(m + m_e)$, as shown in Fig. 8.7a. The force of reaction (this is the thrust) between the vehicle and the exhaust is an internal force in such a system, and the resultant external force is applied only on the vehicle. The appropriate use of equation (2.8a) here is

$$\sum F = \frac{d}{dt}(mv + m_ev_e) = m\frac{dv}{dt} + \dot{m}v + \dot{m}_ev_e \tag{8.32a}$$

$$\sum F = m\frac{dv}{dt} + \dot{m}(v - v_e) = m\frac{dv}{dt} - \dot{m}_e(v - v_e) \tag{8.32b}$$

Equation (8.32b) is the model for the general situation described. The term $\dot{m}_e(v - v_e)$ represents the force necessary to decelerate the exhaust stream of particles from a velocity v to a lower velocity v_e. It is also the thrust (see Fig. 8.7b) that acts on both the vehicle and the exhaust stream. For Practice Problem 2.7,

$$\sum F = PA \tag{8.32c}$$

and

$$\frac{dv}{dt} = 0 \tag{8.32d}$$

Thus $T = -PA$, and

$$\dot{m}_e(v - v_e) = -PA \tag{8.32e}$$

giving

$$v_e = \frac{1}{\dot{m}_e}PA + v \tag{8.32f}$$

Physically, equations (8.32e) and (8.32f) imply that the exhaust for this vehicle should be directed ahead of the vehicle in order to keep its velocity constant during this period.

Solution b: Equations (8.32b) and (8.32c) are still applicable to the present problem. That is,

$$m\frac{dv}{dt} = PA - \dot{m}(v - v_e)$$

Further, if the displacement of the vehicle is y, then $dy/dt = v$, and

$$m\frac{d^2y}{dt^2} + \dot{m}\frac{dy}{dt} = PA + \dot{m}v_e = 2\dot{m}\frac{dy}{dt} + \dot{m}(v_e - 2v_o - kA) \qquad \text{(8.33a)}$$

Since $\dot{m}$ is a constant, then the vehicle mass at any time after the initial point is given by $m = m_0 + \dot{m}t$. Hence the model to solve is

$$(m_0 - \dot{m}_e t)\frac{d^2y}{dt^2} + \dot{m}_e\frac{dy}{dy} = \dot{m}(v_e - 2v_o - kA) = F_0 \qquad \text{(8.33b)}$$

The Laplace transform solution of equation (8.33b) is facilitated by the multiplication by t^n property (Table 8.5, item 5). Thus

$$m_0[s^2Y - sy(0) - \dot{y}(0)] - \dot{m}_e(-1)\frac{d}{ds}[s^2Y - sy(0) - \dot{y}(0)] + \dot{m}_e[sY - y(0)] = \frac{F_0}{s}$$

Using $y(0) = 0$, we have

$$\dot{y}(0) = v_0; \qquad m_0 s^2 Y - m_0 v_0 + \dot{m}_e\left(2sY + s^2\frac{dY}{ds}\right) + \dot{m}_e sY = \frac{F_0}{s}$$

or

$$\dot{m}_e s^2\frac{dY}{ds} + (\dot{m}_e 3s + m_0 s^2)Y = m_0 v_0 + \frac{F_0}{s}$$

First solving the differential equation in s, we obtain

$$\frac{dY}{ds} + \left(\frac{m_0}{\dot{m}_e} + \frac{3}{s}\right)Y = \frac{m_0 v_0}{\dot{m}_e s^2} + \frac{F_0}{\dot{m}_e s^3} \qquad \text{(8.33c)}$$

Recall from mathematics that there are several techniques for solving any first-order ordinary differential equation. For the present problem the integrating factor (see [1], for example) is given by

$$\exp\left[\int\left(\frac{m_0}{\dot{m}_e} + \frac{3}{s}\right)ds\right] = \exp\left[\frac{m_0 s}{\dot{m}_e} + 3\ln s\right] = s^3\exp\left[\frac{m_0 s}{\dot{m}_e}\right]$$

so that equation (8.33c) becomes

$$\frac{d}{ds}(s^3 e^{m_0 s/\dot{m}_e}Y) = \left(\frac{m_0 v_0}{\dot{m}_e s^2} + \frac{F_0}{\dot{m}_e s^3}\right)s^3 e^{m_0 s/\dot{m}_e} \qquad \text{(8.33d)}$$

Integrating, we have

$$s^3 e^{m_0 s/\dot{m}_e}Y = \int\left(\frac{m_0 v_0}{\dot{m}_e} + \frac{F_0}{\dot{m}_e}\right)e^{m_0 s/\dot{m}_e}\, ds$$

$$= \int\left(\frac{m_0 v_0}{\dot{m}_e}se^{m_0 s/\dot{m}_e}\, ds + \frac{F_0}{\dot{m}_e}e^{m_0 s/\dot{m}_e}\, ds\right)$$

or

$$Y = \frac{1}{s^3} e^{-m_0 s/\dot{m}_e} \left(\frac{\dot{m}_e v_0}{m_0} e^{m_0 s/\dot{m}_e} \left[\frac{m_0 s}{\dot{m}_e} - 1 \right] + \frac{F_0}{m_0} e^{m_0 s/\dot{m}_e} + C \right)$$

where C is a constant of integration, that is, C is not a function of s.

$$Y = \frac{\dot{m}_e v_0}{m_0} \left[\frac{m_0}{\dot{m}_e s^2} - \frac{1}{s^3} \right] + \frac{F_0}{m_0 s^3} + \frac{C}{s^3} e^{-m_0 s/\dot{m}_e}$$

$$Y = \frac{v_0}{s^2} + \left(\frac{F_0}{m_0} - \frac{\dot{m}_e v_0}{m_0} \right) \frac{1}{s^3} + \frac{C}{s^3} e^{-m_0 s/\dot{m}_e} \tag{8.33e}$$

To find the inverse Laplace transform of (8.33e), we note for the last term that since C is not a function of s,

$$\mathcal{L}^{-1} \left[\frac{C}{s^3} \right] = \frac{C}{2} t^2 \quad \text{and} \quad \mathcal{L}^{-1} \left[\frac{C}{s^3} e^{-m_0 s/\dot{m}_e} \right] = \frac{C}{2} (t - m_0/\dot{m}_e)^2$$

Hence

$$y(t) = v_0 t + \left(\frac{F_0 - \dot{m}_e v_0}{2m_0} \right) t^2 + \frac{C}{2} (t - m_0/\dot{m}_e)^2 \tag{8.33f}$$

To determine C, we apply the initial condition $y(0) = 0$ to get $C = 0$. Hence the solution is

$$y(t) = v_0 t + \left(\frac{F_0 - \dot{m}_e v_0}{2m_0} \right) t^2 \tag{8.33g}$$

Substitution of equation (8.33g) in equation (8.33b) will verify the answer. Putting back the value of F_0, we obtain the final solution:

$$y(t) = v_0 t + \frac{\dot{m}}{2m_0} (v_e - v_0 - kA) t^2 \tag{8.33h}$$

It is interesting to note that the solution (8.33h) also includes equation (8.32f), the answer obtained for part (a). To see this, obtain the velocity v by differentiating equation (8.33h) with respect to time. Then setting $v = \text{constant} = v_0$, solve for v_e.

....

Application to Partial Differential Equations

The Laplace transform is often used to solve boundary-value problems, that is, partial differential equations that are subject to conditions prescribed at the physical boundaries of the system. In Sections 2.4 and 4.3 it was demonstrated that such problems arise directly when a distributed-parameter model is assumed for dynamic physical systems. In general, the Laplace transformation of a partial differential equation in two variables results in an ordinary differential equation in one of the variables, which can then be solved by the usual methods. However, the resulting algebraic function of the Laplace parameter s usually requires the application of the complex inversion formula, equation (8.20d), before the final solution can be obtained for even very simple physical problems. The reader interested in pursuing the subject should refer to reference [9] or other texts on the application of Laplace transforms.

Example 8.20	A Simple Boundary-Value Problem in PDE

Consider the longitudinal vibration of a thin rod (under zero body force), equation (4.34) repeated:

$$\frac{\partial^2 u}{\partial t^2} = \frac{E}{\rho}\frac{\partial^2 u}{\partial x^2} = a^2\frac{\partial^2 u}{\partial x^2} \tag{8.34a}$$

If the rod is initially at rest but with a spatial configuration given by

$$u(x, 0) = \sin\left(\frac{\pi x}{L} - \pi\right)$$

where L is the length of the rod, determine the vibration of the rod $u(x, t)$, if the ends of the rod remain fixed.

Solution: Taking the Laplace transform of equation (8.34a) with respect to the time variable, we obtain

$$s^2 U(x, s) - su(x, 0) - \left.\frac{\partial u(x, t)}{\partial t}\right|_{t=0} = s^2 U(x, s) - su(x, 0) = a^2\frac{d^2}{dx^2}U(x, s)$$

The velocity term is zero because the rod is initially at rest, and in the Laplace domain, the partial derivatives with respect to x become ordinary derivatives. Thus we have a second-order ODE for $U(x, s)$:

$$\frac{d^2}{dx^2}U(x, s) - \left(\frac{s}{a}\right)^2 U(x, s) = -\frac{s}{a^2}\sin\left(\frac{\pi x}{L} - \pi\right) \tag{8.34b}$$

The free response is given by the homogeneous solution

$$U_H(x, s) = C_1 e^{sx/a} + C_2 e^{-sx/a}$$

where C_1 and C_2 can be functions of s. Applying the *boundary conditions*, we obtain

$$x = 0: \qquad C_1 + C_2 = \mathscr{L}[u(0, t)] = 0$$
$$x = L: \qquad C_1 e^{sL/a} + C_2 e^{-sL/a} = \mathscr{L}[u(L, t)] = 0$$

the solution of which is $C_1 = C_2 = 0$. Note that the boundary values are Laplace transformed. Thus the free response of (8.34b) is zero. Using the method of undetermined coefficients, we can try for the forced response and hence the total solution of (8.34b),

$$U(x, s) = C(s)\cos\left(\frac{\pi x}{L} + \phi(s)\right)$$

Substituting in equation (8.34b) (the sine function can be replaced by the equivalent cosine term), we obtain

$$-C(s)(\pi/L)^2\cos\left(\frac{\pi x}{L} + \phi(s)\right) - \left(\frac{s}{a}\right)^2 C(s)\cos\left(\frac{\pi x}{L} + \phi(s)\right) \tag{8.34c}$$

$$= -\left(\frac{s}{a^2}\right)\cos\left(\frac{\pi x}{L} - \frac{3\pi}{2}\right)$$

the solution of which is

$$C(s) = \frac{s}{a^2\pi^2/L^2 + s^2} \quad \text{and} \quad \phi(s) = -3\pi/2$$

Note that we also could have applied the phasor transform to equation (8.34b) to arrive at this result. Thus solution of the vibration problem in the Laplace domain is

$$U(x, s) = \frac{s}{a^2\pi^2/L^2 + s^2}\cos\left(\frac{\pi x}{L} - \frac{3\pi}{2}\right) \tag{8.34d}$$

Fortunately, the inverse Laplace transform of equation (8.34d) can be found directly. The result (see item 7 of Table 8.4) is

$$u(x, t) = \cos\left(\frac{\pi x}{L} - \frac{3\pi}{2}\right)\cos\left(\frac{a\pi}{L}t\right)$$

8.3 LAPLACE DOMAIN SOLUTION OF THE VECTOR STATE EQUATION

The Laplace transform formula equation (8.20c) and the linearity and derivative properties (items 1 and 3 in Table 8.5) are also applicable to vector functions. Consider the general nth-order linear state model, with $t_0 = 0$, equation (6.49) repeated:

State:

$$\frac{d}{dt}\mathbf{X} = \mathbf{AX} + \mathbf{BU}; \qquad \mathbf{X}(0) = \mathbf{X}_0 \tag{8.35a}$$

Output:

$$\mathbf{Y} = \mathbf{CX} + \mathbf{DU} \tag{8.35b}$$

Using the above transform properties and the rules of matrix algebra, we obtain

$$s\mathbf{X}(s) - \mathbf{X}_0 = \mathbf{AX}(s) + \mathbf{BU}(s) \Rightarrow (s\mathbf{I} - \mathbf{A})\mathbf{X}(s) = \mathbf{X}_0 + \mathbf{BU}(s)$$

Assuming that $(s\mathbf{I} - \mathbf{A})^{-1}$ exists, then we have

$$\mathbf{X}(s) = (s\mathbf{I} - \mathbf{A})^{-1}\mathbf{X}_0 + (s\mathbf{I} - \mathbf{A})^{-1}\mathbf{BU}(s) \tag{8.36a}$$

Also,

$$\mathbf{Y}(s) = \mathbf{CX}(s) + \mathbf{DU}(s) \tag{8.36b}$$

The inverse Laplace transform of $\mathbf{X}(s)$ yields $\mathbf{X}(t)$, the solution of the system (8.35a).

| Example 8.21 | Solution of a State Model by Laplace Transformation |

Find by the Laplace transform method, the solution to

$$\frac{d}{dt}\begin{bmatrix} x_1 \\ x_2 \end{bmatrix} = \begin{bmatrix} -1 & 0 \\ 1 & -2 \end{bmatrix}\begin{bmatrix} x_1 \\ x_2 \end{bmatrix} + \begin{bmatrix} 1 \\ 0 \end{bmatrix}2U_s(t); \qquad \mathbf{X}(0) = \begin{bmatrix} 2 \\ 3 \end{bmatrix}$$

Solution:

$$(s\mathbf{I} - \mathbf{A})^{-1} = \begin{bmatrix} (s+1) & 0 \\ -1 & (s+2) \end{bmatrix}^{-1}; \qquad \det(s\mathbf{I} - \mathbf{A}) = (s+1)(s+2)$$

$$(s\mathbf{I} - \mathbf{A})^{-1} = \frac{1}{(s+1)(s+2)}\begin{bmatrix} (s+2) & 0 \\ 1 & (s+1) \end{bmatrix}$$

From equation (8.36a),

$$\mathbf{X}(s) = \frac{1}{(s+1)(s+2)}\begin{bmatrix} 2(s+2) \\ 2+3(s+1) \end{bmatrix} + \frac{1}{(s+1)(s+2)}\begin{bmatrix} (s+2) \\ 1 \end{bmatrix}\frac{2}{s}$$

$$\mathbf{X}(s) = \begin{bmatrix} \dfrac{2(s+2)+2(s+2)/s}{(s+1)(s+2)} \\[3mm] \dfrac{2+3(s+1)+2/s}{(s+1)(s+2)} \end{bmatrix} = \begin{bmatrix} \dfrac{2}{s+1} + \dfrac{2}{s(s+1)} \\[3mm] \dfrac{3s^2+5s+2}{s(s+1)(s+2)} \end{bmatrix}$$

$\mathscr{L}^{-1}\mathbf{X}(s)$ is found by an *element-by-element inversion* process. Using Tables 8.4 and 8.5 and the partial fractions expansion, we obtain

$$\mathscr{L}^{-1}\left[\frac{2}{s+1}\right] = 2e^{-t}; \qquad \mathscr{L}^{-1}\left[\frac{2}{s(s+1)}\right] = \int_0^t 2e^{-\tau}\,d\tau = 2(1 - e^{-t})$$

$$\mathscr{L}^{-1}\left[\frac{3s^2+5s+2}{s(s+1)(s+2)}\right] = \mathscr{L}^{-1}\left[\frac{1}{s} + \frac{0}{s+1} + \frac{2}{s+2}\right] = 1 + 2e^{-2t}$$

Thus

$$\mathbf{X}(t) = \begin{bmatrix} 2e^{-t} + 2(1 - e^{-t}) \\ 1 + 2e^{-2t} \end{bmatrix} = \begin{bmatrix} 2 \\ 1 + 2e^{-2t} \end{bmatrix}$$

The State Transition Matrix Revisited

The preceding solution agrees with the result obtained using the expansion of the matrix exponential $e^{\mathbf{A}t}$. The Laplace transform solution of equation (8.35) or (6.49) offers yet another interpretation or means of finding the solution matrix. Consider the inverse Laplace transform of equation (8.36a):

$$\mathbf{X}(t) = \mathscr{L}^{-1}[(s\mathbf{I} - \mathbf{A})^{-1}]\mathbf{X}_0 + \mathscr{L}^{-1}[(s\mathbf{I} - \mathbf{A})^{-1}\mathbf{B}U(s)] \qquad \textbf{(8.37a)}$$

The application of this equation was demonstrated in Example 8.21. However, comparing, term by term, equation (8.37a) with the time-domain solution, equation (6.58), which is repeated here as (8.37b),

$$\mathbf{X}(t) = e^{\mathbf{A}t}\mathbf{X}_0 + \int_0^t e^{\mathbf{A}(t-\tau)}\mathbf{B}U(\tau)\,d\tau \qquad \textbf{(8.37b)}$$

we see that

$$e^{\mathbf{A}t} = \mathcal{L}^{-1}[(s\mathbf{I} - \mathbf{A})^{-1}] \tag{8.38a}$$

and

$$\int_0^t e^{\mathbf{A}(t-\tau)}\mathbf{B}\mathbf{U}(\tau)\, d\tau = \mathcal{L}^{-1}[(s\mathbf{I} - \mathbf{A})^{-1}\mathbf{B}\mathbf{U}(s)] \tag{8.38b}$$

Thus the solution matrix can be found using equation (8.38a). Further, equation (8.38b) yields the forced response of the system (8.35a). We also note that equation (8.38b) agrees with the convolution property, stated for scalar functions in equation (8.29c). Here, the inverse Laplace transform of the product of two Laplace domain matrix functions, $(s\mathbf{I} - \mathbf{A})^{-1}$ and $\mathbf{B}\mathbf{U}(s)$, is given by the convolution integral of the time-domain functions $e^{\mathbf{A}t}$ and $\mathbf{B}\mathbf{U}(t)$. Also, equation (8.38a) is a generalization of the scalar Laplace transform pair (item 4, Table 8.4) $e^{at} = \mathcal{L}^{-1}[(s - a)^{-1}]$.

Example 8.22   The Solution Matrix

Consider again the problem of Example 6.13. That is, find $e^{\mathbf{A}t}$ when $\mathbf{A} = \begin{bmatrix} 1 & 1 \\ 0 & 1 \end{bmatrix}$.

Solution: Using equation (8.38a), we have

$$e^{\mathbf{A}t} = \mathcal{L}^{-1}\begin{bmatrix} (s-1) & -1 \\ 0 & (s-1) \end{bmatrix}^{-1} = \mathcal{L}^{-1}\left(\frac{1}{(s-1)^2}\begin{bmatrix} (s-1) & -1 \\ 0 & (s-1) \end{bmatrix}\right)$$

$$= \mathcal{L}^{-1}\begin{bmatrix} \dfrac{1}{(s-1)} & \dfrac{1}{(s-1)^2} \\ 0 & \dfrac{1}{(s-1)} \end{bmatrix}$$

Using Table 8.4, we obtain as in Example 6.13,

$$e^{\mathbf{A}t} = \begin{bmatrix} e^t & te^t \\ 0 & e^t \end{bmatrix}$$

The Matrix Transfer Function

The Laplace domain input-output relation for the system of equations (8.35a) and (8.35b) is easily obtained. From equations (8.36a) and (8.36b), assuming zero initial state, we have

$$\mathbf{Y}(s) = \mathbf{G}(s)\mathbf{U}(s) = \mathbf{C}(s\mathbf{I} - \mathbf{A})^{-1}\mathbf{B}\mathbf{U}(s) + \mathbf{D}\mathbf{U}(s) \tag{8.39a}$$

so that the matrix transfer function is

$$\mathbf{G}(s) = \mathbf{C}(s\mathbf{I} - \mathbf{A})^{-1}\mathbf{B} + \mathbf{D} \tag{8.39b}$$

Note that the matrix $\mathbf{D}$ represents direct coupling between input and output. Also, the transfer function (8.39b) does not make any reference to the state of the system, hence the view of the transfer function as a "black box" model.

Because $\mathbf{G}(s)$ is a matrix, the rules for matrix multiplication apply to the definition (8.39a). In other words, $\mathbf{Y}(s) = \mathbf{G}(s)\mathbf{U}(s)$ and not $\mathbf{U}(s)\mathbf{G}(s)$. Similarly, given two

Figure 8.8

Transfer Function for a Cascaded System

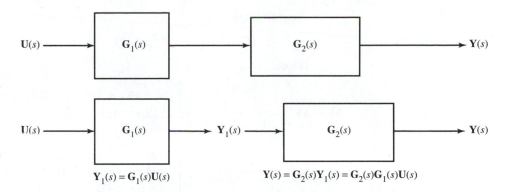

$$\mathbf{Y}_1(s) = \mathbf{G}_1(s)\mathbf{U}(s) \qquad \mathbf{Y}(s) = \mathbf{G}_2(s)\mathbf{Y}_1(s) = \mathbf{G}_2(s)\mathbf{G}_1(s)\mathbf{U}(s)$$

cascaded systems with transfer functions $\mathbf{G}_1(s)$ for the first system (input side) and $\mathbf{G}_2(s)$ for the second system (overall output side), the overall system output $\mathbf{Y}(s)$ is given by (see Fig. 8.8)

$$\mathbf{Y}(s) = \mathbf{G}_2(s)[\mathbf{G}_1(s)\mathbf{U}(s)] = \mathbf{G}_2(s)\mathbf{G}_1(s)\mathbf{U}(s) \tag{8.39c}$$

where the term in brackets represents the input to the second system. In Section 8.2 it was shown that the denominator of the transfer function equal to zero yields the system's characteristic equation. This is also true for the general matrix transfer function with no direct coupling between input and output. We can see this in the definition of the inverse of a matrix. From Appendix B,

$$(s\mathbf{I} - \mathbf{A})^{-1} = \frac{1}{\det(s\mathbf{I} - \mathbf{A})}\text{Adjoint}[(s\mathbf{I} - \mathbf{A})] \tag{8.39d}$$

which gives

$$\mathbf{G}(s) = \frac{1}{\det(s\mathbf{I} - \mathbf{A})}\text{Adjoint}[(s\mathbf{I} - \mathbf{A})]\mathbf{B} \tag{8.39e}$$

The denominator of every term in $\mathbf{G}(s)$ is thus $\det(s\mathbf{I} - \mathbf{A})$, and according to equation (6.68), the characteristic equation for the system is also $\det(s\mathbf{I} - \mathbf{A}) = 0$.

Example 8.23   Matrix Transfer Function and Forced Response

Find the transfer function for

$$\frac{d}{dt}\mathbf{X} = \begin{bmatrix} -3 & 2 & 0 \\ -1 & -4 & 1 \\ -2 & 0 & -1 \end{bmatrix}\mathbf{X} + \begin{bmatrix} 2 \\ 0 \\ 2 \end{bmatrix}\mathbf{U}; \qquad \mathbf{Y} = \begin{bmatrix} y_1 \\ y_2 \end{bmatrix} = \begin{bmatrix} 1 & 0 & 0 \\ 0 & 0 & 1 \end{bmatrix}\mathbf{X}$$

Obtain the forced response (output) when the input is a unit impulse.

Solution:

$$(sI - A) = \begin{bmatrix} (s + 3) & -2 & 0 \\ 1 & (s + 4) & -1 \\ 2 & 0 & (s + 1) \end{bmatrix}$$

From Example 6.18,

$$\det(s\mathbf{I} - \mathbf{A}) = (s + 3)^2(s + 2)$$

Hence

$$(s\mathbf{I} - \mathbf{A})^{-1} = \frac{1}{(s + 3)^2(s + 2)} \begin{bmatrix} \{(s + 4)(s + 1)\} & \{-(s + 1) - 2\} & \{-2(s + 4)\} \\ \{2(s + 1)\} & \{(s + 3)(s + 1)\} & \{-4\} \\ \{2\} & \{s + 3\} & \{(s + 3)(s + 4) + 2\} \end{bmatrix}$$

$$(s\mathbf{I} - \mathbf{A})^{-1} = \frac{1}{(s + 3)^2(s + 2)} \begin{bmatrix} \{(s + 4)(s + 1)\} & \{2(s + 1)\} & \{2\} \\ \{-(s + 1) - 2\} & \{(s + 3)(s + 1)\} & \{s + 3\} \\ \{-2(s + 4)\} & \{-4\} & \{(s + 3)(s + 4) + 2\} \end{bmatrix}$$

$$(s\mathbf{I} - \mathbf{A})^{-1}\mathbf{B} = \frac{1}{(s + 3)^2(s + 2)} \begin{bmatrix} \{2(s + 4)(s + 1) + 4\} \\ \{-2(s + 1) - 4 + 2(s + 3)\} \\ \{-4(s + 2) + 2(s + 3)(s + 4) + 4\} \end{bmatrix}$$

$$\mathbf{C}(s\mathbf{I} - \mathbf{A})^{-1}\mathbf{B} = \frac{1}{(s + 3)^2(s + 2)} \begin{bmatrix} \{2(s + 4)(s + 1) + 4\} \\ \{-4(s + 4) + 2(s + 3)(s + 4) + 4\} \end{bmatrix}$$

Thus

$$\mathbf{G}(s) = \begin{bmatrix} \left\{ \dfrac{2(s^2 + 5s + 6)}{(s + 3)(s + 3)(s + 2)} \right\} \\ \left\{ \dfrac{2(s^2 + 5s + 6)}{(s + 3)(s + 3)(s + 2)} \right\} \end{bmatrix}$$

and $\mathbf{Y}(s) = \mathbf{G}(s)\mathbf{U}(s)$.

For $\mathbf{U}(t) = \delta(t)$, $\mathbf{U}(s) = 1$, so that $\mathbf{Y}(s) = \mathbf{G}(s)$.

$$\mathbf{Y}(t) = \mathscr{L}^{-1}[\mathbf{Y}(s)] = \mathscr{L}^{-1} \begin{bmatrix} \dfrac{2(s^2 + 5s + 6)}{(s + 3)^2(s + 2)} \\ \dfrac{2(s^2 + 5s + 6)}{(s + 3)^2(s + 2)} \end{bmatrix} = \mathscr{L}^{-1} \begin{bmatrix} \dfrac{2}{s + 3} \\ \dfrac{2}{s + 3} \end{bmatrix} = \begin{bmatrix} 2e^{-3t} \\ 2e^{-3t} \end{bmatrix}$$

8.4 Z-DOMAIN SOLUTION OF DISCRETE-TIME SYSTEMS

The *z-transform* plays a role in the analysis and design of discrete-time systems similar to that of the Laplace transform in continuous-time systems. However, as we shall see, the *z*-transform can also be viewed as a special case or a subset of the Laplace transform. The discussion in this section takes cognizance of the background already developed in the transform analysis of continuous time systems and is therefore brief and oriented toward the application of the *z*-transform, especially to the analysis of the more general discrete-time vector state equation. There are two major distinguishing attributes of the discrete-time system model of equations (6.90a) and (6.90b): The dynamic operator is the time delay (written in time-advance format), and the signals **X**, **U**, and **Y** represent sampled data (see Section 2.4). Recall also the discussion of sampled data under the discrete Fourier transform in Section 8.2.

**Introducing the
z-Transform**

A representation of sampled data in terms of a time-delayed sequence of values results in a useful transform for discrete-time systems. Consider the sampled function $(f = f(kT+), k = 0, 1, 2, \ldots; T = \text{sampling period})$ illustrated in Fig. 8.9a. The notation $f(kT+)$ is used to emphasize that at $t = kT$ the sample value is defined immediately *after* and not directly *at* the instant. Note that if a continuous signal $f(t)$ is sampled with a sampling period T, the result is also $f(kT)$, and the definition $f = f(kT+)$ allows for jump discontinuities in $f(t)$ at $t = kT$. The z-transform is defined by

$$\mathscr{Z}[f(kT)] = F(z) = \sum_{k=0}^{\infty} f_k z^{-k} \qquad (8.40)$$

where the parameter z denotes the time-advance operator, so that z^{-k} represents a time delay by kT time units. The magnitude of z is limited to those values for which the series (8.40) is convergent. The definition (8.40) is easily interpreted if it is written in expanded format:

$$F(z) = \mathscr{Z}[f(kT)] = f_0 + z^{-1}f_1 + z^{-2}f_2 + z^{-3}f_3 + \ldots \qquad (8.41)$$

Then as indicated in Fig. 8.9b, the sampled function $f(kT)$ can be viewed as the sum of delayed samples f_k, $k = 0, 1, 2, \ldots$ so that $z^{-k}f_k$ is the z-transform of a delayed (by k sample periods) sample, and $F(z)$ is simply a transform (or a representation) of $f(kT)$ as the sum of an infinite series of delayed samples.

Actually, as will be shown later, $z^{-k}f_k$ is an impulse of strength f_k occurring at time $= kT$, so that $F(z)$ is really a sum of delayed impulses. Armed with the definition (8.40) or (8.41), we can calculate the z-transform of various sampled functions. Table 8.6 shows z-transform pairs for some of the functions whose Laplace transforms were

Figure 8.9

**Interpreting Sampled Data
and the z-transform**

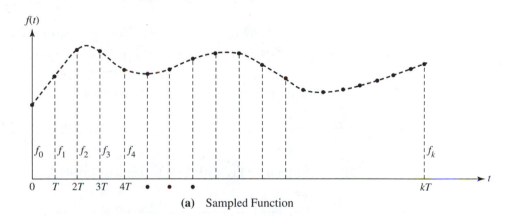

(a)   Sampled Function

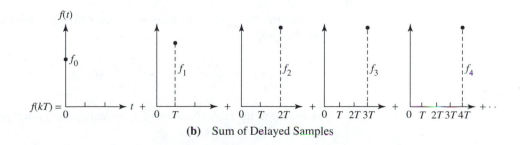

(b)   Sum of Delayed Samples

Table 8.6 Some Common z-Transform Pairs

$F(z) = \sum_{\kappa=0}^{\infty} f_k z^{-k}$	$f(kT)$	$f(t)$	CONVERGENCE $\|z\| > R$
1. $\dfrac{z}{z-1}$	1	1 or $U_s(t)$	1
2. $\dfrac{Tz}{(z-1)^2}$	kT	t	1
3. $\dfrac{T^2 z(z+1)}{(z-1)^3}$	$(kT)^2$	t^2	1
4. $\dfrac{z}{z-c}, c = e^{-aT}$	$e^{-akT} = (e^{-aT})^k$	e^{-at}	$\|e^{-aT}\|$
5. $\dfrac{cTz}{(z-c)^2}, c = e^{-aT}$	$kTe^{-akT} = kT(e^{-aT})^k$	te^{-at}	$\|e^{-aT}\|$
6. $\dfrac{z \sin \omega T}{z^2 - 2z \cos \omega T + 1}$	$\sin(k\omega T)$	$\sin(\omega t)$	1
7. $\dfrac{z^2 - z \cos \omega T}{z^2 - 2z \cos \omega T + 1}$	$\cos(k\omega T)$	$\cos(\omega t)$	1
8. $\dfrac{cz \sin \omega T}{z^2 - 2cz \cos \omega T + c^2}, c = e^{-aT}$	$(e^{-aT})^k \sin(k\omega T)$	$e^{-at}\sin(\omega t)$	$\|e^{-aT}\|$
9. $\dfrac{z^2 - cz \cos \omega T}{z^2 - 2cz \cos \omega T + c^2}, c = e^{-aT}$	$(e^{-aT})^k \cos(k\omega T)$	$e^{-at}\cos(\omega t)$	$\|e^{-aT}\|$
10. 1	$\delta(kT)$	$\delta(t)$	0

given in Table 8.4. To demonstrate the derivation of the z-transform pairs, we consider the entry for $f(t) = t$. From equation (8.41),

$$\mathscr{Z}[f(kT)] = 0 + z^{-1}T + z^{-2}(2T) + z^{-3}(3T) + \ldots \tag{8.42a}$$

$$\mathscr{Z}[f(kT)] = \mathscr{Z}[kT] = z^{-1}T[1 + 2z^{-1} + 3z^{-2} + \ldots] \tag{8.42b}$$

Recall from pure mathematics the power series

$$1 + 2r + 3r^2 + \ldots + nr^{n-1} = \frac{1 - r^n(n+1) + nr^{n+1}}{(1-r)^2} \tag{8.42c}$$

The limit of this series as n becomes arbitrarily large is given by

$$\lim_{n \to \infty} \left(\frac{1 - r^n(n+1) + nr^{n+1}}{(1-r)^2} \right) = \frac{1}{(1-r)^2} \text{ for } |r| < 1 \tag{8.42d}$$

Comparing the term in brackets in equation (8.42b) with the power series (8.42c), we see that with $r = z^{-1}$,

$$\mathscr{Z}[kT] = \frac{z^{-1}T}{(1 - z^{-1})^2} = \frac{Tz}{(z-1)^2} \quad \text{for } |z^{-1}| < 1 \tag{8.42e}$$

as shown in the table. Note that z must satisfy the convergence condition $|z| > 1$ for the transform to exist.

Properties of the z-Transform

The z-transform and the Laplace transform are related through the time-shift property of Laplace transforms. From Table 8.5, the Laplace operator for time advance T is e^{sT}. Thus we can define

$$z = e^{sT} \tag{8.43a}$$

This equation implies that the parameter z is, like s, a complex number. Further, if z is the time-shift operator as given by equation (8.43a), then z^{-k} must represent a delayed unit impulse, since

$$\mathcal{L}[\delta(t - kT)] = e^{-skT}\mathcal{L}[\delta(t)] = e^{-skT} = z^{-k} \tag{8.43b}$$

The z-transform also has properties that parallel those of the Laplace transform. The linearity property follows immediately from the definition (8.40):

$$\mathcal{Z}[c_1 f_1(kT) + c_2 f_2(kT)] = c_1 F_1(z) + c_2 F_2(z) \tag{8.44a}$$

Also the discrete-time equivalent of continuous-time differentiation is time advance, and the z-transform of a time-shifted (advanced by T) signal from (8.41) is

$$\mathcal{Z}[f(t + T)] = f(T) + z^{-1}f(2T) + z^{-2}f(3T) + \ldots$$

$$= z[-f_0 + f_0 + z^{-1}f_1 + z^{-2}f_2 + z^{-3}f_3 + \ldots]$$

That is,

$$\mathcal{Z}[f(t + T)] = z[F(z) - f_0] = z[F(z) - f(0+)] \tag{8.44b}$$

A few other properties of the z-transform, include the following:

Final value:

$$\operatorname*{Lim}_{k \to \infty} f(kT) = \operatorname*{Lim}_{z \to 1}(1 - z^{-1})F(z) \tag{8.44c}$$

if $f(\infty)$ exists, and Convolution:

$$\mathcal{Z}\left[\sum_{n=0}^{\infty} f_{k-n}g_n\right] = F(z)G(z), \quad \text{where } f_{k-n} = 0 \text{ for } n > k \tag{8.44d}$$

Pulse Transfer Functions and Recurrence Solutions

The linearity and time-advance properties facilitate the z-transformation of the discrete-time system model given by equations (6.90a) and (6.90b), that is,

$$z\mathbf{X}(z) - z\mathbf{X}_0 = \mathbf{P}\mathbf{X}(z) + \mathbf{Q}\mathbf{U}(z) \tag{8.45a}$$

and

$$\mathbf{Y}(z) = \mathbf{C}\mathbf{X}(z) + \mathbf{D}\mathbf{U}(z) \tag{8.45b}$$

Solving for the z-domain state and output vectors (assuming that $(z\mathbf{I} - \mathbf{P})^{-1}$ exists), we obtain

$$\mathbf{X}(z) = (z\mathbf{I} - \mathbf{P})^{-1}z\mathbf{X}_0 + (z\mathbf{I} - \mathbf{P})^{-1}\mathbf{Q}\mathbf{U}(z) \tag{8.46a}$$

For zero initial state, we have

$$\mathbf{Y}(z) = [\mathbf{C}(z\mathbf{I} - \mathbf{P})^{-1}\mathbf{Q} + \mathbf{D}]\mathbf{U}(z) = \mathbf{G}(z)\mathbf{U}(z) \tag{8.46b}$$

The z-domain transfer function

$$\mathbf{G}(z) = \mathbf{C}(z\mathbf{I} - \mathbf{P})^{-1}\mathbf{Q} + \mathbf{D} \tag{8.47}$$

is known as the *pulse transfer function*. Note that as in the continuous-time and s-domain cases, the overall denominator of the transfer function for no direct coupling between input and output when set equal to zero gives the characteristic equation of the system (equation (6.89) in this case).

| Example 8.24 | Pulse Transfer Function of Savings-Only Bank |

Determine for the savings-only commercial bank of Example 4.9 the pulse transfer function between money in the bank and the net payments into the bank and the overhead expenditure by the bank. The sampling time is a quarter. Assume that $\alpha_1 = 0.075$, $\alpha_2 = 0.70$, and $\alpha_3 = 0.1$.

Solution: The state model for the bank, equations (4.48a) through (4.48d), was put in vector state format in Section 6.4. The result is

$$\begin{bmatrix} x_{1,k+1} \\ x_{2,k+1} \end{bmatrix} = \begin{bmatrix} 1 + \alpha_1 & 0 \\ (1 - \alpha_1)\alpha_2 & 1 + \alpha_2\alpha_3 \end{bmatrix}\begin{bmatrix} x_{1,k} \\ x_{2,k} \end{bmatrix} + \begin{bmatrix} 1 & 0 \\ 0 & -\alpha_2 \end{bmatrix}\begin{bmatrix} u_{1,k} \\ u_{2,k} \end{bmatrix}$$

$$\begin{bmatrix} y_{1,k} \\ y_{2,k} \end{bmatrix} = \begin{bmatrix} (1 - \alpha_1)(1 - \alpha_2) & (1 - \alpha_2)\alpha_3 \\ 0 & \alpha_3 \end{bmatrix}\begin{bmatrix} x_{1,k} \\ x_{2,k} \end{bmatrix} + \begin{bmatrix} 0 & -(1 - \alpha_2) \\ 0 & 0 \end{bmatrix}\begin{bmatrix} u_{1,k} \\ u_{2,k} \end{bmatrix}$$

where x_1 and x_2 represent the amount of savings and investments, respectively; u_1 and u_2 are, respectively, the net payments into the account and the overhead expenditure. The outputs, money in the bank and profits, are given by y_1 and y_2, respectively. For the given values of α_1, α_2, and α_3, we see that

$$\mathbf{P} = \begin{bmatrix} 1.075 & 0 \\ 0.6475 & 1.07 \end{bmatrix}; \qquad \mathbf{Q} = \begin{bmatrix} 1 & 0 \\ 0 & -0.7 \end{bmatrix}; \qquad \mathbf{C} = \begin{bmatrix} 0.2775 & 0.03 \\ 0 & 0.1 \end{bmatrix};$$

$$\mathbf{D} = \begin{bmatrix} 0 & -0.3 \\ 0 & 0 \end{bmatrix}$$

From equation (8.47),

$$\mathbf{G}(z) = \begin{bmatrix} .2775 & 0.03 \\ 0 & .1 \end{bmatrix}\begin{bmatrix} (z - 1.075) & 0 \\ -0.6475 & (z - 1.07) \end{bmatrix}^{-1}\begin{bmatrix} 1 & 0 \\ 0 & -0.7 \end{bmatrix} + \begin{bmatrix} 0 & -0.3 \\ 0 & 0 \end{bmatrix}$$

$$\mathbf{G}(z) = \frac{1}{(z - 1.075)(z - 1.07)}\begin{bmatrix} 0.2775 & 0.03 \\ 0 & 0.1 \end{bmatrix}\begin{bmatrix} (z - 1.07) & 0 \\ 0.6475 & (z - 1.075) \end{bmatrix}$$

$$\begin{bmatrix} 1 & 0 \\ 0 & -0.7 \end{bmatrix} + \begin{bmatrix} 0 & -0.3 \\ 0 & 0 \end{bmatrix}$$

$$\mathbf{G}(z) = \begin{bmatrix} \left\{\dfrac{0.2775(z - 1.07) + 0.019425}{(z - 1.075)(z - 1.07)}\right\} & \left\{\dfrac{-0.021}{(z - 1.07)} - 0.3\right\} \\ \left\{\dfrac{0.06475}{(z - 1.075)(z - 1.07)}\right\} & \left\{\dfrac{-0.07}{(z - 1.07)}\right\} \end{bmatrix}$$

This is the matrix transfer function for both outputs y_1 and y_2. For just money in the bank at the beginning of each quarter ($y_{1,k}$), the pulse transfer function from above is

$$\mathbf{G}(z) = \left[\left\{ \frac{0.2775(z - 1.07) + 0.019425}{(z - 1.075)(z - 1.07)} \right\} \quad \left\{ \frac{-0.021}{(z - 1.07)} - 0.3 \right\} \right]$$

Note that the same result could have been obtained by considering y_1 as the only output, that is, using $\mathbf{C} = [0.2775 \quad 0.03]$.

....

Given the simplicity of the time-domain (recurrence) solution of the discrete-time model (6.90), a z-domain approach to the solution for the system state is not common. However, as we shall see, the design of dynamic systems (especially for automatic control) is often carried out in the Laplace or z-domains. For discrete-time systems, the result of a z-domain design is usually a pulse transfer function relating the desired output to the input, in the z-domain. Consider the general scalar pulse transfer function, which is a ratio of two polynomials in z:

$$G(z) = \frac{Y(z)}{U(z)} = \frac{b_0 + b_1 z^{-1} + b_2 z^{-2} + \ldots + b_m z^{-m}}{1 + a_1 z^{-1} + a_2 z^{-2} + \ldots + a_n z^{-n}}, \quad n \geq m \qquad \textbf{(8.48a)}$$

The z-domain output is given by

$$(1 + a_1 z^{-1} + a_2 z^{-2} + \ldots + a_n z^{-n}) Y(z) \qquad \textbf{(8.48b)}$$
$$= (b_0 + b_1 z^{-1} + b_2 z^{-2} + \ldots + b_m z^{-m}) U(z)$$

Recalling the interpretation of the z-transform given in Fig. 8.9, we see that equation (8.48b) represents the following time-domain expression:

$$y(k) + a_1 y(k - 1) + a_2 y(k - 2) + \ldots + a_n y(k - n)$$
$$= b_0 u(k) + b_1 u(k - 1) + b_2 u(k - 2) + \ldots + b_m u(k - m) \qquad \textbf{(8.49a)}$$

which yields the recurrence solution

$$y(k) = b_0 u(k) + b_1 u(k - 1) + b_2 u(k - 2) + \ldots$$
$$+ b_m u(k - m) - a_1 y(k - 1) - a_2 y(k - 2) - \ldots - a_n y(k - n) \qquad \textbf{(8.49b)}$$

Note that with the condition $n \geq m$, the current output depends only on present and past inputs (and past outputs), which is our definition for a dynamic system. Also note that the discrete-time-domain solution for a system represented by a pulse transfer function is rather straightforward and simple.

Z-Domain to Discrete-Time Domain

The preceding development also suggests a rather simple approach to finding the values of the inverse z-transform at the sampling instant. Consider the general z-domain function (which is analogous to equation (8.30a) for the Laplace domain):

$$F(z) = \frac{b_m + b_{m-1} z + b_{m-2} z^2 + \ldots + b_0 z^m}{a_n + a_{n-1} z + a_{n-2} z^2 + \ldots + a_0 z^n}, \quad n \geq m \qquad \textbf{(8.50a)}$$

$F(z)$ can be expanded into partial fractions using the technique developed for Laplace domain functions, and by table lookup (using, for example, Table 8.6) the inverse z-transform for various fractions and hence for the entire function can be established. However, if in equation (8.50a) we divide (by long division) the numerator of $F(z)$ by the denominator, we obtain a new function of z,

$$F(z) = f_0 + f_1 z^{-1} + f_2 z^{-2} + \ldots + f_3 z^{-3} \ldots \tag{8.50b}$$

whose coefficients are, by the definition of the z-transform (equation (8.41)), the value sequence of the time-domain-sampled function,

$$f_k = [f_0, f_1, f_2, f_3, \ldots] = \mathcal{Z}^{-1}[F(z)] \tag{8.50c}$$

Example 8.25   Application of z-Transforms

Consider the following second-order discrete-time system:

$$\mathbf{X}_{k+1} = \begin{bmatrix} 0 & 1 \\ -1 & 2 \end{bmatrix} \mathbf{X}_k + \begin{bmatrix} 1 \\ 3 \end{bmatrix} u_k; \qquad y_k = [1 \quad 0]\mathbf{X}_k, \quad \text{and} \quad u_k = 1 \text{ for all } k \geq 0$$

a. Find $G(z)$.
b. Obtain $y(kT)$ by inverse z-transformation of $Y(z)$.
c. Find y_k, $k = 0, 1, 2$, and 3, using the result from (a) and the the recurrence model given by equations (8.48) and (8.49).
d. Repeat (c), but use instead the time-domain sequence given by equations (8.50).
e. Obtain a closed-form solution for y_k by assuming that the sequence of values of y_k can be represented by a low-order polynomial in k.

Solution a:

$$G(z) = [1 \quad 0]\begin{bmatrix} z & -1 \\ 1 & z-2 \end{bmatrix}^{-1}\begin{bmatrix} 1 \\ 3 \end{bmatrix} = \frac{1}{z^2 - 2z + 1}[1 \quad 0]\begin{bmatrix} z-2 & 1 \\ -1 & z \end{bmatrix}\begin{bmatrix} 1 \\ 3 \end{bmatrix}$$

$$G(z) = \left[\left\{\frac{z-2}{z^2-2z+1}\right\} \quad \left\{\frac{1}{z^2-2z+1}\right\}\right]\begin{bmatrix} 1 \\ 3 \end{bmatrix} = \frac{z+1}{z^2-2z+1}$$

Solution b: $Y(z) = G(z)U(z)$. From Table 8.6, item 1,

$$U(z) = \frac{z}{z-1}$$

Hence

$$Y(z) = \left(\frac{z+1}{z^2-2z+1}\right)\left(\frac{z}{z-1}\right) = \frac{z(z+1)}{(z-1)^3}$$

Again from Table 8.6, item 3,

$$y_k = \mathcal{Z}^{-1}[Y(z)] = \left(\frac{t}{T}\right)^2 = k^2$$

where T is the sampling period.

Solution c:

$$G(z) = \frac{Y(z)}{U(z)} = \frac{z+1}{z^2 - 2z + 1} = \frac{0 + z^{-1} + z^{-2}}{1 - 2z^{-1} + z^{-2}}$$

Hence

$$(1 - 2z^{-1} + z^{-2})Y(z) = (0 + z^{-1} + z^{-2})U(z)$$

By equations (8.49a) and (8.49b),

$$y(k) = 2y(k-1) + u(k-1) - y(k-2) + u(k-2)$$

Substituting for k and $u(k)$ and noting that by definition, $y(0) = 0$, we obtain

$$y(1) = 0 + 1 - 0 + 0 = 1; \quad y(2) = 2 + 1 - 0 + 1 = 4;$$
$$y(3) = 8 + 1 - 1 + 1 = 9$$

Solution d:

$$Y(z) = \frac{z^2 + z}{z^3 - 3z^2 + 3z - 1}$$

By long division,

$$
\begin{array}{r}
z^{-1} + 4z^{-2} + 9z^{-3} \\
z^3 - 3z^2 + 3z - 1 \overline{)z^2 \quad + \quad z \quad\quad\quad\quad\quad} \\
\underline{z^2 \quad - 3z + 3 - z^{-1}} \\
4z - 3 + z^{-1} \\
\underline{4z - 12 + 12z^{-1} - 4z^{-2}} \\
9 - 11z^{-1} + 4z^{-2} \\
\underline{9 - 27z^{-1} + 27z^{-2} - 9z^{-3}} \\
\vdots
\end{array}
$$

$$Y(z) = 0 + (1)z^{-1} + (4)z^{-2} + (9)z^{-3} + \ldots$$

By equation (8.50c),

$$y_k = y(kT) = [0, 1, 4, 9, \ldots]$$

Solution e: We note that the recurrence equation obtained in (c) is a nonhomogeneous second-order recurrence relation with constant coefficients. Hence we try (method of undetermined coefficients):

$$y_k = c_0 + c_1 k + c_2 k^2,$$
$$k = 0 \Rightarrow 0 = c_0; \quad k = 1 \Rightarrow 1 = c_1 + c_2; \quad k = 2 \Rightarrow 4 = 2c_1 + 4c_2$$

Hence

$$c_1 = 0; \quad c_2 = 1; \quad y_k = k^2$$

Check:

$$k = 3, \quad y_k = 9; \quad k = 4, \quad y_k = 2(9) + 1 - 4 + 1 = 16 = k^2$$

8.5 PRACTICE PROBLEMS

8.1 Deduce the Fourier transforms of the functions illustrated in Fig. 8.10. *Hint:* Consider operations involving properties of Fourier transforms that can be performed on some elementary functions in Table 8.1 to yield the illustrated functions.

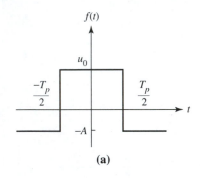

(a)

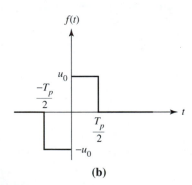

(b)

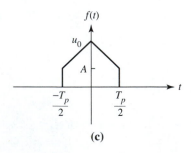

(c)

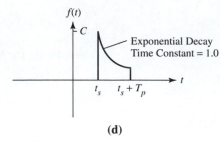

(d)

Figure 8.10 (Problem 8.1)

8.2 Find the Fourier transform of the following functions and plot their amplitude, phase, and power density spectrums:

(a) $2e^{(1.5-3t)}$

(b) $\dfrac{\sqrt{\pi}}{2} e^{-(t/4)^2}$

(c)* $\dfrac{\sin t}{t}$

(d) $\dfrac{1}{1 + t^2}$

8.3 Compute, using Fourier transforms, $f(t)*g(t)$ if $f(t) = e^{-t^2}$ and

(a) $g(t) = \begin{cases} 1 & |t| \le 1/2 \\ 0 & \text{otherwise} \end{cases}$

(b) $g(t) = \begin{cases} 1 - 2|t| & |t| \le 1/2 \\ 0 & \text{otherwise} \end{cases}$

8.4 Given a system with the frequency transfer function

$$G(\omega) = \frac{1}{(4 - \omega^2) + j2\omega}$$

determine the forced response when the input is

(a) A unit step.

(b) An impulse of strength 3.

(c) A unit pulse of duration 2 time units occurring at $t = 0$.

(d) The function $\sin 3t$.

8.5 Repeat Example 8.2 for the parallel resonant circuit shown in Fig. 7.8a. For part (d) of the problem, take $i_{\text{in}}(t) = \delta(t) + e^{-3t^2}$, and retain the given parameter values.

8.6 For the following system models, determine (1) the Fourier domain transfer function, (2) the forced response due to a unit impulse, and (3) the forced response due to the specified input:

(a) $\dfrac{d^2y}{dt^2} + \dfrac{dy}{dt} + y = \cos 2t + \sin 2t$

(b) $\dfrac{d^3y}{dt^3} + 2\dfrac{d^2y}{dt^2} + 2\dfrac{dy}{dt} + y = 3 \sin 2t$

8.7 Determine the Fourier domain transfer function for the LC network of Fig. 3.19 by

(a) Fourier transforming the state model derived in Example 3.13, equations (3.38a) through (3.38e).

(b)* Combining the transforms of the components and or subsystems of the circuit.

8.8 Use the Fourier transform method to find the transfer function of each of the following filters.

(a) The seismic filter of Fig. 8.11a

(b)* The LC low-pass filter of Fig. 8.4

(c) The double-tuned band-pass filter of Fig. 8.11b

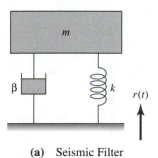

(a) Seismic Filter

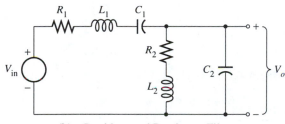

(b) Double-tuned Band-pass Filter

Figure 8.11 (Problem 8.8)

8.9 The transfer function of a filter is

$$G(\omega) = 4\dfrac{\sin^2(a\omega/2)}{a\omega^2}$$

Determine and sketch the output when the input is the signal $e^{-t}U_s(t)$, and $a = 1$.

8.10 Design and realize with passive circuit elements a Butterworth-type low-pass filter having the following specifications:

(a) Cutoff frequency of 500 Hz and an attenuation of 0.25 or less at twice the cutoff frequency

(b) Cutoff frequency of 1 kHz and an attenuation of 0.1 or less at 2.5 times the cutoff frequency

8.11 Repeat Example 8.4 part (b) for $N = 4$ and 6.

8.12 Determine the Fourier series coefficients for the periodic function $f(t) = 2 \sin(3\pi t)$, according to

(a) An exponential Fourier series, equations (8.13a) and (8.13b).

(b) A discrete-time Fourier series approximation, equations (8.14a) and (8.14b), for $N = 3$ and $N = 7$.

8.13 Repeat Example 8.5 using Mathcad.

8.14 Use MATLAB FFT functions to obtain the DFT for a sequence of samples of a unit pulse of width 5 s at $t = 0$ for two cases: $T = 1$, $N = 16$ and $T = 0.1$, $N = 160$. Plot the amplitude spectrum for each case and compare with the spectrum for the Fourier integral transform.

8.15 Compute the DFT of a sequence of N samples of $f(t) = 2 \sin 3\pi t$ defined on $0 \le t \le 2/3$, for (i) $T = 0.22$, $N = 3$ and (ii) $T = 0.06$, $N = 12$

(a) Using MATLAB.

(b) Using Mathcad.

8.16 Compute using a software package the DFT of a sequence of 100 samples at sampling time interval of 0.1 of the function $f(t) = \cos\!\left(\frac{\pi}{2} + \right)\!\left[1 - \cos\!\left(\frac{\pi}{4}t\right)\right]$.

8.17 Use FFT to determine the amplitude spectrum of the function given in Problem 8.2. Explain your choice of sampling time interval and number of samples in each case. Compare graphically your result with that of Problem 8.2.

8.18 Repeat Problem 8.3 using FFT. Your result should be a plot of $f(t)*g(t)$. Use $N = 100$, $T = 0.05$.

8.19* Compute and plot the output of the filter of Problem 8.9 using an FFT program. Determine an adequate sampling time interval and number of samples.

8.20 Using the result already established in Example 8.6 for $n = 1$, that is,

$$\mathscr{L}\left[\dfrac{d}{dt}(A \cos(\omega t + \phi)e^{-\sigma t})\right] = sz$$

where

$$z = \mathscr{L}\left[(A \cos(\omega t + \phi)e^{-\sigma t})\right] \quad \text{and} \quad s = -\sigma + j\omega$$

complete by *induction* the proof that

$$\mathscr{L}\left[\dfrac{d^n}{dt^n}(A \cos(\omega t + \phi)e^{-\sigma t})\right] = s^n z$$

as follows: Suppose for some particular n,

$$\mathscr{L}\left[\frac{d^n}{dt^n}(A\cos(\omega t + \phi)e^{-\sigma t})\right] = \mathscr{L}[A_n\cos(\omega t + \phi_n)e^{-\sigma t}]$$

show that

$$\mathscr{L}\left[\frac{d^{n+1}}{dt^{n+1}}(A\cos(\omega t + \phi)e^{-\sigma t})\right] = s\mathscr{L}[A_n\cos(\omega t + \phi_n)e^{-\sigma t}]$$

Since the result has been demonstrated for $n = 1$, the above implies that it is also true for $n = 2$, and 3, and 4, . . . , and indeed for all $n \geq 1$.

8.21 Repeat Problem 7.38 using Laplace transforms.

8.22 Determine if the following functions are Laplace transformable, and if so, derive the range of the Laplace transform parameter s over which the Laplace transform exists:

(a) e^{at} (b) $\cos(\omega t)$ (c) u_0^{at} (Hint: u_0^{at} can be replaced by an exponential function   (d) $e^{(t^2 - t)}$

8.23 Prove the following Laplace transform pairs by direct integration of equation (8.20c):

(a) $\mathscr{L}[t] = \dfrac{1}{s^2}$ (b) $\mathscr{L}[\cos \omega t] = \dfrac{s}{s^2 + \omega^2}$

(c)* $\mathscr{L}[\sin \omega t] = \dfrac{\omega}{s^2 + \omega^2}$

(d) $\mathscr{L}[U_p(t - t_s)] = \dfrac{e^{-t_s s}(1 - e^{-T_p s})}{s}$

8.24 Derive the Laplace transform of the indicated time functions (refer to Fig. 8.12):

(a)* $2e^{(1.5 - 3t)}$ (b) $e^{-3t} + 3te^{-3t}$ (c) $t^3 + 3t^2 + 2t + 1$

(d) the function shown   (e) $(1 + te^{-t})^2$ (f) $te^{-2t}\cos 3t$

(g)* see figure   (h) $\sin t \sin 5t$ (i) as illustrated

(j) $e^{-t}\sin^2 t$ (k) $\sin(2t)\cos(3t)$ (l) the function shown

(m) $\sin^2(3t)$

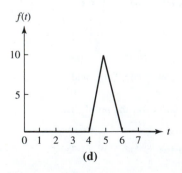

(d)

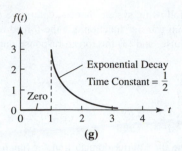

(g)

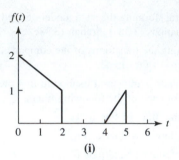

(i)

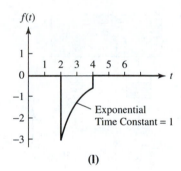

(l)

Figure 8.12 (Problem 8.24)

8.25 Find $\mathscr{L}[f(t)]$ if

(a) $f(t) = 2t^3 + t^2\cos t$ (b) $f(t) = \dfrac{e^{(4 - 3t)} - e^{-t}}{t - 2}$

(c) $f(t) = \begin{cases} \cos t & 0 < t < \pi \\ \sin 2t & t > \pi \end{cases}$

(d) $f(t) = \begin{cases} (t - 1)^3 & t > 1 \\ 0 & 0 < t < 1 \end{cases}$

In Problems 8.26 through 8.30, find the function of time $f(t)$ whose Laplace transform $F(s)$ is given.

8.26 (a) $\dfrac{e^{-s}}{s}$ (b) $\dfrac{e^{-2s}}{s^2}$ (c) $\dfrac{e^s - e^{-2s}}{s}$

(d) $\dfrac{e^{-3s} - e^{-5s}}{s}$ (e) $\dfrac{4e^{-0.5s}}{(s + 2)^2}$

8.27 (a) $\dfrac{3}{s+5}$ **(b)** $\dfrac{2s+7}{s+3}$ **(c)** $\ln\left(\dfrac{s+2}{s+3}\right)$

(d)* $3\ln\left(1+\dfrac{2}{s}\right)$ **(e)** $\ln\left(1+\dfrac{3}{s+3}\right)$

8.28 (a) $\dfrac{3}{s^2+4}$ **(b)** $\dfrac{5}{s^2+5}$ **(c)** $\dfrac{4s}{s^2+4}$ **(d)** $\dfrac{2(s+1)}{s^2+9}$

8.29 (a) $\dfrac{2}{s^2+6s+9}$ **(b)** $\dfrac{1}{s^2+10s+34}$

(c) $\dfrac{5}{s^2+2s+17}$ **(d)** $\dfrac{s+1}{s^2+s+1}$ **(e)*** $\dfrac{s+1}{s^2+2s+2}$

(f) $\dfrac{s+1}{s^2+2s+3}$ **(g)** $\dfrac{s+6}{s^2+2s+9}$ **(h)** $\dfrac{3s-8}{s^2-4s+20}$

8.30 (a) $\dfrac{s^2+4s-5}{[(s+2)^2+9]^2}$ **(b)** $\dfrac{6s}{s^4+18s^2+81}$

(c) $\dfrac{s^2-3}{s^4+6s^2+9}$ **(d)** $\dfrac{s^2-16}{s^4+32s^2+256}$

8.31 If

$$\mathcal{L}[f''(t)] = \ln\left(\dfrac{s}{s+3}\right)$$

and $f(0) = -3$, $f'(0) = 1$, find $F(s)$.

8.32 Show using appropriate properties of Laplace transforms from Table 8.5 that $\mathcal{L}^{-1}\left[\dfrac{1}{s^2}F(s)\right] = \int_0^t \int_0^T f(\lambda)d\lambda dT$.

8.33 Use Laplace transforms to obtain the transfer function for the oscillating tram of Problem 7.30. What is the condition for resonance?

8.34 Solve by Laplace transforms the following differential equations:

(a) (i) $\dfrac{d^2y}{dt^2} + \dfrac{dy}{dt} + y = 3\sin 2t;\ y(0) = \dot{y}(0) = 0$

(ii) Use the solution to (i) to find the solution to

$\dfrac{d^2y}{dt^2} + \dfrac{dy}{dt} + y = \cos 2t + \sin 2t;\ y(0) = \dot{y}(0) = 0$

(b) $\dfrac{d^3y}{dt^3} + 2\dfrac{d^2y}{dt^2} + 2\dfrac{dy}{dt} + y = \delta(t);\ y(0) = 1,\ \dot{y}(0)$

$= \ddot{y}(0) = 0$

(c) $\dfrac{d^2y}{dt^2} + 3\dfrac{dy}{dt} + 2y = u(t);$ system is initially at rest

where $u(t)$ is as illustrated in (i) Problem 8.24d; (ii) Problem 8.24g; (iii) Problem 8.24i; and (iv) Problem 8.24l.

8.35* A second-order system is described by

$$\dfrac{d^2y}{dt^2} + 4\dfrac{dy}{dt} + 3y = f(t)$$

Find the forced response using Laplace transforms if $f(t)$ is as given in Fig. 8.6.

8.36 A first-order system is described by

$$\dfrac{dx}{dt} + ax = u(t);\qquad x(0) = x_0$$

If $u(t)$ is a unit step, find the value of a such that the final value of x is 5.

8.37 Find the initial and final values of the system shown under the prescribed inputs:

(a)* $G(s) = \dfrac{s+2}{s(s+1)},\ u(t) = 2e^{-2t}$

(b) $G(s) = \dfrac{s}{(s+1)^2},\ u(t) = 3e^{-2t}$

(c) $G(s) = \dfrac{1}{(s+1)(s+2)(s+3)(s+4)(s+5)(s+6)},$

$u(t) = \delta(t)$

(d) $G(s) = \dfrac{3s+1}{5s+1},\ u(t) = 2\delta(t)$

8.38 Find the output $y(t)$ for the transfer functions and inputs given in Problem 8.37 (a) by the convolution theorem and (b) by another technique.

8.39 Use partial fractions or other methods to evaluate the inverse Laplace transforms of the following functions of s:

(a) $\dfrac{s+2}{(s+1)(s+3)}$ **(b)** $\dfrac{1}{s(s+1)(s+3)}$

(c) $\dfrac{5s+7}{(s-2)(s+1)(s+3)}$ **(d)** $\dfrac{s+1}{(s+2)(s+3)^2}$

(e)* $\dfrac{s^2-2}{s^3-4s^2+s+6}$ **(f)** $\dfrac{s+1}{(s+1)^3(s+2)}$

(g) $\dfrac{3s^2-13s-7}{(s+1)(s-2)^3}$ **(h)** $\dfrac{s^2+s+1}{(s^2+2s+2)(s^2+6s+9)}$

(i) $\dfrac{2s^3-s^2-1}{(s+1)^2(s^2+1)^2}$ **(j)** $\dfrac{e^{-s}(2s-3)}{s(s+1)(s+3)^2}$

8.40 Using the appropriate function(s) of your math software, repeat (i) Problem 8.39(a), (ii) Problem 8.39(b), (iii) Problem 8.39(c), (iv) Problem 8.39(d), (v) Problem 8.39(e), (vi) Problem 8.39(f), (vii) Problem 8.39(g), (viii) Problem 8.39(h), (ix) Problem 8.39(i), (x) Problem 8.39(j)

8.41 If a system's response to the input $u_1(t) = 2\delta(t)$ is $y_1(t) = e^{-t} - e^{-5t}$, find its response ($y_2(t)$) to the input $u_2(t) = 4e^{-2t}$ by (a) convolution in the time domain and (b) Laplace transforms.

8.42 What is the order of each system and what are the eigenvalues of the systems whose transfer functions were given in Problem 8.37?

8.43 The response of a linear system to an impulse input of strength 3.0 is $y_1(t) = 1 - e^{-2t}$. If this system is subjected to a forcing input $u(t) = e^{-3t}\sin 2t$, find   **(a)** the early value of the response $y(0+)$,   **(b)** the final value of the response $y(\infty)$, and   **(c)** the eigenvalues of the system.

8.44 If the response of a linear system to the input $u(t) = 5 \sin 2t$ is $2 - 3 \sin 2t - 5e^{-3t}$, find the transfer function for the system.

8.45* **(a)** Which of the following systems whose transfer functions are given will respond fastest, and why?

$$G_1(s) = \frac{s+2}{s(s+1)\left(s+\frac{1}{5}\right)} \quad G_2(s) = \frac{5}{s^3 + 4s^2 + 5s + 2}$$

$$G_3(s) = \frac{3s+1}{(s^2 + 6s + 10)(s+4)}$$

(b) What will be the amplitude of the steady-state output of the first system if it is subject to a sinusoidal input of unit amplitude and frequency of 5 rad/s.

(c) What is the differential equation describing the second system?

8.46 The equation describing the vertical ascent of a rocket (see [9]) of total mass m that is subject to aerodynamic resistance R is

$$m\frac{d^2y}{dt^2} + R + \frac{d^2}{dt^2}(mh) = PA - \dot{m}u - mg$$

where y is the vertical displacement of the rocket (exhaust nozzle), h is the distance from the exhaust nozzle of the center of mass, P is the average pressure across the nozzle exit area A, and u is the relative velocity of the expelled gases. Consider a rocket ascending near the surface of the earth with an axially burning solid fuel such that $h = H - am$ (H = constant) and whose nozzle exhausts at atmospheric pressure ($P = 0$). Obtain an expression for the vertical displacement y as a function of time since firing if the initial total mass is m_0, the relative velocity of the expelled gases is constant, the aerodynamic resistance can be approximated during this period by

$$R = b\frac{dy}{dt}, \quad (b = \text{constant})$$

and the rate of fuel consumption $-\dot{m}$ is a constant.

8.47 Solve the following ordinary differential equation with variable coefficients, using Laplace transforms:

$$t\frac{d^2y}{dt^2} + (1+t)\frac{dy}{dt} + y = 0; \quad y(0) = 1, \quad y'(0) = -1$$

8.48 Find by Laplace transformation the state transition matrix and the free response of the following systems:

(a) System of Problem 6.36

(b)* System of Problem 6.37

In Problems 8.49 through 8.52 repeat the indicated problems by using Laplace transforms.

8.49 Problem 6.32

8.50* Problem 6.34

8.51 Problem 6.36

8.52* Problem 6.37

8.53 Solve by Laplace transformation the system

$$\frac{d}{dt}\mathbf{X} = \begin{bmatrix} -3 & -7 \\ 0 & -4 \end{bmatrix}\mathbf{X} + \begin{bmatrix} 1 & 1 \\ 1 & 0 \end{bmatrix}\begin{bmatrix} 1 \\ t \end{bmatrix}; \quad \mathbf{X}(0) = \begin{bmatrix} 1 \\ 1 \end{bmatrix}$$

8.54 Derive $\mathbf{G}(s)$ for the systems given by the following $\mathbf{A}$, $\mathbf{B}$, $\mathbf{C}$, and $\mathbf{D}$ matrices:

(a) $\mathbf{A} = \begin{bmatrix} 0 & 1 \\ -6 & -5 \end{bmatrix}$, $\mathbf{B} = \begin{bmatrix} 0 & 1 \\ 1 & 1 \end{bmatrix}$, $\mathbf{C} = \begin{bmatrix} 1 & 1 \end{bmatrix}$,

$\mathbf{D} = \begin{bmatrix} 0 \\ 1 \end{bmatrix}$

(b) $\mathbf{A} = \begin{bmatrix} 2 & 4 \\ 8 & 6 \end{bmatrix}$, $\mathbf{B} = \begin{bmatrix} 1 \\ 2 \end{bmatrix}$, $\mathbf{C} = \begin{bmatrix} 1 & 0 \end{bmatrix}$, $\mathbf{D} = 0$

(c) $\mathbf{A} = \begin{bmatrix} -1 & -1 & 1 \\ 0 & -2 & 1 \\ -2 & 0 & 1 \end{bmatrix}$, $\mathbf{B} = \begin{bmatrix} 1 & 1 \\ 0 & 0 \\ 1 & 0 \end{bmatrix}$,

$\mathbf{C} = \begin{bmatrix} 1 & 0 & -1 \\ -1 & 2 & 1 \end{bmatrix}$, $\mathbf{D} = \begin{bmatrix} 0 & 1 \\ 1 & 0 \end{bmatrix}$

8.55 Obtain the transfer function between the average daily temperature and the daily price of cantaloupes for the buyers' group prediction model of Example 4.10. What is the condition for the daily price of cantaloupes to continue to fall? By referring to the development in Example 4.10, can you interpret your answer in everyday terms of supply and demand?

8.56* Find the transfer function between the thermometer reading and the ambient temperature in Problem 4.3.

8.57 Derive $G(s)$ for the linearized dual pendulum model of Problem 6.28. Take the output as θ_1.

8.58* Obtain the transfer function between V_2 and V_1 for the lead compensation network of Problem 3.38.

8.59 Find the output response of the systems with the given transfer function $\mathbf{G}(s)$ and input $\mathbf{U}(t)$:

(a) $\mathbf{G}(s) = \left[\left\{ \dfrac{s+1}{s^2 + 3s + 2} \right\} \quad \left\{ \dfrac{9}{s^2 + 3s + 2} \right\} \right]$,

$\mathbf{U}(t) = \begin{bmatrix} te^{-t} \\ 1 \\ 3 \end{bmatrix}$

(b) $\mathbf{G}(s) = \begin{bmatrix} \dfrac{1}{(s+3)^2} \\ \dfrac{s}{(s+3)^2} \end{bmatrix}$, $\mathbf{U}(t) = U_p(t), T_p = 2$

8.60 Derive the z-transform of

(a) $f(kT) = 1$ **(b)** $f(kT) = (e^{-aT})^k$

8.61 Determine the z-transform and the region of convergence of

(a) $f(kT) = \begin{cases} 2^{kT} & 0 \le k \le 4 \\ 0 & \text{otherwise} \end{cases}$ **(b)** $f(kT) = e^{j\omega kT}$

8.62 Determine the z-transform of the sequence

(a)* $f(kT) = a^{kT}\cos(\omega kT)$ **(b)** $f(kT) = a^{kT}\sin(\omega kT)$

8.63 Find using (i) partial fractions expansion and (ii) long division, the inverse z-transform of

(a) $F(z) = \dfrac{z^2}{(z-1)^2}$ **(b)** $F(z) = \dfrac{3z}{(z-1)(z-2)}$

(c)* $F(z) = \dfrac{b}{z-a}$ **(d)** $F(z) = \dfrac{1}{1-0.25z^{-2}}$

assume that $f_k = 0$, for all $k < 0$.

8.64 Find $f(kT)$ if $F(z)$ is given by   **(a)** $\dfrac{z+2}{(z-2)^2}$

(b) $\dfrac{(z^2+1)(z^2-1)}{z^4+3.2}$

8.65 Find the final value of the sampled function $f(kT) = kT(kT - T)$.

8.66 Determine the final value of the function whose z-transform is

$$F(z) = \frac{z^3 + 2z^2 + z + 1}{z^3 + z^2 - 5z + 3}$$

8.67 Determine $\mathbf{P}^k$ for the following matrices, using z-transforms

(a) $\mathbf{P} = \begin{bmatrix} 1 & 1 \\ 0 & 1 \end{bmatrix}$ **(b)** $\mathbf{P} = \begin{bmatrix} 0 & -0.5 \\ 0.5 & 0 \end{bmatrix}$

8.68 The difference equation describing a system is

$$y(k) + 2y(k-1) - 5y(k-2) = 3u(k)$$

What is the pulse transfer function between the output $y(k)$ and the input $u(k)$?

8.69* Find by z-transforms the solution to the discrete-time system

$$\mathbf{X}_{k+1} = \begin{bmatrix} \dfrac{1}{4} & \dfrac{1}{8} \\ 1 & \dfrac{1}{2} \end{bmatrix} \mathbf{X}_k + \begin{bmatrix} 1 \\ 4 \end{bmatrix} u_k; \qquad \mathbf{X}_0 = \begin{bmatrix} 0 \\ 1 \end{bmatrix}, u_k = \text{unit step}$$

8.70 Solve by z-transformation the following system:

$$\mathbf{X}_{k+1} = \begin{bmatrix} 0 & -\dfrac{1}{2} & 0 \\ -1 & 0 & 0 \\ 1 & -\dfrac{1}{4} & 0 \end{bmatrix} \mathbf{X}_k + \begin{bmatrix} 0 & 1 \\ 1 & 0 \\ 0 & 1 \end{bmatrix} \mathbf{U}_k;$$

$$\mathbf{X}_0 = \begin{bmatrix} 0 \\ 2 \\ 0 \end{bmatrix}, \mathbf{U}_k = \begin{bmatrix} 1 \\ 1 \\ 2 \end{bmatrix}$$

8.71* Find the corresponding z-domain input-output relation $G(z)$, given that for an output $y_k = 1 + k$ for $k = 0, 1, 2,$ $\dots$, the input is a unit step.

8.72 Repeat Problem 8.71 for $y_k = 2ke^{-6k}$.

8.73 A system has the pulse transfer function

$$G(z) = \frac{2 + z^{-1}}{1 + 3z^{-1} + z^{-2}}$$

Find the response when the input is $u(kT) = \sin(kT)$.

8.74 In Problem 4.25, suppose the city government kills a fixed number of adult female dogs each year as a control measure. Derive the pulse transfer function between the number of rabid dogs in the city and the number of dogs exterminated each year.

8.75 Given the system

$$x_{1,k+1} = 1.5x_{1,k} + 3u_k; \qquad x_{1,0} = 0$$

$$x_{2,k+1} = 7.5x_{1,k} - x_{2,k} + 9u_k; \qquad x_{2,0} = 1$$

(a) Find the transfer function.

(b) Find the response if the input is (i) a unit step; (ii) the sequence

$$u_k = \begin{cases} 0, 0.5, 1.0, 1.5, 2 & k = 0, 1, \dots, 4 \\ 0 & \text{otherwise} \end{cases}$$

8.76 Find the pulse transfer function for the system

$$\mathbf{X}_{k+1} = \begin{bmatrix} -5 & -6 & 3 \\ 1 & 0 & -1 \\ -1 & 1 & 1 \end{bmatrix} \mathbf{X}_k + \begin{bmatrix} 1 & 0 \\ 0 & 1 \\ 1 & 1 \end{bmatrix} \mathbf{U}_k;$$

$$\mathbf{Y}_k = \begin{bmatrix} 1 & 1 & 0 \\ 1 & 0 & 1 \end{bmatrix} \mathbf{X}_k$$

Chapter

9

Representation of System Dynamics

A N INTUITIVE PREREQUISITE FOR EFFECTIVE CONTROL is knowledge of the system to be controlled. The more complete and better understood this information is, the more effective the control job that can be done. The development of knowledge of systems of interest was our main focus in Part I. In the last three chapters of Part II, we examined such information analytically. In this chapter, we shall discover ways of representing dynamic behavior that complement and enhance this knowledge, both in the transform domain and in state space.

First, we shall consider structural or graphical representations that take advantage of the analytical simplification that occurs when dynamic behavior is examined in the transform domain. Specifically, Laplace domain and *z*-domain operational block diagrams will be introduced, followed by *frequency response diagrams.* Next, we shall explore the relations between these transfer function formulations, which dominate what is generally called "classical" control theory, and the state space description, which we emphasized in most of Part I and Chapter 6 and which characterizes much of what is sometimes called *modern* control. The conversion between one description and the other, which will also facilitate the application of control techniques developed for one formulation in the other description, is discussed. Finally, the idea of the state space and state trajectories, introduced briefly in Section 3.1, is considered in more detail here. Some manifestations of the structure of dynamic systems that are revealed by such representations in state space are examined: The concept of *observability* deals with the ability to completely measure the state of a system, which, in turn, implies a form of knowledge of the system. *Controllability* is a complementary concept as well as an intuitive prerequisite for system control. It deals with the feasibility of control, that is, the circumstances under which any system is *controllable.* System *stability,* to which we have already alluded in earlier chapters, is considered formally and in more detail in the next chapter.

9.1 OPERATIONAL BLOCK DIAGRAMS AND RELATED ALGEBRA

The block diagram that was introduced in Section 1.3 can represent the physical model of a system. In Chapters 7 and 8, the transfer function was shown for linear systems to be, in fact, a system model, provided zero initial state is assumed or only the forced response is considered. The combination of these two ideas can be used advantageously

472

to represent system dynamics. The behavior of a dynamic system subject to some input can usually be better visualized in terms of the individual responses of the physical components constituting the system. Block diagrams of such physical units, in which the cause-effect relationships are specified by the transfer functions of these components, can be systematically interconnected into an operational block diagram that presents in one picture the detailed physical interactions in the system, the inputs and outputs, and the relative behavior (such as response speeds) of various parts of the system. For a complex problem, such a symbolic representation can be indispensable to our understanding of the *dynamics of the system.*

Figure 9.1 illustrates a component block diagram and two examples of operational block diagrams. The first example, Fig. 9.1b, represents a *unilaterally* coupled or cascade system. Here all effects are in one direction: A subsystem can influence components downstream of it but cannot influence those subsystems that are upstream of it. For instance, the input $U_2(s)$ has no effect on subsystem 1, represented by $G_1(s)$, etc. In the second example, Fig. 9.1c, which shows a *bilaterally* coupled system, the coupling is general. Input $U_2(s)$, for example, not only affects subsystems 2 and 3 but also component 1 through subsystem 4 and the feedback line. In either case, we can routinely solve for any output of interest. This algebraic manipulation of a dynamic system is one of the advantages of the transfer function representation.

Consider an interest in the overall output $Y_3(s)$. For the first case (Fig. 9.1b), we have

$$Y_3(s) = G_3(s)Y_2(s) = G_3(s)G_2(s)E_2(s)$$

$$E_2(s) = Y_1(s) + U_2(s) = G_1(s)U_1(s) + U_2(s)$$

so that

$$Y_3(s) = G_3(s)G_2(s)[G_1(s)U_1(s) + U_2(s)] \qquad (9.1)$$

For the second case (Fig. 9.1c), we still have

$$Y_3(s) = G_3(s)G_2(s)E_2(s)$$

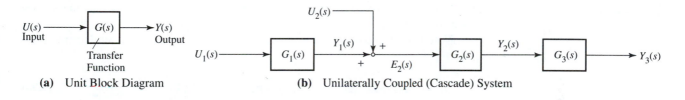

(a)   Unit Block Diagram

(b)   Unilaterally Coupled (Cascade) System

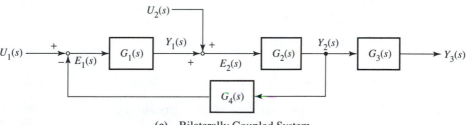

(c)   Bilaterally Coupled System

Figure 9.1

Interconnecting Operational Block Diagrams

However,

$$E_2(s) = Y_1(s) + U_2(s) = G_1(s)E_1(s) + U_2(s) \quad \text{and} \quad E_1(s) = U_1(s) - G_4(s)Y_2(s)$$

so that

$$E_2(s) = G_1(s)U_1(s) - G_1(s)G_4(s)Y_2(s) + U_2(s)$$

But

$$Y_2(s) = G_2(s)E_2(s)$$

hence

$$E_2(s) = G_1(s)U_1(s) - G_1(s)G_4(s)G_2(s)E_2(s) + U_2(s)$$

or

$$[1 + G_1(s)G_4(s)G_2(s)]E_2(s) = G_1(s)U_1(s) + U_2(s)$$

giving

$$E_2(s) = \frac{G_1(s)U_1(s) + U_2(s)}{1 + G_1(s)G_4(s)G_2(s)}$$

and

$$Y_3(s) = \frac{G_3(s)G_2(s)[G_1(s)U_1(s) + U_2(s)]}{1 + G_1(s)G_4(s)G_2(s)} \tag{9.2}$$

Canonical Scalar Feedback

The $[1 + P(s)]$ term is expected in the denominator any time there is negative feedback ($[1 - P(s)]$ for positive feedback). To see this, consider the canonical feedback loop shown in Fig. 9.2a.

Note that the general feedback loop shown in Fig. 9.2b can be reduced to loop (a) with

$$G(s) = G_N(s)G_{N-1}(s) \ . \ . \ . \ G_1(s) \tag{9.3a}$$

and

$$H(s) = H_m(s)H_{m-1}(s) \ . \ . \ . \ H_1(s) \tag{9.3b}$$

Solving for $Y(s)$ in Fig. 9.2a, we have

$$Y(s) = G(s)E(s) = G(s)U(s) - G(s)H(s)Y(s)$$

$$[1 + G(s)H(s)]Y(s) = G(s)U(s)$$

$$\frac{Y(s)}{U(s)} = G_{FL}(s) = \frac{G(s)}{1 + G(s)H(s)} \tag{9.4}$$

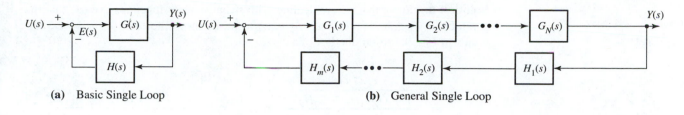

(a) Basic Single Loop

(b) General Single Loop

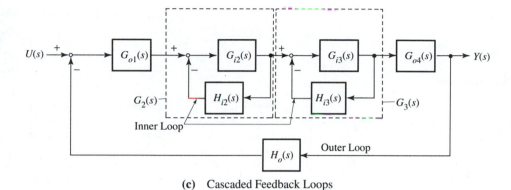

(c) Cascaded Feedback Loops

Figure 9.2

Analysis of Scalar Feedback Loops

This result also demonstrates that the individual transfer functions $G_1(s) \ldots G_N(s)$ may themselves represent internal (inner) feedback loops as shown in Fig. 9.2c. Thus a complex block diagram can be analyzed quickly by recognizing the various feedback loops within it.

| **Example 9.1** | Block Diagram Reduction |

a. Reduce the block diagram shown in Fig. 9.3 to the canonical form of Fig. 9.2a and find C/R.

b. Solve for C/R directly from the block diagram.

Figure 9.3

The Block Diagram

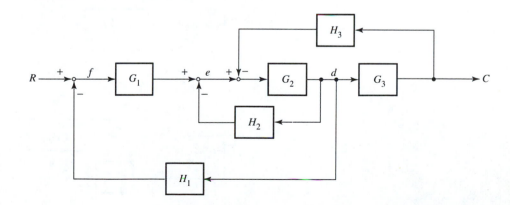

Solution a: Noting that we are dealing with scalar transfer functions, we see that the given block diagram is progressively equivalent, as shown, to each of the block diagrams in Fig. 9.4. From the last one,

$$\frac{C}{R} = \frac{\dfrac{G_3 G_2 G_1}{1 + G_2 G_1 H_1 + G_2 H_2}}{1 + \dfrac{G_3 G_2 H_3}{1 + G_2 G_1 H_1 + G_2 H_2}} \quad \text{or} \quad \frac{C}{R} = \frac{G_3 G_2 G_1}{1 + G_2 G_1 H_1 + G_2 H_2 + G_3 G_2 H_3}$$

Figure 9.4

Equivalent Block Diagrams

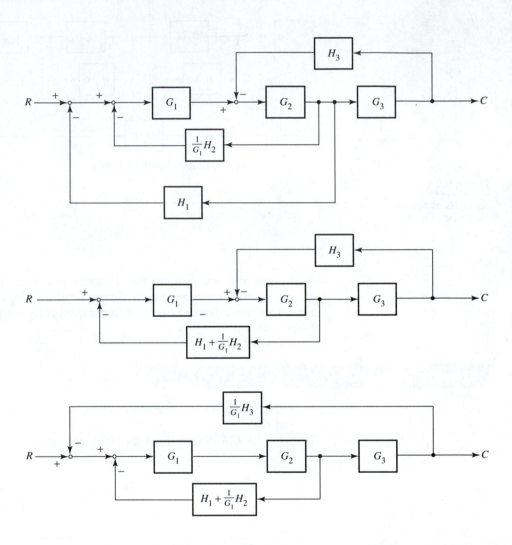

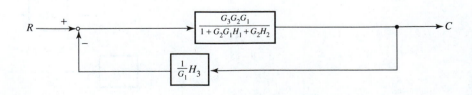

Solution b: With the notation (d, e, f) in Fig. 9.3:

$$f = R - H_1 d \tag{9.5a}$$

$$e = G_1 f - H_2 d \tag{9.5b}$$

$$d = G_2(e - H_3 C) \tag{9.5c}$$

$$C = G_3 d \tag{9.5d}$$

Substituting (9.5c) in (9.5d) we get

$$C = G_3 G_2 e - G_3 G_2 H_3 C \tag{9.5e}$$

Substituting (9.5a) in (9.5b), we have

$$e = G_1 R - G_1 H_1 d - H_2 d = G_1 R - (G_1 H_1 + H_2)d \tag{9.5f}$$

Substituting (9.5c) in (9.5f), we obtain

$$e = G_1 R - (G_1 H_1 + H_2)G_2(e - H_3 C)$$

which gives

$$e = \frac{G_1 R - (G_1 H_1 + H_2)G_2 H_3 C}{1 + G_1 G_2 H_1 + G_2 H_2} \tag{9.5g}$$

Thus equation (9.5e) becomes

$$C = \frac{G_3 G_2 G_1 R}{1 + G_1 G_2 H_1 + G_2 H_2} + \frac{G_3 G_2(G_1 H_1 + H_2)G_2 H_3 C}{1 + G_1 G_2 H_1 + G_2 H_2} - G_3 G_2 H_3 C$$

which yields

$$\frac{C}{R} = \frac{\dfrac{G_3 G_2 G_1}{1 + G_1 G_2 H_1 + G_2 H_2}}{1 + G_3 G_2 H_3 - \dfrac{G_3 G_2(G_1 H_1 + H_2)G_2 H_3}{1 + G_1 G_2 H_1 + G_2 H_2}}$$

or

$$\frac{C}{R} = \frac{G_3 G_2 G_1}{(1 + G_1 G_2 H_1 + G_2 H_2)(1 + G_3 G_2 H_3) - G_3 G_2(G_1 H_1 + H_2)G_2 H_3}$$

$$= \frac{G_3 G_2 G_1}{1 + G_1 G_2 H_1 + G_2 H_2 + G_3 G_2 H_3}$$

Application to Mixed Component Systems

The operational block diagram is particularly useful in describing systems with mixed behavior components. The mixing can be sectorial such as when mechanical and electrical components are interconnected in the same system, or it can be functional, such as when the same system has both time-delay (dead time) and time-integrating elements. The state space models of both types of mixed systems were considered generally in Chapter 5. We consider here an example illustrating the application of block diagrams to functionally mixed component systems.

Self-Excited Chatter   In Section 7.3 we gave machine tool chatter as an example of self-excited vibration. The mechanism for self-excited regenerative chatter has been described in reference [11] for single-point turning. Consider the schematic of a *plunge* cut on a lathe and the exaggerated chip thickness variation shown in Fig. 9.5. As a result of machine deflection due to cutting forces, the cutting tool leaves marks (undulations) on the workpiece. Each such undulation left on a previous revolution returns to act on the tool in the next revolution, by causing a change in the cutting force and consequently inducing further deflection. Such compounding of disturbance (exciting forces) in phase on successive revolutions builds to violent chatter.

A simplified single-point model is obtained by assuming a linear relationship between the cutting force and the chip thickness:

$$F = K_c U_c \tag{9.6a}$$

The proportionality constant K_c is called the *cutting stiffness*. The machine dynamics is represented by its "directional" transfer function:

$$\frac{Y_m}{F} = \frac{G_m(s)}{K_m} \tag{9.6b}$$

which relates the deflection of the machine tool Y_m in the chip thickness direction to a force F in the cutting force direction. Such a description can be developed analytically by the methods of Part I and Chapter 7 or experimentally by frequency response identification (Section 9.2). The contribution of the material from previous revolutions to the desired chip thickness (current total infeed; see Fig. 9.5) is given by the Laplace domain time-shift property

$$d_m = \mu_w e^{-T_w s} \tag{9.6c}$$

where T_w is the current period for one revolution of the workpiece, and μ_w is a scale factor. The machine deflection subtracts from the total infeed, whereas the material from previous revolutions adds to it. The operational block diagram representation of Fig. 9.6 is easily constructed from the preceding developments.

Finally, the transfer function relating the total amount of workpiece removed δ_w to the desired chip thickness U_0 is obtained as follows: From equation (9.4),

$$\frac{Y_m(s)}{U_0(s)} = \frac{(K_c/K_m)G_m(s)}{1 + (1 - \mu_w e^{-T_w s})(K_c/K_m)G_m(s)}$$

But

$$\delta_w(s) = (1 - \mu_w e^{-T_w s})Y_m(s)$$

hence

$$\frac{\delta_w(s)}{U_0(s)} = \frac{(1 - \mu_w e^{-T_w s})(K_c/K_m)G_m(s)}{1 + (1 - \mu_w e^{-T_w s})(K_c/K_m)G_m(s)} \tag{9.7}$$

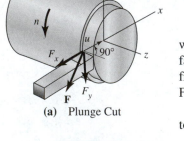

(a)   Plunge Cut

(b)   Chip Thickness

Figure 9.5

Plunge Cut on Lathe and Chip Thickness Variations

Figure 9.6

Single-Point-Turning Block Diagram Representation

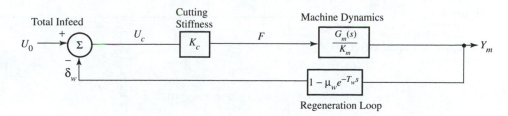

A peculiar behavior found in a few systems, especially process engineering units, is the net result of two opposing simultaneous effects that however, occur at different rates. If the faster effect is such that the overall response is initially in a direction opposite the eventual response direction, the behavior is termed *reverse reaction* or *inverse response*. A typical example is the response of the liquid level in a steam drum boiler following a step increase in the flowrate of the cold feedwater. From the physical model of Fig. 9.7a we deduce that the opposing effects are caused by the addition of heat and the addition of cold water *(feedwater)*. The heat input causes a general reduction in density of the liquid and, more significantly, an increase in the amount of the vapor phase in equilibrium with the liquid. Both result in an increase in the volume of the liquid or a rise in the liquid level in the boiler, which follows a *first-order-integral* (pure capacitive) sense:

Application to Reverse Reaction Processes

$$y(t) = K \int^t u(t)\, dt \Rightarrow Y(s) = \left(\frac{K}{s}\right) U(s); \quad \text{(first-order-integral behavior)} \tag{9.8a}$$

The speed of the first-order-integral response is characterized by the gain K. If we normalize the transfer function given by equation (9.8a), that is,

$$G(s) = \frac{1}{(1/K)s} = \frac{1}{\tau s}; \quad \text{(first-order-integral behavior)} \tag{9.8b}$$

we see that the speed of response can be measured in terms of the time quantity $(1/K)$. The smaller this characteristic time, the faster the response. Using $y_1(t)$ to represent the contribution to the liquid level by heat addition following the change in flowrate of cold feedwater $u(t)$, we take

$$Y_1(s) = \left(\frac{K_1}{s}\right) U(s) \tag{9.8c}$$

The addition of feedwater to the boiler causes the temperature to drop. This leads to a decrease of the liquid level of the boiling water. The major mechanism is the decrease in the volume of the vapor bubbles. The overall process is a *first-order* type of behavior:

$$\frac{dy}{dt} = -\frac{1}{T}y + \frac{K}{T}u(t) \Rightarrow Y(s) = \left(\frac{K}{Ts+1}\right) U(s); \quad \text{(first-order behavior)} \tag{9.9a}$$

As we already know, the speed of response of a first-order system can be measured in terms of the time constant T. For the normalized form,

$$G(s) = \frac{1}{(T/K)s + 1/K} \tag{9.9b}$$

Figure 9.7

Physical Model and Inverse
Response of Liquid Level in
Drum Boiler

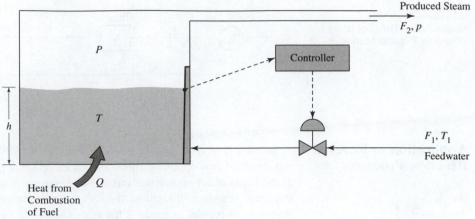

(a)   Physical Model of Liquid Level in Drum Boiler System

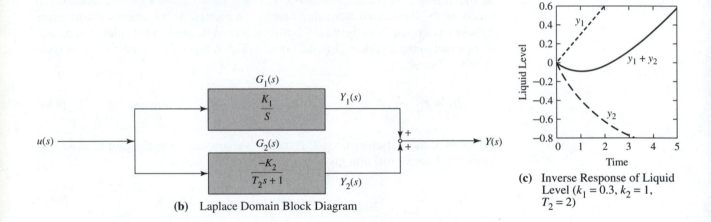

(b)   Laplace Domain Block Diagram

(c)   Inverse Response of Liquid
Level ($k_1 = 0.3$, $k_2 = 1$,
$T_2 = 2$)

the time constant is still T, and the quantity (T/K) is a measure of response speed. As before, the smaller this characteristic time, the faster the response. Using $y_2(t)$ to represent the contribution to the liquid level by cold water added as a result of the increase in feedwater flowrate, we take

$$Y_2(s) = -\left(\frac{K_2}{T_2 s + 1}\right)U(s) \tag{9.9c}$$

where the negative sign accounts for the fact that this is a reduction in liquid level. The net effect of feedwater flow on the liquid level is given by the sum of $y_1(t)$ and $y_2(t)$. Since in practice the heat supply is usually maintained at a constant rate corresponding to a desired rate of steam production, the rate of increase of $|y_2(t)|$ could exceed that of $|y_1(t)|$ initially. That is,

$$\frac{T_2}{K_2} < \frac{1}{K_1} \quad \text{or} \quad K_1 T_2 < K_2 \tag{9.10a}$$

and the resultant initial response of the liquid level is *reverse* or *inverse*. Note that if condition (9.10a) is not the case, an inverse type response will not occur.

Given the algebraic nature of interactions in the Laplace domain, the block diagram representing the system model for this process is easily constructed, as shown in Fig. 9.7b. The overall transfer function can be obtained by executing the indicated block diagram algebra:

$$\frac{Y(s)}{U(s)} = G_1(s) - G_2(s) = G(s) = \left(\frac{K_1}{s} - \frac{K_2}{T_2 s + 1} \right) \tag{9.10b}$$

$$= \frac{(K_1 T_2 - K_2)s + K_1}{s(T_2 s + 1)}$$

Observe from equation (9.10b) that when condition (9.10a) is satisfied, the transfer function has a *positive zero* (that is, values of s at which $|G(s)|$ become zero):

$$s = \frac{-K_1}{K_1 T_2 - K_2}; \quad \text{positive when equation (9.10a) is satisfied}$$

This is a general result, that is, *when a system possesses an inverse response, a root of the numerator of its transfer function has a positive real part.*

A simulation of the block diagram of Fig. 9.7b, that is, solution of the system model given by equation (9.10b) (this can be accomplished by first converting the transfer function into an equivalent vector state equation (see Section 9.3) and solving this equation by the methods of Chapter 6*; see **On the Net**) is shown in Fig. 9.7c. The input u is a unit step. The peculiar nature of a reverse reaction process is evident from this result. With such unusual early behavior, processes with inverse response can be difficult to control satisfactorily if, in some cases, special attention is not paid to the transient behavior.

Inverse response can also occur as a result of similar opposing effects between other first- or second-order systems, including those with dead time (see [12] and Practice Problems 9.14 and 9.39).

z-Transform Block Diagrams

The preceding discussion of s-domain or Laplace transform block diagrams applies to z-transform block diagrams as well. The primary difference is that the transfer functions of s are replaced by transfer functions of z; consequently, the interpretations of the implied block dynamics are different. Consider the basic dynamic operator in discrete-time systems, the delay. As shown in Fig. 9.8a, n unit delays in cascade (unilaterally coupled) are additive in the discrete-time domain. Thus with $u(k)$ as the input to the first unit delay, the output from the last unit delay, or the output of the implied system, is $y(k) = u(k - n)$. For the z-domain block diagram shown in Fig. 9.8b, the usual block diagram algebra gives the overall transfer function as the product of the transfer functions of each of the blocks in cascade, so that

$$Y(z) = [G(z)]U(z) = [G_n(z)G_{n-1}(z) \cdots G_3(z)G_2(z)G_1(z)]U(z) \tag{9.11a}$$

$$= [z^{-1} \cdot z^{-1} \cdots z^{-1} \cdot z^{-1} \cdot z^{-1}]U(z) = [z^{-n}]U(z)$$

---

*Although MATLAB's *step* command, available with the Signals and Systems Tool Box, will compute and automatically plot the step response of an LTI system, given its transfer function as a ratio of two polynomials like equation (9.10b), it is not recommended for this inverse process.

Figure 9.8

Unilaterally Coupled System of Unit Delays

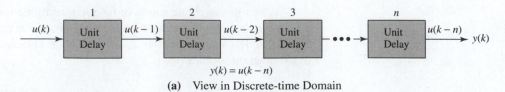

$$y(k) = u(k - n)$$

(a)   View in Discrete-time Domain

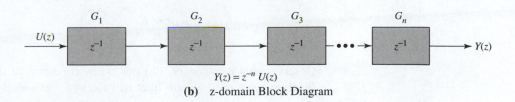

$$Y(z) = z^{-n} U(z)$$

(b)   z-domain Block Diagram

But according to the time-shift property, equation (8.44b), for zero initial conditions (which is implied for transfer function representations), the z-transform of a time-*delayed* signal is

$$\mathcal{Z}[u(t - T)] = z^{-1}U(z) \qquad \text{(9.11b)}$$

By repeated application of (9.11b) we see that

$$\mathcal{Z}[u(t - nT)] = z^{-n}U(z) \qquad \text{(9.11c)}$$

which is the z-transform of $y(k)$, that is, $Y(z)$.

Example 9.2   z-Domain Block Diagram Synthesis

Recall from Section 8.4 that z^{-k} represents a delayed unit impulse, and as shown in Fig. 9.9a, a sampled unit step signal is in fact a sequence of unit impulses. Assuming that a signal generator that can produce a unit impulse

$$u(k) = \begin{cases} 0 & k < 0 \\ 1 & k = 0 \\ 0 & k > 0 \end{cases}$$

is available, draw a block diagram of a system that will produce the pattern of delayed signals represented by the sampled unit step.

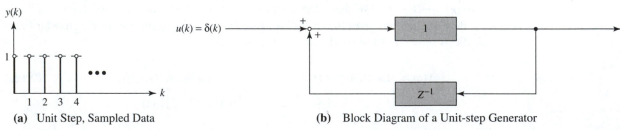

(a)   Unit Step, Sampled Data

(b)   Block Diagram of a Unit-step Generator

Figure 9.9

Representation of a Sampled Unit-Step Signal

Solution: We note from Fig. 9.9a that the unit step function is a repetition of the unit impulse input but with successive inputs delayed by progressively higher numbers of discrete-time units. This can be accomplished by feeding back the impulse output after delaying it by one time unit. The block diagram is shown in Fig. 9.9b. It can be seen that such a delayed feedback signal on passing once again around the loop results in an output impulse that is delayed by two time units, and so on. The block diagram algebra is as before for a basic feedback loop. However, the feedback is positive (additive), so that there is a minus sign instead of a plus sign in the denominator term in equation (9.4). That is,

$$G(z) = \frac{1}{1 - z^{-1}} \tag{9.12}$$

Since the z-transform of the unit impulse is 1 (entry 10 of Table 8.6), the z-transform of the input is $U(z) = 1$. It follows that the z-transform of the output of the circuit synthesized above is

$$Y(z) = G(z)U(z) = \frac{1}{1 - z^{-1}} = \frac{z}{z - 1}$$

This is the z-transform of the unit step (entry 1 of Table 8.6).

9.2 IDENTIFICATION AND FREQUENCY RESPONSE

Frequency response analysis is often used in (a) identification and (b) design of dynamic systems. Some of the design aspects were introduced as part of the phasor and Fourier transform analyses in Chapters 7 and 8. We saw that the frequency spectrum of the output of a linear system is equal to the product of the spectrum of the input and the spectrum of the system's frequency transfer function or its unit impulse response. We were able to influence some characteristics of the output spectrum such as the resonant frequency and amplitude and other specifications of damping behavior by making proper choices of system parameters. A "sketch" of impulse and step response identification of dynamic systems was also given in Section 8.2. Although our ultimate goal is control system design, we shall briefly consider the frequency response technique and the elementary aspects of frequency response identification and computation. The actual practice of design in the frequency domain is covered in Chapter 12.

Frequency Domain Identification

In the frequency response approach to system identification, the steady-state response of the system to (sinusoidal) disturbances at various frequencies (the sinusoidal steady state) is analyzed with the hope of deducing the system dynamics. An advantage of this (black box) approach is that the equations (model) of the system need not be known a priori (recall the discussion of impulse and step response identification). In addition, frequency response testing is normally not as upsetting to a process at the step or impulse input methods. Because only the steady-state input-output relation or (frequency domain) transfer function is sought, the system that is subject to a frequency response test must be *asymptotically stable* (that is, all eigenvalues are on the left half-complex plane, excluding the imaginary axis). The implication is that the system must be such that its free response dies out with time, so that only the steady-state response remains. For a *marginally stable* system (that is, a stable system with eigenvalues on the imaginary axis), any initial disturbance will persist (no steady state) unless a precise initial condition that can exactly cancel out the initial disturbance is imposed. Even then, this is difficult to accomplish in practice, given the often low precision in our knowledge of or our ability to influence the initial state.

As we saw in Section 7.4, for a particular sinusoidal input frequency, the steady-state response can be represented by the phasor transfer function $G(j\omega)$. For sinusoidal or periodic inputs of many frequencies, the frequency response is still represented by $G(j\omega)$ or the Fourier transfer function, where ω takes on values covering the frequency range of interest. Thus the frequency domain transfer function is $G(j\omega)$. As we discussed in Section 8.2 (Transfer Functions, Impulse Response, Convolution), the frequency domain transfer function can be obtained from the Laplace domain transfer function by substituting $s = j\omega$ in the latter. Finally, the amplitude (magnitude) ratio and phase shift at each frequency are given, as before, by

$$M = |G(j\omega)| \quad \text{and} \quad \phi = \angle G(j\omega) \tag{9.13}$$

Since $G(j\omega)$ is a complex number, the frequency response is completely described by M and ϕ, when these are given for all relevant ω, that is, the amplitude and phase spectra are given. In identification (see Example 9.4), the amplitude and phase data are used to reconstruct the transfer function of the system.

Bode and Nyquist Diagrams

The actual analysis of the output data is generally done using frequency response (Bode and Nyquist) diagrams. The *Bode* diagram is a semilogarithmic format for displaying frequency response data: The amplitude ratio M, called the *gain,* is plotted in decibel units (the linear scale) versus the logarithm of the frequency, while the phase shift ϕ is plotted in degrees (another linear scale) versus $\log(\omega)$. (Logarithms to the base 10 are implied unless otherwise stated). The decibel unit is defined as

$$20 \log_{10} M = \text{decibels (db) or gain in db} \tag{9.14}$$

so that the decibel scale converts a logarithmic scale for the gain into a linear scale. Note that on an actual logarithmic abscissa scale, ω is plotted in order to effect $\log(\omega)$.

To simplify (manual) plotting and analysis of Bode diagrams, the general transfer function

$$G(s) = \frac{B(s)}{A(s)} = \frac{b_0 s^m + b_1 s^{m-1} + \cdots + b_{m-1}s + b_m}{a_0 s^n + a_1 s^{n-1} + \cdots + a_{n-1}s + a_n}, \quad a_0 \neq 0 \tag{9.15a}$$

is represented by a combination (product) of canonical forms (constant k, integrator $s^{\pm 1}$, first-order $(Ts + 1)^{\pm 1}$, and second-order $(T^2 s^2 + 2\zeta Ts + 1)^{\pm 1}$):

$$G(s) = \frac{k \prod_{i=1}^{p} (T_i s + 1) \prod_{l=1}^{r} (T_l^2 s^2 + 2\zeta_l s + 1)}{s^q \prod_{h=1}^{\gamma} (T_h s + 1) \prod_{\eta=1}^{\mu} (T_\eta^2 s^2 + 2\zeta T_\eta s + 1)} \tag{9.15b}$$

where $p + 2r = m$ is the order of the numerator polynomial, and $q + \gamma + 2\mu = n$ is the order of the denominator polynomial. Each product term in equation (9.15b) is, in general, a complex number ($s = j\omega$). Because the logarithm of a product is the sum of the logarithms of each term, and the angle of a product of complex numbers is the sum of the angles of each complex number, the Bode plot for a transfer function when put in the form (9.15b) reduces to summation (and subtraction) of gains (in decibels) and phases (in degrees) of individual canonical terms. That is the simplification. If the transfer function cannot be put in canonical form, the definition for M and ϕ can still be used, or the computer can always be utilized as discussed under Frequency Response Computation.

Bode Diagrams for the Canonical Forms

We develop next the Bode diagrams for each of the canonical transfer functions represented in equation (9.15b). These will then serve as building blocks for constructing the diagrams for arbitrary transfer functions.

1. Pure Gain

$$G(j\omega) = k; \qquad k > 0 \tag{9.16a}$$

The gain and phase are

$$M = k \quad \text{and} \quad \phi = 0 \quad \text{for all } \omega \neq 0 \tag{9.16b}$$

Thus

$$\text{Gain (db)} = 20 \log(k) = \text{constant} \tag{9.16c}$$

The gain and phase Bode plots are horizontal lines, as shown in Fig. 9.10a.

2. Integrator
We consider first the pure integrator and the complementary pure derivative term.
(i) Given $G(s) = 1/s$, then

$$G(j\omega) = \frac{1}{j\omega} \tag{9.17a}$$

$$M = \frac{1}{\omega} \quad \text{and} \quad \phi = -90° \quad \text{for all } \omega \neq 0 \tag{9.17b}$$

$$\therefore \text{ Gain (db)} = 20 \log(k) = 20 \log(1) - 20 \log(\omega) \tag{9.17c}$$

Comparing (9.17c) with the equation of a straight line,

$$y = I + Sx \tag{9.17d}$$

where I is the y-intercept when $x = 0$ and S is the slope, we see that equation (9.17c) describes a straight line with $y = $ gain (db), $x = \log(\omega)$, $I = 20 \log(1) = 0$, and $S = -20$. This value of the slope is described as a slope of -20 db/decade (a decade

Figure 9.10

**Bode Plots for Canonical
Transfer Function Forms**

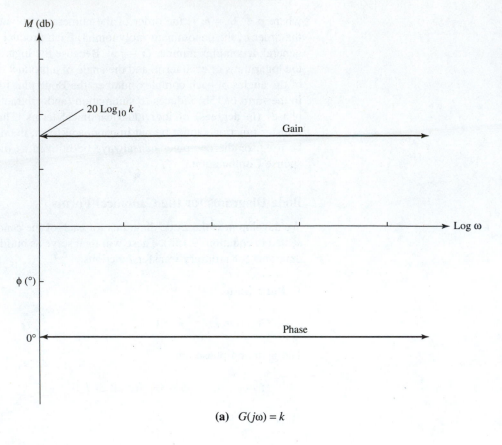

(a) $G(j\omega) = k$

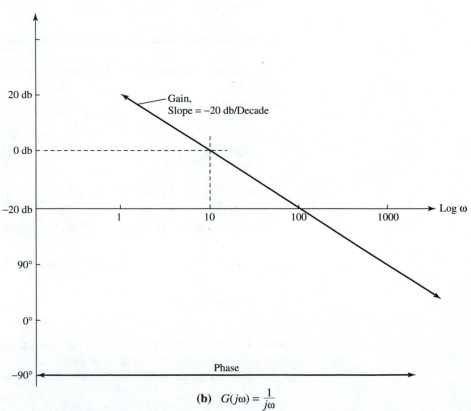

(b) $G(j\omega) = \dfrac{1}{j\omega}$

Figure 9.10 *continued*

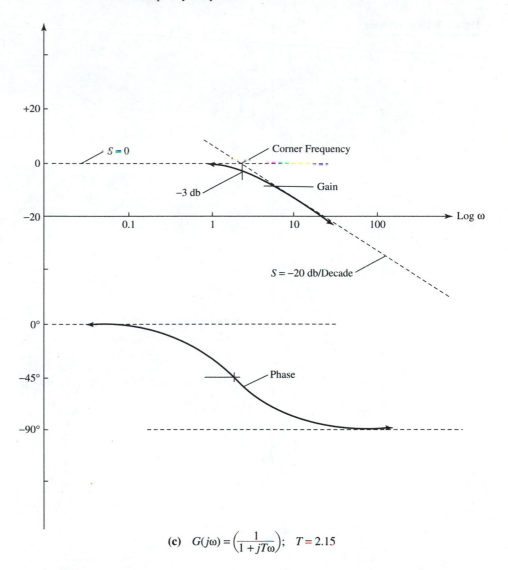

(c) $G(j\omega) = \left(\dfrac{1}{1 + jT\omega}\right)$; $T = 2.15$

is 10, and y decreases by 20 db when ω increases by 10, since $\log(\omega)$ will increase by 1). The Bode plot for the pure integrator is shown in Fig. 9.10b.

(ii) for the pure derivative term,

$$G(s) = s \Rightarrow G(j\omega) = j\omega \tag{9.17e}$$

$$M = \omega, \quad \phi = +90° \quad \text{for all } \omega \neq 0 \tag{9.17f}$$

The Bode plot is also a straight line with

$$S = +20 \text{ db/decade} \quad \text{and} \quad \text{gain (db)} = 0 \quad \text{at } \omega = 1 \tag{9.17g}$$

(iii) For the general integral (or derivative) term,

$$G(s) = s^{\pm q} \tag{9.17h}$$

Figure 9.10 *continued*

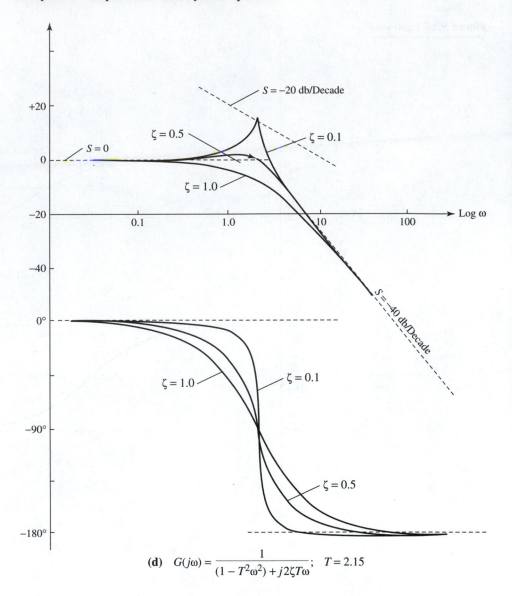

(d) $G(j\omega) = \dfrac{1}{(1 - T^2\omega^2) + j2\zeta T\omega}$; $T = 2.15$

Noting that

$$s^{\pm q} = \prod_{i=1}^{q}(s^{\pm 1})$$

we see that the Bode plot for this general case is still a straight line, and for all ω,

$$\phi = \pm(90q)^0 \quad \text{and} \quad S = \pm(20q) \text{ db/decade} \tag{9.17i}$$

3. First-Order Canonical Form
 (i)

$$G(s) = \frac{1}{Ts + 1} \tag{9.18a}$$

Equation (9.9a) shows that this is the main factor in the transfer function for the first-order state equation. Continuing, we have

$$G(j\omega) = \frac{1}{1 + jT\omega}$$

so that

$$M = \frac{1}{\sqrt{1 + (T\omega)^2}}; \qquad \psi = -\tan^{-1}(T\omega) \tag{9.18b}$$

We sketch the Bode plot to the first-order term by considering its asymptotic behavior: For $T\omega \ll 1$,

$$G(j\omega) \Rightarrow 1; \qquad M \Rightarrow 1$$

and gain (db) $\Rightarrow 0$, while $\phi \Rightarrow 0^0$.
For $T\omega \gg 1$,

$$G(j\omega) \Rightarrow \frac{1}{jT\omega}; \qquad M \Rightarrow \frac{1}{T\omega}$$

and gain (db) $\Rightarrow$ straight line with $S = -20$ db/decade, while $\phi \Rightarrow -90°$ (see pure integrator above).

The transition between the two regions, $T\omega = 1$, defines the *corner* or *break* or *cutoff frequency:*

$$\omega = \frac{1}{T} = \omega_c \tag{9.18c}$$

$$M = \frac{\sqrt{2}}{2} \Rightarrow gain\ (db) = -3.0103\ db \cong -3\ db \tag{9.18d}$$

$$\phi = -\tan^{-1}(1) = -45° \tag{9.18e}$$

As shown in Fig. 9.10c, the first-order plot is sketched by locating the corner frequency, drawing the two asymptotes $y = 0$ db and $y = -20 \log(\omega)$ to meet at the corner frequency, and sketching a fair curve that passes through -3 db at $\omega = 1/T$ and that is asymptotic to these lines.
(ii) The plot for $G(j\omega) = (1 + jT\omega)$ is similar, except that for $T\omega \gg 1$, $S = +20$ db/decade, $\phi \Rightarrow +90°$, and at $\omega = 1/T$, gain (db) $= +3$ db and $\phi = +45°$.

4. Second-Order Canonical Form

$$G(s) = \frac{1}{T^2 s^2 + 2\zeta T s + 1} \Rightarrow G(j\omega) = \frac{1}{(1 - T^2\omega^2) + j2\zeta T\omega} \tag{9.19a}$$

We consider again asymptotic behavior:
For $T\omega \ll 1$,

$$G(j\omega) \Rightarrow 1; \qquad M \Rightarrow 1 \text{ and gain (db)} \Rightarrow 0, \text{ while } \phi \Rightarrow 0^0.$$

For $T\omega \gg 1$, $(T\omega)^2 \gg T\omega$,

$$G(j\omega) \Rightarrow -\frac{1}{T^2\omega^2}$$

gain (db) $\Rightarrow$ straight line with $S = -40$ db/decade, while $\phi \Rightarrow -180°$.
At the corner frequency $T\omega = 1$,

$$G(j\omega) = \frac{1}{j2\zeta T\omega} \quad \text{and} \quad M = \frac{1}{2\zeta T\omega}$$

is a function of ζ.

However, the slope at this point is -20 db/decade. With respect to this -20 db/decade tangent, at the corner frequency, the gain curve is convex (and not concave or inflected), since

$$\frac{1}{2\zeta T\omega} > \frac{1}{\sqrt{(1 - T^2\omega^2)^2 + (2\zeta T\omega)^2}} \quad \text{for all } \omega \neq 0 \tag{9.19b}$$

The phase here is $-90°$. Figure 9.10d shows the Bode plot for the second-order canonical form for a number of values of ζ.

5. Dead Time   Although dead time was not specifically included in equation (9.15b) as one of the canonical forms, it is a distinct feature of dynamic systems that often needs to be considered, as was demonstrated in the plunge cut on a lathe problem. Systems whose transfer functions contain dead time are easily analyzed in the Bode format:

$$G(s) = e^{-sT}; \qquad G(j\omega) = e^{-j\omega T} \tag{9.20a}$$

The magnitude is $M = 1$, so that

$$\text{gain (db)} = 0 \text{ db} \quad \text{for all } \omega \neq 0 \tag{9.20b}$$

and the phase is

$$\phi = -\omega T \text{ rad} = -180 \, \omega T/\pi \text{ deg} \tag{9.20c}$$

Minimum Phase Systems   Except for dead time, a clear pattern of asymptotes and corner frequency values should have been evident in the foregoing analysis of canonical forms. This information, which is summarized in Table 9.1, also suggests a one-to-one correspondence between the phase of a system and the gain. In other words, either the phase information or the amplitude data are sufficient to completely describe the dynamic system. This deduction is true provided the system is *minimum phase*. A minimum phase system is one that satisfies the condition that all the *poles* and *zeros* of the transfer function are either negative real, or zero, or conjugate complex with negative (or zero) real part. For example,

$$G(s) = \frac{T_1 s - 1}{T_2 s + 1}$$

Table 9.1

Summary of Bode Diagram Results for Canonical Forms

ELEMENT	$\omega << \dfrac{1}{T}$	$\omega = \dfrac{1}{T}$	$\omega >> \dfrac{1}{T}$
k	$20 \log_{10} k$ [db]	$20 \log_{10} k$ [db]	$20 \log_{10} k$ [db]
	$\phi = 0°$	$\phi = 0°$	$\phi = 0°$
s^n	$n \cdot 20$ [db/decade]	$n \cdot 20$ [db/decade]	$n \cdot 20$ [db/decade]
	$\phi \longrightarrow n \cdot 90°$	$\phi = n \cdot 90°$	$\phi \longrightarrow n \cdot 90°$
$(Ts + 1)^{\pm 1}$	0 [db]	$-\pm 3$ [db]	± 20 [db/decade]
	$\phi \longrightarrow 0°$	$\phi = +45°$	$\phi \longrightarrow \pm 90°$
$(T^2 s^2 + 2\zeta\, T_5 + 1)^{\pm 1}$	0 [db]	depends on ζ, ± 20 [db/decade]	± 40 [db/decade]
	$\phi \longrightarrow 0°$	$\phi = \pm 90°$	$\phi \longrightarrow \pm 180°$
Dead time: e^{-Ts}	0 [db]	0 [db]	0 [db]
	$\phi = -\omega T$ [rad]	$\phi = -1$ [rad]	$\phi = -\omega T$ [rad]

has a zero, $s = 1/T_1$ which is positive real. Thus this system is not minimum phase, although it is a stable system. Also, dead time e^{-Ts} is not minimum phase; $\exp(-Ts)$ is zero when the real part of s tends to $+\infty$. In terms of the polynomial representation of the transfer function, equation (9.15a), the system can be identified from the magnitude information alone as long as all the coefficients of s in the numerator polynomial have the same sign (see [13]). Obviously, care should be exercised in plotting non–minimum phase systems, whereas in the identification of systems from frequency response data, the possibility of a non–minimum phase system should not be overlooked.

The application of these results in constructing and analyzing Bode diagrams is best illustrated by examples.

Example 9.3 Manual Construction of Bode Diagram

Develop the Bode diagram for the system

$$G(s) = \frac{5(0.1s + 1)}{s(s^2 + 2s + 1)}$$

Solution: Letting $G(s) = G_1(s)G_2(s)G_3(s)G_4(s)$, we obtain the canonical forms:

$G_1(s) = 5$

$G_2(s) = 0.1s + 1; \qquad T = 0.1, \quad \omega_c = 10$

$G_3(s) = \dfrac{1}{s}$

$G_4(s) = \dfrac{1}{s^2 + 2s + 1}; \qquad T = 1, \quad \omega_c = 1, \quad \zeta = 1$

Step 1: The two corner frequencies $\omega_c = 1$ and $\omega_c = 10$ give rise to the three regions: $\omega <<; 1 < \omega < 10;$ and $\omega >> 10$.

Step 2: We obtain and "add" the asymptotes in each region and compute the asymptotic values and the actual values at the corner frequencies. The "addition" is done graphically, but an analytical approach that can be tabulated as shown in Table 9.2 is neater.

Table 9.2 Addition of Asymptotes and Corner Values

REGION	QUANTITY	$G_1(s)$	$G_2(s)$	$G_3(s)$	$G_4(s)$	$G(s)$
$\omega \ll 1$	Gain asymptote	0 db/decade	0 db/decade	−20 db/decade	0 db/decade	−20 db/decade
	Phase asymptote	0°	0°	−90°	0°	−90°
$\omega = 1$	Gain	13.98 db	0.043 db	0 db	−6.02 db	8 db
	Phase	0°	5.71°	−90°	−90°	−174.3°
$1 < \omega < 10$	Gain asymptote	0 db/decade	0 db/decade	−20 db/decade	−40 db/decade	−60 db/decade
	Phase asymptote	0°	0°	−90°	−180°	−270°
$\omega = 10$	Gain	13.98 db	+3 db	−20 db	−40.1 db	−43.1 db
	Phase	0°	+45°	−90°	−168.6°	−213.6°
$\omega \gg 10$	Gain asymptote	0 db/decade	+20 db/decade	−20 db/decade	−40 db/decade	−40 db/decade
	Phase asymptote	0°	+90°	−90°	−180°	−180°

Step 3: These data, that is, the asymptotes and corner values, are sketched as shown in Fig. 9.11.

Step 4: As an aid to the final sketching of the curves (Fig. 9.11), actual values of gain and phase can be computed at a few frequencies, as needed. This procedure is illustrated in Table 9.3.

Example 9.4 — Identification from Bode Diagram

Identify the system transfer function from the Bode (gain and phase) plot shown in Fig. 9.12. Assume that the beginning and ending slopes of the gain plot can be extrapolated.

Solution: The gain plot begins with an initial slope of −40 db/decade, which suggests a term

$$G_1(s) = \frac{1}{s^2}$$

Figure 9.11

Bode Diagram for
$$G(s) = \frac{5(0.1s + 1)}{s(s^2 + 2s + 1)}$$

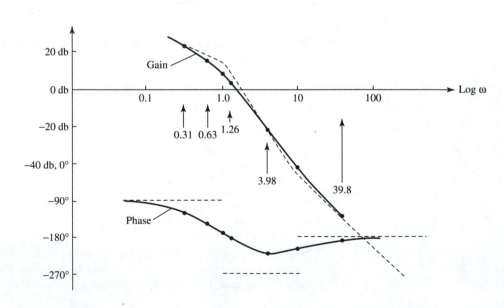

Table 9.3

Some Data Points

ω (rad/s)	GAIN (db)	PHASE (°)
0.31	23.2	−123
0.63	15.1	−150
1.26	3.8	−186
3.98	−22.0	−220
39.90	−69.8	−191

in the transfer function. However, the projected slope does not intersect the 0 db line at $\omega = 1$ but intersects the $\omega = 1$ line at about 26 db. This suggests a pure gain term

$$G_2(s) = k$$

that when "added" to $G_1(s)$ shifts the entire curve upward by 26 db.

Thus $20 \log(k) = 26$, or $k \approx 20$. Further, the slope changes to -60 db/decade with a bend-over point at $\omega = 0.2$ as shown. This suggests the "addition" of a first-order term:

$$G_3(s) = \frac{1}{T_3 s + 1}$$

with corner frequency

$$\omega_c = \frac{1}{T_3} = 0.2, \quad \text{or} \quad T_3 = 5$$

There is yet another change in slope to -20 db/decade with a bend-over point at $\omega = 5$. This requires a second-order term

$$G_4(s) = T_4^2 s^2 + 2\zeta T_4 s + 1$$

which will add $+40$ db/decade to the slope beyond the corner frequency

$$5T_4 = 1 \quad \text{or} \quad T_4 = 0.2$$

Note that this second-order term is the mirror image of the second-order canonical form shown in Fig. 9.10 (that is, a numerator term instead of the denominator function).

Figure 9.12

Frequency Response Identification: Minimum Phase System

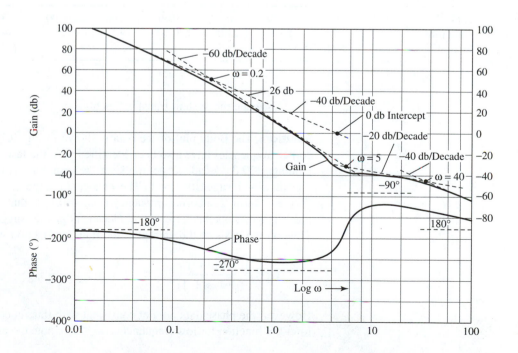

Keeping this in mind, and comparing the curves in Fig. 9.10 with the region around $\omega = 5$ in Fig. 9.12, we see that ζ must be less than 0.5. We can determine the actual value of ζ from the contribution of the second-order term at the corner frequency. From Fig. 9.12, the curve is 6 dB below the intersection of the two asymptotes at $\omega = 5$. That is, assuming that any other term with corner frequency greater than 5 (the next change of slope is at $\omega = 40$) contributes very little to the gain at $\omega = 5$, then

$$20 \log(2\zeta T_4 \omega) = -6, \quad \text{or} \quad \zeta = \frac{10^{-6/20}}{(2)(0.2)(5)} = 0.25$$

Finally, there is the further change of slope from -20 db decade to -40 db/decade at $\omega = 40$. Hence the final term is the transfer function is

$$G_5(s) = \frac{1}{0.025s + 1}$$

The system transfer function as identified is thus:

$$G(s) = \frac{20(0.04s^2 + 0.1s + 1)}{s^2(5s + 1)(0.025s + 1)}$$

This must be correct, since the phase plot is also consistent (see the phase asymptotes: $-180°$, $-270°$, $-90°$, and $-180°$). The system is minimum phase.

Example 9.5   Identification from Experimental Data

The data from a frequency response test, after reduction, were as follows:

Table 9.4 Experimental Data

ω (s^{-1})	0.01	0.04	0.08	0.20	0.60	2.00	4.00	8.00	10.00	20.00
Gain (db)	40.0	29.6	24.1	21.0	20.1	19.4	17.9	14.5	13.0	7.7
Phase (°)	−84.7	−69.8	−54.4	−34.6	−33.5	−82.0	−155	−288	−351	−649

Estimate the system transfer function.

Solution: The Bode plot of the data is shown in Fig. 9.13. The gain asymptotes and corner frequencies have also been identified as in the last example. We see that the beginning -20 db/decade slope implies a $1/s$ term, but the 0 db crossing of this asymptote at $\omega = 1$ suggests $k = 1$. The change in slope to a horizontal line requires a $(T_1 s + 1)$ term. The corner frequency is $\omega_c = 0.1$, so that $T_1 = 10$. The gain curve further changes slope back to -20 db/decade at $\omega = 5$, suggesting the term $1/(0.2s + 1)$ Thus on the basis of the gain plot, the transfer function would be

$$G(s) = \frac{10s + 1}{s(0.2s + 1)}$$

However, the phase data are not consistent with this function. Although the phase follows the function for low frequencies ($\omega < 1.0$; note the phase asymptotes required by

Figure 9.13

Frequency Response Identification: Non–Minimum Phase System

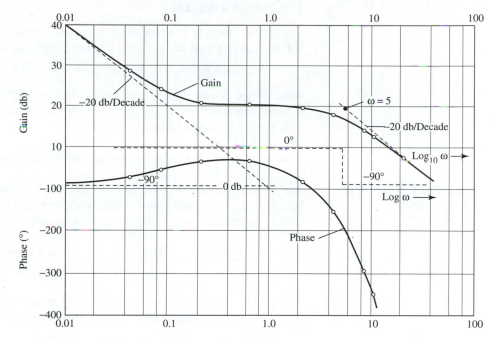

the function above: $-90°$, $0°$, and $-90°$), it deviates from the function progressively by larger and larger angles as the frequency gets bigger. This is similar to the behavior of dead time e^{-Ts}, which will add to the phase the angle $(-180\omega T/\pi)°$ while leaving the gain unaffected. To determine T and explore the presence of dead time, we consider the data at $\omega = 10$ and $\omega = 20$. The angle of $G(s)$ is

$$\angle G(j\omega) = -90° + \tan^{-1}(10\omega) - \tan^{-1}(0.2\omega) \Rightarrow \angle G(j10) = -64°$$

$$\text{and } \angle G(j20) = -76.3°$$

The measured angles are, however, $-351°$ at $\omega = 10$ and $-649°$ at $\omega = 20$. Thus the contribution of the dead time at $\omega = 20$ is

$$-\frac{(180)(20)T}{\pi} = -649 + 76.3 = -572.7°$$

hence $T = 0.5$

As a check, at $\omega = 10$, the angle contributed by the dead time is

$$\frac{(180)(10)(0.5)}{\pi} = -286.5°$$

Hence the total angle is $-286.5 - 64 = -350.5°$, which agrees with the measurement.

The system transfer function is thus

$$G(s) = \frac{e^{-0.5s}(10s + 1)}{s(0.2s + 1)}$$

The Nyquist Diagram

The *Nyquist* diagram is a plot of $G(j\omega)$ in the complex plane, for $\omega = 0$ to ∞. Recall that we can represent the complex number $z = G(j\omega)$ in two ways:

$$(1)\ z = \text{Real}[G(j\omega)] + j\text{Img}[G(j\omega)]; \qquad (2)\ z = Me^{j\phi}$$

where M and ϕ were defined previously. Due to the second form, the Nyquist diagram is sometimes referred to as the *polar plot*. No matter which representation of the complex number is used, a locus of points in the complex plane, known as the *Nyquist locus*, results when ω varies from 0 to ∞. An arrow on the locus is used to indicate the direction of increasing ω. Some examples of Nyquist diagrams are shown in Fig. 9.14.

Frequency Response Computation

The problem in frequency response computation is to compute the frequency response data (generate Bode or Nyquist plots) for a given model (transfer function) of the system. Consider equation (9.15a) again. Substituting $s = j\omega$ in $G(s)$, gives us

$$G(j\omega) = \frac{(b_m - b_{m-2}\omega^2 + b_{m-4}\omega^4 - \cdots) + j(b_{m-1}\omega - b_{m-3}\omega^3 + b_{m-5}\omega^5 - \cdots)}{(a_n - a_{n-2}\omega^2 + a_{n-4}\omega^4 - \cdots) + j(a_{n-1}\omega - a_{n-3}\omega^3 + a_{n-5}\omega^5 - \cdots)} = \frac{BR + jBI}{AR + jAI} \quad \textbf{(9.21)}$$

BR, BI, AR, and AI can be computed (for any n and m) at each frequency ω according to the pattern that is discernible for each term in equation (9.21). For the Nyquist plot we can write

$$\frac{BR + jBI}{AR + jAI} = X(\omega) + jY(\omega) \qquad\qquad \textbf{(9.22a)}$$

Figure 9.14

Examples of Nyquist Plots

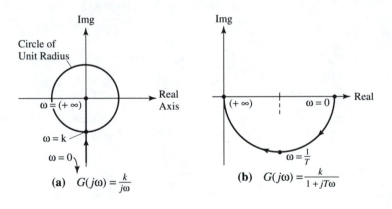

(a) $G(j\omega) = \dfrac{k}{j\omega}$

(b) $G(j\omega) = \dfrac{k}{1 + jT\omega}$

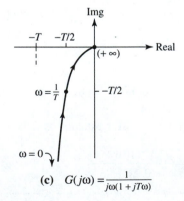

(c) $G(j\omega) = \dfrac{1}{j\omega(1 + jT\omega)}$

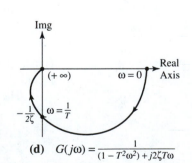

(d) $G(j\omega) = \dfrac{1}{(1 - T^2\omega^2) + j2\zeta T\omega}$

where, using "*" for multiplication, we have

$$X(\omega) = \left[\frac{BR\mathord{*}AR + BI\mathord{*}AI}{AR^2 + AI^2} \right] \tag{9.22b}$$

$$Y(\omega) = \left[\frac{BI\mathord{*}AR - BR\mathord{*}AI}{AR^2 + AI^2} \right] \tag{9.22c}$$

The Nyquist diagram is then a plot of $Y(\omega)$ versus $X(\omega)$ in the complex plane, for all $0 < \omega < \infty$. For the Bode diagram, the gain (db) and the phase are plotted versus the log (ω), where

$$\text{Gain (db)} = 20 \log |G(j\omega)| = 20 \log \sqrt{X(\omega)^2 + Y(\omega)^2} \tag{9.23a}$$

$$= 20 \log \sqrt{\left(\frac{BR^2 + BI^2}{AR^2 + AI^2} \right)}$$

$$\text{Phase} = \angle G(j\omega) = \tan^{-1} \left[\frac{Y(\omega)}{X(\omega)} \right] \tag{9.23b}$$

Dead time e^{-sT} can also be handled within this framework by slightly modifying $G(j\omega)$.

Consider

$$G(s) = e^{-sT} G_0(s) \tag{9.23c}$$

Substituting $s = j\omega$, and using $e^{-j\omega T} = \cos(\omega T) - j\sin(\omega T)$, we obtain

$$G(j\omega) = [\cos(\omega T) - j \sin(\omega T)] \left[\frac{(BR)_0 + j(BI)_0}{(AR)_0 + j(AI)_0} \right] \tag{9.23d}$$

where $(BR)_0$, $(BI)_0$, $(AR)_0$, and $(AI)_0$ are as defined previously (equation (9.21), but with $G_0(j\omega)$ replacing $G(j\omega)$). Equation (9.23d) multiplied out gives

$$G(j\omega) = \frac{BR + jBI}{AR + jAI} = X(\omega) + jY(\omega) \tag{9.23e}$$

as before, except for the modification

$$BR = (BR)_0 \cos(\omega T) + (BI)_0 \sin(\omega T)$$

$$BI = (BI)_0 \cos(\omega T) - (BR)_0 \sin(\omega T)$$

$$AR = (AR)_0 \tag{9.23f}$$

and

$$AI = (AI)_0$$

The particular algorithm used in a computational software package can of course differ in some details from this presentation. However, these equations will be quite useful if your math software does not include explicit Bode and Nyquist functions.

Example 9.6	Frequency Response Computation by Math Software

Compute and plot using MATLAB, the Nyquist and bode diagrams for a system with the open-loop transfer function

$$G(s) = \frac{1}{0.1s^4 + 1.02s^3 + 0.3s^2 + s}$$

Solution:

```
% fig9_15.m
% Script to generate Figure 9.15
% Fig. 9.15a is the Nyquist Diagram
% plotted within manually scaled axes
% Fig. 9.15b is the Bode Diagram
% plotted automatically by the bode (num,den,w) function
%
% TRANSFER FUNCTION AND FREQUENCY RANGE DATA
num = [1];                              % numerator polynomial coefficients
den = [0.1 1.02 0.3 1 0];              % denominator polynomial coefficients
w = logspace (−2,4,240);               % frequency vector equal logarithmic spacing

% NYQUIST DIAGRAM
[re,im,w] = nyquist (num,den,w);       % generate Nyquist diagram data
axis ([−6 2 −12 4]);                   % scale axes
plot (re,im);                          % plot generated data
title ('(a) Nyquist Diagram');
ylabel ('Imaginary Axis');                      xlabel ('Real Axis');
text (−2.5,2.03,'>');                  % indicate direction for increasing frequency
hold on; plot ([−6 2], [0 0], ':');
plot ([0 0], [−12 4], ':');            % show axes
%
hold off; pause;                       % press any key to continue
axis; axis;                            % restore automatic axes scaling
%
% BODE DIAGRAM
bode(num,den,w);                       % generate data and plot Bode diagram
title('(b) Bode Diagram');
```

MATLAB includes explicit Bode and Nyquist M-files in its Signals and Systems Toolbox. These functions can be used to calculate the frequency response of continuous-time, linear, time-invariant systems. Script file **(fig9_15.m)** invokes the appropriate frequency response functions to generate both the Nyquist and Bode diagrams for the given transfer function.

The output produced by the script is shown in Fig. 9.15. Note that the form of the Nyquist function used simply calculates the data that were then plotted within the script. This was done to prevent the distortion that would otherwise have resulted from automatic axes scaling for this particular transfer function. Also observe that the polynomial coefficients are ordered in decreasing power of *s*. The MATLAB **nyquist** and

Figure 9.15

Computation of Nyquist and Bode Plots by MATLAB

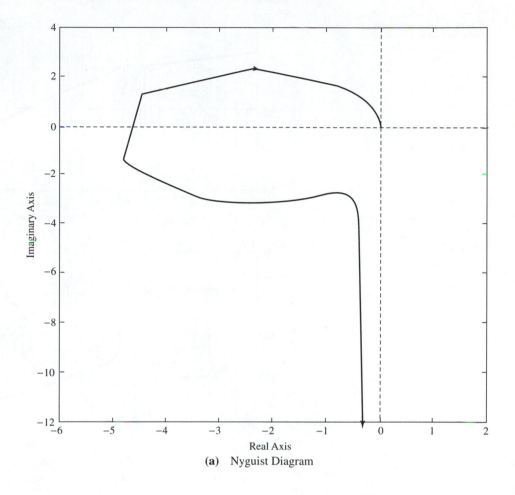

(a) Nyguist Diagram

bode M-files automatically add $\pm 360°$ to the phase at the $\pm 180°$ transition points so that the indicated phase is continuous above or below $\pm 180°$.

A further constraint that is perhaps evident in the Nyquist plot is the problem of frequency increments. Because of the highly nonlinear nature of these functions, equations (9.22) and (9.23), very small (but fixed) increments in ω may be necessary in order to generate a smooth curve. The resulting number of program steps required to cover a fairly reasonable range of frequencies may be too large to be practicable. In Example 9.6 this problem was ameliorated somewhat by incrementing $\log(\omega)$ rather than ω. Thus with a fixed increment in $\log(\omega)$ of only 0.025, a frequency range of 0.01 to 10,000 rad/s could be covered in 240 steps. In spite of this, the new fairly sharp corners on the Nyquist plot suggest that an even more complex (variable increment) frequency net could be more appropriate in many cases.

9.3 RELATIONS BETWEEN TRANSFER FUNCTIONS AND STATE MODELS

The results of the preceding sections suggest that a transfer function representation of a dynamic system can be derived directly. This is particularly true for experimentally obtained data with which, as discussed in Section 8.2, an impulse or step response

Figure 9.15 *continued*

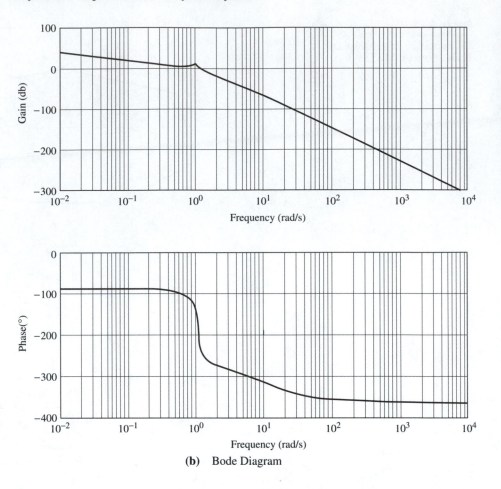

(b)   Bode Diagram

identification or, as demonstrated in Section 9.2, a frequency response identification can be used to derive the linear system transfer function. We saw in Chapter 8 a definite procedure for obtaining the Laplace or z-domain transfer function from the time-domain state description (equations (8.39b) and (8.47)). This process of going from the state equation to the transfer function is unique; unfortunately, the reverse process of finding the state representation for a given transfer function is not. Rather, it is possible to have more than one state space structure for a specified transfer function. Recall from Section 3.1 that the same system can be described with different sets of state variables. Indeed, given any state representation, one can add any number of unobservable (unmeasurable) state variables to obtain new state descriptions, all of which have the same transfer function. Thus the concept of *minimal realizations* of the transfer function refers to those state models (equations (6.49a) and (6.49b)) derived from the transfer function but whose **A** matrices have the least dimensions. It can be shown (see [3]) that a realization of a given transfer function matrix $\mathbf{G}(s)$ is minimal if and only if the state model (equations (6.49a) and (6.49b)) is completely controllable and observable. *Observability* and *controllability* are discussed in Section 9.4, and we will not be concerned here with the rigorous procedures for obtaining a minimal realization. However, the particular signal flow graph technique discussed next is conceptually simple, practically useful, and often results in a minimal realization. Also, although we shall consider mainly *single-input, single-output* (SISO) systems, the results also apply to the relation between one output component and one input acting alone in a *multiple-input, multiple-output* (MIMO) system.

Signal Flow Graphs

Table 9.5 defines the basic elements of the signal flow graph. Note that we have given both the state and transfer function models of the integrator element. We consider first the related problem of constructing a signal flow graph from a state space description. Given the state model

$$\dot{\mathbf{X}} = \mathbf{A}\mathbf{X} + \mathbf{B}\mathbf{U} \tag{9.24a}$$

or

$$\dot{x}_1 = a_{11}x_1 + a_{12}x_2 + \ldots + a_{1n}x_n + b_{11}u_1 + b_{12}u_2 + \ldots + h_{1,}u_r$$

$$\dot{x}_2 = a_{21}x_1 + a_{22}x_2 + \ldots + a_{2n}x_n + b_{21}u_1 + b_{22}u_2 + \ldots + b_{2r}u_r \tag{9.24b}$$

$$\vdots$$

$$\dot{x}_n = a_{n1}x_1 + a_{n2}x_2 + \ldots + a_{nn}x_n + b_{n1}u_1 + b_{n2}u_2 + \ldots + b_{nr}u_r$$

and

$$\mathbf{Y} = \mathbf{C}\mathbf{X} \tag{9.25a}$$

or

$$y_1 = c_{11}x_1 + c_{12}x_2 + \ldots + c_{1n}x_n$$

$$\vdots \tag{9.25b}$$

$$y_m = c_{m1}x_1 + c_{m2}x_2 + \ldots + c_{mn}x_n$$

These equations are already in the signal flow graph state format (see Table 9.5). To construct the graph, we begin with the integrators. There are as many integrators as the order of the system and these are numbered from **right to left.** To interconnect the integrators we observe the following two rules, which also apply to the construction of the signal flow graph from transfer function models:

1. The inputs of the system can be connected only to the inputs of integrators. The gains (transmittances) on these connecting links are fixed by (fix) the rows of **B**.
2. The outputs of the integrators are the state variables, and apart from being connectable to the inputs of integrators where the transmittances of the links are fixed by (fix) the rows of **A**, they can also be connected only to the outputs of the system, in which case the links are determined by (determine) the columns of the **C** matrix.

Table 9.5

Some Basic Symbols of Signal Flow Graphs

ELEMENT	SIGNAL FLOW GRAPH	MODEL
Integrator	$\frac{1}{s}$ $W(s)\!\circ\!\!\longrightarrow\!\!\circ X(s)$	State: $\frac{d}{dt}x = w$ Transfer function: $X(s) = \frac{1}{s}W(s)$
Time delay	z^{-1} $W(z)\!\circ\!\!\longrightarrow\!\!\circ X(z)$	State: $x_{k+1} = w_k$ Transfer function: $X(z) = z^{-1}W(z)$
Summer		$X = G_1W_1 + G_2W_2 + \cdots + G_nW_n$

| **Example 9.7** | Signal Flow Graph from State Equation |

Construct a signal flow graph for the system

$$\dot{\mathbf{X}} = \begin{bmatrix} -10 & -12 & 6 \\ 2 & 0 & -2 \\ -1 & 3 & 1 \end{bmatrix} \mathbf{X} + \begin{bmatrix} 1 & 0 \\ 0 & 1 \\ 1 & 1 \end{bmatrix} \mathbf{U}; \qquad \mathbf{Y} = \begin{bmatrix} 1 & 1 & 0 \\ 1 & 0 & 1 \end{bmatrix} \mathbf{X}$$

Solution:

Step 1: Draw the integrators and label the state variables.

Step 2: Connect up the graph according to the two rules and the definitions in Table 9.5.

The inverse problem of writing the state model, given the signal flow graph, is straightforward. We simply write the state equation for each integrator, using the definitions in Table 9.5. This is easily verified with the graph of Example 9.7 (Fig. 9.16).

From Transfer Function to State Model

We return now to the task of transforming a transfer function representation into a state space description. The following example illustrates the procedure:

$$G(s) = \frac{b_1 s + b_0}{s^2 + a_1 s + a_0} \tag{9.26}$$

Step 1: Convert the transfer function into integrator format by dividing by the highest power of s in the denominator

$$G(s) = \frac{b_1/s + b_0/s^2}{1 + a_1/s + a_0/s^2} = \frac{Y(s)}{U(s)} \tag{9.27a}$$

Figure 9.16

Signal Flow Graph from State Equation

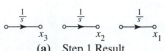

(a)   Step 1 Result

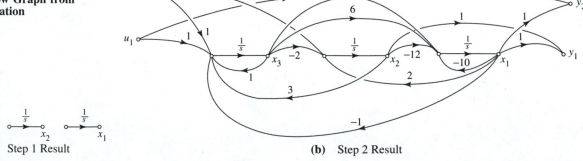

(b)   Step 2 Result

Step 2: Write the output equation, and group the integrators by order:

$$Y(s) = \left[\frac{b_0 U(s) - a_0 Y(s)}{s^2} \right] + \left[\frac{b_2 U(s) - a_1 Y(s)}{s} \right] \qquad (9.27b)$$

Step 3: Draw as many integrators as the highest order integration. The rightmost integrator corresponds to the lowest-order term, and its output is x_1. Proceed in this order from right to left linking the integrators with a 1 gain (see Fig. 9.17a).

Step 4: Connect up according to the two rules given previously and the definitions in Table 9.5. Connect x_1 to y by a 1 link, that is, let $y = x_1$.

Step 5: From the eventual signal flow graph, write the state model as discussed above:

$$\dot{x}_1 = -a_1 x_1 + x_2 + b_1 u$$
$$\dot{x}_2 = -a_0 x_1 + b_0 u \qquad (9.28)$$
$$y = x_1$$

The signal flow graph technique for s-domain transfer functions applies to the transformation of z-domain transfer functions to discrete-time state models, provided we replace $1/s$ with $1/z$.

Example 9.8   **z-Domain Block Diagram to State Model**

Find the discrete-time model for the control system whose z-domain block diagram is shown in Fig. 9.18.

Solution:

$$\frac{Y(z)}{R(z)} = G(z) = \frac{0.5kz/[(z-1)(z-0.5)]}{1 + 0.5kz/[(z-1)(z-0.5)]} = \frac{0.5kz}{(z-1)(z-0.5) + 0.5kz}$$

$$= \frac{0.5kz}{z^2 + (0.5k - 1.5)z + 0.5}$$

Step 1:

$$G(z) = \frac{0.5kz^{-1}}{1 + (0.5k - 1.5)z^{-1} + 0.5z^{-2}} = \frac{Y(z)}{R(z)}$$

Figure 9.17

Signal Flow from Transfer Function

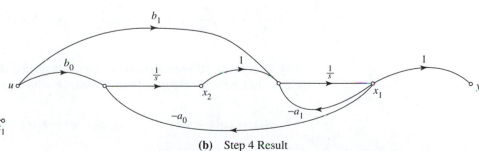

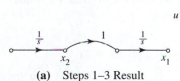

(a)   Steps 1–3 Result

(b)   Step 4 Result

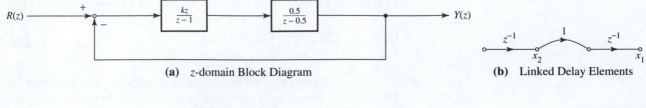

(a) *z*-domain Block Diagram   **(b)**   Linked Delay Elements

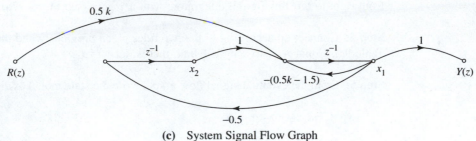

(c)   System Signal Flow Graph

Figure 9.18

**z-Domain Block Diagram and
Signal Flow Graph**

Step 2:

$$Y(z) = z^{-2}[-0.5Y(z)] + z^{-1}[0.5kR(z) - (0.5k - 1.5)Y(z)]$$

Step 3: See Fig. 9.18b.

Step 4: See Fig. 9.18c.

Step 5: The state model is

$$\mathbf{X}_{k+1} = \begin{bmatrix} -(0.5k - 1.5) & 1 \\ -.5 & 0 \end{bmatrix}\mathbf{X}_k + \begin{bmatrix} 0.5k \\ 0 \end{bmatrix}r_k; \qquad y_k = [1 \quad 0]\mathbf{X}_k$$

The Companion Form   A special case in the transformation between transfer functions and state models is the
representation of the system by an ordinary differential equation such as equation
(6.15), repeated here,

$$\frac{d^n y}{dt^n} + a_{n-1}\frac{d^{n-1}y}{dt^{n-1}} + \ldots + a_1\frac{dy}{dt} + a_0 y = b_0 u(t) \tag{9.29}$$

or by an ordinary difference equation, such as equation (8.49a), but without an instantaneous link between input and output, that is, with $b_0 = 0$:

$$y(k) + a_1 y(k - 1) + \ldots + a_n y(k - n) = b_1 u(k - 1) + b_2 u(k - 2)$$
$$+ \ldots + b_m u(k - m) \tag{9.30}$$

We note from our discussions in Sections 8.2 and 8.4 that the transfer functions corresponding to equations (9.29) and (9.30) are, respectively,

$$G(s) = \frac{b_0}{s^n + a_{n-1}s^{n-1} + \ldots + a_1 s + a_0} = \frac{Y(s)}{U(s)} \qquad (9.31)$$

and

$$G(z) = \frac{b_1 z^{-1} + b_2 z^{-2} + \ldots + b_m z^{-m}}{1 + a_1 z^{-1} + a_2 z^{-2} + \ldots + a_n z^{-n}} = \frac{Y(z)}{U(z)} \qquad (9.32)$$

where $m \leq n$ for *realizable* systems. Note that these are scalar transfer functions.

The companion form of the state model, that is, the equivalent system of n first-order ODEs or vector differential equations (in n variables) can be developed as follows. Consider as an example the third-order ODE

$$\frac{d^3 y}{dt^3} + a_2 \frac{d^2 y}{dt^2} + a_1 \frac{dy}{dt} + a_0 y = b_0 u(t) \qquad (9.23a)$$

We can rewrite equation (9.33a) as (9.33b):

$$\frac{d^3 y}{dt^3} = (-a_0)y + (-a_1)\frac{dy}{dt} + (-a_2)\frac{d^2 y}{dt^2} + (b_0)u(t) \qquad (9.33b)$$

We then let

$$x_1 = y, \; x_2 = \frac{dx_1}{dt} = \frac{dy}{dt}, \qquad x_3 = \frac{dx_2}{dt} = \frac{d^2 x_1}{dt^2} = \frac{d^2 y}{dt^2}$$

and

$$\frac{dx_3}{dt} = \frac{d^2 x_2}{dt^2} = \frac{d^3 x_2}{dt^3} = \frac{d^3 y}{dt^3}$$

Substituting equation (9.33b) for (dx_3/dt), we obtain the following set of first-order equations:

$$\dot{x}_1 = x_2$$
$$\dot{x}_2 = x_3$$
$$\dot{x}_3 = -a_0 x_1 - a_1 x_2 - a_2 x_3 + b_0 u$$

From this result we can see that the state representation for equation (9.29) or (9.31) is

$$\dot{\mathbf{X}} = \mathbf{A}\mathbf{X} + \mathbf{B}u; \qquad y = \mathbf{C}\mathbf{X} \qquad (9.34a)$$

where **A** is in the companion form

$$\mathbf{A} = \begin{bmatrix} 0 & & & \\ 0 & & & \\ \cdot & & \mathbf{I} & \\ \cdot & & & \\ \cdot & & & \\ 0 & & & \\ \hline -a_0 & -a_1 & \cdots & -a_{n-1} \end{bmatrix}; \qquad \mathbf{B} = \begin{bmatrix} 0 \\ 0 \\ \cdot \\ \cdot \\ b_0 \end{bmatrix}; \tag{9.34b}$$

$$\mathbf{C} = \begin{bmatrix} 1 & 0 & 0 & \cdots & 0 \end{bmatrix}$$

According to reference [14], the discrete-time formulation for equation (9.30) or (9.32) is

$$\hat{\mathbf{X}}_k = \hat{\mathbf{P}}\hat{\mathbf{X}}_{k-1} + \hat{\mathbf{Q}}u_{k-1}; \qquad y_k = \hat{\mathbf{C}}\hat{\mathbf{X}}_k \tag{9.35a}$$

where $\hat{\mathbf{P}}$ is of a companion matrix form,

$$\hat{\mathbf{P}} = \begin{bmatrix} 0 & & & \\ 0 & & & \\ \cdot & & \mathbf{I} & \\ \cdot & & & \\ 0 & & & \\ \hline -a_n & -a_{n-1} & \cdots & -a_1 \end{bmatrix} \tag{9.35b}$$

$$\hat{\mathbf{C}} = \begin{bmatrix} 1 & 0 & 0 & \cdots & 0 \end{bmatrix} \tag{9.35c}$$

and $\hat{\mathbf{Q}}$ is given by

$$\hat{\mathbf{Q}} = \begin{bmatrix} 1 & 0 & 0 & \cdots & 0 \\ a_1 & 1 & 0 & & 0 \\ a_2 & & 1 & \ddots & \cdot \\ \cdot & & & \ddots & 0 \\ \cdot & & & & \\ a_{n-1} & a_{n-2} & \cdots & a_1 & 1 \end{bmatrix}^{-1} \begin{bmatrix} b_1 \\ \cdot \\ \cdot \\ b_m \\ 0 \\ \cdot \\ \cdot \\ 0 \end{bmatrix} \tag{9.35d}$$

Linear Transformation to Companion Form

Reference [14] also gives the transformation matrix (recall Application of Linear Transformations, Section 6.3) to convert the general single-input discrete-time system model

$$\mathbf{X}_{k+1} = \mathbf{P}\mathbf{X}_k + \mathbf{Q}u_k; \qquad y_k = \mathbf{C}\mathbf{X}_k \tag{9.36a}$$

to companion form

$$\hat{\mathbf{X}}_{k+1} = \hat{\mathbf{P}}\hat{\mathbf{X}}_k + \hat{\mathbf{Q}}u_k; \qquad y_k = \hat{\mathbf{C}}\hat{\mathbf{X}}_k \tag{9.36b}$$

provided the system (9.36a) is observable. The inverse information matrix is

$$
\mathbf{T}^{-1} = \begin{bmatrix} \mathbf{C} \\ \mathbf{CP} \\ \mathbf{CP}^2 \\ \vdots \\ \mathbf{CP}^{n-1} \end{bmatrix}
\tag{9.37a}
$$

so that, as in equations (6.78c)–(6.78e),

$$
\hat{\mathbf{P}} = \mathbf{T}^{-1}\mathbf{PT} = \left[\begin{array}{cccc|c} 0 & & & & \\ 0 & & & & \\ \vdots & & \mathbf{I} & & \\ \vdots & & & & \\ 0 & & & & \\ \hline -a_n & -a_{n-1} & \cdots & -a_1 \end{array}\right]; \qquad \hat{\mathbf{Q}} = \mathbf{T}^{-1}\mathbf{Q};
\tag{9.37b}
$$

$$
\hat{\mathbf{C}} = \mathbf{CT}
$$

and

$$
\hat{\mathbf{X}} = \mathbf{T}^{-1}\mathbf{X}
\tag{9.37c}
$$

As we shall see in Section 9.4, the existence of $\mathbf{T} = [\mathbf{T}^{-1}]$ is the condition for observability of the discrete-time system (9.36a). Finally, we observe from equation (9.35d) that

$$
\begin{bmatrix} b_1 \\ b_2 \\ b_3 \\ \vdots \\ \vdots \\ b_n \end{bmatrix} = \begin{bmatrix} 1 & 0 & 0 & \cdots & 0 \\ a_1 & 1 & 0 & & 0 \\ a_2 & & 1 & \ddots & \vdots \\ \vdots & & & \ddots & 0 \\ a_{n-1} & a_{n-2} & \cdots & a_1 & 1 \end{bmatrix} \hat{\mathbf{Q}}
\tag{9.38a}
$$

where $a_1, a_2, \ldots a_{n-1}$ are obtained from (9.37b). Hence

$$
G(z) = \frac{b_1 z^{-1} + b_2 z^{-2} + \ldots + b_n z^{-n}}{1 + a_1 z^{-1} + a_2 z^{-2} + \ldots + a_n z^{-n}}
\tag{9.38b}
$$

Example 9.9   Pulse Transfer Function to State Model

Derive a state model for the system with the following pulse transfer function:

$$
G(z) = \frac{z^{-1} - 4z^{-2}}{1 + 4z^{-1}\,pl\,5z^{-2} - 2z^{-3}}
$$

Solution: According to equations (9.35b) and (9.35c),

$$
\hat{\mathbf{P}} = \begin{bmatrix} 0 & 1 & 0 \\ 0 & 0 & 1 \\ 2 & -5 & -4 \end{bmatrix}; \qquad \hat{\mathbf{C}} = [1 \quad 0 \quad 0]
$$

From equation (9.35d),

$$
\hat{\mathbf{Q}} = \begin{bmatrix} 1 & 0 & 0 \\ 4 & 1 & 0 \\ 5 & 4 & 1 \end{bmatrix}^{-1} \begin{bmatrix} 1 \\ -4 \\ 0 \end{bmatrix} = \begin{bmatrix} 1 & 0 & 0 \\ -4 & 1 & 0 \\ 11 & -4 & 1 \end{bmatrix} \begin{bmatrix} 1 \\ -4 \\ 0 \end{bmatrix} = \begin{bmatrix} 1 \\ -8 \\ 27 \end{bmatrix}
$$

so that the state model, as given by equation (9.35a), is

$$
\begin{aligned}
x_1(k) &= x_2(k-1) + u(k-1) \\
x_2(k) &= x_3(k-1) - 8u(k-1) \\
x_3(k) &= 2x_1(k-1) - 5x_2(k-1) - 4x_3(k-1) + 27u(k-1) \\
y(k) &= x_1(k)
\end{aligned}
$$

Example 9.10   Linear Transformation to Companion Form

Transform the following discrete-time state model to companion form and write the corresponding transfer function:

$$
\mathbf{X}_{k+1} = \begin{bmatrix} 1 & 1 \\ 0 & 1 \end{bmatrix}\mathbf{X}_k + \begin{bmatrix} 2 \\ 1 \end{bmatrix}u_k; \qquad y_k = [2 \quad -1]\mathbf{X}_k
$$

Solution: The inverse transformation matrix, from equation (9.37a), is

$$
\mathbf{T}^{-1} = \begin{bmatrix} 2 & -1 \\ 2 & 1 \end{bmatrix}
$$

so that

$$
\mathbf{T} = \begin{bmatrix} 2 & -1 \\ 2 & 1 \end{bmatrix}^{-1} = \begin{bmatrix} 1/4 & 1/4 \\ -1/2 & 1/2 \end{bmatrix}
$$

From equation (9.37b),

$$
\hat{\mathbf{P}} = \begin{bmatrix} 2 & -1 \\ 2 & 2 \end{bmatrix}\begin{bmatrix} 1 & 1 \\ 0 & 1 \end{bmatrix}\begin{bmatrix} 1/4 & 1/4 \\ -1/2 & 1/2 \end{bmatrix} = \begin{bmatrix} 0 & 1 \\ -1 & 2 \end{bmatrix} = \begin{bmatrix} 0 & 1 \\ -a_2 & -a_1 \end{bmatrix};
$$

$$
\hat{\mathbf{Q}} = \begin{bmatrix} 2 & -1 \\ 2 & 1 \end{bmatrix}\begin{bmatrix} 2 \\ 1 \end{bmatrix} = \begin{bmatrix} 3 \\ 5 \end{bmatrix}
$$

and

$$
\hat{\mathbf{C}} = [2 \quad -1]\begin{bmatrix} 1/4 & 1/4 \\ -1/2 & 1/2 \end{bmatrix} = [1 \quad 0]
$$

as required.

Thus, the companion form, from equation (9.36b), is

$$\hat{\mathbf{X}}_{k+1} = \begin{bmatrix} 0 & 1 \\ -1 & 2 \end{bmatrix} \hat{\mathbf{X}}_k + \begin{bmatrix} 3 \\ 5 \end{bmatrix} u_k; \qquad y_k = [1 \quad 0]\hat{\mathbf{X}}_k$$

From equation (9.38a),

$$\begin{bmatrix} b_1 \\ b_2 \end{bmatrix} = \begin{bmatrix} 1 & 0 \\ -2 & 2 \end{bmatrix} \begin{bmatrix} 3 \\ 5 \end{bmatrix} = \begin{bmatrix} 3 \\ -1 \end{bmatrix}$$

so that

$$G(z) = \frac{3z^{-1} - z^{-2}}{1 - 2z^{-1} + z^{-2}}$$

9.4 CONCEPTS IN STATE SPACE

The question of observability of a system featured prominently in the last section. Such concepts and others to be discussed here are most vividly illustrated in state space. Because direct solution of the system model is not required, state space methods have been used effectively in the analysis and design of nonlinear dynamic systems. However, our objective here is merely to introduce the subject, especially as an illustration of another useful representation of system dynamics. It should be noted that the state space representation can perhaps be generated more easily by simulation of the solution of the full differential equation, including nonlinear problems, on a digital computer. Indeed, the production of most of the sketches shown in the figures here was aided by digital computer simulation.

State Trajectories for Second-Order Systems

The state space was defined in Section 3.1 as the (n-dimensional) mathematical space whose coordinates are the state variables. When the state variables are related according to $\dot{x}_{j-1} = x_j; j = 2, 3, \ldots, n$, the state space is known as the *phase space*. If, in particular, $n = 2$, then we have a state plane *(phase plane)*. The "state" of a system at any time is specified by the value of the state variables. This specification represents a "point" in the state space. For any given initial state, the succession of states projected on the state space gives a *trajectory* or locus of "points" along which the system will "move" from that initial state. Different initial states yield different trajectories. The family of trajectories yields an instantaneous picture of the stability behavior of the system; changes in the speed of motion with time and from point to point along the trajectory can be observed clearly; and if the system is under control, the effectiveness of the control is apparent. The "speed of motion" is determined by which *mode(s)* dominate the response at any time. In general, the fast modes dominate the early motion, while the slow modes (after the fast ones have decayed sufficiently) determine the later motion. The system modes, as we saw in Section 6.3, can be represented in the state space by a set of linearly independent vectors called eigenvectors. If these vectors also *span* the state space (this is possible for distinct eigenvalues; see Appendix B for further explanation of span), then any trajectory in the state space can be expressed as a function of these eigenvectors. Thus we also find in the state space technique a confirmation of the result that the system response is made up of a combination of the modes.

The Isocline Method

We shall illustrate some of the preceding ideas with a second-order system to which we shall apply the *isocline* technique of constructing state trajectories. Consider the system of Example 6.15:

$$\dot{\mathbf{X}} = \begin{bmatrix} -1 & 0 \\ 1 & -2 \end{bmatrix} \mathbf{X}; \qquad \mathbf{X}(0) = \begin{bmatrix} 2 \\ 3 \end{bmatrix} \tag{9.39a}$$

The eigenvalues were found to be -1 and -2, so that the modes are e^{-t} (slow) and e^{-2t} (fast). The corresponding eigenvectors were also computed (in Example 6.16),

$$\mathbf{V}_1 = \begin{bmatrix} 2 \\ 2 \end{bmatrix} \quad \text{and} \quad \mathbf{V}_2 = \begin{bmatrix} 0 \\ 1 \end{bmatrix} \tag{9.39b}$$

and are indicated in the state space as shown in Fig. 9.19a.

Figure 9.19

State Plane Trajectories and Isoclines

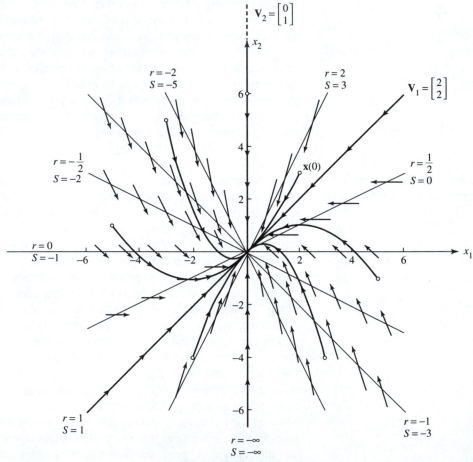

(a) Eigenvectors in State Space

(b) The Isocline Method

A quick graphical technique for constructing second-order system trajectories is to find a family of curves with the same slope in the state plane, called *isoclines*. The trajectories are then drawn tangent to these curves. Continuing with the example, we see that the slope is

$$S = \frac{dx_2}{dx_1} = \frac{\dot{x}_2}{\dot{x}_1} = \frac{x_1 - 2x_2}{-x_1} \tag{9.40a}$$

Introducing the parameter $r = x_2/x_1$ to identify each isocline, we have

$$S = -1 + 2r \tag{9.40b}$$

Now we consider various values of r: The x_1 axis corresponds to $r = 0$ (since $x_2 = 0$), hence $S = -1$. This family of curves is a family of straight lines of -1 slope, as shown in Fig. 9.19b, sketched over the x_1-axis as short strokes of slope $= -1$. The arrows on these strokes have been drawn to coincide with the direction of motion. The same direction of motion is also indicated by the arrows on the state trajectories. The direction of motion may be obtained from the state equation. Take, for example, $r = -1$. The slope is $S = -3$, as sketched. But

$$\frac{x_2}{x_1} = -1 \Rightarrow x_2 = -x_1 \tag{9.40c}$$

Substituting equation (9.40c) in one of the state equations (the second one, for instance), we get

$$\dot{x}_2 = x_1 - 2x_2 = -x_2 - 2x_2 = -3x_2 \tag{9.40d}$$

Equation (9.40d) indicates that x_2 decreases with time when x_2 is positive and increases with time when x_2 is negative. The arrows on $r = -1$ are simply drawn to reflect this information. Actually, for this particular problem, the first state equation will determine the direction of motion for most of the isoclines. Using the procedure described, we sketch other isoclines corresponding to a sufficient number of values of r (sufficient to yield, on inspection, the pattern of trajectories), as shown. Finally, the trajectories (shown as solid lines) are sketched, starting from various arbitrary initial states and using the slopes as a guide. The trajectory corresponding to the initial state given in equation (9.39a) is also identified.

Some important observations can be made from Fig. 9.19b:

1. None of the trajectories cross one another. It follows that the solution to the differential equation (9.39a) as illustrated in the state plane is unique.
2. Any motion starting on an eigenvector ($\mathbf{X}_0 = [0 \quad 6]^T$ or $\mathbf{X}_0 = [6 \quad 6]^T$, for example) remains on that eigenvector (to the origin).
3. For this example, all trajectories converge to the origin (in other cases, all trajectories may, for instance, diverge from the origin).
4. Near the origin (the terminal state) all trajectories are tangent to one of the eigenvectors—the one representing the slow mode ($r = 1$). This agrees with what we said previously about fast and slow modes: The slow modes dominate the final response of a stable system.

The Case of a Controlled System   The first observation depends somewhat on the fact that a free system was considered. Introducing a forcing input implies a means of forcing the system to move out of one trajectory into another. (The preceding analysis will not hold in this case unless the forcing inputs are functions of the state variables or can be treated essentially as extra state variables).

Example 9.11   Trajectories for Proportional Feedback Control

Consider the system of Fig. 9.19 under control such that

$$\dot{x}_1 = -x_1 + u$$
$$\dot{x}_2 = x_1 - 2x_2 \qquad\qquad\qquad (9.41)$$
$$y = x_2$$

Determine the phase plane trajectories for proportional feedback control with a setpoint of 4. That is, $u = k(4 - y) = k(4 - x_2)$. Let $k = 6$.

Solution: The controlled system is

$$\dot{x}_1 = -x_1 + kx_2 + 4k \qquad\qquad\qquad (9.42a)$$
$$\dot{x}_2 = x_1 - 2x_2 \qquad\qquad\qquad (9.42b)$$

The new equilibrium point obtained by setting the LHS to zero is

$$x_{1e} = \frac{8k}{2 + k}, x_{2e} = \frac{4k}{2 + k} \qquad\qquad\qquad (9.43)$$

For $k = 6$, $x_{1e} = 6$, $x_{2e} = 3$. Note that $x_{2e} = 3$ is different from the desired value or setpoint of 4. The difference, as shown in Fig. 9.20, is known as *offset* (see Section 11.1). From equation (9.43), the effect can be reduced to zero for this type of control, only in the limit that $k \longrightarrow \infty$.

The transformation

$$\hat{x}_1 = x_1 - 6; \qquad \hat{x}_2 = x_2 - 3$$

gives a coordinate system with origin at the equilibrium point. From equations (9.42a) and (9.42b), the new system of equations is

$$\frac{d}{dt}\hat{x}_1 = -\hat{x}_1 - 6\hat{x}_2$$

$$\frac{d}{dt}\hat{x}_2 = \hat{x}_1 - 2\hat{x}_2$$

We define

$$r = \frac{\hat{x}_2}{\hat{x}_1} \quad \text{and} \quad S = \frac{d\hat{x}_2/dt}{d\hat{x}_1/dt}$$

As before,

$$S = \frac{2r - 1}{6r + 1}$$

Figure 9.20

Trajectories for Proportional Feedback Control

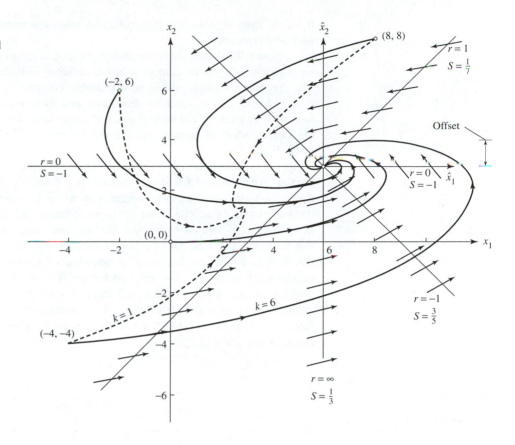

The trajectories can be sketched in the new coordinate system according to the previous procedure. The original coordinates, x_1 and x_2, can still be used as shown by observing that the origin of the new coordinates is at the point $(6, 3)$ of the original coordinates.

We can clearly see the behavior of the controlled system from Fig. 9.20. It is evident, for instance, that the system is stable but that this controller cannot quite drive the system to the desired final value (there is instead an offset). Also, by comparing the new trajectories with the uncontrolled system trajectories of Fig. 9.19 we can see that the control has indeed transformed the system behavior. For example, a new value of the controller parameter k will result in a new set of trajectories, which could intersect the old ones, so that a general control could move the system across trajectory lines. A few trajectories for $k = 1$ are shown by dashed lines in Fig. 9.20. Note the new equilibrium state and corresponding larger offset.

Application to Some Nonlinear Control Systems

As another illustration of controlled system behavior as well as the application of the above methods to nonlinear systems, let us consider the phenomenon of saturation in the proportional controller just discussed. As we mentioned in Section 6.1, *saturation* is a common characteristic of engineering actuators. A valve or piston-cylinder that reaches the end of its travel while the controlling input is still finite is experiencing saturation. Electronic amplifiers saturate, an open tank saturates when it begins to overflow, and even digital computers have maximum and minimum numbers that they can normally represent.

Figure 9.21 shows a (time domain) block diagram of the controlled system of Example 9.11 but with saturation in the actuator isolated from the controller block. The figure also shows other nonlinearities common with physical actuators and *on-off control systems* such as *dead zone* and *hysteresis*. On-off or *bang-bang* type control systems are a consequence of a discontinuous or switching control law (recall the classification of control systems summarized in Table 1.3). A familiar on-off controller is the space heater control system that is discussed at the beginning of Chapter 11. Dead zone or *dead band,* which in hydraulic actuators can result from overlap in valves and friction in pistons, is a threshold over which an input change causes no output change. A nonlinearity in which two output values are obtained for the same input value, one while the input is increasing and the other while the input is decreasing, is termed hysteresis. Sticking and backlash in mechanisms often give rise to hysteresis. *Coulomb friction* is another very common nonlinearity, although this fact is hidden by the usual assumption of viscous friction made in order to obtain a linear model. The relay behavior (Fig. 9.21c) that represents the actual physical device used in the on-off control system—an electromagnetic relay, for example—is also a fair description of the relation between the Coulomb frictional force and the relative velocity when the normal force between the sliding surfaces is constant (equation (2.22)).

Figure 9.21

Examples of Typical Actuator Nonlinearities

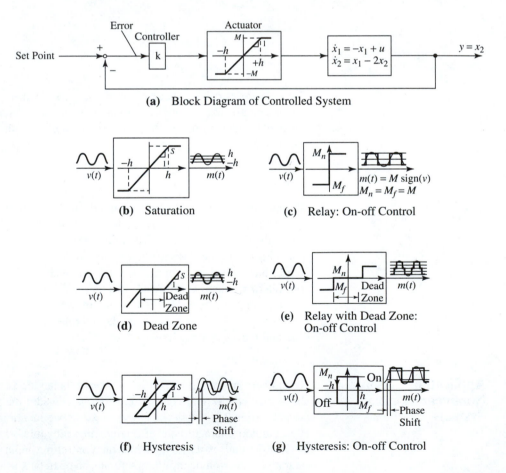

(a)   Block Diagram of Controlled System

(b)   Saturation

(c)   Relay: On-off Control

(d)   Dead Zone

(e)   Relay with Dead Zone: On-off Control

(f)   Hysteresis

(g)   Hysteresis: On-off Control

Returning to the problem of Example 9.11, we note that depending on the value of x_2, there are three possible expressions for u that give three sets of trajectory patterns that apply in the respective ranges of x_2. That is, for a setpoint of 4 and $k = 6$, we have

$$4 - x_2 \geq \frac{M}{6}; \qquad u = M \tag{9.44a}$$

$$\frac{-M}{6} < 4 - x_2 < \frac{M}{6}; \qquad u = 6(4 - x_2) \tag{9.44b}$$

$$4 - x_2 \leq \frac{-M}{6}; \qquad u = -M \tag{9.44c}$$

If we suppose $M = 9$, then the boundary values of x_2 or the switching lines from one trajectory pattern to the other are $x_2 = 2.5$ and $x_2 = 5.5$. The trajectories for the middle region (9.44b) are exactly as previously developed in Example 9.11 and sketched in Fig. 9.20. For the first region (9.44a), that is, $x_2 \leq 2.5$, the first two of equations (9.41) become

$$\dot{x}_1 = -x_1 + 9 \tag{9.45a}$$

$$\dot{x}_2 = x_1 - 2x_2 \tag{9.45b}$$

The equilibrium point for the new system is (9,4.5) and a transformation to the coordinates

$$\hat{x}_1 = x_1 - 9 \quad \text{and} \quad \hat{x}_2 = x_2 - 4.5$$

gives

$$\frac{d}{dt}\hat{x}_1 = -\hat{x}_1$$

$$\frac{d}{dt}\hat{x}_2 = \hat{x}_1 - 2\hat{x}_2$$

which is precisely the original uncontrolled system (9.39a). Thus the trajectories for this region are the same as those shown in Fig. 9.19 except that they are shifted to an origin at (9,4.5). Similarly, for the third region, $x_2 \geq 5.5$, the origin is $(-9,4.5)$ and the trajectory pattern is again the same as in Fig. 9.20. The final result is shown in Fig. 9.22a. The controlled system behavior including the switching between trajectory patterns is again clear. In Fig. 9.22b is shown the situation for a more severe saturation, that is, with $M = 4.8$. Here the trajectories converge to the equilibrium point corresponding to the upper saturation limit, which is outside the active control region (between the two switching lines) and independent of the controller parameter k.

Finally, we consider on-off control with hysteresis (Fig. 9.21g) of the system (9.41) such that $M_n = 4.8$ ("on"), $M_f = -4.8$ ("off"), and $h = 1.5$. We again assume feedback control of x_2 with the desired value, this time equal to 0.5. Under these conditions, the switching function (input to the on-off control element) is $v(t) = 0.5 - x_2$, and the desired state is ($x_{1d} = 1.0$, $x_{2d} = 0.5$). During the on period, $u = 4.8$, so that equations (9.41a) and (9.41b) give the trajectory pattern of Fig. 9.20 but with the origin at (4.8,2.4), as we discussed previously. These are the on trajectories. The dominant

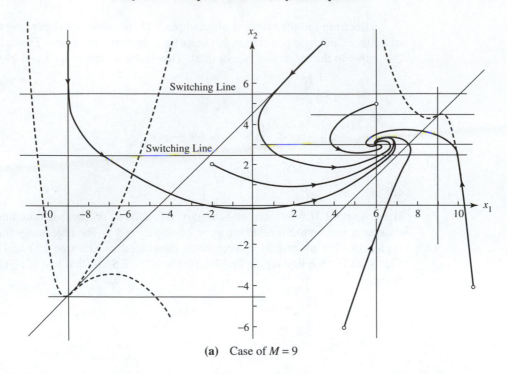

(a)   Case of $M = 9$

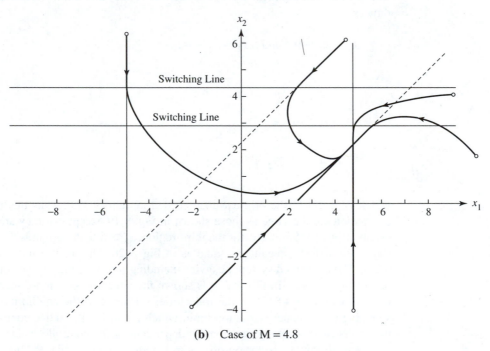

(b)   Case of M = 4.8

Figure 9.22

Trajectories for Proportional Control with Saturation

mode in this trajectory pattern is the 45° line through this origin (see Fig. 9.23). Similarly, when the switch is off, $u = -4.8$, so that again the off tracjectory pattern is the same as that of the trajectories of the controlled system but with origin at $(-4.8, 2.4)$. The dominant mode for this case is also shown in Fig. 9.23.

Figure 9.23

On-off Control with Hysteresis and Limit Cycling

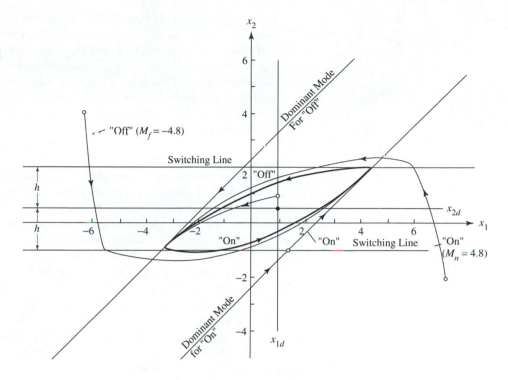

Let us consider now an initial state of $(1.4, -1)$ that is on the dominant on trajectory. If the switch is on at this point, the system will remain on this trajectory until the switching line $x_2 = 2$, that is, $(0.5 - x_2) = -h$, is reached. The controller then switches to off and the system subsequently follows an off trajectory until the lower switching line, $x_2 = -1$ or $(0.5 - x_2) = h$, is reached. The controller then switches back to on, and the system next follows an on trajectory that passes through this point. From Fig. 9.23 we see that for any initial state, all trajectories quickly converge to a closed loop formed by an on and an off trajectory around the desired state. This includes initial states such as $(1,1)$ within the loop. Such a loop that represents a steady-state oscillation of the system is known as a *limit cycle*. The limit cycle in this example is said to be "stable" in the sense that all trajectories, whether from within or without the loop, converge to the limit cycle (from which they cannot subsequently escape). Limit cycles are fairly common phenomena in nonlinear control systems, and the state space representation (such as Fig. 9.23) gives a very useful analytical tool particularly for systems of low order. Note that with this controller neither the desired state nor, indeed, any state within the limit cycle can be reached from any state outside or on the limit cycle.

Controllability and Observability of Dynamic Systems

The preceding examples illustrate the important idea of forcing inputs (control) and their capacity to move the system about in state space from trajectory to trajectory.

Meaning of Controllability

A system is said to be *controllable* if it can be moved from any state $\mathbf{X}(t_0)$ at $t = t_0$ to any other desired state $\mathbf{X}(t_1)$ in a finite time interval ($\tau = t_1 - t_0$) by applying a piecewise continuous input $\mathbf{U}(t)$, where t is contained in $[t_0, t_1]$.

By a suitable change of variables in the system, we can also speak of controllability in terms of being able to drive the system to the origin from any initial state. More vividly, in state space, *controllability implies reachability of all state space.* A system is uncontrollable if there is a section of state space to which it cannot be driven (or from which it cannot escape). This will be the case when some state variables are not affected by the control inputs. In other words, *a system is uncontrollable if not every state variable is affected by the control inputs.* Consider the Example of Fig. 9.19, and suppose the control system is such that x_1 unaffected by the inputs. Then for any initial state on the eigenvector $(\mathbf{V}_2)$ trajectory $r = -\infty$, the control will never be able to move the system off this trajectory!

Note that controllability as defined here depends only on the physical structure of the system as long as there are no severe constraints on the forcing inputs applied to the system. Thus in Fig. 9.20, although the proportional controller with a modest value of the parameter k is unable to drive the system precisely to any desired final state, a sufficiently large value of k or another type of controller can overcome this deficiency. Similarly, the results shown in Fig. 9.23 do not imply that a suitable on-off control that can move the system to any set point cannot be found. Indeed, as we shall see below, the system considered in these examples is completely controllable. However, the lesson in these results is that practical limitations in the controller such as saturation and hysteresis could cause a system not to be *completely controllable* (unable to reach some regions of state space) even though structurally they should be fully controllable. This is particularly true for linear systems, which are almost never completely controllable when the control vector is restricted to a compact set (*bounded-* as opposed to *infinite-control;* see [15]).

The Diagonal Case   It is not practical, in general, to determine from mere inspection of the system equations or diagrams whether the control reaches all state variables because of the coupling among these variables. However, as we saw in Section 6.3, the (modal) state variables are completely decoupled in a diagonal system. It follows that one way to explore controllability in a nondiagonal system is to apply a linear transformation to diagonal form as discussed in Chapter 6.

Consider the system of Example 6.16:

$$\dot{\mathbf{X}} = \begin{bmatrix} -3 & 4 & 4 \\ 1 & -3 & -1 \\ -1 & 2 & 0 \end{bmatrix} \mathbf{X} + \mathbf{BU}; \qquad y = \mathbf{CX} \tag{9.46}$$

With

$$\mathbf{B} = \begin{bmatrix} 2 \\ -\dfrac{1}{2} \\ 1 \end{bmatrix} \text{ and } \mathbf{C} = \begin{bmatrix} -1 & 1 & 0 \end{bmatrix} \tag{9.47a}$$

The transformation matrix and its inverse were found to be

$$\mathbf{T} = \begin{bmatrix} 1 & 0 & 1 \\ -1 & 1 & \dfrac{1}{2} \\ 1 & -1 & 0 \end{bmatrix}; \qquad \mathbf{T}^{-1} = \begin{bmatrix} 1 & -2 & -2 \\ 1 & -2 & -3 \\ 0 & 2 & 2 \end{bmatrix} \tag{9.47b}$$

so that the diagonalized system was given as

$$\frac{d}{dt}\hat{x}_1 = -3\hat{x}_1 + u \tag{9.47c}$$

$$\frac{d}{dt}\hat{x}_2 = -2\hat{x}_2$$

$$\frac{d}{dt}\hat{x}_3 = -\hat{x}_3 + u$$

$$y = \hat{\mathbf{C}}\hat{\mathbf{X}}$$

We see that the modal state variable $\hat{x}_2$ is not affected by the input, so that this system with the **B** matrix as in equation (9.47a) is uncontrollable. Note that in equation (9.47c), the **B** matrix is

$$\hat{\mathbf{B}} = \begin{bmatrix} 1 \\ 0 \\ 1 \end{bmatrix} \tag{9.47d}$$

so that the mode corresponding to an all-zero row in $\hat{\mathbf{B}}$ will not be affected by the control input(s). Consider now a new **B** matrix,

$$\mathbf{B} = \begin{bmatrix} 1 \\ 2 \\ 0 \end{bmatrix} \tag{9.48a}$$

This represents a new system, since the **B** matrix is part of the system structure. The corresponding $\hat{\mathbf{B}}$ matrix in the diagonal form is, by equation (6.78d),

$$\hat{\mathbf{B}} = \mathbf{T}^{-1}\mathbf{B} = \begin{bmatrix} 1 & -2 & -2 \\ 1 & -2 & -3 \\ 0 & 2 & 2 \end{bmatrix}\begin{bmatrix} 1 \\ 2 \\ 0 \end{bmatrix} = \begin{bmatrix} -3 \\ -3 \\ 4 \end{bmatrix} \tag{9.48b}$$

so that the new system is controllable. In general, *a linear system is controllable if no row of its $\hat{\mathbf{B}}$ matrix contains all zeros.*

Modal Domain State Space and Observability

In the state space with the modal variables as the coordinates, Fig. 9.24, we can see that for the uncontrollable system, $\hat{\mathbf{B}}u$, which represents a vector in the $(\hat{x}_1, \hat{x}_3)$ plane, is always perpendicular to $\hat{x}_2$. Consequently $\hat{x}_2$ cannot be influenced by this control. In contrast, $\hat{\mathbf{B}}u$ for the controllable system is not normal to any modal state coordinate.

Observability   We have also shown in Fig. 9.24 the measurement vector $(\hat{\mathbf{C}})^T$ (designated $\mathbf{C}_1$), where $\hat{\mathbf{C}}$ is given by equation (6.78e) as

$$\hat{\mathbf{C}} = \mathbf{CT} = \begin{bmatrix} -1 & 1 & 0 \end{bmatrix}\mathbf{T} = \begin{bmatrix} -2 & 1 & -\frac{1}{2} \end{bmatrix} \tag{9.49a}$$

Figure 9.24

Controllability and Observability Concepts in Model Domain State Space

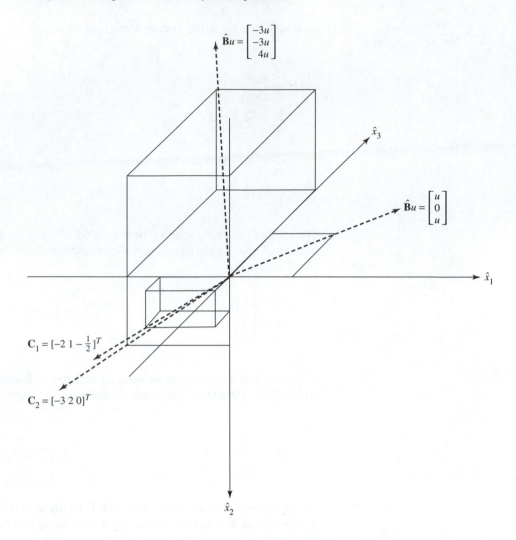

$$\hat{\mathbf{B}}u = \begin{bmatrix} -3u \\ -3u \\ 4u \end{bmatrix}$$

$$\hat{\mathbf{B}}u = \begin{bmatrix} u \\ 0 \\ u \end{bmatrix}$$

$\hat{x}_3$

$\hat{x}_1$

$\mathbf{C}_1 = [-2 \; 1 \; -\tfrac{1}{2}]^T$

$\mathbf{C}_2 = [-3 \; 2 \; 0]^T$

$\hat{x}_2$

The measurement $y = \hat{\mathbf{C}}\hat{\mathbf{X}}$ can be viewed as the projection of the state vector $\hat{\mathbf{X}}$ on the vector $(\hat{\mathbf{C}})^T$. We can see that for $\hat{\mathbf{C}}$ as given by equation (9.49a), all the modal state variables can be projected on the measurement vector. Suppose on the other hand that $\mathbf{C} = [-1 \quad 2 \quad 0]$, so that

$$\hat{\mathbf{C}} = [-1 \quad 2 \quad 0]\mathbf{T} = [-3 \quad 2 \quad 0] \tag{9.49b}$$

$(\hat{\mathbf{C}})^T$ (designated $\mathbf{C}_2$) is also shown in Fig. 9.24. This vector lies in the $(\hat{x}_1, \hat{x}_2)$ plane and is orthogonal to $\hat{x}_3$, so that $\hat{x}_3$ has no projection on $\mathbf{C}_2$. The third mode is not *observable* (measurable) in this case. *A system is observable if all its states are derivable (directly or indirectly) from the output vector of the system.* Hence, from these considerations, a linear system is observable if no column of the (diagonal form) $\hat{\mathbf{C}}$ matrix is all zero. Further, the number of all-zero columns in the $\hat{\mathbf{C}}$ matrix gives the number of *unobservable modes*.

The preceding conditions, which are necessary and sufficient for the controllability and observability of linear systems in the diagonal form (modal domain), are further illustrated in Fig. 9.25 using signal flow graphs.

Figure 9.25

Illustration with Modal Domain Signal Flow

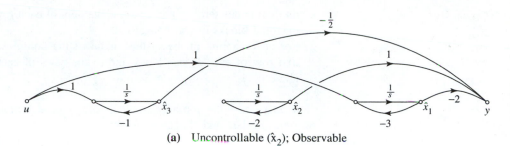

(a) Uncontrollable ($\hat{x}_2$); Observable

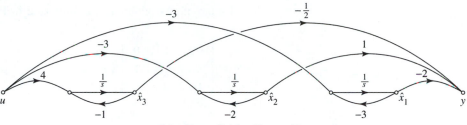

(b) Controllable; Observable

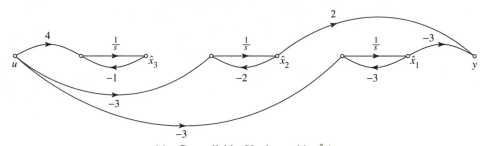

(c) Controllable; Unobservable ($\hat{x}_3$)

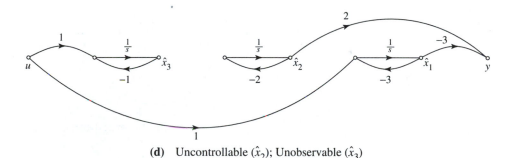

(d) Uncontrollable ($\hat{x}_2$); Unobservable ($\hat{x}_3$)

The Jordan Form

Recall that the diagonal block is a special case of the Jordan canonical form. By observing carefully the structure of the Jordan block, equation (6.81b), we can deduce the following conditions for controllability and observability:

1. A linear system in Jordan canonical form is controllable if at least one element of $\hat{\mathbf{B}}$ in the row corresponding to the bottom row of each Jordan block (the rest depend on this) is not zero, and at least one element in each row of $\hat{\mathbf{B}}$ corresponding to the diagonal blocks is also nonzero.

2. For observability of the system, none of the columns of $\hat{\mathbf{C}}$ corresponding to the first row of each Jordan block and to the rows of each diagonal block should be all zero.

Note that since the Jordan canonical form (see equation (6.81a)) can be made up of pure Jordan and diagonal blocks, the necessary and sufficient conditions for controllability and observability of linear systems in this form consist of specifications for the Jordan blocks and the previously derived conditions for the diagonal blocks.

The Jordan block conditions are illustrated in Fig. 9.26 with the system

$$\frac{d}{dt}\hat{\mathbf{X}} = \begin{bmatrix} -2 & 1 & 0 \\ 0 & -2 & 1 \\ 0 & 0 & -2 \end{bmatrix} \hat{\mathbf{X}} + \hat{\mathbf{B}}u; \qquad y = \hat{\mathbf{C}}\hat{\mathbf{X}} \qquad \text{(9.50a)}$$

Figure 9.26

Controllability/observability in a Jordan Block

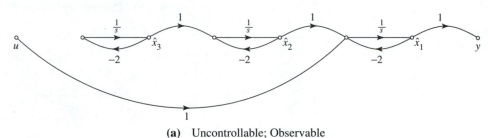

(a) Uncontrollable; Observable

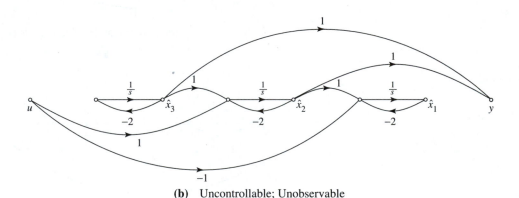

(b) Uncontrollable; Unobservable

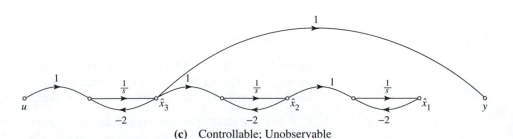

(c) Controllable; Unobservable

for the following cases:

a. $\quad \hat{\mathbf{B}} = \begin{bmatrix} 1 \\ 0 \\ 0 \end{bmatrix}; \qquad \hat{\mathbf{C}} = \begin{bmatrix} 1 & 0 & 0 \end{bmatrix}$ (9.50b)

b. $\quad \hat{\mathbf{B}} = \begin{bmatrix} 1 \\ -1 \\ 0 \end{bmatrix}; \qquad \hat{\mathbf{C}} = \begin{bmatrix} 0 & 1 & 1 \end{bmatrix}$ (9.50c)

c. $\quad \hat{\mathbf{B}} = \begin{bmatrix} 0 \\ 0 \\ 1 \end{bmatrix}; \qquad \hat{\mathbf{C}} = \begin{bmatrix} 0 & 0 & 1 \end{bmatrix}$ (9.50d)

The General Case

The controllability and observability criteria based on the canonical transformations may require substantial computational effort, since the system eigenvalues and eigenvectors all have to be computed prior to transformation. The necessary and sufficient conditions for the controllability/observability of the general linear system, equation (6.49), can be derived on the basis of the expansion of the state transition matrix $e^{\mathbf{A}t}$, reference [13]. Consider the solution to equation (6.49a):

$$\mathbf{X}(t_1) = e^{\mathbf{A}(t_1 - t_0)}\mathbf{X}(t_0) + \int_{t_0}^{t_1} e^{\mathbf{A}(t_1 - \tau)}\mathbf{B}\mathbf{U}(\tau)\, d\tau \tag{9.51a}$$

Changing the definitions of the limits of integration, we have

$$\mathbf{X}(t_0) = e^{\mathbf{A}(t_0 - t_1)}\mathbf{X}(t_1) + \int_{t_1}^{t_0} e^{\mathbf{A}(t_0 - \tau)}\mathbf{B}\mathbf{U}(\tau)\, d\tau$$

We can always define $\mathbf{X}(t_1) = 0$ (by a change of variables in the original system, for instance) and take

$$\mathbf{X}(t_0) = -\int_{t_0}^{t_1} e^{\mathbf{A}(t_0 - \tau)}\mathbf{B}\mathbf{U}(\tau)\, d\tau \tag{9.51b}$$

Now, $e^{\mathbf{A}t}$ is given by equation (6.59) and also by the Cayley-Hamilton theorem and Sylvester formula (see [3] and Appendix B, equation (B4.4d)):

$$e^{\mathbf{A}t} = \sum_{i=0}^{n-1} \gamma_i(t)\mathbf{A}^i \tag{9.51c}$$

Substituting (9.51c) in (9.51b), we obtain

$$\mathbf{X}(t_0) = -\int_{t_0}^{t_1} \sum_{i=0}^{n-1} \gamma_i(t_0 - \tau)\mathbf{A}^i\mathbf{B}\mathbf{U}(\tau)\, d\tau \tag{9.51d}$$

If we define the $n \times (n \times r)$ matrix

$$\mathbf{M} = \begin{bmatrix} \mathbf{B}; & \mathbf{A}\mathbf{B}; & \mathbf{A}^2\mathbf{B}; & \cdots & \mathbf{A}^{n-1}\mathbf{B} \end{bmatrix} \tag{9.52}$$

and the $(n \times r)$ vector

$$\mathbf{W} = \begin{bmatrix} \mathbf{W}_0 \\ \mathbf{W}_1 \\ \cdot \\ \cdot \\ \mathbf{W}_{n-1} \end{bmatrix} = -\int_{t_0}^{t_1} \gamma_i(t_0 - \tau)\mathbf{U}(\tau)\, d\tau, \quad i = 0, \dots, n-1 \tag{9.53a}$$

where each $\mathbf{W}_i$ is an r-vector, equation (9.51d) reduces to

$$\mathbf{X}(t_0) = \mathbf{MW} \tag{9.53b}$$

The linear system (6.49a) and (6.49b) is controllable if the system can be driven, in finite time, to $\mathbf{X}(t_0)$ from any $\mathbf{X}(t_1)$, that is, if n independent scalar equations satisfy the matrix equation (9.53b)—in other words, if equation (9.53b) can be solved for an independent set of state variables $\mathbf{X}(t_0)$. The condition for this (that is, *consistency* of the nonhomogeneous system (9.53b); see Appendix B) is

$$\text{rank } \mathbf{M} = n \tag{9.54}$$

Further, if $\mathbf{M}$ is a square matrix, the condition (9.54) is equivalent to requiring that the determinant of $\mathbf{M}$ be nonzero. In general, the rank of $\mathbf{M}$ is equal to the number of controllable modes of the system.

Consider the system of Example 9.11, equations (9.41),

$$\mathbf{A} = \begin{bmatrix} -1 & 0 \\ 1 & -2 \end{bmatrix}; \qquad \mathbf{B} = \begin{bmatrix} 1 \\ 0 \end{bmatrix}$$

so that

$$[\mathbf{B} \quad \mathbf{AB}] = \begin{bmatrix} 1 & -1 \\ 0 & 1 \end{bmatrix}$$

The determinant is 1. Since $[\mathbf{B} \quad \mathbf{AB}]$ is square, its rank is 2, so that the system is completely controllable.

Observability   For the case of observability, it is sufficient to consider the free system, since the input vector is (considered) always measurable. Thus $\mathbf{X}(t) = e^{\mathbf{A}(t-t_0)}\mathbf{X}(t_0)$ and

$$\mathbf{Y}(t) = \mathbf{C}e^{\mathbf{A}(t-t_0)}\mathbf{X}(t_0) \tag{9.55a}$$

Using equations (9.51c) and (9.55a), we obtain

$$\mathbf{Y}(t) = \sum_{i=0}^{n-1} \gamma_i(t - t_0)\mathbf{CA}^i\mathbf{X}(t_0) = \sum_{i=0}^{n-1} \mathbf{CA}^i\gamma_i(t - t_0)\mathbf{X}(t_0) \tag{9.55b}$$

If we define then $n \times (n \times m)$ matrix

$$\mathbf{L} = [\mathbf{C}^T; \quad \mathbf{A}^T\mathbf{C}^T; \quad (\mathbf{A}^T)^2\mathbf{C}^T; \quad \cdots \quad (\mathbf{A}^T)^{n-1}\mathbf{C}^T] \tag{9.56}$$

and

$$\mathbf{V} = [\gamma_0; \quad \gamma_1; \quad \cdots; \quad \gamma_{n-1}] \tag{9.57a}$$

equation (9.55b) becomes

$$\mathbf{Y}(t) = \mathbf{VL}^T\mathbf{X}(t_0)$$

The derivation of all elements of the state vector $\mathbf{X}(t_0)$ from the measurement $\mathbf{Y}(t)$, that is, the observability of the system (6.49a) and (6.49b) requires that

$$\text{rank } \mathbf{L} = n \tag{9.58}$$

In general, the rank of $\mathbf{L}$ gives the number of observable modes of (6.49a) and (6.49b).

The conditions (9.54) and (9.58) are also the criteria for the controllability and observability, respectively, of the linear discrete-time system (6.90a) and (6.90b) provided that in equations (9.52) and (9.56), $\mathbf{A}$ is replaced by $\mathbf{P}$, and $\mathbf{B}$ is replaced by $\mathbf{Q}$. Observe that for $m = 1$ (single output), $\mathbf{L}$ is square and for the discrete-time case is equal to the transpose of $\mathbf{T}^{-1}$ given in equation (9.37a), so that the condition for the existence of $\mathbf{T}$ is indeed the observability of the system (9.36a).

Example 9.12 Controllability and Observability of a Linear System

Examine the controllability and observability of the system

$$\dot{\mathbf{X}} = \begin{bmatrix} 1 & -1 \\ -1 & 1 \end{bmatrix}\mathbf{X} + \begin{bmatrix} 2 \\ 2 \end{bmatrix}u; \qquad y = [1 \quad 1]\mathbf{X}$$

a. By transforming the system into canonical form.
b. By the general criteria.

Solution a: The eigenvalues can be shown to be 0 and 2. From equation (6.67b), the eigenvectors are

$$\mathbf{V}_1 = \begin{bmatrix} 1 \\ 1 \end{bmatrix} \quad \text{and} \quad \mathbf{V}_2 = \begin{bmatrix} 1 \\ -1 \end{bmatrix}$$

so that the transformation matrix and inverse are

$$\mathbf{T} = \begin{bmatrix} 1 & 1 \\ 1 & -1 \end{bmatrix} \quad \text{and} \quad \mathbf{T}^{-1} = \begin{bmatrix} 1/2 & 1/2 \\ 1/2 & -1/2 \end{bmatrix}$$

This gives

$$\hat{\mathbf{B}} = \mathbf{T}^{-1}\mathbf{B} = \begin{bmatrix} 2 \\ 0 \end{bmatrix} \quad \text{and} \quad \hat{\mathbf{C}} = \mathbf{CT} = [2 \quad 0]$$

so that the system is both uncontrollable and unobservable. One of the system modes (the same one in this case) is neither controllable nor observable.

Solution b:

$$\mathbf{M} = [\mathbf{B}; \quad \mathbf{AB}] = \begin{bmatrix} 2 & 0 \\ 2 & 0 \end{bmatrix} \quad \text{and} \quad \mathbf{L} = [\mathbf{C}^T \quad \mathbf{A}^T\mathbf{C}^T] = \begin{bmatrix} 1 & 0 \\ 1 & 0 \end{bmatrix}$$

The matrices

$$\begin{bmatrix} 2 & 0 \\ 2 & 0 \end{bmatrix} \quad \text{and} \quad \begin{bmatrix} 1 & 0 \\ 1 & 0 \end{bmatrix}$$

are equivalent (see Appendix B) and have rank 1. Also, the determinant of each matrix is zero. Thus only one mode is controllable or observable, and the system is neither controllable nor observable.

Local Controllability of Nonlinear Systems

Although there are some controllability theories for nonlinear systems, such criteria are always for specific types or classes of nonlinear systems. For instance, reference [15] gives the sufficient conditions for complete controllability of *bilinear* systems (nonlinear systems with the general form

$$\dot{\mathbf{X}} = \mathbf{AX} + \sum_{k=1}^{r} \mathbf{N}_k u_k \mathbf{X} + \mathbf{BU}$$

where, as before, $\mathbf{X}$ and $\mathbf{U}$ are the n-dimensional state vector and the r-dimensional control vector, respectively; $\mathbf{A}$ and $\mathbf{B}$ are, respectively, $n \times n$ and $n \times r$ matrices of constants; and $\mathbf{N}_k$ ($k = 1, \ldots , r$) is an $n \times n$ constant matrix). In general, the foregoing developments can be applied to the determination of the controllability of a nonlinear system within the vicinity of an equilibrium point. According to Lee and Markus, reference [16], the nonlinear system

$$\dot{\mathbf{X}} = \mathbf{f}(\mathbf{X}, \mathbf{U}, t) \tag{9.59a}$$

is controllable if for some $\tilde{\mathbf{U}}(\mathbf{X})$, the system

$$\dot{\mathbf{X}} = \mathbf{f}(\mathbf{X}, \tilde{\mathbf{U}}(\mathbf{X})) \tag{9.59b}$$

is asymptotically stable, and the linear approximation of (9.59a) in a neighborhood of the origin (recall Linearization of the Vector State Equation, Section 6.3),

$$\dot{\mathbf{X}} = \mathbf{AX} + \mathbf{BU} \tag{9.60a}$$

where $\mathbf{A}$ and $\mathbf{B}$ are given by equations (6.50b), repeated here,

$$\mathbf{A} = \frac{\partial \mathbf{f}}{\partial \mathbf{X}}(\mathbf{0}, \mathbf{0}); \qquad \mathbf{B} = \frac{\partial \mathbf{f}}{\partial \mathbf{U}}(\mathbf{0}, \mathbf{0}) \tag{9.60b}$$

is controllable. Asymptotic stability of linear systems is discussed in Section 10.1. Stability of nonlinear systems can be investigated via a state space representation, as in the preceding discussion or by simulation or other techniques. The conditions under which (9.59b) can be made stable and the determination of an appropriate $\tilde{\mathbf{U}}(\mathbf{X})$ (stability control) are beyond the scope of the present discussion. In equation (9.60b), it is assumed that the final state (the desired state, or the equilibrium state) $\mathbf{X}_d$, of (9.59a) is the origin. Where this is not the case, then the change of variables $\hat{\mathbf{X}} = (\mathbf{X} - \mathbf{X}_d)$ should be made in the system description, equation (9.59a).

| **Example 9.13** | Controllability of a Moist Material Drying Process |

Consider the moist product drying model described in Example 4.4, equation (4.20), repeated here:

$$\dot{x}_1 = \left(-\tfrac{1}{2}m\right)x_2$$

$$\dot{x}_2 = (8\lambda^2 m)x_1 - (5\lambda)x_2 - (8\lambda^2 m)q_e \qquad (9.61)$$

$$\dot{x}_3 = \frac{1}{C_e}\left[\left(\frac{c_a + w_2 c_v}{1 + w_2}\right)\dot{m}_2(T_2 - x_3) + \left(\frac{c_w - c_v}{2}\right)x_2 x_3 - \left(\frac{H}{2}\right)x_2\right]$$

Determine the feasibility of controlling this system within the neighborhood of an operating point (x_{1d}, x_{2d}, x_{3d}) and nominal moist air flow $\dot{m}_d$, using only the equilibrium moisture content q_e and the humid air flow in the drying chamber $\dot{m}_2$ as control inputs. Assume that all other parameters are constants.

Solution: The system (9.61) was linearized in Example 6.12. With the transformation

$$\hat{\mathbf{X}} = \mathbf{X} - \mathbf{X}_d; \qquad \mathbf{U} = \begin{bmatrix} q_e - x_{1d} \\ \dot{m}_2 - \dot{m}_d \end{bmatrix}$$

equations (9.60a) and (9.60b) yield (see equations (6.55) and (6.56))

$$\frac{d}{dt}\hat{\mathbf{X}} = \mathbf{A}\hat{\mathbf{X}} + \mathbf{B}\mathbf{U}$$

where

$$\mathbf{A} = \begin{bmatrix} 0 & (-1/2m) & 0 \\ (8\lambda^2 m) & (-5\lambda) & 0 \\ 0 & \left(\dfrac{(c_w - c_v)x_{3d} - H}{2C_e}\right) & \left(\dfrac{(c_w - c_v)}{2C_e}x_{2d} - \dfrac{(c_a + w_2 c_v)}{C_e(1 + w_2)}\dot{m}_d\right) \end{bmatrix}$$

$$= \begin{bmatrix} 0 & a_1 & 0 \\ a_2 & a_3 & 0 \\ 0 & a_4 & a_5 \end{bmatrix}$$

$$\mathbf{B} = \begin{bmatrix} 0 & 0 \\ (-8\lambda^2 m) & 0 \\ 0 & \dfrac{(c_a + w_2 c_v)(T_2 - x_{3d})}{C_e(1 + w_2)} \end{bmatrix} = \begin{bmatrix} 0 & 0 \\ b_1 & 0 \\ 0 & b_2 \end{bmatrix}$$

If we assume the operating point (new origin) of the nonlinear model is asymptotically stable, the suggested controls will be feasible provided rank $[\mathbf{B} \quad \mathbf{AB} \quad \mathbf{A^2B}] = 3$. But

$$[\mathbf{B} \quad \mathbf{AB} \quad \mathbf{A^2B}] = \begin{bmatrix} 0 & 0 & a_1 b_1 & 0 & a_1 a_3 b_1 & 0 \\ b_1 & 0 & a_3 b_1 & 0 & (a_1 a_2 + a_3^2) & 0 \\ 0 & b_2 & a_4 b_1 & a_5 b_2 & (a_3 a_4 + a_4 a_5) & a_5^2 \end{bmatrix}$$

which is equivalent to ($\Rightarrow$) (see Appendix B on computation of rank)

$$\begin{bmatrix} a_1b_1 & 0 & 0 & a_1a_3b_1 & 0 \\ a_3b_1 & b_1 & 0 & (a_1a_2 + a_3^2) & 0 \\ a_4b_1 & 0 & b_2 & (a_3a_4 + a_4a_5) & a_5^2 \end{bmatrix} \Rightarrow \begin{bmatrix} a_1b_1 & 0 & 0 & 0 & 0 \\ 0 & b_1 & 0 & 0 & 0 \\ 0 & 0 & b_2 & 0 & 0 \end{bmatrix} \Rightarrow \begin{bmatrix} 1 & 0 & 0 \\ 0 & 1 & 0 \\ 0 & 0 & 1 \end{bmatrix}$$

so that the rank is indeed 3.

On the Net

- Fig 9_7.m and boiler.m (MATLAB): source file for Fig. 9.7c
- SC11.MCD (Mathcad): computation of frequency response by Mathcad
- Fig9_15.m (MATLAB): source file for Fig. 9.15, Example 9.6
- Fig10_1.m (MATLAB): phase plane trajectories illustrating system stability; source files for Fig. 9.19b
- Auxiliary exercise: Practice Problems 9.19, (9.18, 9.23, 9.24, 9.25, 9.27 can be done with computer assistance), 9.39.

9.5 PRACTICE PROBLEMS

9.1 Find the closed-loop eigenvalues for the system whose block diagram is shown in Fig. 9.27.

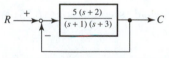

Figure 9.27 (Problem 9.1)

9.2 Reduce the block diagram shown in Fig. 9.28 to canonical form and find the expression for the output. If $k = 10$, what is the equivalent time constant for the overall system?

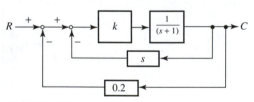

Figure 9.28 (Problem 9.2)

9.3* Find the effective time constant of the subsystem shown in Fig. 9.29 (in dashed lines). Determine the overall transfer function (C/R) for the diagram.

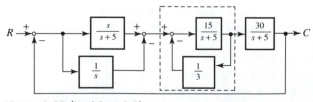

Figure 9.29 (Problem 9.3)

In Problems 9.4 to 9.7, find the overall transfer function (*C/R*) for the block diagram by **(a)** direct algebra and **(b)** by first reducing the system to canonical form.

9.4 The block diagram shown in Fig. 9.30.

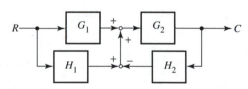

Figure 9.30 (Problem 9.4)

9.5 The given block diagram in Fig. 9.31.

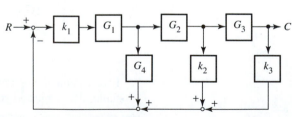

Figure 9.31 (Problem 9.5)

9.6* The feedforward system shown in Fig. 9.32.

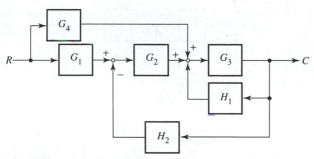

Figure 9.32 (Problem 9.6)

9.7 The given cascade system in Fig. 9.33.

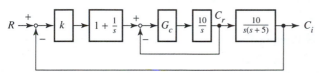

Figure 9.33 (Problem 9.7)

9.8 Find the following transfer functions for the block diagram shown in Fig. 9.34.

(a) C_r/R

(b) C_i/R

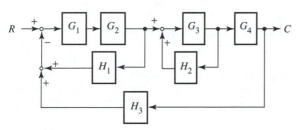

Figure 9.34 (Problem 9.8)

9.9 Determine for the given system of Fig. 9.35, C/R_2 by setting $R_1 = 0$.

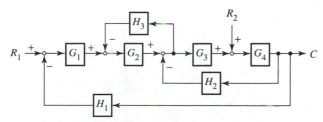

Figure 9.35 (Problem 9.9)

9.10 Draw the operational diagrams for the walking beam and hydraulic amplifier of Example 5.6. Combine these block diagrams into an overall representation of the hydraulic servomotor shown in a box in Fig. 5.21. What is the transfer function between y_2 and y_p?

9.11(a) Draw the operational diagram for each component of the room temperature control system shown in Fig. 5.23 (and in Fig. 5.24), using the relations given in Example 5.7. For the bimetallic element and lever include a block relating $V_{s,k}$ to $(\theta_d - \theta_k)$, and represent this relation with a small sketch of $V_{s,k}$ versus $(\theta_d - \theta_k)$ placed inside the box. Similarly indicate the relation between $i_{c,k}$ and $V_{s,k}$.

9.11(b) Combine the block diagrams from Problem 9.11a into an overall representation of the temperature control system of Fig. 5.23. Identify the output, the control inputs, and the disturbance inputs.

9.12* For small wheel angles ϕ, the rate of change of the heading of a car traveling at steady speed v is approximately proportional to ϕ:

$$\frac{d}{dt} h = \frac{v}{L} \phi$$

Construct a block diagram representation of the process between the sensing of the car's current deviation from desired heading ($h_d - h$) by the driver, his or her reaction in turning the steering wheel θ, and the subsequent heading of the car. Assume that the angle of turn is proportional to the deviation in heading and that the driver has a finite reaction time. Also, the wheel angle is proportional to θ but with some backlash (See Fig. 9.36.)

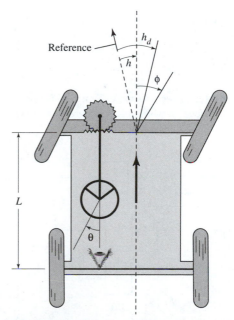

Figure 9.36 (Problem 9.12)

9.13 Draw the operational block diagram of the simple rolling mill described in Practice Problem 5.17, which includes the effect of the time of travel of the material between the rolling wheels and the sensing wheels. Assume that the rolling wheel reduces the material thickness instantaneously and that the hydraulic amplifier spool valve position (refer to Example 5.6) is the difference between the thumbscrew input and the walking beam input.

9.14 Consider reverse reaction systems arising out of two opposing processes as indicated. Determine in each case the condition for an inverse response to occur and the corresponding positive zero.

(a)* Two first-order processes,

$$G(s) = \frac{K_1}{\tau_1 s + 1} - \frac{K_2}{\tau_2 s + 1}$$

(b) Two first-order processes with dead time,

$$G(s) = \frac{K_1 e^{-T_1 s}}{\tau_1 s + 1} - \frac{K_2 e^{-T_2 s}}{\tau_2 s + 1}$$

(c) Second- and first-order processes,

$$G(s) = \frac{K_1}{T_1^2 s^2 + 2\zeta T_1 s + 1} - \frac{K_2}{T_2 s + 1}$$

(d) Two second-order processes,

$$G(s) = \frac{K_1}{T_1^2 s^2 + 2\zeta_1 T_1 s + 1} - \frac{K_2}{T_2^2 s^2 + 2\zeta_2 T_2 s + 1}$$

(e) Two second-order processes with dead time,

$$G(s) = \frac{K_1 e^{-T_1 s}}{\tau_1^2 s^2 + 2\zeta_1 \tau_1 s + 1} - \frac{K_2 e^{-T_2 s}}{\tau_2^2 s^2 + 2\zeta_2 \tau_2 s + 1}$$

9.15(a), (b), (c) Determine the overall transfer function $Y(z)/U(z)$ for each system whose block diagram is shown in Fig. 9.37.

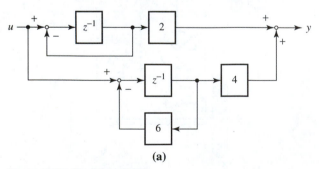

(a)

Figure 9.37 (Problem 9.15)

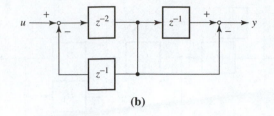

(b)

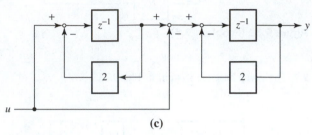

(c)

Figure 9.37 continued (Problem 9.15)

9.16(a), (b) Using the unit impulse generator of Example 9.2, draw a block diagram of a system that will produce the output pattern shown in Fig. 9.38.

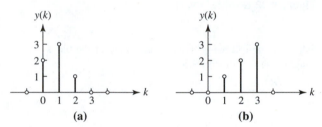

(a)   **(b)**

Figure 9.38 (Problem 9.16)

9.17(a), (b)* A signal generator produces a signal given by

$$u_k = \begin{cases} 0, & k < 0 \\ k, & 0 \le k \le 3 \\ 0, & k > 3 \end{cases}$$

Draw a block diagram of the system that will produce each of the output patterns shown in Fig. 9.39, using this signal source as the basic unit.

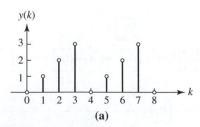

(a)

Figure 9.39 (Problem 9.17)

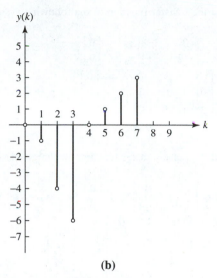

(b)

Figure 9.39 *continued* (**Problem 9.17**)

9.18 Sketch the Nyquist and Bode diagrams for

(a) $G_1(s) = \dfrac{1 - 5s}{(1 + 5)}$ and $G_2(s) = \dfrac{1 + 5s}{(1 + s)}$

Comment on the differences if any between the Bode diagrams for the two systems.

(b) $G_2(s) = \dfrac{1 + s}{s}$

9.19 Plot, using available computational software or manual means, the Bode diagram for

(a) $G(s) = \dfrac{5(1 + 0.5s)}{(1 + s)(1 + 0.2s)}$ **(b)** $G(s) = \dfrac{2}{s^2(2 + s)}$

(c) $G(s) = \dfrac{4e^{-0.5s}}{(1 + s)(1 + 10s)}$

(d) $G(s) = \dfrac{(1 + 2s)(2 - 3s)}{(2 + s)(1 + 3s)}$

9.20(a), (b), (c) Identify the system transfer function from the asymptotic gain characteristics shown in Fig. 9.40. Assume minimum phase systems.

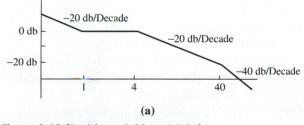

(a)

Figure 9.40 (Problems 9.20 and 9.21)

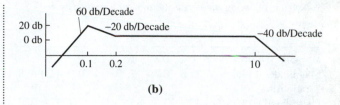

(b)

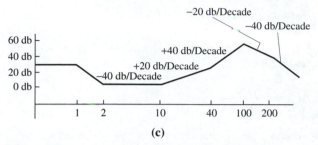

(c)

Figure 9.40 *continued* (**Problems 9.20 and 9.21**)

9.21 Construct the phase diagram for the minimum phase system with the given gain asymptotes in Fig. 9.40.

9.22* A minimum phase system has the Bode diagram phase shift asymptotes shown in Fig. 9.41. If the static gain is 1, identify the system transfer function, $G(s) = k[?]$.

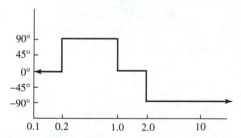

Figure 9.41 (Problem 9.22)

9.23 From the data obtained in a frequency response test, the following values were deduced for the input-output amplitude ratios at the corner frequencies:

ω (rad/s)	0.1	0.5	5.0	50
$\lvert G \rvert$ (db)	+48	+20	0	−40

(a) Sketch the Bode amplitude plot. (Assume that starting and ending slopes can be extrapolated.)

(b) Estimate the system transfer function.

(c) Sketch the Bode phase shift plot. What are your assumptions?

9.24 The following data was obtained in a frequency response test:

ω (rad/s)	0.1	0.2	0.4	0.8	1	2	4	6
Magnitude ratio (db)	14.0	13.6	11.6	8.30	7.20	1.90	−4.20	−8
Phase lag (°)	13	25	45	75	85	125	175	220

Estimate the transfer function for the system. *Hint:* Consider the presence of a delay.

9.25* A first-order system has the following frequency transfer function:

$$G(j\omega) = \frac{j\omega K}{1 + j\omega K}$$

where K is an adjustable parameter and ω is in rad/s.

(a) Sketch the Bode diagram (gain in db, and phase $\omega = 0.1$ to 10) for $K = 1$.

(b) Sketch the Nyquist diagram ($\omega = 0$ to ∞; indicate direction of increasing ω) for $K = 1$.

(c) Find a value of K that will permit this system to be used effectively as a filter in a piece of electronic equipment to suppress a 0.143-Hz low-frequency hum. Assume that a low-frequency noise can be considered suppressed if the ratio of the output noise amplitude to the input noise is less than 0.25.

9.26 Show that the Nyquist plot for the transfer function shown in Fig. 9.14b should indeed be a semicircle.

9.27 The transfer function for a pollution-monitoring instrument is given by

$$G(s) = \frac{e^{-0.3s}}{(1 + 0.1s)(1 + s)}$$

(a) Determine the frequency response of the instrument in the Bode format.

(b) What is the bandwidth of the instrument?

9.28 Convert the following nth-order ordinary differential equation into a vector differential equation ($\dot{\mathbf{X}} = \mathbf{AX} + \mathbf{B}u$) with n state variables. That is, specify $\mathbf{X}$, $\mathbf{A}$, $\mathbf{B}$, and u for

(a) $\dfrac{d^3y}{dt^3} + 6\dfrac{d^2y}{dt^2} + 11\dfrac{dy}{dt} + 5y = f$

(b) $\dfrac{d^3y}{dt^3} + 4\dfrac{d^2y}{dt^2} + 5\dfrac{dy}{dt} + 2y = 3\sin 3t$

(c) $\dfrac{d^4y}{dt^4} + 10\dfrac{d^3y}{dt^3} + 35\dfrac{d^2y}{dt^2} + 50\dfrac{dy}{dt} + 24 = \cos t$

9.29* Convert the third-order ODE of Problem 6.24(ii) into a system of three first-order equations. Indicate the resulting $\mathbf{A}$ and $\mathbf{B}$ matrices.

9.30 Find the transfer function of the following systems:

(a) $\begin{bmatrix} \dot{x}_1 \\ \dot{x}_2 \end{bmatrix} = \begin{bmatrix} 0 & 1 \\ -2 & -3 \end{bmatrix} \begin{bmatrix} x_1 \\ x_2 \end{bmatrix} + \begin{bmatrix} 0 \\ 9 \end{bmatrix} u; \qquad y = x_1$

(b) $\dot{\mathbf{X}} = \begin{bmatrix} 0 & 1 & 0 \\ 0 & 0 & 1 \\ -4 & -6 & -4 \end{bmatrix} \mathbf{X} + \begin{bmatrix} 0 \\ 0 \\ 1 \end{bmatrix} u;$

$y = \begin{bmatrix} 1 & 0 & 0 \end{bmatrix}\mathbf{X}$

9.31 Find the pulse transfer function for the given system.

(a) $\mathbf{X}_{k+1} = \begin{bmatrix} 0 & 1 \\ -0.5 & 1.5 \end{bmatrix} \mathbf{X}_k + \begin{bmatrix} 0.1 \\ 0 \end{bmatrix} u_k; \quad y_k = \begin{bmatrix} 1 & 0 \end{bmatrix}\mathbf{X}_k$

(b)* $\mathbf{X}_{k+1} = \begin{bmatrix} 0 & 1 & 0 \\ 0 & 0 & 1 \\ -\frac{1}{4} & \frac{1}{2} & -\frac{1}{4} \end{bmatrix} \mathbf{X}_k + \begin{bmatrix} 1 \\ 0 \\ -1 \end{bmatrix} u_k;$

$y_k = \begin{bmatrix} 1 & 0 & 0 \end{bmatrix}\mathbf{X}_k$

9.32 Sketch the signal flow graph for the system with the transfer function

(a) $\dfrac{1 - 0.5s}{(1 + s)(1 + 0.5s)}$ **(b)** $\dfrac{2s + 7}{s^2 - 5s + 6}$

(c)* $\dfrac{s^2 + 4s + 2}{s(s^3 + 3s^2 + 3s + 1)}$

9.33 Sketch the signal flow graph for the system with the pulse transfer function

(a) $\dfrac{z^2 + z + 1}{z^2 - z - 0.5}$ **(b)** $\dfrac{z^2 - 3}{(z^2 + 1)(z + 2)}$

(c)* $\dfrac{0.5z^2}{z^3 - 2z^2 + z - 0.5}$

9.34 Draw the signal flow graph for the system of

(a) Problem 6.32. **(b)** Problem 6.35.

(c) Problem 8.69; let $y_k = x_{1k}$. **(d)** Problem 8.76.

9.35 Find the state representation for the system with the transfer function given in

(a) Problem 9.32a. **(b)** Problem 9.32b. **(c)*** Problem 9.32c.

Determine in each case whether the system is asymptotically stable.

9.36 Find the companion form state representation for the system with the transfer function given in

(a) Problem 9.33a. **(b)** Problem 9.33b. **(c)*** Problem 9.33c.

In Problem 9.37 and Problem 9.38, obtain the state model for the given signal flow graphs.

9.37(a), (b), (c) The continuous-time system graphs of Fig. 9.42.

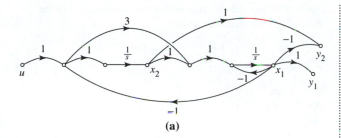

(a)

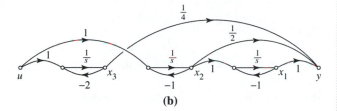

(b)

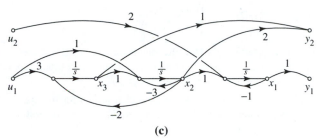

(c)

Figure 9.42 (Problem 9.37)

9.38(a), (b), (c) The discrete-time system graphs of Fig. 9.43.

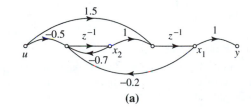

(a)

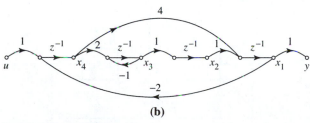

(b)

Figure 9.43 (Problem 9.38)

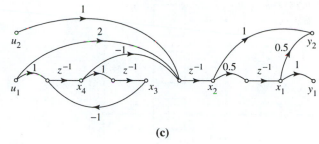

(c)

Figure 9.43 continued (Problem 9.38)

9.39 From the conditions for occurrence of inverse response determined in Problem 9.14, select reasonable values of the parameters in the process transfer functions, and simulate the overall response of the following systems, by first converting the transfer function into an equivalent vector state equation and solving this equation using a computational mathematics package.

(a)* The system of Problem 9.14a

(b) The system of Problem 9.14b

(c) The system of Problem 9.14c

In Problems 9.40 and 9.41, transform the given discrete-time state model into companion form and write the corresponding transfer function

9.40 The second-order system

$$\mathbf{X}_{k+1} = \begin{bmatrix} 1/4 & 1/8 \\ 1 & 1/2 \end{bmatrix}\mathbf{X}_k + \begin{bmatrix} 1 \\ 4 \end{bmatrix}u_k; \qquad y_k = [1 \quad 0]\mathbf{X}_k$$

9.41(a) Problem 6.42a, with $y_k = [1 \quad 1]\mathbf{X}_k$

9.41(b)* Problem 8.70, with $y_k = [1 \quad 2 \quad 1]\mathbf{X}_k$

In Problems 9.42 to 9.44, for the linear systems with the given **A** matrices,

(a) Obtain the characteristic equation, modes, eigenvectors, and trajectory pattern in the state space.

(b) Indicate whether the system is stable, asymptotically stable, or unstable from consideration of the trajectory pattern.

(c) Sketch the complex plane location of the roots of the characteristic equation obtained in (a). What is the relation between root placement in the complex plane and stability?

9.42 $\mathbf{A} = \begin{bmatrix} -2 & -1 \\ -2 & -3 \end{bmatrix}$

9.43 $\mathbf{A} = \begin{bmatrix} 2 & 1 \\ 1 & 3 \end{bmatrix}$

9.44 $\mathbf{A} = \begin{bmatrix} -1 & 2 \\ -5 & 1 \end{bmatrix}$

9.45(a), (b)* Determine from only the given block diagram in Fig. 9.44 whether the system is controllable and observable.

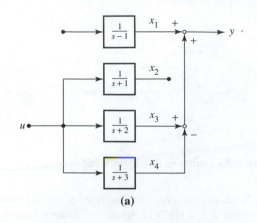

(a)

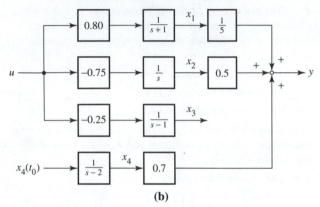

(b)

Figure 9.44 (Problem 9.45)

9.46 Is the following second-order discrete-time system

$$\mathbf{X}_{k+1} = \begin{bmatrix} -1/2 & 0 \\ 1 & -2 \end{bmatrix} \mathbf{X}_k + \begin{bmatrix} 0 \\ 1 \end{bmatrix} u_k; \qquad \mathbf{X}_0 = \begin{bmatrix} 2 \\ 2 \end{bmatrix}$$

(a) Stable?   **(b)** Controllable? Justify your answers.

In Problems 9.47 to 9.50, determine whether the system is

(a) Controllable.   **(b)** Observable.

9.47 The third-order continuous-time system

$$\dot{\mathbf{X}} = \begin{bmatrix} -1 & 0 & 0 \\ 0 & -2 & 0 \\ 0 & 0 & -3 \end{bmatrix} \mathbf{X} + \begin{bmatrix} 1 \\ 2 \\ 0 \end{bmatrix} u; \qquad y = \begin{bmatrix} 1 & 1 & 1 \end{bmatrix} \mathbf{X}$$

9.48 The system of Problem 8.69, with $y_k = x_{1k}$

9.49* The system given in Problem 8.76

9.50 The system of

(a) Problem 9.37a. **(b)** Problem 9.37c. **(c)** Problem 9.38c.

9.51 Determine the observability of the linear second-order system the triple {**A,B,C**}

$$\mathbf{A} = \begin{bmatrix} -1 & 1 \\ 0 & 1 \end{bmatrix}; \qquad \mathbf{B} = \begin{bmatrix} 0 \\ 1 \end{bmatrix}; \qquad \mathbf{C} = \begin{bmatrix} -2 & 1 \end{bmatrix}$$

9.52 Ascertain the controllability and observability of the following multivariable system

$$\dot{\mathbf{X}} = \begin{bmatrix} 3 & 2 & 7 \\ 1 & 5 & 6 \\ -4 & -1 & -3 \end{bmatrix} \mathbf{X} + \begin{bmatrix} 1 & 1 \\ 0 & 2 \\ 1 & 0 \end{bmatrix} \mathbf{U}; \qquad \mathbf{Y} = \begin{bmatrix} 1 & 0 & 1 \\ 0 & 1 & 1 \end{bmatrix} \mathbf{X}$$

9.53* For the discrete-time system with the z-domain block diagram of Fig. 9.45, if

$$G_p(z) = \frac{Y(z)}{M(z) + aV(z)} = \left(\frac{1-c}{a} \right) \frac{z^{-2}}{1 - cz^{-1}} \quad \text{and}$$

$$D(z) = \frac{M(z)}{E(z)} = \left(\frac{a}{1-c} \right) \frac{1 - cz^{-1}}{1 - z^{-2}}$$

(a) Obtain the signal flow graphs for the plant and the controller.

(b) Combine the results from (a) and determine the signal flow graph for the open-loop system

$$G_0(z) = G_p(z)D(z) = \frac{Y(z)}{E(z)}$$

Indicate using dashed lines the reference input and feedback signals that would make the signal flow graph represent the overall closed-loop system.

(c) Determine the discrete-time domain model for the open-loop system using the result from (b). That is, determine **P** and **Q** in

$$\mathbf{X}_{k+1} = \mathbf{P}\mathbf{X}_k + \mathbf{Q} \begin{bmatrix} e_k \\ v_k \end{bmatrix}$$

(d) Determine the controllability of the open-loop system by diagonalizing the model in (c), that is, by finding $\hat{\mathbf{P}}$ and $\hat{\mathbf{Q}}$ in

$$\hat{\mathbf{X}}_{k+1} = \hat{\mathbf{P}}\hat{\mathbf{X}}_k + \hat{\mathbf{Q}} \begin{bmatrix} e_k \\ v_k \end{bmatrix}$$

where $\hat{\mathbf{P}}$ is diagonal

(e) Sketch the signal flow graph for the diagonalized (open-loop) system obtained in (d). Indicate, as in (b), the overall closed-loop system.

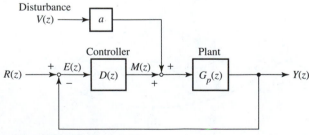

Figure 9.45 (Problem 9.53)

IN THE LAST CHAPTER, TWO IMPORTANT ASPECTS of dynamic behavior, controllability and observability, were introduced. In this chapter, we consider another significant aspect of behavior that is determined by the structure of a system, namely, the *stability* of dynamic systems. Because of this structural dependence, stability behavior can be investigated in state space, as was evident in Example 9.11. This viewpoint will be considered further in Section 10.1. As is the case for all other forms of dynamic behavior, stability is determined by the eigenvalues of the system. In Section 10.2 we explore the dependence of stability on the values of the roots of the system's characteristic equation. We shall introduce a technique that examines the implications of a graphical display of the values of the characteristic roots as well as methods for investigating stability directly on the system's characteristic (polynomial) equation. Since most of these procedures are first explained for continuous-time systems, similar techniques for discrete-time systems are introduced in Section 10.3. In Section 10.4 we consider stability concepts in the frequency domain. Again, both graphical interpretations and analytical measures are investigated. Finally, a few concepts relative to the stability of nonlinear systems are introduced in Section 10.5.

Another implication of the structural basis of stability is that considerations for control and stability necessarily go together. The introduction of control in a system implies a change in the structure of the system that immediately has an impact on the system's stability behavior. A previously stable system can become unstable depending on the control law implemented. In general, a controlled system that has become unstable can be deemed to be out of control. Yet for some systems, especially certain nonlinear systems such as some *bilinear systems,* the ability to introduce *adaptive* or *variable-structure* control in which, depending on the values taken by the control variables, the system can pass through stable and unstable states, implies that the system can be steered throughout the required region of state space, thus allowing for *complete controllability* without *infinite control* (recall that complete controllability of most linear systems is not feasible with *bounded* control). Some of the techniques introduced in Sections 10.2 and 10.3 are directly applicable to closed-loop systems, and most of the discussion in Sections 10.4 and 10.5 is on the stability behavior and stability measures of systems under feedback control. The effect of control on stability is a major factor in the design of control systems considered in Part III.

10.1 STABILITY CONCEPTS IN STATE SPACE

We observed in Section 9.4 that for the example second-order system, equation (9.39a), all trajectories in the state plane converged to the origin and were tangent to the eigenvector corresponding to the slow mode, near the origin. This implies that the origin is an equilibrium point for this system. In the *Lyapunov concept of stability* (see [3] and [17] and below for more details), we determine whether a specific origin (not the system, as such) is stable or not (a given system may have more than one equilibrium point with different stability characteristics). The origin is *stable* if for small disturbances away from the origin the system always remains within a finite region surrounding the origin. If in addition the system always returns ultimately to the origin, the origin is said to be *asymptotically stable.* The *system* (that is, with respect to a specific origin) is *unstable* within a region surrounding the equilibrium point if there is always some small disturbance within this region for which the system moves out of the region. In the trajectory pattern of Fig. 9.19 (also shown in Fig. 10.1a), we saw that no matter the disturbance away from the origin to any point in the state space, the resulting (initial) state is on a trajectory that converges to the origin. Thus this *system* is (asymptotically) stable or has an asymptotically stable equilibrium point. Another asymptotically stable trajectory pattern is shown in Fig. 10.1b. In this case the system is oscillatory, but the origin is nonetheless (asymptotically) stable. Figures 10.1c and 10.1d illustrate second-order trajectories for stable (but not asymptotically stable) *systems.* In the first case (Fig. 10.1c) the trajectories converge to a line, whereas in the second case (Fig. 10.1d) the region to which the system always returns after an initial disturbance is an ellipse. Figures 10.1e through 10.1h illustrate more or less corresponding trajectory patterns of unstable second-order *systems.* The system of Fig. 10.1f is oscillatory, but its origin is unstable.

Another aspect of the equilibrium point of Fig. 10.1a is that it is *globally stable,* since the stability is not limited to "small" disturbances within finite regions of the state space. Every initial state in the entire state space is on a "stable" trajectory. This is generally true for stable linear systems. Alternatively, the system is said to be *stable in the large.* Many nonlinear systems exhibit *local stability,* that is, they behave in a stable manner in disjoint regions of state space. Within each region the behavior of the system can be similar to any of the trajectory patterns in Figs. 10.1a to 10.1d. However, given a sufficiently large disturbance, the system can escape from one stability region and begin on a trajectory that can be part of another stability region. These other aspects of stability are illustrated in Fig. 10.2 using the intuitive model of a ball and hills and valleys. This model also explains the distinction between the stable behaviors shown in Figs. 10.1a through 10.1d. Suppose in Fig. 10.2 that the valley surfaces are frictionless. Then, in principle, once the ball is disturbed within a valley, it will not come to rest at the lowest point in the valley but rather will oscillate indefinitely, with some fixed amplitude, about the equilibrium point. Such an equilibrium point is *stable in the sense of Lyapunov.* In the case of the "infinite" valley in Fig. 10.2a, the stability is global. This is the situation illustrated in state space in Fig. 10.1d. No matter the magnitude of the disturbance, the ball cannot escape from the valley; only the (fixed) amplitude of the ensuing oscillation (the size of the ellipse in Fig. 10.1d) is affected. However, as shown in Fig. 10.2c, for "finite" valleys, a sufficiently large disturbance can allow the ball to escape one valley and possibly be captured by another valley. This is local stability. Similarly, the (peaks of the) hills in Figs. 10.2b and 10.2c illustrate, respectively, global and local instability. If on the other hand there is some friction on the valley surfaces that gives rise to any amount of damping of the ball motion, the ball will always come to rest at an equilibrium point. This is asymptotic stability. Evidently, depending on the amount of damping and the shape of the valley, the ball can come to rest without oscillation (as implied by Fig. 10.1a) or with some oscillation (as implied by Fig. 10.1b).

Figure 10.1

Some Stability Concepts in State Space

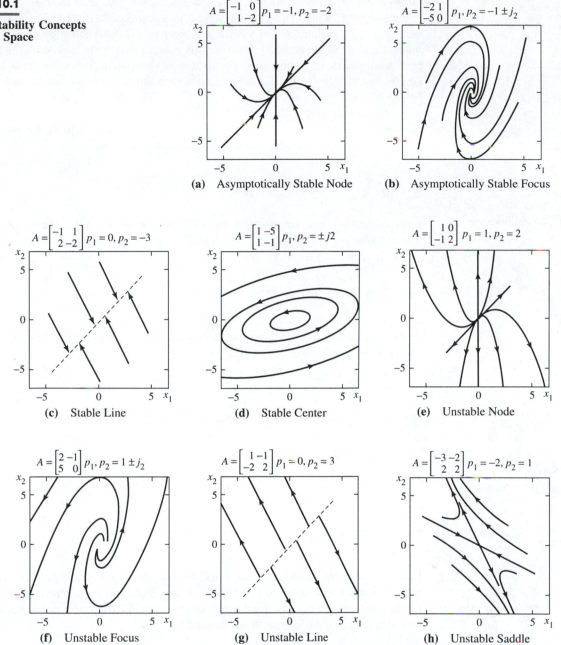

(a) Asymptotically Stable Node

(b) Asymptotically Stable Focus

(c) Stable Line

(d) Stable Center

(e) Unstable Node

(f) Unstable Focus

(g) Unstable Line

(h) Unstable Saddle

In engineering practice, asymptotic stability is preferred over mere stability, since more precise design is possible, and the eventual position of the system is assured.

Equilibrium or Critical Point

The concepts of equilibrium and disturbance are central to stability. In the foregoing discussion, disturbance was taken into account by considering the state of the system following a disturbance as the initial state for the subsequent response. For most engineering components (uncontrolled) movements away from the equilibrium state (insta-

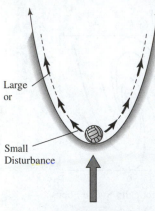

Large
or

Small
Disturbance

Stable in the Large

(a)   Global Stability

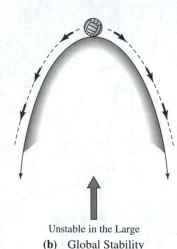

Unstable in the Large

(b)   Global Stability

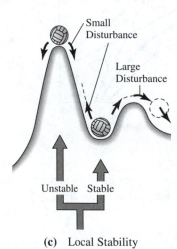

Small
Disturbance

Large
Disturbance

Unstable   Stable

(c)   Local Stability

Figure 10.2

Some Aspects of Stability

bility) result in large strains that these systems may not be able to sustain. There are many examples: A high-rise building subject to an earthquake disturbance near its resonant frequency, an airplane or a missile subject to the disturbance of an unstable air mass, a reactor or boiler suddenly losing its coolant supply. In each case the result is poor performance or, even worse, destruction of the system. It is clear from these examples that stability of the equilibrium state is an important aspect of dynamic system behavior.

As another practical matter, the equilibrium point is not necessarily the origin. However, by a suitable change of variables, we can always transfer the equilibrium point to the origin. Consider the following system:

$$\frac{d}{dt}\mathbf{X}_g = \mathbf{f}(\mathbf{X}_g, t) = \mathbf{f}(\mathbf{X}_g(t)) \tag{10.1a}$$

Since the elements of the vector $\mathbf{f}$ do not depend explicitly on t, the system is free and stationary or *autonomous*. The equilibrium point, $\mathbf{X}_e$, is defined as the solution to

$$\frac{d}{dt}\mathbf{X}_g = \mathbf{0} = \mathbf{f}(\mathbf{X}_e) \tag{10.1b}$$

To investigate stability, we make a transformation of coordinates from the original state vector $\mathbf{X}_g$ to a new state vector $\mathbf{X}$:

$$\mathbf{X} = \mathbf{X}_g - \mathbf{X}_e \tag{10.1c}$$

and deal with the new system whose equilibrium point is the origin:

$$\frac{d}{dt}\mathbf{X} = \mathbf{f}(\mathbf{X}); \qquad \mathbf{f}(\mathbf{0}) = \mathbf{0} \tag{10.1d}$$

or if the system is linear,

$$\frac{d}{dt}\mathbf{X} = \mathbf{A}\mathbf{X} \tag{10.1e}$$

In the forthcoming material we shall assume that all the necessary conditions are satisfied, so that the solution to (10.1d) or (10.1e) exists and is unique for given initial conditions (see [3]). Further, for the nonlinear system (10.1d), our conclusions are valid within a finite region in state space defined by

$$\|\mathbf{X}\| < R \tag{10.2a}$$

where R is some finite, real, and positive number, and the symbol $\|\ \ \|$ stands for *norm,* and the implied norm of $\mathbf{X}$ in equation (10.2a) is the Euclidean norm (see Appendix B):

$$\|\mathbf{X}\| = (x_1^2 + x_2^2 + \cdots + x_n^2)^{1/2} \tag{10.2b}$$

From the definition (10.2b), the region in state space (10.2a) is, in general, a *hypersphere* of radius R. In the phase plane (two-dimensional state space), it is a circle of radius R.

Finally, although the general problem of stability concerns deviations of the motion of a system, whether free or under some forcing, about some fixed response of the system, it can be shown (see [17]) that a stability problem of this general kind can be reduced to the simpler problem of the stability about a fixed origin of a free system.

Stability in the Sense of Lyapunov

Formally an equilibrium state is said to be *stable* (in the sense of Lyapunov) if for any positive scalar ϵ there exists a positive scalar δ such that $\|\mathbf{X}(t_0)\| < \delta$ implies $\|\mathbf{X}(t)\| < \epsilon$, $t \geq t_0$. In other words (see the phase plane illustration, Fig. 10.3a), any trajectory starting at any point $\mathbf{X}_0 = \mathbf{X}(t_0)$ inside the hyperspherical region of radius δ always remains in the hyperspherical region of radius ϵ. An equilibrium point is *asymptotically stable* if (1) it is *stable* and (2) $\mathbf{X}(t) \longrightarrow \mathbf{0}$ as $t \longrightarrow \infty$, that is, every trajectory starting inside the region δ converges to the origin as time increases indefinitely. Finally, an equilibrium state is *unstable* if it is not stable: There exists an $0 < \epsilon < R$ such that for every $\epsilon > \delta > 0$ there is an $\mathbf{X}(t_0)$ with $\|\mathbf{X}(t_0)\| < \delta$ and $\|\mathbf{X}(t_1)\| \geq \epsilon$ for some $t_1 > t_0$. The equilibrium point is *completely unstable* if this condition holds for every tracjectory starting inside the region δ.

Lyapunov Method of Stability Analysis

We already know that the stability characteristics of a system can be determined by solution of the system equation. The trajectory patterns in Fig. 10.1 were obtained by solving the respective autonomous linear system equations for various initial states (MATLAB **ode23** (see script file **fig10_1.m; On the Net**) was employed in this case). The *direct method of Lyapunov* or the *Lyapunov second method* is a technique for ascertaining the stability (in the sense of Lyapunov) of a system without necessarily solving the system equations. The specific stability tests are given by the following theorems (see [17]):

Lyapunov Stability Theorem   If there exists a *Lyapunov function V(**X**)* in some region S around the origin, then the origin is stable for all $\mathbf{X}(t_0)$ contained in S.

Lyapunov Asymptotic Stability Theorem   If there exists a *Lyapunov function V(**X**)* in some region S around the origin, and if in addition, $-dV(\mathbf{X})/dt$ along a trajectory is *positive definite* in S, then the origin is asymptotically stable.

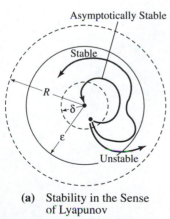

(a) Stability in the Sense of Lyapunov

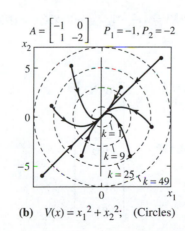

(b) $V(x) = x_1^2 + x_2^2$;   (Circles)

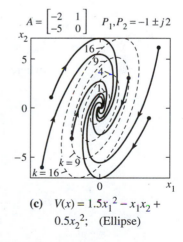

(c) $V(x) = 1.5x_1^2 - x_1 x_2 + 0.5x_2^2$;   (Ellipse)

Figure 10.3

Definition of Stability Illustrated in the Phase Plane

What Is a Lyapunov Function?   A *candidate* scalar function $V(\mathbf{X})$ becomes a *Lyapunov function* if $V(\mathbf{X})$ is *positive definite* and $dV(\mathbf{X})/dt$ along a trajectory is *negative definite* or *negative semidefinite* in a region S.

A scalar function $V(\mathbf{X})$ is *positive definite* (see Appendix B) if

1. $V(\mathbf{0}) = 0$;
2. $V(\mathbf{X})$ is positive throughout some region of state space S outside the origin, that is, $V(\mathbf{X}) > 0$, $\mathbf{X} \in S$; $\mathbf{X} \neq \mathbf{0}$;
3. $V(\mathbf{X})$ is continuous throughout S, and
4. $V(\mathbf{X})$ has continuous first partial derivatives with respect to x_i, that is,

$$\frac{\partial V(\mathbf{X})}{\partial x_i}, \quad i = 1, \ldots, n$$

are continuous.

If condition (2) is weakened to $V(\mathbf{X}) \geq 0$ rather than $V(\mathbf{X}) > 0$, then $V(\mathbf{X})$ is *positive semidefinite*. Similarly, by reversing the inequality sign in condition (2), that is, $V(\mathbf{X}) < 0$ or $V(\mathbf{X}) \leq 0$, $\mathbf{X} \in S$; $\mathbf{X} \neq \mathbf{0}$, we obtain the definition for $V(\mathbf{X})$ *negative definite* or *negative semidefinite*.

Finally, by the time derivative $dV(\mathbf{X})/dt$ along a trajectory of a system (equation (10.1d), for example), we mean

$$\frac{d}{dt} V(\mathbf{X}) = \frac{\partial V(\mathbf{X})}{\partial x_2} \frac{dx_1}{dt} + \frac{\partial V(\mathbf{X})}{\partial x_2} \frac{dx_2}{dt} + \cdots + \frac{\partial V(\mathbf{X})}{\partial x_n} \frac{dx_n}{dt} \qquad \textbf{(10.3a)}$$

$$= [\dot{x}_1; \quad \dot{x}_2; \quad \cdots \quad \dot{x}_n] \begin{bmatrix} \dfrac{\partial V(\mathbf{X})}{\partial x_1} \\[6pt] \dfrac{\partial V(\mathbf{X})}{\partial x_2} \\ \vdots \\ \dfrac{\partial V(\mathbf{X})}{\partial x_n} \end{bmatrix}$$

or

$$\frac{d}{dt} V(\mathbf{X}) = \left(\frac{d\mathbf{X}}{dt} \right)^T \cdot \mathbf{grad}(V(\mathbf{X})) = (\mathbf{f}(\mathbf{X}))^T \cdot \mathbf{grad}(V(\mathbf{X})) \qquad \textbf{(10.3b)}$$

where,

$$\mathbf{grad}(V(\mathbf{X})) = \begin{bmatrix} \dfrac{\partial V(\mathbf{X})}{\partial x_1} \\[6pt] \dfrac{\partial V(\mathbf{X})}{\partial x_2} \\ \vdots \\ \dfrac{\partial V(\mathbf{X})}{\partial x_n} \end{bmatrix} \qquad \textbf{(10.3c)}$$

is presumed to exist, since $V(\mathbf{X})$ has continuous first partial derivatives with respect to all x_i.

To see the relation between the Lyapunov function and stability, consider a situation where $V(\mathbf{X}) = k$ represents in state space a hypersphere of radius $k^{1/2}$ or a hyperellipsoid (of, say, major axis proportional to $k^{1/2}$). Then imagine, as illustrated in Figs. 10.3b and 10.3c, the region S of state space covered with many such closed contours (surfaces) of decreasing sizes, $k_m, k_{m-1}, \ldots, k_1, k_0$, the smallest ($k_0 = 0$) being the origin (recall that $V(\mathbf{0}) = 0$). Suppose a state trajectory in the same region crosses all these surfaces (contours) in a direction going from the larger-size contour to the smaller-size contour. Then along such a trajectory, $V(\mathbf{X})$ is monotonously decreasing ($dV(\mathbf{X})/dt$ is negative definite along the trajectory), and the trajectory must therefore converge to the origin (the smallest $V(\mathbf{X})$); that is, the system is asymptotically stable.

Example 10.1   Determining Stability by an Arbitrary Test Function

Using the test function $V(\mathbf{X}) = x_1^2 + x_2^2$, ascertain the stability of the second-order linear and stationary systems shown in state space in Figs. 10.1a and 10.1b. Note that if, for example, these systems represent mechanical systems with x_1 and x_2 viewed, respectively, as displacement and velocity, then $V(\mathbf{X})$ is an energy-type function, since x_1^2 and x_2^2 are proportional to the potential and kinetic energies in the system, respectively.

Solution: We note that $V(\mathbf{0}) = 0$ and that $V(\mathbf{X})$ is positive throughout all state space (that is, for all values of x_1 and x_2 outside the origin). Also by equation (10.3c),

$$\mathbf{grad}(V(\mathbf{X})) = \begin{bmatrix} 2x_1 \\ 2x_2 \end{bmatrix}$$

For the case of Fig. 10.1a,

$$\mathbf{A} = \begin{bmatrix} -1 & 0 \\ 1 & -2 \end{bmatrix}$$

and by equation (10.3b),

$$\frac{d}{dt} V(\mathbf{X}) = \begin{bmatrix} -x_1 \\ x_1 - 2x_2 \end{bmatrix}^T \begin{bmatrix} 2x_1 \\ 2x_2 \end{bmatrix} = -2x_1^2 + 2x_1x_2 - 4x_2^2$$

$$= -(x_1 - x_2)^2 - x_1^2 - 3x_2^2$$

which is negative **definite** for all values of x_1 and x_2 outside the origin. Thus $V(\mathbf{X})$ is a Lyapunov function for the system of Fig. 10.1a, and this system is asymptotically stable, as we already know. In the phase plane, the contours $x_1^2 + x_2^2 = k$ are circles around the origin. A few of these circles (dashed lines) superimposed on the trajectory pattern of Fig. 10.1a, are shown in Fig. 10.3b to illustrate the present development.

For the case of Fig. 10.1b,

$$\mathbf{A} = \begin{bmatrix} -2 & 1 \\ -5 & 0 \end{bmatrix}$$

and by equation (10.3b),

$$\frac{d}{dt} V(\mathbf{X}) = \begin{bmatrix} -2x_1 + x_2 \\ -5x_1 \end{bmatrix}^T \begin{bmatrix} 2x_1 \\ 2x_2 \end{bmatrix} = -4x_1^2 - 8x_1 x_2$$

which is not negative definite or even negative semidefinite (consider, for example, $x_1 = 1$ and $x_2 = -1$, which gives $\dot{V}(\mathbf{X}) = 4$). Thus, using the present test function, we cannot reach any conclusion on the stability of the origin for this system, although we already know, from Fig. 10.1b, that the origin is asymptotically stable.

The second case in the preceding example illustrates one of the difficulties in using the Lyapunov direct method. The Lyapunov stability theorems given above provide only sufficiency conditions for stability. If a candidate Lyapunov function fails to satisfy the requirements of the stability theorem, no valid conclusions can be reached, since another choice of a test function may well meet the stability conditions. The second apparent difficulty in the application of this method is that it provides no explicit clue for the construction of suitable $V(\mathbf{X})$ for the general system (10.1d). However, a formal method exists for generating appropriate Lyapunov functions for linear, stationary, lumped-parameter systems (both continuous-time and discrete-time systems). The resulting test for *asymptotic stability* constitutes both necessary and sufficiency conditions (see [17]). We illustrate this procedure next using the inconclusive case of Example 10.1. For a linear, stationary discrete-time system, simply replace the $\mathbf{A}$-matrix in the procedure with the system's $\mathbf{P}$-matrix.

Example 10.2   Generating Lyapunov Functions for Linear, Stationary Systems

Consider again the autonomous second-order linear system:

$$\dot{\mathbf{X}} = \begin{bmatrix} -2 & 1 \\ -5 & 0 \end{bmatrix} \mathbf{X}$$

Determine whether the origin of this system is asymptotically stable, using as a Lyapunov function the following quadratic form:

$$V(\mathbf{X}) = \mathbf{X}^T \mathbf{E} \mathbf{X} \tag{10.4a}$$

Solution: Note that $V(\mathbf{X})$ in the quadratic form (10.4a) is positive definite if (see Appendix B) the coefficients of $\mathbf{E}$ satisfy the Sylvester condition. That is, if

$$e_{11} > 0; \quad \begin{vmatrix} e_{11} & e_{12} \\ e_{21} & e_{22} \end{vmatrix} > 0; \quad \cdots; \quad \begin{vmatrix} e_{11} & e_{12} & \cdots & e_{1n} \\ e_{21} & e_{22} & \cdots & e_{2n} \\ \vdots & \vdots & & \vdots \\ e_{n1} & e_{n2} & \cdots & e_{nn} \end{vmatrix} > 0 \tag{10.4b}$$

Further, since $x_i x_j = x_j x_i$, subsequent computations can be simplified without any loss of generality by letting $\mathbf{E}$ be symmetric. That is,

$$e_{ij} = e_{ji} \tag{10.4c}$$

Continuing, we differentiate (10.4a) with respect to time,

$$\dot{V}(\mathbf{X}) = \dot{\mathbf{X}}^T \mathbf{E} \mathbf{X} + \mathbf{X}^T \mathbf{E} \dot{\mathbf{X}}$$

Substituting the equation for the linear stationary system, equation (10.1e), we obtain

$$\dot{V}(\mathbf{X}) = (\mathbf{AX})^T \mathbf{E} \mathbf{X} + \mathbf{X}^T \mathbf{E} \mathbf{AX}$$

Using $(\mathbf{AX})^T = \mathbf{X}^T \mathbf{A}^T$, we obtain

$$\dot{V}(\mathbf{X}) = \mathbf{X}^T (\mathbf{A}^T \mathbf{E} + \mathbf{E} \mathbf{A})\mathbf{X}$$

Defining the matrix $\mathbf{W}$ by

$$\dot{V}(\mathbf{X}) = -\mathbf{X}^T \mathbf{W} \mathbf{X} \tag{10.4d}$$

so that $\dot{V}(\mathbf{X})$ is negative definite when $\mathbf{W}$ is positive definite, requires

$$-\mathbf{W} = \mathbf{A}^T \mathbf{E} + \mathbf{E} \mathbf{A} \tag{10.4e}$$

Since any positive definite $\mathbf{W}$ will do, the identity matrix is usually the convenient choice. We therefore proceed with the following two steps, which constitute the method:

1. Solve for $\mathbf{E}$ using

$$\mathbf{A}^T \mathbf{E} + \mathbf{E} \mathbf{A} = -\mathbf{I} \tag{10.4f}$$

$$\begin{bmatrix} -2 & 1 \\ -5 & 0 \end{bmatrix}^T \begin{bmatrix} e_{11} & e_{12} \\ e_{12} & e_{22} \end{bmatrix} + \begin{bmatrix} e_{11} & e_{12} \\ e_{12} & e_{22} \end{bmatrix} \begin{bmatrix} -2 & 1 \\ -5 & 0 \end{bmatrix} = \begin{bmatrix} -1 & 0 \\ 0 & -1 \end{bmatrix}$$

or

$$\begin{bmatrix} (-4e_{11} - 10e_{12}) & (e_{11} - 2e_{12} - 5e_{22}) \\ (e_{11} - 2e_{12} - 5e_{22}) & (2e_{12}) \end{bmatrix} = \begin{bmatrix} -1 & 0 \\ 0 & -1 \end{bmatrix}$$

which has the solution $e_{11} = 1.5$, $e_{12} = -0.5$, and $e_{22} = 0.5$.

2. Test for the positive definiteness of $\mathbf{E}$ (using the Sylvester condition (10.4b)). If $\mathbf{E}$ is positive definite, then the origin is *asymptotically stable*. Otherwise (since this technique provides necessary and sufficiency conditions), the origin is *not asymptotically stable*.

Applying (10.4b), $e_{11} = 1.5 > 0$, we have

$$\begin{vmatrix} e_{11} & e_{12} \\ e_{12} & e_{22} \end{vmatrix} = e_{11}e_{22} - e_{12}^2 = 0.75 - 0.25 = 0.5 > 0$$

Thus $\mathbf{E}$ is positive definite, $V(\mathbf{X}) = \mathbf{X}^T \mathbf{E} \mathbf{X}$ is a Lyapunov function, and the origin of the given linear, stationary system is asymptotically stable. Figure 10.3c shows the tracjectory pattern for this system (Fig. 10.1a) with some contours of constant $V(\mathbf{X})$ superimposed. In general, when $V(\mathbf{X})$ is given by the quadratic form, equation (10.4a) with $\mathbf{E}$ positive definite, $V(\mathbf{X}) = constant$ represents a hyperellipsoid in state space (ellipse in phase plane).

10.2 STABILITY AND EIGENVALUE PLACEMENT

The eigenvalues of a system determine its stability and other behavior. It follows that the most straightforward means of investigating a linear system's stability is to examine the possible values of its characteristic roots. One way to accomplish this is to look at the placement of the eigenvalues in the complex plane. We did this in Chapters 6 and 7, in connection with second-order systems. We found that, in general, eigenvalues in the right half of the complex plane (RHP) cause instability, whereas roots in the left half of the complex plane (LPH) are stable. Root placement directly on the imaginary axis causes marginal or neutral stability; although if the multiplicity is more than 1, instability results. Figure 10.4 illustrates these ideas. Observe also that the larger the imaginary component of the eigenvalue, the larger the frequency of oscillation, and the larger and more negative (hence the smaller) the real component of the root, the shorter the decay time. Of course, this complex plane picture portrays not only the relative stability of the system but other aspects of its dynamic behavior as well. Another way of exploring the nature of the eigenvalues of a linear system is to deduce this information from the characteristic polynomial itself. This method does not require determination of the actual values of the characteristic roots. The *root locus* method and the *Routh stability criterion,* which are considered in this section are, respectively, examples of the first and second approaches to stability analysis based on the system's eigenvalues.

The Root Locus Technique

The eigenvalues of a system are functions of the system's parameters. In general, one is thus interested in the placement of the roots in the complex plane as some parameter of the system is varied. The result is a sequence of root positions along which the eigenvalues will "move" as this parameter changes. This sequence of points is known as the *root locus.* Once the root locus is available, the values of the parameter that give the desired quality and stability of system behavior can be chosen. The root locus technique is one of the fundamental approaches to classical feedback controller design.

Figure 10.4

Relative Stability and Eigenvalue Placement

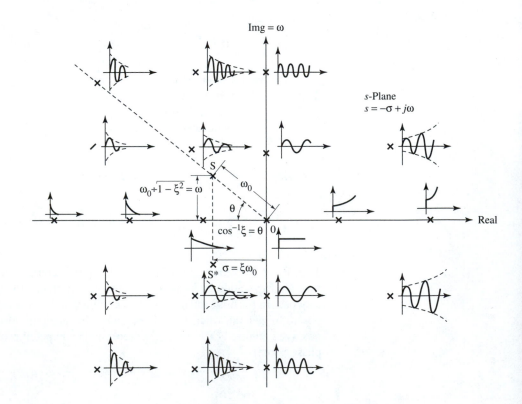

Graphical Construction of Root Loci

We construct the root locus according to the following procedure:

a. We put the system characteristic equation in the form

$$1 + kG_o(s) = 0 \Rightarrow G_o(s) = -\frac{1}{k}, \quad k > 0 \qquad \text{(10.5a)}$$

where k is the (design) parameter of interest and $G_o(s)$ must be in factored form (for manual plotting),

$$G_o(s) = \frac{(s - z_1)(s - z_2) \cdots (s - z_m)}{(s - p_1)(s - p_2) \cdots (s - p_n)}, \quad n \geq m \qquad \text{(10.5b)}$$

where $z_1, z_2, \ldots, z_m$ and $p_1, p_2, \ldots, p_n$ are known as the open-loop zeros and open-loop poles, respectively. These terms reflect what happens to $G_o(s)$ when s takes on the corresponding values—$G_o(s)$ goes to zero for the *zeros* and to infinity for the *poles*.

Consider, for example, the antiaircraft gun position control system illustrated in Fig. 10.5a. The antiaircraft gun follows the motion of an airplane as shown. An automatic control system positions the elevation angle θ of the muzzle in proportion to an input voltage command v that is generated by a radar tracking unit (not shown). The block diagram of a simplified model of the servomechanism is shown in Fig. 10.5b. The control law is *PI,* that is *proportional-plus-integral* control (recall Table 1.3; *a proportional-plus-integral-plus-derivative* (PID) control law was introduced in Example 3.19). We shall assume, for this exercise, that the PI controller is constrained to $T_i = 1$ by some (physical) considerations. With this constraint, the closed-loop transfer function for the overall system is

$$G(s) = \frac{\dfrac{10}{s(s + 5)}\left(\dfrac{10}{s + 10}\right)k_c\left(\dfrac{s + 1}{s}\right)}{1 + \dfrac{10}{s(s + 5)}\left(\dfrac{10}{s + 10}\right)k_c\left(\dfrac{s + 1}{s}\right)}$$

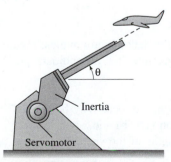

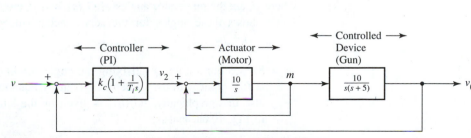

(a) Schematic Diagram of Physical System

(b) Block Diagram of Control System

Figure 10.5

Antiaircraft Servo Example

This gives the characteristic equation,

$$1 + 100k_c\frac{(s + 1)}{s^2(s + 5)(s + 10)} = 0 \tag{10.5c}$$

Comparing equation (10.5c) with (10.5a) and (10.5b), we can set $k = 100k_c$ and

$$G_o(s) = \frac{(s + 1)}{(s)(s)(s + 5)(s + 10)}$$

in which $m = 1$, $z_1 = -1$, $n = 4$, $p_1 = p_2 = 0$, $p_3 = -5$, and $p_4 = -10$.

b. the points on the root locus must satisfy the *angle condition*

$$\angle G_o(s) = -180° \pm 360° \, N; \quad N = \text{an integer} \tag{10.6a}$$

c. We scale the root locus using the *magnitude condition*

$$|G_o(s)| = \frac{1}{k} \tag{10.6b}$$

These two requirements follow directly from equation (10.5a) if k is restricted to only positive real values. In general, $G_o(s)$ is a complex number that in polar format has a magnitude and an angle. By equation (10.5a), $G_o(s)$ must equal a negative real number, that is, it must lie on the negative real axis. This is the statement of equation (10.6a), which is the equation with which the locus is generated. The graphical solution of (10.6a) for all s is facilitated by the following four steps and five rules. The rules permit the rest of the root locus to be quickly sketched after only a few points are determined on it:

> **Step 1** Locate in the s-plane the poles and zeros of $G_o(s)$ (distinguished by different markings, for example, $\times$ and $\bigcirc$, respectively; see Fig. 10.6a and note that there is a double pole at $s = 0$).
>
> **Step 2** By trial and error, find a point q that satisfies the angle condition (10.6a). Note that for graphical construction of the root locus, equation (10.6a) can be written

$$\sum_{j=1}^{m}\alpha_j - \sum_{i=1}^{n}\beta_i = -180° \pm 360° \, N \tag{10.7a}$$

where α_j are the numerator angles of $G_o(s)$, and β_i are the denominator angles. Figure 10.6a shows these angles for the servomechanism example, for a candidate point $q = (-\sigma) + j\omega$.

> **Step 3** For a point q satisfying the angle condition, the locus may be labeled at this point with the value of k given by equation (10.6b) and $s = q$. Note also that graphically, $|G_o(q)|$ is given by the lengths of the vectors indicated in Fig. 10.6a, that is,

$$|G_o(q)| = \frac{|q - z_1|}{|q|^2|q - p_3||q - p_4|} \tag{10.7b}$$

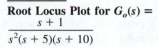

Figure 10.6

Root Locus Plot for $G_o(s) =$
$$\frac{s + 1}{s^2(s + 5)(s + 10)}$$

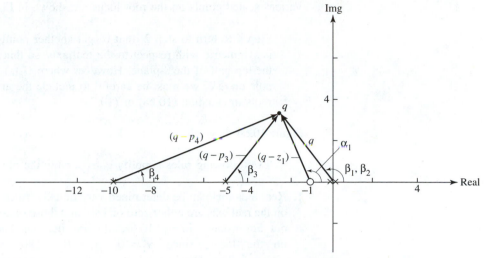

(a) Poles, Zeros, and Angle Condition

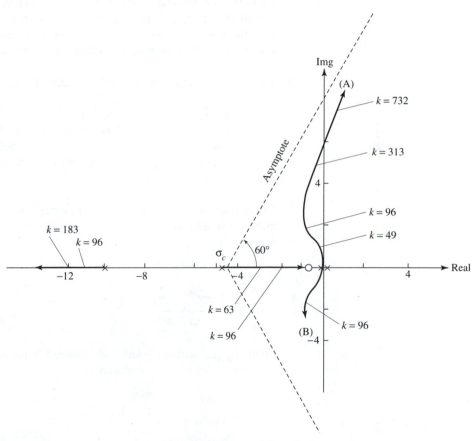

(b) Asymptotes and Root Loci

Various scaled points on the root locus are shown in Fig. 10.6b.

> **Step 4** Return to step 2 (that is, get another point q). Note that the root locus is symmetric with respect to the real axis, so that we need to investigate only the top half of the s-plane. However where $G_o(s)$ has a complex-conjugate pole or zero, we must be careful to include the angle contributions from both roots in equation (10.6a) or (10.7a).

Figure 10.6b shows the eventual plot of the root locus for the given characteristic equation.

The following rules simplify considerably the plotting of a root locus:

1. **Real axis loci** can be determined very quickly, since the angles of poles and zeros on the real axis are either zero or 180° and those of complex-conjugate roots cancel out. For instance, in Fig. 10.6a, all points between 0 and -1 on the real axis are not on the locus, since $\alpha_1 = 0$, $\beta_1 = \beta_2 = 180°$, and $\beta_3 = \beta_4 = 0°$, so that $\angle G_o = -360°$. On the other hand, all points between -1 and -5 are on a root locus, since $\alpha_1 - \beta_1 - \beta_2 - \beta_3 - \beta_4 = 180 - 180 - 180 - 0 - 0 = -180°$. The rest of the real axis loci shown by heavy lines in Fig. 10.6b are similarly determined. The arrows on the loci point in the direction of increasing k; see rule 2.
2. **The number of branches** is equal to the number of poles of $G_o(s)$. All branches start with $k = 0$ at the pole and end with $k \longrightarrow \infty$ at a zero.

To see this, note from equations (10.5a) and (10.5b) that

$$(s - p_1)(s - p_2) \cdot \cdot \cdot (s - p_n) + k(s - z_1)(s - z_2) \cdot \cdot \cdot (s - z_m) = 0 \qquad \textbf{(10.7c)}$$

so that when $k = 0$, the roots are the poles, and when $k \longrightarrow \infty$, the roots are the zeros. For the example (see Fig. 10.6b), there are four branches ($n = 4$), three of which approach theoretical zeros at infinite distances from the origin. The fourth branch ends at the zero at -1 on the real axis.

3. **Asymptotes** The branches approach straight-line asymptotes at large distances from the origin. The center of the asymptotes on the real axis for $(n - m) \geq 2$ is given by

$$(-\sigma_c) = \frac{\displaystyle\sum_{i=1}^{n} p_i - \sum_{j=1}^{m} z_j}{(n - m)} \qquad \textbf{(10.8a)}$$

where the real parts of p and z are implied in equation (10.8a). The angles the asymptotes make with the real axis are

$$\gamma = \pm \frac{180° \, L}{(n - m)}; \qquad L = 1, 3, 5, \ldots \qquad \textbf{(10.8b)}$$

For our example, there are four branches, with center at

$$(-\sigma_c) = \frac{(0 + 0 - 5 - 10) - (-1)}{3} = \frac{-14}{3} = -4.67$$

and angles of

$$\gamma = \pm\frac{180°}{3}, \pm\frac{540°}{3} = \pm60°, \pm180°$$

Note that two of the branches are on the real axis. The $\pm60°$ asymptotes are shown in Fig. 10.6b.

4. **Breakaway or break-in points** The point σ_b at which the locus leaves or enters the real axis may be found by solving for the value of s for which

$$\frac{d}{ds}\left[\frac{1}{G_o(s)}\right] = 0 \tag{10.8c}$$

or

$$\sum_{i=1}^{n}\left(\frac{1}{\sigma_b - p_i}\right) = \sum_{j=1}^{m}\left(\frac{1}{\sigma_b - z_j}\right) \tag{10.8d}$$

An alternative procedure is given by rule 5. In the present case, the *breakaway point* is by default at the origin. Equation (10.8c) or (10.8d) will also yield $s = 0$ as one of the solutions.

5. **Angle of departure/arrival** The angle of departure/arrival of a locus from/to a complex conjugate pole/zero may be found by selecting trial points very close to the pole/zero and applying the angle condition. Similarly, the breakaway or break-in point can be located using trial points at very small vertical distances from the real axis.

The shape, near the origin, of the root locus branches that start at the pole at the origin was determined for the present example by the application of rule 5. After determining a few other points on the top branch, using step 2, we sketched the balance of the entire root locus, shown in Fig. 10.6b, by drawing a smooth curve through the available points.

Summary of Manual Approach   To fix ideas, let us summarize the manual generation of the root locus shown in Fig. 10.6.

1. *Put characteristic equation in the form $1 + kG_0(s) = 0$:* See equation (10.5c); $k = 100k_c$.
2. *Indicate the poles and zeros on the root locus:* See Fig. 10.6. There is one zero at $z_1 = -1 + j0$, and there are four poles at $p_1 = p_2 = 0$, $p_3 = -5 + j0$, and $p_4 = -10 + j0$.
3. *Real axis loci:* As explained previously, only the points between $p_3 = -5$ and $z_1 = -1$, and those between $p_4 = -10$ and a theoretical zero at $z_2 = -\infty$ satisfy the angle condition. Because each set of points begins at a pole and ends at a zero, each set is a complete root locus branch.
4. *Number of branches:* With four poles, four branches are expected, but two were found in rule 3. Hence the remaining two must originate at the two (double) poles

at the origin. The zeros for these branches must be theoretical ones at infinite distances from the origin. Since the two branches (*A* and *B*) are mirror images of each other, we need only establish one—(*A*).

5. *Asymptotes:* Equation (10.8a) gives the center of the asymptotes at -4.67, and equation (10.8b) gives the four angles $\pm 60°$, $\pm 180°$. As explained before, branches *A* and *B* must be asymptotic to the lines drawn from -4.67 at angles of $\pm 60°$ (see Fig. 10.6).

6. *Breakaway/break-in points:* Branches *A* and *B* break away at the origin because their poles are coincident at the origin. There is no break-in condition in this example.

7. *Angle of departure/arrival:* The angle at which branch *A* leaves the origin can be established by finding one or two points on this branch close to the origin. For example, let $s = -\sigma + j0.5$ in angle condition (10.6a):

$$\tan^{-1}\left(\frac{0.5}{1-\sigma}\right) - 2\tan^{-1}\left(\frac{0.5}{-\sigma}\right) - \tan^{-1}\left(\frac{0.5}{5-\sigma}\right) - \tan^{-1}\left(\frac{0.5}{10-\sigma}\right) = -\pi + 2N\pi$$

By trial and error find $\sigma = 0.089$. The point $-0.089 + j0.5$ establishes a departure angle only slightly greater than $90°$.

8. *Establish shape of branch with additional points:* From the established departure angle of branch *A*, this branch must cross the imaginary axis to proceed asymptotic to the $60°$ line. To find the imaginary axis crossing, let $s = j\omega$ in equation (10.6a), and find $\omega = 5.92$. Two other points between the imaginary axis crossing and the point $-0.089 + j0.5$ should establish the shape of branch *A*. Using an approach similar to the one used in (7), establish the points $-0.88 + j1.87$ and $-0.50 + j4.59$, for example. Finally, sketch branch *A* as a fair curve through the established points and asymptotic to the $60°$ line.

9. *Scale the root locus:* Compute *k* at selected points on the root locus, using equation (10.6b). For example at $s = -0.88 + j1.87$,

$$k = \frac{|s|^2|s+5||s+10|}{|s+1|} = 96.014$$

and at $-0.50 + j4.59$, $k = 313.132$. These values and some others are shown in Fig. 10.6. At $s = j5.92$, $k = 525.651$.

The root locus shows the stability behavior of the system relative to the parameter $k = 100k_c$. For example, the system is oscillatory for all $k_c > 0$, since a pair of complex-conjugate eigenvalues will always be located on branches *A* and *B*. For all $k_c > 5.256$, the system is unstable. The system is marginally stable at approximately $k_c = 5.256$, and for all $0 < k_c < 5.256$ the system is asymptotically stable.

Parameters in the Root Locus   Although the proportional gain k_c (actually $100k_c$) was the subject (varied) parameter in the foregoing illustration, the characteristic equation (10.5a) is neither restricted to any type of controller nor, in fact, to a controlled system. The variable parameter in the root locus can be any quantity in the characteristic equation that by algebraic arrangement can be put in the form of equation (10.5a). Suppose, for example, the subject parameter in the antiaircraft servo system is the integral constant T_i and not the proportional gain k_c. The system characteristic equation can still be put in the necessary root locus form, by taking advantage of the zero RHS of the characteristic equation:

$$1 + k_c \left[\frac{10}{s(s+5)} \right]\left[\frac{10}{s+10} \right]\left[1 + \frac{1}{T_i s} \right] = 0 \Rightarrow$$

$$1 + \frac{100 k_c}{s(s+5)(s+10)} + \frac{100 k_c}{T_i s^2 (s+5)(s+10)} = 0$$

or

$$\frac{s(s+5)(s+10) + 100 k_c}{s(s+5)(s+10)} + \frac{100 k_c}{T_i s^2 (s+5)(s+10)} = 0$$

which gives

$$1 + \frac{100 k_c s(s+5)(s+10)}{T_i s^2 (s+5)(s+10)[s(s+5)(s+10) + 100 k_c]} = 0$$

which gives finally,

$$1 + \left(\frac{1}{T_i} \right) \frac{100 k_c}{s[s(s+5)(s+10) + 100 k_c]} = 0 \qquad (10.9a)$$

Figure 10.7

Root Locus Plot for $G_o(s) = \dfrac{236}{s[s(s+5)(s+10) + 236]}$

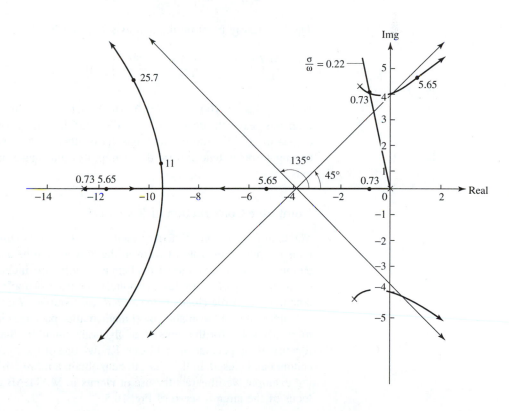

Thus

$$G_o(s) = \frac{100k_c}{s[s(s + 5)(s + 10) + 100k_c]} \qquad \text{(10.9b)}$$

and the root locus parameter is $k = (1/T_i)$.

The root locus plot of the system (10.9a) with k_c fixed at 2.36 is shown in Fig. 10.7. The poles of $G_o(s)$ (there are no zeros; $m = 0$) as determined by computer (see the *root* or *roots* function of your computational mathematics software) are

$$p_1 = 0; \quad p_2 = -12.51112844; \quad p_3, p_4 = -1.2443578 \pm j4.161074982$$

Another illustration of manual plotting of root locus is the following: The real axis loci are as shown because the angle condition is satisfied only between -12.5112844 and 0:

$$-(\beta_1 + \beta_2) = -(180 + 0) = -180°$$

p_3 and p_4 contribute a net angle of zero along the real axis. There are four branches, since $n = 4$. The center of asymptotes on the real axis is

$$\frac{-0 - 12.5112844 - 2(1.2443578)}{4} = -3.75$$

The angles of the asymptotes are

$$\pm\frac{180}{4}; \ \pm\frac{3(180)}{4} = \pm45°; \pm135°$$

The breakaway point on the real axis is given by

$$\frac{d}{ds}\left(\frac{s^4 + 15s^3 + 50s^2 + 236s}{236}\right) = 0$$

or $4s^3 + 45s^2 + 100s + 236 = 0$, which yields the point -9.235. Thus two root locus branches leave the real axis at $s = -9.235$ and tend toward asymptotes that make angles of $\pm135°$ with the real axis. Two other branches originating at the complex-conjugate poles tend toward the asymptotes that make angles of $\pm45°$ with the real axis.

Computer Construction of Root Loci

Manual construction of the root locus as described is still very tedious for many systems in spite of the rules. However, the process can be automated for computer application. Some computational mathematics software include such *root locus* routines especially as part of a relevant applications pack or toolbox. MATLAB, for example, contains the M-file **rlocus** in the *Signals and Systems Toolbox*. Where a formal root locus function is not available, most mathematics packages include a *root* function to numerically solve for the roots of an algebraic equation. Such a function can be applied repetitively to generate a root locus. Knowledge of the rules of the manual plotting procedure can be used, in this case, to help obtain a more efficient solution. In the following example we illustrate the use of **rlocus** in MATLAB to generate and plot the root locus of the aircraft servo of Fig. 10.5.

| Example 10.3 | Finding and Plotting Root Locus by Computer |

Find and plot, by computer, the locus of the roots of

$$1 + k\frac{(s + 1)}{s^2(s + 5)(s + 10)} = 0$$

using the **rlocus** function M-file in MATLAB.

Solution: The MATLAB function rlocus(*num,den*) calculates and plots the locus of roots of the characteristic equation,

$$G(s) = 1 + kG_0(s) = 1 + k\frac{num(s)}{den(s)} = 0$$

with the parameter (open-loop gain factor) k automatically determined. The vectors *num* and *den* specify, respectively, the numerator and denominator coefficients of $G_o(s)$ in descending powers of s. For the present problem, $num(s) = s + 1$, so that the vector $num = [1 \quad 1]$,

$$den(s) = s^2(s + 5)(s + 10) = s^4 + 15s^3 + 50s^2 + 0s + 0$$

so that the vector $den = [1 \quad 15 \quad 50 \quad 0 \quad 0]$.

The script file used to generate the root locus shown in Fig. 10.8 is as follows:

```
% fig10_8.m
% Script to generate root locus shown in Fig. 10.8
%
num = [1 1];                    % numerator coefficients
den = [1 15 50 0 0];           % denominator coefficients
rlocus(num,den);               % find and plot root locus
title('Root Locus by MATLAB');
```

An advantage of computer construction of the root locus that should be evident from the foregoing is that the poles and zeros of $G_o(s)$ need not be known a priori.

The Routh Stability Test

The Routh stability criterion, reference [18], allows one to determine the asymptotic stability from a system's characteristic equation without first having to find the eigenvalues of the system. The test, which consists of two conditions, is applied as follows:

1. Put the characteristic equation of interest into the polynomial form:

$$a_0s^n + a_1s^{n-1} + \ldots + a_{n-1}s + a_n = 0; \qquad a_0 > 0 \qquad \textbf{(10.10)}$$

2. The necessary condition for stability is

$$a_i > 0; \qquad i = 0, 1, 2, \ldots, n \qquad \textbf{(10.11a)}$$

This is only a necessary condition. If it is violated, the system is unstable. If it is satisfied, the system may or may not be asymptotically stable.

Figure 10.8

Root Locus Plot by MATLAB for $G_o(s) = \dfrac{(s + 1)}{s^2(s + 5)(s + 10)}$

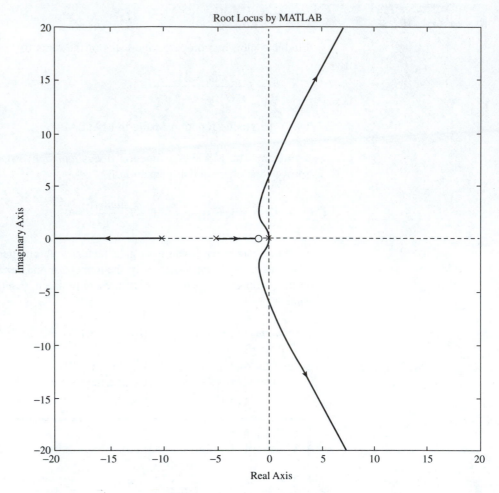

3. The sufficient condition for asymptotic stability (that is, for systems that satisfy the necessary condition) is that if all the numerical values in the first column of the following array:

$$
\begin{array}{cccccc}
a_0 & a_2 & a_4 & a_6 & \cdot & \cdot & \cdot \\
a_1 & a_3 & a_5 & a_7 & \cdot & \cdot & \cdot \\
b_1 & b_2 & b_3 & \cdot & \cdot & \cdot \\
c_1 & c_2 & c_3 & \cdot & \cdot & \cdot \\
\cdot & \cdot & & & & \\
\cdot & \cdot & & & & \\
\cdot & \cdot & & & & \\
\end{array}
$$

are positive, that is,

$$a_0 > 0,\ a_1 > 0,\ b_1 > 0,\ c_1 > 0,\ \text{etc.} \tag{10.11b}$$

then the system is asymptotically stable. Otherwise,

(i) The system is not asymptotically stable, and

(ii) The number of sign changes in the first column is equal to the number of roots in the right half of the complex plane (that is, the unstable roots). This second conclusion is also true for systems that fail the necessary condition.

Construct the array as follows:

a. Arrange the coefficients of the characteristic polynomial according to the pattern

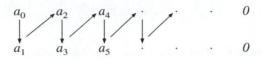

The final elements are zeros.

b. Compute subsequent rows of coefficients according to

$$b_1 = a_2 - \frac{a_0 a_3}{a_1}; \qquad b_2 = a_4 - \frac{a_0 a_5}{a_1}; \qquad b_3 = a_6 - \frac{a_0 a_7}{a_1}; \qquad \text{etc.} \qquad \textbf{(10.11c)}$$

$$c_1 = a_3 - \frac{a_1 b_2}{b_1}; \qquad c_2 = a_5 - \frac{a_1 b_3}{b_1}; \qquad \text{etc.} \qquad \textbf{(10.11d)}$$

Alternatively (and particularly for computer application), if we let the element in the ith row and jth column be $r(i, j)$, then for $i = 3, 4, \ldots, n + 1$,

$$r(i, j) = r(i - 2, j + 1) - \frac{r(i - 2, 1)r(i - 1, j + 1)}{r(i - 1, 1)} \qquad \textbf{(10.11e)}$$

(j) terminates after computing $r(i, j)$, for which $r(i - 2, j + 1)$ is the terminal element of the $(i - 2)$ row. Since this element is zero, the resulting $r(i, j)$ also zero. Note that in this scheme, $a_0 = r(1, 1)$.

Example 10.4	Resolution of Asymptotic Stability by the Routh Test

Determine whether the system specified by the following characteristic equation is asymptotically stable:

$$s^5 + 7s^4 + 18s^3 + 23s^2 + 17s + 6 = 0$$

Solution: The necessary condition is satisfied because all the coefficients are greater than zero. From the array

$i = 1$:	1	18	17	0
$i = 2$:	7	23	6	0
$i = 3$:	$14\frac{5}{7}$	$16\frac{1}{7}$	0	
$i = 4$:	15.32	6	0	
$i = 5$:	10.38	0		
$i = 6$:	6	0		

The sufficient condition is also satisfied, since all the first column elements are positive. Hence the system is asymptotically stable. Note that there are $(n + 1)$ rows in the array, where n is the order of the characteristic polynomial.

Example 10.5	Number of Unstable Roots

Determine the number of unstable roots of the system with the characteristic equation

$$s^4 + 3s^3 + s^2 + 6s - 1 = 0$$

Solution: The system is unstable because it fails the necessary condition of the Routh test—the coefficient (-1) is not positive. However, the Routh array can still be used to find the number of unstable roots. Applying steps (a) and (b) of the array construction procedure, we obtain the array

$$
\begin{array}{rrrr}
1 & 1 & -1 & 0 \\
3 & 6 & 0 & 0 \\
-1 & -1 & 0 & \\
3 & 0 & 0 & \\
-1 & 0 & & \\
\end{array}
$$

Since there are three sign changes in the first column of this array: $(+3/-1)$, $(-1/+3)$ and $(+3/-1)$, the system has three unstable roots.

....

More on the Routh Test

There can be a few complications in the Routh test for some systems. Some examples of these situations, their implications, and procedures for handling them are as follows:

1. If a zero appears in the first column of a row of otherwise nonzero elements:
   a. The normal method of constructing the array will break down.
   b. The system has at least one *unstable* root.
   c. The Routh array can still be found by one of two other methods:
      i. In agreement with physical problems in which parameters are not known precisely, replace the zero with a small positive number ϵ, and algebraically carry on as normal.
      ii. Make a change of variables $s = 1/p$ in the characteristic equation and construct a new array for the characteristic equation in p.
2. If an entire row is zero:
   a. The normal method will break down.
   b. The system has at least one pair of *equal and opposite roots* (real, imaginary, or complex).
   c. The number of such all-zero rows equals the *multiplicity* of the equal and opposite roots.
   d. Proceed with the construction of the Routh array by replacing any all-zero row i with a row consisting of the coefficients of a polynomial obtained by differentiating an auxiliary polynomial equation that is constructed as follows:
      i. The highest power in the auxiliary polynomial is $n - i + 2$.
      ii. Succeeding powers in the polynomial decrease by 2.
      iii. The coefficients of the polynomial are the elements of the $(i - 1)$th row.
   e. The roots of the auxiliary polynomial equation are the equal and opposite roots of the system that caused the all-zero row.

Example 10.6 Marginally Stable System

Use the Routh criterion to determine the number of roots that lie on the right half of the complex plane for the system with the following characteristic equation:

$$s^5 + s^4 + 3s^3 + 3s^2 + 2s + 26 = 0$$

Also, identify the roots that are located symmetrically about the origin.

Solution: Part of the Routh array is as follows:

$$i = 1: \quad 1 \quad 3 \quad 2 \quad 0$$
$$i = 2: \quad 1 \quad 3 \quad 2 \quad 0$$
$$i = 3: \quad 0 \quad 0 \quad 0$$

The third row ($i = 3$) is a row of all zeros. To proceed with the Routh array, we find the highest power of the auxiliary polynomial equation: $n - i + 2 = 5 - 3 + 2 = 4$. The coefficients are 1, 3, 2, 0, so that the auxiliary characteristic equation is

$$s^4 + 3s^2 + 2 = 0 \ (= p^2 + 3p + 2; \qquad p = s^2)$$

The roots of this characteristic equation are

$$s = +j, -j, +\sqrt{2}j, \text{ and } -\sqrt{2}j$$

where $j = \sqrt{-1}$. These are the roots of the original system that are symmetrically located about the origin. (Observe that the original characteristic polynomial is divisible by the auxiliary polynomial, and the factor is $(s + 1)$). Finally, differentiating the auxiliary equation with respect to s, we get

$$4s^3 + 6s = 0$$

Replacing row 3 in the Routh array with the coefficients 4, 6, 0 and proceeding with the normal array construction, we have

$$i = 1: \quad 1 \quad\quad 3 \quad 2 \quad 0$$
$$i = 2: \quad 1 \quad\quad 3 \quad 2 \quad 0$$
$$i = 3: \quad 4 \quad\quad 6 \quad 0$$
$$i = 4: \quad 1.5 \quad 2 \quad 0$$
$$i = 5: \quad \frac{2}{3} \quad 0$$
$$i = 6: \quad 2 \quad\quad 0$$

Because there are no sign changes in the first column, there are no roots in the right-half s-plane. Indeed, the fifth root is given by $(s + 1) = 0$, or $s = -1$.

....

Some Application Concepts

The most straightforward application of the Routh array is the determination of the range of values of a design parameter for which a system of interest is asymptotically stable. Often the characteristic equation of a system under design (say, for purposes of control) may include one or more design parameters that can be carried through to the Routh array. Application of the foregoing Routh stability criteria can then determine limiting values of the parameters for which the stability conditions are satisfied. See, for example, Practice Problems 10.29 and 10.31. Other applications of the Routh array are possible through the use of various transformations; see [19]. The transformation $s = -p$ interchanges the RHP and LHP and therefore allows the determination of the number of roots n_L in the left-half complex plane (the *unstable roots* of the characteristic equation in p). Since an nth order system has n roots, and the Routh array for the characteristic polynomial (in s) gives the number of roots in the RHP n_R, the number of purely imaginary roots can be determined to be $n - n_R - n_L$.

Specification of Speed of Response or Decay Rate   If all the roots of a system under design fall to the left of the line $s = -\sigma$, the speed of response of the system will be faster than that of an exponential decay with time constant $T = 1/\sigma$. The transformation $s = (p - \sigma)$ maps the imaginary axis to the line $s = -\sigma$, so that the Routh array constructed with the characteristic polynomial in p will allow this design specification of speed of response to be met.

Example 10.7   Application on Speed of Response

Consider again the armature-controlled DC motor actuator of Example 8.6. Suppose we wish to determine the spring constant for the valve actuator shaft k (see Fig. 8.5) so that the overall system will have a desired speed of response. For the given parameter values, the characteristic equation given by equation (e) of Example 8.6 can be written as

$$s^3 + 90s^2 + (K + 62000)s + 50K = 0$$

where $K = k/J = 1000k$ is now to be determined. Find the range of values of k such that the decay rate of the controlled system is faster than that of e^{-5t}.

Solution: The desired maximum time constant is $T = \dfrac{1}{5}$. Thus the transformation is $s = (p - 5)$. Substituting this value in the characteristic equation, we obtain

$$p^3 + 75p^2 + (K + 61175) + (45K - 307875) = 0$$

The necessary condition for the stability of the transformed system is given by $45K - 307875 > 0$, which yields $K > 6842$, or $k > 6.842$. The Routh array is

1	$K + 61175$	0
75	$45K - 307875$	0
$0.4K + 65280$	0	
$45K - 307875$	0	

which yields no further condition on K. Thus the design for a minimum speed of response of time constant $= 0.2$ is $k > 6.842$ N $\cdot$ m.

10.3 STABILITY OF DISCRETE-TIME SYSTEMS

We saw in Section 6.4 that the relative stability of linear discrete-time systems is determined by the magnitude of the eigenvalue compared with 1. In general, stability is assured if the magnitude of the eigenvalue is less than 1. Roots with magnitudes greater than 1 result in instability. Eigenvalues of single multiplicity with magnitudes exactly equal to 1 give rise to marginally stable modes. The response pattern for roots on the real axis was illustrated in Fig. 6.22. Negative real eigenvalues give oscillatory modes, whereas real and positive roots are nonoscillatory. Further, as shown in Fig. 10.9 (see also Practice Problem 6.45), eigenvalues with imaginary components (these occur in conjugate pairs), in general, yield oscillatory modes. For unstable roots, the rate of divergence increases with the magnitude of the eigenvalue, while the rate of convergence increases with decreasing magnitude of stable roots.

Based on this understanding of the relative stability of discrete-time systems as a function of eigenvalue placement in the complex plane, the stability of discrete-time systems can be investigated directly by the root locus technique, provided the characteristic equation can be put in the appropriate form.

Example 10.8	Direct Application of Root Locus Method

Determine the stability of the SISO discrete-time system with the following pulse transfer function:

$$G(z) = \frac{1}{z^2 + 2z - 1}$$

If the system is subject to integral feedback control,

$$U(z) = -\left(\frac{kz}{z - 1}\right)Y(z)$$

with unit feedback gain, sketch the (transfer function) block diagram for the controlled system and explore, using the *root locus technique,* its stability behavior.

Figure 10.9

Stability and Discrete-time System Root Placement

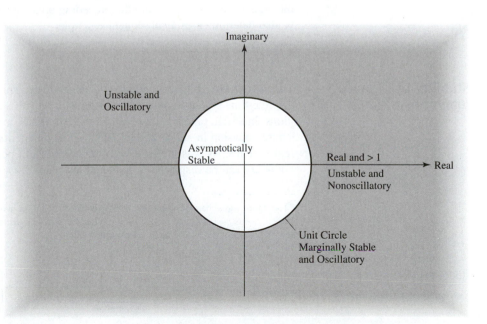

Solution: The characteristic equation of the uncontrolled system is

$$z^2 + 2z - 1 = 0$$

which has the roots $-1 + \sqrt{2}$ and $-1 - \sqrt{2}$, or 0.4142 and -2.4142. Thus one of the eigenvalues lies outside the unit circle in the z-plane, and the system is therefore unstable.

The closed-loop system block diagram is as shown in Fig. 10.10, and from equation (9.4), the closed-loop transfer function is

$$G_{CL}(z) = \frac{kz/[(z-1)(z^2 + 2z - 1)]}{1 + kz/[(z-1)(z^2 + 2z - 1)]} = \frac{kz}{(z-1)(z^2 + 2z - 1) + kz}$$

Thus the controlled system characteristic equation is

$$(z - 1)(z^2 + 2z - 1) + kz = 0$$

or

$$1 + k\frac{z}{(z-1)(z^2 + 2z - 1)} = 0$$

The root locus plot for this characteristic equation with k as the adjustable parameter is shown in Fig. 10.11.* The unit circle is also shown on the same scale. It should be evident that there is no value of k for which all the eigenvalues of this system lie inside the unit circle. The value of k (between 3.5 and 5.0) that places the closed-loop system eigenvalues on the unit circle (marginal stability) can be found by substituting one of the eigenvalues ($z = -1$) into the preceding characteristic equation. The result, $k = 4$, is also determined in a subsequent example.

....

Stability by Transformation to s-Plane

The *imaginary axis* for continuous-time systems, and the *unit circle* (Fig. 10.9) for discrete-time systems constitute *stability boundaries* separating stable and unstable regions. In design practice, these boundaries should be avoided because of the lack of absolute precision in actual design data and the nonstationarity (no matter how small) of real systems. As was the case with the equilibrium point, the location of the stability boundary can be displaced in the complex plane by a suitable transformation of vari-

---

*The figure was generated using the MATLAB **rlocus** function invoked with left-hand arguments [**r,k**] = rlocus(*num,den*) and plotted using plot(**r**,'-'). The gain vector **k** was then available for scaling the root locus (see script M-file **fig10_11.m; On the Net**).

Figure 10.10
_____

System Block Diagram

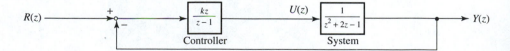

Figure 10.11

Root Locus Plot for

$$G_o(z) = \frac{z}{(z - 1)(z^2 + 2z - 1)}$$

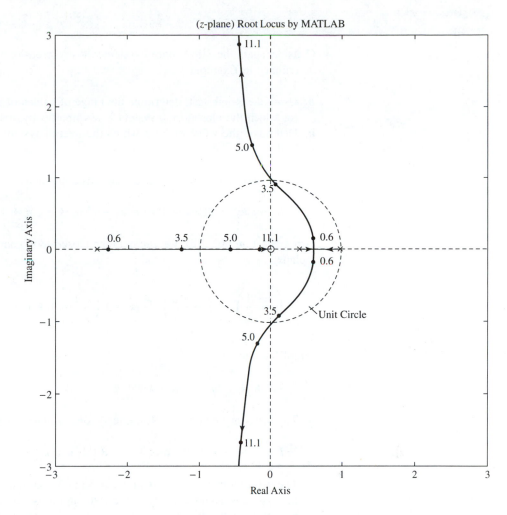

(z-plane) Root Locus by MATLAB

ables. This powerful design technique, which has already been introduced with respect to the application of the Routh test to design for a specified speed of response, will be further illustrated in this section. In particular, the unit circle of the z-plane can be transformed to the vertical imaginary axis of the s-plane using the transformation

$$s = \frac{z + 1}{z - 1} \tag{10.12a}$$

which gives

$$z = \frac{s + 1}{s - 1} \tag{10.12b}$$

which is the transformation mapping the inside of the unit circle in the z-plane into the left-half s-plane. This means that design techniques for investigating the stability of linear continuous-time systems based on the system characteristic equation or function, such as the root locus method and the previously mentioned Routh test (see Section 10.2), can be conveniently applied to linear discrete-time systems if we express the characteristic function in terms of s, using equation (10.12b).

Example 10.9   Application of s-Plane Methods

Consider again the SISO control system whose closed-loop characteristic equation was determined in Example 10.8.

a. Using the Routh test, determine the range of values of the controller gain k, if any, for which the closed-loop system is asymptotically stable.

b. Determine the value of k for which the overall system will operate at the stability limit.

Solution: From Example 10.8, the closed-loop characteristic equation is

$$(z - 1)(z^2 + 2z - 1) + kz = z^3 + z^2 + (k - 3)z + 1 = 0$$

Using equation (10.12b), we can map the closed-loop characteristic equation onto the s-plane:

$$\left(\frac{s+1}{s-1}\right)^3 + \left(\frac{s+1}{s-1}\right)^2 + (k-3)\left(\frac{s+1}{s-1}\right) + 1 = 0$$

Simplifying we have

$$ks^3 + (4 - k)s^2 + (8 - k)s + k - 4 = 0$$

a. The necessary condition (Routh test) for asymptotic stability is

$$k > 0, (4 - k) > 0 \quad \text{or} \quad 4 > k, 8 > k, \text{ and } k > 4$$

The two requirements $4 > k$ and $k > 4$ imply that there is no value of k for which the system is asymptotically stable. Although it is not necessary to proceed further, note that the Routh array for this characteristic equation is

$i = 1$:	k	$8 - k$	0
$i = 2$:	$4 - k$	$k - 4$	0
$i = 3$:	8	0	
$i = 4$:	$k - 4$	0	

and the necessary condition that the first-column elements be positive also yields the same conclusion previously reached.

b. Operation at the stability limit implies that the closed-loop eigenvalues all lie on the unit circle in the z-plane. In the s-plane, the stability limit implies that the roots lie on the imaginary axis. On the imaginary axis, $s = j\omega$. Hence

c. $k - 4 - (4 - k)\omega^2 + j\omega[(8 - k) - k\omega^2] = 0$

This equation must be satisfied by both the real and imaginary parts, that is,

$$\omega[(8 - k) - k\omega^2] = 0 \quad \text{and} \quad (k - 4)(\omega^2 + 1) = 0$$

From which, $\omega = 0$ and ± 1, and $k = 4$

The solution is thus $k = 4$. As a check, note that the three roots in the s-plane are $s = 0$, $s = +j$, and $s = -j$. From equation (10.12b), the corresponding closed-loop eigenvalues are $z = -1$, $z = +j$, and $z = -j$, which are all on the unit circle. Further, the characteristic equation for these roots is

$$(z - j)(z + j)(z + 1) = z^3 + z^2 + z + 1$$

which is the same as the controlled system characteristic equation above, with $k = 4$.

....

Jury's Inners Stability Test

In this section we consider the following discrete-time system characteristic equation:

$$F(z) = d_0 z^n + d_1 z^{n-1} + \ldots + d_{n-1} z + d_n = 0 \tag{10.13}$$

Like the Routh test for continuous-time systems, the *Jury test* for asymptotic stability of a discrete-time system (see [2]) applies directly to equation (10.13). The method consists of two parts, as follows:

(1) Necessity Test Substitute $z = 1$ and $z = -1$ in $F(z)$ and require thus

$$F(1) > 0 \text{ and } (-1)^n F(-1) > 0 \tag{10.14a}$$

This test ensures that all real eigenvalues are within the ± 1 stability range.

(2) Sufficiency Test Construct the $(n - 1) \times (n - 1)$ triangular matrices $\mathbf{V}$ and $\mathbf{W}$, using the coefficients of $F(z)$ thus

$$\mathbf{V} = \begin{bmatrix} d_0 & d_1 & d_2 & \cdots & d_{n-2} \\ 0 & d_0 & d_1 & \cdots & d_{n-3} \\ 0 & 0 & d_0 & \cdots & d_{n-4} \\ \vdots & & & & \vdots \\ 0 & 0 & 0 & \cdots & d_0 \end{bmatrix}; \tag{10.14b}$$

$$\mathbf{W} = \begin{bmatrix} d_2 & d_3 & d_4 & \cdots & d_{n-1} & d_n \\ d_3 & d_4 & d_5 & \cdots & d_n & 0 \\ d_4 & d_5 & \cdots & d_n & 0 & 0 \\ d_5 & \cdots & d_n & 0 & 0 & 0 \\ \vdots & & 0 & & \vdots & \vdots \\ d_n & 0 & \cdots & 0 & 0 & 0 \end{bmatrix}$$

and obtain the two matrices $\mathbf{H}_1$ and $\mathbf{H}_2$ representing their sum and difference:

$$\mathbf{H}_1 = \mathbf{V} + \mathbf{W}; \qquad \mathbf{H}_2 = \mathbf{V} - \mathbf{W} \tag{10.14c}$$

Then $\mathbf{H}_1$ and $\mathbf{H}_2$ must be *positive innerwise* for the system to be asymptotically stable. That is, the determinants of the matrices, starting from the center element(s) and proceeding outward to the entire matrix, must be positive. If n is odd, the center element is a 2×2 matrix, whereas it is a scalar quantity when n is even.

| **Example 10.10** | Application of Jury's Test |

Determine the asymptotic stability of the discrete-time system with the fourth-order z-domain characteristic equation

$$5z^4 - 6.2z^3 + 4.4z^2 - 5.8z + 2.8 = 0$$

Solution: The necessary conditions, equation (10.14a), are satisfied:

$$5 - 6.2 + 4.4 - 5.8 + 2.8 = 0.2 > 0 \quad \text{and}$$

$$5 + 6.2 + 4.4 + 5.8 + 2.8 = 24.2 > 0$$

For the sufficiency test, we compute from equations (10.14b) and (10.14c),

$$\mathbf{H}_1 = \begin{bmatrix} 5 & -6.2 & 4.4 \\ 0 & 5 & -6.2 \\ 0 & 0 & 5 \end{bmatrix} + \begin{bmatrix} 4.4 & -5.8 & 2.8 \\ -5.8 & 2.8 & 0 \\ 2.8 & 0 & 0 \end{bmatrix} = \begin{bmatrix} 9.4 & -12.0 & 7.2 \\ -5.8 & 7.8 & -6.2 \\ 2.8 & 0 & 5 \end{bmatrix}$$

and

$$\mathbf{H}_2 = \begin{bmatrix} 0. & -0.4 & 1.6 \\ 5.8 & 2.2 & -6.2 \\ -2.8 & 0 & 5 \end{bmatrix}$$

so that the conditions to be satisfied are

$$7.8 > 0 \quad \text{and} \quad 2.8[-12.0(-6.2) - 7.2(7.8)]5[9.4(7.8) - 12.0(5.8)]$$

$$= 69.672 > 0$$

$$2.2 > 0 \quad \text{and} \quad -2.8[-6.2(-0.4) - 2.2(1.6)] + 5[0.6(2.2) + 5.8(0.4)]$$

$$= 21.112 > 0$$

where the 3×3 determinants were computed by cofactor expansion along the bottom row of each matrix (see Appendix B). Because the sufficiency conditions are also met, the discrete-time system is asymptotically stable. An application of the Jury test in design is given in Example 14.1.

....

10.4 STABILITY IN THE FREQUENCY DOMAIN

In Section 9.2 we introduced frequency response diagrams and frequency domain identification. The frequency response diagram could be used in system identification because the diagram is simply a form of representation of the system dynamics. Here we would examine a system's frequency response to deduce its stability behavior. We would also consider specifications of some performance measures for relative stability. However, whereas similar time domain or transient response specifications such as percent overshoot, settling time, and degree of damping are based on the closed-loop or overall system characteristics, some frequency domain performance specifications are associated with the so called *open-loop design technique,* while others are based on the closed-loop system characteristics. Open-loop frequency domain design involves the determination of the closed-loop controller using the system's open-loop transfer function or characteristics. The Bode and Nyquist diagrams, introduced in Section 9.2, are examples of (well-established and popular) open-loop design tools. On the other hand, closed-loop frequency domain design requires the closed-loop characteristics. However, the common practice is still to obtain first the closed-loop frequency response using the open-loop frequency response data (diagrams).

The Closed-Loop Frequency Response

Consider the closed-loop control system (frequency domain) block diagram shown in Fig. 10.12. We define the **open-loop** transfer function as the *measured output* transfer function

$$G_o(j\omega) = G_p(j\omega)G_c(j\omega)H(j\omega) \tag{10.15a}$$

The **closed-loop** transfer function is, according to equation (9.4),

$$G_{CL}(j\omega) = \frac{G_p(j\omega)G_c(j\omega)}{1 + G_p(j\omega)G_c(j\omega)H(j\omega)} \tag{10.15b}$$

In terms of the open-loop transfer function (10.15a), the closed-loop transfer function is

$$G_{CL}(j\omega) = \frac{G_p(j\omega)G_c(j\omega)}{1 + G_o(j\omega)} \tag{10.15c}$$

On Bode Diagram   Equation (10.15c) suggests that given $H(j\omega)$, $G_{CL}(j\omega)$ can be approximated from the Bode plot of the open-loop frequency response $G_o(j\omega)$:

$$\text{If } |G_o(j\omega)| \gg 1, \text{ then } G_{CL}(j\omega) \longrightarrow \frac{G_p(j\omega)G_c(j\omega)}{G_o(j\omega)} = \frac{1}{H(j\omega)}$$

and

$$\text{If } |G_o(j\omega)| \ll 1, \text{ then } G_{CL}(j\omega) \longrightarrow G_p(j\omega)G_c(j\omega) = \frac{G_o(j\omega)}{H(j\omega)}$$

In particular, if a *unity feedback system* ($H(j\omega) = 1$) can be assumed, then the closed-loop frequency response is a function of the open-loop frequency response only:

$$G_{CL}(j\omega) = \frac{G_o(j\omega)}{1 + G_o(j\omega)} \tag{10.15d}$$

Figure 10.12

Reference Structure for Closed-loop Frequency Response Analysis

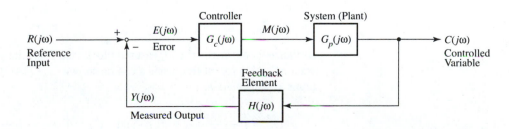

For many practical systems, the open-loop gain at low frequencies is usually greater than 1 and generally increases as frequency decreases, whereas at high frequencies the open-loop gain is often less than 1 and normally decreases as the frequency increases. Thus, as shown in Fig. 10.13a (gain plot), for a unity feedback system ($H(j\omega) = 1$), the approximate closed-loop frequency response is given by the 0 db line for low frequencies, up to the frequency at which the open-loop gain is 1 (0 db), and thereafter it is given by the open-loop frequency response. The forcing frequency ω_g at which the (open-loop) gain is 0 db is known as the *gain crossover frequency:*

$$|G_o(j\omega_g)| = 1 \quad \text{or} \quad 20 \log |G_o(j\omega_g)| = 0 \tag{10.16}$$

The corresponding approximate closed-loop phase plot, assuming a minimum phase system, will be 0° for low frequencies up to ω_g, after which it will follow the $\angle[G_o(j\omega)]$ line.

On Nyquist Diagram. The closed-loop frequency response for unity feedback systems can also be obtained (exactly in this case) from the open-loop Nyquist locus. Consider the typical Nyquist plot $G_o(j\omega)$ shown in Fig. 10.13b. At any point P on this locus the closed-loop phasor, equation (10.15d) repeated here,

$$G_{\text{CL}}(j\omega) = \frac{G_o(j\omega)}{1 + G_o(j\omega)} \tag{10.17a}$$

is given by

$$G_{\text{CL}}(j\omega) = \frac{\text{phasor OP}}{\text{phasor CP}} \tag{10.17b}$$

since "vectorially" (recall the rectangular or vector-type representation of complex numbers) **OP** = **OC** + **CP**, and point C is such that **OC** = $-1 + 0j$ and **OP** is of course $G_o(j\omega)$, so that **CP** = $1 + G_o(j\omega)$. The closed-loop frequency response (10.17a) can as usual be represented by two quantities, a magnitude and a phase:

$$M = |G_{\text{CL}}(j\omega)| = \frac{\text{OP}}{\text{CP}} \tag{10.17c}$$

$$\phi = \angle[G_{\text{CL}}(j\omega)] = \angle(\text{CPO}) \tag{10.17d}$$

To see the last result, note from Fig. 10.13b and equation (10.17b) that, measuring clockwise,

$$\phi = [\angle(\text{DOP}) - \angle(\text{DCP})]$$

But by geometry,

$$\angle(\text{DOP}) = \angle(\text{DCP}) + \angle(\text{CPO})$$

so that ϕ is negative when measured clockwise from the phasor CP. The loci of constant M and ϕ for all frequencies can be shown (see [14]) to be circles on the Nyquist plane with centers at

$$C_M = \left(-\frac{M^2}{M^2 - 1} + j0\right) \quad \text{and} \quad C_\phi = \left(-\frac{1}{2} + j\frac{1}{2 \tan \phi}\right) \tag{10.18a}$$

Figure 10.13

From Open-loop to Closed-loop Frequency Response

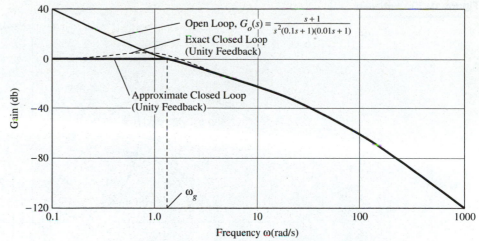

(a)   Bode Plot Approximation of Closed-loop Frequency Response

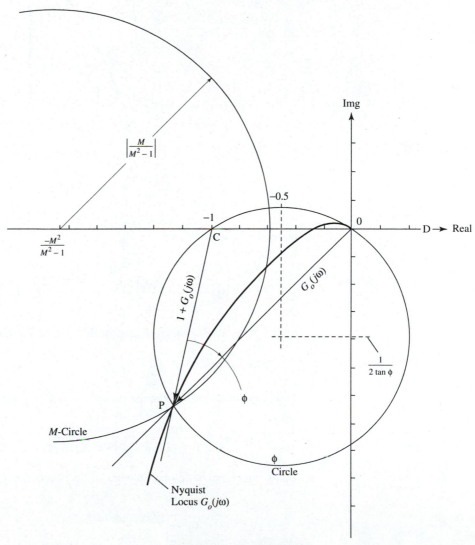

(b)   Phasors, M and ϕ Circles

Figure 10.13 *continued*

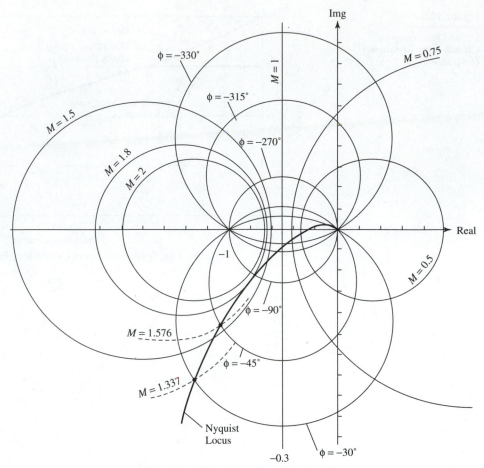

(c)   Closed-loop Data from Nyquist Plot

respectively, and radii of

$$R_M = \left| \frac{M}{M^2 - 1} \right| \quad \text{and} \quad R_\phi = \frac{\sqrt{\tan^2\phi + 1}}{|2\tan\phi|} \tag{10.18b}$$

respectively. To demonstrate this for the case of the magnitude, let

$$G_o(j\omega) = X + jY \tag{10.19a}$$

Thus any locus on the Nyquist plane is a plot of Y versus X. Proceeding, we have

$$G_{CL}(j\omega) = \frac{G_o(j\omega)}{1 + G_o(j\omega)} = \frac{X + jY}{(1 + X) + jY} \tag{10.19b}$$

Using the complex number algebra of Section 6.2, we obtain

$$M^2 = \frac{X^2 + Y^2}{(1 + X)^2 + Y^2} \tag{10.19c}$$

so that

$$M^2[(1 + X)^2 + Y^2] = X^2 + Y^2$$

Expanding the expression, we get

$$(M^2 - 1)X^2 + (M^2 - 1)Y^2 + 2XM^2 = -M^2$$

Dividing by $(M^2 - 1)$, we have

$$X^2 + Y^2 + 2X\left(\frac{M^2}{M^2 - 1}\right) = -\frac{M^2}{M^2 - 1}$$

Finally, completing the square on the LHS by adding $[M^2/(M^2 - 1)]^2$ to both sides, we obtain

$$\left(X + \frac{M^2}{M^2 - 1}\right)^2 + Y^2 = \left(\frac{M}{M^2 - 1}\right)^2 \tag{10.19d}$$

This is the equation of a circle with radius and center as previously stated. The M- and ϕ-circles (as they are called) that pass through the point P on the Nyquist locus are shown in Fig. 10.13b. These circles correspond to the frequency of $G_o(j\omega)$ at P. At any other frequency (any other point), a pair of intersecting M- and ϕ-circles can also be determined as shown in Fig. 10.13c (see $M = 1.337$, $\phi = -30°$; $M = 1.576$, $\phi = -45°$; for example), so that the closed-loop frequency response (that is, values of M and ϕ as functions of frequency) can be generated from the open-loop Nyquist locus. Observe in Fig. 10.13c that the phase angle above the real axis is 180° less than the phase below the real axis, for the same ϕ-circle. Also the line $(-0.5 \pm j\omega)$ is coincident with the $(M = 1)$-circle and is a line of symmetry.

Bandwidth and Resonance Peak

Bandwidth was defined in Section 7.2 as the range of frequencies (of the input) over which the system response is satisfactory. A common quantification of bandwidth that is similar to equation (7.30c) is that *bandwidth is the frequency range over which the (overall or closed-loop output over input) magnitude ratio does not differ by more than −3 db from the magnitude ratio at a specified frequency.* Recall that $20 \log(\sqrt{2}/2) = -3$, so that if M_{max} is viewed as the (desired magnitude ratio at the specified frequency, then equation (7.30c) also represents the current definition of bandwidth. When we apply this new definition of bandwidth to the approximate closed-loop frequency response illustrated by Fig. 10.13a, it is evident that the *approximate closed-loop bandwidth is the gain crossover frequency* ω_g. Also, from the reciprocal relation between the equivalent time constant and the corner frequency, equation (9.18c), we conclude that, in general, the larger the bandwidth, the faster the speed of response, so that *bandwidth is primarily a measure of speed of response.* However, the relationship has been re-examined here in the context that the bandwidth of the closed-loop system is an approximate function of the open-loop frequency response.

Resonance Peak   The behavior of systems considered in Section 7.2 was also characterized by a peak amplitude of vibration (*resonance peak* M_p) and the corresponding *frequency at resonance* ω_R, which is the input frequency at which the peak amplitude occurs (not necessarily the *resonant,* or *natural, frequency*). We are concerned presently with these quantities for a closed-loop control system, and with the *resonance peak as a measure of the relative stability* of such a system. The resonance peak for a unity feedback system M_p can be determined directly from the Nyquist diagram of the open-loop transfer function as the largest value of M of the M-circle(s) tangent to the Nyquist locus. Note that more than one M-circle can be tangent to $G_o(j\omega)$; however, the one with the largest value of M yields M_p. In Fig. 10.13c, the largest value of M of an M-circle tangent to the Nyquist locus appears to be $M_p = 1.8$. The value of the frequency at this tangent point is ω_R.

Frequency domain design for specified bandwidth or resonance peak is discussed in Chapter 12.

Relative Stability and Gain and Phase Margins

Due to the inherent delays in a system, there is usually a phase lag between the system input and output. The phase shift, as can be seen in the Bode diagrams of Section 9.2, is normally a function of frequency. Also, the output-over-input amplitude ratio is often a decreasing function of the forcing frequency. For a feedback control system with a sinusoidal reference input, the output and hence the feedback signal will normally lag behind the input (see Fig. 10.14; the output and input frequencies are assumed to be the same or very close to each other). However, at some forcing frequency $\omega = \omega_p$, this phase lag (which is due to the delays inherent in the controlled system) will approach 180° ($\phi = -180°$), so that the feedback signal will add to the error input instead of subtracting from it (see Fig. 10.14c). If in addition the (output over error input, that is, the *open-loop*) amplitude ratio at this same frequency is greater than or equal to 1 (gain db ≥ 0 db), then the subsequent output amplitude will continue to increase, resulting ultimately in very large amplitude oscillations or instability. These observations can be further explained by noting that the amplitude of the response of the closed-loop sys-

Figure 10.14

Gain, Phase Shift, and Instability

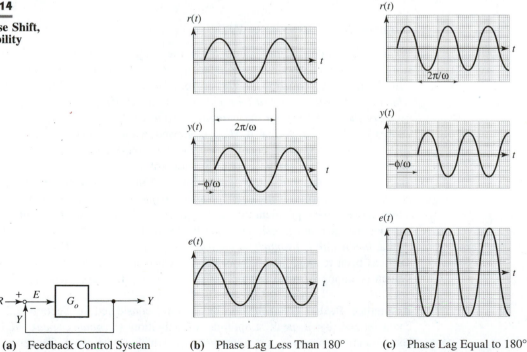

(a)   Feedback Control System   **(b)**   Phase Lag Less Than 180°   **(c)**   Phase Lag Equal to 180°

tem would become arbitrarily large if the denominator of the closed-loop frequency transfer function, equation (10.15c), should become zero:

$$1 + G_o(j\omega) = 0 \qquad\qquad\qquad\qquad\qquad \textbf{(10.20a)}$$

Hence instability occurs when

$$G_o(j\omega) = -1 \qquad\qquad\qquad\qquad\qquad \textbf{(10.20b)}$$

That is, *the system is at the stability limit for any frequency at which the open-loop phase lag is 180° and the open-loop gain is simultaneously 1.* Consequently, one of the usual objectives in the open-loop method of controller design is to separate by as large a frequency range as possible the points of occurrence of the critical (open-loop) phase of $-180°$ and (open-loop) gain of 1 (0 db). This is often accomplished by specifying either the *gain margin* or the *phase margin*

The **gain margin** is the factor by which the gain differs from (is smaller than) 1 (0 db) at the forcing frequency for which the phase is $-180°$

$$\text{Gain margin} = \frac{1}{|G_o(j\omega_p)|}; \qquad (>1)$$

or

$$\text{Gain margin (db)} = -20 \log|G_o(j\omega_p)|; \qquad (>0) \qquad\qquad \textbf{(10.21a)}$$

where

$$\angle[G_o(j\omega_p)] = -180° \qquad\qquad\qquad\qquad \textbf{(10.21b)}$$

and ω_p, the frequency at which the phase is $-180°$, is called the *phase crossover frequency*.

The **phase margin** is the amount by which the phase differs from (is greater than) $-180°$ at the forcing frequency for which the gain is 0 db:

$$\text{Phase margin} = \angle[G_o(j\omega_g)] - (-180°); \qquad (>0) \qquad\qquad \textbf{(10.22)}$$

where ω_g, the frequency at which the gain is 0 db, is the *gain crossover frequency* that was defined in equation (10.16).

The gain and phase margins are thus (open-loop) measures of (closed-loop) relative stability. Figure 10.15 illustrates these specifications in Bode and Nyquist formats. Observe in Fig. 10.15a that the phase crossover frequency and the gain crossover frequency coincide (and become the frequency of sustained constant-amplitude oscillation ω_p) at the stability limit (shown by the gain curve in dashed lines). The stability limit curve in the Nyquist format (see Fig. 10.15b) passes through the $(-1 + j0)$ point. Also, at the stability limit, the gain margin (in db) and the phase margin are simultaneously zero. For the same phase data in Fig. 10.15a, a (parallel) gain curve above the stability limit curve (that is, for $k > 97.9$) will result in an unstable closed-loop system, for which the concepts of gain and phase margins are meaningless (the gain and phase margins are defined to be positive for a stable system). *A gain margin of at least 8 db and a phase margin of at least 30° have been found in practice to give reasonably good closed-loop performance.*

Figure 10.15

Definition of Gain and Phase Margins

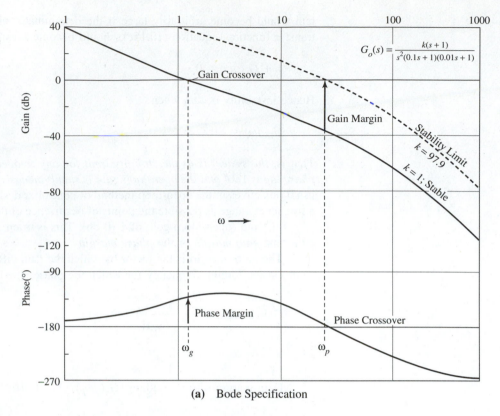

$$G_o(s) = \frac{k(s + 1)}{s^2(0.1s + 1)(0.01s + 1)}$$

(a)   Bode Specification

Example 10.11   Analysis of Gain and Phase Margins

Given the open-loop transfer function

$$G(j\omega) = \frac{ke^{-j\,\pi/8}}{j\omega(1 + j\omega)}$$

if the gain is adjusted to provide a phase margin of 45°, what is the gain margin? What is the approximate bandwidth of the closed-loop system?

Solution: The problem can be solved either graphically or analytically. The graphical solution is left as an exercise (see Practice Problem 10.41). The open-loop transfer function can be written as

$$\frac{ke^{-j\,\pi/8}}{j\omega(1 + j\omega)} = \frac{ke^{-j\,\pi/8}}{\omega\sqrt{1 + \omega^2}\,e^{j[\pi/2 + \tan^{-1}(\omega)]}}$$

so that

$$\angle G(j\omega) = -\frac{\pi}{8} - \frac{\pi}{2} - \tan^{-1}(\omega)$$

Figure 10.15 *continued*

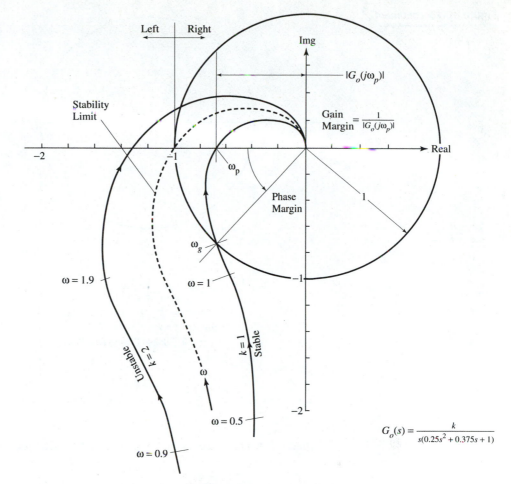

(b) Nyquist Specification

From equation (10.22),

Phase margin $= \angle G(j\omega_g) + 180°$

where ω_g is the gain crossover frequency:

$$-\frac{\pi}{8} - \frac{\pi}{2} - \tan^{-1}(\omega_g) + \pi = \frac{\pi}{4} \Rightarrow \tan^{-1}(\omega_g) = \frac{\pi}{8}$$

so $\omega_g = 0.4142$.

By definition of the gain crossover frequency,

$$|G(j\omega_g)| = 1 = \frac{k}{\omega_g\sqrt{1 + \omega_g^2}} \Rightarrow k = 0.448$$

From equation (10.21a),

Gain margin $= -20 \log|G(j\omega_p)|$

Figure 10.15 *continued*

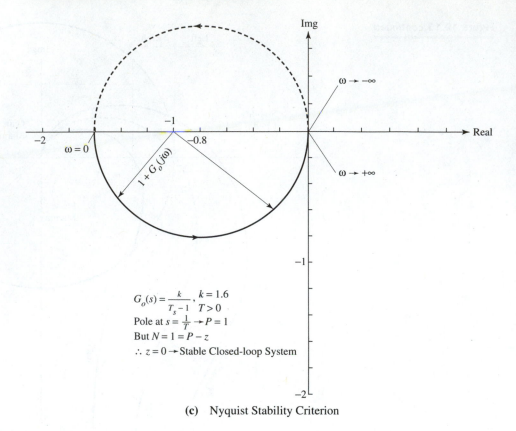

$$G_o(s) = \frac{k}{T_s - 1}, \quad \begin{matrix} k = 1.6 \\ T > 0 \end{matrix}$$

Pole at $s = \frac{1}{T} \rightarrow P = 1$

But $N = 1 = P - z$

$\therefore z = 0 \rightarrow$ Stable Closed-loop System

(c) Nyquist Stability Criterion

where ω_p = phase crossover frequency, that is, $\angle G(j\omega_p) = -180°$ or

$$-\frac{\pi}{8} - \frac{\pi}{2} - \tan^{-1}(\omega_p) = -\pi$$

or

$$\tan^{-1}(\omega_p) = \frac{3\pi}{8} \Rightarrow \omega_p = 2.4142$$

The gain margin is therefore given by

$$\text{Gain margin} = -20 \log\left(\frac{0.448}{2.4142\sqrt{1 + 2.4142^2}}\right) = 22.97 \text{ db}$$

The approximate closed-loop system bandwidth is the gain crossover frequency: $\omega_g = 0.4142 \text{ s}^{-1}$.

....

Nyquist Stability Criterion

To understand the Nyquist specification shown in Fig. 10.15b, recall that increasing the open-loop gain above the stability limit value will result in an unstable closed-loop system. In the Nyquist plane, as it is illustrated in Fig. 10.15b, raising or lowering the open-loop gain relative to the critical value will result in a shift of the Nyquist plot of

$G_o(j\omega)$ to the left or to the right, respectively, of the $(-1 + j0)$ point. Assuming that the open-loop response $G_o(s)$ is stable (that is, there are no poles of $G_o(s)$ in the RHP) and that the power of s in the denominator of $G_o(s)$ is greater than or equal to the power of s in the numerator, (one form of) the **Nyquist stability criterion,** under these conditions, is that

> The closed-loop system is unstable if the point $(-1 + 0j)$ lies to the right-hand side of the plot of $G_o(j\omega)$ while proceeding along the Nyquist locus in the direction of increasing ω.

Since many engineering systems normally meet the conditions under which this criterion is applicable, it is a convenient form of the Nyquist stability theorem. However, a more general statement of the stability criterion, which does not require a stable open-loop as a precondition, is that

> A system is stable if and only if the net number of counterclockwise rotations of the "vector" from $(-1 + 0j)$ to a point on the curve of $G_o(j\omega)$ (that is, the vector $1 + G_o(j\omega)$) as ω varies from $-\infty$ to $+\infty$ is equal to the number of poles of $G_o(j\omega)$ in the right-half complex plane. Note that, in this regard, the locus of the "vector" (actually, phasor) $G_o(j\omega)$ as ω goes from $-\infty$ to 0 is the mirror image about the real axis of the locus for ω varying from $+\infty$ to 0, hence the number of rotations of the $1 + G_o(j\omega)$ vector as ω goes from $-\infty$ to $+\infty$ is twice the number of rotations of the same vector for ω varying from 0 to $+\infty$.

The theoretical basis of the Nyquist criterion is beyond the scope of this book; however, fairly simple but informal proofs of the theory are given in references [3] and [17]. We shall concentrate here on the application of the theory. Figure 10.15c illustrates the generalized Nyquist criterion. If the number of poles of $G_o(s)$ in the RHP is P, and the number of roots of the closed-loop system (zeros of $1 + G_o(s)$) in the RHP is Z, then the number of net counterclockwise rotations N of the vector from $(-1 + 0j)$ to a point on the curve $G_o(j\omega)$ as ω is varied from $-\infty$ to $+\infty$ is equal to $P - Z$. That is,

$$N = P - Z \tag{10.23a}$$

or

$$2N_0 = P - Z \tag{10.23b}$$

where N_0 is in this case the number of counterclockwise rotations of the $1 + G_o(j\omega)$ vector as ω varies from 0 to $+\infty$. If in equation (10.23a) $N = P$, as required by the Nyquist criterion, then $Z = 0$, that is, there are no roots of the closed-loop system in the RHP, which is the usual condition for stability. In terms of the transfer function representation of equation (9.15a)

$$G_o(s) = \frac{B(s)}{A(s)}$$

so that P is the number of unstable roots (roots in the RHP) of the polynomial $A(s)$, and Z is the number of unstable roots of the polynomial $B(s) + A(s)$, that is, the unstable roots of the closed-loop characteristic equation. In the example shown in Fig. 10.15c, $A(s) = Ts - 1$, so that $P = 1$ (the unstable root is $s = 1/T$). The number of counterclockwise rotations of the vector from $(-1 + j0)$ is one, so that $Z = 0$. The closed-loop characteristic equation $(k + Ts - 1 = 0)$ has the root $s = (1 - k)/T$, so that the system is indeed stable for all $k > 1$.

In equations (10.23a) and (10.23b), if $P = 0$, then $N = 0$ is required for stability. Thus another useful statement of the convenient form of the Nyquist theorem is as follows:

> If $G_o(s)$ has no poles in the RHP, then the closed-loop system is (asymptotically) stable if and only if the Nyquist locus of $G_o(s)$ does not encircle the $(-1 + j0)$ point as ω passes from $-\infty$ to $+\infty$.

Figure 10.16a illustrates this criterion for an arbitrary $G_o(s)$ that is assumed to have no unstable poles. Observe in Fig. 10.16a that if the region to the right of the locus is shaded as it is traversed, with ω going from 0 to $+\infty$, stability is conveniently determined according to whether or not the $(-1 + j0)$ point falls within the shaded region. The rest of Fig. 10.16 shows further illustrations of this criterion along with the application of the general theory. Note that the number of encirclements of the $(-1 + j0)$ point is a signed quantity. N increases in the positive sense for counterclockwise rotations of the $1 + G_o(j\omega)$ vector as ω varies from 0 to $+\infty$. For clockwise rotations, N is negative. The indication of *type* shown in these figures is determined according to a representation of the open-loop transfer function in the form of equation (9.15a). The power of s in the denominator q determines the type ($q = 0$ for type 0, $q = 1$ for type 1, etc.).

10.5 STABILITY AND NONLINEAR SYSTEMS

The application of the state space (including phase space or phase plane) method to the analysis of some common nonlinear systems was introduced very briefly in Section 9.4. When combined with digital computer simulation (discussed in Section 6.4), this approach represents the most general means of analysis and design of nonlinear systems. This is so because digital computer solutions are flexible, so that there are few if any restrictions on the type of nonlinear system or on the nature of the design problem. Digital computer simulation is also a powerful technique for analysis of nonlinear systems because it is effective in handling most of the special problems that arise from nonlinearity. These include the absence of general mathematical solutions for many nonlinear relations; the nonapplicability of the principle of superposition; the dependence of the response (including stability behavior) of many nonlinear systems on the initial conditions and on the nature of the input variables; and the introduction of unusual behavior characteristics that are peculiar to nonlinear systems, such as *limit cycles, hysteresis,* and *self-excitation.* Because of these problems, most of the other techniques of analysis for nonlinear systems are either approximate or restricted to specific system types.

Some of the methods considered in Sections 9.4 and 10.1 with respect to representation of system dynamics and stability behavior can be applied to certain nonlinear systems. The *linearization* approach used previously to determine local controllability of nonlinear systems can also be used to extend stability concepts for linear systems to neighborhoods of operating or equilibrium points of nonlinear systems in which the deviation from linear behavior can be presumed to be small. An extension of this concept is the *piecewise-linear* approach, in which entire nonlinear systems can be approximated by several linear regions. Given the previous detailed treatment of linear systems, these approaches will not be covered here. The *Lyapunov second method* introduced in Section 10.1 is applicable to both linear and nonlinear systems provided an appropriate Lyapunov function can be generated (see Practice Problem 10.8). Although certain classes of nonlinear problems lend themselves to formal procedures for constructing suitable Lyapunov functions (see [17], for example), we will not pursue that topic here.

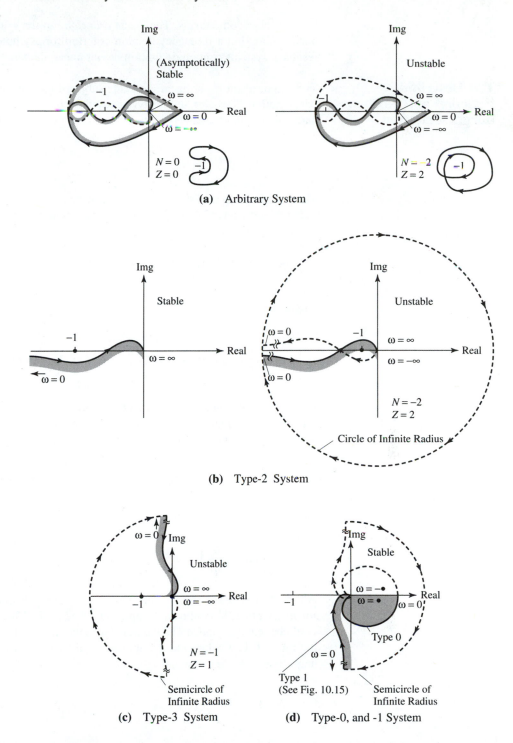

(a) Arbitrary System

(b) Type-2 System

(c) Type-3 System

(d) Type-0, and -1 System

In line with the discussions in the last section, we shall consider first, in an abridged manner, an approximate frequency domain analysis of the stability of systems that include nonlinear elements. The *describing function* is essentially an approximate frequency transfer function of a nonlinear component that permits the extension of some of the powerful transfer-function methods for linear systems to systems containing nonlinear components. The *circle criterion,* on the other hand, is not an approximate method but a frequency domain criterion for asymptotic stability of single-loop feedback systems containing time-invariant linear elements and nonlinear components.

The Describing Function and Kochenburger Criterion

The output of most nonlinear elements subject to a sinusoidal input is usually a periodic function. However, unlike the situation for linear components, this output contains harmonics of the fundamental frequency. Consider the input sinusoid

$$u = A \sin(\omega t) \tag{10.24a}$$

The periodic output, $y = y(u) = f(\omega t)$, can be described by a Fourier series (see Section 7.4, equations (7.42) and (7.43); let $t_0 = 0$ and $\Gamma = 2\pi/\omega$),

$$y = \frac{1}{2}p_0 + \sum_{n=1}^{\infty} [p_n \cos(n\omega t) + q_n \sin(n\omega t)] \tag{10.24b}$$

where

$$p_n = \frac{1}{\pi} \int_0^{2\pi} y \cos(n\omega t) d(\omega t); \qquad q_n = \frac{1}{\pi} \int_0^{2\pi} y \sin(n\omega t) d(\omega t) \tag{10.24c}$$

Recall that if the output is assumed to be *zero mean,* then the term p_0 vanishes. An approximate technique due to Kochenburger, reference [20], is to represent y (approximately) by only the fundamental or first component:

$$y \cong p_1 \cos(\omega t) + q_1 \sin(\omega t) = \sqrt{p_1^2 + q_1^2} \sin(\omega t + \phi) \tag{10.24d}$$

where

$$\phi = \tan^{-1}\left(\frac{p_1}{q_1}\right) \tag{10.24e}$$

Although it does not apply to all nonlinear systems, this approximation is reasonable for many practical nonlinear components. The *describing function* that *describes* the nonlinear element, much like the transfer function of a linear system, is then defined as the ratio of the fundamental component of the output of the nonlinear element to the amplitude of the input signal. That is,

$$N(A) = \frac{1}{A}(q_1 + jp_1) \tag{10.25}$$

or

$$|N| = \frac{\text{amplitude of } y}{\text{amplitude of } u} = \frac{\sqrt{p_1^2 + q_1^2}}{A} \tag{10.26a}$$

$$\angle N = (\text{phase of } y) - (\text{phase of } u) = \tan^{-1}\left(\frac{p_1}{q_1}\right) \tag{10.26b}$$

Table 10.1 gives the describing functions for the actuator nonlinearities shown in Fig. 9.21. A more extensive table of describing functions is given in reference [21]. Note that for the *single-valued* nonlinearities,

$$|N| = \left(\frac{q_1}{A}\right); \qquad \angle N = 0 \tag{10.26c}$$

We illustrate next the evaluation of describing functions, using the case of *dead zone* (the second entry in the table).

Table 10.1 Describing Functions for Some Nonlinearities

Characteristic	Describing Function	Characteristic	Describing Function
Saturation	$\angle N = 0$ $\|N\| = \frac{2M}{\pi h} \sin^{-1}\left(\frac{h}{A}\right) +$ $\frac{2M}{\pi A}\sqrt{1 - \left(\frac{h}{A}\right)^2}; A \geq h$ $\|N\| = \frac{M}{h}; A < h$	On/Off Element	$\angle N = 0$ $\|N\| = \frac{4M}{\pi A}$
Dead Zone	$\angle N = 0$ $\|N\| = k\left[1 - \frac{2}{\pi}\sin^{-1}\left(\frac{h}{A}\right)\right.$ $\left. \frac{-2h}{\pi A}\sqrt{1 - \left(\frac{h}{A}\right)^2}\right]; A \geq h$ $\|N\| = 0; A < h$	On/Off + Dead Zone	$\angle N = 0$ $\|N\| = \frac{4M}{\pi A}\sqrt{1 - \left(\frac{h}{A}\right)^2}; A \geq h$ $\|N\| = 0; A < h$
Saturation with Hysteresis	For $A \geq M + h$: $N = \left\{\frac{1}{2} + \frac{1}{\pi}\left[\sin^{-1}\left(\frac{M+h}{A}\right)\right.\right.$ $- \sqrt{1 - \left(\frac{M-h}{A}\right)^2}$ $+ \frac{M+h}{A}\sqrt{1 - \left(\frac{M+h}{A}\right)^2}$ $\left.\left.+ \frac{M-h}{A}\sqrt{1 - \left(\frac{M-h}{A}\right)^2}\right]\right\}$ $-j\left\{\frac{4Mh}{\pi A^2}\right\}$ For $A < M + h$ (Hysteresis Alone): $N = \left\{\frac{1}{2} + \frac{1}{\pi}\left[\sin^{-1}\left(\frac{A-2h}{A}\right)\right.\right.$ $\left.\left.+ \left(\frac{A-2h}{A}\right)\sqrt{1 - \left(\frac{A-2h}{A}\right)^2}\right]\right\}$ $-j\left\{\frac{4h}{\pi A^2}(A - 2h)\right\}$	On/Off with Hysteresis	$\angle N = -\sin^{-1}\left(\frac{h}{A}\right)$ $\|N\| = \frac{4M}{\pi A}$

Example 10.12   Describing Function for Dead Zone

Obtain the describing function for a dead zone component having dead bandwidth of $2h$ and slope of k.

Solution: Figure 10.17 shows the periodic output y for a sinusoidal input u and identifies some parameters relevant to the present solution. Note that there will be no output if half the dead bandwidth is more than the amplitude of the input, $h > A$. The dead zone function can be described in terms of y and u by

$$y = \begin{cases} 0 & -h < u < h \quad \text{or} \quad h > A \\ k(u - h) & u \geq h \quad \text{and} \quad h \leq A \\ k(u + h) & u \leq -h \quad \text{and} \quad h \leq A \end{cases}$$

In terms of the variable ωt, and over the period $-\pi$ to π, the function is

$$y(\omega t) = \begin{cases} 0 & -\beta < \omega t < \beta \quad \text{or} \quad h > A \\ k(A \sin \omega t - h) & \beta \leq \omega t \leq \pi - \beta \quad \text{and} \quad h \leq A \\ k(A \sin \omega t + h) & -\pi + \beta \leq \omega t \leq -\beta \quad \text{and} \quad h \leq A \end{cases} \qquad \textbf{(10.27a)}$$

Figure 10.17

Fundamental Response of Dead Zone Element to Sinusoidal Input

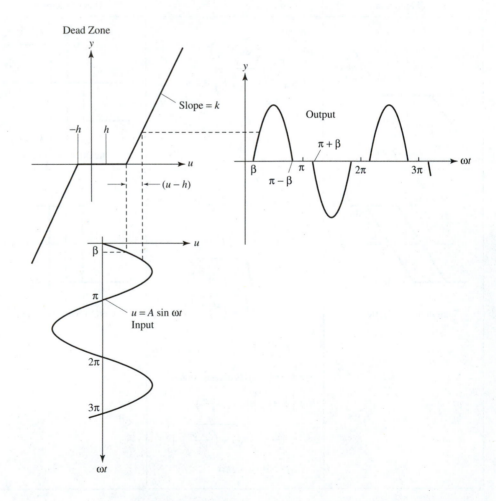

Note that

$$\beta = \sin^{-1}\left(\frac{h}{A}\right) \tag{10.27b}$$

As can be seen in Fig. 10.17, $y(\omega t)$ is an odd (zero-point symmetry) function. Hence (see Effects of Symmetry, Section 7.4) $p_1 = 0$, and equation (10.24c) for q_1 should be

$$q_1 = \begin{cases} \dfrac{2}{\pi}\displaystyle\int_0^\pi y\,\sin(\omega t)\,d(\omega t) & A \geq h \\ 0 & A < h \end{cases}$$

Substituting for y from (10.27a) and noting from Fig. 10.17 that $y(\omega t) = 0$ for $\pi - \beta < \omega t \leq \pi$, we have

$$\frac{2}{\pi}\int_\beta^{\pi-\beta} k(A\,\sin(\omega t) - h)\sin(\omega t)\,d(\omega t) =$$

$$q_1 = \begin{cases} \dfrac{2kA}{\pi}\left[\left(\dfrac{\omega t}{2} - \dfrac{\sin(2\omega t)}{4}\right)\Bigg|_\beta^{\pi-\beta} + \dfrac{h}{A}\cos(\omega t)|_\beta^{\pi-\beta}\right] & A \geq h \\ 0 & A < h \end{cases}$$

Using equation (10.27b) and relations such as

$$\sin(2\omega t) = 2\,\sin(\omega t)\cos(\omega t) \quad \text{and} \quad \cos(\beta) = \sqrt{1 - \sin^2\beta} = \sqrt{1 - \left(\frac{h}{A}\right)^2}$$

we obtain

$$q_1 = \begin{cases} Ak\left[1 - \dfrac{2}{\pi}\sin^{-1}\left(\dfrac{h}{A}\right) - \dfrac{2h}{\pi A}\sqrt{1 - \left(\dfrac{h}{A}\right)^2}\right] & A \geq h \\ 0 & A < h \end{cases} \tag{10.27c}$$

Applying equations (10.26a) through (10.26c), we find

$$|N| = \begin{cases} k\left[1 - \dfrac{2}{\pi}\sin^{-1}\left(\dfrac{h}{A}\right) - \dfrac{2h}{\pi A}\sqrt{1 - \left(\dfrac{h}{A}\right)^2}\right] & A \geq h \\ 0 & A < h \end{cases} \tag{10.28a}$$

and

$$\angle N = 0 \tag{10.28b}$$

which is the same as the entry in the table.

....

To utilize the describing function in stability analysis, we need to partition the system with the nonlinearity so that it can be described by the combined function $N(A)G(j\omega)$, where $G(j\omega)$ represents the linear components, while the nonlinear elements are **combined** in $N(A)$. A unity feedback system with such an open-loop description is shown in Fig. 10.18a. We can see that the stability analysis can be carried out as before (see Fig. 10.15) if the open-loop transfer function $G_o(j\omega)$ is replaced by $N(A)G(j\omega)$. Unfor-

Figure 10.18

Describing Function and
Nonlinear System and
Stability

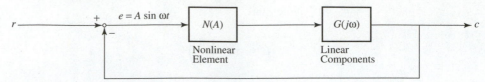

(a)   Unity Feedback Control System with Nonlinear Element

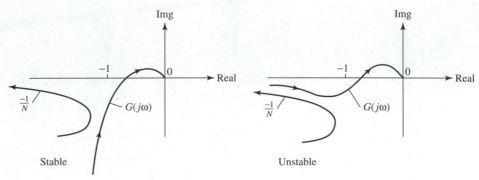

(b)   Kochenberger Stability Criterion

tunately, $N(A)G(j\omega)$ consists of several Bode or Nyquist loci, one for each value of A. However, the stability limit is still given by equation (10.20b), that is,

$$N(A)G(j\omega) = -1$$

or

$$G(j\omega) = \frac{-1}{N(A)} \qquad \text{(10.29)}$$

Put in this form, separate loci can be plotted in the Nyquist plane for $G(j\omega)$ and for the negative inverse of $N(A)$, that is, $-1/N(A)$ (for all A), and instead of requiring that for stability the $N(A)G(j\omega)$ "locus" must pass the $(-1, j0)$ point on its left, we now require that the $G(j\omega)$ locus pass the $-1/N(A)$ locus on its left as ω increases. This stability criterion is illustrated in Fig. 10.18b. Observe that for the unstable case the system would have been stable according to the Nyquist stability criterion if it did not have the nonlinearity.

Application of the Root Locus Technique

For single-valued nonlinearities for which the phase of the describing function is zero, the root locus technique can also be used to investigate system stability. In this case the closed-loop characteristic equation should be put in such a form that the variable parameter is $N(A)$ or an algebraic function of $N(A)$. Consider again the aircraft servo example, with an inherent nonlinearity N in the actuator, such that the block diagram of Fig. 10.5b can be modified as shown in Fig. 10.19a.

Using the methods of Section 9.1, we can obtain an equivalent block diagram such as given ultimately in Fig. 10.19b. The open-loop transfer function for this system is

$$G_o(s) = N\frac{100k_c(1 + T_i s)}{s^2(s + 5)T_i s}\left[\frac{10k_c(1 + T_i s) + T_i s^2(s + 5)}{10k_c(1 + T_i s)}\right] \qquad \text{(10.30a)}$$

Figure 10.19

Root Locus and Describing Function Stability Analysis

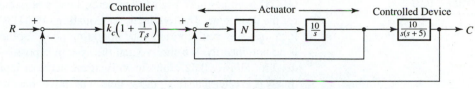

(a) Servomechanism with Nonlinear Actuator

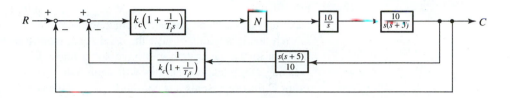

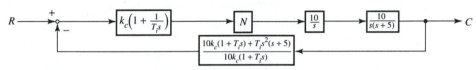

(b) Equivalent Block Diagrams

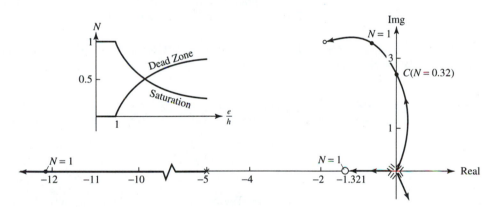

(c) Root Locus and Limit Cycle

If we let

$$G_o(s) = NG(s) \tag{10.30b}$$

then

$$G(s) = 10\left[\frac{10k_c(1 + T_i s) + T_i s^2(s + 5)}{T_i s^3(s + 5)}\right] \tag{10.30c}$$

and the characteristic equation is

$$1 + NG(s) = 0 \tag{10.31}$$

Suppose N represents a dead zone (Table 10.1) with $h = 0.5$, say, and slope $k = 1$. To investigate the effect of the nonlinearity via the root locus technique we can fix k_c and T_i at some reasonable values, for example, $k_c = 2$ and $T_i = 1$, so that N in equation

(10.31) becomes the variable parameter. This is possible because the phase shift is zero for the dead zone. The root locus for equation (10.31) is shown in Fig. 10.19c. The root locations at $N = 1$ represents the operating point for the choice of k_c and T_i when there is no nonlinearity. We observe that the system is positively damped at this point and indeed for all $N > 0.32$. Consequently, if the actuator input signal e (Fig. 10.19a) is such that N is greater than 0.32, the system will move in a direction in which e decreases due to the damping. From the describing function of the dead zone shown in Table 10.1 and in the insert in Fig. 10.19c, N will decrease with e, and the system will progress to the critical point C. Any decrease of N beyond C results in unstable conjugate-complex roots that cause e and hence N to increase, so that the system again returns to C. If e is such that it is within or near the dead zone, then N is zero or close to zero, and from the root locus, the system will have some unstable roots. The resulting large and increasing amplitude oscillations will again cause the operating point to migrate to C. Point C thus represents a stable *limit cycle*. This is the same phenomenon that was illustrated in the phase plane in Fig. 9.23. The frequency of the sustained oscillation can be obtained from the root locus plot and is about 2.8 rad/s.

As a practical matter, equation (10.31) and the root locus plot of Fig. 10.19c apply to any other single-valued nonlinearity in the actuator. However, the interpretation of the root locus can be different. If N represents saturation (Table 10.1, $M/h = 1$), for instance, from the insert in Fig. 10.19c, we see that N decreases as e increases, so that it will take a large e for the roots of the system to migrate into the RHP. Moreover, once the system becomes unstable, further or consequent increases in e will result in even more migration of the roots further into the RHP, so that the system remains unstable. Thus no stable limit cycle can develop at point C.

Stability of Limit Cycle of a Nonlinear System

To illustrate the frequency domain analysis of a system with a nonlinear component, using describing functions, we present the following example.

Example 10.13   Limit Cycles and the Nyquist Locus

Determine the stability behavior of the third-order system shown in Fig. 10.20a, which is subject to a nonlinearity much like *saturation with hysteresis* (Table 10.1) but whose describing function was determined experimentally as shown in Table 10.2.

Solution: The Nyquist locus of the linear part of the forward loop,

$$G(s) = \frac{1}{s(0.25s^2 + 0.375s + 1)}$$

with $s = j\omega$, is shown in Fig. 10.20b with some values of ω indicated. The locus of $-1/N$ is also shown on the same scales and coordinates. Note that the angles of $-1/N$ are 180° less than those of $1/N$, and the magnitude is simply $1/|N|$. The values marked on the curve are however those of A, that is, the amplitude of the error signal e. We observe that the pair of loci intersect at two points, C and D. According to the Kochenberger stability criterion shown in Fig. 10.18, these points must represent stability limits for the system. To assist in interpreting the meaning of these limits of stability, we have also shown on the same Nyquist plane the loci for $G(-0.1 + j\omega)$ and $G(0.1 + j\omega)$, which represent slightly *positively damped* and *negatively damped* operations, respectively, of the same system, where 0.1 is the damping factor.

Consider now operation at point C. The frequency of oscillation here is 1.83 rad/s. If the amplitude of oscillation increases due to some disturbance, the operation

Figure 10.20

Nyquist Locus and Describing Function Stability Analysis

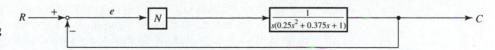

(a) Control System with Nonlinear Element

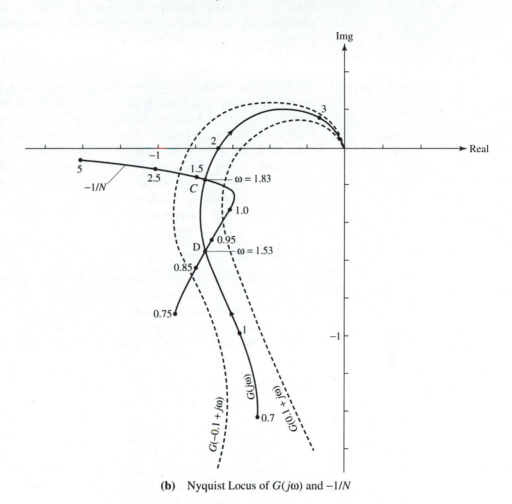

(b) Nyquist Locus of $G(j\omega)$ and $-1/N$

will migrate along the $-1/N$ locus toward the region of positive damping, which will cause the operation point to return to C. Similarly, a decrease in amplitude at C will result in movement in the direction of negative damping, which also will cause a return of the working point to C. Point C is therefore a *stable limit cycle* just like point C in Fig. 10.19c. On the other hand, if point D represents the system operation, an increase in A will move the working point in the direction of negative damping, which will

Table 10.2 Describing Function Data for a Nonlinear Element

A	5.000	2.500	1.500	1.000	0.950	0.850	1.750
$|N|$	0.713	0.993	1.230	1.470	1.180	0.976	0.794
$\angle N$ (°)	−3.270	−6.840	−11.310	−28.070	−34.440	−38.660	−44.360

cause further increases in the oscillating amplitude, so that the operating point will progressively move to C. If the error amplitude remains small enough so that the working point is always to the left of D, then the system will remain completely stable. However, such stability is not *global*, since a sufficiently large disturbance could force the operation to move to D and subsequently to point C. Point D (where the operating frequency is 1.53 rad/s) therefore represents an *unstable limit cycle*. It is unstable in the sense that any disturbance at operating point D will cause the operation to diverge from point D.

....

Circle Criterion

The describing-function-in-the-frequency-domain approach discussed in the foregoing is often limited by practical considerations to the investigation of the existence/nonexistence of limit circles and the stability of limit circles. Further, the describing function itself is only an approximate representation of the nonlinearity and may not apply to some nonlinear components. The *circle criterion* is also a Nyquist-type frequency domain stability criterion that, nevertheless, requires no approximation for the nonlinearity and provides an explicit condition for *asymptotic stability*. However, as presented here, this criterion does not include a necessary condition and is only a sufficient condition. The ensuing discussion follows the development in reference [17].

System Configuration

The circle criterion test for asymptotic stability applies to a system partitioned into a nonlinear component and a linear part, as shown in Fig. 10.21a (*v* in this figure represents external disturbances that are assumed to enter the system at the indicated point). This situation is similar to Fig. 10.18a, which was used for the describing function approach, except that this time *N* is the actual nonlinear relation.

Sufficiency Condition

The circle criterion may be stated with respect to Fig. 10.21 as follows:

If the nonlinearity *N*(Fig. 10.21a) lies inside a sector defined by the lines of slope α and β (Fig. 10.21b), and if the frequency response $G(j\omega)$ of the linear part of the system (defined in Fig. 10.21a) does not touch or encircle the *disk* passing through the points $-1/\alpha$ and $-1/\beta$ in the Nyquist plane (Fig. 10.21c), then the closed-loop response is bounded, in the sense that the loop variable *e* or *m* (Fig. 10.21a) is *square integrable* for forcing input *r* or *v* (Fig. 10.21a) that is square integrable.

A function $u(t)$ is square integrable in the interval $0 \leq t \leq \infty$ if $\int_0^\infty u^2 \, dt$ is bounded for all initial states.

The center and radius of the disk or *circle* shown in Fig. 10.21c are given by

$$C_d = \frac{-(\alpha + \beta)}{2\alpha\beta} + j0; \qquad R_d = \frac{(\beta - \alpha)}{2\alpha\beta}; \quad \text{(center and radius of disk)} \tag{10.32}$$

Note that the circle criterion, as stated, uses the Nyquist plot $G(j\omega)$ of the linear part. A special case of the circle criterion occurs when $\alpha = 0$ and $\beta = K$, and the disk of the circle criterion (see equation (10.32)) degenerates into a vertical line (circle of infinite radius) passing through the point $-1/K$. The resulting criterion shown in Fig. 10.21d,

Figure 10.21

Circle Criterion for Asymptotic Stability

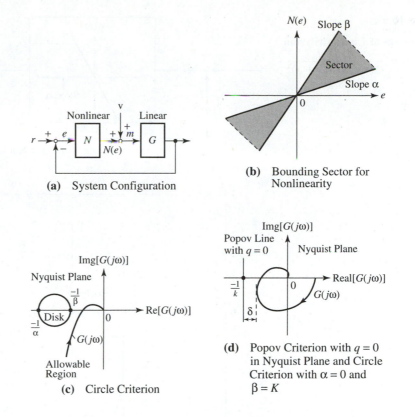

(a) System Configuration

(b) Bounding Sector for Nonlinearity

(c) Circle Criterion

(d) Popov Criterion with $q = 0$ in Nyquist Plane and Circle Criterion with $\alpha = 0$ and $\beta = K$

is equivalent to the *Popov criterion* (not discussed here; see [17]). Another special case is the situation where α is negative (the lower line in Fig. 10.21b is a line of negative slope). The origin of the complex plane coordinates is then inside the disk, and the entire $G(j\omega)$ plot must remain *strictly* inside the circle to satisfy the condition for asymptotic stability in this case. Finally, if $\alpha = \beta = C$, then the nonlinear part becomes a linear time-invariant system that is simply a gain C, and the test disk degenerates into a point at $-1/C$. The circle criterion in this case is, as should be expected, the same as the Nyquist stability test, provided $G(s)$ has no poles in the RHP. To see this, note that the magnitude C of the gain element can be incorporated in the open-loop transfer function, that is, $G_o(s) = CG(s)$, leaving a unit gain for N. The critical point then becomes the $(-1 + j0)$ point, as it is in the Nyquist criterion.

The circle test is valid for stationary and time-varying nonlinearities and even *dynamic* nonlinearities that satisfy the condition

$$0 \leq \frac{N(e)}{e} \leq \beta$$

Example 10.14 Application of Circle Criterion

Consider the control system with a nonlinear component shown in Fig. 10.22a. For nonlinearities like *saturation* and *dead zone*, α (Fig. 10.21b) necessarily has to be zero. However, the value of β depends on the slope of the proportional region of the nonlinearity. Determine the asymptotic stability limit of β using the circle criterion.

Figure 10.22

Illustration of Circle Criterion

for $G(s) = \dfrac{2}{(20s + 1)(10s + 1)^2}$

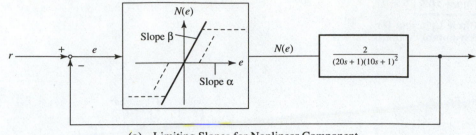

(a)   Limiting Slopes for Nonlinear Component

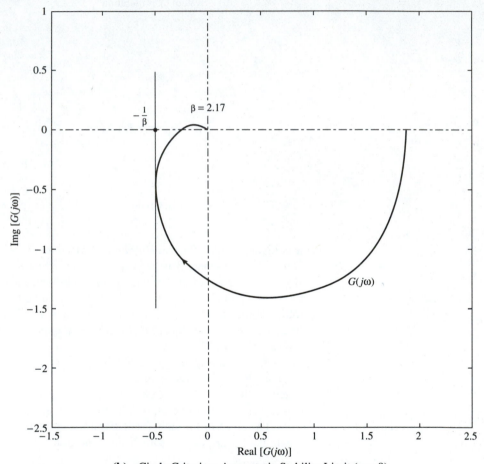

(b)   Circle Criterion: Asymptotic Stability Limit ($\alpha = 0$)

Solution: The Nyquist locus of $G(s)$ shown in Fig. 10.22b was generated using MAT-LAB (see script M-file **fig10_22.m; On the Net**) by invoking the **nyquist** function with left-hand arguments: $[re, \ im, \ w] = \text{nyquist}(num, \ den)$, where $num = [2]$, and $den = [2000 \ \ 500 \ \ 40 \ \ 1]$, followed by $\text{plot}(re, im)$. As discussed earlier, with $\alpha = 0$, the test *circle* in the circle criterion is a vertical line passing through $-1/\beta$. The limiting value of β is given by the vertical line that is tangent to the Nyquist locus while lying to the left of it. This line was determined graphically by plotting on a large scale the Nyquist locus in the neighborhood of the suspected tangency point (plot not shown). The tangent line was found to cross the real axis at $-1/\beta = -1/2.17$, so that the solution is $\beta = 2.17$.

....

Hyperstability

Stability theorems such as the circle criterion that apply to whole classes of nonlinearities without requiring approximations of the nonlinear behavior determine what is often termed *absolute stability* of the system having the nonlinear component. A generalization of absolute stability by Popov known as *hyperstability* finds significant application in modern control theory, especially in adaptive control (see Section 13.3). The reader interested in hyperstability theory should refer to the book by Popov, reference [22], which contains an extensive treatment of this subject.

O n   t h e   N e t

- fig10_1.m and stable.m (MATLAB): source file for Fig. 10.1
- fig10_3.m and Lyapunov.m (MATLAB): source file for Figs. 10.3b and 10.3c
- SC12.MCD (Mathcad): source file for example 10.3, using *root* function of Mathcad
- fig10_8.m (MATLAB): source file for Fig. 10.8, example 10.3
- fig10_11.m (MATLAB): source file for Fig. 10.11
- fig10_22.m (MATLAB): source file for Fig. 10.22
- Auxiliary exercise: Practice Problems 10.4, 10.8, 10.24 through 10.27

10.6 PRACTICE PROBLEMS

10.1 (a) Explain the difference between stability and asymptotic stability.

10.1 (b) Determine whether the system with the characteristic equation

(a) $s(s^2 + 4)(s - 1) = 0$;   **(b)** $s(s^2 + 4)(s + 1) = 0$

is merely stable, unstable, or asymptotically stable.

10.2 Comment on the stability behavior of each of the systems whose phase plane trajectories are shown in Fig. 10.1.

10.3* Which of the following systems whose transfer functions are given responds faster and why?

$$G_A(s) = \frac{1 + 11s}{12s^2 + s + 1}; \qquad G_B(s) = \frac{3 + 110s}{1200\,s^2 + 10s + 1}$$

10.4 Determine the relative stability of the systems of Problem 10.3 by obtaining an autonomous state equation for each transfer function via the signal flow graph, making any necessary change of variables so that the equilibrium point is at the origin, and using a computational math software product to obtain and plot the phase plane trajectories for a number of properly chosen but different initial states.

10.5 Using the energy function $V(\mathbf{X}) = x_1^2 + x_2^2$ as a test function, ascertain the stability of the equilibrium point of the following autonomous stationary systems:

(a) System of Fig. 10.1c   **(b)** System of Fig. 10.1d

(c) System of Fig. 10.1e   **(d)** System of Fig. 10.1f

(e) System of Fig. 10.1g   **(f)** System of Fig. 10.1h

10.6 Repeat Problem 10.5 using as Lyapunov function the quadratic form of equation (10.4a).

10.7 Ascertain by the Lyapunov second method the stability of the origin of

(a) The discrete-time system

$$\mathbf{X}_{k+1} = \begin{bmatrix} 1 & -1/2 \\ -2 & 1 \end{bmatrix}\mathbf{X}_k$$

(b) The continuous-time system with

$$\mathbf{A} = \begin{bmatrix} 1 & -1 & 0 \\ -1 & 1 & -1 \\ 0 & -1 & 1 \end{bmatrix}$$

(c) The discrete-time system with

$$\mathbf{P} = \begin{bmatrix} 0 & -1/2 & 0 \\ -1 & 0 & 0 \\ 1 & -1/4 & 0 \end{bmatrix}$$

10.8 (a)* A second-order system is given by

$$\dot{x}_1 = x_2$$

$$\dot{x}_2 = -ax_1 - bx_2 - x_1^2 x_2; \qquad a > 0, \quad b > 0$$

(i) Is this a linear or a nonlinear system? If it is nonlinear, obtain by linearization an approximate linear model that is valid in the neighborhood of the equilibrium point.

(ii) Determine the conditions on a and b for the system to be asymptotically stable in a small neighborhood of the equilibrium point.

(iii) Using as candidate Lyapunov function

$$V(\mathbf{X}) = \frac{1}{2}(ax_1^2 + x_2^2)$$

determine whether the origin is asymptotically stable, stable, or unstable.

(iv) For the case $a = b = 1$, use math software to obtain a family of trajectories starting at different initial states in the phase plane. Ascertain the stability of the origin of the system from the plot of the pattern of trajectories. Also, indicate on the same scale a family of curves representing the function $V(\mathbf{X}) = $ constant, where $V(\mathbf{X})$ is as given in (iii).

10.8 (b) Repeat Problem 10.8a for the system

$$\dot{x}_1 = x_2$$
$$\dot{x}_2 = -ax_1 - bx_2 - x_2^3; \qquad a > 0, \quad b > 0$$

10.9 Plot and label the eigenvalues of each of the systems shown in Fig. 10.1 on the complex plane (the eigenvalues were indicated on the figure). Discuss the correlation between root placement in the complex plane and stability of equilibrium point and general behavior of each system as given by the pattern of trajectories in Fig. 10.1

10.10 Determine the stability of the following system from the values of the eigenvalues:

(a) System with the transfer function

$$G(s) = \frac{s + 2}{s(s + 1)(s + \frac{1}{5})}$$

(b) Discrete-time system of Problem 10.7a

(c) Continuous-time system of Problem 10.7b

(d) Discrete-time system of Problem 10.7c

10.11 (a) Find the range of values of the controller gain k_c for which the closed-loop system of Fig. 10.23 is asymptotically stable. What is the value of k_c at the stability limit?

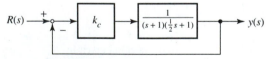

Figure 10.23 (Problem 10.11)

10.11 (b) Given the plant transfer function

$$G_p(s) = \frac{1}{(s + 1)(s + 2)(s + 3)}$$

find a proportional feedback control k such that the closed-loop system is at the stability limit.

10.12* A rabbit-fox environment is described by the following model:

$$\dot{x}_1 = 2x_1 - 3x_2 \quad \text{and} \quad \dot{x}_2 = 2x_1 - x_2,$$

where x_1 is the rabbit population and x_2 is the fox population. A decision is made to control the rabbit-fox environment by introducing linear feedback in the form of a disease that is fatal to rabbits but does not affect foxes, thereby reducing the rate of growth of the rabbit population by an amount kx_1

(a) Draw a Laplace domain block diagram of the feedback control system. (Assume a zero setpoint on the rabbit population.)

(b) Find the smallest value of k that prevents the population explosion from occurring (assuming arbitrary initial population sizes).

10.13 Is the system whose block diagram is given in Fig. 10.24 stable?

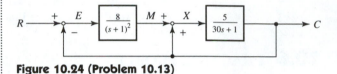

Figure 10.24 (Problem 10.13)

10.14* Consider the antiaircraft gun servomechanism of Fig. 10.5a. If the equation representing the servomotor-inertia system is

$$T\frac{d^2\theta}{dt^2} + \frac{d\theta}{dt} = c_1 v_1$$

where v_1 is the voltage applied to the motor (c_1 is a constant), θ is measured by an electrical device: v_θ (in volts) $= c_2\theta$, and the controller uses proportional action: $v_1 = k_c(v - v_\theta)$.

(a) Draw the block diagram (in the s-domain) for this system and identify the output and the setpoint (desired value).

(b) Obtain the system's transfer function.

(c) Find the "steady-state" angle $\theta(\infty)$ for a step change in v.

(d) Find the undamped natural frequency (cycles per second) and damping ratio of the complete system.

(e) Derive an expression for k_c that will give the fastest response to a step change in v without "overshoot" of the steady-state $\theta(\infty)$.

10.15 Given the block diagram shown in Fig. 10.25,

(a) Obtain the **A**, **B**, and **C** matrices for the state model.

(b) What are the eigenvalues of the system?

(c) What are the possible modes of response of this system?

(d) If the forcing input is kept constant at $u = d$, obtain in terms of d the equilibrium state of the system.

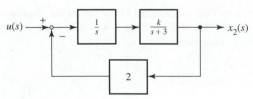

Figure 10.25 (Problem 10.15)

10.16* For the given block diagram in Fig. 10.26, find the transfer function $G(s) = c/r$ if the symbol Ⓞ stands for

(a) Subtraction.

(b) Addition.

(c) Which of the systems represented by (a) and (b) is stable? Explain your answer.

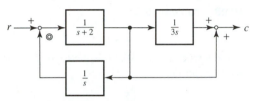

Figure 10.26 (Problem 10.16)

10.17 The dynamics of a small spacecraft attempting to land on a certain planet is given by

$$\dot{x}_1 = -x_2 \quad \text{and} \quad \dot{x}_2 = x_1 + u$$

where x_1 is the position above the planet, x_2 is the velocity toward the planet, and u represents the rocket's thrust.

(a) Is this system asymptotically stable?

(b) In order to land the spacecraft smoothly it is necessary to regulate the thrust using negative feedback of the spacecraft velocity, that is, $u = -k_c x_2$.

Find a range of value of k_c that will work.

Note: Landing smoothly implies the asymptotic reduction of the position and velocity of the spacecraft to zero, with no oscillations.

10.18 Find k_1, k_2 such that the eigenvalues of the system with the characteristic equation

$$z^2 + k_1 z + 1 - k_2 = 0$$

are $-\frac{1}{2}$ and $\frac{1}{4}$.

10.19* A discrete-time system is described by

$$\mathbf{X}_{k+1} = \begin{bmatrix} 1 & -2 \\ -1/2 & 1 \end{bmatrix} \mathbf{X}_k + \begin{bmatrix} 1 \\ 0 \end{bmatrix} u_k; \quad y_k = [1 \quad -1]\mathbf{X}_k$$

(a) Determine whether this system is stable.

(b) If a feedback controller, $u_k = -ky_k + u_{k-1}$, is applied to the system, determine the range of k for which the controlled system is asymptotically stable.

10.20 Consider the system of Problem 8.65. If it is placed under proportional feedback control, $u_k = -ky_k$, with $y_k = [-1/4 \quad 1]\mathbf{X}_k$, determine the value of k at the stability limit of the controlled system.

10.21 Sketch (manually) the root locus for the characteristic equation

(a) $1 + k \left[\dfrac{(s+2)^2}{s(s+4)(s+3)} \right] = 0$

(Darken your lines, indicate the directions of increasing k, and label at least one value of k). *Hints:* (i) There is a double zero. (ii) The root locus does not leave the real axis!

(b) $1 + \dfrac{2}{bs^3 + s^2 + 3s} = 0;$ b is the design parameter

(c) $1 + k\dfrac{s^2 - 4s + 8}{s^2 + 4s + 8} = 0$

10.22 Consider the closed-loop characteristic equation of Example 10.7. Put this equation in the form of equation (10.5a) and plot the root locus with k as the adjustable parameter. Indicate on the same scale the vertical line $s = -5$, which demarcates the complex plane with respect to those eigenvalues whose real parts are less than -5. From the intersection of this line with the root locus, verify the design for speed of response of Example 10.7. Explain your result.

10.23 Sketch (manually) the root locus for a proportional feedback control system with the open-loop transfer function

(a) $G_o(s) = \dfrac{k_c}{(s+1)(s+2)}$

(b) $G_o(s) = \dfrac{k_c}{(s+1)^3}$

(c)* $G_o(s) = \dfrac{(s+2)^2}{s(s+6)(s+3)}$

10.24 Repeat Problem 10.21 using appropriate computer software.

10.25 Repeat Problem 10.23 using appropriate computer software.

10.26* Sketch the root locus for the open-loop transfer function

$$kG_0(s) = \dfrac{k(s+1)}{s^2(s+3)(s+6)}, \quad \text{for } 0 \le k \le \infty$$

For what value of k do purely imaginary roots exist? What are the values of these roots?

10.27 Explain how the root locus technique can be used to find the eigenvalues of any system.

(a) Plot the root locus for the characteristic equation

$$1 + \dfrac{k(1-s)}{s(1+s)} = 0$$

where k is a variable parameter. *Hint:* Modify the rules to apply for $(1 - s)$ or, alternatively, let $(1 - s) = -1(s - 1)$.

(b) Use a computational mathematics package to generate and plot the root locus.

10.28 For the system specified by the given characteristic equation, determine (i) the stability of the system, (ii) the number of unstable roots if the system is not stable, and (iii) the number of roots on the imaginary axis, if any:

(a) $7s^5 - 5s^4 + 11s^3 - 9s^2 - 3s + 1 = 0$

(b) $8s^4 + 4s^3 + 2s^2 + 4s = 0$

(c)* $s^6 + 2s^5 + 8s^4 + 6s^3 + 17s^2 + 4s + 10 = 0$

(d) $s^5 + 7s^4 + 18s^3 + 23s^2 + 17s + 6 = 0$

(e) $s^5 + 2s^4 + 4s^3 + 8s^2 + 3s + 2 = 0$

(f) $s^4 + s^3 - s - 1 = 0$

10.29 Consider the given Laplace domain block diagram in Fig. 10.27. Find, using the Routh criterion, the range of k for which the overall system is asymptotically stable.

(a) The control system shown

(b) The closed-loop system shown

(c)* The control system shown

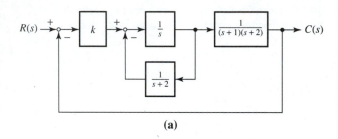

(a)

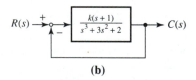

(b)

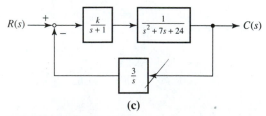

(c)

Figure 10.27 (Problem 10.29)

10.30 Determine the stability of the following characteristic polynomials. Where the system is unstable, what is the number of unstable roots?

(a) $s^5 + s^4 + 2s^3 + s^2 + s + k$

(b) $s^3 + 3s^2 - 6s + 10$

10.31* The parameters k and T are adjustable in the feedback control system shown in Fig. 10.28. Determine and sketch on a set of kT-axes the allowable design domain that will give a closed-loop response with all time constants less than or equal to 1 time unit.

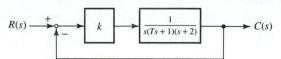

Figure 10.28 (Problem 10.31)

10.32 With reference to Fig. 10.29, is the following control system stable or unstable?

(a) $G_1 = \dfrac{1}{s+1};$ $G_2 = \dfrac{15}{s+5};$ $H = \dfrac{3}{s+3}$

(b) $G_1 = \dfrac{1}{s+1};$ $G_2 = \dfrac{145}{(s+5)(s+2)};$ $H = 1$

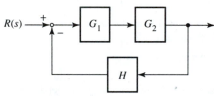

Figure 10.29 (Problem 10.32)

10.33 Generate and plot the root locus of the system of Problem 10.19b in the s-plane after first transforming the characteristic equation from a function of z to a function of s. Indicate on the plot the value of k obtained in Problem 10.19b and comment on your results.

10.34 Generate and plot in the z-plane the root locus of the system of Problem 10.19b, with k as the adjustable parameter. Indicate on the plot the unit circle and the value of k obtained in Problem 10.19b, and comment on the results.

10.35 What is the characteristic equation of the system of Problem 10.7c?

(a) Using the transformation equations (10.12a) and (10.12b) and the Routh test, determine the stability of the system.

(b) Determine the stability of the system using Jury's inners test.

10.36 Using Jury's inners stability test, ascertain the asymptotic stability of the discrete-time system with the following characteristic equation:

(a) $5z^5 - 3z^4 + 9z^3 - 7z^2 - z + 1 = 0$

(b)* $3z^5 + 9z^4 + 5z^3 + 11z^2 + 7z - 1 = 0$

(c) $5z^6 + 7z^5 + 11z^4 + 4z^3 + z^2 = 0$

10.37 Use Jury's inners stability test to verify the result of Example 10.8 that there is no value of k for which all the eigenvalues of the closed-loop system lie inside the unit circle.

10.38 Sketch the Bode and Nyquist diagrams for the plant

$$G_p(s) = \frac{10}{(1 + 2s)^2(1 + s)}$$

Determine on each diagram the gain margin and phase margin.

10.39 The transfer function of a linear system is

$$G(s) = \frac{s + 2}{s(s + 1)(s + 0.2)}$$

Find the gain margin, the phase margin, and the bandwidth for this system.

10.40 Repeat Problem 10.39 for the plant with the transfer function

(a)* $G_p(s) = \dfrac{10}{s(1 + s)\left(1 + \dfrac{1}{4}s\right)}$

(b) $G_p(s) = \dfrac{10}{\left(1 + \dfrac{10}{3}s\right)(1 + s)^2}$

10.41 (a) Solve the problem of Example 10.10 graphically according to the following procedure:

1. Plot the Bode gain and phase diagrams for $G(j\omega)$ with $k = 1$.
2. Identify the frequency ω_g at which the phase is $-180° + 45° = -135°$.
3. Determine the value of k that will give a parallel gain curve shifted vertically so that it crosses the 0 db line at $\omega = \omega_g$. This makes the phase margin 45°.
4. Using this parallel gain curve and the previous phase plot, determine the gain margin as defined in Fig. 10.15a.

10.41 (b) Repeat Problem 10.41a using the Nyquist diagram with an equivalent procedure.

10.42 Consider proportional control of a plant with the transfer function

$$G_p(s) = \frac{0.5}{(s + 1)^2(s + 5)}$$

(a) Sketch the Bode diagram for the $G_p(j\omega)$.

(b) Determine approximately the value of the proportional feedback control k_c that will give a unity feedback system with a gain margin of 8 db.

10.43 For the given open-loop polar plot in Fig. 10.30, determine by measurement the gain and phase margins for the feedback control system.

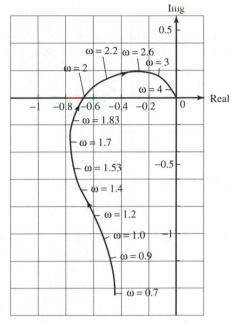

Figure 10.30 (Problems 10.43 and 10.47)

10.44* Repeat Problem 10.43 for the Nyquist diagram shown in Fig. 10.31.

10.45 (a) What is meant by resonance in the response of a controlled system?

10.45 (b) Show that the ϕ-contours in the Nyquist plane are circles.

10.46 Given the open-loop frequency response data of Problem 9.24, obtain the closed-loop frequency response

(a) Approximately on a Bode diagram.

(b) Exactly via the Nyquist diagram.

(c) What is the bandwidth of the closed-loop system?

(d) What is the resonant peak M_p and its frequency of occurrence?

10.47 For the polar plot of the open-loop transfer function of a unity feedback system given by Fig. 10.30 determine approximately

(a) The resonant peak M_p and the frequency of occurrence of the peak amplitude.

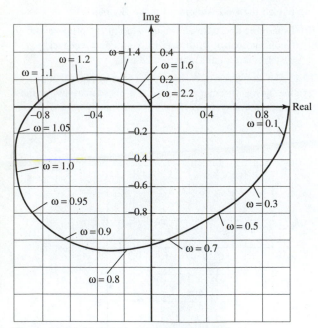

Figure 10.31 (Problems 10.44 and 10.48)

(b) The Nyquist and Bode plots of the closed-loop system. Can you identify the results in (a) on the Bode diagram?

(c) The bandwidth of the closed-loop system.

10.48 Repeat Problem 10.47 for the Nyquist diagram given in Fig. 10.31.

10.49 Determine the stability of the closed-loop systems whose open-loop Nyquist plots are shown in Fig. 10.32. Indicate in each case the number of roots Z of the closed-loop system in the right half-plane.

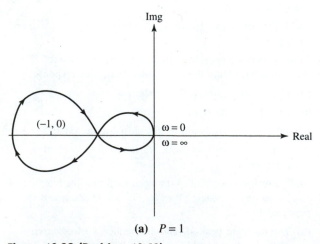

(a) $P = 1$

Figure 10.32 (Problem 10.49)

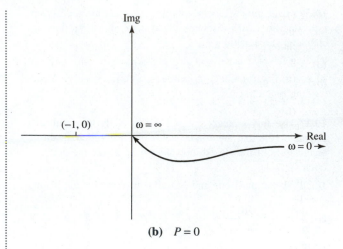

(b) $P = 0$

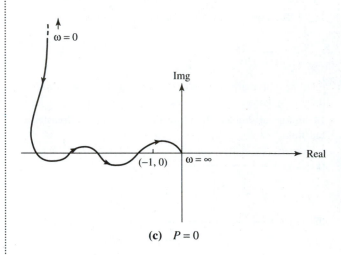

(c) $P = 0$

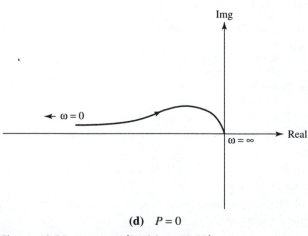

(d) $P = 0$

Figure 10.32 continued (Problem 10.49)

10.50 Using Example 10.12, for dead zone, as a guide, derive the describing function for each of the remaining nonlinear characteristics shown in Table 10.1. Sketch the describing function (that is, plot N versus e/h) in each case.

10.51 (a)* Determine and sketch (N/S versus e/h) the describing function for saturation with an arbitrary slope S.

10.51 (b) Suppose that in Fig. 10.18a the nonlinear element represents saturation of unknown slope S, and

$$G(s) = \frac{2.5}{s(1 + 0.25s)(1 + s)}$$

Find the value of S for which the closed-loop system will remain completely stable.

10.52 For the given system in Fig. 10.33, determine the range of values of e for which the system is stable when the nonlinear element is

(a) Saturation ($h = 1$, slope $S = 5$).

(b) Dead zone ($h = 1$, slope $S = 5$).

(c) On-off element ($M = 10$).

(d)* On-off element with dead zone ($M = 10$, $h = 1$).

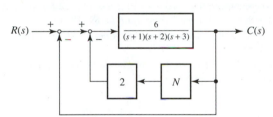

Figure 10.33 (Problem 10.52)

10.53 Repeat Problem 10.52 for the system shown in Fig. 10.34. Determine any stable and unstable limit cycles.

10.54 Provide an interpretation of Fig. 10.19 when N represents another single-valued nonlinearity such as

(a)* An on-off element.

(b) An on-off element with dead zone.

Does a stability limit cycle develop? If so, what is the frequency of the limit cycle?

10.55 Determine the stability behavior of the system of Fig. 10.20a when N represents

(a) An on-off element with hysteresis element, $M = 10$ and $h = 1$.

(b) A saturation-with-hysteresis element, $M = 4$, $h = 0.5$, and slope $= 1$.

10.56 In Fig. 10.21a, let

$$G(s) = \frac{s + 2}{(s + 0.5)^2}$$

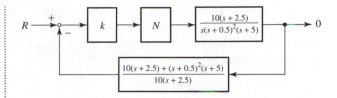

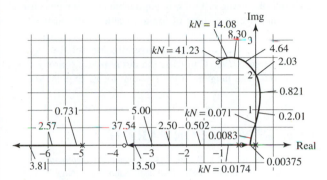

Figure 10.34 (Problem 10.53)

Determine by the circle criterion the largest slope β of $N(e)$ for asymptotic stability.

10.57 Repeat Problem 10.56 for

$$G(s) = \frac{s + 2}{(s + 0.5)^2(s + 1)}$$

10.58 Given the control system shown in Fig. 10.35 if $\alpha = 5.0$, obtain the asymptotic stability limit value of β by

(a) The circle test.

(b) Repeat (a) for the case $\alpha = -5$.

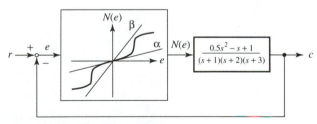

Figure 10.35 (Problem 10.58)

10.59 (a)* For the system of Example 10.14, suppose the nonlinearities are such that a negative lower bound slope is necessary. Determine by the circle criterion the asymptotic stability limit values of α and β under this condition.

10.59 (b) Recall that in the circle criterion, the condition $\alpha = \beta = C$ implies that the "nonlinear" component is just a gain element. For the system of Example 10.14, determine the largest possible proportional gain C that will leave the feedback system asymptotically stable under this condition.

10.60 Consider the nonlinear system of Fig. 10.21a with an unstable linear part,

$$G(s) = \frac{1}{(s-1)(s+2)(s+3)}$$

Note that the circle criterion is applicable even for this case. Let the upper bound slope of the allowable nonlinearities be given by $\beta = 50$. Determine the smallest possible lower bound slope α that will yield an overall system that is asymptotically stable. *Hint:* Construct the circle passing through the point -0.02 and tangent to the Nyquist locus. Explain your solution.

REFERENCES FOR PART II

1. Boyce, W. E., and DiPrima, R. C., *Elementary Differential Equations and Boundary Value Problems,* John Wiley, New York, 1969.

2. Auslander, D. M., Takahashi, Y., and Rabins, M. J., *Introducing Systems and Control,* McGraw-Hill, New York, 1974.

3. Barnett, S., *Introduction to Mathematical Control Theory,* Clarendon Press, Oxford, 1975.

4. Betts, K. S., "Math Packages Multiply," *Mechanical Engineering,* 112, no. 8 (August 1990): 32–36.

5. Dahlquist, G., and Bjorck, A., *Numerical Methods,* trans. Ned Anderson, Prentice-Hall, Englewood Cliffs, N.J., 1974.

6. Hutton, D. V., *Applied Mechanical Vibrations,* McGraw-Hill, New York, 1981.

7. Bohn, E. V., *The Transform Analysis of Linear Systems,* Addison-Wesley, Reading, Mass., 1963.

8. Poularikas, A. D., and Seeley, S., *Signals and Systems,* 2d ed., PWS-KENT, Boston, Mass. 1991.

9. Spiegel, M. R., *Theory and Problems of Laplace Transforms,* McGraw-Hill, New York, 1965.

10. Meriam, J. L., and Kraige, L. G., *Engineering Mechanics,* Vol. 2, *Dynamics,* John Wiley, New York, 1992.

11. Maddux, K., et al., "Design of Chatter Free Machine Tools," ASME paper No. 75-DET-5.

12. Stephanopoulos, G., *Chemical Process Control,* Prentice-Hall, Englewood Cliffs, N.J., 1984.

13. Graupe, D., *Identification of Systems,* Robert E Krieger, New York, 1975.

14. Lee, R. C. K., *Optimal Estimation, Identification and Control,* M.I.T. Press, Cambridge, 1964.

15. Rink, R. E., and Mohler, R. R., "Completely Controllable Bilinear Systems," *SIAM J. Control,* 6, no. 3 (1968): 477–487.

16. Lee, E. B., and Markus, L., "Optimal Control for Nonlinear Processes," *Arch. Rational Mech. Anal. J.,* 8 (1961): 36–58.

17. Takahashi, Y., Rabins, M. J., and Auslander, D. M., *Control and Dynamic Systems,* Addison-Wesley, Reading, Mass., 1972.

18. Routh, E. J., *Dynamics of a System of Rigid Bodies,* 3d ed., Macmillan, New York, 1877.

19. Takahashi, T., *Mathematics of Automatic Control,* Rinehart and Winston, New York, 1966.

20. Kochenburger, R., "Frequency-Response Method for Analysis of a Relay Servomechanism," *Trans. A.I.E.E.,* 69 (1950):

21. Gibson, J. E., *Nonlinear Automatic Control,* McGraw-Hill, New York, 1963.

22. Popov, V. M., *Hyperstability of Control Systems,* Springer-Verlag, New York, 1973.

System Design

W̲E ARRIVE, FINALLY, AT THE LAST STAGE of the dynamic system investigation: *design*. As we mentioned in Chapter 1, our focus is on *design for automatic controls*. Two intuitive prerequisites for automatic control discussed in the last part of the text are knowledge of systems of interest and feasibility of control. It is also necessary to specify the quality of control in the design of control systems. The design of a controller depends on the performance objectives for the control system. For a given specification of control quality, various controllers can be designed, each of which attempts to satisfy well that specification. Thus a fairly wide range of design methods and types of control systems (recall the classification of control systems in Table 1.3) will be featured in Part III. We begin in Chapters 11 and 12 with the structurally simple and easily visualized *single-input, single-output* (SISO) plant controllers and progress to the fully *multiple-input, multiple-output* (MIMO) *multivariable* control systems introduced in Chapter 13. Chapter 14 covers a similar range of topics for *discrete-time control systems,* while Chapters 11 to 13 concentrate on continuous-time systems. We conclude in Chapter 15 with topics relating to the implementation of control system designs using the microcomputer. Controller hardware, as a component of an overall control system, influences system behavior and should, in general, be taken into consideration in the design stage.

Chapters 11 and 12 deal mostly with problems of *classical feedback controller design* and *performance specifications* that can be solved directly using some of the analytical techniques already developed, especially in the last chapter. *Time domain* methods are used in Chapter 11, and Chapter 12 concentrates on *frequency domain* techniques. Many of the control problems in Chapters 13 and 14, however, require more general solution methods or a *modern control* approach. In addition to multivariable systems, *feedforward* and *cascade* control systems are also featured in Chapter 13. Chapter 14 briefly explains the ideas of *supervisory, hierarchical,* and *numerical control* systems and discusses in some detail *direct digital control* and *finite-time settling* design. In Chapter 15, we discuss sensing and actuation concepts and hardware and deal with concerns for interfacing the microcomputer with components of the controlled system that are external to the computer. These lead to an illustration of a complete realization of a computer control system.

Finally, it is not enough for a controller to simply satisfy the performance specifications. There is the need, sometimes, to answer the question, How good is a good controller? Concerns for high performance and maintaining good performance or stability in the face of incomplete knowledge of or changes in the system dynamics and environmental inputs lead to such ideas as *observer* design and *adaptive* and H_∞ *control* systems—advanced concepts briefly introduced in Chapter 13. The concept of *optimal control,* which results from the search for the "best controller," is another advanced topic discussed briefly in Chapters 13 (*maximum principle* approach) and 14 (*dynamic programming* approach). A general design approach that is applicable in all these cases and is dispersed throughout Part III is digital computer simulation. In today's complex problems, it is unusual that an analytical design will be implemented directly as a result of analysis. Some form of simulation of the system and the design is often necessary not only to verify the analysis but to acquire more insight into the system behavior and to *fine tune* the design.

Chapter

11

Introducing Automatic Control Systems Design

WHAT IS AUTOMATIC CONTROL? The meaning of *automatic closed-loop control* was explained in Section 1.2. The word *control* is self-explanatory and implies among other concepts, manipulation, regulation, and tracking. The word *automatic* is used here to imply *not manual*. Consider, for example, a single-pipe steam heater system, such as may be found in some older homes. Usually, the only way to maintain a comfortable temperature range in a room equipped with such a heater (see Fig. 11.1a) is for the occupants to periodically open or close the steam valve, according to whether they are feeling cold or very warm. This is manual control. In contrast, for the warm-air furnace system illustrated in Fig. 11.1b, all that is required is for the room *thermostat* to be set to the desired room temperature, and the *burner valve* is automatically turned on and off in order to keep the room temperature close to the set value. This is automatic control. In this chapter (refer to Table 1.3) we consider the automatic control of SISO *continuous-time plants* (nature of loop and object of control), using also *continuous controllers* (nature of controller), designed using primarily *classical methods* in the *time* (also implies *Laplace*) *domain* (nature of analytical technique).

11.1 STAGE FOUR: DESIGN

In stage four of a dynamic investigation we seek the combination of parameter values and, where possible, even the arrangement of components that yield the desired solution. For design for automatic control, the satisfaction of the *goal* of the study is usually measured by a set of *performance specifications* that will be introduced shortly. As was shown in Fig. 1.12, the design process is *iterative*. The automatic control system design that is acceptable with respect to the performance specifications must also be *workable* or practicable. Sometimes, as in the case of *optimal* control, we require that it also be the *best* workable solution possible for the given specifications. Where such requirements are not met, we iterate through the dynamic study, as suggested by Fig. 1.12, starting again at *physical modeling* or *model construction*. The underlying assumption, of course, is that a physical system will normally always admit of a practical or optimal solution, but this is not strictly true, in practice, for every single case.

The design process is also *interactive,* especially with respect to the intelligence or knowledge available from *model solution*. All relevant controlled system behaviors determined through model solution are normally brought to bear on the design process.

Figure 11.1

Manual and Automatic Space Heating Controls

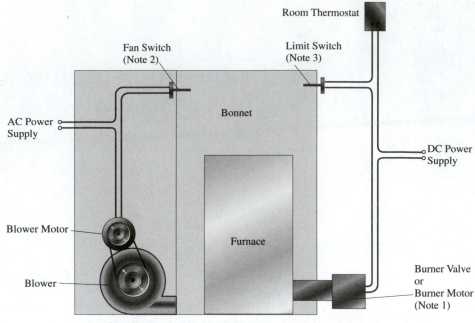

(b)   Warm-air Furnace System; Arrangement of Controls

Notes: 1. The burner motor or valve is directly connected to the room thermostat.
2. The fan switch turns on the blower whenever the air temperature inside the bonnet is high.
3. The limit switch is a safety device to shut off the burner should the bonnet air temperature become excessive.

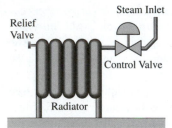

(a)   One-pipe Space Heating Element

This factor implies a minor but very significant iterative loop (not shown in Fig. 1.12) between the *design* and *solve model* stages. Although not apparent in design using well-established and closed-form techniques, this iterative loop is quite evident in design by computer simulation.

Continuous-Time Single-Loop Feedback Control

The concept of *feedback control* was introduced in Section 1.2. Figure 11.2a shows the single-loop feedback control structure in the *s*-domain block diagram format. As explained in Section 1.2, sometimes an auxiliary block (**actuator**) can be interposed between the **controller** and the **plant** for clarity or emphasis, but the actuator can also be incorporated in either the controller or the **system.** The object of control is variously termed the *system, plant,* or **process.** The control system is *single-loop* in the sense that only one output variable is fed back.

Feedback control or *closed-loop control* results when the controller output is governed by the difference between actual and desired system behavior (see [1]). The *controlled variable* represents the actual behavior of the system, which can differ from the desired behavior as given by the *setpoint (desired value, reference input,* etc.). The controlled variable is affected by the controller output—the *manipulated variable*—as well as by disturbances or upsets. Anything that affects the actual behavior of the system from outside the feedback loop is considered a *disturbance.* As a practical matter, if the nature of a disturbance is understood, the first line of action should be to attempt

Figure 11.2

Feedback Control Structure

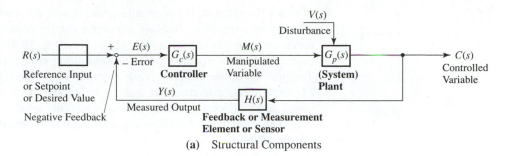

(a)   Structural Components

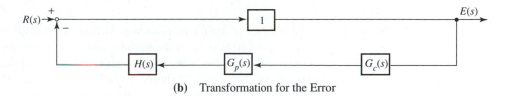

(b)   Transformation for the Error

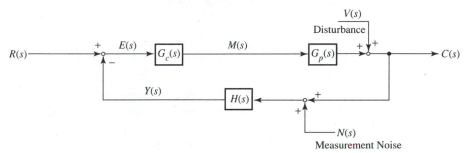

(c)   Model for Disturbance and Measurement Noise Inputs

to eliminate it at the source (see Section 13.1). However, this is not often possible. The *error* is the result of some form of a comparison (represented symbolically in Fig. 11.2a by a small circle with signs on the inputs) between the desired value and the measurement of the actual behavior of the system. The measurement or **feedback element** is a means for converting the controlled variable into a form that is easily utilized in the controller. For instance, a thermocouple can be used to convert a temperature variable into a voltage signal. In Fig. 11.2a the outline of an **input element** is indicated following the setpoint, but no transfer function is shown (a 1 is implied). However, input elements are necessary to condition the setpoint for the controller just as the feedback element conditions the controlled variable for the controller. For example, the setting of a thermostat knob to a temperature position can be converted into a differential voltage signal with a potentiometer that has its own characteristics. Thus without a specific indication of the input component, a proper interpretation of Fig. 11.2a is that the setpoint is being recognized after the input element.

The controller is the implementation of the algorithm or rule (*control law* or *control action*) for driving the error to zero or a small value. There are various forms of control elements, as shown in Table 1.3 (nature of controller). In continuous controllers, the manipulated variable is a continuous function of the error. This is in contrast to discontinuous or on-off control components, of which the room thermostat (Fig. 11.1b) is a typical example. A controller such as (again) many temperature controllers in which the setpoint, although adjustable, is held constant for long periods of time is often called a *regulator. Tracking* refers to a control situation in which the setpoint pattern is continuously changing. An example (see Fig. 10.5) is the radar-generated plane-position reference input to the muzzle-positioning control of an antiaircraft gun system. The position of the plane, which is continuously changing (indeed, the plane may be executing evasive maneuvers), has to be tracked (with necessary compensation for speed, heading, etc.) by the gun in order to score a hit.

Our objective in the following discussion is to present some common procedures for designing the controller $[G_c(s)]$. But before doing this, let us examine a major implication of the feedback structure.

Stability and Sensitivity   The question of stability, which was discussed generally in the last chapter, is particularly important now, since the introduction of certain types of control can cause an otherwise stable system to become unstable and thereby defeat the purpose of the control. The overall transfer function for the feedback control system of Fig. 11.2a, that is, the closed-loop transfer function, is according to equations (9.3) and (9.4) or equation (10.15b), repeated here,

$$G_{CL}(s) = \frac{C(s)}{R(s)} = \frac{G_p(s)G_c(s)}{1 + G_p(s)G_c(s)H(s)} \tag{11.1}$$

In terms of the measured output transfer function, which we define as the open-loop transfer function, equation (10.15a), repeated,

$$G_o(s) = G_p(s)G_c(s)H(s) \tag{11.2a}$$

the closed-loop transfer function (equation (10.15c)) and characteristic equation are

$$G_{CL}(s) = \frac{G_p(s)G_c(s)}{1 + G_o(s)} \tag{11.2b}$$

$$1 + G_o(s) = 0 \text{ or } 1 + G_p(s)G_c(s)H(s) = 0 \tag{11.3}$$

Equation (11.3) is totally different from the original (uncontrolled) system characteristic equation,

$$[\text{Denominator of } G_p(s)] = 0 \tag{11.4a}$$

or the open-loop system characteristic equation,

$$[\text{Denominator of } G_o(s)] = 0 \tag{11.4b}$$

although they are all related. The implications of this difference for the stability of the overall system is obvious. *Although the original (uncontrolled) system may be stable, the feedback system, if designed without regard to stability, may not be stable!* In this respect, note also that the closed-loop characteristic equation (11.3) is affected by the measurement element $H(s)$. Although in theory very little attention is paid to this component, in practice the dynamics of the sensors in a feedback control system may be all-important!

Sensitivity

The dependence of the characteristics of a system on those of one of its components or parameters is measured by the corresponding *sensitivity*. The differential sensitivity of a system's overall transfer function $O(s)$ with respect to the quantity $I(s)$ or the element with transfer function $I(s)$ is given by

$$S_I^O(s) = \frac{dO(s)/O(s)}{dI(s)/I(s)} = \frac{I(s)}{O(s)} \frac{dO(s)}{dI(s)}; \quad \text{(sensitivity)} \tag{11.5a}$$

For the system's characteristics, which are represented by $O(s)$, to be insensitive to variations in the quantity $I(s)$, the sensitivity $S_I^O(s)$ needs to be zero or as small as possible. Note that in equation (11.5a) the sensitivity is a function of s and is not necessarily fixed. Putting $s = 0$ gives the *static* sensitivity, while $s = j\omega$ gives the *dynamic* (or frequency dependence of the) sensitivity. The sensitivity with respect to a particular component can therefore be high over certain frequencies and yet insignificant over another range of frequencies. Similarly, performance under transient and steady-state conditions can differ significantly. The correlation between transient response and frequency response is considered in Section 12.1.

What is the sensitivity of the closed-loop system of Fig. 11.2a with respect to the feedback element? According to equation (11.5a),

$$S_H^{G_{CL}}(s) = \frac{H(s)}{G_{CL}(s)} \frac{dG_{CL}(s)}{dH(s)}$$

But from equation (11.1),

$$G_{CL}(s) = \frac{G_p(s)G_c(s)}{1 + G_p(s)G_c(s)H(s)}$$

so that

$$\frac{dG_{CL}(s)}{dH(s)} = \frac{-[G_p(s)G_c(s)]^2}{[1 + G_p(s)G_c(s)H(s)]^2} \quad \text{and} \quad \frac{H(s)}{G_{CL}(s)} = \frac{[1 + G_p(s)G_c(s)H(s)]H(s)}{G_p(s)G_c(s)}$$

Thus

$$S_H^{G_{CL}}(s) = \frac{-G_p(s)G_c(s)H(s)}{1 + G_p(s)G_c(s)H(s)} = \frac{-G_o(s)}{1 + G_o(s)} \tag{11.5b}$$

We can see in equation (11.5b) a confirmation of the importance of the feedback element to the closed-loop system dynamics. For instance, for frequencies for which the *open-loop gain* $|G_o(s)| \gg 1$, $S_H^{G_{CL}}(s) \longrightarrow -1$, which represents significant sensitivity. A similar result is obtained, as should be expected, for the sensitivity of the closed-loop characteristics with respect to the *input* element (shown in outline in Fig. 11.2a). Suppose the transfer function for the input element is $I(s)$. For this case the closed-loop transfer function, equation (11.1), should be amended to

$$G_{CL}(s) = \frac{C(s)}{R(s)} = \frac{G_p(s)G_c(s)I(s)}{1 + G_p(s)G_c(s)H(s)} \tag{11.5c}$$

and

$$S_I^{G_{CL}}(s) = \frac{I(s)}{G_{CL}(s)} \frac{dG_{CL}(s)}{dI(s)} \qquad (11.5d)$$

$$= \frac{I(s)[1 + G_p(s)G_c(s)H(s)]}{G_p(s)G_c(s)I(s)} \frac{G_p(s)G_c(s)}{1 + G_p(s)G_c(s)H(s)} = 1$$

The closed-loop system is *very* sensitive to the input element characteristics at all frequencies!

Feedback Control Objective and Open-Loop Gain

The sensitivity with respect to the forward-loop components, $G(s) = G_p(s)G_c(s)$, is

$$S_G^{G_{CL}}(s) = \frac{G(s)}{G_{CL}(s)} \frac{dG_{CL}(s)}{dG(s)}$$

$$= \frac{G(s)[1 + G(s)H(s)]}{G(s)} \left[\frac{1}{1 + G(s)H(s)} - \frac{G(s)H(s)}{(1 + G(s)H(s))^2} \right]$$

$$S_G^{G_{CL}}(s) = \frac{1}{1 + G(s)H(s)} = \frac{1}{1 + G_o(s)} \qquad (11.5e)$$

For the same range of frequencies as above, for which $|G_o(s)| \gg 1$, the sensitivity of the overall system with respect to $G(s) = G_p(s)G_c(s)$ tends to $1/G_o(s)$, which also implies a strong dependence of the closed-loop characteristics on the forward-loop elements. The control system however will become insensitive to variations in the forward-loop parameters in the limit that the open-loop gain

$$|G_o(s)| \longrightarrow \infty \text{ or } |S_G^{G_{CL}}| \longrightarrow 0.$$

As a practical matter, the parameters of the plant $G_p(s)$ are seldom completely stationary or completely known, although the controller $G_c(s)$ could have been designed assuming fixed (nominal) values for the plant parameters. It follows that since $G_o(s) = G_p(s)G_c(s)H(s)$, errors due to such lack of knowledge or variations in plant parameters can be minimized if the controller gain is sufficiently high: $|G_c(s)| \longrightarrow \infty$. In other words, we can replace the requirement on the open-loop gain, $|G_o(s)| \longrightarrow \infty$, with one on the controller gain, $|G_c(s)| \longrightarrow \infty$. As shown next, this conclusion is also reached for the *overall* feedback control objective, and it therefore represents a major advantage of the feedback control structure relative to open-loop configurations.

Feedback Control Objective   Our fundamental *objective* in feedback control is correspondence between the measurement of the controlled variable and the setpoint:

$$Y(s) = R(s) \text{ or } E(s) = 0, \quad \text{for all } R(s) \text{ and } V(s); \text{ (overall objective)} \qquad (11.6a)$$

To find $E(s)$, note that Fig. 11.2a can be transformed as shown in Fig. 11.2b, so that

$$E(s) = \left[\frac{1}{1 + G_o(s)} \right] R(s) = S_G^{G_{CL}}(s)R(s) \qquad (11.6b)$$

$E(s)$ will vanish for all $R(s)$ when again the *open-loop gain* $|G_o(s)| \longrightarrow \infty$. If we consider the plant $G_p(s)$ and measurement $H(s)$ transfer functions as fixed design inputs, then $|G_o(s)| \longrightarrow \infty$ implies $|G_c(s)| \longrightarrow \infty$, as before. *Most engineering systems become unstable in this limit.* This conclusion can be obtained from the root locus plots already constructed in Chapter 10 (Figs. 10.6, 10.7, and 10.8) by observing the trend in each case whereby some roots migrated toward the RHP as the adjustable parameter k became arbitrarily large. The *gain* of the control action $M(s)$ relative to the error $E(s)$, that is, $|G_c(s)|$, is often referred to as the *tightness* of the loop. Thus there is the risk of loss of control (instability) in a sufficiently tight loop. As a result, *feedback controller design is generally a compromise between control quality and relative stability.*

As a further demonstration of conflict between performance objectives, consider the requirement for minimizing the propagation of measurement noise. For the model given by Fig. 11.2c, it is left as an exercise (see Practice Problem 11.3c) for the reader to show that the gain of the (controlled) system output relative to noise input before the feedback sensor, as indicated, is

$$|S_{C/N}(s)| = |S_G^{G_{CL}}(s) - 1| \tag{11.6c}$$

However, $|S_{C/N}| \longrightarrow 0$, which is needed to minimize the effect of measurement noise, is in conflict with $|S_G^{G_{CL}}| \longrightarrow 0$, required to minimize the effect of variations in forward-loop parameters or, for that matter, to reject disturbances that enter the plant (see Practice Problem 11.3d).

Time Response Performance and Design

We can deduce from the preceding discussion the following elements of feedback control objective:

1. **Determination of degree of system stability.** Not only is stability necessary, the relative stability is important.
2. **Determination of transient response.** Dynamic and static performance can also differ.
3. **Determination of steady-state performance.**

These objectives form the basis for evaluating the performance of feedback controllers, and the basic goal of control system design is to meet the performance specifications. In general, the measures of the performance of dynamic systems are

1. **Relative stability**
2. **Speed of response**
3. **System accuracy or permissible error**

These performance measures are associated, in the order given, with the listed performance objectives of feedback control systems. In practice, the actual specification depends on the type of prototype input to which the system is subjected. As was shown in Part II, the inputs that have been used for identification of deterministic systems include the *step, ramp, pulse, impulse,* and *sinusoidal* input of many frequencies (frequency response). Step response and frequency response are easier to realize. Step response specifications, in particular, and time response specifications, in general, are considered here (frequency domain design is the subject of Chapter 12).

Figure 11.3

Typical Step Response
and Specifications

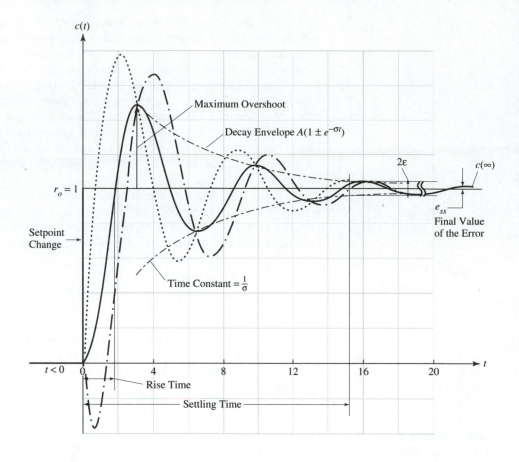

Step Response Specifications

For feedback control system design, we consider the *unit-step response* from zero initial state without loss of generality. The solid curve (the other two curves are explained below) in Fig. 11.3 shows such a typical step response (for an underdamped second-order system, for example, see Fig. 7.7b). The response is characterized by certain quantities shown on the figure that can be used to specify performance (see Section 7.1, Underdamped Second-Order Transient Response Characteristics, for the derivation of the expressions):

1. *Rise time* $\left[t_r = \frac{1}{\omega}\left(\frac{\pi}{2} - \phi\right); \quad \tan \phi = -\sigma/\omega\right]$, equation (7.16b): The time for the response to first reach the desired value (that is, $r(t) = 1$). It is a specification of the speed of response.
2. *Percent overshoot* $[100e^{-\sigma\pi/\omega}]$, equation (7.16f): The amount, as a percentage, by which the maximum value of the response exceeds the setpoint (sometimes defined relative to the steady-state value rather than to the setpoint). It is a specification that actually embodies all three performance measures, especially relative stability and system accuracy.
3. *Settling time* $\left[t_{ss} = \frac{-1}{\sigma}\ln\left(\epsilon\sqrt{1 - \zeta^2}\right)\right]$, equation (7.16g): The time it takes the response to settle within some prescribed range of values, usually given as a percentage of the setpoint (for example, $\pm\epsilon$, $\epsilon = 5\%$). It is determined by the rate of decay (decay envelope of the response) and is therefore a specification of the speed of response and relative stability.

4. *Final value of the error,* or the steady-state error: Specifies the permissible error and is the difference between the desired value and the final value of the response.

The steady-state error is a general time domain specification for system *(static)* accuracy or permissible error that is not limited to the step-input response. For the SISO feedback loop of Fig. 11.2a,

$$e_{ss} = \underset{t \to \infty}{\text{Limit}}[r(t) - c(t)] = r(\infty) - c(\infty) \tag{11.7a}$$

Also, by the final-value theorem (we assume that all necessary conditions are met in each case),

$$e_{ss} = \underset{t \to \infty}{\text{Limit}}[e(t)] = \underset{s \to 0}{\text{Limit}}[sE(s)] \tag{11.7b}$$

From equation (11.6b),

$$e_{ss} = \underset{s \to 0}{\text{Limit}}\left[\frac{sR(s)}{1 + G_o(s)}\right] \tag{11.7c}$$

Effect of Zeros on Second-Order System Step Response

The solid curve in Fig. 11.3, on which the preceding step response specifications are based, is for a second-order system, equation (7.2), with the *normalized* transfer function

$$\frac{C(s)}{R(s)} = G_{CL}(s) = \frac{\omega_0^2}{s^2 + 2\zeta\omega_0 s + \omega_0^2}; \quad \text{(for Fig. 11.3, } \omega_0 = 1 \text{ and } \zeta = 0.2\text{)} \tag{11.7d}$$

A more general second-order (normalized) transfer function, especially considering the capacity of controllers to alter the system structure (see example 11.7) is*

$$\frac{C_z(s)}{R(s)} = G_{CL}(s) = \frac{(\omega_0^2/b)(s + b)}{s^2 + 2\zeta\omega_0 s + \omega_0^2} \tag{11.7e}$$

$$= \left[\frac{\omega_0^2}{s^2 + 2\zeta\omega_0 s + \omega_0^2} + \frac{(\omega_0^2/b)s}{s^2 + 2\zeta\omega_0 s + \omega_0^2}\right]$$

Equation (11.7e) shows that the general second-order system now admits a zero, $s = -b$, in addition to the usual two poles (eigenvalues). It also shows, in the expanded form, that the new transfer function is the sum of the old one and another second-order system with the same characteristic equation as the old one. We can conclude therefore that the system with a finite zero and the one without a zero are both characterized by the same response modes, but the combination of these modes in the respective step responses differs. The eigenvalues and hence the decay rate $\sigma = \zeta\omega_0$, and the damped frequency $\omega = \omega_0\sqrt{1 - \zeta^2}$, remain the same. For stable systems, as assumed, this further implies that the final value of the error (the steady-state response) will remain unchanged. However the transient response measures, rise time, % overshoot, and settling time, are all expected to change.

---

*The most general and realizable second-order transfer function will allow for a numerator polynomial that can go up to second order (that is, with up to two zeros), but such a system is rare in practice.

The expanded form in equation (11.7e) suggests that the changes in these specifications are small for large $|b|$, that is, the farther away the zero is from the origin of the complex plane, the less the effect (see also the following results). The more dramatic effect on the transient response occurs when b is negative. This occurrence gives a positive zero and, as discussed in Section 9.1, an *inverse response*. This is the situation illustrated by the dash-dot curve in Fig. 11.3, which represents the system (11.7d) with a positive zero $s = 1$ ($b = -1$) added. The dotted curve illustrates (11.7d) with a negative zero $s = -b = -1$.

To quantify the anticipated effects, consider the solution of (11.7e) for $c_z(t)$ under zero initial conditions and $r(t) = 1$ (unit step). Recalling the solution of (11.7d) for $c(t)$ under the same conditions, equation (7.16a) and the derivative property of Laplace transforms, we take

$$c_z(t) = c(t) + \frac{1}{b}\frac{d}{dt}c(t) \tag{11.7f}$$

$$= 1 - \frac{1}{\cos\phi}\left[\left(\frac{b-\sigma}{b}\right)\cos(\omega t + \phi) + \sin(\omega t + \phi)\right]e^{-\sigma t}$$

The cosine and sine terms can be combined to yield

$$c_z(t) = 1 - \frac{M_z}{\cos\phi}[\cos(\omega t + \phi + \phi_z)]e^{-\sigma t} \tag{11.7g}$$

where

$$M_z = \frac{1}{|b|}\sqrt{\omega^2 + (b-\sigma)^2} \quad \text{and} \quad \phi_z = \tan^{-1}\left(\frac{\omega}{b-\sigma}\right) \tag{11.7h}$$

By the same techniques used in Section 7.1, we can determine the corresponding rise time, % overshoot, and settling time to be

Rise time:

$$t_{rz} = \frac{1}{\omega}\left(\frac{\pi}{2} - \phi - \phi_z\right) = \frac{1}{\omega}\left(\frac{\pi}{2} - \phi\right) - \frac{\phi_z}{\omega} = t_r - \frac{\phi_z}{\omega} \tag{11.7i}$$

Time to peak:

$$t_{pz} = \frac{\pi}{\omega} - \frac{\phi_z}{\omega} = t_p - \frac{\phi_z}{\omega} \tag{11.7j}$$

% Overshoot:

$$100M_z e^{-\sigma\left(\frac{\pi}{\omega} - \frac{\phi_z}{\omega}\right)} = \left(M_z e^{\sigma\frac{\phi_z}{\omega}}\right)100e^{-\pi\frac{\sigma}{\omega}} \tag{11.7k}$$

$\pm\epsilon$ Settling time:

$$t_{ssz} = -\frac{1}{\sigma}\ln\left(\frac{\epsilon\sqrt{1-\zeta^2}}{M_z}\right) = -\frac{1}{\sigma}\ln(\epsilon\sqrt{1-\zeta^2}) + \frac{\ln(M_z)}{\sigma} \tag{11.7l}$$

$$= t_{ss} + \frac{\ln(M_z)}{\sigma}$$

That is, relative to the second-order system without a zero, for the same system with a finite zero (at $s = -b$): (1) the rise time and the time to peak are both *reduced* by ϕ_z/ω, (2) the % overshoot is *amplified* by the factor $M_z e^{\delta\phi_z/\omega}$, and (3) the settling time is *increased* by $\ln(M_z)/\sigma$.

These results can be verified in Fig. 11.3 by computing these changes using the parameter values $\omega_0 = 1$, $\zeta = 0.2$, and $b = 1$ (or $b = -1$ for the inverse response case). Practice Problem 11.12 also illustrates the application of these relations to a second-order system with a negative zero.

When these step response specifications are applied generally to a system under design, the assumption is that the overall behavior of the system is second order or can be approximated as such. Proportional control is intuitively the simplest control law. Let us begin by exploring it.

Example 11.1 Proportional Controller via Time Response Specifications

An X-ray scanning device is used to inspect printed circuit boards and wafer chips mounted on a screw-driven X-Y table, as shown in Fig. 11.4a. The position of the table is controlled by a PC-based controller. Figure 11.4b shows the block diagram for proportional control ($G_c(s) = k_c$) of one of the axes of the table. $G_p(s)$ represents the motor and table dynamics, with the ratio of inertia to damping coefficient $J/\beta = \frac{1}{4}$. Design the proportional controller (find the range of k_c) such that

1. The percent overshoot for a step input is less than 40%.
2. The final value of the error for a ramp input $r(t) = u_0 t$ is less than 10% of the magnitude (slope) of the ramp u_0.

Figure 11.4

Circuit Board Scanning Control System

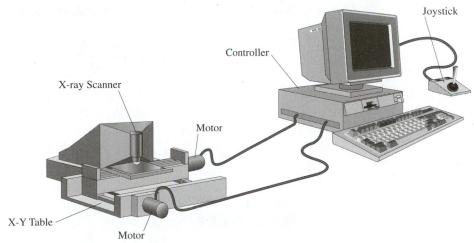

(a) Example of X-Y Point-to-point Control System (Operator Has Option to Manually Override System Using Joystick.)

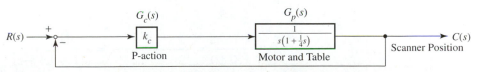

(b) Block Diagram of Proportional Control System

Solution: The closed-loop transfer function is

$$G_{CL}(s) = \frac{\dfrac{k_c}{s\left(1 + \frac{1}{4}s\right)}}{1 + \dfrac{k_c}{s\left(1 + \frac{1}{4}s\right)}} = \frac{k_c}{s\left(1 + \frac{1}{4}s\right) + k_c};\ \ \text{(no zero)}$$

The closed-loop characteristic equation is therefore

$$\frac{1}{4}s^2 + s + k_c = 0 \quad \text{or} \quad s^2 + 4s + 4k_c = 0$$

From the canonical second-order characteristic equation (7.4a),

$$s^2 + 2\zeta\omega_0 s + \omega_0^2 = 0$$

We see that

$$\omega_0 = 2\sqrt{k_c} \quad \text{and} \quad 2\zeta\omega_0 = 4$$

so that

$$\zeta = \frac{2}{2\sqrt{k_c}} = \frac{1}{\sqrt{k_c}}$$

Now, from equation (11.7k), 40% overshoot requires

$$M_z e^{\phi_z/\omega} e^{-\pi\sigma/\omega} = \exp\left(\frac{-\pi\zeta}{\sqrt{1 - \zeta^2}}\right) \leq 0.4$$

or

$$\frac{-\pi\zeta}{\sqrt{1 - \zeta^2}} \leq \ln(0.4) = -0.9163 \Rightarrow \frac{\pi^2\zeta^2}{1 - \zeta^2} \geq 0.8396 \Rightarrow \zeta^2 \geq 0.0784$$

That is, $1/k_c \geq 0.0784$ or $k_c \leq 12.755$.
Next, we note that

$$G_o(s) = G_p(s)G_c(s)H(s) = \frac{k_c}{s\left(1 + \frac{1}{4}s\right)} \quad \text{and} \quad E(s) = \frac{1}{1 + G_o(s)}R(s)$$

With

$$\frac{1}{1 + G_o(s)} = \frac{1}{1 + \dfrac{k_c}{s\left(1 + \frac{1}{4}s\right)}} = \frac{s\left(1 + \frac{1}{4}s\right)}{s\left(1 + \frac{1}{4}s\right) + k_c}$$

given

$$r(t) = u_0 t, \qquad R(s) = \mathscr{L}[u_0 t] = u_0/s^2$$

by the final-value theorem, or equation (11.7c),

$$e_{ss} = \underset{s \to 0}{\text{Limit}} \left[\frac{s.s\left(1 + \frac{1}{4}s\right)u_0}{s^2[s\left(1 + \frac{1}{4}s\right) + k_c]} \right] = \frac{u_0}{k_c} \leq 0.1u_0$$

That is, $1/k_c \leq 0.1$ or $k_c \geq 10$. Hence the solution is $10 \leq k_c \leq 12.755$.

Static Accuracy and System Type

Besides serving as a simple illustration of the application of time response specifica-tions and what is meant by controller design, something else was established in the pre-ceding example. For the ramp input, the proportional controller had a steady-state error $e_{ss} = u_0/k_c$ that would remain finite except in the limit $k_c \longrightarrow \infty$. Such an error, called *offset,* was discovered in Example 9.11—another proportional control system. Is this an accident of only *proportional* control or does it also have something to do with the *type* of system and the type of input? For that matter, what is the steady-state error perfor-mance for all the fundamental single-mode controls—*proportional* control, *integral* control, and *derivative* control—for some general class of dynamic systems?

Consider again a general representation of the (overall) open-loop transfer func-tion $G_p(s)G_c(s)H(s) = G_o(s)$ in the form of equation (9.15a):

$$G_o(s) = \frac{K(b_0 + b_1 s + \cdots + b_m s^m)}{s^q(a_0 + a_1 s + \cdots + a_{n-q} s^{n-q})}; \tag{11.8a}$$

$$m \geq 0, \quad q \geq 0, \quad (n - q) \geq 0, \quad n \geq m, \quad \frac{b_0}{a_0} \neq 0$$

where $n \geq m$ is the condition for realizability (present state does not depend on future events) of the transfer function, and b_0/a_0 is assumed to be finite and nonzero. The clas-sification of systems represented as in equation (11.8a) into *types* was introduced in Section 10.4 in connection with the Nyquist stability criterion (Fig. 10.16): $q = 0$ for a *type*-0 system, $q = 1$ for a *type*-1 system, and so on. Substituting equation (11.8a) into equation (11.7c), we obtain

$$e_{ss} = \underset{s \to 0}{\text{Limit}} \left[\frac{sR(s)}{1 + \dfrac{Kb_0/a_0}{s^q}} \right] \tag{11.8b}$$

Step (Position) Input   Consider the case of step input $r(t) = u_0 U_s(t)$ to the closed-loop system. Substituting $R(s) = u_0/s$ in equation (11.8b), we find for type 0, 1, 2, . . . systems,

$$e_{ss} = \begin{cases} \dfrac{u_0}{1 + Kb_0/a_0} & 0 \quad \text{type 0} \\ 0 & \text{type 1, 2, . . .} \end{cases} ; \quad \text{(step input)} \tag{11.8c}$$

Thus type-0 systems subject to a steplike input always have an offset. This is the case no matter the type of controller, provided the overall open-loop system is type 0. In Example 9.11, for instance, as the reader can verify,

$$G_o = \frac{X_2(s)}{U(s)} = \frac{k_c}{(s + 1)(s + 2)} = \frac{k_c}{2 + 3s + s^2}$$

so that $b_0/a_0 = \frac{1}{2}$, $K = k_c$, $m = 0$, $n = 2$, and $q = 0$, that is, a type-0 system. From equation (11.8c), $e_{ss} = 2u_0/(2 + k_c)$, and if the magnitude of the step is 4, then the steady-state error will be $8/(2 + k_c)$. From equation (9.43), the offset for the control system of Example 9.11 for a setpoint of 4 is

$$4 - \frac{4k}{2 + k} = \frac{8}{2 + k}$$

which agrees with the result from equation (11.8c).

Ramp (Velocity) Input   For a ramp input, $r(t) = u_0 t$ and $R(s) = u_0/s^2$, equation (11.8b) yields

$$e_{ss} = \begin{cases} \infty & \text{Type } 0 \\ \dfrac{u_0}{Kb_0/a_0} & \text{Type } 1 \\ 0 & \text{Type } 2, 3, \ldots \end{cases} ; \quad \text{(ramp input)} \qquad \textbf{(11.8d)}$$

The steady-state error is infinite for a type-0 subject to a ramp input under feedback control! For a type-1 system, an offset results for a ramp input, whereas type-2 and higher systems yield zero static error.

The proportional control system of Example 11.1, with

$$G_0(s) = \frac{k_c}{s\left(1 + \frac{1}{4}s\right)}; \qquad K = k_c, \frac{b_0}{a_0} = 1$$

is a type-1 system. According to equation (11.8d), an offset is expected, and its value is

$$e_{ss} = \frac{u_0}{k_c}$$

as found in the example.

Static Accuracy and Mode of Control

To explore the relation, if any, between the mode of control (proportional, integral, or derivative) and steady-state error, let us separate $G_c(s)$ from $G_o(s)$ and consider the class of the balance of the open loop system $G_p(s)H(s) = G_{PH}(s)$, which can be represented by equation (11.8a), that is,

$$G_{PH}(s) = \frac{K(b_0 + b_1 s + \cdots + b_m s^m)}{s^q(a_0 + a_1 s + \cdots + a_{n-q} s^{n-q})}; \qquad \textbf{(11.9a)}$$

$$m \geq 0, \quad q \geq 0, \quad (n - q) \geq 0 \, n \geq m, \quad \frac{b_0}{a_0} \neq 0$$

where, in this case, q can be viewed as the number of *pure integrations* in the transfer function. As before, $n \geq m$ is required for realizability, and b_0/a_0 is finite and not zero.

Proportional Control (P-action)   For *proportional* mode control, $G_c(s)$ is just a proportional gain k_c. Thus substituting this value and equation (11.9a) into equation (11.7c), we obtain for a step input, $r(t) = u_0 U_S(t)$, $R(s) = u_0/s$,

$$e_{ss} = \begin{cases} \dfrac{u_0}{1 + k_c K b_0/a_0} & q = 0 \\ 0 & q = 1, 2, \ldots \end{cases} \quad ; \quad \text{(P-action, step input)} \tag{11.9b}$$

and for a ramp input, $r(t) = u_0 t$ and $R(s) = u_0/s^2$,

$$e_{ss} = \begin{cases} \infty & q = 0 \\ \dfrac{u_0}{k_c K b_0/a_0} & q = 1 \\ 0 & q = 2, 3, \ldots \end{cases} \quad ; \quad \text{(P-action, ramp input)} \tag{11.9c}$$

Hence proportional control of the class of systems (11.9a) with no pure integration always has a steady-state error (offset) for a steplike input. The static error is infinite for the same type of system under a ramp input, while systems with one pure integration have a finite offset under proportional control and a ramp input. Again, for the control system of Example 9.11,

$$G_{PH} = \frac{1}{(s+1)(s+2)} = \frac{1}{2 + 3s + s^2}; \qquad K = 1, \quad \frac{b_0}{a_0} = \frac{1}{2}, \quad q = 0$$

so that for a step input with $u_0 = 4$, equation (11.9b) gives $e_{ss} = 8/(2 + k_c)$, as before. Also, for the system of Example 11.1,

$$G_{PH}(s) = \frac{1}{s\left(1 + \frac{1}{4}s\right)}; \qquad K = 1, \quad \frac{b_0}{a_0} = 1, \quad q = 1$$

and equation (11.9c) predicts an offset for a ramp input that is equal to u_0/k_c, as before.

Note that in each case the offset can be reduced by increasing the proportional controller gain k_c. However, this is at the risk that some systems will become unstable, as we mentioned earlier.

Integral Control (I-action) $G_c(s)$ for pure *integral* control is $G_c(s) = 1/(T_i s)$, where T_i is a constant (controller parameter). From equations (11.9a) and (11.7c), the steady-state error in this case is

$$e_{ss} = \underset{s \to 0}{\text{Limit}} \left[\frac{sR(s)}{1 + \dfrac{K b_0/a_0}{T_i s^{q+1}}} \right] \tag{11.10a}$$

According to equation (11.10a), for a step input, the steady-state error is zero no matter the number of pure integrations in the system:

$$e_{ss} = 0; \quad \text{(I-action, step input)} \tag{11.10b}$$

That is, there is no offset with pure integral control for a steplike input. The reason can be explained physically as follows: The integrator will continue to accumulate the error and generate a correcting input (manipulated variable) as long as there is a finite error. Consequently, the error is eventually reduced to zero.

It is unfortunate that this "stubbornness" in the integral controller can create a problem known as *reset windup* in some applications. In practice the integrator may become saturated before the error is reduced to zero, and as long as the sign of the error does not change, it will continue to hold the controller output near the saturation value even though the controlled variable is close to the desired value. The consequence of this can be a large offset in the response. Another problem with integral control, as will be explained later, is its tendency to *destabilize* the system through its influence on the closed-loop eigenvalues. In addition, I-action slows down the control, since integration is, in general, a slow process.

The situation is different for a ramp input. Because the magnitude of the input is increasing continuously while the integral action, as a dynamic process, involves some delay, an offset should be expected for systems with no pure integration. According to equation (11.10a), we have for a ramp input

$$e_{ss} = \begin{cases} \dfrac{u_0 T_i}{K b_0 / a_0} \neq 0 & q = 0 \\ 0 & q = 1, 2, \ldots \end{cases} \quad ; \quad \text{(I-action, ramp input)} \qquad \text{(11.10c)}$$

Derivative Control (D-action)   Finally, for *derivative* control, we shall analyze the transfer function $G_c(s) = T_d s$, although this form (pure differentiation) is not physically realizable. An example of a realizable approximate differentiator was given in Practice Problem 3.50 (see also Practice Problems 11.25 and 11.32, and Section 12.3). From equations (11.9a) and (11.7c),

$$e_{ss} = \underset{s \to 0}{\text{Limit}} \left[\frac{s R(s)}{1 + \dfrac{T_d K b_0 / a_0}{s^{q-1}}} \right] \qquad \text{(11.11a)}$$

and for step and ramp inputs, equation (11.11a) gives the results

$$e_{ss} = \begin{cases} u_0 & q = 0 \\ \dfrac{u_0}{1 + T_d K b_0 / a_0} & q = 1 \\ 0 & q = 2, 3, \ldots \end{cases} \quad ; \quad \text{(D-action, step input)} \qquad \text{(11.11b)}$$

$$e_{ss} = \begin{cases} \infty & q = 0 \text{ and } 1 \\ \dfrac{u_0}{T_d K b_0 / a_0} & q = 2 \\ 0 & q = 3, 4, \ldots \end{cases} \quad ; \quad \text{(D-action, ramp input)} \qquad \text{(11.11c)}$$

Thus, unless the system includes sufficient pure integration, pure derivative action is more or less a "nonstarter" as far as steady-state performance is concerned. This could have been anticipated, since for a system with a finite final value, at steady state nothing is changing anymore, and the differentiator will have no output even though the error is not zero.

11.2 CLASSICAL FEEDBACK CONTROLLERS

In general, derivative action is normally combined with proportional control to give a *proportional-plus-derivative control* (PD-control). This combination is effective particularly for undamped or lightly damped systems, since the negative feedback of the derivative of the controlled variable is somewhat akin to the addition of (viscous) damping to the system. D-action thus has a stabilizing influence on some systems. Also, derivative control is generally used to speed up control action. In practice, various combinations of the basic single-mode controllers are used in order to capitalize on the relative advantages of each. Much of the control theory developed between the late 1950s and early 1970s in the frequency and Laplace domains, the so-called *classical control theory*, was based on such combinations of the basic control modes. Table 11.1 summarizes the usual classical controllers along with their theoretical transfer functions. The realizable forms of these controllers will be considered where this is necessary.

One-, Two- and Three-Mode Process Controllers

We shall generally refer to controllers derived from one or more of the three basic control modes as *one-* or *two-* or *three-mode* controllers. For instance, the PID controller in Table 11.1, is a three-mode controller. The designation *three-term* (etc.) controller can also be used. The original application of such controllers, which were usually implemented using pneumatic, hydraulic, fluidic, or analog-electronic computing devices, was mostly in the process and power industries. The physical realization of the values of the parameters k_c, T_i, and T_d in such controllers was usually by means of adjustable knobs on the body of the computing device (controller). The procedure for selecting optimal values of these parameters, which was thus called *tuning*, was often a trial-and-error process. However, some systematic approaches and tuning rules of thumb also exist.

Ziegler-Nichols Rules

One such systematic approach, which was developed by Ziegler and Nichols, references [2–4], consists of practical rules of thumb for the design of most of the classical controllers in Table 11.1 when the plant **unit-step response** is *sigmoidal* (see Fig. 11.5a), or the plant **exhibits a stability limit under proportional control** (see Fig. 11.5b). Such responses *(plant signature)* are typical of many industrial processes, and the Ziegler-Nichols rules, which were developed from experiments on various industrial processes, have proved quite useful in control system practice, especially when the

Table 11.1

Some Classical Feedback Controllers

CONTROL	SYMBOL	TRANSFER FUNCTION[a]
Proportional	P	k_c
Integral	I	$\dfrac{1}{T_i s}$
Proportional-plus-derivative	PD	$k_c(1 + T_d s)$
Proportional-plus-integral	PI	$k_c\left(1 + \dfrac{1}{T_i s}\right)$
Proportional-plus-integral-plus-derivative	PID	$k_c\left(1 + \dfrac{1}{T_i s} + T_d s\right)$

[a]The derivative forms are not realizable, as shown.

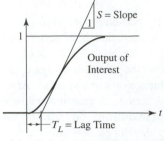

(a)   Unit-step-input Response

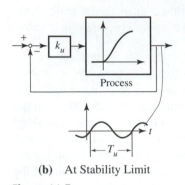

(b)   At Stability Limit

Figure 11.5

Parameters for Sigmoidal Plant and Process with Stability Limit

tuning criterion is the same as or similar to that used in the original experiments of Ziegler—*minimization of integrated absolute error* (IAE):

$$\text{IAE (integrated absolute error)} = \int_0^\infty |e(t)| \, dt \qquad \textbf{(11.12a)}$$

Tuning criteria or *performance indices* such as the IAE constitute a category of (time response) *performance measures* (more complex than those already introduced). Some of the more common criteria will be discussed in the next section.

There are thus two approaches to controller design via the Ziegler-Nichols rules. In the first method, the parameters T_L (lag time) and S (slope of the tangent at the inflection point), shown in Fig. 10.5a, are measured on the record of an actual unit-step input response experiment performed on the plant. From these two quantities, the controller settings are determined as follows:

P-control: $\quad k_c = \dfrac{1}{T_L S}$

PI-control: $\quad k_c = \dfrac{0.9}{T_L S} \quad \text{and} \quad T_i = 3.3 T_L \qquad \textbf{(11.12b)}$

PID-control: $\quad k_c = \dfrac{1.2}{T_L S}, \quad T_i = 2 T_L \quad \text{and} \quad T_d = 0.5 T_L$

In the second approach, the process is placed under proportional feedback control and the controller gain is adjusted (increased) until the stability limit of the process is reached (this presumes that the original plant is stable). The value of the gain at this point k_u and the period of the ensuing (constant amplitude) oscillation T_u are both recorded. The actual controller settings are again given by two quantities:

P-action: $\quad k_c = 0.5 k_u$

PI-action: $\quad k_c = 0.45 k_u \quad \text{and} \quad T_i = 0.83 T_u \qquad \textbf{(11.12c)}$

PID-action: $\quad k_c = 0.6 k_u, \quad T_i = 0.5 T_u \quad \text{and} \quad T_d = 0.125 T_u$

Obviously the second method cannot be applied if the plant fails to exhibit a stability limit under proportional control. However, the experiment to determine the controller parameters can be simulated using a model of the process (if it exists) rather than being performed on an actual plant.

Example 11.2   Application of Ziegler-Nichols Rules

Determine by Ziegler-Nichols rules the controller settings for P-, PI-, and PID-control of a servomotor actuator whose behavior is similar to the DC motor actuator of Fig. 8.5, except that the equivalent parameter values are ($J = \beta = L_a = k = 0.01$, $R_a = 0.04$, and $K_T = 0.01732$). Assume unity feedback.

Solution: For the given motor and actuator parameter values, the transfer function from Example 8.6 is

$$\frac{Y(s)}{V_a(s)} = \frac{b_0}{s^3 + a_2 s^2 + a_1 s + a_0} = \frac{173.2}{s^3 + 5s^2 + 8s + 4}$$

Assuming for this exercise that $b_0 = 173.2$ is factored into the controller, we can simply take

$$G_p(s) = \frac{1}{s^3 + 5s^2 + 8s + 4}$$

Note that (from the methods of Chapter 9) this plant is controllable. The unit-step input response of the plant without control is shown in Fig. 11.6.* We note that the response pattern is sigmoidal.

First method: If for the purposes of this example we assume that this step response was recorded in an actual experiment on the physical plant, then we can proceed with the first design approach of Ziegler and Nichols by measuring the quantities T_L and S on the unit-step response curve. When this is done (see insert in Fig. 11.6a), the following values are obtained: $T_L = 0.5$, and $S = 0.102$. Thus according to equation (11.12b), the following parameter values are suggested for the three controllers (note that the k_c values are actually for $b_0 k_c$):

P-control: $\quad k_c = \dfrac{1}{[T_L S]} = \dfrac{1}{[(0.5)(0.102)]} = 19.61$

PI-control: $\quad k_c = \dfrac{0.9}{[T_L S]} = 17.65 \quad \text{and} \quad T_i = 3.3 T_L = 1.65$

PID-control: $\quad k_c = \dfrac{1.2}{[T_L S]} = 23.53, \quad T_i = 2 T_L = 1 \quad \text{and} \quad T_d = 0.5 T_L = 0.25$

Second method: We assume the plant will have a stability limit under proportional control. This will be confirmed by the analysis. From equations (11.2a) and (11.2b), the closed-loop transfer function under proportional control ($H(s) = 1$ for unity feedback) is

$$G_{\text{CL}}(s) = \frac{k_u}{s^3 + 5s^2 + 8s + 4 + k_u}$$

so that the characteristic equation is

$$s^3 + 5s^2 + 8s + k_u + 4 = 0$$

The eigenvalues of a system at the stability limit lie on the imaginary axis in the s-plane. This condition is given by $s = j\omega$, where ω is the circular frequency of (constant-amplitude) oscillation. Thus putting $s = j\omega$ in the characteristic equation, we obtain

$$(k_u + 4 - 5\omega^2) + j(8\omega - \omega^3) = 0$$

The real and imaginary parts of the complex number on the LHS must each be zero to satisfy the equation. That is,

$$8 - \omega^2 = 0 \Rightarrow \omega = 2\sqrt{2}$$

---

*This figure can be generated with MATLAB's **step** (*num, den*) function by providing the relevant overall transfer function's numerator and denominator polynomial coefficients (see the script M-file **fig11_6a.m; On the Net**).

Figure 11.6

Simulation of System Response with and without Control

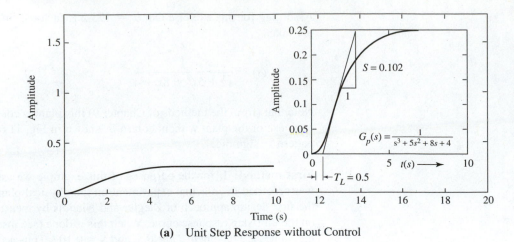

$$G_p(s) = \frac{1}{s^3 + 5s^2 + 8s + 4}$$

$S = 0.102$

$T_L = 0.5$

(a) Unit Step Response without Control

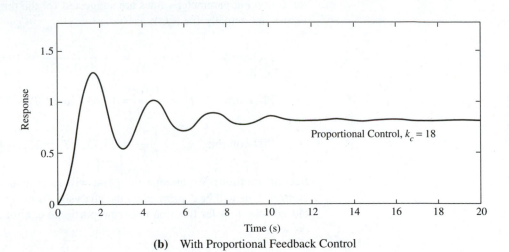

Proportional Control, $k_c = 18$

(b) With Proportional Feedback Control

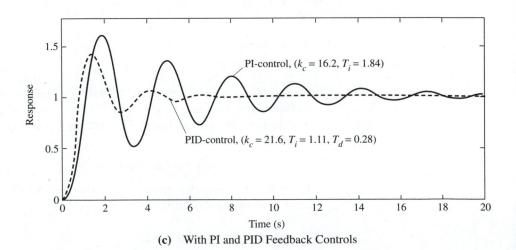

PI-control, $(k_c = 16.2, T_i = 1.84)$

PID-control, $(k_c = 21.6, T_i = 1.11, T_d = 0.28)$

(c) With PI and PID Feedback Controls

and

$$k_u + 4 - 5\omega^2 = k_u + 4 - 40 = 0 \Rightarrow k_u = 36$$

The period of oscillation at the stability limit is

$$T_u = \frac{2\pi}{\omega} = \frac{\pi}{\sqrt{2}}$$

Hence the controller design (see equation (11.12c) is

P-control: $k_c = 0.5k_u = (0.5)(36) = 18$

PI-control: $k_c = 0.45k_u = 16.2$ and $T_i = 0.83T_u = (0.83)\left(\pi/\sqrt{2}\right) = 1.84$

PID-control: $k_c = 0.6k_u = 21.6$, $T_i = 0.5T_u = 1.11$,   and

$$T_d = 0.125T_u = 0.28$$

The behavior of the closed-loop system $G_{CL}(s)$ for the last three controllers is shown in Figs. 11.6b (proportional control) and 11.6c (PI- and PID-control). Note the considerable improvement in the response pattern (relative to the uncontrolled system), particularly with respect to the offset.

As a practical matter, the controller parameter values given by the Ziegler-Nichols rules should be viewed as starting points in the controller design. Further adjustments of the controller (*fine tuning*), preferably carried out on the actual (physical) plant with the controller implemented and in place, are still required for a given system.

Controller Selection and Tuning

In Example 11.2 we carried out the design of three different controllers for the same system. Which of the three controllers should be selected for the system? In general, the choice of controller is a trade-off between the desired response pattern and the simplicity of the control algorithm, which can determine controller cost as well as tuning complexity. The characterization of the desired response pattern can vary with application, system, and design practitioner. Ideally, the tuning criterion used to determine the optimum controller parameter values should also be used to select the type of controller, provided it adequately characterizes the desired response pattern. However, it is more likely that a combination of performance criteria will be necessary to characterize the desired system behavior. Table 11.2 summarizes the values of the previously introduced category of performance measures, which are based on a few points on the response curve as well as the value of the IAE (the Ziegler-Nichols tuning criterion), which is an integral response criterion, as determined from the behavior of the system of Example 11.2 (Figs. 11.6b and 11.6c) under each of the three controllers.

Clearly, if the desired system behavior is best measured by the IAE, then the PID-controller should be selected for this system. A similar criterion is the *decay ratio*. In the process industries, a step response pattern having one-quarter decay ratio *(quarter-decay response)* is often considered a reasonable trade-off between a fast rise time (quick response) and an acceptable settling time (adequate stability) for continu-

Table 11.2

System Performance under P-, PI-, and PID-controllers

PERFORMANCE MEASURE	P-CONTROL	PI-CONTROL	PID-CONTROL
Offset	0.18	0	0
Rise time (s)	1.16	1.09	0.80
% Overshoot	54.3[a]	60.0	42.2
Decay ratio[b]	0.39[a]	0.58	0.16
5% Settling time (s)	8.8[a]	14.5	4.4
IAE (at 30 s)[c]	6.104[d]	3.174	1.062

[a]Based on the final value of the response rather than the reference value.
[b]Ratio of successive response amplitudes (recall the *logarithmic decrement* equation (7.12h)). Although it was not included in the earlier discussion of performance specifications, the *decay ratio* is a measure of transient response.
[c]As defined in equation (11.12a), the IAE should be computed to $t \to \infty$. However, for this problem, full accuracy is essentially obtained by $t = 30$ s, for all three controllers.
[d]With the error defined relative to the reference value. If the error for the IAE (not for the controller) is redefined in terms of the final value, the corresponding IAE is only 1.594.

ous controllers (see [5]); The PID-controller would also be selected from Table 11.2 on this basis. However, there are applications in which a small offset is completely tolerable. It could be that the final value of the response need not be very accurate (a velocity input situation, for instance), or the application can permit translations of the reference value to accommodate the offset. Given such situations, the rather moderate performance of the P-controller shown in Table 11.2 (note that the IAE of 6.104 is based on the reference value; on the basis of the final value, the IAE is only 1.594), coupled with the ease of tuning (only one parameter is to be determined) could recommend proportional control for this system. Finally, it would appear that for this particular system, the PI-controller is unlikely to represent the best choice of type of controller except perhaps on the basis of steady-state error and relative ease of tuning only.

Controller Tuning

The problem of determining adequate controller parameter values is actually an *optimization* problem. Specifically, given the dynamics of the object of control, represented by a number of variables: $d_1, d_2, d_3, \ldots$, and the controller algorithm, containing a number of adjustable parameters: $c_1, c_2, c_3, \ldots$, we want to determine, in terms of the system variables, the controller parameters that minimize an error criterion: $J = \text{function}(c_1, c_2, c_3, \ldots , d_1, d_2, d_3, \ldots .)$ subject to the response of the controlled system to a given input. Depending on the number of variables involved, the nature of the system dynamics, and the complexity of the error function J, the resulting optimization problem can indeed be formidable.

For the classical controllers of Table 11.1, for example, the number of controller parameters is at most three, that is, for the PID algorithm. A small number of system variables generally implies a limitation on the number or type of systems that can be modeled by that set of variables. A model of very wide applicability is likely to involve a large number of variables. The two variables of the first method of Ziegler and Nichols, lag time T_L and slope S, can define a *first-order-lag-plus-dead-time* model (see Fig. 11.7a), which fortunately is a simple model that does represent many *self-regulating* industrial processes:

$$G_m(s) = \frac{Ke^{-T_L s}}{Ts + 1}$$

(11.13a)

Figure 11.7

Comparison of Integral
Performance Criteria
for PI-control of a
First-order-lag-plus-dead-time

Process: $G_p(s) = \dfrac{e^{-0.27s}}{Ts + 1}$

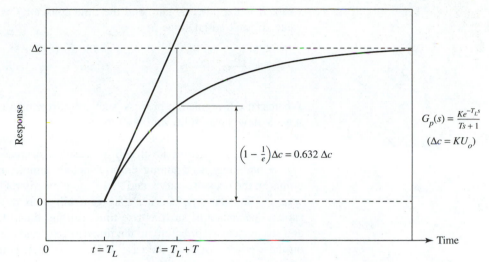

$G_p(s) = \dfrac{Ke^{-T_L s}}{Ts + 1}$

$(\Delta c = KU_o)$

$\left(1 - \dfrac{1}{e}\right)\Delta c = 0.632\,\Delta c$

(a) Step Input Response of a First-order-lag-plus-dead-time Process

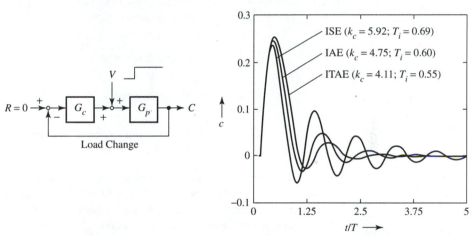

Load Change

ISE $(k_c = 5.92; T_i = 0.69)$

IAE $(k_c = 4.75; T_i = 0.60)$

ITAE $(k_c = 4.11; T_i = 0.55)$

(b) Optimal Responses: Unit Step Disturbance at Manipulated Input

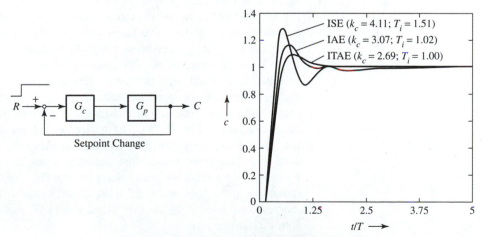

Setpoint Change

ISE $(k_c = 4.11; T_i = 1.51)$

IAE $(k_c = 3.07; T_i = 1.02)$

ITAE $(k_c = 2.69; T_i = 1.00)$

(c) Optimal Responses: Unit Step Setpoint Change

where K is the process gain, and the time constant T is given by the 100% response value ΔC and initial slope S:

$$T = \frac{\Delta C}{S} \tag{11.13b}$$

Whereas a self-regulating process, such as most temperature processes, is one that will seek a new level following a step change in the manipulated input, a *non-self-regulating process* (a tank-level control process for example), will not seek a new steady value. Rather, the level will increase (or decrease) indefinitely.

Commonly used tuning criteria include simple measures based on just a few points on the response curve, and more complex criteria that are based on the entire response. The latter type of criteria are generally more difficult to evaluate. The time response specifications such as rise time, settling time, decay ratio, percent overshoot, and steady-state error belong to the first category, and controller tuning or design based on these criteria, such as the case of Example 11.1, is usually straightforward. In the second category are a number of *integral criteria,* of which the most popular (in terms of usage) are the *integral absolute value of the error* (IAE), defined previously by equation (11.12a), the *integral of the square of the error* (ISE), and the *integral of time multiplied by the absolute value of the error* (ITAE):

$$\text{IAE (integrated absolute error)} = \int_0^\infty |e(t)|\, dt \tag{11.14a}$$

$$\text{ISE (integrated square error)} = \int_0^\infty [e(t)]^2\, dt \tag{11.14b}$$

$$\text{ITAE (integrated time multiplied by absolute error)} = \int_0^\infty t\,|e(t)|\, dt \tag{11.14c}$$

To some extent, each performance index emphasizes different aspects of the system response. For example, for the type of self-regulating processes under consideration, the ISE penalizes large errors and overshoots more than the IAE, while the IAE is in general a more sensitive error function. Thus overshoots of responses based on the ISE should in general be lower than those of responses obtained using the IAE. The ISE, however, tends to give long settling times. The ITAE is apparently the most sensitive of the three criteria. As a result of the presence of the t (time) product term in equation (11.14c), the ITAE also weighs more heavily errors that occur late in the response while essentially ignoring those that occur very early in the response. This suggests that the ITAE can allow large overshoots but yield relatively short settling times.

These attributes of the above tuning criteria are confirmed by the result shown in Fig. 11.7b, which is a computer simulation of the optimal responses, based on each of the three performance criteria, of a first-order-lag-plus-dead-time control object subject to a unit-step disturbance v at the manipulated input (*load change;* see block diagram in Fig. 11.7b). The parameters of the control object (see Fig. 11.7a) are defined relative to the time constant to make the model more general, and the controller parameter values that optimize each response were determined for a PI-control algorithm (two parameters) to limit the problem complexity. It can be seen that the response that is optimal with respect to the IAE has a higher overshoot than the response relative to the ISE. Also, the ITAE has the largest overshoot and the shortest settling time.

With respect to setpoint changes (see Fig. 11.7c), the observation relative to the sensitivity to the error and the settling time of each of the three tuning criteria is essentially upheld. However, other attributes of these integral criteria compare differently. Nonetheless, the results of Fig. 11.7 can provide a good qualitative basis for choosing a tuning criterion for most processes.

Computer-Assisted Design of a Nonlinear System

In the absence of specific guidelines or rules of thumb, such as those of Ziegler and Nichols, the computational capacity necessary to apply an integral performance criterion or similar performance index is substantial. This is especially true for nonlinear systems, for which an analytical solution for the optimal parameter values is unlikely to be available. Such situations call for either full-fledged *computer-aided design* tools or the application of the computer to assist in the control system design process.

We can describe the general process of system design as the arrangement of components and the determination of component parameter values to achieve a specific goal. A process of this nature can be carried out in three interrelated parts:

1. **Formation of system models:** which includes the mathematical description and solution of component behavior and interactions.
2. **Realization:** that is, association of each element of the model with an actual item of hardware.
3. **Optimization:** which involves mostly the determination of parameter combinations that yield the best system performance.

Computer-assisted system design is the application of the computer to some parts of the design process, especially those involving the formulation and solution of system models. Examples of such applications include the direct simulation of a system model in order to determine from the results the effect of a particular realization or optimization, and the approximation of component behavior that is available in an inconvenient form (such as performance curves and tables of properties) with appropriate and convenient mathematical or numerical functions. We have already seen in this text a number of these applications and some computer-assisted control system design. Digital computer simulation of dynamic systems was introduced in Section 6.5. In Section 9.2, frequency response computation, especially by math software, which is the essential element in frequency response simulation, was considered. Finally, in Section 9.4 as well as in Chapter 10, the computer was used in a number of applications involving the analysis and design of nonlinear as well as linear systems. Our objective here is merely to illustrate some of the computer modeling or programming details underlying such unified executions of model solution and control system design.

Example 11.3   Design of a Control System Having a Nonlinear Actuator

Consider the servomechanism (similar to the antiaircraft gun positioning example of Fig. 10.5) but with a nonlinear actuator, as shown in Fig. 11.8a. Let the actuator behavior (which includes saturation) be given by the following experimentally determined data (Table 11.3):

Develop a computational mathematics software program to (interactively) design (tune) the proportional-plus-integral controller shown, such that for a step setpoint in-

Figure 11.8

Servomechanism with Nonlinear Actuator

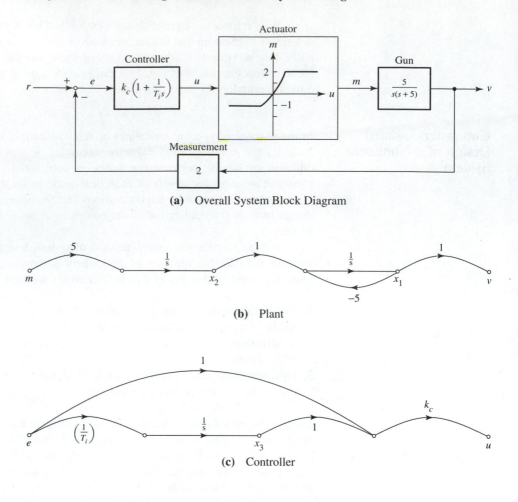

(a)   Overall System Block Diagram

(b)   Plant

(c)   Controller

put ($r(t) = 2$) the integral of the absolute value of the error (IAE), equation (11.14a), is minimized. Assume starting values of k_c and T_i of 2.35 and 0.90, respectively. To reduce the optimization effort, let T_i be fixed at the starting value.

Solution: We consider here a MATLAB solution (for Mathcad, see **SC13.MCD; On the Net**).

1. **Approximation of functions:** The behavior of the actuator is given in a form that is not convenient for the simulation of the overall system. The problem of approximating a function f with a convenient function f^* is of two types. In the first case, called *interpolation,* the approximation f^* is required to be exact at (pass through) all the data points. This implies that f^* contains as many unknowns as there are data values. In the second case, known as *smoothing,* f^* is required to pass as close as possible to the data points, according to some measure of the closeness between f^* and f. In this case f^* contains fewer unknowns than the number of data points, and the problem is said to be *overdetermined.* The reader who is not familiar with the subject of approximation of functions should refer to texts on numerical methods, such as reference [6].

In general, interpolation is often applied to problems in which f and hence the data values are sufficiently well established. For example, f is a known infinite se-

Table 11.3

u	m
≤ -0.6	-1.00
-0.3	-0.45
0.0	0.00
0.3	0.50
0.6	1.10
≥ 0.9	2.00

ries, while f^* is a truncated series with only a few terms. Smoothing, on the other hand, is useful for problems in which there is some uncertainty or scatter in the data. Another factor that sometimes recommends smoothing for our present type of endeavor is the need to minimize the amount of computation at any stage of the simulation. Smoothing facilitates this through the reduction in the number of unknowns relative to the available data. By far the most popular smoothing technique is the *method of least squares*. Here the measure of closeness between f^* and f, which is made as small as possible, is the sum of the squares of the deviation $(f^* - f)$.

Most computational mathematics software, such as Mathcad and MATLAB, contains functions for *correlation* and *linear regression* as well as interpolation functions. For the present example, the experimental data for the actuator is assumed to be a candidate for interpolation (the case of smoothing is given as an exercise in Practice Problem 11.24). MATLAB's **spline** function can be used to approximate the active region of the actuator characteristics $(-0.6 \le u \le 0.9)$; see function M-file **picad.m** listing. Given the data vectors **vu** and **vm** from Table 11.3, **spline** evaluates m directly from **vu, vm,** and u.

2. **Conversion of plant and controller transfer functions to state models:** The gun transfer function implies

$$(s^2 + 5s)v = 5m \Rightarrow v = \frac{5m}{s^2} - \frac{5v}{s}$$

which has the signal flow graph (see Section 9.3) shown in Fig. 11.8b. The state model is thus

$$\dot{x}_1 = -5x_1 + x_2 \tag{11.15a}$$

$$\dot{x}_2 = 5m \tag{11.15b}$$

$$v = x_1 \tag{11.15c}$$

Similarly, the controller transfer function gives the signal flow graph shown in Fig. 11.8c and the state model

$$e = r - 2v = r - 2x_1 \tag{11.15d}$$

$$\dot{x}_3 = \frac{1}{T_i} e \tag{11.15e}$$

$$u = k_c(e + x_3) \tag{11.15f}$$

3. **Actuator model:** The actuator model is given by the cubic spline interpolation function of the data for the active region of the actuator plus the representation of the saturation, which is accomplished in MATLAB with an explicit **if-elseif-else** construct.

4. **Tuning criterion:** Equations (11.15a) through (11.15f) plus the actuator description represent the overall system dynamic model. However, to design the proportional controller, we need to monitor the IAE. In a full-fledged computer-aided design exercise, the determination of the optimal controller parameters is normally accom-

plished automatically by the software. However, some measure of designer control of the simulation process is sometimes useful. Let $\dot{x}_4 = |e(t)|$, then $x_4(t) = \int_0^t |e(t)|\, dt$, so that

$$\lim_{t \to \infty}[x_4(t)] = \text{IAE} \tag{11.15g}$$

But as a practical matter, we can define

$$\text{IAE} = x_4(t_f) \tag{11.15h}$$

where t_f is a time determined to be sufficiently large and after which the changes in x_4 are negligible.

5. **Computer model: ode45** (or **ode23**) can be used as the basis for the computer model. In addition, an overall loop controlled by the designer can be introduced via the **while** statement to simulate interactive tuning of the controller.

   The interactive tuning of k_c works by displaying the current value of k_c and the resulting IAE. The designer is expected to determine from examination of all previous k_c,IAE pairs whether the optimum k_c has been found. If this is the case, a negative value is entered for k_c to terminate the program, otherwise a new k_c, whose choice is guided by the trend deduced from the existing k_c,IAE pairs, is entered to continue the program run. For some other tuning criterion, the time response of the controlled variables can instead be displayed for a qualitative evaluation.

6. **Program execution:** Observe in script M-file **fig 11_9.m** that zero initial conditions are assumed, as is normal in control system simulation. The execution of the computer model shown in the M-files **picad.m** and **fig11_9.m** occurs in two stages. First, an adequate t_f is determined. Beginning with an initial estimate of t_f and for a fixed reasonable value of k_c, a value of t_f is sought such that for values of time above it, the computed value of IAE remains essentially the same. For the present problem, $t_f = 6.4$ was found for convergence of the IAE to three decimal places. The second stage of the execution was carried out as discussed under item 5. The k_c,IAE pairs that were generated are given in Table 11.4 in the same sequence of generation or search pattern.

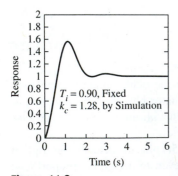

Figure 11.9

Performance of Minimum IAE PI-controller

```
function xdot = picad(t,x)
iT = 1/.9; r = 2;                % Note: Kc is a global variable
% Approximation of Functions, Interpolation Data:
vu = [−.6 −.3 0 .3 .6 .9]; vm = [−1 −.45 0 .5 1.1 2];
%
xdot = zeros(4,1);
xdot(1) = −5 .* x(1) + x(2);
e = r − 2 .* x(1); u = Kc* (e + x(3));
if u < −.6
        m = −1;                  % saturation, lower limit
elseif u > .9
        m = 2;                   % saturation, upper limit
else
        m = spline(vu,vm,u);     % cubic spline interpolation
end;
xdot(2) = 5*m;
xdot(3) = iT*e;
xdot(4) = abs(e);
```

```
% Script to generate Figure 11.9
% Uses function M-file, picad.m
%
global Kc; Ks = 2.35;                    % Kc is global
t0 = 0; tf = 6.4;                        % time interval
x0 = [0 0 0 0]';                         % zero initial conditions
disp('Note: This is a long running program!');
while Ks > 0                             % interactive design loop
Kc = Ks;
disp('   Please wait');
[t,x] = ode45('picad',t0,tf,x0);
%
IL = length (x(:,4)); disp('   IAE = '); disp(x(IL,4));  % monitor performance
Ks = input('enter Kc, a negative value to quit: ');      % Designer Interaction
end;
axis([0 6 0 2]);
plot(t,x(:,1));
text(1.4,.6,'Ti = 0.90, fixed');
text(1.4,.5,'Kc = 1.28, by simulation');
ylabel('Response'); xlabel('Time (sec)');
```

Table 11.4

Interactive Tuning Results (MATLAB)

s/n	k_c	$x_4(6.4)$	s/n	k_c	$x_4(6.4)$	s/n	k_c	$x_4(6.4)$
1	2.35	1.6835	6	1.40	1.6582	11	1.28	1.6568
2	2.50	1.6862	7	1.30	1.6569	12	1.27	1.6568
3	2.00	1.6759	8	1.20	1.6584	13	1.26	1.6570
4	1.50	1.6607	9	1.25	1.6572			
5	1.00	1.7028	10	1.29	1.6568			

From this result, the optimum choice of k_c is 1.28. The controlled system performance for this value of the proportional gain is shown in Fig. 11.9.

11.3 ROOT LOCUS AND ROUTH TEST DESIGN

The performance specifications of control systems can also be given directly in terms of the permissible values of the eigenvalues of the overall system. Such specifications can include decay rates for dominant (slowest) modes, effective damping in the system, and optimum frequencies for oscillatory modes. As was mentioned in Section 10.2, the root locus method permits the investigation of the s-plane placement of the system eigenvalues as a function of some design parameter. Thus, given the root locus and the interpretation of the control objectives in terms of desired system eigenvalues, the controller design can be fixed. Similarly, the Routh test is carried out on the system characteristic equation of which the eigenvalues are the solution. Specifications relative to allowable values of the roots of the characteristic equation can therefore be examined by the Routh test by carrying through to the Routh array any parameters whose values are to be determined. As was illustrated in Section 10.2, certain transformations on the characteristic equation can be employed to expand such applications of the Routh array.

Design Using the Root Locus Method

Root locus design, in general, involves the interpretation of design specifications in the complex plane. The root locus as a complex plane display of the characteristic equation (10.5a), repeated here,

$$1 + kG_o(s) = 0 \tag{11.16}$$

can be employed to directly determine the adjustable parameter k that could be part of a controller under design or, for that matter, part of any design problem, including other than control system design. Such direct use of the root locus method in system design was illustrated in Practice Problems 10.22, 10.33, 10.34, and 10.52 through 10.54. Further practice is provided by Practice Problems 11.27 through 11.32.

Example 11.4   Root Locus and Design

Consider an electromagnetic relay–based servo system, the core of which is shown in Fig. 11.10a. Compared with the device shown in Fig. 5.13 and modeled in Example 5.4, the manipulative input $V_{\text{in}} = V_{CE} - V_{CC}$. Suppose for design relative to a nominal operating configuration, the assumptions

$$L(\theta) \approx L(\theta_0) = L$$

and

$$\frac{L(\theta)i}{\angle_a\theta} \approx \frac{L(\theta_0)i_0}{\angle_a\theta_0} = K_B$$

suggested in that example are valid, then the electromagnetic relay model, equation (5.49), becomes the linear system

$$\frac{d}{dt}\theta = \omega$$

$$\frac{d}{dt}\omega = \frac{3}{m_d\angle_a^2}\left[(m_a g\angle_G - k\angle_s^2)\theta - \beta\omega - \left(\frac{1}{2}\angle_a K_B\right)i\right] = a_{21}\theta - a_{22}\omega - a_{23}i$$

$$\frac{d}{dt}i = \frac{1}{L}[(\angle_a K_B)\omega - R_m i - V_{\text{in}}] = a_{32}\omega - a_{33}i - \frac{V_{\text{in}}}{L}$$

which has the open-loop transfer function

$$\frac{\theta(s)}{(a_{23}/L)V_{\text{in}}(s)} = G_0(s) = \frac{1}{s^3 + (a_{22} + a_{33})s^2 + (a_{22}a_{32} + a_{23}a_{32} - a_{21})s - a_{21}a_{33}}$$

If the present relay parameters have such values that

$$G_0(s) = \frac{1}{s^3 + 16s^2 + 44s - 160}$$

for a unit feedback proportional control system,

a. Sketch the root locus with the proportional gain k as the adjustable parameter.

b. What value of k will give a closed-loop behavior with maximum damping for all time (transient and steady state)?

Figure 11.10

**Design of an Electromagnetic
Relay–based Servomechanism**

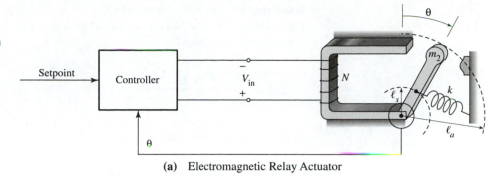

(a)   Electromagnetic Relay Actuator

Root Locus by MATLAB

(b)   Root Locus Plot for $G_o(s) = \dfrac{1}{(s-2)(s+8)(s+10)}$

Solution a: The root locus plot is shown in Fig. 11.10b (see script M-file **fig11_10b.m; On the Net**). For manual plotting, note the following: The closed-loop characteristic equation is

$$1 + \frac{k}{s^3 + 16s^2 + 44s - 160} = 1 + \frac{k}{(s - 2)(s + 8)(s + 10)} = 0$$

Hence there are three poles: 2, -8, -10, and no zero. The center of gravity is

$$-\sigma = \frac{-10 - 8 + 2}{3} = -5.33$$

The angles of the asymptote are $\pm 180/3 = \pm 60°$. The breakaway point is given:

$$\frac{d}{ds}\left(\frac{\text{denominator of } G}{\text{numerator of } G}\right) = 0 = \frac{d}{ds}(0.1s + 1)(s^2 + 6s - 16) = 0$$

$$(0.1s + 1)(2s + 6) + 0.1(s^2 + 6s - 16) = 0 = 0.3s^2 + 3.2s + 4.4$$

$$s = \frac{-3.2}{0.6} \pm \sqrt{\left(\frac{3.2}{0.6}\right)^2 - \frac{4.4}{0.3}} = -5.33 \pm 3.71 = -1.62 \text{ or } -9.04$$

That is, the breakaway point is -1.62.

Solution b: Observing the distribution of values of k, we conclude that the largest value of k for which there is no root with smaller real part (more damping) is the value of k at the breakaway point. Now,

$$k = |s - 2||s + 8||s + 10|$$

At $s = -1.62$,

$$k = |-1.62 - 2||-1.62 + 8||-1.62 + 10| = (3.62)(6.38)(8.38) = 193.54$$

is the solution.

....

Relative Effects of I- and D-Action

Inasmuch as the transfer function $G_o(s)$ can be represented, in general, by a ratio of numerator and denominator polynomials, equation (11.16) also implies that a root locus plot reflects the relative influences of the poles and zeros of $G_o(s)$. Because these quantities can depend on the controller parameters for a control system, the qualitative effect of a particular control action on performance can be deduced by examining its influence on the placement of open-loop poles and zeros. This aspect of controller design by the root locus technique is explored next.

As further illustration of the relative effects on performance of integral and derivative controls, some of which were discussed in Section 11.1, consider again the antiaircraft servomechanism of Fig. 10.5a. The block diagram of the control system is shown again in Fig. 11.11a, for PID-action. The corresponding closed-loop characteristic equation is

$$1 + (100k_c)\frac{\left(1 + \dfrac{1}{T_i s} + T_d s\right)}{s(s + 5)(s + 10)} = 0 \tag{11.17a}$$

With only proportional control $\left(\dfrac{1}{T_i} = T_d = 0\right)$, the resulting characteristic equation,

$$1 + (100k_c)\frac{1}{s(s + 5)(s + 10)} = 0 \tag{11.17b}$$

has the root locus shown in Fig. 11.11b. The root locus for the same system with proportional-plus-integral control ($T_d = 0$, but with $T_i = 1$ as before), that is, for the characteristic equation

$$1 + (100k_c)\frac{(s + 1)}{s^2(s + 5)(s + 10)} = 0 \tag{11.17c}$$

Figure 11.11

Relative Effects of I- and D-action

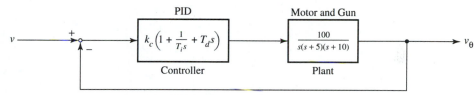

(a)   Block Diagram of Control System

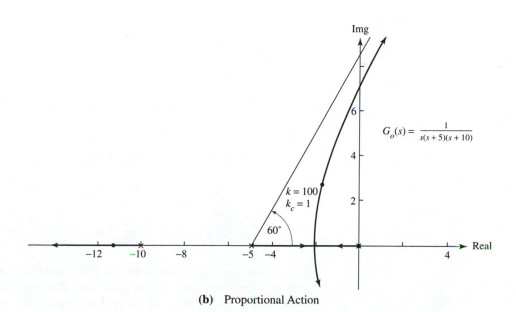

(b)   Proportional Action

Figure 11.11 *continued*

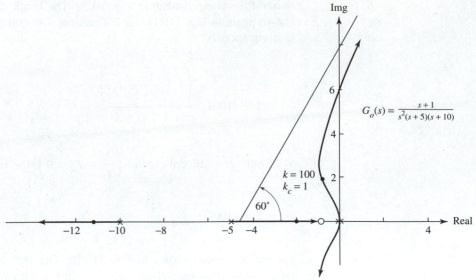

(c)   Proportional-plus-integral Action

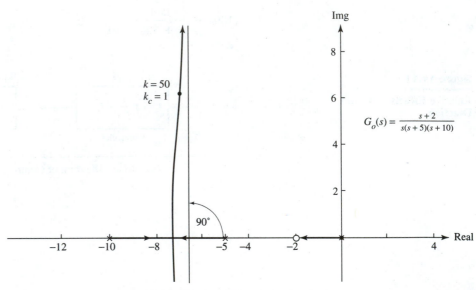

(d)   Proportional-plus-derivative Action

is repeated as Fig. 11.11c, while Fig. 11.11d shows the root locus for proportional-plus-derivative action $\left(\dfrac{1}{T_i} = 0 \text{ but with } T_d = 0.5\right)$, that is, for the characteristic equation,

$$1 + \left(\frac{100 k_c}{2}\right)\frac{(s+2)}{s(s+5)(s+10)} = 0, \quad \text{where } T_d = 0.5 \qquad \textbf{(11.17d)}$$

The construction of these root loci is left as a practice exercise for the reader. From a study of these plots it should now be evident that *integral action introduces a pole at* $s = 0$, which "pulls" the root locus branches toward the imaginary axis (stability boundary), notwithstanding the zero (in this case at $s = -1$) that is also added. Thus for a given controller setting, say, $k_c = 1$, the system will *tend to be less stable and to be slower*. In contrast, *derivative control adds a zero*, which "pushes" the root locus branches away from the stability boundary and further into the LHP, *thus speeding up response*. Note that this effect of an added zero agrees with the analytically determined results in Section 7.1 for the step response of second-order systems. Also, comparing the "operating point" $k_c = 1$ in Figs. 11.11b and 11.11d by referring to the relations between second-order eigenvalue components shown in Fig. 7.2, we see that *the damping in the system is also increased by D-action*. These developments provide further clues on when to include or leave out particular modes in the controller design. We should also remember, however, that derivative control provides no relief from offsets for most systems and that the actual effect of a realizable D-controller will be slightly different from that of Fig. 11.11d (see Practice Problem 11.33).

Application to Dead Time and Other Nonlinear Systems

The application of the root locus technique to systems including single-valued nonlinearities was discussed in Section 10.5. This approach can be used for the design of nonlinear control systems of a similar nature given appropriate interpretations of the design specifications in the complex plane (see Practice Problems 11.46 and 11.47). We consider here the specific case of nonlinearity due to a transportation lag.

Dead time or *transportation lag* is a common behavior compartment in many industrial processes. Given a pure delay term e^{-sT} in the transfer function, substitution of $s = -\sigma + j\omega$ gives us

$$e^{-sT} = e^{\sigma T} e^{-j\omega T} \tag{11.18a}$$

and using Euler's formula, equation (6.41c), we obtain

$$e^{-sT} = e^{\sigma T}(\cos \omega T - j \sin \omega T) \tag{11.18b}$$

so that the contribution of e^{-sT} to the real and imaginary parts or $G_o(s)$ can be determined.

To illustrate the corresponding modifications for manual plotting of the root locus, we consider the following characteristic equation:

$$1 + G_o(s) = 1 + ke^{-sT} = 0 \tag{11.18c}$$

From equation (11.18b), the real and imaginary parts of the LHS of (11.18c) must each be zero:

$$1 + ke^{\sigma T}\cos \omega T = 0 \tag{11.18d}$$

$$ke^{\sigma T}\sin \omega T = 0 \tag{11.18e}$$

Equation (11.18e) gives $\omega T = N\pi$, $N = 0, 1, 2, \ldots$, but equation (11.18d) requires that $\cos \omega T$ be negative, so that the solutions for ωT and k are

$$\omega T = \pm\pi, \pm3\pi, \pm5\pi, \ldots \tag{11.18f}$$

$$k = e^{-\sigma T} \tag{11.18g}$$

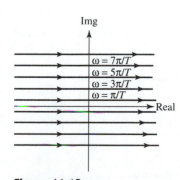

Figure 11.12

Root Locus Plot for $[1 + ke^{-st} = 0]$

Note that from equation (11.18a), $-\omega T$ is the angle of $G_o(s)$, and equation (11.18f) satisfies the angle condition and is hence the equation of the root loci. These are shown in Fig. 11.12. There are an infinite number of branches, all parallel to the real axis.

Equation (11.18g) gives the scaling of the loci. For a transfer function combining the dead time with other elements these branches become asymptotes of the root loci.

| Example 11.5 | Root Locus of a Process with Dead Time |

Consider a proportional temperature controller for a hot water heater, with remote sensing of the hot water temperature as shown in Fig. 11.13a. Develop a simple model for this system and plot the root locus for the corresponding closed-loop system characteristic equation.

Figure 11.13

Root Locus Plot for
$[1 + RC_s + k_c e^{-sT} = 0]$

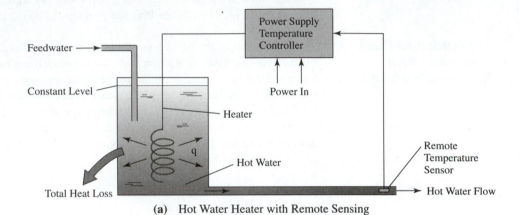

(a)   Hot Water Heater with Remote Sensing

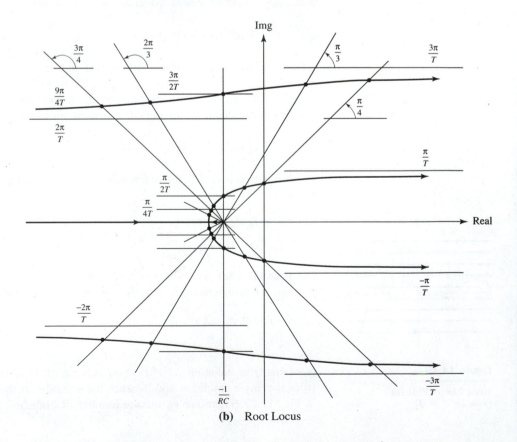

(b)   Root Locus

Solution: A single-lump model of the hot water heater gives

$$c\frac{d\theta}{dt} = q - \frac{\theta}{R} \tag{11.19a}$$

where θ is the temperature of the water in the heater relative to the ambient temperature, and θ/R is an approximation for all heat losses, including those due to the inlet and outlet flows. For a constant-level water heater, the pressure drop from the water heater to the remote sensor is essentially fixed, so that the velocity of the flowing fluid v is also constant. The sensed temperature (relative to the ambient temperature) can therefore be represented by $\theta_{sensor}(t) = \theta(t - T); \; T = L/v$.

For proportional feedback control of the heater, we take

$$Rq = k_c[\theta_r - \theta(t - T)] \tag{11.19b}$$

where θ_r is the setpoint. Substituting equation (11.19b) in equation (11.19a) and taking the Laplace transform (with zero initial conditions), we have

$$\frac{\theta(s)}{\theta_r(s)} = \frac{k_c}{1 + RCs + k_c e^{-sT}} \tag{11.19c}$$

Thus the closed-loop characteristic equation is $1 + RCs + k_c e^{-sT} = 0$. To put this in the form of equation (10.5a) or (11.16), we can divide by $(1 + RCs)$ to get

$$1 + \frac{(k_c/RC)e^{-sT}}{s + 1/RC} = 0 \tag{11.20a}$$

so that $k = k_c/RC$, and

$$G_o(s) = \frac{e^{-Ts}}{s + 1/RC}$$

Now, from equation (11.18a),

$$\angle G_o(s) = -\omega T - \angle\left[s - \left(\frac{-1}{RC}\right)\right] \quad \text{or} \quad \angle G_o(s) = -\omega T - \beta \tag{11.20b}$$

where the angles in equation (11.20b) are in radians. At very large distances from the origin, in the RHP, β tends to zero, so that in this region, considering the angle condition (10.6a) or (10.7a), the root loci are asymptotic to the horizontal lines $\omega T = \pm\pi$, $\pm 3\pi, \pm 5\pi, \ldots$.

In the LHP, at considerable distances from the origin, β tends to π, so that the root loci in this region are asymptotic to the lines $\omega T = 0, \pm 2\pi, \pm 4\pi, \ldots$.

These asymptotes are similar to the horizontal lines of Fig. 11.12. Consider now the vertical line through the pole, $s = -1/RC$. Along this line, $\beta = \pm\pi/2$, so that the intersection of the vertical line and all horizontal lines, $\omega T = \pm\pi/2, \pm 5\pi/2, \ldots$, are all on a root locus. Similarly, the radial lines originating from the pole and making angles of $\pm 3\pi/4$ with the real axis intersect the horizontal lines $\omega T = \pm\pi/4$, $\pm 9\pi/4, \ldots$.

Proceeding in this manner with other radial lines, we can easily sketch the root loci (see Fig. 11.13b). The root locus around the real axis breakaway point is determined in the same way by using radial lines that are closer and closer to the real axis. Finally, the scaling equation, using $s = -\sigma + j\omega$, is

$$k = e^{-\sigma T}\sqrt{\left(\frac{1}{RC} - \sigma\right)^2 + \omega^2} \qquad (11.20c)$$

Example 11.6   Root Locus Design for a Process with Dead Time

With the loci of the controlled system eigenvalues having been obtained as a function of a design parameter, the root locus design implies selection of the parameter values that will result in a system that meets the performance specifications. Consider again the temperature controller of Example 11.5. Suppose we desire that the speed of response of the controlled system be at least as fast as the response of the heater, which is given by the heater's time constant RC. Determine the proportional controller gain k_c that will meet this specification. For a quantitative solution, let $T = 10\pi$ (31.4 s) and $RC = 40/3 = 13.3$ s.

Solution: The specification requires that all closed-loop eigenvalues lie to the left of the vertical line $s = -1/RC$ in Fig. 11.13. This figure is sketched again in Fig. 11.14a but with the values of the proportional gain k_c (= RCk) indicated at various points on the root loci. A particular value of $k_c = 0.009$ is shown on all the branches, and it can be seen that this represents a workable design. For this value of k_c, all the system eigenvalues have real parts that are less than $-1/RC = -0.075$ (actually, less than $-1.13/RC = -0.08475$). Other designs are also feasible. In particular, note that $k_c = 0.014798$ will give the fastest possible response for the controlled system. The smallest negative real part of the eigenvalues in this case is about $-1.424/RC$, or -0.10683. Figure 11.14b shows a simulation of the uncontrolled and controlled systems. Although there is some improvement in response speed, steady-state error (offset) is clearly not a factor in the present design.

Second-Order Dominance   The modes corresponding to those eigenvalues that are closest to the imaginary axis dominate the response of a stable dynamic system. This is true because the response modes corresponding to eigenvalues located farther in the LHP die out relatively quickly (see Fig. 10.4). The root loci of many engineering systems, such as the servomechanism discussed earlier, or the system of Example 11.4, are often characterized by a pair of "dominant" branches near the imaginary axis. *The pair of complex-conjugate eigenvalues located on these branches therefore dominate the response of the system and cause it to behave much like a second-order system.* Such behavior is said to be *second-order dominant,* and the controller performance of second-order dominant systems may be specified in terms of an equivalent desirable second-order response. Recall that given the eigenvalues of an oscillatory second-order system, the frequency of oscillation, the damping ratio, the decay rate, etc., are completely determined (see Fig. 7.2). Indeed, the wealth of our graphical and/or analytical experience with second-order systems can be brought to bear on the design of controllers for systems whose closed-loop characteristics can be approximated by a second-order response.

Figure 11.14

Root Locus Design for

$$\left[1 + \frac{40}{3}s + k_c e^{-10ms} = 0 \right]$$

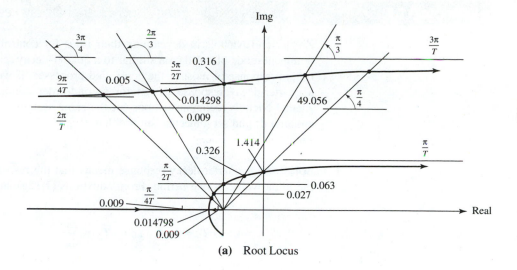

(a) Root Locus

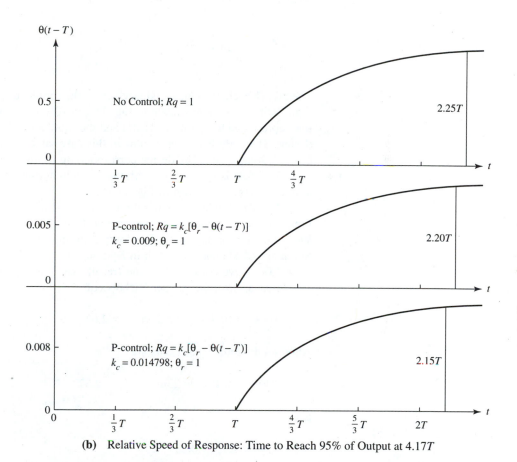

(b) Relative Speed of Response: Time to Reach 95% of Output at 4.17T

Example 11.7	Second-Order Dominance and Quarter-Decay Response

Ziegler and Nichols in developing their rules for controller settings observed that a dominant mode of oscillation similar to a quarter-decay response characterized the optimum response of most of the controlled processes. If we interpret this to mean that a quarter-decay response design for second-order dominant systems will duplicate Ziegler-Nichols recommendations for the same system, determine this design for the antiaircraft gun servomechanism problem (Fig. 10.5b) and compare the result with the Ziegler-Nichols rule.

Solution: A quarter-decay response means that the response decays to one-fourth of the initial value in one period. From equations (7.12g) and (7.12c),

$$e^{\sigma T_d} = 4 \Rightarrow \sigma T_d = \ln 4, \quad \text{and} \quad T_d = \frac{2\pi}{\omega}$$

so that

$$\frac{\sigma}{\omega} = \frac{1}{2\pi} \ln 4 = 0.22 \tag{11.21}$$

The branches close to the imaginary axis in the root locus plot for the servomechanism, Fig. 10.6b, are sketched again in Fig. 11.15a. The point of intersection of the straight line represented by equation (11.21) and the root locus (the dominant branch) gives the design. There are two intersections in this case, at $k = 18.5$ and at $k = 205.7$ (that is, $k_c = 0.19$ and 2.06). The design corresponding to the more stable and faster response (see Fig. 10.4) is $k_c = 2.06$. The transient unit-step-input response of the control system as designed is shown in Fig. 11.15b.

For the Ziegler-Nichols rule, we first find the proportional gain at the stability limit. This value (that is, k_u) can be read off Fig. 10.6 or calculated as in Example 11.2. The result by the calculation performed during the development of Fig. 10.6b (see Summary of Manual Approach in Section 10.2) was $k_u = 525.65$, with $\omega_u = 5.92 \Rightarrow T_u = 1.06$. If we consider P-action (recall that T_i was predetermined to be 1.0 for the servomechanism), the Ziegler-Nichols equation (11.12c) gives

$$100k_c = 0.5k_u = 262.8 \Rightarrow k_c = 2.63$$

Also, for PI-action the results are

$$100k_c = 0.45k_u = 236.5 \Rightarrow k_c = 2.37$$

and

$$T_i = 0.83T_u = 0.88$$

The Ziegler-Nichols recommendation for this problem, $k_c = 2.63$ (or $k_c = 2.37$ for PI-control), is evidently not close to the quarter-decay response design $k_c = 2.06$ (see Practice Problem 11.39).

Figure 11.15

**Quarter-decay Response
and Dominant Branch**

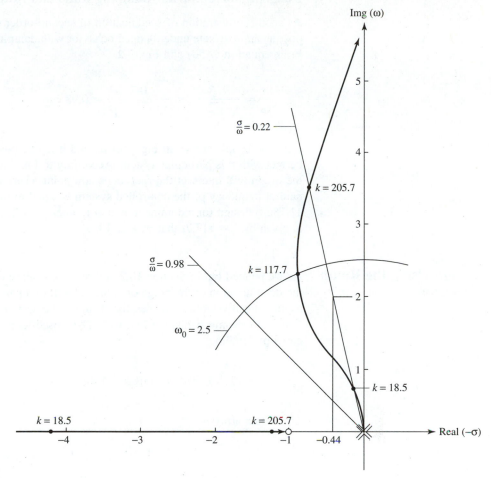

(a)   Root Locus

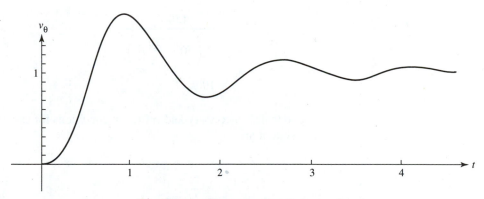

(b)   Transient Response, Step Reference Input

Designing for Dominant Damping Ratio and Natural Frequency

As another illustration of specification of second-order dominant characteristics, suppose an approximate underdamped behavior with damping ratio equal to 0.7 is desired. From equation (7.5d) and Fig. 7.2,

$$\frac{\sigma}{\omega} = \frac{\zeta}{\sqrt{1 - \zeta^2}} = \frac{0.7}{\sqrt{1 - 0.7^2}} = 0.98$$

This line is also shown in Fig. 10.15a, and it is obvious that this specification cannot be met with this particular system. According to Fig. 7.2, a circular arc with center at the origin will intersect the root locus at a point where the approximate characteristic natural frequency of the controlled system is equal to the radius of the arc. From Fig. 11.15a, a design for a dominant natural frequency of 2.5 rad/s for the servomechanism is given by $k = 117.7$, that is, k_c 1.18, ...

Design Using the Routh Criterion

As we pointed out in Section 10.2, a direct application of the Routh array in design is the determination of the range of values of a design parameter for which the system is asymptotically stable. Consider, for example, the design of the PI-controller for the antiaircraft servomechanism of Fig. 10.5. The closed-loop characteristic equation is, from equation (10.9a),

$$s^4 + 15s^3 + 50s^2 + 100k_c s + 100k_c/T_i = 0 \tag{11.22}$$

The Routh array is

1	50	$100k_c/T_i$	0
15	$100k_c$	0	0
$50 - \dfrac{20}{3}k_c$	$100k_c/T_i$		
$100k_c - \dfrac{1500k_c}{T_i\left(50 - \dfrac{20}{3}k_c\right)}$	0		
$100k_c/T_i$	0		

so that the necessary and sufficient conditions for asymptotic stability of the controlled system are

$$k_c > 0: \qquad \frac{k_c}{T_i} > 0$$

which by the first condition implies $T_i > 0$;

$$5 - \frac{20}{3}k_c > 0 \Rightarrow k_c < 7.5$$

and

$$100k_c - \frac{1500k_c}{T_i\left(50 - \frac{20}{3}k_c\right)} > 0 \Rightarrow k_c < 7.5 - \frac{9}{4T_i}$$

The allowable design domain is thus given by

$$T_i > 0 \quad \text{and} \quad 0 < k_c < 7.5 - \frac{9}{4T_i} \tag{11.23}$$

which is shown on a set of k_c, T_i axes in Fig. 11.16. A particular choice of k_c and T_i that are within the allowable design domain can be made by tuning (that is using a performance criterion).

Example 11.7	Design for Specified Speed of Response or Decay Rate

Figure 11.17a shows a numerically controlled (NC) milling machine having two axes powered by servomotors. Assuming the common ballscrew-nut feed drives for each axis, a simple model of the machine can be given by $G_m(s)$ (see Fig. 11.17b), which relates the screw or slide velocity ω to the drive torque T in terms of the equivalent inertia J and damping coefficient β. A typical control structure for such machine tools is cascade control (see Section 13.1), which consists of an inner (velocity) control loop and an outer position or, in this case force, control loop, as illustrated in Fig. 11.17b. Control of the feed force in milling is useful to ensure quality of the finished cut and to manage cutting tool wear. The indicated first-order model $G_f(s)$ is often assumed to relate the cutting force to the feed rate. Finally, if the servomotor dynamics is represented by a simple gain (as suggested in Example 2.2), then the motor model can be

Figure 11.16

Allowable Design Domain for Servo Controller

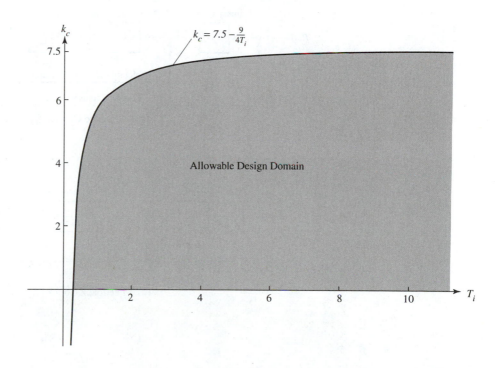

Figure 11.17

**Control of Feed Force
in NC Milling**

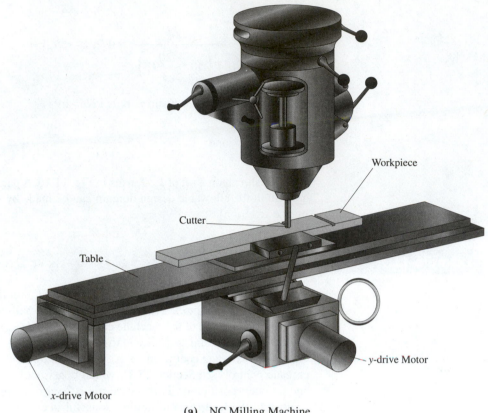

(a)   NC Milling Machine

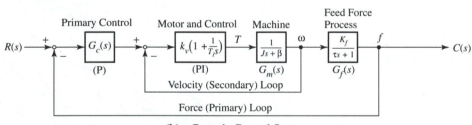

(b)   Cascade Control System

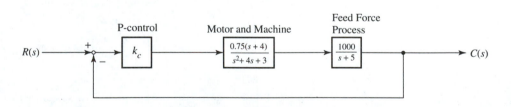

(c)   Feedback Control System

incorporated in the velocity controller transfer function, which for this example is PI-action.

Suppose the machine and feed force process parameter values are $J = 0.02$ kg · m², $\beta = 0.065$ N · m · s, $K_f = 200$ N · s, and $\tau = 0.2$ s, and the velocity loop has been independently tuned so that the inner loop PI-controller parameters are determined to be $k_v = 0.015$ and $T_i = 0.25$ (the tuning of a less-damped machine tool system using step response criteria is given as an exercise in Practice Problem 11.12). Determine the range of values of proportional gain k_c, as the outer loop (force) controller, such that the decay rate of the overall controlled system is faster than that of e^{-2t}.

Solution: The block diagram of Fig. 11.17b is redrawn in Fig. 11.17c, with the known parameter values substituted and the inner loop replaced by its closed-loop transfer function. Note that the motor-machine system is second order with a negative zero that was incorporated into the system by the controller. For a response speed or decay rate that is faster than e^{-2t}, all the controlled system eigenvalues must have real parts that are less than -2. As we discussed in Section 10.2, this requires the transformation $s = (p - 2)$ in the system's characteristic equation.

The closed-loop characteristic equation of the feedback control system of Fig. 11.17c is

$$s^3 + 9s^2 + (23 + k)s + 15 + 4k = 0$$

where $k = 750k_c$. Substituting $s = p - 2$, we obtain

$$p^3 + 3p^2 + (k - 1)p + (2k - 3) = 0$$

The necessary condition for the asymptotic stability of the transformed system is $k - 1 > 0$ and $2k - 3 > 0$. This yields $k > 1.5$. The Routh array,

1	$k - 1$	0
3	$2k - 3$	0
$k/3$	0	
$2k - 3$	0	

produces no additional condition on k. Thus the desired design is $750k_c > 1.5$, or $k_c > 0.002$

....

O n   t h e   N e t

- fig11_3.m and zeroef.m (MATLAB): source file for Fig. 11.3
- fig11_6a.m (MATLAB): source file for Fig. 11.6a
- fig11_6b.m and pro11_6.m (MATLAB): source file for Fig. 11.6b
- fig11_6c.m, ppi11_6.m and pid11_6.m (MATLAB): source file for Fig. 11.6c
- F11_7B.MCD (Mathcad): simulation of optimal response, Fig. 11.7b
- F11_7C.MCD (Mathcad): simulation of optimal response, Fig. 11.7c
- SC13.MCD (Mathcad): design of control system with nonlinear actuator, Example 11.3
- fig11_9.m and picad.m (MATLAB): design of control system with nonlinear actuator, Example 11.3
- fig11_10.m (MATLAB): source file for Fig. 11.10b, Example 11.4
- Auxiliary exercise: Practice Problems 11.12, 11.23 through 11.26.

11.4 PRACTICE PROBLEMS

11.1 Explain feedback control. Give two reasons why the stability of a feedback control system is an important design consideration.

11.2 Given generally the input-output transfer function

$$G(s) = \frac{b_0 + b_1 s + b_2 s^2 + \cdots + b_m s^m}{a_0 + a_1 s + a_2 s^2 + \cdots + a_n s^n}; \quad m \geq 0, \quad n > 0$$

what is

(a) The realizability condition on $G(s)$?

(b) The stability condition on $G(s)$?

(c) The condition that for a unit step input the output will settle down to a nonzero constant value of 1.0?

(d) The condition for $G(s)$ to be minimal phase?

11.3 Determine, and comment on your result, the sensitivity of the closed-loop system of Fig. 11.2a with respect to

(a) The controller $G_c(s)$

(b) The plant $G_p(s)$
Show by appropriate transformation of the block diagram of Fig. 11.2c that the transfer function between the output $C(s)$ and

(c) The measurement noise $N(s)$ is $S_{C/N}(s) = S_G^{G_{CL}}(s) - 1$.

(d) The disturbance input $V(s)$ is $S_{C/V}(s) = S_G^{G_{CL}}(s)$.

11.4* What is the sensitivity of the proportional control system of Example 11.1, Fig. 11.4, to small changes in k_c if the nominal value of k_c is 11.4?

11.5 From the given unit step response of a second-order plant (see Fig. 11.18), determine

(a) The damped circular frequency.

(b) The damping ratio.

(c) The percent overshoot.

(d) The decay ratio.

(e) The rise time.

(f) The settling time for 5% tolerance on the steady-state error.

11.6 The unit step responses of the components of a proportional feedback control system are indicated in Fig. 11.19. Using analyses correct to two decimal places, determine the overall (Laplace domain) transfer function for the system.

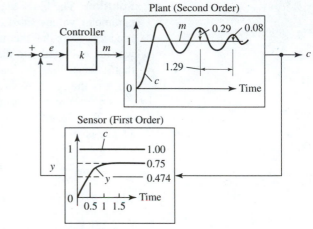

Figure 11.19 (Problem 11.6)

11.7(a) A first-order system is described by

$$\frac{dx}{dt} + ax = u(t); \quad x(0) = 0$$

If the input is a unit step, find the value of a such that the final value of the response x is 5.

11.7(b) The specification on a particular second-order liquid-level control system is that the overshoot of the step response will not exceed 20%.

(i) What is the corresponding limiting value of the damping ratio?

(ii) For an undamped natural frequency of 0.1 rad/s, determine the 5% settling time of the output after a step input occurs.

11.8 Determine the range of k_c for the servomechanism shown in Fig. 11.20 for which the step response percent overshoot is 15% or less.

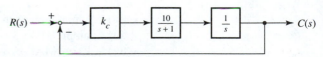

Figure 11.20 (Problem 11.8)

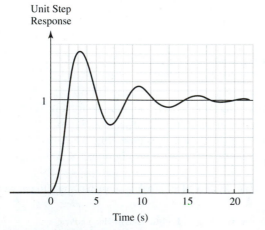

Figure 11.18 (Problem 11.5)

11.9* For the control system whose block diagram is shown in Fig. 11.21, it is desired that the system should have the following step response characteristics:

(1) Damping ratio of 0.75

(2) Steady-state error of 7.5% of the step magnitude

Determine the values of k and T that satisfy these conditions.

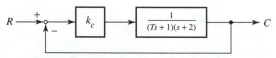

Figure 11.21 (Problem 11.9)

11.10 For the control system whose block diagram is shown in Fig. 11.22, it is desired that this system have the following characteristics:

(1) Undamped natural frequency $\omega_n = 5$

(2) Steady-state error for a ramp input $r(t) = At$ of 5% of the slope A

(a) Determine the values of k and T that satisfy these conditions,

(b) Find the corresponding 5% settling time (step-response characteristic).

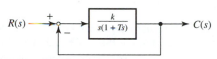

Figure 11.22 (Problem 11.10)

11.11 The control system shown in Fig. 11.23 is to be designed for the following step response specifications:

(1) Percent overshoot of 25%

(2) Natural (circular) frequency of 5 units

Determine the values of k and T that satisfy these conditions.

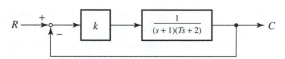

Figure 11.23 (Problems 11.11 and 11.28)

11.12* Consider the independent tuning of the velocity loop of the NC machine of Example 11.7. Suppose the machine parameter values, this time, are $J = 0.05$ kg $\cdot$ m^2 and $\beta = 0.01$ N $\cdot$ m $\cdot$ s.

(a) Show that the closed-loop transfer function for the velocity control system is

$$G_{CL}(s) = \frac{20k_v(s + 1/T_i)}{s^2 + (0.2 + 20k_v)s + 20k_v/T_i}$$

(b) Using, if necessary, computer assistance to solve any nonlinear system of equations, determine k_v and T_i so that the velocity loop will have a unit-step response having a rise time of 0.15 s and a 5% settling down time of 0.5 s.

(c) Repeat (b) for a step response specification of 10% overshoot and settling time of 1.75 s.

11.13 Given the system shown in Fig. 11.24,

(a) What is the open-loop response, that is, with the feedback loop disconnected, for a step change in $r(t)$ of height r_0 from a point at which all the initial conditions are zero?

(b) What is the closed-loop response for the same change as proposed in (a)?

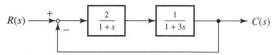

Figure 11.24 (Problem 11.13)

11.14 Consider the feedback control structure of Fig. 11.2a, but with the disturbance input affecting the plant as a load change (see block diagram in Fig. 11.7b).

(a) Obtain the transfer function between $C(s)$ and $V(s)$ for a zero setpoint.

(b) Obtain the sensitivity of the closed-loop system to $V(s)$.

(c) Investigate the steady-state "immunity" to disturbance of the feedback structure by computing $c(\infty)$ for a unit step disturbance for the elementary controllers—proportional, integral, and derivative controller, assuming that $H(s) = 1$ and $G_p(s)$ is of the general form

$$G(s) = \frac{b_0 + b_1 s + b_2 s^2 + \cdots + b_m s^m}{a_0 + a_1 s + a_2 s^2 + \cdots + a_n s^n}; \quad n \geq m;$$

$$\frac{b_0}{a_0} \neq 0$$

What do you conclude?

11.15* Consider the closed-loop control system of Example 11.2.

(a) Determine the type of system represented by the open-loop transfer function for each of the three controllers investigated in the example.

(b) Verify analytically the values of the offset under each controller as listed in Table 11.2.

11.16 The Ziegler-Nichols approximation (fit) for a sigmoidal unit-step response (Fig. 11.5a) can be described as a first-order-lag-plus-dead-time model.

(a) Using the parameters T_L and S, defined in Fig. 11.5a, obtain the transfer function for the approximate system on the basis of the above description.

(b) Given a first-order-lag-plus-dead-time process with the following transfer function:

$$G_p(s) = \frac{5e^{-0.4s}}{2s + 1}$$

obtain a preliminary design for PID-control of the process using the first method of Ziegler and Nichols.

11.17 The result of a step input test on a certain industrial process is as shown in Fig. 11.25. If, however, the magnitude of the step was 5, determine the preliminary feedback controller design for the plant using Ziegler-Nichols rules, assuming

(a) P-action

(b) PI-action

(c) PID-action

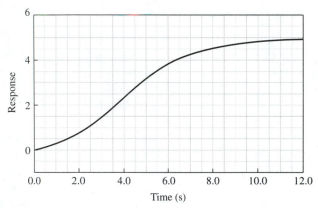

Figure 11.25 (Problem 11.17)

11.18 Given a plant with the transfer function

$$G_p(s) = \frac{10}{(s + 2)(s + 1)^3}$$

(a) Determine, using Ziegler-Nichols rules, the controller settings k_c and T_i for PI-control of the plant.

(b) Obtain preliminary controller settings for PID-control and P-control.

(c) Use computer simulation to investigate system performance, as in Table 11.2, under the three controllers determined in (a) and (b).

11.19* Repeat Problem 11.18 for the plant

$$G_p(s) = \frac{10}{(s + 10)(s^2 + 2s + 1)}$$

11.20 Given the step response shown in Fig. 11.6a, recommend the parameter values for PID-control of the plant using the first method of Ziegler and Nichols.

11.21(a) Under what conditions are the Ziegler-Nichols rules most appropriate?

11.21(b) Given the open-loop transfer function

$$G_o(s) = \frac{k}{(s + 10)(s^2 + 6s - 16)}$$

determine the value of k for the closed-loop system using Ziegler-Nichols rules.

(a) What are the corresponding settings for PI-action:

$$\left[G_c(s) = k_c\left(1 + \frac{1}{T_i s}\right) \right]$$

11.22* Given the plant with the transfer function

$$G_p(s) = \frac{40}{(s + 4)(s^2 + 4s + 5)}$$

determine using Ziegler-Nichols rules the controller settings k_c and T_i for PI-control of the plant, that is, for

$$G_c(s) = k_c\left(1 + \frac{1}{T_i s}\right)$$

11.23(a) Using computer simulation, obtain a comparison (similar to Fig. 11.7b) of the ISE, IAE, and ITAE for proportional control of the first-order-lag-plus-dead-time process used for Fig. 11.7, subject to a unit step load change.

11.23(b) Repeat (a) for a unit-step setpoint change.

11.24 Repeat the computer-assisted control system design of Example 11.3 by fixing k_c at 1.28, while T_i is determined from the simulation. Plot the resulting optimal unit step input response.

11.25 Repeat the computer-assisted control system design of Example 11.3 by using the method of least squares to fit the active region of the actuator characteristics with a second-degree polynomial.

11.26 Consider proportional-plus-derivative (PD) control of the undamped pendulum system shown in Fig. 11.26. Considering a very simple actuator model, $G_a(s) = 0.007$ rad/V and the measurement gain $H(s) = 5$ V/rad, determine the minimum value of k_c and maximum value of T_d for the controller that will give a 10% steady-state error (step input) and a decay rate for the overall controlled system that is faster than that of e^{-5t}.

(a)* Assuming the ideal PD-controller:

$$G_c(s) = k_c(1 + T_d s)$$

(b)* Assuming the realizable PD-controller:

$$G_c(s) = k_c\left[1 + \left(\frac{T_d s}{1 + \tau s}\right)\right]$$

Take $\tau = 0.1T_d$.

(c) Investigate the effectiveness of using the ideal derivative operator to design the realizable but approximate differentiator by sketching and comparing the unit step responses of the ideal system and the realizable system, using in both cases the values of k_c and T_d determined for the ideal control system.

(d) Repeat (c) for $\tau = 0.3T_d$ and $\tau = 0.05T_d$. What can you conclude from these results about the effect of τ in the approximate differentiator?

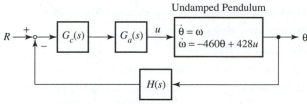

Figure 11.26 (Problem 11.26)

11.27 Sketch the root locus for the open-loop transfer function

$$kG_o(s) = \frac{k(s+1)}{s^2(s+3)(s+6)}, \quad \text{for } 0 \le k \le \infty$$

For what value of k do purely imaginary roots exist? What are the values of these roots?

11.28* Given the feedback loop shown in Fig. 11.23,

(a) Find the characteristic equation.

(b) Find the offset for a unit-step input.

(c) Sketch the root locus for k assuming $T = 1$.

(d) Determine the value of k for a closed-loop operation with a circular frequency of 2 units, assuming $T = 1$.

11.29(a) Under what conditions are the Ziegler-Nichols rules most appropriate?

11.29(b) Given

$$\text{Plant } G_p(s) = \frac{1}{(s+1)(s+2)(s+3)}; \quad \text{Control } G_c(s) = k$$

obtain a value for k that will give a satisfactory (not just for stability alone) feedback response pattern for a step change in the setpoint.

(c) Sketch the root locus for the system of (b) with k as the design parameter.

(d) Although the plant is third order, certain values of k cause it to behave like a second-order system. Explain this using the sketch in (c).

(e) If satisfactory response is considered a second-order-type pattern with a natural frequency of 0.27 Hz, specify an approximate value of k that will give this type of response.

11.30 The root locus and the block diagram for a proportional feedback control system are as shown in Fig. 11.27. It is desired that the overall system operate with a damping ratio of $\sqrt{2}/2$.

(a) Indicate approximately on the root locus the locations corresponding to this operating point.

(b) Calculate exactly the value of k that will cause the system to operate with this damping ratio.

(c) Determine the minimum value of k such that the closed-loop response is convergent within 2 time units.

(d) Find the value of k that will maximize the damping in the closed-loop system response.

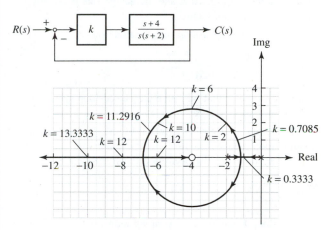

Figure 11.27 (Problem 11.30)

11.31 Consider the system

$$\dot{\mathbf{X}} = \begin{bmatrix} -4 & 1 \\ -20 & 0 \end{bmatrix}\mathbf{X} + \begin{bmatrix} 1 \\ 0 \end{bmatrix}u; \quad y = x_1$$

Plot the root locus for a proportional feedback control, $u = -kx_1$. Determine from this plot the controller gain for which the controlled system will have a damping ratio of 0.5. Verify this result by direct analysis of the closed-loop characteristic equation.

11.32 Consider again the system of Fig. 11.17c. Determine by the root locus technique the minimum value of k such that the maximum closed-loop time constant will be $\frac{1}{2}$.

11.33* Repeat the example of Fig. 11.11d but with the derivative part of the PD-controller $T_d s$ replaced by the approximate but realizable differentiator $T_d s/(1 + \tau s)$. Assume $\tau = 0.1T_d$ and keep the value of T_d in the example. Compare the new root locus and the "operating point" of $k_c = 1$ with the results of Fig. 11.11d.

11.34 Consider the use of the machine tool chatter model of Fig. 9.6 in machine design. Suppose

$$G_m(s) = \frac{10^5}{s^2 + 10s + 10^5}$$

Set up the overall system characteristic equation to allow determination by the root locus method of the machine parameter (k_c/k_m) that will prevent chattering instability at a given workpiece rotating speed. Putting $s = -\sigma + j\omega$ in your solution, determine the angle and magnitude conditions for this problem.

11.35 Obtain the fastest possible PI-controller for the hot water heater with remote sensing system of Example 11.6 (Fig. 11.13a). Let the integral controller parameter T_i be predetermined to be $3.3T$, where T and RC have the same values as in Fig. 11.14.

11.36 Determine T_i for quarter-decay response for the servo-mechanism of Fig. 10.5 with k_c fixed at 2.36. *Hint:* The root locus is shown in Fig. 10.7, and the design line is already indicated on the root locus.

11.37* Given the open-loop transfer function

$$G_o(s) = \frac{ke^{-0.5s}}{(s + 1)(s + 0.5)}$$

find the value of k that will give the closed-loop system an approximate quarter-decay response.

11.38 Explain briefly what is meant by the *dominant roots* of a single-loop feedback control system. What value of k will give the system with the open-loop transfer function

$$G_o(s) = \frac{k}{(s + 1)^3}$$

a closed-loop behavior with damping ratio of 0.5?

11.39 It can be argued that the quarter-decay response result of Fig. 11.15 is not directly comparable with Ziegler-Nichols rules for PI-control, since the parameter that was fixed did not have the same value. Derive the transfer function for the servomechanism with $T_i = 0.88$, and obtain the quarter-decay response design for k_c. Do you get a better match between the result and Ziegler-Nichols?

11.40 Given the open-loop transfer function

$$G_o(s) = \frac{k}{(s + 10)(s^2 + 6s - 16)}$$

(a) Determine the value of k that will give an approximate closed-loop second-order behavior of circular frequency of 2 rad/s.

(b) Determine the value of k for the closed-loop system that will give a quarter-decay response.

(c) Determine the value of k for the closed-loop system using Ziegler-Nichols.

11.41 Consider the proportional control of a plant with the transfer function

$$G_p(s) = \frac{2}{(s + 2)^2(s + 5)}$$

(a) Find approximately the value of the proportional gain that will give the closed-loop system a dominant second-order behavior with damping ratio of $\sqrt{2}/2$.

(b) If the plant is placed under proportional-plus-integral control, determine by Ziegler-Nichols rules the controller settings k_c and T_i.

11.42* Repeat Problem 11.41 for a plant with the transfer function

$$G_p(s) = \frac{0.5}{(s + 1)^2(s + 5)}$$

11.43 Given the feedback loop shown in Fig. 11.28,

(a) Find the characteristic equation.

(b) Sketch the root locus.

(c) Find the offset for a unit-step input.

(d) Replace the proportional action k_c, with integral control $[1/(T_i s)]$ and compute the characteristic equation. Use the Routh test to determine if the new system is unconditionally stable or the conditions on T_i for which the new system is stable.

(e) Find the offset for a unit-step input into the new feedback system.

(f) Can you choose T_i so that this offset is zero? If so, what is the value of T_i? Explain your answer.

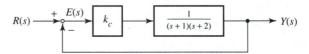

Figure 11.28 (Problem 11.43)

11.44 Consider the feedback control system shown in Fig. 11.29, in which the parameters k and T are adjustable.

(a) Determine and sketch the allowable design domain on a set of kT-axes that will give a closed-loop response that is convergent within about 3 time units.

(b) Derive, in terms of k and T, the expression for the offset in the system, if any, for a unit-step input.

Figure 11.29 (Problem 11.44)

11.45* In Problem 11.26(b) use the Routh test and math software to establish values for τ and T_d closest to those found in the problem. How does the ratio τ/T_d compare with $\tau = 0.1T_d$?

11.46 For the system shown in Fig. 11.30, find the value of k_c that will cause the closed-loop system to oscillate continuously as a linear system (small-amplitude oscillation). What is the period of the oscillation?

11.47* If in Problem 11.46 the nonlinear block is replaced by an on/off element with hysteresis ($M = 5$, $h = 0.5$), determine the stability of the resulting system for $k_c = 1.0$. Discuss any limit cycles. For what value of k_c is the system unconditionally stable?

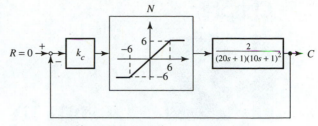

Figure 11.30 (Problem 11.46)

12

Design in the Frequency Domain

THIS CHAPTER CONTINUES WITH THE DESIGN of continuous-time control systems, but in the frequency domain. Some aspects of frequency domain design, including most of the performance measures employed in this chapter, have already been introduced as part of the phasor transform analysis of Chapter 7 and the stability analysis in the frequency domain of Chapter 10. Thus we begin directly in Section 12.1 with control system design for specified performance. Frequency domain performance specifications such as bandwidth, resonance peak, and gain and phase margins are considered. The correlation between these performance measures and the time domain specifications employed in Chapter 11 are explored, especially for systems with second-order-type behavior, so that such control systems can be designed in the frequency domain even though the desired performance is specified in the time domain, or vice versa. In Section 12.2 we introduce the traditional design approach of *system compensation,* a method of design in which relevant networks or *compensators* are introduced into the system in order to *reshape* its frequency response. The problem of *pole-zero cancellation,* which is often associated with this technique, is briefly highlighted, and the design of common compensators such as *lead, lag,* and *lag-lead* compensators is illustrated. Finally, we consider in Section 12.3 the design of compensators representing the classical mode controllers introduced in Chapter 11, as well as examples of nonlinear control system compensation.

12.1 DESIGN FOR SPECIFIED PERFORMANCE

In Section 11.1 we introduced general performance measures for control systems design—relative stability, system accuracy, and speed of response—and explored such time domain or transient response specifications of these measures as percent overshoot, settling time, and degree of damping. In this section we consider control system design based on similar frequency domain specifications of performance introduced in Part II of the text: *bandwidth, resonance peak, frequency at resonance, gain margin,* and *phase margin.* By determining the relationships between the frequency response and transient response, we can utilize the same frequency domain design techniques, although the performance is specified in the time domain. However, as we pointed out in Section 10.4, frequency domain methods are generally of two types: open-loop design techniques, and other methods based on the closed-loop system characteristics. In

Section 11.1 we showed that most feedback control objectives can be reduced to requirements on the open-loop sensitivity. We shall consider here mostly open-loop frequency domain design that involves the determination of the closed-loop controller using the system's open-loop transfer function or characteristics.

Design for Given Frequency/Bandwidth and Resonance Peak

We illustrate in the following examples control systems design for specified bandwidth and for given resonance characteristics. In addition to direct analysis, we shall employ the Bode and Nyquist diagrams as our *open-loop design* tools.

| **Example 12.1** | High-Pass Filter |

A first-order filter has the following frequency transfer function:

$$G(j\omega) = \frac{j\omega T}{1 + j\omega T}$$

where T is an adjustable parameter and ω is in radians per second.

a. Sketch the Bode diagram (gain in decibels and phase in degrees) for $0.1 \le \omega \le 10$, and $T = 1$.
b. Find a value of T if the filter is to be used in a piece of electronic equipment to suppress a 0.143-Hz low-frequency hum. Assume that useful signals in the equipment have considerably higher frequencies and that the low-frequency noise can be considered suppressed if the ratio of the output noise amplitude to the input noise amplitude is less than 0.25.

Solution a: Note that the implied noise control system is an *open-loop* system. Alternatively, $G(j\omega)$ is the *overall* system transfer function. The Bode diagram shown in Fig. 12.1 is easily sketched using the manual method of Section 9.2 or a computational math software package (Fig. 12.1 was generated using MATLAB. See script M-file **fig12_1.m; On the Net**). The system is a *high-pass* filter in the sense that input signals of frequencies higher than the cutoff frequency ($\omega = 1/T = 1$, in this case) are passed through without attenuation, while signals of lower frequency are attenuated.

Solution b: The frequency in the attenuation region at which the gain is 0.25 or $20 \log_{10}(0.25) = -12$ dB is indicated in the gain plot of Fig. 12.1 as $\omega = 0.2582$. This location will correspond to the noise frequency of $\omega = 2\pi(0.143) = 0.8985$ if the design corner frequency is shifted to the right by an amount such that the new corner frequency is $(0.8985/0.2582 = 3.48)$ times the old one. Note that subtraction on the logarithmic frequency scale implies division of the actual frequencies. The Bode diagram resulting from this change in corner frequency is shown as the dashed line in Fig. 12.1. The corresponding value of T is $1/3.48 = 0.287$. Thus the design is $T < 0.287$.

| **Example 12.2** | Design for Specified Closed-Loop Bandwidth |

A device for producing special sound track uses an armature-controlled DC motor to rotate a plastic disk (with cut grooves) against a scratch point, as shown in Fig. 12.2a. An audio pickup feeds back the produced sound, which is compared with the desired angular vibration. The load torque T_L in this arrangement is essentially a disturbance,

Figure 12.1

**Bode Diagram and Design
of a High-pass Filter**

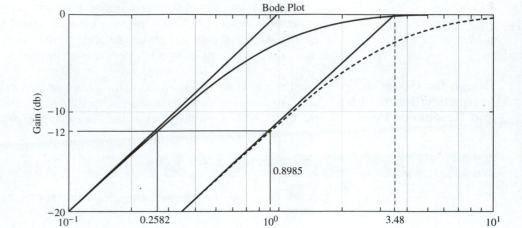

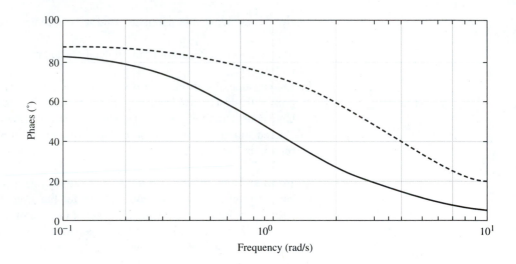

and the transfer function between the armature voltage $V_a = k_D u$ and the disk angle y is, according to equations (5.52a through 5.52c),

$$\frac{Y(s)}{V_a(s)} = \frac{K_b}{s[JL_a s^2 + (JR_a + \beta L_a)s + (\beta R_a + K_b^2)]}$$

$$\Rightarrow \frac{Y(s)}{U(s)} = G_p(s) = \frac{0.5}{s(0.25s^2 + 0.375s + 1)}$$

Determine a proportional controller k_c such that the closed-loop bandwidth is about 1.5 rad/s (0.24 Hz). Assume a unity feedback control system.

Solution: In general, the closed-loop bandwidth for most practical systems is (recall Fig. 10.13a) approximately the open-loop gain crossover frequency. Thus we determine the value of k_c that will yield a gain crossover frequency equal to the given bandwidth. Such a design should be considered preliminary or a first attempt, since the bandwidth definition is an approximate one.

Figure 12.2

Control of a Vibration-producing Device

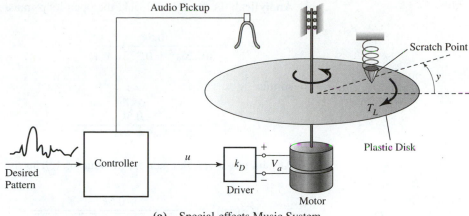

(a) Special-effects Music System

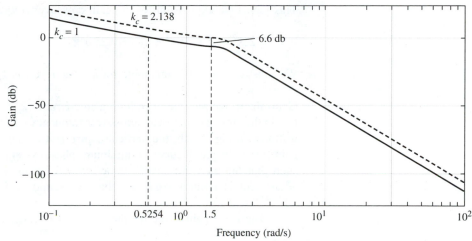

(b) Open-loop Bode Gain Plot and Design

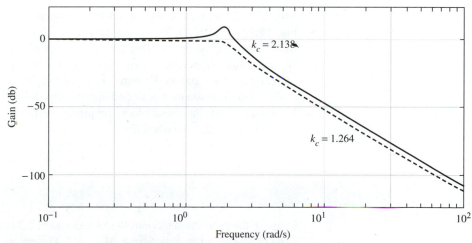

(c) Closed-loop Frequency Response for $k_c = 2.138$ and 1.264

Analytical: For unity feedback, the open-loop transfer function is

$$G_o(s) = \frac{0.5k_c}{s(0.25s^2 + 0.375s + 1)}$$

so that

$$G_o(j\omega) = \frac{0.5k_c}{-0.375\omega^2 + j(\omega - 0.25\omega^3)}$$

At gain crossover, the gain has a value of 1. That is,

$$|G_o(j\omega)| = \frac{0.5k_c}{\sqrt{0.375^2\omega^4 + (\omega - 0.25\omega^3)^2}} = 1$$

This yields the expression for the gain crossover frequency in terms of k_c:

$$0.25k_c^2 = 0.375^2\omega^4 + (\omega - 0.25\omega^3)^2$$

We substitute $\omega = 1.5$ and solve for k_c. The result is $k_c = 2.138$.

Graphical: We plot the open-loop gain curve for $k_c = 1$, as shown in Fig. 12.2b, and obtain the corresponding gain crossover frequency ($\omega_g = 0.5254$). We determine, as illustrated in Fig. 12.2b, the vertical adjustment of the gain curve, that is, the parallel gain curve (the phase curve for minimum-phase systems is not affected by this process) such that the gain crossover frequency will be about 1.5 rad/s. Since the required adjustment is around +6.6 db, the corresponding k_c, that is the design, is $k_c = 10^{(6.6/20)} = 2.138$.

The Bode gain curve for the closed-loop transfer function,

$$G_{CL}(s) = \frac{0.5k_c}{s(.25s^2 + .375s + 1) + 0.5k_c}$$

with $k_c = 2.138$, that is, the closed-loop frequency response as designed, is shown in Fig. 12.2c (solid line). The given closed-loop bandwidth is only approximately realized. Depending on the detailed definition of the bandwidth, the present design may not be acceptable. Also, the closed-loop behavior is similar to that of an underdamped second-order system (see Example 12.6) and suggests that the gain k_c can be reduced to diminish the resonance peak and improve the design. Such a situation is represented, for example, by the closed-loop frequency response for $k_c = 1.264$, which is also shown in Fig. 12.2c (dashed line).

Example 12.3   Design for Specified Resonance Peak

Suppose the proportional controller of Example 12.2 is to be designed instead for a desired closed-loop resonance peak, $M_p = 2.2$. Determine the value of k_c that will satisfy this specification.

Figure 12.3

**Design for Specified
Resonance Peak**

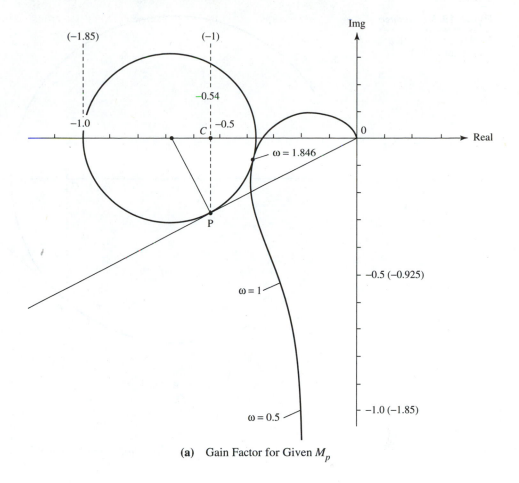

(a)   Gain Factor for Given M_p

Solution: Following the development of the relation between the resonance peak and
M-circles in Section 10.4, we proceed with the design as follows:

1. Construct the Nyquist locus of $G_o(s)$ for a particular value of $k_c = K$, for example,
    $K = 1$. That is, plot

    $$G_K(s) = \frac{0.5K}{s(0.25s^2 + 0.375s + 1)}; \qquad s = j\omega, \quad K = 1$$

    The Nyquist plot is shown in Fig. 12.3a.
2. Construct the tangent line OP from the origin at an angle θ (see Fig. 12.3b) given
    by

    $$\theta = \sin^{-1}\left(\frac{1}{M_p}\right) = \sin^{-1}\left(\frac{1}{2.2}\right) \approx 27°$$

3. By trial and error draw a circle with center on the negative real axis and tangent to
    both the $G_K(j\omega)$ locus and the line drawn at angle $\theta = 27°$ (see Fig. 12.3a).
4. From the tangent point P, between the circle and the line, erect a perpendicular to
    the negative real axis to meet the real axis at C.

Figure 12.3 *continued*

**Design for Specified
Resonance Peak**

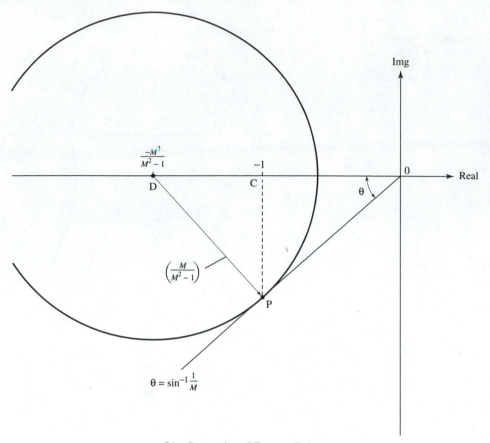

(b)   Properties of Tangent Point

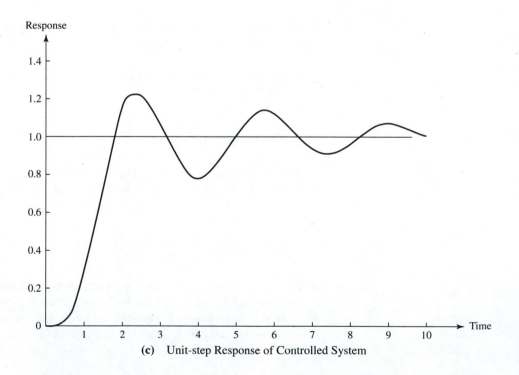

(c)   Unit-step Response of Controlled System

5. The scale value at point C should be -1, so that the ratio of -1 to the actual scale value (-0.54) is the amount by which K should be changed in order for the circle to become the M_p circle. That is,

$$k_c = \left(\frac{-1}{-0.54}\right)K = (1.85)(1) = 1.85$$

Explanation: The properties of the tangent point P to an M-circle are illustrated in Fig. 12.3b (see also Fig. 10.13b). From the right triangle DPO,

$$\sin\theta = \left(\frac{\dfrac{M}{M^2-1}}{M^2-1}\Bigg/ M^2\right) = \frac{1}{M}$$

so that

$$\theta = \sin^{-1}\left(\frac{1}{M}\right)$$

Also

$$\overline{DC} = \left(\frac{M}{M^2-1}\right)\cos(90° - \theta) = \left(\frac{M}{M^2-1}\right)\sin\theta = \frac{1}{M^2-1}$$

so that

$$\overline{CO} = \left(\frac{M^2}{M^2-1}\right) - \frac{1}{M^2-1} = 1$$

That is, C is the $(-1, j0)$ point.

Step 5 can be explained as follows: Given the Nyquist plot for $G_o(j\omega)$, the Nyquist plot for $FG_o(j\omega)$, where F is a constant factor, can be obtained simply by multiplying the real and imaginary axes scales of $G_o(j\omega)$ by F. (The new scale is shown in parenthesis in Fig. 12.3a.) In step 5, F was determined as the factor that will make point C the $(-1, j0)$ point.

Notes:

The use of an arbitrary factor K in step 1 allows the application of this design technique to an existing Nyquist locus where the old gain factor is other than 1. For instance, if we interpret the 0.5 factor in $G_p(s)$ as an existing gain factor, then $K = 0.5$ in Fig. 12.3a, and the new gain factor for the specified M_p is $FK = (1.85)(0.5) = 0.925$.

The frequency at resonance corresponding to the given M_p can be read off the Nyquist locus. The value in this case is $\omega_R = 1.846$.

Figure 12.3c shows the unit-step response of the controlled system using this design.

Design Based on Gain- and Phase-Margin Criteria

We shall now illustrate the design for specified gain or phase margin with the simple case of proportional control. The various performance specifications can, however, apply to more complex controllers, as we shall see. For now, recall that gain margin and phase margin are measures of closed-loop relative stability but are determined using the open-loop frequency response.

| Example 12.4 | Design for Specified Gain or Phase Margin |

Assuming a unity feedback control system, determine a "good" proportional controller for a plant with the transfer function

$$G_p(s) = \frac{2}{(s + 2)^2(s + 5)}$$

Solution for Fixed Gain Margin: We illustrate first a graphical solution for a desirable gain margin that is similar to the approach used in Example 12.2. The Bode plot of $G_p(s)$, that is,

$$G_0(s) = \frac{2k_c}{(s + 2)^2(s + 5)}, \quad \text{with } k_c = 1$$

is shown in Fig. 12.4. From these data, the current gain margin is about 39.8 db, while the gain plot is everywhere below 0 db. For proportional control, a large gain is desirable to reduce offset. Thus we can raise the gain plot sufficiently to give a desirable gain margin as long as the resulting phase margin is acceptable. The raising of the gain plot can be accomplished by merely changing the gain scale (see inside scale in Fig. 12.4).

If we fix the gain margin at 9.8 db (recall from Section 10.4 that gain margins of at least 8 db and phase margins of at least 30° were recommended as reasonable for

Figure 12.4

Gain-factor Design for Given Gain Margin

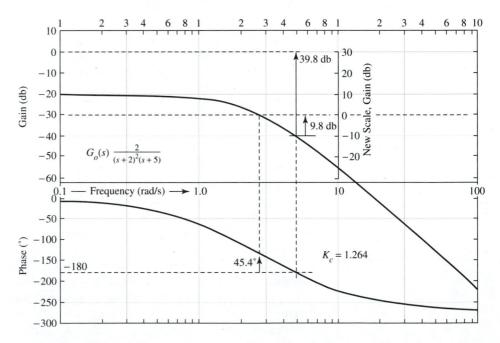

good closed-loop performance), the required change in scale is $39.8 - 9.8 = 30$ db, as shown. On the basis of the new scale the phase margin is about $45.4°$, which is acceptable. The design is complete and is given by $20 \log_{10}(k_c) = 30 \longrightarrow k_c = 31.62$.

Solution for Fixed Phase Margin: We consider another solution that can be obtained by fixing the phase margin at a desirable value while accepting the resulting gain margin if it is reasonable. Using an analytical approach this time, let us pick a desirable phase margin of $45°$. The required gain crossover frequency can be found by solving

$$\angle G_o(j\omega) = -180 + 45 = -135°$$

But

$$\angle G_o(j\omega) = -2 \tan^{-1}\left(\frac{\omega}{2}\right) - \tan^{-1}\left(\frac{\omega}{5}\right)$$

The value of ω that yields a phase angle of $-135°$ can be found by an iterative solution or by a nonlinear equations solution function of a computational math package, such as MATLAB's **fzero** function or Mathcad's **root** or **find** functions. For example, for MATLAB, we can first define and save the function M-file **fazangle.m**,

```
function a = fazangle(w)
a = (135*pi/180) - 2*atan(w/2) - atan(w/5);
```

and then obtain

```
wg = fzero('fazangle',2.7)
wg = 2.6893
```

The current gain at this frequency is found by evaluating the magnitude of the open-loop transfer function with $k_c = 1$ and $\omega = wg = 2.6893$

$$|G_o(j\omega)| = \frac{2k_c}{(4 + \omega^2)\sqrt{25 + \omega^2}} = 0.0314$$

Thus the required gain factor so that at this frequency the gain will be 1 is

$$k_c = \frac{1}{0.0314} = 31.9$$

Note that the phase crossover frequency can also be found by the same technique used in obtaining wg above. The result is $\omega_p = 4.899$, and by equation (10.21a), the gain margin is

$$-20 \log |G_o(j\omega_p)| = -20 \log \left[\frac{2(31.9)}{(4 + 4.899^2)\sqrt{25 + 4.899^2}} \right] = 9.75$$

which is acceptable.

....

Correlation between Transient and Frequency Response

Performance specifications for frequency domain design of control systems can be given in terms of time response performance, such as those considered in Chapter 11, provided adequate relationships exist between performance measures in the frequency and time domains. In general, exact relationships can be established among most of the performance measures for canonical first- and second-order linear systems. From these, approximate correlations between closed-loop frequency response and time domain response can be deduced for control system design.

First-Order Behavior

Consider the first-order canonical (closed-loop) transfer function of the form (equation (9.18a))

$$G_{CL}(s) = \frac{K}{Ts + 1} \Rightarrow G_{CL}(j\omega) = \frac{K}{1 + j\omega T} \tag{12.1a}$$

The transient response of a system with this transfer function is characterized by the *time constant T*, which is a measure of speed of response. As shown in Section 9.2, the frequency response of the system (12.1a) is characterized by the *corner* or *break* or *cutoff frequency* ω_c, which is related to the time constant by equation (9.18c), repeated here:

$$\omega_c = \frac{1}{T} \tag{12.1b}$$

In Example 12.1 the desired corner frequency of 3.48 rad/s corresponds to a specification of speed of response or time constant of $T = 0.287$ s. Depending on the application, the cutoff frequency can be interpreted in terms of bandwidth. For a high-pass filter, such as in Example 12.1, the range of frequencies greater than the corner frequency (that is, the high or "pass" frequencies) is normally viewed as the bandwidth. For a low-pass filter (see Fig. 12.5a), the bandwidth is usually the range of frequencies less than the cutoff frequency. As shown in Fig. 12.5c, a *band-pass filter* passes frequencies within a specified range (band). Hence it has two cut-off frequencies, and the bandwidth is typically the frequency range between the low and the high corner frequencies.

Figure 12.5

Frequency Response (Ideal Behavior) of Filters and Resonator

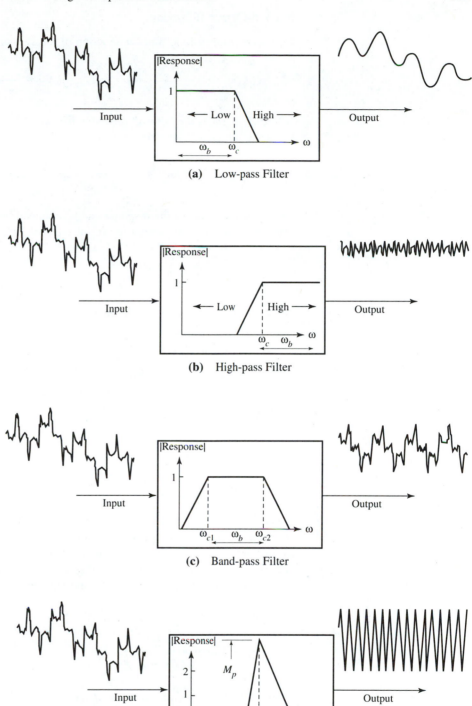

(a) Low-pass Filter

(b) High-pass Filter

(c) Band-pass Filter

(d) Resonator

Second-Order Behavior

Consider a closed-loop transfer function of the canonical form (equation (9.19a))

$$G_{CL}(s) = \frac{K}{T^2 s^2 + 2\zeta Ts + 1} \Rightarrow G_{CL}(j\omega) = \frac{K}{(1 - \omega^2/\omega_0^2) + j2\zeta\omega/\omega_0} \tag{12.2a}$$

where $\omega_0 = 1/T$ is the *natural frequency*. The closed-loop magnitude ratio (or the magnitude of the closed-loop response for unit input magnitude) is

$$M(\omega) = |G_{CL}(j\omega)| = \frac{K}{\sqrt{(1 - \omega^2/\omega_0^2) + 4\zeta^2\omega^2/\omega_0^2}} \tag{12.2b}$$

The peak amplitude of the response can be obtained by differentiating (12.2b) with respect to frequency and setting the result to zero. This requires that

$$2\zeta^2 - \left(1 - \frac{\omega^2}{\omega_0^2}\right) = 0$$

which gives the frequency at resonance:

$$\omega_R = \omega_0\sqrt{1 - 2\zeta^2} \tag{12.2c}$$

Substituting (12.2c) in (12.2b), we obtain the resonance peak:

$$M_p = \frac{K}{2\zeta\sqrt{1 - \zeta^2}} \tag{12.2d}$$

Note that equation (12.2c) requires that for resonance conditions to obtain ($M_p > 1$),

$$\zeta < 0.7071 \tag{12.2e}$$

Equation (12.2d) gives the correlation between the transient response characteristic *damping ratio* and the frequency response performance measure *resonance peak*. Similar performance measures in the time domain are the *peak value of the response Y_p* or the *overshoot O_p*, both of which can also be correlated with the damping ratio. According to equations (7.16d) and (7.16e), and equation (12.2a),

$$Y_p = K + O_p \tag{12.2f}$$

$$O_p = K \exp\left(\frac{-\pi\zeta}{\sqrt{1 - \zeta^2}}\right) \tag{12.2g}$$

Figure 12.6a shows a plot of both M_p and Y_p versus the damping ratio for a practical range of design values of M_p, and $K = 1$. The correlation between M_p and Y_p given by equations (12.2d) and (12.2f) is shown in Fig. 12.6b. Such a relationship can be used to convert a transient response performance specification of Y_p or overshoot for a second-order or approximate-second-order control system to an equivalent frequency domain specification of resonance peak, or vice versa.

Bandwidth and Phase Margin   The relationships between the frequency domain performance measures—bandwidth and phase margin—and the underdamped second-

Figure 12.6

Correlation Between Second-order Transient and Frequency Response

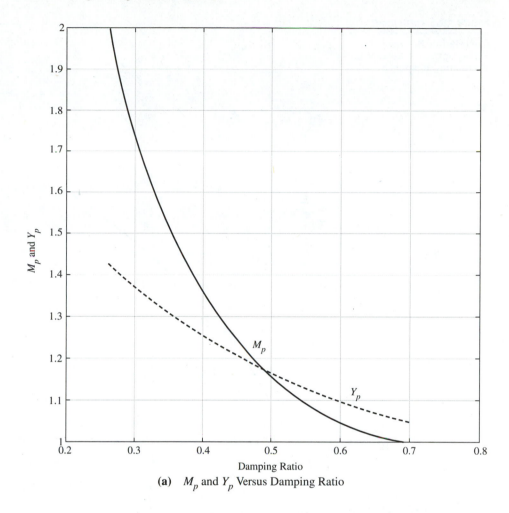

(a) M_p and Y_p Versus Damping Ratio

order transient response characteristics—damping ratio and natural frequency—can also be established. Recall that the bandwidth was defined in Section 10.4 as that frequency range over which the magnitude ratio does not differ by more than -3 db from the magnitude ratio at a specified frequency. Further, considering the desirable or specified frequency as zero (static) or low frequency, the (unity feedback) closed-loop bandwidth was approximately equal to the open-loop gain crossover frequency ω_g.

For a second-order closed-loop system (12.2a), the consistent definition of bandwidth is given by

$$M(\omega) = |G_{\mathrm{CL}}(j\omega)| > \left(\frac{\sqrt{2}}{2}\right)K \tag{12.3a}$$

since the static or low-frequency amplitude ratio is equal to K. Substituting equation (12.2b) into (12.3a), we have

$$\frac{K}{\sqrt{(1 - \omega^2/\omega_0^2)^2 + 4\zeta^2\omega^2/\omega_0^2}} > \left(\frac{\sqrt{2}}{2}\right)K \Rightarrow \left(1 - \frac{\omega^2}{\omega_0^2}\right)^2 + 4\zeta^2\frac{\omega^2}{\omega_0^2} < 2$$

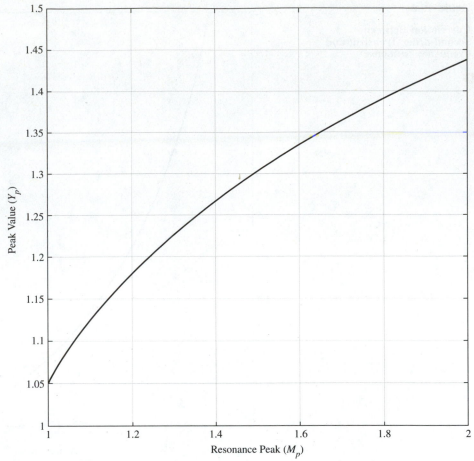

(b)   Peak Value Versus Resonance Peak

This yields the expression for the bandwidth ω_b:

$$\frac{\omega_b^4}{\omega_0^4} + (4\zeta^2 - 2)\frac{\omega_b^2}{\omega_0^2} - 1 = 0$$

or

$$\frac{\omega_b^2}{\omega_0^2} = (1 - 2\zeta^2) + \sqrt{2 - 4\zeta^2(1 - \zeta^2)} \qquad \textbf{(12.3b)}$$

Note that equation (12.3b) is different from the open-loop gain crossover frequency ratio. According to equation (10.22), the phase margin γ is given by

$$\gamma = \angle G_o(j\omega_c) + 180°; \qquad |G_o(j\omega_c)| = 1 \qquad \textbf{(12.3c)}$$

For a unity feedback system, the open-loop and closed-loop transfer functions are related by

$$G_{\text{CL}}(j\omega) = \frac{G_o(j\omega)}{1 + G_o(j\omega)}$$

Figure 12.6 *continued*

**Correlation Between
Second-order Transient and
Frequency Response**

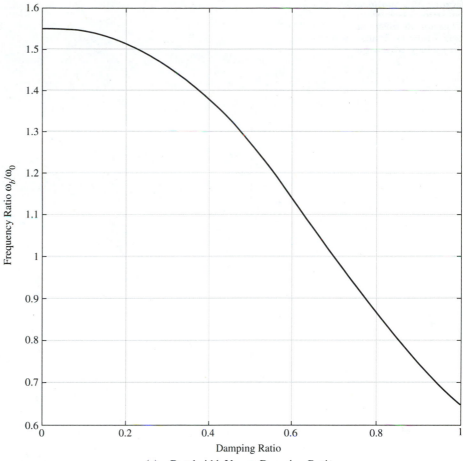

(c)   Bandwidth Versus Damping Ratio

Substituting for $G_{CL}(j\omega)$ from (12.2a) and solving for $G_o(j\omega)$, we obtain

$$G_o(j\omega) = \frac{K}{1 - K - \omega^2/\omega_0^2 + j2\zeta\omega/\omega_0} \tag{12.3d}$$

Substituting (12.3d) into equation (12.3c) gives us

$$\gamma = 180° - \tan^{-1}\left(\frac{2\zeta\omega_c/\omega_0}{1 - K - \omega_c^2/\omega_0^2}\right); \tag{12.3e}$$

$$\frac{\omega_c^2}{\omega_0^2} = (1 - K) - 2\zeta^2 + \sqrt{[(1 - K) - 2\zeta^2]^2 + 2K - 1}$$

so that the phase margin is a function of the damping ratio only. Note that in equation (12.3d) the real part of the denominator is negative for $K = 1$, so that the arctangent in equation (12.3e) is, for this case, an angle between 90° and 180°.

The correlations given by equations (12.3b) and (12.3e) are also shown in Figs. 12.6c and 12.6d, respectively, for the underdamped range of damping ratios $0 < \zeta < 1$, and $K = 1$.

Figure 12.6 *continued*

**Correlation Between
Second-order Transient and
Frequency Response**

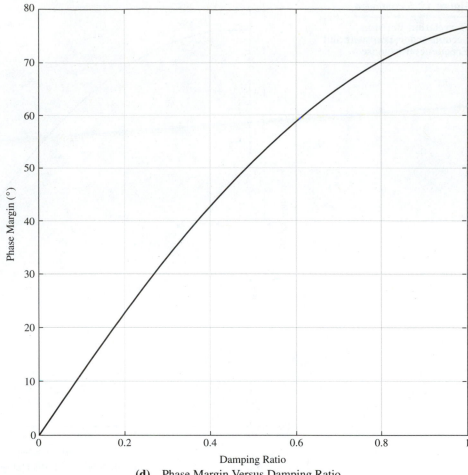

(d)   Phase Margin Versus Damping Ratio

Example 12.5   Design Using Correlation between ζ and M_p

Consider pure integral control, $G_c(s) = k/s$, of the first-order temperature process

$$G_p(s) = \frac{1}{(10s + 1)}$$

(see equation (11.19a) with $RC = 10$, for example). Determine the parameter k such that the closed-loop resonance peak is $M_p = 1.4$. Based on the correlation between transient and frequency response, what are the corresponding frequency at resonance, percent overshoot for a step input, phase margin, and closed-loop bandwidth?

Solution: Assuming unity feedback, the closed-loop transfer function is

$$G_{\mathrm{CL}}(s) = \frac{G_p(s)G_c(s)}{1 + G_p(s)G_c(s)} = \frac{k}{s(10s + 1) + k} = \frac{1}{\dfrac{10}{k}s^2 + \dfrac{1}{k}s + 1} \qquad \textbf{(12.4a)}$$

This is in the form of equation (12.2a) with $K = 1$.
   Thus

$$|G_{\text{CL}}(j\omega)| = \frac{1}{\sqrt{(1 - \omega^2/\omega_0^2)^2 + (1/10k)\omega^2/\omega_0^2}}; \qquad \omega_0 = \sqrt{\frac{k}{10}} \qquad \text{(12.4b)}$$

Since the closed-loop resonance peak is specified, the correlation (12.2d) can be applied. That is, the damping ratio can be read off Fig. 12.6a or calculated from

$$\frac{1}{2\zeta\sqrt{1 - \zeta^2}} = 1.4$$

The result is $\zeta = 0.3874$. However, since the closed-loop transfer function is second order, the damping ratio is also given (see equation (12.2a)) by equation (12.4a):

$$4\zeta^2 = \frac{1}{10k} \qquad\qquad\qquad\qquad \text{(12.4c)}$$

Substituting $\zeta = 0.3874$ into equation (12.4c), we obtain the solution $k = 0.1666$.
   From equation (12.4b), the natural frequency is $\omega_0 = 0.129$ rad/s, and equation (12.2c) gives the frequency at resonance as $\omega_R = 0.108$ rad/s.
   The open-loop and closed-loop frequency responses of the controlled system with the design value of k are shown in Figs. 12.7a and 12.7b, respectively. It can be seen from Fig. 12.7b that the specified resonance peak of 1.4 (2.92 db) is satisfied, and the frequency at resonance is as predicted.
   The peak value of the response can be read off Fig. 12.6a or Fig. 12.6b, so that the overshoot is given by equation (12.2f). Alternatively, equation (12.2g) can be used, since ζ is available. Either way, the result is $O_p = 0.267$.
   Equation (12.3e) gives the phase margin

$$\frac{\omega_c}{\omega_0} = \sqrt{-2\zeta^2 + \sqrt{4\zeta^4 + 1}} = 0.8625 \quad \text{and}$$

$$\gamma = 180° - \tan^{-1}\left[\frac{2(.3874)}{-0.8625}\right] = 180 - 138 = 42°$$

which agrees with Fig. 12.7a. The approximate closed-loop bandwidth is given by equation (12.3b) or Fig. 12.6c. The result $\omega_b = 0.179$ rad/s is indicated in Fig. 12.7b, and it satisfies the definition of bandwidth given in equation (12.3a). The open-loop gain crossover frequency $\omega_g = 0.1113$, which is shown in Fig. 12.7a, would yield a smaller bandwidth in this case.

Example 12.6   *Second-Order Approximations*

In Chapter 11 we introduced *second-order dominant* behavior, in which the characteristics of a higher-order system are approximated by those of a second-order system whose eigenvalues are the pair of dominant complex-conjugate roots possessed by the

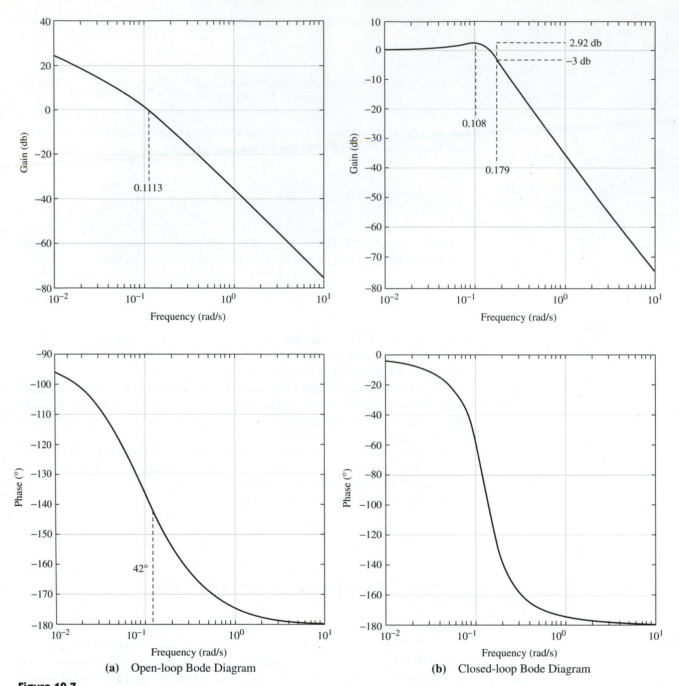

(a) Open-loop Bode Diagram   **(b)** Closed-loop Bode Diagram

Figure 12.7

Open-loop and Closed-loop Frequency Response of Controlled System

system. The approach presumes knowledge of the eigenvalues or the use of a design technique such as the root locus method. For transfer functions of the form

$$G(s) = \frac{b_0}{a_0 + a_1 s + a_2 s^2 + \cdots + a_n s^n} \tag{12.5a}$$

the following two analytical approximations due to Gustafson (see [7]) can be useful:

$$G_{a1}(s) = \frac{b_0}{a_0 + a_1 s + a_2 s^2} \tag{12.5b}$$

$$G_{a2}(s) = \frac{b_0}{r(n+1, 1) + r(n, 1)s + r(n-1, 1)s^2} \tag{12.5c}$$

where n is the order of the original system, and the denominator coefficients in equation (12.5c) are the last three first-column elements of the Routh array for the denominator polynomial of the original transfer function (12.5a) (see Section 10.2, equation (10.11e)).

Consider again the design for the specified closed-loop bandwidth problem of Example 12.2. Using the transient response correlations developed earlier and each of the preceding second-order approximations, obtain the values of the proportional controller k_c that yield approximately the desired ω_b, and compare your results with the solution of Example 12.2.

Solution: Recall that the closed-loop transfer function for the controlled system is

$$G_{CL}(s) = \frac{0.5k_c}{0.25s^3 + 0.375s^2 + s + 0.5k_c}$$

Thus the system is third order. The second-order approximation corresponding to (12.5b) is

$$G_{a1}(s) = \frac{0.5k_c}{0.375s^2 + s + 0.5k_c} = \frac{1}{\dfrac{0.75}{k_c}s^2 + \dfrac{2}{k_c}s + 1} \tag{12.6a}$$

Note that the approximate transfer function has been put in the form of equation (12.2a). For the approximation equation (12.5c), the Routh array is

$$
\begin{array}{lcc}
i = 1 & 0.25 & 1 \\[4pt]
i = 2 & 0.375 & 0.5k_c \\[4pt]
i = 3 & \left(1 - \dfrac{k_c}{3}\right) & 0 \\[4pt]
i = 4 & 0.5k_c & 0
\end{array}
$$

so that the second-order approximation is

$$G_{a1}(s) = \frac{0.5k_c}{0.375s^2 + \left(1 - \dfrac{k_c}{3}\right)s + 0.5k_c} \tag{12.6b}$$

$$= \frac{1}{\dfrac{0.75}{k_c}s^2 + 2\left(\dfrac{1}{k_c} - \dfrac{1}{3}\right)s + 1}$$

Assuming that these second-order approximations are adequate, we see that the relevant correlation for the present design effort is equation (12.3b):

$$\omega_b^2 = \omega_0^2\left[(1 - 2\zeta^2) + \sqrt{2 - 4\zeta^2(1 - \zeta^2)}\right] = 1.5^2 \tag{12.6c}$$

We consider now the first case of equations (12.5b) and (12.6a). The second-order characteristic denominator polynomial of (12.6a) gives

$$2\zeta T = \frac{2}{k_c}; \qquad T = \sqrt{\frac{0.75}{k_c}}$$

Hence

$$\omega_0^2 = \frac{k_c}{0.75} \tag{12.6d}$$

$$\zeta^2 = \frac{1}{0.75k_c} \tag{12.6e}$$

Equations (12.6c), (12.6d), and (12.6e) can be solved for k_c (using, for example, MATLAB's **fzero** function). The result is $k_c = 2.146$ ($\omega_0 = 1.692$ and $\zeta = 0.6213$).

For the second case of equations (12.5c) and (12.6b), the natural frequency is still given by equation (12.6d), while the square of the damping ratio is

$$\zeta^2 = \frac{k_c}{0.75}\left(\frac{1}{k_c} - \frac{1}{3}\right)^2 \tag{12.6f}$$

Again, a simultaneous solution of the nonlinear set of equations (12.6c), (12.6d), and (12.6f) gives $k_c = 1.264$ ($\omega_0 = 1.298$ and $\zeta = 0.5943$).

Considering the results of Example 12.2, we see that the solution given by the approximation equation (12.5b), $k_c = 2.146$, is closer to the analytical or graphical solution ($k_c = 2.138$). However, the solution due to the second approximation equation (12.5c), $k_c = 1.264$, satisfies better the bandwidth specification (see the closed-loop response for $k_c = 1.264$ in Fig. 12.2b). In general, the adequacy of an approximation depends on the specific application. how well the approximation fits the given behavior should be ascertained by simulation, and, if possible, additional performance criteria other than the design specification should be examined. For example, the approximation (12.5c) gave rise to equation (12.6b) that includes a stability condition: $k_c < 3$, whereas the approximation (12.5b) resulted in equation (12.6a) that does not include a stability condition on k_c and could therefore lead to a design solution for a controlled system that is also unstable (see Practice Problem 12.16).

12.2 DESIGN BY FREQUENCY DOMAIN COMPENSATION

Feedback control system design using Bode and Nyquist diagrams entails shaping and reshaping the open-loop frequency response until the closed-loop system specifications are satisfied. This is usually accomplished through the introduction of appropriate control circuits, known as *compensation networks,* in the forward or feedback loop of the control system (see Fig. 12.8). Compensation is a two-stage process: (1) Specific components or networks are added to the system in order to change its overall structure, and (2) these components are adjusted in order to satisfy performance specifications. There are a large number of different compensation networks. Most of them were developed to meet the peculiar requirements of specific systems. Our discussion here is limited to some of the more general and common compensating elements and their methods of design (adjustment). Most of these techniques of compensation, although considered for linear systems, are also applicable to nonlinear systems, provided the compensation is being used to modify the effective transfer functions of the linear components in the control loop. However, before proceeding with the discussion of compensation schemes, we draw the reader's attention to the potential problem of *pole-zero cancellation.* Although the addition of new elements may change the overall behavior of a system, it does not necessarily physically alter the existing components of the system.

The Problem of Pole-Zero Cancellation

The problem of pole-zero cancellation occurs when a zero (numerator) term in the transfer function of the controller exactly cancels a pole (denominator) term in the plant transfer function of a control system that is under analysis. This event is problematic, since such cancellation is not "physical," Rather, some mode of the system can become

Figure 12.8

Some Possible Locations of Compensation Networks

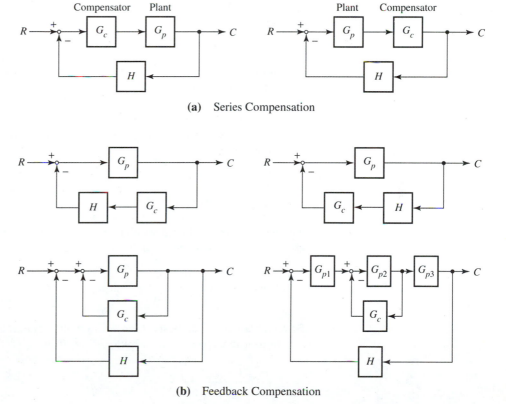

(a)   Series Compensation

(b)   Feedback Compensation

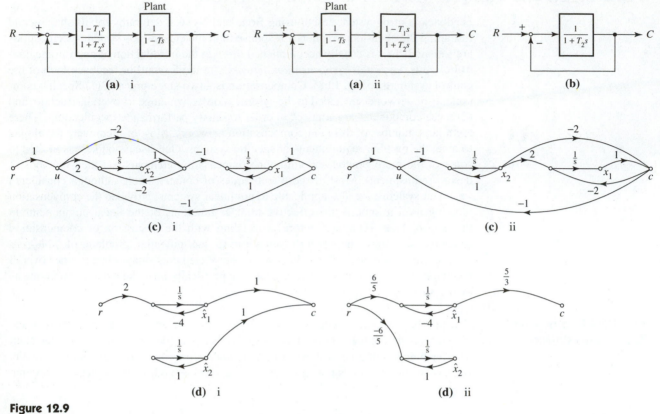

Figure 12.9

Series Compensation with
Pole-zero Cancellation

uncontrollable or unobservable. Although pole-zero cancellation is possible with most controllers, including those discussed in the previous chapter, the nature of compensation networks makes them more susceptible to the phenomenon.

Consider, for example, the first-order plant shown in Fig. 12.9a. This control object has a pole $s = 1/T$; $T > 0$, in the RHP and is consequently unstable. A stabilizing series compensation

$$G_c(s) = \frac{(1 - T_1 s)}{(1 + T_2 s)}$$

with the parameters chosen such that $T_1 = T$ will theoretically cancel the unstable pole, giving the reduced system (shown in Fig. 12.9b), which is stable:

$$\text{Open-loop pole} = \frac{-1}{T_2}; \quad \text{closed-loop pole} = \frac{-2}{T_2}, \quad T_2 > 0$$

However a careful analysis gives a different picture. The closed-loop characteristic equation of the system (Fig. 12.9(a)i, $T_1 = T$) is

$$1 + \frac{1 - T_s}{1 + T_2 s} \frac{1}{1 - Ts} = 0$$

If we refrain from canceling the $(1 - Ts)$ terms, the closed-loop characteristic equation reduces to

$$(1 + T_2s)(1 - Ts) + 1 - Ts = (1 - Ts)(2 + T_2s) = 0 \qquad \textbf{(12.7)}$$

We see that in addition to the new pole $(s = -2/T_2)$, the old (unstable) pole $(s = 1/T)$ is still present. Equation (12.7) represents the correct procedure in handling this type of system. Further insight into the problem can be gained from the signal flow graphs, Fig. 12.9c, constructed directly from the block diagrams of Fig. 12.9a. To simplify the presentation, we have assumed that $T_1 = T = 1$ and $T_2 = \frac{1}{2}$. In Fig. 12.9(c)i, which represents the case in which the compensating element is positioned before the plant, the closed-loop state equation is easily deduced:

$$\dot{\mathbf{X}} = \begin{bmatrix} -1 & -1 \\ -6 & -2 \end{bmatrix} \mathbf{X} + \begin{bmatrix} 2 \\ 6 \end{bmatrix} r; \qquad c = [1 \quad 0]\mathbf{X} \qquad \textbf{(12.8a)}$$

The eigenvalues are -4 and 1, as given also by equation (12.7), and the transformation matrix to diagonal form is

$$\mathbf{T} = \begin{bmatrix} 1 & 1 \\ 3 & -2 \end{bmatrix}.$$

This yields the modal domain representation

$$\frac{d}{dt}\hat{\mathbf{X}} = \begin{bmatrix} -4 & 0 \\ 0 & 1 \end{bmatrix} \hat{\mathbf{X}} + \begin{bmatrix} 2 \\ 0 \end{bmatrix} r; \qquad c = [1 \quad 1]\hat{\mathbf{X}} \qquad \textbf{(12.8b)}$$

whose signal flow graph is shown in Fig. 12.9(d)i. We see that the system is uncontrollable, and the uncontrollable mode is the unstable one. The corresponding results for the other position of the series compensator are given in Figs. 12.9(c)ii and 12.9(d)ii and described by the following equations:

$$\dot{\mathbf{X}} = \begin{bmatrix} -2 & 6 \\ 1 & -1 \end{bmatrix} \mathbf{X} + \begin{bmatrix} 0 \\ -1 \end{bmatrix} r; \qquad c = [1 \quad -2]\mathbf{X} \qquad \textbf{(12.9a)}$$

$$\frac{d}{dt}\hat{\mathbf{X}} = \begin{bmatrix} -4 & 0 \\ 0 & 1 \end{bmatrix} \hat{\mathbf{X}} + \begin{bmatrix} 6/5 \\ -6/5 \end{bmatrix} r; \qquad c = [5/3 \quad 0]\hat{\mathbf{X}}; \qquad \textbf{(12.9b)}$$

$$\mathbf{T} = \begin{bmatrix} 1 & 1 \\ -1/3 & 1/2 \end{bmatrix}$$

This time, the "canceled" mode, although controllable, is unobservable. The controlled system is, however, still unstable. In general, pole-zero cancellation is an ineffective means of dealing with an undesirable pole.

Gain-Factor Compensation

The simplest compensating element is a proportional or gain factor, that is, an amplifier. On a Bode diagram, the magnitude plot can be shifted up or down, that is, the proportional factor can be increased or decreased without affecting the phase angle plot (assuming the system is minimum phase). This is the basis of gain-factor compensation via the Bode diagram, and as we shall see, it is also the basis for the effective use of a gain factor in conjunction with other compensation networks that affect both the magnitude and phase. The proportional controllers that we designed in Section 12.1 are all forward-loop *gain compensators,* and the Bode diagram technique was used in

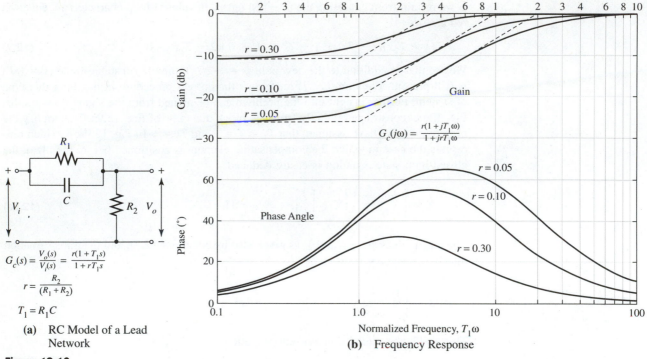

$$G_c(s) = \frac{V_o(s)}{V_i(s)} = \frac{r(1 + T_1 s)}{1 + rT_1 s}$$

$$r = \frac{R_2}{(R_1 + R_2)}$$

$$T_1 = R_1 C$$

(a)   RC Model of a Lead Network

Figure 12.10

Lead Compensator

many of the examples in that section. Example 12.3 illustrates a basic design method for specified resonance peak using the Nyquist diagram.

Lead and Lag Compensation

Lead Compensation   A compensation network with the transfer function

$$G_c(s) = \frac{r(1 + T_1 s)}{1 + rT_1 s}; \quad r < 1; \quad \text{(lead compensator)} \tag{12.10a}$$

is often used as a series compensator to increase, generally, the phase of the controlled system. Such a compensation network may be realized in an electrical unit by the *RC* circuit shown in Fig. 12.10a (see also Practice Problem 3.38). A lead network may also be realized with mechanical and/or other components. The parameter r is the lead ratio, and T_1 can be viewed as the network time constant.

Figure 12.10b shows the Bode plot for the *lead compensator* for three values of r covering the typical range of this parameter ($0.05 \leq r \leq 0.30$). From this figure we can see that "addition" (recall that on a logarithmic scale, multiplication of factors is achieved by adding their logarithms) of this compensator will in general lower the overall *low-frequency* gain and raise the overall *midfrequency* phase angle. The terms *low-frequency* and *midfrequency* are relative to the corner frequency given by the time constant T_1. Because of the effect on the phase, series lead compensation is normally used to improve the phase margin of a system. This in turn implies that lead compensation can improve the relative stability of the system. Since the gain is also affected by the lead compensator, it is common to apply gain compensation after the addition of a lead network. The combined effect on the system gain of the lead element and the gain compensator can also be used to increase the controlled system bandwidth and therefore the speed of response (see Bandwidth and Resonance Peak, Section 10.4, and Design for Specified Closed-Loop Bandwidth, Example 12.2).

Another view of lead compensation is in terms of an approximate (but realizable) proportional-plus-derivative controller. If we write the transfer function (12.10a) as

$$G_c = \frac{k(1 + T_1 s)}{1 + T_2 s}$$

where $T_2 = rT_1$ and $k = r$, we see that for the given range of values of r, $T_2 \ll T_1$, and $G_c(s) \approx k(1 + T_1 s)$. Thus according to the discussion of relative effects of I- and D-action in Section 11.3 (see Fig. 11.11), lead compensation can increase system damping (improve relative stability) and speed up response.

Adjusting Lead Compensators   As is illustrated in the following example, the design of a lead controller can be facilitated by knowledge of the precise frequency ω_m at which the peak phase advance occurs. From equation (12.10a), the lead network phase angle is

$$\theta = \tan^{-1}(\omega T_1) - \tan^{-1}(\omega r T_1) \tag{12.10b}$$

Setting the derivative of θ with respect to ω, to zero

$$\frac{d}{d\omega} [\tan^{-1}(\omega T_1) - \tan^{-1}(\omega r T_1)] = 0$$

and noting that

$$\frac{d}{d\omega} (\tan^{-1} x) = \frac{dx}{d\omega} \bigg/ (x^2 + 1)$$

we find

$$\omega_m = \frac{1}{T_1 \sqrt{r}} \tag{12.10c}$$

The corresponding phase θ_m is

$$\theta_m = \tan^{-1}\left(\frac{1}{\sqrt{r}}\right) - \tan^{-1}(\sqrt{r})$$

Or since

$$\tan^{-1} x = \frac{\pi}{2} - \tan^{-1}\left(\frac{1}{x}\right)$$

$$\theta_m = 90° - 2 \tan^{-1} \sqrt{r} \tag{12.10d}$$

From equation (12.10a), the gain of the lead compensator at $\omega = \omega_m$ is

$$M = \sqrt{r} \tag{12.10e}$$

Example 12.7   Series Lead Compensation

A quality parameter in plastic extrusion is the melt temperature in the die. In a simplified plant model, the die temperature T is related to the screw displacement θ by a first-order process with time constant $\tau = 1$:

$$\frac{T(s)}{\theta(s)} = \frac{k_e}{\tau s + 1} = \frac{10}{s + 1}$$

while the screw is modeled as a well damped second-order system:

$$\frac{\theta(s)}{U(s)} = \frac{1}{Js^2 + \beta s + k_r} = \frac{1}{0.0625s^2 + 0.5s + 1} = \frac{1}{\left(1 + \frac{1}{4}s\right)^2}$$

where u is the drive torque. Thus the overall plant model is

$$G_p(s) = \frac{T(s)}{U(s)} = \frac{10}{(1 + s)\left(1 + \frac{1}{4}s\right)^2}$$

a. Design a (series) lead compensator for this system that will give an overall phase margin of 45° while keeping the bandwidth the same as the uncompensated system.
b. Repeat the design, assuming a hardware limitation in the lead ratio of $r \leq 0.1$.

Solution a: From the Bode plot of $G_p(j\omega)$, Fig. 12.11, the uncompensated phase margin is about 7.3° and the gain margin is about 1.94 dB, both of which are quite low. The gain crossover frequency, and hence the bandwidth, is about 4.4 rad/s. A lead compensator will raise the midfrequency phase and hence the phase margin. At $\omega = 4.4$, we need a phase lead of $45° - 7.3° = 37.7°$. Thus we should choose the parameters r and

Figure 12.11

Illustration of Lead Compensation

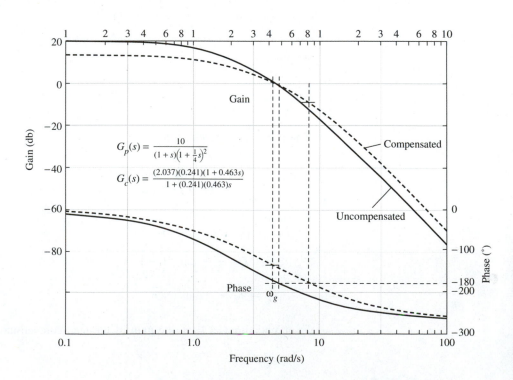

$$G_p(s) = \frac{10}{(1 + s)\left(1 + \frac{1}{4}s\right)^2}$$

$$G_c(s) = \frac{(2.037)(0.241)(1 + 0.463s)}{1 + (0.241)(0.463)s}$$

T_1 of the network so that the phase angle is 37.7° at $\omega = 4.4$ rad/s. We can select r by letting $\phi_m = 37.7°$ in equation (12.10d). That is,

$$\tan^{-1}\sqrt{r} = \frac{90° - 37.7°}{2} = 26.15°, \quad \text{or } r = 0.241$$

Since this is within the practical range of r, we shall use this value. From equation (12.10c), we let

$$\frac{1}{T_1\sqrt{r}} = \omega_m = 4.4$$

For $r = 0.241$, we obtain

$$T_1 = 0.463$$

The lead compensator also lowers the low-frequency gain. In particular, at the gain crossover frequency $\omega = 4.4$, the gain will be lowered by a factor of

$$\left| \frac{r(1 + j4.4T_1)}{1 + j4.4rT_1} \right| = \sqrt{r} = 0.4904$$

Thus in addition to the lead compensator, a gain compensator $K = 1/0.4909 = 2.037$ is required. This will maintain the same gain crossover frequency and hence, approximately, the same bandwidth. The final controller design is

$$G_c(s) = \frac{Kr(1 + T_1 s)}{1 + rT_1 s}; \qquad K = 2.037, \quad r = 0.241, \quad T_1 = 0.463$$

The Bode plot of $G_o(s) = G_p(s)G_c(s)$ is shown by the dashed curve in Fig. 12.11, and it is clear that this is an acceptable design. The new gain margin is approximately 10 dB. The phase margin is of course 45°.

Solution b: Since r is limited to $r \le 0.1$, and equation (12.10d) gives $r = 0.241$, we should choose $r = 0.1$. This gives $\phi_m = 54.9°$. However, from equation (12.10b), at $\omega = 4.4$, the phase lead is

$$\phi(4.4) = \tan^{-1}(4.4T_1) - \tan^{-1}(4.4rT_1)$$

If we require $\phi(4.4) = 37.7°$, then (from Fig. 12.10, or by graphical solution, or trial and error, or nonlinear equation solution algorithm.) $T_1 = 0.212$. The required gain compensation is

$$K = \left| \frac{1 + j4.4(0.1)(0.212)}{0.1(1 + j4.4(0.212))} \right| = 7.372$$

Thus the new design is $K = 7.372$, $r = 0.1$, and $T_1 = 0.212$. The corresponding gain margin is about 18.9 db.

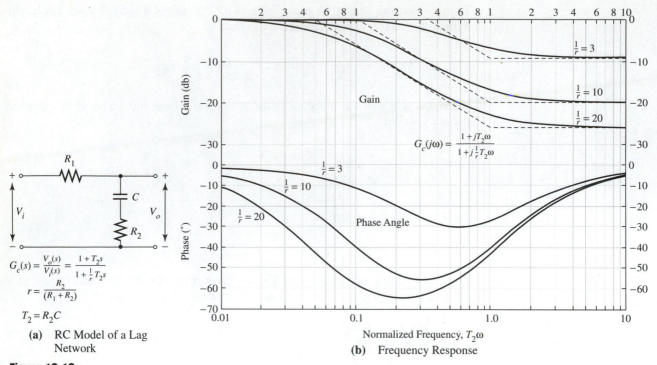

$$G_c(s) = \frac{V_o(s)}{V_i(s)} = \frac{1 + T_2 s}{1 + \frac{1}{r} T_2 s}$$

$$r = \frac{R_2}{(R_1 + R_2)}$$

$$T_2 = R_2 C$$

(a)   RC Model of a Lag Network

Figure 12.12

Lag Compensator

Lag Compensation

The transfer function for a *lag compensator* can be written as

$$G_c(s) = \frac{1 + T_2 s}{1 + \frac{1}{r} T_2 s}, \quad (1/r) > 1; \text{ (lag compensator)} \tag{12.11a}$$

An *RC* circuit realization, for example, is shown in Fig. 12.12a, while the frequency response is given in Bode format in Fig. 12.12b for lag ratios $(1/r)$ covering the typical range $(3 \le 1/r \le 20)$. From this figure we can deduce that the main effect of lag compensation is to decrease the high-frequency gain (increase the attenuation), whereas it decreases the phase angle in the low- to midfrequency region (increases the phase lag). Also, a lag compensator can decrease the system bandwidth and the phase and/or gain margins and generally can cause a system to be more sluggish. Further, for the given range of values of $(1/r)$, we can see that the lag compensator is "like" an approximate proportional-plus-integral controller:

$$G_c(s) \approx r \left(\frac{1 + T_2 s}{T_2 s} \right)$$

so that lag compensation is generally used to improve steady-state performance (permissible error or system accuracy). However, it does not have the same capacity as true PI-action to actually drive the error completely to zero in all cases.

Adjustment   The design of a lag compensator is similar to that of a lead compensator and normally also requires gain compensation. From equation (12.11a) we note that the phase lag is

$$\phi = \tan^{-1}\left(\frac{1}{r}T_2\omega\right) - \tan^{-1}(T_2\omega) \tag{12.11b}$$

The maximum phase lag (minimum phase shift) occurs at

$$\omega_m = \frac{1}{T_2\sqrt{1/r}} \tag{12.11c}$$

and the corresponding phase lag is

$$\phi_m = 2\tan^{-1}\sqrt{1/r} - 90° \tag{12.11d}$$

Example 12.8   Feedback Lag Compensation

Consider feedback-loop compensation of a robot arm–positioning system using a lag compensator, as shown by the block diagram of Fig. 12.13a. Determine the design of the lag compensator and the series gain factor (amplifier) for the proper functioning of the robot, according to the following specifications:

  i. Steady-state velocity (ramp) input error less than 5% of the magnitude (slope) of the input,
 ii. Phase margin of $35 \pm 2.5°$
iii. Gain crossover frequency of 1 rad/s.

Solution: We consider first the gain compensation that will satisfy the error specification. We shall then determine the balance of the control system, if possible, to meet the balance of the specifications. It is expected that some fine tuning will normally be required, although this task will not be undertaken here.

For the transfer function of the lag compensator given by equation (12.11a), the open-loop transfer function for the control system is

$$G_o(s) = \frac{k(1 + T_2 s)}{s(1 + .5s)(1 + .05s)\left(1 + \frac{1}{r}T_2 s\right)} \tag{12.12a}$$

According to equation (11.7c), for a velocity input of magnitude u_0 the steady-state error is

$$e_{ss} = \underset{s\to 0}{\text{Limit}}\left[\frac{u_0/s}{1 + G_o(s)}\right] = \underset{s\to 0}{\text{Limit}}\left[\frac{u_0}{s + \dfrac{k(1 + T_2 s)}{(1 + .5s)(1 + .05s)\left(1 + \frac{1}{r}T_2 s\right)}}\right] = \frac{u_0}{k}$$

Substituting the steady-state error specification, we obtain

$$\frac{u_0}{k} \le 0.05 u_0 \Rightarrow k \ge 20$$

Figure 12.13

Block Diagram of Control System and Bode Diagrams of Uncompensated and Compensated Systems

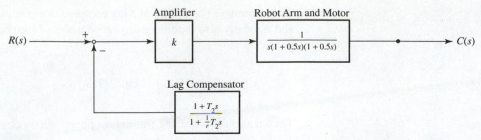

(a)   Block Diagram of Robot Arm–Positioning Control System

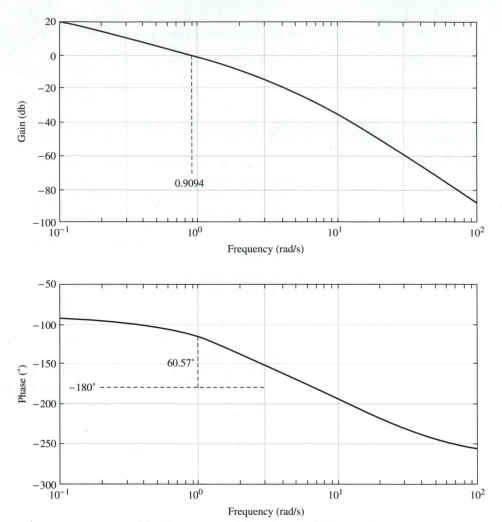

(b)   Uncompensated Open-loop Bode Diagram

Let us allow a little design margin by taking $k = 21$.

Figure 12.13b shows the Bode diagram for the uncompensated open-loop transfer with $k = 1$. That is, for the transfer function

$$G_{ou}(s) = \frac{k}{s(1 + .5s)(1 + .05s)} = \frac{k}{0.025s^3 + 0.55s^2 + s} \tag{12.12b}$$

Figure 12.13 *continued*

Block Diagram of Control System and Bode Diagrams of Uncompensated and Compensated Systems

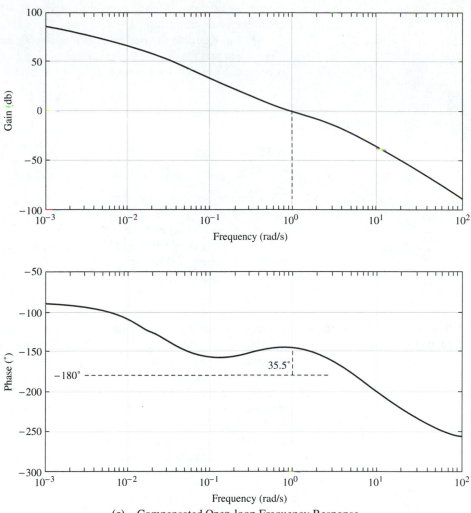

(c) Compensated Open-loop Frequency Response

the uncompensated system gain crossover frequency is $\omega_{gu} = 0.9094$. However, the desired gain crossover frequency is $\omega_g = 1.0$. Without regard to the error- and phase-margin specifications, the gain adjustment necessary to give a gain crossover frequency of 1.0 is $k = 1.2531$. This is obtained by substituting $s = j\omega$ and $\omega = 1$ in equation (12.12b) and solving for k so that $|G_{ou}(j\omega)| = 1$. Since the required gain factor for the steady-state error constraint is $k = 21$, the compensated gain curve will need to be attenuated by a factor of about

$$\frac{21}{1.2531} \approx 16.76$$

or lowered by about 24.5 dB at the frequency $\omega = 1$ in order to maintain the required gain crossover frequency. This attenuation is to be accomplished with the lag compensator.

Also, from Fig. 12.13b the uncompensated phase margin for a gain crossover frequency of 1 rad/s, that is, $180° + \angle G_{ou}(\omega = 1)$, is $60.57°$. Thus to meet the phase-margin specification, the lag compensator should also introduce a phase lag of about 25° at the same frequency $\omega = 1$.

Let us examine the Bode diagram of the lag compensator shown in Fig. 12.12b. It is evident that for the attenuation and phase lag required, a lag ratio close to $1/r = 20$ is needed. Consider, for instance, $1/r = 19$ at $T_2\omega = 2$,

$$20 \log |G_c(j\omega)| = 20 \log\left[\sqrt{\frac{1 + 2^2}{1 + 4(19^2)}}\right] = -24.61$$

and

$$\angle G_c(j\omega) = \tan^{-1}(2) - \tan^{-1}(38) = -25.06°$$

Thus we can select the lag compensator with lag ratio $1/r = 19$ and time constant $T_2 = 2$ (so that $T_2\omega = 2$ when $\omega = 1$). The open-loop Bode plot of the transfer function equation (12.12a) with these parameter values for the lag compensator and the gain factor $k = 21$ is shown in Fig. 12.13c. The gain crossover frequency is essentially 1.0 rad/s, as required, and the other specifications are also satisfied.

Lag-Lead and Cascade Compensation

Given the somewhat divergent effects of lead and lag compensators, a combination of both compensators can retain the advantages of both controllers while minimizing their disadvantages. Such a combined controller can be implemented by connecting the lag and lead networks in *cascade* and designing them sequentially, that is, lag compensation followed by lead compensation. This technique is the same as has been applied thus far with respect to lead-gain compensation and lag-gain compensation and is used in Example 12.9. An alternative approach is to have a single network in which both lag and lead compensation are combined. An example of an *RC* circuit realization of *lag-lead* compensation was given in Practice Problem 7.16. One possible three-parameter transfer function for this network is

$$G_c(s) = \frac{(1 + T_1 s)(1 + T_2 s)}{T_1 T_2 s^2 + \left(T_1 + \dfrac{1}{r} T_2\right)s + 1}; \quad \text{(lag-lead compensator)} \tag{12.13}$$

where T_1, T_2, and r have the same meanings as in Figs. 12.10a and 12.12a.

Example 12.9   Series Lag-Lead Compensation

Consider again the extrusion system of Example 12.7. Design a (series) lag-lead compensation such that in addition to an overall phase margin of 45°, the closed-loop unit-step input steady-state error is less than or equal to 1% and the gain crossover frequency is about 5 rad/s.

Solution: If we assume unity feedback, the open-loop transfer function of the compensated system is

$$G_o(s) = \frac{10}{(1 + s)\left(1 + \dfrac{1}{4}s\right)^2} \frac{(1 + T_2 s)}{\left(1 + \dfrac{1}{r_2}T_2 s\right)} \frac{kr_1(1 + T_1 s)}{(1 + r_1 T_1 s)}$$

where a gain compensation k has also been added in anticipation of the need for it. According to equation (11.7c), the steady-state error for a unit-step input is

$$e_{ss} = \underset{s \to 0}{\text{Limit}}\left[\frac{1}{1 + G_0(s)}\right] = \underset{s \to 0}{\text{Limit}}\left[\frac{(1 + s)\left(1 + \dfrac{1}{4}s\right)^2\left(1 + \dfrac{1}{r_2}T_2s\right)(1 + r_1T_1s)}{(1 + s)\left(1 + \dfrac{1}{4}s\right)^2\left(1 + \dfrac{1}{r_2}T_2s\right)(1 + r_1T_1s) + 10kr_1(1 + T_2s)(1 + T_1s)}\right]$$

$$e_{ss} = \frac{1}{1 + 10kr_1}$$

and for 1% error,

$$\frac{1}{1 + 10kr_1} = \frac{1}{100}$$

so that $kr_1 \sim 10$.

Thus a gain increase of about 20 db will be required. Note that (as was the case in the last example) the lag compensation does not affect the steady-state error for this system. The open-loop Bode plot of $G_p(j\omega)$ was given in Fig. 12.11 and is repeated in Fig. 12.14a. From this figure we see that if the 0-db line is lowered by 20 db (this is equivalent to an amplification of 10), the gain crossover frequency increases from about 4.4 to about 11.2 rad/s. The resulting phase margin is about $-46°$. An excessive amount of phase lead (91°) would be required to bring the phase margin to 45° at this gain crossover frequency. However, the lag compensator can be positioned to reduce the gain crossover frequency from around 11.2 to about 5 rad/s with only a small reduction in the phase at 5 rad/s, which is presently about $-183°$. The "high-frequency" gain attenuation necessary to accomplish this is about 17.5 db (Fig. 12.14a), and from Fig. 12.12b, a lag ratio $3 < 1/r_2 < 10$ can be chosen for this purpose. Also, to minimize the phase lag, the corner frequency $\omega_c = 1/T_2$ can be fixed at 0.5 rad/s, (that is, $T_2 = 2$), so that from Fig. 12.12b, the phase reduction at 5 rad/s will be only about 5°. However looking ahead, we see that the combination of gain and lead compensation will introduce a gain factor at $\omega = 5$ given by equation (12.10a) as

$$M = (kr_1)\left|\frac{1 + j5T_1}{1 + j5r_1T_1}\right| = (kr_1)M_1$$

Since (kr_1) is fixed by the steady-state error specification, we see that the balance factor

$$M_1 = \left|\frac{1 + j5T_1}{1 + j5r_1T_1}\right| \tag{12.14a}$$

is actually greater than 1 for any $r_1 < 1$. Thus $1/r_2$ should be chosen to give a gain attenuation greater than 17.5 db to compensate for this magnification.

If we choose $1/r_2 = 15$, the attenuation at 5 rad/s according to equation (12.11a) is 23.5 db, giving the balance factor

$$20 \log M_1 = 23.5 - 17.5 = 6 \text{ db} \tag{12.14b}$$

whereas the actual phase reduction is, according to equation (12.11b), 5.33°. The lag-compensated system using the design $T_2 = 2$ and $1/r_2 = 15$ is shown by the dashed line

Figure 12.14

**Illustration of
Lag-lead Compensation**

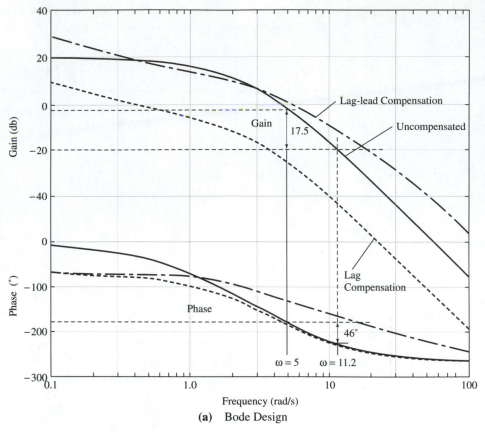

(a) Bode Design

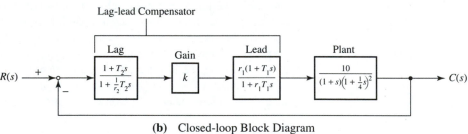

(b) Closed-loop Block Diagram

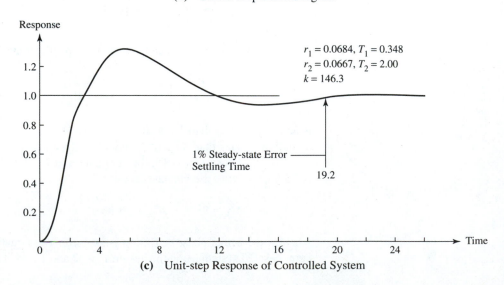

(c) Unit-step Response of Controlled System

(short strokes) in Fig. 12.14a. The required phase lead at the design gain crossover frequency is thus

$$\phi = (183 + 5.33) + 45 - 180 = 53.33°$$

From equation (12.10b), this requires

$$\tan^{-1}(5T_1) - \tan^{-1}(5r_1T_1) = 53.33° \tag{12.15}$$

Solving equations (12.14a and 12.14b) and (12.15) simultaneously (by trial and error or nonlinear equation solution function, for instance), we obtain the design $T_1 = 0.348$, and $r_1 = 0.0684$. Also $k = 10/r_1 = 146.3$. The Bode diagram for the overall lag-lead-compensated system is shown in Fig. 12.14a (long dashes). The corresponding closed-loop block diagram is shown in Fig. 12.14b. Observe that the gain compensator (amplifier network) has been placed between the lag and lead compensators, where it will prevent the loading of the first circuit by the second. The load-effect isolation *(unilateral coupling)* property of amplifiers was discussed in Section 3.4. Finally, Fig. 12.14c shows a simulation of the unit-step input response of the compensated system. It is evident that the specification of 1% steady-state error is also achieved.

12.3 CLASSICAL MODE CONTROLLERS AND NONLINEAR EXAMPLES

Cascade compensation as just illustrated can be applied to other networks, including two or more of the same compensator. For instance, two lead networks can be cascaded with necessary isolation to achieve an overall phase lead that is greater than the capacity of a single lead compensator. An open-loop frequency response design of the classical PID feedback controller can be accomplished with the same technique by considering the D-, I- and P-controllers as compensators in cascade and "adding" them sequentially to the process. A derivative controller in the *minor* (inner) feedback loop (see Fig. 12.8b, lower block diagrams) is often used as a means of stabilizing (increasing the damping in) a control system. The applications of PID-action to a process with dead-time nonlinearity, and minor-loop rate feedback to another system having a saturation-with-hysteresis-type nonlinear component, respectively, are illustrated in this section. The solution of the latter design example is accomplished by computer-assisted design in the Nyquist plane.

Proportional, Reset, and Rate Compensation

Figure 12.15 shows the Bode diagrams for the integral or *reset* control mode (actually proportional-plus-integral control),

$$G_c(j\omega) = 1 + \frac{1}{j\omega T_i} = \frac{(1 + j\omega T_i)}{j\omega T_i} \tag{12.16a}$$

and the (realizable) derivative or *rate* control mode,

$$G_c(j\omega) = \frac{(1 + j\omega T_d)}{(1 + j\omega \tau)} \tag{12.16b}$$

The realization of such an approximate differentiator using an op-amp circuit was considered in Example 3.19 (see Fig. 3.32). It can be shown, and the reader should verify,

Figure 12.15

**Control-mode
Response Curves**

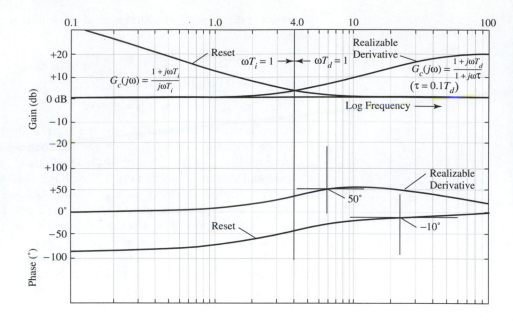

that the transfer function representing the model for that approximate differentiator circuit or equation (3.49a) is

$$\frac{-V_{\text{out}}(s)}{V_{\text{in}}(s)} = \left(\frac{R_f}{R_i}\right)\left[\frac{1 + Ts}{1 + \tau s}\right]$$

(12.16c)

The Bode diagrams of Fig. 12.15 illustrate the relative effects of the classical mode controllers on process dynamics. Note that the derivative curve is applicable to the case where $\tau = 0.1T_d$. In general, the derivative mode raises the high-frequency process phase and gain but has no effect on the low-frequency phase and gain. The high and low frequency are relative to the corner frequency given by $\omega T_d = 1$. In the case of the reset mode, the corner frequency is given by $\omega T_i = 1$, and reset raises the low-frequency gain but does not affect the high-frequency gain. Finally, the overall process phase is lowered by reset.

Although the detailed addition of these control modes to a process can be guided in an arbitrary way by their anticipated effects, the practice described in reference [8] is to "add" the *derivative mode* by aligning the $+50°$ phase point of the controller and the $-180°$ phase point of the open-loop process Bode curve. This is followed by adding the *reset mode* with the $-10°$ phase point of the reset curve aligned with the $-170°$ phase point of the process-plus-derivative curve obtained previously. As an aid in graphically positioning these controller curves or analytically applying them to the process, note that according to equation (12.16b), for $\tau = 0.1T_d$, the $-50°$ phase shift occurs for the derivative mode at

$$\tan^{-1}(\omega T_d) - \tan^{-1}(0.1\omega T_d) = 50° \Rightarrow \omega T_d = 1.713$$

(12.17a)

and according to equation (12.16a), the $-10°$ phase shift occurs for the reset at

$$\tan^{-1}(\omega T_i) - 90° = -10° \Rightarrow \omega T_i = 5.67$$

(12.17b)

The corner frequencies corresponding to equations (12.17a) and (12.17b) may provide more convenient reference points in the case of a graphical analysis. Finally, the *proportional band* is determined to satisfy the gain- or phase-margin criterion. Note that the eventual PID-controller implied by this technique is the combination *(second-canonical form)*

$$G_c(s) = k_c(1 + T_d s)\left(1 + \frac{1}{T_i s}\right) \tag{12.18a}$$

as opposed to the *first-canonical form* given in Table 11.1. The realizable representation of (12.18a) is

$$G_c(s) = k_c\left(\frac{1 + T_d s}{1 + \tau s}\right)\left(1 + \frac{1}{T_i s}\right), \quad T_d \gg \tau \tag{12.18b}$$

Example 12.10   PID Compensation of Process with Dead-Time Nonlinearity

Design a cascade PID-controller for the hot water heater with remote sensing system of Fig. 11.13a using a desired gain margin of 10 dB. Assume that the hot water heater system is adequately described by equations (11.19a) and (11.19b), and treat the quantity q/C as the manipulated variable. Let the heater time constant RC be equal to $40/3$ time units, and the transportation lag or dead time T be equal to 10π time units.

Solution: According to equations (11.19a) and (11.19b), the plant model is

$$\theta_{\text{sensor}}(s) = e^{-Ts}\theta(s) = \left(\frac{e^{-Ts}}{s + \frac{1}{RC}}\right)\frac{q}{C} = \left(\frac{[RC]e^{-Ts}}{[RC]s + 1}\right)\frac{q}{C} \tag{12.19a}$$

so that the plant transfer function is

$$G_p(s) = \frac{(40/3)e^{-10\pi s}}{1 + (40/3)s} \tag{12.19b}$$

We shall utilize a graphical approach in this example. The analytical design is given as an exercise in Practice Problem 12.27. We also assume in the following that $\tau = 0.1T_d$ for the derivative controller.

    The dash-dot lines in Figure 12.16 show the open-loop Bode diagram of the uncompensated system, that is, of $G_p(j\omega)$. The $-180°$ phase shift occurs at $\omega = 0.076$ (uncompensated phase crossover frequency), and the gain here is 19.5 db, so that the uncompensated "gain margin" is -19.5 db. To add the derivative mode with its 50° phase frequency aligned with the process $-180°$ phase frequency, we let $\omega T_d = 1.713$ at $\omega = 0.076$. This gives $T_d = 22.5$ and $\tau = 2.25$

    Note that the corresponding corner frequency is $\omega = 1/22.5 = 0.0444$. The derivative mode phase curve as positioned is shown (by short dashed lines) in Fig. 12.16 on the same frequency scale as the process curve but with a different phase angle scale, as indicated. The derivative gain curve is not shown. The resulting process-plus-derivative Bode phase (only) curve is shown by long dashed lines in the figure.

    The $-170°$ phase frequency on this process-plus-derivative curve is approximately $\omega = 0.097$. Thus to add the reset mode with its $-10°$ phase frequency aligned

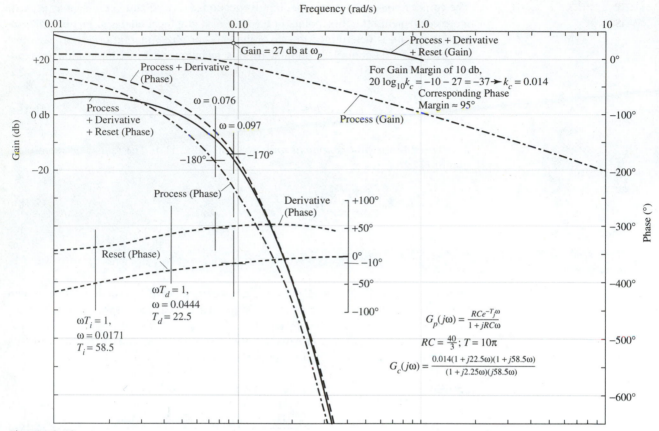

Figure 12.16

Design of PID-compensator for the Hot Water Heater with Remote Sensing System

with this frequency we let $\omega T_i = 5.67$ at $\omega = 0.097$, which gives approximately $T_i = 58.5$

The reset mode phase curve is shown as positioned by short dashed lines in Fig. 12.16, on the same scales as the derivative curve. The resulting process-plus-derivative-plus-reset Bode (gain plus phase) diagram is shown by solid lines in Fig. 12.16. From these curves the (compensated) phase crossover frequency is about 0.095 rad/s, and the gain at phase crossover is approximately 27 db. It follows that to achieve a 10-db gain margin a proportional band is required to lower the gain curve by $10 - (-27) = 37$ db. That is,

$$20 \log_{10} k_c = -37 \Rightarrow k_c = 0.014$$

With this value of the gain compensation, the system phase margin as designed is 95°. The final controller is therefore

$$G_c(s) = \frac{0.014(1 + 22.5s)(1 + 58.5s)}{(1 + 2.25s)(58.5s)}$$

Rate Feedback Compensation

To appreciate the stabilizing effect of rate feedback, consider a servomechanism with a minor-loop *rate* (velocity) feedback in addition to a major-loop *proportional* (position) feedback, as shown in Fig. 12.17a. This combination is equivalent to a proportional-plus-derivative controller, whose stabilizing effect was demonstrated in Section 11.3 (see Fig. 11.11). It was established, although using a linear plant, that such a controller introduces a zero that speeds up the system response and increases the damping in the system. Finally, note that the combination of derivative and proportional feedback can also arise normally in a servomechanism as velocity and position feedback sensors.

| Example 12.11 | Nonlinear System Compensation via the Nyquist Locus |

Consider again the third-order system with an actuator having a saturation-with-hysteresis–like nonlinearity of Example 10.13. The describing function for the nonlinearity is given by Table 10.2. Recall from Section 10.5 that the feedback system can exhibit both stable and unstable limit cycling, but the system can be stabilized (Kochenburger criterion) by reducing the gain.

a. Obtain such a stabilizing proportional control k such that the gain margin of the linear part is 8 db.

b. Suppose the servo system is subjected to (a realizable) rate feedback in parallel with the original position feedback, as shown in Fig. 12.17a. Obtain the design of the rate and gain compensators that will eliminate the oscillation and satisfy the 8-db gain-margin specification for the linear part and otherwise perform as well as or better than the proportional controller obtained in (a). Assume that the rate compensator is of the form

$$G_c(s) = \frac{1 + Ts}{1 + \tau s}, \quad \text{with } \tau = 0.05T$$

Solution a: We first organize the describing function data for use with our computational mathematics software. To construct the Nyquist locus $-1/N$ of the nonlinear part, we note that

$$N = |N|e^{j\phi} \Rightarrow -\frac{1}{N} = \frac{-1}{|N|e^{j\phi}} = \frac{1}{|N|}e^{j(\pi - \phi)} \tag{12.20a}$$

$$= \frac{1}{|N|}[\cos(\pi - \phi) + j\sin(\pi - \phi)]$$

Table 12.1 Describing Function and Nyquist Locus Data

A	5.00	2.50	1.50	1.00	0.95	0.85	0.75
$\|N\|$	0.713	0.993	1.230	1.470	1.180	0.976	0.794
$\phi\ (°)$	-3.27	-6.84	-11.31	-28.07	-34.44	-38.66	-44.36
$R = (1/|N|)\cos(180 - \phi)$	-1.400	-1.000	-0.797	-0.600	-0.699	-0.800	-0.900
$I = (1/|N|)\sin(180 - \phi)$	-0.080	-0.120	-0.159	-0.320	-0.479	-0.640	-0.880

Thus from Table 10.2 we can generate the real and imaginary parts of the describing function of the nonlinearity, as shown in Table 12.1. Such data are easily utilized by a function like MATLAB's **nyquist** (see script M-file **fig12_17.m**) to plot the $-1/N$ locus.

The open-loop transfer function of the linear part under proportional control is

$$G_{ou}(s) = \frac{k}{s(0.25s^2 + 0.375s + 1)} \Rightarrow \tag{12.20b}$$

$$G_{ou}(j\omega) = \frac{k}{-0.375\omega^2 + j(\omega - 0.25\omega^3)}$$

The Nyquist loci of the uncompensated linear part, that is, $G_{ou}(j\omega)$ with $k = 1$, and $-1/N$, which are shown in Fig. 10.20b, are repeated in Fig. 12.17b (solid lines) using MATLAB.

From equation (12.20b), the phase crossover frequency is given by

$$\angle G_{ou}(j\omega) = -\tan^{-1}\left(\frac{\omega - 0.25\omega^3}{-0.375\omega^2}\right) = -180° \Rightarrow \omega_p = 2$$

so that for a gain margin of 8 db,

$$\frac{1}{|G_{ou}(j2)|} = \frac{0.375(4)}{k} = 10^{(8/20)} \Rightarrow k = 0.597$$

Figure 12.17b also shows (dashed line) the plot of $G_{ou}(j\omega)$ with this value of k. Because this locus does not intersect the $-1/N$ curve and passes to the right of it as frequency increases, a proportional controller design for the specified gain margin is feasible, and the solution is $k = 0.597$.

The phase margin corresponding to this solution is $\gamma = 74.95°$.

Solution b: For the given form of rate feedback and the block diagram of Fig. 12.17a, the compensated open-loop transfer function (linear part) is

$$G_{oc}(s) = \frac{k}{s(0.25s^2 + 0.375s + 1)}\left[1 + \left(\frac{1 + Ts}{1 + \tau s}\right)\right] \tag{12.20c}$$

$$= \frac{(T + \tau)ks + 2k}{0.25\tau s^4 + (0.375\tau + 0.25)s^3 + (\tau + 0.375)s^2 + s}$$

A MATLAB script M-file to interactively determine the design parameters k and T ($\tau = 0.05T$), **fig12_17.m,** is shown in the box. The search relies on visual cues as follows: The locus of $G_{oc}(j\omega)$ must cross the real axis at the same point as $G_{ou}(j\omega)$ for $k = 0.597$ (this guarantees that the gain margin is 8 db), and it must pass the $-1/N$ curve on the right and must not intersect it (this assures that there will be no sustained oscillations, that is, no possibility of limit cycling). Further, a better performance relative to the proportional controller under these circumstances implies a less sluggish response or lower phase margin. Since the phase margin is given, on the Nyquist plane, by the angle between the negative real axis and the line from the origin to the point of intersection of the unit circle and the open-loop locus, the

Figure 12.17

Rate Feedback Compensation of a System Having Saturation-with-hysteresis– type Nonlinearity

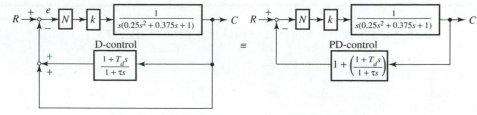

(a)   Rate Feedback Compensation of Nonlinear System

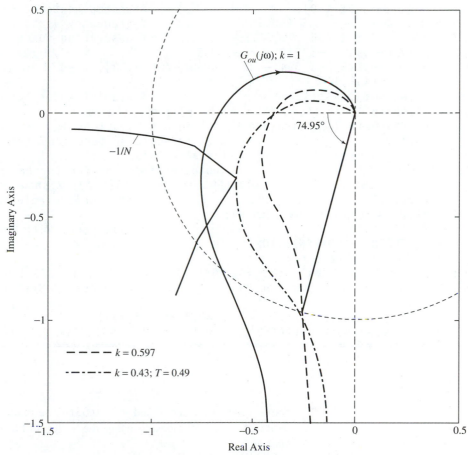

(b)   Gain Reduction and Rate Feedback Solutions

visual cue with respect to performance is that the compensated open-loop locus should intersect the unit circle to the left of the intersection point of the design locus for proportional control.

Note that the script M-file declares the vector P whose elements are $P(1) = k$ and $P(2) = T$ to be a global variable. Thus these values can be varied between successive runs of the program without changing the script file. Following a fair trial, our solution (to two decimal places) as shown in Fig. 12.17b (dash-dot line) is $k = 0.43$ and $T = 0.49$.

Note that this solution does not allow much of a safety margin with respect to potential limit cycling. However, the phase margin is $\gamma = 73.5°$, which represents a slightly improved performance. This value is obtained from equation (12.20c) using the

```
% fig12_17.m
% Script to generate Nyquist Plots of Figure 12.17b
%
[x,y] = meshdom(−1:.1:1,−1:.1:1);
V = x .^2 + y .^2;
global P;
if isempty(P)
      P = input ('Enter parameters [K T] ');
end;
% TRANSFER FUNCTION AND FREQUENCY RANGE DATA
den = [0.25 .375 1 0];                      % uncompensated denominator polynomial
w = logspace (−.5,3.5,250);                 % frequency vector, equal log spacing
A = [5 2.5 1.5 1 .95 .85 .75];              % describing funct. frequency
R = [−1.4 −.9999 −.7972 −.6 −.6989 −.8 −.9];    % −1/N real part
I = [−.08 −.1199 −.1594 −.32 −.4793 −.64 −.88]; % −1/N imaginary part
% NYQUIST DIAGRAMS
[re,im,w] = nyquist(1,den,w);               % Nyquist Locus for Uncompensated System
axis('square'); axis([−1.5 .5 −1.5 .5]); plot(re,im);
title('(b) Gain Reduction and Rate Feedback Solutions'); hold on;
plot(R,I);                       % plot (describing function) locus of −1/N
ylabel('Imaginary Axis'); xlabel('Real Axis');
text(−.4,.14,'>');              % indicate direction for increasing frequency
plot([−1.5 .5],[0 0],':'); plot([0 0],[−1.5 .5],':'); % show axes
contour(V,[1],−1:.1:1,−1:.1:1,':');         % show unit circle for phase margin
[rk,ik,w] = nyquist(.597,den,w);
plot(rk,ik,'−');                            % Plot Locus With Reduced Gain
% RATE FEEDBACK COMPENSATION
K = P(1); T = P(2); t = .05*T;
num = [K*(T + t) 2*K]; dec = [.25*t .375*t + .25 t + .375 1 0];
[rc,ic,w] = nyquist(num,dec,w);             % Compensated System Nyquist Locus
plot(rc,ic,'−.');
text(−1.44,−1.2,sprintf('− − − K = %4.3f',.597));
text(−1.44,−1.4,sprintf('−.−.− K = %3.2f, T = %3.2f',K,T)); hold off;
```

above values of k and T and $\tau = 0.05T$. The reader may wish to undertake an experiment for a better design using the script M-file **fig12_17.m; On the Net.** Good *hunting*.

....

On the Net

- fig12_1.m (MATLAB): source file for Fig. 12.1, Example 12.1
- fig12_2.m (MATLAB): source file for Figs. 12.2b and 12.2c, Example 12.2
- fazangle.m (MATLAB): function file, Example 12.4
- fig12_7.m (MATLAB): source file for Fig. 12.7, Example 12.5
- fig12_13.m (MATLAB): source file for Figs. 12.13b and 12.13c, Example 12.8
- fig12_17.m (MATLAB): interactive design of nonlinear system compensation, source file for Fig. 12.17, Example 12.11
- Auxiliary exercise: Computer assistance can be employed in essentially all practice problems of this chapter; however, the more specific cases include Practice Problems 12.1 through 12.5, 12.8 through 12.11, 12.27, and 12.32 through 12.34.

12.4 PRACTICE PROBLEMS

12.1 Repeat the problem of Example 12.1 if the noise to be suppressed has a frequency of 0.71 Hz, and the noise is considered suppressed when the noise output over input amplitude ratio is less than 0.5.

12.2 Consider the electrical filter circuit shown in Fig. 12.18. The resistance R is variable.

(a) Obtain the transfer function between the output and input voltages $V_o(s)/V_i(s)$.

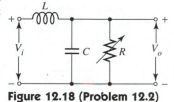

Figure 12.18 (Problem 12.2)

(b) Plot the Bode diagram for the following set of values: $L = 25$ mH, $C = 10$ μF, and $R = 50$. What type of filter does this circuit represent with these parameter values, and what is the bandwidth?

(c) Determine a setting for the variable resistance R such that the cutoff frequency or effective bandwidth is 220 Hz.

(d) What is the approximate first-order transfer function for the circuit as designed in (c)?

12.3* The mechanical speed control system shown in Fig. 12.19, uses a highly damped flywheel whose moment of inertia is adjustable in fixed units by removing or adding the indicated cylindrical segments.

(a) Obtain the transfer function between the shaft speed and the input torque $\omega(s)/T(s)$.

(b) Plot the Bode diagram for $J = 0.25$ and $\beta = 0.1$. What type of filter does this system represent with these parameter values, and what is the bandwidth?

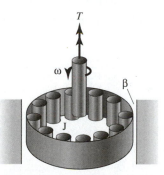

(c) If each cylinder changes **Figure 12.19 (Problem 12.3)** the flywheel moment of inertia by 0.05 unit, determine a setting for the moment of inertia J such that the cutoff frequency or effective bandwidth is 1 rad/s.

12.4 The mechanical device shown in Fig. 12.20 is used as a displacement filter.

(a) Obtain the transfer function between the displacement $X_2(s)/X_1(s)$.

(b) Plot the Bode diagram for $k_1 = 1$, $k_2 = 5$, and $\beta = 0.01$. What type of filter does this device

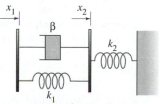

Figure 12.20 (Problem 12.4)

represent with these parameter values, and what is the bandwidth?

(c) Determine a value for the viscous damper parameter β such that the device bandwidth will be 50 Hz.

12.5 The RLC circuit shown in Fig. 12.21 is to be used as a filter.

(a) Obtain the transfer function between the output and input voltages $V_o(s)/V_i(s)$.

(b) Plot the Bode diagram for the following set of values: $R_1 = 40$ Ω, $L_1 = 5.0$ mH, and $C_1 = 20$ μF. What is the peak amplitude ratio? If bandwidth for

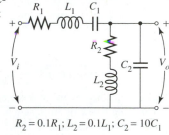

$R_2 = 0.1R_1; L_2 = 0.1L_1; C_2 = 10C_1$

Figure 12.21 (Problem 12.5)

this circuit is defined as the frequency range over which the amplitude ratio does not differ by more than 3 dB from the peak amplitude ratio, what is the bandwidth, and what type of filter does this circuit represent with these parameter values?

(c) Suppose the capacitor C_1 represents a variable condenser. Determine (if you wish, by computer assistance) a good setting for C_1 if satisfactory input voltage frequencies are in the range 50–150 Hz.

(d) What is an approximate second-order transfer function for the circuit as designed in (c)?

12.6 Determine the gain-factor compensation so that the closed-loop system will have approximately the specified bandwidth for each of the following plants:

(a) $G_p(s) = \dfrac{1}{s(1 + 0.2s)}$; $\qquad \omega_b = 0.8$ rad/s

(b) $G_p(s) = \dfrac{1}{s(1 + 0.25s)^2}$; $\qquad \omega_b = 0.5$ Hz

(c)* $G_p(s) = \dfrac{10}{(1 + s)(1 + 0.25s)^2}$; $\qquad \omega_b = 2.0$ rad/s

12.7 Determine the gain-factor compensation so that the closed-loop system will have a peak value of $M_p = 1.5$ for each of the following plants:

(a) $G_p(s) = \dfrac{1}{s(1 + 0.2s)}$ $\qquad$ **(b)** $G_p(s) = \dfrac{1}{s(1 + 0.25s)^2}$

(c) $G_p(s) = \dfrac{10}{(1 + s)(1 + 0.25s)^2}$

12.8 Consider proportional control of a plant with the transfer function

$$G_p(s) = \frac{0.5}{(s + 1)^2(s + 5)}$$

(a) Sketch the Bode diagram for the plant, $G_p(j\omega)$.

(b) Determine approximately the value of the proportional feedback control k that will give the feedback system a gain margin of 8 db.

12.9 Design a series gain-factor compensation for the following systems so that the overall phase margin is 40° or more and the overall gain margin is 8 db or more:

(a) $G_p(s) = \dfrac{10}{(1 + 2s)^2(1 + s)}$

(b)* $G_p(s) = \dfrac{10}{(s + 1)^2(0.5s + 1)}$

12.10 Consider control of the pressure on the face of the plasticizing screw of an injection molding machine. The process model is similar to that for melt temperature in plastic extrusion (see Example 12.7), except for a much shorter time constant:

$$G_p(s) = \dfrac{2}{(s + 2)^2(s + 5)}$$

Design a series gain compensation for the process such that the phase margin is 45°. What is the resulting gain margin?

12.11 Consider integral control $G_c(s) = [1/(T_i s)]$ of the plant

$$G_p(s) = \dfrac{10}{(s + 10)(s^2 + 0.2s + 1)}$$

Assuming unit feedback gain, determine the value of T_i that satisfies both the gain- and phase-margin criteria.

12.12 Show from measurement on the unit-step response of the controlled system shown in Fig. 12.3c that the design specification of $M_p = 2.2$ in Example 12.3 is satisfied.

12.13 Repeat Problem 12.7(a) using the correlation between M_p and ζ. What are the corresponding frequency at resonance, percent overshoot for a step input, closed-loop bandwidth, and phase margin?

12.14 Determine by a frequency domain method the range of k_c for the servomechanism shown in Fig. 12.22 such that the step response percent overshoot is 15% or less.

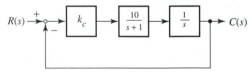

Figure 12.22 (Problem 12.14)

12.15* Design using the correlation between ζ and ω_b, the proportional control system of Example 12.5 such that $\omega_b = 0.2$. What are the corresponding resonance peak, frequency at resonance, percent overshoot for a step input, and phase margin?

12.16 Consider again the design for the specified resonance peak problem of Example 12.3. Using the transient and frequency response correlations and the second-order approximations of Gustafson, equations (12.5a) through (12.5c), obtain the proportional gain k_c that will yield approximately the given M_p. Compare the result for each approximation with the value from Example 12.3. Show that the result for the approximation (12.5b) gives rise to an unstable/nonminimum phase system.

12.17 Given a unity feedback control of the electromagnetic relay actuator of Fig. 11.10, with the open-loop transfer function

$$G_o(s) = \dfrac{k}{(s + 10)(s^2 + 6s - 16)}$$

use the second-order approximation equation (12.5b) for the closed-loop system and the correlation between transient and frequency response to

(a) Determine the value of k that will give an approximate closed-loop second-order behavior of circular frequency of 2 rad/s.

(b) Determine the value of k for the closed-loop system that will give a quarter-decay response.

Compare your results, if any, with those obtained in Practice Problem 11.40.

12.18 Repeat Problem 12.17 using the second-order approximation equation (12.5c).

12.19* For proportional control of a plant with the transfer function

$$G_p(s) = \dfrac{0.5}{(s + 1)^2(s + 5)}$$

find approximately, by frequency domain methods, the value of the proportional gain that will give the closed-loop system an approximate second-order behavior with damping ratio of $\sqrt{2}/2$.

12.20 For the system

$$\dfrac{d}{dt} x = \begin{bmatrix} 1 & -2 \\ 0 & -1 \end{bmatrix} x + \begin{bmatrix} 1 \\ 0 \end{bmatrix} u; \qquad y = [1 \quad 0]x$$

(a) Determine the eigenvalues and whether the system is stable.

(b) Discuss fully the possibilities and implications—especially with respect to pole-zero cancellation, stability, observability and controllability—of feedback control of the system by utilizing the compensator

$$G_c(s) = \dfrac{(s - 1)}{(s + 1)}$$

at each of the first four locations illustrated in Fig. 12.8. Take $H = 1$ in each case.

12.21 Design a series lead compensation for the vibration-producing device with transfer function

$$G_p(s) = \frac{1}{s(1 + 0.5s)(1 + 0.2s)}$$

such that the phase margin is $40 \pm 3°$ and the gain crossover frequency is greater than or equal to 1.0 rad/s.

12.22* Design a series lead-lead compensation for the system

$$G_p(s) = \frac{40}{s^2(s + 4)}$$

such that the phase margin is greater than 45°, and the gain margin is greater than 6 dB. Assume that the lead ratio for each lead compensator is limited to $0.05 \leq r \leq 0.25$.

12.23 Repeat Problem 12.21 using series lag compensation instead.

12.24 Design a series lag compensation for the system

$$G_p(s) = \frac{100}{s(s^2 + 3s + 50)}$$

such that the phase margin is about 40°, and the gain crossover frequency is reduced to about half the uncompensated value.

12.25 Design a lag-lead compensation for the system of Problem 12.21 such that in addition to the specifications there, the steady-state error for a unit-step input is less than or equal to 5%.

12.26 Consider position control rather than feed force control for an NC milling machine (see Fig. 11.17) with PI-control of the velocity loop such that the overall open-loop transfer function is

$$G_p(s) = \frac{5(0.1s + 1)}{s(s^2 + 2s + 1)}$$

Design a lag-lead compensation for the system such that the bandwidth is approximately the same as the uncompensated system, the phase margin is about 45°, and the open-loop gain at $\omega = 0.1$ rad/s is less than 15 db.

12.27 Repeat Example 12.10 using an analytical-plus-computational-mathematics-software-assisted approach. Your answers for k_c, T_d, and T_i should be close to but not necessarily equal to the results obtained in the example.

12.28* Consider the pollution monitoring instrument of Problem 9.27:

$$G_p(s) = \frac{e^{-0.3s}}{(1 + 0.1s)(1 + s)}$$

Design a cascade PID compensation, equation (12.18b), for this system so that the final bandwidth of the instrument is at least 1.0 rad/s. Let $\tau = 0.1T_d$, and pay attention to the eventual gain and phase margins.

12.29 For the thermal process control system shown in Fig. 12.23, find the controller settings k_c, T_i, T_d using the Bode diagram and the method described in Section 12.3 (Proportional, Reset, and Rate Compensation). Assume a desired overall gain margin of 5 db.

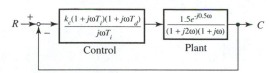

Figure 12.23 (Problem 12.29)

12.30 Consider again the robot arm–positioning system of Example 12.8. Instead of the lag compensator, determine, for the same specifications, the design of a realizable rate feedback in parallel with a unit position feedback. Assume that a series gain compensator is also applied and that $\tau = 0.05T$ for the derivative controller.

12.31 Repeat Example 12.11 using a feedback lead compensator instead of rate-plus-proportional feedback.

12.32 Consider again the third-order vibration-producing servomechanism of Example 12.11. Suppose the nonlinear component is an on/off-with-hysteresis element (see Table 10.1 for the describing function) with $M/h = 1$. Obtain the design of the realizable rate feedback such that the control system is stable without cycling for frequencies up to 1000 rad/s, and the gain margin for the linear part is as large as possible, while the phase margin is as small as possible. Note that for the locus of $-1/N$, it is necessary to consider only $-1 < h/A < 1$.

12.33* A servo system with backlash or mechanical hysteresis (assume the describing function is given by the entry in Table 10.1 for hysteresis alone) in the actuator has a linear part with the transfer function

$$G(s) = \frac{1 + s}{s^2(1 + 0.75s)}$$

Determine the design of a rate feedback controller in parallel with a unit position feedback such that the phase margin is 30°. Assume that $\tau = 0.05T$ and that no proportional band is allowed. Note that under these circumstances only the T parameter of the derivative controller is to be determined. Note also that the parameter for the nonlinearity is h/A, and the relevant values are $0 < h/A < 1$ (take $0.05 \leq h/A \leq 0.95$).

12.34 Obtain a feedback lead compensator that will meet the same phase-margin specification for the servo system of Problem 12.33.

13

Multiloop and Other Control Configurations

THE LAST TWO CHAPTERS FOCUSED EXCLUSIVELY on SISO systems. In this chapter we introduce the design of MIMO and other multiloop systems. Whereas the transfer function representation of system dynamics was sufficient for the design of SISO systems, for MIMO systems both transfer functions and state space models are useful in describing the behavior of control objects. Also, for many MIMO control systems, additional knowledge of the plant needs to be available in the controller, in the form of complete state measurement or otherwise as an algorithm—that is, an *observer*—that can estimate any unknown state. The design of *state observers* as well as the concept of *adaptive control,* which goes beyond concern for unmeasured state to accommodate other aspects of the lack of complete knowledge of the plant dynamics and inputs, are also introduced in this chapter. The chapter concludes with a brief exploration of the idea of *optimal control* and a short discussion of the H_∞ *control* problem for continuous-time systems.

We begin in Section 13.1 with *feedforward* and *cascade control* configurations that are applicable to both SISO and MIMO systems. However, by approaching these subjects from the SISO point of view, we ensure a gradual transition from the design techniques of previous chapters to the problems of this chapter. Also for *multivariable control systems,* which is the subject of Section 13.2, the *decoupling control* method, which retains much of the SISO design ideas, is considered first. The complex problem of *state vector feedback control* of multivariable systems is explored gradually in terms of simpler components: ideal state vector feedback and *eigenvalue assignment, scalar controlling input,* scalar controlling input plus integral action, and extension to *vector controlling input.* The design of state observers and adaptive control concepts is considered next, in Section 13.3, for deterministic systems. The design of both an ideal *asymptotic* observer, which estimates the entire state vector, and the *supplemental observer,* which only estimates the unmeasurable state variables, is discussed. Although there are two major approaches to the subject of adaptive control: *model reference adaptive control* (MRAC) *systems* and *self-tuning regulators* (STR), our brief introduction of this subject in Section 13.3 emphasizes MRAC, which can be illustrated fairly well within the scope of this text. Similarly, in Section 13.4, only deterministic optimal control theory is discussed, including *the linear quadratic regulator,* which is a basic feedback control law. Finally, some underlying principles of H_∞ optimal control are briefly outlined, and the concept is illustrated with an example problem.

13.1 FEEDFORWARD AND CASCADE CONFIGURATIONS

Although the SISO control system design developed in Chapters 11 and 12 was based on the canonical feedback structure (recall Fig. 11.2a), in automatic control practice, special control system configurations are sometimes used to deal more effectively with such practical problems as gross mismatch in the speed of response of distinct parts or processes in a controlled system, undesirable actuator characteristics, frequent occurrence of disturbances of large magnitudes, or very large amounts of process lags and/or dead times or to take advantage of the knowledge of the entry point into the controlled system of load disturbances or even the knowledge of the detailed nature of the disturbances themselves. One such control configuration, the *feedforward* control system, is not a closed-loop structure at all but rather an example of an *open-loop* control system as explained in Section 1.2. The other control configuration considered here, *cascade* control system (not to be confused with cascade or sequential *compensation* of the last chapter), is essentially a type of multiloop feedback structure. Although the design of feedforward and cascade control systems is illustrated here from the now-familiar SISO point of view, these are general concepts, applicable as well to MIMO systems. In fact, given that any of the feedback loops in a SISO cascade configuration incorporates some dynamic behavior, the (measured) output for such a loop effectively constitutes an independent state variable, so that an alternative interpretation of the resulting structure as a *multi-output,* that is, MIMO, system is appropriate.

Introducing Feedforward Control Systems

The object of feedback control is to keep the controlled variable close to the desired value in the presence of disturbances. Since no knowledge of the disturbance is presumed, the feedback controller merely tries, within the limits of its design and the available power in the actuators, to correct the errors introduced in the controlled variable by the disturbances. One can therefore see how the performance of a feedback controller could deteriorate in the presence of frequent (relative to the speed of signal propagation through the closed-loop system for instance) disturbances of large magnitudes. A *feedforward control system* is one that predicts the manipulative inputs that will keep the controlled variables at their desired values when *measured* disturbances enter the process (see [5]). Thus whereas feedback control determines the manipulative inputs from measurement of the controlled variables, feedforward control relies on measurement of the disturbances (see Fig. 13.1).

If the measurement of disturbances is complete and accurate and if the feedforward controller, as a predictive device, is designed on the basis of complete and accurate knowledge of the controlled process, then in terms of immunity to disturbances, feedforward control should be superior to feedback control. The performance of a feedforward controller should also be better than that of a feedback controller in terms of the speed and effectiveness of corrective action for deviations of the controlled variables from the desired values. Whereas in feedforward control the disturbance measurements and the process model are used to anticipate deviations in the controlled variables and to take corrective action, in feedback control such deviations must first occur and secondly must be detected after the conditions responsible for the deviation have propagated through all the time-delaying processes in the control loop.

Transportation Lag and Time Lag

One way to represent the amount of delay time in a control loop is to consider the *dynamic* (integral and delay) processes or equivalent *first-order lags* and *dead times* in the

Figure 13.1

Feedback and Feedforward Control Systems

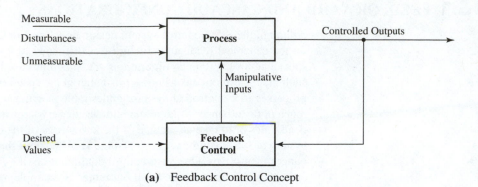

(a)   Feedback Control Concept

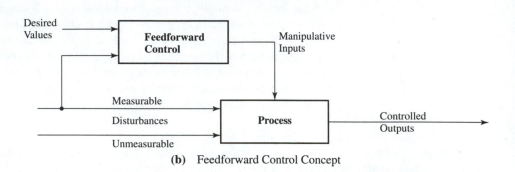

(b)   Feedforward Control Concept

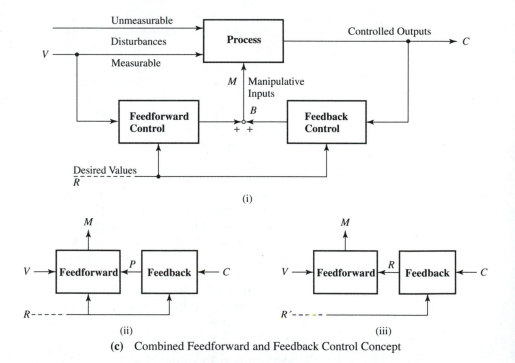

(i)

(ii)

(iii)

(c)   Combined Feedforward and Feedback Control Concept

loop. Dead time or *transportation lag,* whose model is given by the Laplace transform (time-shift property)

$$G(s) = e^{-sT}; \qquad T = \text{delay time}; \quad \text{(dead time or transportation lag)} \tag{13.1}$$

is a common phenomenon that has characterized the behavior of some processes in our earlier examples. Essentially all physical processes involve some pure- or transport-time delay between input and output. However, this behavior is nonlinear, as we have shown in previous discussions.

A single *time lag,* that is, a first-order lag, is represented by the first-order state model or the transfer function (equation (9.18a) or (12.1a) revisited)

$$G(s) = \frac{1}{Ts + 1}; \qquad T = \text{time constant}; \quad \text{(first-order lag)} \tag{13.2}$$

In general, higher-order (integral) dynamic processes can be viewed in terms of the *equivalent* number of first-order processes. Thus the time lag is inherent in most dynamic system behavior, and signal propagation through a time lag is delayed in the sense that it takes a finite time (several time constants) for the output of a first-order lag to reach its full value for a given input.

Other Advantages and Disadvantages of Feedforward Control

Figure 13.1 reveals yet another advantage as well as one major disadvantage of feedforward control. The control loop in Fig. 13.1b is open-loop, so that if the process and controller are ordinarily stable, instability cannot occur in the overall system. Recall that a stable open-loop system can become unstable under (closed-loop) feedback control. The major shortcoming of feedforward control illustrated in Fig. 13.1 is the possibility, in practice, that some significant disturbances will not be measurable and can then act on the process without an attendant corrective action. Since this is the forte of feedback controllers, a common solution to this problem is to combine, in one system, feedforward and feedback controls. The resulting control system is, of course, more expensive, and the cost must be justified with respect to the relative improvement in performance over using either of the controllers alone.

Finally, the requirement for a very accurate process model (even gradual changes in plant parameters with time must be predictable) is a disadvantage of feedforward control. Because of this requirement, exclusive feedforward controllers tend to be sophisticated and expensive, and as a result, they are often associated with systems of complex dynamic behavior where the cost of control can be justified. Again, in this case, a not-so-sophisticated feedforward controller can be combined with a feedback controller to provide mutual compensation for the shortcomings of the individual controllers. Figure 13.1c shows three possible implementations of the combined feedforward and feedback control concept. In the first case, c(i), the feedback controller can be used to bias the output of the feedforward controller, that is, the manipulated variables. In the second case, c(ii), the output of the feedback controller can be used to adjust (some parameter in) the feedforward controller. In the third case, c(iii), the feedback controller output is the setpoint for the feedforward controller. These combined controllers are further developed as exercises in Practice Problems 13.2 and 13.3.

Comparative Design Examples of Feedforward Control

The preceding discussion is illustrated next by two example problems. In the first problem, feedforward and feedback controllers are designed and compared for a two-tank level control system (see Fig. 13.2a), which represents a system that can be described quite accurately mathematically and for which the significant disturbance, in this case the supply line pressure or flow fluctuations, is completely measurable. In the second problem, the hot-water heater with remote sensing system of the last two chapters is re-examined for feedforward control. Recall (Example 11.6, Fig. 11.14b) that the performance of (proportional) feedback control for this problem was particularly bad, especially with respect to steady-state offset and speed of response, and that a PID controller was designed for the same system in Example 12.10. The factors of long delays in the control loop and inaccurate process model are also illustrated by this second example.

| **Example 13.1** | Two-Tank Level-Control System |

Consider the control of the level h_2 in the two-water-tank system shown in Fig. 13.2a. The flow through a centrifugal pump (or the capacity) can be regulated either by changing the pump (impeller) speed or by *throttling* (ideally, a constant-enthalpy pressure re-

Figure 13.2

Two-tank Level Control System

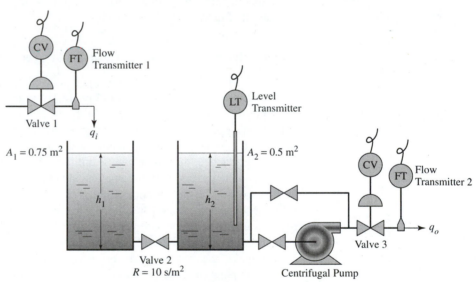

(a)   Two-tank Level Control System Components

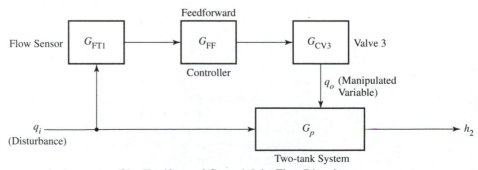

(b)   Feedforward Control, Inlet Flow Disturbance

Figure 13.2

**Two-tank Level
Control System**

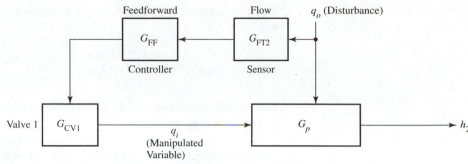

(c) Feedforward Control, Outlet Flow (Demand) Disturbance

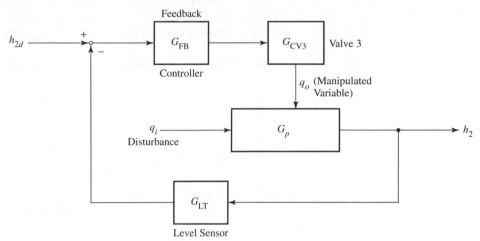

(d) Feedforward Control, Inlet Flow Disturbance

duction usually accomplished in a valve or constriction) in the discharge piping. The latter mode of regulation is assumed, and the outlet flow q_o is considered a system input. The water supply inlet flow q_i is another system input. Note that the outflow is limited to 0.002 m³/s maximum.

a. Assuming the component parameters given in Fig. 13.2a, show that the matrix transfer function between h_2 and the system inputs q_i and q_o is

$$G_p(s) = \left[\frac{R}{(1 + T_1 s)(1 + T_2 s) - 1} \quad \frac{-R(1 + T_1 s)}{(1 + T_1 s)(1 + T_2 s) - 1} \right] \quad \textbf{(13.3a)}$$

$$= \left[\frac{0.8}{s(1 + 3s)} \quad \frac{-0.8(1 + 7.5s)}{s(1 + 3s)} \right]$$

b. Design a feedforward controller, as illustrated in Fig. 13.2b, to maintain the tank level h_2 at the desired value, assuming inlet flow q_i is the disturbance input. Ignore the dynamics of the flow transducers (transmitters) and control valves and take

$$G_{CV3}(s)G_{FT1}(s) = 1.0$$

c. Design a feedback controller for h_2, as illustrated in Fig. 13.2d, by assuming simply that

$$G_{CV3}(s)G_{FB}(s) = k_c \quad \text{and} \quad G_{LT}(s) = 1.0$$

Determine the minimum value of k_c such that the largest time constant in the closed-loop system behavior is 8 s.

d. Simulate the response of the controlled system using the controllers designed in (b) and (c). Assume the initial state $h_1 = h_2 = 1$ m, and the setpoint $h_{2d} = 1$. Let the inlet flow disturbance be given by a pulse train of magnitude 0.001 and a width of 10 s.

e. Explore the design of an exact feedforward controller for the system configuration shown in Fig. 13.2c in which the outlet flow (water demand by users for instance) is the disturbance input, and the inlet flow q_i is the manipulated variable.

Solution a: The vector state equation for the two-tank system is

$$\frac{d}{dt}\begin{bmatrix} h_1 \\ h_2 \end{bmatrix} = \begin{bmatrix} -1/T_1 & 1/T_1 \\ 1/T_2 & -1/T_2 \end{bmatrix}\begin{bmatrix} h_1 \\ h_2 \end{bmatrix} + \begin{bmatrix} 1/A_1 & 0 \\ 0 & -1/A_2 \end{bmatrix}\begin{bmatrix} q_i \\ q_o \end{bmatrix} \tag{13.3b}$$

$$y = h_2 = \begin{bmatrix} 0 & 1 \end{bmatrix}\begin{bmatrix} h_1 \\ h_2 \end{bmatrix}$$

where $T_1 = RA_1$, and $T_2 = RA_2$.

Solving for the matrix transfer function, from equation (8.39b), we obtain

$$\mathbf{Y}(s) = \mathbf{G}_p(s)\begin{bmatrix} Q_i(s) \\ Q_o(s) \end{bmatrix}; \qquad \mathbf{G}_p(s) = \mathbf{C}(s\mathbf{I} - \mathbf{A})^{-1}\mathbf{B}$$

$$\mathbf{G}_p(s) = \begin{bmatrix} \dfrac{R}{(1 + T_1 s)(1 + T_2 s) - 1} & \dfrac{-R(1 + T_1 s)}{(1 + T_1 s)(1 + T_2 s) - 1} \end{bmatrix} = \begin{bmatrix} G_1(s) & G_2(s) \end{bmatrix}$$

Substituting the values for R, A_1, and A_2 from Fig. 13.2a, we obtain equation (13.3a), where

$$G_1(s) = \frac{0.8}{s(1 + 3s)} \quad \text{and} \quad G_2(s) = \frac{-0.8(1 + 7.5s)}{s(1 + 3s)}$$

Solution b: The feedforward control (Fig. 13.2b) is

$$Q_o(s) = G_{CV3}(s)G_{FF}(s)G_{FT1}(s)Q_i(s) \tag{13.4a}$$

Thus the controlled output is

$$H_2(s) = [G_1(s) + G_2(s)G_{CV3}(s)G_{FF}(s)G_{FT1}(s)]Q_i(s)$$

Complete immunity from the disturbance q_i requires

$$G_1(s) + G_2(s)G_{CV3}(s)G_{FF}(s)G_{FT1}(s) = 0 \tag{13.4b}$$

so that the "exact" feedforward controller is

$$G_{FF}(s) = \frac{-G_1(s)}{G_2(s)G_{CV3}(s)G_{FT1}(s)} = \frac{1}{1 + 7.5s} \tag{13.4c}$$

which is *realizable*. For simulation, equations (13.4c) and (13.4a) give

$$\frac{d}{dt} q_o = -\frac{q_o}{7.5} + \frac{q_i}{7.5} \tag{13.4d}$$

Solution c: The system model to use in determining the feedback controller design is

$$\frac{d}{dt} h_1 = -\frac{1}{T_1} h_1 + \frac{1}{T_1} h_2 \tag{13.5a}$$

$$\frac{d}{dt} h_2 = \frac{1}{T_2} h_1 - \frac{1}{T_2} h_2 - \frac{1}{A_2} q_o \tag{13.5b}$$

$$q_o = -k_c(h_{2d} - h_2) \tag{13.5c}$$

Note that in equation (13.5c) the outflow is defined to be positive for the negative feedback already implied in equation (13.5b). Also, there is the physical constraint (nonlinearity) on the manipulated variable, $0.002 \geq q_o \geq 0$, which however is ignored in the following determination of the minimum k_c. Using equations (13.3a) and (13.5c), we obtain the closed-loop transfer function for h_2:

$$H_2(s) = \frac{0.8k_c(1 + 7.5s)H_{2d}(s)}{s(1 + 3s) + 0.8k_c(1 + 7.5s)}$$

so that the closed-loop characteristic equation is

$$s^2 + \left(\frac{1}{3} + 2k_c\right)s + \frac{0.8}{3}k_c = 0$$

and the closed-loop eigenvalues are

$$-\left(\frac{1}{6} + k_c\right) \pm \sqrt{\left(\frac{1}{6} + k_c\right)^2 - \frac{0.8}{3}k_c}$$

For a maximum time constant of 8, the minimum decay rate is $\frac{1}{8} = 0.125$, so that

$$-\left(\frac{1}{6} + k_c\right) + \sqrt{\left(\frac{1}{6} + k_c\right)^2 - \frac{0.8}{3}k_c} = -0.125$$

Solving for k_c, we find $k_c = 1.5625$.

Solution d: Figure 13.3 shows the simulation of the response of the two-tank control system. The feedforward controller can maintain the level of tank 2 exactly at the initial value (setpoint) provided the initial state is an equilibrium state. A nonequilibrium initial state will give rise to an additional free (uncontrolled) response. This lack of

Figure 13.3

Simulation of Two-tank Level Control

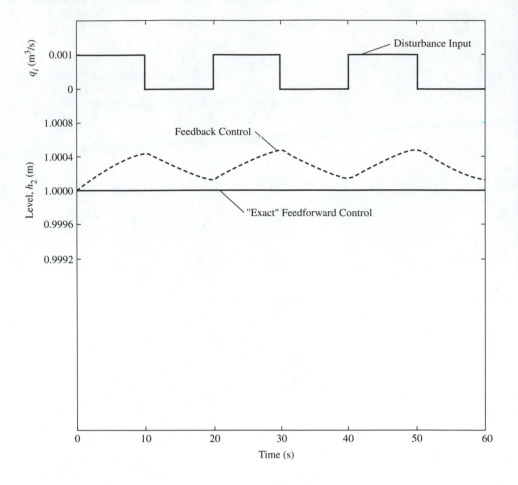

control is due to not including the setpoint in the design of the feedforward controller (see Example 13.2 for a feedforward controller design that makes use of the setpoint). Although the feedback controller is not able to keep the controlled variable exactly equal to the desired value, the magnitude of the error is small, so that feedforward control is not significantly better than feedback control for the conditions of this problem.

Solution e: The feedforward control is

$$Q_i(s) = G_{CV1}(s)G_{FF}(s)G_{FT2}(s)Q_o(s)$$

which gives

$$H_2(s) = [G_1(s)G_{CV1}(s)G_{FF}(s)G_{FT2}(s) + G_2(s)]Q_o(s) \qquad \textbf{(13.6a)}$$

so that for complete immunity of h_2 from the disturbance,

$$G_{FF}(s) = \frac{-G_2(s)}{G_1(s)G_{CV1}(s)G_{FT2}(s)} \qquad \textbf{(13.6b)}$$

If $G_{CV1}(s)G_{FT2}(s) = 1$ (constant), as before, than $G_{FF}(s) = (1 + 7.5s)$, which is not realizable. Thus an exact feedforward control may not always be feasible for a given problem.

Example 13.2	Hot-Water Heater with Remote Sensing

Design a simple feedforward controller for the hot-water heater with remote sensing system of Example 11.5 (Fig. 11.13a). Assume that the disturbance input is the ambient temperature θ_a, so that the hot-water heater model (equations (11.19a) and (11.19b)) becomes

$$\frac{d\theta}{dt} = \frac{q}{C} - \frac{\theta - \theta_a}{RC} \tag{13.7a}$$

$$\theta_{\text{sensor}}(t) = \theta(t - T) \tag{13.7b}$$

Compare the performance of the feedforward controller with that of the PID feedback controller designed in Example 12.10. Consider also the controller performance when the design model is inaccurate due to an unmeasured (disturbance) feedwater input. That is,

$$\frac{d\theta}{dt} = \frac{q}{C} - \frac{\theta - \theta_a}{RC} - \frac{W}{C}(\theta - \theta_a)$$

or

$$\frac{d\theta}{dt} = \frac{q}{C} - \left(\frac{1 + RW}{RC}\right)(\theta - \theta_a) \tag{13.7c}$$

where W is the product of the mass flowrate and specific heat of the feedwater. Let $RW = 0.25$.

Solution: A very simple and surprisingly effective feedforward controller can be obtained by considering the steady-state behavior of the how-water system. That is, from equation (13.7a),

$$\frac{q}{C} - \frac{\theta - \theta_a}{RC} = 0 \tag{13.8a}$$

Solving for the manipulated variable (q/C) and substituting the setpoint θ_d for the hot-water temperature θ, we obtain the control

$$\left(\frac{q}{C}\right) = \left(\frac{1}{RC}\right)(\theta_d - \theta_a) \tag{13.8b}$$

Note that this design assumes measurement of θ_a but ignores the effect of the time delay in the remote sensing line, whereas the design of the PID-controller (see Example 12.10 and Practice Problem 12.27) shown in Fig. 12.16 includes the effect of the time delay.

The performance of the feedforward controller is shown in Fig. 13.4a. The settling time is within about 4 sensing line delay times. However, a large offset results

Figure 13.4

Response of Hot-water Heater with Remote Sensing

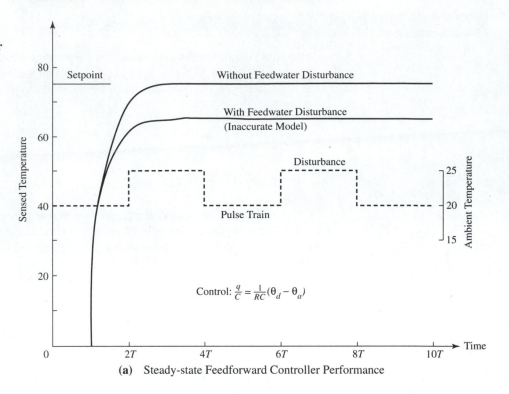

(a) Steady-state Feedforward Controller Performance

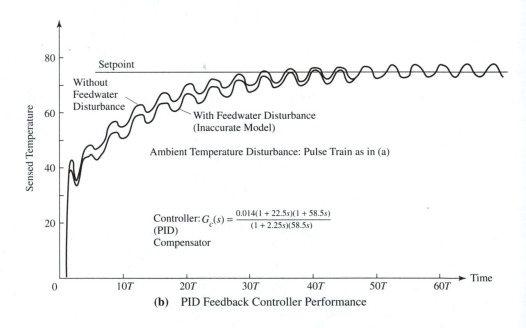

(b) PID Feedback Controller Performance

when the model used in the controller design is inaccurate. The response of the system under the PID feedback controller is simulated in Fig. 13.4b. This result suggests that the feedback controller performs better than the feedforward controller under the condition of an inaccurate process model or unmeasured disturbance. However, the settling time is in this case quite long—over 50 times the sensing line delay time—and the deleterious effect of frequent large-amplitude disturbances (sustained ambient temperature oscillations) is also quite evident in the rather large fluctuations in the value of the controlled variable. Although this slow response might be acceptable in certain processes such as a domestic hot-water heater system, it could be bad for others.

Cascade Control Systems

In a *cascade control system* such as is shown in Fig. 13.5a, a primary variable C_2 is held closer to its desired value by *interlocking* a *primary controller* with a controller for a related *secondary variable* C_1 (see [9]). Usually the interlocking of the primary and secondary variables is accomplished, as shown in the figure, by making the output of the primary (outer) or *master controller* the setpoint for the secondary (inner) or *slave controller.* It is easy to deduce the two major advantages and hence application of cascade control by observing the controller structures shown in Figs. 13.5a and 13.5b.

Mismatch in Response Speeds

Consider first the problem of mismatch in the speeds of response of different parts of the system. Suppose the dynamics of process I are faster than those of process II. In single-loop feedback control system practice, illustrated by Fig. 13.5b, this problem can manifest itself by premature (that is prior to C_2's coming close to the desired value) saturation in process I actuators, thereby rendering the overall control system erratic and not very effective. With cascade control it is possible to make the dynamics of the inner loop enclosing process I be even faster, that is, to approach a *static* gain. Note that the fastest process, in this context, is a static one, since in that case the present output is determined immediately by the present input. The result is the apparent elimination of process I dynamics from the overall system and hence better performance of the total system. We can demonstrate this conclusion by looking at the closed-loop transfer function of the system output C_2 relative to the setpoint R, under cascade control. This result is

$$\frac{C_2}{R} = \frac{G_{p2}G_3G_{cm}}{1 + G_{p2}G_3G_{cm}H_2} \tag{13.9a}$$

where G_3 is the transfer function for the secondary loop by itself:

$$G_3 = \frac{G_{p1}G_{cs}}{1 + G_{p1}G_{cs}H_1} \tag{13.9b}$$

Ordinarily, the speed of the secondary loop can be increased by increasing its open-loop gain. Since this must imply $|G_{cs}| \gg 1$, equation (13.9b) approaches

$$G_3 \approx \frac{1}{H_1} \tag{13.9c}$$

so that the closed-loop transfer function, equation (13.9a), approaches equation (13.9d),

$$\frac{C_2}{R} \approx \frac{G_{p2}(1/H_1)G_{cm}}{1 + G_{p2}(1/H_1)G_{cm}H_2} \tag{13.9d}$$

which does not depend on process I dynamics (G_{p1}).

Figure 13.5

**Cascade Control and
Single-loop Feedback Control**

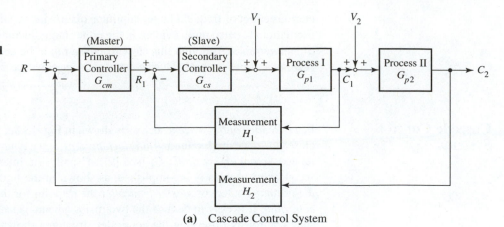

(a)   Cascade Control System

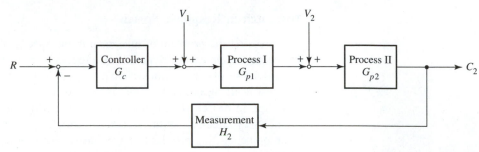

(b)   Single-loop Feedback Control System

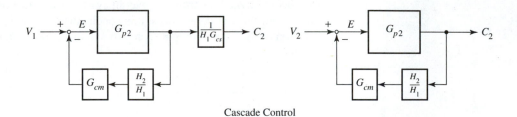

Cascade Control

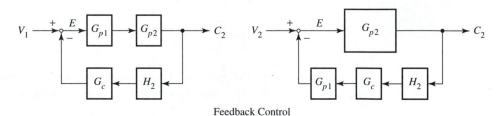

Feedback Control

(c)   Effect of Load Disturbance

Point of Entry of Major Disturbances

The second major application of cascade control occurs when most of the significant
disturbances enter the system through the secondary loop. Under single-loop feedback
control system design (Fig. 13.5b), such disturbances remain "uncompensated" and
free to upset the system. The controller merely tries to correct the errors introduced in

the controlled variable C_2 by these load upsets. However, with the cascade control system shown in Fig. 13.5a, the secondary loop can be sufficiently *desensitized* to the disturbance inputs to free the primary controller to do a better job of maintaining the output C_2 near the desired value. It can be shown that even where the major disturbances enter the outer loop (such as V_2 in Fig. 13.5), cascade control can still provide some improvement in performance.

Consider the transfer function relating the controlled output C_2 to the disturbance input V_1. For the cascade control system we have

$$\frac{C_2}{V_1} = \frac{G_{p2}G_{p1}}{1 + G_{cs}(G_{p1}H_1 + G_{p2}G_{p1}G_{cm}H_2)} \tag{13.10a}$$

and using equations (13.9b) and (13.9c), we have

$$\frac{C_2}{V_1}_{[cascade]} \approx \left[\frac{G_{p2}}{1 + G_{p2}G_{cm}(H_2/H_1)} \right] \frac{1}{H_1 G_{cs}} \tag{13.10b}$$

For the single-loop (feedback control) system,

$$\frac{C_2}{V_1}_{[single-loop]} = \frac{G_{p2}G_{p1}}{1 + G_{p2}G_{p1}G_cH_2} \tag{13.10c}$$

Similarly, the transfer function between the output C_2 and the disturbance input V_2 is given by

$$\frac{C_2}{V_2}_{[cascade]} = \frac{G_{p2}}{1 + G_{p2}G_3G_{cm}H_2} \approx \frac{G_{p2}}{1 + G_{p2}G_{cm}(H_2/H_1)} \tag{13.11a}$$

$$\frac{C_2}{V_2}_{[single-loop]} = \frac{G_{p2}}{1 + G_{p2}G_{p1}G_cH_2} \tag{13.11b}$$

The block diagrams representing equations (13.10b), (13.10c), (13.11a), and (13.11b) are shown in Fig. 13.5c. It is evident that the single-loop control circuits incorporate more process time lags (and delays) in the feedback path, so that changes in C_2 due to a disturbance input are delayed more, thereby maintaining the error E at a larger value. Further, for the case of V_1 (that is, the disturbance input falls within the inner loop), the cascade control system effectively contains the extended subunit $1/(H_1 G_{cs})$, which can greatly attenuate the effect of V_1 on C_2. Complete immunity of the control system to the disturbances entering the secondary loop requires zero coupling between V_1 and C_2, or $C_2/V_1 = 0$. From equation (13.10b), this situation is approached in the limit that

$$|G_{cs}| \longrightarrow \infty \tag{13.12}$$

This requirement on G_{cs} is also consistent with the solution for the case of a mismatch in the speeds of response of the inner and outer loops. In practice, G_{cs} is a simple controller such as a proportional gain of a sufficiently large and feasible magnitude.

Example 13.3

Undamped Servomechanism with Speed Mismatch

Consider the undamped pendulum system of Fig. 13.6a. the actuator characteristics are given by the block diagram in Fig. 13.6b, the open-loop step response data shown in Fig. 13.6c, and a physical limitation of actuator travel to $+0.52$ and -0.52 in. from the neutral (vertical) position. Design and simulate a single-loop PID-controller for the pendulum system that will move the pendulum from an initial state to a final specified state in minimum time. Let the performance criterion be the settling time with 10% tolerance on the final value of the response. Assume that the "PID"-controller is of the form

$$G_c(s) = \left[\frac{1}{T_i s} + \left(\frac{Ts + 1}{\tau s + 1} \right) \right] k_c$$

Note that the proportional part of this controller is essentially a gain compensator.

Figure 13.6

Undamped Pendulum Servomechanism

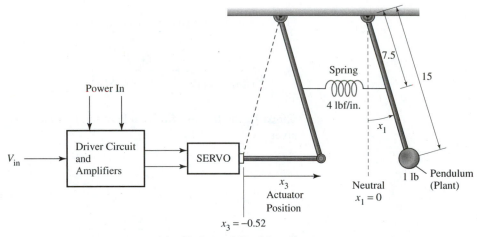

(a)   Undamped Pendulum System

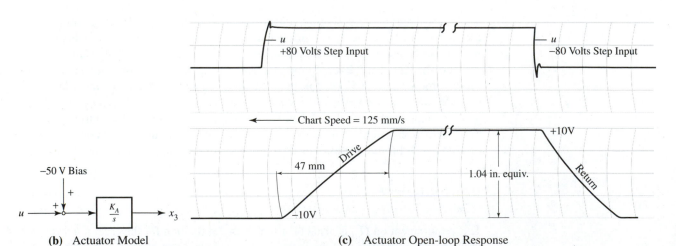

(b)   Actuator Model

(c)   Actuator Open-loop Response

Solution: The pendulum system model (see Section 3.2) is as follows:

$$\dot{x}_1 = x_2 \tag{13.13a}$$

$$\dot{x}_2 = -411.73x_1 + 25.73x_3 \tag{13.13b}$$

where x_1 is the pendulum position in radians, and x_3 is the actuator position in inches. This model gives a natural system frequency of $\sqrt{411.73}/2\pi = 3.229$ Hz.

From the given actuator open-loop step-input-test response data (see Fig. 13.6c), it is evident that the actuator has different drive and return speeds. Using the drive slope for an approximate model, we see that the actuator covered 1.04 in. $(0.52 - (-0.52) = 1.04)$ in 0.376 s ([47 mm]/[125 mm/s] = 0.376 s) under a step input of 80 V. Hence (see block diagram of Fig. 13.6b),

$$(80 - 50)K_A = \frac{1.04}{0.376} = 2.766 \text{ in./s} \tag{13.14a}$$

$$K_A = \frac{(2.766 \text{ in./s})}{(30 \text{ V})} = 0.092 \text{ in./V-s} \tag{13.14b}$$

If we consider the pendulum subsystem alone, the angular position of this harmonic oscillator is given by

$$x_{1\text{free}}(t) = A \cos\left(\sqrt{411.73}t\right)$$

where the pendulum is assumed to be initially at rest, and A depends on the initial position. The corresponding horizontal linear velocity of the pendulum is

$$x_{2\text{free}}(t) = -A\sqrt{411.73} \sin\left(\sqrt{411.73}t\right) \text{ rad/s}$$

which gives approximately

$$x_{2\text{free}}(t) \approx -304.37 \, A \sin\left(\sqrt{411.73}t\right) \text{ in./s} \tag{13.15}$$

The magnitude of this velocity varies with the pendulum position and can range from a very small (zero) value to a large value, possibly several times the actuator velocity (of 2.766 in./s). This mismatch in the speeds of the actuator and pendulum subsystems recommends cascade control for this problem.

Cascade control: Figure 13.7a shows the block diagram of a simple proportional feedback control system for the actuator as the inner or secondary loop of the overall cascade control system. Practical considerations such as the measurement or feedback element (potentiometer) and isolating amplifiers have been included. Figures 13.7b and 13.7c show the computer simulation of the step-input and ramp-input responses of the actuator under this control. These results illustrate the computer-assisted design of the proportional controller. Ideally, the closed-(slave) loop behavior should approach a static gain, so that the step-input response in Fig. 13.7b should approach a step (this re-

Figure 13.7

Secondary (Actuator) Control Loop

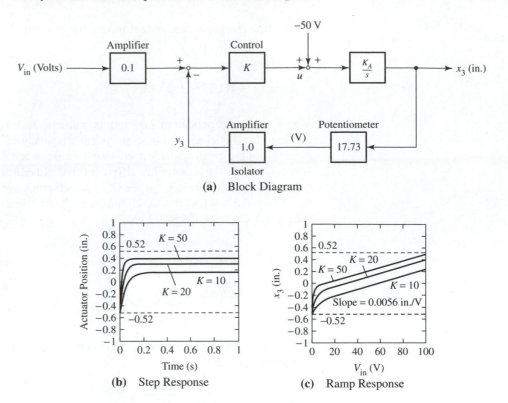

(a)   Block Diagram

(b)   Step Response

(c)   Ramp Response

quires $K \longrightarrow \infty$), while the ramp-input results of Fig. 13.7c should approach a straight line of slope (see equation (13.9c))

$$\frac{x_3}{V_{in}} = \frac{(0.1)}{(17.73)(1)} = 0.00564 \text{ in./V}$$

Although Fig. 13.7b suggests using a very high proportional control gain, Fig. 13.7c indicates that a smaller gain could be adequate given other possible practical considerations, such as saturation and limited range of trouble-free travel in the actuator. At any rate, there is a limit to how much the overall controlled system can be made insensitive to the inner-loop dynamics.

   If we assume a proper choice of K has been made and the ideal behavior of the actuator under control can be presumed, then Fig. 13.8a shows the block diagram representation of the overall controlled system with the actuator dynamics reduced to a mere gain $K_A = 0.0056$ in./V. Figure 13.8b gives the simulation of the step-input response of this controlled system for various choices of the primary controller parameters. The desired final state in all cases is $x_1 = 0.02$ rad. The initial state was $x_1(0) = -0.03$ rad. Apparently, considering the system performance specification, the "best" combination of primary controller parameters is given by settings I:

$$k_c = 2, \quad T_i = 0.01, \quad \tau = 0.01, \quad \text{and } T = 1.01$$

This result is based on a limited and ad hoc search and is only illustrative. The formal determination of optimal control for continuous-time systems is one of the subjects introduced in Section 13.4.

Figure 13.8

Performance of Overall Controlled System

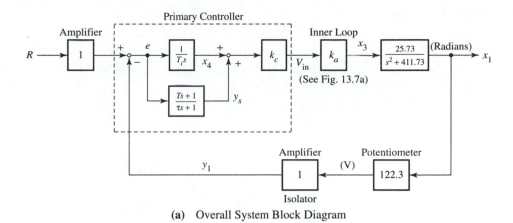

(a) Overall System Block Diagram

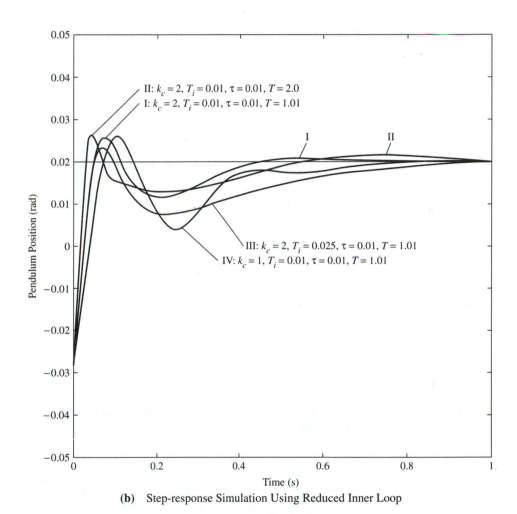

(b) Step-response Simulation Using Reduced Inner Loop

Figure 13.8c shows the response of the original system under this master controller but with the full actuator model restored, that is, according to Fig. 13.7a and the choice $K = 50$. Although the response obtained is not as "good" (particularly with respect to the overshoot) as that given by the reduced model in Fig. 13.8b, about the same

Figure 13.8 *continued*

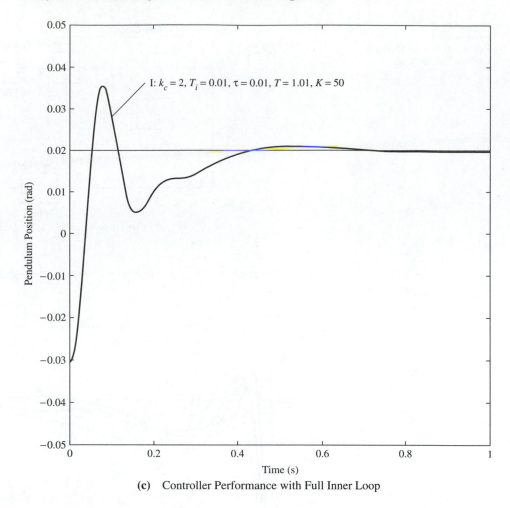

I: $k_c = 2$, $T_i = 0.01$, $\tau = 0.01$, $T = 1.01$, $K = 50$

(c)   Controller Performance with Full Inner Loop

settling time of around 0.5 s is attained in each case. In practice, this design could serve as the starting point for fine-tuning the controller.

Tuning the Cascade Controller

Note that the eventual (physical) inner-loop proportional control gain K can be determined during the realization of the primary controller. The design process adopted in this example can be viewed as one possible solution to the difficult problem of tuning the multiple (and dependent) controllers in a cascade control arrangement. The process has the following steps:

1. By simulating the secondary loop only, determine the controller design that gives a good "reduced" secondary loop behavior.
2. Using the reduced secondary loop model, determine by simulation of the overall system, or otherwise, the optimum primary loop controller settings.
3. Implement the master controller determined in (2) and tune the slave controller to give a good response (in terms of the simulation result in (1) or in terms of setpoint-following ability).
4. Freeze the slave controller settings and then fine-tune the master controller.

13.2 MULTIVARIABLE CONTROL SYSTEMS

Control systems having more than one input and/or output are known as *multivariable control systems*. For instance, most cascade control systems that can be designed from the scalar-variable, single-loop point of view, as was done in the last section, are actually also multivariable systems. This is true because the secondary variable for the non-trivial cases is usually an independent state variable of the system. Hence the secondary and primary variables can represent two system outputs. Consider the oscillating pendulum system shown in Figs. 13.7a and 13.8a. The secondary variable x_3 is given by

$$\frac{d}{dt} x_3 = K_A(u - 50) \tag{13.16a}$$

$$u = K[0.1k_c(x_4 + y_5) - 17.73x_3] \tag{13.16b}$$

$$\frac{d}{dt} x_4 = \left(\frac{1}{T_i}\right)(R - 122.3x_1) \tag{13.16c}$$

$$y_5 = x_5 + \left(\frac{T}{\tau}\right)(R - 122.3x_1) \tag{13.16d}$$

and

$$\frac{d}{dt} x_5 = \left(\frac{1}{\tau}\right)(R - 122.3x_1 - y_5)$$

or

$$\frac{d}{dt} x_5 = \left(\frac{1}{\tau}\right)\left[122.3\left(\frac{T}{\tau} - 1\right)x_1 - x_5 - \left(\frac{T}{\tau} - 1\right)R\right] \tag{13.16e}$$

Equation (13.16b) represents multivariable control. Indeed, combining equations (13.16b) and (13.16d) we get

$$u = K\left[0.1k_c\left(x_4 + x_5 + \frac{T}{\tau}R - 122.3\frac{T}{\tau}x_1\right) - 17.73x_3\right]$$

or what is in effect *state vector feedback* (to be discussed shortly):

$$u = \left[0.1Kk_c\left(\frac{T}{\tau}\right)R\right] - [k_1x_1 + k_3x_3 + k_4x_4 + k_5x_5] \tag{13.17}$$

where $k_1 = 12.23Kk_cT/\tau$, $k_3 = 17.73K$, and $k_4 = k_5 = -0.1Kk_c$.

In equation (13.17), the two state variables x_4 and x_5 were introduced by the control (recall that feedback control alters the system's dynamic structure), whereas x_1 and x_3 are the original system "outputs."

The Concept of Decoupling Control

Perhaps a logical first step in our study of multivariable control systems would be to try to apply to the new type of control system the design methods (see Chapters 11, 12, and the earlier parts of this chapter) with which we are already familiar. As shown in Fig. 13.9a, by replacing scalar variables with vector variables and scalar transfer functions with matrix transfer functions, we can represent linear multivariable or MIMO

Figure 13.9

Multivariable Control System

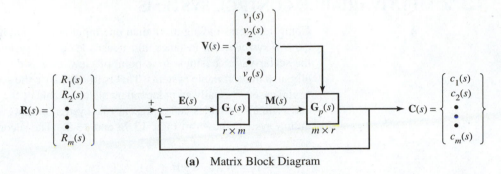

(a) Matrix Block Diagram

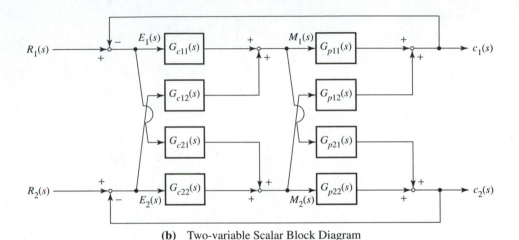

(b) Two-variable Scalar Block Diagram

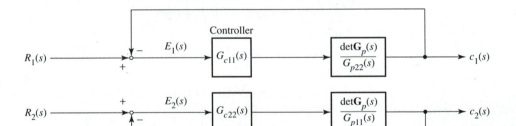

(c) Decoupled Two-variable Control System

control systems in the same block-diagram form as single-variable, single-loop or SISO feedback control systems (see [10]). The block-diagram algebra is also essentially the same provided we take care to preserve the order of multiplication of matrices, since the multiplication of matrices is, in general, noncommutative.

We obtain the closed-loop matrix transfer function $\mathbf{G}_{CL}(s)$ for Fig. 13.9a as follows:

$$\mathbf{C}(s) = \mathbf{G}_p(s)\mathbf{M}(s) = \mathbf{G}_p(s)\mathbf{G}_c(s)\mathbf{E}(s)$$

where

$$\mathbf{E}(s) = \mathbf{R}(s) - \mathbf{C}(s)$$

If we define the open-loop matrix transfer function

$$\mathbf{G}_o(s) = \mathbf{G}_p(s)\mathbf{G}_c(s) \tag{13.18a}$$

then

$$\mathbf{C}(s) = \mathbf{G}_o(s)[\mathbf{R}(s) - \mathbf{C}(s)] \quad \text{or} \quad [\mathbf{I} + \mathbf{G}_o(s)]\mathbf{C}(s) = \mathbf{G}_o(s)\mathbf{R}(s)$$

so that

$$\mathbf{G}_{\mathrm{CL}}(s) = [\mathbf{I} + \mathbf{G}_o(s)]^{-1}\mathbf{G}_o(s) \tag{13.18b}$$

Consequently, the characteristic equation of the closed-loop system is equation (6.68), repeat here:

$$\det[\mathbf{I} + \mathbf{G}_o(s)] = 0 \tag{13.18c}$$

Note the apparent similarity between equations (13.18b) and (13.18c) and the corresponding scalar equations (11.2b) and (11.3), respectively. The design by classical techniques of the controller $\mathbf{G}_c(s)$ from equations (13.18a) through (13.18c) is however not straightforward due to the complexity of the scalar components of these equations.

Consider a two-input ($r = 2$), two-output ($m = 2$) multivariable plant. Let

$$\mathbf{G}_p(s) = \begin{bmatrix} G_{p11}(s) & G_{p12}(s) \\ G_{p21}(s) & G_{p22}(s) \end{bmatrix} \quad \text{and} \quad \mathbf{G}_c(s) = \begin{bmatrix} G_{c11}(s) & G_{c12}(s) \\ G_{c21}(s) & G_{c22}(s) \end{bmatrix}$$

Then from equation (13.18a),

$$\mathbf{G}_o(s) = \begin{bmatrix} G_{11}(s) & G_{12}(s) \\ G_{21}(s) & G_{22}(s) \end{bmatrix} \tag{13.19a}$$

$$= \begin{bmatrix} \{G_{p11}G_{c11} + G_{p12}G_{c21}\} & \{G_{p11}G_{c12} + G_{p12}G_{c22}\} \\ \{G_{p21}G_{c11} + G_{p22}G_{c21}\} & \{G_{p21}G_{c12} + G_{p22}G_{c22}\} \end{bmatrix}$$

and equation (13.18c) becomes

$$[1 + G_{p11}G_{c11} + G_{p12}G_{c21}][1 + G_{p21}G_{c12} + G_{p22}G_{c22}]$$
$$- [G_{p11}G_{c12} + G_{p12}G_{c22}][G_{p21}G_{c11} + G_{p22}G_{c21}] = 0 \tag{13.19b}$$

which is quite complex. Further insight into this problem is revealed by the scalar block-diagram representation of the two-variable control system in Fig. 13.9b. Here the complexity in the scalar equations is manifested in the (cross) *coupling* between the two feedback loops. Recall that this was also the problem in tuning cascade controllers.

Ideal Independent Control

The concept of *decoupling control* is to **simplify the design problem** by severing the coupling between the control loops (using elements of the controller known as *decouplers*) to achieve independent control. *Ideal independent control* can be defined as a situation in which the ith output $c_i(t)$ is a function of the ith reference input $R_i(t)$ **only**. This is possible if all the off-diagonal elements of the $\mathbf{G}_o(s)$ matrix are zero. To see this, note that if $\mathbf{G}_o(s)$ is diagonal, then so is $\mathbf{I} + \mathbf{G}_o$ and $[\mathbf{I} + \mathbf{G}_o]^{-1}$ and hence $[\mathbf{I} + \mathbf{G}_o]^{-1}\mathbf{G}_o$. By equation (13.18b), $\mathbf{G}_{CL}(s)$ is therefore diagonal, and the closed-loop system is *decoupled*. Setting the off-diagonal elements of $\mathbf{G}_o(s)$ to zero gives the design for the decouplers or *compensators*. From equation (13.19a), for example,

$$G_{p11}(s)G_{c12}(s) + G_{p12}(s)G_{c22}(s) = 0 \quad \text{and} \quad G_{p21}(s)G_{c11}(s) + G_{p22}(s)G_{c21}(s) = 0$$

give

$$G_{c12}(s) = \frac{-G_{p12}(s)G_{c22}(s)}{G_{p11}(s)} \tag{13.20a}$$

$$G_{c21}(s) = \frac{-G_{p21}(s)G_{c11}(s)}{G_{p22}(s)} \tag{13.20b}$$

Substituting equations (13.20a) and (13.20b) into equation (13.19a), we obtain the decoupled open-loop transfer function elements

$$G_{11}(s) = \frac{det\mathbf{G}_p(s)G_{c11}(s)}{G_{p22}(s)} \tag{13.20c}$$

$$G_{22}(s) = \frac{det\mathbf{G}_p(s)G_{c22}(s)}{G_{p11}(s)} \tag{13.20d}$$

The choice of $G_{c12}(s)$ and $G_{c21}(s)$ in equations (13.20a) through (13.20d) as the decouplers leaves $G_{c11}(s)$ and $G_{c22}(s)$ as the main controllers, which can then be designed by our previous (scalar) techniques. The scalar block diagram of the decoupled second-order example is shown in Fig. 13.9c. Obviously, given our discussion in Section 12.2 about pole-zero cancellation, the true effectiveness of the decoupling control system can be established properly only by a **complete** simulation of the control system or from an actual implementation.

Example 13.4	Control of a Continuous Stirred Tank Reactor

Consider again the well-mixed CSTR of Example 4.7. The schematic description in Fig. 4.11 is repeated as Fig. 13.10. The reaction is first order: that is, A (reactant) $\longrightarrow$ B (product), and the rate of disappearance of A (creation of product) $= kx_1$. To simplify the present model, assume that the rate constant k is independent of temperature and that the densities and specific heats of reacting mixtures are constant and the same in the feed stream and in the reactor.

Figure 13.10

Stirred Chemical Reactor

w = Reactor Feedrate, kg/s ($w = 5$)

v = Feed Temperature, K($v = 300$)

u_1 = Concentration of Reactant (A) in Feed Stream, kg/m^3

u_2 = Rate of Heat Input kW

x_1 = Reactor Concentration (A), kg/m^3

x_2 = Reactor Temperature, K

ρ = Density of Reacting Mixture, kg/m^3 ($\rho = 1200$)

c = Specific Heat of Reacting Mixture, kJ/kg·K ($c = 3$)

H = Heat of (Endothermic) Reaction, kJ/kg (of A Consumed) ($H = 2000$)

k = Reaction Rate Constant, kg/m^3·sec ($k = 0.02$)

V = Reactor Volume, m^3 ($V = 5$)

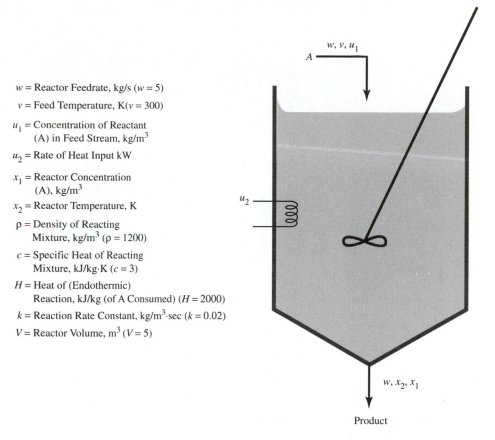

a. Derive a multivariable model for the reactor in which the state variables are x_1 and x_2; the manipulated inputs are u_1 and u_2; and the disturbance input is v, the feed temperature.

b. Design a decoupling multivariable control system for the reactor consisting of appropriate (decoupling) compensators and P- and PI-controllers for the reactor concentration and temperature, respectively. Assume that the desired reactor performance is such that following a setpoint change, all outputs converge to within 5% of the reference values in 135 s.

c. Simulate the transient behavior of the overall control system subject to the initial state $x_1 = 0$ kg/m^3 and $x_2 = 295$ K, and the corresponding reference state, $R_1 = 60$ kg/m^3 and $R_2 = 350$ K.

Solution a: Under the stated assumptions, the simplified solution developed in Example 4.7 is adequate. The state model is thus given by equations (4.32a) through (4.32c), repeated here:

$$\frac{d}{dt}x_1 = -kx_1 + \frac{w}{\rho V}(u_1 - x_1) \tag{13.21a}$$

$$\frac{d}{dt}x_2 = -\frac{Hk}{\rho c}x_1 + \frac{w}{\rho V}(v - x_2) + \frac{u_2}{\rho Vc} \tag{13.21b}$$

$$c_1 = x_1 \text{ and } c_2 = x_2 \tag{13.21c}$$

These results give the vector state equation (13.22a) and output equation (13.22b):

$$\frac{d}{dt}\begin{bmatrix} x_1 \\ x_2 \end{bmatrix} = \begin{bmatrix} -\left(k + \dfrac{w}{\rho V}\right) & 0 \\ -\left(\dfrac{Hk}{\rho c}\right) & -\left(\dfrac{w}{\rho V}\right) \end{bmatrix}\begin{bmatrix} x_1 \\ x_2 \end{bmatrix}$$

$$+ \begin{bmatrix} \left(\dfrac{w}{\rho V}\right) & 0 \\ 0 & \left(\dfrac{1}{\rho Vc}\right) \end{bmatrix}\begin{bmatrix} u_1 \\ u_2 \end{bmatrix} + \begin{bmatrix} 0 \\ \left(\dfrac{w}{\rho V}\right) \end{bmatrix} v \tag{13.22a}$$

$$\begin{bmatrix} c_1 \\ c_2 \end{bmatrix} = \begin{bmatrix} x_1 \\ x_2 \end{bmatrix} \tag{13.22b}$$

If we ignore the disturbance input, the plant transfer function is given by equation (8.39b):

$$\mathbf{G}_p(s) = (s\mathbf{I} - \mathbf{A})^{-1}\mathbf{B} = \begin{bmatrix} G_{p11}(s) & G_{p12}(s) \\ G_{p21}(s) & G_{p22}(s) \end{bmatrix}$$

where

$$G_{p11}(s) = \frac{w}{w + \rho V(s + k)} \tag{13.23a}$$

$$G_{p12}(s) = 0 \tag{13.23b}$$

$$G_{p21}(s) = \frac{HVkw}{[w + \rho V(s + k)](\rho Vcs + wc)} \tag{13.23c}$$

$$G_{p22}(s) = \frac{1}{\rho Vcs + wc} \tag{13.23d}$$

Solution b: The controllers to be designed are $G_{c11}(s) = k_{c1}$ and $G_{c22}(s) = k_{c2}[1 + 1/(T_i s)]$. According to (13.20a), since $G_{p12}(s) = 0$, only one compensator is needed. That is, from equation (13.20b),

$$G_{c21}(s) = \frac{-k_{c1}HVkw}{w + \rho V(s + k)} \tag{13.24a}$$

Note that this decoupler is realizable. From equation (13.19a), the decoupled open-loop transfer function is

$$\mathbf{G}_o(s) = \begin{bmatrix} \dfrac{wk_{c1}}{w + \rho V(s + k)} & 0 \\ 0 & \dfrac{k_{c2}[1 + 1/(T_i s)]}{\rho Vcs + wc} \end{bmatrix}$$

so that the decoupled closed-loop relations are

$$\frac{C_1(s)}{R_1(s)} = \frac{\dfrac{wk_{c1}}{w + \rho V(s + k)}}{1 + \dfrac{wk_{c1}}{w + \rho V(s + k)}} \quad \text{and} \quad \frac{C_2(s)}{R_2(s)} = \frac{\dfrac{k_{c2}[1 + 1/(T_i s)]}{\rho Vcs + wc}}{1 + \dfrac{k_{c2}[1 + 1/(T_i s)]}{\rho Vcs + wc}}$$

The corresponding closed-loop characteristic equations (one for each decoupled loop) are thus

$$\rho Vs + wk_{c1} + w + \rho Vk = 0 \quad \text{or} \quad 6000s + 5k_{c1} + 5 + 120 = 0 \qquad \textbf{(13.24b)}$$

and

$$\rho VcT_i s^2 + (wc + k_{c2})T_i s + k_{c2} = 0 \quad \text{or}$$

$$18000 T_i s^2 + (15 + k_{c2})T_i s + k_{c2} = 0 \qquad \textbf{(13.24c)}$$

If we interpret the controlled system performance specification to imply that all closed-loop time constants must be less than 45 s, (that is, assuming sufficient (95%) convergence in three time constants; note that $1 - e^{-3} = 0.95$), equation (13.24b) gives

$$\frac{5k_{c1} + 125}{6000} > \frac{1}{45} \quad \text{or} \quad k_{c1} > 1.667 \qquad \textbf{(13.24d)}$$

Recognizing that proportional control is subject to offset, and noting from equation (13.21a) that the reactor concentration loop is a simple first-order system, we can obtain further conditions on k_{c1} by looking at the equilibrium concentration. Substituting $u_1 = k_{c1}(R_1 - x_1)$ into equation (13.21a) and setting the left-hand side to zero, we obtain

$$0 = -\left[k + \frac{w}{\rho V}(1 + k_{c1})\right]x_1 + \frac{w}{\rho V}k_{c1}R_1 \quad \text{or} \quad x_1 = \left[\frac{k_{c1}}{\left(1 + \dfrac{\rho Vk}{w} + k_{c1}\right)}\right]R_1$$

so that to minimize the difference between x_1 and R_1 at equilibrium,

$$k_{c1} \gg 1 + \frac{\rho Vk}{w} = 25$$

For a 5% error in the final value of x_1,

$$k_{c1} > \frac{(0.95)(25)}{(0.05)} = 475 \qquad \textbf{(13.24e)}$$

To find k_{c2} and T_i, we can use the Routh array method as in Example 11.7. Putting $s = (p - 1/45)$ in equation (13.24c), we obtain the resulting characteristic polynomial equation in p:

$$(18000 T_i)p^2 + (k_{c2} - 785)T_i p + [(385 - k_{c2})/45]T_i + k_{c2} = 0$$

and the Routh array is

$$18000T_i \qquad \left(\frac{385 - k_{c2}}{45}\right)T_i + k_{c2}$$

$$(k_{c2} - 785)T_i \qquad\qquad 0$$

$$\left(\frac{385 - k_{c2}}{45}\right)T_i + k_{c2} \qquad 0$$

The stability conditions are thus

$$k_{c2} > 785 \quad \text{and} \quad T_i < \frac{45k_{c2}}{k_{c2} - 385} \tag{13.24f}$$

Solution c: The equations to simulate are still those of the **total (coupled) system.** That is (see Fig. 13.9b), equations (13.21a) and (13.21b) with the following control equations:

$$u_1 = k_{c1}(R_1 - x_1) \tag{13.25a}$$

and

$$U_2(s) = G_{c21}(s)E_1(s) + G_{c22}(s)E_2(s)$$

Substituting for $G_{c21}(s)$ from equation (13.24a) and for $G_{c22}(s)$ (that is PI-action), we take

$$U_2(s) = \frac{-1000k_{c1}}{6000s + 125}E_1(s) + \frac{k_{c2}}{T_i s}E_2(s) + k_{c2}E_2(s) \tag{13.25b}$$

We then transform equation (13.25b) to state equation format (see Signal Flow Graphs, Section 9.3):

$$u_2 = x_3 + x_4 + k_{c2}(R_2 - x_2) \tag{13.25c}$$

$$\frac{d}{dt}x_3 = -\frac{k_{c1}}{6}(R_1 - x_1) - 0.02083x_3 \tag{13.25d}$$

$$\frac{d}{dt}x_4 = \frac{k_{c2}}{T_i}(R_2 - x_2) \tag{13.25e}$$

Figure 13.11 shows the simulation result for the following choice of controller settings that satisfy equations (13.24e) through (13.24f): $k_{c1} = 500$, $k_{c2} = 786$, $T_i = 74$.

The performance, as can be determined from the figure, is fairly good.

....

State Vector Feedback and Eigenvalue Assignment

For many systems, decoupling multivariable control is generally not as satisfactory as Example 13.4 would suggest. A major reason is that this design approach neglects the system's internal state, with the result that system controllability/observability can be lost due to pole-zero cancellation in the decoupling compensator (see Section 12.2). Also, since the decouplers are more or less fixed by the plant dynamics, these controllers can be **unrealizable** or even **unstable.** Another possible disadvantage of *design simplification through diagonalization of the open-loop transfer function* is that the interaction that is lost between the control loops could, if retained, make for better or more effective control (see [11]).

Figure 13.11

Decoupling Multivariable Controller Performance

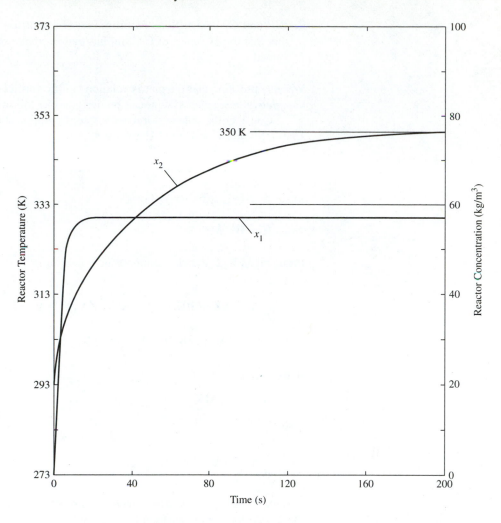

A multivariable control system design technique that retains the interaction between controlled variables is the *state vector feedback* (SVF) method. As the name implies, in this technique each control input is a function of the state vector. This is a direct approach to multivariable control system design that, in the general case, can be an ambitious and impractical undertaking. We reduce the problem complexity by imposing certain restrictions on the system.

Ideal Case

Consider a linear nth-order plant, equations (13.26a) and (13.26b):

$$\frac{d}{dt}\mathbf{X} = \mathbf{AX} + \mathbf{BU} \tag{13.26a}$$

$$\mathbf{Y} = \mathbf{CX} \tag{13.26b}$$

with the following restrictions:

1. The control vector $\mathbf{U}$ and the output vector $\mathbf{Y}$ are both n-vectors. That is, there are as many control variables and output variables as there are state variables.

2. $\mathbf{B}^{-1}$ and $\mathbf{C}^{-1}$ exist. Note that from (1) these matrices are ($n \times n$) square. Note further that the existence of $\mathbf{C}^{-1}$ implies that *complete state vector measurement* is assumed.

We also presume that the matrix relations in the forthcoming discussion are sparse. A *completely decoupled* system can be designed as follows:

Consider the state vector feedback configuration shown in Fig. 13.12. Assume for simplicity that $\mathbf{C} = \mathbf{I}$. Find $\mathbf{H}$ such that

$$\mathbf{BH} = \mathbf{M} \text{ is diagonal} \tag{13.27a}$$

Since $\mathbf{B}^{-1}$ exists, fixing $\mathbf{M}$ gives

$$\mathbf{H} = \mathbf{B}^{-1}\mathbf{M} \tag{13.27b}$$

From Fig. 13.12, the closed-loop state equation is

$$\frac{d}{dt}\mathbf{X} = \mathbf{AX} + \mathbf{BH}[\mathbf{G}_c(\mathbf{r} - \mathbf{X}) - \mathbf{KX}] = \mathbf{AX} + \mathbf{M}[\mathbf{G}_c(\mathbf{r} - \mathbf{X}) - \mathbf{KX}]$$

$$= (\mathbf{A} - \mathbf{MK})\mathbf{X} + \mathbf{MG}_c(\mathbf{r} - \mathbf{X})$$

If we define

$$\mathbf{N} = \mathbf{A} - \mathbf{MK} \tag{13.27c}$$

then

$$\frac{d}{dt}\mathbf{X} = \mathbf{NX} + \mathbf{MG}_c(\mathbf{r} - \mathbf{X}) \tag{13.27d}$$

Further, if we choose $\mathbf{N}$ to be diagonal and require a diagonal controller matrix $\mathbf{G}_c$, then the closed-loop state equation (13.27c) reduces to

$$\frac{d}{dt}\mathbf{X} = \mathbf{A}_d\mathbf{X} + \mathbf{MG}_c\mathbf{r} \tag{13.27e}$$

where

$$\mathbf{A}_d = (\mathbf{N} - \mathbf{MG}_c) \tag{13.27f}$$

is diagonal.

Figure 13.12

State Vector Feedback Multivariable Control System: A State Space (Vector Controlling Input) Approach

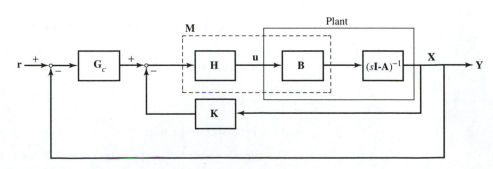

$\mathbf{A}_d$ can be viewed as the *desired closed-loop* **A**-*matrix,* and since $\mathbf{MG}_c$ is also diagonal, then the system is completely decoupled. Considering equation (13.27d), we can describe each control loop in the decoupled system by the Laplace domain expression

$$sx_i = N_{ii}x_i + M_{ii}G_{cii}(r_i - x_i) \tag{13.28a}$$

which gives an ideal independent control, since each output x_i depends only on the *i*th input r_i. The *i*th-loop transfer function is thus

$$\frac{x_i}{r_i} = \frac{M_{ii}G_{cii}}{s - N_{ii} + M_{ii}G_{cii}} \tag{13.28b}$$

so that the characteristic equation is

$$s - N_{ii} + M_{ii}G_{cii} = 0 \tag{13.28c}$$

Given the desired closed-loop eigenvalues (note that the algorithm $\mathbf{G}_c$ is assumed to be fixed a priori), we can find the N_{ii}'s using equation (13.28c), and from these we can determine the state vector feedback gain **K** by equation (13.27c). This is the *eigenvalue assignment* design technique. Recall that the behavior of a dynamic system is completely determined by the eigenvalues. Thus by specifying the closed-loop system eigenvalues, we can achieve any design goal that is compatible with the technique.

Example 13.5   State Vector Feedback Control of a CSTR

Consider again the well-stirred chemical reactor of Example 13.4. Suppose the controller $\mathbf{G}_c$ in Fig. 13.12 is simply the integral controller

$$\mathbf{G}_c = \begin{bmatrix} 1/(T_1s) & 0 \\ 0 & 1/(T_2s) \end{bmatrix}$$

Determine the state vector feedback control **K** that will satisfy the performance objectives.

Solution: From equation (13.22a), we note that **B** is diagonal and invertible. Hence we can simply choose $\mathbf{H} = \mathbf{I}$, that is,

$$\mathbf{M} = \mathbf{B} = \begin{bmatrix} w/(\rho V) & 0 \\ 0 & 1/(\rho Vc) \end{bmatrix} = \begin{bmatrix} 1/1200 & 0 \\ 0 & 1/18000 \end{bmatrix}$$

The characteristic equations for the two loops are

$$s - N_{11} + \frac{w}{\rho V}\left(\frac{1}{T_1s}\right) = 0 \Rightarrow s^2 - N_{11}s + \frac{w}{\rho VT_1} = 0 \tag{13.29a}$$

and

$$s - N_{22} + \frac{1}{\rho Vc}\left(\frac{1}{T_2s}\right) = 0 \Rightarrow s^2 - N_{22}s + \frac{1}{\rho VcT_2} = 0 \tag{13.29b}$$

Following Example 13.4, we can require that each of the real parts of the closed-loop eigenvalues be less than $-1/45$. However, to assure faster recovery of the system, especially the reactor temperature, from unbalance, we can select the real parts $-\sigma_1 = -1/20$ for the concentration loop and $-\sigma_2 = -1/10$ for the temperature loop. For a quarter-decay-type response (for instance) we can choose

$$\omega_1 = \frac{\sigma_1}{0.22} = \frac{1}{4.4}, \quad \text{and} \quad \omega_2 = \frac{\sigma_2}{0.22} = \frac{1}{2.2}$$

The eigenvalue assignment is then

$$s_1 = -\frac{1}{20} \pm \frac{j}{4.4} \quad \text{and} \quad s_2 = -\frac{1}{10} \pm \frac{j}{2.2}$$

The corresponding characteristic equations are given by

$$s^2 + 2\sigma s + \sigma^2 + \omega^2 = 0$$

or

$$s^2 + 0.1s + 0.054 = 0 \tag{13.30a}$$

and

$$s^2 + 0.2s + 0.2166 = 0 \tag{13.30b}$$

Figure 13.13

Controller Performance for the Chemical Reactor

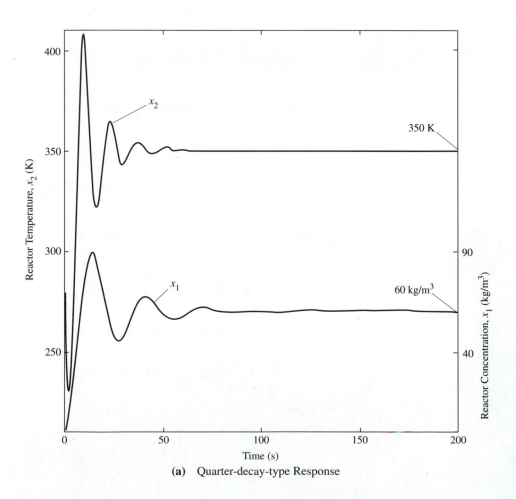

(a) Quarter-decay-type Response

Comparing equations (13.29) and (13.30), we can deduce

$$N_{11} = -0.1, \quad T_1 = \frac{w}{0.054\rho V} = 0.0154, \quad \text{and}$$

$$N_{22} = -0.2, \quad T_2 = \frac{1}{0.2166\rho V c} = 2.565(10^{-4})$$

From equations (13.22a) and (13.27c), the state vector feedback gains are given by

$$\begin{bmatrix} -0.02083 & 0 \\ -0.01111 & -0.00083 \end{bmatrix} - \begin{bmatrix} k_{11}/1200 & k_{12}/1200 \\ k_{21}/18000 & k_{22}/18000 \end{bmatrix} = \begin{bmatrix} -0.1 & 0 \\ 0 & -0.2 \end{bmatrix}$$

which yields

$$k_{11} = 95, \; k_{12} = 0, \; k_{21} = -200, \; k_{22} = 3585$$

The simulation results for the system of Fig. 13.12 using these data are shown in Fig. 13.13a. Again, a good response is obtained. The rather severe transient performance is due to the lack of sufficient damping in a quarter-decay-type behavior and the absence, in the present simulation, of any constraints on the controls. Some limitations on the magnitude of the control inputs is expected in practice. Figure 13.13b shows the response for a design requiring a damping ratio of $\sqrt{2}/2$ in each of the control loops but retaining the same decay rates as before. Note that for this value of damping ratio (see

Figure 13.13

Controller Performance for the Chemical Reactor

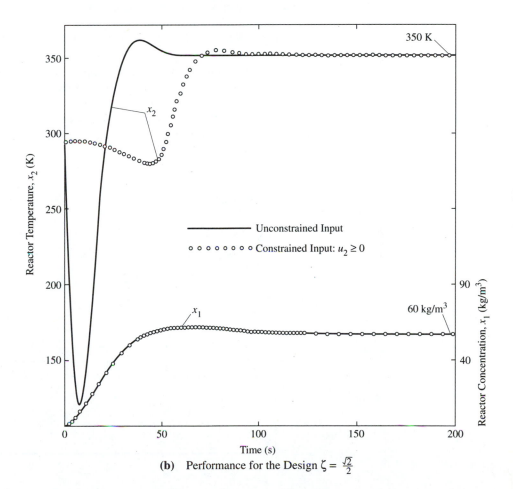

(b) Performance for the Design $\zeta = \frac{\sqrt{2}}{2}$

equation (7.5d)), $\omega = \sigma$, so that the characteristic equations (13.30a) and (13.30b) become

$$s^2 + 0.1s + 0.005 = 0 \quad \text{and} \quad s^2 + 0.2s + 0.02 = 0$$

Thus only T_1 and T_2 change for the new eigenvalues ($T_1 = 0.1667$ and $T_2 = 0.00278$). The "circle" line-type curves in Fig. 13.13b constitute the corresponding reactor response when the heat input is constrained to $u_2 \geq 0$. Observe that the temperature response pattern in both figures is *reverse reaction* (see Section 9.1)—that is, the process step-response pattern exhibits an initial negative dip for a positive step input. The power of the eigenvalue assignment technique is clearly demonstrated by these results.

Assuming that all the constraints imposed on the "ideal" multivariable control system are satisfied, we note that there are still a few loose ends in the procedure. In particular, the detailed specifications of **H** and $\mathbf{G}_c$ are not included as such in the analysis. However, as just demonstrated, the method has some possibilities, especially for simple systems in which the matrix relations are sparse.

Scalar Controlling Input and Integral Action

In many practical control problems the number of manipulated variables for an nth-order plant is often less than n. We consider first the case of a (single) *scalar controlling input* and simple proportional state vector feedback control:

$$u = -\mathbf{KX}; \qquad \mathbf{K} = [k_1 \quad k_2 \quad \cdots \quad k_n] \tag{13.31a}$$

We shall still retain the constraint of complete state vector measurement (that is, complete observability of the system). Although the limitation on **B** is relaxed, **B** is now a column vector. Figure 13.14 shows the structure of such a multivariable control system. Note that $\mathbf{C} = \mathbf{I}$ (that is, $\mathbf{Y} = \mathbf{X}$) has been assumed to simplify analysis and that there is no loss in generality from this assumption. Finally, observe that in this case,

$$\mathbf{G}_c = \mathbf{K} = [k_1 \quad k_2 \quad \cdots \quad k_n] \tag{13.31b}$$

Substituting equation (13.31a) in the plant equation, we have

$$\frac{d}{dt}\mathbf{X} = \mathbf{AX} + \mathbf{B}u \tag{13.32a}$$

Figure 13.14

State Vector Feedback with Scalar Controlling Input

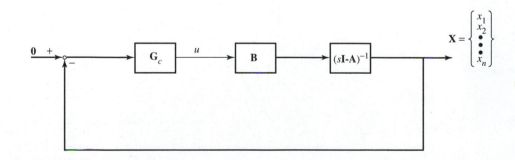

we obtain

$$\frac{d}{dt}\mathbf{X} = \mathbf{AX} - \mathbf{BKX} = [\mathbf{A} - \mathbf{BK}]\mathbf{X} \tag{13.32b}$$

The closed-loop state equation (13.32b) gives the closed-loop characteristic equation

$$\det[s\mathbf{I} - (\mathbf{A} - \mathbf{BK})] = 0 \tag{13.32c}$$

or factoring $(s\mathbf{I} - \mathbf{A})$ and noting that the determinant of a product is the product of the determinants (see Appendix B, equation (B2.2d)), we obtain

$$\det(s\mathbf{I} - \mathbf{A})\det[\mathbf{I} + (s\mathbf{I} - \mathbf{A})^{-1}\mathbf{BK}] = 0 \tag{13.32d}$$

Equation (13.32d) can be reduced further using the result, equation (B2.2e) of Appendix B,

$$\det(\mathbf{I} + \mathbf{HK}) = 1 + \mathbf{KH} \tag{13.32e}$$

where $\mathbf{I} = n \times n$ identity matrix, $\mathbf{H} = n$-column vector, and $\mathbf{K} = n$-row vector. Thus defining

$$\mathbf{H}(s) = (s\mathbf{I} - \mathbf{A})^{-1}\mathbf{B} \tag{13.32f}$$

and substituting in equation (13.32d), we get the closed-loop characteristic equation

$$\det_{\mathrm{CL}}(s) = 0 = \det(s)(1 + \mathbf{KH}) = 0 \tag{13.33a}$$

where $\det(s)$ is the open-loop characteristic polynomial, that is,

$$\det(s) = \det(s\mathbf{I} - \mathbf{A}) = s^n + a_{n-1}s^{n-1} + \ldots + a_1 s + a_0 \tag{13.33b}$$

The design process is essentially complete at this point. Given the desired closed-loop characteristic equation, we can find the state vector feedback gain $\mathbf{K}$.

Reference [10] gives a direct method for determining $\mathbf{K}$ from the coefficients of the open-loop characteristic equation and those of the desired closed-loop characteristic polynomial. This approach is useful for systems whose model is given in Laplace domain transfer function form. Also, a significant result from that development is that in addition to the observability constraint, the original system has to be controllable for the present method to be applicable. When these conditions are met, the resulting controlled system is also both observable and controllable.

Example 13.6 Level Control of Three-Tank System

For the three-tank system shown in Fig. 13.15a, determine a state vector feedback proportional control that will yield a closed-loop system with the desired eigenvalues -6, -4, and -3. Simulate the response of the controlled system from an initial state $x_1(0) = x_2(0) = x_3(0) = 1$ to the desired final state $x_1 = 2$, $x_2 = 1.5$, and $x_3 = 1.0$.

Figure 13.15

State Vector Feedback
Control of Three-tank System

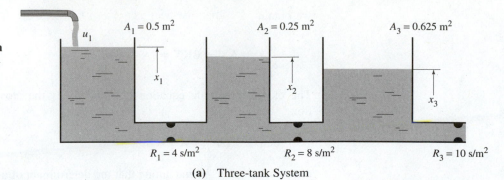

(a) Three-tank System

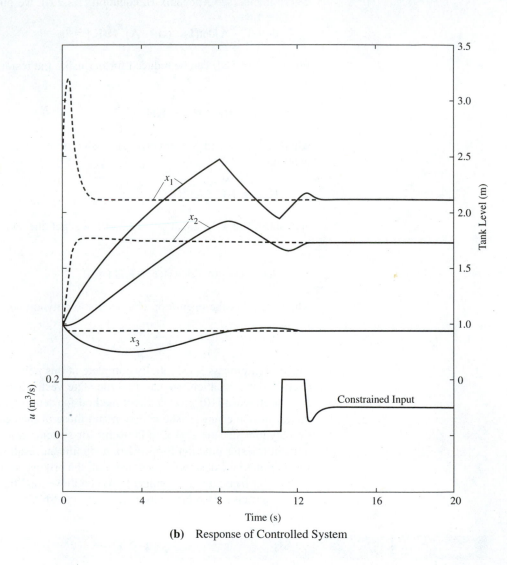

(b) Response of Controlled System

Solution: Using the parameter values given in Fig. 13.15a, we obtain the state model of the system:

$$\dot{\mathbf{X}} = \begin{bmatrix} -0.5 & 0.5 & 0 \\ 1 & -1.5 & 0.5 \\ 0 & 0.2 & -0.36 \end{bmatrix} \mathbf{X} + \begin{bmatrix} 2 \\ 0 \\ 0 \end{bmatrix} u = \mathbf{AX} + \mathbf{B}u \qquad \textbf{(13.34a)}$$

$$\det(s) = \det(s\mathbf{I} - \mathbf{A})$$
$$= (s + 0.5)(s + 1.5)(s + 0.36) - 0.1(s + 0.5) - 0.5(s + 0.36) \qquad \textbf{(13.34b)}$$

and

$$(s\mathbf{I} - \mathbf{A})^{-1} = \frac{1}{\det(s)} \begin{bmatrix} (s + 1.5)(s + 0.36) - 0.1 & 0.5(s + 0.36) & 0.25 \\ (s + 0.36) & (s + 0.5)(s + 0.36) & 0.5(s + 0.5) \\ 0.2 & 0.2(s + 0.5) & (s + 0.5)(s + 1.5) - 0.5 \end{bmatrix}$$

where, again,

$$\det(s) = \det(s\mathbf{I} - \mathbf{A})$$
$$= (s + 0.5)(s + 1.5)(s + 0.36) - 0.1(s + 0.5) - 0.5(s + 0.36)$$
$$= s^3 + 2.36s^2 + 0.87s + 0.04$$

From equation (13.32f),

$$\mathbf{H}(s) = (s\mathbf{I} - \mathbf{A})^{-1}\mathbf{B} = \frac{1}{\det(s)} \begin{bmatrix} 2\{(s + 1.5)(s + 0.36) - 0.1\} \\ 2(s + 0.36) \\ 0.4 \end{bmatrix}$$

so that

$$\det_{CL}(s) = \det(s) + 2k_1[(s + 1.5)(s + 0.36) - 0.1]$$
$$+ 2k_2(s + 0.36) + 0.4k_3 \qquad \textbf{(13.34c)}$$

which gives the closed-loop characteristic

$$s^3 + (2.36 + 2k_1)s^2 + (0.87 + 3.72k_1 + 2k_2)s$$
$$+ (0.04 + 0.88k_1 + 0.72k_2 + 0.4k_3) = 0 \qquad \textbf{(13.34d)}$$

The desired closed-loop characteristic equation is

$$(s + 6)(s + 4)(s + 3) = s^3 + 13s^2 + 54s + 72 = 0 \qquad \textbf{(13.34e)}$$

Matching coefficients in equations (13.34d) and (13.34e), we have

$$2.36 + 2k_1 = 13 \Rightarrow k_1 = 5.32; \qquad 0.87 + 3.72k_1 + 2k_2 = 54 \Rightarrow k_2 = 16.67$$

and

$$0.04 + 0.88k_1 + 0.72k_2 + 0.4k_3 = 72 \Rightarrow k_3 = 138.19$$

The scalar controlling input is thus

$$u = 5.32(2 - x_1) + 16.67(1.5 - x_2) + 138.19(1 - x_3) \tag{13.34f}$$

The simulation of equations (13.34a) and (13.34f), subject to the given initial state, is shown (in broken lines) in Fig. 13.15b. The solid lines represent the response for a constrained input: $0 \le u \le 0.2$. This manipulated variable u for this case is also indicated in the figure.

Adding Integral Action

Note from Fig. 13.15b that the response of the controlled system was subject to offset (actually, no state variable converged to its actual desired value). From our experience, such a situation calls for some form of integral control. Fortunately, I-action is easily incorporated in the state vector feedback (scalar controlling input) procedure.

Consider, for example, the general second-order system

$$\dot{x}_1 = a_{11}x_1 + a_{12}x_2 + b_1 u \tag{13.35a}$$

$$\dot{x}_2 = a_{21}x_1 + a_{22}x_2 + b_2 u$$

Suppose proportional state vector feedback control plus integral control of x_1 is desired. That is, (still assuming zero reference state),

$$u = -\left[k_1 x_1 + k_2 x_2 + \frac{1}{T_i} \int x_1 \, dt \right] \tag{13.35b}$$

We can define a third state variable,

$$\dot{x}_3 = x_1 \tag{13.35c}$$

so that the manipulated variable, equation (13.35b), becomes

$$u = -[k_1 x_1 + k_2 x_2 + k_3 x_3] = -\mathbf{KX} \tag{13.35d}$$

where

$$k_3 = \frac{1}{T_i} \tag{13.35e}$$

"Proportional" state vector feedback controller design can then be carried out as usual but with the third-order system given by equations (13.35a) and (13.35c). That is with,

$$\mathbf{A} = \begin{bmatrix} a_{11} & a_{12} & 0 \\ a_{21} & a_{22} & 0 \\ 1 & 0 & 0 \end{bmatrix} \quad \text{and} \quad \mathbf{B} = \begin{bmatrix} b_1 \\ b_2 \\ 0 \end{bmatrix}$$

The application of this procedure to the three-tank level-control system of Example 13.6 is left as an exercise (see Practice Problem 13.15). Another application is given by Example 13.9.

Extension to a Vector Controlling Input

The state vector feedback (scalar controlling input) method can also be applied approximately to a system with more than one manipulated variable. For example, for a third-order, two-input system,

$$\dot{\mathbf{X}} = \mathbf{AX} + \mathbf{QU}; \qquad \mathbf{QU} = \begin{bmatrix} q_{11} & q_{12} \\ q_{21} & q_{22} \\ q_{31} & q_{32} \end{bmatrix} \begin{bmatrix} u_1 \\ u_2 \end{bmatrix} \tag{13.36a}$$

A scalar controlling input u can be defined by its distribution into the vector $\mathbf{U}$ using appropriate weighting factors:

$$u_1(t) = w_1 u(t) \tag{13.36b}$$

$$u_2(t) = w_2 u(t)$$

so that

$$\mathbf{QU} = \begin{bmatrix} q_{11}w_1 + q_{12}w_2 \\ q_{21}w_1 + q_{22}w_2 \\ q_{31}w_1 + q_{32}w_2 \end{bmatrix} u = \mathbf{B}u \tag{13.36c}$$

Equation (13.36a) then becomes

$$\dot{\mathbf{X}} = \mathbf{AX} + \mathbf{B}u; \qquad \mathbf{B}u = \begin{bmatrix} b_1 \\ b_2 \\ b_3 \end{bmatrix} u = \begin{bmatrix} q_{11}w_1 + q_{12}w_2 \\ q_{21}w_1 + q_{22}w_2 \\ q_{31}w_1 + q_{32}w_2 \end{bmatrix} u \tag{13.36d}$$

which can then be used for design, as before.

Example 13.7   Vector Control of Three-Tank System

Suppose the middle tank in the three-tank system of Example 13.6 also has a flow control input, as shown in Fig. 13.16a. Determine a state vector feedback control that will give the same closed-loop eigenvalues as before but utilizing both flow inputs.

Solution: The new system state equation (see Fig. 13.16a) is

$$\dot{\mathbf{X}} = \begin{bmatrix} -0.5 & 0.5 & 0 \\ 1 & -1.5 & 0.5 \\ 0 & 0.2 & -0.36 \end{bmatrix} \mathbf{X} + \begin{bmatrix} 2 & 0 \\ 0 & 4 \\ 0 & 0 \end{bmatrix} \begin{bmatrix} u_1 \\ u_2 \end{bmatrix} \tag{13.37}$$

Given the distribution weights w_1 and w_2, equation (13.36d) gives us

$$\mathbf{B} = \begin{bmatrix} 2w_1 \\ 4w_2 \\ 0 \end{bmatrix}$$

Figure 13.16

**State Vector Feedback for
Vector Controlling Input**

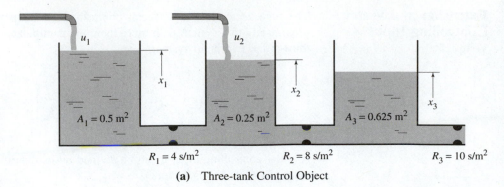

(a)   Three-tank Control Object

(b)   Response for Unconstrained Inputs

Figure 13.16

State Vector Feedback for Vector Controlling Input

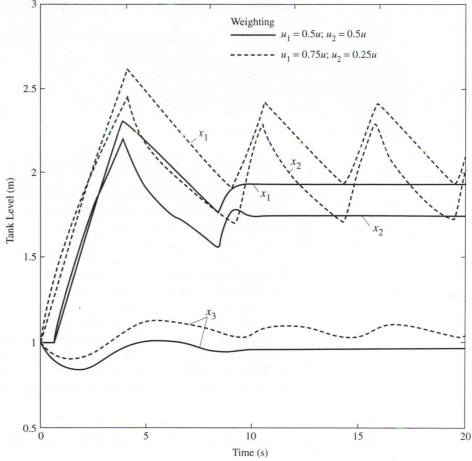

(c) Response for Constrained Inputs: $0 \leq u_1 \leq 0.2$; $0 \leq u_2 \leq 0.2$

and from equation (13.32f),

$$\mathbf{H}(s) = \frac{1}{\det(s)} \begin{bmatrix} 2w_1\{(s + 1.5)(s + 0.36) - 0.1\} + 4w_2\{0.5(s + 0.36)\} \\ 2w_1(s + 0.36) + 4w_2(s + 0.5)(s + 0.36) \\ 2w_1(0.2) + 4w_2\{0.2(s + 0.5)\} \end{bmatrix}$$

Note that $\det(s)$ and $(s\mathbf{I} - \mathbf{A})^{-1}$ are the same as in Example 13.6. The closed-loop characteristic equation is thus

$$\det(s) + 2w_1k_1[(s + 1.5)(s + 0.36) - 0.1] + 2w_2k_1(s + 0.36)$$

$$+ 2w_1k_2(s + 0.36) + 4w_2k_2(s + 0.5)(s + 0.36) + 0.4w_1k_3$$

$$+ 0.8w_2k_3(s + 0.5) = 0$$

which reduces to

$$s^3 + (2.36 + 2w_1k_1 + 4w_2k_2)s^2$$

$$+ (0.87 + 3.72w_1k_1 + 2w_2k_1 + 2w_1k_2 + 3.44w_2k_2 + 0.8w_2k_3)s$$

$$+ (0.04 + 0.88w_1k_1 + 0.72[w_2k_1 + w_1k_2 + w_2k_2] + 0.4w_1k_3 + 0.4w_2k_3) = 0$$

so that if, for example, $w_1 = w_2 = 0.5$ (equal weighting of the inputs), then, as before,

$$2.36 + k_1 + 2k_2 = 13, \quad 0.87 + 2.86k_1 + 2.72k_2 + 0.4k_3 = 54$$

and

$$0.04 + 0.8k_1 + 0.72k_2 + 0.4k_3 = 72,$$

which gives the design

$$k_1 = -27.8, \quad k_2 = 19.22, \quad k_3 = 200.9$$

The response of the system with this control is shown in Fig. 13.16b (solid line). Note that the equations simulated are (13.37), (13.36b), and (13.31a).

The determination of practical weighting factors in control systems design is often a difficult problem, requiring much insight and perhaps some experimentation. The broken curve in Fig. 13.16b represents the response due to the weighting $w_1 = 0.75$, $w_2 = 0.25$, which gives the feedback gains $k_1 = 31.77$, $k_2 = -37$, $k_3 = 179.8$. Although this weighting appears to give a faster response, the performance in terms of steady-state offset is not as good as that for the even distribution of the input (see Practice Problem 13.18 for the performance when I-action is added). Figure 13.16c shows the effect of constraints on the inputs while retaining the present design.

13.3 INTRODUCING STATE OBSERVERS AND ADAPTIVE CONTROL

The state vector feedback multivariable control systems considered thus far has required complete state vector measurement as a constraint that must be satisfied by the system. Also, we know that achieving an optimal control requires a lot of information about the system to be controlled. However, in many practical control situations, the state vector is not completely available for direct measurement, and the exact structure of many physical systems is often obscure or uncertain. There is also the physical reality that practical systems are seldom stationary, and significant parameter variations as well as environmental variations are quite common. The process of discovering a good model for a given system is the subject of model *identification,* (see [10]). In cases the basic structure of the system can be determined but the various model parameters are the unknown, the term *parameter estimation* is more appropriate. The frequency response identification discussed in Chapter 9 falls in this category. The problem of making good estimates of the state variables from the available data is the subject of *state estimation.* An *observer* is a (state-estimator) system driven by the available outputs and inputs of an original system with the objective of reconstructing the original system's state vector. Thus the observer can, in principle, be utilized as a subsystem in a state vector feedback multivariable control scheme in which complete state measurement is otherwise not feasible.

Identification or parameter estimation is also a critical aspect of *adaptive control,* which deals with the more complex problem of automatic control of a system in the presence of uncertainties, structural perturbations, and environmental variations (see [12]). The general concept of an adaptive solution is to design a controller that is self-adjusting in response to variations in the plant and its inputs. Some example of problems requiring adaptive control include the following:

1. *Flight control:* The flight environment of missiles and supersonic aircraft can vary very quickly and widely due to the very high flight speeds of these vehicles and large altitude changes and/or long distances that are usually involved. Also, the aerodynamic characteristics of flight vehicles not only change with the flight environment (for example, air density and wind speed and direction) but are functions of the vehicle velocity (Mach number) as well. In the case of missiles, the inertia properties also vary widely with time due to the rapid depletion of the fuel, which constitutes the bulk of the vehicle's mass.

2. *Ship control:* Although ships move quite slowly relative to airborne vehicles, the effect of changes in the operating environment, for example, wind and waves, and changes in the dynamic characteristics such as draft, speed, and course on a ship's steering and fuel economy are nonetheless considerable. This is so because of the massive structure of most ships, especially tankers.

3. *Industrial drives and motion control:* Electrical drive parameters are seldom well known and can vary rapidly and widely in some applications. Their low-speed characteristics are very often nonlinear and substantially different from the high-speed characteristics. Many industrial drives are subject to very large load variations. The drive moment of inertia in most motion control systems, such as machine tool drives, is usually a function of the load (work), which is variable and not normally measurable.

4. *Process control:* Many process control systems are actually composed of several interacting processes, and their dynamic behavior is often complex, nonlinear, and poorly understood. Most chemical processes include time-varying systems or systems with variable time constants and time delays, and/or large time delays. Also, environmental disturbances such as variations in the feed stream properties, or the feedstock of batch processes, are quite common.

There are two major forms of adaptive control systems. An adaptive control scheme employing continuous plant identification followed by control action that is adaptive is the principle of the *self-tuning regulator* (STR). On the other hand, the adaptation of the control action in *model reference adaptive control* (MRAC) makes the output of the "unknown" plant approach asymptotically the output of a reference model that is part of the control system.

Design of State Observers

Our discussion of *state observers* will be restricted to simple, continuous-time, linear, time-invariant, deterministic systems. We shall first consider an observer for the entire state vector without regard to the number of state variables that can actually be measurable. The principle of such a state estimator, the *asymptotic* observer, is illustrated in Fig. 13.17 for a plant described by the state space model

$$\frac{d}{dt}\mathbf{X} = \mathbf{A}\mathbf{X} + \mathbf{B}\mathbf{U}; \qquad \mathbf{Y} = \mathbf{C}\mathbf{X} \tag{13.38a}$$

where $\mathbf{A}$, $\mathbf{B}$, and $\mathbf{C}$ are the plant parameters, $\mathbf{X}$ is an n-vector of state variables, and $\mathbf{Y}$ is an m-vector of output variables. The observer is essentially a model of the plant but with an additional "control" input, in this case a (proportional) feedback control input with the plant output vector as the reference *(tracking)* data. The observer parameters, which are not necessarily equal to the plant parameters, are represented by $\hat{\mathbf{A}}$, $\hat{\mathbf{B}}$, and $\hat{\mathbf{C}}$, and the estimated state and output vectors are $\hat{\mathbf{X}}$ and $\hat{\mathbf{Y}}$, respectively.

Figure 13.17

The Asymptotic Observer

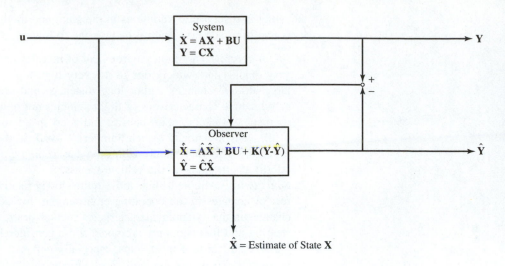

$\hat{\mathbf{X}}$ = Estimate of State $\mathbf{X}$

The Ideal Case

If we assume exact measurement of the inputs and outputs of the plant, a simple *ideal* observer for the system (13.38a), that is, one in which the plant structure and parameters are known completely and accurately, $\hat{\mathbf{A}} = \mathbf{A}$, $\hat{\mathbf{B}} = \mathbf{B}$, and $\hat{\mathbf{C}} = \mathbf{C}$, is given by

$$\frac{d}{dt}\hat{\mathbf{X}} = \mathbf{A}\hat{\mathbf{X}} + \mathbf{B}\mathbf{U} + \mathbf{K}(\mathbf{Y} - \hat{\mathbf{Y}}); \qquad \hat{\mathbf{Y}} = \mathbf{C}\hat{\mathbf{X}} \tag{13.38b}$$

This model can be implemented on a digital (micro) or analog computer. As was pointed out previously, the model utilizes *feedback* of the measurement error $(\mathbf{Y} - \hat{\mathbf{Y}})$, which is proportional to the actual error in the estimate of state $(\mathbf{X} - \hat{\mathbf{X}})$, to compensate for inaccuracies introduced not only by lack of knowledge of the actual initial state of the system $(\hat{\mathbf{X}}(0) \neq \mathbf{X}(0))$ but by the practical problem of imperfect knowledge of the system structure, that is, of $\mathbf{A}$, $\mathbf{B}$, and $\mathbf{C}$. In equation (13.38b), $\mathbf{K}$ is the $(n \times m)$-dimensional feedback matrix.

Subtracting equation (13.38b) from (13.38a), we obtain

$$\frac{d}{dt}(\mathbf{X} - \hat{\mathbf{X}}) = \mathbf{A}(\mathbf{X} - \hat{\mathbf{X}}) - \mathbf{K}(\mathbf{Y} - \hat{\mathbf{Y}}) = \mathbf{A}(\mathbf{X} - \hat{\mathbf{X}}) - \mathbf{K}\mathbf{C}(\mathbf{X} - \hat{\mathbf{X}})$$

Thus the error,

$$\mathbf{E} = \mathbf{X} - \hat{\mathbf{X}} \tag{13.38c}$$

obeys the equation

$$\frac{d}{dt}\mathbf{E} = (\mathbf{A} - \mathbf{K}\mathbf{C})\mathbf{E} = \mathbf{A}*\mathbf{E} \tag{13.38d}$$

If we choose the feedback gain $\mathbf{K}$ so that the system (13.38d) is asymptotically stable, that is, the eigenvalues of $\mathbf{A}*$ are sufficiently small (large and negative), then $\mathbf{E}(t)$ tends to zero as time increases, so that the estimated state converges *asymptotically* to the actual state

$$\hat{\mathbf{X}}(t) \longrightarrow \mathbf{X}(t) \tag{13.38e}$$

as time increases.

Consider, for instance, the oscillating pendulum of Example 13.3. For simplicity, let the parameters be normalized so that the model is

$$\frac{d}{dt}\begin{bmatrix} x_1 \\ x_2 \end{bmatrix} = \begin{bmatrix} 0 & 1 \\ -1 & 0 \end{bmatrix}\begin{bmatrix} x_1 \\ x_2 \end{bmatrix} + \begin{bmatrix} 0 \\ 1 \end{bmatrix}u; \tag{13.39a}$$

$$y = \begin{bmatrix} 1 & 0 \end{bmatrix}\begin{bmatrix} x_1 \\ x_2 \end{bmatrix} \Rightarrow \mathbf{A} = \begin{bmatrix} 0 & 1 \\ -1 & 0 \end{bmatrix}; \quad \mathbf{C} = \begin{bmatrix} 1 & 0 \end{bmatrix}$$

We shall assume, as is often the case, that the angular velocity x_2 is not directly measurable. However, both the angular displacement x_1 and the actuator displacement u can be measured exactly. The purpose of the observer is therefore to estimate the (angular) velocity. Let

$$\mathbf{K} = \begin{bmatrix} k_1 \\ k_2 \end{bmatrix}$$

then

$$\mathbf{A}^* = \mathbf{A} - \mathbf{KC} = \begin{bmatrix} -k_1 & 1 \\ -1 - k_2 & 0 \end{bmatrix}$$

and

$$\det(s\mathbf{I} - \mathbf{A}^*) = \begin{vmatrix} s + k_1 & -1 \\ 1 + k_2 & s \end{vmatrix} = s(s + k_1) + 1 + k_2$$

Hence the characteristic equation is

$$s^2 + k_1 s + (1 + k_2) = 0$$

If we specify the eigenvalues of the error system as -1 and -2, then the desired characteristic equation is

$$(s + 1)(s + 2) = s^2 + 3s + 2 = 0$$

so that $k_1 = 3$ and $k_2 = 1$.

Note that from equation (13.38d),

$$\mathscr{L}[\mathbf{E}(t)] = [s\mathbf{I} - (\mathbf{A} - \mathbf{KC})]^{-1}\mathbf{E}(0) = \frac{1}{(s + 1)(s + 2)}\begin{bmatrix} s & 1 \\ -2 & s + 3 \end{bmatrix}\mathbf{E}(0)$$

In the absence of other information, the best estimate of the initial state $\mathbf{X}(0)$ is

$$\hat{\mathbf{X}}(0) = \mathbf{0} \tag{13.39b}$$

Hence $\mathbf{E}(0) = \mathbf{X}(0)$, and

$$\underset{t \to \infty}{\text{Limit}}\mathbf{E}(t) = \underset{s \to 0}{\text{Limit}}s\mathbf{E}(s) = \underset{s \to 0}{\text{Limit}}\frac{s}{(s+1)(s+2)}\begin{bmatrix} s & 1 \\ -2 & s+3 \end{bmatrix}\mathbf{X}(0) = \mathbf{0}$$

as desired.

Note that the observer, as designed, is given by equations (13.38b) and (13.39b) and the result,

$$\mathbf{K} = \begin{bmatrix} 3 \\ 1 \end{bmatrix} \tag{13.39c}$$

The Supplemental Observer

The observer we just constructed gave the estimate of the entire n-state, although some of the state variables are measurable. This could be an unnecessary expenditure of time and computational resources, especially for large-(n) order systems. The *supplemental observer* merely provides the unmeasurable state, that is, it supplements the measurement. Consider the system (13.38a) with $\mathbf{Y}$ an m-vector and $\mathbf{X}$ an n-vector, such that (n–m) state variables are not measured or are not measurable. Let us define the supplementary (n–m) output vector

$$\mathbf{W} = \mathbf{K}\mathbf{X} \tag{13.40a}$$

such that

$$\mathbf{X} = \begin{bmatrix} \mathbf{C} \\ \hline \mathbf{K} \end{bmatrix}^{-1} \begin{bmatrix} \mathbf{Y} \\ \hline \mathbf{W} \end{bmatrix} = \mathbf{M}^{-1} \begin{bmatrix} \mathbf{Y} \\ \hline \mathbf{W} \end{bmatrix} \tag{13.40b}$$

$\mathbf{K}$ is (n–m) $\times$ n, and $\mathbf{C}$ is, of course, $m \times n$. Equation (13.40b) defines the matrix

$$\mathbf{M} = \begin{bmatrix} \mathbf{C} \\ \hline \mathbf{K} \end{bmatrix} \tag{13.40c}$$

whose inverse must exist for this construction to work—that is, the estimation of $\mathbf{X}$ by $\mathbf{Y}$ and $\hat{\mathbf{W}}$. Note that $\mathbf{Y}$ is assumed to be known exactly, whereas $\hat{\mathbf{W}}$ is the estimate of $\mathbf{W}$.

The supplemental observer algorithm is derived as follows:
From equation (13.40b),

$$\begin{bmatrix} \mathbf{Y} \\ \hline \mathbf{W} \end{bmatrix} = \mathbf{M}\mathbf{X}$$

so that

$$\frac{d}{dt}\begin{bmatrix} \mathbf{Y} \\ \hline \mathbf{W} \end{bmatrix} = \mathbf{M}\frac{d}{dt}\mathbf{X} = \mathbf{M}\mathbf{A}\mathbf{X} + \mathbf{M}\mathbf{B}\mathbf{U}$$

Using equation (13.40b) again, we obtain

$$\frac{d}{dt}\begin{bmatrix} \mathbf{Y} \\ \hline \mathbf{W} \end{bmatrix} = (\mathbf{M}\mathbf{A}\mathbf{M}^{-1})\begin{bmatrix} \mathbf{Y} \\ \hline \mathbf{W} \end{bmatrix} + (\mathbf{M}\mathbf{B})\mathbf{U} \tag{13.40d}$$

If we partition the matrices in equation (13.40d) as follows:

$$\mathbf{MAM}^{-1} = \left[\begin{array}{c|c} & \\ \hline \mathbf{H}_y & \mathbf{H}_w \end{array}\right] \begin{array}{l} \}m \\ \}(n-m) \end{array} \quad \text{and} \quad \mathbf{MB} = \left[\begin{array}{c} \\ \hline \mathbf{B}_w \end{array}\right] \begin{array}{l} \}n \\ \}(n-m) \end{array}$$
$$\underbrace{}_{m} \ \underbrace{}_{(n-m)}$$

then the supplemental observer equation (note that there are matrices in the blank partitions but they are not of present interest) is given by

$$\frac{d}{dt}\hat{\mathbf{W}} = \mathbf{H}_y\mathbf{Y} + \mathbf{H}_w\hat{\mathbf{W}} + \mathbf{B}_w\mathbf{U}; \qquad \hat{\mathbf{W}}(0) = \mathbf{0}$$

or

$$\frac{d}{dt}\hat{\mathbf{W}} = \mathbf{H}_w\hat{\mathbf{W}} + [\mathbf{H}_y\mathbf{Y} + \mathbf{B}_w\mathbf{U}]; \qquad \hat{\mathbf{W}}(0) = \mathbf{0}$$

where zero initial values for the estimated output variables are again necessary because we presume the absence of any information a priori. As shown by equation (13.40e), the supplemental observer is driven only by $\mathbf{U}$ and $\mathbf{Y}$. For $\hat{\mathbf{W}}$ to converge rapidly to $\mathbf{W}$ or close to $\mathbf{W}$, all transients in $\hat{\mathbf{W}}$ must die out quickly. This again requires an asymptotically stable $\mathbf{H}_w$.

Thus the conditions for the design of $\mathbf{K}$ are

1. $\left[\dfrac{\mathbf{C}}{\mathbf{K}}\right]^{-1}$ must exist.
2. The eigenvalues of $\mathbf{H}_w$ must be sufficiently small.

Note again that the estimation of the unavailable state variables in $\mathbf{X}$ is given by $\mathbf{Y}$ and $\hat{\mathbf{W}}$ and equation (13.40b).

Example 13.8   Velocity Observer for Servomechanism

Design a supplemental observer to estimate the velocity of the pendulum system equation (13.39a). Using simulation, compare the performance of this state estimator with the full observer obtained earlier (equations (13.38b), (13.39b), and (13.39c)). Let the system input in all cases be $u = te^{-0.2t}$.

Solution: Let

$$\mathbf{K} = [k_1 \quad k_2]$$

then

$$\mathbf{M} = \left[\frac{\mathbf{C}}{\mathbf{K}}\right] = \left[\begin{array}{c|c} 1 & 0 \\ \hline k_1 & k_2 \end{array}\right]$$

$\det \mathbf{M} = k_2$, so that $\mathbf{M}^{-1}$ will exist for any $k_2 \neq 0$. This is the first condition on $\mathbf{K}$. The observer equation can be derived directly, in this simple case, from the estimation

$$W = \mathbf{KX} = k_1 x_1 + k_2 x_2$$

which gives

$$\frac{d}{dt}\hat{W} = k_1\frac{d}{dt}x_1 + k_2\frac{d}{dt}x_2 \quad \text{and} \quad x_2 = \frac{W - k_1x_1}{k_2}$$

Using the available knowledge of the plant structure, equation (13.39a), we have

$$\frac{d}{dt}\hat{W} = k_1x_2 - k_2x_1 + k_2u$$

Substituting for x_2 and $y = x_1$, we obtain

$$\frac{d}{dt}\hat{W} = \left(\frac{k_1}{k_2}\right)\hat{W} - \left(\frac{k_1^2 + k_2^2}{k_2}\right)y + k_2u; \qquad \hat{W}(0) = 0 \qquad \text{(13.41a)}$$

Thus

$$H_w = \frac{k_1}{k_2}; \qquad H_y = -\left(\frac{k_1^2 + k_2^2}{k_2}\right); \qquad B_w = k_2$$

The estimation of x_2 is

$$\hat{x}_2 = \frac{\hat{W} - k_1y}{k_2} \qquad \text{(13.41b)}$$

The second condition on **K** will be satisfied by any $(k_1/k_2) < 0$.

The design $k_1 = -1$, $k_2 = 1$, for example, meets the two conditions. Note that the equations simulated are as follows:

1. **System:** equation (13.39a) with the initial state $x_1(0) = 0$, and $x_2(0) = 2$
2. **Full observer:** $\hat{x}_1 = \hat{x}_2 + \{3\}(y - \hat{x}_1); \qquad \hat{x}_1(0) = 0$
   $\hat{x}_2 = -\hat{x}_1 + u + \{1\}(y - \hat{x}_1); \qquad \hat{x}_2(0) = 0$
3. **Supplemental observer:** equations (13.41a) and (13.41b) with the initial state $\hat{W}(0) = 0$.

To fix ideas, let us also derive the supplemental observer equations using the general design procedure outlined earlier. The plant model, equation (13.39a), gives

$$\mathbf{A} = \begin{bmatrix} 0 & 1 \\ -1 & 0 \end{bmatrix}; \qquad \mathbf{B} = \begin{bmatrix} 0 \\ 1 \end{bmatrix}; \qquad \mathbf{C} = \begin{bmatrix} 1 & 0 \end{bmatrix}$$

Equation (13.40a) gives the supplementary output:

$$W = \begin{bmatrix} k_1 & k_2 \end{bmatrix}\begin{bmatrix} x_1 \\ x_2 \end{bmatrix} = k_1x_1 + k_2x_2$$

From equation (13.40c),

$$\mathbf{M} = \begin{bmatrix} 1 & 0 \\ k_1 & k_2 \end{bmatrix}$$

so that

$$\mathbf{M}^{-1} = \frac{1}{k_2}\begin{bmatrix} k_2 & 0 \\ -k_1 & 1 \end{bmatrix}; \qquad k_2 \neq 0$$

Further,

$$\mathbf{MAM}^{-1} = \begin{bmatrix} 0 & 1/k_2 \\ -k_2 - k_1^2/k_2 & k_1/k_2 \end{bmatrix} \quad \text{and} \quad \mathbf{MB} = \begin{bmatrix} 0 \\ k_2 \end{bmatrix}$$

Equation (13.40d) with these matrices appropriately partitioned is

$$\frac{d}{dt}\begin{bmatrix} y \\ W \end{bmatrix} = \begin{bmatrix} 0 & 1/k_2 \\ -k_2 - k_1^2/k_2 & k_1/k_2 \end{bmatrix}\begin{bmatrix} y \\ W \end{bmatrix} + \begin{bmatrix} 0 \\ k_2 \end{bmatrix}u$$

Figure 13.18

Performance of Full and Supplemental Observers

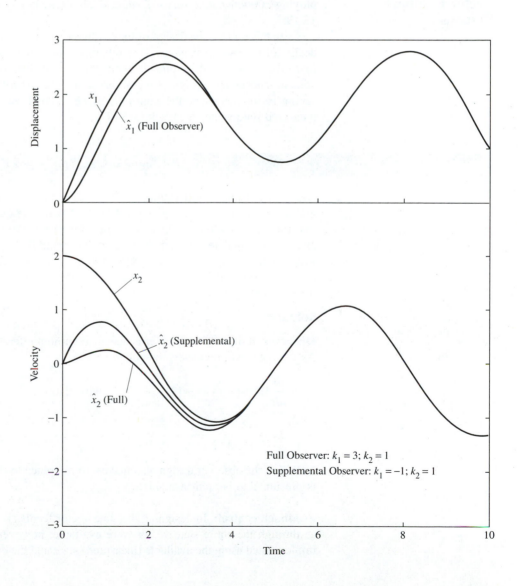

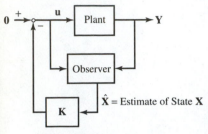

Figure 13.19

**Estimated State Vector
Feedback Control**

so that

$$H_y = -k_2 - \frac{k_1^2}{k_2}; \qquad H_w = \frac{k_1}{k_2}; \qquad B_w = k_2$$

as before. Finally, substituting these results into equation (13.40e), we obtain the supplemental observer equation (13.41a), whereas equation (13.40b) yields the estimation of x_2, equation (13.41b).

Figure 13.18 shows the performance of both the full observer and the supplemental observer. The responses are remarkable even for the modest (with respect to speed of convergence) eigenvalues chosen. Both the full observer and the supplemental observer are able to converge exactly to the system state within about five time units.

Application to State Vector Feedback Systems

Observers such as just discussed can be applied in a closed-loop control system by employing *estimated state vector feedback*. The basic configuration is illustrated in Fig. 13.19.

In this process the design of the feedback gains can always be separated from the design of the estimator gains. This separation or noninteracting property, which was first investigated for deterministic estimators by Luenberger (see [13]) allows for convenient independent design of the observer and the controller. Note that where the observer is supplemental, the control will be a combination of estimated state and measured state vector feedback (MSVF).

Example 13.9   Estimated and Measured State Vector Feedback Control

Design the combination estimated and measured state vector feedback (with integral action) control system given in Fig. 13.20a for the oscillating pendulum system. Let the desired closed-loop eigenvalues for the controlled system be -1.5, -2, and -2.5. Simulate the performance of the controlled system subject to the initial state $x_1(0) = 0$, $x_2(0) = 2$; zero reference input $r = 0$, and a disturbance input (see Fig. 13.20a) $v = 0.2$. Use the supplemental observer design derived in Example 13.8 for the same system.

Solution:

Observer design: The observer is designed without reference to the controller design. As we obtained previously, the equations are

$$\frac{d}{dt}\hat{W} = \{-1\}\hat{W} - \{2\}y + u; \qquad \hat{W}(0) = 0 \tag{13.42a}$$

$$\hat{x}_2 = \frac{\hat{W} - \{-1\}y}{\{1\}} \tag{13.42b}$$

Note that the observer design also makes no reference to the disturbance input, which is presumed to be unmeasurable.

Feedback control: In designing the state vector feedback control system, we proceed as through the entire state vector were available; however, the eventual controller is implemented using the available (measured) state and the estimated state (from the ob-

Figure 13.20

Estimated and Measured State Vector Feedback Control

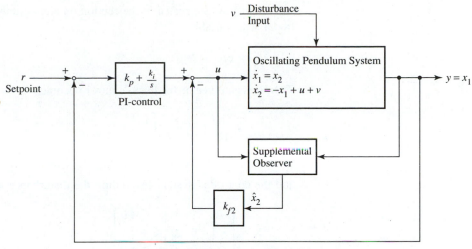

(a) Overall System Scalar Block Diagram

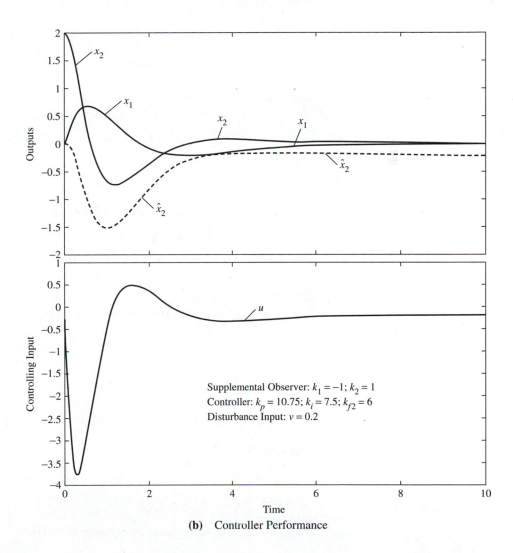

(b) Controller Performance

server unit). As suggested in the section on adding integral action, we define the additional state variable

$$\dot{x}_3 = x_1 \tag{13.42c}$$

Then the state vector feedback control becomes

$$u = -[k_p \quad k_{f2} \quad k_i] \begin{bmatrix} x_1 \\ x_2 \\ x_3 \end{bmatrix} = -\mathbf{KX}$$

and the controlled system (excluding the disturbance input) becomes

$$\dot{\mathbf{X}} = \begin{bmatrix} 0 & 1 & 0 \\ -1 & 0 & 0 \\ 1 & 0 & 0 \end{bmatrix} \mathbf{X} + \begin{bmatrix} 0 \\ 1 \\ 0 \end{bmatrix} u$$

From equation (13.32f),

$$\mathbf{H}(s) = (s\mathbf{I} - \mathbf{A})^{-1}\mathbf{B} = \frac{1}{\det(s)} \begin{bmatrix} s^2 & s & 0 \\ -s & s^2 & 0 \\ s & 1 & s^2+1 \end{bmatrix} \begin{bmatrix} 0 \\ 1 \\ 0 \end{bmatrix} = \frac{1}{\det(s)} \begin{bmatrix} s \\ s^2 \\ 1 \end{bmatrix}$$

where $\det(s) = s(s^2 + 1)$.

From equation (13.33a) the closed-loop characteristic equation is

$$s^3 + s + k_p s + k_{f2}s^2 + k_i = 0, \text{ or } s^3 + k_{f2}s^2 + (1 + k_p)s + k_i = 0 \tag{13.42d}$$

The desired closed-loop characteristic equation is

$$(s + 1.5)(s + 2)(s + 2.5) = s^3 + 6s^2 + 11.75s + 7.5 = 0 \tag{13.42e}$$

Comparing equations (13.42d) and (13.42e), we see that

$$k_{f2} = 6; \qquad k_p = 10.75; \qquad k_i = 7.5 \tag{13.42f}$$

Simulation: The equations to simulate are the following:

System:

$$\dot{x}_1 = x_2; \qquad x_1(0) = 0 \tag{13.43a}$$

$$\dot{x}_2 = -x_1 + u + v; \qquad x_2(0) = 2 \tag{13.43b}$$

$$v = 0.2 \tag{13.43c}$$

$$y = x_1 \tag{13.43d}$$

Observer: equations (13.42a) and (13.42b)

Control:

$$\dot{x}_3 = y \tag{13.43e}$$

$$u = -[k_p y + k_{f2}\hat{x}_2 + k_i x_3] \tag{13.43f}$$

where k_p, k_{f2}, and k_i are given by equation (13.42f)

The simulation results are shown in Fig. 13.20b. The control is able to drive the actual system state to zero, as desired, within about seven time units, even with the steady offset in the estimated pendulum velocity due to the unmeasured disturbance. An observer design that can eliminate this offset is explored in Practice Problem 13.23

Adaptive Control Concepts

The estimated and measured state vector feedback control scheme shown in Fig. 13.20a looks structurally similar to the *self-tuning regulator* (STR) form of adaptive control shown in Fig. 13.21a. Both utilize *feedforward* of the plant input and *feedback* of the plant output to estimate *on-line* process parameters (in the case of the STR) or process state variables (as in Fig. 13.20a). However, there is a major difference in the application of the estimated quantities. Whereas in Fig. 13.20a the estimated values are merely inputs to a predetermined controller, in the STR the estimated parameters are used to adjust the controller. Specifically, the STR includes an on-line solution for an optimal control system if the plant parameters are known. This *regulator design* is then applied "adaptively" to the unknown plant using recursively estimated values of the plant parameters. In general, analytical development of STR systems, which are typically discrete-time or mixed discrete- and continuous-time systems, requires some knowledge of recursive parameter estimation and stochastic control theory, which is not considered in this book.

In contrast, *model reference adaptive control* (MRAC) has been applied to mostly deterministic systems and can be used with both continuous-time and discrete-time systems. According to reference [12], the basic functions common to most adaptive control systems are

1. Identification of unknown parameters or measurement of an index of performance,
2. Decision on the control strategy, and
3. On-line modification of the parameters of the controller or the (plant) input signal.

Again, an estimated and measured state feedback control system lacks the last function, although it includes the first two. In the model reference adaptive system (MRAS), the desired index of performance is given by a *reference model,* and the comparison between the desired and measured indices of performance is normally obtained directly by comparing the output of the reference model with that of the plant. As shown in Figs. 13.21b and 13.21c, two basic types of MRAC systems are distinguished by the use of the result of this comparison either to modify the controller parameters (*parameter-adaptive* MRAS) or to generate an auxiliary plant input (*signal-synthesis* MRAS).

There are three structural classifications of MRAC systems. The most commonly used structure, the *parallel* structure (Fig. 13.22a), is also represented by Figs. 13.21b and 13.21c. The *series* MRAS and the *series-parallel* MRAS are shown in Figs. 13.22b and 13.22c, respectively. Note that there are two alternative series-parallel structures: In one case the reference model is in parallel with another reference model, and in the other case the reference model is in parallel with another *adjustable system.*

To describe analytically the MRAC design problem, we consider the case of parallel MRAS. By *system structure* we mean the nature of the (mathematical) description of the real plant—ordinary differential equations or partial differential equations or difference equations, linear or nonlinear, order of the system, etc. As stated

Figure 13.21

**Adaptive Control Systems:
Self-tuning Regulator and
Model Reference Adaptive
Control Systems**

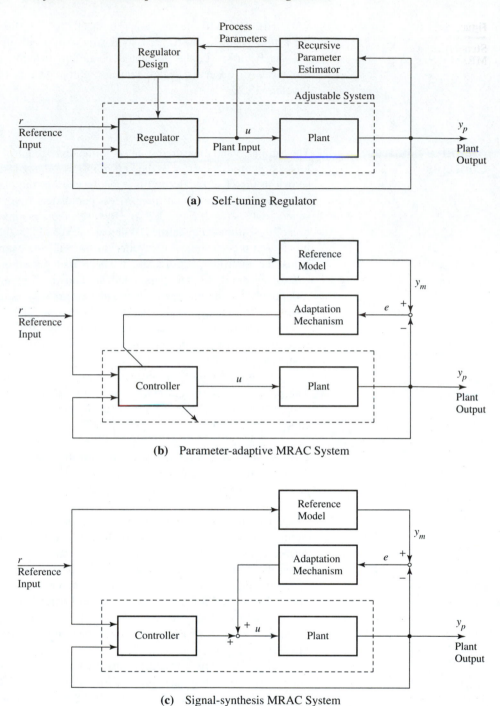

(a) Self-tuning Regulator

(b) Parameter-adaptive MRAC System

(c) Signal-synthesis MRAC System

earlier, both the structure of the plant and the plant parameters can be unknown. Nonetheless, a broad state variable description for each of the subunits (assuming that all state vectors are accessible) can be undertaken as follows (subscripts m and p are used to distinguish variables and parameters of the model and the adjustable system or plant, respectively):

Figure 13.22

Structural Classification of MRAC Systems

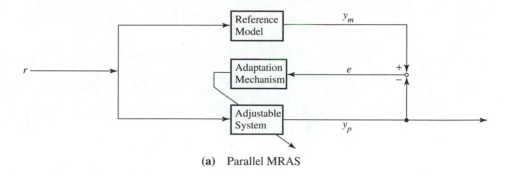

(a) Parallel MRAS

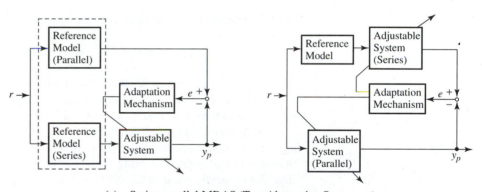

(b) Series MRAS

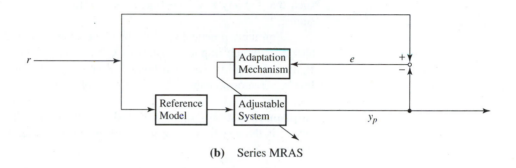

(c) Series-parallel MRAS (Two Alternative Structures)

Reference model:

$$\frac{d}{dt}\mathbf{X}_m = \mathbf{A}_m\mathbf{X}_m + \mathbf{B}_m\mathbf{r}; \qquad \mathbf{X}_m(0) = \mathbf{X}_{m-0} \tag{13.44a}$$

The reference model is assumed to be stable and completely controllable.

Adjustable system—parameter adaptation: The adjustable system parameters are functions of time and the generalized state error, which is defined by

$$\mathbf{e} = \mathbf{X}_m - \mathbf{X}_p \tag{13.44b}$$

where $\mathbf{X}_p$ is the adjustable system state vector, so that

$$\frac{d}{dt}\mathbf{X}_p = \mathbf{A}_p(\mathbf{e}, t)\mathbf{X}_p + \mathbf{B}_p(\mathbf{e}, t)\mathbf{r}; \tag{13.44c}$$

$$\mathbf{X}_p(0) = \mathbf{X}_{p-0}, \quad \mathbf{A}_p(0) = \mathbf{A}_{p-0}, \quad \mathbf{B}_p(0) = \mathbf{B}_{p-0}$$

The design objective is to determine an adaptation law for adjusting $\mathbf{A}_p(\mathbf{e}, t)$ and $\mathbf{B}_p(\mathbf{e}, t)$:

$$\mathbf{A}_p(\mathbf{e}, t) = \mathbf{A}_p(\mathbf{e}, \tau, t) + \mathbf{A}_p(0); \qquad 0 \le \tau \le t \tag{13.44d}$$

$$\mathbf{B}_p(\mathbf{e}, t) = \mathbf{B}_p(\mathbf{e}, \tau, t) + \mathbf{B}_p(0); \qquad 0 \le \tau \le t \tag{13.44e}$$

such that for any $\mathbf{r}$, $\mathbf{e} \longrightarrow \mathbf{0}$ as $t \longrightarrow \infty$. In addition, the entire system is to remain stable.

For many practical systems the plant "characteristic" parameters $\mathbf{A}_p$ are often not alterable (which implies $\mathbf{A}_p(\mathbf{e}, t) = \mathbf{A}_p(0) =$ fixed), whereas plant control parameters $\mathbf{B}_p$ (which can involve such things as physical placement of actuators, power supply levels, actuator characteristics—for example, valve trim) are sometimes alterable.

Adjustable system—signal synthesis: In the case of a signal-synthesis MRAC system, it is the auxiliary input that depends on the generalized state error,

$$\frac{d}{dt}\mathbf{X}_p = \mathbf{A}_p\mathbf{X}_p + \mathbf{B}_p[\mathbf{r} + \mathbf{u}_a(\mathbf{e}, t)]; \qquad \mathbf{X}_p(0) = \mathbf{X}_{p-0}, \quad \mathbf{u}_a(0) = \mathbf{u}_{a-0} \tag{13.44f}$$

and this dependence is determined by the adaptation law

$$\mathbf{u}_a(\mathbf{e}, t) = \mathbf{u}_a(\mathbf{e}, \tau, t) + \mathbf{u}_a(0); \qquad 0 \le \tau \le t \tag{13.44g}$$

where $\mathbf{u}_a(\mathbf{e}, \tau, t)$ is the functional relationship between $\mathbf{u}_a(\mathbf{e}, t)$ and $\mathbf{e}$ during $0 \le \tau \le t$. Again, the design objective is to determine the adaptation law such that for any $\mathbf{r}$ the error is reduced to zero asymptotically while the overall system remains stable.

Reference [12] gives a detailed overview of the design techniques for determining the adaptation law. Because MRAC systems behave like nonlinear systems as well as time-varying systems, stability is an especially important consideration in their design. Consequently, most of the successful design techniques such as the *Lyapunov* approach and the *hyperstability* approach are based on relevant stability criteria. A good survey of these techniques and, in particular, the application of the hyperstability concept, can be found in reference [14]. Because we developed the background on the Lyapunov method of stability analysis in Chapter 10, let us briefly explore an example of a Lyapunov MRAS design scheme for a limited problem.

Signal-Synthesis MRAS Design Illustration

Specifically, we shall consider signal-synthesis MRAC of a plant that can be described in companion form (recall equation (9.34b)):

$$\mathbf{A}_p = \begin{bmatrix} 0 & & & \\ 0 & & & \\ \cdot & & \mathbf{I} & \\ \cdot & & & \\ 0 & & & \\ \hline -a_{p_0} & -a_{p_1} & \cdots & -a_{p_{n-1}} \end{bmatrix}; \qquad \mathbf{B}_p = \begin{bmatrix} 0 \\ 0 \\ \cdot \\ \cdot \\ b_p \end{bmatrix} \tag{13.45}$$

Note that the implied control objects are also single-input systems. According to equations (13.44a), (13.44b), and (13.44f), the differential equation for the error is

$$\dot{\mathbf{e}} = \mathbf{A}_m\mathbf{X}_m + \mathbf{B}_m\mathbf{r} - \mathbf{A}_p\mathbf{X}_p - \mathbf{B}_p[\mathbf{r} + \mathbf{u}_a]$$

$$= \mathbf{A}_m\mathbf{X}_m - \mathbf{A}_m\mathbf{X}_p + \mathbf{A}_m\mathbf{X}_p - \mathbf{A}_p\mathbf{X}_p + (\mathbf{B}_m - \mathbf{B}_p)\mathbf{r} - \mathbf{B}_p\mathbf{u}_a$$

$$= \mathbf{A}_m\mathbf{e} + (\mathbf{A}_m - \mathbf{A}_p)\mathbf{X}_p + (\mathbf{B}_m - \mathbf{B}_p)\mathbf{r} - \mathbf{B}_p\mathbf{u}_a$$

or

$$\dot{\mathbf{e}} = \mathbf{A}_m\mathbf{e} + \mathbf{f} \tag{13.46a}$$

where

$$\mathbf{f} = (\mathbf{A}_m - \mathbf{A}_p)\mathbf{X}_p + (\mathbf{B}_m - \mathbf{B}_p)\mathbf{r} - \mathbf{B}_p\mathbf{u}_a \tag{13.46b}$$

The design objective is to adjust $\mathbf{f}$ in such a way that $\mathbf{e}$ tends to zero as time increases, that is, the error system is asymptotically stable. Consider a positive definite ("Lyapunov") function for the error system

$$\mathbf{V} = \mathbf{e}^T\mathbf{Q}\mathbf{e}$$

From equation (13.46a), the time derivative of $\mathbf{V}$ is

$$\dot{\mathbf{V}} = \dot{\mathbf{e}}^T\mathbf{Q}\mathbf{e} + \mathbf{e}^T\mathbf{Q}\dot{\mathbf{e}} = (\mathbf{e}^T\mathbf{A}_m^T + \mathbf{f}^T)\mathbf{Q}\mathbf{e} + \mathbf{e}^T\mathbf{Q}(\mathbf{A}_m\mathbf{e} + \mathbf{f})$$

$$= \mathbf{e}^T(\mathbf{A}_m^T\mathbf{Q} + \mathbf{Q}\mathbf{A}_m)\mathbf{e} + 2\mathbf{e}^T\mathbf{Q}\mathbf{f}$$

If, as in Chapter 10, we define the matrix $\mathbf{W}$ by

$$\dot{\mathbf{V}} = -\mathbf{e}^T\mathbf{W}\mathbf{e} + 2\mathbf{e}^T\mathbf{Q}\mathbf{f} \tag{13.46c}$$

so that

$$-\mathbf{W} = \mathbf{A}_m^T\mathbf{Q} + \mathbf{Q}\mathbf{A}_m \tag{13.46d}$$

then $\dot{\mathbf{V}}$ is negative definite when $\mathbf{W}$ is positive definite and

$$\mathbf{e}^T\mathbf{Q}\mathbf{f} < 0; \qquad \mathbf{e} \neq 0 \tag{13.46e}$$

Now, according to the Lyapunov stability analysis of Chapter 10, for any positive definite $\mathbf{W}$, a positive definite $\mathbf{Q}$ that satisfies equation (13.46d) can always be found if $\mathbf{A}_m$ is an asymptotically stable matrix. This is part of the already-stated presumption that the reference model is stable and completely controllable. Next, if we let $\mathbf{A}_m$ be of companion form as well, then we can see from equation (13.46b) that all rows of $\mathbf{f}$ will be zero except the last or nth row f_n, which we can write as

$$f_n = \hat{f}_n - b_p u_a \tag{13.46f}$$

where $\hat{f}_n$ is the nth and nonzero row of

$$(\mathbf{A}_m - \mathbf{A}_p)\mathbf{X}_p + (\mathbf{B}_m - \mathbf{B}_p)\mathbf{r}$$

We also have the following simplification:

$$\mathbf{e}^T\mathbf{Q}\mathbf{f} = \mathbf{e}^T q_n' f_n = \mathbf{e}^T q_n'(\hat{f}_n - b_p u_a) \qquad (13.46\text{g})$$

where q_n' is the nth row of $\mathbf{Q}^T$ or the nth column of $\mathbf{Q}$. Assuming that $b_p > 0$, we can see that the following *switching control* law satisfies the requirement (13.46e):

$$|u_a| > \frac{1}{b_p}|\hat{f}_n| \qquad (13.46\text{h})$$

$$\text{sign}\,(u_a) = \text{sign}\,(\mathbf{e}^T q_n')$$

Example 13.10   Signal-Synthesis MRAC of Undamped Oscillator

a. Using the Lyapunov scheme just described, determine a signal-synthesis model reference adaptive control for the undamped pendulum system

$$\dot{x}_1 = x_2 \qquad (13.47\text{a})$$

$$\dot{x}_2 = -\omega^2 x_1 + u + v$$

where v is an unknown disturbance input and ω^2 varies with time in an obscure manner. Assume that the desired reference model behavior is characterized by the eigenvalues -2 and -2.5. Let $v = 0.2$ and

$$\omega^2 = \omega_0^2 + 0.5\sin(4t); \qquad \omega_0^2 = 1$$

and simulate the response of the controlled system for zero setpoint and the initial conditions $x_1(0) = 0$ and $x_2(0) = 2$.

b. Suppose we desire to include PI-feedback control on the pendulum position as in Example 13.9. Can the present Lyapunov MRAC scheme be applied to the resulting system? If so, design the signal-synthesis MRAS for the same desired closed-loop eigenvalues and PI-control parameters used in Example 13.9, and simulate the response for the same pendulum parameter values and initial conditions given in (a). Compare the results with that of the corresponding measured state vector feedback controller determined in Example 13.9.

Solution a: The adjustable system is given by (13.47a). Hence

$$\mathbf{A}_p = \begin{bmatrix} 0 & 1 \\ -\omega^2 & 0 \end{bmatrix}; \qquad \mathbf{B}_p = \begin{bmatrix} 0 \\ 1 \end{bmatrix}$$

Note that since the setpoint is zero, $u = u_a$ in this case.

Reference model: Let

$$\mathbf{A}_m = \begin{bmatrix} 0 & 1 \\ a_1 & a_2 \end{bmatrix} \quad \text{and} \quad \mathbf{B}_m = \mathbf{B}_p = \begin{bmatrix} 0 \\ 1 \end{bmatrix}$$

The reference model is given by equation (13.44a), that is,

$$\dot{x}_{m_1} = x_{m_2} \tag{13.47b}$$

$$\dot{x}_{m_2} = a_1 x_{m_1} + a_2 x_{m_2}$$

The desired stability behavior requires that

$$\det(s\mathbf{I} - \mathbf{A}_m) = s^2 - a_2 s - a_1 = (s + 2)(s + 2.5) = s^2 + 4.5s + 5$$

that is

$$a_1 = -5 \quad \text{and} \quad a_2 = -4.5$$

Signal synthesis: Let

$$\mathbf{Q} = \begin{bmatrix} q_{11} & q_{12} \\ q_{12} & q_{22} \end{bmatrix}$$

Note that

$$\mathbf{A}_m^T \mathbf{Q} + \mathbf{Q}\mathbf{A}_m = \begin{bmatrix} 2a_1 q_{12} & a_1 q_{22} + q_{11} + a_2 q_{12} \\ a_1 q_{22} + q_{11} + a_2 q_{12} & 2(q_{12} + a_2 q_{22}) \end{bmatrix}$$

If we simply take $\mathbf{W} = \mathbf{I}$, then equation (13.46d) yields the following set of algebraic equations:

$$2a_1 q_{12} = -1$$

$$a_1 q_{22} + q_{11} + a_2 q_{12} = 0$$

$$2(q_{12} + a_2 q_{22}) = -1$$

which has the solution $q_{11} = 1.1167$, $q_{12} = 0.1$, and $q_{22} = 0.1333$. The reader can verify that with these elements $\mathbf{Q}$ is indeed positive definite.

According to equations (13.46b) and (13.46f),

$$\hat{f}_n = (\mathbf{A}_m - \mathbf{A}_p)\mathbf{X}_p = (a_1 + \omega_0^2)x_1 + a_2 x_2$$

Note that we used the nominal value of the natural frequency (that is, ω_0), since the actual value of ω is assumed to be unknown. Similarly, we ignored the disturbance v. Finally,

$$\mathbf{e}^T q_n' = e_1 q_{12} + e_2 q_{22} = q_{12}(x_{m_1} - x_1) + q_{22}(x_{m_2} - x_2)$$

From equation (13.46h), with $b_p = 1$, the control can be taken as

$$u_a = \text{sign}(\mathbf{e}^T q_n')[|\hat{f}_n| + \epsilon] \tag{13.47c}$$

where ϵ is a small positive number. The value of ϵ does not affect the controller performance as long as this value is relatively small. For the simulations, $\epsilon = 0.01$ was used.

Equations (13.47a) through (13.47c) are the simulation equations (see script M-file **fig13_23.m; On the Net**). In general, the initial states of the model and the plant will not be coincident. The model initial state can be taken at the equilibrium or desired state, but this is not necessary. The following values were used here: $x_{m_1}(0) = 0$, and $x_{m_2}(0) = 0.5$. Figure 13.23a shows the results of the simulation. The control is remarkably effective. Recall that this is a switching control action.

Figure 13.23

Signal-synthesis MRAC of Undamped Pendulum Servomechanism

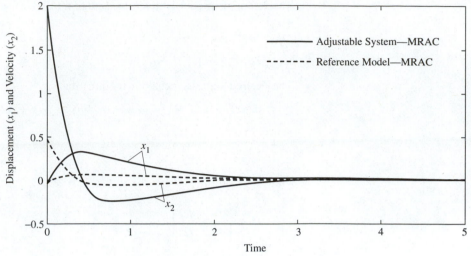

(a)   Performance of Signal-synthesis MRAC

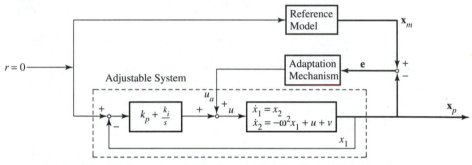

(b)   Configuration of Signal-synthesis MRAC with PI-feedback

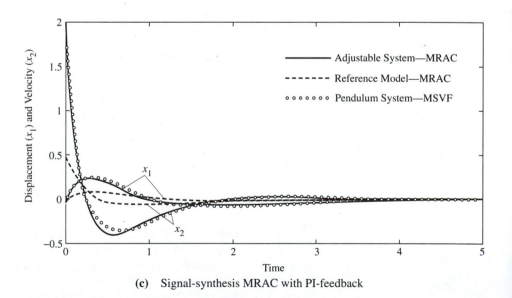

(c)   Signal-synthesis MRAC with PI-feedback

Solution b: Although the Lyapunov scheme introduced here does not specifically include an explicit feedback control loop, it can be applied to the present problem if the adjustable system is defined to incorporate the feedback control, and the resulting system is of the necessary form. The configuration of the MRAC system in this case is shown in Fig. 13.23b.

Adjustable system: With the auxiliary variable for the integral action defined as before, the plant equation becomes

$$\dot{x}_1 = x_2; \qquad x_1(0) = 0$$
$$\dot{x}_2 = -(k_p + \omega^2)x_1 - k_i x_3 + u_a + v; \qquad x_2(0) = 2 \qquad \textbf{(13.48a)}$$
$$\dot{x}_3 = x_1; \qquad x_3(0) = 0$$

By a simple rearrangement of the variables, the above equation can be put in companion form. If we let $x_{p_3} = x_2$, $x_{p_2} = x_1$ and $x_{p_1} = x_3$, in equation (13.48a), the adjustable system is

$$\dot{x}_{p_1} = x_{p_2}; \qquad x_{p_1}(0) = 0$$
$$\dot{x}_{p_2} = x_{p_3}; \qquad x_{p_2}(0) = 0 \qquad \textbf{(13.48b)}$$
$$\dot{x}_{p_3} = -k_i x_{p_1} - (k_p + \omega^2)x_{p_2} + u_a + v; \qquad x_{p_3}(0) = 2$$

which is in companion form:

$$\mathbf{A}_p = \begin{bmatrix} 0 & 1 & 0 \\ 0 & 0 & 1 \\ -k_i & -(k_p + \omega^2) & 0 \end{bmatrix}; \qquad \mathbf{B}_p = \begin{bmatrix} 0 \\ 0 \\ 1 \end{bmatrix}$$

Reference model: Let the reference model be given by equation (13.44a) with $\mathbf{r} = \mathbf{0}$, and

$$\mathbf{A}_m = \begin{bmatrix} 0 & 1 & 0 \\ 0 & 0 & 1 \\ a_1 & a_2 & a_3 \end{bmatrix}; \qquad \mathbf{B}_m = \mathbf{B}_p = \begin{bmatrix} 0 \\ 0 \\ 1 \end{bmatrix}; \qquad \mathbf{X}_m(0) = \begin{bmatrix} 0 \\ 0 \\ 0.5 \end{bmatrix}$$

For the desired closed-loop eigenvalues -1.5, -2, and -2.5,

$$\det(s\mathbf{I} - \mathbf{A}_m) = s^3 - a_3 s^2 - a_2 s - a_1 = (s + 1.5)(s + 2)(s + 2.5)$$
$$= s^3 + 6s^2 + 11.75s + 7.5$$

so that $a_1 = -7.5$, $a_2 = -11.75$, and $a_3 = -6$.

Control synthesis: If we let

$$\mathbf{Q} = \begin{bmatrix} q_{11} & q_{12} & q_{13} \\ q_{12} & q_{22} & q_{23} \\ q_{13} & q_{23} & q_{33} \end{bmatrix} \quad \text{and} \quad \mathbf{W} = \begin{bmatrix} 1 & 0 & 0 \\ 0 & 1 & 0 \\ 0 & 0 & 1 \end{bmatrix}$$

equation (13.46d) yields the positive definite solution

$$q_{11} = 1.8726, \quad q_{12} = 1.2065, \quad q_{13} = 0.0667, \quad q_{22} = 2.0684, \quad q_{23} = 0.1452,$$
$$q_{33} = 0.1075$$

Also

$$\hat{f}_n = (\mathbf{A}_m - \mathbf{A}_p)\mathbf{X}_p = (a_1 + k_i)x_{p_1} + (a_2 + k_p + \omega_0^2)x_{p_2} + a_3 x_{p_3}$$

And

$$\mathbf{e}^T q_n' = e_1 q_{13} + e_2 q_{23} + e_3 q_{33}$$

$$= q_{13}(x_{m_1} - x_{p_1}) + q_{23}(x_{m_2} - x_{p_2}) + q_{33}(x_{m_3} - x_{p_3})$$

The control is again given by equation (13.47c). Figure 13.23c shows (solid lines) the simulation results for this (switching) control and (circles) the corresponding results for the measured state vector feedback control (MSVF) obtained in Example 13.9:

$$u = -[k_p x_1 + k_{f2} x_2 + k_i x_3] = -[10.75 x_1 + 6 x_2 + 7.5 x_3]$$

or in the present formulation

$$u_a = -k_{f2} x_{p_3} = -6 x_{p_3}$$

Note that to be consistent with the Lyapunov MRAC scheme, x_2 or x_{p_3} is assumed to be measurable (no observer is used). The performance of the two controllers under these conditions is very close. Recall that the underlying design criterion utilized for both controllers was *eigenvalue assignment* and the same desired closed-loop characteristics were used. Finally, observe, by comparing Figs. 13.23a and 13.23c, that the addition of a feedback loop to the basic signal-synthesis MRAC did not result in any improvement in performance for this system.

13.4 INTRODUCING CONTINUOUS-TIME OPTIMAL CONTROL AND THE H_∞ CONTROL CONCEPT

We saw in the last example that for the oscillating pendulum servomechanism, state vector feedback control and signal-synthesis model reference adaptive control (one a continuous controller and the other a discontinuous controller) both satisfied well the performance specification implied by the assigned eigenvalues. For a particular performance index, such as the settling time, for example, is there a controller that would be more satisfactory (yield a shorter settling time) than either of these two controllers? Indeed, what is (the description of) the controller that will yield the *shortest* settling time possible for this system? The first question has to do with the existence of an optimal control for a particular system and index of performance. Although this question can sometimes be answered by a detailed analysis of the system (see [15], for example), the common practice of engineers, which is sustained by experience, is to assume that an optimum solution normally exists and when necessary to attempt to discover that solution. A successful search for an optimal solution of course confirms its existence; however, a given unsuccessful search is not conclusive evidence of the lack of existence of an optimum solution.

The second question is the subject of *optimal control system design*. This is a different design problem from the control system design we have undertaken thus far. In particular, whereas in our previous work the form of the controller (for example, PI-, PID-, MSVF-control) was generally known in advance and only the controller parameters were to be determined, in optimal control system design, the solution consists of both the form of the controller and the controller parameter values. Indeed, depending on the problem, the optimal control can be open-loop or closed-loop. A practical problem that arises from this loose nature of the optimal design is the difficulty of implementing the resulting controller, especially if it should turn out to be quite complex. However, the continuing availability of faster and cheaper microcomputers has mostly eliminated this problem.

Another consideration or problem that is of increasing interest in control systems design is the capacity of a feedback controller, even if optimal, to remain satisfactory in the face of unforseen disturbances, model errors such as those due to neglected or ignored system dynamics and nonlinearities or lack of knowledge, errors in parameter identification and state measurement, and variations in the plant and its environment. These factors, in effect, render *uncertain* any model of the system that is used to design the controller. The problem of control system design that purposefully incorporates model uncertainty in general classes of (SISO and MIMO) systems is the subject of *robust control* theory. *Robustness* implies the system's ability to maintain performance in the face of uncertainties. For a feedback control system, we can view the *performance* as an assessment of how well the system deals with unexpected or undesirable inputs and the *robustness* as an assessment of how resilient the system is in the face of internal changes in its behavior (see [16]). A particularly useful index of robustness as well as performance for general SISO and MIMO systems is the H$_\infty$ norm. This chapter concludes with a brief description of the H$_\infty$ optimal control problem.

Nature of the Optimal Control Problem

We have already carried out controller optimization (or *tuning*) in this text. In particular, we used *parameter optimization* to come up with the "best" design for a given controller. That is, given the controller structure (say PI-action), we determine the controller parameter values (k_c and T_i) that minimized a given performance index (IAE) for example). This was the approach used in Example 11.3 to develop an optimal PI-controller for the nonlinear servomechanism. Such a control law is referred to as a *specific optimal control law* because the controller configuration was fixed.† The fundamental optimization problem considered here is to determine, for a particular system, which of all possible controller configurations results in the lowest value of a given performance index. Specifically, the optimal control problem for continuous-time systems can be stated as follows:

Objective:

Determine a control function $\mathbf{U}(t) = \mathbf{U}^*$ that will act on the system over the interval $t_0 \le t \le t_f$ in such a way as to **minimize** the *performance index:*

$$J(\mathbf{U}) = \int_{t_0}^{t_f} G(\mathbf{X}(t), \mathbf{U}(t), t)\, dt \tag{13.49a}$$

subject to the following *constraints.*

Plant dynamics:

$$\dot{\mathbf{X}}(t) = \mathbf{f}(\mathbf{X}(t), \mathbf{U}(t), t); \qquad t \in [t_0, t_f] \tag{13.49b}$$

---

†Since one of the controller parameters in that example (T_i) was prescribed and was not adjusted, the resulting design was *suboptimal* rather than *optimal.*

Admissible controls:

$$\mathbf{U}(t) \in \mathbf{U}_a; \quad \text{for all } t \in [t_0, t_f] \tag{13.49c}$$

Admissible trajectory:

$$\mathbf{X}(t) \in \mathbf{X}_a; \qquad \text{for all } t \in [t_0, t_f] \tag{13.49d}$$

In the preceding problem statement, the *performance index* is a formulation (the integral form is generally adopted) of the requirements on the plant that are to be met in the best possible manner. As stated in equation (13.49a), optimum performance is achieved when the performance index J is a minimum. J is in this sense commonly termed a *cost function,* since it is the general practice of engineers and others to minimize cost. However, an optimization problem can also be stated in terms of *maximization* of a performance criterion, but there is no loss in generality in equation (13.49a) because the minimization of J is equivalent to the maximization of $-J$.

The optimal control problem stated above is a *constrained optimization problem.* Although the requirement on the system dynamics is obvious, it is nonetheless a constraint. The optimal control is to be determined for a plant that behaves in a **specific manner** during the *control interval* $[t_0, t_f]$. Further, there are physical limitations on the plant behavior that are often the result of physical limitations on the plant components—a piston that can travel only so far, a drive motor with so much holding torque, a process fluid that reacts with the material of the containment when its temperature exceeds a certain value, and the like. These physical limitations give rise to restrictions on the values of the control variables as well as the state variables. An *admissible control* is a control that satisfies the *control constraints* during the control interval. Similarly, a state trajectory that satisfies the state variable constraints during the control interval can be called the *admissible trajectory.*

Some Typical Optimal Control Problems

The performance index as a mathematical expression of the desired qualities of the controlled system can be as varied as there are different plants, applications, and design personnel. However, most engineering situations fit or can be made to fit into a few categories of control problems that are defined by the format of their performance indices and some of which have been investigated extensively. By utilizing one or a combination of the following forms of performance criteria, the control system designer can often take advantage of available solutions or methodologies.

Minimum-Time Problem   The objective of the *minimum-time problem* is to transfer the system from a specified initial state $\mathbf{X}(t_0) = \mathbf{X}_0$ at time t_0, to a final state $\mathbf{X}(t_f)$ in minimum time. This objective is expressed by the performance index $J = t_f - t_0$ or in the integral form,

$$J = \int_{t_0}^{t_f} dt \tag{13.50a}$$

The final state $\mathbf{X}(t_f)$ can be specified directly, or a target set $\mathbf{S}_f$ containing the final state $\mathbf{X}(t_f) \in \mathbf{S}_f$ can be given.

Minimum-Fuel-Consumption Problem   The performance criterion for the *minimum-fuel-consumption problem* can be written as

$$J = \int_{t_0}^{t_f} (w_1 |u_1(t)| + w_2 |u_2(t)| + \ldots) \, dt \tag{13.50b}$$

where $w_1, w_2, \ldots$ are appropriate nonnegative weighting factors for the relevant control variables. This equation is applicable to certain control systems such as those in rocket-propelled spacecraft where the magnitude of the control action can be associated with fuel- (the thrust produced by a rocket engine is proportional to the propellant flowrate) or other critical material-consumption.

Minimum-Energy-Consumption Problem   Recall from our discussion of the properties of the Fourier transform that the energy content of a general signal $f(t)$ is proportional to the time integral of the square of the signal, equation (8.5b):

$$E = \int_{-\infty}^{\infty} f(t)^2 \, dt$$

For multiple control inputs, the performance index for the *minimum-energy-consumption problem* is

$$J = \int_{t_0}^{t_f} (\mathbf{U}(t)^T \mathbf{U}(t)) \, dt$$

More generally, the performance criterion can be written as

$$J = \int_{t_0}^{t_f} (\mathbf{U}^T(t) \mathbf{W} \mathbf{U}(t)) \, dt \tag{13.50c}$$

where $\mathbf{W}$ is a positive definite diagonal matrix of weighting factors.

Regulator Problem   The general performance index that is often used to capture the objectives of the *regulator problem* is

$$J = \frac{1}{2} \mathbf{X}^T(t_f) \mathbf{H} \mathbf{X}(t_f) + \frac{1}{2} \int_{t_0}^{t_f} (\mathbf{X}^T(t) \mathbf{Q} \mathbf{X}(t) + \mathbf{U}^T(t) \mathbf{R} \mathbf{U}(t)) \, dt \tag{13.50d}$$

where $\mathbf{R}$ is a real, symmetric, positive definite, constant matrix, and $\mathbf{H}$ and $\mathbf{Q}$ are real, symmetric, positive semidefinite, constant matrices,* and the general factor of $\frac{1}{2}$, which does not influence the optimal control solution, is included to facilitate analysis. To understand equation (13.50d), note first, that it can be viewed as a combination of (three) performance indices: $J = \frac{1}{2}(J_1 + J_2 + J_3)$, where

$$J_1 = \mathbf{X}^T(t_f) \mathbf{H} \mathbf{X}(t_f); \qquad J_2 = \int_{t_0}^{t_f} (\mathbf{X}^T(t) \mathbf{Q} \mathbf{X}(t)) \, dt; \qquad J_3 = \int_{t_0}^{t_f} (\mathbf{U}^T(t) \mathbf{R} \mathbf{U}(t)) \, dt$$

---

*The element of $\mathbf{Q}$ or $\mathbf{R}$ can be functions of time if it is necessary to vary such weighting factors during the control interval.

Next, recall from our discussion in Section 11.1 that the setpoint or desired state is **constant** during *regulation*. Thus without loss of generality, equation (13.50d) presumes that the system coordinates have been transformed, if necessary, so that the origin is the desired state. Under these circumstances, $\mathbf{X}(t)$ also represents the error signal $\mathbf{e}(t)$, and J_2 implies a preference for the *integral square error* (ISE) as a good measure of transient performance. Finally, J_1 represents an objective to minimize the deviation of the final state of the system from the desired or reference state, and J_3 expresses the desire that the magnitudes of the relevant control variables remain within physical bounds during the control interval.

Note that the weighting factors in $\mathbf{Q}$ (also $\mathbf{H}$) and $\mathbf{R}$ have somewhat divergent connotations relative to the expenditure of control effort. Consider the simple case where $\mathbf{Q} = \mathrm{diag}[q_1, q_2, \ldots]$ and $\mathbf{R} = \mathrm{diag}[r_1, r_2, \ldots]$. Whereas a large value of the weighting factor r_i relative to other weighting factors in $\mathbf{R}$ implies greater penalty for the expenditure of the control effort u_i, a relatively large q_i implies the diversion and use of relatively more control effort to regulate x_i.

Equation (13.50d) also presumes the capacity to express the control goal in terms of the system's internal state. If it is necessary to use the system's output instead, the *output regulator problem* can be obtained by substituting the output vector $\mathbf{Y}$ for the state vector $\mathbf{X}$ in equation (13.50d).

Tracking Problem   Using again an ISE-type measure of the tracking objective of controlling a system in such a way that during the control interval the system's tracjectory $\mathbf{X}(t)$ is as close as possible to a desired trajectory $\mathbf{r}(t)$, we can write a general performance index for the *tracking problem* as

$$J = \frac{1}{2}\mathbf{e}^T(t_f)\mathbf{H}\mathbf{e}(t_f) + \frac{1}{2}\int_{t_0}^{t_f}(\mathbf{e}^T(t)\mathbf{Q}\mathbf{e}(t) + \mathbf{U}^T(t)\mathbf{R}\mathbf{U}(t))\,dt; \tag{13.50e}$$

$$\mathbf{e}(t) = \mathbf{r}(t) - \mathbf{X}(t)$$

This is another combined performance index, and the significance of the individual terms is similar to those of corresponding terms in the regulator problem. Also, for output tracking, the error vector can be given by $\mathbf{e}(t) = \mathbf{r}(t) - \mathbf{Y}(t)$, where $\mathbf{r}(t)$ is in this case the desired *output* trajectory.

Some Basic Concepts of Calculus of Variations

Two analytical techniques are often used to solve optimal control problems such as those introduced in the preceding discussion: the *calculus of variations* approach with the *maximum principle* by L. S. Pontryagin (see [17]) as a fundamental theorem, and the theory of *dynamic programming* by R. E. Bellman (see [18]) with the *principle of optimality* as a fundamental rule. Although dynamic programming is a "natural" basis for discrete-time optimal control, the maximum principle is an elegant approach in the continuous-time domain (see [18]).

Before we attempt the solution of some illustrative continuous-time optimal control problems, let us examine some basic concepts of calculus of variations that can underpin such a solution. Our discussion would be brief and cursory. The reader interested in more detailed developments of the subjects introduced here should refer to texts on optimal control methods and theory, such as references [15] and [19].

Functionals and Their Extremization

A *functional* $J(x)$ is a rule that assigns to each function x in a certain class (*domain of the functional*) $\mathcal{D}$, a unique real number. By this definition, the performance indices we considered earlier, equations (13.50a) through (13.50e), are all functionals, and an example of the domain of a functional is the class of continuous functions of time t that are of concern to us. Note that unlike a **function** whose domain can be a subset of real numbers, the domain of a **functional** is a class of functions. It is important to keep this difference in mind, since some of the results of extremization of functionals are similar to what obtains in the minimization or maximization of functions.

Calculus of variations is concerned with finding functions that maximize or minimize (extremize) a given functional. To establish some of the required results of calculus of variations, consider the simpler problem of functionals of a single function. In particular, consider the scalar function $x(t) \in \mathscr{D}$, where $\mathscr{D}$ is the class of functions with continuous first derivatives, and the functional

$$J(x) = \int_{t_0}^{t_f} g(x, \dot{x}, t) \, dt \tag{13.51a}$$

where g has continuous first and second partial derivatives with respect to all its arguments. The problem of determining the conditions on the trajectory $x(t)$ such that along it $J(x)$ is extremum can be broken up into four cases, which can be analyzed separately as follows: (1) *fixed end-points* problem; (2) *specified terminal time, free terminal state* problem; (3) *free terminal time, fixed terminal state* problem; and (4) *free terminal time, free terminal state* problem. We shall examine the first problem and merely state the results for the others. For the last problem, we would assume some restriction on the terminal point in terms of a relationship between the final time and state: $x(t_f) = \phi(t_f)$:

Case of fixed end-points: In this problem the function or trajectory $x(t)$ connecting two given points, $(t_0, x(t_0))$ and $(t_f, x(t_f))$, along which $J(x)$ is extremum is sought. Consider for a given (fixed) t the change in J due to a small change in x, that is, a variation in x, δx:

$$\Delta J(x, \delta x) = J(x + \delta x) - J(x) = \int_{t_0}^{t_f} g(x + \delta x, \dot{x} + \delta \dot{x}, t) \, dt - \int_{t_0}^{t_f} g(x, \dot{x}, t) \, dt$$

Expansion of the RHS in a Taylor series about $(x, \dot{x})$,

$$\Delta J(x, \delta x) = \int_{t_0}^{t_f} \left[g(x, \dot{x}, t) + \left(\frac{\partial}{\partial x} g(x, \dot{x}, t) \right) \delta x + \left(\frac{\partial}{\partial \dot{x}} g(x, \dot{x}, t) \right) \delta \dot{x} + \text{higher variations} \right] dt - \int_{t_0}^{t_f} g(x, \dot{x}, t) \, dt$$

gives the *first variation:*

$$\delta J(x, \delta x) = \int_{t_0}^{t_f} \left[\left(\frac{\partial}{\partial x} g(x, \dot{x}, t) \right) \delta x + \left(\frac{\partial}{\partial \dot{x}} g(x, \dot{x}, t) \right) \delta \dot{x} \right] dt$$

Substituting

$$\delta x(t) = \int_{t_0}^{t_f} \delta \dot{x}(t) \, dt + \delta x(t_0)$$

and integrating by parts, we obtain

$$\delta J(x, \delta x) = \int_{t_0}^{t_f} \left(\frac{\partial}{\partial x} g(x, \dot{x}, t) \right) \delta x \, dt + \frac{\partial}{\partial \dot{x}} g(x, \dot{x}, t) \, \delta x \Big|_{t_0}^{t_f} \tag{13.51b}$$

$$- \int_{t_0}^{t_f} \frac{d}{dt} \left(\frac{\partial}{\partial \dot{x}} g(x, \dot{x}, t) \right) \delta x \, dt$$

Since $(t_0, x(t_0))$ and $(t_f, x(t_f))$ are given, $\delta x(t_f) = \delta x(t_f) = 0$. Hence

$$\delta J(x, \delta x) = \int_{t_0}^{t_f} \left[\frac{\partial}{\partial x} g(x, \dot{x}, t) - \frac{d}{dt}\left(\frac{\partial}{\partial \dot{x}} g(x, \dot{x}, t) \right) \right] \delta x \, dt$$

If $x = x^*$ is the function for which $J(x^*)$ is extremum, then just as for functions,

$$\delta J(x^*, \delta x) = 0 \qquad\qquad\qquad \text{(13.51c)}$$

is a *(fundamental) necessary condition (of variational calculus)*. Thus

$$\int_{t_0}^{t_f} \left[\frac{\partial}{\partial x} g(x^*, \dot{x}^*, t) - \frac{d}{dt}\left(\frac{\partial}{\partial \dot{x}} g(x^*, \dot{x}^*, t) \right) \right] \delta x \, dt = 0$$

Since δx is arbitrary in the interval (t_0, t_f), the necessary condition reduces to

$$\frac{\partial}{\partial x} g(x^*, \dot{x}^*, t) - \frac{d}{dt}\left(\frac{\partial}{\partial \dot{x}} g(x^*, \dot{x}^*, t) \right) = 0; \quad \text{(Euler-Lagrange)} \qquad \text{(13.51d)}$$

which is known as the *Euler-Lagrange equation*. For the fixed end-points case, equation (13.51d) must be solved subject to those given end points: $x(t_0)$ and $x(t_f)$. This is generally a nonlinear and time-varying second-order ODE problem. It is a *two-point boundary value problem* because the boundary conditions are specified at two points rather than at one point *(initial-value problem)*. The Euler-Lagrange equation can be shown to apply to the other cases except that the two-point boundary conditions are different in each case.

Case of fixed t_f, free $x(t_f)$ The first boundary condition is of course $x(t_0)$ specified. This also means that $\delta x(t_0) = 0$. Using this fact and the Euler-Lagrange equation (13.51d), we see from equation (13.51b) that

$$\left. \frac{\partial}{\partial \dot{x}} g(x^*, \dot{x}^*, t) \delta x \right|_{t_f} = 0$$

Since $x(t_f)$ is free, $\delta x(t_f)$ must be arbitrary, so that the second boundary condition is

$$\left. \frac{\partial}{\partial \dot{x}} g(x^*, \dot{x}^*, t) \right|_{t_f} = 0 \qquad\qquad\qquad \text{(13.51e)}$$

Case of free t_f fixed $x(t_f)$: The second boundary condition for this case can be shown to be

$$\left[g(x^*, \dot{x}^*, t) - \dot{x}^* \frac{\partial}{\partial \dot{x}} g(x^*, \dot{x}^*, t) \right]\Bigg|_{t_f} = 0 \qquad\qquad \text{(13.51f)}$$

The first boundary condition is the same for all four cases.

Case of free t_f free $x(t_f) = \phi(t_f)$: The second end condition is given by

$$\left[\left(\frac{\partial}{\partial \dot{x}} g(x^*, \dot{x}^*, t) \right)(\dot{\phi} - \dot{x}^*) \right]\Bigg|_{t_f} + g(x^*, \dot{x}^*, t)|_{t_f} = 0; \quad \text{(transversality condition)} \quad \text{(13.51g)}$$

which is called the *transversality condition*.

Extension to Functionals of More Than One Independent Function and Constrained Extremization

Given the functional of n independent functions,

$$J(\mathbf{X}) = \int_{t_0}^{t_f} G(\mathbf{X}, \dot{\mathbf{X}}, t)\, dt; \qquad \mathbf{X}(t) = \begin{bmatrix} x_1(t) \\ x_2(t) \\ \cdot \\ \cdot \\ \cdot \\ x_n(t) \end{bmatrix} \tag{13.52a}$$

where $x_i(t)$ is an independent function with continuous first derivatives, and G has continuous first and second derivatives with respect to all its arguments, the conditions on the trajectory $\mathbf{X} = \mathbf{X}^*(t)$, along which $J(\mathbf{X})$ is extremum, follow directly from the single-function problem. That is,

$$\frac{\partial}{\partial \mathbf{X}} G(\mathbf{X}^*, \dot{\mathbf{X}}^*, t) - \frac{d}{dt}\left(\frac{\partial}{\partial \dot{\mathbf{X}}} G(\mathbf{X}^*, \dot{\mathbf{X}}^*, t) \right) = \mathbf{0}; \quad \text{(Euler-Langrange)} \tag{13.52b}$$

where

$$\frac{\partial}{\partial \mathbf{Z}} G = \begin{bmatrix} \dfrac{\partial G}{\partial z_1} \\[6pt] \dfrac{\partial G}{\partial z_2} \\ \cdot \\ \cdot \\ \cdot \\ \dfrac{\partial G}{\partial z_n} \end{bmatrix} \tag{13.52c}$$

and the previous boundary conditions are transformed in the same way. Equation (13.52b) consists of n simultaneous Euler-Lagrange equations, each generally a nonlinear, second-order ODE.

Constrained Extremization   If the functions $\mathbf{X}(t)$ are also subject to an ℓ-vector equality constraint,

$$\varphi(\mathbf{X}, \dot{\mathbf{X}}, t) = \mathbf{0}; \qquad t \in [t_0, t_f] \tag{13.53a}$$

the new problem can be solved as before, by considering an *augmented functional* formed by *adjoining* the constraint to J:

$$J_a(\mathbf{X}, \lambda) = \int_{t_0}^{t_f} \{ G(\mathbf{X}, \dot{\mathbf{X}}, t) + \lambda^T(t)[\varphi(\mathbf{X}, \dot{\mathbf{X}}, t)] \}\, dt \tag{13.53b}$$

where the functions $\lambda_i(t)$, $i = 1, 2, \ldots, \ell$, are known as *Lagrange multipliers*. Note that equation (13.53a) reduces the number of components of $\mathbf{X}$ that are independent. Suppose $\mathbf{X}$ has (n) component functions, with the constraint (13.53a), the number of independent components of $\mathbf{X}$ reduces to $(n - \ell)$. If we define the *augmented integrand function*

$$G_a(\mathbf{X}, \dot{\mathbf{X}}, \lambda, t) = G(\mathbf{X}, \dot{\mathbf{X}}, t) + \lambda^T \varphi(\mathbf{X}, \dot{\mathbf{X}}, t) \tag{13.53c}$$

so that

$$J_a(\mathbf{X}, \lambda) = \int_{t_0}^{t_f} G_a(\mathbf{X}, \dot{\mathbf{X}}, \lambda, t)\, dt \tag{13.53d}$$

we can see that the necessary conditions for the constrained problem are

$$\varphi(\mathbf{X}^*, \dot{\mathbf{X}}^*, t) = \mathbf{0}; \qquad t \in [t_0, t_f] \tag{13.53e}$$

and the Euler-Lagrange condition

$$\frac{\partial}{\partial \mathbf{X}} G_a(\mathbf{X}^*, \dot{\mathbf{X}}^*, \lambda^*, t) - \frac{d}{dt}\left(\frac{\partial}{\partial \dot{\mathbf{X}}} G_a(\mathbf{X}^*, \dot{\mathbf{X}}^*, \lambda^*, t) \right) = \mathbf{0} \tag{13.53f}$$

Note from the definition (13.53c), that equation (13.53f) is the same as

$$\frac{\partial}{\partial \mathbf{X}} G(\mathbf{X}^*, \dot{\mathbf{X}}^*, t) + \left(\frac{\partial}{\partial \mathbf{X}} \varphi(\mathbf{X}^*, \dot{\mathbf{X}}^*, t) \right)^T \lambda^*(t) \tag{13.53g}$$

$$- \frac{d}{dt}\left[\left(\frac{\partial}{\partial \dot{\mathbf{X}}} G(\mathbf{X}^*, \dot{\mathbf{X}}^*, t) \right) + \left(\frac{\partial}{\partial \dot{\mathbf{X}}} \varphi(\mathbf{X}^*, \dot{\mathbf{X}}^*, t) \right)^T \lambda^*(t) \right] = \mathbf{0}$$

where

$$\frac{\partial \varphi}{\partial \mathbf{X}} = \begin{bmatrix} \dfrac{\partial \varphi_1}{\partial x_1} & \dfrac{\partial \varphi_1}{\partial x_2} & \cdots & \dfrac{\partial \varphi_1}{\partial x_n} \\[2mm] \dfrac{\partial \varphi_2}{\partial x_1} & \dfrac{\partial \varphi_2}{\partial x_2} & \cdots & \dfrac{\partial \varphi_2}{\partial x_n} \\[1mm] \vdots & \vdots & \cdots & \vdots \\[1mm] \dfrac{\partial \varphi_l}{\partial x_1} & \dfrac{\partial \varphi_l}{\partial x_2} & \cdots & \dfrac{\partial \varphi_l}{\partial x_n} \end{bmatrix} \tag{13.53h}$$

Application to Control Problems

The typical control problems described previously, equations (13.50a) through (13.50e), can all be represented by a general performance index of the form

$$J = H(\mathbf{X}(t_f), t_f) + \int_{t_0}^{t_f} G(\mathbf{X}(t), \mathbf{U}(t), t) \, dt \tag{13.54a}$$

where $\mathbf{X}$ is an n-vector and $\mathbf{U}$ is an r-vector. The problem of discovering a control $\mathbf{U}(t) = \mathbf{U}^*(t)$ that minimizes J subject to the plant dynamics constraint equation (13.49b), repeated here:

$$\dot{\mathbf{X}}(t) = \mathbf{f}(\mathbf{X}(t), \mathbf{U}(t), t); \qquad t \in [t_0, t_f] \tag{13.54b}$$

but with the admissible control and admissible state trajectories allowed to be *unbounded,* can be solved by application of the constrained minimization results obtained previously.

First, if we assume that H is a differentiable function, then we can write

$$H(\mathbf{X}(t_f), t_f) = \int_{t_0}^{t_f}\left(\frac{d}{dt} H(\mathbf{X}(t), t) \right) dt + H(\mathbf{X}(t_0), t_0)$$

$$= \int_{t_0}^{t_f}\left(\left(\frac{\partial}{\partial \mathbf{X}} H(\mathbf{X}, t) \right)^T \dot{\mathbf{X}} + \frac{\partial}{\partial t} H(\mathbf{X}, t) \right) dt + H(\mathbf{X}(t_0), t_0)$$

so that equation (13.54a) becomes

$$J = \int_{t_0}^{t_f} \left(G(\mathbf{X}, \mathbf{U}, t) + \left(\frac{\partial}{\partial \mathbf{X}} H(\mathbf{X}, t) \right)^T \dot{\mathbf{X}} + \frac{\partial}{\partial t} H(\mathbf{X}, t) \right) dt + H(\mathbf{X}(t_0), t_0)$$

However, since $\mathbf{X}(t_0)$ and t_0 are fixed, minimization of the performance index is equivalent to the minimization of the integral performance index

$$J = \int_{t_0}^{t_f} \left(G(\mathbf{X}, \mathbf{U}, t) + \left(\frac{\partial}{\partial \mathbf{X}} H(\mathbf{X}, t) \right)^T \dot{\mathbf{X}} + \frac{\partial}{\partial t} H(\mathbf{X}, t) \right) dt \qquad (13.54\text{c})$$

Second, the plant dynamics constraint (13.54b) is the same as the n-vector equality constraint

$$\mathbf{f}(\mathbf{X}, \mathbf{U}, t) - \dot{\mathbf{X}} = \mathbf{0}; \qquad t \in [t_0, t_f] \qquad (13.54\text{d})$$

We can therefore form the augmented functional

$$J_a = \int_{t_0}^{t_f} \left(G(\mathbf{X}, \mathbf{U}, t) + \left(\frac{\partial}{\partial \mathbf{X}} H(\mathbf{X}, t) \right)^T \dot{\mathbf{X}} + \frac{\partial}{\partial t} H(\mathbf{X}, t) + \lambda^T [\mathbf{f}(\mathbf{X}, \mathbf{U}, t) - \dot{\mathbf{X}}] \right) dt \quad (13.54\text{e})$$

or by defining the augmented integrand function, we have

$$G_a(\mathbf{X}, \dot{\mathbf{X}}, \mathbf{U}, \lambda, t) = G(\mathbf{X}, \mathbf{U}, t) + \lambda^T [\mathbf{f}(\mathbf{X}, \mathbf{U}, t) - \dot{\mathbf{X}}] \qquad (13.54\text{f})$$

$$+ \left(\frac{\partial}{\partial \mathbf{X}} H(\mathbf{X}, t) \right)^T \dot{\mathbf{X}} + \frac{\partial}{\partial t} H(\mathbf{X}, t)$$

$$J_a = \int_{t_0}^{t_f} G_a(\mathbf{X}, \dot{\mathbf{X}}, \mathbf{U}, \lambda, t) \, dt$$

which is the same as equation (13.53d).* It follows that the necessary conditions for determination of the optimal control are

$$\mathbf{f}(\mathbf{X}^*, \mathbf{U}^*, t) - \dot{\mathbf{X}}^* = \mathbf{0}; \qquad t \in [t_0, t_f] \qquad (13.54\text{g})$$

$$\frac{\partial}{\partial \mathbf{X}} G_a(\mathbf{X}^*, \dot{\mathbf{X}}^*, \mathbf{U}^*, \lambda^*, t) - \frac{d}{dt} \left(\frac{\partial}{\partial \dot{\mathbf{X}}} G_a(\mathbf{X}^*, \dot{\mathbf{X}}^*, \mathbf{U}^*, \lambda^*, t) \right) = 0 \qquad (13.54\text{h})$$

However, from the definition (13.54f), the condition (13.54h) can be shown to be equivalent to the following two equations:

$$\dot{\lambda}^*(t) = -\frac{\partial}{\partial \mathbf{X}} [G(\mathbf{X}^*, \mathbf{U}^*, t) + \lambda^{*T} \mathbf{f}(\mathbf{X}^*, \mathbf{U}^*, t)] \qquad (13.54\text{i})$$

$$\frac{\partial}{\partial \mathbf{U}} [G(\mathbf{X}^*, \mathbf{U}^*, t) + \lambda^{*T} \mathbf{f}(\mathbf{X}^*, \mathbf{U}^*, t)] = \mathbf{0} \qquad (13.54\text{j})$$

---

*Note that in this case the net number of independent components is $(n + r) - n = r$, that is, $\mathbf{X}$ and $\mathbf{U}$ components less constraint components.

We introduce the *Hamiltonian,* which is defined as

$$\mathfrak{h}(\mathbf{X}, \mathbf{U}, \lambda, t) = G(\mathbf{X}, \mathbf{U}, t) + \lambda^T \mathbf{f}(\mathbf{X}, \mathbf{U}, t) \tag{13.55a}$$

Equations (13.54i) and (13.54j) then become

$$\dot{\lambda} = -\frac{\partial}{\partial \mathbf{X}} \mathfrak{h}(\mathbf{X}, \mathbf{U}, \lambda, t) \tag{13.55b}$$

$$\mathbf{0} = \frac{\partial}{\partial \mathbf{U}} \mathfrak{h}(\mathbf{X}, \mathbf{U}, \lambda, t) \tag{13.55c}$$

Further, the system dynamics constraint can also be written in terms of the Hamiltonian:

$$\dot{\mathbf{X}} = \frac{\partial}{\partial \lambda} \mathfrak{h}(\mathbf{X}, \mathbf{U}, \lambda, t) \tag{13.55d}$$

Because of the similarity in form between equations (13.55b) and (13.55d), the Lagrange multiplier $\lambda(t)$ is usually referred to as the *co-state vector* in control applications. To obtain a solution to the optimal control problem, we solve equations (13.55a) through (13.55d) simultaneously subject to the boundary conditions. Note that $(2n + r)$ equations are involved, of which $2n$ are first-order differential equations ((13.55b) and (13.55d)) and r are algebraic equations (equation (13.55c)). The boundary conditions consist of n conditions on the initial state $\mathbf{X}(t_0) = \mathbf{X}_0$ and n (or $(n + 1)$ if the terminal time t_f is free) boundary conditions, which depend on the end conditions of the problem, as discussed earlier. Note that when t_f is free, t_f becomes an unknown variable in addition to the $2n$ constants of integration, making a total of $(2n + 1)$ unknowns. The boundary conditions for the four cases of end conditions considered previously are given in Table 13.1. These boundary conditions can be obtained by substituting the augmented integrand function equation (13.54f) into equations (13.51e) through (13.51g), adjusted for vectors.

A procedure for solving an optimal control problem using the results given by equations (13.55a) through (13.55d) and boundary conditions such as those of Table 13.1 is illustrated in the following example.

Table 13.1 Boundary Conditions for Different Optimal Control Problem End Points

PROBLEM END CONDITION	BOUNDARY CONDITIONS
t_f is fixed and $\mathbf{X}(t_f)$ is also fixed	$\mathbf{X}(t_0) = \mathbf{X}_0$ $\mathbf{X}(t_f) = \mathbf{X}_f$
t_f is fixed and $\mathbf{X}(t_f)$ is free	$\mathbf{X}(t_0) = \mathbf{X}_0$ $\left[\dfrac{\partial}{\partial \mathbf{X}} H(\mathbf{X}, t) - \lambda(t) \right]\Big\|_{t_f} = \mathbf{0}$
t_f is free and $\mathbf{X}(t_f)$ is fixed	$\mathbf{X}(t_0) = \mathbf{X}_0,\ \mathbf{X}(t_f) = \mathbf{X}_f,$ and $\left[\mathfrak{h}(\mathbf{X}, \mathbf{U}^*, \lambda, t) + \dfrac{\partial}{\partial \mathbf{X}} H(\mathbf{X}, t) \right]\Big\|_{t_f} = 0$
t_f and $\mathbf{X}(t_f)$ are free but related by $\mathbf{X}(t_f) = \phi(t_f)$	$\mathbf{X}(t_0) = \mathbf{X}_0,\ \mathbf{X}(t_f) = \phi(t_f),$ and $\left[\mathfrak{h}(\mathbf{X}, \mathbf{U}^*, \lambda, t) + \dfrac{\partial}{\partial \mathbf{X}} H(\mathbf{X}, t) \right]\Big\|_{t_f} + \left[\dfrac{\partial}{\partial \mathbf{X}} H(\mathbf{X}, t) - \lambda(t) \right]^T \dot{\phi}(t)\Big\|_{t_f} = 0$

| Example 13.11 | Optimal Control of Undamped Oscillator |

Consider again the undamped pendulum system of Example 13.10. Determine a control that will drive the system from an initial state $x_1(0) = 0$ and $x_2(0) = 2$ to the final state $x_1(t_f) = x_2(t_f) = 0$ in minimum time and with minimum control effort. Simulate the response of this control for the nominal plant as well as for a variable plant with disturbance input as described in Example 13.10.

Solution: Note that the specified end condition is of the type t_f is free and $\mathbf{X}(t_f)$ is fixed.

The *given* (nominal) *plant* is

$$\dot{x}_1 = x_2$$
$$\dot{x}_2 = -x_1 + u$$

thus

$$\mathbf{f}(\mathbf{X}, \mathbf{U}, t) = \begin{bmatrix} x_2 \\ -x_1 + u \end{bmatrix}$$

The *performance index* can be taken as

$$J = \int_0^{t_f} \left(1 + \frac{1}{2} u^2 \right) dt$$

This is a combination of the *minimum-time* and *minimum-energy-consumption* problems and should capture the stated objective of the present problem. Hence

$$H(\mathbf{X}(t_f), t_f) = 0 \quad \text{and} \quad G(\mathbf{X}, u, t) = 1 + \frac{1}{2} u^2$$

Step 1: *Form the Hamiltonian* (equation (13.55a)):

$$\mathfrak{H}(\mathbf{X}, u, \lambda, t) = 1 + \frac{1}{2} u^2 + \lambda_1 x_2 + \lambda_2(-x_1 + u)$$

Step 2: *Solve the control equation* (13.55c) *to obtain* $\mathbf{U}^*$ *in terms of* $\mathbf{X}$, λ, *and* t:

$$\frac{\partial}{\partial u} \mathfrak{H}(\mathbf{X}, u, \lambda, t) = 0 = u + \lambda_2$$

thus

$$u^* = -\lambda_2$$

Note that

$$\frac{\partial^2}{\partial u^2} \mathfrak{H}(\mathbf{X}, u, \lambda, t) = 1$$

so that the Hamiltonian is indeed minimized by u^*.

Step 3: *Substitute* **U*** *in the Hamiltonian to obtain* $\mathfrak{h}(\mathbf{X}, \mathbf{U}^*, \lambda, t)$:

$$\mathfrak{h}(\mathbf{X}, \mathbf{U}^*, \lambda, t) = 1 + \lambda_1 x_2 - \lambda_2 x_1 - \frac{1}{2}\lambda_2^2$$

Step 4: *Using* $\mathfrak{h}(\mathbf{X}, \mathbf{U}^*, \lambda, t)$, *solve the state and co-state equations* (13.55d) *and* (13.55b) *subject to the given boundary conditions* (Table 13.1):

$$
\begin{aligned}
\dot{x}_1 &= x_2 \\
\dot{x}_2 &= -x_1 - \lambda_2
\end{aligned}
\text{; (state equations)}
$$

$$
\begin{aligned}
\dot{\lambda}_1 &= \lambda_2 \\
\dot{\lambda}_2 &= -\lambda_1
\end{aligned}
\text{; (co-state equations)}
$$

We can utilize Laplace transforms to solve these first-order equations. If we let

$$\mathbf{Z} = \begin{bmatrix} x_1 \\ x_2 \\ \lambda_1 \\ \lambda_2 \end{bmatrix}$$

then the equation to solve is $\dot{\mathbf{Z}} = \mathbf{AZ}$, where

$$\mathbf{A} = \begin{bmatrix} 0 & 1 & 0 & 0 \\ -1 & 0 & 0 & -1 \\ 0 & 0 & 0 & 1 \\ 0 & 0 & -1 & 0 \end{bmatrix}$$

The solution (see equation (8.37a)) is

$$\mathbf{Z}(t) = \mathscr{L}^{-1}[(s\mathbf{I} - \mathbf{A})^{-1}]\mathbf{Z}(0)$$

where $\mathbf{Z}(0)$ includes two of the given boundary conditions:

$$\mathbf{Z}(0) = \begin{bmatrix} x_1(0) \\ x_2(0) \\ \lambda_1(0) \\ \lambda_2(0) \end{bmatrix} = \begin{bmatrix} 0 \\ 2 \\ \lambda_1(0) \\ \lambda_2(0) \end{bmatrix}$$

Carrying out the indicated matrix operations and inverse Laplace transformations (see Tables 8.4 and 8.5), we obtain the following result:

$$x_1(t) = 2\sin t + \frac{1}{2}\lambda_1(0)(\sin t - t\cos t) - \frac{1}{2}\lambda_2(0)t\sin t$$

$$x_2(t) = 2\cos t + \frac{1}{2}\lambda_1(0)t\sin t - \frac{1}{2}\lambda_2(0)(\sin t + t\cos t)$$

$$\lambda_1(t) = \lambda_1(0)\cos t + \lambda_2(0)\sin t$$

$$\lambda_2(t) = -\lambda_1(0)\sin t + \lambda_2(0)\cos t$$

Applying the given end-point boundary conditions, we get two equations with three unknowns:

$$0 = 2 \sin t_f + \frac{1}{2}\lambda_1(0)(\sin t_f - t_f\cos t_f) - \frac{1}{2}\lambda_2(0)t_f\sin t_f \tag{a}$$

$$0 = 2 \cos t_f + \frac{1}{2}\lambda_1(0)t_f\sin t_f - \frac{1}{2}\lambda_2(0)(\sin t_f + t_f\cos t_f) \tag{b}$$

The unknowns are $\lambda_1(0)$, $\lambda_2(0)$, and t_f. We obtain the third equation or fifth boundary condition from (see Table 13.1)

$$\left[\mathfrak{h}(\mathbf{X}, \mathbf{U}^*, \lambda, t) + \frac{\partial}{\partial \mathbf{X}}H(\mathbf{X}, t)\right]\Bigg|_{t_f} = 0 = 1 + \lambda_1(t_f)x_2(t_f) - \lambda_2(t_f)x_1(t_f) - \frac{1}{2}\lambda_2^2(t_f)$$

$$= 1 - \frac{1}{2}\lambda_2^2(t_f)$$

which gives

$$\lambda_2(t_f) = \pm\sqrt{2}$$

or using the negative root (which yields a meaningful solution), we obtain the third equation,

$$-\sqrt{2} = -\lambda_1(0)\sin t_f + \lambda_2(0)\cos t_f \tag{c}$$

These three equations, (a) through (c) (including the constraint $t_f > 0$), can be solved simultaneously using, for example, a Mathcad *Given-Find* solve block or MATLAB's **fsolve** function. The solution is $\lambda_1(0) = 0.014$, $\lambda_2(0) = 1.434$, and $t_f = 2.963$.

Step 5: *Obtain the optimal control by substituting the result of step 4 into the expression for* **U*** *obtained in step 2.*
    The optimal control is

$$u^*(t) = -\lambda_2(t) = 0.014 \sin t - 1.434 \cos t; \qquad 0 \le t \le 2.963$$

Note that this is not a feedback control. Figure 13.24a shows the simulation of the system's response under the optimal control law above, with the same law applied for periods $t > 2.963$. Obviously a different control law is required for $t > 2.963$ if the state achieved at $t = t_f$ is to be maintained subsequently. This is another problem, but the solution can be deduced from the system's dynamic equations to be $u(t) = 0$; $t > 2.963$. The effect of the combined control law is shown in Fig. 13.24b. Although the performance is remarkable, the dependence of the present control on the system having the precise initial state, parameters, and operating environment is an obvious disadvantage. This is demonstrated by the dashed curves in Fig. 13.24b, which show the response for the same combined control law but applied to the system with the frequency perturbation $\omega^2 = 1 + 0.5 \sin(4t)$ and disturbance input $v = 0.2$, described in Example 13.10. Not only is the desired final state not achieved at t_f, but this information is not available to the controller, and the control will ordinarily be switched to the second law at $t = t_f$, as if the required final state had been attained.

Figure 13.24

Optimal Control of
Undamped Oscillator

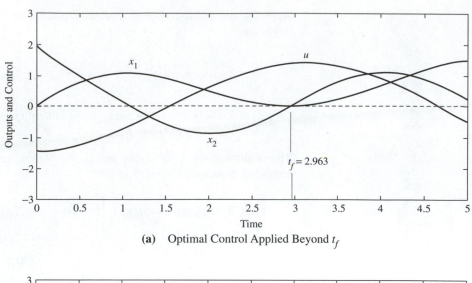

(a)   Optimal Control Applied Beyond t_f

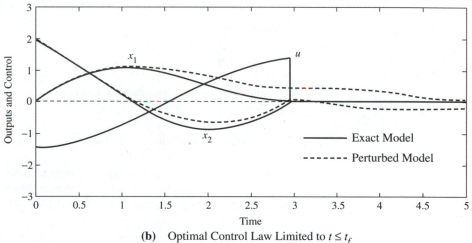

(b)   Optimal Control Law Limited to $t \leq t_f$

The Maximum (Minimum) Principle and Time-Optimal Control

We now examine the full optimal control problem as stated in equations (13.49a) through (13.49d). The difference between the control problem of equations (13.49) and control problems such as the one solved in Example 13.11 is the complete absence in the latter of control variable constraints (equation (13.49c)) and perhaps the exclusion of more general state variable constraints (*equality*-type state variable constraints, see equation (13.53a) were included): equation (13.49d).

Control Constraints and Minimum Principle

The Hamiltonian approach outlined in the last example is applicable to problems in which the control is unbounded. This is a requirement of the variational calculus used to develop the method. Equation (13.55c), for example, requires the partial differentiation of $\natural$ with respect to $\mathbf{U}$, which for arbitrary $\mathbf{X}$, λ and t is feasible only for unconstrained $\mathbf{U}$. Consider now an optimal control problem having the control variable constraints, equation (13.49c) repeated:

$$\mathbf{U}(t) \in \mathbf{U}_a; \quad \text{for all } t \in [t_0, t_f] \tag{13.56a}$$

Equation (13.55c) is not applicable in this case. However, according to the work of Pontryagin (see [17]), the optimal control **U*** can still be chosen to *minimize* the Hamiltonian. Because the Hamiltonian was defined in Pontryagin's original work with a sign opposite of the one used here, his result was stated in terms of *maximizing* $\mathfrak{h}$; hence it is referred to as *Pontryagin's maximum principle*. With respect to the present application, however, the appropriate terminology would be Pontryagin's minimum principle.

Given, then, an optimal control problem with control variable constraints (13.56a), we apply the *minimum principle* in place of equation (13.55c). Specifically, we determine **U*** by minimizing $\mathfrak{h}(\mathbf{X}, \mathbf{U}, \lambda, t)$ with respect to the admissible controls while treating other variables as constant, that is,

$$\mathfrak{h}(\mathbf{X}, \mathbf{U}^*, \lambda, t) = \min_{\mathbf{U}(t) \in \mathbf{U}_a} \mathfrak{h}(\mathbf{X}, \mathbf{U}, \lambda, t) \tag{13.56b}$$

Equation (13.55c) then becomes a special case of equation (13.56b), applicable when the control is unbounded.

State Variable Inequality Constraints

The Hamiltonian technique included a handle for state variable *equality constraints* by using Lagrange multipliers to adjoin the constraints to the performance index. A mechanism for handling state variable *inequality constraints* will round out the present examination of the full optimal control problem of equations (13.49). The usual approach to inequality constraints is to convert them into equivalent equality constraints. In particular, we shall use an approach that incorporates the inequality constraints into expanded *plant dynamics (equality)* constraints. Consider the *l*-vector inequality constraints on state variables:

$$\varphi(\mathbf{X}(t), t) \leq 0 \tag{13.56c}$$

We assume that the components of φ are continuously differentiable in **X** and t. A technique to convert (13.56c) into equality constraints is thus: Let a new $((n + 1)$th) state variable be given by

$$\dot{x}_{n+1} = f_{n+1}$$
$$= -[\varphi_1(\mathbf{X}, t)]^2 U_s(\varphi_1) - [\varphi_2(\mathbf{X}, t)]^2 U_s(\varphi_2) - \cdots - [\varphi_l(\mathbf{X}, t)]^2 U_s(\varphi_l) \tag{13.56d}$$

where $U_s(\varphi_i)$ is a *strict* unit-step function as defined in equation (6.21), repeated here:

$$U_s(\varphi_i(\mathbf{X}, t)) = \begin{cases} 0, & \varphi_i(\mathbf{X}, t) \leq 0 \\ 1, & \varphi_i(\mathbf{X}, t) > 0 \end{cases}; \quad i = 1, 2, \ldots, l \tag{13.56e}$$

Applying the definition (13.56e) to equation (13.56d), we can see that $\dot{x}_{n+1}(t) \leq 0$ for all t and that $\dot{x}_{n+1}(t) = 0$ only at those times when all the constraints (13.56c) are satisfied. If we now specify the boundary conditions for (13.56d) to be

$$x_{n+1}(t_0) = x_{n+1}(t_f) = 0 \tag{13.56f}$$

then

$$x_{n+1}(t) = 0 \text{ for all } t \in [t_0, t_f]$$

is the solution to (13.56d) if $\dot{x}_{n+1}(t) = 0$ for all $t \in [t_0, t_f]$. Since this is indeed the case when the inequality constraints (13.56c) are satisfied for all $t \in [t_0, t_f]$, it follows that successful solution of the optimal control problem, amended with the $(n + 1)$th state equation (13.56d) and additional boundary conditions (13.56f), implies satisfaction of the constraints (13.56c). We shall again illustrate a procedure for solving optimal control problems (including state variable inequality constraints) using Pontryagin's minimum principle with an example.

Example 13.12	Application of Pontryagin's Minimum Principle

Develop completely the boundary value problem that should be solved to obtain the optimal control of the undamped oscillator of Example 13.11 if the controller is subject to saturation,

$$|u(t)| \leq 1 \tag{13.57a}$$

and the state variables are also subject to the constraint

$$x_1(t) \geq 0 \tag{13.57b}$$

$$-0.5 \leq x_2(t) \leq 2 \tag{13.57c}$$

Assume that the performance index, the initial state, and specified final state of the plant remain the same as in Example 13.11.

Solution: We first formally indicate what is given and then outline the steps in the procedure.

Given—*plant equations* (13.54b):

$$\dot{x}_1 = f_1 = x_2$$
$$\dot{x}_2 = f_2 = -x_1 + u$$

Given—*performance index* equation (13.54a):

$$J = \int_0^{t_f} \left(1 + \frac{1}{2}u^2\right) dt \Rightarrow H(\mathbf{X}(t_f), t_f) = 0 \quad \text{and} \quad G(\mathbf{X}, u, t) = 1 + \frac{1}{2}u^2$$

Given—*control variable constraints,* equation (13.56a):

$$|u(t)| \leq 1, \quad \text{for all } t \in [0, t_f]$$

Given—*state variable constraints,* equation (13.56c): The given state variable constraints, equations (13.57b) and (13.57c), can be interpreted in the form of the inequality constraints equation (13.56c), as follows:

$$\varphi_1(\mathbf{X}, t) = -x_1(t) \leq 0$$
$$\varphi_2(\mathbf{X}, t) = [-x_2(t) - 0.5] \leq 0$$
$$\varphi_3(\mathbf{X}, t) = [x_2(t) - 2] \leq 0$$

Note that the constraint on x_2 required two equations, hence $\ell = 3$. We now outline the procedure:

Step 1: *Form the Hamiltonian* (equation 13.55a), including equation (13.56d)).
First, from equation (13.56d),

$$f_{n+1} = -[x_1]^2 U_s(-x_1) - [x_2 + 0.5]^2 U_s(-x_2 - 0.5) - [x_2 - 2]^2 U_s(x_2 - 2)$$

Next, equation (13.55a) gives

$$\mathfrak{h}(\mathbf{X}, u, \lambda, t) = 1 + \frac{1}{2}u^2 + \lambda_1 x_2 + \lambda_2(-x_1 + u)$$

$$-\lambda_3([x_1]^2 U_s(-x_1) + [x_2 + 0.5]^2 U_s(-x_2 - 0.5)$$

$$+ [x_2 - 2]^2 U_s(x_2 - 2))$$

Step 2: *Minimize $\mathfrak{h}(\mathbf{X}, \mathbf{U}, \lambda, t)$ with respect to all admissible control vectors to obtain $\mathbf{U}^*$ in terms of $\mathbf{X}, \lambda$, and t.*
Consider only the terms of $\mathfrak{h}(\mathbf{X}, \mathbf{U}, \lambda, t)$ that contain $u(t)$, abbreviated $\mathfrak{h}{:}u$:

$$\mathfrak{h}{:}u = \frac{1}{2}u^2 + \lambda_2 u \qquad\qquad\text{(13.57d)}$$

When $|u(t)| \le 1$,

$$\frac{\partial}{\partial u}\mathfrak{h}(\mathbf{X}, u, \lambda, t) = 0 \Rightarrow u^*(t) = -\lambda_2(t)$$

Thus as long as $|\lambda_2(t)| \le 1$,

$$u^*(t) = -\lambda_2(t)$$

However, when $|\lambda_2(t)| > 1$, equation (13.57d) suggests that the following control minimizes $\mathfrak{h}(\mathbf{X}, \mathbf{U}, \lambda, t)$:

$$u^*(t) = \begin{cases} -1, & \text{for } \lambda_2(t) > 1 \\ +1, & \text{for } \lambda_2(t) < -1 \end{cases}$$

Thus summarizing, we have

$$u^*(t) = \begin{cases} -1; & \lambda_2(t) > 1 \\ -\lambda_2(t); & \lambda_2(t) \le 1 \\ 1; & \lambda_2(t) < -1 \end{cases} \qquad\qquad\text{(13.57e)}$$

Step 3: *Substitute $\mathbf{U}^*$ in the Hamiltonian to obtain $\mathfrak{h}(\mathbf{X}, \mathbf{U}^*, \lambda, t)$, equation (13.56b):*

$$\mathfrak{h}(\mathbf{X}, u^*, \lambda, t) = 1 + \frac{1}{2}u^{*2} + \lambda_1 x_2 + \lambda_2(-x_1 + u^*)$$

$$-\lambda_3([x_1]^2 U_s(-x_1) + [x_2 + 0.5]^2 U_s(-x_2 - 0.5)$$

$$+ [x_2 - 2]^2 U_s(x_2 - 2))$$

where u^* is given by equation (13.57e).

Step 4: *Using $\mathfrak{h}(\mathbf{X}, \mathbf{U}^*, \lambda, t)$, solve the state and co-state equations (13.55d) and (13.55b) subject to the given boundary conditions (equation (13.56f) and Table 13.1):*

$$\dot{x}_1 = x_2$$

$$\dot{x}_2 = -x_1 + u^* \qquad\qquad\qquad \text{; (state equations)}$$

$$\dot{x}_3 = -[x_1]^2 U_s(-x_1) - [x_2 + 0.5]^2 U_s(-x_2 - 0.5) - [x_2 - 2]^2 U_s(x_2 - 2)$$

$$\dot{\lambda}_1 = \lambda_2 + 2x_1\lambda_3 U_s(-x_1)$$

$$\dot{\lambda}_2 = -\lambda_1 + 2\lambda_3[x_2 + 0.5]U_s(-x_2 - 0.5)$$

$$\qquad\qquad + 2\lambda_3[x_2 - 2]U_s(x_2 - 2); \quad \text{(co-state equations)}$$

$$\dot{\lambda}_3 = 0$$

The boundary conditions are

$$x_1(0) = 0, \quad x_2(0) = 2, \quad x_1(t_f) = 0, \quad x_2(t_f) = 0$$

including, from equation (13.56f),

$$x_3(0) = 0, \quad x_3(t_f) = 0$$

and from Table 13.1

$$\left[\mathfrak{h}(\mathbf{X}, \mathbf{U}^*, \lambda, t) + \frac{\partial}{\partial \mathbf{X}} H(\mathbf{X}, t)\right]\bigg|_{t_f} = 0$$

or

$$1 + \frac{1}{2}[u^*(t_f)]^2 + \lambda_2(t_f)u^*(t_f) = 0$$

where $u^*(t)$ is given by equation (13.57e).

The state and co-state equations and the above boundary conditions define the boundary value problem. To complete the outline of the procedure described here, we give the last step, which is the same as in Example 13.11:

Step 5: *Obtain the optimal control by substituting the result of step 4 into the expression for $\mathbf{U}^*$ obtained in step 2.*

The boundary value problem obtained in this example is nonlinear, and a closed-form solution is not available. However, a numerical solution could be obtained and then fit with an analytical function. Such a control would, strictly speaking, be considered *suboptimal*.

Time-Optimal Control

The development in Example 13.12 suggests that when there are limits on the control action, the optimal control is for a good part of the time at the controller limit in one direction or the other. For the *minimum-time problem,* which is one of the categories of optimal control problems we introduced as closely approximating some real design objectives, we find that for most practical control variable constraints or nonlinearities, such as *saturation* and *hysteresis* (recall Fig. 9.21), the optimal control is a *discontinuous* or *switching control law* (on-off or bang-bang control, for example). We illustrate with the following example how the switching law can be discovered (much like the state space investigation carried out in Section 9.4) without solving the complete boundary value problem. Some computational and practical difficulties that may be encountered in implementing such a controller will also be discussed in the example.

| Example 13.13 | Minimum-Time Control of Undamped Servomechanism |

Suppose we remove the state variable constraints on the undamped oscillator of Example 13.12 and instead of the performance index stated in that problem, seek to minimize the time to drive the system from the given initial state to the given final state. Derive an optimal control solution to this problem and simulate the response of the controlled system.

Solution:

Given—plant dynamics:

$$\dot{x}_1 = f_1 = x_2$$
$$\dot{x}_2 = f_2 = -x_1 + u$$

Given—performance index:

$$J = \int_0^{t_f} (1)\, dt \Rightarrow H(\mathbf{X}(t_f), t_f) = 0 \quad \text{and} \quad G(\mathbf{X}, u, t) = 1$$

Given—control variable constraints:

$$|u(t)| \le 1, \quad \text{for all } t \in [0, t_f]$$

Step 1: *The Hamiltonian is*

$$\mathfrak{h}(\mathbf{X}, u, \lambda, t) = 1 + \lambda_1 x_2 - \lambda_2 x_1 + \lambda_2 u$$

Step 2: To obtain $\mathbf{U}^*$, consider the terms of $\mathfrak{h}(\mathbf{X}, \mathbf{U}, \lambda, t)$ that contain $u(t)$: $\mathfrak{h}{:}u = \lambda_2 u$. With $|u(t)| \le 1$, the Hamiltonian will be minimized if

$$u^*(t) = \begin{cases} -1, & \text{if } \lambda_2(t) > 0 \\ +1, & \text{if } \lambda_2(t) < 0 \end{cases}$$

or

$$u^*(t) = -\text{sign}(\lambda_2(t))$$

Step 3: *Substitute $\mathbf{U}^*$ in the Hamiltonian:*

$$\mathfrak{h}(\mathbf{X}, u^*, \lambda, t) = 1 + \lambda_1 x_2 - \lambda_2 x_1 - \lambda_2\, \text{sign}(\lambda_2)$$

Step 4: *Solve the state and co-state equations object to the boundary conditions:*

$$\begin{aligned} \dot{x}_1 &= x_2 \\ \dot{x}_2 &= -x_1 + u^*;\ u^* = -\text{sign}(\lambda_2) \end{aligned}; \text{ (state equations)}$$

$$\begin{aligned} \dot{\lambda}_1 &= \lambda_2 \\ \dot{\lambda}_2 &= -\lambda_1 \end{aligned}; \text{ (co-state equations)}$$

Let us consider the set of state trajectories that result for each value of the control:

Case of $u^* = -1$ **Case of $u^* = 1$**

$$\dot{x}_1 = x_2$$ $$\dot{x}_1 = x_2$$

$$\dot{x}_2 = -x_1 - 1$$ $$\dot{x}_2 = -x_1 + 1$$

solving, we obtain

$$x_1 = -A \cos t + B \sin t - 1$$ $$x_1 = -A \cos t + B \sin t + 1$$

$$x_2 = A \sin t + B \cos t$$ $$x_2 = A \sin t + B \cos t$$

where A and B are constants that depend on the initial conditions. For arbitrary initial values

$$x_1(0) = x_{1-0}, \text{ and } x_2(0) = x_{2-0},$$

$$x_1(t) = (x_{1-0} + 1)\cos t + x_{2-0}\sin t - 1 \qquad x_1(t) = -(1 - x_{1-0})\cos t$$
$$+ x_{2-0}\sin t + 1$$

$$x_2(t) = -(x_{1-0} + 1)\sin t + x_{2-0}\cos t \qquad x_2(t) = (1 - x_{1-0})\sin t + x_{2-0}\cos t$$

Figure 13.25

Time-optimal Control of Undamped Servomechanism

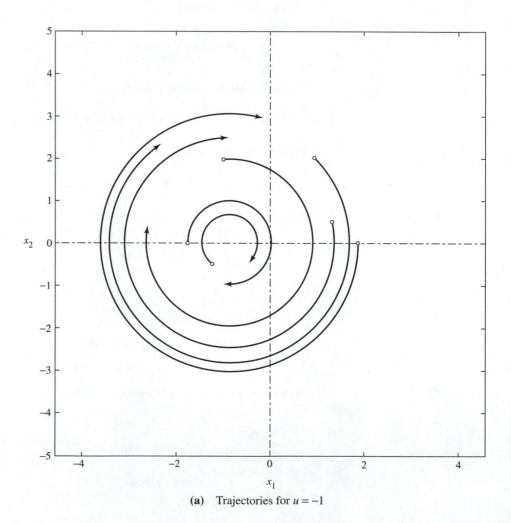

(a)   Trajectories for $u = -1$

Figure 13.25 *continued*

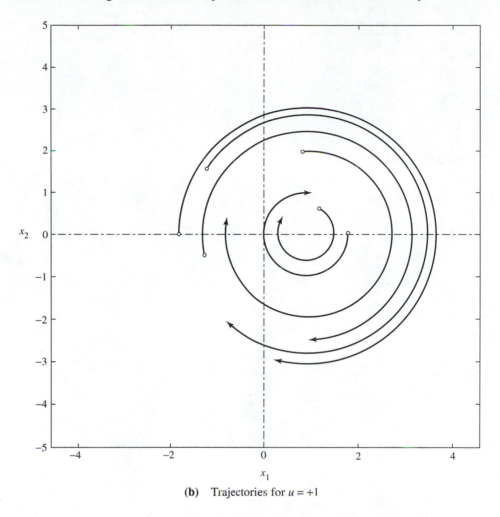

(b) Trajectories for $u = +1$

The sets of trajectories described by the preceding equations can be obtained simply by computer solution of these equations starting from different initial states.† The results shown in Figs. 13.25a and 13.25b indicate that the trajectories are sets of *circles*. Imagine the two sets of trajectories combined in the state space (see Fig. 13.25c). There is one trajectory in each set of circles (recall that without active control trajectories in the same set will not cross one another) that passes through the origin, the desired terminal state. Any motion starting on such a trajectory with the consistent value of u^* will continue along the trajectory until it reaches the origin. Motion proceeding otherwise along a trajectory of the family of circles corresponding to one value of u^* can, on reaching a trajectory of the other family that goes directly to the origin, be transferred to that trajectory by switching to the corresponding value of u^*. Our problem is to determine precisely at what point in time or at what point in state space to perform the switching so that the motion will subsequently continue to the origin.

Of course, one approach is to solve the full boundary-value problem to determine the time function $\lambda_2(t)$ from which $u^*(t) = -\text{sign}(\lambda_2(t))$ is the switching function. This is an *open-loop* type of control, as we have observed before. A usually much less formidable approach is to seek the *switching curve*—the locus of points in state space

---

†See script M-file **fig13_25; On the Net.**

Figure 13.25 *continued*

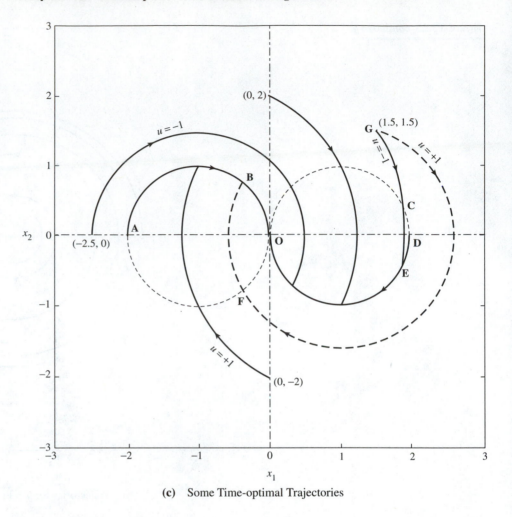

(c)   Some Time-optimal Trajectories

along which the control must be switched from one value consistent with the previous trajectory to another value corresponding to the desired new trajectory—in other words, a parametric representation of the switching function sign $(\lambda_2(t))$ in terms of the state variables. The resulting *switching control law* is of a *feedback* type, since it depends on the state variables.

Continuing with the idea of a parametric curve in x_1 and x_2, and in this case, circular trajectories, we have

For $u^* = -1$

$$(x_1 + 1)^2 = (x_{1-0} + 1)^2\cos^2 t + x_{2-0}^2\sin^2 t + 2(x_{1-0} + 1)x_{2-0}\sin t \cos t$$

and

$$x_2^2 = (x_{1-0} + 1)^2\sin^2 t + x_{2-0}^2\cos^2 t - 2(x_{1-0} + 1)x_{2-0}\sin t \cos t$$

so that

$$(x_1 + 1)^2 + x_2^2 = (x_{1-0} + 1)^2 + x_{2-0}^2 = C^2$$

which is the equation of a circle with center at $x_1 = -1, x_2 = 0$. Proceeding in the same fashion, we have

For $u^* = 1$

$$(x_1 - 1)^2 + x_2^2 = (1 - x_{1-0})^2 + x_{2-0}^2 = D^2$$

which also represents a circle but with center at $x_1 = 1$, $x_2 = 0$. If these curves must pass through the origin, then the admissible initial states on each curve must be such that $C = 1$ and $D = 1$. Thus the switching curve is composed of the pair of circles of unit radii:

$$(x_1 + 1)^2 + x_2^2 = 1; \quad \text{corresponding to the control, } u^* = -1$$

and

$$(x_1 - 1)^2 + x_2^2 = 1; \quad \text{corresponding to the control, } u^* = 1$$

These circles can be recognized in Figs. 13.25a and 13.25b as the two trajectories, one from each family of circles, that intersect the origin. Figure 13.25c shows the actual switching curve along with a few *controlled* trajectories obtained by simulating the response of the controlled system with different initial states. Note that although each initial state is simultaneously on a trajectory of each family of circles, a judicious choice of the initial value of u^* is necessary to assure the path of least travel time. Consider, for example, the trajectory starting at G (1.5, 1.5). For the initial control $u = 1$, motion will proceed on a trajectory (dashed line) that is a circle with center at (1, 0). This trajectory first intersects the "left" unit circle at F. If the control is switched at this point to $u = -1$, the system will move clockwise along this unit circle to the origin, that is, the path FABO. However, if switching is deferred until this unit circle is intersected again at B, the path to the origin FBO will be shorter and faster. Thus only the top part of the "left" circle is part of the switching curve. Similarly, if the initial control at (1.5, 1.5) is $u = -1$, the alternative paths to the origin, GCDEO with switching at C and GEO with switching at E, suggest that only the lower part of the "right" unit circle is on the switching curve. Thus for an initial state (1.5, 1.5) the judicious initial control is $u = -1$, and the optimal trajectory is the solid line GEO.

The true switching curve for this problem is the curve ABOED, which is indicated by a solid line in Fig. 13.25c, and the switching control law is

$$u^*(t) = \begin{cases} -1; & (x_1(t) + 1)^2 + x_2^2(t) = 1 \quad \text{and} \quad x_2(t) \geq 0 \\ +1; & (x_1(t) - 1)^2 + x_2^2(t) = 1 \quad \text{and} \quad x_2(t) \leq 0 \end{cases}$$

Close examination of Fig. 13.25c (on magnified scales) will reveal that following the switch in the control value at the switching curve, the subsequent path followed in each case is a circular arc of radius slightly greater or less than unity. This is a numerical necessity. Because the state of the system is computed at discrete points in time, a tolerance band must be imposed on the switching curve that is consistent with the degree of time discretization, the computational accuracy, and the speed with which the system is moving in the vicinity of the switching point. If the band is too narrow, the switching point can be missed altogether. If the band is too wide, the switching can occur too far from the switching curve, so that the subsequent motion will be on a trajectory that does not pass sufficiently close to the origin. The corresponding sources of this problem in a real system are errors in measurement of the state and the accuracy of the description of the system by the model with which the switching curve was derived.

Figure 13.25 *continued*

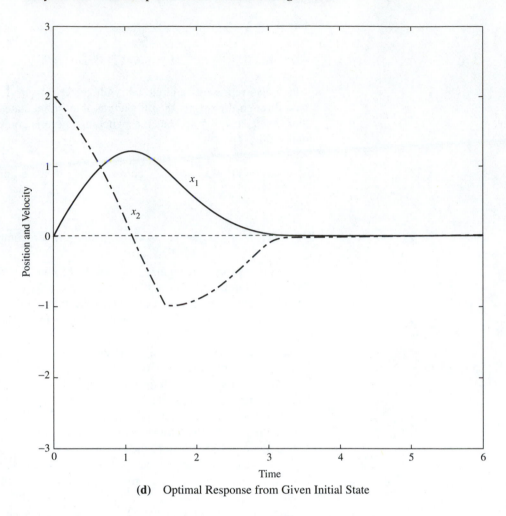

(d)   Optimal Response from Given Initial State

Also, if a digital computer is used to realize the controller, then the relative *sampling rate* will be an additional factor.

The resolution of this problem using the differential equation solvers **ode23** and **ode45** in MATLAB (3.5, Student Edition) is complicated by the *explicit* nature of these solution schemes and the automatic step size feature. The following approach was used to obtain the responses shown in Fig. 13.25c (see script M-file **fig13_25.m, On the Net**): **ode23** was used with a specified desired accuracy of the solution. The integration time was set at 6.0, which was considered a sufficient period for the system to be driven to the origin from any of the initial states considered. The present undamped pendulum system represents a process of moderate speed. Since ordinarily the automatic step size that would be determined for this process and the desired integration time could be larger than is necessary for a reasonable switching curve tolerance band, the step size was indirectly adjusted by adding a faster but decoupled process to the system equations. Indeed, the *stiff* problem (see Example 6.21),

$$\dot{x}_3 = a(\sin t - x_3); \qquad a = 100$$

was used. Note that the value of a controls the speed of the fast component of this stiff system, while speed of the slow component is comparable to that of the pendulum system. A good ("SINGULARITY LIKELY"-free) desired accuracy of the solution under these circumstances was determined at 0.2E-04 (the default accuracy of solution for **ode23** is 1.0E-03). The function M-file **timeopt.m (On the Net)** containing the differential equations, including the simulation of the controller, is as follows:

```
function xdot = timeopt (t,x)
ep1 = .0068; ep2 = .029;
xdot = zeros(3,1);
% CONTROL
if (x(2) >= 0) & (abs((x(1)+1)*(x(1)+1)+x(2)*x(2)-1)
 < ep1),u_s = -1;
end;
if (x(2) <= 0) & (abs((x(1)-1)*(x(1)-1)+x(2)*x(2)-1)
< ep1),
u_s = 1;
end;
if (abs(x(1)) < ep2) & (abs(x(2)) < ep2),
u_s = 0;
end;
% SYSTEM
xdot(1) = x(2);
xdot(2) = -x(1) + u_s;
xdot(3) = 100*(sin(t) - x(3));
```

The width of the tolerance band on the switching curve that is represented in the program by the quantity $ep1$ was determined as the smallest value for which the switching curve was not "missed" for all initial states considered.

The next problem is what happens when the system reaches the origin. If the control is left at its final value, the system will not come to rest at the origin but will proceed to move around the unit circle corresponding to that control. The solution illustrated in function M-file **timeopt.m** is to switch the control off ($u = 0$) when the system is close enough to the origin. Again, having a tolerance region *(circle)* about the origin is both a numerical and a practical necessity. The radius of the *tolerance circle* is represented by $ep2$ in this function M-file. With $ep1$ fixed at 0.0068, the minimum $ep2$ for which the arrival at the origin was detected for all initial states considered was 0.029. The initial state $(-2.5, 0)$ happened to be controlling for both $ep1$ and $ep2$. Obviously, larger values for these quantities may be required for a less specific design (see Practice Problem 13.30).

Figure 13.25d shows the optimal response of the controlled system from the given initial state $\mathbf{X}(0) = [0 \quad 2]^T$ and confirms the effectiveness of the time-optimal controller.

Depending on the detailed objective, in practice the solution adopted here for the terminal motion may not be the best, although it is likely to be the cheapest since turning the control off is still part of the general *discontinuous* nature of the controller. However, $u = 0$ implies that the system is no longer under control and will eventually drift out of the tolerance circle, where the controller will be reactivated. Thus the system will not actually come to rest but will oscillate about the origin with an amplitude determined by the radius of the tolerance circle (limit cycling). An alternative approach to a completely discontinuous control system is to change to a *linear* (feedback) type control (for example, $u = -k_1 x_1 - k_2 x_2$) once the system enters the tolerance region about the final state. This *saturation* type of control, as opposed to the earlier *relay* (actually *relay-with-dead-zone;* see Fig. 9.21) type, would be able to bring the system to rest at the origin but could represent a more expensive controller.

....

Optimal Linear Quadratic Regulator

The results of Example 13.12 point out the difficulties in obtaining closed-form optimal control laws for general problems, whereas Example 13.11 suggests that where a closed-form solution is obtained, it may not be of the advantageous feedback controller configuration. It can, however, be shown that for a **linear system** and **unbounded control,** the *quadratic* regulator problem (including also the tracking problem; performance indices (13.50d) and (13.50e)) has a **state vector feedback controller as optimal.** We outline the main results next without a formal derivation.

The *linear quadratic regulator* (LQR) refers to the following optimal control problem:

Minimize the *quadratic* performance index, equation (13.50d), repeated here:

$$J = \frac{1}{2}\mathbf{X}^T(t_f)\mathbf{H}\mathbf{X}(t_f) + \frac{1}{2}\int_{t_0}^{t_f} (\mathbf{X}^T(t)\mathbf{Q}(t)\mathbf{X}(t) + \mathbf{U}^T(t)\mathbf{R}(t)\mathbf{U}(t))\,dt \qquad \textbf{(13.58a)}$$

where $\mathbf{H}$ and $\mathbf{Q}(t)$ are real, symmetric, positive semidefinite $n \times n$ matrices; $\mathbf{R}(t)$ is a real, symmetric, positive definite $r \times r$ matrix; and $t \in [t_0, t_f]$ is a specified control interval.

Subject to *linear* plant dynamics and initial state

$$\dot{\mathbf{X}}(t) = \mathbf{A}(t)\mathbf{X}(t) + \mathbf{B}(t)\mathbf{U}(t); \qquad \mathbf{X}(t_0) = \mathbf{X}_0 \qquad \textbf{(13.58b)}$$

where $\mathbf{X}$ is an n-state vector, $\mathbf{U}$ is an r-control input vector, and $\mathbf{A}$ and $\mathbf{B}$ are $n \times n$ and $n \times r$ real matrices, respectively.

The Hamiltonian for this problem, equation (13.55a), is

$$\mathfrak{h}(\mathbf{X}, \mathbf{U}, \lambda, t) = \frac{1}{2}(\mathbf{X}^T(t)\mathbf{Q}(t)\mathbf{X}(t) + \mathbf{U}^T(t)\mathbf{R}(t)\mathbf{U}(t)) + \lambda^T(t)[\mathbf{A}(t)\mathbf{X}(t) + \mathbf{B}(t)\mathbf{U}(t)]$$

and the control equation (13.55c) yields

$$\frac{\partial}{\partial u}\mathfrak{h}(\mathbf{X}, \mathbf{U}, \lambda, t) = \mathbf{R}(t)\mathbf{U}(t) + \mathbf{B}^T(t)\lambda(t) = 0$$

or

$$\mathbf{U}*(t) = -\mathbf{R}(t)^{-1}\mathbf{B}^T(t)\lambda(t) \qquad \textbf{(13.59a)}$$

Substituting equation (13.59a) into the co-state and state equations (13.55b) and (13.55c), we obtain

$$\dot{\lambda}(t) = -\mathbf{Q}(t)\mathbf{X}(t) - \mathbf{A}^T(t)\lambda(t) \qquad \textbf{(13.59b)}$$

$$\dot{\mathbf{X}}(t) = \mathbf{A}(t)\mathbf{X}(t) - \mathbf{B}(t)\mathbf{R}(t)^{-1}\mathbf{B}^T(t)\lambda(t) \qquad \textbf{(13.59c)}$$

If we let

$$\lambda(t) = \mathbf{S}(t)\mathbf{X}(t) \tag{13.59d}$$

where **S** is an $n \times n$ time-varying, symmetric, positive definite matrix, then substitution in (13.59b) gives

$$\dot{\mathbf{S}}(t)\mathbf{X}(t) + \mathbf{S}(t)\dot{\mathbf{X}}(t) = -\mathbf{Q}(t)\mathbf{X}(t) - \mathbf{A}^T(t)\mathbf{S}(t)\mathbf{X}(t)$$

and using (13.59c) to eliminate $\dot{\mathbf{X}}$, gives us

$$\dot{\mathbf{S}}(t)\mathbf{X}(t) + \mathbf{S}(t)[\mathbf{A}(t)\mathbf{X}(t) - \mathbf{B}(t)\mathbf{R}(t)^{-1}\mathbf{B}^T(t)\mathbf{S}(t)\mathbf{X}(t)] = -\mathbf{Q}(t)\mathbf{X}(t)$$
$$- \mathbf{A}^T(t)\mathbf{S}(t)\mathbf{X}(t)$$

or

$$\dot{\mathbf{S}}(t) + \mathbf{Q}(t) - \mathbf{S}(t)\mathbf{B}(t)\mathbf{R}(t)^{-1}\mathbf{B}^T(t)\mathbf{S}(t) + \mathbf{S}(t)\mathbf{A}(t) + \mathbf{A}^T(t)\mathbf{S}(t)]\mathbf{X}(t) = \mathbf{0}$$

which can be true for an arbitrary $\mathbf{X}(t)$ only if

$$\dot{\mathbf{S}}(t) + \mathbf{Q}(t) - \mathbf{S}(t)\mathbf{B}(t)\mathbf{R}(t)^{-1}\mathbf{B}^T(t)\mathbf{S}(t) + \mathbf{S}(t)\mathbf{A}(t) + \mathbf{A}^T(t)\mathbf{S}(t) = \mathbf{0} \tag{13.60a}$$

Equation (13.60a) is referred to as the *matrix Riccati equation.* It is a matrix differential equation representing a system of $n(n + 1)/2$ (not n^2 because of the symmetry of **S**) first-order, nonlinear, time-varying ordinary differential equations. The boundary condition for equation (13.60a) is (consider, for example, equation (13.59d) and the second entry in Table 13.1)

$$\mathbf{S}(t_f) = \mathbf{H} \tag{13.60b}$$

If a symmetric positive definite solution $\mathbf{S}(t)$ of equations (13.60a) and (13.60b) can be found, then from equations (13.59a) and (13.59d), the optimal control being sought is

$$\mathbf{U}^*(t) = -\mathbf{R}(t)^{-1}\mathbf{B}^T(t)\mathbf{S}(t)\mathbf{X}(t) = -\mathbf{K}(t)\mathbf{X}(t);$$
$$\mathbf{K}(t) = \mathbf{R}(t)^{-1}\mathbf{B}^T(t)\mathbf{S}(t) \tag{13.60c}$$

which is a (time-varying) *state vector feedback* control law. Note that in this development the plant, equation (13.58b), can be *either time-varying or stationary.*

Infinite-Time (Stationary) Linear State Regulator

Consider specifically a *stationary* plant:

$$\mathbf{A}(t) = \mathbf{A}, \mathbf{B}(t) = \mathbf{B}$$

The performance index of equation (13.58a) refers to a *finite-time problem,* since t_f is finite. If we allow the final time to be arbitrarily large (approach infinity), the final state $\mathbf{X}(t_f)$ should approach the equilibrium state, which, as usual, is presumed to be the origin $\mathbf{0}$. Note that the idea of an equilibrium state anticipates that the *controlled* system is asymptotically stable. The *terminal state constraint* in J is thus no longer necessary and the *infinite-time (time-invariant) linear state regulator* problem is

Minimize the *quadratic* performance index,

$$J = \frac{1}{2}\int_{t_0}^{\infty} (\mathbf{X}^T(t)\mathbf{Q}\mathbf{X}(t) + \mathbf{U}^T(t)\mathbf{R}\mathbf{U}(t))\, dt \tag{13.61a}$$

where $\mathbf{Q}$ is an $n \times n$ real, symmetric, positive semidefinite, constant matrix; $\mathbf{R}$ is an $r \times r$ real, symmetric, positive definite, constant matrix; and

$$t \in [t_0, t_f]$$

is a specified control interval.

Subject to *linear stationary* plant dynamics and initial state,

$$\dot{\mathbf{X}}(t) = \mathbf{A}\mathbf{X}(t) + \mathbf{B}\mathbf{U}(t); \qquad \mathbf{X}(t_0) = \mathbf{X}_0 \tag{13.61b}$$

The previous formulations for the finite terminal time case are still valid except that an additional condition must be imposed on the system to guarantee a bounded performance index. This condition is that

The pair $[\mathbf{A}, \mathbf{B}]$ must be completely controllable. $\tag{13.61c}$

Note that the linear regulator problem does not require that the original plant be stable. Whereas for finite t_f, J can still be finite even when an uncontrollable state is unstable, for infinite t_f, an unstable state that is also not controllable will eventually cause J to become unbounded, hence the condition (13.61c).

It can be shown that for the performance index (13.61a) and the assumption $\mathbf{X}(t \longrightarrow \infty) = \mathbf{0}$, the matrix $\mathbf{S}(t)$ must be constant, so that $\dot{\mathbf{S}}(t) = \mathbf{0}$. Hence equation (13.60a) reduces to a nonlinear algebraic matrix equation, the *reduced* (or *algebraic*) *matrix Riccati equation:*

$$\mathbf{S}\mathbf{A} + \mathbf{A}^T\mathbf{S} - \mathbf{S}\mathbf{B}\mathbf{R}^{-1}\mathbf{B}^T\mathbf{S} + \mathbf{Q} = \mathbf{0} \tag{13.62a}$$

The optimal control is in this case a (time-invariant) state feedback control:

$$\mathbf{U}^*(t) = -\mathbf{R}^{-1}\mathbf{B}^T\mathbf{S}\mathbf{X}(t) = -\mathbf{K}\mathbf{X}(t); \qquad \mathbf{K} = \mathbf{R}^{-1}\mathbf{B}^T\mathbf{S} \tag{13.62b}$$

A symmetric positive definite solution $\mathbf{S}$ of equation (13.62a) always exists (see [15]) if the *controlled system* is asymptotically stable, that is, if $(\mathbf{A} - \mathbf{B}\mathbf{R}^{-1}\mathbf{B}^T\mathbf{S}) = (\mathbf{A} - \mathbf{B}\mathbf{K})$ is a stable matrix; and the condition (13.61c) is necessary to assure such asymptotic stability of the feedback system.

The Riccati equation appears in other control fields such as *stochastic* control, and its solution is of fundamental importance in control system design. Although a number of special numerical schemes are available for solving the Riccati equation, we shall illustrate here only MATLAB's **lqr** function, which is applicable to the solution of the linear quadratic regulator problem equations (13.61a) through (13.61c) and the associated matrix Riccati equation (13.62a), and control (13.62b).

Example 13.14	Linear Quadratic Regulator Design

Given again the low-order pendulum system

$$\dot{x}_1 = x_2$$

$$\dot{x}_2 = -x_1 + u$$

Determine the optimal control u that will minimize the performance index (13.61a) with

$$\mathbf{Q} = \begin{bmatrix} 1 & 0 \\ 0 & 1 \end{bmatrix} \quad \text{and} \quad \mathbf{R} = R = 1$$

a. By direct solution of equations (13.62a) and (13.62b).
b. By use of MATLAB's *lqr* function.

Solution a: We note that

$$\mathbf{A} = \begin{bmatrix} 0 & 1 \\ -1 & 0 \end{bmatrix}; \quad \mathbf{A}^T = \begin{bmatrix} 0 & -1 \\ 1 & 0 \end{bmatrix}; \quad \mathbf{R}^{-1} = 1$$

Also

$$[\mathbf{B} \quad \mathbf{AB}] = \begin{bmatrix} 0 & 1 \\ 1 & 0 \end{bmatrix} \quad \text{and} \quad \det \begin{bmatrix} 0 & 1 \\ 1 & 0 \end{bmatrix} = -1$$

so that the given system is completely controllable.
Let

$$\mathbf{S} = \begin{bmatrix} s_{11} & s_{12} \\ s_{21} & s_{22} \end{bmatrix}$$

then

$$\mathbf{SB} = \begin{bmatrix} s_{12} \\ s_{22} \end{bmatrix} \quad \text{and} \quad \mathbf{B}^T\mathbf{S} = [s_{21} \quad s_{22}]$$

Applying equation (13.62a), we obtain

$$\mathbf{SA} + \mathbf{A}^T\mathbf{S} - \mathbf{SBR}^{-1}\mathbf{B}^T\mathbf{S} + \mathbf{Q} = \begin{bmatrix} -s_{12} & s_{11} \\ -s_{22} & s_{21} \end{bmatrix} + \begin{bmatrix} -s_{21} & -s_{22} \\ s_{11} & s_{12} \end{bmatrix}$$

$$- \begin{bmatrix} s_{12}s_{21} & s_{12}s_{22} \\ s_{22}s_{21} & s_{22}^2 \end{bmatrix} + \begin{bmatrix} 1 & 0 \\ 0 & 1 \end{bmatrix} = \mathbf{0}$$

which yields the four algebraic equations

$$-s_{12} - s_{21} - s_{12}s_{21} + 1 = 0$$

$$s_{11} - s_{22} - s_{12}s_{22} = 0$$

$$-s_{22} + s_{11} - s_{22}s_{21} = 0$$

$$s_{21} + s_{12} - s_{22}^2 + 1 = 0$$

The middle two equations imply that $s_{12} = s_{21}$, so that **S** is indeed symmetric. The first equation then gives

$$s_{21}^2 + 2s_{21} - 1 = 0$$

which has the solution

$$s_{21} = s_{12} = -1 \pm \sqrt{2} \tag{a}$$

The last equation and one of the middle equations then yield

$$s_{22} = \pm\sqrt{2s_{21} + 1} \tag{b}$$

$$s_{11} = s_{22}(1 + s_{21}) \tag{c}$$

We can determine the correct sign in equations (a) and (b) from the requirement that the controlled system be stable. We consider now equation (13.62b). The optimal feedback gain matrix is

$$\mathbf{K} = \mathbf{R}^{-1}\mathbf{B}^T\mathbf{S} = [s_{21} \quad s_{22}]$$

so that $u = -s_{21}x_1 - s_{22}x_2$. Substituting for u in the plant equation, we obtain

$$\begin{matrix} \dot{x}_1 = x_2 \\ \dot{x}_2 = -x_1 - s_{21}x_1 - s_{22}x_2 \end{matrix} \longrightarrow \ddot{x}_2 + s_{22}\dot{x}_2 + (1 + s_{21})x_2 = 0$$

We see that the controlled system characteristic equation is

$$p^2 + s_{22}p + (1 + s_{21}) = 0 \tag{d}$$

and for stability,

$$1 + s_{21} > 0 \quad \text{or } s_{21} > -1, \quad \text{and } s_{22} > 0$$

The proper signs in equations (a) and (b) are thus determined, that is, $s_{21} = -1 + \sqrt{2} = 0.4142$, and $s_{22} = \sqrt{(2s_{21} + 1)} = 1.3522$. Also, from equation (c), $s_{11} = 1.9123$. Hence

$$\mathbf{S} = \begin{bmatrix} 1.9123 & 0.4142 \\ 0.4142 & 1.3522 \end{bmatrix} \quad \text{and} \quad \mathbf{K} = [0.4142 \quad 1.3522]$$

Note that from equation (d), the eigenvalues of the controlled system are

$$p_1, p_2 = -\frac{s_{22}}{2} \pm \sqrt{\frac{s_{22}^2}{4} - 1 - s_{21}} = -0.6761 + j0.9783, -0.6761 - j0.9783$$

Solution b: MATLAB's *lqr* function, which is available in the *Signals and Systems Toolbox,* can be applied to this problem as shown in the following script M-file, **exm13_14.m.** The command [K, S, E] = lqr(A, B, Q, R) returns the optimal feedback gain matrix K = **K** in equation (13.62b); the unique positive definite solution S = **S** of the matrix Riccati equation (13.62a); and the closed-loop system eigenvalues, E = eig(A − BK), all associated with the time-invariant linear quadratic regulator problem (13.61). Note that this command will not give a solution for a case where the condition (13.61c) is not met.

Script M-file **exm13_14.m**

```
% Exm13_14.m
clear; clg;
A = [0 1; -1 0];
B = [0 1]';
Q = [1 0; 0 1];
R = 1;
[K,S,E] = lqr(A,B,Q,R);
disp('K ='); disp(K);
disp('S ='); disp(S);
disp('E ='); disp(E);
```

When the above script is executed, the following output is obtained:

K =

      0.4142   1.3522

S =

      1.9123   0.4142

      0.4142   1.3522

E =

      $-0.6761 + 0.9783i$

      $-0.6761 - 0.9783i$

which is in agreement with the previous solution.

Output Regulator Problem and Nonzero Setpoint

In the *output regulator problem* we seek controls $\mathbf{U}(t)$: $t \in [t_0, t_f]$ that minimize the quadratic performance index

$$J = \frac{1}{2}\mathbf{Y}^T(t_f)\hat{\mathbf{H}}\mathbf{Y}(t_f) + \frac{1}{2}\int_{t_0}^{t_f}(\mathbf{Y}^T(t)\hat{\mathbf{Q}}(t)\mathbf{Y}(t) + \mathbf{U}^T(t)\mathbf{R}(t)\mathbf{U}(t))\,dt \qquad \textbf{(13.63a)}$$

subject to the plant dynamics equation (13.58b) and the output equation

$$\mathbf{Y}(t) = \mathbf{C}\mathbf{X}(t) \qquad\qquad \textbf{(13.63b)}$$

If we substitute equation (13.63b) into the performance index, we obtain

$$J = \frac{1}{2}\mathbf{X}^T(t_f)\mathbf{C}^T\hat{\mathbf{H}}\mathbf{C}\mathbf{X}(t_f) + \frac{1}{2}\int_{t_0}^{t_f}(\mathbf{X}^T\mathbf{C}^T(t)\hat{\mathbf{Q}}(t)\mathbf{C}\mathbf{X}(t) + \mathbf{U}^T(t)\mathbf{R}(t)\mathbf{U}(t))\,dt$$

which is the same as the *state* regulator performance index provided we set

$$\mathbf{H} = \mathbf{C}^T\hat{\mathbf{H}}\mathbf{C} \qquad\qquad \textbf{(13.63c)}$$

$$\mathbf{Q}(t) = \mathbf{C}^T\hat{\mathbf{Q}}(t)\mathbf{C} \qquad\qquad \textbf{(13.63d)}$$

Thus the solution of the output regulator problem is the same as equations (13.60) or (13.62) with the substitutions (13.63c) and (13.63d). Note that if $\hat{\mathbf{H}}$ and $\hat{\mathbf{Q}}$ are symmetric positive semidefinite, as required, then $\mathbf{H}$ and $\mathbf{Q}$ as given by equations (13.63c) and (13.63d), respectively, are also positive semidefinite provided $\mathbf{C}$ is not zero, as must be presumed.

Nonzero Setpoint   The preceding solutions to the linear quadratic regulator problem assume that the desired final state is the equilibrium point or the origin. However, there are situations in practice in which this assumption cannot be sustained. Consider the time-invariant system, for example. Suppose the desired operating point or the setpoint is $\mathbf{X}_d$. Substituting the following transformations

$$\hat{\mathbf{X}}(t) = \mathbf{X}(t) - \mathbf{X}_d \tag{13.64a}$$

$$\hat{\mathbf{U}}(t) = \mathbf{U}(t) - \mathbf{U}_d \tag{13.64b}$$

where $\mathbf{U}_d$ is to be determined, into the state equation (13.61b) gives us

$$\frac{d}{dt}\hat{\mathbf{X}}(t) = \mathbf{A}\hat{\mathbf{X}}(t) + \mathbf{B}\hat{\mathbf{U}}(t) + \mathbf{A}\mathbf{X}_d + \mathbf{B}\mathbf{U}_d \tag{13.64c}$$

For the equilibrium point of the system (13.64c) to be the origin, we require

$$\mathbf{A}\mathbf{X}_d + \mathbf{B}\mathbf{U}_d = 0 \tag{13.64d}$$

$\mathbf{U}_d$ can thus be viewed as the static control required to maintain the state at the point $\mathbf{X}_d$ different from the origin. Unfortunately, for an arbitrary $\mathbf{X}_d$, equation (13.64d) cannot be solved for $\mathbf{U}_d$ in many cases. However, solutions are more likely to be found if the dimension of $\mathbf{U}$ is the same as that of $\mathbf{X}$ or if the application allows some choice in the value of $\mathbf{X}_d$. Assuming for now that a solution of (13.64d) exists, then we can solve the regulator problem for the system

$$\frac{d}{dt}\hat{\mathbf{X}}(t) = \mathbf{A}\hat{\mathbf{X}}(t) + \mathbf{B}\hat{\mathbf{U}}(t) \tag{13.64e}$$

to obtain the control

$$\hat{\mathbf{U}}(t) = -\mathbf{K}\hat{\mathbf{X}}(t) \tag{13.64f}$$

In terms of the original system variables, the control to be applied is

$$\mathbf{U}(t) = \hat{\mathbf{U}}(t) + \mathbf{U}_d = -\mathbf{K}\hat{\mathbf{X}}(t) + \mathbf{U}_d = -\mathbf{K}\mathbf{X}(t) + \mathbf{K}\mathbf{X}_d + \mathbf{U}_d$$

or

$$\mathbf{U}(t) = -\mathbf{K}\mathbf{X}(t) + \mathbf{U}_{sp} \tag{13.65a}$$

where

$$\mathbf{U}_{sp} = \mathbf{K}\mathbf{X}_d + \mathbf{U}_d$$

If in particular we have an output regulator problem or a choice in the values of significant elements of $\mathbf{X}_d$ that can be expressed by the output (setpoint) equation

$$\mathbf{Y}_{sp} = \mathbf{C}\mathbf{X}_d \tag{13.65b}$$

then we can determine $\mathbf{U}_{sp}$ as follows: We substitute equation (13.65a) into the state equation (13.61b) to get

$$\dot{\mathbf{X}} = (\mathbf{A} - \mathbf{BK})\mathbf{X} + \mathbf{BU}_{sp}$$

For an asymptotic steady-state value

$$\mathbf{X}(t \longrightarrow \infty) = \mathbf{X}_d$$

we must have

$$\mathbf{0} = (\mathbf{A} - \mathbf{BK})\mathbf{X}_d + \mathbf{BU}_{sp} \text{ or } \mathbf{X}_d = -(\mathbf{A} - \mathbf{BK})^{-1}\mathbf{BU}_{sp}$$

Substituting equation (13.65b), we obtain

$$\mathbf{Y}_{sp} = -\mathbf{C}(\mathbf{A} - \mathbf{BK})^{-1}\mathbf{BU}_{sp} \tag{13.65c}$$

Generally, as before, a solution of (13.65c) is not guaranteed. However, if $\mathbf{U}$ and $\mathbf{Y}$ have the same dimensions and $\mathbf{C}(\mathbf{A} - \mathbf{BK})^{-1}\mathbf{B}$ is not singular, a *unique* solution of (13.65c) for $\mathbf{U}_{sp}$ always exists. Finally, note that these output regulator results presume the system is completely observable.

Example 13.15   Output Regulator with Nonzero Setpoint

The third-order system

$$\dot{\mathbf{X}} = \mathbf{AX} + \mathbf{B}u; \qquad \mathbf{A} = \begin{bmatrix} -3 & 4 & 4 \\ 1 & -3 & -1 \\ -1 & 2 & 0 \end{bmatrix}, \quad \mathbf{B} = \begin{bmatrix} 1 \\ 2 \\ 0 \end{bmatrix}; \qquad y = \mathbf{CX};$$

$$\mathbf{C} = \begin{bmatrix} -1 & 1 & 0 \end{bmatrix}$$

was shown in Section 9.4 to be controllable and observable. Determine the optimal control for output regulation of this system about the setpoint $y_{sp} = 1.5$ if the performance index is

$$J = \frac{1}{2} \int_{t_0}^{\infty} (y^2(t) + u^2(t))\, dt$$

Simulate the response of the controlled system from a zero initial state.

Solution: From the given performance index, $\mathbf{R} = 1$ and $\hat{\mathbf{Q}} = 1$, so that from equation (13.63d),

$$\mathbf{Q} = \begin{bmatrix} -1 \\ 1 \\ 0 \end{bmatrix} \begin{bmatrix} -1 & 1 & 0 \end{bmatrix} = \begin{bmatrix} 1 & -1 & 0 \\ -1 & 1 & 0 \\ 0 & 0 & 0 \end{bmatrix}$$

which is symmetric positive semidefinite.

With this value of $\mathbf{Q}$ and the given values of $\mathbf{A}$ and $\mathbf{B}$, we can solve, using MATLAB, for instance, the matrix Riccati equation (13.62a) and feedback gain matrix equation (13.62b). The result (see script M-file **fig13_26.m; On the Net**) is

K =

−0.0041	0.3648	0.5360

S =

0.1161	−0.0601	0.0915
−0.0601	0.2125	0.2223
0.0915	0.2223	0.4747

E =

$$-2.6403 + 0.7177i$$

$$-2.6403 - 0.7177i$$

$$-1.4449$$

We can also solve equation (13.65c) to yield $u_{sp} = -1.8028$. Thus the optimal control law is

$$u = 0.0041x_1 - 0.3648x_2 - 0.5360x_3 - 1.8028$$

Figure 13.26 shows the response of the system under this controller. The output converges very quickly to the desired value. The behavior of the individual state variables, the internal state of the system, is also shown in Fig. 13.26. Note that the resulting steady-state or equilibrium state corresponding to the specified desired output y_{sp} is $x_1 = -3$, $x_2 = -1.5$, and $x_3 = -0.5$.

....

Performance Robustness and the H_∞ Norm

An advantage of feedback control that was demonstrated for SISO systems is that under certain conditions (for example, when inputs cover only a certain frequency range) it also provides some degree of immunity to plant uncertainties from various sources. In other words, SISO feedback control is *robust* under certain conditions. A fundamental problem in control theory and practice is the design of controllers that perform well not just for a single plant and under known inputs but rather for a family of plants, whether SISO or MIMO, under various types of inputs and disturbances. H_∞ control is an example of a formulation of such a problem. Our efforts in this section are limited, however, only to explaining the H_∞ norm and developing a typical statement of the H_∞ optimal control problem.

Another Look at Performance and Robustness in SISO Systems

In Section 11.1 we saw that for a SISO system the basic objective of feedback control, the ability to follow a reference input with negligible error, is determined by (see equation (11.6b) the *loop sensitivity S*:

$$S = S_G^{G_{CL}} = \frac{1}{1 + G_o(s)} = \frac{1}{T(s)} \tag{13.66a}$$

Figure 13.26

Response of Output Regulator with Nonzero Setpoint

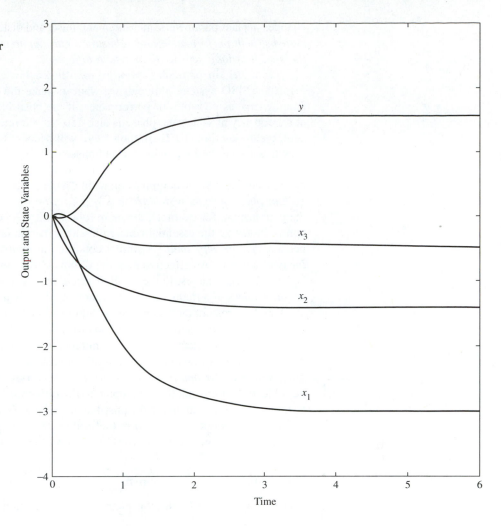

where $T(s) = 1 + G_o(s)$ is termed the *return difference*. Also, for most systems, and over a reasonable range of frequencies, this objective, and hence a major aspect of the *performance* of the feedback system, is determined by the *open-loop gain* $|G_o(s)|$ (or, more generally, the return difference, which we want to make as large as possible). Moreover, both the loop sensitivity and the open-loop gain (and the return difference) are functions of frequency. Consequently, in Chapter 12 we designed SISO feedback control systems on the basis of frequency performance specifications, by manipulating (shaping and reshaping) either a semilogarithmic display (Bode design) or a complex plane (polar) display (Nyquist design) of the open-loop gain versus frequency function.

We also saw in Section 11.1 that the stability of the SISO feedback system (recall that the closed-loop characteristic equation is given by $T(s) = 0$) as well as the degree of its immunity to changes in plant or forward-loop parameters (see equation (11.5e)), its ability to minimize the effect of measurement noise (see equation (11.6c)), and its capacity to reject disturbances that enter the plant (see Practice Problem 11.3d) all are also determined by the loop sensitivity (or the return difference) and, in some of these cases, in a manner that is in conflict with the main requirements for good performance. As we explained at the beginning of Section 13.4, *robustness* deals with the resilience of a feedback control system in the face of such changes in internal behavior. That is, robustness deals with maintenance of stability and minimization of the effects of changes in plant parameters, measurement noise, and disturbance inputs. Thus *for SISO systems, both performance and robustness can apparently be evaluated by the same tools such as loop sensitivity or return difference.*

Consider, in particular, *stability robustness.* For a minimum phase (and unity feedback) SISO system, although we want to make the return difference or the loop gain as large as possible for performance, if the open-loop gain should become unity at a frequency at which the phase is also 180° (if the return difference should become zero, recall equation (10.21a)), the system will become unstable. Thus for stability we seek to separate the loop gain of 1 and phase of 180° by as large a frequency range as possible, in other words, we try to keep the return difference from coming close to zero at any frequency. We accomplished this in Chapters 10 and 12 by specifying either the *gain or phase margin, which are thus specifications of stability robustness.* The higher the gain margin, for example, the more robust (with respect to stability) the SISO system is. Note that the essential concern in either margin is that the return difference not become *zero or very small.* A similar condition in multivariable systems is the *singularity* of a matrix. An effective measure of how close a matrix comes to being singular is the ratio of the largest to the smallest *singular values* of the matrix (see below). The singular values of a frequency matrix function are known as *principal gains* (see below also), and it turns out that the notion of "multivariable system gains" based on such singular values exhibits similar correlations with performance and robustness as do SISO system gains. For instance, there are a number of theorems (see [20] for a typical one) that define the gain and phase margins of a multivariable system on the basis of the singular values of the *multivariable return difference* (see equation (13.66d)). However, our objective here is not the development of *multivariable frequency response analysis* or multivariable design using Nyquist-like techniques (see [16], for example). Our objective is to introduce the H$_\infty$ norm (defined next) as an appropriate measure of performance and robustness for both SISO and MIMO systems.

Principal Gains and H$_\infty$ Norm

Consider again the closed-loop matrix transfer function, equation (13.18b), repeated here:

$$\mathbf{G}_{CL}(s) = [\mathbf{I} + \mathbf{G}_o(s)]^{-1}\mathbf{G}_o(s) = \mathbf{S}(s)\mathbf{G}_o(s) = \mathbf{T}(s)^{-1}\mathbf{G}_o(s)$$
$$\doteq \mathbf{S}(s)\mathbf{G}_p(s)\mathbf{G}_c(s) \tag{13.66b}$$

where

$$\mathbf{S}(s) = [\mathbf{I} + \mathbf{G}_p(s)\mathbf{G}_c(s)]^{-1} \tag{13.66c}$$

is the (multivariable) loop sensitivity and

$$\mathbf{T}(s) = \mathbf{I} + \mathbf{G}_p(s)\mathbf{G}_c(s) \tag{13.66d}$$

is the (multivariable) *return difference.* Although the matrix $\mathbf{G} = \mathbf{G}_{CL}(s)$, or for that matter any matrix $\mathbf{G}$, does not have a unique gain (for example in the multiple-input context, a different gain will be associated with each input direction), we can consider the idea of a bounded range of gains such as given by the *induced matrix norm* (see Appendix B, equation (B3.6a)):

$$\|\mathbf{G}\| = \sup_{w \neq 0} \frac{\|\mathbf{G}w\|}{\|\mathbf{w}\|} \tag{13.67a}$$

where $\|\mathbf{w}\|$ stands for the (vector) *norm* or "length" of $\mathbf{w}$, and (13.67a) implies the largest value of the ratio of the "length" of $\mathbf{z} = \mathbf{G}\mathbf{w}$ to the "length" of $\mathbf{w}$. In particular, if we define "length" of a vector in terms of the Euclidean norm,

$$\|\mathbf{w}\|_s = \sqrt{(\mathbf{w}^H\mathbf{w})} \tag{13.67b}$$

where $\mathbf{w}^H = \mathbf{w}^{*T}$ is the transpose of the complex conjugate of $\mathbf{w}$, then it can be shown (see [21], for example, and also Appendix B, equation (B3.6f)) that (13.67a) becomes

$$\|\mathbf{G}\|_s = \sqrt{|\text{maximum eigenvalue of } \mathbf{G}^H\mathbf{G}|} \tag{13.67c}$$

The positive square roots of the eigenvalues of $\mathbf{G}^H\mathbf{G}$ (or $\mathbf{G}^T\mathbf{G}$ if $\mathbf{G}$ is real and not complex valued) are called the *singular values* of $\mathbf{G}$, so that the norm of $\mathbf{G}$, equation (13.67c), is the largest singular value.

If $\mathbf{G} = \mathbf{G}(s) = \mathbf{G}(j\omega)$, then the singular values of $\mathbf{G}$ are functions of frequency ω and are known as *principal gains* of $\mathbf{G}(s)$. If we denote the largest and smallest principal gains by $\overline{\sigma}$ and $\underline{\sigma}$, respectively, then

$$\overline{\sigma}(\mathbf{G}(j\omega)) = \|\mathbf{G}(j\omega)\|_s \tag{13.67d}$$

We can consider the "gain" of a multivariable system as a frequency function that is bounded above and below by $\overline{\sigma}(\omega)$ and $\underline{\sigma}(\omega)$, respectively. Note that singular values or principal gains can be computed with respect to the multivariable system specification that is appropriate—closed-loop transfer function $\sigma(\mathbf{G}_{CL}(j\omega))$, open-loop transfer function $\sigma(\mathbf{G}_o(j\omega))$, loop sensitivity $\sigma(\mathbf{S}(j\omega))$, return difference $\sigma(\mathbf{T}(j\omega))$, etc. The computation of a few principal gains of a multivariable system is illustrated in the following example. By using a decoupled system, we can also determine the corresponding SISO specifications for each decoupled loop.

Example 13.16	Singular Values of a Decoupled Multivariable System

Consider, as the model of an actual multivariable system, the decoupled open-loop transfer function

$$\mathbf{G}_o(s) = \begin{bmatrix} \dfrac{2.5k_{c1}}{6s + 0.125} & 0 \\[2ex] 0 & \dfrac{k_{c2}[1 + 1/(T_i s)]}{5.2s + 3.1} \end{bmatrix}$$

Assuming the following values of the controller parameters; $k_{c1} = 1$, $k_{c2} = 8$, and $T_i = 0.775$, determine the largest and smallest principal gains of the closed-loop transfer function and the loop sensitivity, and show that for such a decoupled multivariable system, these principal gains are the same as the closed-loop gain and loop sensitivity-gain of the two constituent (decoupled) SISO loops.

Solution: For the given controller parameter values, the open-loop matrix transfer function is

$$\mathbf{G}_o(s) = \begin{bmatrix} \dfrac{2.5}{6s + 0.125} & 0 \\[2ex] 0 & \dfrac{8s + 10.3226}{s(5.2s + 3.1)} \end{bmatrix}$$

If we assume unity feedback loops or the multivariable system configuration of Fig. 13.9a, the closed-loop matrix transfer function and the multivariable loop sensitivity are given by equations (13.66b) and (13.66c), respectively, that is,

$$
\mathbf{S}(s) = [\mathbf{I} + \mathbf{G}_o(s)]^{-1} =
\begin{bmatrix}
1 + \dfrac{2.5}{6s + 0.125} & 0 \\[4mm]
0 & 1 + \dfrac{8s + 10.3226}{s(5.2s + 3.1)}
\end{bmatrix}^{-1}
$$

$$
=
\begin{bmatrix}
\dfrac{6s + 0.125}{6s + 2.625} & 0 \\[4mm]
0 & \dfrac{s(5.2s + 3.1)}{5.2s^2 + 11.1s + 10.3226}
\end{bmatrix}
$$

and

$$
\mathbf{G}_{CL}(s) = \mathbf{S}(s)\mathbf{G}_o(s) =
\begin{bmatrix}
\dfrac{2.5}{6s + 2.625} & 0 \\[4mm]
0 & \dfrac{8s + 10.3226}{5.2s^2 + 11.1s + 10.3226}
\end{bmatrix}
$$

Singular values: We consider first the closed-loop transfer function. If we substitute $s = j\omega$, then

$$
\mathbf{G}_{CL}^H(j\omega) = \mathbf{G}_{CL}^T(-j\omega)
$$

so that

$$
\mathbf{G}_{CL}^H(j\omega)\mathbf{G}_{CL}(j\omega) =
\begin{bmatrix}
\dfrac{2.5}{-j6\omega + 2.625} & 0 \\[4mm]
0 & \dfrac{-j8\omega + 10.3226}{-5.2\omega^2 - j11.1\omega + 10.3226}
\end{bmatrix}
\begin{bmatrix}
\dfrac{2.5}{j6\omega + 2.625} & 0 \\[4mm]
0 & \dfrac{j8\omega + 10.3226}{-5.2\omega^2 + j11.1\omega + 10.3226}
\end{bmatrix}
$$

$$
\mathbf{G}_{CL}^H(j\omega)\mathbf{G}_{CL}(j\omega) =
\begin{bmatrix}
\dfrac{6.25}{36\omega^2 + 6.8906} & 0 \\[4mm]
0 & \dfrac{64\omega^2 + 106.556}{27.04\omega^4 + 15.855\omega^2 + 106.556}
\end{bmatrix}
$$

The positive square roots of the eigenvalues of this matrix (which can be obtained by solving equation (6.68) with $\mathbf{A} = \mathbf{G}_{CL}^H(j\omega)\mathbf{G}_{CL}(j\omega)$) and hence the singular values are

$$
\sigma_1(\mathbf{G}_{CL}(j\omega)) = \sqrt{\dfrac{6.25}{36\omega^2 + 6.8906}} = \underline{\sigma}(\mathbf{G}_{CL}(j\omega))
$$

$$
\sigma_2(\mathbf{G}_{CL}(j\omega)) = \sqrt{\dfrac{64\omega^2 + 106.556}{27.04\omega^4 + 15.855\omega^2 + 106.556}} = \overline{\sigma}(\mathbf{G}_{CL}(j\omega))
$$

These are also the principal values (see solid curves, Fig. 13.27), as indicated. The principal gains of the loop sensitivity can be computed in a similar manner. The result (see dashed curves, Fig. 13.27) is

$$\sigma_1(\mathbf{S}_{CL}(j\omega)) = \sqrt{\frac{36\omega^2 + 0.0156}{36\omega^2 + 6.8906}} = \overline{\sigma}(\mathbf{S}(j\omega))$$

$$\sigma_2(\mathbf{S}(j\omega)) = \sqrt{\frac{27.04\omega^4 + 9.61\omega^2}{27.04\omega^4 + 15.855\omega^2 + 106.556}} = \underline{\sigma}(\mathbf{S}(j\omega))$$

SISO closed-loop gain and sensitivity: From the given matrix open-loop transfer function, the individual unity feedback closed-loop transfer functions are

$$G_1(s) = \frac{\dfrac{2.5}{6s + 0.125}}{1 + \dfrac{2.5}{6s + 0.125}} = \frac{2.5}{6s + 2.625} \quad \text{and}$$

$$G_2(s) = \frac{\dfrac{8s + 10.3226}{s(5.2s + 3.1)}}{1 + \dfrac{8s + 10.3226}{s(5.2s + 3.1)}} = \frac{8s + 10.3226}{5.2s^2 + 11.1s + 10.3226}$$

Figure 13.27

Principal Gains of a Decoupled Control System

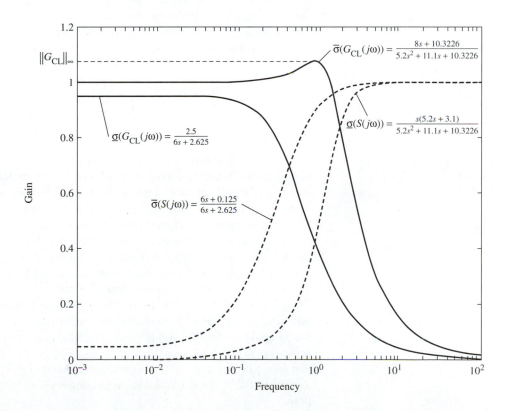

so that the closed-loop (frequency) gains are

$$|G_1(j\omega)| = \sqrt{\frac{6.25}{36\omega^2 + 6.8906}} = \underline{\sigma}\,(\mathbf{G}_{CL}(j\omega))$$

$$|G_2(j\omega)| = \sqrt{\frac{64\omega^2 + 106.556}{27.04\omega^4 + 15.855\omega^2 + 106.556}} = \overline{\sigma}(\mathbf{G}_{CL}(j\omega))$$

Similarly, the individual loop sensitivities are

$$S_1(s) = \frac{1}{1 + \dfrac{2.5}{6s + 0.125}} = \frac{6s + 0.125}{6s + 2.625}$$

$$S_2(s) = \frac{1}{1 + \dfrac{8s + 10.3226}{s(5.2s + 3.1)}} = \frac{s(5.2s + 3.1)}{5.2s^2 + 11.1s + 10.3226}$$

Hence the frequency domain loop sensitivity gains are also the same as the principal values:

$$|S_1(j\omega)| = \overline{\sigma}(\mathbf{S}(j\omega)) \quad \text{and} \quad |S_2(j\omega)| = \underline{\sigma}(\mathbf{S}(j\omega))$$

The "matrix" transfer function *gains* just illustrated are defined at individual frequencies. It is desirable for design of optimal robust controllers to use a single number (a performance index) to measure the matrix "gain." One such norm on a transfer function $\mathbf{G}$ that is associated with its principal gains is the *operator norm* (see [16]), with the definition

$$\|\mathbf{G}\|_\infty = \sup_\omega \overline{\sigma}(\mathbf{G}(j\omega)) \tag{13.68}$$

H_∞ Norm H_∞ denotes the set of all *proper* ($\lim_{s \to \infty} G(s)$ exists) transfer functions G for which $\|G\|_\infty < \infty$ and that are also exponentially stable. Thus the H_∞ norm of a transfer matrix $\mathbf{G}$ is the maximum over all frequencies of its largest singular value. The meaning of the H_∞ norm of a given physical system depends on the characteristic represented by $\mathbf{G}$. In general, and recalling the definitions (13.67a) and (13.67b), the H_∞ norm can be viewed as equal to the maximum over all disturbances (that is, inputs) $\mathbf{w}$ (not equal to zero) of the quotient of the amount of energy coming out of the system and the amount of energy going into the system. Graphically, equation (13.68) implies that $\|\mathbf{G}\|_\infty$ is the peak value that can be read off a plot of $\overline{\sigma}(\mathbf{G}(j\omega))$ versus frequency. The H_∞ norm on the closed-loop transfer function of the system of Example 13.16, $\|\mathbf{G}_{CL}\|_\infty$, is indicated in Fig. 13.27. The correlation between the present global (MIMO and SISO) measures of performance and robustness and our earlier SISO characteristics is again evident from this result. Recalling the definition in Section 10.3 of the *resonance peak*—which was also one of our measures for relative stability—we can see that for a SISO system, $\|G\|_\infty$ is the same as the system's M_p (the largest distance from a point on the Nyquist contour to the origin) when G is the closed-loop transfer function.

An H$_\infty$ Control
Problem and a Solution

The essential idea in H_∞ optimal control is to design a controller that will optimize performance (stability or some other goal for the system) for the worst external input condition. Bad external input conditions are the manifestations of *perturbations* of the system or equivalently of *uncertainty* (of parameters, ignored dynamics, neglected nonlinearities, etc.) in our description of the system. One method of handling the problem of robustness is *to treat the uncertainty as additional input(s) to the system.**

Modeling Uncertainty

Figure 13.28 illustrates our model of an uncertain system. We consider the overall system (see Fig. 13.28a) as having two sets of inputs and two sets of outputs. The set of all manipulable (control) variables is represented by **u**, **w** represents the set of all other *exogenous* inputs such as disturbances and measurement noise, and **y** is the set of measured signals which are available to the controller. Any other outputs (real or otherwise theoretical) that may be of interest are represented by **z**. In particular, **z** represents those outputs of the system whose dependence on **w** we want to minimize presumably by using the measurements **y** to choose the inputs **u**. In Fig. 13.28b the subsystem Σ_K represents all the uncertainties in the system or the "uncertain" part of the plant such that if the transfer matrix Σ_K is zero, then the *nominal* plant model from **u** to **y** is recovered. We assume that the control (compensation) is known, so that it is combined with the "certain" part of the plant to give the *nominal plant.*

---

*The stochastic optimal linear regulator, the *linear-quadratic-Gaussian* (LQG) controller design (stochastic systems are outside the scope of this book) treats such inputs as *white noise* additive to the output measurement, but this approach is not suitable for some types of uncertainty such as parameter uncertainty.

Figure 13.28

Representation of an
Inaccurately Known Plant
Under Feedback Control

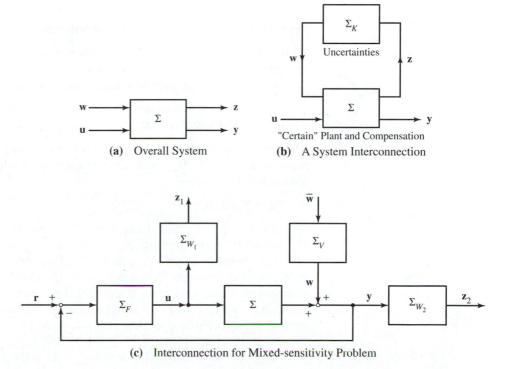

(a)   Overall System

(b)   A System Interconnection

(c)   Interconnection for Mixed-sensitivity Problem

To further explain the interconnection in Fig. 13.28b, consider again the oscillator of Example 13.10 with some slight damping. Let the nominal plant be given by

$$\dot{x}_1 = x_2$$
$$\dot{x}_2 = -\omega^2 x_1 - \beta x_2 + bu$$
$$= -ax_1 - 0.1x_2 + bu; \quad \text{(nominal plant: } a = a_0 = 1, b = b_0 = 1)$$

$$y = x_1$$

(13.69a)

Suppose the effect of the disturbance input and the frequency parameter variation is uncertainty in the values of ω^2 (that is, a) and b, and our knowledge of this effect is limited only to information on the range in which the actual values of these parameters lie: $a \in [a_0 - \Delta a, a_0 + \Delta a]$, $b \in [b_0 - \Delta b, b_0 + \Delta b]$. We can rewrite the second equation (13.69a) as follows:

$$\dot{x}_2 = -(a_0 + a - a_0)x_1 - \beta x_2 + (b_0 + b - b_0)u$$
$$= -a_0 x_1 - \beta x_2 + b_0 u + (a_0 - a)x_1 + (b - b_0)u$$

and thus deduce the following interconnection (structure):

$$\Sigma: \begin{cases} \dot{x}_1 = x_2 \\ \dot{x}_2 = -a_0 x_1 - \beta x_2 + b_0 u + w \\ y = x_1 \\ \mathbf{z} = \begin{bmatrix} x_1 \\ u \end{bmatrix} \end{cases}$$

(13.69b)

$$\Sigma_K: \{ w = [(a_0 - a) \quad (b - b_0)]\mathbf{z}$$

(13.69c)

The system Σ_K is static (equation (13.69c) involves no dynamics) in this case.

Formulation of an H_∞ Control Problem and a Solution

With reference to the model of Fig. 13.28b, a regular full-information H_∞ control problem (see [22]) can be described as follows:

Consider a linear, time-invariant, finite-dimensional system Σ,

$$\Sigma: \begin{cases} \dot{\mathbf{x}} = \mathbf{Ax} + \mathbf{Bu} + \mathbf{Ew} \\ \mathbf{z} = \mathbf{Cx} + \mathbf{D}_1\mathbf{u} + \mathbf{D}_2\mathbf{w} \end{cases}$$

(13.70a)

where $\mathbf{x}$ is an n-vector, $\mathbf{u}$ is an r-vector, $\mathbf{w}$ is a q-vector, and $\mathbf{z}$ is an m-vector, and the matrices $\mathbf{A}$, $\mathbf{B}$, $\mathbf{C}$, $\mathbf{D}_1$, $\mathbf{D}_2$, and $\mathbf{E}$ are of the appropriate dimensions. Determine a static feedback law,

$$\mathbf{u} = \mathbf{F}_1\mathbf{x} + \mathbf{F}_2\mathbf{w}$$

(13.70b)

such that the resulting closed-loop system is *internally stable* and its transfer matrix $\mathbf{G}_{wz}$ has an H_∞ norm that is strictly less than some a priori given bound γ.

The closed-loop system is called internally stable if with $\mathbf{w} = \mathbf{0}$ and for all initial states, the state converges to zero as $t \longrightarrow \infty$. Given (13.70a) and (13.70b), the closed-loop system is internally stable if and only if the matrix $(\mathbf{A} + \mathbf{BF}_1)$ is asymptotically stable. Also for this system, it is left as an exercise (see Practice Problem 13.53) to show that the closed-loop transfer matrix is given by

$$\mathbf{G}_{wz}(s) = (\mathbf{C} + \mathbf{D}_1\mathbf{F}_1)(s\mathbf{I} - \mathbf{A} - \mathbf{BF}_1)^{-1}(\mathbf{E} + \mathbf{BF}_2) + (\mathbf{D}_2 + \mathbf{D}_1\mathbf{F}_2)$$

(13.70c)

Note that the problem statement is in terms of the H_∞ norm of the closed-loop system's being less than some value, whereas our ultimate goal is to *minimize* the H_∞ norm. It is understood that in principle the minimum H_∞ norm can be obtained by some search procedure over admissible control laws. Note further that in terms of the compensation, the given full-information H_∞ control problem statement is not the most general. The *static state* feedback considered is less general than *static* (output) feedback, which in turn is less general than *dynamic* output feedback.

The solution of the H_∞ control problem that is illustrated next is for a further restricted feedback law:

$$\mathbf{u} = \mathbf{Fx} \tag{13.70d}$$

However, relative to the control law (13.70b), this restriction is not a significant limitation, since it is in tune with control system practice in which a feedback law that depends on a measurement of the disturbance is not normally applied.

Result.   It can be shown for the system (13.70a) and $\gamma > 0$ (see [22]) that assuming the subsystem

$$\dot{\mathbf{x}} = \mathbf{Ax} + \mathbf{Bu} \tag{13.71a}$$

$$\mathbf{z} = \mathbf{Cx} + \mathbf{D}_1\mathbf{u}$$

has no *invariant zeros* on the imaginary axis, and $\mathbf{D}_1$ is *injective;* that if

$$\mathbf{D}_2^T\mathbf{D}_2 < \gamma^2\mathbf{I} \tag{13.71b}$$

and a positive semidefinite solution $\mathbf{P}$ of the following algebraic Riccati equation:

$$0 = \mathbf{A}^T\mathbf{P} + \mathbf{PA} + \mathbf{C}^T\mathbf{C} \tag{13.71c}$$

$$- \begin{bmatrix} \mathbf{B}^T\mathbf{P} + \mathbf{D}_1^T\mathbf{C} \\ \mathbf{E}^T\mathbf{P} + \mathbf{D}_2^T\mathbf{C} \end{bmatrix}^T \begin{bmatrix} \mathbf{D}_1^T\mathbf{D}_1 & \mathbf{D}_1^T\mathbf{D}_2 \\ \mathbf{D}_2^T\mathbf{D}_1 & \mathbf{D}_2^T\mathbf{D}_2 - \gamma^2\mathbf{I} \end{bmatrix}^{-1} \begin{bmatrix} \mathbf{B}^T\mathbf{P} + \mathbf{D}_1^T\mathbf{C} \\ \mathbf{E}^T\mathbf{P} + \mathbf{D}_2^T\mathbf{C} \end{bmatrix}$$

exists such that the closed-loop matrix

$$\mathbf{A}_{\text{CL}} = \mathbf{A} - [\mathbf{B} \quad \mathbf{E}] \begin{bmatrix} \mathbf{D}_1^T\mathbf{D}_1 & \mathbf{D}_1^T\mathbf{D}_2 \\ \mathbf{D}_2^T\mathbf{D}_1 & \mathbf{D}_2^T\mathbf{D}_2 - \gamma^2\mathbf{I} \end{bmatrix}^{-1} \begin{bmatrix} \mathbf{B}^T\mathbf{P} + \mathbf{D}_1^T\mathbf{C} \\ \mathbf{E}^T\mathbf{P} + \mathbf{D}_2^T\mathbf{C} \end{bmatrix} \tag{13.71d}$$

is asymptotically stable, then the static feedback law (13.70d), with $\mathbf{F}$ given by

$$\mathbf{F} = -[\mathbf{D}_1^T(\mathbf{I} - \gamma^{-2}\mathbf{D}_2\mathbf{D}_2^T)^{-1}\mathbf{D}_1]^{-1}[\mathbf{D}_1^T\mathbf{C} + \mathbf{B}^T\mathbf{P} + \mathbf{D}_1^T\mathbf{D}_2(\gamma^2\mathbf{I} - \mathbf{D}_2^T\mathbf{D}_2)^{-1}(\mathbf{D}_2^T\mathbf{C} + \mathbf{E}^T\mathbf{P})] \tag{13.71e}$$

exists such that when it is applied to the system (13.70a), the resulting closed-loop system is internally stable, and the closed-loop operator $\mathbf{G}_{wz}$ (equation (13.70c), note that $\mathbf{F}_2 = \mathbf{0}$ in this case) has H_∞ norm less than γ, i.e., $\|\mathbf{G}_{wz}\|_\infty < \gamma$.

For the meaning of invariant zeros of the system (13.71a) and their determination, see Appendix B and Example 13.17. $\mathbf{D}_1$ injective means that (in equation (13.70a)) no two inputs $\mathbf{u}$ have the same effect on the output $\mathbf{z}$ unless they are the same inputs.

The Mixed-Sensitivity Problem.   Figure 13.28c shows an interconnection that is typical of a special type of H_∞ control problem known as the *mixed-sensitivity problem*. The term Σ_F represents the precompensator that is to be designed. Suppose the transfer matrices $\mathbf{G}_F$ and $\mathbf{G}$ are associated with the subsystems Σ_F and Σ, respectively. Assume for the time being that unity transfer matrices are associated with all the other subsystems shown in Fig. 13.28c. Then the transfer matrix from $\tilde{\mathbf{w}}$ to $\mathbf{z}_1$ can be determined to be the "closed-loop" transfer matrix

$$\mathbf{G}_{\mathrm{CL}} = \mathbf{G}_F(\mathbf{I} + \mathbf{GG}_F)^{-1} \tag{13.72a}$$

The same expression also describes the transfer matrix from $\mathbf{r}$ to $\mathbf{u}$, and it can be termed the *control sensitivity function*. In practice the control inputs are often limited physically, so that a bound on the control sensitivity function is desirable. Similarly, the transfer matrix from $\tilde{\mathbf{w}}$ to $\mathbf{z}_2$ under the assumptions is

$$\mathbf{S} = (\mathbf{I} + \mathbf{G}_F\mathbf{G})^{-1} \tag{13.72b}$$

which also expresses the transfer matrix from $\mathbf{r}$ to $(\mathbf{r} - \mathbf{y})$ or the *sensitivity function,* as before. $\mathbf{S}$, as we saw in Section 11.1, needs to be made small for good *tracking* performance. Also, as we found in that section, a trade-off is often necessary in these generally conflicting requirements on the two sensitivities $\mathbf{G}_{\mathrm{CL}}$ and $\mathbf{S}$.

In addition to bandwidth limitations on actuators and sensors, we usually require optimal *tracking* performance only up to some frequency. Furthermore, our system models are typically accurate over some limited bandwidth. The H_∞ norm, on the other hand, provides a *uniform bound over all frequencies,* so that very conservative results can ordinarily be obtained for specific problems. In Fig. 13.28c the systems Σ_{W_1}, Σ_{W_2}, and Σ_V are weights that can be chosen in such a way that regulation is concentrated over frequencies of interest rather than uniformly over all frequencies. If we now associate the transfer matrices $\mathbf{G}_{W_1}$, $\mathbf{G}_{W_2}$, and $\mathbf{G}_V$, respectively, with these systems, then the *mixed* sensitivities are $\mathbf{G}_{W_1}\mathbf{G}_{CL}\mathbf{G}_V$ and $\mathbf{G}_{W_2}\mathbf{S}\mathbf{G}_V$, or the transfer matrix from the disturbance $\tilde{\mathbf{w}}$ to $\mathbf{z}_1$ and $\mathbf{z}_2$ whose H_∞ norm is to be minimized is

$$\mathbf{G}_{wz} = \begin{bmatrix} \mathbf{G}_{W_1}\mathbf{G}_{CL}\mathbf{G}_V \\ \mathbf{G}_{W_2}\mathbf{S}\mathbf{G}_V \end{bmatrix} \tag{13.72c}$$

Example 13.17   Static State Feedback H∞ Control of Underdamped Oscillator

Consider the slightly damped oscillator described in equations (13.69a) through (13.69c). It is desired to track a reference position signal θ_d using a controller that will yield a *robustly stable* system with respect to uncertainties in the parameters a and b, as described in equations (13.69a) through (13.69c). It is further desired to take into account possible controller gain and bandwidth limitations due to physical restrictions on actuator capacity.

Using the results given in equations (13.71a) through (13.71e) design a static state feedback control (13.70d) that will minimize the *weighted integrated tracking error* and the *weighted control input,* assuming the following first-order weights:

$$W_{\text{tracking error}}(s) = \frac{1 + 0.02s}{1 + 0.2s}$$

$$W_{\text{control input}}(s) = 0.2\left(\frac{1 + 0.1s}{1 + 0.02s}\right)$$

Elaboration: A first-order weight of the form

$$W(s) = \epsilon\left(\frac{1 + \tau s}{1 + Ts}\right) \tag{13.73}$$

represents a low-pass filter if $\tau << T$ or a high-pass filter if $\tau >> T$. The given weight for the tracking error is a low-pass filter with a cutoff frequency of 5. This implies an interest in tracking, particularly well, low-frequency signals. The integration of the tracking error also implies a similar interest, since it is intended to guarantee zero tracking error at steady state *(zero frequency)*. On the other hand, the given weight on the control input is a high-pass filter with a corner frequency of 50. This limits the controller to higher operating speeds, where the corresponding actuation forces are presumably within the capacity of the actuator.

The quantity ϵ can be used to express the relative importance of the goals represented by the weights on different variables. For the present system, $\epsilon = 0.2$ in the control input weight suggests, in terms of the gains of these circuits, that equal importance is placed on both the tracking and control input goals. A value of $\epsilon = 0.1$ could imply that the tracking goal is emphasized twice as much as the control input goal.

Solution: Figure 13.29a shows a representation of this control system using a structure that is essentially equivalent to a *mixed-sensitivity* problem. However, the problem solution is on the basis of equations (13.71a) through (13.71e).

The various subsystem state models are as follows:

$$\text{Basic plant: } \begin{cases} \dot{x}_1 = x_2 \\ \dot{x}_2 = -x_1 - 0.1x_2 + u + w \\ w = \begin{cases} 0, & \text{nominal plant} \\ -\Delta a(t)x_1 + \Delta b(t)u, & \text{uncertain plant} \end{cases} \end{cases}$$

$$\text{Control weight: } \begin{cases} \dot{x}_3 = -\dfrac{1}{T_1}x_3 + \dfrac{1}{T_1}\left(1 - \dfrac{\tau_1}{T_1}\right)u = -50x_3 - 200u \\ z_1 = \epsilon x_3 + \epsilon\dfrac{\tau_1}{T_1}u = \epsilon x_3 + \epsilon 5u \end{cases}$$

Integrated error: $\dot{x}_4 = x_1$ (that is, using $\theta_d = 0$ for design purposes)

$$\text{Error weight: } \begin{cases} \dot{x}_5 = -\dfrac{1}{T_2}x_5 + \dfrac{1}{T_2}\left(1 - \dfrac{\tau_2}{T_2}\right)x_4 = -5x_5 + 4.5x_4 \\ z_2 = x_5 + \dfrac{\tau_2}{T_2}x_4 = x_5 + 0.1x_4 \end{cases}$$

Comparing these results with equation (13.70a), we can establish that

$$\mathbf{A} = \begin{bmatrix} 0 & 1 & 0 & 0 & 0 \\ -1 & -0.1 & 0 & 0 & 0 \\ 0 & 0 & -50 & 0 & 0 \\ 1 & 0 & 0 & 0 & 0 \\ 0 & 0 & 0 & 4.5 & -5 \end{bmatrix}; \quad \mathbf{B} = \begin{bmatrix} 0 \\ 1 \\ -200 \\ 0 \\ 0 \end{bmatrix};$$

$$\mathbf{C} = \begin{bmatrix} 0 & 0 & \epsilon & 0 & 0 \\ 0 & 0 & 0 & 0.1 & 1 \end{bmatrix}$$

$$\mathbf{D}_1 = \begin{bmatrix} 5\epsilon \\ 0 \end{bmatrix}; \quad \mathbf{D}_2 = \mathbf{0}; \quad \mathbf{E} = \begin{bmatrix} 0 \\ 1 \\ 0 \\ 0 \\ 0 \end{bmatrix}$$

Figure 13.29

Structure and Performance of H_∞ Controller

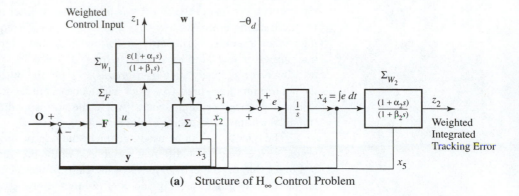

(a)   Structure of H_∞ Control Problem

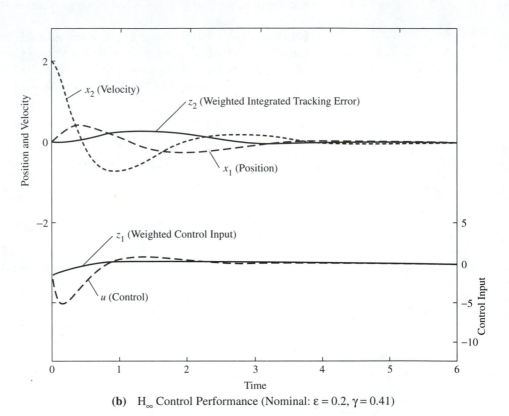

(b) H_∞ Control Performance (Nominal: $\varepsilon = 0.2$, $\gamma = 0.41$)

Note that equation (13.71b) is satisfied: $\mathbf{D}_2^T\mathbf{D}_2 = \mathbf{0} < \gamma^2\mathbf{I}$. Also, $\mathbf{D}_1$ is injective, and the system $(\mathbf{A},\mathbf{B},\mathbf{C},\mathbf{D}_1)$ has no invariant zeros on the imaginary axis (see the demonstration in Appendix B).

    With $\epsilon = 0.2$, the minimum positive γ (to two decimal places) for which a positive semidefinite solution $\mathbf{P}$ of equation (13.71c) was obtained was $\gamma = 0.41$. The solution is

$$\mathbf{P} = \begin{bmatrix} 0.6079 & 0.2764 & -0.0045 & 0.7308 & 0.0199 \\ 0.2764 & 0.1908 & -0.0033 & 0.2952 & 0.0029 \\ -0.0045 & -0.0033 & 5.8449\text{E-}5 & -0.0047 & -3.3398\text{E-}5 \\ 0.7308 & 0.2952 & -0.0047 & 1.1535 & 0.1087 \\ 0.0199 & 0.0029 & -3.3398\text{E-}5 & 0.1087 & 0.1 \end{bmatrix}$$

Figure 13.29 *continued*

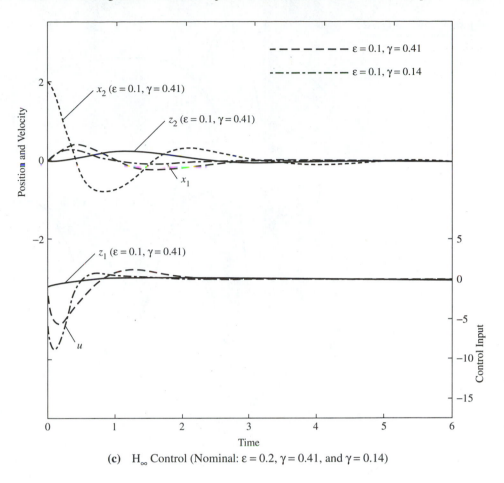

(c)   H$_\infty$ Control (Nominal: $\varepsilon = 0.2$, $\gamma = 0.41$, and $\gamma = 0.14$)

which is positive definite, and the corresponding feedback matrix, equation (13.71e), is

$$\mathbf{F} = [-1.1764 \quad -0.8508 \quad -0.1850 \quad -1.2352 \quad -0.0096]$$

The nominal system performance under the static state feedback equation (13.70d) with **F** as given ($\varepsilon = 0.2$ and $\gamma = 0.41$) is shown in Fig. 13.29b.* Both the weighted and unweighted outputs are shown. The effect of the relative weight factor ε is illustrated by Fig. 13.29c. With $\varepsilon = 0.1$, which implies tighter control on the integrated tracking error, better steady-state error performance can be observed (compare the solid- and dashed-line curves in Fig. 13.29c with the corresponding curves in Fig. 13.29b), although this is at the expense of some loss of control of the magnitude of the control input (especially early in the response history). The result (solid- and dashed-line curves) shown in Fig. 13.29c is based on a solution for **P** using $\varepsilon = 0.1$ and $\gamma = 0.41$. The feedback matrix corresponding to this solution is

$$\mathbf{F} = [-1.7108 \quad -0.9428 \quad -0.1842 \quad -2.0540 \quad -0.0299]$$

However, with $\varepsilon = 0.1$, it is possible to obtain a positive definite solution of equation (13.71c) for γ as small as 0.14. The feedback matrix for this case, that is, $\varepsilon = 0.1$ and $\gamma = 0.14$, is

$$\mathbf{F} = [-3.5824 \quad -2.1800 \quad -0.1626 \quad -4.0280 \quad -0.0448]$$

*See Script M-file **fig13_29.m; On the Net.**

Figure 13.29 *continued*

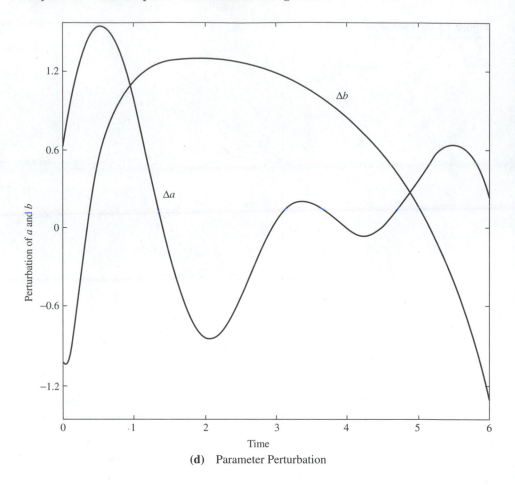

(d)   Parameter Perturbation

This value of **F** represents an improved design, since our ultimate objective is the minimization of γ. Figure 13.29c also shows the results (the none-weighted position x_1 and control input u only) for this case $\epsilon = 0.1$ and $\gamma = 0.14$ (see the dash-dot lines) and confirms the improvement in overall performance for this design relative to the case of $\epsilon = 0.1$ and $\gamma = 0.41$.

The (original case of $\epsilon = 0.2$, $\gamma = 0.41$) performance of the H_∞ controller when plant parameters are uncertain is shown in Fig. 13.29e. Although we can observe some deterioration in the response (compare Figs. 13.29e and 13.29b) relative to our goals of integrated tracking error and especially weighted control input, the performance of the present H_∞ controller is remarkable, considering, as shown in Fig. 13.29d, the rather severe parameter variations it had to overcome (the value of the parameters a and b vary by more than 50% within the solution interval).

Figure 13.29e also compares the performance of the H_∞ controller with that of the signal-synthesis model reference adaptive controller designed in Example 13.10 (part (a), that is, without PI-feedback). Recall that the oscillator model used for this MRAS design did not include a damping term. Thus the entire parameter b represents uncertainty due to *ignored dynamics* for the MRAC design. Considering this model uncertainty and the severity of the parameter perturbations already alluded to, the performance of this signal-synthesis MRAC is quite *robust* although not as good as that of H_∞ control, especially relative to the control input performance measure.

Figure 13.29 *continued*

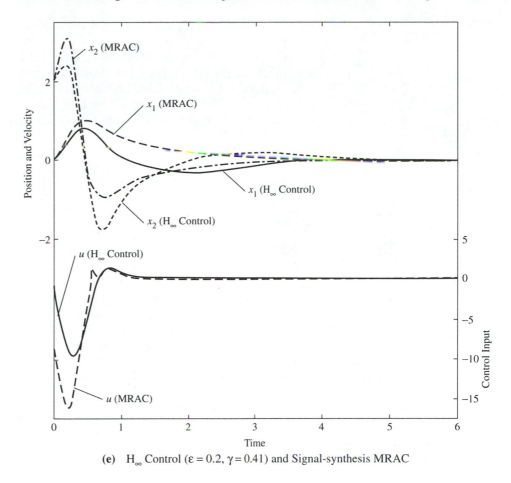

(e)   H$_\infty$ Control ($\varepsilon = 0.2$, $\gamma = 0.41$) and Signal-synthesis MRAC

On the Net

- fig13_3.m, feedfor.m and feedbac.m (MATLAB): source file for Fig. 13.3, Example 13.1
- F13_4a.MCD (Mathcad): feedforward control of hotwater heater system, Example 13.2
- fig13_7.m, act13_7s.m and act13_7r.m (MATLAB): source file for Figs. 13.7b and 13.7c, Example 13.3
- fig13_8.m, pid13_8b.m and pid13_8c.m (MATLAB): source file for Figs. 13.8b and 13.8c, Example 13.3
- fig13_11.m and decoup.m (MATLAB): source file for Fig. 13.11, Example 13.4
- fig13_13.m, chemrea.m and chemreac.m (MATLAB): source file for Fig. 13.13, Example 13.5
- fig13_15.m, tank3ul.m and tank3ulc.m (MATLAB): source file for Fig. 13.15b, Example 13.6
- fig13_16.m, tank3u2.m and tank3u2c.m (MATLAB): source file for Figs. 13.16b and 13.16c, Example 13.7
- fig13_18.m and observa.m (MATLAB): source file for Fig. 13.18, Example 13.8
- fig13_20.m and esticon.m (MATLAB): source file for Fig. 13.20b, Example 13.9
- fig13_23.m, penMRAC1.m, penMRAC2.m and msvfcon.m (MATLAB): source file for Figs. 13.23a and 13.23c, Example 13.10
- fig13_24.m, opt1.m, opt2.m and opt3.m (MATLAB): source file for Fig. 13.24, Example 13.11

- SC14.MCD (Mathcad): time-optimal control of undamped servomechanism, Example 13.13
- fig13_25.m and timeopt.m (MATLAB): source file for Fig. 13.25, Example 13.13
- exm13_14.m (MATLAB): solution of matrix Riccati equation, Example 13.14
- fig13_26.m and outreg.m (MATLAB): source file for Fig. 13.26, Example 13.15
- fig13_27.m (MATLAB): source file for Fig. 13.27, Example 13.16
- F13_29E.MCD (Mathcad): H_∞ controller and signal-synthesis MRAC performance, Example 13.17
- fig13_29.m, hinfcnom.m, hinficon.m, and hinfMRAC.m (MATLAB): source file for Figs. 13.29b through 13.29e, example 13.17
- Auxiliary exercise: Practice Problems 13.2, 13.15, 13.16, 13.18, 13.23, 13.25, 13.27, 13.30, 13.45, 13.46, 13.48, 13.51, and 13.54.

13.5 PRACTICE PROBLEMS

13.1 Consider the thermal system of Fig. 13.30 with material transport

$$C_1\dot{x}_1 = -wx_1 - w(x_1 - x_2) + wv$$

$$C_2\dot{x}_2 = w(x_1 - x_2) + u$$

where C_1, C_2 = heat capacitances, w = flowrate times specific heat, x_1, x_2 are temperatures, and v = feed temperature. Let $C_1 = C_2 = 1$ and $w = 1$. Explore the design of an exact feedforward controller that can maintain the temperature $y = x_1$ at the desired value assuming

(a) u is the manipulated variable, and the feed temperature is a measured disturbance.

(b) v is the manipulated variable, and u is the measured disturbance.

(c) As in case (a) but obtain an approximate feedforward control based on the steady-state behavior of the system.

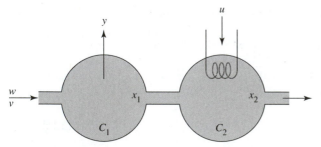

Figure 13.30 (Problem 13.1)

13.2 Consider again the two-tank level-control system of Example 13.1.

(a) Simulate the response of the system for both the feedforward controller, equation (13.4e), and the feedback controller, equation (13.5c), for a *nonequilibrium* initial state $h_1(0) = 1$ and $h_2(0) = 0.99$, and verify that for the same

setpoint and input-flow disturbance given in the example, the feedback controller gives a better transient (or early) response.

(b) Consider a *combined* feedforward and feedback control for the problem, as shown in Fig. 13.31. This combination of controllers is of the type illustrated in Fig. 13.1c(i). Show that for the same performance specifications considered in Example 13.1, the design of the independent feedforward and feedback controllers obtained in that example remains adequate for the combined controller. Simulate the performance of this controller for the initial conditions and setpoint given in (a), and confirm that this combined controller performs better than either of the individual controllers acting alone.

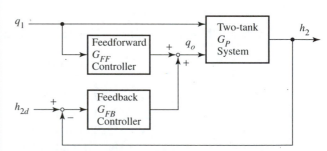

Figure 13.31 (Problem 13.2(b))

13.3* Consider an in-line heater, as shown in Fig. 13.32a, in which a process liquid is heated by steam. The heat release by the steam q is proportional to the steam flowrate u, which is controlled by a valve as shown. A simple first-order model for this process, assuming the proportionality factor between the q and u is unity, is

$$\rho c V \frac{d}{dt} T_o = \rho c w (T_i - T_o) + u$$

(a) Considering the liquid flowrate w and inlet temperature T_i as disturbance inputs, the following feedforward control for the product temperature T_o, has been proposed:

$$u = (\rho c)w(T_{SL} - T_i)$$

where T_{SL} is the setpoint or desired value of the product temperature. Draw a schematic diagram of this control system and explain or derive the proposed control.

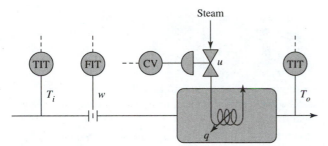

Figure 13.32a (Problem 13.3(a))

(b) Suppose a combined feedforward and feedback control of the type illustrated in Fig. 13.1c(iii) is to be applied to this system, as shown by the block diagram in Fig. 13.32b, that is, the feedforward controller setpoint T_{SL} is supplied by the feedback controller, and T_{SP} is the effective product temperature setpoint. For the same feedforward controller given in (a), determine an expression for a proportional feedback controller $G_{FB} = k_c$, such that the closed-loop time constant is 0.5 or less.

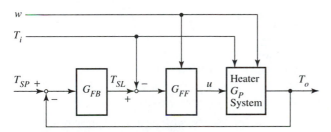

Figure 13.32b (Problem 13.3(b))

(c) Suppose the liquid inlet temperature is found to vary very little, and the desired value of the product temperature is relatively fixed. The feedforward control in (a) can in this case be taken as

$$u = [(\rho c)(T_{SL} - T_{IN})]w = k_r w$$

Fig. 13.32c shows a block diagram of a combined feedforward and feedback control for this process, which is of the type illustrated in Fig. 13.1c(ii), that is, the feedback controller is used to adjust the parameter k_r. The feedforward control system here can be considered a

ratio controller with k_r as the *ratio constant.* Determine for this case and the same specification in (b), $G_{FB} = k_c$.

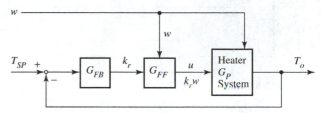

Figure 13.32c (Problem 13.3(c))

13.4 Design the cascade control system shown in Fig. 13.33 such that the overall system behavior is similar to a second-order response with a damping ratio of 0.25 and so that all the eigenvalues have real parts that are less than -1.

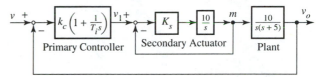

Figure 13.33 (Problem 13.4)

13.5 Consider the stirred chemical reactor of Fig. 13.10, equations (13.21a) through (13.21c), with no external heat input ($u_2 = 0$) and with exothermic reaction ($-H = 2000$). Let all other quantities remain as given in the figure.

(a) Draw a Laplace domain block diagram of the overall control system in the format of the cascade control system of Fig. 13.5a, in which the reactor concentration subsystem, equation (13.21a), is the inner process (I), while the temperature subsystem, equation (13.21b), is the outer process (II), and the feed temperature remains the disturbance input.

(b) Design the cascade control system represented in (a) using a proportional secondary controller and a PI-main controller, and using the same performance criteria as in Example 13.4.

13.6 (a) What do you understand by the term

(i) Multivariable control system?

(ii) Ideal independent control of a multivariable system?

13.6 (b) How does the technique of eigenvalue assignment meet the ultimate design objective in a multivariable control system?

13.7 What is the closed-loop characteristic equation of the multivariable system in Fig. 13.34? Note that all the transfer functions are matrices.

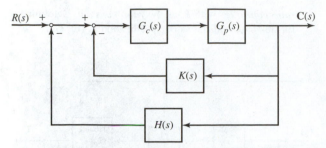

Figure 13.34 (Problem 13.7)

13.8* Consider the two mixing heaters as shown in Fig. 13.35. Design the decouplers or compensators G_{c12} and G_{c21} for this system when the controllers G_{c11} and G_{c22} are P-action: k_{11}, k_{22}. Assume that the heaters are insulated and do not lose heat to the environment. Also, no heat is lost in the outlet flow from the last heater. Let all the quantities C_1, C_2, and w be equal to 1. The feed temperature v is the disturbance input. Take C = thermal capacitance and w = mass flowrate times specific heat.

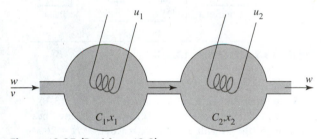

Figure 13.35 (Problem 13.8)

13.9 Consider the state vector feedback multivariable control system shown in Fig. 13.36. If the plant model is

$$\dot{\mathbf{X}} = \begin{bmatrix} -3 & 2 \\ 4 & -5 \end{bmatrix} \mathbf{X} + \begin{bmatrix} 1 & 0 \\ 0 & 1 \end{bmatrix} \begin{bmatrix} u_1 \\ u_2 \end{bmatrix}$$

and the controller is I-action,

$$\mathbf{G}_c(s) = \begin{bmatrix} 1/(T_1 s) & 0 \\ 0 & 1/(T_2 s) \end{bmatrix}$$

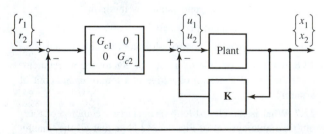

Figure 13.36 (Problem 13.9)

Show that by proper choice of

$$\mathbf{K} = \begin{bmatrix} k_{11} & k_{12} \\ k_{21} & k_{22} \end{bmatrix}$$

an independent control is possible. Determine $\mathbf{K}$ and T_1, T_2 if all the desired closed-loop eigenvalues must be -3.

13.10 Design the state vector feedback control shown in Fig. 13.12 for the system of Problem 13.8, using

$$\mathbf{G}_c = \begin{bmatrix} \dfrac{g_1}{s+1} & 0 \\ 0 & \dfrac{g_2}{s+2} \end{bmatrix}$$

where g_1 and g_2 are constants. Let all closed-loop decay rates be greater than or equal to 10 and the damping ratio be about 0.5.

13.11* Recall the rabbit-fox environment of Problem 10.12. Suppose, in order to control the environment, linear feedback is introduced in the form of a disease that is fatal to rabbits but whose effect on foxes is to increase their appetite for rabbits. That is, the net effect of the disease on the rabbit-fox environment is the reduction of the rate of growth of the rabbit population only by an amount $(k_1 x_1 + k_2 x_2)$,

(a) Draw a Laplace domain block diagram of this multivariable feedback control system.

(b) Determine the values of k_1 and k_2 that will yield a stable environment with the overall characteristic equation

$$s^2 + s + 7 = 0.$$

13.12 Design a state vector feedback control for the system

$$\dot{\mathbf{X}} = \begin{bmatrix} -3 & 2 \\ 4 & -5 \end{bmatrix} x + \begin{bmatrix} 1 \\ 0 \end{bmatrix} u$$

such that the desired closed-loop behavior is represented by the characteristic equation $s^2 + 10s + 50 = 0$.

13.13 Repeat Problem 13.12 with integral action on x_2. Let the additional closed-loop eigenvalue be -2.

13.14 Design the state vector feedback system shown in Fig. 13.37 such that all the closed-loop system eigenvalues are equal to -2.

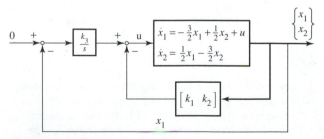

Figure 13.37 (Problem 13.14)

13.15 Consider again the three-tank level-control system of Example 13.6 (Fig. 13.15a). Obtain a state vector feedback proportional-plus-integral control design for this system such that three of the desired closed-loop system eigenvalues are the same as in the example (-6, -4 and -3), while the remaining three may be chosen in the range $-6 < \sigma < -3$ (ideally with actual values that simplify the design process). Simulate the response of the system for the same initial and final states given in Example 13.6 and compare your controller performance with those of the example.

13.16* Repeat Problem 13.13 assuming that there are two inputs such that

$$\dot{\mathbf{X}} = \mathbf{AX} + \begin{bmatrix} u_1 \\ u_2 \end{bmatrix}$$

where $\mathbf{A}$ is the same as in that problem. For the determination of the weighting for the distribution of the inputs, simulate the zero-setpoint response of the system from the initial state $[-1 \quad -1 \quad 0]^T$ and minimize both the overshoot and settling time of x_1 and x_2.

13.17 Design a direct proportional state vector feedback control for the system of Problem 13.8 using a distributed scalar controlling input. Let the desired closed-loop eigenvalues be $-p_1$ and $-p_2$, where p_1 and p_2 are positive constants. Explore the possibilities of the following weighting combinations:

(a) $w_1 = 0.5$, $w_2 = 0.5$

(b) $w_1 = 1.0$, $w_2 = 0$

(c) $w_1 = 0$, $w_2 = 1$

13.18 Repeat Problem 13.15 assuming the middle tank in the three-tank system of Fig. 13.15a also has a flow control input. Let the control action be distributed between the two flow inputs using the same weighting factors considered in Example 13.7. Compare your results with those of Example 13.7.

13.19* Given

$$\dot{\mathbf{X}} = \begin{bmatrix} 0 & 1 & 0 \\ 0 & 0 & 1 \\ 1 & -1 & -2 \end{bmatrix} \mathbf{X} + \begin{bmatrix} 0 \\ 1 \\ 0 \end{bmatrix} u; \qquad y = [0 \quad 1 \quad 0]\mathbf{X}$$

find a realizable (single-loop) linear feedback control that will make all the characteristic roots of the closed-loop system equal to -1. *Note:* The controller does not necessarily have to be one of the traditional types. Let the problem determine the form of the controller.

13.20 What is the purpose of an observer, and why is one needed in some control systems?

13.21 Is an observer required for complete state vector feedback control of the following system?

$$\dot{\mathbf{X}} = \begin{bmatrix} -1 & 1 \\ 0 & 1 \end{bmatrix} \mathbf{X} + \begin{bmatrix} 0 \\ 1 \end{bmatrix} u; \qquad y = [-2 \quad 1]\mathbf{X}$$

If so, design a supplemental observer for this system. Limit the magnitude of the observer gains to 10 or less.

13.22 For the system

$$\dot{\mathbf{X}} = \begin{bmatrix} -2 & 0 \\ 1 & -3 \end{bmatrix} \mathbf{X} + \begin{bmatrix} 1 \\ 0 \end{bmatrix} u; \qquad y = [0 \quad 1]\mathbf{X}$$

design

(a) A full observer. Give the equations for the estimation and determine the conditions on the elements of the feedback vector $\mathbf{k}$ such that the observer in the absence of disturbances will converge asymptotically to the state without oscillation.

(b) A supplemental observer. Give the conditions on k_1 and k_2

13.23* In Example 13.9, where there was a disturbance input, the estimated state $\hat{x}_2$ was offset from the actual state x_2. Explore the possibility of eliminating this offset under such conditions by using proportional-plus-integral action on the feedback of the error in the estimated state of a full observer. Determine a set of values k_{1p}, k_{1i}, k_{2p}, k_{2i} for

$$\mathbf{K} = \begin{bmatrix} k_{1p} + k_{1i}/s \\ k_{2p} + k_{2i}/s \end{bmatrix}$$

such that for a step-input disturbance, the error system is asymptotically stable with speed faster than that of e^{-t}, and the offset is indeed zero. *Hint:* Use the Routh test, then compute the transfer function $\mathbf{E}(s)/V(s)$, and use the final-value theorem.

13.24 Design the estimated and measured state vector feedback for the system of Problem 13.22 such that the resulting closed-loop system is critically damped with a minimum time constant of 1.

13.25 Determine a signal-synthesis model reference adaptive control for the plant

$$\dot{x}_1 = x_2$$

$$\dot{x}_2 = -10x_1 - 11x_2 + u$$

such that the reference model is characterized by the eigenvalues -2 and -3. Simulate the response of the controlled system for zero setpoint and initial state $[0 \quad -1]^T$ assuming that the plant output is subject to an unmodeled dead time $e^{-0.2s}$. In other words, the measurement of $\mathbf{X}$ used in the controller is $\mathbf{X}_m(t) = \mathbf{X}(t - 0.2)$.

13.26* Design a signal-synthesis model reference adaptive control for the system

$$\dot{x}_1 = x_2$$

$$\dot{x}_2 = x_3$$

$$\dot{x}_3 = -x_1 - 2x_2 - x_3 + 2u$$

assuming that all states are measurable and the desired behavior of the system can be represented by the third-order characteristic equation $s^3 + 16s^2 + 44s + 40 = 0$.

13.27 Consider the third-order servomotor actuator of Example 11.2 with the transfer function

$$G_p(s) = \frac{1}{s^3 + 5s^2 + 8s + 4}$$

The *proportional* control determined for this plant using Ziegler-Nichols rules ($k_c = 19.61$) gives a closed-loop system with the characteristic equation $s^3 + 5s^2 + 8s + 23.61 = 0$.

Assuming that all internal states are accessible, design a signal-synthesis MRAC for this system that will give the same desired closed-loop behavior. Simulate the response of the system under both the MRAC and the P-action of Example 11.2, and comment on your results.

13.28 Consider the first-order object $\dot{x} = -x + u$. Determine the value of u that minimizes

$$J = \int_0^{t_f} \left(1 + \frac{1}{2}u^2 \right) dt$$

subject to

(a) $x(0) = 1$, $x(t_f = 1) = 0$

(b) $x(0) = 1$, $t_f = 1$, $x(t_f)$ is free

(c) $x(0) = 1$, $x(t_f) = 0$, t_f is free

(d) $x(0) = 1$, $x(t_f) = 1 - t_f$

13.29* Repeat, using the performance index

$$J = \int_0^{t_f} (x^2 + u^2) \, dt$$

(a) Problem 13.28a

(b) Problem 13.28b

13.30 Consider again Example 13.13. Using MATLAB and the M-files **timeopt.m** and **fig13_25.m** on the net, explore using simulation the dependence of the tolerance bands ep_1 and ep_2 on the admissible region for system motion in state space by finding one initial state only within the square region $(-2.5, -2.5) \le (x_1, x_2) \le (2.5, 2.5)$ for which the required tolerance bands—such that the switching curve is not missed ep_1, and arrival at the origin is not undetected ep_2—are larger than the values determined in Example 13.13.

13.31 Given the plant

$$\dot{x}_1 = x_2$$
$$\dot{x}_2 = u$$

it is desired to transfer the system from the initial state $x_1(0) = 2$, $x_2(0) = 0$ to a final state on the unit circle, $x_1^2 + x_2^2 = 1$, in minimum time.

(a) Formulate completely the two-point boundary-value problem whose solution will yield the necessary control.

(b) Solve the problem formulated in (a)

13.32 For the plant given in Problem 13.31, find the control $u(t)$ that will minimize

$$J = x_1(t_f) + \frac{1}{2} \int_0^{t_f} u^2 \, dt$$

subject to $x_1(0) = 0$, $x_2(0) = 0$.

13.33 For the same plant in Problem 13.31, determine the control $u(t)$ that will transfer this system from an arbitrary initial state to the origin in minimum time if the admissible controls are $|u(t)| \le 1$.

13.34* Find the optimal control to take the following system:

$$\dot{x}_1 = -x_1 + u$$
$$\dot{x}_2 = u$$

from an arbitrary initial state to the origin in minimum time, with the control constraint $|u| \le 1$.

13.35 Determine for the second-order plant

$$\dot{x}_1 = x_2$$
$$\dot{x}_2 = -2x_1 - x_2 + u$$

a control $u(t)$ that will take the system from an arbitrary initial state to the final state $\mathbf{x}(t_f) = \mathbf{0}$ in the shortest possible time.

13.36 Consider the following *minimum energy* problem: *Minimize*

$$J = \frac{1}{2} \int_0^1 u^2 \, dt$$

subject to

$$\dot{x}_1 = x_2 \qquad \mathbf{X}(0) = \begin{bmatrix} 1 \\ 0 \end{bmatrix} \qquad \mathbf{X}(1) = \begin{bmatrix} 0 \\ 0 \end{bmatrix}$$
$$\dot{x}_2 = u$$

and the control constraint $|u| \le 6$.

(a) Using the minimum principle, formulate the TPBVP for the optimal control.

(b) Solve the problem by reasoning out the possibilities and the terminal boundary condition.

13.37 Given the system

$$\dot{x}_1 = -x_1 + x_2 + u_1; \qquad x_1(0) = 1; \qquad x_1(1) = 0$$
$$\dot{x}_2 = x_1 - x_2 + u_2; \qquad x_2(0) = 0; \qquad x_2(1) = 0$$

determine the optimal control $\mathbf{U}^*$ that minimizes the performance index

$$J = \frac{1}{2} \int_0^1 (u_1^2 + u_2^2) \, dt$$

13.38* Consider the following *minimum fuel* problem:

$$Minimize \quad J = \int_0^1 |u(t)| \, dt$$

subject to $\quad \begin{matrix} \dot{x}_1 = x_2 \\ \dot{x}_2 = u \end{matrix} \quad \mathbf{X}(0) = \begin{bmatrix} 0 \\ 0 \end{bmatrix} \quad \mathbf{X}(1) = \begin{bmatrix} 1 \\ 0 \end{bmatrix}$

and the control constraint $|u| \le 4$.

(a) Using the minimum principle, formulate the TPBVP for the optimal control.

(b) Solve the problem by reasoning out the possibilities and the terminal boundary condition.

13.39 Given the first-order system and performance index

$$\dot{x} = -3x + 4u; \qquad x(0) = x_0; \, x(\infty) = 0:$$

$$J = \frac{1}{2} \int_0^\infty (x^2 + u^2) \, dt$$

(a) Determine and solve the Euler-Lagrange TPBVP for the optimal control $u^*(t)$ and optimal trajectory $x^*(t)$.

(b) Show using your results from (a) that the optimal control can be realized with a feedback law, $u^*(x) = -kx^*$, that is independent of the initial condition. Confirm this result by solving the algebraic Riccati equation for this problem.

(c) Starting with the feedback control law $u = -kx$, determine by parameter optimization the value of k that optimizes the performance index for this specific controller and show that the result agrees with the optimal control theory.

13.40 Solve as an Euler-Lagrange TPBVP the optimal linear quadratic regulator

$$Minimize \quad J = \frac{1}{2} \int_0^1 (x^2 + u^2) \, dt \quad subject \ to \quad \dot{x} = -x + u$$

Verify your answer using the relevant Riccati equation.

13.41 Repeat Problems 13.39a and 13.39b using the following second-order system and performance index:

$$\begin{matrix} \dot{x}_1 = x_2; \, x_1(0) = x_{1-0} \\ \dot{x}_2 = -x_2 + u; \, x_2(0) = x_{2-0} \end{matrix} \quad ; \quad J = \frac{1}{2} \int_0^\infty (x_1^2 + 4u^2) \, dt$$

13.42* Repeat Problems 13.39a and 13.39b using the following second-order system and performance index:

$$\dot{x}_1 = x_2; \, x_1(0) = 1$$

$$\dot{x}_2 = -2x_1 - 3x_2 + u; \, x_2(0) = 0;$$

$$J = \frac{1}{2} \int_0^\infty (5x_1^2 + 5x_2^2 + u^2) \, dt$$

13.43 Consider the following *tracking* problem. Given the servomechanism

$$\dot{x}_1 = -0.1x_1 + x_2$$

$$\dot{x}_2 = u$$

it is desired to drive the system from the initial state $x_1(0) = 1$, $x_2(0) = 0$ to the origin while tracking the straight line $x_1 = -x_2$.

(a) Formulate this problem as a linear quadratic regulator problem. You may consider that the path is given by $x_1 + x_2 = 0$, so that the integral of $(x_1 + x_2)^2$ can be minimized.

(b) Solve the problem formulated in (a).

13.44 Solve the following optimal linear quadratic regulator problem:

$$Minimize \quad J = \frac{1}{2} \int_0^\infty (\mathbf{X}^T \mathbf{Q} \mathbf{X} + \mathbf{U}^T \mathbf{R} \mathbf{U}) \, dt$$

subject to $\quad \begin{matrix} \dot{x}_1 = x_2 + u_2 \\ \dot{x}_2 = -4x_1 - 4x_2 + u_2 \end{matrix}$

where

$$\mathbf{Q} = \begin{bmatrix} 2 & -1 \\ -1 & 36 \end{bmatrix} \quad and \quad \mathbf{R} = \begin{bmatrix} 1 & 0 \\ 0 & 1 \end{bmatrix}$$

13.45 Recall the simulation of optimal responses of a first-order-lag-plus-dead-time control object subject to PI-control in Fig. 11.7. In normalized time coordinates, the control object can be described by

$$G_p(s) = \frac{e^{-0.2s}}{s + 1}$$

In the process industries, an often-used linear approximation of dead time is the first-order *Padé* approximation:

$$e^{-T_d s} \approx \frac{1 - \dfrac{T_d}{2} s}{1 + \dfrac{T_d}{2} s}$$

(a) Show that with this approximation this control object can be described by the state model

$$\dot{x}_1 = -\frac{2}{T_d} x_1 + \frac{4}{T_d} x_2$$

$$\dot{x}_2 = -x_2 + u$$

$$y = x_1 - x_2$$

where x_2 is the output of the first-order lag component

(b) Obtain the linear quadratic *output* regulator for this second-order system for $\hat{\mathbf{Q}} = 1$, $\mathbf{R} = 1$, and zero set-point. Simulate the response of the controlled system from the initial state $x_1 = 0$, $x_2 = 1$.

(c) Consider the possibility of PI-action on y, that is,

$$u = -k_c\left(y + \frac{1}{T_i}\int y\, dt\right)$$

Let $\dot{x}_3 = y$, and solve the zero setpoint third-order linear quadratic output regulator problem. (You may have to solve the Riccati equation by longhand if your math software is unable to do so. You can verify that this system is controllable). Show that the optimal feedback gain for x_3 is zero, so that this form of PI-action on the given output does not admit of an optimal (output regulator) solution except the same P-action solution of (b). Simulate the response of the controlled system from the initial state $x_1 = 0$, $x_2 = 1$, $x_3 = 0$, and compare the results with the unit-setpoint-change ISE performance shown in Fig. 11.7.

13.46* Repeat Problem 13.45 using the second-order *Padé* approximation to dead time:

$$e^{-T_d s} = \frac{T_d^2 s^2 - 6T_d s + 12}{T_d^2 s^2 + 6T_d s + 12}$$

(a) The state model in this case is

$$\dot{x}_1 = -\frac{6}{T_d}x_1 + x_2 - \frac{12}{T_d}x_3$$

$$\dot{x}_2 = -\frac{12}{T_d^2}x_1$$

$$\dot{x}_3 = -x_3 + u$$

$$y = x_1 + x_3$$

where x_3 is the output of the first-order lag component

(b) Use the initial state $x_1 = 0$, $x_2 = 0$, $x_3 = -1$. Also, establish that the nonzero optimal feedback gain is the same as for the first-order approximation in Problem 13.45.

(c) In lieu of a fourth-order model consider directly the use of the third-order optimal feedback gains obtained in Problem 13.45. That is, let $k_1 = k_1$ of Problem 13.45, $k_2 = k_3$ of Problem 13.45, and $k_3 = k_2$ of Problem 13.45. Show that the resulting controlled system performance is the best in the series.

13.47 Consider the undamped servomechanism of Example 13.3 with the overall system block diagram of Fig. 13.8a. Referring to this diagram and equations (13.16c) through (13.16e), and assuming the actuator inner loop can be approximated by the ideal behavior $k_a = 0.0056$ in./V,

(a) Formulate but do not solve the fourth-order output regulator problem that can represent this system, that is, determine **A**, **B**, **C**, and **Q** matrices and the correspondence between the feedback gains **K** and the PID-controller parameters of Example 13.3. Take $\hat{Q} = 1$ and $R = 1$. *Hint:* It may be necessary to consider a two-input system (one being V_{IN}) to ensure that the **A** and **B** matrices do not contain any undetermined parameters.

(b) Show that the resulting [**A**, **B**] system is completely controllable.

(c) Can you solve the corresponding algebraic Riccati equation (by longhand or software)? If so, determine the predicted optimal values of the PID-controller parameters. Note that the values in Example 13.3 are for an essentially minimum time objective.

13.48 Recall the continuous stirred tank reactor of Example 13.4:

$$\begin{bmatrix}\dot{x}_1 \\ \dot{x}_2\end{bmatrix} = \begin{bmatrix} -k + \dfrac{w}{\rho V} & 0 \\[2ex] -\dfrac{Hk}{\rho c} & -\dfrac{w}{\rho V} \end{bmatrix}\mathbf{X} + \begin{bmatrix} \dfrac{w}{\rho V} & 0 \\[2ex] 0 & \dfrac{1}{\rho Vc} \end{bmatrix}\begin{bmatrix}u_1 \\ u_2\end{bmatrix}$$

$$+ \begin{bmatrix} 0 \\[1ex] \dfrac{w}{\rho V} \end{bmatrix}v; \quad \mathbf{X}(0) = \begin{bmatrix}0 \\ 295\end{bmatrix}; \quad \mathbf{X}_{\text{desired}} = \begin{bmatrix}60 \\ 350\end{bmatrix}$$

(a) Develop completely the boundary-value problem that should be solved to obtain the optimal control for the objective that following a setpoint change, all outputs should converge to the reference values in 135 s.

(b) Reformulate the problem as a nonzero setpoint linear state regulator problem and solve the problem for

$$\mathbf{Q} = \mathbf{R} = \begin{bmatrix}1 & 0 \\ 0 & 1\end{bmatrix}$$

that is, determine **K** and $\mathbf{U}_{sp}$. Simulate this control system (including the disturbance), and compare the response with that obtained in Example 13.4. Explore the possibility of improving the performance with a different **Q**.

13.49* Consider the three-tank level-control system of Example 13.6:

$$\dot{\mathbf{X}} = \begin{bmatrix} -0.5 & 0.5 & 0 \\ 1 & -1.5 & 0.5 \\ 0 & 0.2 & -0.36 \end{bmatrix}\mathbf{X} + \begin{bmatrix}2 \\ 0 \\ 0\end{bmatrix}u; \quad \mathbf{X}(0) = \begin{bmatrix}1 \\ 1 \\ 1\end{bmatrix};$$

$$\mathbf{X}_{\text{desired}} = \begin{bmatrix}2.0 \\ x_{2d} \\ x_{3d}\end{bmatrix}$$

where x_{2d} and x_{3d} are to be determined.

(a) For a nonzero setpoint optimal linear state regulator formulation with $\mathbf{Q} = \mathbf{I}_3$ and $\mathbf{R} = 1$, obtain the optimal feedback gains **K**, the "setpoint input" u_{sp}, x_{2d}, and x_{3d}, and the closed-loop eigenvalues, and compare these results with the design in Example 13.6 using the present closed-loop eigenvalues as the desired eigenvalues.

(b) Formulate completely the TPBVP whose solution will yield the necessary control for the above state regulator problem when the input is constrained to $0 \le u \le 0.2$.

13.50 Repeat problem 13.49 for the three-tank (vector) level control system of Example 13.7:

$$\dot{X} = \begin{bmatrix} -0.5 & 0.5 & 0 \\ 1 & -1.5 & 0.5 \\ 0 & 0.2 & -0.36 \end{bmatrix} X + \begin{bmatrix} 2 & 0 \\ 0 & 4 \\ 0 & 0 \end{bmatrix} \begin{bmatrix} u_1 \\ u_2 \end{bmatrix};$$

$$X(0) = \begin{bmatrix} 1 \\ 1 \\ 1 \end{bmatrix}; \qquad X_{desired} = \begin{bmatrix} 2.0 \\ x_{2d} \\ x_{3d} \end{bmatrix}$$

The corresponding R and input constraints are $R = I_2$, $0 \le u_1 \le 0.2$ and $0 \le u_2 \le 0.2$.

13.51 Determine and plot as a function of frequency the largest and smallest principal gains for the closed-loop transfer matrix and the loop sensitivity matrix of the following multivariable systems. Identify in each case the H_∞ norm for the closed-loop transfer matrix.

(a) The stirred chemical reactor decoupling control system of Example 13.4

  (i) Using the decoupled system equations and the controller parameter values determined in that example.

  (ii) Using the total (coupled) system equations (13.25a) through (13.25e) with the controller parameter values determined in the example.

(b) The system of Problem 13.9 with the output $Y = X$.

(c) The multivariable system of Fig. 13.37, Problem 13.14. Assume any small values of k_1, k_2, and k_3 such that the overall system is asymptotically stable.

13.52 Discuss the similarities and differences, if any, between a typical *robust* control system and an *adaptive* control system.

13.53 Show that for the system (13.70a) with the control (13.70b) the closed-loop transfer matrix between z and w is given by equation (13,70c). *Hint:* Substitute u, that is, equation (13.70b), into equation (13.70a), and determine the matrices A^*, B^*, C^*, and D^* such that

$$\dot{x} = A^*X + B^*w$$

$$z = C^*x + D^*w$$

13.54 Rework completely the problem of Example 13.17, including all necessary simulations and comparison with signal-synthesis MRAC, for the case of no restriction on the control input (that is, $\epsilon = 0$). What is the minimum γ (determined by simulation or otherwise) for which a static state feedback control that minimizes the weighted integrated tracking error is obtained?

14

Discrete-Time Control Systems

THE APPLICATION OF DIGITAL COMPUTERS TO THE CONTROL of processes and systems is now commonplace. This is true not only for the so called *smart products* that include an *embedded* small computer as one of their components, and the *process* industries, which pioneered the use of large computers in a *supervisory* role, but for *power* industries and *motion control* systems, which for a long time traditionally relied on analog-type controls. The applications range from simple toys and household appliances to automobiles, industrial robots, aircraft autopilots, and chemical plants. A major motivation for the expanding use of computers is the continuing development of faster, smaller, and cheaper microprocessors. The use of digital computers in control has also grown as a result of the increased use of modern control theory (such as discussed in Chapter 13), which requires sophisticated signal processing, a forte of digital computers.

Digital computers are discrete-time systems, and their presence in a control loop imparts a discrete-time nature to the overall system (see Nature of Plant, Table 1.3). Our development of the subject of discrete-time control will somewhat parallel the approach used for continuous-time control. Following a brief introduction in Section 14.1 of discrete-time control concepts—supervisory, *hierarchical* and *distributed* control, *sequence* control, *numerical* control, and *direct digital control* (see Table 1.3, Nature of Controller), we consider in Section 14.2 the application of classical or conventional feedback control to discrete-time systems or single-loop (SISO) digital controllers along with the problem of performance specification and parameter tuning for such systems. Also discussed in this section is the analysis and design of systems containing both discrete-time signals and continuous-time signals—the so-called sampled-data systems. Next, discrete-time state space–based control systems or modern control, as exemplified by the *finite-time settling* state vector feedback controller and observer design, are considered in Section 14.3. This leads to the introduction in Section 14.4 of the discrete-time optimal control problem via the *dynamic programming* approach. This material complements our discussion in Chapter 13 of continuous-time optimal control. As was the case in previous chapters of Part III, aspects of design by digital simulation are dispersed throughout the present chapter, including its application in parameter tuning. Computer simulation assistance in this task is especially important due to the complication that is sometimes introduced by the dependence of digital controller parameters on the *sampling period* and sampling conditions.

14.1 DIGITAL COMPUTERS IN CONTROL LOOPS

A digital computer can be applied to process control in a number of different modes. At the lowest level of individual process loops, the computer can be employed in the direct determination and application of manipulated inputs necessary to maintain process variables at their desired values. This is the *direct digital control* (DDC) mode. At the highest level, the computer can be used in a *supervisory* sense to coordinate the individual process loops by using knowledge of their interaction and measured response data to determine loop setpoints that collectively guarantee optimum performance or behavior in a specific way of the overall process. At midlevels are other modes of use such as *data logging, sequencing,* and *numerical control.*

Supervisory Control

A typical process control environment may include several interacting process loops. The optimum operating strategy is determined by the combined effect of the current values of the individual loop variables, operational constraints (for example, equipment and safety limitations), product specifications, and even external inputs such as raw material composition and cost. A supervisory control system in which the individual loop setpoints are calculated from a master program stored in computer memory can be very effective in realizing the performance goals of the overall process. The supervisory control program represents a model of the process that can be determined by prior knowledge or from tests or experiments performed *on-line* on the process, or both. As for digital computer software, it must be adequate for the tasks to be performed and consistent with the capacity of the computer. The ability of a digital computer to collect large quantities of data, analyze such data, and make logical decisions based on the results is what recommends it for the supervisory control function.

Although the supervisory computer determines the optimum operating strategy, it interacts with the process indirectly by merely providing the setpoints for the constituent process loops. The direct implementation of the decisions is mostly left to the loop controllers, which can be all analog controllers, in which case the supervisory computer is the only digital computer in the system, or all digital controllers, or a combination of analog and digital controllers. However, provisions can be made for the supervisory computer to assume direct control of a critical loop in the event of failure of that loop's controller. Figure 14.1a illustrates a supervisory control configuration having only direct digital control (DDC) loops.

Some Forms of Supervisory Control

There are many forms of supervisory control. In terms of the nature of the control strategy, the types of supervisory control range from very simple *ratio control* to complex *program control* strategies. Other forms of supervisory control include *sequence control* and *numerical control,* which are discussed below as being also bona fide modes of application of the digital computer in process control, and *cascade control* which we explored in Chapter 13 for continuous-time control systems. It is possible for different process loops to be subject to different forms of supervisory control within the same overall system.

Ratio Control   In ratio control the supervisory computer sets the loop controller (parameter) to a value that has a fixed relationship with the measured value of a process variable. *Ratio control* is an example of the type of implementation of combined feedforward and feedback control concepts illustrated in Fig. 13.1c(ii) for continuous-time control, in which the output of the feedback controller is used to adjust (some parameter in) the feedforward controller (see also Practice Problem 13.3c, which deals with

Figure 14.1

Examples of Supervisory and Distributed Controls and Data Logging Configuration

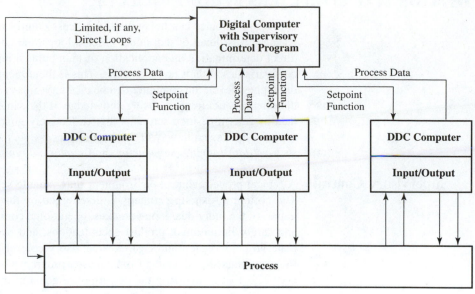

(a)   A Supervisory Control System

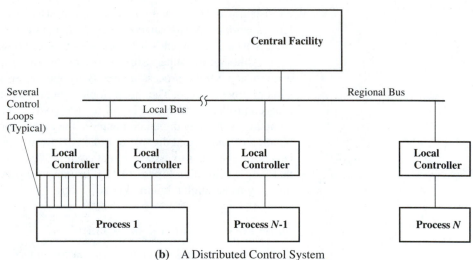

(b)   A Distributed Control System

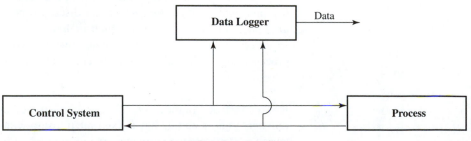

(c)   A Data Logging System

ratio control of feedwater flow through a control valve). In our present context consider the supervisory computer in place of the feedback controller, and replace the feedforward controller with the process loop controller.

Cascade Control   The *cascade control* process illustrated in Fig. 13.5a for multiloop continuous-time control also describes the same form of supervisory computer control. In this case the supervisory computer is the *primary* or *master* controller, and the loop controller is the *secondary* or *slave* controller. A typical supervisory control system can thus contain several secondary or process loops in parallel with one master or supervisory controller. The master controller is used to alter the setpoints of the individual slave controllers that control secondary process variables in order to achieve overall system goals or performance specifications on primary variables. A digital computer is more suited for master control functions because of its flexibility, speed, capacity for low-cost implementation of sophisticated control laws via software, and the ease with which such a controller implementation can be changed if necessary.

Program Control   Program control refers to a procedure or phase of control action that is usually applied for a purpose that is different or distinct from the routine operation of the process. The duration of the procedure can be very brief or, as in the case of start-up or shutdown of large nuclear power plants, it can last for a number of days. A supervisory control program is used to determine, over the required period of the time, the necessary pattern and sequence of changes of setpoints. In addition to optimal and safe start-up and shutdown of complex processes, program control is also used to quickly and cost effectively move a process from one set of operating conditions to another set in order to begin producing a new product with the same plant. Program control makes possible *flexible manufacturing systems* (FMS)—computer-controlled manufacturing processes that adapt automatically to random changes in product design, models, or styles.

Hierarchical and Distributed Controls

A *hierarchical* control system is a multilevel control configuration similar to the *pyramidal* structure in the management of organizations. The highest levels of the hierarchy are typically concerned with overall system performance, product and process specifications, safety, profitability, and other longer term goals and changes. Components at these levels receive information both from the lower levels and from external sources and generate decisions that affect the goals and even the structure (an adaptive control action, for instance) of lower levels. Components at the lowest levels deal with much shorter term goals and changes. They utilize input information from direct measurements on the process and setpoint and other loop-specific short-term objectives sent from higher levels and generate corrective actions on the process and performance feedback to upper levels. Supervisory control implies a hierarchical configuration even when there are only two control levels. The supervisory controller is in principle the highest level of a hierarchical control system.

Distributed Control Systems

The hierarchical structure also applies to distributed control systems. Figure 14.1b illustrates a distributed control system consisting of a network of local control stations coordinated by a central facility. Computers at the central facility are used for overall control of the system and such other tasks as data logging (described below), status display, diagnosis, and trend analysis. The network can be nested, that is, a node on the regional bus can connect to a local bus linking several local controllers in the vicinity. Each local controller, in turn, can be controlling a number of process loops, as shown. The local controllers are capable of independent operation, thus making for a highly reliable system. The nested network structure also makes it easy to expand or modify the system. An intranet or internet-type regional communication link can facilitate integrated control of processes at plant sites separated by long distances.

Data Logging

The ability of digital computers to collect and organize large quantities of data also recommends them for use as data loggers. In *data logging,* a digital computer is used to *log* (mostly analog) input signals for subsequent data processing. As shown in Fig. 14.1c, the data logger is not normally an active part of the digital control loop. However the data generated by the logging of relevant process variables can be processed *off-line* or *on-line* by computer to deduce subsequent optimum operating strategy for the plant. In most cases, the logged data are part of a required operating record, such as in nuclear power plants, or the essential output, as in automated laboratory and meteorological data acquisition systems.

Sequence and Numerical Control Systems

Sequence control and *numerical control* systems are generally characterized by an overall fixed order in which actions take place and a limited functional dependence between the overall control instructions and the resultant values of process variables. These characteristics limit the field of processes or systems to which sequence and numerical controls are normally applied.

Sequence Control

Two examples of familiar processes that utilize *sequence control* are washing machine cycles and traffic control systems. Other examples include machine tools, packaging equipment, and automatic test and analysis machines. The control actions in a sequence control system take place in a predetermined and fixed sequence, with the end of one action signaling the beginning of the next action. Although feedback of the completion of an action can be used, there is typically no comparison of desired and actual *values* of quantities. However, a sequence controller may also involve delays to allow for the operating time of system actuators and error checking to determine component failures as well as other safety and operator interaction features.

Numerical Control

Numerical control is typically applied to machine tools and is essentially the operation of such machines by coded instructions. The control uses predetermined instructions, often coded in a symbolic program or other high-level language, to control the sequence of manufacturing operations performed by the machine tool. The name *numerical control* (NC) reflects the nature of the instructions, which consist mainly of *numerical* information such as tool position or displacement or associated direction and/or rotating speed. *Computer numerical control* (CNC) results when the coded instructions are resident and are issued by a computer. Typical manufacturing operations to which numerical control has been applied include milling, turning, drilling, boring, punching, riveting, grinding, welding, flame cutting, garment knitting, and wire winding. In such NC machine tools, the coded instructions are ultimately translated into electromechanical movements of the relevant tool, and the ultimate servomechanism can be analog or digital.

CNC machine tools make it possible to directly produce a part from the same computer used to design it, without an intervening stage of part drawings and other manual inputs, since the *computer aided design* (CAD) process can provide both the necessary part surface information to drive the machine tool and the design data required to access (on-line) the applicable *computer aided manufacturing* (CAM) database.

Direct Digital Control Systems

In DDC the control laws are implemented in a digital computer as computer programs that accept inputs representing controlled variables and other measurements on the process, setpoints, and even operator's instructions and generate computer outputs corresponding to manipulated variables or signals that are sent directly to process actuators (e.g., piston displacements or valve positions) and operator's information (see Fig. 14.2). The term *direct digital control* derives from the fact that the decisions of the computer are applied directly to the process (see [5]).

An advantage of software realization of control laws over analog computer or other hardware implementation is the absence of restrictive computing hardware constraints. For example, the value of the output voltage of an electronic operational amplifier circuit is limited to the range of the power supply voltage, and the availability of such circuit components as resistors and capacitors in discrete values limits the range of parameters values that can be obtained from combinations of these components. On the other hand, processing speed can be a problem peculiar to software controllers. For real-time control to be possible, the digital computer must process information fast enough to ensure that computation of the manipulated variables is concluded within the sampling period, no matter how complex the control algorithm or how short the sampling time. Fortunately, the processing capacity and speed of digital computers continue to increase, so that a large and increasing number of processes can be placed under DDC. A further advantage of DDC is that it permits fast and low-cost implementation of even more sophisticated control systems, since the associated control law can simply be *programmed*.

Figure 14.2

Direct Digital Control System

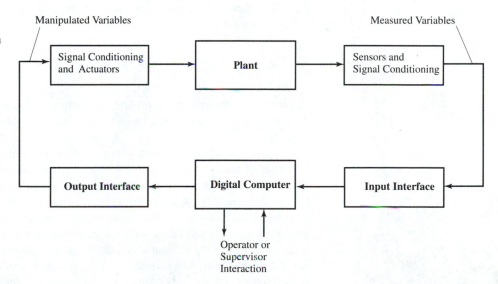

14.2 SINGLE-LOOP DIGITAL CONTROLLERS

There are two general approaches to the development of digital control algorithms for a given control object. The first approach, which is complementary to the practice developed in Chapter 11 for analog controllers, is to give the form of the algorithm and then determine and adjust its parameters (that is, tune the algorithm) to obtain the desired closed-loop response. This is the method considered first. The second approach, which is discussed subsequently, is to give the desired closed-loop transfer function or model and then determine from this and the plant model the form and parameters of the control algorithm.

Two-Term (PI) and Three-Term (PID) Control

The simplest control action that can be implemented using a digital computer is the *on-off* discontinuous control. Although it is easy and relatively inexpensive to accomplish, this type of control does not take into account the magnitude of the error between the desired and actual values of the controlled variables and hence does not meet the requirements of a good number of processes. Two of the most common digital control algorithms in use are the discrete-time equivalents of the classical three-term and two-term single-loop feedback controllers: the proportional-plus-integral-plus-derivative (PID) control law and the simpler proportional-plus-integral (PI) control algorithm.

The PID Algorithm

It is sufficient to consider the PID algorithm, since the result can be reduced to the PI function. The PI-controller can be obtained in the following by setting $T_d = 0$ in the PID algorithm. Consider the classical PID control law:

$$\frac{M(s)}{E(s)} = G_c(s) = k_c \left(1 + \frac{1}{T_i s} + T_d s \right)$$

where $M(s)$ and $E(s)$ are, respectively, the (Laplace domain) manipulated and error variables. The continuous-time model for this law is

$$m(t) = k_c \left[e(t) + \frac{1}{T_i} \int_0^t e(t)\, dt + T_d \frac{d}{dt} e(t) \right] + m(0) \tag{14.1}$$

Euler's method, equation (6.94c), applied to equation (14.1) gives, in terms of *sampled* data, the algorithm

$$m_k = k_c \left[e_k + \frac{T}{T_i} \sum_{i=0}^{k} e_i + \frac{T_d}{T}(e_k - e_{k-1}) \right] + m_B \tag{14.2a}$$

where T is the sampling interval and m_B is the initial actuator (position) value, that is, the value of the manipulated variable when the control loop is placed on automatic control.

Equation (14.2a) is known as the *position algorithm* because the value calculated is the actual manipulated variable, which in many cases represents an actuator (valve, piston, etc.) position. The necessity to initialize the position algorithm, that is, to somehow supply the initial actuator position m_B to the computer, can be a disadvantage for this controller. Knowledge of m_B is necessary to ensure a smooth transition ("bumpless transfer") from manual or no control to automatic control. Another possible disadvantage is that should the computer fail, the resulting zero output from the computer could imply zero position, to which the actuator would respond, perhaps, with undesirable consequences. On the other hand, for the *velocity algorithm* in which the computed output is the change in the manipulated variable,

$$\Delta m_k = m_k - m_{k-1}$$

a zero output implies that the actuator should hold the previous position, which, under many circumstances, would represent a safe ("fail-safe") response.

The PID velocity algorithm can be obtained by writing equation (14.2a) for m_{k-1} and subtracting the result from m_k, which gives

$$\Delta m_k = k_c\left[(e_k - e_{k-1}) + \frac{T}{T_i}e_k + \frac{T_d}{T}(e_k - 2e_{k-1} + e_{k-2})\right] \qquad \textbf{(14.2b)}$$

It is assumed that the computer output for the velocity algorithm goes to an integrating-type actuator (such as a stepping motor), where the following integration is performed:

$$m_k = \Delta m_k + m_{k-1} = m_0 + \sum_{i=1}^{k}\Delta m_k \qquad \textbf{(14.2c)}$$

As a practical matter, the position of most actuators will not follow equation (14.2c) exactly, so that some loss of control quality is expected in the final control element.

An alternative set of PID algorithms is obtained by applying the trapezoidal rule equation (6.99) to equation (14.1), instead of Euler's method. The resulting position algorithm is

$$m_k = k_c\left[e_k + \frac{T}{T_i}\sum_{i=1}^{k}\frac{1}{2}(e_i + e_{i-1}) + \frac{T_d}{T}(e_k - e_{k-1})\right] + m_B \qquad \textbf{(14.3a)}$$

and the corresponding velocity algorithm is

$$\Delta m_k = k_c\left[(e_k - e_{k-1}) + \frac{T}{2T_i}(e_k + e_{k-1}) + \frac{T_d}{T}(e_k - 2e_{k-1} + e_{k-2})\right] \qquad \textbf{(14.3b)}$$

Since the preceding algorithms are all in software, many of the limitations typical of the corresponding analog PID hardware can be eliminated through programming: *Reset windup* can be overcome for the position algorithm by placing an upper limit on the summation term in equations (14.2a) and (14.3a). Alternatively, the summation can be terminated whenever the control element saturates and resumed when the manipulated variable returns to within the control range. This "antiwindup" feature is already inherent in the velocity algorithm. *Proportional and derivative kicks* due to setpoint changes can be reduced by ramping setpoints to their new values or eliminated altogether as explained below. However, the controller can become sluggish with respect to load changes. This situation can be compensated for by storing in the computer two sets of controller parameters: one applicable to a period of time following a setpoint change (recall *program control*), and the second set for all other times.

Canonical Velocity Algorithm

The PI and PID DDC algorithm preferred in practice is the velocity form. If we substitute for the error $e_k = r_k - y_k$ (see Fig. 14.3a or 14.3d) in equations (14.2b) and (14.3b) and assume

$$r_k = r_{k-1} = r_{k-2} \qquad \textbf{(14.4a)}$$

we obtain for equation (14.2b),

$$\Delta m_k = k_c \left[(y_{k-1} - y_k) + \frac{T}{T_i}(r_k - y_k) + \frac{T_d}{T}(2y_{k-1} - y_{k-2} - y_k) \right] \qquad \textbf{(14.4b)}$$

and for equation (14.3b),

$$\Delta m_k = k_c \left[(y_{k-1} - y_k) + \frac{T}{T_i}\left(r_k - \frac{1}{2}[y_{k-1} + y_k] \right) \right.$$
$$\left. + \frac{T_d}{T}(2y_{k-1} - y_{k-2} - y_k) \right] \qquad \textbf{(14.4c)}$$

In terms of independent P, I, and D adjustments we can write the canonical algorithm as follows:

$$\Delta m_k = k_p(y_{k-1} - y_k) + k_I(r_k - y_k)$$
$$+ k_D(2y_{k-1} - y_{k-2} - y_k); \quad \text{(canonical velocity)} \qquad \textbf{(14.5a)}$$

so that for the velocity algorithm derived from Euler's method, equation (14.4b), we have

$$k_p = k_c; \ k_I = \frac{k_c T}{T_i}; \qquad k_D = \frac{k_c T_d}{T}; \quad \text{(Euler)} \qquad \textbf{(14.5b)}$$

while for the algorithm derived from the trapezoidal rule, equation (14.4c),

$$k_p = k_c - \frac{1}{2}k_I; \qquad k_I = \frac{k_c T}{T_i}; \qquad k_D = \frac{k_c T_d}{T}; \quad \text{(trapezoidal)} \qquad \textbf{(14.5c)}$$

Thus the canonical algorithm, equation (14.5a), describes both (Euler and trapezoidal) velocity forms of the PID algorithm. The absence of the setpoint r in both the proportional and derivative terms of the velocity algorithm (equation (14.5a), for instance) and hence the elimination of proportional and derivative kicks (large, sudden changes in output following setpoint changes) is due to the assumption (14.4a). Because only the integral term now contains the setpoint, the simplification of the controller given by equation (14.4a) *cannot be used if I-action is not included in the algorithm.*

Sampled-Data Systems and Parameter Tuning   The quantitative determination of a given digital control algorithm requires the identification of the object of control. If we adopt the point of view that most processes or plants are continuous-time systems, then a digital control loop on such a process will involve both discrete-time data and continuous-time signals. Such a system having both discrete and continuous signals is often referred to as a *sampled-data system.* The equivalent object of control for sampled-data systems $G_p(z)$ is determined in equation (14.6c).

Sampled-Data System

A single-variable digital process control loop with a continuous-time process, a SISO sampled-data system, will have the configuration shown in Fig. 14.3a. The actuation and measurement subsystems can be considered incorporated in the plant, so that the structure shown in the figure is a general one. The continuous plant output $y(t)$ must be converted to discrete-time data by *sampling* at discrete points in time. Physically, sampling is normally accomplished using *analog-to-digital converter* (ADC) hardware and a digital computer sampling program, as will be discussed in the next chapter. For now, we represent this process with the *sampler* block shown in Fig. 14.3a and assume sampling is performed at equal intervals of time or *sampling period T.*

Figure 14.3b shows the relation between the continuous function $y(t)$ and the sampled function y_k, which is a discrete-time approximation of the continuous-time function. The characteristics of this approximation such as *bandwidth* and *aliasing error* were discussed in Section 8.2 in connection with the discrete-time Fourier series and the discrete Fourier transform. The sampled function is also a *time-delayed sequence* of values or numbers $f(kT)$, $k = 0, 1, \ldots$, and as discussed in Section 8.4, it has a z-transform $Y(z)$, given by equation (8.40). We shall also assume that the set-point input is available a priori as sampled data having the z-transform $R(z)$. Continuing around the control loop in Fig. 14.3a, we see that the manipulated variable as a digital computer output is also a sampled function or a sequence of values with the z-transform $M(z)$. The ultimate control input to the continuous-time process, which itself must be a continuous function, is derived from the manipulated variable by the process described next.

Zero-Order Hold

A series of numbers $m_k = m(kT)$, $k = 0, 1, 2, \ldots$, is a limited description of a continuous function $m(t)$. In particular, the values at times other than the sampling instants are undefined and must be approximated, usually by some interpolation technique. The simplest such reconstruction of $m(t)$ that uses the least information is a zero-slope first-order (or again Euler) approximation illustrated in Fig. 14.3c. This operation is referred to as *zero-order hold,* that is, an operation in which the value of each sampled signal is "held" constant (zero slope) over the sampling period of T time units. The result is the *staircase* pattern signal shown in Fig. 14.3c. Physically, the conversion of the discrete-time data into a continuous-time signal is usually accomplished with *digital-to-analog converter* (DAC) hardware (also discussed in the next chapter). Although without additional *signal conditioning,* the DAC output, especially for relatively long sampling periods, may not quite represent a zero-order held function, we shall always assume the conditions of Fig. 14.3a.

The zero-order hold is represented in Fig. 14.3a by its Laplace domain transfer function. To understand this transfer function, consider the impulse input response of a zero-order hold. In Section 8.4 it was shown that the sampled function m_k, $k = 0, 1, 2, \ldots$, can be viewed as an impulse train, with the impulse at $t = kT$ having the strength $m(kT)$ (see equations (8.41) and (8.43b)). A unit impulse can also be viewed as an impulse train consisting of an initial impulse of unit strength followed by impulses of zero strength. The staircase pattern that will result when such "sampled data" are passed through a zero-order hold is a unit *pulse* of width equal to the sampling rate. Since the Laplace domain transfer function of a linear system is the same as the Laplace transform of the system's unit-impulse response (equation (8.28b)), the transfer function of the zero-order hold is the same as the Laplace transform of a unit pulse. From Table 8.4, item 14,

$$\mathscr{L}[\text{zero-order hold}] = \frac{1 - e^{-Ts}}{s} \tag{14.6a}$$

Figure 14.3

Single-loop Digital Control Structure

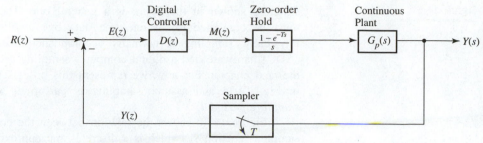

(a)   Continuous-time Process (Sampled-data System)

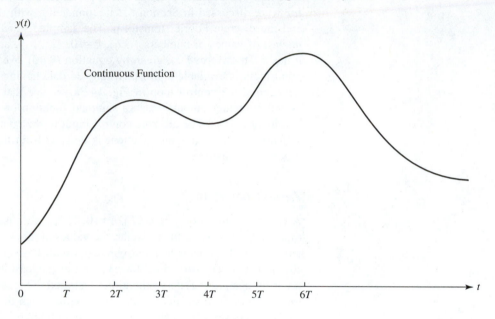

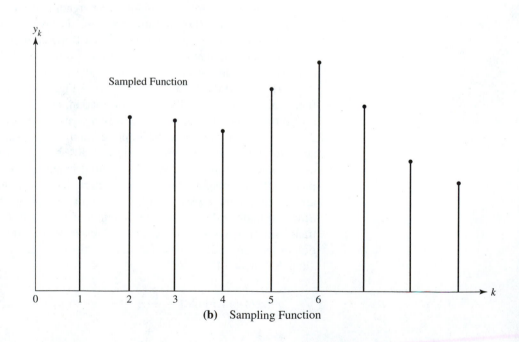

(b)   Sampling Function

Figure 14.3 continued

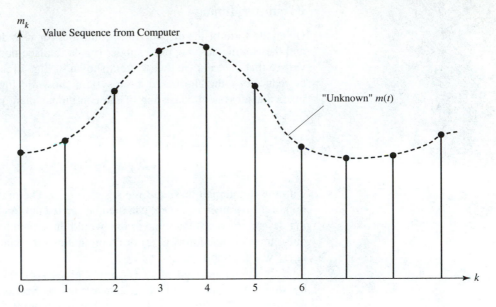

Value Sequence from Computer

"Unknown" $m(t)$

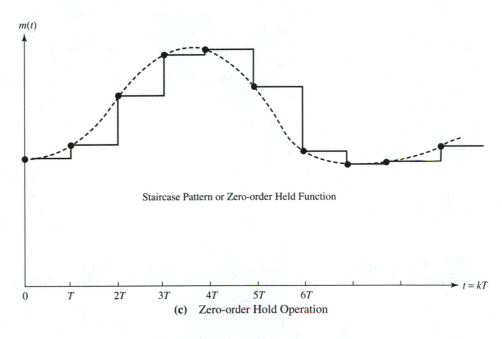

Staircase Pattern or Zero-order Held Function

(c) Zero-order Hold Operation

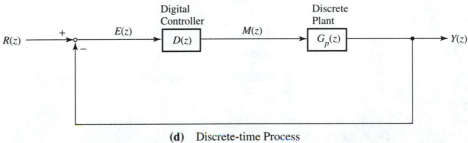

(d) Discrete-time Process

Parameter Tuning

Given the form of the discrete-time controller (or its z-domain transfer function) $D(z)$ and the continuous-time plant model (or its Laplace-domain transfer function) $G_p(s)$, we see that the sampled-data system shown in Fig. 14.3a is also equivalent to, or can be reduced to, the all-discrete-time system shown in Fig. 14.3d, provided we can replace the subsystem consisting of the continuous plant plus zero-order hold,

$$G(s) = \frac{1 - e^{-Ts}}{s} G_p(s) \tag{14.6b}$$

with the counterpart discrete-time system $G_p(z)$. The design of $D(z)$ or the determination of the parameters of $D(z)$ can then be carried out. As Fig. 14.3d implies, the resulting design process is the same as one would undertake for a discrete-time plant whose pulse transfer function is $G_p(z)$ or a continuous-time plant whose discrete-time model $G_p(z)$ was available a priori.

It should be evident that various aspects of this solution can be facilitated by some of the discrete-time system results of Sections 6.4–6.5, Section 8.4, and also Section 9.3. For example, using the methods of Section 9.3, we can obtain a linear state model for $G(s)$, equation (14.6b), and convert it to the discrete-time equivalent using the results of Section 6.5. From the developments in Section 8.4 we can then derive the pulse transfer function $G_p(z)$ of this discrete-time state model. This approach will be illustrated in Example 14.8 in connection with optimal controller design for a continuous-time process using discrete-time dynamic programming.

We consider here (see example 14.1) the z-domain design approach. Using the interpretation, equation (8.43a), $z = e^{sT}$, we can define

$$G_p(z) = (1 - z^{-1})\mathcal{Z}\{\mathcal{L}^{-1}[G_p(s)/s]\} \tag{14.6c}$$

We can also interpret $\mathcal{L}^{-1}[G(s)/s]$ as the (time domain) unit-step response of $G(s)$, if this is convenient. Further, note that the design of a digital controller normally involves an additional parameter relative to its analog counterpart. This parameter is the sampling period T, which is introduced by the zero-order hold.

Example 14.1	Computer-Assisted Parameter Tuning

Consider, as shown in Fig. 14.4a, the direct digital PI-control of the antiaircraft servo system of Fig. 10.5. Assuming the velocity form of the PI algorithm of equation (14.2b), determine by simulation the approximate values of k_c and T_i that will minimize the integrated squared error (ISE). Note that the actuator is an integrating type and is therefore compatible with the "velocity" output of the controller.

Solution:

1. Pulse transfer function: The pulse transfer function of the actuator-gun system is given by equation (14.6c):

$$G_p(z) = (1 - z^{-1})\mathcal{Z}\left\{\mathcal{L}^{-1}\left[\frac{100}{s^2(s + 5)(s + 10)}\right]\right\}$$

Figure 14.4

Design of Direct Digital Control Servomechanism

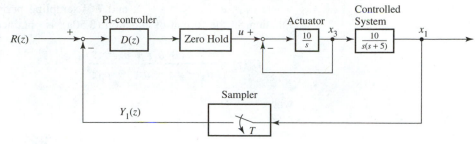

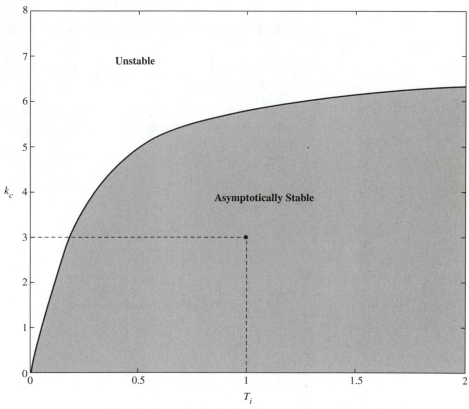

(a) Block Diagram of Direct Digital PI-controlled Servomechanism

(b) Allowable PI-controller Settings for Sampling Period = 0.2

Using partial fraction expansions (Section 8.2), we obtain

$$G_p(z) = (1 - z^{-1})\mathcal{Z}\left\{\mathcal{L}^{-1}\left[\frac{-3}{s} + \frac{10}{s^2} + \frac{4}{s+5} - \frac{1}{s+10}\right]\right\}$$

$$= (1 - z^{-1})\mathcal{Z}\left\{\frac{1}{5}(-3 + 10t + 4e^{-5t} - e^{-10t})\right\}$$

Looking up the z-transforms from Table 8.6, we get

$$G_p(z) = \frac{z-1}{5z}\left[\frac{-3z}{z-1} + \frac{10Tz}{(z-1)^2} + \frac{4z}{z-c_1} - \frac{z}{z-c_2}\right]$$

where $c_1 = e^{-5T}$, $c_2 = e^{-10T}$, and T = sampling period. The resulting pulse transfer function, in the format of equation (8.50a), is obtained after some manipulation:

$$G_p(z) = \frac{b_2 + b_1 z + b_0 z^2}{a_3 + a_2 z + a_1 z^2 + a_0 z^3} \tag{14.7a}$$

where

$$a_0 = 5; \quad a_1 = -5(1 + c_1 + c_2); \quad a_2 = 5(c_1 + c_2 + c_1 c_2); \quad a_3 = -5c_1 c_2$$

$$b_0 = 4c_1 - c_2 - 3 + 10T; \quad b_1 = [(5 - 10T)c_2 - (10T + 5)c_1 - 3c_1 c_2 + 3]$$

$$b_2 = (3 + 10T)c_1 c_2 - 4c_2 + c_1$$

The PI-control velocity algorithm, equation (14.2b) with $T_d = 0$, is

$$m_k = k_c \left[(e_k - e_{k-1}) + \frac{T}{T_i} e_k \right] = k_c \left[\left(1 + \frac{T}{T_i} \right) e_k - e_{k-1} \right] \tag{14.7b}$$

This is directly the controller output, since the integration equation (14.2c) is accomplished in this case in the actuator. Taking the z-transform of equation (14.7b), we obtain

$$M(z) = k_c \left[\left(1 + \frac{T}{T_i} \right) E(z) - z^{-1} E(z) \right]$$

so that

$$D(z) = k_c \left[\left(1 + \frac{T}{T_i} \right) - z^{-1} \right]$$

From Fig. 14.4a or Fig. 14.3d, the closed-loop pulse transfer function is given by

$$G_{CL}(z) = \frac{G_p(z)D(z)}{1 + G_p(z)D(z)} \tag{14.8a}$$

so that the closed-loop characteristic equation in the z-domain is

$$1 + G_p(z)D(z) = 0 \tag{14.8b}$$

substituting for $G_p(z)$ and $D(z)$ in equation (14.8b), we get

$$1 + k_c \left(\frac{k_i z - 1}{z} \right) \left(\frac{b_2 + b_1 z + b_0 z^2}{a_3 + a_2 z + a_1 z^2 + a_0 z^3} \right) = 0 \tag{14.9a}$$

where

$$k_i = 1 + \frac{T}{T_i} \tag{14.9b}$$

Equation (14.9a) further reduces to the form, equation (10.13) repeated,

$$F(z) = d_0 z^n + d_1 z^{n-1} + \ldots + d_{n-1} z + d_n = 0 \tag{14.10}$$

where for the present case, $n = 4$, and

$$d_0 = a_0 = 5; \qquad d_1 = a_1 + k_c k_i b_0; \qquad d_2 = a_2 + k_c(k_i b_1 - b_0)$$

$$d_3 = a_3 + k_c(k_i b_2 - b_1); \qquad d_4 = -k_c b_2$$

2. Application of Jury's inners stability test: In the tradition of continuous-time system design, we can obtain initial estimates for the controller parameters by requiring the overall controlled system to be (asymptotically) stable. The Jury test for asymptotic stability (see Section 10.3) applies directly to equation (14.10) and can be used for this purpose.

The necessary conditions, equation (10.14a), are

$$5 + d_1 + d_2 + d_3 + d_4 > 0 \tag{14.11a}$$

$$5 - d_1 + d_2 - d_3 + d_4 > 0 \tag{14.11b}$$

For the sufficiency test, equations (10.14b) and (10.14c) give

$$H_1 = \begin{bmatrix} 5 & d_1 & d_2 \\ 0 & 5 & d_1 \\ 0 & 0 & 5 \end{bmatrix} + \begin{bmatrix} d_2 & d_3 & d_4 \\ d_3 & d_4 & 0 \\ d_4 & 0 & 0 \end{bmatrix} = \begin{bmatrix} (5 + d_2) & (d_1 + d_3) & (d_2 + d_4) \\ d_3 & (5 + d_4) & d_1 \\ d_4 & 0 & 5 \end{bmatrix}$$

and

$$\mathbf{H}_2 = \begin{bmatrix} (5 - d_2) & (d_1 - d_3) & (d_2 - d_4) \\ -d_3 & (5 - d_4) & d_1 \\ -d_4 & 0 & 5 \end{bmatrix}$$

so that the conditions to be satisfied are

$$5 + d_4 > 0 \tag{14.11c}$$

$$5 - d_4 > 0 \tag{14.11d}$$

$$5(5 + d_2)(5 + d_4) - (d_1 + d_3)(5d_3 - d_1 d_4) - d_4(d_2 + d_4) > 0 \tag{14.11e}$$

$$5(5 - d_2)(5 - d_4) - (d_1 - d_3)(d_1 d_4 - 5d_3) + d_4(d_2 - d_4)(5 - d_4) > 0 \tag{14.11f}$$

3. Selecting the sampling period: The sampling time T is involved in the stability conditions, along with other controller parameters k_c and T_i. However, the normal practice is to determine the sampling interval independently, from other considerations. Unfortunately, there are no general quantitative criteria for selecting the sampling period except perhaps the *sampling theorem* where it is applicable. In general, the control-loop performance improves as the sampling time is decreased. However, the hardware computing time generally also increases with decreasing sampling interval. The choice of sampling period is thus often a compromise between these two conflicting requirements.

The qualitative implication of the sampling theorem (equation (6.96)) is that the sampling interval must be small enough to ensure that essential information is not lost. With this requirement as a guide, sampling periods should normally not be greater than the shortest process and or input dead time, the shortest time constant, half the shortest oscillating period, etc. For the present problem, considering the shortest time constant of the controlled system in Fig. 14.4a, we can select $T = 0.2$ s.*

---

*Actually, considering the actuator, we should select $T \leq 0.1$. However, simulations suggest $T = 0.2$ is a good start.

4. Initial parameter estimates: With T fixed at this value of the time constant, equations (14.11) were coded on the computer (see script M-file **fig14_14b.m; On the Net**) and solved for pairs of trial k_c and T_i for which the controlled system was not asymptotically stable. The result is shown in Fig. 14.4b using k_c and T_i coordinates. From these data, a middle-of-the-range choice of the initial parameters, $k_c = 3.0$ and $T_i = 1.0$, was made.

5. Controller design by simulation: The following script and function M-files, **exmp14_1.m** and **e14_1RHf.m,** illustrate the computer model used to determine by trial and error the "optimum" digital controller parameters k_c and T_i (recall the interactive tuning in Example 11.3) for the selected sampling time. The following observations are noteworthy:

a. The computer model simulates simultaneously a continuous-time and a discrete-time system. The step size for the continuous-time system simulation ΔT is assumed to be an integer ($COUNT$) multiple of the sampling interval T. For direct control of the step size, the implicit trapezoidal algorithm **TRAP** was used rather than **ode23** or **ode45**.

b. The continuous plant model used in the simulation (see Section 9.3) is

$$\frac{d}{dt}x_1 = x_2 - 5x_1$$

$$\frac{d}{dt}x_2 = 10x_3$$

$$\frac{d}{dt}x_3 = 10(u - x_3) = 10[k_c(k_i x_7 - x_8) - x_3]$$

where $u = k_c(k_i x_7 - x_8)$ (see equation (14.7b)) is the zero-order held output of the PI-controller (explained below). The tuning criterion, the minimization of the integrated squared error (ISE), can be approximated as follows:

$$\frac{d}{dt}x_4 = x_7(t)^2; \qquad ISE \approx x_4(t_f)$$

where t_f is the final or maximum time in the simulation.

c. Sampling with zero-order hold is simulated by creating a "staircase pattern" time function or "zero-order held" function of sampling instants x_6 that is used to time and "hold" the sampling of the error $x_7 = r - x_1(x_6)$.

d. The time-delayed signal needed in the control law $e(t - T)$ is obtained by simply defining another variable, which is the zero-order held error x_7 delayed by $COUNT$ sampling instants:

$$x_8^{<k>} = x_7^{<k-COUNT>}$$

Note that for the first $COUNT$ instants or for T seconds, x_8 is necessarily zero.

d. Because the manipulated variable (control) u, is computed in the above scheme using zero-order held samples of the error, the zero-order hold of the manipulated variable is automatically simulated.

```
% exmp14_1.m: Example 14.1, Solution Continued
% Note: script violates MATLAB Student Version element limit for matrices
% Uses function M-file, e14_1RHf.m (user defined right-hand functions)
% This program uses the if function of MATLAB to realize zero-order
% held sampled data. Time delayed signals are created by defining
% additional variables whose current values are the appropriately delayed
% values of other variables. The MATLAB version of the implicit Euler-
% predictor-trapezoidal-corrector module TRAP, is used to solve model
%
% PROVIDE INITIAL CONDITIONS AND PARAMETERS
global k_c k_i;
COUNT = 10; r = 1; k_c = 2.5; Ti = 0.46;
n = 7; nl = 5;                  % give size of state vector
f = zeros(nl,1); fw = zeros(nl,1);
t0 = 0; tf = 6; T = 0.2; dT = T / COUNT;
h = 0.5*dT; TP = T - h; Np1 = ((tf - t0)/dT) + 1;
while k_c > 0                   % interactive design loop
k_i = 1 + (T / Ti);
x(:,1) = [0 0 0 0 t0 0 r 0]';% provide initial state
%
% MAIN (time) LOOP
for k = 1:Np1,
% PREDICTOR
[fw(1),fw(2),fw(3),fw(4),fw(5)] = e14_1RHf(x(:,k));
x(:,k+1) = x(:,k);             % simply use previous value
% TRAPEZIODAL CORRECTOR
for L = 1:7,                   % Uses 7 iterations in the corrector
[f(1),f(2),f(3),f(4),f(5)] = e14_1RHf(x(:,k+1));
for j = 1:nl,
x(j,k+1) = x(j,k) + h * (fw(j) + f(j));
end
if (x(5,k) - x(6,k)) < TP,
   x(6,k+1) = x(6,k); x(7,k+1) = x(7,k);
else                          % x6 is defined as a "staircase sampling time"
   x(6,k+1) = x(5,k); x(7,k+1) = r - x(1,k);
end                           % x7 is defined as a zero-order held error
if k <= COUNT,
   x(8,k+1) = x(8,k);
else                          % x8 is defined as a delayed zero-order held error
   x(8,k+1) = x(7,k-COUNT+1);
end
end
end
end
% DISPLAY RESULTS
disp('kc,Ti,ISE = '); disp([k_c Ti x(4,Np1)]);      % monitor performance
k_c = input ('enter kc, negative to quit: ');       % designer interaction
Ti = input('enter Ti also: ');
end;
```

6. Results: The variable t_f was chosen equal to 6. $COUNT = 10$, that is $\Delta T = 0.02$, was used. The search for optimal controller parameter values was conducted by simply changing the values of k_c and T_i in the program and observing the resulting value of ISE relative to previous ISE values. With a resolution in the trial values of the parameters

```
function [f1, f2, f3, f4, f5] = e14_1RHf(x)
f1 = x(2) - 5.* x(1);
f2 = 10 .* x(3);
f3 = 10 .* (k_c .* (k_i .* x(7) - x(8)) - x(3));
f4 = x(7) .* x(7);
f5 = 1;                    % x5 is defined as time t
```

of 0.1 for k_c and 0.01 for T_i, the "optimal" setting $k_c = 2.5$ and $T_i = 0.46$ was found. Figure 14.6b shows the performance of the controlled system (both $x_1(t)$ and $u(t)$) for this digital controller setting. The search for the optimum controller setting was restricted to only those pairs of values of k_c and T_i that satisfy the stability conditions (Fig. 14.4b). Such *constrained optimization* is necessary because the computer model is an approximation of the system model, and its stability behavior can be somewhat different from that of the actual system.

Recommendations by Takahashi, Chan, and Auslander

The Ziegler-Nichols rules for analog controllers, equations (11.12b) and (11.12c), apply to processes that are characterized by the S-shaped step input response (Fig. 11.5a) or by a stability limit under proportional control (the *ultimate sensitivity* method). The process dynamics of such systems were represented in the Ziegler-Nichols formulas by two parameters, the tangent slope S and the time lag T_L, as defined in Fig. 11.5a, or the stability limit proportional controller gain k_u and the period of sustained oscillation T_u, as shown in Fig. 11.5b. Takahashi, Chan, and Auslander, (see [23] for details) developed similar rules, expressed in terms of these same parameters, for (P, PI, and PID) direct digital control of similar processes. However, unlike the Ziegler-Nichols rules, which were deduced from experiments on actual plants, the Takahashi, Chan, and Auslander recommendations shown in Table 14.1 were derived from computer simulation of the control of two representative but very simple models of the actual processes:

1. **A delayed integrator**

$$G_p(s) = \frac{Se^{-T_L s}}{s} \tag{14.12a}$$

   for non-self-regulating processes (see Fig. 14.5a) and
2. **A first-order-lag-plus-dead-time model**

$$G_p(s) = \frac{Se^{-T_L s}}{1 + s} \tag{14.12b}$$

   for self-regulating processes (Fig. 14.5b). Self-regulating and non-self-regulating processes were explained in the discussion of controller tuning in Section 11.2.

The tuning criterion used by Takahashi et al. [23] was the minimization of the integral of the square error (ISE). Finally, the controller parameters shown in Table 14.1 apply to the canonical form of the controller such as equation (14.5a). Specific settings for other forms of the control algorithm can be determined by comparing such algorithms with equation (14.5a).

Table 14.1 Takahashi, Chan, and Auslander DDC Parameter Tuning Rules

TYPE	TRANSIENT RESPONSE METHOD[a]			ULTIMATE SENSITIVITY METHOD[b]		
	K_P	K_I	K_D	K_P	K_I	K_D
P	$\dfrac{1}{S(T_L + T)}$			$\dfrac{1}{2}k_u$		
PI	$\dfrac{0.9}{S\left(T_L + \dfrac{T}{2}\right)} - \dfrac{K_I}{2}$	$\dfrac{0.27T}{S\left(T_L + \dfrac{T}{2}\right)^2}$		$0.45k_u - \dfrac{K_I}{2}$	$0.54\dfrac{k_u}{T_u}T$	
		Except $T_L/T \longrightarrow 0$		Decrease the value if $T_L = T/4$		
PID	$\dfrac{1.2}{S(T_L + T)} - \dfrac{K_I}{2}$	$\dfrac{0.6T}{S\left(T_L + \dfrac{T}{2}\right)^2}$	$\dfrac{0.6}{ST}$ or $\dfrac{0.5}{ST}$ [c]	$0.6k_u - \dfrac{K_I}{2}$	$1.2\dfrac{k_u}{T_u}T$	$\dfrac{3k_uT_u}{40T}$
		Except $T_L/T \longrightarrow 0$		Not recommended if $T_L/T = 1/4$		

[a]S and T_L are the two parameters (defined in Figs. 11.5 and 14.5) representing process dynamics.
[b]k_u and T_u are the stability limit gain and resulting cycling period, respectively, for discrete-time proportional control of the process.
[c]The lower value is recommended when T_L/T is close to an integer number: 0, 1 or 2.

Example 14.2   Tuning by Takahashi, Chan, and Auslander Rules

Determine the PI-controller settings for the servomechanism of Example 14.1 (Fig. 14.4a) using the Takahashi, Chan, and Auslander recommendations. Compare these settings with the parameter values determined by simulation in Example 14.1.

Solution: Following Takahashi et al. [23], we can directly determine the two dynamic process parameters by simulation, in this case, of the direct digital proportional control of the plant. This is easily accomplished using the computer model of script M-file **exmp14_1.m*** with the necessary modification of the control: $u = k_u(x_7 - x_8)$ instead of $u = k_c(k_ix_7 - x_8)$. A value of the proportional gain k_u resulting in sustained oscillation of the controlled process was determined by trial and error as $k_u = 6.97$. A segment of the output for this control is shown in Fig. 14.6a, from which (that is, the sampled output) we can deduce $T_u = 4T = 0.8$, approximately. As indicated in reference [23], other simple oscillatory patterns of such period as $2T$ or $6T$ can be obtained for some problems.

From Table 14.1, the PI-DDC parameters are given by

$$K_I = 0.54k_u\frac{T}{T_u} = 0.54\frac{(6.97)(0.2)}{0.8} = 0.941$$

$$K_P = 0.45k_u - \frac{K_I}{2} = 0.45(6.97) - 0.5(0.941) = 2.666$$

For the Euler velocity algorithm, equation (14.5b) gives $k_c = 2.666$ and $T_i = 0.567$. The system performance for this control is shown in Fig. 14.6b along with the results of Example 14.1.† In general, both results are quite close. Although the ISE for the

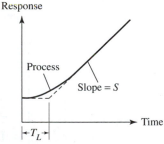

Response

Process

Slope = S

Time

$\mid\leftarrow T_L\rightarrow\mid$

(a) Non-self-regulating Process

Response

Process

$0.63\Delta y$

Δy

Time

$\mid\leftarrow T_L\rightarrow\mid\leftarrow\tau\rightarrow\mid$

(b) Self-regulating Process

Figure 14.5

Simple Models for Common Processes

*The Mathcad version **SC15.MCD** was used for the results shown here. See document **FIG14_6a.MCD**; On the Net.

†See Mathcad document **FIG14_6b.MCD**; On the Net.

Figure 14.6

**Design and Performance of a
Servomechanism Under DDC**

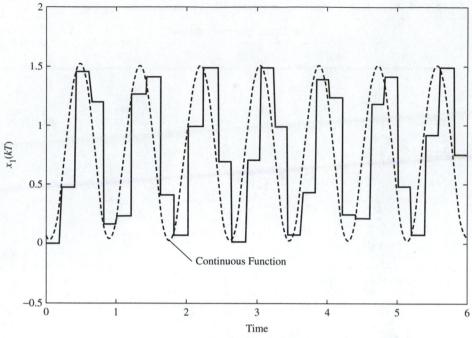

(a)   Stability Limit Under Proportional DDC: $k_u = 6.97$, $T_u = 4T$

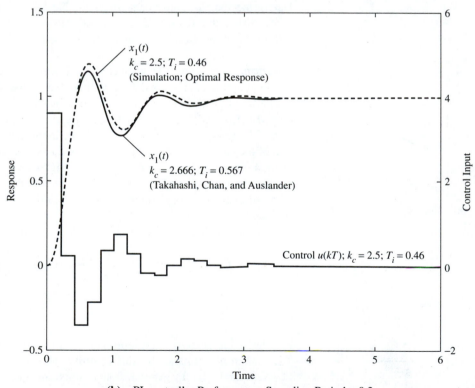

(b)   PI-controller Performance, Sampling Period = 0.2

present design is a little higher than that for the design obtained in Example 14.1, the present result has a smaller overshoot. However, the settling time is also slightly more than that of the optimal solution. In the final analysis, fine-tuning on the actual process is essential for either design.

....

Minimal Response Algorithms in the *z*-Domain

As we alluded to earlier, implementation of digital control laws in software implies that the flexibility and sophistication of the digital computer is also at the disposal of the control system designer. Consequently, complex and even nonlinear control algorithms are possible for digital computer control. These algorithms include state vector feedback and state observer systems, introduced in Chapter 13, for continuous-time systems, optimal control systems, also considered in Chapter 13 for continuous-time systems, and briefly explored here for discrete-time systems, in Section 14.4, *finite-time settling* controllers, considered in the next section, and many others.

Deadbeat Response

A *deadbeat* or *minimal response* algorithm reference [5], is one that satisfies the following performance specifications:

1. The rise time is a minimum.
2. The settling time is finite.
3. The steady-state error is zero.

Specifications (2) and (3) are remarkable considering the asymptotic behavior that generally characterized all our previous (continuous-time) controllers. Indeed, specifications (2) and (3) imply that a minimal response controller can, in principle, bring the object of control to exactly the desired state in a finite number of sampling periods! As we shall see, such precise control requires considerable knowledge of the system under control. Also, in some cases we should expect practical problems with the hardware implementation of the control scheme. If the sampling period is too short or the desired settling time is not long enough, the magnitude of the required control inputs may be beyond the capacity of the system actuators. Alternatively, the system may take longer than expected to reach the desired final state.

z-Domain Design

Consider the sampled data control system shown in Fig. 14.3d. The closed-loop pulse transfer function, equation (14.8a) repeated, is

$$G_{CL}(z) = \frac{G_p(z)D(z)}{1 + G_p(z)D(z)} \tag{14.13a}$$

If the desired closed-loop performance characteristics can be used to specify $G_{CL}(z)$, then the digital control law can be determined from equation (14.13a) as

$$D(z) = \frac{1}{G_p(z)} \left[\frac{G_{CL}(z)}{1 - G_{CL}(z)} \right] \tag{14.13b}$$

provided such a controller is realizable (see Practice Problem 13.19 for a similar technique applied to a continuous-time control system).

A realizable pulse transfer function is one whose output does not depend on any future input or output. It follows from (14.13b) that care should be taken in specifying $G_{CL}(z)$. Further, since

$$G_{CL}(z) = \frac{Y(z)}{R(z)} \tag{14.13c}$$

the feasibility of this approach can also depend on the nature of the reference input $R(z)$. Nonetheless, many different specifications of $G_{CL}(z)$ are possible for different design methods. Two examples follow.

Finite-Time Settling Control

A digital control system is said to be *finite-time settling* when its closed-loop transfer function for a reference input is a finite polynomial in z^{-1} (see [11]). This condition implies that all closed-loop poles are at the origin $z = 0$. Consider, for instance, the finite polynomial

$$G_{CL}(z) = 1 + 2z^{-1} + 3z^{-2} + 4z^{-3}$$

Factoring $z^{-3} = (1/z^3)$ gives us

$$G_{CL}(z) = \frac{z^3 + 2z^2 + 3z + 4}{z^3}$$

which has the characteristic equation $z^3 = 0$ or three poles, each $z = 0$.

One simple specification of a finite-time settling control system that satisfies the minimal response criteria (1) to (3) for a step reference input is

$$G_{CL}(z) = z^{-N} \tag{14.14}$$

where N is the number of sampling instants after which the error becomes and remains zero. Consider, for example, the unit-step reference input

$$r_k = 1, k = 0, 1, 2, \ldots \quad \text{and} \quad R(z) = 1 + z^{-1} + z^{-2} + \ldots = \frac{1}{1 - z^{-1}}$$

From equations (14.13c) and (14.14), the z-domain output is

$$Y(z) = \frac{z^{-N}}{1 - z^{-1}} = z^{-N}(1 + z^{-1} + z^{-2} + \ldots)$$

that is, a delayed unit step.

In other words, after N sampling instants, the output is exactly 1—the same as the reference input. The choice of N depends on $G_p(z)$ and other practical considerations.

Example 14.3   Minimal Response Design of System with Dead Time

Design a minimal response controller for the hot-water heater with remote sensing system of Fig. 11.13a that for a step reference input will drive the sensed temperature to the desired value in the smallest number of sampling instants.

Solution: For a plant output $y = \theta_{\text{sensor}}$ and input $u = q/C$, the plant Laplace domain transfer function, equation (11.20b), is

$$G_p(s) = \frac{e^{-10\pi s}}{s + 3/40}$$

From equation (14.6c), the pulse transfer function for the plant with zero-order hold is

$$G_p(z) = (1 - z^{-1})\mathcal{Z}\left\{ \mathcal{L}^{-1}\left[\frac{e^{-10\pi s}}{s(s + 3/40)} \right] \right\} \tag{14.15a}$$

Assuming, for simplicity (see Example 14.4 for the general case), that the sampling period T is chosen to be an integer fraction of the dead time, $T = 10\pi/n$, $n =$ integer, then using the interpretation $z = e^{ST}$, we can write

$$G_p(z) = z^{-n}(1 - z^{-1})\mathcal{Z}\left\{ \mathcal{L}^{-1}\left[\frac{1}{s(s + 3/40)} \right] \right\}$$

Applying entry 4 of Laplace transforms Table 8.4 and the integral property, Table 8.5, we have

$$G_p(z) = z^{-n}(1 - z^{-1})\mathcal{Z}\left\{ \frac{40}{3}(1 - e^{-3t/40}) \right\}$$

Using z-transform Table 8.6, we obtain

$$G_p(z) = \left(\frac{40}{3} \right) z^{-n}(1 - z^{-1})\left(\frac{z}{z - 1} - \frac{z}{z - c} \right)$$

where $c = e^{-3T/40}$.

Finally,

$$G_p(z) = \left(\frac{40}{3} \right) z^{-n}\left(\frac{1 - c}{z - c} \right) = \left(\frac{40}{3} \right)(1 - c)\frac{z^{-(n+1)}}{1 - cz^{-1}} \tag{14.15b}$$

Since no matter the input to the plant, there will be no response for time nT, a realizable $G_{\text{CL}}(z)$ must also contain a time delay of at least nT. It follows that the smallest possible number of sample periods after which the sensed temperature will become equal to the reference value is $N = n + 1$. We thus specify

$$G_{\text{CL}}(z) = z^{-(n+1)} \tag{14.16a}$$

Substituting equations (14.16a) and (14.15b) into equation (14.13b), we obtain the controller

$$D(z) = \left(\frac{3/40}{1 - c} \right)\left[\frac{1 - cz^{-1}}{1 - z^{-(n+1)}} \right] \tag{14.16b}$$

The corresponding discrete-time domain expression for the manipulated variable is

$$m_k = m_{k-N} + \left(\frac{3/40}{1 - c} \right)[e_k - ce_{k-1}] \tag{14.16c}$$

Figure 14.7 shows a simulation of this design for two values of n: $n = 3$ and $n = 1$. The initial conditions and setpoint are the same as in Fig. 13.4, which gives the performance of two continuous-time controllers: a feedforward and a feedback controller. Although there is no disturbance input in the present simulation, the performance of the digital controller appears impressive, especially for the case $n = 3$, which represents a speci-

Figure 14.7

Minimal Response Control of Hot-water Heater with Remote Sensing

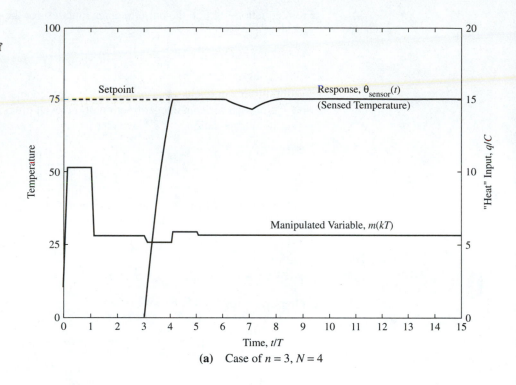

(a)   Case of $n = 3$, $N = 4$

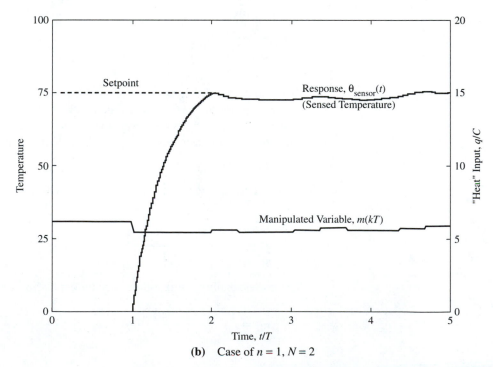

(b)   Case of $n = 1$, $N = 2$

fication of time to reach the desired value of $\left(1 + \frac{1}{3}\right)$ times the dead time as compared with twice the dead time, or the case $n = 1$. However, as we discussed earlier, such rapid convergence can be at a cost. Figure 14.7 shows that the first peak of the manipulated variable is much higher for the case $n = 3$. This peak value, which is given by equation (14.16c):

$$m_{\text{peak}} = \left(\frac{3/40}{1 - c}\right)r_0; \qquad r_0 = \text{setpoint}$$

can in practice exceed the capacity of the system actuator. It is thus not sufficient to evaluate the performance of a control algorithm by examining only the (controlled variable) response.

....

The Ringing Problem

Another practical problem that can be revealed by examination of the controller output is the demand for excessive actuator movements (with respect to time), or vibration, that the actuator will not be able to follow or that can damage the actuator or other downstream components. The *ringing* problem and a controller design method due to Dahlin (see [28]) are illustrated by the following example.

Example 14.4   Dahlin's Method and Actuator Ringing

Design a digital controller for the hot-water heater with remote sensing system of Fig. 11.13a, assuming that the sampling period for the control system is fixed at $T = 20$. Let the desired closed-loop behavior be like the response of a first-order lag with dead time:

$$G_{\text{CL}}(s) = \frac{e^{-L_d s}}{\tau s + 1} \tag{14.17a}$$

as suggested by Dahlin, reference [28], rather than a deadbeat response as in Example 14.3.

Solution:

Pulse transfer function for plant with dead time: First, we determine the plant pulse transfer function, equation (14.15a). Let the plant dead time $L_p = 10\pi$ be given by

$$L_p = nT + L; \qquad 0 < L < T \tag{14.17b}$$

so that n is the largest integer number of sampling periods T in L_p. For the present problem, $L_p \sim 31.4159$, so that for $T = 20$, for example, $n = 1$ and $L \sim 11.4159$. Putting (14.17b) in (14.15a), we obtain

$$G_p(z) = (1 - z^{-1})Z\left\{\mathscr{L}^{-1}\left[\frac{e^{-nTs}e^{-Ls}}{s(s + a)}\right]\right\}$$

where $a = 3/40$. Proceeding, we have

$$G_p(z) = z^{-n}(1 - z^{-1})Z\left\{\mathscr{L}^{-1}\left[\frac{e^{-Ls}}{s(s + a)}\right]\right\} \tag{14.17c}$$

By the combination of entry 4 of Laplace transform Table 8.4 and the integral and time shift properties, both of Table 8.5, equation (14.17c) becomes

$$G_p(z) = \frac{1}{a} z^{-n}(1 - z^{-1}) \mathcal{Z}\{[1 - e^{-a(t-L)}]U_s(t - L)\}$$

By equation (8.40),

$$G_p(z) = \frac{1}{a} z^{-n}(1 - z^{-1}) \sum_{k=0}^{\infty} [1 - e^{-a(kT-L)}]U_s(kT - L)z^{-k}$$

or

$$G_p(z) = \frac{1}{a} z^{-n}(1 - z^{-1})\{[1 - e^{-a(T-L)}]z^{-1} + [1 - e^{-a(2T-L)}]z^{-2}$$
$$+ [1 - e^{-a(3T-L)}]z^{-3} + \ldots\}$$

Substituting

$$b = \frac{(T - L)}{T} \tag{14.17d}$$

we obtain

$$G_p(z) = \frac{1}{a} z^{-n}(1 - z^{-1})\{[z^{-1} + z^{-2} + z^{-3} + \ldots]$$
$$- [e^{-abT}z^{-1} + e^{-abT}e^{-aT}z^{-2} + e^{-abT}e^{-2aT}z^{-3} + \ldots]\}$$

or

$$G_p(z) = \frac{1}{a} z^{-(n+1)}(1 - z^{-1})\{[1 + z^{-1} + z^{-2} + \ldots] \tag{14.17e}$$
$$- e^{-abT}[1 + e^{-aT}z^{-1} + e^{-2aT}z^{-2} + \ldots]\}$$

Finally, recognizing the convergent power series in equation (14.17e), we have

$$G_p(z) = \frac{1}{a} z^{-(n+1)}(1 - z^{-1})\left\{\left[\frac{1}{1 - z^{-1}}\right] - e^{-abT}\left[\frac{1}{1 - z^{-1}e^{-aT}}\right]\right\} \tag{14.17f}$$
$$= \frac{z^{-(n+1)}[(1 - e^{-abT}) + (e^{-abT} - e^{-aT})z^{-1}]}{a(1 - e^{-aT}z^{-1})}$$

If we define

$$c_1 = e^{-aT}; \qquad c_2 = e^{-abT} - e^{-aT}; \qquad c_3 = 1 - e^{-abT} \tag{14.18a}$$

equation (14.17f) becomes

$$G_p(z) = \frac{z^{-(n+1)}(c_3 + c_2 z^{-1})}{a(1 - c_1 z^{-1})} \tag{14.18b}$$

Dahlin's method: The pulse transfer function for the desired closed-loop (Laplace domain) transfer function, equation (14.17a) plus zero hold, is given by equation (14.6c). We choose L_d to be an integral number N of the sampling period T, so that

$$G_{CL}(z) = (1 - z^{-1})\mathcal{Z}\left\{\mathcal{L}^{-1}\left[\frac{e^{-NTs}}{s(\tau s + 1)}\right]\right\} \tag{14.19a}$$

$$= \frac{(1 - e^{-T/\tau})z^{-(N+1)}}{1 - e^{-T/\tau}z^{-1}}; \qquad L_d = NT$$

If we substitute equation (14.19a) in equation (14.13b), the control algorithm is

$$D(z) = \frac{1}{G_p(z)}\left[\frac{(1 - e^{-T/\tau})z^{-(N+1)}}{1 - e^{-T/\tau}z^{-1} - (1 - e^{-T/\tau})z^{-(N+1)}}\right] \tag{14.19b}$$

For the present plant, represented by equation (14.18b), the apparent time delay resulting from the dead time is nT (see equation (14.17c)). Hence N should be chosen equal to n. Substituting this result and equation (14.18b) into equation (14.19b), we get

$$D(z) = \frac{ac_5(1 - c_1 z^{-1})}{(c_3 + c_2 z^{-1})[1 - c_4 z^{-1} - c_5 z^{-(n+1)}]} \tag{14.19c}$$

where $c_4 = e^{-T/\tau}$ and $c_5 = 1 - c_4$.

Simulation results: With N fixed, Dahlin's algorithm still contains one free parameter, the time constant τ, which can be used as a tuning parameter. This can be an advantage of this design technique. Figure 14.8a shows the performance of Dahlin's algorithm on the hot-water heater system for three values of τ and again for no disturbance input. Apparently, control becomes tighter as the quantity τ/T gets smaller. Thus with a proper choice of τ (such as $\tau/T = 1.3$ in Fig. 14.8a), a satisfactory design that is also not too strenuous for the process can be determined. (Observe the lower magnitudes of the manipulated variable in Fig. 14.8a relative to Fig. 14.7a.)

....

The manipulated variable pattern shown in Fig. 14.8a reveals, however, that the algorithm *rings* noticeably for $\tau/T = 1.3$. By examining the controller transfer function, we can identify the predominant cause of the ringing and perhaps eliminate it if it is a problem. According to Fig. 6.22, a pole with a value $-1 \leq p < 0$ will give rise to a (stable) ringing mode. For $p = -1$, the ringing amplitude is constant, while for $-1 < p < 0$ the amplitude decays with time. The actual controller response pattern will of course depend on the input sequence and the combined effect of all the poles and even zeros of the controller transfer function. However, in general, ringing can be suspected when the controller transfer function includes *ringing poles,* and the severity of any such ringing will generally depend on how close the magnitude of the ringing pole is to 1. For the present controller, equation (14.19c), the poles (for $n = 1$) are

$$p = \frac{-c_2}{c_3} = -0.6365; \qquad p = 0.5c_4 + \sqrt{(0.25c_4^2 + c_5)} = +1.011$$

and

$$P = 0.5c_4 - \sqrt{(0.25c_4^2 + c_5)} = -0.5476$$

Figure 14.8

Illustration of Dahlin's Method and Ringing Phenomenon

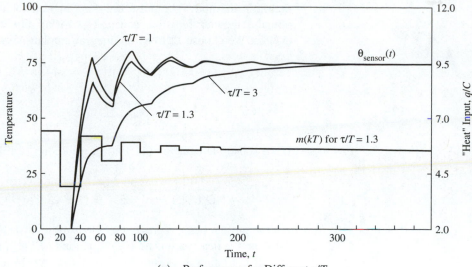

(a)   Performance for Different τ/T

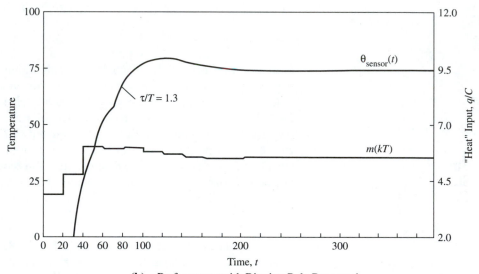

(b)   Performance with Ringing Pole Removed

so that there are two poles that can cause ringing. Figure 14.8b shows the performance of the controller obtained by eliminating the "more severe" ringing pole ($z = -0.6365$) in the manner suggested by Dahlin (see [28]), that is, by replacing the factor $(c_3 + c_2 z^{-1})$ with its steady-state gain $(c_3 + c_2)$, which is obtained by letting z approach 1 in the factor:

$$D(z) = \left(\frac{a c_5}{c_3 + c_2} \right) \frac{(1 - c_1 z^{-1})}{1 - c_4 z^{-1} - c_5 z^{-2}} \tag{14.20}$$

The controller output shown in Fig. 14.8b confirms the elimination of the ringing without much degradation of the controller performance.

14.3 DISCRETE-TIME STATE SPACE DESIGN

The digital controller as determined by equation (14.13b) is essentially a compensation network and is susceptible to *pole-zero cancellation*, which can cause uncontrollability or unobservability, as explained in Section 12.2. Consider the overall open-loop consisting of the digital controller and the plant (see Fig. 14.3d). For the hot-water heater problem, with $n = 1$, equations (14.15b) and (14.16b) give

$$G_0(z) = G_p(z)D(z) = \left[\left(\frac{1-c}{a} \right) \frac{z^{-2}}{1 - cz^{-1}} \right] \left[\left(\frac{a}{1-c} \right) \frac{1 - cz^{-1}}{1 - z^{-2}} \right] 3 \qquad (14.21)$$

The pole-zero cancellation involving the term $(1 - cz^{-1})$ is evident. The signal flow graph of the control system based on equation (14.21) is developed in Figs. 14.9a to 14.9c. Note that in the plant signal flow graph the ambient temperature disturbance input $v(t) = \theta_a(t)$ has been included to give the overall plant input:

$$\frac{q}{C} + \frac{\theta_a}{RC} = m(t) + av(t)$$

(see equation (13.7a)).

The signal flow graph in Fig. 14.9d, obtained after diagonalization of the open loop (see Practice Problem 9.53), indicates two uncontrollable (but stable) modes driven by the disturbance input. If these modes had been unstable, the performance of the algorithm, equation (14.16c), would have been unsatisfactory. As was the case for continuous-time systems, this type of attention to the system's internal state and interaction between controlled variables lead, in general, to state space (multivariable control system) design methods and, in particular, to state vector feedback control.

Finite-Time Settling State Vector Feedback Control

The control algorithm for finite-time settling to be developed next requires for its feasibility the controllability of the open-loop system, which guarantees that the closed-loop system will also be controllable (see Practice Problem 14.17). Further, since the control technique is state vector feedback, integral action can be added routinely, as was the case with continuous-time state vector feedback control (Section 13.2), to eliminate any offset in the controlled variable. The only problem is the necessity for complete state measurement. However, as we did in Chapter 13, we can incorporate an observer in the controller for systems where the complete state vector is not available. We shall find that such systems must however be observable, thus reinforcing again the notion that more precise control requires more information about the system.

Consider the general nth order linear and stationary discrete-time system

$$\mathbf{X}_{k+1} = \mathbf{P}\mathbf{X}_k + \mathbf{Q}\mathbf{U}_k \qquad (14.22a)$$

where $\mathbf{U}_k$ is an r-vector, $\mathbf{X}_k$ is an n-vector, $\mathbf{P}$ and $\mathbf{Q}$ are $(n \times n)$ and $(n \times r)$ matrices, respectively. We seek a control law,

$$\mathbf{U}_k = \mathbf{K}\mathbf{X}_k \qquad (14.22b)$$

(with $\mathbf{K}$ an $r \times n$ matrix) that can bring the system (14.22a) from an arbitrary initial state $\mathbf{X}_0$ to a final state at the origin $\mathbf{X}_N = \mathbf{0}$ in a finite number of at most N sampling periods.

Figure 14.9

Minimal Response and Controllability of Hot-water Heater System

$$G_p(z) = \frac{Y(z)}{M(z) + aV(z)} = \left(\frac{1-c}{a}\right)\frac{z^{-2}}{1-cz^{-1}}$$

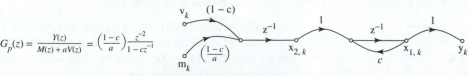

(a) Plant Signal Flow Graph

$$D(z) = \frac{M(z)}{E(z)} = \left(\frac{a}{1-c}\right)\frac{1-cz^{-1}}{1-z^{-2}}$$

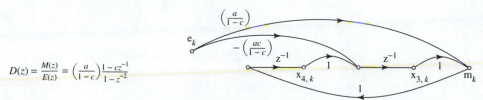

(b) Minimal Response Controller Signal Flow Graph

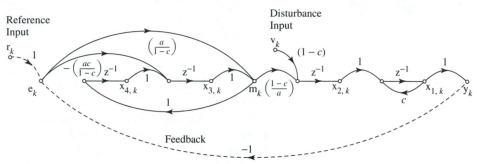

(c) Overall Control System Non-diagonalized Signal Flow Graph

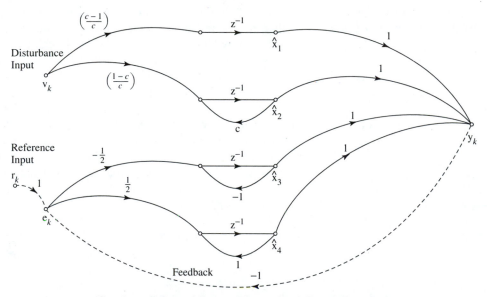

(d) Overall Control System Diagonalized Signal Flow Graph

Case of *n/r* Equal to an Integer

Although the finite-time settling control technique is a special case of the eigenvalue assignment method in which all the desired closed-loop eigenvalues are zero, we can for this general case seek the control law directly by computing the sequence of states from the initial state $\mathbf{X}_0$ to the final state $\mathbf{X}_N$:

$$\mathbf{X}_1 = \mathbf{PX}_0 + \mathbf{QU}_0 = (\mathbf{P} - \mathbf{QK})\mathbf{X}_0$$

$$\mathbf{X}_2 = (\mathbf{P} - \mathbf{QK})\mathbf{X}_1 = \left\{\mathbf{P}^2 - [\mathbf{PQ};\quad \mathbf{Q}]\begin{bmatrix}\mathbf{K} \\ \mathbf{K}(\mathbf{P} - \mathbf{QK})\end{bmatrix}\right\}\mathbf{X}_0$$

$$\vdots$$

$$\mathbf{X}_N = \left\{\mathbf{P}^N - [\mathbf{P}^{N-1}\mathbf{Q};\quad \cdots;\quad \mathbf{PQ};\quad \mathbf{Q}]\begin{bmatrix}\mathbf{K} \\ \mathbf{K}(\mathbf{P} - \mathbf{QK}) \\ \vdots \\ \mathbf{K}(\mathbf{P} - \mathbf{QK})^{N-1}\end{bmatrix}\right\}\mathbf{X}_0 = \mathbf{0}$$

The solution for $\mathbf{X}_N$ will exist for any arbitrary initial state $\mathbf{X}_0$ if

$$\mathbf{P}^N = [\mathbf{P}^{N-1}\mathbf{Q};\quad \cdots;\quad \mathbf{PQ};\quad \mathbf{Q}]\begin{bmatrix}\mathbf{K} \\ \mathbf{K}(\mathbf{P} - \mathbf{QK}) \\ \vdots \\ \mathbf{K}(\mathbf{P} - \mathbf{QK})^{N-1}\end{bmatrix} = \mathbf{MW} \qquad (14.22c)$$

where $\mathbf{M}$ is the $(n \times Nr)$ matrix

$$\mathbf{M} = [\mathbf{P}^{N-1}\mathbf{Q};\quad \cdots;\quad \mathbf{PQ};\quad \mathbf{Q}] \qquad (14.22d)$$

Since $\mathbf{P}^N$ is an $(n \times n)$ matrix, a unique solution of equation (14.22c) for $\mathbf{W}$ is possible if $\mathbf{M}$ is square, that is, $Nr = n$, or

$$\frac{n}{r} = N \qquad (14.23a)$$

and further, $\mathbf{M}^{-1}$ exists, that is,

$$\det\mathbf{M} \neq 0 \qquad (14.23b)$$

If the conditions (14.23a) and (14.23b) are satisfied, then

$$\begin{bmatrix}\mathbf{K} \\ \mathbf{K}(\mathbf{P} - \mathbf{QK}) \\ \vdots \\ \mathbf{K}(\mathbf{P} - \mathbf{QK})^{N-1}\end{bmatrix} = \mathbf{M}^{-1}\mathbf{P}^N \qquad (14.23c)$$

so that $\mathbf{K}$ is given by the first r rows of $\mathbf{M}^{-1}\mathbf{P}^N$. In summary, since $\mathbf{M}$ is the controllability matrix, recall equation (9.52), the condition for the solution (14.23c) is (1) $n/r = N =$ integer and (2) the system (14.22a) is controllable.

Case of n/r Not Equal to an Integer

If n/r is not equal to an integer, then we can find the integer number $\Delta n < r$ such that

$$\frac{(n + \Delta n)}{r} = N = \text{integer} \tag{14.24a}$$

For example, if $n = 4$ and $r = 3$, let $\Delta n = 2$ and $N = 2$. We then introduce Δn additional but fictitious state variables $x^*_{n+1}, x^*_{n+2}, \ldots, x^*_{n+\Delta n}$, which are not coupled to the real state variables, for example,

$$x^*_{n+i,k+1} = x^*_{n+i,k}; \qquad i = 1, 2, \ldots, \Delta n \tag{14.24b}$$

and which have zero initial state,

$$x^*_{n+i,0} = 0; \qquad i = 1, 2, \ldots, \Delta n \tag{14.24c}$$

The $(n + \Delta n) \times (n + \Delta n)$ **P**-matrix and expanded **Q**-matrix for the augmented system are thus

$$\mathbf{P}^* = \left[\begin{array}{c|c} \mathbf{P}_{(n \times n)} & \mathbf{0} \\ \hline \mathbf{0} & \mathbf{I}_{(\Delta n \times \Delta n)} \end{array} \right] \tag{14.25a}$$

$$\mathbf{Q}^* = \left[\begin{array}{c} \mathbf{Q}_{(n \times r)} \\ \hline \mathbf{\Delta Q}_{(\Delta n \times r)} \end{array} \right] \tag{14.25b}$$

The elements of $\mathbf{\Delta Q}$ must be carefully chosen in such a way that the Δn additional state variables are controllable. The design is then determined as before with $\mathbf{P}^*$, $\mathbf{Q}^*$, and $(n + \Delta n)$ replacing $\mathbf{P}$, $\mathbf{Q}$, and n, respectively, in equations (14.22) and (14.23).

| **Example 14.5** | State Vector Feedback Control of Hot-Water Heater |

Determine the finite-time settling state vector feedback control algorithm for the hot-water heater with remote sensing system, assuming that the z-domain plant transfer function is given by equation (14.15b) with $n = 1$ and that all plant state variables are available as measured outputs. Let the desired steady-state error in the sensed hot-water temperature be zero.

Solution: Figure 14.10a shows the typical state vector feedback configuration with I-action on one of the state variables. Integral control of the sensed hot-water temperature $x_{1,k}$ is necessary to ensure zero steady-state error. Note from Table 8.6 that

$$Z\left\{ \mathscr{L}^{-1}\left[\frac{1}{s} \right] \right\} = Z\{U_s(t)\} = \frac{z}{z - 1} \tag{14.26a}$$

Figure 14.10

Finite-time Settling State Vector Feedback Control of Hot-water Heater with Remote Sensing

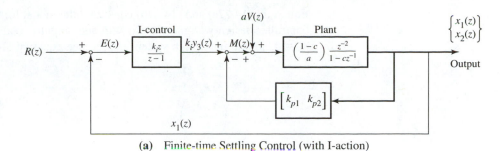

(a) Finite-time Settling Control (with I-action)

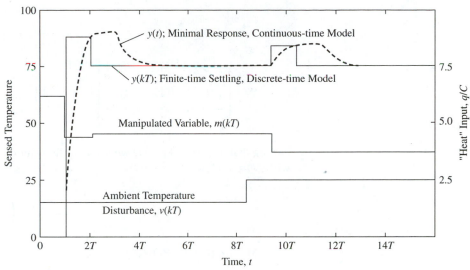

(b) Finite-time Settling Controller Performance Under Disturbance Input

hence the entry for the I-control block. Also, the corresponding discrete-time model for the I-control block is

$$x_{3,k+1} = x_{3,k} + e_k \tag{14.26b}$$
$$y_{3,k} = x_{3,k} + e_k$$

From the plant signal flow graph, Fig. 14.9a, the plant state model is

$$x_{1,k+1} = cx_{1,k} + x_{2,k} \tag{14.27a}$$
$$x_{2,k+1} = \left(\frac{1-c}{a}\right)m_k + (1-c)v_k \tag{14.27b}$$

From the control system block diagram, Fig. 14.10a, and equation (14.26b),

$$x_{3,k+1} = x_{3,k} + r_k - x_{1,k} \tag{14.27c}$$

and

$$m_k = -[(k_{p1} + k_i)x_{1,k} + k_{p2}x_{2,k} - k_i(x_{3,k} + r_k)] \tag{14.27d}$$

Equations (14.27a) and (14.27b) constitute the model for simulation. However, for the controller design we ignore the disturbance input v^k and assume zero reference input $r_k = 0$, so that the design system model (see equation (14.22a)) is

$$
x_{1,k+1} = cx_{1,k} + x_{2,k}
$$

$$
x_{2,k+1} = \left(\frac{1-c}{a}\right)m_k \tag{14.28a}
$$

$$
x_{3,k+1} = -x_{1,k} + x_{3,k}
$$

$$
m_k = -[k_1 x_{1,k} + k_2 x_{2,k} + k_3 x_{3,k}]
$$

Thus

$$
\mathbf{P} = \begin{bmatrix} c & 1 & 0 \\ 0 & 0 & 0 \\ -1 & 0 & 1 \end{bmatrix}; \qquad \mathbf{Q} = \begin{bmatrix} 0 \\ \left(\dfrac{1-c}{a}\right) \\ 0 \end{bmatrix} \tag{14.28b}
$$

Noting that $n = 3$, and $r = 1$, we see that the solution, equation (14.23c), is applicable with $N = 3$. Now,

$$
\mathbf{M}^{-1} = [\mathbf{P}^2\mathbf{Q}; \quad \mathbf{PQ}; \quad \mathbf{Q}]^{-1} = \begin{bmatrix} \dfrac{c(1-c)}{a} & \left(\dfrac{1-c}{a}\right) & 0 \\ 0 & 0 & \left(\dfrac{1-c}{a}\right) \\ \left(\dfrac{1-c}{-a}\right) & 0 & 0 \end{bmatrix}^{-1}
$$

$$
= \begin{bmatrix} 0 & 0 & \left(\dfrac{-a}{1-c}\right) \\ \left(\dfrac{a}{1-c}\right) & 0 & \left(\dfrac{ca}{1-c}\right) \\ 0 & \left(\dfrac{a}{1-c}\right) & 0 \end{bmatrix}
$$

so that $\mathbf{M}^{-1}$ exists (the system (14.28) is controllable), and

$$
\mathbf{M}^{-1}\mathbf{P}^N = \mathbf{M}^{-1}\begin{bmatrix} c^3 & c^2 & 0 \\ 0 & 0 & 0 \\ -(c^2 + c + 1) & -(c+1) & 1 \end{bmatrix}
$$

$$
= \begin{bmatrix} \dfrac{a(c^3 + c + 1)}{1-c} & \dfrac{a(c+1)}{1-c} & \dfrac{-a}{1-c} \\ \dfrac{-ac(c+1)}{1-c} & \dfrac{-ac}{1-c} & \dfrac{ac}{1-c} \\ 0 & 0 & 0 \end{bmatrix}
$$

Thus

$$k_{p1} + k_i = \frac{a(c^2 + c + 1)}{1 - c}; \qquad k_{p2} = \frac{a(c + 1)}{1 - c}; \qquad k_i = \frac{a}{1 - c}$$

Figure 14.10b shows the simulation results for the system of equations (14.27a) and (14.27b), using the controller parameters determined above and subject to a step change in ambient temperature or a low-frequency disturbance input. As can be seen from the dashed line, which represents the performance of the z-domain minimal response controller of Example 14.3, the present finite-time settling state vector feedback controller using complete knowledge of the plant state responds faster and without loss of control under such a disturbance input. However, note that the minimal response simulation is based on the continuous-time plant, whereas the discrete-time model of the plant was used in simulating the present finite-time settling response.

State Vector Feedback with Eigenvalue Assignment Method

For a low-order, simple system such as the hot-water heater (that is, the discrete-time model), the finite-time settling controller design can be obtained perhaps more easily via the *eigenvalue assignment* approach. Consider again the system (14.28a). If we substitute for m_k in the second equation using the last equation, we obtain the closed-loop system $\mathbf{P}$ matrix

$$\mathbf{P}_{CL} = \begin{bmatrix} c & 1 & 0 \\ -bk_1 & -bk_2 & -bk_3 \\ -1 & 0 & 1 \end{bmatrix}$$

where $b = (1 - c)/a$. The closed-loop characteristic equation is given by $\det(z\mathbf{I} - \mathbf{P}_{CL}) = 0$, or substituting $\mathbf{P}_{CL}$, we obtain

$$z^3 + [bk_2 - 1 - c]z^2 + [bk_1 - c(bk_2 - 1) - bk_2]z + [cbk_2 - bk_1 - bk_3] = 0$$

For finite-time settling, all the closed-loop eigenvalues should be zero, that is, the desired closed-loop characteristic equation should be $z^3 = 0$. Hence we must require

$$bk_2 - 1 - c = 0 \Rightarrow k_2 = \frac{a(c + 1)}{1 - c};$$

$$bk_1 - c(bk_2 - 1) - bk_2 = 0 \Rightarrow k_1 = \frac{a(c^2 + c + 1)}{1 - c}$$

and

$$cbk_2 - bk_1 - bk_3 = 0 \Rightarrow k_3 = \frac{-a}{1 - c} = -k_i$$

which is the same result we obtained in Example 14.5.

This development is applicable to the design of other types of discrete-time state vector feedback control systems such as the ones explored with respect to continuous-time control systems: ideal state vector feedback control, state vector feedback with scalar controlling input, and state vector feedback with vector controlling input derived from distributed scalar input. The only difference is that values assigned to the eigenvalues will not necessarily be all zero.

Finite-Time Settling Observer

In Example 14.5 the hot-water heater internal state $x_{2,k}$, which was presumed to be a known output in the finite-time settling design, is actually not available. This limitation can be removed by incorporating an observer in the control system. In particular, we can consider, in place of the asymptotic observer we discussed for continuous-time systems, a *finite-time settling observer*. Given the system

$$\mathbf{X}_{k+1} = \mathbf{P}\mathbf{X}_k + \mathbf{Q}\mathbf{U}_k; \qquad \mathbf{X}_k = \mathbf{X}_0 \text{ at } k = 0 \tag{14.29a}$$

the output is

$$\mathbf{Y}_k = \mathbf{C}\mathbf{X}_k \tag{14.29b}$$

where as before $\mathbf{P}$ and $\mathbf{Q}$ are $(n \times n)$ and $(n \times r)$, respectively, and $\mathbf{C}$ is $(m \times n)$, that is, there are m outputs. Let the model or *auxiliary observer equation* be

$$\hat{\mathbf{X}}_{k+1}^{0} = \hat{\mathbf{P}}\hat{\mathbf{X}}_k + \hat{\mathbf{Q}}\mathbf{U}_k; \qquad \hat{\mathbf{X}}_0^0 = \mathbf{0} \tag{14.30a}$$

the output is

$$\hat{\mathbf{Y}}_{k+1}^{0} = \hat{\mathbf{C}}\hat{\mathbf{X}}_{k+1}^{0} \tag{14.30b}$$

where $\hat{\mathbf{X}}$ is the estimate of $\mathbf{X}$, and $\hat{\mathbf{X}}_k^0$ is an *a priori* estimate of $\mathbf{X}_k$. Then, given the weight vector $\mathbf{f}$, the main observer equation is

$$\hat{\mathbf{X}}_{k+1} = \hat{\mathbf{X}}_{k+1}^{0} + \mathbf{f}(\mathbf{Y}_{k+1} - \hat{\mathbf{C}}\hat{\mathbf{X}}_{k+1}^{0}) \tag{14.30c}$$

where

$$\hat{\mathbf{X}}_0 = \mathbf{f}\mathbf{Y}_0 = \mathbf{f}\mathbf{C}\mathbf{X}_0 \tag{14.30d}$$

since for lack of knowledge, $\hat{\mathbf{X}}_0^0 = \mathbf{0}$.

Although $\mathbf{f}$ can be determined by other techniques, our present interest is in the observer gain for finite-time settling, that is, $\hat{\mathbf{X}}_k - \mathbf{X}_k = \mathbf{0}$, for all $k \geq N - 1$ (see [24]). We shall also restrict our attention to the case of an accurate model, that is, $\hat{\mathbf{P}} = \mathbf{P}$, $\hat{\mathbf{Q}} = \mathbf{Q}$, and $\hat{\mathbf{C}} = \mathbf{C}$.

First, suppose n/m **= integer.** Starting from equation (14.30d), we know the error

$$\mathbf{e}_k = \mathbf{X}_k - \hat{\mathbf{X}}_k \tag{14.31a}$$

at $k = 0$:

$$\mathbf{e}_0 = \mathbf{X}_0 - \hat{\mathbf{X}}_0 = (\mathbf{I} - \mathbf{f}\mathbf{C})\mathbf{X}_0 \tag{14.31b}$$

Let

$$\mathbf{e}_{k+1}^{0} = \mathbf{X}_{k+1} - \hat{\mathbf{X}}_{k+1}^{0} \tag{14.31c}$$

Then subtracting equation (14.30a) from equation (14.29a), we have

$$\mathbf{e}_{k+1}^{0} = \mathbf{P}\mathbf{e}_k \tag{14.31d}$$

From the main observer equation (14.30c),

$$\hat{\mathbf{X}}_{k+1} - \hat{\mathbf{X}}_{k+1}^{0} = \mathbf{f}\mathbf{C}(\mathbf{X}_{k+1} - \hat{\mathbf{X}}_{k+1}^{0}) = \mathbf{f}\mathbf{C}\hat{\mathbf{e}}_{k+1}^{0} \tag{14.31e}$$

We can also write

$$\hat{\mathbf{X}}_{k+1} - \hat{\mathbf{X}}_{k+1}^{\,0} = (\hat{\mathbf{X}}_{k+1} - \mathbf{X}_{k+1}) + (\mathbf{X}_{k+1} - \hat{\mathbf{X}}_{k+1}^{\,0})$$

so that from equations (14.31a) and (14.31c), equation (14.31e) becomes

$$\mathbf{e}_{k+1}^{\,0} - \mathbf{e}_{k+1} = \mathbf{fCe}_{k+1}^{\,0}.$$

Applying equation (14.31d) we obtain

$$\mathbf{Pe}_k - \mathbf{e}_{k+1} = \mathbf{fCPe}_k$$

which gives the recursive relation for the error:

$$\mathbf{e}_{k+1} = (\mathbf{I} - \mathbf{fC})\mathbf{Pe}_k. \tag{14.32a}$$

Computing the sequence $\mathbf{e}_i$, $\mathbf{i} = 1, 2, \ldots$, and using equation (14.31b), we get

$$\mathbf{e}_1 = (\mathbf{I} - \mathbf{fC})\mathbf{Pe}_0 = (\mathbf{I} - \mathbf{fC})\mathbf{P}(\mathbf{I} - \mathbf{fC})\mathbf{X}_0 = \left\{ \mathbf{P} - [\mathbf{f}; \quad (\mathbf{I} - \mathbf{fC})\mathbf{Pf}] \begin{bmatrix} \mathbf{CP} \\ \mathbf{C} \end{bmatrix} \right\} \mathbf{X}_0$$

$$\mathbf{e}_2 = (\mathbf{I} - \mathbf{fC})\mathbf{Pe}_1 = \left\{ \mathbf{P} - [\mathbf{f}; \quad (\mathbf{I} - \mathbf{fC})\mathbf{Pf}; \quad (\mathbf{I} - \mathbf{fC})^2\mathbf{P}^2\mathbf{f}] \begin{bmatrix} \mathbf{CP}^2 \\ \mathbf{CP} \\ \mathbf{C} \end{bmatrix} \right\} \mathbf{X}_0$$

$$\vdots$$

$$\mathbf{e}_{N-1} = \left\{ \mathbf{P} - [\mathbf{f}; \quad (\mathbf{I} - \mathbf{fC})\mathbf{Pf}; \quad \cdots; \quad (\mathbf{I} - \mathbf{fC})^{N-1}\mathbf{P}^{N-1}\mathbf{f}] \begin{bmatrix} \mathbf{CP}^{N-1} \\ \vdots \\ \mathbf{CP} \\ \mathbf{C} \end{bmatrix} \right\} \mathbf{X}_0 \tag{14.32b}$$

It can be shown that equation (14.32b) is satisfied when

$$\mathbf{f} = \mathbf{P}^{N-1} \begin{bmatrix} \mathbf{CP}^{N-1} \\ \vdots \\ \mathbf{CP} \\ \mathbf{C} \end{bmatrix}^{-1} \begin{bmatrix} \mathbf{I}_{(m \times n)} \\ \mathbf{0} \end{bmatrix}$$

or

$$\mathbf{f} = \mathbf{P}^{N-1}\mathbf{L}^{-1} \begin{bmatrix} \mathbf{I}_{(m \times n)} \\ \mathbf{0} \end{bmatrix} \tag{14.33a}$$

where $\mathbf{L}$ is the $(Nm \times n)$ matrix

$$\mathbf{L} = \begin{bmatrix} \mathbf{CP}^{N-1} \\ \vdots \\ \mathbf{CP} \\ \mathbf{C} \end{bmatrix} \tag{14.33b}$$

Since $\mathbf{L}^{-1}$ must exist for this observer design to be feasible, $\mathbf{L}$ must be square, that is $Nm = n$. Further, $\mathbf{L}$ is the observability matrix, (equation (9.56)), so that the conditions for the design are (1) $n/m = N =$ integer, and (2) the system (14.29) is observable.

If n/m **is not an integer,** then as in the case of the finite-time settling controller, we can find Δn such that $(n + \Delta n)/m = N =$ integer.

The Δn additional fictitious state variables and the expanded $\mathbf{P}$-matrix are the same as given in equations (14.24b), (14.24c), and (14.25a). However, the expanded $\mathbf{C}$-matrix is

$$\mathbf{C}^* = \left[\mathbf{C}_{(m \times n)} \mid \Delta\mathbf{C}_{(m \times \Delta n)} \right] \tag{14.33c}$$

where $\Delta\mathbf{C}$ must be carefully chosen to ensure that $\mathbf{L}^{-1}$ (based on $\mathbf{C}^*$ and $\mathbf{P}^*$) exists. The weight factor $\mathbf{f}$ is then given by equation (14.33a) with the proper substitution of $\mathbf{P}^*$ and $\mathbf{C}^*$ for $\mathbf{P}$ and $\mathbf{C}$, respectively.

Example 14.6　Application of Finite-Time Settling Observer

Consider again the hot-water heater with remote sensing system of Example 14.5. Design a finite-time settling observer that can be used to estimate the internal state $x_{2,k}$, assuming the sensed temperature $x_{1,k}$ is the only available output. Using this observer with the finite-time settling state vector feedback controller designed in Example 14.5, simulate the response of the actual continuous-time plant, subject to the same ambient temperature disturbance as in Fig. 14.10b.

Solution: The discrete-time system is given by equations (14.27a) through (14.27d) with the output equation

$$y_k = x_{1,k} \tag{14.34a}$$

Thus $n = 2$ and $m = 1$, so that $N = 2$. Ignoring the disturbance input, we have

$$\hat{\mathbf{P}} = \mathbf{P} = \begin{bmatrix} c & 1 \\ 0 & 0 \end{bmatrix}; \qquad \hat{\mathbf{Q}} = \mathbf{Q} = \begin{bmatrix} 0 \\ \dfrac{1-c}{a} \end{bmatrix}; \qquad \hat{\mathbf{C}} = \mathbf{C} = [1 \quad 0]$$

Thus

$$\mathbf{L} = \begin{bmatrix} c & 1 \\ 1 & -c \end{bmatrix}$$

$\det(\mathbf{L}) = -1$, so that $\mathbf{L}^{-1}$ exists and is

$$\mathbf{L}^{-1} = \begin{bmatrix} 0 & 1 \\ 1 & -c \end{bmatrix}$$

From equation (14.33a) the observer gains are

$$\mathbf{f} = \begin{bmatrix} c & 1 \\ 0 & 0 \end{bmatrix}\begin{bmatrix} 0 & 1 \\ 1 & -c \end{bmatrix}\begin{bmatrix} 1 \\ 0 \end{bmatrix} = \begin{bmatrix} 1 \\ 0 \end{bmatrix} \tag{14.34b}$$

The observer algorithm is therefore as follows:

1. Obtain m_k and y_k [for $k = 0$ only, initialize $\hat{\mathbf{X}}_0^0$ by equation (14.30d), that is, $\hat{x}_{1,0}^0 = y_0$　and　$\hat{x}_{2,0}^0 = 0$]

2. Compute $\hat{\mathbf{X}}^0_{k+1}$ by equation (14.30c), that is,

$$\hat{x}_{1,k} = \hat{x}_{1,k}^{\,0} + y_k - \hat{x}_{1,k}^{\,0} = y_k \tag{14.35a}$$

$$\hat{x}_{2,k} = \hat{x}_{2,k}^{\,0} + 0 \tag{14.35b}$$

3. Compute $\hat{x}_{k+1}^{\,0}$ by equation (14.30a), that is,

$$\hat{x}_{1,k+1}^{\quad\;\,0} = c\hat{x}_{1,k} + \hat{x}_{2,k} \tag{14.35c}$$

$$\hat{x}_{2,k+1}^{\quad\;\,0} = \left(\frac{1-c}{a}\right)m_k \tag{14.35d}$$

4. Set $k = k + 1$ and return to step 1.

Note from equation (14.35a) that the estimation is exact, as should be expected for $x_{1,k}$, which is the measurement. Thus equation (14.35c) is not needed.

   For the plant, the continuous-time or actual plant is simulated, that is,

$$\frac{dx}{dt} = m(kT) + a(v - x) \tag{14.36a}$$

$$y(t) = x(t - 10\pi)$$

where T is the sampling period, which for this problem is equal to the dead time: $T = 10\pi$. y_k is thus the sample of $y(t)$ at $t = kT$, while $m(kT)$ is the zero-order held (staircase pattern) signal of the discrete-time control input, equation (14.27d) of Example 14.5:

$$m_k = -[(k_{p1} + k_i)y_k + k_{p2}\hat{x}_{2,k} - k_i(x_{3,k} + r_k)] \tag{14.36b}$$

where the estimated value $\hat{x}_{2,k}$, from equations (14.35a) through (14.35d), has replaced the unavailable state $x_{2,k}$, and the additional state due to the integral control $x_{3,k}$ is given by equation (14.27c), repeated here:

$$x_{3,k+1} = r_k - y_k + x_{3,k}; \qquad x_{3,0} = 0 \tag{14.36c}$$

Figure 14.11a shows the general configuration for finite-time settling state vector feedback control using a finite-time settling observer. If we assume, as shown in the figure, that all m-outputs are subject to integral action, then the total system order for the controller is $n + m$. The total number of sampling steps for the system output vector to settle to the setpoint value is the sum of the number of steps for the observer to converge and the number of steps for the controller to settle. That is, let $(n + \Delta n_0)/m = N_0$ for the observer and $(n + m + \Delta n_c)/r = N_c$ for the controller.

   Then the number of steps needed for $\mathbf{Y} = \mathbf{r}$ is

$$(N_0 - 1) + N_c$$

For the present problem, $n = 2$, $\Delta n_0 = \Delta n_c = 0$, and $m = r = 1$. Hence it should take $(2 - 1) + 3$, or 4, steps for the output to converge to the setpoint. This result is confirmed by the simulation of the hot-water heater response shown in Fig. 14.11b. Following each step change in the disturbance input, the sensed temperature (output) takes four sampling periods to reach and remain at the setpoint value.

Figure 14.11

**Finite-time Settling Control
Via Finite-time
Settling Observer**

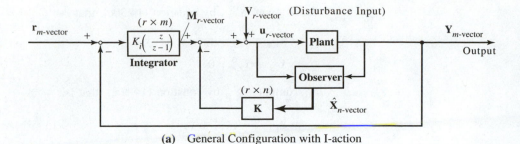

(a)   General Configuration with I-action

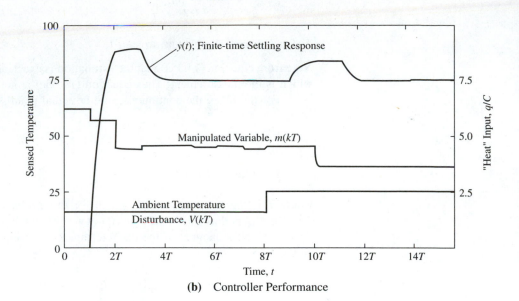

(b)   Controller Performance

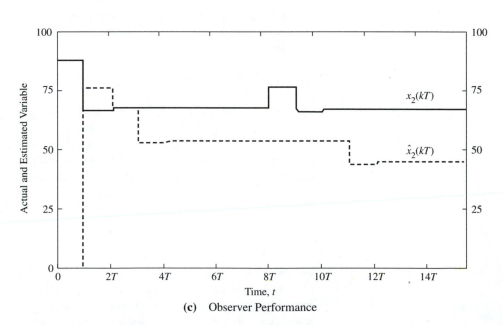

(c)   Observer Performance

Referring to Fig. 14.10b, observe that the performance of the minimal response controller is practically the same as the performance of the finite-time settling controller (recall the specification, equation (14.16a)). The only problem is the potential for loss of controllability (already discussed) of the z-domain design technique. The present result also suggests that the finite-time settling controller performance can be less satisfactory for high-frequency disturbance inputs so that the technique is perhaps best suited for step disturbance inputs such as load changes (see [24]).

Figure 14.11c shows that the finite-time settling observer is convergent within the same four steps; however, the estimated state (broken line) is being compared with the state as determined by the discrete-time model of the continuous-time plant (solid line). Finally, as would be expected for a proportional-gain observer (see Practice Problem 13.23 on eliminating a similar offset for a continuous-time controller), the present finite-time settling observer also exhibits a bias or offset whose value depends on the disturbance input.

14.4 INTRODUCING DISCRETE-TIME OPTIMAL CONTROL

As we alluded to at the beginning of this chapter, one of the consequences of the growth (and continuing expansion) of the computational capabilities of microcomputers is the increased application of advanced control algorithms. *Optimal control* is an advanced control concept in which we seek a high-performance control algorithm that can compel the controlled system to behave in the "best possible way." The calculus of variations approach to determining the optimal control law was developed in Section 13.4. In this section we consider a technique for developing the optimal control algorithm for discrete-time systems based on the theory of dynamic programming.

Some relevant aspects of the theory of dynamic programming are first deduced to provide a good framework for the subsequent development of the basic relations for discrete optimal control. Although the resulting computational algorithm is quite general, its implementation is restricted in practice to those problems for which the computer memory (storage) and/or speed requirements are within the capacity of the computer (controller) hardware. However closed-form and hence compact solutions are also feasible for some special cases, including the case of linear systems with quadratic performance indices for which (as we saw for continuous-time systems) the proportional state vector feedback controller is optimal. This solution (see [25]) will also be considered here but without a rigorous derivation.

Elements of Dynamic Programming

The theory of dynamic programming is the study of *multistage decision processes* [18], and many engineering systems (or processes) are multistage processes or can be approximated as such.

Multistage Process

Let $\mathbf{x}$ be a state of a system or a point in a set or space $\mathbf{X}$. If $\mathbf{x}_0$ is the initial state, and $\mathbf{x}_1$ is the state one time unit later, and $\mathbf{T}(\mathbf{x}_0)$ is a transformation with the property that the transformed state $\mathbf{x}_1 = \mathbf{T}(\mathbf{x}_0)$ belongs to $\mathbf{X}$ for all $\mathbf{x}_0$ in $\mathbf{X}$, then the set of successive states (vectors) $[\mathbf{x}_0, \mathbf{x}_1, \mathbf{x}_2, \ldots, \mathbf{x}_k, \ldots]$ such that

$$\mathbf{x}_{k+1} = \mathbf{T}(\mathbf{x}_k) \tag{14.37a}$$

for all $k = 0, 1, 2, \ldots$, is called a *multistage process*. An example is the free linear discrete-time system

$$\mathbf{x}_{k+1} = \mathbf{P}\mathbf{x}_k \tag{14.37b}$$

where $\mathbf{x}_k$ is an n-vector and $\mathbf{P}$ is an $(n \times n)$-matrix.

Recurrence Relations   Consider, for example, the function

$$J = \sum_{i=0}^{N} h(\mathbf{x}_i) \tag{14.38a}$$

where $h(\mathbf{x})$ is a prescribed function. Given the transformation $\mathbf{T}$ whereby $\mathbf{x}_i$ can be obtained from $\mathbf{x}_{i-1}$, J is completely determined by the initial state $\mathbf{x}_0$ and the number of stages N, since for all $\mathbf{x}_0$ in $\mathbf{X}$, we can write

$$J_N(\mathbf{x}_0) = \sum_{i=0}^{N} h(\mathbf{x}_i) = h(\mathbf{x}_0) + h(\mathbf{T}(\mathbf{x}_0)) + h(\mathbf{T}^2(\mathbf{x}_0)) + \ldots + h(\mathbf{T}^N(\mathbf{x}_0)) \tag{14.38b}$$

where $\mathbf{T}^2(\mathbf{x}_0) = \mathbf{T}(\mathbf{x}_1) = \mathbf{T}(\mathbf{T}(\mathbf{x}_0))$, $\mathbf{T}^3(\mathit{x}_0) = \mathbf{T}(\mathbf{x}_2) = \mathbf{T}(\mathbf{T}(\mathbf{x}_1)) = \mathbf{T}(\mathbf{T}(\mathbf{T}(\mathbf{x}_0)))$, etc. The dependence of J on the initial state $\mathbf{x}_0$ and on N is expressed by writing J as $J_N(\mathbf{x}_0)$, and N is allowed to take on the values $N = 0, 1, 2, \ldots$. For $N \geq 1$, the partial sum

$$h(\mathbf{T}(\mathbf{x}_0)) + h(\mathbf{T}^2(\mathbf{x}_0)) + \cdots + h(\mathbf{T}^N(\mathbf{x}_0)) = h(\mathbf{x}_1) + h(\mathbf{T}(\mathbf{x}_1)) \tag{14.38c}$$
$$+ h(\mathbf{T}^2(\mathbf{x}_1)) + \cdots + h(\mathbf{T}^{N-1}(\mathbf{x}_1))$$

is equal to $J_{N-1}(\mathbf{x}_1) = J_{N-1}(\mathbf{T}(\mathbf{x}_0))$ by the definition (14.38b). Consequently, J or $J_N(\mathbf{x}_0)$ satisfies the *recurrence relation*

$$J_N(\mathbf{X}_0) = h(\mathbf{x}_0) + J_{N-1}(\mathbf{T}(\mathbf{x}_0)); \qquad N \geq 1 \tag{14.38d}$$

with $J_0(\mathbf{x}_0) = h(\mathbf{x}_0)$.

Consider as another example,

$$J_N(\mathbf{x}_0) = \prod_{i=0}^{N} h(\mathbf{x}_i) \tag{14.39a}$$

That is,

$$J_N(\mathbf{x}_0) = h(\mathbf{x}_0)h(\mathbf{x}_1)h(\mathbf{x}_2) \cdots h(\mathbf{x}_N) = h(\mathbf{x}_0)h(\mathbf{T}(\mathbf{x}_0))h(\mathbf{T}^2(\mathbf{x}_0)) \cdots h(\mathbf{T}^N(\mathbf{x}_0))$$

But

$$h(\mathbf{T}(\mathbf{x}_0))h(\mathbf{T}^2(\mathbf{x}_0)) \cdots h(\mathbf{T}^N(\mathit{x}_0)) = h(\mathbf{x}_1)h(\mathbf{T}(\mathbf{x}_1)) \cdots h(\mathbf{T}^{N-1}(\mathbf{x}_1))$$
$$= J_{N-1}(\mathbf{x}_1) = J_{N-1}(\mathbf{T}(\mathbf{x}_0))$$

Hence the recurrence relation is

$$J_N(\mathbf{X}_0) = h(\mathbf{x}_0)J_{N-1}(\mathbf{T}(\mathbf{x}_0)) \tag{14.39b}$$

Multistage Decision Processes

Suppose a multistage process can be sufficiently influenced so that the transformation $\mathbf{T}$ now depends on another vector as well: $\mathbf{T} = \mathbf{T}(\mathbf{x}, \mathbf{u})$, where the value of $\mathbf{u}$ can be chosen from a set of allowable vectors $\mathbf{U}$. Let $\mathbf{u}_k$ be the choice at the kth stage, so that $\mathbf{x}_1 = \mathbf{T}(\mathbf{x}_0, \mathbf{u}_0)$, and generally,

$$\mathbf{x}_{k+1} = \mathbf{T}(\mathbf{x}_k, \mathbf{u}_k) \tag{14.40a}$$

then the vector $\mathbf{u}_k$ is called the decision vector or decision variable, the choice of $\mathbf{u}_k$ is called a decision, and the set of successive states $[\mathbf{x}_0, \mathbf{x}_1, \mathbf{x}_2, \ldots, \mathbf{x}_k, \ldots]$ is now called a *multistage decision process*. An example is the forced linear discrete-time system

$$\mathbf{x}_{k+1} = \mathbf{P}\mathbf{x}_k + \mathbf{Q}\mathbf{u}_k \tag{14.40b}$$

where $\mathbf{x}_k$ is an n-vector, $\mathbf{P}$ is an $(n \times n)$-matrix, $\mathbf{u}_k$ is an r-vector, and $\mathbf{Q}$ is an $(n \times r)$-matrix. Another example is the general discrete-time state vector system

$$\mathbf{x}_{k+1} = \mathbf{f}(\mathbf{x}_k, \mathbf{u}_k); \qquad k \in [0, N-1] \tag{14.40c}$$

Optimal Policy Our concern here is with multistage decision processes in which the $\mathbf{u}_k$ are chosen so as to minimize a given scalar function of the state and decision variables:

$$J = J(\mathbf{x}_0, \mathbf{x}_1, \mathbf{x}_2, \ldots; \mathbf{u}_0, \mathbf{u}_1, \ldots) \tag{14.41a}$$

If the values of $\mathbf{u}$ are chosen with respect to scalar criteria J that depend only on the past and present stages of the process, then the decision at stage k or the function

$$\mathbf{u}_k = \mathbf{u}_k(\mathbf{x}_0, \mathbf{x}_1, \mathbf{x}_2, \ldots, \mathbf{x}_k; \mathbf{u}_0, \mathbf{u}_1, \ldots, \mathbf{u}_{k-1})$$

is called the *policy function* or simply the *policy*. An *optimal policy* is a policy that minimizes the criterion function J (also called the *cost function* or, if it is being maximized, *return function*). For criterion functions like (14.38a) or (14.39a), we saw that $J = J_N(\mathbf{x}_0)$. It follows from equation (14.40a) that for optimal control, the policies can in general be further restricted to functions of only the current state and the stage of the process, that is,

$$\mathbf{u}_k = \mathbf{u}(\mathbf{x}_k, k) = \mathbf{u}_k(\mathbf{x}_k) \tag{14.41b}$$

and the desired behavior of the controlled system can be judged based on the criterion (function) that must attain its minimum value when such behavior is the "best possible." The scalar criteria in this case are, as before, termed *performance measures* or *performance indices* (see below).

Principle of Optimality

For policies that depend only on the current state, "an optimal policy has the property that whatever the initial state and initial decision are, the remaining decisions must constitute an optimal policy with regard to the state resulting from the first decision." This is the *Principle of Optimality,* reference [18]. Consider a discrete and deterministic type N-stage decision process, that is, the set of vectors $[\mathbf{x}_0, \mathbf{x}_1, \mathbf{x}_2, \ldots, \mathbf{x}_N; \mathbf{u}_0, \mathbf{u}_1, \ldots, \mathbf{u}_{N-1}]$, with $\mathbf{x}_{k+1} = \mathbf{T}(\mathbf{x}_k, \mathbf{u}_k)$ for each k. Starting in state $\mathbf{x}_0$ with an optimal sequence of decisions, $\mathbf{u}_0^0, \mathbf{u}_1^0, \ldots, \mathbf{u}_{N-1}^0$ such that $\mathbf{x}_1 = \mathbf{T}(\mathbf{x}_0, \mathbf{u}_0^0)$, $\mathbf{x}_2 = \mathbf{T}(\mathbf{x}_1, \mathbf{u}_1^0)$, etc., $\mathbf{u}_1^0, \mathbf{u}_2^0, \ldots, \mathbf{u}_{N-1}^0$ must constitute an optimal sequence of decisions for the $(N - 1)$ stage process starting in state $\mathbf{x}_1$.

For example, let the criterion function for a multistage decision process be

$$J(\mathbf{x}_0, \mathbf{x}_1, \ldots; \mathbf{u}_0, \mathbf{u}_1, \ldots) = \sum_{i=0}^{N} h(\mathbf{x}_i, \mathbf{u}_i) \tag{14.42a}$$

Let the minimum value of J depend only on the initial state $\mathbf{x}_0$ and the number of stages N. Then we can denote this minimum value of J, which must have been obtained starting in state $\mathbf{x}_0$ with an optimal policy, by $J_N^0(\mathbf{x}_0)$. From the principle of optimality and the preceding developments we can conclude that no matter the initial decision $\mathbf{u}_0$, for $N \geq 1$,

$$h(\mathbf{x}_0, \mathbf{u}_0) + [h(\mathbf{x}_1, \mathbf{u}_1) + \cdots + h(\mathbf{x}_N, \mathbf{u}_N)] = h(\mathbf{x}_0, \mathbf{u}_0) + J_{N-1}(\mathbf{T}(\mathbf{x}_0, \mathbf{u}_0)) \tag{14.42b}$$

and since this relation holds for all initial decisions $\mathbf{u}_0$, we must minimize (14.42b) over $\mathbf{u}_0$ to obtain $J_N^0(\mathbf{x}_0)$, so that

$$J_N^0(\mathbf{x}_0) = \min_{\mathbf{u}_0}[h(\mathbf{x}_0, \mathbf{u}_0) + J_{N-1}^0(\mathbf{T}(\mathbf{x}_0, \mathbf{u}_0))]; \qquad N \geq 1 \tag{14.43a}$$

with

$$J_0^0(\mathbf{x}_0) = \min_{\mathbf{u}_0}[h(\mathbf{x}_0, \mathbf{u}_0)] \tag{14.43b}$$

Thus the problem of determining a sequence of decisions that minimize a criterion function has been converted into that of solving a functional equation (14.43a).

The Discrete-Time Optimal Control Problem

Consider the multistage decision process, equation (14.40c), repeated here:

$$\mathbf{x}_{k+1} = \mathbf{f}(\mathbf{x}_k, \mathbf{u}_k); \qquad \mathbf{x}_0 = \mathbf{x}(0) \tag{14.44a}$$

where $\mathbf{x}_k$ is an n-state vector, $\mathbf{u}_k$ is an r-control vector, k is the time index, and $\mathbf{f}$ is an n-function vector (possibly a nonlinear state transformation vector). We seek the optimal control sequence $[\mathbf{u}_0^0, \mathbf{u}_1^0, \ldots, \mathbf{u}_{N-1}^0]$ that minimizes the general performance functional

$$J = H(\mathbf{x}_N) + \sum_{i=0}^{N-1} h(\mathbf{x}_i, \mathbf{u}_i) \tag{14.44b}$$

such that at each (time) stage, $\mathbf{u}_k$ can be assumed to be restricted as in equation (14.41b) and

$$\mathbf{u}_k \in \mathbf{U} \tag{14.44c}$$

and

$$\mathbf{x}_k \in \mathbf{X} \tag{14.44d}$$

where H and h are scalar-valued functions, and $\mathbf{U}$ and $\mathbf{X}$ are sets of allowable vectors.

The preceding constitutes a *discrete-time optimal control problem*, and from our earlier discussion we can deduce that both the optimal control sequence

$[\mathbf{u}_0^0, \mathbf{u}_1^0, \ldots, \mathbf{u}_{N-1}^0]$ and the minimum value of the cost functional are both functions of the initial state, so that the recurrence relation equation (14.43a) is applicable.

Typical Performance Measures

Note that equation (14.44b) is our discrete-time analog of the general continuous-time optimal control performance index, equation (13.54a). As was the case in Section 13.4, there are different performance functionals for specific optimal control problems. Following reference [26], we can identify some common forms of performance measures that can apply to typical discrete-time optimal control problems:

1. **Minimum-time problems** seek, as before, to minimize the time used to transfer the system from an initial state $\mathbf{x}_0$ to a specified state $\mathbf{x}_N$. For example,

$$J = \sum_{i=0}^{N-1} T_i \tag{14.45a}$$

where T_i is the sampling period between stages i and $i + 1$.

2. **Terminal control** describes the situation in which the system is to be transferred to a final state $\mathbf{x}_N$ as close as possible to some desired state $\mathbf{r}_N$. For instance, if the error at the instant (stage) k is

$$\mathbf{e}_k = \mathbf{x}_k - \mathbf{r}_k \tag{14.45b}$$

then we can have

$$J = \mathbf{e}_N^T \mathbf{W} \mathbf{e}_N \tag{14.45c}$$

where $\mathbf{W}$ is an appropriate real, symmetric positive definite (SPD) or symmetric positive semidefinite (SPSD) matrix (see Appendix B) of weighting factors.

3. **Minimum effort** implies, as in the continuous-time case (minimum fuel consumption), that the control effort to be expended in attaining the final state should be minimized. A possible type of discrete-time performance index in this case is

$$J = \sum_{i=0}^{N-1} \sum_{j=1}^{r} w_j |u_{i,j}| \tag{14.45d}$$

where $w_j, j = 1, 2, \ldots, r$, are weighting factors, and u_{ij} is the jth element of the control vector at stage i. Another applicable performance measure is

$$J = \sum_{i=0}^{N=1} \mathbf{u}_i^T \mathbf{W} \mathbf{u}_i \tag{14.45e}$$

where $\mathbf{W}$ is a real SPD matrix of weighting factors.

4. **Tracking problems,** unlike terminal control problems, require the system state $\mathbf{x}_k$ to follow as closely as possible the desired state $\mathbf{r}_k$ for all $0 \le k \le N$. Thus a typical performance index is

$$J = \sum_{i=0}^{N} \mathbf{e}_i^T \mathbf{W} \mathbf{e}_i \tag{14.45f}$$

5. **Quadratic performance indices.** Measures of performance such as (14.45c), (14.45e), and (14.45f) are, as before, quadratic criteria. The general quadratic cost functional is a combination of terminal control, minimum effort control, and tracking. An example (see [25]) is

$$J = \frac{1}{2}\mathbf{x}_N^T \mathbf{H} \mathbf{x}_N + \frac{1}{2}\sum_{i=0}^{N-1}\{\mathbf{x}_i^T \mathbf{W}_i \mathbf{x}_i + 2\mathbf{u}_i^T \mathbf{M}_i \mathbf{x}_i + \mathbf{u}_i^T \mathbf{R}_i \mathbf{u}_i\} \tag{14.45g}$$

where N is specified, $\mathbf{H}$ and $\mathbf{W}_i$ are SPSD, $\mathbf{R}_i$ is SPD, and $\mathbf{W}_i - \mathbf{M}_i^T \mathbf{R}_i^{-1} \mathbf{M}_i$ must be positive semidefinite

Computing a Solution

Consider the optimal control problem equations (14.44a) through (14.44d), defined for every $k < N$, that is,

$$\text{Minimize} \quad J_k = H(\mathbf{x}_N) + \sum_{i=k}^{N-1} h(\mathbf{x}_i, \mathbf{u}_i) \tag{14.46a}$$

subject to equations (14.44a), (14.44c), and (14.44d)

Evidently the minimum value of J_k depends only on the starting state $\mathbf{x}_k$ and the number of stages $(N - k)$ or simply k, since N is fixed in this scheme. Denoting this minimum value of J_k by $J_k^0(\mathbf{x}_k)$ and noting from the principle of optimality that J_k must be minimized with respect to the starting decision $\mathbf{u}_k$, we obtain as in equations (14.43a) and (14.43b), but with a backward running index, the recursive algorithm

$$J_k^0(\mathbf{x}_k) = \min_{\mathbf{u}_k}[h(\mathbf{x}_k, \mathbf{u}_k) + J_{k+1}^0(\mathbf{f}(\mathbf{x}_k, \mathbf{u}_k))]; \qquad k \le (N - 1) \tag{14.46b}$$

with

$$J_N^0(\mathbf{x}_N) = H(\mathbf{x}_N) \tag{14.46c}$$

Note that the eventual minimum cost $J_N^0(\mathbf{x}_0)$ (original nomenclature) is now replaced by $J_0^0(\mathbf{x}_0)$ (new nomenclature, with backward running index k). Equations (14.46a) through (14.46c) suggest that given $J_N^0(\mathbf{x}_N)$ we can calculate both the cost functions and the control functions recurrently as follows:

1. Let $k = N - 1$. Considering equation (14.46b), we need to either compute and store in the computer the values of $h(\mathbf{x}_k, \mathbf{u}_k)$, $\mathbf{f}(\mathbf{x}_k, \mathbf{u}_k)$, and $J_{k+1}^0(\mathbf{f}(\mathbf{x}_k, \mathbf{u}_k))$ (equal to $H(\mathbf{x}_N)$ for $k = N - 1$ only) or store the formulas for calculating them, for each $\mathbf{x}_k$ in $\mathbf{X}$. In practice, the state space is discretized into a finite set of values $\{\mathbf{x}_k^{<m>}\}$, $m = 1, 2, \ldots, M$, whereas $\mathbf{u}_k$ can be restricted to a finite set: $\{\mathbf{u}_k^{<j>}\}$, $j = 1, 2, \ldots, R$. Thus $J_k^0(\mathbf{x}_k)$, for instance, will be given by $\{J_k^0(\mathbf{x}_k^{<m>})\}$, $m = 1, 2, \ldots, M$, where, if necessary, some values of $J_k^0(\mathbf{x}_k)$ can be obtained by interpolation or extrapolation (the values of $\mathbf{f}(\mathbf{x}_k, \mathbf{u}_k) = \mathbf{x}_{k+1}$ may not fall exactly on the state grid even though the values of $\mathbf{x}_k$ are on the grid). Similarly, values of $h(\mathbf{x}_k, \mathbf{u}_k)$, for example, will consist of $h(\mathbf{x}_k^{<m>}, \mathbf{u}_k^{<j>})$, $m = 1, \ldots, M; j = 1, \ldots, R$. It is easy to see how computer *high-speed* memory capacity can become a problem in many cases.

   To determine $J_k^0(\mathbf{x}_k)$, equation (14.46b), we evaluate for each m, the expression $[h(\mathbf{x}_k^{<m>}, \mathbf{u}_k) + J_{k+1}^0(\mathbf{f}(\mathbf{x}_k^{<m>}, \mathbf{u}_k))]$, for $\mathbf{u}_k = \mathbf{u}_k^{<1>}, \mathbf{u}_k^{<2>}, \ldots, \mathbf{u}_k^{<R>}$, and choose the value of $\mathbf{u}_k$, that is, $\mathbf{u}_k^0(\mathbf{x}_k^{<m>})$, that yield(s) the minimum cost, that is $J_k^0(\mathbf{x}_k^m)$. Repeating this procedure at $k = N - 1$, for $m = 1, 2, \ldots, M$, yields the cost function $J_{N-1}^0(\mathbf{x}_{N-1})$ and the control function $\mathbf{u}_{N-1}^0(\mathbf{x}_{N-1})$.

2. We let $k = N - 2$ and similarly obtain $J^0_{N-2}(\mathbf{x}_{N-2})$ and $\mathbf{u}^0_{N-2}(\mathbf{x}_{N-2})$. Thus, recurrently, we compute the sequences $\{J^0_{N-L}(\boldsymbol{x}_{N-L})\}$ and $\{\boldsymbol{u}^0_{N-L}(\boldsymbol{x}_{N-L})\}$ $L = 1, 2, \ldots, N$.

Note that at each stage the same procedure, equation (14.46b), is used to compute $J^0_k(\mathbf{x}_k)$ given $J^0_{k+1}(\mathbf{x}_{k+1})$. Hence the set of computer instructions is small and easily written in a high-level computer language or computational math software procedure. Indeed, $h(\mathbf{x}_k, \mathbf{u}_k)$ and $\mathbf{f}(\mathbf{x}_k, \mathbf{u}_k)$, if not determined using stored formula, need to be computed only once at $k = N - 1$ and stored for subsequent use. The storage capacity problem for the dynamic programming algorithm is also somewhat diminished, since to determine $J^0_k(\mathbf{x}_k)$ we need only $J^0_{k+1}(\mathbf{x}_{k+1})$. Thus $J^0_{k+1}(\mathbf{x}_{k+1})$. can be "washed out" of the computer high-speed memory once $J^0_k(\mathbf{x}_k)$ is computed. Finally, the constraints on permissible values of the state and/or policy functions serve only to reduce the numbers of $\mathbf{x}$- and $\mathbf{u}$-values over which to search at each stage and hence to make the application of the dynamic programming technique easier. This is an advantage of the method over other optimization techniques (for example, the calculus of variations approach for continuous-time systems), which tend to become more complicated when constraints are introduced.

Next, we illustrate the details of this procedure, first with a very simple example (one state variable and one input system), which is solved manually. This example is followed by another, more involved (multivariable system) example for which a pair of computational mathematics procedures will be developed using the experience of the first example.

Example 14.7 — Formulation and Computation of a Simple Problem

The annual performance of a high-yield investment account may be described (from the owner's viewpoint) as in Example 4.9 (see equation (4.48a)) by

$$x_{1,k+1} = (1 + a)x_{1,k} + u_{1,k}$$

where $x_{1,k}$ is the amount in the account, a is the rate of return or interest rate for annual compounding (assume 100%), and $u_{1,k}$ is the net annual investment or payment into the account (deposits less withdrawals). If it is desired to accumulate as close to 2000 (money units) as possible, by the end of the third year while expending minimum (payments) effort, formulate the problem and determine by dynamic programming the optimal net payments (investment) policy for

a. An opening balance of zero.
b. An opening balance of 200.

Assume that

  (i) The maximum deposit the owner can contribute a year is 400.
 (ii) The maximum withdrawal that is allowed per year in this type of account is 200.
(iii) Minimum investment effort can be interpreted in this case as

$$Minimize \sum_{i=0}^{N-1} x_{1,i}u_{1,i}$$

where i is a time index for 1-year investment periods and N is the maturity time (3 years).

Solution—Problem Formulation: The state equation (for $a = 1$) is

$$x_{1,k+1} = 2x_{1,k} + u_{1,k}; \qquad \text{(a) } x_{1,0} = 0; \qquad \text{(b) } x_{1,0} = 200 \tag{14.47a}$$

Note that although this is a SISO system, we have retained a state-vector and input-vector index in the nomenclature for the variables, so that all the potentially applicable indices are represented in this problem solution. From equation (14.47a),

$$\mathbf{f}(\mathbf{x}_k, \mathbf{u}_k) = f_1(\mathbf{x}_k, \mathbf{u}_k) = 2x_{1,k} + u_{1,k} \tag{14.47b}$$

In addition to a minimized investment effort, the accumulated amount after 3 years is to be as close as possible to 2000. Thus the performance index is a combination of terminal control and minimum effort types, with the latter as specified previously. With no further information to weight the relative importance of the different components of the overall performance measure, we can simply let (see equations (14.45b) and (14.45c))

$$J = [(x_{1,3} - 2000)^2 + \sum_{i=0}^{2} x_{1,i} u_{1,i}] 10^{-3} \tag{14.47c}$$

The factor 10^{-3} is just for numerical convenience and can be removed. Hence comparing equations (14.47c) and (14.44b), $N = 3$ as already indicated, we have

$$H(\mathbf{x}_N) = 0.001(x_{1,N} - 2000)^2 \tag{14.47d}$$

$$h(\mathbf{x}_k, \mathbf{u}_k) = 0.001(x_{1,k} \cdot u_{1,k}) \tag{14.47e}$$

Solution—Computation: In order to retain the main object of this example, which is to illustrate the computation of optimal control by dynamic programming, we consider a gross discretization of the state space and allowable control as follows:

$$x_{1,k}^{<m>} = 200(m - 1); \qquad m = 1, \ldots, 11$$

$$u_{1,k}^{<j>} = 200(j - 2); \qquad j = 1, \ldots, 4$$

We can set up the computation scheme as shown in Table 14.2. Note that for a multivariable system, each cell under the state $(\mathbf{x}_k)$ column will correspond to a single node in the discretized state space, that is, a single combination of values of the state variables. Similarly, each input column corresponds to one combination of allowable values of input variables. As explained previously, the entries $\mathbf{f}(\mathbf{x}_k, \mathbf{u}_k)$ and $h(\mathbf{x}_k, \mathbf{u}_k)$ are computed once at $k = (N - 1) = 2$ and saved for use at other stages, whereas the only entry required and shown in the table for $k = N = 3$ is $J_N^0(\mathbf{x}_N) = H(x_{1,3}^{<m>}, m = 1, 2, \ldots, M)$. Note the dual notation in Table 14.2 $J(\mathbf{x}_k, \mathbf{u}_k)$ represents equation (14.46b) before a minimum is selected as well as the selected minimum $J_k^0(\mathbf{x}_k)$, which is shown in highlighted (bold) characters.

The other values in Table 14.2 can be explained as follows: Consider stage $k = (N - 1) = 2$. Suppose the state is $\mathbf{x}_2 = x_{1,k}^{<5>} = 800$, that is, $m = 5$. If the control is $\mathbf{u}_2 = -200$, that is, $u_{1,k}^{<1>}$, then by equation (14.47b), the subsequent state $\mathbf{x}_3 = x_{1,3} = \mathbf{f}(\mathbf{x}_2, \mathbf{u}_2) = 2(800) - 200 = 1400$. Equation (14.47e) gives $h(\mathbf{x}_2, \mathbf{u}_2) = 0.001(800)(-200) = -160$, and $J_{k+1}^0(\mathbf{f}(\mathbf{x}_k, \mathbf{u}_k))$ or $J^0(k + 1) = J_3^0(1400) = 360$ is read off the (previously computed) column for $k = 3$ and row for $\mathbf{x}_k = 1400 = x_{1,k}^{<8>}$. Finally, by equation (14.46b), $J_2(\mathbf{x}^{<5>}, \mathbf{u}^{<1>}) = -160 + 360 = 200$. Continuing in a similar manner for the same $\mathbf{x}_2$ (that is, $m = 5$), if we let the control be $\mathbf{u}_2 = 0$, (that is, $j = 2$), we obtain $\mathbf{x}_3 = 1600$, $h(\mathbf{x}_2^{<5>}, \mathbf{u}_2^{<2>}) = 0$, $J_3^0(1600) = 160$, and $J_2(\mathbf{x}^{<5>}, \mathbf{u}^{<2>}) = 160$. For $\mathbf{u}_2 = 200$, or $j = 3$, we find $\mathbf{x}_3 = 1800$, $h(\mathbf{x}_2^{<5>}, \mathbf{u}_2^{<3>}) = 160$, $J_3^0(1800) = 40$, and $J_2(\mathbf{x}^{<5>}, \mathbf{u}^{<3>}) = 200$; and for $j = 4$

Table 14.2 Dynamic Programming Computation Scheme

STATE x_k		$k = 3$	$u_k^{<j>}, k = 2$				$u_k^{<j>}, k = 1$				$u_k^{<j>}, k = 0$			
			-200	0	200	400	-200	0	200	400	-200	0	200	400
2000	$f(x_k, u_k)$		3800	4000	4200	4400								
	$h(x_k, u_k)$		-400	0	400	800								
$(m = 11)$	$J^0(k+1)$		3240	4000	4840	5760								
	$J(x_k, u_k)$	**0**	**2840**	4000	5240	6560								
1800	$f(x_k, u_k)$		3400	3600	3800	4000								
	$h(x_k, u_k)$		-360	0	360	720								
$(m = 10)$	$J^0(k+1)$		1960	2560	3240	4000								
	$J(x_k, u_k)$	**40**	**1600**	2560	3600	4720								
1600	$f(x_k, u_k)$		3000	3200	3400	3600								
	$h(x_k, u_k)$		-320	0	320	640								
$(m = 9)$	$J^0(k+1)$		1000	1440	1960	2560								
	$J(x_k, u_k)$	**160**	**680**	1440	2280	3200								
1400	$f(x_k, u_k)$		2600	2800	3000	3200								
	$h(x_k, u_k)$		-280	0	280	560								
$(m = 8)$	$J^0(k+1)$		360	640	1000	1440								
	$J(x_k, u_k)$	**360**	**80**	640	1280	2000								
1200	$f(x_k, u_k)$		2200	2400	2600	2800								
	$h(x_k, u_k)$		-240	0	240	480								
$(m = 7)$	$J^0(k+1)$		40	160	360	640								
	$J(x_k, u_k)$	**640**	**-200**	160	600	1120								
1000	$f(x_k, u_k)$		1800	2000	2200	2400								
	$h(x_k, u_k)$		-200	0	200	400								
$(m = 6)$	$J^0(k+1)$		40	0	40	160								
	$J(x_k, u_k)$	**1000**	**-160**	0	240	560								
800	$f(x_k, u_k)$		1400	1600	1800	2000								
	$h(x_k, u_k)$		-160	0	160	320								
$(m = 5)$	$J^0(k+1)$		360	160	40	0	80	680	1600	2840				
	$J(x_k, u_k)$	**1440**	200	**160**	200	320	**-80**	680	1760	3160				
600	$f(x_k, u_k)$		1000	1200	1400	1600								
	$h(x_k, u_k)$		-120	0	120	240								
$(m = 4)$	$J^0(k+1)$		1000	640	360	160	-160	-200	80	680				
	$J(x_k, u_k)$	**1960**	880	640	480	**400**	**-280**	-200	200	920				
400	$f(x_k, u_k)$		600	800	1000	1200								
	$h(x_k, u_k)$		-80	0	80	160								
$(m = 3)$	$J^0(k+1)$		1960	1440	1000	640	400	160	-160	-200				
	$J(x_k, u_k)$	**2560**	1880	1440	1080	**800**	320	160	**-80**	-40				
200	$f(x_k, u_k)$		200	400	600	800								
	$h(x_k, u_k)$		-40	0	40	80								
$(m = 2)$	$J^0(k+1)$		3240	2560	1960	1440	1520	800	400	160	240	-80	-280	-80
	$J(x_k, u_k)$	**3240**	3200	2560	2000	**1520**	1480	800	440	**240**	200	-80	**-240**	0
0	$f(x_k, u_k)$		-200	0	200	400								
	$h(x_k, u_k)$		0	0	0	0								
$(m = 1)$	$J^0(k+1)$			4000	3240	2560		2560	1520	800		800	240	-80
	$J(x_k, u_k)$	**4000**	4000	4000	3240	**2560**		2560	1520	**800**		800	240	**-80**

($\mathbf{u}_2 = 400$), the results are $\mathbf{x}_3 = 2000$, $h(\mathbf{x}_2^{<5>}, \mathbf{u}_2^{<4>}) = 320$, $J_3^0(1600) = 0$, and $J_2(\mathbf{x}^{<5>}, \mathbf{u}^{<4>}) = 320$. The actual value of the cost function given by equation (14.46b) for $m = 5$ at $k = 2$ is $\min(j_2(\mathbf{x}^{<5>}, \mathbf{u}^{<j>}); J = 1, 2, \ldots, 4)$. That is, $J_2^0(\mathbf{x}_2 = 800) = \min(200, 160, 200, 320) = 160$. This quantity is identified in the table by bold characters, as explained previously. The corresponding $\mathbf{u}_2^0(\mathbf{x}_2 = 800) = 0$.

The quantities for $\mathbf{x}_k > 2000$ represent extrapolations (no interpolations were necessary for the present discretization). However, the "extrapolated" values were not determined by some approximation technique but by calculation using the same equations for $\mathbf{x}_k \le 2000$. Since no approximation was used, it was necessary to compute such extrapolated quantities at $k = 2$ if they were required at some $k < 2$. For example, at $k = 1$, if $\mathbf{x}_1 = 800$, then for $\mathbf{u}_1 = 0$, say, $\mathbf{x}_2 = 1600$, so that $J_2^0(1600)$ is required. But at $k = 2$ and $\mathbf{x}_2 = 1600$, $\mathbf{x}_3 > 2000$ for all $\mathbf{u}_2$.

The preceding argument holds as long as the extrapolated states are within the allowable set. In principle, at stage k, any combination of input variables that results in a subsequent state is outside the set of allowable states $\mathbf{X}$ must be excluded from consideration in the determination of the cost. Although the present problem does not specifically indicate the set $\mathbf{X}$, if we interpret the desire to accumulate *as nearly as possible* 2000 money units to imply a quantity between 1800 and 2000 (note that such considerations also help to determine an adequate state discretization), then the admissible states can be specified as $\mathbf{X} = [0, 2000]$, and any $\mathbf{u}_k$ for which $\mathbf{x}_{k+1} > 2000$ will be excluded from consideration at stage k. This implies that all the quantities within the shaded area of Table 14.2 need not be computed. As can be observed from Table 14.2, this will, for the present problem, result in some vacant cells—allowable states for which there are no feasible control inputs. In the computer procedure developed subsequently, we shall assume that the allowable state and input sets are such that there is always at least one feasible input combination for every allowable node on the state grid. This assumption will be satisfied, for example, for the present problem under the condition of calculating instead of approximating the extrapolated values, if we take the allowable state set to be $\mathbf{X} = [0, 14600]$.

Returning to the explanation of the values in Table 14.2, we consider next stage $k = 1$. Suppose again $\mathbf{x}_1 = 800$, then the set of values for $f(\mathbf{x}_1, \mathbf{u}_1)$ and $h(\mathbf{x}_1, \mathbf{u}_1)$, $\mathbf{u}_1 = -200, 0, 200, 400$, are exactly the same as for $k = 2$ and do not have to be recalculated. For instance, if $\mathbf{u}_1 = -200$, then $\mathbf{x}_2 = 1400$ and $h(\mathbf{x}_1, \mathbf{u}_1) = -160$ (see the $k = 2$ column, row $\mathbf{x}_k = 800$). $J_{k+1}^0(\mathbf{x}_2)$ is obtained from the $k = 2$ column, row $\mathbf{x}_k = 1400$, that is, 80. Consequently, $J_1(\mathbf{x}_1, \mathbf{u}_1) = -160 + 80 = -80$. Similarly for $\mathbf{u}_1 = 0, 200$, and 400, we obtain $J_1(\mathbf{x}_1, \mathbf{u}_1) = 680, 1760$, and 3160, respectively, so that $J_1^0(\mathbf{x}_1 = 800) = -80$.

The procedure for $k = 0$ is exactly the same as for $k = 1$. However, since the initial state is specified, only the cells for the initial states of interest were completed for $k = 0$. As a result, it was necessary to complete the table only for $k = 1$, up to $\mathbf{x}_1 = 800$. Entries above this point would have required the extrapolation of the $k = 2$ column beyond $\mathbf{x}_2 = 2000$, and so on. However, for a general automatic procedure, as developed next, the computations should be carried out for all feasible states at $k = 0$ to obtain a truly general solution.

Solution—Optimal Control and Cost: Table 14.2 is actually the solution of the optimal control problem. However, the essential solution—the optimal control and cost for each allowable state at each initial stage, can be extracted from this solution, as shown in Table 14.3 The values in shaded cells in Table 14.3 are the additional quantities that will be available from Table 14.2 if, as we discussed earlier, the allowable states are extended sufficiently beyond $\mathbf{X} = [0, 2000]$ so that no cell for $0 \le \mathbf{x}_k \le 2000$ is completely empty.

Table 14.3 Optimal Control and Minimum Cost

x(k)	k = 0		k = 1		k = 2		k = 3
	$u^0(0)$	$J_0^0(x(0))$	$u^0(1)$	$J_1^0(x(1))$	$u^0(2)$	$J_2^0(x(2))$	$J_3^0(x(3))$
2000	−200	156120	−200	28000	−200	2840	0
1800	−200	118640	−200	20120	−200	1600	40
1600	−200	86280	−200	13520	−200	680	160
1400	−200	59040	−200	8200	−200	80	360
1200	−200	36920	−200	4160	−200	−200	640
1000	−200	19920	−200	1400	−200	−160	1000
800	−200	8040	−200	−80	0	160	1440
600	−200	1280	−200	−280	400	400	1960
400	−200	−360	200	−80	400	800	2560
200	200	−240	400	240	400	1520	3240
0	400	−80	400	800	400	2560	4000

Case (a):

Suppose $\mathbf{x}_0 = 0$. From Table 14.3 the minimum cost is −80, that is, $J_0^0(\mathbf{x}_0 = 0) = -80$, and the initial decision (optimal control) is $\mathbf{u}_0^0 = 400$. The balance of the optimal trajectory is as follows: By equation (14.47a) or using $f(\mathbf{x}_0, \mathbf{u}_0)$ from Table 14.2, $\mathbf{x}_1 = 400$. Using Table 14.3 again, we get $\mathbf{u}_1^0 = 200$. With this control, equation (14.47a) or Table 14.2 yields the next state, $\mathbf{x}_2 = 1000$, and this time Table 14.3 gives $\mathbf{u}_2^0 = -200$. The final state is $\mathbf{x}_3 = 1800$. Thus the optimal policy for a *new account* is to make an initial deposit of 400 followed by another deposit of 200 at the beginning of the next year and a withdrawal of 200 in the third year. The total "cost" for this strategy, which is the minimum relative to all other strategies for a new account, is $J = -80$.

Case (b):

If $\mathbf{x}_0 = 200$, the minimum cost is −240, and the optimal policy is $\mathbf{u}_0^0 = 200 \Rightarrow \mathbf{x}_1 = 600$; $\mathbf{u}_1^0 = -200 \Rightarrow \mathbf{x}_2 = 1000$; and $\mathbf{u}_2^0 = -200 \Rightarrow \mathbf{x}_3 = 1800$. In other words, with an initial balance of 200, the optimal policy is to make an initial deposit of 200 followed by withdrawals of 200 each in the following 2 years.

As we alluded to earlier, note that in both cases the error in the final state is within the resolution of the state observation determined by the discretization of the **x**-values.

The storage memory problem notwithstanding, the flexibility of digital computation of optimal control by dynamic programming recommends the application of the method to sampled data systems or even to continuous-time systems, where the procedure can be used to approximate the equivalent continuous-time optimal controller. This is especially true when it is possible to deduce from the dynamic programming results a closed-form expression (algorithm) for the optimal control. In this regard, note that dynamic programming results such as Table 14.3 are in fact representations of the optimal control as a function of the state. We now consider an example illustrating discrete-time optimal control of a multivariable (continuous-time) plant.

Example 14.8   Continuous to Discrete-Time Multivariable Example

Recall the continuous stirred tank reactor of Example 13.5, with the following second-order, two-input state model:

$$\frac{d}{dt}\begin{bmatrix} x_1 \\ x_2 \end{bmatrix} = \begin{bmatrix} -\left(k + \dfrac{w}{\rho V}\right) & 0 \\ -\left(\dfrac{Hk}{\rho c}\right) & -\left(\dfrac{w}{\rho V}\right) \end{bmatrix}\begin{bmatrix} x_1 \\ x_2 \end{bmatrix} + \begin{bmatrix} \left(\dfrac{w}{\rho V}\right) & 0 \\ 0 & \left(\dfrac{1}{\rho V c}\right) \end{bmatrix}\begin{bmatrix} u_1 \\ u_2 \end{bmatrix} \quad \textbf{(14.48a)}$$

$$= \begin{bmatrix} -2.0833 \cdot 10^{-2} & 0 \\ -1.1111 \cdot 10^{-2} & -8.3333 \cdot 10^{-4} \end{bmatrix}\begin{bmatrix} x_1 \\ x_2 \end{bmatrix}$$

$$+ \begin{bmatrix} 8.333 \cdot 10^{-4} & 0 \\ 0 & 5.556 \cdot 10^{-5} \end{bmatrix}\begin{bmatrix} u_1 \\ u_2 \end{bmatrix} = \mathbf{AX} + \mathbf{BU}$$

where the disturbance input term has been excluded from the present application.

We desire to drive this system from any initial state, such as ($x_1(0) = 0$, $x_2(0) = 295$), to within 5% of a final state ($\mathbf{x}_s = [60 \quad 350]^T$) in about 45 s, using a discrete-time optimal controller, subject to the following quadratic performance index:

$$J = \frac{1}{2}\Delta\mathbf{x}_N^T\mathbf{H}\Delta\mathbf{x}_N + \frac{1}{2}\sum_{i=0}^{N-1}(\Delta\mathbf{x}_i^T\mathbf{W}\Delta\mathbf{x}_i + 2\Delta\mathbf{u}_i^T\mathbf{M}\Delta\mathbf{x}_i + \Delta\mathbf{u}_i^T\mathbf{R}\Delta\mathbf{u}_i); \quad \textbf{(14.48b)}$$

$$\Delta\mathbf{x}_i = \mathbf{x}_i - \mathbf{x}_s; \qquad \Delta\mathbf{u}_i = \mathbf{u}_i - \mathbf{u}_s$$

where $\mathbf{x}_i$ is the *sampled* state vector or the state vector in the approximate discrete-time state model, $\mathbf{u}_i$ is the digital controller output, $\mathbf{u}_s$ is the steady-state input necessary to maintain a nonzero setpoint $\mathbf{x}_s$ (see Output Regulator Problem and Nonzero Setpoint, Section 13.4), and

$$\mathbf{H} = \begin{bmatrix} 1 & 0 \\ 0 & 1 \end{bmatrix}, \quad \mathbf{W} = \begin{bmatrix} 1 & 0 \\ 0 & 1 \end{bmatrix}, \quad \mathbf{M} = \begin{bmatrix} 0.75 & 0 \\ 0 & 0.25 \end{bmatrix}, \quad \mathbf{R} = \begin{bmatrix} 1.0 & 0 \\ 0 & 0.5 \end{bmatrix} \textbf{(14.48c)}$$

a. Obtain an approximate but adequate discrete-time state model of the continuous-time plant, as discussed in Section 6.5, using a sampling period of 5 s.

b. Develop computational mathematics programs to compute the optimal control of the equivalent discrete-time plant obtained in (a) by discrete-time dynamic programming

c. Generate the plant's response subject to this control.

Assume constraints on the state and control as follows:

$$0 \leq x_1 \leq 70; \qquad 295 \leq x_2 \leq 355 \quad \textbf{(14.48d)}$$

$$-2000 \leq u_1 \leq 19000; \qquad 17300 \leq u_2 \leq 107900 \quad \textbf{(14.48e)}$$

Solution a: As developed in Appendix B, equations (B4.3e) through (B4.3h) the discrete-time state model is approximately

$$\hat{\mathbf{x}}_{k+1} = \hat{\mathbf{P}}\hat{\mathbf{x}}_k + \hat{\mathbf{Q}}\hat{\mathbf{u}}_k; \qquad \hat{\mathbf{x}}_0 = \begin{bmatrix} 0 \\ 295 \end{bmatrix}$$

where

$$\hat{\mathbf{P}} = \mathbf{I} + \mathbf{A}T + \mathbf{A}^2\frac{T^2}{2!} + \mathbf{A}^3\frac{T^3}{3!} + \cdots; \qquad \hat{\mathbf{Q}} = T\left(\mathbf{I} + \mathbf{A}\frac{T}{2!} + \mathbf{A}^2\frac{T^2}{3!} + \cdots\right)\mathbf{B}$$

But from equation (14.48a),

$$\mathbf{A} = \begin{bmatrix} -2.0833 \cdot 10^2 & 0 \\ -1.1111 \cdot 10^{-2} & -8.3333 \cdot 10^{-4} \end{bmatrix}; \qquad \mathbf{A}^2 = 10^{-3}\begin{bmatrix} 0.4340 & 0 \\ 0.2407 & 0.0007 \end{bmatrix};$$

$$\mathbf{A}^3 = 10^{-5}\begin{bmatrix} -0.9042 & 0 \\ -0.5023 & -0.0001 \end{bmatrix}$$

so that with $T = 5$ and a desired accuracy in the results implied by the 5% final-value error, the approximation (B4.3i) will be adequate. That is,

$$\hat{\mathbf{P}} = \mathbf{I} + \mathbf{A}T = \begin{bmatrix} 0.8958 & 0 \\ -0.0556 & 0.9958 \end{bmatrix}; \qquad \hat{\mathbf{Q}} = T\mathbf{B} = 10^{-3}\begin{bmatrix} 4.1665 & 0 \\ 0 & 0.2778 \end{bmatrix}$$

However, for numerical convenience, we make the following change of variables:

$$\hat{\mathbf{x}} = 10\mathbf{x} \quad \text{and} \quad \hat{\mathbf{u}} = 10^4\mathbf{u}$$

and consider instead the discrete-time state model

$$\mathbf{x}_{k+1} = \mathbf{P}\mathbf{x}_k + \mathbf{Q}\mathbf{u}_k$$

or

$$\begin{bmatrix} x_1 \\ x_2 \end{bmatrix}_{k+1} = \begin{bmatrix} 0.8958 & 0 \\ -0.0556 & 0.9958 \end{bmatrix}\begin{bmatrix} x_1 \\ x_2 \end{bmatrix}_k + \begin{bmatrix} 4.1665 & 0 \\ 0 & 0.2778 \end{bmatrix}\begin{bmatrix} u_1 \\ u_2 \end{bmatrix}_k; \qquad \textbf{(14.49a)}$$

$$\begin{bmatrix} x_1 \\ x_2 \end{bmatrix}_0 = \begin{bmatrix} 0 \\ 29.5 \end{bmatrix}$$

The corresponding desired final state is

$$\mathbf{x}_s = \begin{bmatrix} 6 \\ 35 \end{bmatrix} \qquad\qquad\qquad \textbf{(14.49b)}$$

At steady state, a controlled discrete-time system satisfies

$$\begin{aligned} \mathbf{x}_s &= \mathbf{P}\mathbf{x}_s + \mathbf{Q}\mathbf{u}_s & \mathbf{x}_s &= (\mathbf{I} - \mathbf{P})^{-1}\mathbf{Q}\mathbf{u}_s \\ & \quad\text{or} \\ \mathbf{y}_s &= \mathbf{r}_d = \mathbf{C}\mathbf{x}_s & \mathbf{u}_s &= [\mathbf{C}(\mathbf{I} - \mathbf{P})^{-1}\mathbf{Q}]^{-1}\mathbf{r}_d \end{aligned} \qquad \textbf{(14.49c)}$$

Substituting $\mathbf{P}$, $\mathbf{Q}$, and $\mathbf{r}_d = \mathbf{x}_s$, from (14.49a) and (14.49b), and $\mathbf{C} = \mathbf{I}$, we obtain

$$\mathbf{u}_s = \begin{bmatrix} 0.15 \\ 1.73 \end{bmatrix} \tag{14.49d}$$

Solution b: A MATLAB S3.5 solution is described here. (A corresponding Mathcad 3.1 solution pair, **dynap.mcd** and **optcon.mcd,** developed by the author is also available **On the Net.**) The script M-file that computes the optimal control and minimum cost **dynap.m** is listed in the box. The number of state variables $n = 2$, inputs $r = 2$, and stages $N = 45/5 = 9$. We took $M = 11$ and $R = 7$, respectively, to be small but reasonable numbers of *quantization* levels of the state and input:

$$x_{1,k}^{<1,2,\ldots,11>} = 0, .7, 1.4, \ldots, 7; \tag{14.50a}$$

$$x_{2,k}^{<1,2,\ldots,11>} = 29.50, 30.05, 30.60, \ldots, 35.00$$

$$u_{1,k}^{<1,2,\ldots,7>} = -.2, .15, .50, \ldots, 1.9; \tag{14.50b}$$

$$u_{2,k}^{<1,2,\ldots,7>} = 1.73, 3.24, 4.75, \ldots, 10.79$$

This means that there are, in this case, (11×11), or 121, discrete nodes in the admissible state space and (7×7), or 49, combinations of permissible inputs (see **dynap.m** listing; the Mathcad program **dynap.mcd** actually uses one-dimensional representations of these combinations and has functions for converting from the two-dimensional to the one-dimensional indices and vice versa).

Note the relationship between the ranges of state variables and inputs and the constraint sets:

$$X_{1,1} = 0 \le x_{1,k} \le X_{1,2} = 7, \qquad X_{2,1} = 29.5 \le x_{2,k} \le X_{2,2} = 35.5 \tag{14.51a}$$

$$-0.2 \le u_{1,k} \le 1.9; \qquad 1.73 \le u_{2,k} \le 10.79 \tag{14.51b}$$

Every state on the state grid has at least one input combination for which the corresponding next state is within the constraint set. To show this, we express the requirement that the minimum and maximum values of the "next" state variables be within the constraint set and verify that each of the following resulting conditions can be satisfied by at least one allowable input value:

$$x_{1,k+1}(\min) = 0.8958x_{1,k}(\min) + 4.1665u_{1,k} = 4.1665u_{1,k} \ge 0$$

$$x_{2,k+1}(\min) = -0.0555x_{1,k}(\max) + 0.9958x_{2,k}(\min) + 0.2778u_{2,k}$$

$$= 28.9869 + 0.2778u_{2,k} \ge 29.5$$

$$x_{1,k+1}(\max) = 0.8958x_{1,k}(\max) + 4.1665u_{1,k} = 6.2706 + 4.1665u_{1,k} \le 7$$

$$x_{2,k+1}(\max) = -0.0555x_{1,k}(\min) + 0.9958x_{2,k}(\max) + 0.2778u_{2,k}$$

$$= 34.853 + 0.2778u_{2,k} \le 35.5$$

Initialization of minimum cost terms: Script M-file **dynap.m** uses the (perhaps time-consuming) strategy of recomputing the terms $h(\mathbf{x}_k, \mathbf{u}_k)$ and $\mathbf{f}(\mathbf{x}_k, \mathbf{u}_k)$ at each stage for every state and input node as required (**dynap.mcd** uses the alternative method of initializing and storing all three terms $H(\mathbf{x}_N)$, $h(\mathbf{x}_k, \mathbf{u}_k)$, and $\mathbf{f}(\mathbf{x}_k, \mathbf{u}_k)$). This leaves the minimum cost at N or $H(\mathbf{x}_N)$ as the only precomputed term. A function M-file **termH.m** is used for this purpose. Similarly, $h(\mathbf{x}_k, \mathbf{u}_k)$ and $\mathbf{f}(\mathbf{x}_k, \mathbf{u}_k)$ are obtained as needed by calling the function M-files **hsum.m** and **fkplus1.m**.

Script M-file **dynap.m**

```
% DYNAP.M                       by E.I. Umez-Eronini, 09-08-95. All rights reserved.
% Discrete-Time Optimal Control via Dynamic Programming Script File
% Uses function M-files, termH.m, fkplus1.m, hsum.m & Lookup.m
% The optimal control function (table) is written to the file -
% OPCON.MAT, using SAVE command and is read using LOAD command in
% the companion script M-file, fig14_12.m
%
clear;
% ENTER VARIOUS DATA VALUES
T = 5; N = 9; n = 2; M = 11; r = 2; R = 7;
rd = [6 35]'; us = [.15 1.73]';
x(1,1) = 0; x(2,1) = 29.5; dx = [0.7 .55];
X(1,1) = 0; X(2,1) = 29.5;                  % define state
X(1,2) = 7; X(2,2) = 35.5;                  % constraint set
u(1,1) = -.2; u(2,1) = 1.73; du = [0.35 1.51];
%
% DISCRETIZE AND STORE ADMISSIBLE STATE AND CONTROLS
for i = 1:n,
 for m = 2:M,
  x(i,m) = x(i,m-1)+dx(i);
 end;
end;
for q = 1:r,
 for j = 2:R,
  u(q,j) = u(q,j-1)+du(q);
 end;
end;
%
Lx1 = M*fix(1024/(M*M));                    % demarcation of output array
% INITIALIZE COST FUNCTION
for i = 1:M,
 for m = 1:M,
  Jold(i,m) = termH(x(1,i)-rd(1),x(2,m)-rd(2));
 end;
end;
%
% RECURSIVELY CALCULATE MINIMUM COST AND CONTROL
disp(' Long Running Program, Please Wait!');
for L = 1:N,
 gc_ = (L-1)*M;                             % group count
 for i = 1:M,
  ptr = i+gc_;                              % file pointer
  for m = 1:M,
   xk = [x(1,i) x(2,m)]';
   J0(i,m) = 1E6;                           % a very large number
   for q = 1:R,
    for j = 1:R,
     uk = [u(1,q) u(2,j)]';
     y = fkplus1(xk,uk);                    % get next state
```

continued

```
% use this input only if state belongs to constraint set
    if (y(1)>=X(1,1))&(y(1)<=X(1,2))&(y(2)>=X(2,1))&(y(2)<=X(2,2)),
    Jkpl = hsum(xk-rd,uk-us) + lookup(y,M,x,Jold);
      if Jkpl < J0(i,m),
        J0(i,m) = Jkpl; uk0(1) = uk(1); uk0(2) = uk(2);    % store result
      end;
     end;
    end;                                    % j-loop
   end;                                     % q-loop
   if ptr <= Lx1,
     u01(ptr,m) = uk0(1); u02(ptr,m) = uk0(2);
   else
     px = ptr-Lx1; u11(px,m) = uk0(1); u12(px,m) = uk0(2);    % store result
   end;
  end;                                      % m-loop
 end;                                       % i-loop
 Jold = J0;
end;                                        % next stage, L-loop
save opcon u01 u02 u11 u12;                 % save optimal control function on disk
```

Interpolation/extrapolation, $J_{k+1}^0(\mathbf{x}_{k+1})$: Since $\mathbf{f}(\mathbf{x}_k, \mathbf{u}_k)$ values are, potentially, not on the state nodes, **dynap.m** (as well as **dynap.mcd**) interpolates or extrapolates all $J_{k+1}^0(\mathbf{x}_{k+1})$ values, as long as the corresponding state $\mathbf{x}_{k+1} = \mathbf{f}(\mathbf{x}_k, \mathbf{u}_k)$ is admissible. A first-order *Lagrange*-type two-variable interpolation formula is used. Since interpolation/extrapolation can also be required in the optimal control versus state function (table) during application of control to the plant, the interpolation formula is developed as a general function M-file **Lookup.m**, which requires the data to be interpolated to be passed to it as one of the arguments.

Optimal control function: The optimization result is saved as a disk file consisting of concatenated arrays of $M \times M$ values of each input variable at successive stages. That is, each input value is a function of the $M \times M$ combinations of state variables or state nodes. Because an array in MATLAB S3.5 is limited to 1024 elements, a facility is provided in **dynap.m** to split the optimal control function into an appropriate number of array variables. For $N = 9$, two such array variables ($u01$, $u02$, and $u11$, $u12$, respectively) were required to hold the $11 \times (9 \times 11) = 1089$ elements of each input function.

Subroutines: The listings of the function M-files follow.

Function M-file **termH.m**

```
function H = termH(x,y)
% Computes the terminal term in the cost function
% Returns a scalar value, given the state variables
% Replace below with your own function
H = 0.5*(x*x + y*y);
```

Solution c: Application of the dynamic programming results simply requires retrieval of the control versus state function, setting the initial condition, and subsequently interpolating the control for each state. However, to interpolate the control "table" the

Function M-file **fkplus.m**

```
function y = fkplus1(x,u)
% FKPLUS1 computes the vector y = x(k+1)
%       given x(k) and u(k) and the formula
%       below: x(k+1) = f(x(k),u(k))
%
p11 = 0.8958; p21 = -0.0556; p22 = 0.9958;
q11 = 4.1665; q22 = 0.2778;
y = zeros(2);
y(1) = p11*x(1) + q11*u(1);

y(2) = p21*x(1) + p22*x(2) + q22*u(2);
```

Function M-file **hsum.m**

```
function h = hsum(x,u)
% HSUM computes the integral term h(x,u)
%       of the cost function, given x and u
%       and the scalar valued functions below.
%
xqx = x(1)*x(1) + x(2)*x(2);
umx = .75*u(1)*x(1) + .25*u(2)*x(2);
uru = u(1)*u(1) + 0.5*u(2)*u(2);
h = 0.5*(xqx + uru) + umx;
```

original state space grid has to be restored. At each stage, the *lookup* function is called twice to interpolate for the two input variables. These actions are implemented in the script M-file **fig14_12.m**. Note that in **Dynap.m** the control function is stored with the initial controls ($k = 0$) last. Thus the control data are retrieved bottom-to-top. (**Dynap.mcd** stores the control data in the strict time sequence.)

Figure 14.12 shows the response of the controlled system. Since the simulation interval is greater than the computed control interval (200 versus 45 s), the control table is simply cycled through continuously. This approach is also required when the control objective is not met at the conclusion of the optimization time interval. By repeated cycling through the optimal control data, the desired solution is approached, it is hoped, asymptotically. However, for the present problem, the control objective is accomplished well within the optimization time interval, although it should be noted that it is the response of the equivalent (approximate) discrete-time model that is simulated, not the actual continuous-time plant. Further, the control action will be subjected to zero-order hold prior to being applied to the plant. Nonetheless, the performance of the optimal controller (recall the results for state vector feedback design in Fig. 13.13) is remarkable.

The Discrete-Time Linear Quadratic Problem

For a *linear plant* with a *quadratic performance index* (equation (14.45g)) subject to *no constraints of state and control,* the dynamic programming solution of the optimal control problem—*the linear quadratic* (LQ) problem—is an analytical or closed-form algorithm that is a state- or output-vector feedback control. This result also follows directly from the recursive algorithm, equation (14.46b). We shall examine first the

Function M-file **Lookup.m**

```
function r = lookup(y,M,x,f)
% given the grid points: x(M,M) and
% the values of a function at the grid
% points: f(M,M), LOOKUP uses linear
% interpolation (Lagrange-type) to
% approximate f(y) at a given, possibly
% none-grid, point y. This value is r.
%
% DETERMINE INTERPOLATION POINTS
for k = 1:2,
m = 1;
while (x(k,m)<=y(k)) & (m<M),
m = m+1;
end;
i2(k) = max([2 m]);
m = M;
while (x(k,m)>=y(k)) & (m>1),
m = m-1;
end;
i1(k) = m;
if i1(k) ==i2(k),
i1(k) =i2(k) -1;
end;
end;
% LOOKUP VALUE
d1 = (x(2,i1(2))-x(2,i2(2)))*(x(1,i1(1))-x(1,i2(1)));
d2 = (x(1,i2(1))-x(1,i1(1)))*(x(2,i1(2))-x(2,i2(2)));
c1 = f(i1(1),i1(2))*(y(1)-x(1,i2(1)))*(y(2)-x(2,i2(2)))/d1;
c2 = f(i1(1),i2(2))*(y(1)-x(1,i2(1)))*(y(2)-x(2,i1(2)))/
     (-d1);
c3 = f(i2(1),i1(2))*(y(1)-x(1,i1(1)))*(y(2)-x(2,i2(2)))/d2;
c4 = f(i2(1),i2(2))*(y(1)-x(1,i1(1)))*(y(2)-x(2,i1(2)))/
     (-d2);
r = c1 + c2 + c3 + c4;
```

problem of regulation about the *origin* when complete state vector feedback is feasible and then extend the result to the *time-invariant output regulator* with *nonzero setpoint* problem.

Consider again the discrete-time state model

$$\mathbf{x}_{k+1} = \mathbf{P}_k\mathbf{x}_k + \mathbf{Q}_k\mathbf{u}_k \tag{14.52a}$$

where $\mathbf{x}_0$ is given and $\mathbf{x}_k$ and $\mathbf{u}_k$ are not constrained. Let the performance index to be minimized be given by equation (14.45g), repeated here:

$$J = \frac{1}{2}\mathbf{x}_N^T\mathbf{H}\mathbf{x}_N + \frac{1}{2}\sum_{i=0}^{N-1}\{\mathbf{x}_i^T\mathbf{W}_i\mathbf{x}_i + 2\mathbf{u}_i^T\mathbf{M}_i\mathbf{x}_i + \mathbf{u}_i^T\mathbf{R}_i\mathbf{u}_i\} \tag{14.52b}$$

Note that for regulation about the origin, $\mathbf{x}_N = \mathbf{0}$, so that $\mathbf{H} = \mathbf{0}$. However, we want to retain this terminal term for application to the nonzero setpoint case. For finite $\mathbf{x}_N$, the

Script M-file **Fig14_12.m**

```
% Fig14_12.m; uses function M-files, fkplus1.m and Lookup.m
% Simulates response of a discrete-time plant subject to Optimal Control as
% computed by Dynap.m using Dynamic Programming. The control
% data is retrieved from the disk file OPCON.MAT using the LOAD command
clear;
% RESTORE REQUIRED DATA VALUES
T = 5; N = 9; n = 2; M = 11; x(1,1) = 0; x(2,1) = 29.5; dx = [0.7 .55];
Lx1 = M*fix(1024/(M*M)); fu1 = zeros(M,M); fu2 = zeros(M,M);
t(1) = 0; xk(1,1) = 0; xk(2,1) = 29.5;        % initial state
Nmx = 40; Nm = Nmx+1; Lt = (N-1)*M+1;        % Nmx = simulation stages
% RESTORE DISCRETIZED STATE DATA FOR USE IN INTERPOLATION
for i = 1:n,
 for m = 2:M,
  x(i,m) = x(i,m-1)+dx(i);
 end;
end;
load opcon;                                   % load optimal control
table
for k = 1:Nm,                                 % BEGIN
 k1 = k+1; t(k1) = t(k)+T;                    % time
 if Lt > Lx1,
  pt = Lt-Lx1;
  for i = 1:M,
   for m = 1:M,
    fu1(i,m) = u11(pt,m); fu2(i,m) = u12(pt,m);
   end;
   pt = pt+1;
  end;
 else
   pt = Lt;
   for i = 1:M,
    for m = 1:M,
     fu1(i,m) = u01(pt,m); fu2(i,m) = u02(pt,m);
    end;
    pt = pt+1;
   end;
 end;
 uk(1,k) = lookup([xk(1,k) xk(2,k)]',M,x,fu1);
 uk(2,k) = lookup([xk(1,k) xk(2,k)]',M,x,fu2);
 y = fkplus1([xk(1,k) xk(2,k)]',[uk(1,k) uk(2,k)]');
 xk(1,k1) = y(1); xk(2,k1) = y(2); Lt = Lt-M;
 if Lt <1,
  Lt = (N-1)*M+1;
 end;
end;
axis('square'); axis([0 200 0 40]); plot(t,xk(1,:),'r'); hold on;
plot(t,xk(2,:),'g'); plot(t(1:Nm),uk(1,:),':r');
plot(t(1:Nm),uk(2,:),':g');
ylabel('Response and Control'); xlabel('Time - sec');
title('Fig. 14.12 Optimal control by dynamic programming');
text(5,.4,'u1'); text(30,2,'u2'); text(60,6,'x1'); text(100,35,'x2');
hold off;
```

Figure 14.12

Optimal Control by Dynamic Programming

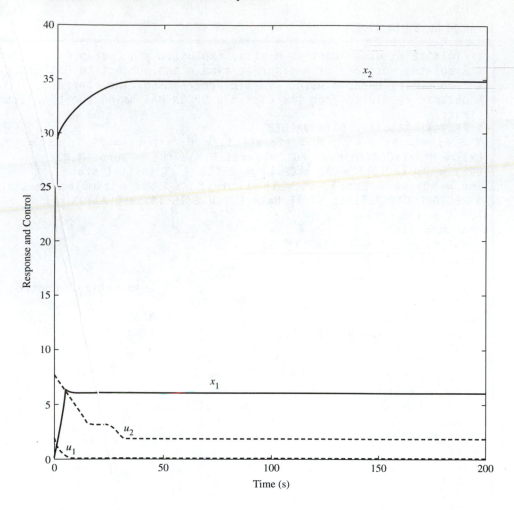

requirement that the minimum J with respect to unconstrained $\mathbf{u}_i$ remain finite leads to the condition that $\mathbf{W}_i - \mathbf{M}_i^T \mathbf{R}_i^{-1} \mathbf{M}_i$ must be *positive semidefinite*.

Applying equations (14.46b) and (14.46c) for $k = N - 1$ to the above cost functional, we have

$$J_{N-1}^0(\mathbf{x}_{N-1}) = \min_{\mathbf{u}_{N-1}} \frac{1}{2} \{ \mathbf{x}_{N-1}^T \mathbf{W}_{N-1} \mathbf{x}_{N-1} + 2\mathbf{u}_{N-1}^T \mathbf{M}_{N-1} \mathbf{x}_{N-1}$$
$$+ \mathbf{u}_{N-1}^T \mathbf{R}_{N-1} \mathbf{u}_{N-1} + \mathbf{x}_N^T \mathbf{H} \mathbf{x}_N \}$$

Substituting equation (14.52a) for $\mathbf{x}_N$, we obtain

$$J_{N-1}^0(\mathbf{x}_{N-1}) = \min_{\mathbf{u}_{N-1}} \frac{1}{2} [\mathbf{x}_{N-1}^T \mathbf{W}_{N-1} \mathbf{x}_{N-1} + 2\mathbf{u}_{N-1}^T \mathbf{M}_{N-1} \mathbf{x}_{N-1} + \mathbf{u}_{N-1}^T \mathbf{R}_{N-1} \mathbf{u}_{N-1}$$
$$+ (\mathbf{P}_{N-1} \mathbf{x}_{N-1} + \mathbf{Q}_{N-1} \mathbf{u}_{N-1})^T \mathbf{S}_N (\mathbf{P}_{N-1} \mathbf{x}_{N-1} + \mathbf{Q}_{N-1} \mathbf{u}_{N-1})]$$
$$= \frac{1}{2} [\mathbf{x}_{N-1}^T (\mathbf{W}_{N-1} + \mathbf{P}_{N-1}^T \mathbf{S}_N \mathbf{P}_{N-1}) \mathbf{x}_{N-1}] + \min_{\mathbf{u}_{N-1}} \frac{1}{2} \{ 2\mathbf{u}_{N-1}^T (\mathbf{M}_{N-1}$$
$$+ \mathbf{Q}_{N-1}^T \mathbf{S}_N \mathbf{P}_{N-1}) \mathbf{x}_{N-1} + \mathbf{u}_{N-1}^T (\mathbf{R}_{N-1} + \mathbf{Q}_{N-1}^T \mathbf{S}_N \mathbf{Q}_{N-1}) \mathbf{u}_{N-1} \}$$

where we have defined

$$\mathbf{S}_N = \mathbf{H} \tag{14.52c}$$

Since $\mathbf{u}_{N-1}$ is unconstrained, the quantity $\{\cdot\}$ can be minimized by taking its partial derivative with respect to $\mathbf{u}_{N-1}$ and setting this quantity to zero. The result is the *state vector feedback control:*

$$\mathbf{u}_{N-1}^0 = -[\mathbf{R}_{N-1} + \mathbf{Q}_{N-1}^T\mathbf{S}_N\mathbf{Q}_{N-1}]^{-1}[\mathbf{M}_{N-1} + \mathbf{Q}_{N-1}^T\mathbf{S}_N\mathbf{P}_{N-1}]\mathbf{x}_{N-1} \tag{14.52d}$$

or

$$\mathbf{u}_k^0 = -[\mathbf{R}_k + \mathbf{Q}_k^T\mathbf{S}_{k+1}\mathbf{Q}_k]^{-1}[\mathbf{M}_k + \mathbf{Q}_k^T\mathbf{S}_{k+1}\mathbf{P}_k]\mathbf{x}_k = -\mathbf{K}_k\mathbf{x}_k \tag{14.53a}$$

To determine $\mathbf{S}_{k+1}$, we substitute equation (14.52d) back into the expression for $J_{N-1}^0(\mathbf{x}_{N-1})$ to get

$$J_{N-1}^0(\mathbf{x}_{N-1}) = \frac{1}{2}\mathbf{x}_{N-1}^T\{\mathbf{W}_{N-1} + \mathbf{P}_{N-1}^T\mathbf{S}_N\mathbf{P}_{N-1} - (\mathbf{M}_{N-1} + \mathbf{Q}_{N-1}^T\mathbf{S}_N\mathbf{P}_{N-1})^T$$
$$(\mathbf{R}_{N-1} + \mathbf{Q}_{N-1}^T\mathbf{S}_N\mathbf{Q}_{N-1})^{-1}(\mathbf{M}_{N-1} + \mathbf{Q}_{N-1}^T\mathbf{S}_N\mathbf{P}_{N-1})\}\mathbf{x}_{N-1}$$

If we define

$$J_{N-1}^0(\mathbf{x}_{N-1}) = \frac{1}{2}\mathbf{x}_{N-1}\mathbf{S}_{N-1}\mathbf{x}_{N-1} \tag{14.53b}$$

we obtain the *discrete matrix Riccati equation,*

$$\mathbf{S}_k = \mathbf{W}_k + \mathbf{P}_k^T\mathbf{S}_{k+1}\mathbf{P}_k$$
$$- (\mathbf{M}_k + \mathbf{Q}_k^T\mathbf{S}_{k+1}\mathbf{P}_k)^T(\mathbf{R}_k + \mathbf{Q}_k^T\mathbf{S}_{k+1}\mathbf{Q}_k)^{-1}(\mathbf{M}_k + \mathbf{Q}_k^T\mathbf{S}_{k+1}\mathbf{P}_k) \tag{14.53c}$$

which can be solved recursively starting with (14.52c)

Stationary Case and Output Regulator with Nonzero Setpoint

If the matrices $\mathbf{W}$, $\mathbf{P}$, $\mathbf{M}$, $\mathbf{Q}$, and $\mathbf{R}$ are not functions of (time) k, and we consider, as in the continuous-time case, the *infinite-time* problem $N \longrightarrow \infty$, then the discrete matrix Riccati equation (14.53c) has a steady-state solution $\mathbf{S}$, and the optimal feedback gain defined in equation (14.53a) is time invariant:

$$\mathbf{K} = [\mathbf{R} + \mathbf{Q}^T\mathbf{S}\mathbf{Q}]^{-1}[\mathbf{M} + \mathbf{Q}^T\mathbf{S}\mathbf{P}] \tag{14.54a}$$

provided the system (14.52a) is *completely* controllable. $\mathbf{S}$ can be obtained from (14.53c) by an iterative solution of the implicit equation

$$\mathbf{S} = \mathbf{W} + \mathbf{P}^T\mathbf{S}\mathbf{P} - (\mathbf{M} + \mathbf{Q}^T\mathbf{S}\mathbf{P})^T(\mathbf{R} + \mathbf{Q}^T\mathbf{S}\mathbf{Q})^{-1}(\mathbf{M} + \mathbf{Q}^T\mathbf{S}\mathbf{P}) \tag{14.54b}$$

Output Regulation about a Nonzero Setpoint

Consider the time-invariant linear discrete-time system (14.52a) with the output equation

$$\mathbf{y}_k = \mathbf{C}\mathbf{x}_k \tag{14.55a}$$

We assume that this system is completely controllable and observable (the complete state vector $\mathbf{x}_k$ can be reconstructed from the measurement $\mathbf{y}_k$). As explained in Section 13.4, regulation of the system about a nonzero reference state $\mathbf{r}_d$ requires a steady-state controlling input $\mathbf{u}_s$ to maintain the final output equal to the reference value $\mathbf{y}_s = \mathbf{r}_d$. Equation (14.49c) gives $\mathbf{u}_s$ in terms of the setpoint, assuming the indicated inverse matrices exist.

Consider then a general quadratic performance index for optimal regulation of the system, or equation (14.48b) repeated:

$$J = \frac{1}{2}\Delta\mathbf{x}_N^T\mathbf{H}\Delta\mathbf{x}_N + \frac{1}{2}\sum_{i=0}^{N-1}(\Delta\mathbf{x}_i^T\mathbf{W}\Delta\mathbf{x}_i + 2\Delta\mathbf{u}_i^T\mathbf{M}\Delta\mathbf{x}_i + \Delta\mathbf{u}_i^T\mathbf{R}\Delta\mathbf{u}_i); \qquad \textbf{(14.55b)}$$

$$\Delta\mathbf{x}_i = \mathbf{x}_i - \mathbf{x}_s; \qquad \Delta\mathbf{u}_i = \mathbf{u}_i - \mathbf{u}_s$$

The preceding results can be shown to apply to the new problem as follows: Write the state equation for the deviation system,

$$\begin{aligned}\Delta\mathbf{x}_{k+1} = \mathbf{x}_{k+1} - \mathbf{x}_s &= \mathbf{P}\mathbf{x}_k + \mathbf{Q}\mathbf{u}_k - \mathbf{x}_s \\ &= \mathbf{P}(\mathbf{x}_k - \mathbf{x}_s) + \mathbf{Q}(\mathbf{u}_k - \mathbf{u}_s) + \mathbf{P}\mathbf{x}_s + \mathbf{Q}\mathbf{u}_s - \mathbf{x}_s \\ &= \mathbf{P}\Delta\mathbf{x}_k + \mathbf{Q}\Delta\mathbf{u}_k + (\mathbf{P}-\mathbf{I})\mathbf{x}_s + \mathbf{Q}\mathbf{u}_s\end{aligned}$$

But by equation (14.49c),

$$(\mathbf{P}-\mathbf{I})\mathbf{x}_s + \mathbf{Q}\mathbf{u}_s = 0$$

so that

$$\Delta\mathbf{x}_{k+1} = \mathbf{P}\Delta\mathbf{x}_k + \mathbf{Q}\Delta\mathbf{u}_k \qquad\qquad\qquad \textbf{(14.55c)}$$

The pair of equations (14.55c) and (14.55b) is now identical with the pair (14.52a) and (14.52b), so that the optimal control is given by $\Delta\mathbf{u}_k^0 = -\mathbf{K}\Delta\mathbf{x}_k$, where $\mathbf{K}$ is given by equations (14.54a) and (14.54b). Further, substituting the definitions of $\Delta\mathbf{u}$ and $\Delta\mathbf{x}$, we get

$$\mathbf{u}_k^0 = -\mathbf{K}\mathbf{x}_k + \mathbf{K}\mathbf{x}_s + \mathbf{u}_s \qquad\qquad\qquad \textbf{(14.55d)}$$

Or using equation (14.49c), we have

$$\begin{aligned}\mathbf{u}_k^0 &= -\mathbf{K}\mathbf{x}_k + [\mathbf{I} + \mathbf{K}(\mathbf{I}-\mathbf{P})^{-1}\mathbf{Q}][\mathbf{C}(\mathbf{I}-\mathbf{P})^{-1}\mathbf{Q}]^{-1}\mathbf{r}_d \\ &= -\mathbf{G}_{FB}\mathbf{x}_k + \mathbf{G}_{FF}\mathbf{r}_d\end{aligned} \qquad \textbf{(14.55e)}$$

The optimal control can be viewed as a combination of feedback and feedforward controls, as suggested in equation (14.55e).

Example 14.9   Linear Quadratic Regulation

Consider again the equivalent linear discrete-time optimal control system of Example 14.8, equation (14.49a). Let $\mathbf{y}_k = \mathbf{x}_k$ and $\mathbf{r}_d = [6 \quad 35]^T$, and suppose the state and input variables are unconstrained. Determine the optimal control and simulate the response of the controlled system for the same time interval as in Example 14.8, assuming the same quadratic performance index given by equations (14.48b) and (14.48c). Compare your results with those of Example 14.8.

Solution: The optimal control problem as stated is an output regulator problem with nonzero setpoint. However, there is complete state measurement $\mathbf{C} = \mathbf{I}$. Thus following the developments above, the optimal control is given by equations (14.54a), (14.54b), and (14.55d). We note that

$$\mathbf{W} - \mathbf{M}^T\mathbf{R}^{-1}\mathbf{M} = \begin{bmatrix} 1 & 0 \\ 0 & 1 \end{bmatrix} - \begin{bmatrix} 0.75 & 0 \\ 0 & 0.25 \end{bmatrix}\begin{bmatrix} 1 & 0 \\ 0 & 2 \end{bmatrix}\begin{bmatrix} 0.75 & 0 \\ 0 & 0.25 \end{bmatrix}$$

$$= \begin{bmatrix} 0.4375 & 0 \\ 0 & 0.8750 \end{bmatrix}$$

is positive definite.

Implementation: We also consider here a MATLAB simulation. The iterative solution of equation (14.54b) is implemented in an independent subroutine or function M-file **dricati.m,** which can return either the unsteady solution $\mathbf{S}_k$ (given $\mathbf{S}_{k+1}$) or the steady-state solution $\mathbf{S}$ (given $\mathbf{H}$), according to the value of one of the arguments passed to it. Also returned by the subroutine are the corresponding optimal feedback gain matrix $\mathbf{K}_k$ (or $\mathbf{K}$) and an indication of whether the steady-state solution converged or not. The subroutine listing is as follows:

Function M-file **dricati.m:**

```
function [s,k,c]=dricati(w,p.m,q,r,H,g)
% Dricati.m recursively obtains the unsteady (g=0) or
% steady (g=1) solution of the Discrete Matrix Riccati Equation
%       S = W + P'SP - (M+Q'SP)'inv(R+Q'SQ)(M+Q'SP)
% It returns the solution s, the feedback gain K such
% that the control law u = -Kx minimizes the cost function
%       J = x(N)'Hx(N)+sum(i)[x(i)'Wx(i)+2u(i)'Mx(i)+u(i)'Ru(i)]
% and c = NUMBER OF ITERATIONS if solution converged or c = 0 if no convergence
nmax = 100;                              % limit on # of iterations
n = length(p); et = .00001;              % n = system order, et = tolerance
s = H;                                   % initial solution, S(N) or S(k+1) = H
pT = p'; qT = q'; L = 1; c = 0; cvg = 0;
while cvg < 1,
  s1 = s; cvg = 1; f = (r + qT*s1*q)\(m+qT*s1*p);
  s = w + pT*s1*p - (m+qT*s1*p)'*f; d = abs(s-s1);
  for i = 1:n,
    for j = 1:n,
      if d(i,j) > et,
        cvg = 0;
      end;
    end;
  end;
  L = L+1;
  if (L <= 2) & (g == 1),
    cvg = 0;
  elseif (L > nmax) | (g == 0),
    cvg = 1;
  end;
end;
k = (r+qT*s*q)\(m+qT*s*p);
if L <= nmax,
  c = 1;
end;
```

The solution for **S** and **K** obtained by the preceding program after 15 iterations is

$$
\mathbf{S} = \begin{bmatrix} 0.7059 & -0.0652 \\ -0.0652 & 1.9908 \end{bmatrix}; \quad \mathbf{K} = \begin{bmatrix} 0.2563 & -0.0134 \\ -0.0423 & 1.2235 \end{bmatrix}
$$

Note that **dricati.m** does not check for singularity of the matrix $(\mathbf{R} + \mathbf{Q}^T\mathbf{SQ})$, so that proper attention should be paid to the matrix arguments passed to the subroutine. A typical situation is the case $\mathbf{M} = \mathbf{R} = \mathbf{H}\,(= \mathbf{S}_{k+1}) = \mathbf{0}$. From equation (14.53c) one should in this case directly take $\mathbf{S}_k = \mathbf{W}_k$ for the unsteady problem or enter **dricati.m** with $\mathbf{H} = \mathbf{W}$ for the steady-state problem.

In the present problem, the steady-state optimal feedback matrix is determined once and then applied at each stage to compute the control according to equation (14.55d). The control is then substituted in equation (14.49a) to obtain the corresponding next state. These operations are implemented in a script M-file **fig14_13.m** as follows:

```
% Fig14_13.m
% Script file to generate Fig. 14.13
% Uses function M-files dricati.m and fkplus1.m
clear;
% ENTER VARIOUS DATA VALUES
T = 5; N1 = 41; i2 = eye(2); H = i2; W = i2;
P = [.8958 0;-.0556 .9958]; Q = [4.1665 0;0 .2778]; C = i2;
R = [1 0;0 .5]; M = [.75 0;0 .25]; rd = [6 35]';
[S,Kfb,cv] = dricati(W,P,M,Q,R,H,1);        % obtain optimum steady state gain
if cv < 1,                                  % abort program if there's no solution
  error ('Riccati equation did not converge');
end;
Kff = (i2+Kfb*inv(i2-P)*Q)*inv(C*inv(i2-P)*Q);  % feedforward gain
% BEGIN COMPUTATION OF CONTROLLED RESPONSE
t(1) = 0; xk(1,1) = 0; xk(2,1) = 29.5;      % initial state
for k = 1:N1,
  k1 = k+1; t(k1) = t(k)+T;                  % time
  x = [xk(1,k) xk(2,k)]'; u0 = -Kfb*x+Kff*rd;  % optimal control
  y = fkplus1(x,u0');                        % compute next state
  uk(1,k) = u0(1); uk(2,k) = u0(2); xk(1,k1) = y(1); xk(2,k1) = y(2);
end;
axis('square'); axis([0 200 0 40]);
plot(t,xk(1,:),'r'); hold on; plot(t,xk(2,:),'g');
plot(t(1:N1),uk(1,:),':r'); plot(t(1:N1),uk(2,:),':g');
ylabel('Response and Control'); xlabel('Time - sec');
title('Fig. 14.13 Linear quadratic regulator solution');
text(5,.4,'u1'); text(30,2,'u2'); text(60,6,'x1'); text(100,35,'x2');
hold off;
```

Results: Figure 14.13 shows the response of the controlled system as well as the optimal control in the same format as that of Fig. 14.12. Although the present results represent an unconstrained optimization, the two figures are very close. This is as anticipated, since the constraint sets in Example 14.8 were chosen to be essentially be-

Figure 14.13

Linear Quadratic Regulator Solution

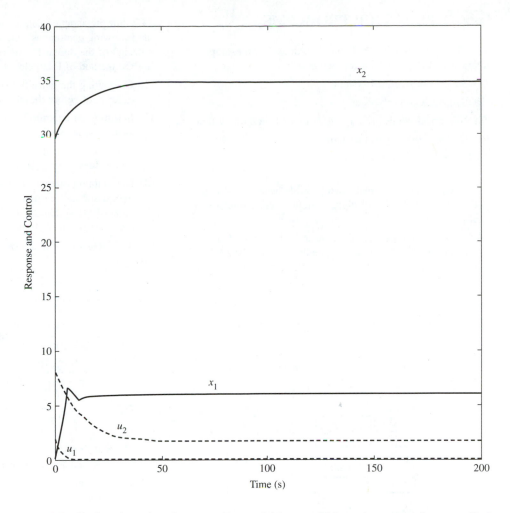

yond the limit values that the respective variables would have in a properly controlled response from the given initial state.

14.5 PRACTICE PROBLEMS

14.1 Using a detailed block diagram, describe on example application of each of the following:

(a) Supervisory control system

(b) Sequence control system

(c) Numerical control (NC) system or CNC control system

(d) Direct digital control system

14.2 Infer at least two advantages of a distributed control system over a centralized system.

14.3 Determine and explain two possible benefits and two potential disadvantages of digital computer control relative to analog control.

14.4 Consider the design of the discrete-time controller shown in Fig. 14.14. Show specifically for load disturbance that with $R(z)$ set to zero

$$Y(z) = \frac{Z\{\mathcal{L}^{-1}[G_p(s)N(s)]\}}{1 + G_p(z)D(z)}$$

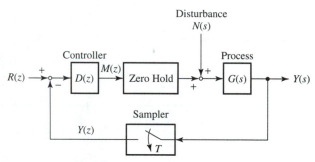

Figure 14.14 (Problem 14.4)

14.5* Consider again the discrete-time system of Problem 10.19. What is the closed-loop pulse transfer function? Determine by simulation or otherwise an adequate sampling period and a value of the controller parameter k that minimizes the IAE.

14.6 In Example 14.1 it was suggested that a sampling period $T = 0.1$ is probably more consistent with the time constants of the servomechanism. Repeat the computer-assisted parameter tuning of the PI-controller of the antiaircraft servomechanism of Example 14.1, and the corresponding tuning by Takahashi, Chan, and Auslander rules of Examples 14.2, using a sampling period $T = 0.1$. Compare your optimum controller settings and performance (ISE) with the results of Examples 14.1 and 14.2. MATLAB users should first determine the optimum performance for $T = 0.2$ for the controller settings of both examples, using script M-file exmp14_1.m or sxmp14_1.m as applicable to their situation.

14.7 For the sampled-data control system (Fig. 14.3a) with the following continuous-time plant transfer functions, obtain a design of the digital PI-controller equation (14.3b) following the method of Example 14.1, that is,

(1) Computing the pulse transfer function assuming a zero-order hold before the plant.

(2) Selecting an appropriate sampling period based on system dynamics.

(3) Using the Jury's inners stability test to establish allowable values of controller parameters.

(4) Determining optimal values of these parameters by hybrid simulation (that is, simulating the continuous parts of the system as such while also simulating the discrete components as discrete-time systems).

Use the integrated square error (ISE) as the tuning criterion:

(a) $G_p(s) = \dfrac{1}{s(1 + 0.2s)}$ **(b)** $G_p(s) = \dfrac{1}{(s + 1)(0.5s + 1)}$

(c) $G_p(s) = \dfrac{s + 4}{(s + 3)(s + 5)}$ **(d)** $G_p(s) = \dfrac{0.5}{(s + 1)^2(s + 5)}$

(e) $G_p(s) = \dfrac{10}{(1 + s)(1 + 0.25s)^2}$

(f) $G_p(s) = \dfrac{1}{(s + 10)(s^2 + 6s - 16)}$

14.8 Repeat Problem 14.7 when applicable, using Takahashi, Chan, and Auslander recommendations, that is,

(1) Select an appropriate sampling period based on system dynamics.

(2) Determine by direct simulation of direct digital proportional control (DDPC) of the plant the stability limit proportional controller gain k_u and the period of sustained oscillation T_u for all observed oscillatory patterns, if any exists.

(3) For the plants exhibiting a stability limit under DDPC, use the parameters above corresponding to the fundamental oscillatory mode and the Takahashi, Chan, and Auslander parameter tuning rules to determine the controller parameters.

Consider both the Euler and Trapezoidal algorithms.

14.9 Using a sampling period of $T = 0.25$, determine the Takashashi, Chan, and Auslander P-, PI-, and PID-parameter tuning recommendations (Table 14.1) for

(a) The system of Example 11.2 using the transient response shown in Fig. 11.6a.

(b) The system whose transient response is given in Problem 11.17.

14.10 Repeat Problem 14.7 for PID-control.

14.11 Repeat Problem 14.8 for PID-control.

14.12* In the z-domain design approach due to Kalman,† an optimum step response is considered one with minimum settling time or overshoot and no steady-state error. Such a response can be obtained, as shown in Fig. 14.15, in a minimum of two sampling periods if the manipulated variable takes on two intermediate values, m_0 and m_1, and a final value m_f (see Fig. 14.3c) that should be equal to the ratio of the desired final output value y_f to the plant gain K.

(a) With reference to the system block diagram in Fig. 14.3d show that for any step input of magnitude r_0

(i) $\dfrac{1}{r_0}Q(z) = \dfrac{M(z)}{R(z)} = \dfrac{1}{r_0}(q_0 + q_1 z^{-1} + q_2 z^{-2})$ (1)

$\qquad = \dfrac{1}{r_0}[m_0 + (m_1 - m_0)z^{-1} + (m_f - m_1)z^{-2}]$

(ii) $\dfrac{1}{r_0}P(z) = \dfrac{Y(z)}{R(z)} = \dfrac{1}{r_0}(p_1 z^{-1} + p_2 z^{-2})$ (2)

$\qquad = \dfrac{1}{r_0}[y_1 z^{-1} + (y_f - y_1)z^{-2}]$

(iii) $G_p(z) = \dfrac{P(z)}{Q(z)}$ (3)

(iv) $D(z) = \dfrac{Q(z)}{1 - P(z)}$ (4)

In addition to the preceding, the following relations should also hold (see figure):

$q_0 + q_1 + q_2 = m_f = \dfrac{y_f}{K}$ (5)

$p_1 + p_2 = y_f$ (6)

In practice, the two relations (5) and (6) may not hold for the coefficients of $G_p(z)$. However, multiplying the numerator and denominator of $G_p(z)$ by y_f/s_p, where s_p is the sum of the numerator coefficients of $G_p(z)$, will ensure that these equations are satisfied. Also, the practical approach following this step is not to solve for the coefficients of $P(z)$ and $Q(z)$ from (3) but to utilize the numerator and denominator of the factored $G_p(z)$ as $r_0 P(z)$ and $r_0 Q(z)$, respectively, in (4) to determine the controller. The only constraint is that the number of intermediate switches in $m(t)$ and hence the order of $Q(z)$ must be greater than or equal to the order of the plant.

(b) Using the technique of (a) and the pulse transfer function for the hot-water heater with remote sensing, for a sample period $T = 20$ (equation (14.18b)), verify that the digital controller transfer function is

$D(z) = \dfrac{0.0965 - 0.0215z^{-1}}{1 - z^{-2}(0.6111 + 0.3889z^{-1})}$

---

†R. E. Kalman (discussion of) "Sampled-Data Processing Techniques for Feedback Control System." A. R. Bergen, and J. R. Ragazzini, *Trans. AIEE*, 73, no. 2, (November 1954): 236–247.

Note that the dead time in the plant transfer function does not affect the application of the method.

(c) Simulate the performance of this controller with the digital computer and compare it with Dahlin's method, Fig. 14.8a.

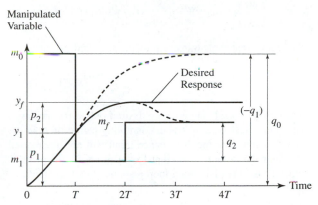

Figure 14.15 (Problem 14.12)

14.13 Design a digital controller for the pollution-monitoring equipment of Practice Problems 9.27 and 12.28, that is,

$G_p(s) = \dfrac{e^{-0.3s}}{(1 + 0.1s)(1 + s)}$

(a) So that the closed-loop system is finite-time settling,

$G_{CL}(z) = z^{-(n+1)}$

where n is the number of sample periods contained in the dead time. Obtain a simulation of your controller for $n = 1$ and $n = 2$.

(b) Using the method of Dahlin. For each value of $n = 1$ and $n = 2$, obtain a simulation of your controller for $\lambda = 1.0, 1.5,$ and 3.0.

(c) Repeat (b) for a sampling period $T = 0.25$. Also, simulate the design for $\lambda = 1.0, 1.5,$ and 3.0.

14.14 Repeat Problem 14.12(b), that is, determine the controller $D(z)$ using the pulse transfer function for the hot-water heater given by equation (14.15b) but with $n = 2$.

14.15 Using Kalman's method as described in Problem 14.12, derive the controller $D(z)$ for a piece of pollution control equipment with the pulse transfer function

$G_p(z) = \dfrac{0.50z^{-1} + 0.19z^{-2}}{1 - 0.40z^{-1} + 0.08z^{-2}}$

using a sampling period $T = 0.3$ and for $y_f = 1$.

14.16 Consider the hot-water heater with remote sensing system of Example 14.3. Let the sampling period be equal to the dead time. That is,

$G_p(s) = \dfrac{e^{-Ts}}{s + 3/40}$

Determine the discrete-time controller $D(z)$ such that for a unit-step disturbance input, the (desired) output is

$$y(0) = 0; \qquad y(T) = 0; \qquad y(2T) = \left(\frac{40}{3}\right)(1 - e^{-3T/40})$$

$$y(3T) = \left(\frac{40}{3}\right)(1 - e^{-6T/40}); \qquad y(nT) = 0 \quad \text{for all } n \geq 4$$

Hint: Show that the z-transform of the specified output is

$$Y(z) = \left(\frac{40}{3}\right)[(1 - e^{-3T/40})z^{-2} + (1 - e^{-6T/40})z^{-3}]$$

Note that the specification of the desired output recognizes that with a dead time of T and a sampling period of T, in the worst possible case the disturbance input will not be detected until after the instant $3T$, so that for $2T \leq t \leq 3T$ the plant output is just the plant open-loop response to a unit-step input.

14.17* For a given open-loop system governed by the equations

$$\mathbf{x}_{k+1} = \mathbf{P}\mathbf{x}_k + \mathbf{Q}\mathbf{u}_k; \qquad \mathbf{y}_k = \mathbf{C}\mathbf{x}_k$$

and the corresponding closed-loop system determined as shown in Fig. 14.16 and which can be described by

$$\mathbf{x}_{k+1} = \mathbf{P}_c\mathbf{x}_k + \mathbf{Q}\mathbf{r}_k; \qquad \mathbf{y}_k = \mathbf{C}\mathbf{x}_k; \qquad \mathbf{P}_c = \mathbf{P} - \mathbf{Q}\mathbf{C}$$

show for a third-order system the following result, which is in fact true for nth order: A closed-loop system is controllable (or uncontrollable) and, observable (or unobservable) if and only if its open-loop system is controllable (or uncontrollable) and observable (or unobservable), and vice versa. That is, show that

$$\text{rank } [\mathbf{Q};\mathbf{PQ};\mathbf{P}^2\mathbf{Q}] = \text{rank } [\mathbf{Q};\mathbf{P}_c\mathbf{Q};\mathbf{P}_c^2\mathbf{Q}] \text{ and rank } \begin{bmatrix} \mathbf{C} \\ \mathbf{CP} \\ \mathbf{CP}^2 \end{bmatrix}$$

$$= \text{rank } \begin{bmatrix} \mathbf{C} \\ \mathbf{CP}_c \\ \mathbf{CP}_c^2 \end{bmatrix}$$

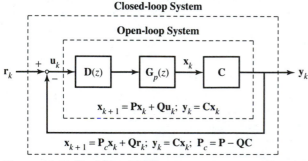

Closed-loop System

Open-loop System

$$\mathbf{x}_{k+1} = \mathbf{P}\mathbf{x}_k + \mathbf{Q}\mathbf{u}_k; \ \mathbf{y}_k = \mathbf{C}\mathbf{x}_k$$

$$\mathbf{x}_{k+1} = \mathbf{P}_c\mathbf{x}_k + \mathbf{Q}\mathbf{r}_k; \ \mathbf{y}_k = \mathbf{C}\mathbf{x}_k; \ \mathbf{P}_c = \mathbf{P} - \mathbf{Q}\mathbf{C}$$

Figure 14.16 (Problem 14.17)

14.18 Assuming that all process variables are available as measured outputs, design a finite-time settling state vector feedback controller, if feasible (if a design is not feasible, explain why), for

(a) The discrete-time system of Practice Problem 6.43.

(b) The discrete-time system of Practice Problem 8.69.

(c)* The discrete-time system of Practice Problem 8.70.

(d) The pollution-monitoring equipment of Practice Problem 9.27, with $T = 0.3$.

14.19 Repeat Problem 14.18 with I-action on the first variable.

14.20 Repeat Problem 14.18 with I-action on the second variable.

14.21 Design a state vector feedback finite-time settling discrete-time controller and observer, where feasible, for the systems represented in the following practice problems. If a design is not feasible in any case, explain why.

(a) 6.36; with I-action and without

(b) 8.76

(c)* 9.31a; with I-action and without

(d) 9.31b

(e) 9.38a; with I-action and without

(f) 9.38b

(g)* 9.27; the pollution-monitoring equipment with $T = 0.3$. Assume that the sensed variable is the only available output.

14.22 A train leaves a station at constant speed such that the distance of the train from the station x_1 can be described by $x_1(k + 1) = x_1(k) + vT$, where v is the speed and T is the sampling period. Suppose a passenger who disembarks at the station forgets his briefcase in the train but remembers this only when the train has reached a distance d from the station. If the passenger decides to give chase on a parallel track, using his car, and if the motion of the car is given by

$$x_2(k + 1) = x_2(k) + Tx_3(k)$$

$$x_3(k + 1) = x_3(k) + Tu(k)$$

where x_2 is the distance of the car from the station, and the car acceleration u is limited to $|u| \leq a$, what will be the optimal sequence of accelerations $u(k)$ so that the car will catch the train in the shortest possible time. Formulate this problem by dynamic programming.

14.23* Consider the first-order object

$$x(k + 1) = x(k) + 0.5u(k)$$

with the initial state $x(0) = 2$ and the end state not specified. Determine the optimal control by dynamic programming for the quadratic performance index

$$J = 2x(N)^2 + \sum_{k=0}^{2}[2x(k)^2 + u(k)^2]$$

Show that the solution is a feedback controller by specifying the feedback gain. Use the discretization x: $(-1.0, -0.5, 0, 0.5, 1.0, 1.5, 2.0, 2.5, 3.0)$ and u: $(-2.0, -1.0, -0.5, 0, 0.5, 1.0, 2.0)$.

14.24 Using Euler's method, convert the following continuous-time *soft-landing problem* into a discrete-time system. Then solve the resulting problem by dynamic programming:

Minimize $\int_0^{t_f}(1 + |u|)\, dt$ *subject to*

1. Dynamics: $\dfrac{d}{dt}x_1 = x_2; \dfrac{d}{dt}x_2 = u$

2. Initial condition: $x_1(0) = 4; x_2(0) = 0$

3. Final condition: $x_1(t_f) = 0; x_2(t_f) = 0$

4. Control constraint: $|u| \le 1$

Display your results using the following sketches: **(a)** x_2 versus x_1; **(b)** x_1 versus t; **(c)** x_2 versus t; and **(d)** u versus t. Use the net x_1: $(0, 1, 2, 3, 4,)$; x_2: $(-2, -1.5, -1.0, -0.5, 0)$; and u: $(-1.0, -0.75, -0.50, -0.25, 0, 0.25, 0.50, 0.75, 1.0)$.

14.25 Given the second-order discrete time system

$$\mathbf{x}_{k+1} = \begin{bmatrix} 0.9688 & 0.2500 \\ -0.250 & 0.9688 \end{bmatrix}\mathbf{x}_k + \begin{bmatrix} 0.0312 \\ 0.2500 \end{bmatrix}u_k; \qquad \mathbf{x}_0 = \begin{bmatrix} 0 \\ 2 \end{bmatrix}$$

(a) Use dynamic programming to determine a control function (table) that will drive this system from the given initial state to a zero final state in minimum time. Use a sampling period of $T = 0.25$ and state and control discretizations of 13 quantization levels each, with $-2 \le x_1 \le 2$, $-2 \le x_2 \le 2$, and $-2 \le u \le 2$. Simulate the response of the plant over the time period $0 \le t \le 6$.

(b) Repeat problem (a) but with the plant controller subject to the constraints $|u_k| \le 1$, $x_{1,k} > 0$, and $-0.5 \le x_2 \le 2$.

(c) For the same sampling period and state and control discretizations in (a), use dynamic programming to deter-

mine the control that minimizes the quadratic performance index

$$J = \frac{1}{2}\sum_{i=0}^{N-1}[\mathbf{x}_i^T\mathbf{W}\mathbf{x}_i + u_i^2]; \qquad \mathbf{W} = \mathbf{I}$$

for $N = 20$, and simulate the response of the controlled system for the same time interval and initial state given in (a).

(d) Repeat (c) by assuming no constraints and determine the optimal feedback gain matrix for infinite time quadratic regulation. Compare the response of the system under the resulting state vector feedback control with the result from (c).

14.26* Consider again the antiaircraft servomechanism of Example 14.1 under DDC. The pulse transfer function of the continuous-time plant with zero-order hold was given in equation (14.7a).

(a) Sketch the z-domain signal flow graph for this pulse transfer function and show that with a sampling period of $T = 0.2$, the corresponding discrete-time state model is

$$\mathbf{x}_{k+1} = \begin{bmatrix} 1.5032 & 1 & 0 \\ -0.5530 & 0 & 1 \\ 0.0498 & 0 & 0 \end{bmatrix}\mathbf{x}_k + \begin{bmatrix} -0.0672 \\ 0.1363 \\ 0.0151 \end{bmatrix}u_k;$$

$$y_k = [1 \quad 0 \quad 0]\mathbf{x}_k$$

(b) For an ISE-type quadratic performance index given by

$$J = \sum_{i=0}^{N-1}\Delta\mathbf{x}_i^T\mathbf{W}\Delta\mathbf{x}_i; \qquad \Delta\mathbf{x}_i = \mathbf{x}_i - \mathbf{x}_s; \qquad \mathbf{W} = \mathbf{I}$$

determine the optimal feedback gain (1×3) matrix $\mathbf{G}_{FB}$ and feedforward (scalar) gain G_{FF} that minimize this cost functional, and simulate the response of the discrete-time system subject to this control and setpoint $r_d = 1$ from a zero initial state over a 6-s time interval. Compare your results with those shown in Fig. 14.6b.

Chapter 15

Realization of Microcomputer Control Systems

I N THIS CONCLUDING CHAPTER WE VERY BRIEFLY CONSIDER some significant aspects of realizing a control system in which a digital computer, specifically a microcomputer or PC, is one of the major components—the *controller*—in the control loop. We are particularly interested in *real-time* computer control, in which the computer must carry out its programmed tasks at regular intervals while simultaneously responding to external (physical or real-world) events occurring at unpredictable times. The computer must be connected somehow to the external events, and program instructions are necessary to direct the interaction between the computer and the external activity. There are thus two major parts to the problem: (1) *hardware interfacing* or connecting the computer to external equipment and (2) *software design* or programming the computer to carry out its control calculations while interacting with external components. The second part has essentially been considered in the preceding chapter and some earlier chapters, except perhaps for necessary modifications to account for the interface between the computer operations and the external events. This aspect and the first part are discussed in this chapter in a manner that is considerably independent of any particular computer hardware or software system, and both parts are integrated and illustrated using a simple example or case study.

We begin in Section 15.1 with the fundamental problem of *interfacing* the computer and the process that we desire to control, that is, instrumentation concepts that tailor real-world physical phenomena into the digital format that is understood by computers and vice versa. Interfacing a computer for *sensing and control* encompasses three fundamental functions: *data acquisition, command delivery,* and *signal conditioning.* Depending on the performance requirements and the nature of the application and associated equipment, either analog interfacing or digital interfacing or both may be required, and different data acquisition and conditioning hardware can be chosen or combined in one system. In Section 15.2 we examine these features of computer measurement and control systems, including the interfacing and signal conditioning requirements of representative types of sensors and actuators. In Section 15.3 we very briefly introduce the practice of general-purpose instrumentation systems that combine all or most of the fundamental functions in convenient and easily programmed microcomputer interface units or boards. Finally, we utilize one such interface in a case study that integrates hardware and software to realize a complete digital computer liquid level/flow process control system.

15.1 INTERFACING WITH EXTERNAL EQUIPMENT

Interfacing involves circuitry and programming necessary to efficiently integrate a microprocessor into an overall system. Although the natural input or output signal for a digital computer is a two-state signal—on or off, 0 or 1—the common measurement or control signal for many sensors and actuators is an analog signal—a continuously variable voltage or current. Typical examples of such equipment include potentiometric and piezoelectric sensors, many electromagnetic actuators such as the electric motor, and, indeed, most of the transducers and other devices modeled in the first part of the text. However, there are also some sensors and actuators whose output or control input signals are similar to or compatible with the two-state or logic signal. Two very good examples of such a sensor and actuator are the pulse-generating tachometer and the stepping motor. Although such equipment generates or accepts two-state signals, a number of operations on the signals, including electrical isolation, filtering, level shifting, *buffering,* and parallel-to-frequency or frequency-to-digital pulse conversion, are still necessary to couple these signals between the computer and the external equipment. Consequently, we explore next the application of two types of sensory and control interfaces (there are also *user-interaction* interfaces (keyboard and CRT display interfaces, for example) and *operational overhead* interfaces (data storage and computer network interfaces, for example) (see [27]) between the digital computer and external equipment: (1) *analog-to-digital* (ADC) and *digital-to-analog* (DAC) converters, which can convert an analog voltage to a number in the computer and vice versa, and (2) *digital input/output* connections. Strictly speaking the latter type of interface is the only means of communication between the computer and external devices, since the ADC, DAC, and even user-interaction interfaces are attached to the computer through a digital input/output (I/O), connection *(port)* of some sort.

It will be helpful, in the following discussions, to keep in mind that within the computer itself, information flow takes place in groups of two-state (binary) signals called words. The binary signal then constitutes a digit or a bit of the word. Different word lengths are in use in small computers—from 4 and 8 bits used in the early calculator processors, some more recent control-oriented microprocessors, and bit-sliced microprocessors; through 16- and 32-bits used in general-purpose machines; to 64 bits, which are available in some recent high-performance microcomputers. Such a word or unit of words is used to represent everything in the computer from instructions for the processor to data for processing. A program written in any language (for example QuickBASIC or C) must eventually be translated to a sequence of such word units (machine code) before it can be executed by the computer. Although not absolutely necessary, such rudimentary knowledge of the internal workings of microcomputers will be presumed in this section.

Digital-to-Analog Conversion

An often necessary control interface component is a circuit to convert a digital word (resulting from a computer operation) to a proportional analog signal (voltage or current usable by some external equipment). After appropriate conditioning (for example, voltage/current amplification), the output of a DAC can be used, for instance, to control the speed of a DC servomotor or manipulate some other *final control element* (actuator). Two general types of DACs are illustrated next with 4-bit circuits; however, 12- and 16-bit devices are common in practice.

Binary-Weighted Resistor DAC.   Figure 15.1 shows the circuit for a simple 4-bit binary-weighted resistor DAC. The switches are digitally controlled so that, as a typical protocol, a switch is closed if the corresponding bit (of the computer output word) is on.

Consider the state shown in Fig. 15.1 with only the least significant bit (LSB) set (the corresponding 4-bit binary word is [0001]. For the particular values of the circuit parameters indicated, we can easily determine, using the operational amplifier circuit analysis rules of Section 3.4, that $V_{out} = -0.5$ V. We can also determine for any state of the switches representing the binary word $[s_3 s_2 s_1 s_0]$, where $s_i = 1$ if bit i is set and $s_i = 0$ otherwise, that

$$V_{out} = -0.5(s_3 \cdot 2^3 + s_2 \cdot 2^2 + s_1 \cdot 2^1 + s_0 \cdot 2^0) \text{ V} \qquad (15.1a)$$

so that the *full-scale output,* representing the binary word [1111] for this particular circuit is

$$V_{FS} = -0.5(8 + 4 + 2 + 1) \text{ V} \quad \text{or} \quad -7.5 \text{ V}$$

Resistor Ladder DAC.   Figure 15.2 illustrates a 4-bit R/2R ladder DAC. Consider the state indicated in which only the most significant bit (MSB) is on (set) while the rest of the bits are off (not set, switches closed to ground). R_1 and R_2 are then in parallel to ground with the equivalent resistance value of R. The (series) sum of this R and R_3, which is $2R$, and the resistor R_4 are also in parallel, and the equivalent resistance value is also R. Continuing in this manner, we see that the resistor network reduces to $R_8 = 2R$, set to MSB as shown, and an overall equivalent resistance of $2R$ set to ground. Analysis of this reduced circuit yields $V_{out} = -V_{ref}$. If only the next most significant bit is on, the equivalent circuit analysis gives

$$V_{out} = -\tfrac{1}{2} V_{ref}$$

For the LSB the result is

$$V_{out} = -\frac{1}{2^3} V_{ref}$$

so that for the 4-bit R/2R ladder DAC circuit, in general,

$$V_{out} = -\frac{1}{2^3}(s_3 \cdot 2^3 + s_2 \cdot 2^2 + s_1 \cdot 2^1 + s_0 \cdot 2^0)V_{ref} \qquad (15.1b)$$

Figure 15.1

Binary-weighted Resistor DAC

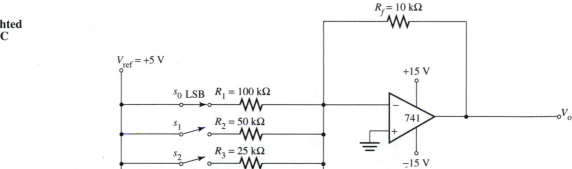

Figure 15.2

R/2R Ladder DAC

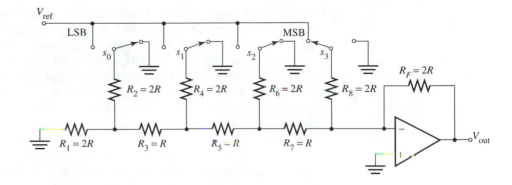

Some DAC Characteristics and Specifications

The performance features of DACs that are important to their proper use include the resolution, the accuracy, and the settling time. Other features include linearity, input coding, and output range. The *resolution* is a measure of the least level of output. The number of bits in the input word determines the possible levels of output. An 8-bit converter has 2^8, or 256, output levels, so that its resolution can be described as "8 bits" or "1 part in 256" or "0.39%." If the 8-bit converter has a range of 0 to $+10$ V (10 V full-scale), say, then the least level of output will be 39 mV. A number of factors, including errors in the op-amp circuits and in the resistor values, can cause a DAC to be inaccurate. The *accuracy* of a DAC is the fractional error in the (voltage) output and is typically specified in terms of the number of least significant bits. An 8-bit DAC with a 10-V full-scale output and $\pm\frac{1}{2}$ LSB accuracy has a maximum error in any output of $(0.5)(0.0039)(10$ V$)$, or 19.5 mV. The *settling time* is the time it takes the converter output to settle with the specified accuracy to the final value following a change in the input word. It follows that the maximum frequency of input word changes at which the DAC can be accurately operated is determined by the output settling time. *Linearity* is a specification of the maximum deviation from a proportional relationship between the binary inputs to a DAC and their analog output voltages, over the input range. It is usually expressed as a percentage of the instrument's full-scale output. The input *word coding* for most DACs is either binary or binary coded decimal (BCD). There are current- and voltage-output DAC models with user-selectable bipolar (for example -10 to $+10$ V) and unipolar (for example, 0 to $+10$ V) *output ranges*. The variety of models also includes *latched* or unlatched outputs. DAC converter boards having latched outputs will maintain the output value indefinitely until instructed to change. The output of DAC boards not having this facility will decay with time (not true zero-order hold). To maintain the output in this case the digital value has to be written repeatedly to the DAC at a prescribed (*refresh*) rate.

Analog-to-Digital Conversion

The outputs of most pressure transducers, temperature sensors, flowmeters, and many measurement elements (sensors) are small currents or voltages that after amplification or other conditioning must be converted to a usable form for a digital computer. The sensory interface component used to convert analog data to corresponding digital data is the *analog-to-digital converter* (ADC). There are many types of ADCs. We illustrate here the operation of only two basic types: the integrating ADC and the successive-approximation ADC.

Integrating-Type ADC   Figure 15.3 shows the block diagram of the *single-slope* ADC, which is an example of an integrating-type ADC. Conversion is initiated after the ramp and counters have been reset to zero, when the analog voltage (assumed positive for this illustration) is applied to the plus input of the comparator (a comparator is essentially an op-amp circuit in open-loop mode; see Section 3.4). The comparator output goes *(logic)* high, since its negative input is (zero) lower than the analog voltage. This starts *(enables)* the ramp and also the AND gate, which causes the clock input to reach the counters. The counters count up the clock pulses while the ramp output grows at a known rate. When the ramp output exceeds the analog voltage, the comparator output goes (logic) low, disabling the AND gate and hence cutting off the clock. The timing and control unit then *latches* the count data (which is proportional to the analog voltage) to the output circuit and resets the counters and the ramp generator for another conversion cycle. If the latched data are passed through decoder/drivers to a display, then the resulting instrument is a simple digital voltmeter.

Successive-Approximation-Type ADC   Figure 15.4 shows a simplified block diagram of a *successive-approximation* ADC. The circuit operates as follows: On the first clock pulse the successive-approximation register (SAR) turns on the MSB to the DAC. If the DAC output is greater than the analog voltage (signified by a logic low output from the comparator, the SAR resets the MSB. If on the other hand the DAC output is less than the analog voltage (logic high output from the comparator) the SAR keeps the MSB set. At the next clock pulse, the SAR sets the next most significant bit and responds to the comparator output as before, that is, leaving the bit set or resetting the bit according to whether the comparator output is high or low. The SAR proceeds in this manner until all the bits have been tried. It then sends out the *end-of-conversion* (EOC) signal indicating that the data output is a valid word. Tying the EOC signal to the *start conversion* (SC) input as shown causes the converter to cycle continuously. For low- to medium-speed applications, the function of the SAR can be performed by a computer program routine, so that analog-to-digital conversion can be accomplished with a DAC and a comparator circuit.

ADC Characteristics and Specifications

The performance characteristics of ADCs are similar to those of DACs. These characteristics include the resolution, the accuracy, the linearity, the conversion time, and the quantization error. *Resolution, accuracy,* and *linearity* have meanings analogous to the definitions of the same quantities for the DAC. Linearity for an ADC is normally ex-

Figure 15.3

Single-slope ADC

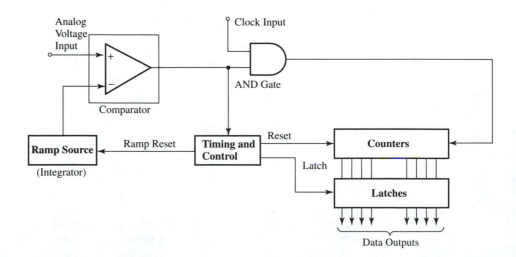

Figure 15.4

Successive-approximation ADC

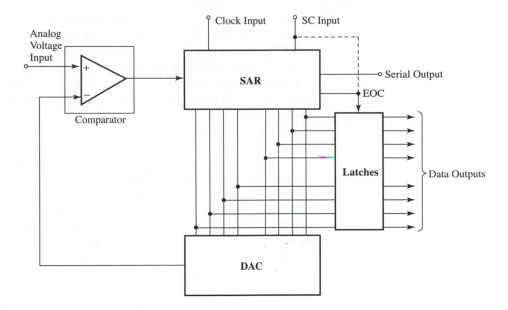

pressed as a percentage of the device's full-scale output or fraction of the LSB. The *conversion time* of an ADC is the time to digitize each sample. This time depends on the type of conversion. A successive-approximation conversion, for example, can be faster than ramp- or slope-type conversion for the same number of bits. The *quantization error* is the difference between the actual analog waveform and its digitized representation. This error is strictly a function of both the sampling period and the irregularity of the input waveform. However, under nominal conditions, the quantization error is ideally $\pm\frac{1}{2}$ LSB, so that this specification is also an indication of the resolution of the converter. The output *word coding* of ADCs, which is user selectable in some cases, includes 2's complement, offset binary, and binary. Similarly, many ADCs are available with selectable input *voltage ranges,* including bipolar and unipolar options. As a rule, the input to the ADC must be relatively constant during the conversion time. Hence depending on how rapidly the input signal changes and the speed of the ADC, an important decision in the application of an ADC is whether or not to use a *sample-and-hold* circuit at the input to the ADC.

Digital Input/Output

Data communication between subsystems of a microcomputer or between a microprocessor and nearby (a few meters distance) external digital signal sources/receptors (devices) is typically implemented in *parallel mode* (that is, all the bits that compose a word are sent at the same time over multiple signal wires). The basic digital interface is thus the *parallel I/O port.* The input to the DAC and the output from the ADC are two-state or three-state *(tristate)* logic signals that are compatible and can be connected to the parallel output and input ports, respectively, of a microcomputer. Three-state logic has the normal logic low and high output states of two-state logic plus a high-impedance (float) output state that in a three-state device is created by a separate "enable" input. *Logic* outputs, in our present context, are low-power signals in which the logic state of the signal is determined by the voltage (or current) level. Different voltage (current) levels are available with different logic technologies or families. For example, for the standard TTL family, the minimum logic "1" output voltage is typically 2.4 V, which guarantees that an output "high" will always be equal to or greater than 2.4 V. The maximum logic "0" output voltage is 0.4 V, which means that an output "low" will always be equal to or less than 0.4 V. The corresponding maximum output currents *(sourcing* and *sinking)* are 0.4 mA and 16 mA, respectively. For a (5-V supply) CMOS (complementary metal oxide on silicon) logic family, the logic "1" minimum output voltage is 4.5 V, and the maximum output current is 0.36 mA at 2.4 V. The logic "0" maximum output voltage is 0.5 V, while the corresponding maximum output current is 0.4 mA at 0.4 V. There are also emitter-coupled logic (ECL) and logic subfamilies such as low-power Schottky TTL.

These and other such detailed characteristics should be known for the digital I/O port in order to properly condition the signals to or from other equipment or, sometimes, the signals transmitted over cables connected between similar ports. Flip-flops, latches, and drivers and receivers or *transceivers* are essential components of digital I/O connections or I/O ports through which the processor communicates with other devices. *Flip-flops, latches,* and *registers,* are memory circuits that store either a 1 or 0 on their outputs. Normally such outputs can change only when a special *clock,* or *strobe,* or *enable* input is pulsed (see [28]). A *driver* is an output device capable of generating from the (internal) input of the part with which it is associated a standardized voltage or current that other parts or devices can use. An input device that can convert a signal from a driver into a signal that is usable by the part with which it is associated is known as a *receiver* (see [27]). Other related components that are sometimes part of computer or peripheral I/O ports include *buffers,* which are similar to drivers but have higher voltage- and current-handling capabilities, and *power transistors,* which are used for handling even higher current levels.

Functional Description of a Parallel Digital I/O Interface

An I/O port, in its simplest form, is realized when means to store the input logic state and timing to indicate when the output logic state is valid are provided. A slightly more detailed functional description of a *parallel digital I/O interface* is shown in Fig. 15.5. The **bus transceivers** are 3-state *buffers* for the data lines to and from the computer I/O channel. An **address decoder** takes address information put out by the processor and determines the component or device being addressed. **Handshaking circuitry** includes

Figure 15.5

Functional Description of a Parallel Digital I/O Interface

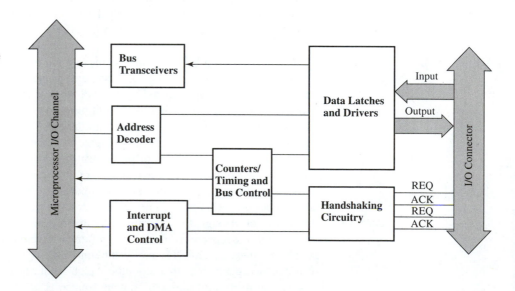

registers, flip-flops, and other logic gates. The control of data transfer between the computer and another device is always split between the two systems, and the series of operations by both systems that establishes this control is known as *handshaking*. A typical handshaking sequence proceeds as follows: A device wishing to transfer data to the processor places *(stores)* the data in the *input latch* and issues a *request* (REQ) for **interrupt** through the handshaking circuit. When the processor *services the interrupt* (see below), it initiates transfer of the data from the **data latches** to the bus transceivers by generating the digital I/O port's address, which is decoded in the *address decoding circuit,* and by issuing the appropriate **timing/control** signals, one of which is a "buffer full" status signal and an *acknowledgment* (ACK) signal to the device. Subsequent timing/control signals cause the data to be read from the transceivers into the processor and a "buffer empty" status signal to be issued. Similarly, the processor can *write* data to an external device by placing the data in the bus transceivers and addressing (REQ) the target device. Further handshaking operations cause the device to initiate the transfer of the data to the output **drivers,** the *driving* of the digital I/O lines with the data value, and acknowledgment (ACK) of received data by the device. Note that the I/O data lines can be as wide (contain as many bits in parallel) as the microcomputer's data word.

Input/Output Control Strategies   *Interrupt* capability in microprocessor-based systems is one way of dealing with random and/or infrequent input signals (see [28]). For example, in the simple case of a *single-level interrupt* system, the processor on receipt of an interrupt signal completes the instruction (if any) in progress and diverts program execution to an *interrupt service routine*. The processor interrupt input must then be *disabled* to allow the service routine to perform the requested task without further interruption. Following such "servicing" of the interrupt, the interrupt input is reenabled, and program execution is returned to the point at which it was first interrupted. In the more typical case of a *multiple interrupt* system, means of automatically determining which task should be executed first *(priority decoding)* is included in the system. Another way of dealing with unscheduled events, and perhaps the simplest I/O control strategy, is *polling*. In this approach the processor on a periodic basis examines the status of each connected device and takes action where necessary. It is assumed that the processor-monitoring cycle is sufficiently fast that no event is missed. The most sophisticated or fastest I/O control method is probably *direct memory access* (DMA). With this control strategy, data can be transferred to/from the computer memory from/to a device without interrupting the processor, which can be doing other things.

Serial Digital I/O

An alternative to data transfer in parallel form is *serial data transmission*. In this approach, data bits are transferred one after the other along a single signal connection. As we alluded to earlier, the normal method of data communication between the microcomputer and remote devices is by parallel I/O. Two common circuits in *serial ports* is the UART *(universal asynchronous receiver/transmitter)* and the USART *(universal synchronous and asynchronous receiver/transmitter,* although the latter is almost always used in the asynchronous mode) that perform the necessary parallel-to-serial and serial-to-parallel data conversions (see Fig. 15.6).

The bits are generated from the output word (usually byte) at the transmit end and then reconstituted into the corresponding input byte at the receiving end, according to a specific rule or *protocol*. The protocol covers both the structure of the message (see example serial data stream in Fig. 15.6) as well as the *handshaking* procedures, if any,

Figure 15.6

Serial Link from
Microcomputer to Remote
Device and an RS-232
Data Structure

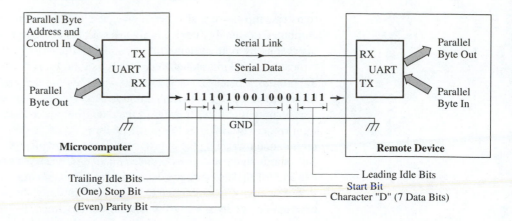

that control the data transfer process. The American Electronics Industry (EIA) standard *RS-232C* (its updates include RS-422 and RS-485) is probably the most widely used protocol system or standard for *bit-serial* data transfers. Some major features of the RS-232C standard include the following:

1. **Signal transmission speed:** Several standard rates of transmission in bits per second *(baud)* are supported, for example: 300, 600, 1200, 2400, 4800, 9600, 19,200, 38,400.

2. **Signal levels:** A voltage signal within -15 V to -3 V corresponds to a logic 1, while a logic 0 is given by voltage signals in the range $+3$ V to $+15$ V. Note that since these *bipolar* voltages are not comparable to the usual digital logic, *line transceivers* (not shown in Fig. 15.6) are necessary for transition between the UARTs and the signal lines. *Current-loop* signals in which a logic 1 is represented by 20 mA and logic 0 by 0 mA are sometimes used to achieve longer-distance transmissions. For transmission over telephone networks, *modems* are used at each end of the link to covert from pulse signals to sine waves of different frequencies (for example, 1270 Hz for logic 1 and 1070 Hz for logic 0).

3. **Code word and error detection:** The *American Standard Code for Information Interchange* (ASCII) is typical. An ASCII code word *(byte)* is a 7-bit (or 8-bit, for the extended ASCII character set) binary number representing each numeral and letter of the alphabet (each character). For example, the 7-bit ASCII code for the character "D" is (1000100) or 44 Hex *(hexadecimal)*. An extra *(parity)* bit is added to make the number of 1's in each code word odd *(odd parity)* or even *(even parity)* and thus to enable single- or odd number-error detection (the occurrence of a single error or an odd number of errors changes the parity of a code word and is therefore detected). The structure of a transmitted message (see Fig. 15.6) hence consists of 8- or 9-bit *characters* (7 or 8 *data bits* and one parity bit) each of which is framed by a *start bit* (which is a logic 0 and the first bit of a transmission sequence) and one or two *stop bits* (which are logic 1 bits added to signal the end of a character). Between each character (message) the line is held at logic 1 *(idle bits)*.

4. **Handshaking signals and pin connections:** A serial communication link can be as simple as a three-line connection—*transmit* (TX), *receive* (RX), and *ground* (GND) (recall Fig. 15.6)—that is just the data connections with no control signals. However, a full complement of handshaking signals includes *request to send* (RTS), *clear to send* (CTS), *data set ready* (DSR), *data carrier detect* (DCD), *data terminal ready* (DTR), and *ring indicator* (RI). Note, however, that only software control

is used in RS-485. RS-232 connectors include 4-pin (RX/TX, TX/RX, DSR/DTR, and GND), 9- and 25-pin configurations. The pinouts for 9-pin (DB9) and 25-pin (DB25) serial ports are shown in Table 15.1.

Serial and Parallel Data Communications   The advantage of parallel data communication over serial communication is that it is a much faster mode of data transfer. However, since parallel data transfer requires more communication lines, it can be more expensive, especially over long distances. Thus serial communication is a common choice for data transfer at low rates or over long distances. Indeed, most microcomputers have one or two serial ports as standard components. However, note that the RS-232 standard defines data communications between two types of serial devices: *data terminal equipment* (DTE) and *data communications equipment* (DCE). Whereas the connection between DTE and DCE devices can be made *straight through,* that is TX pin to TX pin, DTR pin to DTR pin, etc., serial data communications between two DTE devices require a special cable or adapter called a *null modem* cable (a connector that switches the appropriate signal lines). Computers and many instruments operate as DTE devices.

15.2 COMPUTER DATA ACQUISITION AND CONTROL

Although sensory and control interfaces can be combined on one interface board or package, they are nonetheless distinct. Sensory interfaces permit the computer to monitor external (real-world) events through *data acquisition* from relevant process sensors, while control interfaces enable the computer to influence external events by the *application of control* inputs (commands) to relevant process actuators. In this section we examine the features of data acquisition from and control of some representative sensors and actuators, respectively, including considerations for signal capture and conditioning. The illustrations cover both digital- and analog-interface systems.

Pulse Measurements and Commands

We begin with digital interface systems. First, recall that we have both parallel and serial digital I/O but that the parallel data word or byte is basic to the two I/O systems. As a practical matter, for pulse inputs and commands the individual bits of the data word or byte usually carry the sensing or control signals. The variety of such *single-bit* signals commonly encountered in practice is quite small and can be represented by two basic types: *contact closure* and *voltage level* signals (see [29]). Whereas digital input measurements generally cover the two types of signals, pulse output commands are in-

Table 15.1

Serial Interface Port Signals and Connector Pinouts

SIGNAL	NAME	DB9 PIN	DB25 PIN
DCD	Data carrier detect	1	8
RX	Receive data	2	3
TX	Transmit data	3	2
DTR	Data terminal ready	4	20
GND	Signal ground	5	7
DSR	Data set ready	6	6
RTS	Request to send	7	4
CTS	Clear to send	8	5
RI	Ring indicator	9	22

variably voltage level signals. We shall illustrate pulse inputs with the digital interfacing requirements of two rather simple devices: a *switch* or contact and a pulse-generating tachometer. For pulse output we shall revisit the stepping motor.

Contact Closure Type Signals

Contact closure signals represent sensory inputs in which the essential information is the *state* of the contacts (closed or open). Such signals can be *manually* triggered or *automatically* set. Manual inputs include *on-off* or contact switch inputs such as from a *toggle* switch, and *momentary* or *pushbutton* switch inputs such as from a keyboard key or spring-loaded switches in general. Examples of sources of automatically set contact signals include limit or trip switches (such as are found on most machine tools), thermostatic switches (the bimetallic strip type, for instance; recall the temperature control system of Example 5.7 of Section 5.4), and *relay* contacts (see Example 5.4). Whatever the source, the contact information must be converted by the interface into an appropriate voltage level at the digital input port. Figure 15.7a illustrates a simple typical interfacing circuit arrangement for a *normally open* mechanical switch. With this arrangement an open contact results in a logic 0 data input (the resistor clamps the input to (0 V) ground). On contact closure, the supply voltage (+5 V) is connected to the input line, giving rise to a logic 1 input. The resistor is sized to limit the current to the required logic level. Figure 15.7a assumes TTL type conditions.

Pulse Signal Conditioning

In general, the signal-conditioning techniques that are applicable to contact closure signals are also useful for voltage level type signals, although the purposes may differ. Some signal conditioning of pulse signals may be necessary because of the nature of closure of the contact, the duration of the threshold voltage level, the transfer distance, or the need to isolate the computer from remote high voltages, power transients, or noise. For signal transfer over significant distances the usual conditioning is to use a buffer or a power transistor driver. Isolation allows the computer to be interfaced to an external device without an electrical connection. This is often accomplished by electrical relays or optical methods.

Optoisolation   Figure 15.7b illustrates an optical isolation arrangement for digital input to a microcomputer, assuming a *floating* DC source on the switch circuit. The *optoisolator* consists of two parts: a transmitter (*photodiode* or LED (*light emitting diode*)) and a receiver (*phototransistor*) separated by a physical gap. Closure of the switch (which represents the remote pulse input) causes current to flow into the photodiode through a limiting resistor. The light emitted by the photodiode stimulates the phototransistor, which then drives the input signal for further conditioning or transfer to the digital input port. Optical isolators can erect a voltage barrier of several kilovolts. They contain no moving parts and are capable of very high speed operation.

Signal Shaping and Debouncing   Contact closures triggered by mechanical movement (whether manual or automatic) are generally subject to *bounce*—a transient mechanical vibration or an initial period of successive rapid makes-and-breaks of the contact following activation. The resulting *ringing* signal must be filtered or *debounced*. Debouncing techniques, in general, are either *hardware* or *software*. Besides damping and other physical and material design considerations for the contact it-

Figure 15.7

Contact Closure Pulse Signal Interfacing and Some Conditioning Circuits

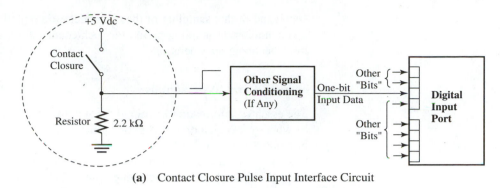

(a) Contact Closure Pulse Input Interface Circuit

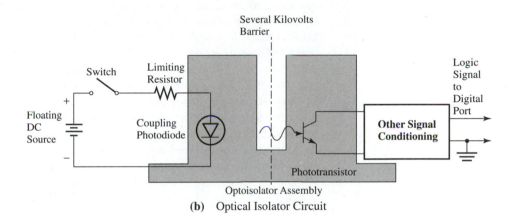

(b) Optical Isolator Circuit

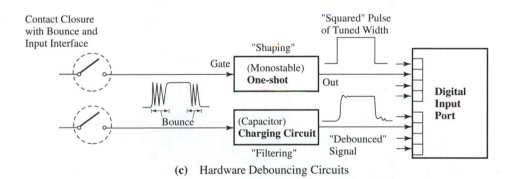

(c) Hardware Debouncing Circuits

self, hardware debouncing implies a shaping or filtering circuit, as shown in Fig. 15.7c. The typical shaping circuit is a type of *latch* circuit—a *one-shot* or an *edge-triggered monostable multivibrator*. Essentially the one-shot output is set on a rising edge of the (gate) input (the very first make-contact signal). The output remains set for a time period predetermined by the circuit constants and is then reset. This period cannot be so short as to capture subsequent contact makes within the bounce or so long as to miss future valid switch inputs. Typical periods are about 10 to 80 ms. As indicated in Fig. 15.7c, filtering can involve a type of capacitor *charging circuit* (an *RC* network) that smooths out the spurious contacts. A simple software debouncing technique is to *program* a sufficient delay period (say, 50 ms) between the detection of the initial contact signal and further sampling of the contact closure input for future contact signals; that is, the minimum sampling period for such contact closure digital inputs is determined by debouncing considerations.

Voltage Level Type Signals

The essential information in voltage level pulse signals is the level of the voltage, that is, whether it is high or low. A good example of such a digital input signal is the output of a pulse-generating tachometer, which is a common means of measuring the angular velocity of rotating equipment. Figure 15.8a is a physical model illustrating the operating principle of the photodetecting type of this speed sensor. An opaque disk with a transparent region or hole or slot (or more than one symmetrically disposed slots) is connected to the rotating shaft or generally caused to rotate at the same angular velocity that is being measured. A photodiode-phototransistor pair is mounted (in a mask or shutter) straddling the disk so that light from the photodiode is blocked by the disk and hence does not reach the phototransistor except when the transparent region registers with the mask. The transistor will thus produce a high electrical output once (or as many times as there are slots on the disk) during each revolution of the disk. The angular speed of the disk is determined from the number of such high output states over a fixed sample period. The direction of rotation can also be determined by comparing the outputs from two photodiode-phototransistor pairs placed at the same circumferential distance but at different angular positions.

The type of signal conditioning that may be required in this type of data acquisition is the same shaping circuit that can be used to debounce contact signals (see Fig.

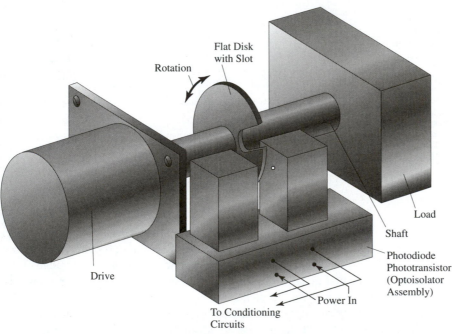

(a)   Arrangement for Shaft-speed Measurement

(b)   Incremental Encoder Disk

Figure 15.8

The Pulse-generating Tachometer

15.7c). Here the objective is to "catch," at the instant of occurrence, the very narrow "spike" signal that is the response of the phototransistor to the momentary light input from the photodiode during passage of the slot without resorting to impossible or extremely short sampling periods. As before, the output from the one shot is a pulse of "tunable" width. Given the speed range of the tachometer and the sampling rate, an adequate pulse width can be selected for the one-shot.

The pulse-generating tachometer is clearly an application of optoisolators. Another application is the (rotary) *position encoder*. There are both *incremental* encoders and *absolute* encoders. If the above disk contains instead a large number of evenly spaced slots at the same circumference, then each pulse output from the optoisolator corresponds to an angular step of the disk. The incremental displacement of the disk can be determined from the number of pulses. An example of the disk of such an *incremental* (rotary) *encoder* is shown in Fig. 15.8b. Observe that there is a second track with a single mark. This is called the *Z channel* or marker, and the single output pulse per revolution associated with this marker is used to establish the zero position. The use of two offset optoisolators on the main channel allows the direction of travel as well as the (angular) distance of travel to be determined. An *absolute encoder* uses several concentric tracks of slots on the same disk, each with its independent optical sensor, to provide a unique position information or code (pattern of light transmission) for each shaft location.

Pulse Outputs and the Stepping Motor

Figure 15.9 illustrates the basic range of interfacing requirements for pulse outputs based primarily on the electrical current needs of the remote devices. At the highest power or current levels, the digital signal may require both a power transistor driver and a high degree of isolation of the computer from the external device. The optoisolator circuit was discussed previously. Note that the ultimate drive or switch circuit for the remote device is not shown and can be considered part of the external equipment. The switching on and off of high-power electric motors and actuation of large electric punch presses or similar equipment are typical situations. At the next lower power level, mains-driven motors, for example, heaters, valves, AC lights, and other alarms can be turned on or off by electrical relays, as depicted. This drive circuit was discussed and analyzed in Examples 3.18 and 5.4. Note that an electrical relay provides some degree of electrical isolation as well. Devices with moderate current requirements, such as small DC resistance lamps, solenoid valves, voice-coil motors, and loudspeakers, can be controlled by an appropriate power transistor–driven digital line, as shown in Fig. 15.9. At low power levels are many digital indicating or display devices such as LEDs and other indicating diodes, and filament and other low-power lamps, which draw currents on the order of or slightly above the nominal capacity of a microcomputer output port. As shown in the figure, a regular current driver or gate that can be part of the specification of the output port often suffices for such devices.

The Stepping Motor

A stepping motor is a good example of a device that can be manipulated directly through a computer's pulse output connection. Its applications include positioning, contouring, and indexing equipment of all kinds, robots, and other manipulators. Stepping motors and their drives were introduced in Section 5.3, and models of some drive circuits were given in Figs. 5.17 and 5.18. Essentially all that is required from the microcomputer are the signals to turn the transistor switches on and off for the various phases of the motor. From our discussion of junction transistors in Section 3.4, we can see that the base current necessary to switch the drive transistors of many (low-power)

Figure 15.9

**Basic Interfaces to
Representative Pulse Output
Control Devices**

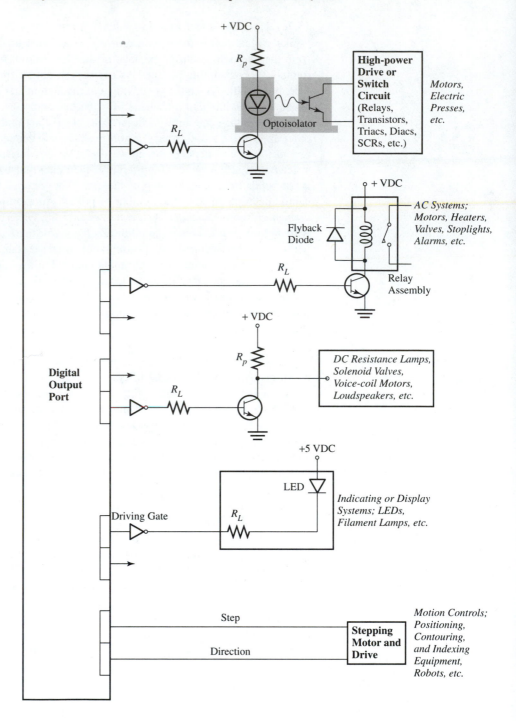

stepping motors is less than the available drive current from a microcomputer's digital
output port, so that direct control of such stepping motors by a computer is feasible. A
very flexible computer programming approach, in this case, is to use a table that contains the pattern of pulse outputs for the desired phase-energizing scheme. A control
program then steps through the table (does table lookup) with appropriate delay in-
between-steps (determined by the desired motor speed or step rate) to drive the motor

accordingly. In this way, complex motions can be executed with the same program merely by changing the pattern in the table or by accessing a different table.

Consider, for example, the two-phase-on coil-energizing sequence shown in Table 5.2 and repeated in Table 15.2. In general, four pulse outputs from the computer will be required to drive the four-phase stepping motor described, although a close examination of the coil-energizing sequence should reveal that the energizing sequences for coils (poles) 1B and 2B are the negation of the sequences for coils 1A and 2A, respectively, so that only two pulse outputs can be used to drive the motor, provided external logic circuits (gates) are provided to generate the other two pulse inputs for the motor from the available two outputs. If we assume that the four pulse outputs are associated with the four LSBs of the computer output data word, then the binary or hexadecimal coding of the output word (all other bits are assumed off except the four of interest) that can be stored in the control program table is as shown in the second and third columns, respectively, of Table 15.2. Stepping through the table in the downward direction will result in motor rotation in a particular direction (counterclockwise, say). Reversing direction in the table (stepping through the values in an upward direction) will cause the motor to rotate in the opposite direction (say, clockwise).

As we alluded to in Section 5.3, stepping motor drives for all but the lowest power motors are often sophisticated devices, most with options for assessing different energizing schemes, motor acceleration and deceleration techniques, step resolutions, pulse current level controls and waveforms, etc. With many such drives, only two basic inputs are required from the computer: a step signal and a direction signal. This situation is shown in Fig. 15.9. The step and direction interface is usually optically isolated and conditioned, as necessary, within the drive, and the desired motor speed is simply encoded in the frequency of the step pulse train output from the computer.

Features of Analog Data Acquisition

Figure 15.10a illustrates the general components for computer analog data acquisition and the relation between the input data, the conditioned data, the quantized data, and the coded digital data. The primary analog data input interface, the ADC, and its characteristics were explored in Section 15.1. As we saw for sensors with pulse outputs, the analog output signals from sensors of physical events require one or more forms of conditioning before they can be useful. Such conditioning includes isolation (of the sensor from downstream devices, including from *common-mode* interference), multiplexing, latching and buffering (as discussed previously), amplification, filtering (to remove noise or unwanted frequency components), level shifting, shaping, linearizing, addition/subtraction of signals, and provision of excitation. The physical components of such conditioning circuits are elements with which we are already familiar from Part I of the text: resistors, capacitors, inductors, transformers, transistors, operational amplifiers, and other integrated circuits. The conditioning circuits themselves are also similar to or the same as many of the systems we modeled in that part of the book and will not be considered much further here. We shall examine instead the functions and

Table 15.2

Two-Phase-On Unipolar Coil Energization

COIL OR POLE				4-BIT CODE				VALUE	
2B	1B	2A	1A	2^3	2^2	2^1	2^0	Hex	(Decimal)
off	off	on	on	0	0	1	1	3	(3)
off	on	on	off	0	1	1	0	6	(6)
on	on	off	off	1	1	0	0	C	(12)
on	off	off	on	1	0	0	1	9	(9)

Figure 15.10

Features of Computer Analog Data Acquisition

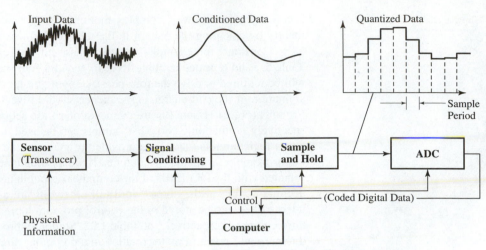

(a) Functional Components of Computer Analog Data Acquisition

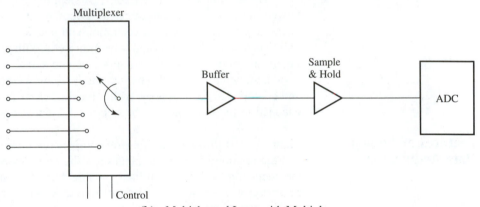

(b) Multichannel Input with Multiplexer

features of the parts of analog data acquisition identified in Fig. 15.10a. Ideally, as shown in the figure, the signal conditioning, the sample and hold, and the analog-to-digital conversion are all coordinated/controlled by the computer. However external *counter-timer* boards and controllers are sometimes used for this purpose.

Sample and Hold

Since a (rapidly) varying input signal will have changed by the time conversion is completed, sensitive data acquisition systems require *sample-and-hold* circuitry to quickly "catch" the signal value and hold this value during the conversion (recall that ADC data acquisition requires that the data be relatively constant during conversion). Very often it is necessary to have several signals share one communication channel to the computer. As shown in Fig. 15.10b, the solution for many multichannel data acquisition products is to employ a single sample-and-hold ADC module and an *input multiplexer* to alternately select (switch) the inputs for sampling. Another possibility is to simultaneously sample and hold all the inputs using multiple sample-and-hold circuits and then employ an *output* multiplexer to select each held signal for conversion by a single ADC. Clearly, sample-and-hold circuits are necessary for multiplexed systems.

Figure 15.10 *continued*

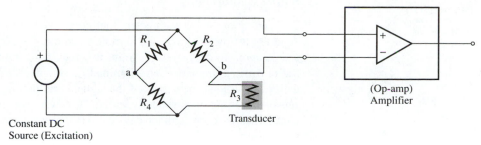

(c) A Bridge Resistance Conditioning Circuit

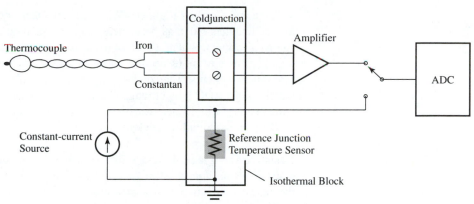

(d) A Method of Thermocouple Cold-junction Compensation

Further, the hold period must extend beyond the conversion time to the time required to transfer other data in the multiplexed set. For fast-changing measurements, employing a sample-and-hold circuit for each measured value before multiplexing is probably the more effective solution for adequate representation or correlation of signals.

Quantization and Sample Period

Because of the discrete-time nature of a microcomputer, its presence in a control system involving analog equipment generally renders the overall arrangement a sampled-data system. Consequently, sampling is perhaps the most important feature of analog data acquisition systems, and the choice of sampling rates is a critical decision in planning a control system. Indeed, we saw in Chapter 14 that the sampling period is always an important design parameter in developing digital control algorithms for sampled-data systems. Again, as is the case for many critical design parameters, the choice of sampling period is often a compromise. Although the frequency of sampling must be sufficiently high to allow acceptable reproduction of the data of interest in the acquired signal and to reflect in some way the manner in which the measured data changes with time, high sampling rates affect the overhead associated with servicing the ADC (recall also the limitation of the ADC conversion time), saving the data (both the data transfer rate to memory and (since high sampling rate normally imply large data volumes) the amount of required memory), reducing the data, and applying the results. It should also be noted that for each of the multiplexing options discussed above, the maximum sampling rate per channel is determined by dividing the maximum sampling rate for the single ADC by the number of channels sampled.

Signal Conditioning

The variety of analog parameters that are sensed in industrial processes is enormous. A partial list includes voltage, current, magnetic flux, displacement, position, velocity, acceleration, force, pressure, strain, flowrate, temperature, density, humidity, concentration, pH, illumination and luminance, sound level, X-rays, and alpha, beta and gamma radiation. The details of the signal conditioning that can be necessary for such sensory information are equally varied. However, as explained earlier, the conditioning circuit elements are familiar devices. Also, a categorization of common analog input signals into a small number of families with related specific signal conditioning requirements is possible. There are essentially four such categories: *thermocouple* signals, *variable resistance* signals, *current* signals, and *voltage* signals.

Voltage and Current Signals   Whereas many sensor outputs are voltage signals (analog tachometers (such as permanent-magnet DC motors) and DVTs (differential voltage transformers, which are often used to sense mechanical movement, for example), a number of sensors (especially industrial ones, such as many pressure cells) are current output systems. The conditioning that is specific to voltage and current signals is matching the ADC's input range. This is usually easily accomplished with op-amp circuits that perform the necessary amplification or attenuation, or both current-to-voltage conversion and amplification/attenuation.

Resistive Devices and Bridge Measurements   A common requirement of variable-resistance transducers is an *excitation* source (voltage or current supply) to convert the resistive function into a measurable signal. However, this group of sensors can further be categorized into those requiring *bridge* measurements and those that do not necessarily require a bridge circuit. *Resistance temperature detectors* (RTDs), *thermistors,* and variable potentiometer devices are some examples of the latter, and as explained next for thermocouple *cold-junction* compensation (see Fig. 15.10d), a constant-current source and perhaps some op-amp amplification or attenuation can be sufficient conditioning for such devices. Variable-resistance sensors such as *strain gages* and *pressure transducers,* which produce very small changes in resistance at full scale, normally require a Wheatstone bridge circuit, for which the standard excitation is a voltage source. The principle of measuring an unknown resistance using a *balanced bridge* was illustrated in Practice Problem 2.22, using Fig. 2.38. In Practice Problem 2.51 the determination of an unknown *strain* in a test gage using a balanced bridge, and the method of temperature compensation of the bridge sensor, were explored for the same bridge circuit of Fig. 2.38. A bridge conditioning circuit that is typical of many strain or pressure data acquisition systems is shown in Fig. 15.10c. The difference between this figure and Fig. 2.38 is the replacement of the current meter between nodes a and b with voltage-sensing lines that feed an op-amp circuit (details not shown). Instead of adjusting R_2 to balance the bridge following a change in R_3, the circuit measures the potential difference between a and b resulting from this change and uses it to determine the change in resistance and hence the transducer output. Temperature compensation can also be realized with this conditioning circuit, as in Practice Problem 2.51.

Thermocouples   Perhaps one of the most demanding analog input signal conditioning requirements is that of a thermocouple sensor. Not only are thermocouple output signals of very low power (less than ± 100 mV for the popular types), thus requiring considerable amplification and shielding from noise, but the relation between the output voltage and the junction temperature is not linear (hence linearization is also required) and is subject to cold-junction error, so that these sensors further require *cold-junction compensation* for accurate operation. Figure 15.10d illustrates a typical approach to conditioning thermocouple measurements. All inputs of the thermocouple (a J-type or iron-constantan thermocouple is illustrated) are applied to terminals that are located on an *isothermal* block to ensure negligible temperature difference between the input connectors (cold junction) and the reference junction sensor. This sensor, which is not a thermocouple, is mounted in the block to measure the cold-junction temperature. The arrangement shown in Fig. 15.10d assumes that an RTD, a thermistor, or other variable-resistance temperature sensor is used for this secondary measurement. Both the measurement of the cold-junction temperature and the thermocouple output (following amplification) are converted in the ADC and sent to the computer, where the compensation and linearization are performed by software. This method of conditioning thermocouple signals thus requires two analog input channels and appropriate software. Thermocouples are popular because they are inexpensive and provide useful measurements over a wide temperature range.

Analog Outputs and Pulse Modulation

Unlike input signals from the outputs of analog sensors, the variety of analog command signals that can be output from a DAC interface is limited to two: voltage and current range signals. The various types of remote analog equipment or actuators generally incorporate drive systems that are designed to operate with one or the other command input. Even in cases where the drive and actuator are clearly physically distinct, it is customary to consider the two as part of the remote equipment. Similarly with microcomputer control, it is usually more cost effective to perform most signal conditioning functions such as filtering, shaping, level shifting, and addition/subtraction of signals in software within the microcomputer. This leaves isolation, signal amplification (voltage or current), and/or wiring circuits to ensure that the DAC output specifications are met as the primary analog output signal conditioning requirements.

Voltage Output   Typical DAC voltage output ranges are 0 to +5 VDC, 0 to +10 VDC, −2.5 to +2.5 VDC, −5 to +5 VDC, and −10 to +10 VDC. The associated current *drain* specifications are typically ±5 mA *load current* and 40 mA maximum *short circuit current*. This implies, as shown in Fig. 15.11a, a minimum load resistance for a given voltage output range. If the external device and its drive meet this condition, then often no other output signal conditioning is required. This is the case with many analog actuators such as DC motors with voltage-input power amplifier drives.

Current Output   DAC current-output specifications are typically the *industry standard* 4–20 mA *current sink*. For proper operation and power dissipation considerations, the voltage across the output circuit is usually limited to a minimum and a maximum condition. Typical specifications are 8 V and 36 V, respectively. Figure 15.11b shows two approaches that are used in practice to connect the process loop in order to meet these requirements. In the first case, a grounded loop supply (say, 24 V) is used with the load *floating*. This approach allows the same supply to power other loops. In the second case, the supply is floating while the load is grounded.

Pulse Control of DC Motors

Although the common approach to computer control of DC motors is by armature or field voltage control (see Section 5.3) through a DAC and an appropriate drive amplifier circuit, it is possible to control the speed of a low-power DC motor by *pulse-width*

Figure 15.11

**Analog Output Connections
and Pulse-width and
Pulse-frequency Modulation**

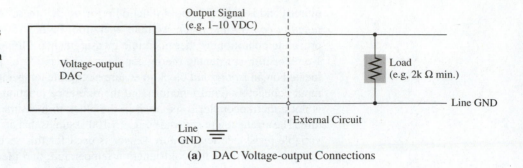

(a) DAC Voltage-output Connections

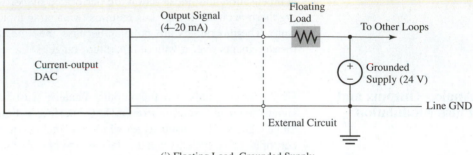

(i) Floating Load, Grounded Supply

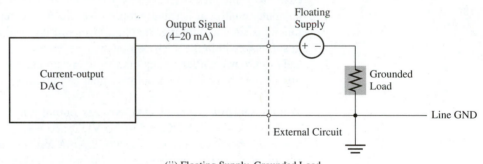

(ii) Floating Supply, Grounded Load

(b) DAC 4–20 mA Current-loop-output Connections

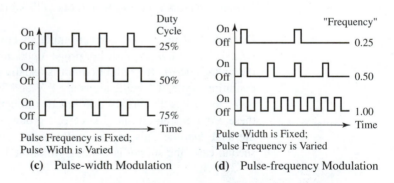

(c) Pulse-width Modulation   **(d)** Pulse-frequency Modulation

modulation (PWM) or *pulse-frequency modulation* (PFM) through a computer pulse
output connection. However, such pulse control of DC motors is still more conveniently or flexibly accomplished using a DAC interface. PWM and PFM are illustrated
in Figs. 15.11c and 15.11d, respectively. Basically, the speed of a DC motor is a func-

tion of the input power and drive characteristics, while the area under an input pulse train is a measure of the average power available from such an input. *Duty cycle* is the (area) percentage of one cycle during which the pulse is on, and PWM is the variation of the duty cycle of a continuous train of pulses. Since the frequency is normally held constant while the on-off time ratio is varied, the duty cycle in PWM is determined by the pulse width. Thus in PWM, power increases with duty cycle. In PFM, power is a function of frequency (the frequency is related to the duty cycle in this case, since the pulse width is fixed). However, the performance of a given motor and drive circuit can be different under PWM and PFM.

15.3 ILLUSTRATION OF A COMPUTER IMPLEMENTATION: PRELIMINARIES

We conclude this chapter by considering a simple but complete example of the implementation of a control system using a real-time microcomputer. A small computer-based data acquisition and control system generally consists of six parts: (1) the physical system whose behavior is to be measured and controlled, (2) the sensors and actuators, (3) the computer system, (4) the interfaces between the computer and the sensors and actuators, (5) signal conditioning hardware (both on the input and output sides), and (6) all relevant software (including signal conditioning and control algorithm). The physical system considered here is a simple liquid level/flow control system (see the schematic diagram, Fig. 15.14) that is typical of the process control industry. The primary sensor, which is described further in Section 15.4, is a pressure transducer. The actuator is a pneumatic (diaphragm-actuated) control valve. Control valves are the most common *final control elements* in process industries. Although valves (short constrictions) in general were considered in some detail in Section 3.3, the physical and operating characteristics of process control valves in particular will be discussed further to their application in systems similar to the one illustrated here.

In addition to the characteristics of sensors and actuators, the nature, especially *bus* structure, of the computer employed in realizing a control system determines to a major extent the type of interface and conditioning hardware as well as software that are applicable. The type of computer used in the present illustration is an IBM PC (AT or compatible). This terminology covers computer systems using the AT or 16-bit ISA bus and includes computers based on the Intel 286- and 386-series of processors, as well some 486-series processors (some systems based on the 486 series do not use the ISA bus but its 32-bit extension, the EISA). Some other microcomputer types include those using the EISA bus, Intel Pentium-series processors, the IBM PS/2 series computers or, more specifically, computers using the *microchannel* bus system, and the *Macintosh* series of computers. Although custom interfaces and conditioning circuits can be designed and built for each application and for a given computer, it is more convenient and sometimes cost effective to select and configure one or more commercially available general interface and signal conditioning instrument modules. Such instrumentation systems are briefly introduced in this section.

Process Control Valves Revisited

Because the control of many industrial processes is based on regulating the flow of some fluid—liquid, gas, vapor, or fluidized solids—in a flow channel of one form or another, valves with adjustable flow openings (note that there are valves that control flow rate without varying the flow opening; an example is the *L-valve,* used in fluidized-bed units) are common equipment in process control systems. The primary characterization of the behavior of such control valves concerns the nature of the dependence of the flow through the valve on the amount of flow opening.

Control Valve Characteristics

Figure 15.12a is a schematic diagram showing the (internal) structural details of a pneumatically actuated control valve. Of particular interest now are the *stem* and the *plug* and the way in which the stem movement combined with the plug shape defines the flow opening. The amount of (opening) movement of the valve stem is referred to as the *valve lift*. Control *valve characteristics* is the term used to describe the relationship between the movement of the valve *stem* and the fluid flow through the valve. As we know from Part I of the text, this relationship is strongly determined by the pressure drop across the valve. We thus speak of the *inherent flow characteristic* of a valve, which is defined as the relationship between the fractional valve lift and the relative flow through the valve at a *constant pressure drop*. The *installed characteristic* is the actual lift-versus-flow characteristic under the system operating conditions. Figure 15.12b illustrates the inherent flow characteristics typical of some commercially available valves.

To describe quantitatively the relationships shown in Fig. 15.12b, we define the *valve gain* as the ratio of the change in flow (ΔQ) to the change in stem position (ΔS), or the slope of the valve characteristic curve:

$$G_v = \frac{\Delta Q}{\Delta S}; \quad \text{(valve gain)} \tag{15.2a}$$

Figure 15.12

A Pneumatically Actuated Valve and Examples of Inherent Flow Characteristics

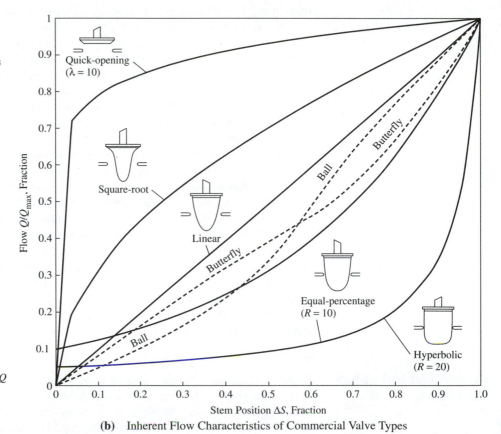

(a) Pneumatically Actuated Valve

(b)   Inherent Flow Characteristics of Commercial Valve Types

Note that ΔS varies from 0 (fully closed) to 1 (fully open). In terms of the valve gain, the valve behavior equation (1.2) (or equation (3.26) with $\alpha = 2$) can be written as

$$Q = G_v \Delta S \left[K \sqrt{\frac{\Delta P}{\rho}} \right] \qquad \text{(15.2b)}$$

where ρ is the density of the fluid and K is a new constant that depends on valve size and the nature of the fluid. The characteristic curves shown in Fig. 15.12b are plots of the relative flow (Q/Q_{max}) versus the relative valve stem position ΔS for a fixed $\left[K \sqrt{(\Delta P/\rho)} \right]$.

Valve Flow Coefficient   Generally, for single-phase liquids (*nonflashing* liquids), equation (15.2b) is traditionally expressed as

$$Q = C_v \sqrt{\frac{\Delta P}{\gamma}}; \quad \text{(C_v; valve flow coefficient)} \qquad \text{(15.3)}$$

where γ is the specific gravity of the liquid. Equation (15.3) defines the *valve flow coefficient C_v*, which is a function of valve size, plug profile (valve type), and stem opening (lift). The valve flow coefficient determines the sizing of control valves. It gives the dependence between the size and form of the valve and the required conditions of pressure drop and flowrate.

Valve Types   Because of the strong dependence of the amount of flow opening and hence the flow through the valve on the plug shape for a given valve lift, the plug profile is used to classify control valves according to type of inherent flow characteristics. In Fig. 15.12b some of the representative plug shapes have been identified with the corresponding flow characteristics or *valve types*. These examples are *linear, quick opening, square root, equal percentage,* and hyperbolic. Other valve characteristic types include *butterfly, ball* (or *circular*), and *sine*.

Linear Control Valve   For a *linear* valve, the valve gain is a constant, independent of valve stem position and is thus equal to the ratio of the maximum flow to the maximum (relative) stem position (which has a value of 1):

$$G_v = \frac{Q_{max}}{\Delta S_{max}} = Q_{max}; \quad \text{(linear valve gain)} \qquad \text{(15.4a)}$$

Hence from equation (15.2b), we can write for the linear valve characteristic,

$$Q = Q_{max} \Delta S \left[K \sqrt{\frac{\Delta P}{\rho}} \right]; \quad \text{(linear valve characteristics)} \qquad \text{(15.4b)}$$

For nonlinear valves in general, the valve gain is a function of stem position, and equation (15.2a) can be rewritten as

$$G_v = \frac{Q_{max}}{\Delta S_{max}} f(\Delta S) = Q_{max} f(\Delta S); \quad \text{(nonlinear valve gain)} \qquad \text{(15.4c)}$$

Expressed this way, the specific types of nonlinear characteristics are determined by the function $f(\Delta S)$ and equation (15.2b). The expressions for some of the previously mentioned nonlinear valve characteristics are as follows:

quick-opening control valve:

$$Q = Q_{max}(\Delta S)^{1/\lambda}\left[K\sqrt{\frac{\Delta P}{\rho}}\right] \tag{15.5a}$$

where λ is a valve parameter of relatively large value such as 10 or 50 (the quick-opening curve in Fig 15.12b is for $\lambda = 10$). Note that equation (15.5a) represents the linear relation if we set $\lambda = 1$. The quick-opening equation also represents the square-root characteristics below, with $\lambda = 2$.

square-root control valve:

$$Q = Q_{max}\sqrt{\Delta S}\left[K\sqrt{\frac{\Delta P}{\rho}}\right] \tag{15.5b}$$

equal-percentage control valve:

$$Q = Q_{max}R^{\Delta S - 1}\left[K\sqrt{\frac{\Delta P}{\rho}}\right] \tag{15.5c}$$

where R is a valve parameter known as the *rangeability:*

$$R = \frac{Q_{max}}{Q_{min}};\quad \text{(rangeability)} \tag{15.5d}$$

Equal-percentage and hyperbolic valves (described next) have finite flows at zero valve stem position, and the rangeability reflects this behavior in the expression for the valve characteristics. The characteristics in Fig. 15.12b are for $R = 10$ (equal percentage) and $R = 20$ (hyperbolic).

hyperbolic control valve:

$$Q = Q_{max}\frac{1}{R - (R - 1)\Delta S}\left[K\sqrt{\frac{\Delta P}{\rho}}\right] \tag{15.5e}$$

Choice of Valve Type   The selection of a valve type for a given application is usually based on experience and comprehensive understanding of the effect a given valve will have on the process. However, one common guiding principle is the idea of *gain matching*—the valve characteristics should compensate for inherent changes in pressure relative to flow in the process. For instance, in a throttling process, which typically involves an orderly reduction of pressure from a stable high value to a constant low value, the inherent piping and/or other system frictional losses are negligible relative to the pressure drop across the valve. For such near-zero inherent pressure changes with flow, a constant valve gain or linear flow characteristics are recommended.

Recall that frictional losses are functions of flowrate (consider, for example, equations (3.28a) and (3.28b) which describe long constrictions). A process with inherent pressure losses that are relatively higher than the valve pressure drop will require low valve gain at low flow (where frictional/pipe losses are small) and high valve gain at high flow (where frictional/pipe losses are large). From Fig. 15.12b, the valve characteristics meeting these requirements are the equal-percentage and hyperbolic characteristics. At the other extreme is an *on-off* flow control process, which requires very rapid changes in flow as the valve is initially being opened or closed. In this case, the matching valve type that offers very high gain at low flow is the quick-opening characteristic. A design parameter that expresses the relative magnitudes of valve pressure drop and inherent system pressure losses is the *valve pressure drop ratio:*

$$P_R = \frac{P_V}{P_V + P_L}; \quad \text{(valve pressure drop ratio)} \tag{15.6}$$

where P_V is the valve pressure drop, and P_L represents the system's frictional losses. Finally, observe generally that the inherent valve flow characteristics shown in Fig. 15.12b all fall between the two extreme types, that is, quick-opening and hyperbolic characteristics.

Valve Actuators and Positioners

A valve actuator is the means for mechanically varying the valve stem position. Control valve actuators of all sorts are available, consistent with the wide range of applications of these valves in process control. *Spring diaphragm* actuators, which are normally operated by air pressure *(pneumatic actuators;* see Fig. 15.12a, for example) are common, simple, and cheap. *Piston actuators,* which are usually operated by some liquid (oil) pressure *(hydraulic actuators)* are more expensive but are smaller, faster and can generate higher forces. *Motor* actuators *(motor-operated valves)* generally use mechanical linkages and members to convert the rotation of a drive motor into the linear motion of the drive stem. There are also various other types of pneumatic, hydraulic, *electromechanical,* and even *electrohydraulic* valve actuators. Another fairly common and cheap actuator is the *vane-type rotary actuator* for rotary-type valves. The operation is typically by electromechanical means, but it can also be hydraulic or pneumatic.

General Actuator Systems   In general, hydraulic actuation systems respond faster than their pneumatic counterparts, are more efficient, and have the advantage of being self-lubricating. However, hydraulic systems require closer tolerances than pneumatic actuators (which can tolerate minor leakages), and (because of significant changes in oil viscosity with temperature) hydraulic actuators are usually restricted to situations where they will operate over a narrow temperature range. Pneumatic actuators are also more appropriate for chemical and petroleum processing environments and are generally cheaper for low operating pressures. Compared with an electric drive, hydraulic actuation is preferred when high power is required, when high-speed operation is desired, and where some intermittent- and low-speed operation will be necessary. On the other hand, electric drives are generally favored where fluid contamination problems exist, when a very clean and quiet environment is desired, where long-distance power transmission is involved, and when certain flammable materials are to be avoided.

Valve Positioners   The control valve illustrated in Fig. 1.10 has a (pneumatic) *valve positioner* mounted on the (pneumatic) actuator. Valve positioners are secondary controllers (controllers typically in cascade configuration) in the control loop between the actual process (or *primary*) controller and the final control element. Although they are often mounted on the valve or its actuator, where they can utilize a mechanical linkage for the *feedback* of the actual stem position to the positioner, this is not a requirement.

Essentially, when a valve positioner is used, the controller output is connected to the positioner instead of having the process controller connected directly to the control valve. The *control law* implemented by the valve positioner is usually a simple one that can be realized by appropriate means such as pneumatic computing elements, analog electronic components, fluidic components, or purely mechanical computing elements like cams and linkages. Valve positioners represent an increase in cost as well complexity of the control valve and the process control system; hence they should not be used unless they are needed. Examples of situations or applications that call for the use of valve positioners include the following: when it is desired to move the control valve very accurately and in small increments; when excessive friction is present (this often introduces *hysteresis* and *repeatability* errors in the relationship between the command signal and the valve stem position that need correction); when other high unbalanced valve forces (including those due to excessive process pressure changes) exist; and whenever pressures or currents or other actuation signals required to operate the valve are higher than (or incompatible with) the corresponding process controller output.

Commercial Interface and Signal-Conditioning Instrument Modules

By *interface instrument modules* we mean integrated and comprehensive packages of interfacing circuits: ADCs, DACs, digital I/O, and other supporting hardware and software, all of which are designed to work directly with a specific computer type or remote instrument. There are basically two kinds of interface instrument modules: *plug-in board* systems and *remote programmable instrument* interfaces. For plug-in boards, the processing power of the computer (PC) is used to manage the connected instruments by running appropriate data acquisition and control software and also to analyze and present results if necessary. In typical remote instrument interfaces, the connected instruments themselves are programmable. Remote instrument data acquisition systems are, in general, capable of higher data collection rates than microcomputer-driven plug-in systems. There are two kinds of data acquisition and control software systems: *integrated software packages* and *driver software with programming tools.* The integrated packages provide ready-to-use capabilities for acquiring data and/or controlling instruments and, in some cases, analyzing data and presenting results, whereas driver software and programming tools permit programming of the data acquisition and control system in a conventional language such as QuickBASIC, Pascal, or C and provide a variety of utilities to assist the programmer.

Signal conditioning instrument modules are, in our present context, circuits that can condition physical signals of a specific family of sensors or of specific power characterics to meet the I/O specifications of the *interface instrument module* with which they are used. Examples include low-level analog voltage (or current) input modules, thermocouple input modules that simultaneously support several thermocouple types and may include isothermal blocks, multichannel analog output modules with switch-selectable output ranges, multichannel isolation amplifier modules, multichannel multiplexer amplifier modules, low-power logic signal-sensing and control modules, high-voltage power control modules, and pulse and frequency input modules.

General Hardware and Software Features

Each interface instrument package offers a set of features and capabilities including numbers of input and output channels, ADC and DAC resolutions and ranges, maximum supported sampling rate, selectable amplifier gains, analog and digital triggering, supported computer operating system(s), compatible user interface and driver programs, any analysis and presentation software, interfaces to other programming languages, and so forth. The computer types that are usually supported by these instrument modules were discussed earlier. The supported remote instruments include *IEEE 488*-bus (essentially synonymous with the HP-IB (Hewlett Packard interface bus and the GPIB (general-purpose interface bus)) or similar industry-standard programmable instruments (examples include many industrial and laboratory instruments, medical equipment, data loggers and recorders), RS-232/422/485 interface instruments such as some remote data loggers and process meters, and *VMEbus* standard or similar modular and card-based instrumentation systems, or another computer! A growing array of interface instrument products of various specifications for a variety of applications are available commercially. Some of the major developers include National Instruments Corp., Keithley Instruments Inc., Omega Engineering Inc., and Hewlett Packard Corp. Other manufacturers include Fluke Corp., Iotech Inc., Validyne Engineering Corp., and Intelligent Instrumentation Inc.

As we alluded to previously, the functionality of available data acquisition and control software packages ranges from drivers for specific devices to integrated packages for developing complete systems. Most integrated packages can acquire data from various plug-in boards, sometimes from different manufacturers—IEEE-488, RS-232, and VXI instruments on their combinations—and include drivers that can be customized for special devices. The data analysis functions featured by many such software packages encompass fast Fourier transform, waveform analysis and signal generation, various types of statistical analysis, curve fitting, and array operations. In addition, some provide interfaces to other analytical software such as Lotus 1-2-3, Excel, and other spreadsheets; Mathcad and other computational software; as well as programming using formal high-level languages such as BASIC and C. Supported operating systems include DOS, Windows, OS/2, and UNIX. Results presentation features include graphical data, file input/output, printer/plotter output, and even network and other communications support. Besides providing tools to synthesize control algorithms, some packages have built-in functions for popular industrial process controllers such as P-, PI-, PID-algorithms. A variety of user interfaces are available with integrated software packages. The goal in most cases is ease of use. *Menu-driven* interfaces guide the user in decision making while simplifying usage. Custom *control panels* are graphical languages that appear to eliminate programming entirely and present the user with familiar hands-on control knobs, meters, and the like. Other *graphical user interfaces* combine several of these features to allow users to build *virtual instruments* with their attendant control panels.

Some examples of integrated software packages include Laboratory Technologies Corp.'s LABTECH NOTEBOOK, LABVIEW, and LABWINDOWS graphical programming languages from National Instruments Corp., VIEWDAC from Keithley Instruments Inc., VIRTUAL ENGINEERING ENVIRONMENT from Hewlett-Packard, and VISUAL DESIGNER from Intelligent Instrumentation Inc. LABTECH ACQUIRE from Laboratory Technologies Corp.; ASYST and DriveLINX from Keithley Instruments Inc.; and National Instruments Corp.'s NI-DAQ, NI-488.2, and NI-VXI are just a few examples of the many driver-level programming languages and utilities.

The Implemented Interface Hardware and Software

We can further describe the hardware and software features of interface instruments by briefly considering the specific data acquisition and control products used in the liquid level/flow control illustration. Our only objective is to explain the **application** of such products to a typical problem. No endorsement of any specific product is intended.

The Keithley System 570

The measurement and control system for the liquid level/flow control system is based on a Keithley System 570 data acquisition work station, which is an example of the plug-in board type of interface instrument module. The System 570 is a member of the Series 500 family of data acquisition instruments from Keithley Instruments Inc. The family is a modular system consisting of an expansion chassis containing a precision power supply and base board with slots for 10 plug-in modules. Various signal conditioning, conversion, and control modules permit the user to tailor the system to specific applications. Some of the available modules for the 570 include analog input, high-level input, low-level input, isolated high-level input, isolated low-level input, strain gauge and RTD, analog-to-digital, high-resolution analog-to-digital, analog output, high-resolution output, current output, high-speed analog output, digital input, digital output, pulse and frequency input, and power control modules. The Series 500 measurement and control system is built around the memory architecture of the IBM PC/XT/AT, 386, 486, PS/2, and 100% compatible computers and is linked to the computer through a bus extension interface that resides in the host computer. Software choices include VIEWDAC and LABTECH NOTEBOOK. Supported high-level languages include BASIC and C.

The Keithley 570 has capacity for 32 single-ended, or 16 differential analog inputs, 2 analog outputs, 16 digital inputs, 16 digital outputs, and 16 relay controls. Their locations on the Keithley 570 motherboard and the numbering of the analog and digital input and output channels are shown in Fig. 15.13, along with the wire connections for the liquid level/flow control application.

Analog outputs are generated by DACs with 12-bit *resolution,* less than 100 ms *settling time* (to 0.01%), and *maximum nonlinearity* of ±0.012%. The range of the analog outputs is controlled by switches in the interface card plugged into the computer. The setting for the present application is −10 V to +10 V. Although the associated maximum current drain specification on an analog output is 4 mA, the Keithley 570 is short-circuit protected and will not be damaged by higher current drains.

Analog inputs are acquired by 12-bit *successive-approximation* ADCs with 25-ms *conversion time,* and 5-ms *sample-and-hold acquisition time.* For the IBM PC–type host computer used in the liquid level/flow control application, this implies that input sampling rates can go as high as 33 kHz. The expected analog input range is controlled by a dip switch on the Keithley 570 motherboard that is also set for −10 V to +10 V inputs for the present application.

LABTECH NOTEBOOK Software

Although a high-level language program (in BASIC, for example) can be written to communicate directly with the Keithley 570 hardware, the software used in the liquid level/flow control system is a menu-driven DOS version of Laboratory Technologies Corp.'s LABTECH NOTEBOOK. New DOS versions and Windows versions of NOTE-BOOK are also available. A key feature of LABTECH NOTEBOOK is its flexibility. Each data channel can be set up separately with different sampling rates and other characteristics. Data acquisition can commence as soon as the program starts, or it can be triggered by an external source such as the pattern of data on an input channel, or by the operator. Data can also be taken from a disk file and utilized by the program as though they came from an external device. The software can operate in *foreground/background* (other PC programs can run while data acquisition and control is in progress). The program is also capable of *real-time* data analysis with the results available for output on a data channel and/or for display, all in real time. There are also interfaces to other analysis, file management, and graphics programs such as Lotus 1-2-3, Symphony, or RS/1.

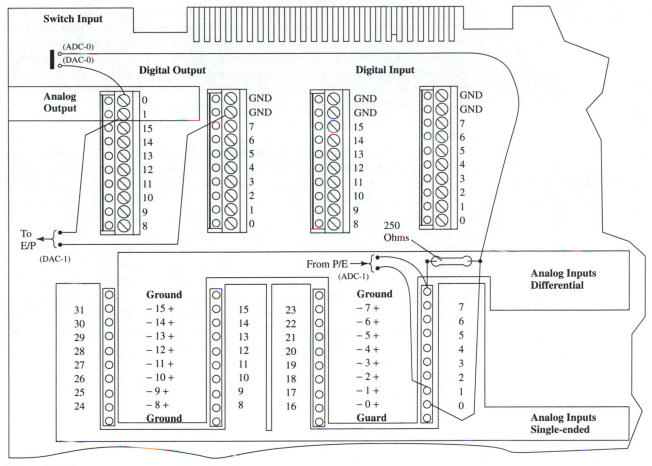

Figure 15.13

**Some Keithley 570
Motherboard Components
and Connections**

In the following description of the liquid level/flow control application, the software details are suppressed in favor of algorithmic statements of the essential objectives of the program. Thus the application can be duplicated on similar equipment using a different software package, or language, or operating system, and so on.

15.4 MICROCOMPUTER REALIZATION OF A LIQUID LEVEL/FLOW CONTROL SYSTEM

The objectives of the liquid level/flow control implementation described in this section are to further illustrate:

1. Data acquisition by a microcomputer
2. The characteristics of liquid level/flow control system components such as flow control valves and pressure transducers
3. Control system design by analysis and digital simulation
4. Microcomputer realization of control systems

The implementation is in three segments: In the first segment, a number of calibration experiments are performed to identify the unknown parameters in the system model. In the second part, optimum parameter values for a PI-feedback controller are determined analytically (and by computer simulation), using the results from the first part. Finally, in the last segment, the main control experiment is carried out, guided by the analytical (and simulation results). The numerical simulation part of this scheme, the subject of Chapters 6 and 14, is actually given in part as an exercise (see Practice Problem 15.16).

Physical Elements and Configuration of Control System

As shown in Fig. 15.14, the physical system consists of a (standpipe) tank with an arrangement of a three-way *(T-ported) plug valve*, a *bypass* line, and a $\frac{1}{2}$-in. *globe valve* (outlet valve) for selecting different tank outflow characteristics, two pressure transducers (E/P and P/E), a pneumatic control valve, and a microcomputer measurement and control workstation. The P/E *(pneumatic-to-electric)* transducer is used to measure the pressure exerted by the liquid in the tank, which is proportional to the liquid level above the sensing point. The E/P *(electric-to-pneumatic)* transducer converts the voltage output from the workstation into a pressure signal to the control valve. The measurement and control system (workstation) consists of a Keithley System 570 instrument module (described in the previous section) hooked up to an IBM PS/2 model 30 microcomputer. Two ADC channels (0 and 1) are used for external data acquisition, while two DAC channels (0 and 1) are used for voltage output. The analog

Figure 15.14

Liquid Level/flow Control System

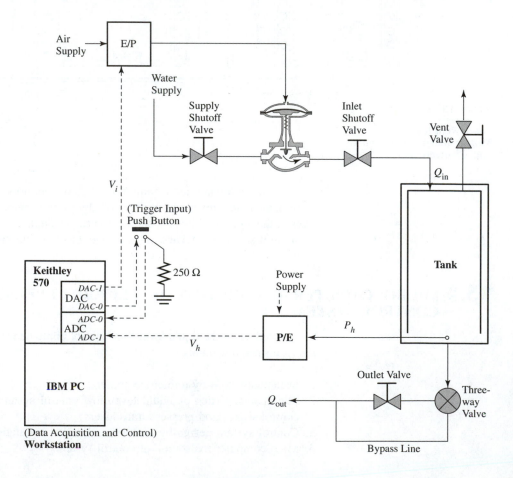

input channel-0 (ADC-0) monitors a push-button switch *(trigger)* input that (recall the contact closure pulse input interface circuit, Fig. 15.7a) is powered by (a 1-V supply) voltage output from the System 570's analog output channel-0 (DAC-0). The tank level voltage signal V_h (leaving the P/E) is acquired on analog input channel-1 (ADC-1), while the command voltage to the E/P V_i is output on analog output channel-1 (DAC-1).

Realization of Object of Control and Other Elements

The object of control, the tank, is actually two standpipes, one inside the other and in communication at the bottom. However, both tanks are considered one equivalent tank with the equivalent liquid level the same as the internal tank liquid level and the corresponding equivalent tank internal cross-sectional area equal to 153.85 cm^2. The *vent valve* for the internal tank is left permanently in the open position so that the top of the equivalent tank is always at atmospheric pressure. The flow into the tank is regulated by the computer control system through voltage commands issued to the E/P, while the level of liquid (water) in the tank is measured by the P/E transducer as described previously. The three positions of the three-way drain valve are OFF (no flow out of the tank), OUTLET (flow is through the outlet valve branch), and BYPASS (flow is through the bypass line). The bypass line actually introduces more resistance to the outflow than the outlet valve at its current opening. The outlet valve opening is fixed in this application.

The E/P transducer is a Moore Products Co. Model 77-16 electric-to-pneumatic transducer. For operation it requires a source of clean, dry, oil-free instrument air at a supply pressure of 20 psig (normal; 30 psig maximum and 18 psig minimum). The specifications include an input range of 4 to 20 mA, DC (16 mA span), output range of 3 to 15 psig, and input impedance of 185 Ω. However, the actual operating ranges depend on the state of calibration of the transducer. The 4–20 mA on 185 Ω gives a voltage input range of 0.74 to 3.7 Vdc. Since the expected maximum current drain on the Keithley 570 analog output is 4 mA, with a 0.74- to 3.7-V output, there is considerable excess current drain by the E/P. This is one of the typical conditions in which a valve positioner or at least a signal conditioning circuit (such as the Keithley series 500 current loop output module, 500-AOM3) is used to boost the output current level. There is also a Model 77-3A E/P with a 0–4 mA, DC input range, as well as voltage output models. However, no such additional or alternative component was used in the present implementation. Although the Keithley 570 would not be damaged by the current levels that obtained in this application, the accuracy could be lowered below rated value by the excess currents.

The P/E transducer is a Moore Model 781P4 pneumatic-to-electric transducer that requires a 3 to 15 psig pressure input signal and has an output specification of 0–10 Vdc into 0–20 mA load, maximum. The transducer also requires an electrical power (115V, 50/60 Hz) supply. Generally, a warm-up period of about 10 minutes should be allowed for the P/E transducer. The pressure input to the P/E transducer comes from a Moore Model 174P *pneumatic pressure transmitter* (see Fig. 15.15) that senses the pressure near the bottom of the tank. The transmitter input range is 10–110 cm water, and it also requires a 20 psig instrument air supply.

The control valve is a Fisher Controls Co. type 513RGS pneumatic *(diaphragm-actuated)* control valve. It is a (diaphragm) pressure-to-open valve. The input range is 3–15 psi, and the corresponding (stem) travel is $\frac{3}{4}$ in. The valve size and rating are $\frac{1}{2}$ in. and 150 gpm, respectively. The flow characteristic is specified as $\frac{1}{4}$ SP M-Flute. This behavior is modeled as a *square-root* characteristic.

Component Calibrations

Although the specifications of the pressure transducers and the control valve are available, and the geometric details of the water tank system are known, calibration of the relevant subsystems is necessary to obtain the true *as-installed* and *operating* conditions and to improve the analytical representations of the behavior of these elements.

Calibration of P/E Transducer and ADC

The P/E transducer measures the level of liquid in the tank (h) as a pressure signal (P_h) and converts this to a voltage signal (V_h). The voltage signal is converted to a digital value (v_h) by the ADC in the Keithley 570, which sends the digital value to the microcomputer. The ranges of the various signals are identified in Fig. 15.15a. Note that the pressure transmitter associated with the P/E transducer has an input range specification of 10–110 cm of water, and the meter scale on the outside of the liquid tank is attached at a distance of about 10 cm above the bottom of the tank. The calibration to be determined (seven data points are used) is the relation between the liquid level in the tank h (cm) and its measurement v_h. This relation is assumed to be of the form

$$h = \text{constant} + \text{slope} \times v_h \tag{15.8a}$$

Note that this relation encompasses both the pressure transmitter and the P/E transducer, and since the calibration data are acquired using the Keithley 570, the calibration also includes the ADC characteristics. The setup of the data acquisition program assumes that the entire calibration process can be completed in less than 20 min (1200 s).

Procedure

1. A well-regulated air supply set at 20 psig is applied to the pressure transmitter and maintained at this pressure level. With the tank inlet shutoff valve closed and the water supply shutoff valve open (see Fig. 15.14), power is applied to the computer, the Keithley 570, and the P/E transducer, which are allowed to warm up. The LABTECH NOTEBOOK was programmed for the following objectives:

   When the program starts, begin to *continuously* (that is, at a sampling rate of 1.0 Hz)

   Output 1.0 V on DAC-0 (this powers the push-button switch)
   Output 3.5 V DAC-1 (this opens the control valve to allow water to enter tank)
   Sample analog input ADC-0 (to detect switch press)
   Sample analog input ADC-1 (tank-level signal)
   Display both analog input signals in windows (for visual cues)

   On each receipt of a switch input (voltage on ADC-0 > 0.9 V)

   Wait for 15 s then record the input value on ADC-1, that is,
   Display the value in a meter window as well as write (append) it to a disk file

2. The data acquisition program is started with the GO command (*normal* start), and the computer opens the inlet flow control valve if it is not already open.

3. With the tank outlet three-way valve set to the OFF (shut) position, the tank inlet shutoff valve is opened to fill the tank to, say, the 70-cm level on the attached scale. This may be better accomplished by overfilling the tank and using the three-way valve to adjust the tank level. The reading of the tank level (voltage input on ADC-1) is traced continuously in the window on the computer screen and can be monitored to determine when the level stabilizes.

Figure 15.15

Liquid-level Data Acquisition System and Calibration

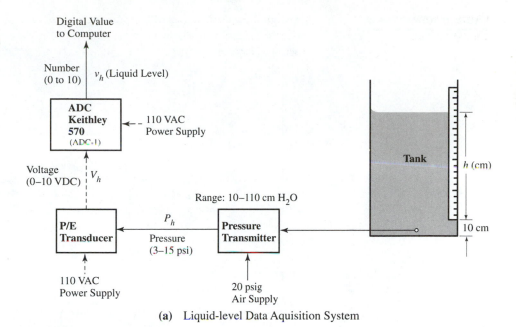

(a) Liquid-level Data Aquisition System

	A	B	C
1	**System Dynamics & Control**		
2	**Liquid Level/Flow Control**		
3	Calibration of P/E & ADC		
4	Date: mm-dd-yy, Time: 18:20:29.39.		
5	Level	Tank Level	Linear Fit
6	volts	h (cm)	h (cm)
7	9.7313	80	80.71214
8	8.7738	70	70.01194
9	7.8456	60	59.63918
10	6.8735	50	48.77582
11	6.1163	40	40.31401
12	5.1832	30	29.88649
13	4.3576	20	20.66029
14			
15		Slope	11.17514
16		Intercept	−28.0365

(b) Spreadsheet Analysis of Tank-level Data

4. After the tank level reading settles, the push button on the front of the Keithley 570 is depressed briefly (held just long enough for the switch input trace in the window to register a jump). On receipt of the switch input, the program waits for about 15 s and then samples the ADC. The sampled value is displayed in the meter (digital window) as well as recorded in the data file **PtoEcal.prn.**

5. Next, by means of the three-way valve, the tank level is adjusted to 60 cm on the scale (note that if the desired level is overshot, the tank can be refilled slightly by opening the tank inlet shutoff valve), and data sampling is triggered again by depression of the push button. In 15 s a sample is taken, and the value in the meter

Figure 15.15 continued

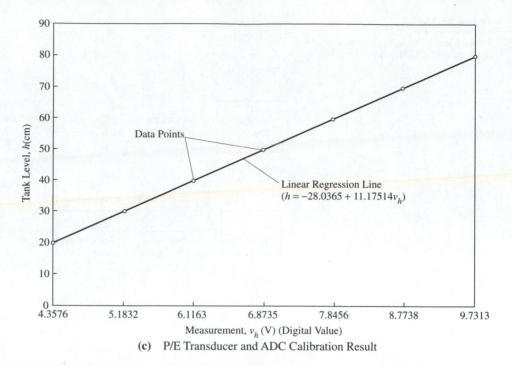

(c) P/E Transducer and ADC Calibration Result

window changes to the reading for the new level. This value is also recorded (*appended* to the old value) in the data file **PtoEcal.prn.**

6. Step 5 is repeated for 50-, 40-, 30-, 20-, and 10-cm tank scale levels, and then the data acquisition program is terminated.

Result   Since **PtoEcal.prn** is an ASCII file, it can be imported into a spreadsheet or other analysis program that reads the ASCII format to determine the relationship between h and v_h. The present result is the first column of the typical spreadsheet segment shown in Fig. 15.15b. The second column in the spreadsheet is the tank level, which is 10 cm higher than the scale readings referred to in the procedure. The corresponding tank-level values determined by a linear regression analysis of the present data are shown in the third column of the spreadsheet. Finally, Fig. 15.15c shows a graphical display of the data and this result, that is the calibration:

$$h = -28.0365 + 11.17514v_h \tag{15.8b}$$

Calibration of Tank Outlet and Bypass Drain Valves

Assuming typical short constriction behavior, we can take the calibration functions as

$$Q_{\text{out}} = Ch^r; \text{ (drain valves)} \tag{15.9}$$

where for the *outlet valve*, $C = C_o$, $r = r_o$, and for the *bypass valve* (line), $C = C_b$, $r = r_b$ are the unknowns. For zero tank inlet flow, the flowrate can be obtained from a measurement of tank level as a function of time, since in this case,

$$Q_{\text{out}} = -A \frac{dh}{dt}; \qquad Q_{\text{in}} = 0 \tag{15.10}$$

where A is the equivalent tank internal cross-sectional area.

Procedure

1. Electrical power and air are applied to the units as shown in Fig. 15.16a, and NOTEBOOK is programmed according to the following objectives:

    When the program starts, begin to *continuously* (at sampling rate of 0.5 Hz)

        Output 3.5 V on DAC-1 (this opens the control valve to allow water to enter tank)

    When data acquisition begins, *continuously* (at sampling rate of 0.5 Hz)

        Sample clock input (read time elapsed)

        Sample analog input ADC-1 (tank-level signal)

Figure 15.16

Liquid Flow Control System and Calibration

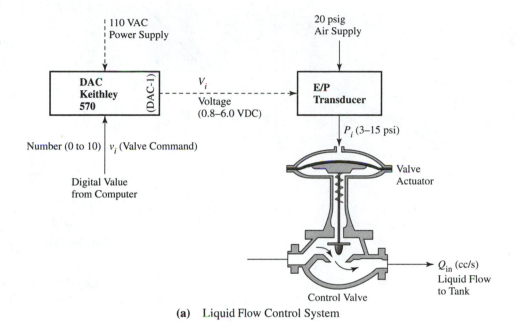

(a) Liquid Flow Control System

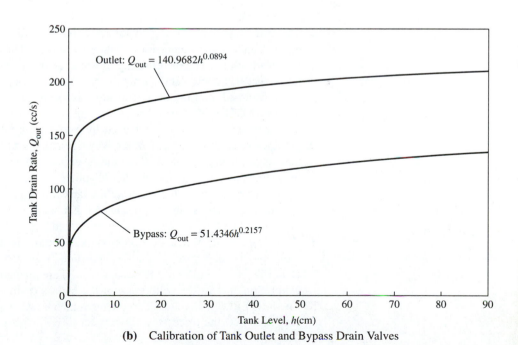

(b) Calibration of Tank Outlet and Bypass Drain Valves

Figure 15.16 *continued*

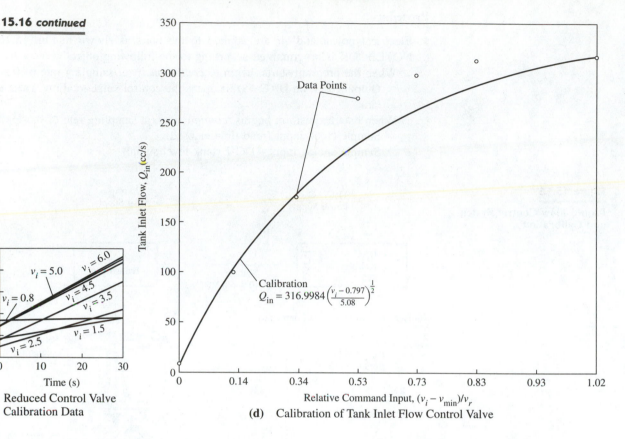

(c) Reduced Control Valve
     Calibration Data

(d)   Calibration of Tank Inlet Flow Control Valve

Display in a window a plot of tank-level reading versus time
Record both the time and tank level readings in a file (overwrite, not append, mode)
Continue procedure for a duration of 120 s, unless data acquisition is terminated earlier.

2. The three-way valve is set to the OFF position, the tank inlet shutoff valve is opened, and the data acquisition program is started. The control valve opens and the tank starts filling. The file **Outlet.prn** simultaneously records the time (channel) and tank-level (channel) data. At about the 75-cm (scale) level the tank inlet shutoff valve is closed, and the data acquisition is terminated using ESC plus any key sequence (NOTEBOOK is returned to its main menu).

3. The acquisition program is reloaded by executing a GO command, and when the graphics screen displays, the three-way valve is set to the OUTLET position. The tank begins to empty, and at about the 70-cm (scale) level, data acquisition is started by hitting any key on the computer keyboard. Note that although data recording to the file **Outlet.prn** occurred during the tank filling process of step 2, the record of the emptying process in the present step is not appended to the previous recording on **Outlet.prn** but is overwritten on it instead.

4. When the tank level nears the 0 cm mark, sampling is stopped as in step 2, and the three-way valve is reset to the OFF position. Then from the File Setup menu, the Data File Name and relevant Header Line(s) are changed to correspond to the next calibration, that is, the calibration of the bypass drain line.

5. The NOTEBOOK is returned to the main menu, and steps 2 and 3 are repeated, except this time, the three-way valve is set to BYPASS in step 3.

Result   Each of the two data files contains two columns of data: time in the first column and the measurement of h, that is, v_h in the second column. From the P/E calibration determined earlier, the v_h values are converted to h values, which are required for determination of the calibration function, equation (15.9). An apparently straightforward approach to finding this function is to differentiate the resulting $h(t)$ numerically and multiply by $-A$ to obtain Q_{out}, as given by equation (15.10). Then from the h and Q_{out} value pairs, the function (15.9) can be fit to the data from each calibration experiment. The problem with this approach, however, is that numerical differentiation is a very noisy process that should be avoided whenever possible.

An alternative approach is to determine the calibration directly from the $h(t)$ data using some nonlinear approximation technique. Consider the following intuitive solution, which can be solved by iteration (or a *goal seek procedure*) in a spreadsheet: Let

$$Q_{out} = -A \frac{dh}{dt} = Ch^r$$

then

$$\frac{dh}{dt} = \left(\frac{-C}{A}\right) h^r$$

which has the solution

$$\left(\frac{1}{1-r}\right)[h^{1-r} - h_0^{1-r}] = \left(\frac{-C}{A}\right)t = \left(\frac{1}{c_2}\right)[h^{c_2} - h_0^{c_2}]$$

where $h_0 = h(t=0)$, and $c_2 = 1 - r$. Solving for t, we obtain

$$t = \left(\frac{-A/C}{c_2}\right)[h^{c_2} - h_0^{c_2}]$$

or if we define

$$z = h^{c_2}; \qquad c_2 = 1 - r, \quad c_1 = \frac{-A/C}{c_2}, \quad c_0 = -c_1 h_0^{c_2} \qquad \text{(15.11a)}$$

then

$$t = c_0 + c_1 z \qquad \text{(15.11b)}$$

Equation (15.11b) describes a straight line with slope equal to c_1 and intercept equal to c_0. If we guess the value of c_2, then the z values can be computed at each data point using this estimate. Equation (15.11b) can then be solved by linear regression for c_0 and c_1. These in turn will give a new value for c_2 from the last equation (15.11a), since h_0 is known. New z values are then computed and the iteration can continue until c_2 converges. With the final c_1 and c_2, the first and second equations (15.11a) give r and C. For the present (outlet and bypass valves) data, the results, which are also shown graphically in Fig. 15.16b, are

$$Q_{out} = 140.9682 h^{0.0894}; \quad \text{(outlet drain valve)} \qquad \text{(15.12a)}$$

$$Q_{out} = 51.4346 h^{0.2157}; \quad \text{(bypass drain valve)} \qquad \text{(15.12b)}$$

Calibration of Flow Control Valve, E/P Transducer, and DAC

The Keithley 570 DAC converts a digital value v_i from the computer into an output voltage. The E/P transducer converts this voltage signal V_i into a control air pressure signal P_i, which acts on the valve actuator to produce a corresponding control valve lift. The tank inlet flow Q_{in} is a function of the control valve lift. Figure 15.16a shows the nominal ranges of the signals. Note that in Fig. 15.16a the range of the command voltage to the control valve actuator is given as 0.8–6 Vdc (not the 0.74–3.7 Vdc suggested by the E/P specifications). The 0.8–6 Vdc range for V_i was determined by experiments of the type described here. To establish the top of the range, higher voltages were sent to the E/P, and the corresponding tank inlet flow rates Q_{in} were measured until the point was reached at which an additional increase in V_i caused essentially no further increase in Q_{in}. The voltage in this case was about 6.0 V (that is, $V_i = 6$) and the corresponding flow was $Q_{in} = 316.9984$ cc/s. The bottom of the voltage range was similarly established around $V_i = 0.8$. Although the E/P will still respond without damage to command inputs outside the above range (but still within the DAC output range), the actuator will be considered *uncontrollable* for such inputs.

For a fixed pressure drop across the control valve, the square-root-type calibration sought (recall equation (15.5b)) is

$$Q_{in} = C_i \left(\frac{V_i - V_{min}}{V_r} \right)^{1/2} \tag{15.13}$$

where C_i, V_r, and V_{min} are unknowns to be determined. Considering the preceding determination of the command voltage range, we can simply establish the unknowns in equation (5.13) by (a) fixing V_{min} at a value only marginally smaller than 0.8, for example, $V_{min} = 0.797$, and (b) requiring that equation (15.13) yield the measured maximum flow at $(V_i - V_{min})/V_r = 1$. The second condition gives $C_i = 316.9984$, and V_r can then be chosen to provide a reasonably good fit between equation (15.13) and the experimental data.

As in the calibration of the outlet and bypass valves, Q_{in} can be obtained from a measurement of tank level as a function of time, since for zero tank outlet flow (drain),

$$Q_{in} = A \frac{dh}{dt}; \qquad Q_{out} = 0 \tag{15.14}$$

However, since Q_{in} is constant for each valve opening (fixed V_i), h versus t is, unlike in the previous calibration, a straight line. These lines for seven values of V_i ranging from 0.8 to 6.0 are used to calibrate the inlet flow control valve subsystem, that is, to determine equation (15.13).

Procedure: Electrical power and 20 psig instrument air supplies are applied to all units as necessary, and the water supply shutoff valve is opened fully.

1. The computer and the Keithley 570 are turned on, and NOTEBOOK is configured essentially as in the calibration of the drain valves. However, the data file is set up also to record simultaneously with the time and tank level, the voltage command (number) output to DAC-1. The Data File Name includes the number "35" as a suffix, corresponding to a desired DAC-1 voltage output to the control valve of 3.5 V.
2. With the three-way valve set to OFF and the tank inlet shutoff valve closed, the NOTEBOOK program is started and allowed to run for about 20 s (monitored on the time display window) before being stopped by entry of an ESC-plus-any key se-

quence. The valve stem will have completely traveled to the position corresponding to the command voltage by this time.

3. The program is reloaded with the GO command, and when the graphics screen appears, the tank inlet shutoff valve is opened **fully** to begin filling the tank. When the tank level reaches around the 5-cm mark, data sampling is initiated by hitting any key on the computer keyboard.

4. The tank level is monitored, and sampling is halted by entering ESC once the tank level reaches about the 60-cm mark. The tank inlet shutoff valve is quickly closed, and the three-way valve is set to BYPASS or OUTLET to drain the tank completely and then reset to OFF.

5. The NOTEBOOK main menu is reentered by hitting another key, and then the CHANNELS setup is adjusted so that the DAC-1 output is changed from 3.5 to 2.5. Next, the FILES setup Data File Name suffix is changed accordingly from 35 to 25. On completion of this task, the NOTEBOOK main menu is reentered, and steps 2 to 4 are repeated.

6. Finally, step 5 is repeated for setpoint values of, say, 1.5, 4.5, and 5.0, and the corresponding data file name suffixes: 15, 45, and 50. We assume, as previously discussed, that a similar process has already been utilized also to record the data for the setpoint values of 6.0 and 0.8.

Result   There are thus seven data files corresponding to voltage-to-valve inputs of 6.0, 5.0, 4.5, 3.5, 2.5, 1.5, and 0.8 V. Each data file is a record of time v_h (that is, measurement of h) and V_i (that is, volts to valve, which is constant). From the time and v_h data and the earlier calibration of the P/E, the v_h values are converted to h values. The plots of the resulting h versus time for each v_i are shown in Fig. 15.16c. The slopes of these lines, which are determined by linear regression, give, according to equation (15.14), the flowrates Q_{in} corresponding to each V_i. Equation (15.13) is then fit to the seven pairs of Q_{in} and V_i values to establish the calibration of the control valve. The choice $V_r = 5.08$ was made for the present application, and the resulting calibration, which is shown in Fig. 15.16b, is

$$Q_{in} = 316.9984 \left(\frac{V_i - 0.797}{5.08} \right)^{1/2} \tag{15.15}$$

Analytical Design and Computer Simulation

An analytical model of the liquid level/flow control system is required to design the controller. This model is simply derived as the combination of the tank description and the calibration of the inlet and drain valves. The resulting model is nonlinear, but it can be linearized about an operating point to facilitate the analytical controller design. However, the controller design through computer simulation can be carried out with the full nonlinear model. This is given as an exercise in Practice Problem 15.16. An alternative approach to a linear model is to fit a linear expression to the *process reaction curve* determined by experiment on the actual plant. The result of this *black box approach* will be used here to validate the analytical linear model.

Analytical Model

The tank behavior is given by equation (1.5c) or (3.32c):

$$\frac{dh}{dt} = \frac{1}{A}(Q_{in} - Q_{out}) \tag{15.16a}$$

where Q_{in} is determined by the control valve calibration or equation (15.15), and Q_{out} is given by the calibration of the drain valves or equations (15.12a) and (15.12b).

Linear Model   For a linearized model, we consider tank drainage through the outlet valve as the normal situation (flow through the bypass line is a disturbance input). Thus the relation between the controlled variable h and the control input V_i is

$$\frac{dh}{dt} = \frac{1}{A}\left[C_i\left(\frac{V_i - V_{min}}{V_r}\right)^{1/2} - C_o h^{r_o}\right] \tag{15.16b}$$

where C_i, V_{min}, V_r, C_o, and r_o were determined previously. We also consider for the purposes of the linear model, two operating or equilibrium points of $h = 30$ cm and $h = 50$ cm. From equation (15.16b), for an equilibrium level h_e,

$$C_i\left(\frac{V_i - V_{min}}{V_r}\right)^{1/2} = C_o h_e^{r_o}$$

and

$$V_e = V_i(h_e) = V_r\left(\frac{C_o}{C_i}h_e^{r_o}\right)^2 + V_{min} \tag{15.16c}$$

Substituting $h_e = 30$ and the values of the calibration constants, we obtain the command voltage necessary to maintain this equilibrium tank level as $V_e = V_i(h_e = 30) = 2.642$. Similarly, $V_i(h_e = 50) = 2.819$. Applying the linearization process described in Section 6.3 to the nonlinear system (15.16b), we obtain

$$\frac{dx}{dt} = ax + bu; \qquad x = h - h_e, \quad u = V_i - V_e \tag{15.17a}$$

$$a = -\frac{C_o r_o}{A}h_e^{r_o - 1}; \qquad b = \frac{C_i}{2A\sqrt{V_r(V_e - V_{min})}} \tag{15.17b}$$

For $h_e = 30$ cm, $a = -0.0037$, $b = 0.336$. For $h_e = 50$ cm, $a = -0.00232$, $b = 0.321$. Note that if the outflow was to go through the bypass line ($C_o = 51.4346$, $r_o = 0.2157$), the corresponding values of the linear model parameter a would be -0.00501 for $h_e = 30$ cm and -0.00335 for $h_e = 50$ cm. The values of b would remain the same as for the outlet valve.

A process reaction curve for an equilibrium level of 30 cm, for example, can be obtained by setting up the data acquisition and control program to acquire and record the tank liquid level as a function of time while sending out to the E/P a constant-voltage output given by $v_i = 2.642$.

With the supply and inlet shutoff valves **fully** open, the tank is filled beyond a desired initial state (say, $h = 60$ cm, which is 50 cm on the scale), and the three-way valve is set to the OUTLET position for normal drainage of the tank. When the tank level reaches the desired initial state, data acquisition is started and sustained until the liquid level comes to equilibrium.

The same experiment simulated with the linearized model simply means solving the equation

$$\frac{dx}{dt} = -0.0037x + 0.336u \qquad\qquad \text{(15.18a)}$$

subject to

$$x(0) = 60 - 30 = 30 \quad \text{and} \quad u = 0$$

that is, the free response from the given initial state. The solution,

$$x(t) = 30e^{-0.0037t} \quad \text{or} \quad h(t) = 30 + 30e^{-0.0037t} \qquad\qquad \text{(15.18b)}$$

is displayed in Fig. 15.17a along with the process reaction curve obtained by the preceding procedure. This result suggests that the present linear model is adequate for the purpose of illustrating the implementation of microcomputer control systems.

Initial PI-Controller Settings

The control algorithm that is implemented in the microcomputer is simply PI-control. We can obtain initial PI-controller settings by design using the linear model. In principle this design should be further investigated using digital computer simulation of the full (nonlinear) system model (as suggested in Practice Problem 15.16) to improve the control system performance, especially with respect to such details as the maintenance of internal variables and external signals within their physical or specified limits and optimization of the choice of sampling period. The ultimate controller design is, of course, still subject to *fine-tuning* after implementation on the physical system.

Consider a PI-controller design that satisfies both a 25% overshoot and a 5% settling time of 90 s for a setpoint $h = 50$ cm. The analytical linear model for this setpoint is

$$\frac{dx}{dt} = -0.00232x + 0.321u; \qquad h = x + 50, \quad V_i = u + 2.819 \qquad \text{(15.19a)}$$

which gives the plant transfer function

$$\frac{X(s)}{U(s)} = G_p(s) = \frac{b}{s - a} \qquad\qquad \text{(15.19b)}$$

where $a = -0.00232$ and $b = 0.321$.

As we alluded to earlier, the computer-controlled liquid level/flow system is a sampled-data system. The control valve command v_i (m_k in the case of the model; see Fig. 15.17b), which is generated in the computer, is a value sequence that is converted into a *staircase* pattern (assumed zero-order held) function V_i (u for the model) by the Keithley 570 DAC. The measurement of the tank level v_h is a sampled function, with

Figure 15.17

Modeling and Design of
Liquid Level/flow
Control System

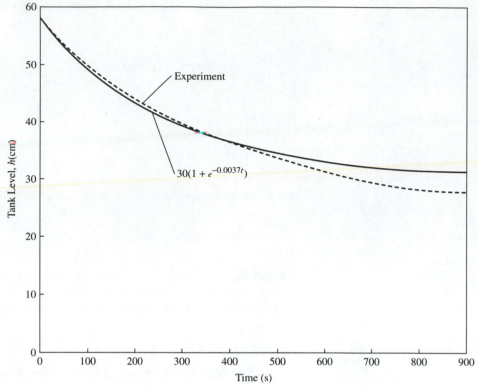

(a) Process Reaction Curve and Response of Linear Model

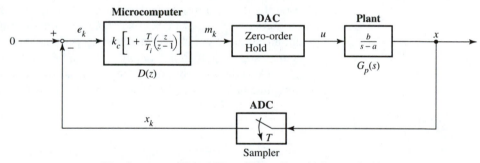

(b) Structure of Liquid Level/Flow Control System Model

the sampling being accomplished by the ADC. Figure 15.17b shows the structure of the control system based on the preceding linear representation of the plant.

Using the z-domain design approach of Chapter 14, we get the model plant pulse transfer function from equation (14.6c):

$$G_p(z) = (1 - z^{-1}) Z\left\{ \mathcal{L}^{-1}\left[\frac{b}{s(s - a)} \right] \right\} \tag{15.19c}$$

$$= (1 - z^{-1}) Z\left\{ \int_0^t b e^{at} dt = \frac{b}{a}(e^{at} - 1) \right\} = \frac{b}{a}\left(\frac{c - 1}{z - c} \right)$$

Figure 15.17 *continued*

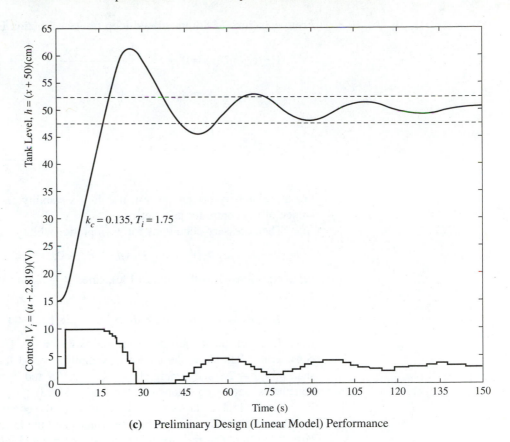

$k_c = 0.135, T_i = 1.75$

(c) Preliminary Design (Linear Model) Performance

where $c = e^{aT}$, and T is the sampling period. The Euler position (we assume we do not have an integrating-type actuator) algorithm, equation (14.2a), adjusted for PI-control, yields

$$m_k = k_c \left[e_k + \frac{T}{T_i} \sum_{i=0}^{k} e_i \right] + m_B \tag{15.20a}$$

Taking $m_B = 0$ for the purposes of this design, and recalling equation (8.50c), we obtain the z-transform of equation (15.20a):

$$M(z) = k_c E(z) + \frac{k_c T}{T_i}(z^{-k} + z^{-k+1} + \cdots + z^{-1} + 1)E(z) \tag{15.20b}$$

$$= k_c \left[1 + \frac{T}{T_i}\left(\frac{z}{z-1} \right) \right] E(z)$$

so that the digital PI-controller transfer function is

$$\frac{M(z)}{E(z)} = D(z) = k_c \left[1 + \frac{T}{T_i}\left(\frac{z}{z-1} \right) \right] = k_c \left(\frac{k_i z - 1}{z - 1} \right); \quad k_i = 1 + \frac{T}{T_i} \tag{15.20c}$$

From equation (14.8a) the closed-loop transfer function is given by

$$1 + G_p(z)D(z) = 0 = z^2 + \left[\frac{k_c b(c-1)k_i}{a} - 2 \right] z + \left[1 - \frac{k_c b(c-1)}{a} \right] \quad \textbf{(15.21a)}$$

$$= d_0 z^2 + d_1 z + d_2$$

where

$$d_0 = 1; \quad d_1 = \frac{k_c b(c-1)k_i}{a} - 2; \quad d_2 = 1 - \frac{k_c b(c-1)}{a} \quad \textbf{(15.21b)}$$

For a preliminary design, we can use Jury's stability test to provide bounds for the choice of the controller parameters.

The necessary conditions for asymptotic stability, equation (10.14a), are

$$1 + d_1 + d_2 > 0 \quad \text{and} \quad 1 - d_1 + d_2 > 0; \qquad \Rightarrow 1 + d_2 > |d_1| \quad \textbf{(15.21c)}$$

From equations (10.14b) and (10.14c), since $\mathbf{V} = d_0$ and $\mathbf{W} = d_2$, the sufficiency conditions are

$$d_0 + d_2 > 0, \quad \text{and} \quad d_0 - d_2 > 0; \qquad \Rightarrow |d_2| < 1 \quad \textbf{(15.21d)}$$

Considering the plant equivalent time constant of $\tau = -1/a \approx 431$ s, we probably have considerable latitude in the choice of sampling period for this system. However, we shall take $T = 2$ s and simply choose $k_c = 0.135$ and $T_i = 1.75$, arbitrary values that nonetheless satisfy equations (15.21c) and (15.21d).

Figure 15.17c shows the step response of the controlled model plant subject to the PI-controller with these parameter values and the initial state $x(0) = -35$, that is, $h(0) = 15$. Note that the model output values for x and u were converted into the corresponding h and V_i values, which are plotted. It is evident from this figure that the overshoot and settling time specifications are well met by the present design. The simulation displayed in Fig. 15.17c was carried out using the same hybrid discrete- and continuous-time procedure (MATLAB script M-file, **exmp14_1.m**) employed in Example 14.1. The present program is given in script M-file **fig15_17.m** (**SC16.MCD** for Mathcad); **On the Net.** In the program, the control valve command v_i is restricted to the range $0 \leq v_i \leq 10$, which is determined by the DAC output range, rather than to the values determined by the E/P calibration or input specifications. Although relaxation of variable constraints of this nature can be removed in the further design by simulation using the full (nonlinear) system model (see Practice Problem 15.16), the present moderate conditions on v_i were imposed so as to be consistent with the main control experiment on the actual plant, which is described next. Restriction of the control valve command to the calibrated range was not implemented in the control experiment because the E/P can sustain the higher input voltages, and apparently, in the absence of a valve positioner or current loop conditioning module in the control valve drive circuit, the actuator response speed can be improved by allowing it to operate within the so-called uncontrollable regions whenever the application demands a sufficiently rapid change in the tank inlet flow.

The Main Control Experiment

The main control experiment duplicates the computer simulation of PI-control shown in Fig. 15.17c but applied to the actual liquid level/flow system. Also, since a weaker performance is anticipated, the experiment and hence data capture are continued for more than the model total simulation time of 150 s. Indeed, the cumulative run time for the experiment is 8 min (480 s), but this includes more than 2 min of additional response time following the application of a disturbance input near $t = 350$ s. The disturbance was created by switching the tank drain from the outlet valve to the bypass line after the response had apparently settled within the specified tolerance of the setpoint. Figure 15.18a shows a comprehensive block diagram of the liquid level/flow control system, including the major structural and operating components and the previously determined calibrations. The result of the PI-control experiment, that is, the tank-level measurement and the actuator command input (the manipulated variable) as functions of time are shown in Fig. 15.18b.

PI-Control and Analog Output Channel Setup

The LABTECH NOTEBOOK analog output channels can implement the PID algorithm. The specific equation implemented is

$$\text{Output} = P \times e(t) + I \times \int e(t)\, dt + D \times \frac{de}{dt} \tag{15.22a}$$

where P is the loop gain, I is the loop reset, D is the loop rate, and $e(t)$ is given by

$$e(t) = \text{loop setpoint} - \text{input channel signal} \tag{15.22b}$$

Although the actual procedure used by NOTEBOOK to compute the integral term can differ from the Euler algorithm on which our preliminary design was based, both are numerical or discrete-time software procedures, and we can compare equation (15.22a) with equation (15.20a) and take

$$\text{Loop gain,} \quad P = k_c \ (= 0.135) \tag{15.22c}$$

$$\text{Loop reset,} \quad I = \frac{k_c}{T_i} \left(= \frac{0.135}{1.75} = 0.077 \right) \tag{15.22d}$$

$$\text{Loop rate,} \quad D = 0 \tag{15.22e}$$

To implement the present preliminary design, the analog output DAC-1 is set up like the last (inlet flow control valve) calibration experiment but modified so that the command voltage output is given by equations (15.22). Note that the existing setup has been P-action. For example, when the NOTEBOOK program objective in the calibration experiment called for output of 3.5 V on DAC-1, this was actually accomplished with a CHANNEL setup having a loop setpoint of 3.5, a loop gain of $P = 1.0$, and an *internally* generated analog input channel signal of 0.

For this experiment, the loop setpoint is the voltage equivalent (consistent with v_h) of the desired tank level, $h = 50$ cm (see Fig. 15.18a). That is,

$$v_h = \frac{1}{11.17514}(h + 28.0365) \Rightarrow \text{loop setpoint} = 6.983 \tag{15.22f}$$

The input channel is ADC-1, which is set up like the last calibration experiment, so that the input channel signal is the measurement of the tank level (that is, the required feedback signal) v_h. Finally, the sampling rate is 0.5 Hz, which gives a sampling period of 2 s, and the duration of the data acquisition interval was changed from 120 to 480 s. Note that according to the program, the input channel signal does not come on until

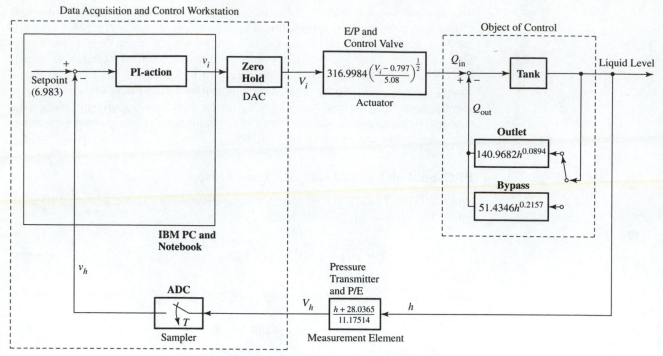

(a)   Structural and Operating Components and Calibration

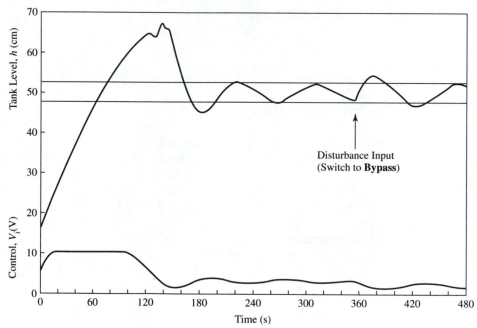

(b)   Performance of Liquid Level/Flow System Subject to Preliminary PI-controller

Figure 15.18

**PI-control Experiment on
Liquid Level/flow
Control System**

data acquisition begins. Prior to this time but after the program has started, a constant voltage that depends on the value of the setpoint is output on DAC-1.

Procedure for the Experiment

Electrical power is applied to all units as usual. The air supply pressure is set at 20 psig, and the water supply shutoff valve is opened fully. The computer and the Keithley workstation are turned on, and NOTEBOOK is set up as in the calibration of the inlet flow control valve with the exceptions already discussed. The Data File Name (say, **LLFPIC50.PRN**) reflects the present experiment—closed-loop PI-control with setpoint of 50 cm.

1. The data acquisition and control program is started and run for about 20 s and then stopped using the ESC key followed by another key sequence. This opens the control valve.
2. Next, the GO command is executed to reload the program. The three-way valve is set to OFF, and the tank inlet shutoff valve is opened fully. The tank starts filling. When the tank level just exceeds the 5-cm mark ($h = 15$ cm), the three-way valve is set to the OUTLET position, and immediately a key on the computer keyboard is pressed to start the data acquisition.
3. The computer screen tank-level and time (meter) displays are monitored, and after the level indication settles well within $\pm 5\%$ of 6.983 (the setpoint) or at time 360 s, whichever comes first, the three-way valve is quickly turned from the OUTLET position to the BYPASS position to apply a disturbance input. The determination of when the tank level comes into the setpoint error band can be facilitated by adding two more traces to the window display, one showing a constant (line) at $1.05(6.983) = 7.332$ and the other indicating the value or line $0.95(6.983) = 6.634$. The program terminates after the 480-s cumulative run time. The tank inlet shutoff valve is then closed, and the tank is drained.

Results and Final Comments

The data recorded in the data file consist of time, tank level v_h, and control valve command V_i, values that can be brought into a spreadsheet or similar analysis program. From the P/E and ADC calibration, the v_h values can be converted to h values, and both h and V_i data can be plotted versus time on the same graph, as shown in Fig. 15.18b. The results show that with the preliminary controller settings, an overshoot of around 29.5% and a settling time of about 225 s are achieved. Thus the specifications are not met, although a substantial degree of control is achieved by this preliminary design. In particular, the system is able to recover from a moderate parameter upset in about 77 s. Following the disturbance input (switching from the outlet drain valve to the bypass line) that was applied at around $t = 352$ s, the response returned within the setpoint error band by $t = 429$ s.

The present controller settings can serve as the starting point for fine-tuning the controller to achieve better performance. Essentially, the controller could be fine-tuned by repeating the preceding PI-control experiment a number of times while adjusting the controller parameters around the preliminary values. However, given the distance of the present performance from the desired behavior, the tuning exercise can be facilitated considerably by first doing some gross tuning. As we have already suggested, the *gross tuning* experiments are best conducted by digital computer simulation using the full (nonlinear) system model. The present results also suggest the use of a more accurate model for further analysis and design of the system given the rather significant mismatch between the behavior of the actual system and the linear model when subject to PI-control. The fundamental problem of the linear model in this situation is that it is specific to a particular operating or equilibrium point (the setpoint), and until the response comes within a small neighborhood of this point, the model remains quite inaccurate. However, we conclude that significant practical aspects of the design of a controller, the implementation of a microcomputer-based control system, and the process of fine-tuning or improving such a control system have been illustrated in this section.

O n   t h e   N e t

- SC16.MCD (Mathcad): source file for Fig. 15.17c
- Fig 15_17.m (MATLAB): design of PI-controller for liquid level/flow control, Fig. 15.17c
- Auxiliary exercise: Practice Problems 15.16 and 15.17.

15.5 PRACTICE PROBLEMS

15.1 In Section 15.2 the optical type of shaft encoder was described. Can you describe two other types of shaft encoders?

15.2 Design a simple interface for a thermistor and your microprocessor.

15.3 A pulse-generating tachometer with signal conditioner such as in Fig. 15.7d is connected to one line of the digital I/O interface to a microcomputer. Assuming the tachometer is to be used to measure angular speeds in the range 30–300 rpm, recommend a sampling period for adequately sampling the tachometer output over this speed range, and the pulse width to be set on the one-short. Explain the factors that influenced your recommendations.

15.4 Using manufacturer catalogs, specification sheets, and other available information, design (that is, specify the components and draw the block diagram for) a computer data acquisition and control system for

(a) A DC motor-speed tachometer and speed controller using armature field voltage.

(b) A DC motor-speed tachometer and speed controller using pulse width modulation.

(c) A water-heater temperature control system utilizing a thermocouple circuit for temperature input and on-off heater control.

(d) An automatic weighing scale utilizing a strain gage bridge circuit for data acquisition. The system should rapidly debounce the weight signal and display the accurate weight.

15.5 What valve characteristic determines the sizing of a control valve?

15.6 Considering the definition of valve gain given in equation (15.2a), can you explain the meaning of *valve sensitivity coefficient?*

15.7 The volume flow rate Q and pressure drop ΔP for flow of water through a control valve are in consistent units $Q = 24$, $\Delta P = 4$. What is the corresponding valve sizing coefficient C_v?

15.8 The flow through a control valve when the stem is 50% open is 5 (rate of flow units). What will be the flow through the same valve when the stem is 75% open, assuming the pressure drop is the same and the valve lift–flow relationship is

(a) Linear.

(b) Quick opening with $\alpha = 10$.

(c) Equal percentage with rangeability = 10.

15.9 What do you consider is the significance in design of the *rangeability* of an equal-percentage or hyperbolic-type control valve?

15.10 Very often, instead of a direct connection between the controller and the valve, a positioner is used. Give three applications (conditions) that call for the use of a valve positioner.

15.11 What is the limiting factor determining the temperature range of a hydraulic actuator?

15.12 Compare further the response of the linear model of the liquid level/flow control system developed in Section 15.4 with the results of the PI-control experiment. In addition to the settling time and overshoot, consider also the rise time, decay ratio, and frequency of oscillation.

15.13 In Section 13.4 various perturbed system uncertainties were mentioned in connection with the H$_\infty$ optimal control, including parameters, ignored dynamics, and neglected nonlinearities. Which of these uncertainties is represented by the disturbance input applied to the liquid level/flow system during the PI-control experiment?

15.14 What are the roles of calibration and simulation in digital computer implementation of control systems?

15.15 Obtain the full nonlinear model relating the tank level h and the command voltage V_i for the liquid level/flow control system illustrated in Section 15.4, using the calibrations determined in that section. Using the nonlinear model and the case in which tank outflow is through the outlet valve, determine the value of the command voltage that will yield an equilibrium tank level of 50 cm.

15.16 Develop a computational software program to simulate the liquid level/flow control system for the nonlinear model obtained in Problem 15.15. Ensure that your computer model has a distinct module (section) for the controller, that it allows the tank exit to be switched between the bypass and the outlet flow valves, and that both the tank level h and the command voltage V_i as well as the error (the difference between the setpoint, in volts, and v_h) can be available as program outputs. Further, the program should permit the application of the limits 0.8 to 6.0 V on the command output v_i. Draw a block diagram of the liquid level/flow control system, indicating in each block the equations solved in your program. Use the program to carry out the following tasks:

(a) Simulate the step-response experiment (process reaction curve) shown in Fig. 15.17a, and compare the model output with the results in the figure.

(b) Simulate on the digital computer, proportional-and-integral feedback control of the level of the nonlinear tank for the setpoint $h = 50$ cm. The scenario should be as follows: The tank is initially at the 15-cm level and open to the outlet valve. After 180 s the tank exit is switched to the bypass line, and the simulation is continued for another 120 s. Determine, using trial and error and, as starting values, the controller settings determined with the linear model in Section 15.4, the value of the PI-controller parameters that generally maximize the speed of response but specifically minimize, for the segment before the disturbance input, the (5%) settling time and the overshoot while the limits on the command voltage are in effect. For your optimum design, plot on the same graph $h(t)$ and $V_i(t)$ versus time, and discuss these results relative to the experimental data shown in Fig. 15.18b.

(c) Repeat (b) with the limits on command voltage removed. Discuss the effect of the limits on the command voltage (dead zone and saturation) on controller performance.

15.17 After carrying out sufficient research on the industry involved to gain an adequate understanding of the process and to gather missing technical details, and following an approach similar to the liquid level/flow control system illustrated in Section 15.4, design for each of the following problems, a microcomputer-based control system with all necessary accessories, including sensors, actuators, and interfacing and signal conditioning components, required to realize the system. Your solution should include

(1) A general description of the process and a technical description/data sheet for each sensor and actuator employed.

(2) Diagram(s) showing detailed data connections between the computer, interface and signal conditioning circuits, and external equipment.

(3) Explanations of the equipment calibrations, if any, useful for implementing the control system, suggestions on the form of the calibration functions (estimates of values of parameters are not necessary), and the analytical model (linear or otherwise) of the process

(4) An overall block diagram of the computer control system as in Fig. 15.18a, with the proposed forms of the calibration functions and plant model entered in the appropriate blocks.

(a) A temperature control system to regulate the reaction in a continuous stirred tank reactor (CSTR). Choose a chemical process that involves no more than two feed streams and one product and illustrates (i) an endothermic reaction; (ii) an exothermic reaction. Note that exothermic processes are, in general, not *self-starting* and require some initial supply of sensible heat input.

(b) End-point position control of a robot arm. Note that varying payloads affect positional accuracy. Use the assumptions of a rigid body and low operating speeds to simplify the robot dynamics. Consider a simple three-degrees-of-freedom spherical coordinate robot.

(c) A temperature (thermal environment) control system for a large office complex. The system should utilize measurements of internal and external temperatures as well as information on the pattern of activity in sections of the building. A single but typical floor of such an office complex can be used to illustrate the solution.

(d) Control of drawing speed in continuous casting of steel. The coordination of various factors such as turndish pouring rate, molten metal temperature, and mold level should be explained.

(e) Control of temperature at the die exit or melt temperature of a single-screw plastics extruder. The melt temperature and output rate are critical to product quality. Important variables that affect the melt temperature include the temperature profile along the extruder (both barrel-wall and die-wall temperatures) and screw speed. Usually the extruder pressure is also an input. Assume a model of two to four heating zones for illustration of the process.

REFERNECES FOR PART THREE

1. Williams, S. B., "Feedback Theory," *Instruments and Control Systems,* vol. 39 (December 1966): 81–84.

2. Ziegler, J. G., and Nichols, N. B., Optimum Settings for Automatic Controllers," *Trans. ASME,* 64, no. 8 (1961): 759.

3. ———, "Process Lags in Automatic Control Circuits," *Trans. ASME,* 65, no. 5 (1943): 433.

4. ———, "Optimum Settings for Controllers," *ISEJ*, (June 1964): 731–734.

5. Smith, C. L., *Digital Computer Process Control,* International Textbook Company, New York, 1972.

6. Dahlquist, G., and Bjorck, A., *Numerical Methods,* Translated by Ned Anderson, Prentice-Hall, Englewood Cliffs, N.J., 1974.

7. Gustafson, R. D., "A Paper and Pencil Control System Design," *Journal of Basic Engineering., ASME Transactions,* 88 (June 1966): 329–36.

8. Anderson, N. A., "Frequency-Response Analysis," *ICS,* (Process Instrumentation), vol. 37 (Jan. 1964): 113–18.

9. Webb, P. U., Reducing Process Disturbances with Cascade Control," *Control, Engineering,* 8, no. 8 (August 1961): 73–76.

10. Friedlander, B., "Estimation Theory and Its Role in Optimal Control," *ASME* publication No. 78-WA/DSC-2.

11. Takahashi, Y., Rabins, M. J., and Auslander, D. M., *Control and Dynamic Systems,* Addison-Wesley, Reading, Mass., 1972.

12. Chalam, V. V., *Adaptive Control Systems (Techniques and Applications),* Marcel Dekker, New York, 1987.

13. Luenberger, O. G., "Observing the State of a Linear System," *IEEE Transactions on Military Electronics,* MIL-8 (April 1964): 74–80.

14. Landau, I. D., *Adaptive Control: The Model Reference Approach,* Marcel Dekker, New York, 1979.

15. Lee, E. B., and Markus L. *Foundations of Optimal Control Theory,* Robert E. Krieger, Malabar, Florida, 1986.

16. Maciejowski, J. M., *Multivariable Feedback Design,* Addison-Wesley, Reading, Mass., 1989.

17. Pontryagin, L. S., et al., *Mathematical Theory of Optimal Processes,* John Wiley, New York, 1962.

18. Bellman, R., and Kalaba, R., *Dynamic Programming and Modern Control Theory,* Academic Press, New York, 1966.

19. Athans, M., and Falb, P. L., *Optimal Control: An Introduction to the Theory and Its Applications,* McGraw-Hill, New York, 1966.

20. Lehtomaki, N. A., Sandell, N. R., Jr., and Athans, M., "Robustness Results on LQG Based Multivariable Control System Designs," *IEEE Trans. on Automatic Control,* AC-26, no. 1 (February 1981): 75–93.

21. Callier, F. M., and Desoer, C. A., Multivariable Feedback Systems, Springer-Verlag, New York, 1982.

22. Stoorvogel, A., *The H∞ Control Problem: A State Space Approach,* Prentice Hall, Englewood Cliffs, N.J., 1992.

23. Takahashi, Y., Chan, C. S., and Auslander, D. M., "Parametereinstellung bei Linearem DDC-Algorithmen," *Regulungstechnik Und Process Datenverarbeitung,* 19, no. 6 (June 1971): 237–44.

24. Takahashi, Y., Tomizuka, M., and Auslander, D. M., "Simple Discrete Control of Industrial Processes (Finite Time Settling Control Algorithm for Single-Loop Digital Controller)," *Trans. ASME, Journal of Dynamic Systems, Measurement and Control,* 97, series no. 4 (December 1975): 354-361.

25. Tomizuka, M., Auslander, D. M., and Takahashi, Y., "A Tutorial Introduction to Discrete Time Optimal Control," ASME publication No. 78-WA/DSC-18.

26. Barnett, S., *Introduction to Mathematical Control Theory,* Clarendon Press, Oxford, 1975.

27. Artwick. B. A., *Microcomputer Interfacing,* Prentice-Hall, Englewood Cliffs, N.J., 1980.

28. Hall, D. V., *Microprocessors and Applications,* McGraw-Hill, New York, 1980.

29. Holland, R. C., *Microcomputers and Their Interfacing,* Pergamon Press, Elmsford, N.Y., 1984.

Appendix A

Selected Constants, Properties, and Conversion Factors

SELECTED CONSTANTS

Table A1 Some Fundamental Constants

NAME	SYMBOL	VALUE—SI	VALUE—U.S. CUSTOMARY
Acceleration of gravity (standard)	g	9.80665 m/s^2	32.174 ft/s^2
Angular velocity of Earth	ω	7.292×10^{-5} rad/s	7.292×10^{-5} rad/s
Angular velocity of Earth-Sun line	Ω	1.991×10^{-7} rad/s	1.991×10^{-7} rad/s
Boltzmann constant	k	1.38066×10^{-23} J/K	1.8331×10^{-23} ft·lbf/R
Diameter of Earth (equatorial)	d	$12,755$ km	7926 mi
Diameter of Earth (mean, that is, of sphere of equal volume)	d	$12,742$ km	7917 mi
Diameter of Earth (polar)	d	$12,713$ km	7900 mi
Electron charge	e	1.60219×10^{-19} C	1.60219×10^{-19} C
Electron rest mass	m_e	9.11×10^{-31} kg	6.2424×10^{-32} lbf·s^2/ft
Escape velocity (Earth)	v	11.18 km/s	6.95 mi/s
Faraday constant	F	$96,485$ C/mol	$96,485$ C/mol
Mass of Earth	m	5.976×10^{24} kg	4.095×10^{23} lbf·s^2/ft
Permeability of free space	μ_0	$4\pi \times 10^{-7}$ H/m	$1.2192\pi \times 10^{-7}$ H/ft
Permittivity of free space	e_0	8.8542×10^{-12} C^2/(J·m)	3.6589×10^{-12} C^2/(ft·lbf·ft)
Speed of light	c	3.00×10^8 m/s	9.8425×10^8 ft/s
Standard atmosphere		101.325 kPa	14.696 psi
Stefan-Boltzmann constant	σ	5.67×10^{-8} W/(m^2·K^4)	4.0788×10^{-8} ft·lbf/(s·ft^2·R^4)
Universal gas constant	R	8.314 J/(mol·K)	1545 ft·lbf/lb·mol·R
Universal gravitational constant	G	6.673×10^{-11} m^3/(kg·s^2)	3.439×10^{-8} ft^4/(lbf·s^4)

Compiled from several sources.

TYPICAL VALUES OF SELECTED PROPERTIES

Table A2 Selected Properties of Some Gases[a]

GAS	CHEMICAL FORMULAR	MOLECULAR WEIGHT	SPECIFIC HEAT[b]	
			c_p [kJ/(kg·K)]	c_v [kJ/(kg·K)]
Air	——	28.97	1.0035	0.7165
Carbon dioxide	CO_2	44.01	0.8418	0.6529
Carbon monoxide	CO	28.01	1.0413	0.7445
Helium	He	4.003	5.1926	3.1156
Hydrogen	H_2	2.016	14.2091	10.0849
Nitrogen	N_2	28.013	1.0416	0.7448
Oxygen	O_2	31.999	0.9216	0.6618
Propane	C_3H_8	44.097	1.6794	1.4909
Steam	H_2O	18.015	1.8723	1.4108

[a]Compiled from several sources.
[b]c_p and c_v are at 300 K.

Table A3 Selected Properties of Some Liquids[a]

LIQUID	DENSITY ρ(kg/m^3)	SPECIFIC HEAT c_p[J/(kg·K)]	BULK MODULUS (GPa)	VISCOSITY $\eta(10^{-4}$ Pa·s)
Ammonia	602	4800		2.31
Freon-12	1310	977		2.69
Mercury	13560	139	24.55	18.23
Oil (engine)	884	1909	1.5	227.36
Water	997	4184	2.21	9.8

[a]Typical values at 300 K; compiled from several sources.

Table A4 Selected Properties of Some Solids[a]

SOLID	DENSITY ρ^b (kg/m³)	YOUNG'S MODULUS E^c (GPa)	POISSON'S RATIO ν^c	SPECIFIC HEAT $c_p{}^b$ [J/(kg·K)]	COEFFICIENT OF THERMAL EXPANSION α^b (10⁻⁶/°C)	THERMAL CONDUCTIVITY k^b [W/(m·K)]	RESISTIVITY ρ^b (10⁻⁸ Ω·m)
Aluminum (alloy 2014)	2730	70	0.33	963	23.2	190	3.7
Brass and bronze	8430	105	0.35	377	20.0	120	6.4
Cast iron (gray)	7210	103	0.20	511	12.1	50.2	67
Concrete (medium strength)	2320	25	0.15	879	9.9	0.93	98
Copper	8910	117	0.35	386	17.6	401	1.7
Glass (0.98 silica)	2190	65		832	8	1.5	
Plastic (ABS)	1060	2		1550	72	0.27	5×10^{20}
Plastic (nylon)	1130	0.24		1670	90	0.34	6×10^{17}
Plastic (Teflon)	2200	0.34		960	126	0.24	1×10^{22}
Rubber (natural)	961	0.00196	0.499	2093	162	0.147	
Sandstone (average values)	2300	40	0.18	921	9	1.3	
Stainless steel (type 304)	7920	193	0.27	502	17.3	15	72
Structural steel (plain carbon)	7860	205	0.29	450	11.7	52	18
Timber (white oak)	590	11		1760		0.16	

[a]Compiled from several sources.

[b]Typical values at 300 K

[c]The modulus of rigidity can be taken as $G = \dfrac{E}{2(1 + v)}$, while the bulk modulus can be taken as $K = \dfrac{E}{3(1 - 2v)}$.

SELECTED CONVERSION FACTORS

Table A5 Conversion Factors from SI to U.S. Customary Units

Acceleration

1 m/s^2	$= 3.2808 \text{ ft/s}^2$
1 m/s^2	$= 39.3701 \text{ in./s}^2$

Area

1 m^2	$= 10.7639 \text{ ft}^2$
1 m^2	$= 1550.0031 \text{ in.}^2$
1 mm^2	$= 1.5500 \times 10^{-3} \text{ in.}^2$

(Area) Moment of Inertia

1 m^4	$= 115.8618 \text{ ft}^4$
1 m^4	$= 2.4025 \times 10^6 \text{ in.}^4$
1 mm^4	$= 2.4025 \times 10^{-6} \text{ in.}^4$

Density

1 kg/m^3	$= 6.2428 \times 10^{-2} \text{ lbm/ft}^3$
1 kg/m^3	$= 3.6127 \times 10^{-5} \text{ lbm/in.}^3$

Energy

1 J	$= 0.7376 \text{ ft·lbf}$
1 J	$= 8.8507 \text{ in.·lbf}$
1 J	$= 9.4783 \times 10^{-4} \text{ Btu}$

Force

1 N	$= 0.2248 \text{ lbf}$

Length

1 m	$= 3.2808 \text{ ft}$
1 m	$= 39.3701 \text{ in.}$
1 mm	$= 3.9370 \times 10^{-2} \text{ in.}$
1 km	$= 0.6214 \text{ mi}$
1 km	$= 0.5400 \text{ (nautical) mi}$

Mass

1 kg	$= 2.2046 \text{ lbm}$
1 kg	$= 6.8522 \times 10^{-2} \text{ slug}$
1 kg	$= 1.1023 \times 10^{-3} \text{ ton}$

Moment or Torque

1 N·m	$= 0.7376 \text{ lbf·ft}$
1 N·m	$= 8.8507 \text{ lbf·in.}$

(Mass) Moment of Inertia

1 kg·m^2	$= 23.7304 \text{ lbm·ft}^2$
1 kg·m^2	$= 0.7376 \text{ lbf·ft·s}^2$
1 kg·m^2	$= 3417.1719 \text{ lbm·in.}^2$
1 kg·mm^2	$= 3.4172 \times 10^{-3} \text{ lbm·in.}^2$

Power

1 W	$= 0.7376 \text{ ft·lbf/s}$
1 W	$= 3.4122 \text{ Btu/h}$
1 W	$= 1.3410 \times 10^{-3} \text{ hp}$

Pressure

$1 \text{ Pa } (= 1 \text{ N/m}^2)$	$= 2.0886 \times 10^{-2} \text{ lbf/ft}^2$
$1 \text{ MPa } (= 10 \text{ bar})$	$= 145.0379 \text{ lbf/in.}^2$

Spring constant

1 N/m	$= 6.8522 \times 10^{-2} \text{ lbf/ft}$
1 N/m	$= 5.7101 \times 10^{-3} \text{ lbf/in.}$

Temperature

$1.8 \times \text{°C} + 32$	$= \text{°F}$
$\text{K} (= \text{°C} + 273.15)$	$= \text{°R}/1.8$

Velocity

1 m/s	$= 3.2808 \text{ ft/s}$
1 m/s	$= 2.2369 \text{ mi/h}$
1 km/h	$= 0.6214 \text{ mi/h}$
1 m/s	$= 1.9438 \text{ knot}$

Volume

1 m^3	$= 35.3147 \text{ ft}^3$
1 m^3	$= 6.1024 \times 10^4 \text{ in.}^3$
$1 \text{ liter } (= 1000 \text{ cm}^3)$	$= 0.2642 \text{ gal}$

Table A6 Conversion Factors from U.S. Customary Units to SI

Acceleration
1 ft/s^2	= 0.3048 m/s^2
1 in./s^2	= 0.0254 m/s^2

Area
1 ft^2	= 9.2903 × 10^{-2} m^2
1 in.^{2}	= 6.4516 × 10^{-4} m^2
1 in.^{2}	= 645.1600 mm^2

(Area) Moment of Inertia
1 ft^4	= 8.6310 × 10^{-3} m^4
1 in.^{4}	= 4.1623 × 10^{-7} m^4
1 in.^{4}	= 4.1623 × 10^5 mm^4

Density
1 lbm/ft^3	= 16.0185 kg/m^3
1 lbm/in.^{3}	= 2.7680 × 10^4 kg/m^3

Energy
1 ft·lbf	= 1.3558 J
1 in.·bf	= 0.1130 J
1 Btu (= 778.16 ft·lbf)	= 1055.0400 J

Force
1 lbf	= 4.4482 N

Length
1 ft	= 0.3048 m
1 in.	= 0.0254 m
1 in.	= 25.4000 mm
1 mi (= 5280 ft)	= 1.6093 km
1 (nautical) mi	= 1.8520 km

Mass
1 lbm	= 0.4536 kg
1 slug (or lbf·s^2/ft)	= 14.5939 kg
1 ton (= 2000 lbm)	= 907.1847 kg

Moment or Torque
1 lbf·ft	= 1.3558 N·m
1 lbf·in.	= 0.1130 N·m

(Mass) Moment of Inertia
1 lbm·ft^2	= 4.2140 × 10^{-2} kg·m^2
1 lbf·ft·s^2	= 1.3558 kg·m^2
1 lbm·in.^{2}	= 2.9264 × 10^{-4} kg·m^2
1 lbm·in.^{2}	= 292.6397 kg·mm^2

Power
1 ft·lbf/s	= 1.3564 W
1 Btu/h	= 0.2931 W
1 hp (= 550 ft·lbf/s)	= 745.6998 W

Pressure
1 lbf/ft^2	= 47.8800 Pa
1 lbf/in.^{2} (or psi)	= 6894.7570 MPa

Spring constant
1 lbf/ft	= 14.5939 N/m
1 lbf/in.	= 175.1268 N/m

Temperature
(°F − 32)/1.8	= °C
°R (= °F + 459.67)	= 1.8 K

Velocity
1 ft/s	= 0.3048 m/s
1 mi/h	= 0.44704 m/s
1 mi/h	= 1.6093 km/h
1 knot (or nautical mi/hr)	= 0.5144 m/s

Volume
1 ft^3	= 2.8317 × 10^{-2} m^3
1 in.^{3}	= 1.6387 × 10^{-5} m^3
1 gal (= 0.13368 ft^3)	= 3.7853 liter

Appendix

B

Some Elements of Linear Algebra

THE OBJECTIVE OF THIS APPENDIX is to provide in one place some of the mathematical concepts that are used at various points in the book but may not have been developed within relevant portions of the text. In particular, we introduce here some significant elements of linear algebra that underpin the analysis of systems of more than one variable. Matrices (and vectors) are especially convenient means of dealing with these systems. Although most of the main results are presented without proofs, the ideas are introduced in a logical order, and in many cases the preceding results that inform a current concept are pointed out, and/or an example is used to illustrate the concept.

B.1 MATRICES: DEFINITIONS

An n by m matrix is an arrangement of objects in n rows and m columns. Consider the system of n linear equations in m variables $x_1, x_2, \ldots, x_m$:

$$
\begin{aligned}
a_{11}x_1 + a_{12}x_2 + \cdots + a_{1m}x_m &= b_1 \\
a_{21}x_1 + a_{22}x_2 + \cdots + a_{2m}x_m &= b_2 \\
&\ \ \vdots \\
a_{n1}x_1 + a_{n2}x_2 + \cdots + a_{nm}x_m &= b_n
\end{aligned}
\tag{B1.1}
$$

A convenient representation of the system (B1.1) for analysis is an arrangement in which the coefficients $a_{11}, a_{12}, \ldots$, are stored (collected) in an $(n \times m)$ array or *matrix* $\mathbf{A}$:

$$
\mathbf{A} =
\begin{bmatrix}
a_{11} & a_{12} & \cdots & a_{1m} \\
a_{21} & a_{22} & \cdots & a_{2m} \\
\vdots & \vdots & & \vdots \\
a_{n1} & a_{n2} & \cdots & a_{nm}
\end{bmatrix}
\tag{B1.2a}
$$

such that

$$
\mathbf{AX} = \mathbf{B}; \qquad
\mathbf{X} =
\begin{bmatrix}
x_1 \\ x_2 \\ \vdots \\ x_m
\end{bmatrix};
\qquad
\mathbf{B} =
\begin{bmatrix}
b_1 \\ b_2 \\ \vdots \\ b_n
\end{bmatrix}
\tag{B1.2b}
$$

X and **B** can be considered single-column ($n \times 1$) matrices or *vectors*, which are defined next. The equivalence between equations (B1.2b) and (B1.1) can be established using the meaning of the product of two matrices, which is explained in Section B.2. the notation $n \times m$ (which is read as *n* by *m*) specifies the number of rows (*n*) and columns (*m*), or the *dimension* of the matrix. For example, the following is a 3×4 matrix:

$$\mathbf{A} = \begin{bmatrix} 1 & 1 & -4 & 0 \\ 1 & -1 & 0 & 2 \\ -2 & 2 & 1 & 0 \end{bmatrix} \tag{B1.2c}$$

Its second row consists of the elements $[1 \quad -1 \quad 0 \quad 2]$, whereas the third column has the objects $[-4 \quad 0 \quad 1]$. A shorthand notation that we shall use in the sequel also to refer to a matrix is $\mathbf{A} = [a_{ij}]$, where the first index $1 \le i \le n$ refers to the rows, and the second index $1 \le j \le m$ refers to the columns. Without the brackets, a_{ij} implies the element of **A** located at the *i*th row and *j*th column. For example, for the given matrix, $a_{31} = -2$, and $a_{22} = -1$.

A matrix with the same number of rows as columns ($n = m$) is called *square*. The objects of matrices considered in this book are either real or complex numbers, but matrix elements can also be functions of real or complex numbers or even specially defined objects.

Vectors A *vector* is an ordered set of objects. Thus each row of a matrix such as **A** of (B1.2c) is a vector or, more specifically, a *row-vector*. Similarly, the same matrix can be viewed as consisting of (column) vectors. Put differently, a vector can be regarded as a matrix having only one column or a matrix with only one row. The order of the objects together with their values implies a direction, while the magnitude of the vector

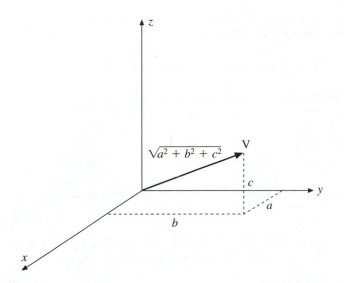

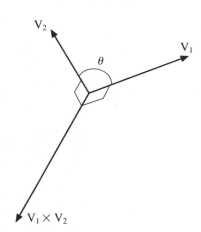

(a) *x-y-z* coordinates of vector (*a, b, c*)

(b) two vectors and their cross product

Figure B1.1

Vectors in 3-space

is determined by the value of the elements only. For example, in geometric three-space, the ordered triple or vector of real numbers $\mathbf{V} = (a, b, c)$ has direction and magnitude (length), as illustrated in Fig. B1.1a. $\mathbf{V}$ in this figure can represent such things as velocity, force, or magnetic flux density. By contrast, entities such as volume, temperature, and magnetic flux, which have only values, are called *scalars*.

Equality, Sum, Scalar Multiple, Dot and Cross Products of Vectors

If $\mathbf{V}_1 = (a_1, b_1, c_1)$ and $\mathbf{V}_2 = (a_2, b_2, c_2)$ are equal, then $a_1 = a_2$, $b_1 = b_2$, and $c_1 = c_2$. This implies that $\mathbf{V}_1$ and $\mathbf{V}_2$ have the same length and the same direction: $\mathbf{V}_1 + \mathbf{V}_2 = (a_1 + a_2, b_1 + b_2, c_1 + c_2)$. A common representation of the vector $\mathbf{V}$ is as the sum of scalar multiples of *unit vectors* in the three coordinate directions: $\mathbf{V} = a\mathbf{i} + b\mathbf{j} + c\mathbf{k}$, where $\mathbf{i} = (1, 0, 0)$, $\mathbf{j} = (0, 1, 0)$, and $\mathbf{k} = (0, 0, 1)$. The *scalar multiple* of a vector is another vector with the same direction (that is, they are parallel) but with the length *scaled* by the scalar value.

The *dot* or *scalar product* of two vectors $\mathbf{V}_1$ and $\mathbf{V}_2$ is a scalar value defined by

$$\mathbf{V}_1 \bullet \mathbf{V}_2 = a_1 a_2 + b_1 b_2 + c_1 c_2 \tag{B1.3a}$$

Geometrically the dot product is the projection of one vector on the other. That is,

$$\mathbf{V}_1 \bullet \mathbf{V}_2 = \|\mathbf{V}_1\| \, \|\mathbf{V}_2\| \cos(\theta) \tag{B1.3b}$$

where $\|\mathbf{V}\|$ is the length or magnitude (or the Euclidean norm, see below) of $\mathbf{V}$, and θ is (as shown in Fig. B1.1b) the angle between the two vectors. It follows from (B1.3b) that the angle between two (nonzero) vectors whose dot product is zero, is 90°, that is, the vectors are perpendicular to each other. Such vectors are said to be *orthogonal*. Another property of the vector dot product is the Cauchy-Schwarz inequality defined (under Vector Norms) below.

The *cross* or *vector product* of two vectors $\mathbf{V}_1 = a_1\mathbf{i} + b_1\mathbf{j} + c_1\mathbf{k}$ and $\mathbf{V}_2 = a_2\mathbf{i} + b_2\mathbf{j} + c_2\mathbf{k}$ is another vector defined by

$$\mathbf{V}_1 \times \mathbf{V}_2 = (b_1 c_1 - b_2 c_1)\mathbf{i} + (a_2 c_1 - a_1 c_2)\mathbf{j} + (a_1 b_2 - a_2 b_1)\mathbf{k} \tag{B1.4a}$$

This product is often computed as a *determinant* (see below):

$$\mathbf{V}_1 \times \mathbf{V}_2 = \begin{vmatrix} \mathbf{i} & \mathbf{j} & \mathbf{k} \\ a_1 & b_1 & c_1 \\ a_2 & b_2 & c_2 \end{vmatrix} \tag{B1.4b}$$

Equation (B1.4b) in turn yields the following property of the cross product:

$$\mathbf{V}_1 \times \mathbf{V}_2 = -\mathbf{V}_2 \times \mathbf{V}_1 \tag{B1.4c}$$

Another property of the cross product, illustrated in Fig. B1.1b, is that $\mathbf{V}_1 \times \mathbf{V}_2$ is perpendicular to both $\mathbf{V}_1$ and $\mathbf{V}_2$. The magnitude of the vector product satisfies

$$\|\mathbf{V}_1 \times \mathbf{V}_2\| = \|\mathbf{V}_1\| \, \|\mathbf{V}_2\| \sin(\theta) \tag{B1.4d}$$

where θ is the angle between $\mathbf{V}_1$ and $\mathbf{V}_2$, and the direction of the cross product is given by the *right-hand rule:* If the fingers of the right hand curl from $\mathbf{V}_1$ to $\mathbf{V}_2$, the thumb points along $\mathbf{V}_1 \times \mathbf{V}_2$. It follows from (B1.4d) that the cross product of two nonzero vectors is equal to zero if and only if the two vectors are parallel (the angle between them is zero or 180°).

Vector Space (R^n)

An *n-vector* is an ordered *n*-tuple $\mathbf{V} = (v_1, v_2, \ldots, v_n)$, that is, $\mathbf{V}$ has n elements or components. The totality of all vectors with n (**real**) components is called *n-space* (and is denoted by R^n; if the components can be complex numbers, for example, the *n*-space is denoted by (C^n)). The set of *n*-vectors $\mathbf{V}_i$ such as R^n that satisfy the following properties:

1. $\mathbf{V}_1 + \mathbf{V}_2 = \mathbf{V}_2 + \mathbf{V}_1$
2. $\mathbf{V}_1 + (\mathbf{V}_2 + \mathbf{V}_3) = (\mathbf{V}_1 + \mathbf{V}_2) + \mathbf{V}_3$
3. $\mathbf{V} + \mathbf{0} = \mathbf{V}$, where $\mathbf{0}$ is the *zero vector* (in R^n, for example, $\mathbf{0} = (0, 0, \ldots, 0)$)
4. $(\alpha + \beta)\mathbf{V} = \alpha\mathbf{V} + \beta\mathbf{V}$, where α, β belong to the (scalar) field F (in R^n α, β are real numbers)
5. $(\alpha\beta)\mathbf{V} = \alpha(\beta\mathbf{V})$
6. $\alpha(\mathbf{V}_1 + \mathbf{V}_2) = \alpha\mathbf{V}_1 + \alpha\mathbf{V}_2$
7. $\alpha\mathbf{0} = \mathbf{0}$.

is called a *vector space*.

Vector Norms

The formula given in Fig. B1.1a for the geometric length or magnitude of the vector is the *Euclidean* norm, which for an *n*-space is defined by

$$\|\mathbf{V}\| = \sqrt{v_1^2 + v_2^2 + \cdots + v_n^2}, \quad \text{where } \mathbf{V} = (v_1, v_2, \ldots, v_n) \tag{B1.5a}$$

Formally, the *norm* of a vector is a function that assigns to every vector in a given space a real number denoted by $\|\mathbf{V}\|$ such that

1. $\|\mathbf{V}\| > 0$ for $\mathbf{V} \neq \mathbf{0}$
2. $\|\mathbf{V}\| = \mathbf{0}$ if and only if $\mathbf{V} = \mathbf{0}$
3. $\|c\mathbf{V}\| = |c| \, \|\mathbf{V}\|$, where c is a scalar and $|c|$ is the absolute value of c.
4. $\|\mathbf{V}_1 + \mathbf{V}_2\| \leq \|\mathbf{V}_1\| + \|\mathbf{V}_2\|$ for all $\mathbf{V}_1$ and $\mathbf{V}_2$; this is known as the *triangle inequality.*
5. $|\mathbf{V}_1 \cdot \mathbf{V}_2| \leq \|\mathbf{V}_1\| \, \|\mathbf{V}_2\|$ for all $\mathbf{V}_1$ and $\mathbf{V}_2$; this is known as the *Cauchy-Schwarz inequality.*

Although there are many vector norms, the Euclidean norm is the most widely used. Two other examples of common vector norms are defined next:

$$\|\mathbf{V}\| = |v_1| + |v_2| + \cdots + |v_n| \tag{B1.5b}$$

$$\|\mathbf{V}\| = \max\{|v_1|, |v_2|, \ldots, |v_n|\} \tag{B1.5c}$$

Span, Linear Independence of Vectors, Basis, and Dimension

The sum of scalar multiples of *n*-vectors,

$$c_1\mathbf{V}_1 + c_2\mathbf{V}_2 + \ldots + c_m\mathbf{V}_m$$

is known as a *linear combination* of the vectors.

Span   Given the *n*-vector space V, the set of all possible *n*-vectors $\mathbf{W}_i$ (for different c_i) that can be expressed as linear combinations of m vectors in V,

$$\mathbf{W} = c_1\mathbf{V}_1 + c_2\mathbf{V}_2 + \cdots + c_m\mathbf{V}_m \tag{B1.6a}$$

where $c_1, c_2, \ldots, c_m$ are scalars (including the all-zero case $\mathbf{W} = \mathbf{0}$), constitute a *subspace* W of V, and the subspace is said to be *spanned* by the vectors $\mathbf{V}_1, \mathbf{V}_2, \ldots, \mathbf{V}_m$.

Linear Independence   The vectors $\mathbf{V}_1, \mathbf{V}_2, \ldots, \mathbf{V}_m$ are *linearly dependent* if and only if there are scalars $c_1, c_2, \ldots, c_m$, not all zero, such that

$$c_1\mathbf{V}_1 + c_2\mathbf{V}_2 + \cdots + c_m\mathbf{V}_m = \mathbf{0} \tag{B1.6b}$$

The vectors $\mathbf{V}_1, \mathbf{V}_2, \ldots, \mathbf{V}_m$ are *linearly independent* if and only if equation (B1.6b) can hold only when $c_1 = c_2 = \ldots = c_m = 0$.

Basis and Dimension   Formally, the vectors $\mathbf{V}_1, \mathbf{V}_2, \ldots, \mathbf{V}_m$ in a subspace form a *basis* for that subspace if (1) they are linearly independent and (2) every vector in the subspace can be written as a linear combination of $\mathbf{V}_1, \mathbf{V}_2, \ldots, \mathbf{V}_m$. It follows that the distinction between a basis and a *spanning set* is that the ordered set of vectors $\{\mathbf{V}_1, \mathbf{V}_2, \ldots, \mathbf{V}_m\}$ is a basis for a vector space V if and only if the $\mathbf{V}$'s span V and are linearly independent. The number of vectors (m) in any basis of a subspace is called the *dimension* of that subspace. Thus the unit vectors $\mathbf{i}$, $\mathbf{j}$, and $\mathbf{k}$, which are easily shown to be linearly independent, form a basis for 3-space, and the dimension of this subspace is three. In general, for any finite value of n, the set of *unit vectors* (n in number)

$$\begin{bmatrix} 1 \\ 0 \\ 0 \\ \cdot \\ \cdot \\ 0 \end{bmatrix}, \begin{bmatrix} 0 \\ 1 \\ 0 \\ \cdot \\ \cdot \\ 0 \end{bmatrix}, \begin{bmatrix} 0 \\ 0 \\ 1 \\ 0 \\ \cdot \\ \cdot \end{bmatrix}, \ldots, \begin{bmatrix} 0 \\ 0 \\ 0 \\ \cdot \\ \cdot \\ 1 \end{bmatrix} \tag{B1.6c}$$

forms a basis for R^n.

Informally, the minimum set of vectors necessary to completely specify a subspace of n-space is the *basis*. A vector (n-) space V is *finite-dimensional* if every vector $\mathbf{V}$ belonging to the space can be written as a linear combination of fixed vectors in the space:

$$\mathbf{V} = c_1\mathbf{V}_1 + c_2\mathbf{V}_2 + \cdots + c_n\mathbf{V}_n \tag{B1.6d}$$

If the vectors $\mathbf{V}_i$ in (B1.6d) are linearly independent (they form a *basis* for V), then the space V is *n-dimensional*.

Other Special Types of Matrices

Null Matrix, Conjugate, Transpose, Symmetric, and Hermitian Matrices

A *null* matrix is one with all its elements equal to zero. Given $\mathbf{A} = [a_{ij}]$, the *conjugate* of $\mathbf{A}$ is the matrix whose elements are the complex conjugates of the elements of $\mathbf{A}$, that is, $\mathbf{A}^* = [a_{ij}^*]$. The *transpose* of $\mathbf{A}$, denoted by $\mathbf{A}^T$, is $[a_{ji}]$. For example, the transpose of the matrix in (B1.2c) is given by

$$\mathbf{A} = \begin{bmatrix} 1 & 1 & -4 & 0 \\ 1 & -1 & 0 & 2 \\ -2 & 2 & 1 & 0 \end{bmatrix}; \qquad \mathbf{A}^T = \begin{bmatrix} 1 & 1 & -2 \\ 1 & -1 & 2 \\ -4 & 0 & 1 \\ 0 & 2 & 0 \end{bmatrix} \tag{B1.7}$$

Note that $(\mathbf{A}^T)^T = [a_{ji}]^T = [a_{ij}] = \mathbf{A}$. An $n \times n$ matrix $\mathbf{A}$ is *symmetric* if it satisfies $\mathbf{A} = \mathbf{A}^T$, that is $[a_{ij}] = [a_{ji}]$. If $\mathbf{A} = -\mathbf{A}^T$ the matrix is said to be *skew-symmetric*. Correspondingly, a complex (square) matrix $\mathbf{H}$ is *Hermitian* if $\mathbf{H}^* = \mathbf{H}^T$ and *skew-Hermitian* if $\mathbf{H}^* = -\mathbf{H}^T$. The eigenvalues (discussed in Section B.3) of a *real* symmetric matrix are real. The eigenvalues of a Hermitian matrix are also real, whereas the eigenvalues of a skew-Hermitian matrix are zero or purely imaginary (have zero real parts).

Triangular, Diagonal, and Identity Matrices

The *main* (or *principal*) *diagonal* elements of an $n \times n$ matrix $\mathbf{A}$ are the elements at the locations $i = j$, that is, the elements $a_{11}, a_{22}, \ldots, a_{nn}$. All the other elements of $\mathbf{A}$ are the *off-diagonal elements*. An $n \times n$ matrix is said to be *triangular* if all the elements **above** *(lower triangular matrix)* or **below** *(upper triangular matrix)* the main diagonal are zero. The following are examples of lower $(\mathbf{L})$ and upper $(\mathbf{U})$ triangular matrices:

$$\mathbf{L} = \begin{bmatrix} -2 & 0 & 0 \\ 1 & 3 & 0 \\ 2 & 5 & -1 \end{bmatrix}; \qquad \mathbf{U} = \begin{bmatrix} -3 & -4 & 1 & 1 \\ 0 & 1 & -4 & 2 \\ 0 & 0 & 0 & -5 \\ 0 & 0 & 0 & 2 \end{bmatrix} \qquad (\text{B1.8a})$$

A *diagonal* matrix is an $n \times n$ matrix with all the off-diagonal elements equal to zero. Such a matrix can be written as $\mathbf{A} = diag[a_{11}, a_{22}, \ldots, a_{nn}]$. For example,

$$diag[-3, -2, -1] = \begin{bmatrix} -3 & 0 & 0 \\ 0 & -2 & 0 \\ 0 & 0 & -1 \end{bmatrix} \qquad (\text{B1.8b})$$

An *identity* matrix, denoted by $\mathbf{I}$ or $\mathbf{I}_n$ (to emphasize its dimension), is a diagonal matrix with the main diagonal elements all equal to 1. For example,

$$\mathbf{I}_3 = \begin{bmatrix} 1 & 0 & 0 \\ 0 & 1 & 0 \\ 0 & 0 & 1 \end{bmatrix} \qquad (\text{B1.8c})$$

The eigenvalues (see Section B.3) of a diagonal matrix are equal to the main diagonal elements. The inverse (defined next) of a diagonal matrix is the diagonal matrix obtained by inverting each element on the main diagonal. It follows that $\mathbf{I}_n^{-1} = \mathbf{I}_n$. The determinant (explained next) of a triangular matrix, including the diagonal or the identity matrix, is just the product of the main diagonal values.

Minor, Cofactor Matrix, Determinant, and Inverse of a Matrix

The *determinant* of the matrix $\mathbf{A}$ is denoted by $|\mathbf{A}|$. For a 1×1 matrix, the determinant is defined as

$$\det[a] = |[a]| = a \qquad (\text{B1.9a})$$

For a 2×2 matrix, the determinant is defined by

$$\det\begin{bmatrix} a & b \\ c & d \end{bmatrix} = \begin{vmatrix} a & b \\ c & d \end{vmatrix} = ad - bc \qquad (\text{B1.9b})$$

Given an $n \times n$ matrix $\mathbf{A} = [a_{ij}]$, the *minor* of the element a_{ij}, denoted by $\mathscr{M}_{ij}$, is the determinant of the $(n - 1) \times (n - 1)$ matrix formed by deleting row i and column j of $\mathbf{A}$. The *cofactor* of a_{ij} is the signed minor, given by $(-1)^{i+j} \mathscr{M}_{ij}$. If $\mathbf{A}$ is 2×2, then the cofactor of each element of $\mathbf{A}$ can be determined using the definition (B1.9a), since the result of deleting an ith row and a jth column of $\mathbf{A}$ is in each case a 1×1 matrix. Similarly, if $\mathbf{A}$ is 3×3, the cofactor of each element of $\mathbf{A}$ can be found by using the definition (B1.9b), since the matrix formed by deleting any *i*th row and *j*th column is in this case a 2 × 2 matrix. For example,

$$\mathbf{A} = \begin{bmatrix} 1 & 0 & 1 \\ -1 & 1 & \frac{1}{2} \\ 1 & -1 & 0 \end{bmatrix}; \qquad (-1)^{1+1}\mathscr{M}_{11} = (1)\begin{vmatrix} 1 & \frac{1}{2} \\ -1 & 0 \end{vmatrix} = \frac{1}{2}, \tag{B1.9c}$$

$$(-1)^{2+3}\mathscr{M}_{23} = (-1)\begin{vmatrix} 1 & 0 \\ 1 & -1 \end{vmatrix} = 1, \ldots$$

Continuing in the manner illustrated in (B1.9c), we have the *cofactor matrix* for **A**, which is formed by replacing each element of **A** by its cofactor, is

$$\text{cofact}[\mathbf{A}] = \begin{bmatrix} \frac{1}{2} & \frac{1}{2} & 0 \\ -1 & -1 & 1 \\ -1 & -\frac{3}{2} & 1 \end{bmatrix} \tag{B1.9d}$$

The *adjoint matrix* is the transpose of the cofactor matrix. For example, for the same **A**,

$$\text{adj}[\mathbf{A}] = [\text{cofact}[\mathbf{A}]]^T = \begin{bmatrix} \frac{1}{2} & -1 & -1 \\ \frac{1}{2} & -1 & -\frac{3}{2} \\ 0 & 1 & 1 \end{bmatrix} \tag{B1.9e}$$

Knowing how to compute the matrix of cofactors for a 3 × 3 matrix means we can compute the same for a 4 × 4 matrix, since when a row and a column are deleted from a 4 × 4 matrix, the result is a 3 × 3 matrix. This argument can thus be continued up to any *n* × *n* matrix.

The *determinant* of an *n* × *n* matrix **A** can determined as the *cofactor expansion* of $|\mathbf{A}|$ by row *I*, which is the sum of the elements of row *I* multiplied by their respective cofactors:

$$|\mathbf{A}| = \sum_{j=1}^{n}(-1)^{I+j}a_{Ij}\mathscr{M}_{Ij} \tag{B1.10a}$$

For example, for the same matrix given in equation (B1.9c), the cofactor expansions by rows 1 and 2 are

$$\begin{vmatrix} 1 & 0 & 1 \\ -1 & 1 & \frac{1}{2} \\ 1 & -1 & 0 \end{vmatrix} = \sum_{j=1}^{3}(-1)^{1+j}a_{1j}\mathscr{M}_{1j} = (1)(1)\begin{vmatrix} 1 & \frac{1}{2} \\ -1 & 0 \end{vmatrix} \tag{B1.10b}$$

$$+ (-1)(0)\begin{vmatrix} -1 & \frac{1}{2} \\ 1 & 0 \end{vmatrix} + (1)(1)\begin{vmatrix} -1 & 1 \\ 1 & -1 \end{vmatrix} = \frac{1}{2}$$

$$= \sum_{j=1}^{3}(-1)^{2+j}a_{2j}M_{2j} = (-1)(-1)\begin{vmatrix} 0 & 1 \\ -1 & 0 \end{vmatrix}$$

$$+ (1)(1)\begin{vmatrix} 1 & 1 \\ 1 & 0 \end{vmatrix} + (-1)(\tfrac{1}{2})\begin{vmatrix} 1 & 0 \\ 1 & -1 \end{vmatrix} = \frac{1}{2}$$

The determinant can also be computed as a cofactor expansion by any column J:

$$|\mathbf{A}| = \sum_{i=1}^{n}(-1)^{i+J}a_{iJ}\mathcal{M}_{iJ} \tag{B1.10c}$$

Obviously, in computing a determinant from the preceding definitions, one would normally choose that row I or column J that simplifies the cofactor expansion. Also, the determinant of a matrix with an all-zero row or an all-zero column is zero. It is easily shown, using the equivalence of the row and column cofactor expansions for the determinant, that for any $n \times n$ matrix $\mathbf{A}$,

$$|\mathbf{A}| = |\mathbf{A}^T| \tag{B1.10d}$$

also

$$|\alpha\mathbf{A}| = |\alpha\mathbf{I}_n|\,|\mathbf{A}| = \alpha^n|\mathbf{A}| \tag{B1.10e}$$

where α is a scalar

The *trace* of an $n \times n$ matrix $\mathbf{A}$, denoted by $\text{Tr}(\mathbf{A})$, is the sum of the diagonal elements of $\mathbf{A}$:

$$\text{Tr}(\mathbf{A}) = \sum_{i=1}^{n}a_{ii} \tag{B1.10f}$$

The *inverse* of an $n \times n$ matrix can be defined in terms of the foregoing as the inverse of the determinant times the transpose of the cofactor matrix:

$$\mathbf{A}^{-1} = \frac{1}{|\mathbf{A}|}[\text{cofact}[\mathbf{A}]]^T = \frac{1}{|\mathbf{A}|}\text{adj}[\mathbf{A}]; \qquad |\mathbf{A}| \neq 0 \tag{B1.11a}$$

Another method of computing the inverse of $\mathbf{A}$ will be described in Section B.2. Obviously, the inverse of a matrix does not exist if the determinant is zero. Such a matrix is said to be *singular.* The 3×3 matrix of (B1.9c) is *nonsingular.* According to equation (B1.11a), its inverse (using (B1.9d) and (B1.10b)) is

$$\begin{bmatrix} 1 & 0 & 1 \\ -1 & 1 & \frac{1}{2} \\ 1 & -1 & 0 \end{bmatrix}^{-1} = \frac{1}{\frac{1}{2}}\begin{bmatrix} \frac{1}{2} & \frac{1}{2} & 0 \\ -1 & -1 & 1 \\ -1 & -\frac{3}{2} & 1 \end{bmatrix}^T = 2\begin{bmatrix} \frac{1}{2} & -1 & -1 \\ \frac{1}{2} & -1 & -\frac{3}{2} \\ 0 & 1 & 1 \end{bmatrix} \tag{B1.11b}$$

$$= \begin{bmatrix} 1 & -2 & -2 \\ 1 & -2 & -3 \\ 0 & 2 & 2 \end{bmatrix}$$

Note that if $\mathbf{A}$ is nonsingular, then $\mathbf{A}^T$ is (by (B1.10d)) also nonsingular, and

$$(\mathbf{A}^T)^{-1} = (\mathbf{A}^{-1})^T \tag{B1.11c}$$

B.2 MATRIX ALGEBRA

Equality, Addition, and Multiplication of Matrices

The matrices $\mathbf{A} = [a_{ij}]$ and $\mathbf{B} = [b_{ij}]$ are *equal* if they have the same dimension, and for each i and j, $a_{ij} = b_{ij}$. If $\mathbf{A}$ and $\mathbf{B}$ have the same dimension, their sum is given by

$$\mathbf{A} + \mathbf{B} = [a_{ij} + b_{ij}] \tag{B2.1a}$$

For example,

$$\begin{bmatrix} 1 & -2 \\ -1 & 2 \\ 1 & 1 \end{bmatrix} + \begin{bmatrix} -3 & 4 \\ 1 & -3 \\ 1 & 2 \end{bmatrix} = \begin{bmatrix} -2 & 2 \\ 0 & -1 \\ 2 & 3 \end{bmatrix} \tag{B2.1b}$$

Note that the transpose of the sum of two matrices is given by $(\mathbf{A} + \mathbf{B})^T = \mathbf{A}^T + \mathbf{B}^T$.

The product of two matrices $\mathbf{A}$ and $\mathbf{B}$ is defined if the number of columns of $\mathbf{A}$ is equal to the number of rows of $\mathbf{B}$. If $\mathbf{A} = [a_{ij}]$ is $n \times r$ and $\mathbf{B} = [b_{ij}]$ is $r \times m$, then the *matrix product* $\mathbf{AB}$ is $n \times m$ and is given by

$$\mathbf{AB} = \left[\sum_{k=1}^{r} a_{ik}b_{kj} \right] \tag{B2.2a}$$

For example,

$$\begin{bmatrix} 1 & -2 \\ -1 & 2 \\ 1 & 1 \end{bmatrix}\begin{bmatrix} 1 & -3 \\ 1 & 2 \end{bmatrix} = \begin{bmatrix} (1)(1) + (-2)(1) & (1)(-3) + (-2)(2) \\ (-1)(1) + (2)(1) & (-1)(-3) + (2)(2) \\ (1)(1) + (1)(1) & (1)(-3) + (1)(2) \end{bmatrix} \tag{B2.2b}$$

$$= \begin{bmatrix} -1 & -7 \\ 1 & 7 \\ 2 & -1 \end{bmatrix}$$

Note that for the preceding example, $\mathbf{BA}$ is not defined, since the number of columns of $\mathbf{B}$ (2) is not equal to the number of rows of $\mathbf{A}$ (3). Also, in cases where both $\mathbf{AB}$ and $\mathbf{BA}$ are defined, $\mathbf{AB}$ is not necessarily equal to $\mathbf{BA}$. Matrices that satisfy the condition $\mathbf{AB} = \mathbf{BA}$ are said to *commute*.

Also note the following properties associated with matrix products:

$$(\mathbf{AB})^T = \mathbf{B}^T\mathbf{A}^T \tag{B2.2c}$$

$$|\mathbf{AB}| = |\mathbf{A}|\,|\mathbf{B}| \tag{B2.2d}$$

$$|\mathbf{I}_n + \mathbf{AB}| = |\mathbf{I}_r + \mathbf{BA}| \tag{B2.2e}$$

where $\mathbf{A} = n \times r$ matrix, $\mathbf{B} = r \times n$ matrix.

Note that if in (B2.2e), $r = 1$, then $|\mathbf{I}_n + \mathbf{AB}| = 1 + \mathbf{BA}$, where $\mathbf{BA}$ is a scalar. If $\mathbf{A}$ is $n \times n$, then

$$\mathbf{AI}_n = \mathbf{I}_n\mathbf{A} = \mathbf{A} \tag{B2.2f}$$

If both $\mathbf{A}$ and $\mathbf{B}$ are $n \times n$ and nonsingular, then

$$(\mathbf{AB})^{-1} = \mathbf{B}^{-1}\mathbf{A}^{-1} \tag{B2.2g}$$

Determination of Rank (and Inverse) of a Matrix

Using the definition for matrix multiplication (B2.2a), we can easily demonstrate the equivalence of the matrix expression (B1.2b) and the linear system of algebraic equations (B1.1). The solution of linear algebraic equations uses operations on the equations that can also be applied to the rows of the matrices $\mathbf{A}$ and $\mathbf{B}$ to solve the matrix system (B1.2b).

Elementary Operations and Equivalent Matrix

The following three *elementary row operations* are defined for an $n \times m$ matrix $\mathbf{A}$:

I Interchange two rows of $\mathbf{A}$.
II Multiply a row of $\mathbf{A}$ by a nonzero number.
III Add a scalar multiple of one row of $\mathbf{A}$ to another row of $\mathbf{A}$.

Elementary *column* operations are defined in a similar way. The matrix formed when an elementary row operation (or a sequence of such operations) are performed on an identity matrix $\mathbf{I}_n$ is called an *elementary matrix* $\mathbf{E}$. If $\mathbf{B}$ is the matrix formed when the same elementary row operation(s) that produced $\mathbf{E}$ are performed on the matrix $\mathbf{A}$, then $\mathbf{B}$ satisfies

$$\mathbf{B} = \mathbf{EA} \tag{B2.3a}$$

Equivalent Matrices Given $n \times m$ matrices $\mathbf{A}$ and $\mathbf{B}$, we say that $\mathbf{A}$ is *row equivalent* to $\mathbf{B}$ if $\mathbf{B}$ can be obtained from $\mathbf{A}$ by a sequence of elementary row operations. If $\mathbf{A}$ is row equivalent to $\mathbf{B}$, then $\mathbf{B}$ is row equivalent to $\mathbf{A}$. If $\mathbf{A}$ is row equivalent to $\mathbf{B}$, and $\mathbf{B}$ is row equivalent to $\mathbf{C}$, then $\mathbf{A}$ is row equivalent to $\mathbf{C}$.

If $\mathbf{A}$ is $n \times n$, elementary operations on $\mathbf{A}$ affect the determinant of the resulting equivalent matrix in predictable ways: Each operation-I changes the sign of the determinant (performing operation I twice yields an equivalent matrix with the same determinant as $\mathbf{A}$); after an operation II, the determinant of the resulting matrix is the (same) scalar multiple of det($\mathbf{A}$); and operation III leaves the determinant unchanged. It follows that the computation of determinants can be facilitated by elementary row (and column) operations provided the effects of these operations on the determinant are taken into account.

Reduced Matrix, Rank, and Inverse

There is always a sequence of elementary row operations on the $n \times m$ matrix $\mathbf{A}$ that will yield an equivalent matrix known as the *reduced matrix* $\mathbf{B} = \mathbf{A}_R$, which is a matrix that satisfies the following four conditions:

1. The first nonzero element (known as the *leading entry*) of any *nonzero row* (row that is not all zero) is 1.
2. All other rows have zero in the column corresponding to the leading entry of each nonzero row.
3. The zero rows, if any, are the lowest rows.
4. The leading entries of succeeding rows must occupy higher (farther to the right) columns.

Consider the following example:

$$\mathbf{A} = \begin{bmatrix} 4 & -2 & 1 & 10 \\ 1 & 0 & 0 & -3 \\ 2 & -3 & 0 & 9 \end{bmatrix};$$

$$\xrightarrow[\text{rows 1 and 2.}]{\text{Interchange}} \begin{bmatrix} 1 & 0 & 0 & -3 \\ 4 & -2 & 1 & 10 \\ 2 & -3 & 0 & 9 \end{bmatrix}; \qquad \xrightarrow[\text{to row 2.}]{\text{Add } -4(\text{row 1})} \begin{bmatrix} 1 & 0 & 0 & -3 \\ 0 & -2 & 1 & 22 \\ 2 & -3 & 0 & 9 \end{bmatrix}$$

$$\xrightarrow[\text{to row 3.}]{\text{Add } -2(\text{row 1})} \begin{bmatrix} 1 & 0 & 0 & -3 \\ 0 & -2 & 1 & 22 \\ 0 & -3 & 0 & 15 \end{bmatrix}; \qquad \xrightarrow[\text{rows 2 and 3.}]{\text{Interchange}} \begin{bmatrix} 1 & 0 & 0 & -3 \\ 0 & -3 & 0 & 15 \\ 0 & -2 & 1 & 22 \end{bmatrix}$$

$$\xrightarrow[\text{row 2 by } -\frac{1}{3}.]{\text{Multiply}} \begin{bmatrix} 1 & 0 & 0 & -3 \\ 0 & 1 & 0 & -5 \\ 0 & -2 & 1 & 22 \end{bmatrix}; \qquad \xrightarrow[\text{to row 3.}]{\text{Add } 2(\text{row 2})} \begin{bmatrix} 1 & 0 & 0 & -3 \\ 0 & 1 & 0 & -5 \\ 0 & 0 & 1 & 12 \end{bmatrix}$$

Rank and Inverse of a Matrix

The *rank* of $\mathbf{A}$ is the number of nonzero rows in $\mathbf{A}_R$. The rank of the preceding 3×4 matrix is 3. If $\mathbf{A}$ is $n \times n$, then rank$(\mathbf{A}) = n$ only if $\mathbf{A}_R = \mathbf{I}_n$.

The inverse of the $n \times n$ matrix $\mathbf{A}$ (if it exists) is also defined as the matrix $\mathbf{B} = \mathbf{A}^{-1}$ such that

$$\mathbf{AB} = \mathbf{BA} = \mathbf{I}_n \tag{B2.3b}$$

From (B2.3a), if it is possible to reduce $\mathbf{A}$ such that $\mathbf{A}_R = \mathbf{I}_n$, then by (B2.3b), $\mathbf{E} = \mathbf{A}^{-1}$. This is another approach to finding the inverse of $\mathbf{A}$. It also follows that $\mathbf{A}$ is nonsingular if $\mathbf{A}_R = \mathbf{I}_n$, or rank$(\mathbf{A}) = n$. The reduction of $\mathbf{A}$ and $\mathbf{I}_n$ can be carried out simultaneously as shown next with the matrix of (B1.11b):

$$
\left[\begin{array}{ccc|ccc}
1 & 0 & 1 & 1 & 0 & 0 \\
-1 & 1 & \frac{1}{2} & 0 & 1 & 0 \\
1 & -1 & 0 & 0 & 0 & 1
\end{array}\right];
\quad \text{Add } 1(\text{row 2}) \text{ to row 3.} \longrightarrow
\left[\begin{array}{ccc|ccc}
1 & 0 & 1 & 1 & 0 & 0 \\
-1 & 1 & \frac{1}{2} & 0 & 1 & 0 \\
0 & 0 & \frac{1}{2} & 0 & 1 & 1
\end{array}\right]
\tag{B2.3c}
$$

$$
\begin{array}{c}
\text{Add } 1(\text{row 1}) \text{ to row 2.} \\
\text{Multiply row 3 by 2.}
\end{array}
\longrightarrow
\left[\begin{array}{ccc|ccc}
1 & 0 & 1 & 1 & 0 & 0 \\
0 & 1 & \frac{3}{2} & 1 & 1 & 0 \\
0 & 0 & 1 & 0 & 2 & 2
\end{array}\right];
\quad
\begin{array}{c}
\text{Add } -1(\text{row 3}) \text{ to row 1.} \\
\text{Add } -\frac{3}{2}(\text{row 3}) \text{ to row 2.}
\end{array}
\longrightarrow
\left[\begin{array}{ccc|ccc}
1 & 0 & 0 & 1 & -2 & -2 \\
0 & 1 & 0 & 1 & -2 & -3 \\
0 & 0 & 1 & 0 & 2 & 2
\end{array}\right]
$$

Thus

$$
\mathbf{A}^{-1} = \begin{bmatrix}
1 & -2 & -2 \\
1 & -2 & -3 \\
0 & 2 & 2
\end{bmatrix}
$$

as before. From the definition (B2.3b), it follows that

$$(\mathbf{A}^{-1})^{-1} = \mathbf{A} \tag{B2.3d}$$

Solution of Homogeneous System of Linear Equations

A *homogeneous* system of linear equations is equation (B1.1) with the right side all equal to zero, or equation (B1.2b) with $\mathbf{B} = \mathbf{0}$, that is,

$$\mathbf{AX} = \mathbf{0} \tag{B2.4a}$$

The preceding system of equations can be solved by the *Gauss-Jordan reduction* or *(complete pivoting) method,* as follows:

1. Find the reduced matrix $\mathbf{A}_R$ and solve the reduced system $\mathbf{A}_R\mathbf{X} = \mathbf{0}$, which has the same solution as (B2.4a).
2. Each nonzero row of $\mathbf{A}_R$ with the leading entry in column j corresponds to an equation containing exactly one *dependent* unknown x_j with coefficient of 1 and perhaps a number of *independent* unknowns.
3. Solve for each dependent unknown as a linear combination of independent unknowns; the independent unknowns have arbitrary values.

Consider the system

$$
\begin{aligned}
x_1 &+ x_2 &- 4x_3 && &+ 3x_5 &= 0 \\
x_1 &- x_2 && &+ 2x_4 &- 5x_5 &= 0 \quad \text{or} \\
-2x_1 &+ 2x_2 && &+ x_4 && &= 0
\end{aligned}
$$ (B2.4b)

$$
\begin{bmatrix}
1 & 1 & -4 & 0 & 3 \\
1 & -1 & 0 & 2 & -5 \\
-2 & 2 & 0 & 1 & 0
\end{bmatrix}
\begin{bmatrix}
x_1 \\ x_2 \\ x_3 \\ x_4 \\ x_5
\end{bmatrix} = \mathbf{0}
$$

Applying the first step of the procedure, we reduce the coefficient matrix:

$$
\begin{bmatrix}
1 & 1 & -4 & 0 & 3 \\
1 & -1 & 0 & 2 & -5 \\
-2 & 2 & 0 & 1 & 0
\end{bmatrix}
\begin{array}{l} \text{Add } -1(\text{row 1}) \text{ to row 2.} \\ \text{Add } 2(\text{row 1}) \text{ to row 3.} \end{array} \longrightarrow
\begin{bmatrix}
1 & 1 & -4 & 0 & 3 \\
0 & -2 & 4 & 2 & -8 \\
0 & 4 & -8 & 1 & 6
\end{bmatrix}
$$

$$
\begin{array}{l} \text{Add } \tfrac{1}{2}(\text{row 2}) \text{ to row 1.} \\ \text{Add } 2(\text{row 2}) \text{ to row 3.} \end{array} \longrightarrow
\begin{bmatrix}
1 & 0 & -2 & 1 & -1 \\
0 & -2 & 4 & 2 & -8 \\
0 & 0 & 0 & 5 & -10
\end{bmatrix}
\begin{array}{l} \text{Multiply row 2 by } -\tfrac{1}{2}. \\ \text{Multiply row 3 by } \tfrac{1}{5}. \end{array} \longrightarrow
\begin{bmatrix}
1 & 0 & -2 & 1 & -1 \\
0 & 1 & -2 & -1 & 4 \\
0 & 0 & 0 & 1 & -2
\end{bmatrix}
$$

$$
\begin{array}{l} \text{Add } -1(\text{row 3}) \text{ to row 1.} \\ \text{Add } 1(\text{row 3}) \text{ to row 2.} \end{array} \longrightarrow
\begin{bmatrix}
1 & 0 & -2 & 0 & 1 \\
0 & 1 & -2 & 0 & 2 \\
0 & 0 & 0 & 1 & -2
\end{bmatrix} = \mathbf{A}_R
$$

From step 2 of the procedure, the dependent unknowns are x_1, x_2, and x_4, and the remaining variables x_3 and x_5 are the independent unknowns, whose values are arbitrary. Finally, step 3 gives the solution: Letting $x_3 = \alpha$ and $x_5 = \beta$, where α and β are arbitrary (constants) values, we have

$$
\begin{aligned}
x_1 &= 2x_3 - x_5 = 2\alpha - \beta \\
x_2 &= 2x_3 - 2x_5 = 2\alpha - 2\beta \quad \text{or} \\
x_4 &= 2x_5 = 2\beta
\end{aligned}
\quad
\begin{bmatrix}
x_1 \\ x_2 \\ x_3 \\ x_4 \\ x_5
\end{bmatrix} =
\begin{bmatrix}
2 & -1 \\
2 & -2 \\
1 & 0 \\
0 & 2 \\
0 & 1
\end{bmatrix}
\begin{bmatrix}
\alpha \\ \beta
\end{bmatrix}
$$

Recall that the rank of $\mathbf{A}$ is equal to the number of nonzero rows of $\mathbf{A}_R$, and by the Gauss-Jordan reduction, the number of dependent unknowns. Thus the condition for the homogeneous system (B2.4a), with $\mathbf{A}$ an $n \times m$ matrix, to have a *nontrivial* solution is $m > \text{rank}(\mathbf{A})$. If $\mathbf{A}$ is square, that is, $n \times n$, the condition for a nontrivial solution, $n > \text{rank}(\mathbf{A})$, implies that $\mathbf{A}$ must be singular, or its determinant must be zero. This is a very useful and important result.

Solution of a Nonhomogeneous System of Linear Equations

Equation (B1.1) or (B1.2b) with $\mathbf{B} \neq \mathbf{0}$ is a *nonhomogeneous* system of linear equations. The general solution is the sum of the solution to the homogeneous problem $(\mathbf{AX} = \mathbf{0})$, say $\mathbf{H}$, and a particular solution of the nonhomogeneous system, say $\mathbf{P}$; that is, $\mathbf{X} = \mathbf{H} + \mathbf{P}$. The particular solution can be found simultaneously with the homogeneous solution by using the complete pivoting method and reducing the *augmented* matrix $[\mathbf{A} | \mathbf{B}]$, which is the $n \times (m + 1)$ matrix obtained by appending $\mathbf{B}$ to the right of $\mathbf{A}$ (that is, assuming $\mathbf{A}$ is $n \times m$). Suppose the reduced matrix is $[\mathbf{A} | \mathbf{B}]_R = [\mathbf{A}_R | \mathbf{C}]$, then the particular solution is the same as a solution to the reduced system $\mathbf{A}_R\mathbf{X} = \mathbf{C}$, while the homogeneous solution is, as before, the solution of $\mathbf{A}_R\mathbf{X} = \mathbf{0}$. It can be shown that the condition for the nonhomogeneous system to have a solution, that is, be *consistent*, is rank($\mathbf{A}$) = rank([$\mathbf{A}|\mathbf{B}$]). Consider the following example with $n = 3$ and $m = 4$:

$$
\begin{aligned}
x_1 &+ x_2 & -4x_3 & & = 3 \\
x_1 &- x_2 & & + 2x_4 & = -5 \\
-2x_1 &+ 2x_2 & & x_4 & = 0
\end{aligned}
\quad \text{so that} \tag{B2.4c}
$$

$$
[\mathbf{A}|\mathbf{B}] =
\begin{bmatrix}
1 & 1 & -4 & 0 & | & 3 \\
1 & -1 & 0 & 2 & | & -5 \\
-2 & 2 & 0 & 1 & | & 0
\end{bmatrix}
$$

The reduced augmented matrix is

$$
[\mathbf{A}_R|\mathbf{C}] =
\begin{bmatrix}
1 & 0 & -2 & 0 & | & 1 \\
0 & 1 & -2 & 0 & | & 2 \\
0 & 0 & 0 & 1 & | & -2
\end{bmatrix}
$$

Since rank($\mathbf{A}$) = rank([$\mathbf{A}|\mathbf{B}$]) = 3, we can proceed to solve for the dependent unknowns (x_1, x_2, x_4) in terms of independent unknown (x_3) and the elements of $\mathbf{C}$, that is,

$$
x_1 = 2x_3 + 1
$$

$$
x_2 = 2x_3 + 2
$$

$$
x_4 = 0x_3 - 2
$$

or letting $x_3 = \gamma$,

$$
\begin{bmatrix}
x_1 \\
x_2 \\
x_3 \\
x_4
\end{bmatrix}
=
\begin{bmatrix}
2 \\
2 \\
1 \\
0
\end{bmatrix}
\gamma +
\begin{bmatrix}
1 \\
2 \\
0 \\
-2
\end{bmatrix}
= \mathbf{H} + \mathbf{P}
$$

If $\mathbf{A}$ is $n \times n$, the condition rank($\mathbf{A}$) = rank([$\mathbf{A}|\mathbf{B}$]) can be met for any $\mathbf{B}$ only if rank($\mathbf{A}$) = n. This implies that the nonhomogeneous system $\mathbf{AX} = \mathbf{B}$, with $\mathbf{A}$ a square ($n \times n$) matrix, is consistent if and only if $\mathbf{A}$ is nonsingular. Of course, in this case, the general solution is the same as the particular solution (the homogeneous solution is the trivial solution) and is also given by

$$
\mathbf{X} = \mathbf{A}^{-1}\mathbf{B}.
$$

B.3 EIGENVALUES AND DIAGONALIZATION

Eigenvalue Problem   An *eigenvalue* of the $n \times n$ matrix $\mathbf{A}$ is a real or complex number p that satisfies

$$
\mathbf{AV} = p\mathbf{V} \tag{B3.1a}
$$

for some nonzero $n \times 1$ matrix or vector $\mathbf{V}$. Any nonzero vector that satisfies (B3.1a) is an *eigenvector* of $\mathbf{A}$ associated with the eigenvalue p. Equation (B3.1a) is known as an eigenvalue problem. Note that $\mathbf{AV}$ is an $n \times 1$ matrix or vector. Since two vectors are equal if their magnitudes and directions are the same, the eigenvalue problem determines the number p such that the vectors $\mathbf{AV}$ and $p\mathbf{V}$ have the same magnitude and direction. The eigenvector is any nonzero vector $\mathbf{V}$ that is parallel to $\mathbf{AV}$, and the eigenvector having a magnitude such that (B3.1a) is satisfied is associated with p.

Thus the same equation (B3.1a) is used to determine the eigenvalues and the associated eigenvectors. Factoring out $\mathbf{V}$, we can write equation (B3.1a) as equation (B3.1b):

$$(p\mathbf{I}_n - \mathbf{A})\mathbf{V} = \mathbf{0} \tag{B3.1b}$$

This is a homogeneous system of linear equations for which (see Section B.2) nontrivial solutions will exist if the determinant of the coefficient matrix is zero:

$$\det(p\mathbf{I}_n - \mathbf{A}) = 0 \tag{B3.1c}$$

Expansion of the determinant gives the *characteristic polynomial,* and equation (B3.1c) is known as the *characteristic equation.* The eigenvalues are the roots (the solutions) of the characteristic equation. Once the eigenvalues are determined by (B3.1c), the eigenvector associated with each eigenvalue can be found by solving (B3.1b). Since equation (B3.1b) is a homogeneous system, the eigenvectors are determined up to the arbitrary constants assigned to the independent unknowns (see Section B.2). A number of techniques are used to fix the eigenvectors, such as the following two:

1. Let the first nonzero element of the vector have unit (1) value.
2. *Normalize* the vector, that is, make its length $\|\mathbf{V}\| = 1$. (The eigenvectors are unit vectors.)

Diagonalization

The properties of diagonal matrices make them convenient and easy to work with, so that there is interest in transforming a nondiagonal square matrix to diagonal form, if possible. An $n \times n$ matrix $\mathbf{A}$ is *diagonalizable* if there exists an $n \times n$ transformation matrix $\mathbf{T}$ such that $\mathbf{T}^{-1}\mathbf{AT}$ is a diagonal matrix.

Condition for Diagonalizability

It can be shown that if the eigenvectors $\mathbf{V}_j, j = 1, \ldots, n$, associated with the eigenvalues of $\mathbf{A}$, $p_j, j = 1, \ldots, n$, are linearly independent, then the $n \times n$ matrix having $\mathbf{V}_j$ as its jth column, that is, $\mathbf{T} = [\mathbf{V}_1, \ldots, \mathbf{V}_n]$, is nonsingular and

$$\mathbf{T}^{-1}\mathbf{AT} = \begin{bmatrix} p_1 & 0 & \cdots & 0 \\ 0 & p_2 & \ddots & 0 \\ \vdots & & \ddots & \vdots \\ 0 & 0 & \cdots & p_n \end{bmatrix} \tag{B3.1d}$$

Thus the diagonal matrix has the eigenvalues (in the same order as the associated eigenvectors) of $\mathbf{A}$ as its main diagonal elements. It can be shown that the associated eigenvectors of an $n \times n$ matrix $\mathbf{A}$ with n *distinct* eigenvalues are linearly independent. Thus such an $\mathbf{A}$ matrix is always diagonalizable ($\mathbf{T}^{-1}$ exists). However, for an $\mathbf{A}$ matrix with nondistinct eigenvalues, it is still possible, in some cases, to find n linearly independent eigenvectors associated with the eigenvalues and hence to diagonalize the $\mathbf{A}$ matrix. Consider, for example, the matrix

$$\mathbf{A} = \begin{bmatrix} -2 & 0 & 3 \\ 1 & -5 & 1 \\ 3 & 0 & -2 \end{bmatrix} \tag{B3.2a}$$

Obtain the eigenvalues and diagonalize the matrix. Equation (B3.1c) yields the characteristic equation

$$
\begin{vmatrix}
p+2 & 0 & -3 \\
-1 & p+5 & -1 \\
-3 & 0 & p+2
\end{vmatrix} = (p+2)(p+5)(p+2) - 3(3(p+5))
$$

$$
= (p+5)((p+2)^2 - 9) = (p+5)^2(p-1) = 0
$$

which gives the eigenvalues $p_1 = p_2 = -5$, and $p_3 = 1$. Although the eigenvalues are not all distinct (-5 is repeated), it may be possible to find corresponding eigenvectors that are linearly independent. Continuing, we substitute $p = -5$ in (B3.1b) to get

$$
\begin{bmatrix}
-3 & 0 & -3 \\
-1 & 0 & -1 \\
-3 & 0 & -3
\end{bmatrix}
\begin{bmatrix}
v_1 \\ v_2 \\ v_3
\end{bmatrix} = \mathbf{0}
$$

which becomes after reduction

$$
\begin{bmatrix}
1 & 0 & 1 \\
0 & 0 & 0 \\
0 & 0 & 0
\end{bmatrix}
\begin{bmatrix}
v_1 \\ v_2 \\ v_3
\end{bmatrix} = \mathbf{0}
$$

The general solution then is

$$
\mathbf{V}_{1,2} = \begin{bmatrix} -\beta \\ \alpha \\ \beta \end{bmatrix}
$$

where α and β are the independent unknowns. Similarly, putting $p = 1$ in (B3.1b) yields the solution

$$
\mathbf{V}_3 = \begin{bmatrix} \gamma \\ \tfrac{1}{3}\gamma \\ \gamma \end{bmatrix}
$$

From the first approach for fixing eigenvalues, the following choice of the (independent) constants: ($\beta = -1$, $\alpha = 0$), ($\beta = 0$, $\alpha = 1$), and $\gamma = 1$ results in the associated eigenvalues

$$
\mathbf{V}_1 = \begin{bmatrix} 1 \\ 0 \\ -1 \end{bmatrix}; \qquad
\mathbf{V}_2 = \begin{bmatrix} 0 \\ 1 \\ 0 \end{bmatrix}; \qquad
\mathbf{V}_3 = \begin{bmatrix} 1 \\ \tfrac{1}{3} \\ 1 \end{bmatrix}
$$

which are easily verified (see Section B.1) to be linearly independent. Thus the transformation matrix and the corresponding diagonal matrix are

$$
\mathbf{T} = \begin{bmatrix}
1 & 0 & 1 \\
0 & 1 & \tfrac{1}{3} \\
-1 & 0 & 1
\end{bmatrix}
\quad \text{and} \quad
\mathbf{T}^{-1}\mathbf{AT} = \begin{bmatrix}
-5 & 0 & 0 \\
0 & -5 & 0 \\
0 & 0 & 1
\end{bmatrix}
\tag{B3.2b}
$$

Orthogonal Diagonalization

An *orthogonal* matrix is an $n \times n$ matrix $\mathbf{A}$ for which $\mathbf{AA}^T = \mathbf{A}^T\mathbf{A} = \mathbf{I}_n$. It follows that the inverse of an orthogonal matrix is its transpose, $\mathbf{A}^{-1} = \mathbf{A}^T$, so that orthogonal matrices are especially convenient to work with. Another property of orthogonal matrices is that an $n \times n$ matrix $\mathbf{T}$ is orthogonal if and only if the *row* vectors of $\mathbf{T}$ are *orthonormal* in n-space or if and only if the *column* vectors of $\mathbf{T}$ are orthonormal. A set of vectors are *orthonormal* if they are orthogonal (recall that this means the dot product of any two of the vectors is zero) and the magnitude (norm) of each vector is 1.

Recall that the eigenvalues of a real symmetric matrix are real. It can be shown that the eigenvectors associated with distinct eigenvalues of a real symmetric matrix are orthogonal, and, indeed, that for any real symmetric $n \times n$ matrix $\mathbf{A}$, there exists an orthogonal matrix that diagonalizes $\mathbf{A}$. For example, obtain an orthogonal diagonalization of the following real symmetric matrix:

$$\mathbf{A} = \begin{bmatrix} 1 & 0 & \sqrt{3}/2 \\ 0 & 2 & 0 \\ \sqrt{3}/2 & 0 & 0 \end{bmatrix}$$

Equation (B3.1c) gives the eigenvalues, $-\frac{1}{2}$, 2, $\frac{3}{2}$, which are all distinct.

Hence the associated eigenvectors are automatically orthogonal. The general solutions of equation (B3.1b) for these eigenvalues are

$$\begin{bmatrix} -\frac{1}{3}\alpha \\ 0 \\ \alpha \end{bmatrix}, \qquad \begin{bmatrix} 0 \\ \gamma \\ 0 \end{bmatrix}, \qquad \begin{bmatrix} \sqrt{3}\beta \\ 0 \\ \beta \end{bmatrix}$$

Fixing the eigenvectors by the second method, so that they are orthonormal (normalized and orthogonal), we get

$$\mathbf{V}_1 = \begin{bmatrix} -\frac{1}{2} \\ 0 \\ \frac{\sqrt{3}}{2} \end{bmatrix}; \qquad \mathbf{V}_2 = \begin{bmatrix} 0 \\ 1 \\ 0 \end{bmatrix}; \qquad \mathbf{B}_3 = \begin{bmatrix} \frac{\sqrt{3}}{2} \\ 0 \\ \frac{1}{2} \end{bmatrix}$$

Thus the orthogonal transformation matrix, its inverse, and the diagonal matrix are

$$\mathbf{T} = \begin{bmatrix} -\frac{1}{2} & 0 & \frac{\sqrt{3}}{2} \\ 0 & 1 & 0 \\ \frac{\sqrt{3}}{2} & 0 & \frac{1}{2} \end{bmatrix}; \qquad \mathbf{T}^{-1} = \mathbf{T}^T = \begin{bmatrix} -\frac{1}{2} & 0 & \frac{\sqrt{3}}{2} \\ 0 & 1 & 0 \\ \frac{\sqrt{3}}{2} & 0 & \frac{1}{2} \end{bmatrix};$$

$$\mathbf{T}^{-1}\mathbf{AT} = \begin{bmatrix} -\frac{1}{2} & 0 & 0 \\ 0 & 2 & 0 \\ 0 & 0 & \frac{3}{2} \end{bmatrix}$$

Quadratic and Definite Forms

A *quadratic form* is an expression

$$v(z) = \sum_{j=1}^{n} \sum_{k=1}^{n} a_{jk} z_j^* z_k \tag{B3.3a}$$

where the z_j's are complex, and each a_{jk} is a complex number. The quadratic form with the z_j's and a_{jk} real is a *real quadratic form* and can be written as (using x in place of z)

$$v(x) = \sum_{j=1}^{n} \sum_{k=1}^{n} a_{jk} x_j x_k = a_{11} x_1^2 + a_{22} x_2^2 + a_{33} x_3^2 + \cdots \tag{B3.3b}$$

$$+ (a_{12} + a_{21}) x_1 x_2 + (a_{13} + a_{31}) x_1 x_2 + \cdots + (a_{23} + a_{32}) x_2 x_3 + \cdots$$

Although all the terms are quadratic, those involving x_j^2 are called *squared* terms, while those involving $x_j x_k$ are called *mixed product* terms. Note that (B3.3b) is only a special case of (B3.3a). For the decomposition (B3.3b) it is customary to assume (without loss of generality) $a_{jk} = a_{kj}$, so that the coefficient matrix $\mathbf{A} = [a_{jk}]$ is symmetric. In terms of the (state) vector $\mathbf{X}$ with the x_j's as components, the quadratic form can be written as

$$V(\mathbf{X}) = \mathbf{X}^T \mathbf{A} \mathbf{X}; \qquad \mathbf{A}^T = \mathbf{A} \tag{B3.3c}$$

In the case of the quadratic form (B3.3a), $V(\mathbf{Z}) = \mathbf{Z}^{*T} \mathbf{A} \mathbf{Z}$, and $\mathbf{A}$ is assumed to be Hermitian ($\mathbf{A}^* = \mathbf{A}^T$). Quadratic forms arise naturally in various applications. For example, the kinetic energy of a system of particles is a quadratic form. In many such applications, an analytical simplification can usually be obtained by transformation (change of variables) to coordinates in which there are no mixed product terms. Such a transformation is given by the *principal axis theorem:*

Principle Axis Theorem   Given the real symmetric matrix $\mathbf{A}$, with eigenvalues γ_1, $\gamma_2, \ldots, \gamma_n$, suppose $\mathbf{T}$ is an orthogonal matrix that diagonalizes $\mathbf{A}$. Then the change of coordinates $\mathbf{X} = \mathbf{T}\mathbf{Y}$ transforms

$$\sum_{j=1}^{n} \sum_{k=1}^{n} a_{jk} x_j x_k$$

to

$$\gamma_1 y_1^2 + \gamma_2 y_2^2 + \cdots + \gamma_n y_n^2$$

Consider again the real symmetric matrix

$$\mathbf{A} = \begin{bmatrix} 1 & 0 & \sqrt{3}/2 \\ 0 & 2 & 0 \\ \sqrt{3}/2 & 0 & 0 \end{bmatrix} \tag{B3.3d}$$

The corresponding quadratic form, equation (B3.3b) or (B3.3c), is $q(x) = x_1^2 + 2x_2^2 + \sqrt{3} x_1 x_2$. However, the eigenvalues of $\mathbf{A}$ are $-\frac{1}{2}, 2, \frac{3}{2}$, and an orthogonal matrix that diagonalizes $\mathbf{A}$ was found previously and is given by

$$\mathbf{T} = \begin{bmatrix} -\frac{1}{2} & 0 & \frac{\sqrt{3}}{2} \\ 0 & 1 & 0 \\ \frac{\sqrt{3}}{2} & 0 & \frac{1}{2} \end{bmatrix}; \qquad \mathbf{T}^{-1} = \mathbf{T}^T = \begin{bmatrix} -\frac{1}{2} & 0 & \frac{\sqrt{3}}{2} \\ 0 & 1 & 0 \\ \frac{\sqrt{3}}{2} & 0 & \frac{1}{2} \end{bmatrix};$$

$$\mathbf{T}^{-1}\mathbf{A}\mathbf{T} = \begin{bmatrix} -\frac{1}{2} & 0 & 0 \\ 0 & 2 & 0 \\ 0 & 0 & \frac{3}{2} \end{bmatrix}$$

In terms of the coordinates y_1, y_2, y_3, is given by $\mathbf{X} = \mathbf{TY} \Rightarrow \mathbf{Y} = \mathbf{T}^{-1}\mathbf{X}$, that is,

$$x_1 = -\frac{1}{2}y_1 + \frac{\sqrt{3}}{2}y_3, \quad x_2 = y_2, \quad x_3 = \frac{\sqrt{3}}{2}y_1 + \frac{1}{2}y_3 \Rightarrow$$

$$y_1 = -\frac{1}{2}x_1 + \frac{\sqrt{3}}{2}x_3, \quad y_2 = x_2, \quad y_3 = \frac{\sqrt{3}}{2}x_1 + \frac{1}{2}x_3$$

the quadratic form given by (B3.3c) is

$$\mathbf{X}^T\mathbf{AX} = \mathbf{Y}^T\mathbf{T}^T\mathbf{ATY} = \mathbf{Y}^T\mathbf{T}^{-1}\mathbf{ATY} = \mathbf{Y}^T \begin{bmatrix} -\frac{1}{2} & 0 & 0 \\ 0 & 2 & 0 \\ 0 & 0 & \frac{3}{2} \end{bmatrix} \mathbf{Y} = -\frac{1}{2}y_1^2 + 2y_2^2 + \frac{3}{2}y_3^2$$

In the case of $\mathbf{A}$ Hermitian, the transformation matrix $\mathbf{T}$ is a *unitary matrix*. A nonsingular matrix $\mathbf{T}$ is unitary if and only if $\mathbf{T}^{*-1} = \mathbf{T}^T$. Note that for a nonsingular complex square matrix,

$$(\mathbf{T}^*)^{-1} = (\mathbf{T}^{-1})^* \tag{B3.3e}$$

Positive (Negative) Definite and Semidefinite

A quadratic form $V(\mathbf{X})$ can be described as having the following *sign* properties:

1. $V(\mathbf{X})$ is *positive (negative) definite* if $V(\mathbf{0}) = 0$ and $V(\mathbf{X}) > 0$ (< 0) whenever $\mathbf{X} \neq \mathbf{0}$.
2. $V(\mathbf{X})$ is *positive (negative) semidefinite* if $V(\mathbf{0}) = 0$ and $V(\mathbf{X}) \geq 0$ (≤ 0) for all values of $\mathbf{X}$, with at least one $\mathbf{X} \neq \mathbf{0}$ such that $V(\mathbf{X}) = 0$.
3. $V(\mathbf{X})$ is *indefinite* if it satisfies neither (1) and (2) and takes both positive and negative values.

These definitions are also applied to scalar functions $V(\mathbf{X})$ that are not necessarily of quadratic form, and definiteness terms are also used to describe the matrix $\mathbf{A}$ that is associated with the form.

One way to determine the sign property of a form is by means of the orthogonal diagonalization (change of coordinates) illustrated earlier. Given an orthogonal transformation matrix $\mathbf{T}$ such that

$$\mathbf{Y}^T(\mathbf{T}^T\mathbf{AT})\mathbf{Y} = \sum_{k=1}^{n} \gamma_k y_k^2 \tag{B3.4a}$$

since $\mathbf{T}$ is nonsingular, $\mathbf{Y} = \mathbf{0}$ if and only if $\mathbf{X} = \mathbf{0}$. Further, $V(\mathbf{X})$ will be (1) positive (negative) definite if and only if all $\gamma_k > 0$ (< 0); (2) positive (negative) semidefinite if and only if all $\gamma_k \geq 0$ (≤ 0), with at least one $\gamma_k = 0$; and (3) indefinite if and only if at least one $\gamma_k > 0$ and one $\gamma_k < 0$.

Applying this approach to a quadratic form with the associated matrix given in (B3.3d), we see that this matrix $\mathbf{A}$ can be described as indefinite.

Sylvester Conditions   Another way to determine the sign property of a quadratic form is by means of the *principal minors* of $\mathbf{A}$. The minor of $\mathbf{A}$, $\mathcal{M}_{ij}$, defined in Section B.1, is a *first* minor. If $\mathbf{A}$ is $n \times n$, then the *order* of the minor $\mathcal{M}_{ij}$ is $n - 1$. (Recall that $\mathcal{M}_{ij}$ is the determinant of the $(n - 1) \times (n - 1)$ matrix formed by deleting row i and column j of $\mathbf{A}$.) We say that $\mathcal{M}_{ij}$ is an $(n - 1)$th-order minor of $\mathbf{A}$. The principal minors $\mathcal{P}_r$ of $\mathbf{A}$ are the rth-order minors whose principal (main) diagonal is part of the principal diagonal of $\mathbf{A}$. An rth-order minor of $\mathbf{A}$ is the determinant obtained by omission of any $(n - r)$ rows and $(n - r)$ columns from $|\mathbf{A}|$. The *leading* principal minors $\mathcal{D}_r$ of $\mathbf{A}$ are:

$$\mathcal{D}_1 = a_{11}, \quad \mathcal{D}_2 = \begin{vmatrix} a_{11} & a_{12} \\ a_{21} & a_{22} \end{vmatrix}, \quad \mathcal{D}_3 = \begin{vmatrix} a_{11} & a_{12} & a_{13} \\ a_{21} & a_{22} & a_{23} \\ a_{31} & a_{32} & a_{33} \end{vmatrix}, \text{ etc.} \quad \text{(B3.4b)}$$

By the *Sylvester conditions,* the quadratic form $V(\mathbf{X})$ is

1. Positive definite if and only if $\mathcal{D}_r > 0$, $r = 1, 2, \ldots, n$.
2. Negative definite if and only if $(-1)^r \mathcal{D}_r > 0$, $r = 1, 2, \ldots, n$.
3. Positive semidefinite if and only if $\det(\mathbf{A}) = 0$ and $\mathcal{P}_r \geq 0$ for all principal minors.
4. Negative semidefinite if and only if $\det(\mathbf{A}) = 0$ and $(-1)^r \mathcal{P}_r \geq 0$ for all principal minors.
5. Indefinite if $V(\mathbf{X})$ satisfies none of the preceding conditions.

Consider, for example, the quadratic form

$$V(\mathbf{X}) = 12x_1^2 - x_2^2 + x_3^2 + 2x_1 x_2 - 6x_1 x_3 + 6x_2 x_3$$

Comparing with equation (B3.3b) with $\mathbf{A}$ symmetric, we obtain the associated matrix

$$\mathbf{A} = \begin{bmatrix} 12 & 1 & -3 \\ 1 & -1 & 3 \\ -3 & 3 & 1 \end{bmatrix}$$

The principal minors of this matrix are

$$\mathcal{D}_1 = 12, -1, 1; \quad \mathcal{D}_2 = \begin{vmatrix} 12 & 1 \\ 1 & -1 \end{vmatrix} = -13; \quad \begin{vmatrix} -1 & 3 \\ 3 & 1 \end{vmatrix} = -10,$$

$$\begin{vmatrix} 12 & -3 \\ -3 & 1 \end{vmatrix} = 3; \quad \mathcal{D}_3 = |\mathbf{A}| = -130$$

It follows by condition 5 that $V(\mathbf{X})$ is indefinite.

Induced Matrix Norm and Singular Value

The concept of norms can be extended to matrices. For example, two simple matrix norms are defined by

$$\|\mathbf{A}\| = \max\left(\sum_{k=1}^{n} |a_{jk}|\right) \quad \text{(B3.5a)}$$

$$\|\mathbf{A}\| = \left(\sum_{i=1}^{n} \sum_{j=1}^{n} |a_{ij}|^2\right)^{1/2} \quad \text{(B3.5b)}$$

These norms satisfy the following properties:

1. $\|\mathbf{A}\| = \|\mathbf{A}^T\|$
2. $\|\mathbf{A} + \mathbf{B}\| \leq \|\mathbf{A}\| + \|\mathbf{B}\|$
3. $\|\mathbf{AB}\| \leq \|\mathbf{A}\| \|\mathbf{B}\|$
4. $\|\mathbf{Ax}\| \leq \|\mathbf{A}\| \|\mathbf{x}\|$

where **A** and **B** are matrices and **x** is a vector.

The *induced matrix norm* may be defined by

$$\|\mathbf{A}\| = \max_{\mathbf{x}}\left(\frac{\|\mathbf{Ax}\|}{\|\mathbf{x}\|}\right) \tag{B3.6a}$$

where $\|\mathbf{x}\|$ is defined as the Euclidean norm given by equation (B1.5a). Using quadratic forms (see above), we express the terms in (B3.6a) as follows:

$$\|\mathbf{x}\| = (\mathbf{x}^T\mathbf{x})^{1/2} \quad \text{and} \quad \|\mathbf{Ax}\| = (\mathbf{x}^T\mathbf{A}^T\mathbf{Ax})^{1/2}$$

so that the induced norm becomes

$$\|\mathbf{A}\| = \max_{\mathbf{x}}\left(\frac{\mathbf{x}^T\mathbf{A}^T\mathbf{Ax}}{\mathbf{x}^T\mathbf{x}}\right)^{1/2} \tag{B3.6b}$$

Further, since $\mathbf{A}^T\mathbf{A}$ is symmetric, there exists an orthogonal diagonalization matrix **T**, and by the principal axis theorem, if we let $\mathbf{x} = \mathbf{Ty}$, then

$$\mathbf{x}^T\mathbf{A}^T\mathbf{Ax} = \mathbf{y}^T\mathbf{T}^T\mathbf{A}^T\mathbf{ATy} = \gamma_1 y_1^2 + \gamma_2 y_2^2 + \cdots + \gamma_n y_n^2 \tag{B3.6c}$$

where $\gamma_1, \gamma_2, \ldots, \gamma_n$ are the eigenvalues of $\mathbf{A}^T\mathbf{A}$. Also, since $\mathbf{T}^T = \mathbf{T}^{-1}$ for an orthogonal matrix,

$$\mathbf{x}^T\mathbf{x} = \mathbf{y}^T\mathbf{T}^T\mathbf{Ty} = \mathbf{y}^T\mathbf{y} = y_1^2 + y_2^2 + \cdots + y_n^2 \tag{B3.6d}$$

The induced norm thus becomes

$$\|\mathbf{A}\| = \max_{\mathbf{y}}\left(\frac{\gamma_1 y_1^2 + \gamma_2 y_2^2 + \cdots + \gamma_n y_n^2}{y_1^2 + y_2^2 + \cdots + y_n^2}\right)^{1/2} \tag{B3.6e}$$

The maximum value of the ratio in (B3.6e) can be shown to be the largest eigenvalue. Hence

$$\|\mathbf{A}\| = |\text{maximum eigenvalue of } \mathbf{A}^T\mathbf{A}|^{1/2}; \quad \text{(induced norm)} \tag{B3.6f}$$

The (positive) square roots of the eigenvalues of $\mathbf{A}^T\mathbf{A}$ are known as the *singular values* of **A**, so that the induced norm of **A** is the largest singular value. This result also applies when **A** is complex, if $\mathbf{A}^T\mathbf{A}$ is replaced by $\mathbf{A}^{*T}\mathbf{A}$.

B.4 FUNCTIONS OF A SQUARE MATRIX

For a square $(n \times n)$ matrix **A** and a positive integer r we define

$$\mathbf{A}^r = \mathbf{AA} \cdots \mathbf{A}, \quad \text{(for a total of } r \text{ factors)} \tag{B4.1}$$

For m a negative integer, $\mathbf{A}^m$ is defined in terms of (B4.1) as follows: Let the positive integer $r = -m$. Then $\mathbf{A}^m = (\mathbf{A}^{-1})^r$, assuming $\mathbf{A}^{-1}$ exists. Similarly, using (B4.1), we can show that $\mathbf{A}^m\mathbf{A}^n = \mathbf{A}^{m+n}$ by letting $r = m + n$. Finally, note that $\mathbf{A}^0$ is interpreted to mean $\mathbf{I}_n$.

The preceding conditions lead to a definition of the *polynomial in* **A** as the matrix

$$p(\mathbf{A}) = c_0\mathbf{I}_n + c_1\mathbf{A} + c_2\mathbf{A}^2 + \cdots + c_r\mathbf{A}^r \tag{B4.2a}$$

where c_i are scalars.

Polynomials in the same matrix commute:

$$p(\mathbf{A})q(\mathbf{A}) = q(\mathbf{A})p(\mathbf{A}) \tag{B4.2c}$$

Given a scalar function defined by the power series

$$p(\gamma) = c_0 + c_1\gamma + c_2\gamma^2 + \cdots = \sum_{k=0}^{\infty} c_k\gamma^k \tag{B4.3a}$$

which is assumed to converge for $|\gamma| < R$, there is an associated matrix function of the square matrix **A** given by the matrix power series

$$p(\mathbf{A}) = c_0\mathbf{I} + c_1\mathbf{A} + c_2\mathbf{A}^2 + c_3\mathbf{A}^3 + \cdots \tag{B4.3b}$$

which is convergent if all the characteristic roots γ_i of **A** satisfy $|\gamma_i| < R$.

For any square matrix **A** with finite elements, it can be shown that the series

$$\phi(\mathbf{A}) \equiv e^{\mathbf{A}} = \mathbf{I} + \mathbf{A} + \frac{1}{2!}\mathbf{A}^2 + \frac{1}{3!}\mathbf{A}^3 + \cdots \tag{B4.3c}$$

converges in the sense that each element of $\phi(\mathbf{A})$ tends to a not-infinite limit as the number of power terms approaches infinity. It follows that the matrix exponential can be expanded as

$$e^{\mathbf{A}t} = \mathbf{I} + \mathbf{A}t + \frac{1}{2!}\mathbf{A}^2t^2 + \frac{1}{3!}\mathbf{A}^3t^3 + \cdots \tag{B4.3d}$$

and from the preceding results,

$$p(\mathbf{A})e^{\mathbf{A}t} = e^{\mathbf{A}t}p(\mathbf{A}) \quad \text{and} \quad e^{-\mathbf{A}t} = [e^{\mathbf{A}t}]^{-1}$$

Conversion of Continuous Vector State Model to Discrete Time

The continuous-time vector state model, equations (6.49a) and (6.49b):

$$\dot{\mathbf{X}} = \mathbf{A}\mathbf{X} + \mathbf{B}\mathbf{U}; \qquad \mathbf{Y} = \mathbf{C}\mathbf{X} + \mathbf{D}\mathbf{U}$$

may be approximated by a discrete-time model, equations (6.90a) and (6.90b):

$$\mathbf{X}_{k+1} = \mathbf{P}\mathbf{X}_k + \mathbf{Q}\mathbf{U}_k; \qquad \mathbf{Y}_k = \mathbf{C}\mathbf{X}_k + \mathbf{D}\mathbf{U}_k$$

From equations (6.64b) and (6.73), the solution of (6.49a) for the time interval $kT \leq t \leq [k+1]T$ is

$$\mathbf{X}([k+1]T) = e^{\mathbf{A}T}\mathbf{X}(kT) + e^{\mathbf{A}T}\int_{kT}^{[k+1]T} e^{-\mathbf{A}\tau}\mathbf{B}\mathbf{U}(\tau)\,d\tau \tag{B4.3e}$$

Assuming that the input vector can be approximated by a *staircase* pattern with sampling period T (recall Fig. 14.3c), $\mathbf{U}(t) \approx \mathbf{U}(k)$, then we can integrate equation (B4.3e), assuming $\mathbf{A}^{-1}$ exists:

$$\mathbf{X}(k+1) = [e^{\mathbf{A}T}]\mathbf{X}(k) + [e^{\mathbf{A}T}(\mathbf{I} - e^{-\mathbf{A}T})\mathbf{A}^{-1}\mathbf{B}]\mathbf{U}(k) \tag{B4.3f}$$

Equation (B4.3f) compared with equation (6.90a) yields the approximate identities

$$\mathbf{P} = e^{\mathbf{A}T} \quad \text{and} \quad \mathbf{Q} = (e^{\mathbf{A}T} - \mathbf{I})\mathbf{A}^{-1}\mathbf{B} \tag{B4.3g}$$

Further, the definition (B4.3d) gives the following result, which does not require existence of $\mathbf{A}^{-1}$:

$$\mathbf{P} = \mathbf{I} + \mathbf{A} + \frac{\mathbf{A}^2 T^2}{2!} + \frac{\mathbf{A}^3 T^3}{3!} + \frac{\mathbf{A}^4 T^4}{4!} + \cdots \tag{B4.3h}$$

$$\mathbf{Q} = T\left(\mathbf{I} + \frac{\mathbf{A}T}{2!} + \frac{\mathbf{A}^2 T^2}{3!} + \frac{\mathbf{A}^3 T^3}{4!} + \cdots\right)\mathbf{B}$$

In general, the number of terms of each expansion in (B4.3h) necessary for full accuracy depends on the elements of $\mathbf{A}$ and the magnitude of T. However, for very small T, the following single-term approximations are sufficient for many problems:

$$\mathbf{P} \approx \mathbf{I} + \mathbf{A}T \quad \text{and} \quad \mathbf{Q} \approx T\mathbf{B} \tag{B4.3i}$$

Also note that if we substitute equation (B4.3i) into the discrete-time vector state model (6.90), we obtain

$$\mathbf{X}_{k+1} = \mathbf{X}_k + T(\mathbf{A}\mathbf{X}_k + \mathbf{B}\mathbf{U}_k) \tag{B4.3j}$$

which is Euler's (numerical) method, equation (6.94c), for the linear time-invariant system (6.49)!

Cayley-Hamilton Theorem and Sylvester's Formula

If we express the characteristic (polynomial) equation of the $(n \times n)$ matrix $\mathbf{A}$ as

$$|s\mathbf{I} - \mathbf{A}| = p_\mathbf{A}(s) = s^n + c_{n-1}s^{n-1} + \cdots + c_1 s + c_0 = 0 \tag{B4.4a}$$

then

$$p_\mathbf{A}(\mathbf{A}) = \mathbf{A}^n + c_{n-1}\mathbf{A}^{n-1} + \cdots + c_1\mathbf{A} + c_0\mathbf{I} = 0 \tag{B4.4b}$$

This is the *Cayley-Hamilton* theorem, which can be stated as: Every square matrix satisfies its own characteristic equation. According to equation (B4.4b), if $\mathbf{A}$ is an $(n \times n)$ matrix, then $\mathbf{A}^n$ can be expressed as a linear combination of $\mathbf{A}^{n-1}, \ldots, \mathbf{A}, \mathbf{I}$, and in turn, $\mathbf{A}^m$ for any $m > n$ is expressible as a linear combination of the same matrices. Consider, for example, the square matrix

$$\mathbf{A} = \begin{bmatrix} -3 & 4 & 4 \\ 1 & -3 & -1 \\ -1 & 2 & 0 \end{bmatrix} \tag{B4.4c}$$

The characteristic equation is

$$|s\mathbf{I} - \mathbf{A}| = s^3 + 6s^2 + 11s + 6 = 0$$

According to the Cayley-Hamilton theorem, the matrix also satisfies

$$\mathbf{A}^3 + 6\mathbf{A}^2 + 11\mathbf{A} + 6\mathbf{I}_3 = 0$$

Thus

$$\mathbf{A}^3 = -6\mathbf{A}^2 - 11\mathbf{A} - 6\mathbf{I}_3$$

$$\mathbf{A}^4 = -6\mathbf{A}^3 - 11\mathbf{A}^2 - 6\mathbf{A} = -6(-6\mathbf{A}^2 - 11\mathbf{A} - 6\mathbf{I}_3) - 11\mathbf{A}^2 - 6\mathbf{A}$$

$$= 25\mathbf{A}^2 + 60\mathbf{A} + 36\mathbf{I}_3$$

$$\mathbf{A}^5 = -6\mathbf{A}^4 - 11\mathbf{A}^3 - 6\mathbf{A}^2 = -90\mathbf{A}^2 - 239\mathbf{A} - 150\mathbf{I}_3$$

and so on.

It further follows from (B4.3c) that $e^{\mathbf{A}t}$ is also expressible as a linear combination in the same matrices, or

$$e^{\mathbf{A}t} = \sum_{i=0}^{n-1} \gamma_i(t)\mathbf{A}^i \tag{B4.4d}$$

Sylvester's Formula

Assuming convergence of the series (B4.3a), it can be shown that if all the characteristic roots γ_i of the $n \times n$ matrix $\mathbf{A}$ are distinct, then the polynomial of $\mathbf{A}$

$$p(\mathbf{A}) = \sum_{k=1}^{n} p(\gamma_k)\mathbf{Z}(\gamma_k) \tag{B4.5a}$$

where

$$\mathbf{F}_k = \frac{(\gamma_1\mathbf{I} - \mathbf{A})(\gamma_2\mathbf{I} - \mathbf{A}) \cdots (\gamma_{k-1}\mathbf{I} - \mathbf{A})(\gamma_{k+1}\mathbf{I} - \mathbf{A}) \cdots (\gamma_n\mathbf{I} - \mathbf{A})}{(\gamma_1 - \gamma_k)(\gamma_2 - \gamma_k) \cdots (\gamma_{k-1} - \gamma_k)(\gamma_{k+1} - \gamma_k) \cdots (\gamma_n - \dot{\gamma}_k)} \tag{B4.5b}$$

$$= \prod_{\substack{j=1 \\ j \neq k}}^{n} [(\gamma_j\mathbf{I} - \mathbf{A})/(\gamma_j - \gamma_k)]$$

Sylvester's formula (B4.5a) also holds when $p(\gamma)$ is a finite series. If $\mathbf{A}$ has repeated characteristic roots, then $p(\mathbf{A})$ is given by

$$p(\mathbf{A}) = \sum_{k=1}^{q} (-1)^{n-\alpha_k} \left[y(\gamma_k)\mathbf{I} - y^{(1)}(\gamma_k)\mathbf{f}(\gamma_k) + \frac{y^{(2)}(\gamma_k)\mathbf{f}^2(\gamma_k)}{2!} + \cdots \right.$$

$$\left. + (-1)^{\alpha_k-1}\frac{y^{(\alpha_k-1)}(\gamma_k)\mathbf{f}^{\alpha_k-1}(\gamma_k)}{(\alpha_k - 1)!} \right] \prod_{\substack{j=1 \\ j \neq k}}^{q} \mathbf{f}^{\alpha_k}(\gamma_k) \tag{B4.5c}$$

where the summation is over q, the number of distinct values of the characteristic roots; α_k is the number of roots having the value γ_k; $y^{(r)}(\gamma_k)$ denotes the rth derivative of y with respect to γ, evaluated at $\gamma = \gamma_k$; and

$$y(\gamma) = \frac{p(\gamma)}{\displaystyle\prod_{\substack{j=1 \\ j \neq k}}^{q} (\gamma - \gamma_j)} \tag{B4.5d}$$

$$\mathbf{f}(\gamma) = \gamma\mathbf{I} - \mathbf{A} \tag{B4.5e}$$

Consider for example,

$$\mathbf{A} = \begin{bmatrix} -3 & 2 & 0 \\ -1 & -4 & 1 \\ -2 & 0 & -1 \end{bmatrix}; \text{ with } |\gamma\mathbf{I} - \mathbf{A}| \tag{B4.5f}$$

$$= \begin{vmatrix} (\gamma + 3) & -2 & 0 \\ 1 & (\gamma + 4) & -1 \\ 2 & 0 & (\gamma + 1) \end{vmatrix} = (\gamma + 3)^2(\gamma + 2)$$

$n = 3$, $q = 2$, $\gamma_1 = -3$, $\alpha_1 = 2$, $\gamma_2 = -2$, and $\alpha_2 = 1$

Equation (B4.5c) then gives

$$p(\mathbf{A}) = (-1)\left[\frac{p(\gamma_1)}{\gamma_1 - \gamma_2}\mathbf{I}_3 - \frac{d}{d\gamma}\left\{\frac{p(\gamma)}{(\gamma - \gamma_2)}\right\}\Bigg|_{\gamma=\gamma_1}(\gamma_1\mathbf{I} - \mathbf{A})\right](\gamma_2\mathbf{I} - \mathbf{A})$$

$$+ \left[\frac{p(\gamma_2)}{(\gamma_2 - \gamma_1)}\right](\gamma_1\mathbf{I} - \mathbf{A})^2$$

$$p(\mathbf{A}) = p(-3)[(\gamma_2\mathbf{I} - \mathbf{A}) - (\gamma_1\mathbf{I} - \mathbf{A})(\gamma_2\mathbf{I} - \mathbf{A})]$$

$$- p^{(1)}(-3)[(\gamma_1\mathbf{I} - \mathbf{A})(\gamma_2\mathbf{I} - \mathbf{A})] + p(-2)(\gamma_1\mathbf{I} - \mathbf{A})^2$$

$$p(\mathbf{A}) = \begin{bmatrix} 3 & 2 & -2 \\ 1 & 2 & -1 \\ 4 & 4 & -3 \end{bmatrix}p(-3) + \begin{bmatrix} 2 & 4 & -2 \\ 0 & 0 & 0 \\ 2 & 4 & -2 \end{bmatrix}p^{(1)}(-3) + \begin{bmatrix} -2 & -2 & 2 \\ -1 & -1 & 1 \\ -4 & -4 & 4 \end{bmatrix}p(-2)$$

Zeros of a Polynomial Matrix

The concept of *poles* and zeros of scalar transfer functions of SISO systems are not easily extended to the transfer function matrices of multivariable systems. The transfer function matrix does not have a clearly distinguishable denominator or numerator to which "poles" or "zeros" can somehow be associated. However, with *factorization methods* the transfer function matrix can be factored into a "numerator matrix" and a "denominator matrix" to which some SISO concepts can be extended. These *factorizing matrices* typically assume one of two forms: matrices whose elements are themselves transfer functions, and *polynomial matrices*—that is, matrices whose elements are polynomials in *s*. Our concern in this brief segment is related to the "zeros" of polynomial matrices. More specifically, we shall outline in a strictly limited way, a process leading to the *invariant zeros* of a dynamic system that are the zeros of a polynomial "system" matrix that will be defined.

The Ring

A *ring* is a nonempty set $\mathcal{R}$ with two operations, $+$ and $\bullet$, such that if a, b, $c \in \mathcal{R}$,

1. $a + (b + c) = (a + b) + c$
2. $a + b = b + a$
3. There exists a 0 in $\mathcal{R}$ such that $a + 0 = 0 + a = a$.
4. $a \bullet (b \bullet c) = (a \bullet b) \bullet c$
5. $a \bullet (b + c) = (a \bullet b) + (a \bullet c)$
6. For every $a \in \mathcal{R}$ there is a $(-a) \in \mathcal{R}$ such that $a + (-a) = 0$.

If for all a, $b \in \mathcal{R}$, $a \bullet b = b \bullet a$, the ring is *commutative*. The ring has an *identity* if there is an element $1 \in \mathcal{R}$ such that $a \bullet 1 = 1 \bullet a = a$.

Most of this appendix has been concerned with *scalar fields,* for example, the field of reals and the complex field. The elements of polynomial matrices are not members of a field but of a *ring,* and certain operations with matrices over, say, a complex field are not applicable for matrices over rings. For example $|\mathbf{A}| \neq 0$ is no longer sufficient for the existence of $\mathbf{A}^{-1}$ as a matrix over the ring $\mathcal{R}$. Because the determinant is now a member of $\mathcal{R}$, the determinant must also have an inverse for the definition of $\mathbf{A}^{-1}$, equation (B1.11a), to have meaning. Formally, a *field* is a commutative ring with identity such that (1) it has at least two elements, and (2) every nonzero element has a reciprocal (is a *unit*). An element $u \in \mathcal{R}$ is called a *unit* of $\mathcal{R}$ if there is a $v \in \mathcal{R}$ such that $uv = vu = 1$. In this case v is called the inverse of u ($v \equiv u^{-1}$).

Smith Form and Zero Polynomial

Let $\mathcal{R}[s]$ denote the ring of polynomials with real coefficients, and $\mathcal{R}^{n \times m}[s]$ the set of all $n \times m$ matrices with components in $\mathcal{R}[s]$. Then an element of $\mathcal{R}^{n \times m}[s]$ is a *polynomial matrix.* A square polynomial matrix ($A(s) \in \mathcal{R}^{n \times m}[s]$) is called *unimodular* if det $A(s)$ is a unit of $\mathcal{R}[s]$. That is, $A(s)$ is invertible over the ring of polynomial matrices. Two polynomial matrices $A(s)$ and $B(s)$ are said to be *unimodularly equivalent* if there exist unimodular matrices $N(s)$ and $M(s)$ such that

$$B(s) = N(s)A(s)M(s) \tag{B4.6}$$

Elementary row (column) operations on polynomial matrices are defined as follows:

  I Interchange any two rows (columns).
 II Multiply a row (column) by a nonzero real or complex number.
III Add a polynomial multiple of a row (column) to another row (column).

The application of a combination of elementary row and column operations on any $n \times m$ polynomial matrix $A(s)$ to produce the unimodularly equivalent $n \times m$ polynomial matrix $B(s)$ can be indicated as in (B4.6) by premultiplying and postmultiplying $A(s)$ by, respectively, ($n \times n$) and ($m \times m$) *unimodular elementary* matrices $N(s)$ and $M(s)$, where N and M represent the row and column operations, respectively. In particular, there is a combination of these operations that will reduce $A(s)$ to the canonical form $B(s) = S(s)$, known as *Smith form:*

$$S(s) = N(s)A(s)M(s); \qquad S(s) = \begin{cases} [\psi(s) \quad \mathbf{0}_{n,m-n}], & m > n \\ \psi(s), & m = n \\ \begin{bmatrix} \psi(s) \\ \mathbf{0}_{n-m,m} \end{bmatrix}, & m < n \end{cases} \tag{B4.7a}$$

where $\psi(s)$ is the diagonal matrix

$$\psi(s) = \left[\begin{array}{cccc|ccc} \varphi_1(s) & 0 & \cdots & 0 & 0 & \cdots & 0 \\ 0 & \ddots & & \vdots & \vdots & & \vdots \\ \vdots & & \ddots & 0 & 0 & \cdots & 0 \\ 0 & \cdots & 0 & \varphi_r(s) & 0 & \cdots & 0 \\ \hline 0 & \cdots & 0 & 0 & 0 & \cdots & 0 \\ \vdots & & & \vdots & \vdots & & \vdots \\ 0 & \cdots & 0 & 0 & 0 & \cdots & 0 \end{array} \right] \tag{B4.7b}$$

with the integer r equal to the *determinantal* (or *normal*) *rank* of $A(s)$ (that is, the order of the highest order minor not identically equal to zero); and $\varphi_i(s)$ are *monic* polynomials (that is, each has the coefficient of its highest power of s equal to 1) with the property that $\varphi_i(s)$ divides $\varphi_{i+1}(s)$ for $i = 1, 2, \ldots, r$, with the $\varphi_i(s)$ being given by

$$\varphi_i(s) = \frac{\mathscr{D}_i(s)}{\mathscr{D}_{i-1}(s)}, \quad i = 1, 2, \ldots, r \tag{B4.7c}$$

where $\mathscr{D}_0(s) = 1$ by definition, and $\mathscr{D}_i(s)$ is the greatest common divisor of minors of order i in $A(s)$, $i = 1, 2, \ldots, r$.

The polynomials $\varphi_i(s)$ are called the *invariant factors* of A, and their product $\varphi_1(s)\varphi_2(s) \ldots \varphi_r(s)$ is called the *zero polynomial* of A.

System Matrix and Invariant Zeros

Consider the linear time-invariant state equation

$$\dot{\mathbf{x}} = \mathbf{A}\mathbf{x} + \mathbf{B}\mathbf{u} \tag{B4.8a}$$

If we assume zero initial conditions, the Laplace transform of (B4.8a) gives

$$s\mathbf{X} = \mathbf{A}\mathbf{X} + \mathbf{B}\mathbf{U} \tag{B4.8b}$$

For the general case where the system output $\mathbf{y}$ depends not only on the state $\mathbf{x}$ but also on the input $\mathbf{u}$ and its derivatives, the Laplace-transformed output can be given by

$$\mathbf{Y} = \mathbf{C}\mathbf{X} + D(s)\mathbf{U} \tag{B4.8c}$$

where $D(s)$ is a polynomial matrix. Equations (B4.8b) and (B4.8c) can be combined into the single equation

$$\begin{bmatrix} s\mathbf{I}_n - \mathbf{A} & -\mathbf{B} \\ \mathbf{C} & D(s) \end{bmatrix} \begin{bmatrix} \mathbf{X} \\ \mathbf{U} \end{bmatrix} = \begin{bmatrix} \mathbf{0} \\ \mathbf{Y} \end{bmatrix} \tag{B4.8d}$$

which defines the polynomial matrix known as the *(Rosenbrock) system matrix:*

$$\Lambda_r(s) = \begin{bmatrix} s\mathbf{I}_n - \mathbf{A} & -\mathbf{B} \\ \mathbf{C} & D(s) \end{bmatrix} \tag{B4.8e}$$

All the mathematical information on the linear system defined by equations (B4.8a) and (B4.8c) or alternatively, in shortened notation, the linear system $(\mathbf{A}, \mathbf{B}, \mathbf{C}, D)$ that is necessary to describe the system's behavior is contained in $\Lambda(s)$. The *invariant zeros* of the linear system $(\mathbf{A}, \mathbf{B}, \mathbf{C}, D)$ are the zeros of $\Lambda(s)$, and the zeros of $\Lambda(s)$ are the roots of the zero polynomial of $\Lambda(s)$.

Consider, for example, the linear system $(\mathbf{A}, \mathbf{B}, \mathbf{C}, \mathbf{D})$ with $\mathbf{A}, \mathbf{B}, \mathbf{C}$, and $\mathbf{D}$ given by

$$\mathbf{A} = \begin{bmatrix} 0 & 1 & 0 & 0 & 0 \\ -1 & -\beta & 0 & 0 & 0 \\ 0 & 0 & -k_1 & 0 & 0 \\ 1 & 0 & 0 & 0 & 0 \\ 0 & 0 & 0 & \lambda_2 & -k_2 \end{bmatrix}; \quad \mathbf{B} = \begin{bmatrix} 0 \\ 1 \\ \lambda_1 \\ 0 \\ 0 \end{bmatrix};$$

$$\mathbf{C} = \begin{bmatrix} 0 & 0 & \epsilon & 0 & 0 \\ 0 & 0 & 0 & \eta_2 & 1 \end{bmatrix}; \quad \mathbf{D} = \begin{bmatrix} \epsilon\eta_1 \\ 0 \end{bmatrix}$$

where λ_1 can be negative, while k_1, k_2, β, ϵ, η_1, η_2, and λ_2 are all positive numbers. We can determine if this system has any invariant zeros on the imaginary axis as follows: By the definition (B4.8e), the system matrix is

$$\Lambda(s) = \begin{bmatrix} s & -1 & 0 & 0 & 0 & 0 \\ 1 & s+\beta & 0 & 0 & 0 & -1 \\ 0 & 0 & s+k_1 & 0 & 0 & -\lambda_1 \\ -1 & 0 & 0 & s & 0 & 0 \\ 0 & 0 & 0 & -\lambda_2 & s+k_2 & 0 \\ 0 & 0 & \epsilon & 0 & 0 & \epsilon\eta_1 \\ 0 & 0 & 0 & \eta_2 & 1 & 0 \end{bmatrix} \tag{B4.9a}$$

This polynomial matrix can be reduced to the canonical Smith form more easily by elementary row and column operations. The following sequence of elementary row and column operations will transform the matrix in (B4.9a) to the result in (B4.9b):

1. Interchange columns 1 and 2.
2. Add $(s+\beta)*$row 1 to row 2.
3. Add $-(s+k_2)*$row 7 to row 5.
4. Add $-\eta_2*$ column 5 to column 4.
5. Add $s*$column 1 to column 2.
6. Add $[-(s+k_1)/\epsilon]*$row 6 to row 3.
7. Add $[-s/\epsilon]*$row 6 to row 4.
8. Add $-\eta_1*$column 3 to column 6.
9. Add $[s(s+\beta)+1]*$row 4 to row 2.
10. Add $-\eta_1 s*$column 2 to column 6.

$$\Lambda_1(s) = \begin{bmatrix} -1 & 0 & 0 & 0 & 0 & 0 \\ 0 & 0 & 0 & 0 & 0 & -\eta_1 s[s(s+\beta)+1]-1 \\ 0 & 0 & 0 & 0 & 0 & -\eta_1(s+k_1)-\lambda_1 \\ 0 & -1 & 0 & 0 & 0 & 0 \\ 0 & 0 & 0 & -\eta_2(s+k_2)-\lambda_2 & 0 & 0 \\ 0 & 0 & \epsilon & 0 & 0 & 0 \\ 0 & 0 & 0 & 0 & 1 & 0 \end{bmatrix} \tag{B4.9b}$$

Note that the row or column number indicated in each of the listed operations applies to the matrix that resulted from the preceding operation.

Next, adding the product of

$$\left[-s^2 + \left(k_1 + \frac{\lambda_1}{\eta_1} - \beta \right)s + (k_1\beta - 1 - k_1^2) + \frac{1}{\eta_1}\left(\lambda_1\beta - 2k_1\lambda_1 - \frac{\lambda_1^2}{\eta_1} \right) \right]$$

and row 3 to row 2 gives us the element

$$\Lambda_2[2, 6] = -(\eta_1 k_1 + \lambda_1)\left[k_1\beta - 1 - k_1^2 + \frac{\lambda_1\beta}{\eta_1} - \frac{2k_1\lambda_1}{\eta_1} - \frac{\lambda_1^2}{\eta_1^2} \right] - 1$$

at (row 2, column 6), which does not depend on s. Row 2 can then be divided by $\Lambda_2[2, 6]$, which is assumed not to be equal to zero, to yield 1 as the element at (row 2, column 6). Finally, we add the product of $[\eta_1(s + k_1) + \lambda_1]$ and row-2 to row-3 and obtain the result

$$
\Lambda_3(s) = \begin{bmatrix} -1 & 0 & 0 & 0 & 0 & 0 \\ 0 & 0 & 0 & 0 & 0 & 1 \\ 0 & 0 & 0 & 0 & 0 & 0 \\ 0 & -1 & 0 & 0 & 0 & 0 \\ 0 & 0 & 0 & -\eta_2(s + k_2) - \lambda_2 & 0 & 0 \\ 0 & 0 & \epsilon & 0 & 0 & 0 \\ 0 & 0 & 0 & 0 & 1 & 0 \end{bmatrix} \qquad \text{(B4.9c)}
$$

Then the following Smith form:

$$
\psi(s) = \begin{bmatrix} 1 & 0 & 0 & 0 & 0 & 0 \\ 0 & 1 & 0 & 0 & 0 & 0 \\ 0 & 0 & 1 & 0 & 0 & 0 \\ 0 & 0 & 0 & 1 & 0 & 0 \\ 0 & 0 & 0 & 0 & 1 & 0 \\ 0 & 0 & 0 & 0 & 0 & -\eta_2(s + k_2) - \lambda_2 \\ 0 & 0 & 0 & 0 & 0 & 0 \end{bmatrix} \qquad \text{(B4.9d)}
$$

can be obtained by the following sequence of operations $(-1*\text{row-1})$, $(-1*\text{row-4})$, $([1/\epsilon]*\text{row-6})$, (interchange rows 2 and 4), (interchange rows 3 and 6), (interchange columns 4 and 6), (interchange "new" row 6 and row 7), and (interchange resulting row 6 and 5).

The invariant zeros are given by the equation

$$
(1)(1)(1)(1)(1)[-\eta_2(s + k_2) - \lambda_2] = 0
$$

which yields

$$
s = -\left(\frac{\lambda_2}{\eta_2} + k_2\right) \qquad \text{(B4.9e)}
$$

Since k_2, η_2, and λ_2 are all positive (nonzero) numbers, the system has no invariant zeros on the imaginary axis.

CHAPTER 1

1.2(a) The relation is static.

1.2(b) Dynamic element

1.4(c) Dynamic

1.4(d) Yes. With visual feedback from the road, driver could make excessive corrections to the vehicle's heading, since the effect of steering wheel input is delayed. These corrections and countercorrections accumulate and at high speed could lead to loss of vehicle control.

1.6 $A = \int_0^t (r_n - r_p - r_s)\, dt + A(0)$; $A(0)$ is initial number of

marketable apples, and t is in days. Alternatively,

$\frac{d}{dt} A = r_n - r_p - r_s$. Relationship $\equiv$ *dynamic;* operator $\equiv$

integrator.

1.10

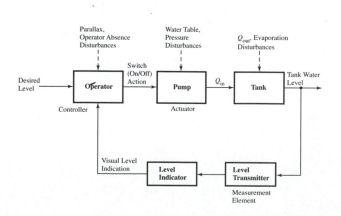

1.16

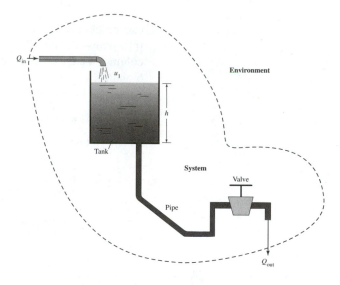

1.19

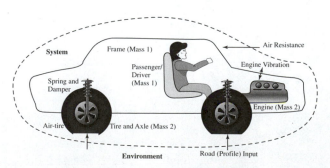

System $\equiv$ sprung mass (car frame, seat, passenger/driver), unsprung mass (engine, tire, and axle), shock absorbers and springs, and compliance and damping of the tires due to the pressurized air. Interactions with environment $\equiv$ air resistance and road (displacement and velocity) input to tires. The engine vibration is an internal interaction.

CHAPTER 2

2.2(b) $F = \left(\dfrac{2\pi\mu rx}{R - r}\right)v$; v = velocity of piston (chamber assumed stationary), and x = penetration.

2.4 $\dfrac{dh}{dt} = -\left(\dfrac{c}{\pi \tan^2 \alpha}\right)h^{-3/2} + \left(\dfrac{h^{-2}}{\pi \tan^2 \alpha}\right)Q_{\text{in}}$; $h \neq 0$.
Dynamic. Outlet diameter must be finite.

2.7 See Example 8.19 solution a.

$T = -PA = \dot{m}_e(v - v_e) = -\dot{m}(v - v_e)$; $v_e = \dfrac{1}{\dot{m}_e}PA + v$

2.10 0.827 mm

2.12 See Fig. 3.17, Example 3.10. Some compromises: water hammer/wave propagation effects ignored, single surge tank is adequate representation of modulating effect, short-term variations in reservoir and tailwater level ignored, dynamics of wicket gate drive ignored. Potential approximations: zero or linear resistance at trash rack, geometric details of tunnel and penstock and surge tank inlet replaced by pipe and surge tank inlet valve (lumped fluid inertia and resistance), uniform-area surge tank, turbine-generator energy converter can be described by a constant power conversion efficiency factor, etc.

2.13 See Fig. 3.39, Problem 3.11. Compromises: drive motor dynamics ignored (treat as velocity input source), structural-details (except inertia) of printhead ignored, paper transport system is independently controlled and its effect is not of interest, vibration of guides and printer frame not a factor. Potential assumptions: no slip between motor pulley and drive cable, mass of pulleys negligible, cable compliance lumped with return spring.

2.15 See Fig. 3.40, Problem 3.12. Compromises: rotor inertia for turbine dynamics, details of drive shaft linkages ignored, wave motion of the sea ignored. Potential assumptions: lumped spring for distributed inertia of drive shaft, mass of drive shaft and propeller lumped into rotary inertia, propeller damping is hydrodynamic.

2.19 0.263 Ω.

2.22 Let $+$ and $-$ nodes of voltage supply be a and c, and left and right nodes of current meter d and b. (1) $V_b = V_d$; (2) $i_{a-d} + i_{c-d} = 0$; and (3) $i_{a-b} + i_{c-b} = 0$.

2.24(d) Take node a between R_1 and R_2, b between R_2 and R_4, c between R_4 and R_5, ground d between R_5 and C; clockwise loop currents i_1 below R_1, i_L above R_3, i_3 below R_3, and i_5 below R_4.

Note method:

$$\frac{(V_s - V_a)}{R_1} - \frac{1}{L}\int (V_a - V_c)\,dt - \frac{(V_a - V_c)}{R_3} - \frac{(V_a - V_b)}{R_2} = 0$$

$$\frac{V_a - V_b}{R_2} - C\frac{dV_b}{dt} - \frac{V_b - V_c}{R_4} = 0;$$

$$\frac{V_b - V_c}{R_4} + \frac{V_a - V_c}{R_3} + \frac{1}{L}\int (V_a - V_c)\,dt - \frac{V_c}{R_5} = 0$$

Loop method:

$$-V_s + R_1 i_1 + R_2(i_1 - i_3) + \frac{1}{C}\int (i_1 - i_5)\,dt = 0;$$

$$R_3(i_L - i_3) + L\frac{di_L}{dt} = 0$$

$$R_4(i_3 - i_5) + R_2(i_3 - i_1) + R_3(i_3 - i_L) = 0;$$

$$-\frac{1}{C}\int (i_1 - i_5)\,dt + R_4(i_5 - i_3) + R_5 i_5 = 0$$

2.29 $f = RC\omega = 2\pi \times 10^{-4}$. For $V_{\text{in}} = \sin \omega t$,

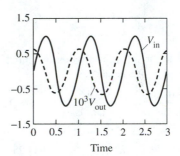

$$V_{out} = \frac{f}{\sqrt{1 + f^2}}\sin(\omega t + \phi); \qquad \phi = \tan^{-1}\left(\frac{1}{f}\right) \approx \frac{\pi}{2}$$

2.34 See Fig. 3.6, Example 3.5. Major assumptions: no slip between cable and pulley, mass of cable is negligible.

2.36 See Example 3.10, solution a.

2.42(a) Let x_1 = position of m_1; x_2 = position of m_2; N_1, N_2 be normal reactions. Let instantaneous contact point between roller and sled = A.
Spring—dynamic:

(tension) $F_s = k(x_1 - x_2)$

Roller—dynamic:

$$m_2 \frac{d}{dt}v_2 = F_s - F_{f2}; \qquad N_2 - m_2 g = 0; \qquad \text{no slip} \rightarrow v_2 = r_2 \omega_2$$

$$\sum M_A = r_2 F_s = I_A \frac{d}{dt}\omega_2; \qquad I_A = \frac{1}{2}m_2 r_2^2 + m_2 r_2^2 = \frac{3}{2}m_2 r_2^2$$

Sled—dynamic:

$$m_1 \frac{d}{dt} v_1 = F_{f2} - F_s; \qquad N_1 - N_2 - m_1 g = 0$$

Major assumptions: roller is a flatlike disk, sled is a point mass hence no moment (all forces concurrent at mass center).

2.44 Let P_1, P_2 = pressure to the left, right of valve; plunger input: $P_1 = F_{in}/A_1$
Valve—static:

$$Q = \frac{1}{R}(P_1 - P_2)$$

Tank—dynamic:

$$A_2 \frac{dx}{dt} = A_2 v = Q$$

Spring—dynamic:

(tension) $F_s = kx;$ $\qquad \dfrac{dx}{dt} = v$

Piston—dynamic:

$$m_2 \frac{dv}{dt} = (P_2 - \rho g x)A_2 - F_s - m_2 g$$

Assumptions: linear valve, no friction on sliding piston, fluid column in smaller tube neglected.

2.47 $\dfrac{dx}{dt} = \dfrac{-0.0025}{\pi \tan^2 \alpha} x + \dfrac{0.01}{\pi \tan^2 \alpha} u; \ x = h - 10 \text{ cm},$

$u = Q_{in} - 5 \text{ cc/s}$

2.51(a) $\dfrac{dR_3}{R_3} = \dfrac{dR_2}{R_2} + \dfrac{dR_4}{R_4}; \ 0.01$ **(a)** $\dfrac{0.0025}{0.0075} = 33\%$

(b) show that $\dfrac{dR_{3-strain}}{R_3} = \dfrac{dR_2}{R_2}$

CHAPTER 3

3.5 Let A = amount present in kilograms and t = time in minutes. $\dfrac{dA}{dt} = -0.0064 A$

3.9 Let x_s = stretch (right-end displacement) of spring.
$\dfrac{dx_s}{dt} = v - \dfrac{k}{\beta} x_s; \ \dfrac{dx}{dt} = v, \ \dfrac{dv}{dt} = -\dfrac{r^2}{J} kx_s;$ output $= x$

3.13 $\dfrac{dx_1}{dt} = v_1; \ \dfrac{dx_2}{dt} = v_2; \ \dfrac{dv_1}{dt} = \dfrac{1}{m_1}[m_1 g + k_1(x_g - x_1) -$

$k_2(x_1 - x_2) + \beta(v_2 - v_1)]; \dfrac{dv_2}{dt} = \dfrac{1}{m_2}[k_2(x_1 - x_2) + m_2 g -$

$\beta(v_2 - v_1)]$; output $x = x_2 - x_1$

3.18 $\dfrac{d\theta}{dt} = \omega; \ \dfrac{d\omega}{dt} = \dfrac{1}{J}\left[\dfrac{k_1}{k_1 + k_2}(\theta_e - \theta) - \beta\omega\right];$ output $= \theta$

3.21 Let y_3, v_3 = downward displacement and velocity of

valve. $\dfrac{dy_3}{dt} = v_3; \ \dfrac{d\theta}{dt} = \omega; \ \dfrac{dv_3}{dt} = \left(\dfrac{L_2^2}{mL_2^2 + J}\right)$

$\left[mg - PA - \dfrac{k_1 L_1}{L_2}\left(\dfrac{L_1}{L_2}y_3 - R\sin\theta\right) - k_2 y_3 - \beta v_3\right];$

output $= y_3$,
where P = combustion chamber pressure is assumed known, A valve face area, damping of push rod is referred to the valve stem, and center of mass of rocker is assumed at the pivot.

3.25 Let x, v = upward displacement and velocity of machine, neglect the mass of the wheels relative to eccentric masses, assume angular acceleration of the wheels α is pre-

scribed. $\dfrac{dx}{dt} = v; \ \dfrac{d\theta}{dt} = \omega, \ \dfrac{d\omega}{dt} = \alpha;$

$\dfrac{dv}{dt} = -2r(\omega^2\cos\theta + \alpha\sin\theta) - 2\dfrac{k}{m}x - g;$ output $= x$.

3.29 Orient fixed X-Y-Z and moving x-y-z coordinates with Y, y axes horizontal from A to B.

$$B_x = \frac{1}{2}\left[mL\left(\frac{mg}{J_{yy}}L\sin\theta\cos\theta - \dot\theta^2\sin\theta\right) + \frac{J_{zz}}{b}\dot\theta p\sin\theta\right];$$

$$A_x = \frac{1}{2}\left[mL\left(\frac{mg}{J_{yy}}L\sin\theta\cos\theta - \dot\theta^2\sin\theta\right) - \frac{J_{zz}}{b}\dot\theta p\sin\theta\right],$$

where $\dfrac{d\theta}{dt} = \dot\theta, \quad \theta(0) = 0;$

$\dfrac{d\dot\theta}{dt} = \dfrac{mg}{J_{yy}}L\sin\theta, \quad \dot\theta(0) = 0; \qquad J_{yy} = \dfrac{1}{4}mr^2 + mL^2;$

$J_{zz} = \dfrac{1}{2}mr^2$

3.33 Let output $= y$, and $A_p = \dfrac{\pi d^2}{4}: \dfrac{dx_1}{dt} = v_2 - \dfrac{k}{R_f A^2}x_1$

3.35 $\dfrac{dy}{dt} = v; \ \dfrac{dv}{dt} = \dfrac{1}{(\rho Ah + A_p m)}$

$$\left[F_{in} - A_p(R_f Av + mg + (k + \rho g)y) + \rho A_p gh - \frac{32\mu hAv}{d^2}\right]$$

3.38 For a choice of ground point at a,

$$\frac{dV_b}{dt} = \frac{1}{C}\left[\frac{V_b}{R_1} + \frac{R_2 + R_L}{R_2 R_L}(V_b + V_1)\right]; V_2 = V_b + V_1$$

3.42(a) (i) $V_{out} = -V_L[\text{switch}(V_{in} + V_L)] + V_{in}[1 - \text{switch}(V_{in} + V_L)]$

(ii) $V_0 = \dfrac{R_d}{R_d + R_c}V_{in} + \dfrac{R_c}{R_d + R_c}(V_d - V_L); \ V_{out} = V_0[\text{switch}(V_{in} + V_L - V_d)] + V_{in}[1 - \text{switch}(V_{in} + V_L - V_d)]$

3.47(a) $i_B = 5.54 \times 10^{-5}$ A; $V_{CE} = 6.68$ V

3.50 $R_f C_1 \dfrac{dV_i}{dt} \approx -\left(R_f C_f \dfrac{dV_0}{dt} + R_f C_f \dfrac{dV_0}{dt} + V_0 \right) \approx -V_0$,

if $1 >> R_f C_f << R_f C_1$

CHAPTER 4

4.3 A = area of bulb, k = conductivity of bath material:

$\dfrac{dT_1}{dt} = \dfrac{hA}{C}(T_2 - T_1)$; $T_1(0) = T_a$; $\dfrac{dT_2}{dt} = \left(\dfrac{-1}{\rho \pi r^2 Lc} \right)$

$\left\{ hA(T_2 - T_1) + \left[\dfrac{k\pi r^2}{w} + h\pi r^2 + \dfrac{2\pi kL}{\ln(\frac{r+w}{r})} \right] (T_2 - T_a) \right\}$;

$T_2(0) = T_0$

4.6 $V \dfrac{d\rho}{dt} = \dot{m}_h + \dot{m}_c - \rho \left(\dfrac{\dot{m}_h}{\rho_h} + \dfrac{\dot{m}_c}{\rho_c} \right)$;

V = volume = constant; $\rho V \dfrac{d\hat{T}}{dt} = \dot{m}_h[c_h(T_h - T_0) - \hat{T} + cT_0] + \dot{m}_c[c_c(T_c - T_0) - \hat{T} + cT_0]$; $c = f(T)$; $\hat{T} = cT$

Let $c = c_c + \dfrac{c_h - c_c}{T_h - T_c}(T - T_c) \rightarrow \left(\dfrac{c_h - c_c}{T_h - T_c} \right)T^2 +$

$\left[c_c - \left(\dfrac{c_h - c_c}{T_h - T_c} \right)T_c \right]T - \hat{T} = 0$

4.9 P_0 = pressure empty tank trapped air, R_v = valve resistance, assume isothermal perfect gas:

$\rho L \dfrac{dQ}{dt} = A(P_p - R_v Q - \rho gh - P_g) - B\left(\dfrac{Q}{A} \right)^\alpha$;

$\dfrac{dh}{dt} = \dfrac{Q}{\pi b^2(H - h)}$; $P_g = \dfrac{P_0 H^2}{(H - h)^2}$

4.13 $\dfrac{dh}{dt} = \dfrac{-1}{AR}(\rho gh - \rho SFr_2^2 \omega^2)$;

$J \dfrac{d\omega}{dt} = \dfrac{\rho^2}{R_v}(gh - SFr_2^2 \omega^2)SFr_2^2 \omega^2$; output = ω.

4.17 $\dfrac{dV}{dt} = \dfrac{1}{\rho}(\rho_1 Q_1 + \rho_2 Q_2 + \rho_3 Q_3) - Q$; $\rho = M_1 c_A + M_2 c_B + M_3 c_C$

$\dfrac{dT}{dt} = \dfrac{1}{\rho c_p V} \left\{ \begin{array}{l} \rho_1 Q_1[c_{p1}(T_1 - T_0) - c_p(T - T_0)] + \\ \rho_2 Q_2[c_{p2}(T_2 - T_0) - c_p(T - T_0)] + \\ \rho_3 Q_3[c_{p3}(T_3 - T_0) - c_p(T - T_0)] + \\ \dfrac{\rho_1 Q_1}{M_1} \Delta \tilde{H}_{(A)} + \dot{q} \end{array} \right\}$

$\dfrac{dc_A}{dt} = \dfrac{1}{V} \left[\dfrac{\rho_1 Q_1}{M_1} - \dfrac{c_A}{\rho}(\rho_1 Q_1 + \rho_2 Q_2 + \rho_3 Q_3) \right]$

$\dfrac{dc_B}{dt} = \dfrac{1}{V} \left[\dfrac{\rho_2 Q_2}{M_2} - \dfrac{c_B}{\rho}(\rho_1 Q_1 + \rho_2 Q_2 + \rho_3 Q_3) \right]$

$\dfrac{dc_C}{dt} = \dfrac{1}{V} \left[\dfrac{\rho_3 Q_3}{M_3} - \dfrac{c_C}{\rho}(\rho_1 Q_1 + \rho_2 Q_2 + \rho_3 Q_3) \right]$

4.20(a) $\dfrac{\partial v}{\partial t} = \dfrac{1}{\rho A} \dfrac{\partial F}{\partial x} + g$; $\dfrac{\partial F}{\partial t} = EA \dfrac{\partial v}{\partial x}$; $v = \dfrac{\partial u}{\partial t}$;

$A = \dfrac{\pi}{4} \left[D^2 - \dfrac{x}{L}(D^2 - d^2) \right]$, or $\dfrac{\partial^2 u}{dt^2} = \dfrac{E}{\rho} \dfrac{\partial^2 u}{\partial x^2} + g$;

IC: $u(x, 0) = 0$, $\dfrac{\partial u}{\partial t}(x, 0) = 0 \Rightarrow \dfrac{\partial F}{\partial x} = -\rho Ag|_{t=0}$;

BC: $u(0, t)$ = given, $F(L, t) = c(u(L, t))^{-1/4}$

4.22 $\dfrac{\partial T}{\partial t} = \dfrac{-1}{\rho Ac} \left[\dfrac{\partial q}{\partial x} + \dfrac{2\pi k}{\ln(r_e/r_i)}(T - T_a) \right]$; $q(x, t) = -kA \dfrac{\partial T}{\partial x}$;

IC: $T(x, 0)$ = given, BC: $q(0, t) = q$, $q(L, t) = h\pi r_i^2(T(L, t) - T_a)$

4.24 $\rho I \dfrac{\partial \omega}{\partial T} = -\dfrac{\partial T}{\partial x} - \alpha|F_w|r_o \text{sign}(\omega)$; $\dfrac{1}{IG} \dfrac{\partial T}{\partial t} = -\dfrac{\partial \omega}{\partial x}$;

$\omega(0, t)$ = given, $T(L, t) = f(F(L, t))$

$\rho A \dfrac{\partial v}{\partial t} = -\dfrac{\partial F}{\partial x} - \dfrac{F_D(x)}{dx} + \rho A \cos(\beta(x))g$; $\dfrac{1}{EA} \dfrac{\partial F}{\partial t} = -\dfrac{\partial v}{\partial x}$;

$F(0,t)$ = given, $v(L, t) = v_B$

$\beta(x)$ = angle of inclination to vertical at x. Can assume zero initial state throughout.

4.26 $\dfrac{dA}{dt} = -cA + u$; A = amount of marketable apples,

$cA = r_P$, input = $u = r_n - r_s$, output = A

4.28 $\dfrac{dx_1}{dt} = (B_1 - M_1)x_1 - a_2 x_1 x_2$; $a_2 = (+)$ constant: mortality rate (rabbits): $M_r = M_1 + a_2 x_2$

$\dfrac{dx_2}{dt} = (B_2 - M_2)x_2 + a_4 x_1 x_2$; $a_4 = (+)$ constant: birthrate (foxes): $B_f = B_2 + a_4 x_1$

CHAPTER 5

5.3 $\dfrac{d\omega_2}{dt} = \dfrac{r_1^2}{(r_1^2 J_2 + r_2^2 J_1)}[T_e - Br_c(\omega_2 - \omega_3)]$;

$\dfrac{d\omega_3}{dt} = \dfrac{Br_c}{J_3}(\omega_2 - \omega_3)$; $\omega_1 = \dfrac{r_2}{r_1}\omega_2$ = output

5.6 See solution for Problem 4.13.

5.9 $\dfrac{di}{dt} = \dfrac{1}{L_m + L_a \cos 2\theta}(V_{in} + 2iL_a\omega_m \sin 2\theta)$; $\dfrac{d\theta}{dt} = \omega_m$;

$T_e = -i^2 L_a \sin 2\theta$

5.11 $\dfrac{dx}{dt} = v$; $\dfrac{dv}{dt} = \dfrac{1}{m}\left(-PA + \dfrac{\epsilon A}{2x^2}V^2 \right)$;

$\dfrac{dV}{dt} = \dfrac{V}{x}v - \dfrac{x}{\epsilon AR}(V + V_s)$; $V_{out} = -V$, where m is the movable diaphragm mass. If m is neglected, then $V_{out} =$

$AR\sqrt{\dfrac{\epsilon}{2P}} \dfrac{dP}{dt} + V_s$.

5.16 $C_1 \dfrac{dT}{dt} = \dfrac{K^2 V_{in}^2}{R} - \dfrac{1}{R_{T2}}(T - T_t) - \dfrac{1}{R_{T1}}(T - T_a);$

$C_2 \dfrac{dT_t}{dt} = \dfrac{1}{R_{T2}}(T - T_t);$ output $= T$, where $T_t =$ thermocouple

junction temperature, and $V_{in} = R_f \left[\dfrac{V_s}{R_1} - \dfrac{\alpha_{AB}}{R_2}(T_t - T_a) \right].$

5.18 $\dfrac{dx}{dt} = \dfrac{QA_1 a^2 P_1^2}{A_1^2 a^2 P_1^2 - k_1 e^2 V_0 P_0}; \ P_1 = \dfrac{1}{a^2 A_1}[F_s ab - k_1 e^2 x];$

$\dfrac{dQ}{dt} = \dfrac{\pi d^2}{\rho \ell}(P_1 - P_2) - \dfrac{32\mu Q}{\rho d^2}; \ P_2 = \dfrac{(k_2 d_2^2 + k_3 d_3^2)\Delta\theta}{A_2 d_p + Q(fh - d_p)/frw},$

where $x =$ piston displacement; Q flow in line, $V_0, P_0 =$ air reference volume, pressure in accumulator; $d_2, d_p,$ $d_3 =$ distance, pivot to spring k_2, brake piston ram, spring k_3; $h =$ transverse distance pivot to drum face; $\Delta\theta =$ shoe angular play. Other assumptions: $w =$ constant, complete power dissipation as friction, laminar flow.

5.20 $A(y_0 - y)\dfrac{dP}{dt} = Q_s P - c_g \dot{m} + PAv; \ Q_s = \dfrac{(P_s - P)}{R_p};$

$m \dfrac{dv}{dt} = -(k_1 + k_2)y - mg; \ \dfrac{dy}{dt} = v;$

$Q_p = c_0 y \, \text{sign}(P_p - \rho_w g h_1) \sqrt{|P_p - \rho_w g h_1|};$

$A_1 \dfrac{dh_1}{dt} = Q_p - \dfrac{\rho g h_1}{R_1}; \ \dfrac{A_2}{\rho_w g} \dfrac{dh_2}{dt} = \dfrac{h_1}{R_1} - \dfrac{h_2}{R_2};$

$\dfrac{b}{d}[\epsilon_a \ell + (\epsilon_w - \epsilon_a)h_2]\dfrac{dV_2}{dt} =$

$i_s - \dfrac{b}{d}(\epsilon_w - \epsilon_a)V_2 \left[\dfrac{\rho_w g}{A_2}\left(\dfrac{h_1}{R_1} - \dfrac{h_2}{R_2} \right) \right]$

$L\dfrac{di}{dt} = KR_f \left(\dfrac{-V_1}{R_1} - \dfrac{V_2}{R_2} \right);$

$P_a = \dfrac{A_s \mu_0 N^2 i^2}{2 A_{rr} x_s^2},$ where $c_g = \dfrac{P}{\rho_a} =$ constant; $c_0 =$ control

valve constant; $L =$ inductance, solenoid; $\ell =$ length, capacitance probe; $b, d =$ probe width of plates, gap.

5.24 $\dfrac{dy}{dt} = v; \ \dfrac{dv}{dt} = -g - \dfrac{k}{m}y - \dfrac{V_{in}^2 \epsilon A}{2my^2};$

$Q = c_0(y_0 - y)\sqrt{\Delta P},$ where $A, y =$ movable plate area, displacement (down); $k, m =$ cantilever spring constant, end mass; $V_{in} =$ input voltage drop; $\Delta P =$ valve pressure drop; $c_0, y_0 =$ valve constants.

5.28 $\dfrac{dy}{dt} = v; \ \dfrac{dv}{dt} = g - \left(\dfrac{V_{in}^2 \epsilon A}{2m} \right)\dfrac{1}{y^2} -$

$\dfrac{E\pi d^2}{4m(r/h)^4} \left[\dfrac{5.33}{1 - v^2}\left(\dfrac{y}{h} \right) + \dfrac{2.60}{1 - v^2}\left(\dfrac{y}{h} \right)^3 \right],$ where $A,$

$m =$ movable plate area, mass; $V_{in} =$ voltage drop; outflow;

$Q = \dfrac{\pi d \Delta P}{c}(H - y), \ \Delta P =$ valve pressure drop.

CHAPTER 6

6.2 $\dfrac{dx}{dt} = -0.45x; \qquad x(0) = 2.773$

6.6 (i) $x(t) = 2.5 - 1.5e^{-t} - 0.5t + [0.5(t - 2) + 1.5e^{-(t-2)} - 1.5]U_S(t-2) + [e^{-(t-4)} + (t - 4) - 1]U_S$
$(t - 4) - (t - 5)U_S(t - 5)$
(ii) $x(t) = 3e^{-t} + 2t - 2$

6.8 (i) $v(t) = 0.08[1 - e^{-0.05t} - (1 - e^{-0.05(t-10)})U_S$
$(t - 10)]$ (iii) $v(t) = 0.04e^{-0.05t}; \ v(0) = 0.04$

6.13 (i) $y(t) = \dfrac{1}{12} + \left(\dfrac{1}{4} - \dfrac{t}{2} \right)e^{-2t} - \dfrac{1}{3}e^{-3t}$

(ii) $y(t) = (t - 1)e^{-2t} + e^{-3t}$

6.18 $1.14; \ -1.5883$ rad (or $-91°$)

6.20 $\dfrac{d}{dt}[\text{Real}\{z\}] = \dfrac{d}{dt}[x(t)];$

$\text{Real}\left\{ \dfrac{d}{dt}[z] \right\} = \text{Real}\{\dot{x}(t) + j\dot{y}(t)\}$

6.23 $(\cos \phi + j \sin \phi)^n = \left(\dfrac{z}{M} \right)^n = (e^{j\phi})^n = e^{jn\phi} =$

$(\cos n\phi + j \sin n\phi)$

6.28 Let $\mathbf{X} = \begin{bmatrix} \theta_1 \\ \theta_2 \\ \omega_1 \\ \omega_2 \end{bmatrix}; \ \mathbf{A} = \begin{bmatrix} 0 & 0 & 1 & 0 \\ 0 & 0 & 0 & 1 \\ -(2 + g) & 1 & 0 & 0 \\ 1 & -g & 0 & 0 \end{bmatrix};$

$\mathbf{B} = \begin{bmatrix} 0 \\ 0 \\ 1 \\ 0 \end{bmatrix}$

6.31 (a) Consider the t^2 term: $\dfrac{A^2}{2} + \dfrac{B^2}{2} + AB \neq$

$\dfrac{(A + B)^2}{2!} = \dfrac{A^2}{2} + \dfrac{B^2}{2} + \dfrac{AB + BA}{2}$ unless $AB = BA$.

6.34 $\begin{bmatrix} x_1(t) \\ x_2(t) \end{bmatrix} = \begin{bmatrix} \dfrac{5}{3}e^{5t} - \dfrac{4}{15}e^{-t} - \dfrac{2}{5}\cos 2t + \dfrac{1}{5}\sin 2t \\ \dfrac{10}{3}e^{5t} + \dfrac{4}{15}e^{-t} + \dfrac{2}{5}\cos 2t - \dfrac{1}{5}\sin 2t \end{bmatrix}$

6.37 $\begin{bmatrix} x_1(t) \\ x_2(t) \\ x_3(t) \end{bmatrix} = \begin{bmatrix} \dfrac{5}{4} - e^{-(t-1)} - \dfrac{15}{4}e^{-2(t-1)} + \dfrac{5}{2}te^{-2(t-1)} \\ \dfrac{15}{4} - 4e^{-(t-1)} - \dfrac{25}{4}e^{-2(t-1)} + \dfrac{5}{2}te^{-2(t-1)} \\ \dfrac{11}{4} - 2e^{-(t-1)} - \dfrac{5}{4}e^{-2(t-1)} + \dfrac{5}{2}te^{-2(t-1)} \end{bmatrix}$

6.42(a) $\begin{bmatrix} x_{1,k} \\ x_{2,k} \end{bmatrix} = \begin{bmatrix} \cos(k\pi/2) + \sin(k\pi/2) \\ 1 + \sin(k\pi/2) - \cos(k\pi/2) \end{bmatrix}$ (b) $\begin{bmatrix} 1 \\ 0 \end{bmatrix}$

6.45(a) $p = \pm j0.5$; the combined mode is $C \cos$
$\left(\dfrac{\pi k}{2} + \phi \right)(0.5)^k$.

6.47 $\left| \dfrac{1}{2}(a^2 T^3 + aT^2 + aT) \right| < 1$

6.54 $y(0.5) = 0.509079$

CHAPTER 7

7.4 (i) $\dot{v}(0+) = 0.5$ m/s^2　**(ii)** $x(t) = 0.5 -$
$\dfrac{\sqrt{3}}{3} \cos\left(\dfrac{\sqrt{3}}{2} t - 0.5236 \right) e^{-t/2}$; F_{d2}(max) = 7.793 N

(iii) $F_s(t) = -175\left[1 - \dfrac{2\sqrt{3}}{3} \cos\left(\dfrac{\sqrt{3}}{2} t - 0.5236 \right) e^{-t/2} \right]$;

$|F_s|_{max} = 203.53$ N

7.8 (i) $y(t) = 2.5 + (y(0) - 2.5)e^{-2t}$

(ii) $y(t) = y(0) + \dfrac{1}{2}\dot{y}(0)\left(1 - e^{-2t} \right)$

7.11 $V_{out}(0) = V_s$; $\dot{V}_{out}(0) = 0$;
$V_{out}(t) = 1.054 V_s \cos(3t - 0.3275)e^{-t}$;

$T_d = \dfrac{2\pi}{3}$; $n = \dfrac{3}{2\pi}$; $\phi = -0.32175$; $f = \dfrac{3}{2\pi}$;

$\zeta = \dfrac{\sqrt{10}}{10} = 0.316$

7.13(b) (i) 6.3　**(ii)** 0.198　**(iii)** 53%　**(iv)** 1.8 s
(v) 14.93 s

7.16 (i) $\dfrac{(1 - \omega^2 R_1 R_2 C_1 C_2) + j\omega(R_1 C_1 + R_2 C_2 + R_1 C_2)}{-\omega^2 R_1 C_1 C_2 + j\omega C_2}$

(ii) $\dfrac{-\omega^2 R_1 C_1 C_2 + j\omega C_2}{(1 - \omega^2 R_1 R_2 C_1 C_2) + j\omega(R_1 C_1 + R_2 C_2 + R_1 C_2)}$

(iii) $\dfrac{1 + j\omega R_2 C_2}{j\omega C_2}$

(iv) $\dfrac{\omega^2 R_1 C_1 C_2 - j\omega C_2}{(1 - \omega^2 R_1 R_2 C_1 C_2) + j\omega(R_1 C_1 + R_2 C_2 + R_1 C_2)}$

7.19 (i) $A = \dfrac{\sqrt{\left(10 + \dfrac{15\omega^2}{4} \right)^2 + \left[\dfrac{-\omega}{4}(30 + 5\omega^2) \right]^2}}{\left(1 - \dfrac{\omega^2}{4} \right)^2 + \dfrac{25\omega^2}{16}}$;

$\phi = \tan^{-1}\left(\dfrac{-\dfrac{\omega}{4}(30 + 5\omega^2)}{10 + \dfrac{15\omega^2}{4}} \right)$

(ii) $A = \dfrac{3}{\omega\sqrt{16 + \omega^2}\sqrt{1 + 9\omega^2}}$;

$\phi = -\dfrac{\pi}{2} - \tan^{-1}\left(\dfrac{\omega}{4} \right) - \tan^{-1}(3\omega)$

7.21 $R = 32.072$ Ω;　　$L = 2.559 \times 10^{-5}$ F

7.25 (i) $\dfrac{d^2\theta}{dt^2} + \left(\dfrac{\beta r^2}{J} \right)\dfrac{d\theta}{dt} + \left(\dfrac{k}{J} \right)\theta = \left(\dfrac{rF_0}{J} \right)\sin(2\pi ft)$

(ii) Base excitation　**(iii)** $G(j\omega) = \dfrac{1}{13000 - \omega^2 + j100\omega}$;

$\omega = 2\pi f$　**(iv)** $\theta(t) = \dfrac{10^{-4}\sin(120\pi nt + \phi)}{\sqrt{(1.3 - 1.44\pi^2 n^2)^2 + 1.44\pi^2 n^2}}$;

$\phi = -\tan^{-1}\left(\dfrac{1.2\pi n}{1.3 - 1.44\pi^2 n^2} \right)$; $n = \dfrac{f}{60}$

7.27 (i) 31.6228 rad/s　**(ii)** $\dfrac{0.002\omega^2}{\sqrt{(1000 - \omega^2)^2}}$　**(iii)** $\dfrac{\pi}{\omega}$

7.32 $d = 0.052$ m, $A = 0.22$ m^2

7.33 $k = 7826$ N/m, $m = 0.196$ kg

7.35(b) 1647 Hz

7.39(a) $f(t) = \dfrac{2\alpha}{\pi}\sum_{n=1}^{\infty} \dfrac{[1 - (-1)^n]}{n}\sin\left(\dfrac{2n\pi}{\Gamma} t \right)$;

$C_n = \dfrac{2\alpha}{n\pi}[1 - (-1)^n]$; $\phi_n = -\dfrac{\pi}{2}$, for all n

7.39(d) $f(t) = \dfrac{b\alpha}{\Gamma} + \dfrac{2\alpha}{\pi}\sum_{n=1}^{\infty} \dfrac{1}{n}\sin\left(\dfrac{nb\pi}{\Gamma} \right)\cos\left(\dfrac{2\pi n}{\Gamma} t \right)$;

$C_0 = \dfrac{2b\alpha}{\Gamma}$, $C_n = \dfrac{2\alpha}{n\pi}\left| \sin\left(\dfrac{nb\pi}{\Gamma} \right) \right|$; $\phi_n = 0$

7.39(h) $f(t) = \dfrac{\alpha}{\Gamma} + \dfrac{2\alpha}{\Gamma}\sum_{n=1}^{\infty} \cos\left(\dfrac{2n\pi}{\Gamma} t \right)$; $C_n = \dfrac{2\alpha}{\Gamma}$; $\phi_n = 0$

7.41(d) $f(t) = \dfrac{4\pi^2}{3} + \sum_{n=1}^{\infty}\left[\dfrac{4}{n^2}\cos(nt) - \dfrac{4\pi}{n}\sin(nt) \right]$

7.47(b) $0.024\cos(t - 1.8158)$

CHAPTER 8

8.2(c) $F(\omega) = \begin{cases} \pi, & |\omega| \le 1 \\ 0, & \text{otherwise} \end{cases}$

8.6(b) $y(t) = e^{-t} + \left[\dfrac{\sqrt{3}}{3}\sin\left(\dfrac{\sqrt{3}}{2} t \right) - \cos\left(\dfrac{\sqrt{3}}{2} t \right) \right] e^{-t/2}$

8.7(b)

$\dfrac{V_{out}}{V_s}(j\omega) = \left[1 + \dfrac{R_s}{R_L} + \omega^4 C_1 L_1 C_2 L_2 - \omega^2\left(C_1 L_1 + C_2 L_2 \right.\right.$
$\left. + C_2 L_1 + C_1 L_2 \dfrac{R_s}{R_L} \right) + j\left\{ \omega\left(R_s(C_1 + C_2) + \dfrac{L_1 + L_2}{R_L} \right)\right.$
$\left.\left. - \omega^3\left(R_s C_1 C_2 L_2 + \dfrac{C_1 L_1 L_2}{R_L} \right) \right\} \right]^{-1}$

8.8(b) $\dfrac{V_{\text{out}}}{V_{\text{in}}}(j\omega) = \Bigg[\left(1 + \dfrac{R_s}{R_L} - \omega^2 LC\right)$

$+ j\omega\left(R_sC + \dfrac{L}{R_L}\right)\Bigg]^{-1}$

8.19 Compare sketch with $f(t) = [t + e^{-(t+1)}]U_S(t+1) - 2[t - 1 + e^{-t}]U_S(t) + [t - 2 + e^{1-t}]U_S(t-1).$

8.23(c) $\displaystyle\int_0^\infty \sin(\omega t)e^{-st}\,dt = -e^{-st}\left(\dfrac{\omega\cos(\omega t) + s\sin(\omega t)}{s^2 + \omega^2}\right)$

$+ \dfrac{\omega}{s^2 + \omega^2}\Bigg|_0^\infty = \dfrac{\omega}{s^2 + \omega^2}$

8.24(a) $\dfrac{2e^{1.5}}{s + 3}$

8.24(g) $\dfrac{3e^{-s}}{s + 2}$

8.27(d) $\dfrac{1 - e^{-2t}}{t}$

8.29(e) $e^{-t}\cos(t)$

8.35 $\left\{\dfrac{5}{4}e^{-(t-1)} - e^{-3(t-1)}\left[\dfrac{11}{12} + \dfrac{3}{2}(t-1)\right] - \dfrac{1}{3}\right\}U_S(t-1)$

8.37(a) Initial value = 0, final value = 2.

8.39(e) $\dfrac{7}{4}e^{3t} - \dfrac{2}{3}e^{2t} - \dfrac{1}{12}e^{-t}$

8.45(i) G_3, dominant time constant smallest

(ii) 0.0422   **(iii)** $\dfrac{d^3y}{dt^3} + 4\dfrac{d^2y}{dt^2} + 5\dfrac{dy}{dt} + 2y = 5u(t)$

8.48(b) $e^{At} = \begin{bmatrix} \{-2te^{-2t} + e^{-t}\} & \{(1+t)e^{-2t} - e^{-t}\} \\ \{4e^{-t} - (4 + 6t)e^{-2t}\} & \{(5 + 3t)e^{-2t} - 4e^{-t}\} \\ \{2e^{-t} - (2 + 2t)e^{-2t}\} & \{(2 + t)e^{-2t} - 2e^{-t}\} \end{bmatrix}$

$\begin{matrix} \{2e^{-t} - (2 + t)e^{-2t}\} \\ \{8e^{-t} - (8 + 3t)e^{-2t}\} \\ \{4e^{-t} - (3 + t)e^{-2t}\} \end{matrix}\Bigg]$

$\mathbf{X}(t) = \begin{bmatrix} (-4 + t)e^{-2(t-1)} + 2e^{-(t-1)} \\ (-10 + 3t)e^{-2(t-1)} + 8e^{-(t-1)} \\ (-3 + t)e^{-2(t-1)} + 4e^{-(t-1)} \end{bmatrix}$

8.50 $\mathbf{X}(t) = \begin{bmatrix} -\dfrac{4}{15}e^{-t} + \dfrac{5}{3}e^{5t} - \dfrac{2}{5}\cos(2t) + \dfrac{1}{5}\sin(2t) \\ \dfrac{4}{15}e^{-t} + \dfrac{10}{3}e^{5t} + \dfrac{2}{5}\cos(2t) - \dfrac{1}{5}\sin(2t) \end{bmatrix}$

8.52 $\mathbf{X}(t) = \begin{bmatrix} \left(\dfrac{-15}{4} + \dfrac{5}{2}t\right)e^{-2(t-1)} - e^{-(t-1)} + \dfrac{5}{4} \\ \left(\dfrac{-25}{4} + \dfrac{15}{2}t\right)e^{-2(t-1)} - 4e^{-(t-1)} + \dfrac{15}{4} \\ \left(\dfrac{-5}{4} + \dfrac{5}{2}t\right)e^{-2(t-1)} - 2e^{-(t-1)} + \dfrac{11}{4} \end{bmatrix}$

8.56 $G(s) =$

$$\dfrac{\left(\dfrac{k\pi r^2}{w} + h\pi r^2 + \dfrac{2\pi kL}{\ln((r+w)/r)}\right)\left(\dfrac{hA/C}{\rho\pi r^2 Lc}\right)}{(s + hA/C)\left[s - \left(\dfrac{k\pi r^2}{w} + h\pi r^2 + \dfrac{2\pi kL}{\ln((r+w)/r)} - hA\right)\left(\dfrac{1}{\rho\pi r^2 Lc}\right)\right] - \dfrac{h^2A^2}{C\rho\pi r^2 Lc}}$$

8.58 $\dfrac{V_2(s)}{V_1(s)} = \dfrac{R_2R_L(R_1Cs - 1)}{R_1R_2R_LCs - (R_1R_2 + R_1R_L + R_2R_L)}$

8.62(a) $F(z) = \dfrac{z^2 - cz\cos(\omega T)}{z^2 - 2cz\cos(\omega T) + c^2}$; $c = a^T$;

$|z| > |a^T|$

8.63(c) $be^{\ln(a)(k-1)}$

8.69 $\begin{bmatrix} x_1(kT) \\ x_2(kT) \end{bmatrix}$

$= \begin{bmatrix} \dfrac{1}{8}[32 - 23e^{\ln(3/4)(k-1)}]U_S(kT - T) \\ e^{\ln(3/4)k} + \left[16 - \dfrac{49}{4}e^{\ln(3/4)(k-1)}\right]U_S(kT - T) \end{bmatrix}$

8.71 $G(z) = \dfrac{z}{z-1}$

CHAPTER 9

9.3 0.1; $\dfrac{C(s)}{R(s)} = \dfrac{90(20 - s)}{90(20 - s) + (s + 10)(s + 5)^2}$

9.6 $\dfrac{C}{R} = \dfrac{(G_1G_2 + G_4)G_3}{1 + (H_1 + H_2G_2)G_3}$

9.12

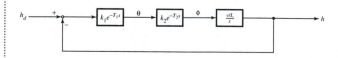

9.14(a) $s = \dfrac{K_2 - K_1}{K_1\tau_2 - K_2\tau_1}$;

If $\dfrac{\tau_1}{\tau_2} > 1$, then $\dfrac{\tau_1}{\tau_2} > \dfrac{K_1}{K_2} > 1$; if $\dfrac{\tau_1}{\tau_2} < 1$, then $\dfrac{\tau_1}{\tau_2} < \dfrac{K_1}{K_2} < 1$.

9.17(b)

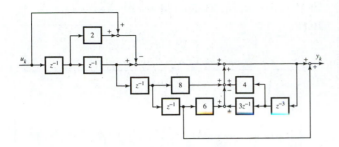

9.22 $G(s) = \dfrac{k(1 + 5s)}{(1 + s)\left(1 + \frac{1}{2}s\right)}$

9.25(a) See example 12.1, part (a)   **(b)** Semicircle ($\omega = 0$ at $(0, 0)$, $\omega \to \infty$ at $(1, 0)$) above real axis, diameter = 1   **(c)** $K < 0.287$; see Example 12.1, part (b)

9.29 $\mathbf{A} = \begin{bmatrix} 0 & 1 & 0 \\ 0 & 0 & 1 \\ -4 & -3 & -2 \end{bmatrix}$; $\mathbf{B} = \begin{bmatrix} 0 \\ 0 \\ \frac{1}{2} \end{bmatrix}$; $u = \sin 2t$

9.31(b) $G(z) = \dfrac{z - \frac{1}{2}}{z^2 - \frac{3}{4}z + \frac{1}{4}}$

9.32(c)

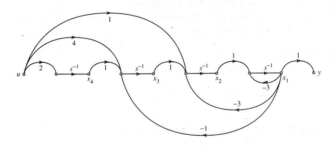

9.33(c)

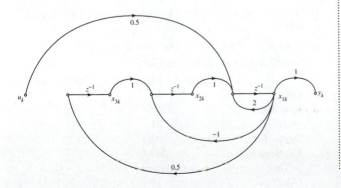

9.35(c) $\dot{x}_1 = -3x_1 + x_2$; $\dot{x}_3 = -x_1 + x_4 + 4u$; $\dot{x}_2 = -3x_1 + x_3 + u$; $\dot{x}_4 = 2u$; $y = x_1$

9.36(c) $\mathbf{X}_{k+1} = \begin{bmatrix} 2 & 1 & 0 \\ -1 & 0 & 1 \\ 0.5 & 0 & 0 \end{bmatrix} \mathbf{X}_k + \begin{bmatrix} 0.5 \\ 0 \\ 0 \end{bmatrix} u_k$;

$y_k = [1 \quad 0 \quad 0] \mathbf{X}_k$

9.39(a) Consider $k_1 = 2$, $k_2 = 1$, $\tau_1 = 4$, $\tau_2 = 1$; the simulation model is

$$\dot{x}_1 = -\frac{1}{\tau_1} x_1 + \frac{k_1}{\tau_1} u; \quad \dot{x}_2 = -\frac{1}{\tau_2} x_2 - \frac{k_2}{\tau_2} u; \quad y = x_1 + x_2$$

9.41(b) $\hat{\mathbf{X}}_{k+1} = \begin{bmatrix} 0 & 1 & 0 \\ 0 & 0 & 1 \\ 0 & \frac{1}{2} & 0 \end{bmatrix} \mathbf{X}_k + \begin{bmatrix} 2 & 2 \\ -\frac{3}{4} & -1 \\ \frac{1}{2} & \frac{3}{4} \end{bmatrix} \begin{bmatrix} u_{1,k} \\ u_{2,k} \end{bmatrix}$;

$\hat{y}_k = [1 \quad 0 \quad 0] \mathbf{X}_k$

$\mathbf{G}(z) =$ $\begin{bmatrix} \dfrac{2z^{-1} - \frac{3}{4}z^{-2} - \frac{1}{2}z^{-3}}{1 - \frac{1}{2}z^{-2}} & \dfrac{2z^{-1} - z^{-2} - \frac{1}{4}z^{-3}}{1 - \frac{1}{2}z^{-2}} \end{bmatrix}$

9.45(b) Not controllable, not observable

9.49(a) (Completely) controllable   **(b)** (Completely) observable

9.53(a) See Figs. 14.9a and 14.9b   **(b)** see Fig. 14.9c

(c) $\mathbf{X}_{k+1} = \begin{bmatrix} c & 1 & 0 & 0 \\ 0 & 0 & \left(\frac{1-c}{a}\right) & 0 \\ 0 & 0 & 0 & 1 \\ 0 & 0 & 1 & 0 \end{bmatrix} \mathbf{X}_k$

$+ \begin{bmatrix} 0 & 0 \\ 1 & (1-c) \\ -\left(\frac{ac}{1-c}\right) & 0 \\ \left(\frac{a}{1-c}\right) & 0 \end{bmatrix} \begin{bmatrix} e_k \\ v_k \end{bmatrix}$ **(d)** System is not controllable if

the disturbance input is not seen as a controlling input. The modal states $\hat{x}_1$ and $\hat{x}_2$ are not affected by e_k although they are subject to v_k. **(e)** see fig. 14.9d

CHAPTER 10

10.3 A responds faster; its rate of decay $\left(\frac{1}{24}\right)$ is larger than B's $\left(\frac{1}{240}\right)$.

10.8(a) **(i)** nonlinear; $\hat{\dot{x}}_1 = \hat{x}_2$; $\hat{\dot{x}}_2 = -a\hat{x}_1 - b\hat{x}_2$
(ii) $b \geq 2\sqrt{a}$ **(iii)** Stable

10.12(a)

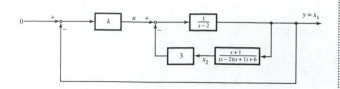

(b) $k = 1$

10.14(a)

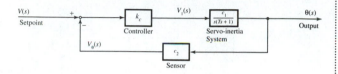

(b) $\dfrac{\theta(s)}{V(s)} = \dfrac{k_c c_1}{s(Ts + 1) + k_c c_1 c_2}$ **(c)** $\theta(\infty) = \dfrac{v_0}{c_2}$

(d) $f_0 = \dfrac{1}{2\pi}\sqrt{\dfrac{k_c c_1 c_2}{T}}; \qquad \zeta = \dfrac{1}{2\sqrt{k_c c_1 c_2 T}}$

(e) $k_c = \dfrac{1}{4Tc_1 c_2}$

10.16(a) $\dfrac{1 + 3s}{3(s^2 + 2s + 1)}$ **(b)** $\dfrac{1 + 3s}{3(s^2 + 2s - 1)}$ **(c)** (a), All

eigenvalues have negative real parts.

10.19(a) Unstable   **(b)** $3 < k < 4$

10.23(c) The root locus does not leave the real axis: one branch is from $(0, 0)$ to $(-2, 0)$, another branch is from $(-3, 0)$ to $(-2, 0)$, and the last branch is from $(-6, 0)$ to $(-\infty, 0)$.

10.26 Locus is like Fig. 10.6 zero: -1, poles: $0, 0, -3, -6$; purely imaginary roots: $\pm j3$, $k = 81$.

10.28(c) (i) Stable but not asymptotically   **(ii)** 0
(iii) Two pairs (or four roots: $\pm j, \pm j\sqrt{2}$

10.29(c) $0 < k < 28$

10.31 $0 < T < 1$, $(1 - T) < k < (2 - 2T)$

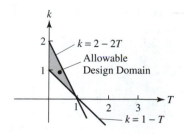

10.36(b) Not asymptotically stable

10.40(a) GM $= -6.02$ db, PM $= -15.01°$, BW ≈ 2.78 rad/s

10.44 GM $= 1.18$ (1.41 db), PM $= 12°$

10.50

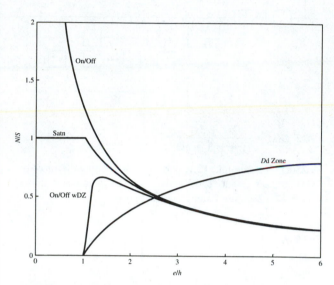

10.51(a)

10.52(d) $e < 1.082$, and $e > 2.614$

10.54(a) Essentially the same as for saturation. N decreases as e/h increases so system will tend to remain stable for operation initially in stable region. If operating point is in unstable region for any reason, system will tend to remain unstable.

10.59(a) $\beta = 2.17$, $\alpha = -0.5$

CHAPTER 11

11.4 $S = 1 - \dfrac{1}{k_c} G_{CL}(s)$; static: $S \approx 0.91$

11.9 $k = 24.667$, $T = 0.018$

11.12(b) $k_v = 0.34$, $T_i = 0.093$ **(c)** $k_v = 0.1$, $T_i = 0.319$; other combinations are possible

11.15 P: type-0, $e_{ss} = 0.18$; PI; type-1, $e_{ss} = 0$; PID: type-1, $e_{ss} = 0$

11.19(a) $k_c = 10.89$, $T_i = 1.138$ **(b)** $k_c = 14.52$, $T_i = 0.6856$, $T_d = 0.1714$; $k_c = 12.10$
(c)

	P-control	PI-control	PID-control
Offset	0.7	0	0
Rise Time	0.57	0.55	0.46
%Overshoot	54	91	57
Decay Ratio	0.78	0.95	0.72
5% Settling Time	6.7	49.4	3.8
IAE	2.123 (at 15)	10.171 (at 60)	0.903 (at 15)

11.22 $k_c = 1.665$, $T_i = 1.138$

11.26(a) $k_c \geq 276.368$, $T_d \leq 0.223$ **(b)** $k_c \geq 276.368$, $T_d \leq 0.201$

11.28(a) $s^2 + \left(\dfrac{T+2}{T}\right)s + \left(\dfrac{k+2}{T}\right) = 0$ **(b)** $\dfrac{2}{2+k}$

(c)

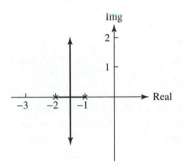

(d) $k = 4.25$

11.33 There is an additional pole of little effect. The added zero is near the old one. Most significant change is the 60° asymptote of dominant branch, relative to 90° in Fig. 11.11d. For sufficiently large k_c, instability is now possible. At operating point $k_c = 1$, there is less damping (and higher frequency) ($s = -4.987 \pm j6.937$ and $-1.147 \pm j0$, compared with $s = -6.901 \pm j5.987$ and $-1.198 \pm j0$)

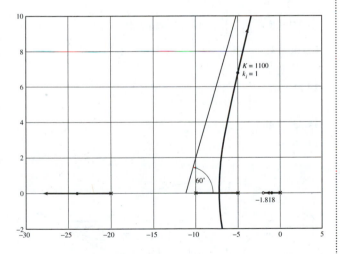

11.37 $k = 1.718$

11.42(a) $k_c = 6.908$ **(a)** $k_c = 64.80$, $T_i = 1.57$

11.45 $k_c \geq 276.368$, $T_d = 0.201$, $\tau = cT_d$, $0 < c \leq 0.09989$

11.47 For $k_c = 1$ system exhibits a stable limit cycle of frequency 0.1154 rad/s; 0.0632.

CHAPTER 12

12.3(a) $G(s) = \dfrac{1}{0.25s + 0.1}$ **(b)** Low-pass, $\omega_c = 0.4$ rad/s

(c) $J = 0.1$

12.6(c) $k_c = 0.2795$

12.9(b) $k_c = 0.316$

12.15 $k = 0.2$; $M_P = 1.512$, $\omega_R = 0.1225$ rad/s, 30.5%, 38.67°

12.19 $k_c = 20.576$

12.22 $\left[\dfrac{0.25(1 + 1.085s)}{1 + 0.25(1.085)s}\right] \rightarrow [1.68] \rightarrow$
$\left[\dfrac{0.25(1 + 1.085s)}{1 + 0.25(1.085)s}\right]$; $GM = 11.179$ db, $PM = 46°$

12.28 $G_c(s) = k_c\dfrac{(1 + 0.377s)(1 + 0.885s)}{(1 + 0.038s)(0.885s)}$;
$0.89 \leq k_c \leq 1.18$

12.33 $T_d = 0.55$

CHAPTER 13

13.3(a) $u(s) = \rho c w(T_{SL} - T_i)$

(b) $k_c \geq \left(\dfrac{2v}{w} - 1\right)$ **(c)** $kc \geq pc\left(\dfrac{2v}{w} - 1\right)$

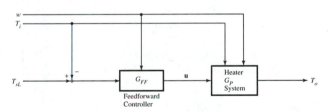

13.8 $G_{c12}(s) = \dfrac{-k_{22}}{s + 1}$; $G_{c21}(s) = \dfrac{-k_{11}}{s + 2}$

13.11(a)

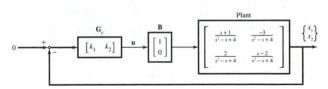

(b) $k_1 = 2$, $k_2 = \frac{1}{2}$

13.16 $w_1 = 0.46$, $w_2 = 0.54$;
$\mathbf{k} = [50.3039 \quad -35.4487 \quad 28.9017]$

13.19 $G_c(s) = \dfrac{s^2 + 2s + 2}{s(s + 2)}$

13.23 $k_{1P} = 3.2$, $k_{1i} = 0$, $k_{2P} = 2.5$, $k_{2i} = 1.5$

13.26 $u_a = \text{sign}(e^T q_n') \left[\left| \frac{1}{2} \hat{f}_n \right| + \epsilon \right]$;

$$e^T q_n' = [0.0125 \quad 0.047 \quad 0.0342] \begin{bmatrix} x_{m1} - x_1 \\ x_{m2} - x_2 \\ x_{m3} - x_3 \end{bmatrix};$$

$$\hat{f}_n = (\mathbf{A}_m - \mathbf{A})\mathbf{X}; \ \mathbf{A}_m = \begin{bmatrix} 0 & 1 & 0 \\ 0 & 0 & 1 \\ -40 & -44 & -16 \end{bmatrix};$$

$$\mathbf{A} = \begin{bmatrix} 0 & 1 & 0 \\ 0 & 0 & 1 \\ -1 & -2 & -1 \end{bmatrix}$$

13.29(a) $u^* = \dfrac{\sqrt{2}}{4} \left\{ \left[\dfrac{(\sqrt{2}+1)e^{-\sqrt{2}} + (\sqrt{2}-1)e^{\sqrt{2}}}{e^{-\sqrt{2}} - e^{\sqrt{2}}} \right] \right.$

$\left[(\sqrt{2}-1)e^{-\sqrt{2}t} + (\sqrt{2}+1)e^{\sqrt{2}t} \right] - e^{-\sqrt{2}t} + e^{\sqrt{2}t} \Big\}$

(b) $u^* = \dfrac{\sqrt{2}}{4} \{ (e^{\sqrt{2}t} - e^{-\sqrt{2}t})$

$- \left[\dfrac{e^{\sqrt{2}} - e^{-\sqrt{2}}}{(\sqrt{2}+1)e^{\sqrt{2}} + (\sqrt{2}-1)e^{-\sqrt{2}}} \right] [(\sqrt{2}-1)e^{-\sqrt{2}t}$

$+ (\sqrt{2}+1)e^{\sqrt{2}t}] \}$

13.34 Initial control: u^*

$= \begin{cases} -1; & (x_1 < 1, x_2 > 0) \text{ and } (x_1 \le 0, x_2 = 0) \\ +1; & (x_1 > -1, x_2 < 0) \text{ and } (x_1 \ge 0, x_2 = 0) \end{cases}$

Switching control: u^*

$= \begin{cases} -1; & e^{x_2(t)} - x_1(t) = 1 \text{ and } x_2(t) \ge 0 \\ +1; & e^{-x_2(t)} + x_1(t) = 1 \text{ and } x_2(t) \le 0 \end{cases}$

13.38(a) $u^*(t) = \begin{cases} -4, & \text{if } \lambda_2(t) \ge 1 \\ 0, & \text{if } -1 < \lambda_2(t) < 1; \\ 4, & \text{if } \lambda_2(t) \le -1 \end{cases}$ $\left. \begin{array}{l} \dot{x}_1 = x_2 \\ \dot{x}_2 = u^* \\ \dot{\lambda}_1 = 0 \\ \dot{\lambda}_2 = -\lambda_1 \end{array} \right\}$,

$x_1(0) = 0, x_2(0) = 0, x_1(1) = 1, x_2(1) = 0$

(b) $u^*(t) = \begin{cases} -4, & t > \frac{1}{2} \\ +4, & t \le \frac{1}{2} \end{cases}$ $\left. \begin{array}{l} x_1(t) = 2t^2 \\ x_2(t) = 4t \end{array} \right\} 0 < t \le \frac{1}{2}$;

$\left. \begin{array}{l} x_1(t) = 4t - 2t^2 - 1 \\ x_2(t) = 4 - 4t \end{array} \right\} \frac{1}{2} < t \le 1$

13.42(a) $\left. \begin{array}{ll} \dot{x}_1 = x_2 & \dot{\lambda}_1 = -5x_1 + 2\lambda_2 \\ \dot{x}_2 = 2x_1 - 3x_2 - \lambda_2, & \dot{\lambda}_2 = -5x_2 - \lambda_1 + 3\lambda_2 \end{array} \right\}$

$x_1(0) = 1, x_2(0) = 0$,

$x_1(\infty) = 0, x_2(\infty) = 0$

$u^*(t) = -e^{-3t}, x_1(t) = 1.5e^{-t} - 0.5e^{-3t}$,

$x_2(t) = -1.5e^{-t} + 1.5e^{-3t}$ **(b)** $u^* = -x_1(t) - x_2(t)$

13.46(b) $\mathbf{k} = [0 \quad 0 \quad 0.4142]$ **(c)** $\mathbf{k}$ is the same as in (b). Overall performance (except for overshoot) is, as in Problem 13.45, better than Fig. 11.7c. Relative to Problem 13.45, ISE and settling time are the same, while overshoot performance is better here than in Problem 13.45.

13.49(a) $\mathbf{k} = [0.9321 \quad 0.335 \quad 0.2613]$, $u_{sp} = 2.7409$, $x_{2d} = 1.6363$, $x_{3d} = 0.9092$; the design of Example 13.6 gives the same feedback gains.
(b) $\dot{x}_1 = -0.5x_1 + 0.5x_2 + 2u^*$, $\dot{\lambda}_1 = -2x_1 + 0.5\lambda_1 - \lambda_2$,
$\dot{x}_2 = x_1 - 1.5x_2 + 0.5x_3$, $\dot{\lambda}_2 = -2x_2 - 0.5\lambda_1 + 1.5\lambda_2 -$
$0.2\lambda_3$, $u^* = \begin{cases} 0.2, & \lambda_1 \le -0.1 \\ 0, & \lambda_1 > -0.1 \end{cases}$ $\dot{x}_3 = 0.2x_2 - 0.36x_3$,
$\dot{\lambda}_3 = -2x_3 - 0.5\lambda_2 + 0.36\lambda_3$, $\mathbf{X}(0)$ and $\mathbf{X}(t_f)$ are as given

CHAPTER 14

14.5 $G_{CL}(z) = \dfrac{k(z - \frac{1}{2})}{z^2 + (k-3)z + (2 - \frac{1}{2}k)}$, $k = 3.186$, independent of T

14.12(c) Compared with Dahlin's method for $\tau = 1.3T$, Kalman's method gives a more oscillatory response but essentially the same steady-state error and settling time. There is some overshoot, although the design assumes none; however, the rise time is short.

14.17 Each system can be obtained from the other by elementary row or column operations.

14.18(c) Depends on $\Delta\mathbf{Q}$: $\Delta\mathbf{Q} = [1 \quad 1]$,

$$\mathbf{K} = \begin{bmatrix} 0 & 0.5 & 0 & 1 \\ -0.25 & -0.625 & 0 & -0.25 \end{bmatrix}; \Delta\mathbf{Q} = [1 \quad 0],$$

$$\mathbf{K} = \begin{bmatrix} \frac{1}{3} & 0 & 0 & \frac{4}{3} \\ -\frac{1}{3} & -\frac{1}{2} & 0 & -\frac{1}{3} \end{bmatrix}; \Delta\mathbf{Q} = [0 \quad 1],$$

$$\mathbf{K} = \begin{bmatrix} -1 & 2 & 0 & 4 \\ 0 & -1 & 0 & -1 \end{bmatrix}$$

14.21(c) $\mathbf{f} = \begin{bmatrix} 1 \\ 1.5 \end{bmatrix}$; without I-action, $\mathbf{k} = [1.5 \quad -3.5]$; with I-action, $k_{p1} = 45$, $k_{p2} = -125$, $k_i = -20$

14.21(g) $\mathbf{f} = \begin{bmatrix} 1 \\ 0.7906 \\ -0.0369 \end{bmatrix}$; $\mathbf{k} = [0 \quad 2.9541 \quad 3.9417]$

14.23 Optimal trajectory: $u^0 = -x$

14.26(a)

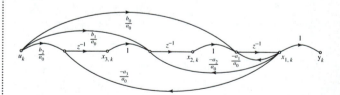

(b) $\mathbf{G}_{FB} = [6.5089 \quad 6.0287 \quad 4.504]$, $G_{FF} = 3.6996$; The response is remarkable. There is no overshoot or oscillation. Settling time is about 1.4 s and compares with 2.5 s in Fig. 14.6.

Index